BotanyLinks

BotanyLinks
www.jbpub.com/botanylinks

The distinctive **BotanyLinks** icon appears next to selected topics in the text, to indicate that additional material is available through Jones and Bartlett's extensive BotanyLinks web site. Upon opening the BotanyLinks home page, click the icon option that corresponds with the text's chapter number and page reference, and you will be connected to a wealth of additional current and reliable information on the Web.

The BotanyLinks icon appears in the text's margin, to indicate the availability of related material on the Web.

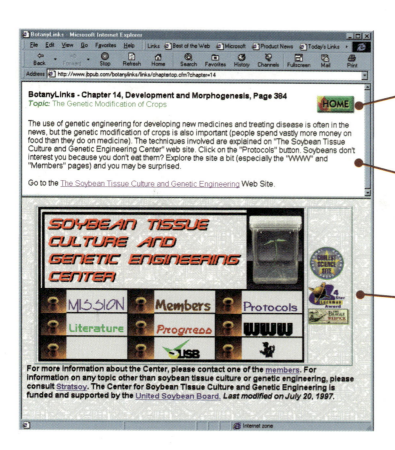

The BotanyLinks icons identify topics that are matched to WWW sites that you can visit from the home page.

Brief, descriptive introductions to each linked site provide context for your explorations.

With a click of your mouse, BotanyLinks takes you to the best and most current sites on the Web.

BotanyLinks .net Questions

The End of Chapter .net Questions invite students to explore a topic in greater depth on the Web, while encouraging the use of active learning skills. A brief overview of the question is provided in the text, while the web site hosts the more detailed and site-specific instructions. In this way, the questions and related web sites can be updated as new information and discoveries come on-line. Clicking on the .net Questions link on the BotanyLinks home page will take students one step further, by giving specific questions to answer on the linked site.

Where appropriate, chapters end with an introduction to the web-based exercises found on-line at BotanyLinks .net Questions.

Detailed site instructions for answering .net Questions appear at the top of the screen.

Students are sent to a carefully chosen web site or given the opportunity to discover their own answers to the .net Questions.

Students can hand in or e-mail answers to their instructor at the instructor's discretion.

PlaBiology Tutor

The Plant ы Tutor Icon indicates topics in the text for which thergnificant interactive content available on the Plant Biolator CD-ROM by Steven E. Sheckler, Stewart A. Hill, and LTaylor of Virginia Polytechnic Institute. Featuring full-color images, dozens of animations, and three full-sxperiment simulations, this CD-ROM provides students w ideal interactive learning aid and instructors with a powlecture enhancement tool.

Plant Biology Tutor CD-ROM

Fourteen Tutorials are grouped into four overall themes covering: ecology, growth and development, reproduction, and systematics and evolution.

The Plant Biology Tutor icon matches topics in the text with topics on the CD-ROM.

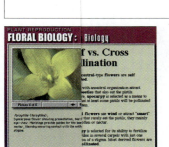

This example one of the six photographic i of this presentation. Any presentatige can be easily enlarged to fult size for closer study.

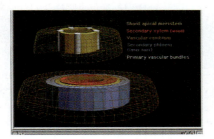

This example shows one of the CD's many 3-dimensional animations, which will aid students in their visualization of difficult biological and botanical concepts.

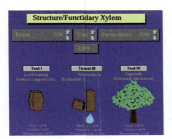

The lab is whdents will explore the changing ons in nature with no-risk re

The Biome Almanac gives students the chance to embark on a one-of-a-kind journey through each of the regions contained within the five biomes.

BOTANY

AN INTRODUCTION TO PLANT BIOLOGY, 2/e

Multimedia Enhanced Edition

JAMES D. MAUSETH

University of Texas, Austin

JONES AND BARTLETT PUBLISHERS

Sudbury, Massachusetts

Boston   Toronto   London   Singapore

World Headquarters
Jones and Bartlett Publishers
40 Tall Pine Drive
Sudbury, MA 01776
978-443-5000
info@jbpub.com
www.jbpub.com

Jones and Bartlett Publishers Canada
P.O. Box 19020
Toronto, ON M5S 1X1
CANADA

Jones and Bartlett Publishers International
Barb House, Barb Mews
London W6 7PA
UK

Chief Executive Officer: Clayton Jones
Chief Operating Officer: Don Jones, Jr.
Publisher: Tom Walker
V.P., Sales and Marketing: Rob McCarry
Senior Managing Editor: Judith H. Hauck
Marketing Director: Rich Pirozzi
Production Manager: Anne Spencer
Manufacturing Director: Jane Bromback
Executive Editor: Brian L. McKean
Project Editor: Kathryn Twombly
Senior Production Editor: Mary Hill
Assistant Production Editor: Ivee Wong

Web Site Design: Andrea Wasik
Cover & Web Walkthrough Design: Anne Spencer
Text Design: Nanci Kappel
Text Artwork: Rolin Graphics, J&R Art Services, Darwin and Vally Hennings
Composition: Progressive Information Technologies
Book Layout: Dorothy Chattin, Claudia Durrell
Cover Manufacture: Coral Graphic Services, Inc.
Book Manufacture: Von Hoffman Press, Inc.

Cover image: © James Randklev/Tony Stone Images.

Frontispiece: *Gazania*, a member of the Compositae family, photographed by the author at the Huntington Botanical Gardens, San Marino, California

Library of Congress Cataloging-in-Publication Data

Mauseth, James D.
    Botany: an introduction to plant biology / James D. Mauseth. —
2/e, Multimedia enhanced ed.
        p.   cm.
    Includes index.
    ISBN: 0-7637-0746-5
    1. Botany.   I. Title.
QK47.M1635   1998   97-42599
580—dc21   CIP

Printed in the United States

02 01 00 99 98   9 8 7 6 5 4 3 2 1

In biology, three topics are so important, so fundamental, that they must permeate every aspect of an introductory botany textbook and should be mentioned or alluded to on every page: they are *evolution by natural selection, analysis of botanical phenomena,* and *diversity of organisms and their components.*

EMPHASIS ON EVOLUTION BY NATURAL SELECTION

Evolution by natural selection is the basis of botany and all other fields of biology. It is possible, without understanding natural selection, to memorize botanical terms, to learn enzyme mechanisms and metabolic pathways, to solve problems in genetics, and so on. But without a knowledge of evolution by natural selection it is not possible to understand biological structures or processes. As instructors, we talk about and teach that the discovery of Darwin and Wallace revolutionized biology, but it can be difficult to explain to students how powerful the concept of natural selection is. In this book, I introduce natural selection in general, nontechnical terms that students new to biology are able to grasp. I believe that students must become familiar with organisms, anatomy, physiology, and genetics before they can truly appreciate the subtleties of evolution, natural selection, selective advantage, and competition. Beginning with Chapter 1, selective advantage and fitness are referred to or discussed in every chapter and on almost every page. It should be impossible for students to forget about evolution for even a few minutes. Natural selection is covered in detail in Chapter 17, after students have become rather sophisticated in their knowledge of plants and botanical phenomena.

ANALYSIS OF BOTANICAL PHENOMENA AND DIVERSITY

Emphasizing natural selection throughout the book has allowed me to incorporate the other two important topics—analysis of botanical phenomena and diversity—into every subject. Diversity includes not only the kingdoms and divisions of organisms, but the

diversity of alternative adaptations, that is, the diversity of ways that plants can be adapted to their environments. There are many types of stems, leaves, and flowers; many types of respiration, storage molecules, and methods of transporting material; many types of pollination, rewards for pollinators, and types of seeds and fruits. Instructors can guide students through an analysis of biological phenomena by emphasizing the diversity of microhabitats and microclimates; the numerous pests and pathogens; the many types of stresses, scarcities, and excesses that plants must face, and by using the principles of evolution by natural selection. There are several types of stem and leaf structures: Which are selectively advantageous under certain conditions and which are adaptive under any conditions? Many types of pollination have evolved, but each type can be analyzed in terms of the selective advantage it offers under particular conditions.

This approach of analyzing diversity of structures and metabolisms on the basis of evolution by natural selection encourages students to participate in biological inquiry. It is so easy to ask "What is the selective advantage of this process compared to that process?" that students will soon realize that not everything in biology is known and understood. Students themselves can think about such questions, rather than leaving them for unknown scientists to answer. Students can create their own hypotheses, their own potential explanations, using natural selection as an analytical tool. Simple observations or thought experiments can often indicate whether hypotheses might be feasible and plausible.

ORGANIZATION

The topics and chapters are organized in a sequence that I find to be the easiest to follow, beginning with the most familiar—structure—and proceeding to the less familiar—metabolism—then finishing with those topics that are probably least familiar to most beginning students—genetics, evolution (in detail), the diversity of organisms, and ecology. Three initial chapters—chemistry, cell structure, and cell division—are not familiar but are so basic they must come first. Because this order may not be preferred by all instructors, sections have been written to be self-contained so they can be taken up in various orders. In courses that emphasize metabolism, the sequence could be Chapters 1, 2, 3, 10 to 15. Courses emphasizing plant diversity could follow a sequence of Chapters 1, 2, (16, 17), 18 to 25. Gene action appears in Chapter 15 of the growth and development section, whereas gene replication is located in Chapter 16 in the genetics section. Although both topics are based on the DNA molecule, it does not seem necessary to cover DNA replication and Mendelian genetics prior to teaching about mRNA, protein synthesis, and gene regulation. These topics, along with the techniques of recombinant DNA analysis, are covered in considerable detail. I believe that beginning students are fully capable of understanding these subjects, which are already becoming central to most fields of biology.

MULTIMEDIA ENHANCED EDITION

BOTANY: An Introduction to Plant Biology, 2/e, Multimedia Enhanced Edition provides students with original web-integrated activities, and direct links to World Wide Web resources. The starting point is BotanyLinks, Jones and Bartlett's extensive botany home page. Students reach BotanyLinks by entering the URL www.jbpub.com/botanylinks into a World Wide Web browser such at Microsoft Internet Explorer or Netscape Navigator.

The leaf icon in the text's margins points to web sites, accessible through BotanyLinks, where further information on similar topics is available. Brief introductions on the BotanyLinks web page place the links in context before the students are linked to the site. Jones and Bartlett monitors the links regularly to ensure that there will always be a working and appropriate site on-line.

The BotanyLinks .net Questions, which are introduced at the end of most chapters, provide students with an opportunity to use the web and their own critical thinking skills to better understand and further explore concepts from the text. The .net Questions send

BotanyLinks
www.jbpub.com/botanylinks

the student to several web sites to help them in their research. The Directory of Organizations assists students in their web research by matching the sites with topics in the text. The unique audiovisual On-Line Glossary is available to aid students in pronouncing botanical terms, and connecting a term with its visual counterpart.

The Plant Biology Tutor sunflower icon in the text's margins identifies topics that are matched to the CD-ROM created by Stephen E. Scheckler, Stewart A. Hill, and David Taylor of the Virginia Polytechnic Institute. This interactive CD is designed to enhance students' understanding of biological and botanical concepts through full-color images, micrographs, animations, tutorials, and full-scale experimental simulations. A table that correlates topics in Plant Biology Tutor with sections in this text is provided on page xxiii, as well as at BotanyLinks.

FEATURES OF THIS EDITION

The following features are used throughout *Botany* to help students learn.

PART OPENERS

Each of the book's four parts is introduced by a brief summary of all the chapters in that part. The opener ties together the main themes and shows how botany is a unified science, not just a body of facts to memorize.

CONCEPTS

Each chapter opens with a section on concepts that will set the stage for the main topics and themes covered in the chapter. It provides an overview and outline of what is to follow.

PLANTS AND PEOPLE

The *Plants and People* boxes are designed to guide students into thinking about plants and how they interrelate with human concerns. Several are oriented toward economics—the importance of plants in providing food, fiber, and medicines. Others discuss the role plants have played in world politics, and others explore the ways in which botanical discoveries have affected the evolution of scientific thought and concepts, such as proving that spontaneous combustion does not occur and that there is no such thing as "vital force" in living creatures. A complete list of the *Plants and People* boxes appears after the detailed table of contents.

BOXES

Boxes elaborate on subjects that, while not essential to the study of botany, help make the material more interesting and understandable. Among the topics are the structure of wood in three dimensions, a key to identifying unknown plants, and resin casting—a new method for studying cell shapes. A complete list of Boxes appears after the detailed table of contents.

ILLUSTRATIONS

The botanical world is full of color, and so is this text. Figure illustrations have been chosen to illuminate points made in the text and to show many of the plants under discussion. Many of the drawings have been redone for this new edition and are both beautiful and botanically accurate. Photographs have been added where our adopters or reviewers have indicated a need for greater variety or multiple views. Extensive labels have been added to many photographs to clarify the features or structures shown. Many of the micrographs are now introduced with either diagrams or low-magnification micrographs to help the reader

understand the orientation of the tissues in the high-magnification illustration. Selected light and electron micrographs are now accompanied by interpretive line drawings to make the photographs more understandable to students.

OTHER FEATURES

Marginal Notes. Marginal notes clarify important points and make connections with other topics. Many notes provide concise summaries of difficult concepts.

Summary. The summary at the end of the chapter consists of numbered points that succinctly list the major topics and concepts discussed.

Important Terms. This feature is new to the second edition and gives a list of terms that should be understood after completing study of the chapter.

Review Questions. Answering these questions will help students master the chapter material. The questions are often open-ended, encouraging students to think about the concepts and synthesize the material, rather than just repeat information.

Glossary. A comprehensive glossary defines major botanical and general biological terms. Each definition is keyed to the chapter where the principal discussion occurs.

Back Endpapers. The endpapers in the back of the book summarize in full-color line drawings the major biological kingdoms of classification and the evolution of plants.

ANCILLARIES

Several ancillaries are available to accompany this text: The *Instructor's Resource Manual* includes the *Instructor's Manual,* the printed *Test Bank* and *Bio-Art.* The *Instructor's Manual,* prepared by Dr. Marshall Sundberg of Louisiana State University, contains chapter outlines, laboratory suggestions, and ideas for fieldwork projects. It also has a wealth of supplemental information that can be included in lectures to provide extra motivation for students. Dr. Ann M. Mickle has prepared a Test Bank of multiple-choice, true/false, matching, and short essay questions. The Test Bank is also available as an *ExaMaster™ Computerized Test Bank* in IBM 5.25″ and 3.5″, as well as Macintosh and Windows. *Bio-Art* consists of approximately 50 illustrations (without labels) reprinted from the book. These can be used as part of a labeling exercise in an exam or can be photocopied and given to students for taking notes. Instructors can add their own labels to customize them to their course.

A set of 150 illustrations from the text have been selected and made available as full-color *Overhead Transparencies.* Each has been chosen because it represents an important complex concept that instructors may wish to discuss with a detailed illustration in lecture.

Plant Biology Tutor CD-ROM by Stephen E. Scheckler, Stewart A. Hill and David Taylor, Virginia Polytechnic Institute contains approximately 2,000 full-color images, dozens of animations, and three full-scale experiment simulations. The fourteen tutorials on this CD-ROM cover four themes: ecology, growth and development, reproduction, and systematics and evolution. With its textual material and colorful animations, *Plant Biology Tutor* provides students with an ideal interactive learning aid and instructors with a powerful lecture enhancement tool.

The Instructor's CD-ROM includes captured images of World Wide Web sites referenced in the text for lecture presentation. It also contains a series of original animations featuring core botanical principles and processes. A Video Resource Library with a full complement of quality videos is available to qualified adopters. Please contact your Jones and Bartlett sales consultant for details.

A Final Word, Just To Students

Plants are of vital importance to all of us in everyday life, providing us not only with food and oxygen, of course, but also with lumber, medicines, fibers, and even protection against high-energy radiation from space. But plants are important to biologists in other ways, more subtle, interesting ways: most of the early discoveries of cell biology, including the discovery of cells themselves, were made in studies of plants, and many enzyme systems were first isolated from plants and were later confirmed to be present in animals as well. Even at present, when thousands of scientists analyze human metabolism and structure in ever greater detail, studies of plant metabolism and structure are essential as a means of truly understanding human biology. For example, when we humans are cold, we generate heat by shivering—a rapid contraction and relaxation of muscle cells that converts mechanical energy to heat energy. But certain plants have an unusual type of respiration (thermogenic respiration) that generates heat more directly and efficiently. Would such a metabolism be beneficial to us? What kinds of changes would have to occur for humans to have thermogenic respiration? Similarly, all organisms use only 20 types of amino acids in building proteins, but some plants have hundreds of types of amino acids, not just 20: if we humans ate such plants, our bodies would try to build our new proteins with those exotic amino acids and we would die. But the plants can somehow tell the difference and are not killed by those amino acids in their cells. How do the plants distinguish between safe and poisonous amino acids? What does our inability to distinguish between them tell us about our protein-synthesizing metabolism—one of our most fundamental metabolisms? By understanding plant biology we can more completely and fully understand and appreciate the details of human biology.

Acknowledgments

This text has benefitted from the generous, conscientious thoughts of many reviewers. They provided numerous suggestions for improving clarity of presentation, or identified illustrative examples that would improve the student's understanding and interest. It has been a pleasure to work with them. I thank them all:

- Vernon Ahmadjian, Clark University
- John Beebe, Calvin College
- Curtis Clark, California State Polytechnic University, Pomona
- Billy G. Cumbie, University of Missouri, Columbia
- Jerry Davis, University of Wisconsin, La Crosse
- Rebecca McBride-DiLiddo, Suffolk University
- John Dubois, Middle Tennessee State University
- Donald S. Emmeluth, Fulton-Montgomery Community College
- Howard Grimes, Washington State University
- James Haynes, State University College at Buffalo
- James C. Hull, Towson State University
- Shelley Jansky, University of Wisconsin, Stevens Point
- Roger M. Knutson, Luther College
- Lillian Miller, Florida Community College at Jacksonville, South Campus
- Louis V. Mingrone, Bloomsburg University
- Rory O'Neil, University of Houston, Downtown
- John Olsen, Rhodes College
- Jerry Pickering, Indiana University of Pennsylvania
- Mary Ann Polasek, Cardinal Stritch College

- Barbara Rafaill, Georgetown College
- Michael Renfroe, James Madison University
- Michael D. Rourke, Bakersfield College
- James L. Seago, Jr., State University of New York at Oswego
- Bruce B. Smith, York College of Pennsylvania
- Garland Upchurch, Southwest Texas State University
- Jack Waber, West Chester University
- James W. Wallace, Western Carolina University
- Peter Webster, University of Massachusetts, Amherst
- Paula S. Williamson, Southwest Texas State University
- Ernest Wilson, Virginia State University

During the course of this project, I have discovered that an author creates only a manuscript; it is the editorial staff that creates an attractive, usable book that actually is effective in transferring the information in the manuscript to the students. I am grateful to the staff of Saunders College Publishing: Ed Murphy for initiating this project and Julie Levin Alexander for supporting and maintaining it through both the first and second editions. Sally Kusch, Senior Project Editor, always knew how to handle every problem, large or small. Lee Marcott, Developmental Editor, had imaginative yet realistic suggestions for the second edition and had a remarkable ability to keep us both working and on schedule. Anne Muldrow, Art Director, and Leslie Ernst, Art Developmental Editor, have greatly improved the art and layout of the second edition, making it easier for students to understand the concepts presented. Carlyn Iverson also contributed to the art program through her sketches and finished illustrations. Sue Howard, the photoresearcher, worked tirelessly to find the best photographs possible. Donna Walker, the copy editor, was quite skillful in simplifying many complex sentences, making them easier to understand. I also want to thank the staff of Jones and Bartlett Publishers for their forward-looking approach in producing this Multimedia Enhanced Edition.

Finally, I would like to express my special appreciation for the help given to me by my family and by Tommy R. Navarre. They always had faith in this project, and they provided me with enthusiasm and confidence whenever I needed it.

James D. Mauseth
Austin, Texas

CONTENTS OVERVIEW

CONTENTS

III   GENETICS AND EVOLUTION   437

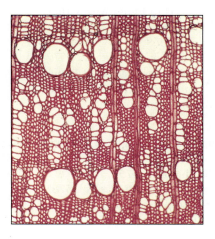

LIST OF PLANTS AND PEOPLE BOXES

LIST OF BOXED READINGS

Correlation Guide Between Mauseth *BOTANY* and *PLANT BIOLOGY TUTOR*

INTRODUCTION TO PLANTS AND BOTANY

1

Flowers are involved in plant reproduction; they produce seeds that perpetuate the species.

CONCEPTS

Botany is the scientific study of plants. This definition requires an understanding of the concepts "plants" and "scientific study." It may surprise you to learn that it is difficult to define precisely what a plant is. Plants have so many types and variations that a simple definition has many exceptions, and a definition that includes all plants and excludes all nonplants may be too complicated to be useful. Also, biologists do not agree about whether certain organisms are indeed plants. Rather than memorizing a terse definition, more is gained by understanding what plants are, what the exceptional or exotic cases are, and why botanists disagree about certain organisms.

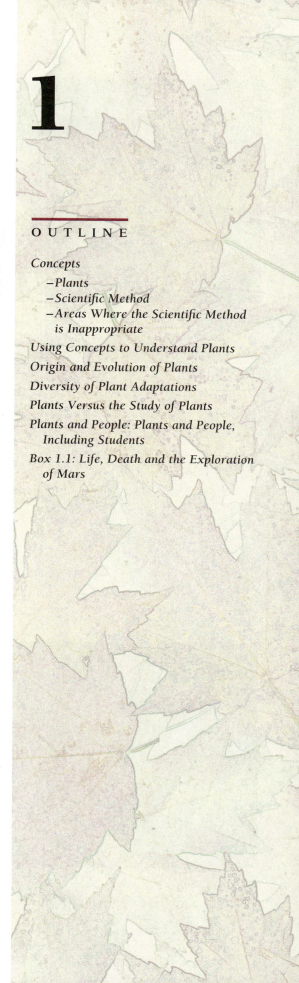

FIGURE 1.1
This morning glory (*Ipomoea*) is obviously a flowering plant. It is a vine with long, slender stems and simple leaves that occur in pairs. It has an extensive root system, not visible here.

FIGURE 1.2
Conifers, like this spruce (*Picea*), produce seeds in cones; the conifers, together with the flowering plants and a few other groups, are known as seed plants.

PLANTS

Your present concept of plants is probably quite accurate: Most plants have green leaves, stems, roots, and flowers (Fig. 1.1). But you can think of exceptions immediately. Conifers such as pine, spruce, and fir have cones rather than flowers (Fig. 1.2), and many cacti and succulents do not appear to have leaves. But both conifers and succulents are obviously plants because they closely resemble organisms that unquestionably are plants. Similarly, ferns and mosses (Figs. 1.3 and 1.4) are easily recognized as plants. Fungi, such as mushrooms (Fig. 1.5) and puffballs, were included in the plant kingdom because they are immobile and produce spores, which function somewhat like seeds. But biologists no longer consider fungi to be plants because recent observations show that fungi differ from plants in many basic biochemical respects.

Algae are more problematical. One group, the green algae (Fig. 1.6), are similar to plants in biochemistry and cell structure, but they also have many significant differences.

FIGURE 1.3
Ferns have several features in common with flowering plants; they have leaves, stems, and roots. However, they never produce seeds, and they have neither flowers nor wood.

FIGURE 1.4
Of all terrestrial plants, mosses have the least in common with flowering plants. They have structures called "leaves" and "stems," but these are not the same as in flowering plants. They have no roots at all.

FIGURE 1.5
Fungi such as these mushrooms are not considered to be plants. They are never green and cannot obtain their energy from sunlight. Also, their tissues and physiology are quite different from those of plants.

FIGURE 1.6
Algae do not look much like plants, but many aspects of their biochemistry and cellular organization are very similar to those of plants. Some of the green algae were the ancestors of land plants; although not considered to be true plants, they are obviously closely related to plants.

Some botanists conclude that it is more useful to include green algae with plants; others exclude them, pointing out that some green algae have more in common with the seaweeds known as red algae and brown algae. Arbitrarily declaring that green algae are or are not plants solves nothing; the important thing is to understand the concepts involved and why disagreement exists.

SCIENTIFIC METHOD

The concept of a scientific study can be understood by examining earlier approaches to studying nature. Until the 15th century, three principal methods for analyzing and explaining the universe and its phenomena were used: religion, metaphysics, and speculative philosophy. In **religious methods,** the universe is assumed either to be created by or to contain deities. The important feature is that the actions of gods cannot be studied: They are either hidden or capricious, changing from day to day and altering natural phenomena. Agricultural studies would be useless because some years crops might flourish or fail because of weather or disease, but in other years crop failure might be due to a god's intervention (a miracle) to reward or punish people. There would be no reason to expect consistent results from experiments. In a religious system, much of the knowledge of the world comes as a revelation from the deity rather than by observation and study of the world. A fundamental principle of all religions is faith: People must believe in the god without physical proof of its existence or actions.

A **metaphysical system** of analyzing the world postulates that in addition to natural forces, there are supernatural, hidden forces that can never be observed or studied. Phenomena that seem unexplainable in terms of the natural processes of physics and chemistry are believed to be controlled by unknown and unknowable forces. Surprisingly, many people still believe in metaphysics without realizing it: Examples of metaphysical forces (if they actually existed) would be luck, bad omens, accurate horoscopes, and reliable methods for picking the winning numbers in a lottery.

Speculative philosophy reached its greatest development with the ancient Greek philosophers. Basically, their method of analyzing the world involved thinking about it logically. They sought to develop logical explanations for simple observations, then followed the logic as far as possible. An example is the philosophical postulation of atoms by

Democritus around 400 BC. From the observation that all objects could be cut or broken into two smaller objects, it follows logically that the two pieces can each be subdivided again into two more, and so on. Finally, some size must be reached at which further subdivision is not possible; objects of that size are atoms. Speculative philosophy did not involve verification; philosophical predictions were made, but no actual experiment or observation was performed to see if they were correct. A problem with this method is that often several alternative conclusions are equally plausible logically; only experimentation reveals which is actually true.

Starting in the 1400s, a new method, called the **scientific method,** began to develop slowly. Several fundamental tenets were established:

1. All accepted information can be derived only from carefully documented and controlled observations or experiments. Claims emanating from deities, priests, prophets, and revelations cannot be accepted automatically; they must be subjected to verification and proof. This separates science from religion.

2. Only phenomena and objects that can be observed and studied are dealt with; claims of supernatural forces that cannot be seen or physically felt or tested must be rejected.

3. All proposed explanations of natural phenomena must be tested and verified; if they cannot be tested by currently available means, they must be viewed with skepticism. Many things cannot yet be explained, but scientists assume that someday we will have instruments to observe and measure the mechanisms behind such phenomena. We do not say that they can never be explained. An example of this is continental drift. In the early 1900s, Alfred Wegener observed that many South American plants and animals resemble those in Africa but are unable to cross the Atlantic Ocean. He hypothesized that Africa and South America had been joined at one time, allowing the plants and animals to spread across the regions, but that the continents later drifted apart. Because there was no way to test and verify this hypothesis then, continental drift was viewed as an interesting but unproven idea. Finally, in the 1960s, it became possible to measure positions accurately enough to establish that South America is indeed moving away from Africa; continental drift was finally proven a half century after it was proposed.

Scientific studies take many forms, but basically they begin with a series of observations, followed by a period of experimentation mixed with further observation and analysis. At some point, a **hypothesis,** or model, is constructed to account for the observations. For example, scientists in the Middle Ages observed that plants never occur in dark caves and grow poorly indoors where light is dim. They hypothesized that plants need light to grow. This can be formally stated as a pair of simple alternative hypotheses: (1) Plants need light to grow. (2) Plants do not need light to grow. The experimental testing may involve the comparison of several plants outdoors, some in light and others heavily shaded, or it may involve several plants indoors, some in the normal gloom and others illuminated by a window or a skylight. Such experiments give results consistent with hypothesis 1; hypothesis 2 would be rejected.

A hypothesis must continue to be tested in various ways. It must be consistent with further observations and experiments, and it must be able to predict the results of future experiments. In this case, the hypothesis predicts that environments with little or no light will have few or no plants. Observations are consistent with these predictions. In a heavy forest, shade is dense at ground level and few plants grow there (Fig. 1.7). Similarly, as light penetrates the ocean, it is absorbed by water until at great depth all light has been absorbed; no plants or algae grow below that depth.

If a hypothesis continues to match observations, we have greater confidence that it is correct, and it may come to be called a **theory.** Occasionally, a hypothesis does not match an observation; that may mean either that the hypothesis must be altered somewhat or that the whole hypothesis has been wrong. For instance, plants such as Indian pipe or *Conopholis* (Fig. 1.8) grow the same with or without light; they do not need light for growth. These are parasitic plants that obtain their energy by drawing nutrients from host plants. Thus our

(a)   (b)

FIGURE 1.7
(a) This aspen forest in Michigan does not have a dense canopy, but it intercepts so much light that few plants survive in the shade. The herb is the bracken fern *Pteridium aquilinum*. (b) Near the aspen forest is an open area with more light; herb growth, in this case a sedge, is much more abundant. *(Courtesy of R. Fulginiti, University of Texas.)*

hypothesis needs only minor modification: All plants except parasitic ones need sunlight for growth. It remains a reasonably accurate predictive model.

> *One of the greatest values of a hypothesis or theory is its power as a predictive model.*

AREAS WHERE THE SCIENTIFIC METHOD IS INAPPROPRIATE

Certain concepts exist for which the scientific method is inappropriate. We all believe that something called morality exists, that it it not right to wantonly kill each other, and that racism and sexism are bad. Science can study, measure, analyze, and describe the factors that cause people to kill each other or to be racist or sexist, and it can predict the outcome of these actions. But science cannot say if such actions are right or wrong, moral or immoral. Consider euthanasia: Many types of incurable cancer cause terrible pain and

FIGURE 1.8
The yellowish flowers pushing out of the pine needle litter constitute almost the entire plant body of this parasitic plant, *Conopholis mexicana*. It is attached to the roots of nearby trees and draws nutrients from them. Like fungi, it cannot obtain its energy from sunlight, but so many other aspects of its anatomy and physiology are like those of ordinary plants that we have no difficulty in recognizing that this is a true plant, not a fungus.

PLANTS & PEOPLE

PLANTS AND PEOPLE, INCLUDING STUDENTS

Plants and people affect each other. Most obvious perhaps are the ways that people benefit from plants: They are the sources of our food, wood, paper, fibers, and medicines. It is difficult to excite students by listing the world production of wheat and lumber in metric tons, but just consider what your life would be like without products such as chocolate, sugar, vanilla, cinnamon, pepper, mahogany, cherry wood, ebony, cotton, linen, roses, orchids, or the paper that examinations are written on. The oxygen we breath comes entirely from plants. Plants affect each of us every day, not simply by keeping us alive but also by providing wonderful sights, textures, and fragrances that enrich our existence.

However, plants and people affect each other in ways that are not readily apparent in our day-to-day lives. Below are a few important topics that you should be aware of. Articles about these appear in the news, and you should think about their importance and how you—as an actual biological organism—interact with the other organisms on this planet.

Biotechnology is a set of laboratory techniques that allow us to alter plants and animals, giving them new traits and characteristics. Farmers have done this for thousands of years with plant breeding and animal husbandry, but biotechnology permits much more rapid, extensive alterations. We must now consider whether such manipulations are safe and worthwhile.

Global warming is caused by a build-up of carbon dioxide in our atmosphere due to the burning of coal, oil, gas, and the trees of forests everywhere (not just tropical rainforests). The carbon dioxide traps heat, preventing Earth from radiating excess energy into space. Global warming could affect the melting of polar ice caps, the circulation of ocean currents, and even the amount and pattern of rainfall. Preserving our forests and planting more trees might help stop and reverse global warming. But the possibility exists that global warming is preventing the occurrence of another ice age. We need more information before deciding whether global warming is harmful.

Desertification is the conversion of ordinary forest or grassland to desert. Accurate measurements are difficult, but it appears that deserts may be spreading as people cut shrubs and trees for firewood and allow goats to eat remaining vegetation. Once an area has been converted to desert, its soil is rapidly eroded away, making recovery difficult. Something as simple as cheap solar cookers might prevent the Sahara desert from spreading farther across Africa.

Habitat loss results when an area is changed so much that a particular species can no longer survive in the area. Significant causes are the construction of highways, housing subdivisions, and shopping malls with enormous parking lots; these eliminate almost all species from an area. But habitats are also lost by logging, farming, mining, damming rivers, and spilling toxic chemicals. As habitat is lost, plants or animals must try to survive on the smaller remaining habitat. Once too little habitat is left, species usually become extinct.

Introduced exotics are organisms that are native to one part of the world but are brought to another part, where they thrive. Examples of introduced exotic animals are Medfly fruit flies in California and zebra mussels in the Great Lakes region. Water hyacinth and kudzu (a vine) were introduced to the United States and are now proliferating and reproducing so vigorously that they are crowding out many plants that normally grow here.

It is simply not realistic to believe that we humans will stop all these activities that have negative impacts on our environment and on the other species with which we share this planet. But we can search for ways to minimize the harm we cause by recycling, conserving resources, and avoiding products that require pollution-causing manufacturing techniques.

Habitat loss is caused by many types of human activity. (a) The simple presence of people is sufficient to frighten away birds and mammals. This in turn affects insects and plants. (b) Even the construction of beautiful parks is habitat destruction.

suffering in their final stages, which may last for months. We have drugs that can arrest breathing so that a person dies painlessly and peacefully. Science developed the drugs and can tell us the metabolic effects of using them, but it cannot tell us if it is right to use them to help a person die and avoid pain. For that answer, it is necessary to turn to a religious or philosophical method of contemplating the world. Biological advances have made us capable of surrogate motherhood, of detecting fetal birth defects early enough to allow a medically safe abortion, and of producing insecticides that protect crops but pollute the environment. These advances have made it more important than ever for us to have well-developed moral and philosophical systems for assessing the appropriateness of various actions.

Finally, we must never forget that we are human beings with emotions. Even if we learn everything that can be known scientifically about plants, a beauty still exists which cannot be analyzed and understood but only felt and appreciated.

USING CONCEPTS TO UNDERSTAND PLANTS

The growth, reproduction, and death of plants—indeed, all aspects of their lives—are governed by a small number of basic principles. Each chapter in this book opens with a section called *Concepts,* which discusses the principles most relevant to the topics in that particular chapter. Here in the first chapter and at the beginning of your study of botany and plants, I want to introduce you briefly to some of these principles and encourage you to use them as you read and think about plants. These concepts will make plant biology more easily understood—the numerous facts, figures, names, and data will be less overwhelming when you realize that they all fit into the patterns governed by a few fundamental concepts.

(a)

1. *Plant metabolism is based on the principles of chemistry and physics.* Some aspects of a plant's growth, development, and response to its environment can seem almost magical and supernatural, but they never are. All the principles you learn in your chemistry or physics classes are completely valid for plants.

2. *Plants must have a means of storing and using information.* After a seed germinates, it grows and develops into a plant, becoming larger and more complex; then it reproduces. All this is possible because the plant is taking in energy and chemical compounds and transforming them into the organic chemical compounds it uses to build more of itself. This requires a complex, carefully controlled metabolism, and there must be a mechanism for storing and using the information that regulates that metabolism. As you may already know, the genes are the primary means of storing this information.

3. *Plants reproduce, passing their genes and information on to their descendants.* Because an individual obtains its genes from its parents, the information it uses to control its metabolism is similar to the information its parent had used; thus, offspring and parents resemble each other. For example, a tomato seed contains genes whose information guides the seed's metabolism into constructing a new tomato plant, but a pea grows into a pea plant because it received different genes and information from its parents (Fig. 1.9).

4. *Genes, and the information they contain, change.* As plants make copies of their genes during reproduction, accidental changes (mutations) occasionally occur, and this causes the affected gene and its information to change. This is quite rare, and most genes (and information) are passed unaltered from parents to offspring. But as mutations occur and change a gene's information, they basically generate new information such that the plant that grows and develops under the control of the mutated gene may be slightly (or significantly) different from its parents. Thus, over time, a gradual evolution occurs in the genes, information, and biology of plants. Consequently, in a large population of many individuals of a species, some variation exists; the individuals are not identical (Fig. 1.10).

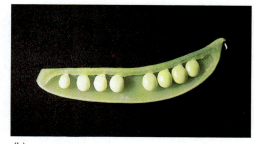

(b)

FIGURE 1.9

(a) The seeds of this tomato (*Lycopersicon*) have received, in the form of genes, the information necessary to produce a new tomato plant, whereas the peas (b) have received from their parents the information for growing into pea plants. Each type of plant differs from other types in the information that it carries.

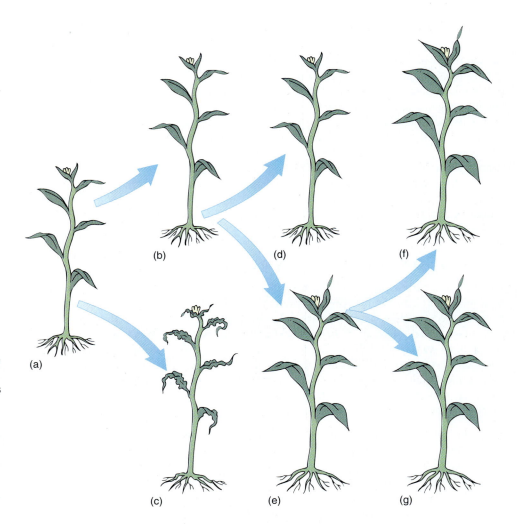

FIGURE 1.10
(a) A plant produces numerous offspring, many of which resemble it strongly (b). Mutations may occur that cause, for instance, leaves to be malformed and poorly shaped for photosynthesis (c); most or all of these mutants die and do not reproduce. The normal plants continue to reproduce (b and d), but another mutation may occur that causes the leaves to be larger and more efficient at photosynthesis (e). These may grow so well that they crowd out the original parental types, and the plant population finally contains only the type with large leaves.

5. *Plants must survive in their own environment.* They must be adapted to the conditions in the area where they live. If they are not adapted to that area's conditions, they grow and reproduce poorly or they die prematurely. Other plants whose genes result in characters that make those plants more suited to live in that area grow and reproduce more successfully and produce more offspring. Also, plants do not exist in isolation: A significant aspect of a plant's environment is the presence of other organisms. Some of the neighboring organisms may be helpful to the plant, others may be harmful, and most perhaps have little effect on it. This concept can be important when trying to understand a plant's structure and metabolism: One type of photosynthetic metabolism and leaf structure may function well if a particular plant always grows in the shade of taller neighbors, whereas a different type of photosynthetic metabolism and leaf structure may be necessary for a plant that grows nearby but in an unshaded area.

6. *Plants are highly integrated organisms.* The structure and metabolism of one part have some impact on all the rest of the plant. When studying the biology of leaves, it is best also to consider how the structure and metabolism of stems, roots, epidermis, and other parts might affect the function of those leaves. Large leaves can absorb more sunlight and energy than can small leaves, but if a plant has large leaves, it may need to have a large root system to absorb water and minerals for the leaves, and it may need wide stems to conduct the water and minerals from the roots to the leaves.

     In addition, keep in mind that structure and metabolism must be integrated: The structure of a cell, tissue, or organ must be compatible with the metabolic function of the same cell, tissue, or organ. For example, if a leaf is fibrous and tough, insects may find it unpalatable and may avoid eating it. But if the leaf is too fibrous, the fibers may block the absorption of sunlight. Such a structure would be incompatible with the function of carrying out photosynthesis.

7. *An individual plant is the temporary result of the interaction of genes and environment.* We must be careful to consider the differences between an individual plant and that plant's species (the group made up of all similar plants). Consider something like a sunflower: An individual plant exists because its parents underwent reproduction and one of their seeds landed in a suitable environment, where the information in the seed's genes interacted with the environment by way of the seed's structure and metabolism. There are two concepts of "sunflower" here: (1) The actual plant that we can observe, measure, cultivate, and enjoy and that interacts with its environment, absorbs resources, responds to changes, attracts pollinators, and resists pathogens (disease-causing organisms). (2) The genetic information that guides all this and that has existed for thousands of years, evolving gradually as it has been passed down through all the ancestors of this particular individual sunflower. This information does not exist just in this one individual but rather in all the currently living sunflower individuals. It will continue to exist in future individuals long after this generation has passed away.

8. *Plants do not have purpose or decision-making capacity.* It is easy for us to speak and write as if plants were capable of thinking and planning. We might say, "Plants produce roots in order to absorb water." But this suggests that the plants are capable of analyzing what they need and deciding what they are going to do. Assuming that plants have human characters such as thought and decision-making capacity is called anthropomorphism, and it should be avoided. Similarly, assuming that processes or structures have a purpose is called teleology, and it too is inaccurate. Consider an alternative way of phrasing the sentence: "Plant roots absorb water," leaving out the phrase "in order to." The reality of the situation is that some of the information in the plant's genes causes the plant to produce roots, which have a structure and metabolism that result in water absorption. Plants have roots because they inherited root genes from their ancestors, not in order to absorb water. Absorbing water is a beneficial result that aids in the survival of the plant, but it is not the result of a decision (anthropomorphism) or purpose (teleology).

ORIGIN AND EVOLUTION OF PLANTS

Life on Earth began about 3.5 billion years ago. At first, living organisms were simple, like present-day bacteria, in both their metabolism and structure. However, over thousands of millions of years cells gradually increased in complexity through evolution by natural selection. The process is easy to understand: As organisms reproduce, their offspring differ slightly from each other in their features—they are not identical. Offspring with features that make them poorly adapted to the habitat probably do not grow well and reproduce poorly if they live long enough to become mature (Fig. 1.11). Offspring with features that cause them to be well-adapted grow well and reproduce abundantly, passing on the beneficial features to their own offspring. This is called **natural selection**. New features come about periodically by mutations, and natural selection determines which new features are eliminated and which are passed on to future generations. Evolution by natural selection is a model consistent with observations of natural organisms, experiments, and theoretical considerations.

As early organisms became more complex, major advances occurred. One was the evolution of the type of photosynthesis that produces oxygen and carbohydrates. This photosynthesis is present in all green plants, but it first arose about 2.8 billion years ago in a bacterium-like organism called a cyanobacterium. Later, cell structure became more efficient as subcellular components evolved. These components, called organelles, are small,

HERMAN®

5-14   © 1985 Universal Press Syndicate

"Another one of nature's mistakes."

FIGURE 1.11
Mutations that produce disadvantageous features usually contribute to the death, sooner or later, of an organism. If the individual cannot undergo reproduction (because it is dead), it cannot pass the mutation on to offspring and the deleterious mutation is eliminated. *(Herman Copyright 1985 Universal Press Syndicate. Reprinted with permission. All rights reserved.)*

BOX 1.1   Life, Death, and the Exploration of Mars

Botany is a subdivison of biology, the study of life. Despite the importance to biology of defining life, no satisfactory definition exists. As we study metabolism, structure, and ecology more closely, we understand many of the processes of life in chemical/physical terms. It is now more difficult to distinguish between biology and chemistry or physics and between living and nonliving. But the lack of a definition for life does not bother biologists; very few short definitions are accurate, and life is such a complex and important subject that a full understanding gained through extensive experience is more useful than a definition.

Although we cannot define life, it is critically important for us to be able to recognize it and to know when it is absent. Many hospitals use artificial ventilators, blood pumps, and drugs to maintain the bodies of victims of accidents or illness. The person's cells are alive, but is the person alive? On a less dramatic scale, how does one recognize whether seeds are alive or dead? A farmer about to spend $100,000 on seed corn wants to be certain that the seed is alive. How do we recognize that coral is alive? It looks like rock but grows slowly—but stalactites are rock and they also grow.

The ability to recognize life or to be certain of its absence is important in space exploration as well. The United States, Russia, Europe, and Japan are organizing a large program for the exploration of Mars, and a search for life will be a key part of the

When we explore Mars and the other planets for life—either currently living forms or signs of extinct organisms—it is critical to understand the characteristics of life. (*Courtesy of NASA*)

experiments. How will we look for life? How will we know if we have found it?

The methods used to search for life on Mars, in seeds, in corals, and in humans vary, but in all cases living beings have all the following characteristics; if even one is missing, the material is not alive:

1. *Metabolism* involving the exchange of energy and matter with the environment must be present. Organisms absorb energy and matter, convert some of it into their own bodies, and excrete the rest. Many nonliving systems also do this: Rivers absorb water from creeks, mix it with mud and boulders, then "excrete" it into oceans.

2. *Nonrandom organization* must be present. All organisms are highly structured, and decay is the process of its molecules returning to a random arrangement. But many nonliving systems also have this feature: Crystals have an orderly arrangement as do many cloud patterns, weather patterns, and ripple patterns in flowing streams.

3. *Growth* must occur. All organisms increase in size from the time they are formed: Fertilized eggs grow into seeds or embryos, and each in turn grows into an adult. At some point growth may cease—we stop getting taller at about 25 years of age. This too is not sufficient to distinguish living from nonliving: Mountains and crystals also grow.

4. A system of *heredity and reproduction* must exist. An organism must produce offspring very similar to itself such that when the individual dies, life persists within its progeny. But fires reproduce and are certainly not alive.

5. *A capacity to respond to the environment* such that metabolism is not adversely affected is necessary. When conditions become dry, an organism can respond by becoming dormant, conserving water, or obtaining water more effectively. But as mountains are raised by geological forces, erosion wears them down, and the faster that the mountains are pushed up, the faster erosion works.

In addition to these absolute requirements of life, two features are almost certainly associated with all forms of life: (1) Organisms *develop,* such that young individuals and old ones have distinctive features; (2) organisms *evolve,* changing with time as the environment changes.

Although these various features are always present in living creatures, no one characteristic is sufficient to be certain that something is living versus inanimate. We have no difficulty being certain that rivers, fire, and crystals are not living, but when we search for life on other planets or even in some exotic habitats here on Earth, deciding whether we have actually discovered life might be quite problematical.

This resurrection plant (*Selaginella lepidophylla*) curls into a ball, dries out, and becomes completely dormant during drought; almost no sign of life can be detected during that period. Yet when rain or dew moistens it again, it uncurls and resumes growth within minutes.

BotanyLinks
www.jbpub.com/botanylinks

completely closed bags of membranes that give each organelle a unique structure and chemistry specialized to a specific function. Division of labor and specialization had come about.

A particularly significant evolutionary step occurred when DNA, the molecule that stores hereditary information, became located in its own organelle—the cell nucleus. Because this step was so important and occurred with so many other fundamental changes in cell metabolism, we classify all cells as prokaryotes if they do not have nuclei (bacteria and cyanobacteria) or as eukaryotes if they do have nuclei (all plants, animals, fungi, and algae; Table 1.1).

By the time nuclei became established, evolution had produced thousands of species of prokaryotes. The newly evolved eukaryotes also diversified. Some acquired an energy-transforming organelle, the mitochondrion, and some acquired chloroplasts, which carry out photosynthesis and convert the energy in sunlight to the chemical energy of carbohydrates. Those with chloroplasts evolved into algae and plants; those without evolved into fungi, protozoans, and animals (Fig. 1.12).

All organisms are classified into five large groups called kingdoms: kingdom Monera, kingdom Plantae, kingdom Animalia, kindgom Myceteae, and kingdom Protista. Parts of the kingdom Protista are closely related to kingdom Plantae because some green algae became adapted to living on land and gradually evolved into the land plants. As a consequence, early land plants resembled algae, but as more mutations occurred and natural selection eliminated less adaptive ones, land plants lost algal characteristics and gained more features suited to surviving on land. Thousands of species arose, but most became extinct, as those that were more fit grew more rapidly, survived longer, and produced more

Prokaryote = "before nuclei"; eukaryote = "having true nuclei."

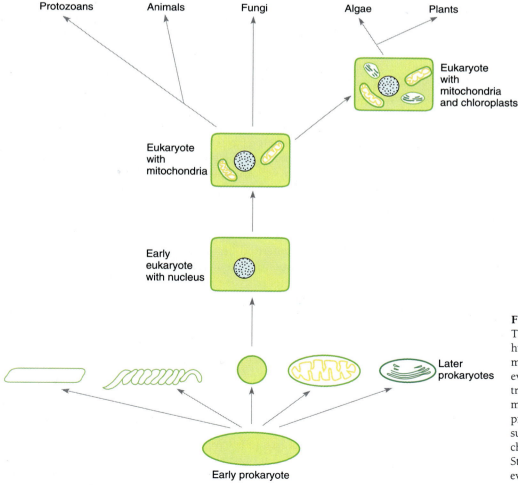

FIGURE 1.12
The earliest cells were simple, but over hundreds of millions of years they became more complex metabolically. Later, some evolved into simple eukaryotic cells with a true nucleus, which later developed mitochondria. Some of these evolved into protozoans and animals. Others evolved such that they have a cell wall made of a chemical called chitin; these became fungi. Still others developed chloroplasts and evolved into plants.

TABLE 1.1   *The Five Kingdoms of Organisms*

Prokaryotes
    Kingdom Monera: bacteria, cyano-
        bacteria
Eukaryotes
    Kingdom Protista: single-cell organisms
        (protozoans, algae); multicellular algae
    Kingdom Myceteae: fungi such as
        mushrooms, puffballs, bread mold
    Kingdom Animalia: animals
    Kingdom Plantae;* plants
        Division Bryophyta: mosses
        Division Pteridophyta: ferns
        Division Coniferophyta: conifers
        Division Anthophyta: flowering plants

* Within kingdom Plantae, many botanists
recognize about 17 divisions; only the four
most familiar are listed here. Many botanists
conclude that algae should be included in
kingdom Plantae.

FIGURE 1.13

This is *Gazania splendens* in the aster family. It has many derived features, especially the bitter, toxic chemicals in its leaves; lettuce is in the same family, but plant geneticists have eliminated the genes that produce the antiherbivore features. As a result, lettuce is sweet both to us and to insects, rabbits, and deer.

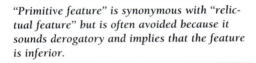

"Primitive feature" is synonymous with "relictual feature" but is often avoided because it sounds derogatory and implies that the feature is inferior.

"Advanced feature" is a commonly used synonym for "derived feature," but it implies that the feature is superior to a relictual one.

offspring. Species that did not become extinct evolved into more species, and so on. The living plants that surround us are the current result of the continuous process of evolution.

Not all organisms evolve at the same rate; some early species were actually so well adapted that they competed successfully against newer species. Algae are so well suited to life in oceans, lakes, and streams that they still thrive even though most features present in modern, living algae must be more or less identical to those present in the ancestral algae that lived more than 1 billion years ago. Features that seem relatively unchanged are **relictual features.** Like the algae, ferns are well-adapted to certain habitats and have not changed much in 150 million years; they too have many relictual features. Modern conifers are similar to early ones that arose about 120 million years ago. The most recently evolved group consists of the flowering plants, which originated about 90 to 100 million years ago with the evolution of several features: flowers; broad, flat, simple leaves; and wood that is efficient for rapid water conduction. The members of the aster family (sunflowers, daisies, dandelions; Fig. 1.13) have many features that evolved recently from features present in ancestral organisms. These are **derived features:** have been derived evolutionarily from ancestral features. One recent (highly derived) feature in the asters is a group of chemical compounds that discourage herbivores from eating the plants.

In the same way that different groups evolve at different rates, various features also evolve at different rates. For instance, the asters are a mixture of the recently derived antiherbivore compounds, less recently derived flowers, and still less recently derived wood. In addition, their bodies are covered with a waxy waterproof layer, the cuticle, that has not changed much since land plants first evolved about 420 million years ago, and their leaves contain chloroplasts and nuclei that are much like those of green algae and are extremely relictual.

DIVERSITY OF PLANT ADAPTATIONS

More than 400,000 species of plants exist today. An unknown number of species, perhaps also several hundred thousand, existed at one time but have become extinct. Virtually all this diversity came about through evolution by natural selection—survival of the fittest.

But the existence of 400,000 types of living plants means that there must be at least 400,000 ways of being fit on today's Earth. For any particular aspect of the environment, many types of adaptation are possible. There is not one single, exclusive, perfect adaptation. Consider plants growing in a climate with freezing winters. Frozen soil is physiologically dry because roots cannot extract water from it. If the winter air is dry and windy, shoots lose water and plants are in danger of dying from dehydration. How can plants adapt to this? Mutations that cause leaves to drop in autumn and mutations that cause bark to develop on all exposed parts of the stem reduce the surface area through which water is lost; such mutations are advantageous in this environment and many species have these adaptations. In other species, the entire shoot system above ground dies, but subterranean bulbs, corms, or tubers persist and produce a new shoot the next spring. Another adaptation is for the plant itself to die but its seeds to live through the harsh conditions. Finally, evergreen species retain both stem and leaves, but the leaves have extra thick cuticles and other modifications that minimize water loss.

This diversity is extremely important, and you must be careful to think in terms of **alternative adaptations,** alternative methods of coping with the environment. The physical and biological world is made up of gradients: It is not simply either hot or cold but rather ranges continuously from hot in some areas at some times to warm, cool, and cold in the same or other areas at other times. Roots face a range of water availability from flooded, waterlogged soil to moist, somewhat dry, dry, and even arid soil (Fig. 1.14). Studying and understanding plants and their survival require that we place the plant in the context of its habitat: What are the significant environmental factors? What predators and pathogens must it protect itself from? What physical stresses must it survive? What are the advantageous, helpful aspects of its habitat? When you think in terms of gradients and ranges of habitat factors, you appreciate the range of responses and adaptations that exists.

PLANTS VERSUS THE STUDY OF PLANTS

Mathematicians, physicists, and chemists study the world and derive interpretations that are considered universally valid. The relationships of numbers to each other are the same everywhere in the Universe; if intelligent beings exist elsewhere, their mathematicians will discover exactly the same mathematical relationships. The same is true for physics and chemistry.

Some aspects of biological knowledge also are universally valid: All metabolic processes, either on Earth or elsewhere in the Universe, can be predicted by fundamental principles of mathematics, physics, and chemistry. Organisms everywhere must take in energy and matter and convert it metabolically into the substance of their own bodies. Thermodynamic principles tell us that this cannot be done with perfect efficiency, so all organisms produce waste heat and waste matter. Furthermore, all organisms must be capable of reproduction and must have a system of heredity.

However, biologists study organisms that exist only on Earth, and we are reasonably certain that if life exists on some other planet, it is not exactly like life here (see Box 1.1). Much of what we study does not have universal truth; therefore, knowing about certain plants in certain habitats does not let us predict precisely the nature of other plants in other habitats. Much of our knowledge is applicable only to a particular set of plants or a certain metabolism. In the biological sciences the fundamental principles have universal validity, but many details are peculiar to the organisms being studied.

As biologists study organisms, we attempt to create a model of an unknown world. Our observations and interpretations constitute a body of knowledge that is both incomplete and inaccurate. Because we do not know everything about all organisms, our knowledge is an incomplete reflection of reality, and at least some of our hypotheses are wrong. In college courses, even introductory botany, you study areas where we are still gathering information and our knowledge is incomplete. Often, not enough data are present to form coherent hypotheses in which we can have great confidence. You must think carefully about the things you read and hear, and analyze whether they seem reasonable and logical and have been verified.

FIGURE 1.14

A mountain provides many types of environments. Higher altitudes are always covered with snow and ice; no plants can survive there. In the lower, warmer altitudes, the steep slopes allow rain to run off rapidly, so the soil is dry. Conifers are well-adapted to these conditions. In the flat valley bottoms, water accumulates as marshes; the roots of sedges and rushes can tolerate the constant moisture, but the roots of conifers would drown. Each type of plant is adapted to specific conditions, even though they grow almost side by side.

As you read, you will deal with two types of information: observations and interpretations. Most observations are reasonably accurate and trustworthy; we usually have to consider only whether the botanist was careful, observed correctly and without error, and reported truthfully. Interpretations are more difficult; they are entirely human constructs based on observations, intuition, previous experience, calculations, and expectations. How can we judge whether an interpretation has any relation to reality? A correct scientific interpretation must make an accurate prediction about the outcome of a future observation or experiment.

Your study of the material in this book, as well as your studies of plants themselves, will be easier if you keep in mind two questions that can be asked about any biological phenomenon:

1. **Are there alternatives to this phenomenon?** Do other structures exist that could perform that same function? Could a different metabolic pathway also occur? If no alternative is possible, why not? If alternatives are possible, do they exist? Did they evolve and then become extinct?

   For example, consider photosynthesis: Do plants have alternative sources of energy other than sunlight? We concluded earlier in the chapter that parasitic plants do not need sunlight; they can grow in dark areas. Also, most seeds germinate underground, in complete darkness. So it seems at first glance that there are alternatives to photosynthesis. But parasitic plants depend on host plants that do carry out photosynthesis, and germinating seeds rely on stored nutrients they obtained from their parent plant, so, at least indirectly, all plants must receive adequate sunlight. On a theoretical basis it seems possible that an insect-trapping plant like a Venus' flytrap might become so efficient that it could live solely by catching animals. But none is known to be that efficient: They catch only enough insects to provide a little extra nitrogen, not enough to provide the plant with all the energy it needs.

2. **What are the consequences?** What are the consequences of a particular feature as opposed to an alternative feature or the absence of the feature? Every feature, structure, or metabolism has consequences for the plant, making it either better or less well adapted. Some may have dramatic, highly significant consequences; others may be close to neutral. When you consider the consequences of a particular feature, you must consider the biology of the plant in its natural habitat as it faces competition, predators, pathogens, and stresses.

   Consider photosynthesis again. A consequence of depending on photosynthesis is that plants with many large leaves can harvest more light than plants with few, small leaves. But a consequence of having large leaves is that they lose more water,

FIGURE 1.15
This tree obtains its energy from sunlight, which is always located in the sky; the plant does not have to hunt for it. Sophisticated sense organs and the power of movement are completely unnecessary; having them would not make the plant more adapted.

they are more easily seen by hungry leaf-eating animals, they are good landing sites for disease-causing fungal spores, and so on. A further consequence of obtaining energy from sunlight is that the light does not need to be hunted the way that animals must hunt for their food. Because the sun rises every morning, plants can just sit in one spot and spread their leaves (Fig. 1.15). They do not need eyes, brains, muscles, or digestive systems to locate, catch, and consume their source of energy. Plants can be very simple and survive, whereas most animals must be complex.

Finally, no matter where you are, plants are readily available for direct observation. You can figure out a great deal by observing a plant and thinking about it (as opposed to just observing it). You will be surprised at how much you already know about plants. As you read, think about how the principles discussed apply to plants you are familiar with. It can be boring to memorize names and terms, but if you think about the material, analyze it, understand it, and see where it is not valid for all plants or all situations, you will find both botany and plants to be enjoyable.

SUMMARY

1. It is difficult to define a plant. It is more important to develop a familiarity with plants and understand how they differ from animals, fungi, protists, and prokaryotes. The differences are presented in later chapters.

2. The scientific method requires that all information be gathered through documented, repeatable observations and experiments. It rejects any concept that can never be examined, and it requires that all hypotheses be tested and be consistent with all relevant observations.

3. Science and religion address completely different kinds of problems. Science cannot solve moral problems; religion cannot explain physical processes.

4. Living organisms have evolved by natural selection. As organisms reproduce, mutations cause some offspring to be less fit, some to be more fit. Those whose features are best suited for the environment grow and reproduce best and leave more offspring than do those that are poorly adapted.

5. For any particular environment, several types of adaptation can be successful.

6. Our knowledge of the world is incomplete and inaccurate; as scientific studies continue, incompleteness diminishes and inaccuracies are corrected.

7. Two simple questions are powerful analytical tools: (1) What are the alternatives? (2) What are the consequences?

REVIEW QUESTIONS

1. How would you distinguish between plants and animals? What characters are important? Be careful to consider unusual plants and animals. Can all animals move? Do they all eat?

2. What are four methods for analyzing nature? Name some advantages and disadvantages of each.

3. What is a hypothesis? A theory? Why is it important that each be able to predict the outcome of a future experiment?

4. Name the five kingdoms, and describe the types of organisms in each. Which are prokaryotes and which are eukaryotes?

5. What are relictual features? Derived features? Which organisms seem to have a larger percentage of relictual features—prokaryotes or eukaryotes? Algae or flowering plants? Amoebas or humans?

6. Plant shoots and roots grow longer only at their tips; the older portions may become wider but not longer. This is called localized growth. Are there any alternatives to localized growth? Do animals grow by localized growth? Do fruits like apples or oranges grow by localized growth?

7. What are the consequences of localized growth? Could a root grow through the soil if all parts of it had to become longer? Could ivy grow up a tree or a building if all parts of the stem and branches had to become longer?

BotanyLinks .net Questions

www.jbpub.com/botanylinks

Visit the .net Questions area of BotanyLinks (http://www.jbpub.com/botanylinks/netquestions) to complete these questions:

1. To study plants and their interactions with the environment, natural areas must be preserved. Where are the natural areas still available? Go to the BotanyLinks home page to complete this exercise.

2. There are many important issues affecting our air, food, water, and environment, but getting information can be difficult. Go to the BotanyLinks home page to find out how to make your opinion matter.

BotanyLinks includes a Directory of Organizations for this chapter.

2

INTRODUCTION TO THE PRINCIPLES OF CHEMISTRY

Chemicals present in soil, air, and water are important for plant life. Fungi such as mushrooms break down dead plant matter, releasing chemicals that can then be used by other plants.

CONCEPTS

Plant metabolism, like that of all other organisms, is based on the fundamental principles of physics and chemistry that govern inanimate matter. All living creatures grow, respond to stimuli, and reproduce; but these processes, however complex, do not rely on any mystical or metaphysical "vital force." The physical and chemical reactions that occur in living cells can be modeled by hypotheses and verified by experimentation and observation. The bodies of plants and other organisms are made of atoms drawn from soil, air, or water, and the energy that drives their metabolism is produced by ordinary chemical reactions.

It is easy to assume that living organisms must possess special properties because they seem too elaborate to be the result of ordinary chemical and physical processes. This was the predominant view until the 1800s, when three major discoveries were made. It was found that biological compounds could be synthesized in the laboratory using inorganic chemicals and ordinary chemical processes. Next, enzymes were extracted from yeast cells, and some steps of fermentation were carried out in vitro (Table 2.1) without the presence of living cells. Finally, Louis Pasteur proved definitively that spontaneous generation does not occur and that there is no such thing as "vital force." These advances, along with the discovery of evolution by natural selection, revolutionized the study of metabolism and showed that analytical techniques and laws of physics, chemistry, and mathematics are sufficient to understand metabolism.

TABLE 2.1 *In Vitro and in Vivo*

in vitro: in glass. This refers to studies performed in test tubes, flasks, Petri dishes, and similar containers in which some aspect of metabolism is manipulated in laboratory conditions and **the metabolic system has been removed from the organism.** Photosynthesis might be studied in vitro by breaking cells open, extracting their pigments and enzymes, and examining how they work by supplying or withholding substances such as carbon dioxide and oxygen or by changing the acidity, temperature, or light.

in vivo: in life. This refers to studies carried out with **intact cells or whole organisms,** whether in natural settings or in a laboratory. Photosynthesis might be studied in vivo by growing whole plants in bright or dim light, low or high levels of carbon dioxide, and other pairs of variables.

ATOMS AND MOLECULES

CHEMICAL BONDS

There are 92 natural elements, each differing from the others by the number of protons in the nuclei of its atoms (Table 2.2). The lightest, hydrogen (H), has only one; the heaviest, uranium (U), has 92. Atomic nuclei also contain neutrons in numbers roughly equal to

TABLE 2.2 *Essential Elements*

Element	Symbol	Atomic Number	Atomic Weight	Amount Needed in Tissues Relative to Molybdenum
Hydrogen	H	1	1	60,000,000
Boron	B	5	10.8	2,000
Carbon	C	6	12	35,000,000
Nitrogen	N	7	14	1,000,000
Oxygen	O	8	16	30,000,000
(Sodium	Na	11	23)	
Magnesium	Mg	12	24.3	80,000
Phosphorus	P	15	31	60,000
Sulfur	S	16	32	30,000
Chlorine	Cl	17	35.4	3,000
Potassium	K	19	39.1	250,000
Calcium	Ca	20	40.1	125,000
Manganese	Mn	25	54.9	1,000
Iron	Fe	26	55.8	2,000
Cobalt	Co	27	58.9	N/A
Copper	Cu	29	63.5	100
Zinc	Zn	30	65.4	300
Molybdenum	Mo	42	95.9	1

These elements are essential for plant life; each carries out one or more vital roles. If any one is missing, a plant cannot survive. Sodium is essential for animals but not for plants. Atomic number corresponds to the number of protons in the atomic nucleus of each; atomic weight is the number of protons plus neutrons in each nucleus.

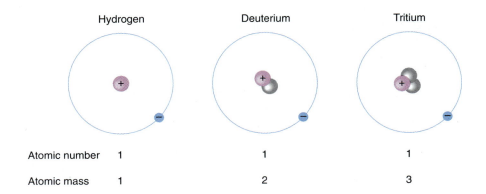

	Hydrogen	Deuterium	Tritium
Atomic number	1	1	1
Atomic mass	1	2	3

FIGURE 2.1
The three isotopes of hydrogen differ in the number of neutrons they contain. Because they all have the same number of protons, they have the same atomic number; they are the same chemical element and have identical properties. Neutrons only cause atoms to move more slowly.

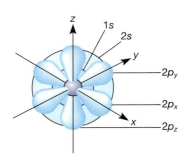

FIGURE 2.2
The lowest energy level for electrons (1s) contains only one orbital; it is drawn here as a sphere centered on the nucleus. The electron may be found anywhere on the sphere. Any orbital can contain either one or two electrons, but no more; hydrogen contains one; helium, with two protons and two electrons, contains two. In all elements with more than two electrons, the first orbital is filled with two electrons and the remaining electrons are placed in other orbitals at higher energy levels. The next level beyond that of hydrogen and helium contains four orbitals: one spherical (2s) and three that are dumbbell shaped (2p). In neon (atomic number 10), all four orbitals in the second energy level are filled with two electrons each, as is the orbital of the first energy level. This is an extremely stable configuration.

protons (Fig. 2.1); neutrons affect only the weight of the atom, not its chemical properties. Around each nucleus are electrons with a negative charge, so each neutralizes the positive charge of a proton. Because of their opposite charges, protons attract electrons, and atoms with equal numbers of protons and electrons are electrically neutral.

Electrical attraction is not the only important factor in determining the number of electrons associated with a nucleus. Electrons fit only into specific orbitals and energy levels around the nucleus, and some arrangements are more stable than others (Fig. 2.2). Helium, neon, and argon, for example, have their outermost energy levels exactly filled with electrons; this is the most stable arrangement possible and also results in equal numbers of electrons and protons, so the atoms are electrically neutral. If one of these atoms loses an electron, it has both an electrical charge (+1) and an unfilled energy level, which is an unstable arrangement. If an atom gains an electron, it again has an unbalanced charge (−1), and the electron is located alone in an orbital of a higher energy level, which is also unstable. Helium, neon, and argon have virtually no tendency to gain or lose electrons or to react with anything; they are called noble gases.

Chlorine, however, even when it is electrically neutral with 17 electrons matching its 17 protons, is not stable. Its outermost energy level (the third level) has its four orbitals full except for one, which contains a single unpaired electron (Table 2.3). Each atom has an extremely strong tendency to absorb one more electron and complete this energy level; this gives the atom a negative charge (Cl^-), which tends to make it slightly unstable, but the new arrangement of 18 electrons is so much more stable that it more than compensates. In plants and animals, chlorine almost always has an extra electron, making it the chloride ion, Cl^-. An atom or molecule that carries a charge is an ion; a negative ion is an *anion*.

Sodium is just the opposite (Table 2.3); when electrically neutral, it has one electron in an orbital by itself. Although losing the electron causes the atom to become the sodium ion Na^+, the remaining electrons are in such stable orbitals with energy levels 1 and 2 exactly filled that this arrangement is favored. A positive ion is a *cation*.

This tendency of electrons to move to the most stable possible configuration, which has been named **electronegativity,** is the driving force behind chemical reactions. The element with the greatest affinity for electrons, fluorine, has an electronegativity of 4.0 (Table 2.3); the noble gases, with no affinity for extra electrons, have electronegativities of 0.0. Only electrons in the highest, partially filled energy level, called **valence electrons,** are involved; they are responsible for forming chemical bonds.

Although an atom such as sodium may have a strong tendency to lose an electron, electrons virtually never fly off into space; instead, they move from one atom to another during a chemical reaction. Imagine a reaction between an atom with low electronegativity, such as sodium, and one with high electronegativity, such as chlorine. Both elements become more stable by the transfer of a valence electron from sodium to chlorine (Fig. 2.3). It is important to emphasize that **by more stable we mean that the atoms have less energy.** This is always the case: A particle is more stable when it has less energy.

When sodium reacts with chlorine, both partners have less energy after the reaction, so energy is liberated to the surroundings. Such an energy-releasing reaction is **exergonic;** if

TABLE 2.3 *Energy Levels and Electronegativity*

Element	Number of Electrons	Number of Electrons for Maximum Stability	Electronegativity
Hydrogen	1	2	2.1
He*	2	2	0.0
Lithium	3	2	1.0
Carbon	6	2 or 10	2.5
Nitrogen	7	2 or 10	3.0
Oxygen	8	10	3.5
Fluorine	9	10	4.0
Neon*	10	10	0.0
Sodium	11	10	1.0
Magnesium	12	10	1.2
Chlorine	17	18	3.0
Argon*	18	18	0.0
Potassium	19	18	0.9
Calcium	20	18	1.0

The first energy level can contain only two electrons, the second level can hold eight, and the third level can hold eight. The noble gases (*) are those elements whose energy levels are exactly full: helium (energy level 1), neon (energy levels 1 and 2), argon (energy levels 1, 2, and 3). Elements that have almost the same number of electrons as the noble gases can become more stable by giving up or taking on electrons so that their energy levels are also exactly full.

energy is released as heat, the reaction is also **exothermic.** Another exergonic reaction is the burning of hydrogen with oxygen; two hydrogen atoms transfer one electron each to an oxygen atom, resulting in water:

$$2H + \tfrac{1}{2}O_2 \longrightarrow H_2O + \text{energy}$$

This reaction can be forced to run backward by adding electrical energy, as in the electrolysis of water to form hydrogen gas and oxygen gas. Because the two products of the reverse reaction have more energy than water (they absorb it from the electrical apparatus), the reverse reaction is **endergonic** (**endothermic** if energy is absorbed in the form of heat). Any endergonic reaction needs a source of energy—an exergonic reaction occurring somewhere. For water electrolysis, the exergonic reaction is the burning of coal at the electricity-generating station. Another endergonic reaction is the formation of carbohydrates in leaves during photosynthesis; the exergonic reactions for this are the thermonuclear reactions in the sun—even though it is 93 million miles away.

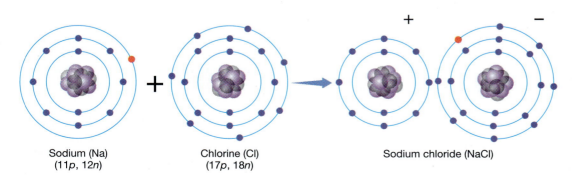

Sodium (Na)
(11*p*, 12*n*)

Chlorine (Cl)
(17*p*, 18*n*)

Sodium chloride (NaCl)

FIGURE 2.3
Transferring its single unpaired valence electron gives sodium a stable arrangement (2 + 8) and gives chlorine the electron it needs to have all its energy levels full (2 + 8 + 8).

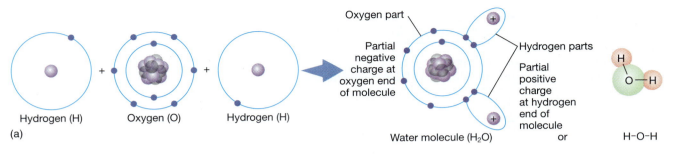

Oxygen part

Partial
negative
charge at
oxygen end
of molecule

Hydrogen parts

Partial
positive
charge
at hydrogen
end of
molecule

Hydrogen (H)   Oxygen (O)   Hydrogen (H)   Water molecule (H_2O)   or   H–O–H

(a)

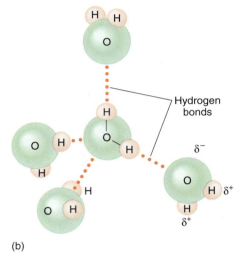

(b)

Hydrogen
bonds

δ^-

δ^+

δ^+

δ *is the Greek letter delta.*

FIGURE 2.4
(a) Oxygen's affinity for electrons is not great enough to pull valence electrons completely away from hydrogen; instead, electrons are shared but spend more time near the oxygen than near the hydrogen. No ions are formed. (b) In a mass of water, all molecules are held weakly to their neighbors by hydrogen bonds. Each bond has little stabilizing effect, but because every molecule has hydrogen bonds, the cumulative effect is significant.

In the sodium/chlorine reaction, a valence electron is actually transferred from one atom to another, converting each into an ion; in a molecule of salt, NaCl, the ions are held together by **ionic bonds**. However, in the hydrogen/oxygen reaction, the transfer is not complete; instead the valence electrons are shared between the nuclei (Fig. 2.4a). Electron sharing has an extra benefit: All electrons are in the most stable orbitals and no build-up of electrical charge occurs. Instead of two H^+ and one O_2^-, the result is just H_2O. A bond in which electrons are shared is described as a **covalent bond**. Whether an ionic bond or a covalent bond forms depends to a large degree on the difference in electronegativities of the two reactants.

WATER

Electrons are often not shared equally; in water, electrons spend more time near the oxygen, which has an electronegativity of 3.5, and less time near the hydrogen, which has an electronegativity of 2.1. Therefore, the oxygen in water has a **partial negative charge** ($\delta-$) and each hydrogen has a **partial positive charge** ($\delta+$). The orbitals are oriented so that both hydrogens and their partial positive charges are on one side of the molecule. The molecule is thus **polar**, with a slightly negative end and a slightly positive end.

Hydrogen Bonding.   When two water molecules come close to each other, the positive charge of one slightly attracts the negative end of the other. This is called **hydrogen bonding**. It is not nearly enough force to pull electrons from one to the other or even enough to cause the two molecules to adhere firmly to each other, but it is sufficient to cause them to adhere slightly (Fig. 2.4b). As a result, water is a rather sticky, viscous substance that can absorb a great deal of energy without warming rapidly and requires a large amount of energy to convert to vapor. Without the stickiness caused by hydrogen bonding, water molecules in plants could not be lifted from the roots to the leaves.

Substances that carry no unbalanced electrical charge, not even a partial one, are **nonpolar** substances, and they do not undergo hydrogen bonding. Consequently, nonpolar substances move easily past each other and flow with little viscosity. When energy is supplied, nothing holds the molecules in place and slows them, so their speed quickly increases, raising their temperature. They also boil and turn to a gas even at a low temperature, with little energy needed. Examples are methane (see Fig. 2.6) and acetylene (see Fig. 2.9).

Water Solubility and Lipid Solubility.   Many other molecules composed of hydrogen bonded to oxygen also have partial charges and undergo hydrogen bonding. When they are placed in water, water molecules surround them and form hydrogen bonds with them (Fig. 2.5a). This permits individual molecules to move into the water: The substance dissolves

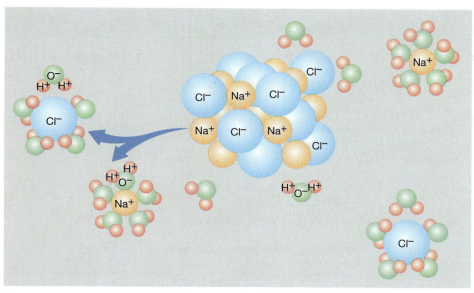

(a)

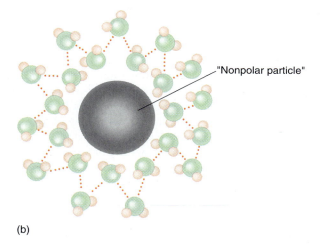

"Nonpolar particle"

(b)

FIGURE 2.5
(a) When a polar material like ordinary table salt, NaCl, is placed in water, the negative ends of water molecules are attracted to the sodium ions and the positive ends are attracted to the chloride ions, forming weak bonds with them. The new bonding is about the same strength as the water-water hydrogen-bonding, so the NaCl molecule can break apart easily. (b) If water molecules were occupying the volume of the nonpolar particle, many (at least 11 in this diagram) new hydrogen-bonds could form, release energy, and stabilize the water.

and is **water soluble.** When a nonpolar substance, which cannot form hydrogen bonds, is placed in water, no interaction occurs between the two types of molecules and the substance does not dissolve. In fact, if a molecule of the substance diffuses into the water, it disrupts the water's own hydrogen bonds—this would be an energy-consuming, endergonic reaction that destabilizes the water (Fig. 2.5b). If the molecule diffuses out of the water, the water molecules can form new hydrogen bonds and release energy, becoming more stable. If the substance is placed in water and agitated violently, it may mix temporarily, but it gradually separates; oil in water is an example of this.

Acids and Bases. When hydrogen combines with oxygen and forms water, the new bonding orbitals have so much less energy than the nonbonding orbitals that they are

Water-soluble substances are hydrophilic (water-loving); those that do not dissolve in water are hydrophobic (water-fearing), also called lipid (fat)-soluble.

TABLE 2.4 *Acids and Bases*		H⁺ Concentration	pH
		H⁺ Concentration	**pH**
Strong acid	Most molecules dissociate to an anion (−) and H⁺. Hydrochloric acid, nitric acid, sulfuric acid	High: $1/10^3$	3
Weak acid	Few molecules dissociate to an anion and H⁺. Acetic acid, citric acid, malic acid	Moderate: $1/10^5$	5
Weak base	Few molecules dissociate to a cation (+) and OH⁻; these combine with H⁺ from water, lowering H⁺ concentration. Asparagine, glutamine, urea	Low: $1/10^9$	9
Strong base	Most molecules dissociate to a cation and OH⁻. Ammonium hydroxide, potassium hydroxide, sodium hydroxide.	Very low: $1/10^{11}$	11

Strong acids and bases are those in which almost all molecules break down and liberate either a proton or a hydroxyl when dissolved in water. In weak acids and bases, many molecules do not break down.

extraordinarily stable. For hydrogen to break away from water requires the input of a large amount of energy, enough to raise electrons out of the low-energy, stable bonding orbitals. In pure water this is so rare that only about one molecule in 10 million breaks down into H⁺ and OH⁻, a proton and a hydroxyl ion, and when they encounter each other, they recombine and form water immediately.

The concentration of H⁺ is known as the **acidity** of a solution; it is measured as pH, which is the negative logarithm of the H⁺ concentration (Table 2.4). In many substances, hydrogen ions are held less tightly than in water molecules, and they give off protons rather easily; any substance that increases the concentration of free protons is an **acid.** Hydrochloric acid, HCl, dissolves in water to give H⁺ and Cl⁻; the extra H⁺ donated to the solution means that HCl is an acid.

A **base** is anything that decreases the concentration of free protons; this is usually accomplished by giving off hydroxyl ions that combine with protons and form water, effectively removing free protons. NaOH breaks down into Na⁺ and OH⁻; the OH⁻ indicates that it is a base.

Acids and bases are important because they contribute protons and hydroxyl ions, which can move onto and off of other compounds present in a solution. Because they carry a + or − charge, when they attach to or detach from a molecule, they affect the charge on the molecule. If a nonpolar, water-insoluble molecule picks up a proton because an acid is present, the nonpolar molecule becomes positively charged and water soluble.

CARBON COMPOUNDS

The concept of life is almost synonymous with the chemistry of carbon. With six protons in its nucleus and an electronegativity of 2.5 (see Table 2.3), carbon has properties that are both unique and essential for life. Carbon can easily and stably exist as a neutral atom with six electrons, or it can form covalent bonds by sharing the valence electrons (usually four) of other atoms. For instance, in methane (CH_4) one carbon atom shares one electron with each of four hydrogen atoms (Fig. 2.6), and in carbon dioxide (CO_2) it shares two electrons with each of two oxygen atoms. In fatty acids, most carbon atoms share two electrons with each of two hydrogens and each of two more carbons (see Fig. 2.21). Because carbon reacts so readily with more carbon, it can form long chains and complex ring structures.

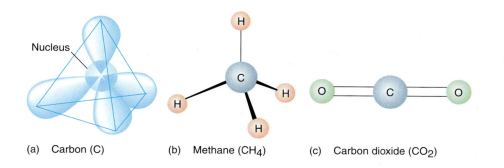

(a)   Carbon (C)   (b)   Methane (CH₄)   (c)   Carbon dioxide (CO₂)

FIGURE 2.6
When an atom shares electrons with another atom, the bonding orbitals have shapes and orientations different from those of isolated atoms, as shown in Figure 2.2. (a) and (b) If carbon bonds to four other atoms, the bonding orbitals point to the four corners of a tetrahedron, which separates each orbital as much as possible from the others. (c) If carbon bonds to only two other atoms, two double-bond orbitals point exactly away from each other.

Carbon can form three types of covalent bonds, depending on how many other atoms it shares electrons with. If there are four other atoms (methane, fatty acids), the carbon is linked by **single bonds,** each bonding orbital containing one electron from the carbon and one from the other atom. These bonding orbitals are arranged in a tetrahedron (Fig. 2.6). In a chain of carbons, the carbon backbone is zigzag, not straight (Fig. 2.7).

Carbon can form a **double bond** by sharing two of its electrons with one other atom that also contributes two electrons. The other two valence electrons of carbon may be in a second double bond or in two single bonds. If two sets of double bonds are present, as in carbon dioxide, the two sets extend in opposite directions, producing a straight molecule (Fig. 2.6). If one double bond and two single bonds are present, the molecule is flat and shaped like a **Y** (see Fig. 2.8). The double bond is extremely rigid and the arms cannot rotate around the carbon. Many organic molecules have carbon-carbon double bonds, and because the double bond cannot rotate, there are two possible forms of

$$-CX=CX-$$

With both Xs on the same side, they are in the *cis* **position** and when on opposite sides in the *trans* **position.** The *cis* form has physical and metabolic properties that differ from those of the *trans* form.

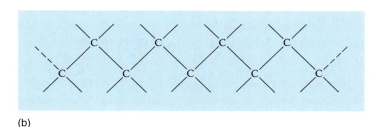

(a)

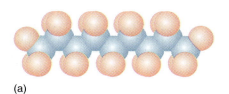

(b)

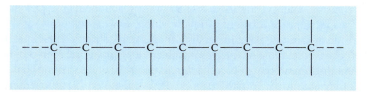

(c)

FIGURE 2.7
(a) A series of carbon atoms bonded to other carbons by single bonds would look something like this, with the blue spheres representing carbon atoms and the red spheres representing other atoms attached to the carbons. Because the bonds are arranged as a tetrahedron, the carbon backbone has a zigzag shape. (b) and (c) are simpler ways of showing the structure in (a).

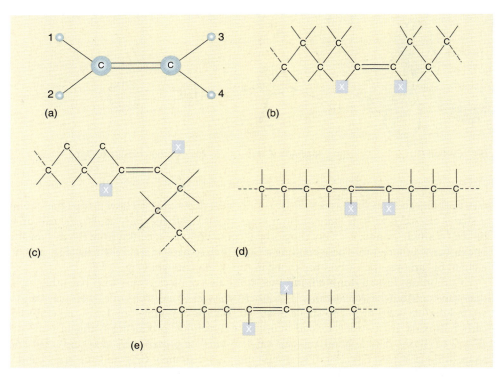

FIGURE 2.8
(a) The carbon-carbon double-bond system is flat: All atoms in this structure are in the same plane. Furthermore, the double bond cannot rotate, so atom 1 cannot change position with atom 2; if the carbon-carbon bond were a single bond instead, such a rotation could occur thousands of times every second. (b) to (e) are simplified ways of presenting carbon chains with one double bond. In (b) and (d), the Xs are in the *cis* position. In (c) and (e) they are in the *trans* position.

FIGURE 2.9
The carbon-carbon triple bond is extremely rare; the extra valence electrons are shown here forming single bonds with hydrogen. This molecule is acetylene.

Rarely, carbon forms a **triple bond** by sharing three of its valence electrons with one other atom (Fig. 2.9). The triple bond is not very stable (its electrons still have a great deal of energy), and triple bonds tend to be broken easily.

MECHANISMS OF CHEMICAL REACTIONS

SECOND-ORDER REACTIONS

One of the easiest chemical reactions to understand is a **second-order reaction**, in which two molecules react to form a third: $A + B \rightarrow AB$. In order for A and B to react, they must be so close that their valence electrons can move between the two sets of orbitals. But as A approaches B, the two sets of electrons repel each other because each set is a negatively charged cloud. A and B cannot simply drift together; they must collide violently enough that electron-cloud repulsion is overcome. The motion of molecules can be speeded by heating them; the faster an atom or molecule moves, the greater is its kinetic energy.

The energy (speed) needed to overcome electron-cloud repulsion and permit chemical reaction is called the **activation energy.** Figure 2.10a shows a reaction diagram of the potential energy of the reactants and products during a reaction. On the left, as A and B approach, they have a certain amount of potential energy that is related not to their speed but to the stability of their electron clouds. As they approach and begin to repel each other, they slow down, and part of the kinetic energy is converted to potential energy. If they have enough kinetic energy (if they are moving rapidly enough), they push very close together, their potential energy is very high, and the electrons rearrange into more stable bonding

PLANTS & PEOPLE

VITAMINS: PLANTS AND HUMAN HEALTH

The many types of organic compounds described in this chapter actually constitute only a small fraction of the large number of diverse organic compounds that plants produce. In addition, organic compounds act as flower pigments, aromas, plant hormones, and defensive compounds that are either irritating or distinctly poisonous. A remarkable aspect of plant metabolism is that during photosynthesis, plants use CO_2 and water to make a single simple organic compound, 3-phosphoglyceraldehyde. Then by altering it chemically with various enzymes, the plants use 3-phosphoglyceraldehyde to construct numerous types of carbohydrates and lipids. By adding simple compounds of nitrogen, phosphorus, sulfur, and so on, all derived from the soil, the plants use the 3-phosphoglyceraldehyde and its derivatives to make amino acids and proteins, nucleotides and nucleic acids, and every other compound in themselves. Actually, to build its entire body and carry out its metabolism, a green plant needs only sunlight, carbon dioxide, water, and a few mineral elements.

In contrast to a plant's well-balanced diet (light, air, water, soil), our well-balanced diet is much more complex, requiring vitamins, essential amino acids, essential fatty acids, and a variety of other compounds. By analyzing the underlying causes of this difference, we can begin to understand certain critical concepts in biology. First, animals cannot photosynthesize because they lack the necessary pigments and enzymes; instead, animals eat plants or other animals as food and then use much of the food's carbohydrate, fat, and protein components for energy production: They respire—oxidize—it and generate energy carriers such as ATP and NADPH $+ H^+$. After animals eat the carbohydrates, fats, and proteins, they digest them to their constituent monomers—the simple sugars, fatty acids, amino acids, and so on. Although it is possible for animals to break these down even further, that would be wasteful. It would require energy and special metabolism to break down the monomers, which would then have to be rebuilt. All plants and animals use the same amino acids, the same nucleotides, and mostly the same fatty acids and sugars. Similarly, as an animal eats a plant, it obtains all the vitamins the plant has produced and uses them in its own metabolism. A mutation in an animal that prevents it from snythesizing a particular amino acid or nucleotide is not deleterious to the animal because it probably gets all that it needs from its food. In fact, it may be slightly advantageous selectively: That animal is not using up resources to make something it gets in its diet, so those extra resources can be used for other needs such as reproduction, defense, growth, and so on.

Over many generations, we would expect natural selection to result in animals with mutations that eliminate synthetic pathways that produce materials the animal already receives reliably in its diet. Consider humans: We have lost the ability to synthesize our own vitamins, several amino acids, and several fatty acids. Even a rather poor diet contains enough of these for humans to lead reasonably healthy lives. In fact, the existence of vitamins was discovered only when long sea voyages required sailors to eat a diet of just dry bread and beef, without any fruits or vegetables (which rotted too quickly). The lack of vitamin C (ascorbic acid) resulted in the deficiency disease scurvy, with symptoms such as bleeding gums, weight loss, and swollen joints. Captain James Cook was one of the first to notice that the disease was cured shortly after the sailors reached land and obtained a diet that included vegetables. He experimented with sauerkraut (the only way to preserve vitamin-rich cabbage) and discovered that it prevented the disease. Later studies showed that citrus fruits were even more effective, and the science of experimental nutrition was initiated. More recently, pellagra was a common deficiency disease in the southern United States. It is caused by a lack of niacin in the diet, and it affected mostly southern blacks who were so poor that they relied almost exclusively on cornmeal for the bulk of their diet; corn is remarkably low in niacin.

The rarity of deficiency diseases shows that a diet must be quite poor to cause noticeable harm to an animal. But do animals need compounds that plants do not make? Yes, with an important example being cholesterol. This is a lipid that plants do not synthesize but that is important to animals. Consequently, all herbivorous animals must be able to make cholesterol, whereas carnivorous animals can obtain it in their food.

The rates at which different organisms have lost the ability to synthesize organic compounds vary greatly. We humans still have the capacity to make almost all the carbohydrates and fats we need, as well as most of the amino acids, but other animals have lost many of their synthetic pathways. Good examples are tapeworms and similar parasites. In their proper habitat (the digestive tract of a mammal), tapeworms receive almost all the nutrients they need. They can synthesize only a few compounds themselves, and almost all their metabolic resources are directed toward growth and reproduction, which occur rapidly. On the other hand, this extreme reliance on the diet means that tapeworms can survive in only a few places.

In summary, as an organism becomes more dependent on its diet, it has a greater potential for growth and reproduction when the environment provides the correct diet but a greater number of deficiency diseases when the environment provides a poor diet.

orbitals, forming either ionic or covalent bonds. If this is an exergonic (energy-liberating) reaction, the valence electrons are more stable (have less energy) in the new bonding orbitals than they did in the old nonbonding orbitals. The liberation of energy lowers the potential energy level. The two electron clouds do not repel each other because of the rearrangement of the valence electrons into bonding orbitals. If *A* and *B* had not had

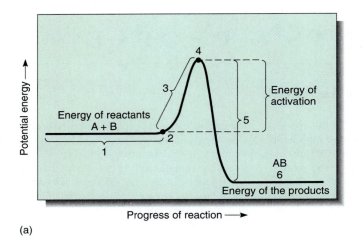

(a)

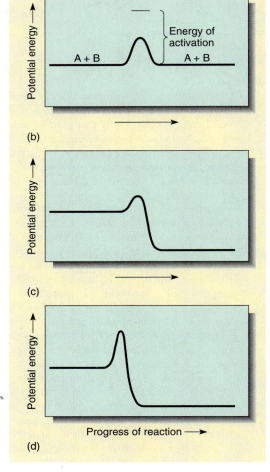

(b)

(c)

(d)

FIGURE 2.10
(a) A potential energy diagram of a second-order reaction. At 1, the two atoms, A and B, are still too far apart to influence each other, but at 2 they begin to repel. During interval 3, they continue to approach because their momentum is great enough to overcome repulsion. At 4, the orbitals rearrange from nonbonding to bonding orbitals, and energy is given up (5) and radiated away from the molecule. Consequently, after the reaction (6) the molecule AB has less potential energy than the two atoms had at 1. (b) If two atoms are moving too slowly, electron cloud repulsion is not overcome and the two atoms bounce off each other. No reaction occurs. (c) A low energy-of-activation barrier. (d) A high energy-of-activation barrier.

enough speed to overcome the activation-energy barrier, they would not have shared electrons, the stabilizing bonding orbitals would not have formed, and the electron clouds of A and B would have repelled each other, flying off in opposite directions (Fig. 2.10b).

The amount of electrical repulsion and therefore the energy of activation needed vary from reaction to reaction. If little repulsion occurs, even slowly moving molecules have enough kinetic energy and the reaction occurs at low temperature (Fig. 2.10c). If the energy-of-activation barrier is very great, the molecules must be moving very rapidly and the reaction occurs only if the substance is heated to a high temperature (Fig. 2.10d). It has nothing to do with how much energy can be liberated during the reaction. Wood can combine with oxygen to produce carbon dioxide and water; this is strongly exergonic and huge amounts of energy are liberated. However, the energy-of-activation barrier is so high that the wood must be heated to several hundred degrees before the reaction can occur.

CATALYSTS

Although there is no way to change the energy of activation of a reaction, it is possible to change the **mechanism** of a reaction. The reaction of A and B to form AB may be difficult because of a high energy of activation, but an alternative set of reactions may be possible, such as

$$A + D \longrightarrow AD$$
$$AD + B \longrightarrow AB + D$$

in which the energies of activation for $A + D$ and for $AD + B$ are both low (Fig. 2.11). If so, a mixture of $A + B + D$ results in the production of AB even at low temperature, whereas a mixture of just $A + B$ does not. A substance such as D that permits a reaction to occur rapidly at low temperature is a **catalyst**; in living organisms, virtually all catalysts are proteins called **enzymes**. Notice that catalysts, enzymes included, emerge from the reaction in the same condition that they entered it; the catalyst itself is not permanently altered by the reaction.

FIRST-ORDER REACTIONS

A **first-order reaction** involves only one molecule, not the collision of two. An example is $AB \rightarrow A + B$, in which a compound breaks down into two parts. First-order reactions also have energies of activation that are overcome as the two parts of the molecule vibrate. As the compound is heated, its component parts vibrate more rapidly, accelerating the reaction. First-order reactions can also be accelerated by catalysts.

REACTION EQUILIBRIA

For every reaction, an opposite back-reaction can also occur; thus, not only can A and B combine to form AB, but AB can break down to form A and B. In any mixture of A and B, both reactions occur. If $A + B \rightarrow AB$ is strongly exergonic and liberates a large amount of energy, then $AB \rightarrow A + B$ is strongly endergonic and requires a large input of energy. In such a set of reactions, almost all the material is ultimately converted to AB because AB is formed more rapidly than it breaks down and is more stable than A and B. If $A + B \rightarrow AB$ is only mildly exergonic, then $AB \rightarrow A + B$ is only mildly endergonic; at equilibrium, A, B, and AB are present in approximately equal quantities, and both reactions occur at about the same rate.

ORGANIC MOLECULES AND POLYMERIC CONSTRUCTION

FUNCTIONAL GROUPS

Millions of carbon compounds are possible, varying not only in the number of carbons they contain but also in the types and numbers of noncarbon atoms, types of bonds, and other factors. This great diversity of carbon compounds actually consists of a small number of families of compounds whose members have similar properties. This is because the properties of a compound are due mostly to the chemical groups, known as **functional groups,** attached to the carbon atoms (Table 2.5). Because carbon compounds can be large, each may have many functional groups of various types, being simultaneously both acids and alcohols, acidic and basic, or lipid soluble in some regions and water soluble in others.

POLYMERIC CONSTRUCTION

A **polymer** is a large compound composed of a number of more or less identical subunits **(monomers).** The simplest example is construction using bricks. The bricks (monomers) are virtually identical, but they can be used to construct many different things: houses, lecture rooms, sidewalks. Simple sugars like glucose are monomers that can be polymerized into starch, cellulose, mucilage, and many other polymers.

Polymeric construction is essential to life for several reasons. First, it reduces the difficulty of construction; to build a brick building, it is only necessary to know how to make bricks and how to assemble them. If a different type of building is needed, only the assembly information needs to be changed, not the mechanism for making bricks. With sugars, for the plant to make starch, it bonds glucose together with a certain type of bond; to make cellulose, it uses a different type of bond. But the mechanism for making glucose does not change.

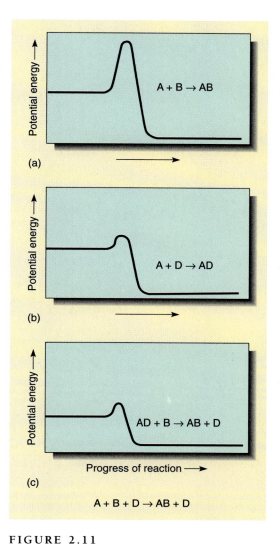

FIGURE 2.11

(a) The activation-energy barrier of this reaction cannot be changed, but a catalyst, D, may exist which provides an alternative reaction mechanism that has low activation-energy barriers in all steps (b) and (c).

TABLE 2.5   *Functional Groups*

	Name	Characteristics
—H	Hydrogen	Low reactivity; lipid solubility
—OH	Hydroxyl	Alcohol group; hydrogen bondings; water solubility
—CH$_3$	Methyl	Low reactivity; lipid solubility
	Carboxyl	Highly reactive; acidic; in water it acts like an acid, giving off a proton: + H$^+$
—NH$_2$	Amino	Highly reactive; basic; in water it acts like a base by absorbing protons —NH$_3$$^+$
	Aldehyde	Moderately reactive; water solubility
	Ketone	Moderately reactive; similar to an aldehyde, but the oxygen is located on an internal carbon rather than a terminal one.
	Phosphate	Highly reactive; when transferred to a compound, that compound usually becomes much more reactive. In water, the hydrogen comes off as a proton, leaving its electron and giving this group a negative charge. It is highly soluble in water.
—SH	Sulfhydryl	Can stabilize the structure of proteins

Polymeric construction allows an organism to have a simple basic metabolism that produces only a few types of monomers. As the physiology changes, only the assembly of monomers changes, not the basic metabolism. For example, as a plant grows, it assembles amino acids into proteins necessary for leaves and stems. At the right time, it begins to assemble the same types of amino acids into different proteins—those necessary for flowers. Even more dramatic, algae have the same sugars and amino acids as flowering plants. During 400 million years of evolution, only the types of proteins formed have changed; the basic mechanism for producing the amino acid monomers is about the same.

Polymeric construction also permits recycling and conservation of resources; once a polymer is no longer needed, it can be depolymerized back to its monomers, which are then used in the construction of a new polymer. All the energy expended in their construction is conserved.

Finally, polymeric construction allows various parts of an organism to work together in construction. As a fertilized egg grows into an embryo in the developing seed, surrounding tissues supply it with sugars, amino acids, and fats. The embryo can quickly assemble these into organelles and grow rapidly. If the tissues supplied the fertilized egg only with carbon, oxygen, nitrogen, and other elements, growth would be much slower.

CARBOHYDRATES

Carbohydrates are defined by two criteria. First, they usually contain only carbon, hydrogen, and oxygen, although a few carbohydrates contain atoms such as nitrogen or sulfur. Second, the ratio of hydrogen to oxygen is close to $2:1$, the same ratio as in water; the generalized chemical formula for carbohydrates is $(CH_2O)_n$.

MONOSACCHARIDES

The simplest carbohydrates are the **monosaccharides,** or simple sugars, glucose being a familiar example. Monosaccharides are small molecules classified by the number of carbon atoms each contains: four- (4C; tetrose), five- (pentose), six- (hexose), seven-carbon sugars, and so on. The pentoses and hexoses are by far the most abundant and most important (Table 2.6; Fig. 2.12).

Various sugars in a class may have the same chemical formula but differ in their atomic arrangements; such molecules are called *isomers.* For example, glucose and fructose are isomers, both having the formula $C_6H_{12}O_6$ but having different chemical structures. Because the functional groups of glucose are arranged differently from those of fructose, the two molecules have distinct shapes and different chemical properties. The differences in chemistry are actually quite slight, but the differences in shape are extremely important. In order for an enzyme to catalyze a reaction, the substrate molecules must fit physically into a very precisely shaped active site. Glucose can enter the active site of certain enzymes, but fructose and other hexoses cannot. Enzymes easily distinguish between isomers by their unique shapes.

Monosaccharides are flexible because all their carbon-carbon bonds are single bonds, not flat, rigid double bonds. When dissolved in water, the molecules flex and rotate around each carbon atom, changing shape thousands of times every second. When one end of a molecule comes close enough to the other end, the two ends may react, forming a closed ring (Fig. 2.12). In glucose the number 1 carbon reacts with the —OH group on the

TABLE 2.6	Common Monosaccharides
4C	Erythrose
5C	Arabinose
	Deoxyribose
	Ribose
	Ribulose
	Xylulose
6C	Fructose
	Galactose
	Glucose
	Mannose
7C	Sedoheptulose

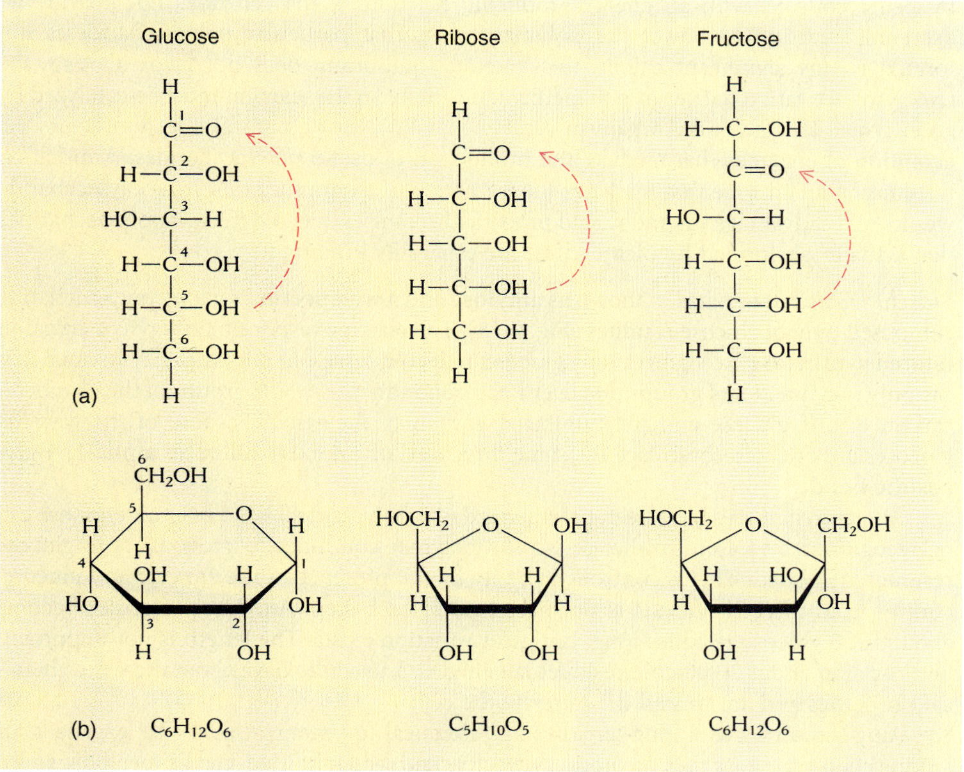

FIGURE 2.12
(a) Ribose is a 5C sugar, a pentose; glucose and fructose are both hexoses (6C). (b) Each sugar can also exist in ring form, but ring formation involves the aldehyde or ketone functional group, which thus does not exist in the ring-form sugars.

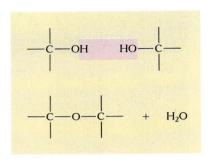

FIGURE 2.13
Because —OH is such a common functional group, many monomers combine by means of dehydration reactions.

number 5 carbon, forming a six-member ring that has oxygen as part of the ring and —CH_2OH as a side group. Ring formation releases energy (it is exergonic), and the ring form is the more stable, more common form for a hexose dissolved in the water of a cell.

Because of ring formation, monosaccharides tend to be rather unreactive, relatively inert molecules, which is ideal for physiological functions such as construction, transport, and energy storage. Monosaccharides can be used to build structures that must be inert, stable, and long lasting. They can be transported from region to region without causing damage by reacting with structures they encounter. Monosaccharides can store energy: They are synthesized in leaves from carbon dioxide and water in an extremely endergonic process—photosynthesis. Once formed, the energy-rich, stable simple sugars can be moved to sites where energy is needed, such as flowers or roots; then they are metabolized, releasing energy at the new location. Animals and fungi use the monosaccharide glucose for transport, but most plants use sucrose, a disaccharide composed of one glucose molecule plus one fructose molecule.

POLYSACCHARIDES

Monosaccharides can act as monomers, reacting with other monosaccharides to form polymers called **polysaccharides.** Extremely short polysaccharides, less then about ten monosaccharides long, are called **oligosaccharides** and are named by the number of sugars they contain: disaccharides such as sucrose, trisaccharides, and so on.

Theoretically, many types of oligosaccharides and polysaccharides are possible. Consider polysaccharides that consist of just glucose: These could be di-, tri-, tetrasaccharides, and so on, each longer and heavier than the previous one. The bond between two monosaccharides results from the interaction of an —OH group on each (Fig. 2.13), so there are numerous possible disaccharides of glucose, each based on the position of the linking bond. During bond formation an entire —OH is removed from one carbon, a hydrogen is removed from the other —OH group, and water is formed; this is a **dehydration reaction.** Reversal of this, breaking the bond by adding water back to it, is **hydrolysis.**

Despite the large number of polysaccharides that are theoretically possible, very few types actually exist. The dehydration reaction is endergonic and must be forced to proceed by being coupled with an energy-producing reaction. If the cell does not provide an exergonic reaction to power the polymerization, that particular polysaccharide is not formed to any significant extent; the reaction equilibrium does not favor it. Also, the energy-of-activation barrier of polymerization is high, so the reaction must be catalyzed by an enzyme. This gives the organism great metabolic control; natural selection favors the evolution of enzymes that mediate the production of useful polysaccharides. A mutation resulting in an enzyme that mediates the formation of a harmful or useless polysaccharide would be disadvantageous and would probably become extinct. Of the few polysaccharides that actually are formed by plants, three are especially important.

Starch.   Starch, technically known as **amylose** and **amylopectin,** is a long polysaccharide composed only of glucose residues (Fig. 2.14). The enzyme responsible for polymerization **(starch synthetase)** recognizes only glucose molecules; these fit onto the enzyme such that the only two functional groups that ever form bonds are the —OH groups of the number 4 carbon of one glucose and the number 1 carbon of the other. Because of the way the glucoses are oriented (both face the same direction), the bond is called an **alpha-1, 4-glycosidic bond.**

After two glucoses have been polymerized, the polymer still has a reactive end that can be recognized by starch synthetase, so the polymer continues to grow as more glucose residues are added. This is controlled simply by the presence of the enzyme and glucose; control from the cell nucleus is not necessary at each step. Amylose molecules become about 1000 glucose residues long, but great variation exists. The length is not important, because after about 20 glucose residues, all amylose molecules have about the same chemical properties and are treated the same by the cell.

Amylose serves as a long-term storage chemical for energy. Sugars are excellent for storing energy because they are not very reactive and, when needed, can be metabolized by

Because of the loss of water, the glucose molecules in the polysaccharide are not really complete glucoses, so the term "glucose residue" is used.

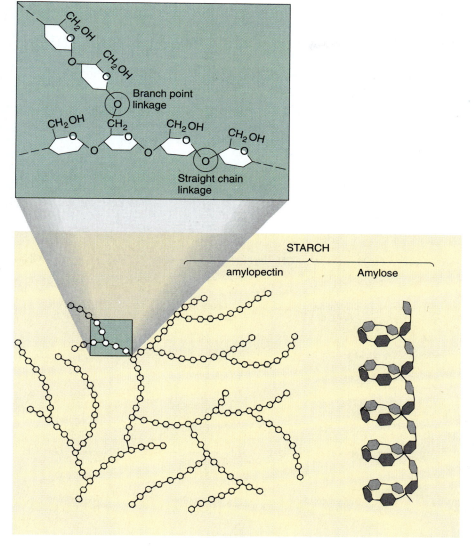

FIGURE 2.14

If glucoses are linked to other glucoses by alpha-1, 4-glycosidic bonds, the result is an unbranched, coiled chain called amylose, the main component of starch. A second enzyme occasionally makes a bond with the number 6 carbon, resulting in an alpha-1, 6-branched-chain amylopectin, also a component of starch.

the protoplasm and mitochondria, and their energy is liberated. However, cells cannot store large amounts of glucose because sugars have a strong tendency to absorb and hold water, causing cells to swell; starches do not do this.

Cells have another enzyme (alpha-1, 6-glucose-starch transglucosylase) that polymerizes glucose units, this time adding a glucose residue onto a number 6 carbon in amylose (Fig. 2.14). This creates a short branch that then grows as though it were a simple amylose molecule, with more glucose residues being added to its number 4 carbon. The second enzyme is not as active, so not every glucose in amylose is affected. Instead a branching occurs only at approximately every 20th glucose. This branched molecule is amylopectin, and its properties differ slightly from those of amylose. The ratio of amylose to amylopectin in starch grains varies from species to species: Potato starch is 78% amylopectin, but starch of wrinkled peas is almost entirely amylose.

Cellulose.   A set of enzymes called **cellulose synthases** polymerizes glucose molecules into the polymer **cellulose** (see Fig. 3.31). The glucose residues have an alternating orientation, and the resulting bond is a **beta-1, 4-glycosidic bond.** This is critically important to the nature of the polymer: Although both amylose and cellulose are unbranched chains of glucose joined at carbons 1 and 4, they are totally distinct chemically and biologically. Cellulose molecules can form large numbers of hydrogen bonds with other cellulose molecules, crystallizing into rigid aggregates that are extremely strong (see Fig. 3.31a). On the

For some enzymes, the spelling is "synthase"; for others it is "synthetase."

cell surface, cellulose molecules hydrogen bond to other polysaccharides and thus become cross-linked into a complex meshwork known as the cell wall (see Chapter 3 for details).

Cellulose is remarkably inert; few organisms have enzymes capable of hydrolyzing the beta-1, 4-glycosidic bond. Wood-rotting fungi and bacteria have cellulose-degrading enzymes, but even termites, cockroaches, and cattle can live on cellulose only because their digestive tracts contain microorganisms with the proper enzymes.

Oligosaccharides.   Many organisms, especially animals but also some plants, attach short chains of sugars onto proteins. These oligosaccharides are most often present on proteins near or at the cell surface and seem to be involved in cell recognition.

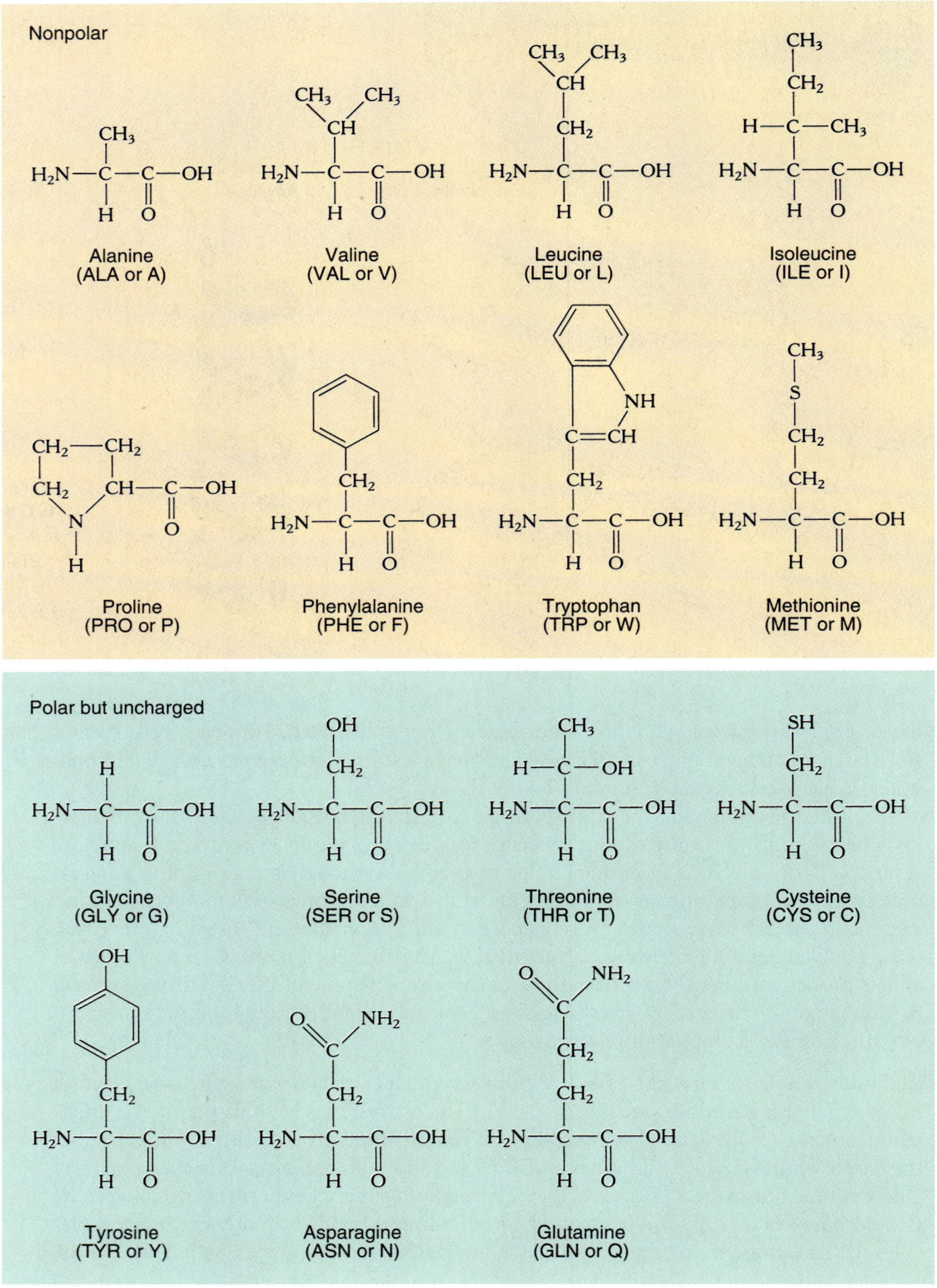

FIGURE 2.15
The amino acids that occur in proteins. Two systems of abbreviations are used for the names of amino acids—an older system that involves three letters for each, and a newer system that uses just one letter each.

Amino Sugars.   In fungi, some sugars contain nitrogen in addition to carbon, hydrogen, and oxygen. These **amino sugars** are polymerized into long chains that can be completely branched and cross-linked. These polymers, known as **chitin,** are part of fungal cell walls.

AMINO ACIDS AND PROTEINS

Proteins are unbranched polymers composed of **amino acids;** they tend to be about 100 to 200 amino acid residues long. Very short proteins, with fewer than about 50 amino acids, are frequently called polypeptides. Twenty amino acids are used for protein synthesis (Fig. 2.15); each consists of one carbon that carries four side groups: (1) —COOH, the carboxyl group that causes it to be an acid; (2) —NH$_2$, the amino group; (3) —H; and (4) a fourth group "R" that differs from one amino acid to another. The R groups are not involved in polymerization; instead, they protrude to the sides of the protein backbone and their properties determine the property of the protein.

The R groups cause amino acids to differ structurally, chemically, and biologically. Some R groups are acidic; others are basic. Nonpolar, hydrophobic R groups give an amino acid a tendency to interact with other nonpolar molecules such as fats and lipids. Polar, hydrophilic R groups cause their amino acids to interact with other polar molecules, avoiding nonpolar ones. Cysteine is unique in having a sulfhydryl in its R group; two sulfhydryls can interact, forming an —S—S— (disulfide) covalent link between two separate cysteines. If the two cysteines are in the same protein, the disulfide link holds the two regions together; if they are in separate proteins, the disulfide links the two proteins.

During protein synthesis, the carboxyl group of one amino acid reacts with the amino group of the next, water is removed, and a **peptide bond** is formed (Fig. 2.16). This sounds similar to the polymerization of monosaccharides and actually is similar if performed in a

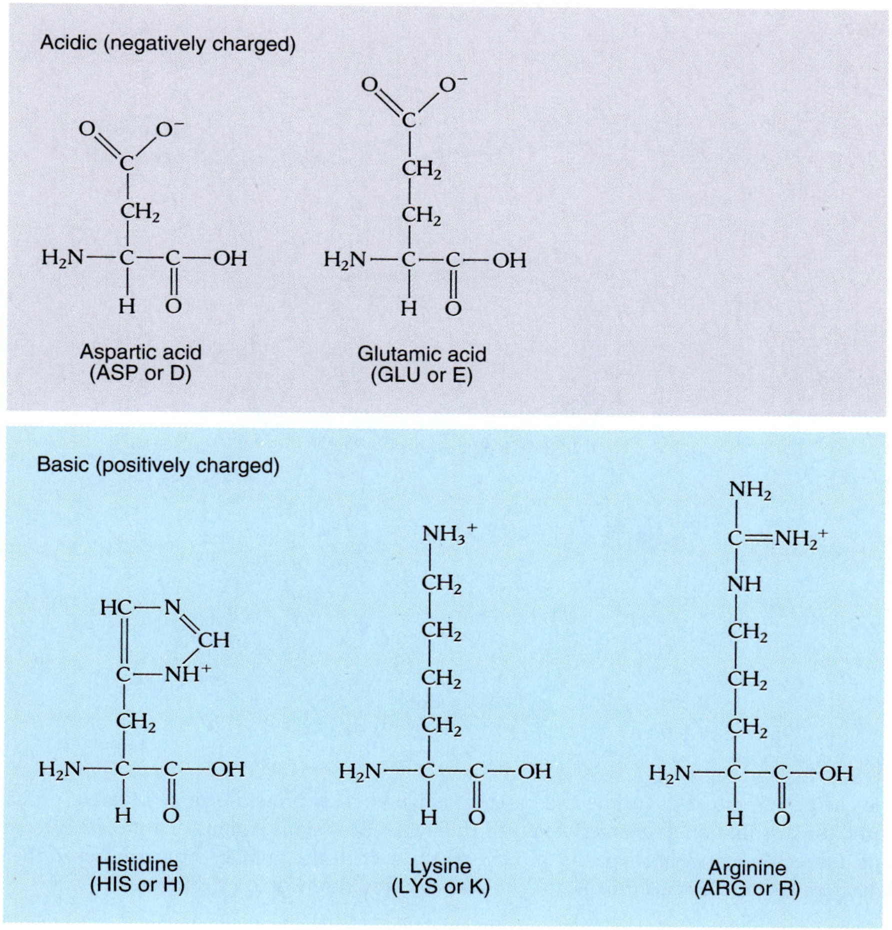

FIGURE 2.15 *(Continued)*

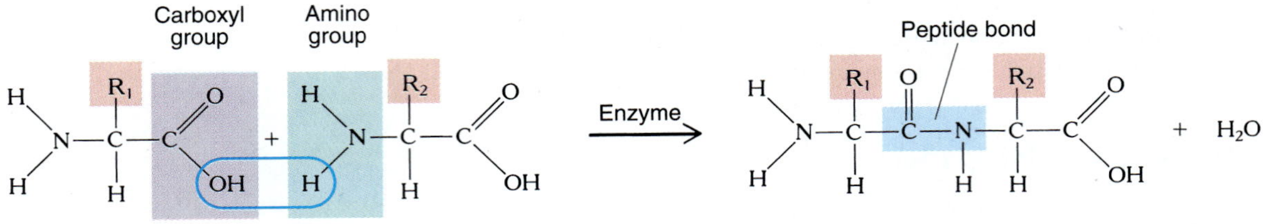

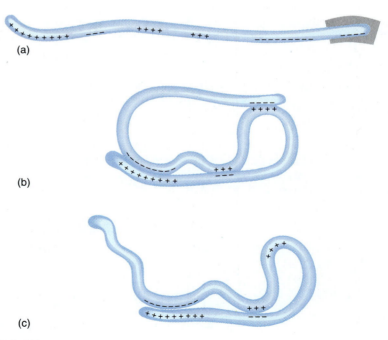

FIGURE 2.16

The peptide bond between two amino acid residues in a protein is formed by the sharing of electrons between a carboxyl carbon and an amino nitrogen. During bond formation, water is removed, so this is a dehydration reaction.

test tube. However, within a cell the two types of polymerization are completely different; the linking of amino acids is mediated by organelles called ribosomes and involves large numbers of enzymes and intermediates. This is necessary because proteins are not as simple as polysaccharides; instead, the sequence and types of amino acids incorporated must be controlled with great precision.

LEVELS OF ORGANIZATION IN PROTEIN STRUCTURE

If the primary structure—the sequence of amino acids—is changed, different regions interact and different shapes result.

Primary Structure.   The amino acid sequence is the protein's **primary structure.** Because each amino acid has a unique R group, the sequence of amino acids produces a particular sequence of R groups. If all R groups were similar chemically, like the side groups in monosaccharides, this sequence would not be very important. But the R groups in proteins are chemically diverse, so a protein can have extremely complex properties. Proteins are flexible (all bonds in the backbone are single bonds), and some regions interact with other regions of the same protein, causing the whole molecule to have a characteristic shape (Fig. 2.17). If the R groups of an entire region are of the correct type, the protein forms a helical structure known as an alpha helix. Alpha helices, which often affect only short regions of protein, are an example of **secondary structures.** Many proteins do not have secondary structure.

FIGURE 2.17

(a) The charged amino acids (acidic and basic) of a hypothetical protein. With a different primary structure, the distribution of charges would be different. (b) Various regions of a protein interact, forming a three-dimensional shape, its tertiary structure. (c) If the primary structure lacked the last four negative amino acids (box in (a)), the tertiary structure would be different.

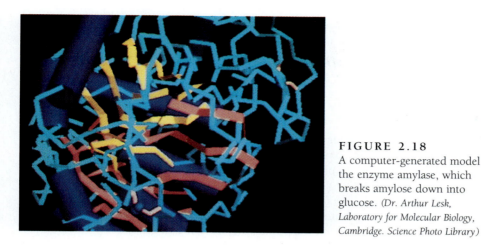

FIGURE 2.18
A computer-generated model of
the enzyme amylase, which
breaks amylose down into
glucose. (*Dr. Arthur Lesk,
Laboratory for Molecular Biology,
Cambridge. Science Photo Library*)

Tertiary Structure.   The physical shape of a protein in its functional mode is its **tertiary structure,** determined largely by primary structure: Positively charged regions attract and bind to negatively charged regions (Fig. 2.17), and hydrophobic R groups interact and form water-free pockets inside the folded protein. Cysteines may link two portions together with disulfide bonds. The protein's overall shape may be globular or fibrous, but the shape and nature of specific sites are usually much more important. Because most plant proteins are enzymes, the active sites must recognize substrates by shape, electrical charge, or hydrophobic properties (Fig. 2.18). For example, the tertiary structure of the enzyme starch synthetase must have a site into which glucose and a growing molecule of amylose can fit. Because glucose is water soluble, the R groups near this active site should be hydrophilic. Without the correct primary structure, the protein does not fold into the proper tertiary structure and a functional active site is not formed.

Tertiary structure is also affected by small molecules and ions that are not part of the protein; their concentration in the protoplasm affects the shape and therefore the activity of many enzymes. Magnesium (Mg^{2+}) and calcium (Ca^{2+}) can interact with negatively charged R groups and alter the tertiary structure; with high levels of magnesium or calcium, two negatively charged regions may become "glued" together, whereas without these two cations the negatively charged regions repel.

Tertiary structure is affected by pH and heat. Protons and hydroxyl ions released by acids and bases interact with charged R groups, changing the way various regions of the protein attract or repel each other. As proteins are heated, the atoms and molecules vibrate more rapidly and unfold. If heated enough, as in cooking an egg, the proteins unfold completely and become denatured. With enzymes, even dilute acids or bases or just mild heating causes enough of a change in shape to distort the active site, and the enzyme cannot function.

Quaternary Structure.   **Quaternary structure** refers to the interaction between two or more separate polypeptides. Quaternary structure is maintained by hydrogen bonding, the interaction of hydrophobic regions, or disulfide bridges. As with secondary structure, not all proteins have this level of organization, but for many it is critical for proper functioning. Many enzymes consist of two or four polypeptides that must be associated to work properly; only when all are aggregated are the active sites completely formed and functional. For example, an enzyme called RuBP carboxylase (discussed in Chapter 10) is a giant enzymatic complex consisting of eight small proteins and eight large ones; the complex is functional and carries out photosynthesis only when all 16 proteins associate with the proper quaternary structure.

A selective advantage of quaternary structure is that it allows **self-assembly** of certain structures; once the individual protein monomers are formed, they automatically associate into the proper structure such as a microtubule or an enzyme complex. No special constructing apparatus or metabolism is needed; however, a cell easily controls self-assembly by altering conditions that affect aggregation of the individual proteins. With a change in the concentration of calcium, magnesium, or protons, the charge of some R groups is altered and the protein takes on a new shape, either promoting or inhibiting self-assembly into the quaternary structure.

NUCLEIC ACIDS

Nucleic acids are polymers composed of monomers called **nucleotides,** but the nucleotides themselves are actually composed of three distinct subunits as well. Each nucleotide is formed by the bonding of (1) a phosphate group, (2) a five-carbon sugar, and (3) a complex ring molecule that contains nitrogen and acts like a base (Fig. 2.19). Nucleic acids contain only five nitrogenous bases that fall into two groups: **Pyrimidines** are molecules composed of a single ring, whereas **purines** consist of two rings (Fig. 2.20). The pyrimidines are cytosine, uracil, and thymine, and the purines are adenine and guanine.

Only two types of five-carbon sugar are ever joined to these nitrogenous bases: **ribose** and **deoxyribose,** which differs from ribose in having an —H group at one site where ribose has —OH. The enzymes involved in joining the sugars to the bases have such precisely shaped active sites that they accurately distinguish ribose from deoxyribose. Only four **ribo**nucleotides (ribose-containing nucleotides) occur because thymine is never attached to ribose. Similarly, uracil is never attached to deoxyribose, so only four **deoxyribo**nucleotides occur.

Details of nucleic acid polymerization are discussed in Chapters 15 and 16, but a few points can be mentioned here. Deoxyribonucleotides and ribonucleotides are never mixed together in the same polymer, so nucleic acid polymers are of just two fundamental types: **deoxyribonucleic acid** (the famous **DNA**) and **ribonucleic acid (RNA).** DNA is found in the nucleus as the main component of chromosomes as well as in plastids and mitochondria. As in proteins, the sequence of monomers is critically important, and any change of sequence can have serious effects. In fact, the **sequence of deoxyribonucleotides** is the genetic information stored in the nucleus. The sequence is used indirectly to guide the polymerization of amino acids into proteins. The primary structure of DNA thus determines the primary structure of proteins and therefore also their secondary, tertiary, and quaternary structures, as well as their function.

Polymers of ribonucleotides, RNA, serve several functions. Some RNA molecules, called messenger RNAs, carry copies of genetic information from the nuclear DNA to ribosomes, the sites of protein synthesis in the protoplasm. Ribosomes are large complexes of enzymes and a second type of RNA, ribosomal RNA. A third class of RNA, transfer RNA, carries amino acids to ribosomes.

LIPIDS

Lipids are fats and oily substances that are extremely hydrophobic and water insoluble. Like carbohydrates, they lack nitrogen and sulfur and consist mainly of carbon, hydrogen, and oxygen. Unlike carbohydrates, they have much more hydrogen than oxygen, often having only two oxygen atoms at one end of a long molecule.

The basic units of many lipids are **fatty acids.** These are long chains containing up to 26 carbon atoms with a carboxyl group at one end. If every carbon atom except the carboxyl carbon carries two hydrogens, the fatty acid is **saturated;** that is, it can hold no more hydrogen (Fig. 2.21; Table 2.7). All carbon-carbon bonds are single, and all parts of the molecule can rotate; in groups, the molecules tend to straighten and crystallize, being stabilized by interactions with closely packed adjacent fatty acids. Because of this tendency to crystallize, saturated fatty acids are solid at room temperature (Fig. 2.22).

If some of the carbons are double bonded to adjacent carbons, the fatty acid is **unsaturated;** the carbon-carbon double bonds are rigid and the molecule has a kink. With several double bonds (polyunsaturated), the molecules are irregular in shape and cannot align well; they have little tendency to crystallize and they remain liquid and oily rather than solid.

Animal fats, such as lard and butter, are usually saturated; plant fats tend to be unsaturated and oily.

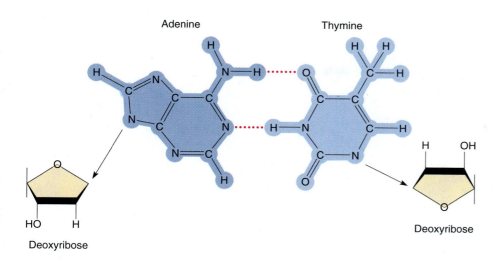

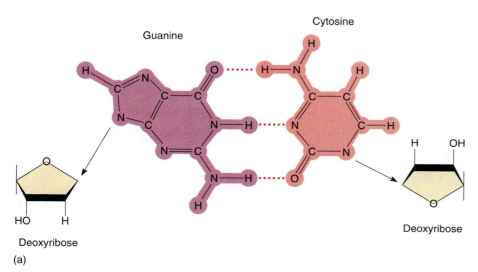

(a)

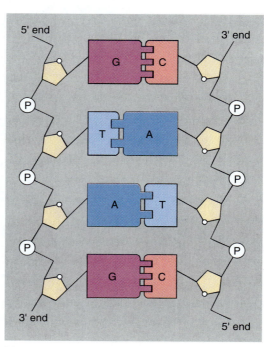

(b)

(c)

FIGURE 2.19

(a) DNA contains four bases; adenine can form two hydrogen-bonds with thymine, and guanine can form three with cytosine. Each base is covalently bound to a sugar, deoxyribose. (b) As the monomers—nucleotides—polymerize into nucleic acid, the sugars are bound to each other by phosphate groups, making a long chain with the bases projecting from the side. If the bases of one nucleic acid complement those of another nucleic acid, the two can form thousands of hydrogen-bonds and adhere to each other, making double-stranded DNA. (c) The two nucleic acids are not straight but rather spiral around a common axis, forming a double helix.

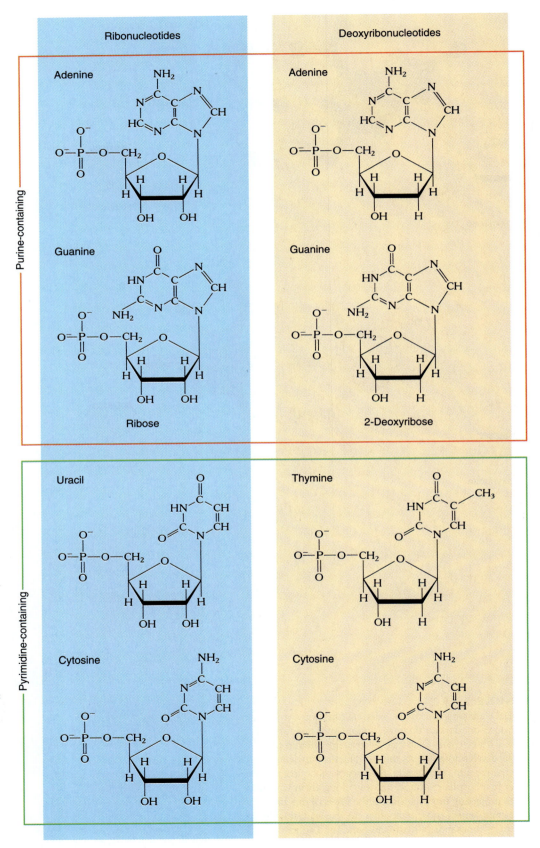

FIGURE 2.20
Purines (adenine and guanine, in the orange rectangle) contain two rings fused together, and pyrimidines (cytosine, uracil and thymine, in the green rectangle) contain a single ring. The sugar ribose is found attached only to adenine, guanine, uracil, and cytosine (blue panel), whereas deoxyribose is attached only to adenine, guanine, thymine, and cytosine (tan panel). Pyrimidines are pie shaped, and a pie can be **cut** (**c**ytosine, **u**racil, **t**hymine).

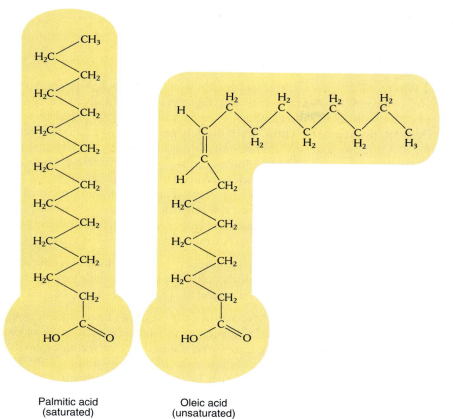

FIGURE 2.21
All fatty acids have a carboxyl group at one end; the rest of the molecule is just carbon and hydrogen. Fatty acids differ in the number of carbon atoms and double bonds they contain. If no double bond is present (that is, the molecule is saturated), the molecule can be straight; if a double bond is present (the molecule is unsaturated), the molecule must be kinked at that point.

FIGURE 2.22
Saturated fats, such as butter or margarine, are so orderly that they can crystallize and require considerable heat to melt, whereas unsaturated fats (oils) are so irregular that they melt even while cool. (*Courtesy R. Tschoepe*)

TABLE 2.7 *Common Fatty Acids*

Number of Carbon Atoms: Number of Double Bonds	Name	Formula
Saturated fatty acids		
4:0	Butyric acid	$CH_3(CH_2)_2COOH$
6:0	Caproic acid	$CH_3(CH_2)_4COOH$
12:0	Lauric acid	$CH_3(CH_2)_{10}COOH$
14:0	Myristic acid	$CH_3(CH_2)_{12}COOH$
16:0	Palmitic acid	$CH_3(CH_2)_{14}COOH$
18:0	Stearic acid	$CH_3(CH_2)_{16}COOH$
26:0	Cerotic acid	$CH_3(CH_2)_{24}COOH$
Unsaturated fatty acids		
Monounsaturated		
16:1	Palmitoleic acid	$CH_3(CH_2)_5CH{=}CH(CH_2)_7COOH$
18:1	Oleic acid	$CH_3(CH_2)_7CH{=}CH(CH_2)_7COOH$
Polyunsaturated		
18:2	Linoleic acid*	$CH_3(CH_2)_3(CH_2CH{=}CH)_2(CH_2)_7COOH$
18:3	Linolenic acid	$CH_3(CH_2CH{=}CH)_3(CH_2)_7COOH$
20:4	Arachidonic acid*	$CH_3(CH_2)_3(CH_2CH{=}CH)_4(CH_2)_3COOH$

* These are essential fatty acids—they are necessary for human growth and development, but they are not synthesized by our bodies. These fatty acids must be present in our diet.

POLYMERS OF FATTY ACIDS

Cutins and Waxes.   Fatty acids tend to polymerize readily with each other, especially when exposed to oxygen. Plants secrete fatty acids through the outer wall of their epidermal cells, and these polymerize when they come into contact with oxygen. If the fatty acids are relatively short, the polymer is **cutin**, but if the fatty acids are longer, the polymer is **wax** (see Fig. 5.23). Cutin and wax are not orderly, linear, well-defined polymers like polysaccharides, proteins, or nucleic acids. Instead, cross-linking between fatty acids can occur almost anywhere, and complex three-dimensional tangles result. Furthermore, the fatty acids involved are mixtures of long-chain and short-chain, saturated and unsaturated, fatty acids. The resulting cutin or wax can be extremely heterogeneous and variable from area to area. Both cutin and wax are waterproof and help reduce water loss from the plant body. They also prevent fungi from invading epidermal cells.

Triglycerides.   **Triglycerides** are composed of three fatty acids combined with one molecule of glycerol (Fig. 2.23a). The three fatty acids within a single triglyceride vary in length and degree of saturation and thus affect the nature of the triglyceride.

Phospholipids.   **Phospholipids** contain glycerol, two fatty acids, and a phosphate group (Fig. 2.23b). This composition is critically important: In triglycerides, all parts of the molecules are hydrophobic, so if they are mixed with water, they coalesce into spherical droplets, the shape that has a minimum surface area exposed to water (see Fig. 2.5b). But the phosphate group of phospholipids is extremely hydrophilic, so these molecules have one end that tends to dissolve in water and one end that repels water. When mixed with water, they form a layer one molecule thick across the top of the water; this is the shape that

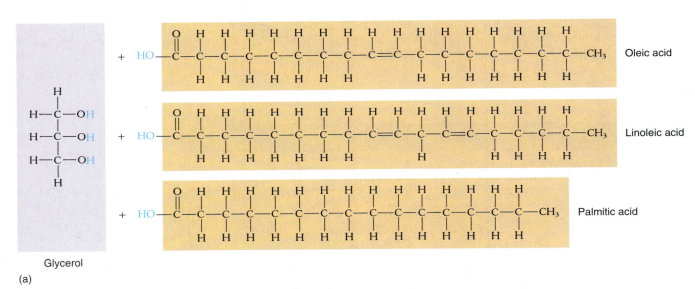

Glycerol

(a)

Oleic acid

Linoleic acid

Palmitic acid

Glycerol

Fatty acid

Fatty acid

Phosphate group

Nonpolar

Polar

Phosphatidic acid

(b)

FIGURE 2.23
(a) In a triglyceride, the carboxyl group on each fatty acid reacts with a hydrogen group on the glycerol in a dehydration reaction. All parts of this molecule are nonpolar. These are extremely hydrophobic: In order for even one of these molecules to diffuse into water, hundreds of water-water hydrogen bonds would have to be broken.
(b) A phospholipid is nonpolar on one end, polar on the other. The polar end can dissolve into water; the nonpolar end can dissolve into fats.

allows both maximum interaction of phosphates with water and minimum interaction of the hydrophobic portion with water (see Fig. 3.7). In cells, phospholipids form two-layered membranes, the hydrophobic layer of one contacting the hydrophobic layer of the other. This dual nature of both repelling and attracting water is exactly the property needed to build biological membranes. Once formed, the membranes stabilize by interacting with proteins that have shapes (tertiary structures) with regions of positive charges that interact with the phosphate groups.

COFACTORS AND CARRIERS

Many enzymes by themselves can convert reactants into products; however, in many cases, small molecules or ions must be present for a reaction to occur. Without ionic charge to help establish the proper tertiary structure, an enzyme may not have a properly formed active site and no catalysis occurs. Such ions are essential to the enzyme's activity and are **cofactors** for the enzyme; examples are ions of magnesium (Mg^{2+}) and iron (Fe^{3+}). Similarly, small organic molecules, called **coenzymes,** carry energy, electrons, or functional groups into the reaction.

ENERGY-CARRYING COENZYMES

The most common energy carrier is **adenosine triphosphate, ATP** (Fig. 2.24). The last two phosphate groups of ATP are attached by **high-energy phosphate bonds** whose bonding orbitals are not very stable. If a phosphate bond is broken and releases a phosphate group (P_i) and adenosine diphosphate (ADP), large amounts of energy are released as electrons rearrange into nonbonding orbitals. ADP can lose a second phosphate group to become adenosine monophosphate (AMP), which also releases large amounts of energy (Fig. 2.25). The phosphate of AMP is attached by a stable, low-energy bond and cannot be used to drive an endergonic reaction. Because the conversions of ATP $\rightarrow$ ADP + P_i and ADP $\rightarrow$ AMP + P_i are both highly exergonic, they can be coupled with endergonic reactions and force them to proceed. ATP can force the reaction $A + B \rightarrow AB$ to proceed by first transferring a phosphate to one of the reactants in an enzyme-mediated phosphorylation: $A + ATP \rightarrow A{-}P + ADP$. The phosphorylated A has more energy than the nonphos-

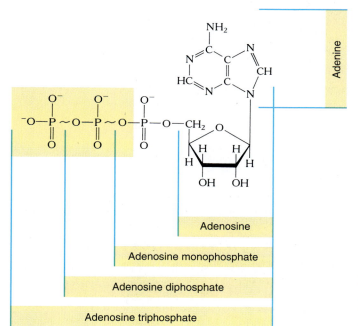

FIGURE 2.24
Adenosine triphosphate. Breaking off the last phosphate to produce ADP results in a more stable set of electron orbitals, and energy is given off. The same is true of removing the second phosphate, but not the third.

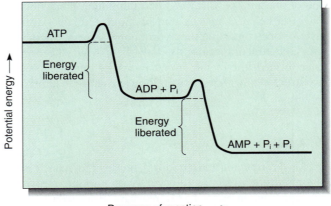

FIGURE 2.25
Dephosphorylation of ATP and ADP is highly exergonic; the liberated energy may be converted to heat or used to drive an endergonic reaction.

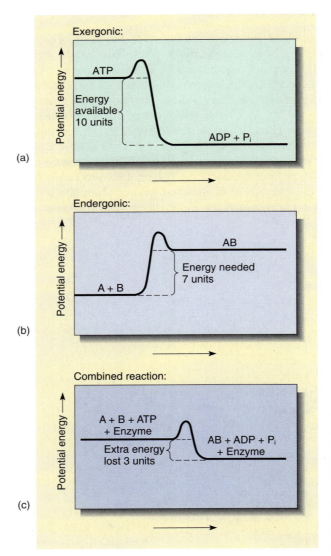

FIGURE 2.26
An exergonic reaction, such as the dephosphorylation of ATP (a), can supply power to an endergonic reaction (b) if it provides enough extra energy and if there is a mechanism, usually an enzyme, that can substitute a new common reaction mechanism for the two old separate mechanisms (c).

phorylated A and the energy-of-activation barrier for the reaction $A—P + B \rightarrow AB + P_i$ is lower than the energy-of-activation barrier for $A + B \rightarrow AB$ (Fig. 2.26).

ATP is such a versatile energy carrier that almost all endergonic reactions are forced to proceed by using the energy carried by ATP. However, ATP is both highly reactive and unstable. It cannot be stored and it cannot be moved from one cell to another; each cell must make its own ATP as it is needed. The majority of the ATP used in daily metabolism is generated by mitochondria, using energy derived from the exergonic breakdown of glucose or lipids. The ATP necessary to build glucose and lipids in the first place is generated in chloroplasts using the energy of sunlight.

ELECTRON CARRIERS

Many metabolic reactions in plants generate molecules that have high electronegativities —strong tendencies to donate electrons. Other reactions involve molecules with low electronegativities, needing electrons to become stabilized. Most molecules of the first type cannot react directly with those of the second type. They occur in separate sites within a cell, or they form at distant times, or the highly electronegative molecules are simply too reactive and too likely to damage other molecules. **Electron carriers** are small molecules that can react with highly electronegative compounds and take the outermost, energetic valence electrons away from them. The carrier itself then becomes reactive, but less so than the first molecule. When it meets the proper enzyme, it is bound to the active site and transfers the electrons to the substrate, activating it. The electron carrier is then released and returns to the site where high-energy electrons are being produced.

Both ATP and electron carriers can be "recharged" and reused repeatedly; they are not used up or destroyed in reactions.

Several electron carriers exist, each with a characteristic electronegativity. Three have only moderate electronegativities: nicotinamide adenine dinucleotide (NAD^+) (Fig. 2.27), flavin adenine dinucleotide (FAD), and flavin mononucleotide (FMN). NAD^+ is capable of transferring electrons between many different metabolic reactions. It picks up two electrons at once, the first of which neutralizes NAD^+ to NAD and the second of which allows NAD to bond with an H^+, forming NADH. (The reactions that form NADH are complex,

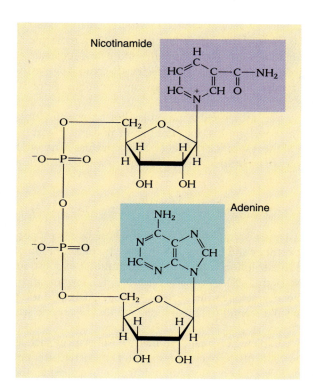

FIGURE 2.27
Nicotinamide adenine dinucleotide (NAD^+): The important area is the site where the positive charge is carried (purple box); this ring can pick up two additional electrons and then deposit them elsewhere. This entire molecule is the "container" for carrying two electrons around a cell. The adenine (green box) is far removed from the site where the electrons are carried, but if the adenine is altered, NAD^+ cannot enter into the proper reactions.

and, to be strictly correct chemically, NADH should always be considered to be associated with a second proton: NADH + H$^+$. Many textbooks do this, but others often simply write either NADH or NADH$_2$. These last two symbols are certainly convenient and make it easier to concentrate on other aspects of the reactions, but to balance chemical equations completely, the term "NADH + H$^+$" must be used.)

Many reactions occurring at many sites produce either ATP or NADH + H$^+$; these diffuse throughout the organelle and cell until by chance they run into enzymes that bind and use them. Later, the ADP and NAD$^+$ are released from the enzyme and again diffuse at random until they return to the recharge sites.

Nicotinamide adenine dinucleotide phosphate (NADP$^+$) is almost identical to NAD$^+$, differing only in having an extra phosphate group. The highly energetic reactions of photosynthesis are used to force electrons onto NADP$^+$, converting it to NADPH + H$^+$, which is used almost exclusively in the chloroplast to make sugars from carbon dioxide and water.

ENZYMES

Enzymes are protein catalysts that accelerate certain chemical reactions by providing alternative mechanisms in which all energy-of-activation barriers are lower than in the original reaction mechanism (see Fig. 2.11). Several important aspects of enzymes can be examined now.

SUBSTRATE SPECIFICITY

The atoms or molecules that an enzyme interacts with are its **substrates,** and these must fit into and be bound by the enzyme's active site if a reaction is to occur. The active site's size, shape, electrical charge, and hydrophobic/hydrophilic nature are determined by the protein's tertiary structure and, if present, its quaternary structure. Many enzymes have a high **substrate specificity;** that is, the active site is so distinctive that only one or two specific substrates fit into it. For example, the enzyme PEP carboxylase binds only to PEP (phosphoenolpyruvate) and to carbon dioxide, so the only reaction it can accelerate is the addition of carbon dioxide to PEP, forming oxaloacetate (see Fig. 10.27b). Other enzymes have low substrate specificity, binding any of several similar substrates. Often this is because only a small part of a substrate fits into the enzyme's active site: As a molecule of amylose grows longer, only the terminal, reactive glucose residue and a new molecule of glucose fit into the active site. The great majority of the polysaccharide chain is not associated with the active site, so the enzyme cannot distinguish an amylose molecule of 50 glucose residues from one of 500 residues.

Some active sites show specificity not for particular molecules but for specific functional groups, being able to bind to carboxyl groups, for example, or phosphate groups, with little or no regard for the nature of the rest of the molecule. Such enzymes carry out a particular type of reaction on many types of molecules, generally having a widespread effect on the metabolism of the cell or organelle.

RATE OF ENZYME ACTION

The binding of a substrate to its enzyme is a second-order reaction whose rate is proportional to the concentrations of the substrates and enzyme. With a given concentration of enzyme, the reaction proceeds faster in a solution with a high concentration of substrate than in one with a low concentration (Fig. 2.28). For many reactions, the enzyme must bind two substrates as well as some coenzyme or carrier such as ATP or NADH. But these bind one at a time to the enzyme, each as a second-order reaction, and the slower one sets the rate for the entire reaction.

Binding to the enzyme is usually complex. For small, symmetrical substances such as magnesium or methane, the ion or molecule fits no matter how it enters the active site. But for a large, complexly shaped molecule, such as a phospholipid or monosaccharide, the molecule may be improperly oriented when it diffuses against the active site. If the wrong

---

The cell has a pool of thousands of molecules of ATP, ADP, NADH, and NAD$^+$, each of which recycles continuously.

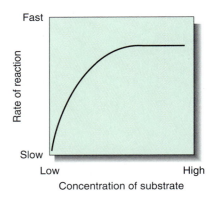

FIGURE 2.28

If a substrate is dilute *(left)*, the rate of reaction is slow because many enzyme molecules spend much of their time empty. As substrate concentration increases *(middle)* the reaction proceeds more rapidly, but at high concentrations *(right)* little or no further increase occurs in reaction rate. The enzymes are already surrounded by substrate, and as quickly as they react with one substrate molecule, another moves into the active site. The active sites spend almost no time empty, so adding more substrate does not increase the rate of reaction.

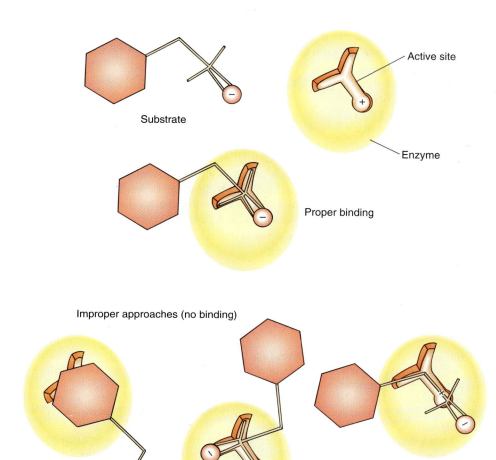

FIGURE 2.29
Usually only part of a substrate fits into the active site, and this part must be oriented properly. The chances are great that when a substrate molecule strikes an enzyme, it either misses the active site or is improperly positioned. The substrate may diffuse against the active site hundreds of times before it actually enters and is bound.

end or group enters (Fig. 2.29), it does not bind to the enzyme but diffuses out again. The requirement for precise binding is greater for enzymes with high substrate specificity; consequently, the rate of binding tends to be slower because only a small fraction of substrate/enzyme collisions have the proper orientation.

CONTROL OF ENZYME ACTIVITY

Cells control the activities of their enzymes by a variety of methods. One is simply producing or not producing the enzyme. Many enzymes necessary for certain specialized types of metabolism are produced only in the cells that need them; for example, enzymes that mediate the production of chlorophyll are produced in leaf cells but not in wood or bark cells.

Once an enzyme is present, its activity can usually be regulated further. Because its tertiary shape is so critical, cellular pH, temperature, and the presence of magnesium and calcium ions are also critical. Slight changes in these factors may alter all proteins within a cell, thereby exerting very broad regulatory power. More precise regulation of specific enzymes is possible by the use of **activators** and **inhibitors,** small molecules that bind to only one specific enzyme and either increase or decrease its activity. An activator may bind to some site, usually not the active site, and thus change the tertiary structure, either causing formation of the active site or improving its characteristics.

An inhibitor impairs the active site when it binds to the enzyme. A **competitive inhibitor** does this by occupying the active site and preventing a substrate from entering: It competes with the substrate for the same site. A **noncompetitive inhibitor** attaches to the enzyme at an area other than the active site; binding alters the enzyme such that (1) the substrate cannot bind, (2) the reaction cannot proceed, or (3) the product cannot be released. In many cases, the inhibitor molecule is either the product of the enzyme itself or is the end-product of the metabolic pathway involving the enzyme. This is **end-product inhibition.** As an enzyme or metabolic pathway functions, a product accumulates and gradually starts to inhibit the enzymes involved in its formation, preventing accumulation of excessive concentrations. If it is used up or secreted and its concentration drops, inhibition is lowered and synthesis begins again.

SUMMARY

1. The metabolism of all organisms is based on universal principles of physics, chemistry, and mathematics.

2. Electronegativity is a measure of the capacity of an element to accept or donate electrons.

3. Electrons, orbitals, and entire atoms are more stable if they have less energy. If a reaction allows valence electrons to move into bonding orbitals that are more stable than the nonbonding orbitals, the product is more stable than the reactants. Energy is given off in this exergonic reaction.

4. In water, hydrogen ions are covalently bonded to oxygen, but water is a polar molecule with partially positive and partially negative ends. This permits hydrogen bonding.

5. Acids increase the concentration of protons in a solution, usually by liberating protons. Bases decrease the proton concentration, usually by liberating hydroxyl ions that bond with protons.

6. Carbon can form single, double, or triple covalent bonds. The single bonds are the most stable, whereas the bonding electrons in double or triple bonds still have considerable energy and can enter chemical reactions relatively easily.

7. If the activation-energy barrier is low, bonding orbitals can be rearranged easily, and even molecules with little energy (low temperature) can react. If orbital rearrangement is difficult, the barrier is high and only high-energy (high-temperature) molecules can react.

8. Catalysts such as enzymes do not change the activation energy of a reaction; rather, they provide an alternative mechanism by which orbitals are more easily rearranged—all activation-energy barriers are lower. Catalysts do not change the total amount of energy liberated or absorbed during a reaction.

9. In all reactions, a reaction equilibrium is established in which there is a mixture of some reactants and some products, some forward reaction and some back-reaction.

10. The physical and chemical properties of organic molecules are most strongly determined by the functional groups present; the backbone often is not so important.

11. Polymeric construction simplifies construction techniques and reduces the amount of information necessary for construction.

12. Three important carbohydrate polymers are starch, cellulose, and oligosaccharides. All are composed of monosaccharides. The type of bond and the presence or absence of branching are important.

13. The primary structure—the amino acid sequence—of proteins determines how the protein folds (secondary and tertiary structure) and how separate proteins aggregate (quaternary structure).

14. Nucleic acids are of two types: deoxyribonucleic acid, a polymer of deoxyribonucleotides, and ribonucleic acid, a polymer of ribonucleotides.

15. Many lipids contain fatty acids, either saturated or unsaturated. A triglyceride consists of glycerol bonded to three fatty acids; a phospholipid consists of a glycerol bonded to two fatty acids and a phosphate group.

16. Many small molecules cycle repeatedly between two states and serve as carriers. ATP carries energy; NADH, NADPH, and FMN carry electrons.

IMPORTANT TERMS

acid
adenosine diphosphate (ADP)
adenosine triphosphate (ATP)
amino acid
amylose
base
catalyst
cellulose
coenzyme
cofactor
covalent bond
dehydration reaction

deoxyribonucleic acid (DNA)
endergonic
endothermic
enzyme
exergonic
exothermic
hydrogen bonding
hydrolysis
lipid
monomer
monosaccharide

nonpolar molecule
nucleic acid
nucleotide
phospholipid
polar molecule
polymer
polysaccharide
protein
ribonucleic acid (RNA)
self-assembly
substrate specificity

REVIEW QUESTIONS

1. Some elements are more stable as ions than as neutral atoms. Why are these elements more stable after having gained or lost electrons? Give examples of biologically important positive ions.

2. Why is it necessary for two reactants to collide vigorously for a reaction to occur? What happens to the shape of valence orbitals if a reaction occurs? What happens if no reaction occurs?

3. An endergonic reaction tends to proceed slowly because it absorbs energy. How do plants force endergonic reactions to occur rapidly? Is ATP involved? Why is ATP such a versatile molecule? Because its breakdown to ADP and P_i is highly exergonic, is its synthesis endergonic?

4. Name five functional groups and give the chemical formula for each. What properties does each functional group give to the molecule to which it is attached?

5. What are carbohydrates, pentoses, and hexoses? Name several hexoses and describe how they differ from each other. How can enzymes distinguish between them?

6. What are the three groups found in every amino acid? What are R groups, and how do they differ from one amino acid to another? Would you agree that because R groups are not involved in forming a peptide bond, they are not really very important?

7. How does a triglyceride differ from a phospholipid? How do these two differ in their ability to dissolve in water? How does this make one especially suitable for the construction of membranes?

BotanyLinks .net Questions

www.jbpub.com/botanylinks

Visit the .net Questions area of BotanyLinks (http://www.jbpub.com/botanylinks/netquestions) to complete this question:

1. The bodies of living organisms contain many chemical compounds, but how many chemical elements? Go to the BotanyLinks home page to investigate.

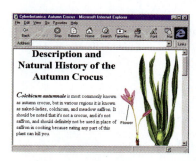

The structure of flowers facilitates sexual reproduction.

I

PLANT STRUCTURE

Plant structure is the physical, material body of a plant, composed of carbohydrates, proteins, lipids, minerals, water, and other components. In terms of both material and energy, it is expensive for an organism to build a body—all the energy and resources used by a tree to construct its body could have been used instead for reproduction, increasing its number of offspring. The theory of evolution by natural selection predicts that organisms must receive enough benefit from their structure to offset the cost of building it: It must be more advantageous selectively to have a body than not to have one. The benefit is that the structure is the framework in which metabolism occurs and the means by which the metabolism interacts with the environment. This can be more easily understood by considering the alternative—organisms with very little structure.

Several organisms have almost no structure whatsoever: viroids, viruses, and mycoplasmas (see Chapters 15 and 19). All are parasites that live, grow, develop, and reproduce only if immersed within the body of a host organism. The host provides a benign environment with enough nutrients for the parasite while also being free of harmful chemicals. The temperature of the host's body is ideal for the parasite, and adequate water is available. Parasites cannot live outside the host-supplied environment. With minimal structure, they have minimal capacity to interact well with the heat, cold, drought, and scarcity of nutrients in a natural environment.

Organisms whose bodies are more structured and complex are able to resist temporary adverse environmental conditions as well as exploit optimal ones. A simple bacterium can absorb mineral nutrients from certain environments. But by constructing a root system, a plant can spread its mineral-absorbing metabolism throughout an extensive, deep volume, growing past dry, rocky, mineral-poor soil and tapping rich, fertile soil. The investment for the plant is great, but sufficient reward is achieved in having an adequate secure source of water and minerals.

As you study the material in this part, think about how the structures discussed facilitate the metabolism that occurs inside them: How do the structures make the metabolism more efficient? What are the selective advantages of the structures? Are alternative structures possible for a particular metabolism? If so, what are the consequences of each?

3

CELL STRUCTURE

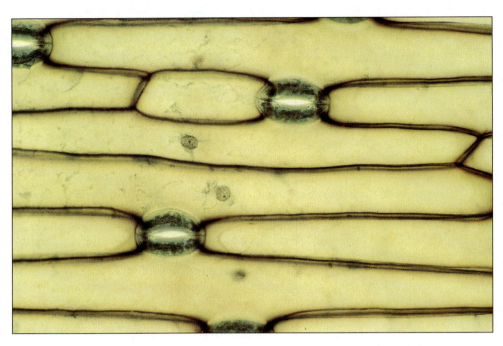

Plant cells have diverse sizes, shapes, and compositions. Their structure and metabolism vary, being correlated to the functions that each particular type of cell carries out.

CONCEPTS

All organisms are composed of small structures called **cells.** In plants, each cell consists of a boxlike cell wall surrounding a mass of protoplasm, which in turn contains its own smaller parts, the **organelles,** such as nuclei, mitochondria, and chloroplasts (Figs. 3.1 to 3.3; Table 3.1). Cells are also the physical framework within which a plant's metabolism occurs. Water and salts are absorbed from soil by root cells, they are transported throughout the plant by cells of the vascular tissues, and the energy of sunlight is used in leaf cells to

TABLE 3.1	Examples of Plant Cell Shapes and Sizes	
Cell	**Shape**	**Dimensions**
Dividing cell in shoot or root	Cube	12 μm × 12 μm × 12 μm
Epidermal cell of lily (Lilium)	Flat, paving stone	45 μm × 143 μm × 15 μm
Photosynthetic cell in leaf of pear (Pyrus)	Short cylinder	7.4 μm diam × 55 μm
Water-conducting vessel cell in oak (Quercus)	Short cylinder	270 μm diam × 225 μm
Fiber cell in hemp (Cannabis)	Long cylinder	20 μm diam × 60,000 μm

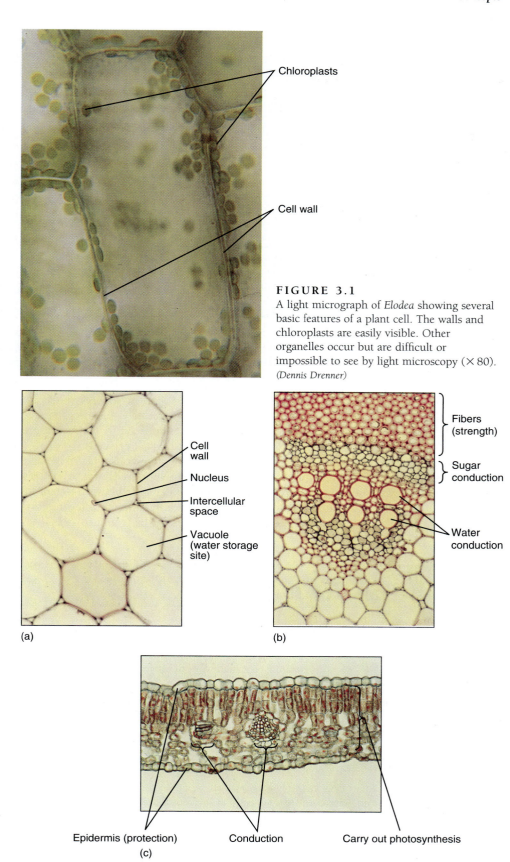

FIGURE 3.1
A light micrograph of *Elodea* showing several basic features of a plant cell. The walls and chloroplasts are easily visible. Other organelles occur but are difficult or impossible to see by light microscopy (× 80).
(Dennis Drenner)

Chloroplasts

Cell wall

Cell wall

Nucleus

Intercellular space

Vacuole (water storage site)

(a)

Fibers (strength)

Sugar conduction

Water conduction

(b)

Epidermis (protection)   Conduction   Carry out photosynthesis
(c)

FIGURE 3.2
Plant metabolism, development, and survival depend on numerous cells working together in a coordinated, integrated fashion. (a) These cells store water in the center of a sunflower stem; they are relatively large and filled mostly with water. The cell walls and a nucleus are visible (× 250). (b) Part of the system that conducts water and nutrients in a sunflower stem. Numerous types of cells occur in specific arrangements that permit efficient conduction. The large red cells in the center conduct water; the small grey cells above them conduct sugars. The cells at the top have thick red walls and provide strength to the stem (× 60). (c) In this transverse section through a leaf of *Ligustrum,* you can see a variety of cells; those in the center carry out photosynthesis (× 150).

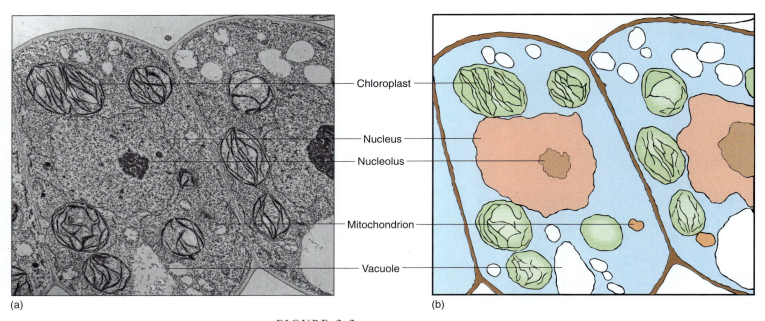

Chloroplast

Nucleus

Nucleolus

Mitochondrion

Vacuole

(a)   (b)

FIGURE 3.3

A micrograph made with a transmission electron microscope of leaf cells. Especially important are the numerous membranes (× 17,000). *(Courtesy of R. Fulginiti, University of Texas)*

convert carbon dioxide and water to carbohydrates. Plant reproduction is also based on cells and cell biology: Some cells in flowers produce pigments or nectar that attract insects which carry pollen between flowers, allowing sperm cells to contact egg cells.

Considering the large number of living organisms and the numerous types of metabolism that must be carried out, one might suspect that there are hundreds of types of cells, but actually just a small number of cell types exists. Most differences between organisms are due to differences in associations of their cells, not in the cells themselves. Regardless of whether a root, stem, leaf, or flower is being constructed, the same basic units—cells—are

---

TABLE 3.2   *Examples of Plant Cell Types, Specializations, and Division of Labor*

Cell Type	Specialization
Cells of shoot/root tips	Cell division; produce new protoplasm
Epidermis	Water retention; cutin and wax are barriers against fungi and insects
Epidermal gland cells	Protection: produce poisons that inhibit animals from harming plants
Green leaf cells	Collect solar energy by photosynthesis
Root epidermal cells	Collect water and minerals
Vascular cells	Transport water, minerals, and organic molecules
Flower cells	Petal cells: pigments that attract pollinators
	Scent cells: fragrances that attract pollinators
	Nectary cells: sugars that attract pollinators
	Stamen cells: indirectly involved in producing sperm cells
	Carpal cells: indirectly involved in producing egg cells
	Fruit cells: produce sugars, aromas, flavorful compounds that attract fruit-eating/seed-dispersing animals

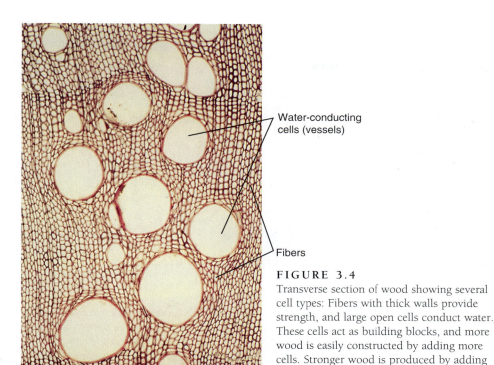

Water-conducting
cells (vessels)

Fibers

FIGURE 3.4
Transverse section of wood showing several
cell types: Fibers with thick walls provide
strength, and large open cells conduct water.
These cells act as building blocks, and more
wood is easily constructed by adding more
cells. Stronger wood is produced by adding
more fiber cells; more conductive wood
results from adding more open cells (× 100).

required (Fig. 3.4). Only the cell associations and minor modifications of the cells themselves change from tissue to tissue or organ to organ.

Although only a few types of cells exist, their differences are important. Any organism composed of more than one cell (a **multicellular organism** rather than a **unicellular** one) always has several types, each specialized for different tasks (see Figs. 3.2 and 3.4; Table 3.2). As a plant develops, the cells in various parts become especially adapted for specific tasks. This **division of labor** allows the whole organism to become more efficient. Unicellular organization has a significant consequence: It does not allow division of labor or specialization. Each cell must perform all tasks—sensing the environment, gathering nutrients, excreting wastes, defense, movement, and reproduction. Because each cell must perform all tasks, it cannot do any one very well. Mutations that make a cell well adapted for protection make it less adapted for other functions and therefore are selectively disadvantageous (Fig. 3.5). The same is true for modifications that improve photosynthesis, reproduction, and so on.

Multicellularity and division of labor result in a more efficient organism, but they have negative consequences as well: As each cell becomes more specialized, it depends more on the others. If a cell evolves toward having thick walls and offering maximum protection, it must rely on other cells of the organism for photosynthesis, mineral absorption, and reproduction. Damage to one part of the organism may result in the death of all cells, even those not initially damaged, whereas in a unicellular organism, a cell dies only if it is damaged directly. Which is more advantageous selectively—a unicellular organism composed of one generalized cell or a multicellular organism composed of specialized ones? The answer is not so obvious: Both types have existed for hundreds of millions of years, so both must be considered highly adaptive and successful.

Like whole individuals, cells have a life span. During their life cycle (**cell cycle**), cell size, shape, and metabolic activities can change dramatically. A cell is "born" as a twin

Selective advantage depends on the environment: In certain habitats unicellular organisms are better adapted, and in others multicellular organisms survive better.

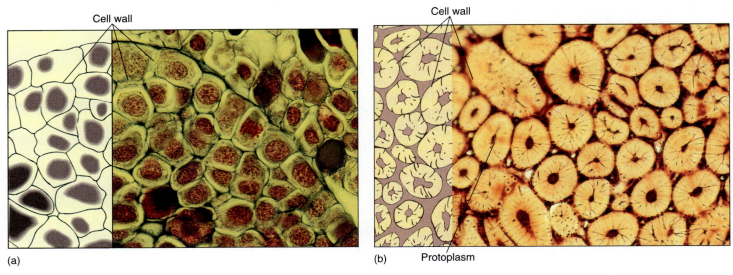

FIGURE 3.5

(a) Cells located at the growing tip of *Pinus* (pine) are specialized for cell growth and division: Their thin cell walls allow sugar, water, and other nutrients to enter the cells easily. But these cells are too soft and weak to be useful as bark or the shell of a nut (× 400). (b) These "stone cells" provide strength in a coconut shell (*Cocos nucifera*) and therefore protect the seed. Almost the entire cell volume is wall. Protoplasm was present when the cells were young, growing and synthesizing their walls, but once the walls were completed, cell metabolism was not necessary and the protoplasts died (remnants are present in the tiny black hole in the center of each cell). Such modifications make it impossible for a stone cell to be a dividing, growing cell as in (a) (× 200).

when its **mother cell** divides, producing two **daughter cells.** Each daughter cell is smaller than the mother cell, and except for unusual cases each grows until it becomes as large as the mother cell was. During this time the cell absorbs water, sugars, amino acids, and other nutrients and assembles them into new, living protoplasm. Once the cell has grown to the proper size, its metabolism shifts as it either prepares to divide or matures and differentiates into a specialized cell.

Both growth and development require a complex and dynamic set of interactions involving all cell parts. That cell metabolism and structure should be complex would not be surprising, but actually they are rather simple and logical. Even the most complex cell has only a small number of parts, each responsible for a discrete, well-defined aspect of cell life.

MEMBRANES

All cells contain at least some membranes, and cells of eukaryotes (plants, animals, fungi, and protista) contain numerous organelles composed of membranes. Membranes perform many important tasks in cell metabolism: They regulate the passage of molecules into and out of cells and organelles; they divide the cell into numerous compartments, each with its own specialized metabolism; and they act as surfaces that hold enzymes. Without membranes, life would be impossible; indeed, many things that cause death in both humans and plants—heat, cold, many poisons, alcohol—are able to kill because they disrupt membranes.

COMPOSITION OF MEMBRANES

All biological membranes are composed of proteins and two layers of phospholipid molecules (see Figs. 2.23 and 3.6). If phospholipids are poured carefully onto the surface of calm water, they automatically form a single layer (a **monolayer**) on the water's surface, with their hydrophilic ends forming hydrogen bonds to water molecules and their hydro-

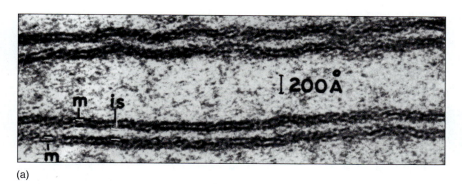

(a)

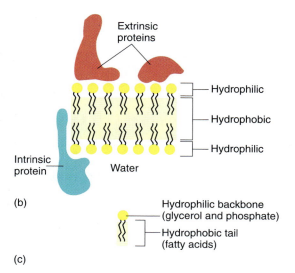

(b)

(c)

FIGURE 3.6

(a) In electron micrographs, membranes appear as two dark lines (proteins) separated by a light region (composed mostly of lipids) ($\times$ 600,000). (b) Pictures such as (a) led scientists to believe that membranes consist of two layers of lipids with proteins located only on the surface. However, many proteins fit partly into the lipid bilayer or pass all the way through it. (c) A phospholipid has three parts: (1) two fatty acids, (2) a backbone of glycerol, and (3) a phosphate group.

phobic fatty acids projecting out of the water (Fig. 3.7). If the water is agitated, the lipid layer doubles over and makes a **bilayer** in which all fatty acids are away from the water and all phosphate groups are in full contact with it. Any break or tear in a bilayer membrane exposes hydrophobic fatty acids to water, so membranes always reseal themselves after a rupture.

The lipid bilayer is a very thin solution. If it contains several types of lipid, they can diffuse laterally throughout the membrane because they are not bonded to each other, but they cannot diffuse vertically from the membrane into the surrounding solution.

All biological membranes contain proteins as well as lipids, usually in a ratio of 60% proteins and 40% lipid (see Fig. 3.6b). Most proteins have large hydrophilic regions, so they associate mostly with the phospholipid phosphates and with water. Many also have large hydrophobic regions that allow them to sink into the membrane and associate with the fatty acids (Fig. 3.8). Variations in hydrophobic and hydrophilic regions mean that various proteins sit entirely on the membrane surface, are partly immersed in it, or span it entirely, with either end projecting out of opposite sides. This allows the protein to have its active site on either or both sides of the membrane or within it. Likewise, a protein may act as a hydrophilic channel that permits small hydrophilic molecules to pass through the membrane. Proteins that are even partially immersed in the lipid bilayer are said to be **intrinsic proteins.** Others, **extrinsic proteins** (also called **peripheral proteins**), are located outside the membrane and merely lie next to it. Although extrinsic proteins may perform important enzymatic functions, they are not an integral part of the membrane's structure.

Some intrinsic proteins contribute to the membrane's fluid nature and, like the lipids, can diffuse laterally. Other proteins interact with adjacent proteins, forming complexes or **domains** (small discrete regions) different from surrounding regions of the membrane. If all membrane components could freely diffuse laterally, the membrane would become homogeneous and no differentiation could occur. But because at least some proteins are bound to their neighbors, membranes are heterogeneous and patchy, and differentiation does take place. Because the membrane is a heterogeneous liquid, it is said to be a **fluid mosaic.**

Some membranes contain a small amount of sugar, usually less than 8%. The sugars occur as short-chain oligosaccharides, each with about 4 to 15 sugar residues. These oligosaccharides are bound to certain intrinsic proteins, converting them into **glycoproteins;** rarely sugars are attached to membrane lipids (**glycolipids**). At present, we think that these polysaccharides make the membrane more distinctive and easy to recognize, an

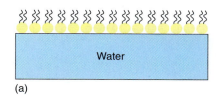

(a)

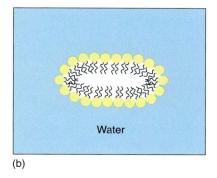

(b)

FIGURE 3.7

(a) In quiet water, phospholipids form a monolayer at the surface; by projecting away from the water, the fatty acids do not disrupt the hydrogen bonding between water molecules. (b) With agitation phospholipids form a bilayer; this arrangement allows the greatest number of hydrogen-bonds to form between phosphates and water while simultaneously minimizing the disruption of water-water hydrogen-bonds by the fatty acids.

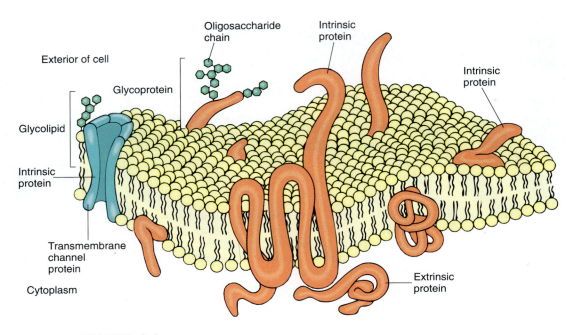

FIGURE 3.8
Intrinsic membrane proteins differ in their size, shape, and location of hydrophobic, lipid-soluble regions; consequently, some sink deep into the membrane, others span it, and some penetrate only one lipid layer.

especially important feature in animals that have an immune system. Specialized protective cells attack and destroy anything they do not recognize as a part of the animal's own body. The glycoproteins and glycolipids occur almost exclusively on the outer surface of the membrane that covers the cell, a position that allows maximum exposure for these polysaccharides. Glycoproteins and glycolipids may be less important in plants.

PROPERTIES OF MEMBRANES

Membranes have several important properties. First, they can grow. Membranes are formed molecule by molecule in certain regions of the cell; then whole pieces of membrane are moved as small bubbles or **vesicles** to different sites in the cell. When the vesicle of preformed membrane arrives at the growing membrane, the two fuse (Fig. 3.9).

In addition to permitting movement of membrane pieces, membrane fusion allows the transport of material. The volume inside the vesicle (the vesicle **lumen**) may be filled with substances that must be accumulated, broken down, or otherwise metabolized at the vesicle's destination. Vesicle movement may also release material to the outside of the cell (Fig. 3.9). This **exocytosis** may be a means to excrete almost anything: wastes, debris, mucilage, proteins, polysaccharides. For example, roots slide through the soil by secreting a slippery, lubricating mucilage that is formed within the root cells, packaged into vesicles that migrate to and fuse with the cell surface membrane, and are then released to the exterior. In many flowers, nectar is secreted from glands by exocytosis. **Endocytosis** is basically the opposite process: A small invagination forms in the outer membrane, and the invagination pinches shut, creating a new vesicle that contains extracellular material. Endocytosis is especially common in algae and other microscopic organisms that take in food particles this way, but it is not known how common it is in plants.

Permeability is an important property of membranes. All biological membranes are **selectively permeable** (also called **differentially permeable**), meaning that certain substances cross the membrane more easily and rapidly than others (Fig. 3.10). Because large regions of a membrane are mostly lipid, membranes are more permeable to hydrophobic

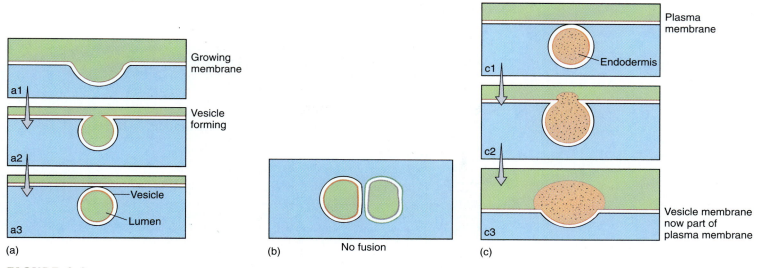

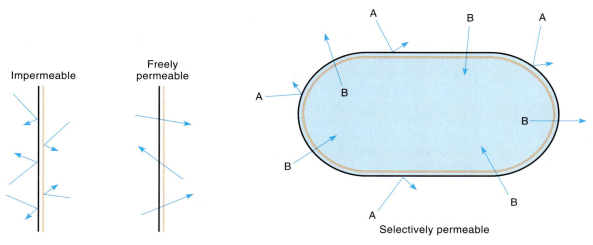

FIGURE 3.9

Numerous organelles produce small vesicles (a) that move through the cell, either remaining discrete for long periods or fusing with other organelles or vesicles. In order to fuse, two membranes must be quite similar in composition. Here, a single membrane bilayer is shown as a double line, one orange and the other black. (b) Membrane dissimilarity prevents two compartments from fusing; for example, the nucleus does not fuse with mitochondria or chloroplasts. (c) When a vesicle fuses with the membrane at the cell surface, its contents are deposited outside the cell and the vesicle membrane becomes part of the cell surface. The vesicle membrane and the plasma membrane have been drawn with the same colors to indicate their similar composition.

substances than to anything that carries an electric charge. However, if charged compounds such as inorganic salts, sugars, and amino acids could not enter cells at all, or if they could enter only slowly, cells would starve.

The movement of charged substances is assisted by large intrinsic proteins that span the membrane and act as hydrophilic channels through it; this is **facilitated diffusion.** Other proteins, called **molecular pumps,** actually bind to a molecule on one side of the

FIGURE 3.10

An impermeable membrane allows nothing to pass through, whereas a freely permeable membrane stops nothing. A selectively permeable membrane allows certain materials to pass through more readily than others, with the result that chemical concentrations on the two sides differ. This membrane is impermeable to substance A but permeable to B.

TABLE 3.3	*Active Transport in Roots of Corn (Zea)*		
	External Concentration	Internal Concentration	Increase
Potassium	1 mM	22 mM	2200%
Calcium	0.2 mM	5 mM	2500%
Chloride	1.2 mM	27 mM	2250%

Mineral nutrients are often present as a dilute solution in soil and must be accumulated by active transport carried out by root cells (the endodermis cells—see Chapter 7). The table shows the ability of endodermis cells to pump nutrients across their membranes from the exterior of the plant to the conducting tissues.

membrane, then by using energy, the protein changes shape and releases the molecule on the other side. By using this active pumping, called **active transport,** cells accumulate substances until the interior concentration of solute far exceeds the exterior concentration. A pump working in the opposite way can actively transport materials out of the cell (Table 3.3). Except for the pumps, the membrane must be impermeable to the molecule; it would do no good to have a molecular pump for a type of molecule that could easily leak back through.

All life depends on the principle of **compartmentalization,** the formation of many compartments, each specialized for a particular process such as producing a particular substance or using a particular precursor. If all the monomers, polymers, sugars, salts, enzymes, and vitamins of an organism were mixed together, some reactions would occur at random but would not be orderly enough to be considered life. Instead, cells are filled with many compartments—the organelles—each surrounded by its own unique selectively permeable membrane. Because the protein channels and pumps are made under the guidance of the nucleus, the cell can control the numbers and types of channels and pumps it makes and inserts into each type of organelle membrane. Organelles thus have membranes with different permeabilities or pumping capacities, depending on the instructions generated by the nucleus.

A **freely permeable membrane,** which allows everything to pass through quickly, would be rather useless for a cell, as would an **impermeable membrane,** one that does not allow anything through at all (Fig. 3.10; Table 3.4).

The last important property of membranes is that they are **dynamic,** constantly changing. Cell membranes are not simply established and left unchanged throughout the life of the cell. Instead, new components are constantly being inserted and old ones removed. If the function of the cell is always the same, the old and new components are similar, but if the cell must change its function, newly inserted components are different from the retracted ones. As the nature of the membrane changes, the nature of the cell changes also.

Your blood is compartmentalized away from your digestive system: The two must work together, but they must not be mixed together.

BASIC CELL TYPES

At the most basic level, both cells and organisms can be classified as either **prokaryotic** or **eukaryotic.** Prokaryotic cells are simpler than eukaryotic ones and are found only in kingdom Monera: bacteria, cyanobacteria, and archaebacteria. It is hypothesized that they represent the most archaic lines of evolution and that eukaryotic cells evolved from them.

Kingdom Monera and prokaryotes were introduced in Chapter 1 (Table 1.1) and are discussed in detail in Chapter 19.

TABLE 3.4 *Summary of Transmembrane Movement*
Impermeable membrane: Nothing passes through; no biological membrane is impermeable to everything.
Freely permeable membrane: Virtually anything can pass through.
Selectively permeable (differentially permeable) membrane: Certain substances pass through rapidly, others pass through slowly.
Facilitated diffusion: The presence of large intrinsic membrane proteins allows hydrophilic, charged molecules to diffuse through the membrane.
Active transport: Large intrinsic membrane proteins bind a molecule and force it through the membrane, consuming energy in the process.
Exocytosis: The fusion of a vesicle with the cell membrane, releasing the vesicle's contents to the cell exterior.
Endocytosis: The invagination of the cell membrane, forming a vesicle that pinches off and carries external material into the cell.

Eukaryotic cells, found in plants, animals, fungi, and protists, are more complex than prokaryotic cells. The most striking difference, the one that gives them their name, is the presence of a true membrane-bounded nucleus in eukaryotic cells. In addition, there are many organelles that allow eukaryotic cells to be more diverse and complex, both morphologically and physiologically.

PLANT CELLS

Although various parts of a plant—roots, wood, bark, leaves, and flower parts—appear to be quite diverse, virtually all their cells have all the following organelles; the exceptions are rare. As each type of cell develops, certain organelles may become modified and more or less abundant, but usually none is lost completely (Fig. 3.11).

PROTOPLASM

All cells, either prokaryotic or eukaryotic, are made of a substance called **protoplasm** (see Fig. 3.3; the protoplasm of a single cell is called its protoplast). This name was given early, when it was thought that protoplasm was a distinct substance like water, oxygen, or iron, and that one of its properties was life itself. We now know that protoplasm is a mass of proteins, lipids, nucleic acids, and water within a cell; except for the wall, everything in the cell is protoplasm, composed of the following organelles.

PLASMA MEMBRANE

The **plasma membrane** (less frequently called the **plasmalemma**) is the membrane that completely covers the surface of the protoplasm (Fig. 3.12). Because it is the outermost surface of the protoplast, it is selectively advantageous for it to be impermeable to harmful materials and permeable to beneficial ones; therefore, it is selectively permeable. There are molecular pumps in the plasma membrane that actively transport needed molecules inward and pump others outward for secretion. Because one side of the plasma membrane faces the external environment and the other side faces the cell, the two sides are quite different, especially in the types of protein they contain. We know very little about the proteins and lipids that make up the plasma membrane because the membrane does not separate cleanly from the protoplast. Almost all attempts to isolate the plasma membrane for study have experienced considerable contamination from other parts of the cell.

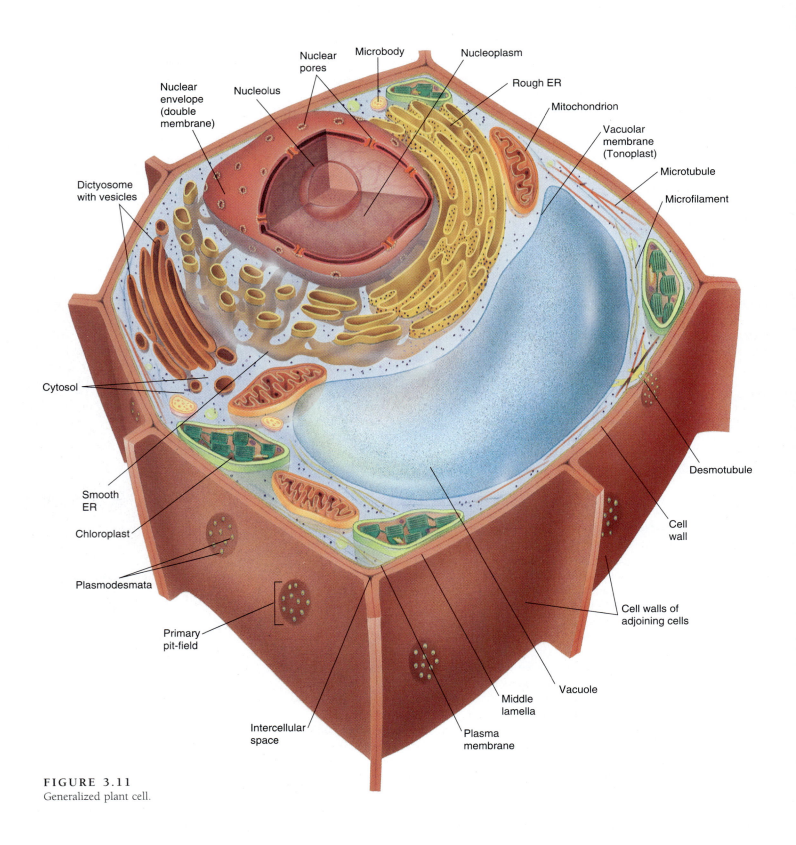

FIGURE 3.11
Generalized plant cell.

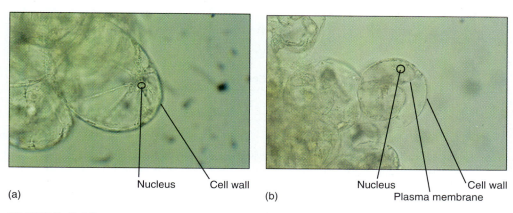

Nucleus   Cell wall

(a)

Nucleus   Cell wall
Plasma membrane

(b)

FIGURE 3.12
(a) A healthy, growing cell. Its protoplast is pressed firmly against the cell wall, so its plasma membrane is not visible (× 160). (b) These cells have been treated with salt to draw water out of the cell, causing the protoplast to shrink. As a result, the plasma membrane is visible (× 160).

NUCLEUS

The **nucleus** (pl.: nuclei; see Fig. 3.11) serves as an archive, or permanent storage place, for the organism's genetic information. The book you are reading is just a brief introduction to plant biology. No one can imagine the number of pages required to store all the information necessary for building and maintaining a single cell. But all that information must be stored in the DNA inside every nucleus, and the storage must be safe and permanent. Information is useless unless it, or copies of it, can be retrieved and used. The nucleus carries out information retrieval by making copies of specific parts of the DNA whenever that information is needed. The copy is not DNA, but a type of ribonucleic acid called messenger RNA. An exciting area of research involves understanding how the nucleus, in response to signals from the rest of the cell, searches its DNA for the needed information and then makes copies of it without accidentally copying other similar information.

The nucleus of a eukaryotic cell is always surrounded by a **nuclear envelope,** which is actually an **outer membrane** and an **inner membrane.** The nuclear envelope separates the nuclear material from the rest of the cell, and it contains numerous small holes, **nuclear pores** (Fig. 3.13), that are involved in the transport of material between the nucleus and the rest of the protoplasm. Nuclear pores have a complex structure and exert control over the movement of materials. If a nucleus is extracted from a cell and placed into water, it swells; this can happen only if the pores prevent material from oozing out as the nucleus absorbs water. Prokaryotes have no nuclear envelope; instead DNA is simply mixed with the rest of the cell contents.

Within the nucleus is a substance called **nucleoplasm.** Like "protoplasm," this name was given before the composition was understood; we now know that nucleoplasm is a complex association of (1) DNA; (2) enzymes and other factors necessary to maintain, repair, and read DNA; (3) histone proteins that support and interact with DNA; (4) several types of RNA; and (5) water and numerous other substances necessary for nuclear metabolism. Nuclear DNA is always closely associated with histones, and this complex of the two is known as chromatin. As a cell ages or its metabolic activities change, so do the nucleus and nucleoplasm. In cells that are undergoing rapid cell division (for example, the cells in root tips, shoot tips, young leaves, and flower buds), the DNA, histones, and duplicating enzymes may dominate the nucleoplasm. But in mature cells that are not dividing, messenger molecules and reading enzymes may be more abundant.

Nuclear
envelope
surface   Nuclear pore

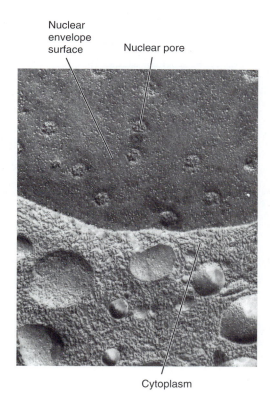

Cytoplasm

FIGURE 3.13
A preparation called a freeze-fracture: The cell is frozen and then tapped to cause it to break. The lipids of the membranes are weak when frozen, so the fracturing often follows the membranes, separating one lipid layer from the other. Here the fracture passed irregularly through the cytoplasm and then entered the nuclear envelope. The nuclear envelope consists of the outer and inner membranes, with a space separating them; the two membranes fuse together at the nuclear pores. Nuclear pores may be distributed uniformly over the nuclear surface or may occur in bands or patches in some species (× 60,000). (*Biophoto Associates*)

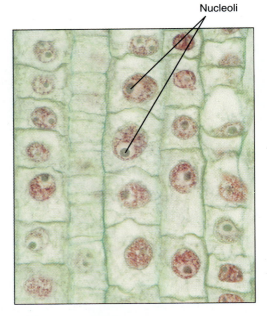

Nucleoli

FIGURE 3.14
Nucleoli usually stain more intensely than the
rest of the nucleus; occasionally they stain a
distinct color. Root tip of hyacinth ($\times$ 250).

*Plant cells always have much more water than
animal cells. Often up to 95% of a plant cell's
weight is water.*

TABLE 3.5   *The Relative Volumes of Organelles in Plant
Cells*

Organelle	Young Cell	Mature Cell
Nucleus	32.4	0.23
Vacuole	4.93	83.3
Mitochondria	5.35	1.01
Plastids	3.72 (proplastids)	3.16 (chloroplasts)
Dictyosomes	0.40	0.04
Hyaloplasm	52.9	12.1
Lipid	0.23	0.00

Young cells tend to be small with large nuclei; as they grow and differentiate, the cell
volume increases and certain organelles become more prominent, depending on the
specialization of the cell. The relative volumes of organelles (the volume of the
organelle expressed as a percentage of the volume of the whole cell) are given here for
a young cell from a shoot tip and for a mature, photosynthetic cell from the outer part
of the stem. The size of the nucleus in both cells is the same, but the mature cell is so
large that the nucleus constitutes only a small part of it, whereas the young cell is so
small that the nucleus occupies one third of its volume. These values are for a cactus,
Echinocereus.

Inside every nucleus is one, two, or rarely several bodies called **nucleoli** (sing.: nucleolus; Fig. 3.14), areas where the components of ribosomes are synthesized and partially
assembled. Each ribosome contains a large amount of ribosomal RNA copied from ribosomal genes in the chromatin.

Nuclei are large, complex organelles, and frequently they occupy a major fraction of
the cell volume (up to 50%; Table 3.5). In certain conducting cells (sieve tube members in
phloem; see Chapter 5), the nucleus breaks down during cell differentiation, and the
mature cell functions for several weeks or months without a nucleus. Our red blood cells
are also enucleate (without a nucleus) at maturity. Both of these cell types are exceptional;
they are enucleate only while performing a very limited type of metabolism, and they die
shortly after losing the nucleus. Several types of plant cells are multinucleate (see Chapter 4).

CENTRAL VACUOLE

Within young, small cells are organelles, **vacuoles,** that have just a single membrane, the
vacuole membrane, also called the **tonoplast.** Vacuoles often appear to be empty (see Fig.
3.11; Table 3.6) because they store mostly water and salts that cannot be preserved for
microscopy. However, they sometimes contain visible crystals, starch, protein bodies, and
various types of granules or fibrous materials in addition to water and salts.

As a cell grows and enlarges, vacuoles expand and merge until there is just one large
central vacuole. Because it contains primarily water and salts, the central vacuole can
expand rapidly, forcing the cell to grow rapidly as well. Animal cells must synthesize
complete protoplasm to grow, but plant cells need only increase the amount of vacuolar
water. Over a long period, plants must produce additional proteins, membranes, and
organelles or they would become almost pure water.

In addition to cell growth, the central vacuole functions in storage of both nutrient
reserves and waste products. In seed cells, vacuoles may be filled with starch or protein that
will be used when the seed germinates, perhaps 10 to 50 years after the material was
deposited in the vacuole. Calcium regulates the activity of many enzymes, and plant cells
keep protoplasmic calcium concentrations at the proper level by moving calcium into the

TABLE 3.6	*Concentrations of Nutrients in Vacuoles Isolated from Barley (Hordeum) Leaf Cells*

Nutrient	Concentration (mM)
Cl^-	56.1
NO_3^-	42.9
HPO_4^- *	64.5
SO_4^{2-}	25.1
K^+	142.2
Na^+	42.2

* Often abbreviated P_i

vacuole, where it reacts with oxalic acid and crystallizes into an inert form. Other nutrients such as potassium may move in and out of the vacuole on a daily basis. The water-soluble pigments in many flowers, fruits, and red beets occur in vacuoles as well.

A system to excrete wastes never evolved in plants; instead, metabolic waste products are pumped across the vacuole membrane and stored permanently in the central vacuole. The tonoplast is otherwise impermeable to these wastes, so they cannot leak back into the cytoplasm where they would be harmful. Holding waste inside forever does not sound like an optimal situation, but it actually may be selectively advantageous: Because most of these compounds are noxious and bitter, they deter animals from eating the plants. Mutations that result in excretion might make the cells taste good, which would be selectively disadvantageous.

The central vacuole is a digestive organelle as well. As organelles age and become impaired, they fuse with the tonoplast and are transported into the central vacuole, where digestive enzymes break them down. The liberated monomers are presumably transported back into the rest of the cell, where they are used again. In animal cells, which do not have central vacuoles, this task is carried out by small vacuoles called lysosomes.

CYTOPLASM

If the nucleus and vacuole are excluded from the protoplasm, the remaining material is referred to as **cytoplasm** and contains the following structures.

MITOCHONDRIA

Cells store energy as highly energetic but fairly unreactive compounds, such as sugars and starches; to utilize the energy, such compounds must be broken down and their energy used to synthesize new compounds that are both highly energetic and very reactive, the most common being adenosine triphosphate (ATP). Because these reactions involve high-energy, reactive compounds, they occur more safely within a discrete organelle than freely in the cytoplasm, where the intermediates might accidentally react with other cell components. **Mitochondria** (sing.: mitochondrion) are the organelles that carry out this cell respiration (Fig. 3.15).

Many steps of respiration are mediated by enzymes bound to mitochondrial membranes. The enzymes are located adjacent to each other so that the product of one reaction (which becomes the reactant in the next reaction) is passed directly to the next enzyme; this controls the highly reactive intermediates (Fig. 3.16). Mitochondrial membranes are folded, forming large sheets or tubes known as **cristae** (sing.: crista). This folding provides

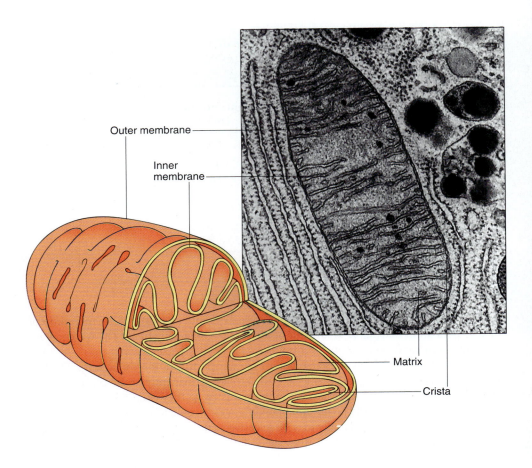

FIGURE 3.15
The inner mitochondrial membrane is folded into platelike cristae, giving it a large surface area; many respiratory enzymes are intrinsic proteins embedded in the crista membrane. The large surface area makes it possible to contain many copies of each enzyme, and more reactions can occur simultaneously.
(Bill Longcore/Photo Researchers)

room for large numbers of enzymes. Reactions that do not involve highly reactive intermediates take place in the liquid **matrix** between the cristae. Around this complex of cristae and matrix is a second membrane, the **outer mitochondrial membrane,** which probably just gives shape and a little rigidity to the mitochondrion. The outer membrane is rather freely permeable, but the **inner mitochondrial membrane,** which forms the cristae, is selectively permeable and has numerous pumps and channels (see Chapter 11).

Mitochondria have their own DNA and ribosomes, both of which are different from those of the rest of the cell. Mitochondrial DNA is a circular molecule and lacks histones; the ribosomes are small and resemble those found in prokaryotes.

FIGURE 3.16
Many enzymes must work together. The product b of enzyme 1 is the reactant of enzyme 2. (a) If the enzymes are in solution, they may not be close together and b must diffuse at random until it encounters enzyme 2; in the meantime, it may accidentally react with something else. (b) If they are located side by side on a membrane, enzyme 1 can pass b directly to enzyme 2, not only speeding up the overall reaction but also eliminating the chance of accidental reactions.

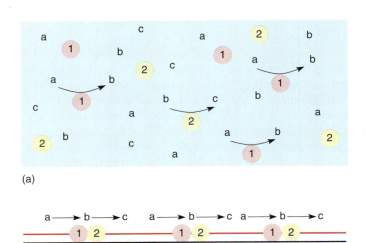

Mitochondria are dynamic organelles; they can grow larger, either as the cell does or as the cell's need for respiration increases. They are often about 1 μm in diameter but up to 5 μm in length, and even much longer in some cases. Mitochondria can also divide into two, increasing the number of mitochondria per cell. Under rare conditions, the mitochondria in some species fuse together, forming a single giant mitochondrion in each cell. Mitochondria occupy between 1% and 25% of the cell volume (see Table 3.5); the higher values are similar to those for animal cells. Because of their ability to divide and fuse, the actual number of mitochondria per cell (usually in the range of 100 to 10,000) may not be as important as their volume.

PLASTIDS

Plastids are a group of dynamic organelles able to perform many functions. One prominent activity is photosynthesis, carried out by the green plastids, chloroplasts. Diverse types of metabolism occur in other classes of plastid: synthesis, storage, and export of specialized lipid molecules; storage of carbohydrates and iron; and formation of colors in some flowers and fruits. Each metabolism is associated with a particular type of plastid; as an organ changes, its plastids may also change, extensively altering their membranes and proteins.

Plastids are found in all plants and algae but never occur in animals, fungi, or prokaryotes.

Like mitochondria, plastids always have an **inner membrane** and an **outer membrane** and an inner fluid called **stroma** (Fig. 3.17). Plastids also have ribosomes and circular DNA that is not associated with histones. Plastids grow and reproduce by pulling apart (see Fig. 4.22); reports of plastid fusion need more study.

Plastids of young, rapidly dividing cells are called **proplastids** and are very simple. The inner membrane has a few folds but little surface area. When exposed to light, proplastids develop into **chloroplasts,** which are green owing to the presence of the photo-

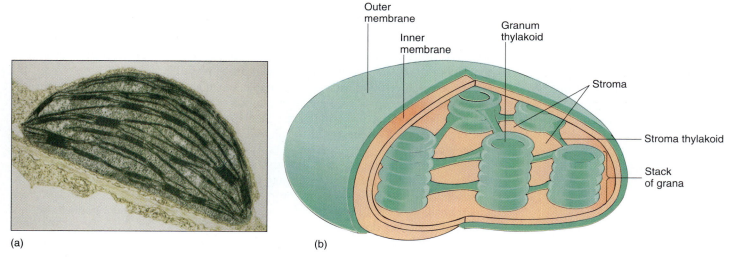

(a) (b)

FIGURE 3.17

(a) All plastids have an outer and an inner membrane, but in chloroplasts the inner membrane is extensive and highly folded. The single membranes are thylakoids, and the multiple membranes are actually stacks of flattened thylakoid vesicles; the stacks are called grana. Like mitochondrial cristae, the extra surface area of the inner membrane provides room for many copies of each enzyme. The photosynthetic pigment chlorophyll is part of the membrane lipid layer (approx. × 30,000). (*Dr. Jeremy Burgess/SPL/Photo Researchers*) (b) The grana are interconnected by thylakoid membranes, and the liquid stroma surrounds the grana. Because the thylakoid membranes are selectively permeable, the concentrations of chemicals inside the thylakoid space differ from those in the stroma. This is essential for photosynthesis.

TABLE 3.7	*Types of Plastids*
Amyoplasts:	Store starch; considered to be leucoplasts
Chloroplasts:	Carry out photosynthesis
Chromoplasts:	Contain abundant colored lipids; in flowers and fruits
Etioplasts:	A specific stage in the transformation of proplastids to chloroplasts; occur when tissues are grown without light
Leucoplasts:	Colorless plastids; synthesize lipids and other materials
Proplastids:	Small, undifferentiated plastids

synthetic pigment **chlorophyll** (Table 3.7). Because many intermediates of photosynthesis are highly reactive, energetic compounds, the controlling enzymes must be incorporated into the membranes, just as in mitochondria. This requires the inner membrane to become more extensive and elaborately folded. Membrane sheets, **thylakoids,** project into the stroma. This increase in membrane area also provides more space for the insertion of the photosynthetic pigments: Chlorophyll has a lipid-soluble tail (see Fig. 10.2), so it is part of the membrane. In certain regions the thylakoids form small baglike vesicles that become stacked together. The stack of vesicles is called a **granum** (pl.: grana). A key feature of photosynthesis is the active transport of protons (H^+) into a small space to build up an electrical charge; the grana vesicles are needed to accumulate these protons from the stroma. The actual conversion of carbon dioxide to carbohydrate occurs in the stroma, catalyzed by enzymes free in solution rather than bound to any of the membranes.

Chloroplasts are larger than mitochondria, about 4 to 6 μm in diameter. An individual leaf cell may contain as many as 50 chloroplasts. When chloroplasts photosynthesize rapidly, they produce sugar faster than the cell can use it, so it is temporarily polymerized into starch grains inside the chloroplasts.

In plant tissues that cannot photosynthesize (roots, bark, wood) proplastids develop into **amyloplasts,** which accumulate sugar and store it as starch for months or years (Fig. 3.18). Each amyloplast produces large starch grains that virtually fill the stroma; few internal membranes are present. In starchy vegetables, amyloplasts constitute the bulk of the tissue. If exposed to light, amyloplasts can be converted to chloroplasts. Nonphotosynthetic parasitic plants such as *Conopholis* (see Fig. 1.8) lack chloroplasts but do have amyloplasts; no plant is ever without some form of plastid.

In some flowers and fruits—for example, tomatoes and yellow squash—bright red, yellow, or orange lipids accumulate in plastids as they differentiate into highly colored **chromoplasts.** An extensive, undulate system of membranes is present, but no grana, and the pigments may be present either as part of the membrane or as discrete droplets, **plastoglobuli** (sing.: plastoglobulus). As fruits ripen, chloroplasts often synthesize large amounts of lipid pigments, alter their thylakoids, and convert to chromoplasts, as when apples and tomatoes change from green to red. Lipid pigments are present in low amounts as small plastoglobuli even in leaf chloroplasts, but they are masked by the abundant green chlorophyll; they can be seen in autumn when cold weather causes chlorophyll, but not lipids, to be broken down and the leaves turn red or yellow.

Many cells have large, unpigmented plastids that have neither chlorophyll nor lipid pigments. These **leucoplasts** may be involved in various types of synthesis: Many types of fats and lipids are synthesized only in plastids and then transported to other organelles and inserted into their membranes. "Leucoplast" is a purely descriptive term—any colorless plastids, including proplastids and amyloplasts, can be considered leucoplasts.

Iron is an essential nutrient for plants and animals and is stored attached to a large protein. Although almost identical to human ferritin, the plant protein is called **phytoferritin.** Ferritin is found throughout animal cytoplasm and nuclei, but phytoferritin is stored almost exclusively in plastids. As much as 80% of the iron in leaves may be found in chloroplasts, and the leucoplasts and amyloplasts of seeds often have especially large amounts.

Examples of long-term starch storage: potatoes, yams, squash, and starchy seeds such as wheat, rye, oats, and corn.

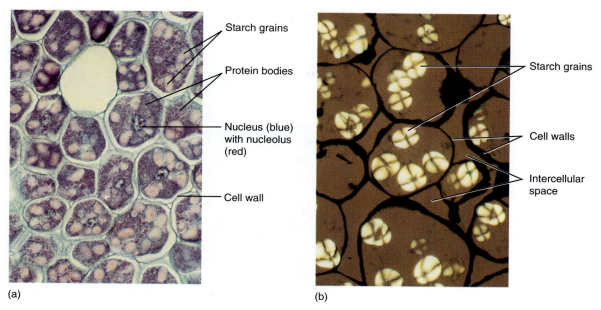

FIGURE 3.18
(a) Cells from a developing bean seed. The large pink bodies are starch grains; the small bluish-red ones are protein bodies. These colors are not natural. The material was stained to make the colorless starch and protein visible. The starch grains occur in plastids called amyloplasts; as the bean germinates, the starch is used as a source of energy. When the starch decreases, the amyloplasts convert to chloroplasts and turn green (× 150). (b) Starch grains that have been photographed in polarized light rather than being stained. Substances that have an orderly, crystalline structure shine brightly in polarized light (× 200).

RIBOSOMES

Immersed in the protoplasm are **ribosomes,** particles responsible for protein synthesis (Fig. 3.19). They are complex aggregates of three molecules of RNA (**ribosomal RNA**) and about 50 types of protein that associate and form two subunits (see Fig. 15.10). Compared with animal cells such as those in the liver and pancreas, most plant cells synthesize little protein and have few ribosomes. However, some do produce large amounts of proteins and are rich in ribosomes, such as the protein-rich seeds of legumes like peas and beans and the cells that secrete the digestive enzymes of insectivorous plants. Each molecule of messenger RNA is long enough for six to ten ribosomes to attach to it and read it simultaneously. The ribosomes are thus bound together by the messenger RNA, forming a cluster called a **polysome.**

Details of ribosome structure and protein synthesis are presented in Chapter 15.

ENDOPLASMIC RETICULUM

A typical plant cell is so small that diffusion, the random movement caused by molecular motion, carries a molecule to many different parts of the cell every second. Diffusion may be the only means by which small molecules like monosaccharides and cofactors move around the cell, but some large molecules such as proteins are carried by the **endoplasmic reticulum (ER),** a system of narrow tubes and sheets of membrane that form a network throughout the cytoplasm (Fig. 3.19).

A large proportion of a cell's ribosomes are attached to the ER, giving it a rough appearance; consequently, this ER is called **rough ER,** or **RER.** As an attached ribosome synthesizes a protein, it passes through the membrane and collects in the lumen. If the

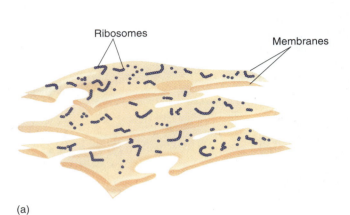

Ribosomes   Membranes

(a)

FIGURE 3.19
(a) Ribosomes only rarely occur free in the cytoplasm; instead, they are usually attached to membranes such as the type shown here, called endoplasmic reticulum. The amount of ribosomes attached is greater in cells that produce abundant protein (such as protein-rich seeds) or less in cells that synthesize little protein (such as the water-storage cells of Fig. 3.2a). (b) Electron micrograph of a cross-section of a hair cell that secretes protein-rich mucilage. The mucilage is synthesized in the endoplasmic reticulum, which is unusually abundant for a plant cell (× 3000). *(Courtesy of R. Fulginiti, University of Texas)*

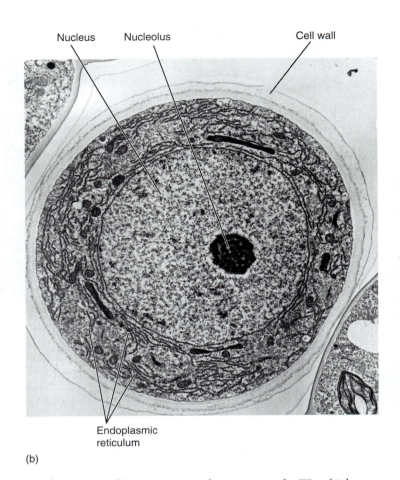

Nucleus   Nucleolus   Cell wall

Endoplasmic
reticulum

(b)

protein is a storage product, as in seeds of legumes, it merely remains in the ER, which may become quite swollen. But if the protein is to be secreted (digestive enzymes, mucilages, adhesive proteins, certain nectars), then its accumulation causes regions of the ER to form vesicles. These detach, move to the plasma membrane, and fuse with it, releasing their contents to the cell's exterior by exocytosis (see Fig. 3.9). In many cases the protein must be modified before export; if so, the ER pinches off only very small vesicles from regions where the ER is close to another organelle, the dictyosome, which carries out the protein modification.

Endoplasmic reticulum that lacks ribosomes is **smooth ER,** or **SER,** and it is involved in lipid synthesis and membrane assembly. The lipids range from simple to extremely complex; as they are produced, the lipids are inserted into the membrane, then vesicles form and pinch off, carrying the new membrane to other parts of the cell. Once the ER-derived vesicles reach the correct organelle, they fuse with it and the vesicles become a new patch of membrane in the organelle. SER is abundant only in cells that produce large amounts of fatty acids (cutin and wax on epidermal cells), oils (palm oil, coconut oil, safflower oil), and fragrances of many flowers.

DICTYOSOMES

Much of the material that is to be secreted by a cell must first be modified by a **dictyosome,** a stack of thin vesicles held together in a flat or curved array (Fig. 3.20). ER vesicles accumulate on one side of the dictyosome, then fuse together and form a wide, thin vesicle called a **cisterna** (pl.: cisternae) that becomes attached to the dictyosome (Fig. 3.21). Soon more ER vesicles gather next to this one and form a new cisterna. The first cisterna becomes embedded more deeply in the dictyosome as more vesicles accumulate on that side, which for obvious reasons is the **forming face.** At the other side, the **maturing face,** vesicles are being released; their contents have been processed. After separation, vesicles can move to

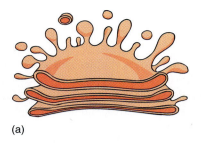

(a)

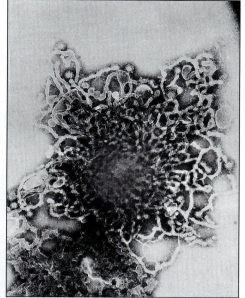

(c)

Vesicles fusing with plasma membrane

Dictyosomes

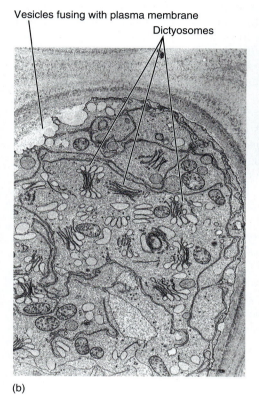

(b)

FIGURE 3.20

(a) Dictyosomes are stacks of flattened vesicles; the edges form a network of tubules and small vesicles. (b) With rapid dictyosome activity, as in these mucilage-secreting cells, the cytoplasm may almost fill with dictyosome vesicles, and fusion of vesicles with the plasma membrane is so abundant that the plasma membrane appears scalloped (×12,000). (c) Face view of dictyosome isolated from a plant cell; the network of peripheral tubules is visible (×60,000). (*b and c, Courtesy of H. Mollenhauer, Texas A&M University*)

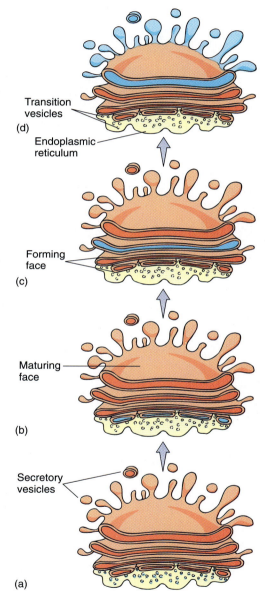

Transition vesicles

Endoplasmic reticulum

(d)

Forming face

(c)

Maturing face

(b)

Secretory vesicles

(a)

FIGURE 3.21

(a) Vesicles derived from endoplasmic reticulum migrate a short distance to a dictyosome-forming face. (b) At the forming face, the vesicles fuse into a new dictyosome vesicle, called either a vesicle or a cisterna. (c) The cisterna "moves through" the dictyosome as more vesicles form on one side while other vesicles are released from the maturing face (d).

the plasma membrane and release their contents. The outer edges of dictyosomes form an interconnected network of curving tubes, and these may absorb the contents from the center of the dictyosome cisterna, then detach and move away. It is not known why some dictyosomes concentrate material in the central cisternae whereas others use the peripheral ones.

Dictyosomes can form large, complex associations. In animal cells that secrete very large amounts of protein, hundreds of dictyosomes associate side by side and form a cup-shaped structure called a **Golgi body** or **Golgi apparatus.** In the Golgi body, the dictyosome's maturing faces are on the inner side of the cup, while the forming faces and associated ER are on the outside. Dictyosomes only rarely aggregate into Golgi bodies in plants, one example being in root hairs: All the dictyosomes are part of a giant Golgi body located at the tip of the hair where growth and cell wall formation occur.

Different types of processing may occur within a dictyosome: modification of the vesicle's membrane or modification of its contents. If the vesicle is to fuse with the plasma membrane after release, the vesicle membrane must be made similar to the plasma membrane. If the vesicle is to remain separate from all the other organelles, acting as a storage vesicle, its membrane must be made unique and incapable of fusion.

The alteration of the vesicle's contents involves the addition of sugars onto proteins, forming glycoproteins. Sugar-containing proteins occur in the plasma membrane, the cell wall, and as storage products in seeds. Strong evidence is accumulating that dictyosomes also polymerize sugars to polysaccharides used in cell wall construction.

The movement of vesicles from ER to dictyosomes and then to other sites was hypothesized on the basis of the presence of numerous vesicles located between ER and dictyosomes and the similarity of the contents of dictyosome cisternae to those of vesicles found fusing with the plasma membrane. This hypothesis predicts that if a secretory cell is given a very brief dose of radioactive sugar or protein precursors, much of the radioactivity is soon found in the ER and later in the dictyosomes. Even later, both are nonradioactive but vesicles at the plasma membrane and external material are radioactive. Experiments have verified both this prediction and the related hypothesis that new membrane is synthesized in the ER and then transported by vesicles to growing organelles. Although organelles appear to be distinct entities when viewed by light or electron microscopy, they are actually highly interrelated by this membrane flow. All membranes of the cell, except the inner membranes of mitochondria and plastids, actually constitute just one extensive system, the **endomembrane system.**

In a few cases, it has been possible to measure the rate of membrane flow. The insectivorous plant *Drosophyllum* has leaves covered with glands that produce a sticky secretion. The fluid contains digestive enzymes that are processed by dictyosomes in the gland's cells. Once an insect has been caught, each gland produces a visible drop of digestive fluid at a rate of 1.3 μm³ per minute. Dictyosome vesicles are about 0.15 μm in radius, so each has a volume of 0.014 μm³, and approximately 100 must fuse with the plasma membrane every minute to deliver the 1.3 μm³ of fluid and enzymes. Each vesicle has a surface area of 0.27 μm², so 27 μm² of vesicle membrane fuses with the plasma membrane every minute during secretion. From calculations of the average cell volume and surface area of the plasma membrane, these secretory cells could double their plasma membrane every 20 seconds. It is important for these cells to be able to retract membrane material from the plasma membrane and recycle it. The mucilage-secreting cells of root tips are more leisurely: Each cell has approximately 800 dictyosomes that together contribute 14 to 26 μm² of membrane per minute to the plasma membrane, which has a surface area of about 1000 μm². Each dictyosome receives a new cisterna every 20 to 40 minutes.

MICROBODIES

Viewed by electron microscopy, cells are seen to contain numerous small, spherical bodies about 0.5 to 1.5 μm in diameter (Fig. 3.22). These are so nondescript that it is not easy to

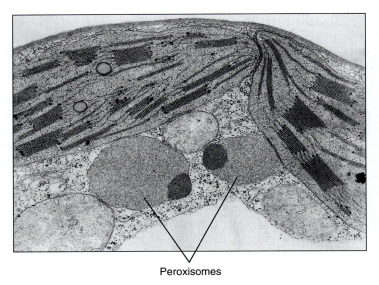

Peroxisomes

FIGURE 3.22
Electron micrograph showing two chloroplasts with two peroxisome-type microbodies next to them. In many plants, photosynthesis is accompanied by a process called photorespiration, in which peroxisomes produce the amino acid glycine (see Chapter 10 for details) (approx. × 66,000). *(BPS)*

tell exactly what they are: ER vesicles, dictyosome vesicles, or distinct organelles. However, by using special chemical reactions and stains, it has been possible to determine the contents and even the types of reactions in these bodies. Some are organelles now called **microbodies** and there are two classes: **peroxisomes** and **glyoxysomes.** Both types isolate reactions that either produce or use the dangerous compound peroxide, H_2O_2. If peroxide were to escape through the microbody membrane, it would damage almost anything it encountered. However, both types of microbody contain the enzyme catalase, which detoxifies peroxide by converting it to water and oxygen.

Peroxisomes are involved in detoxifying certain by-products of photosynthesis and are found closely associated with chloroplasts (see Chapter 10). Glyoxysomes, which occur only in plants, are involved in converting stored fats into sugars. They are important during the germination of fat-rich, oily seeds such as peanut, sunflower, and coconut.

All of the above, except ribosomes, are organelles composed of membranes; the following are nonmembranous organelles.

CYTOSOL

Most of the volume of cytoplasm is a clear substance called **cytosol** or hyaloplasm. It is mostly water, enzymes, and the numerous chemical precursors, intermediates, and products of enzymatic reactions (Table 3.8). Within cytosol are free ribosomes (not attached to RER), as well as skeletal structures—the microtubules and microfilaments.

MICROTUBULES

Microtubules (Fig. 3.23) are the most abundant and easily studied of the structural elements of a cell, and they have many functions. They act as a "cytoskeleton," holding certain regions of the cell surface back while other parts expand. Without them, cells would be just spheres, but by reinforcing specific areas, cell growth and expansion are directed to weaker areas. In other cases, microtubules assemble into arrays like an antenna which either catch vesicles and guide them to specific sites or cover a region, thereby excluding the vesicles

In animals, peroxisomes are abundant in liver and kidney cells, where they break down many foreign compounds that contaminate our food.

TABLE 3.8	*Subunits of the Cell*
Whole cell	
Cell wall	
Protoplasm	
Nucleus	
Vacuole	
Cytoplasm	
All remaining organelles	
Cytosol	

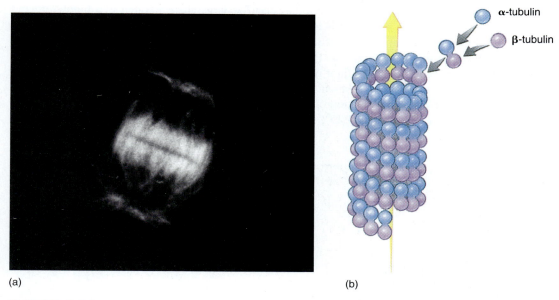

(a) (b)

FIGURE 3.23

(a) Microtubules that are part of a dividing nucleus and are involved in pulling chromosomes to the ends of the cells. A fluorescent dye was used to stain only microtubules, so no other organelles are visible ($\times$ 7000). *(Courtesy of Dr. Kevin C. Vaughn, Southern Weed Science Laboratory)* (b) Alpha and beta tubulin associate into a dimer called tubulin, and dimers aggregate into a microtubule. When no longer needed, microtubules depolymerize back to the monomers, which are recycled to build other microtubules.

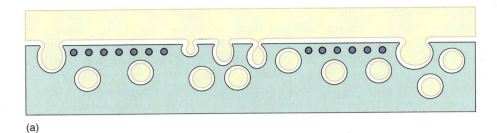

(a)

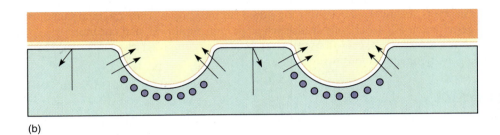

(b)

FIGURE 3.24
Microtubules (depicted here in purple) are often located next to the plasma membrane. (a) They may act as a screen that keeps vesicles away from the plasma membrane, or (b) they may pull the plasma membrane away from the cell wall, allowing material to accumulate there.

(Fig. 3.24). Finally, microtubules are the means of motility for both organelles and whole cells. The framework that moves chromosomes during division of the nucleus is composed of microtubules, and microtubules can attach to and move whole nuclei, mitochondria, and other organelles (see Fig. 4.12).

Microtubules are composed of just two types of globular protein, **alpha tubulin** and **beta tubulin,** which associate as dimers called tubulin that further crystallize into a straight tubule with a diameter of 20 to 25 nm. This is a reversible process; when a microtubule is no longer needed, it depolymerizes back into its component monomers, which disperse into the cytosol until the cell needs to assemble a new microtubule. Control of polymerization and depolymerization is not well-understood. When microtubules occur as individuals or small clusters, new tubulin dimers are added to or removed from one end automatically. In other instances, microtubules occur in large clusters, often in a highly ordered arrangement, and a small body is usually associated with the orderly production of microtubules. For example, when the nucleus undergoes division, an array of microtubules called the spindle is formed in the middle of the cell, and spindle microtubules push and pull the chromosomes to their proper positions.

In all animals and in some fungi and algae, a pair of organelles called **centrioles** is associated with the formation of the spindle. A centriole is made up of nine sets of three short microtubules (Fig. 3.25); the nine triplets are held together by fine protein spokes.

FIGURE 3.25
(a) The two members of a pair of centrioles are always located perpendicular to each other. Here, one is seen in transverse section, the other in longitudinal section (× 100,000). *(Biology Media)* (b) Each part of a centriole contains a circle of nine sets of three microtubules attached to each other by fine filaments along their sides.

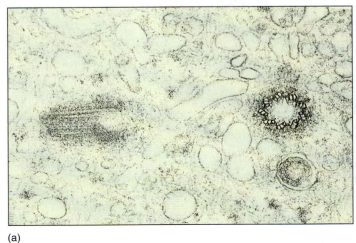

(a)

(b)

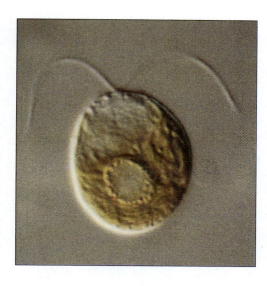

FIGURE 3.26
Flagella occur on many algal and fungal cells, especially the unicellular organisms. This is the alga *Dunalliela.* See also Figure 20.23. Most cells of multicellular organisms are not flagellated, but their sperm cells are (×500).
(Courtesy of G. Thompson; University of Texas)

Centrioles were assumed to be responsible for the organization and polymerization of the spindle microtubules even though plants never have centrioles. At present, however, centrioles are suspected to have nothing to do with spindle formation, even in animals. Rather, it may be that the pair of centrioles may be separated by the growth of the spindle, which pushes them to opposite ends of the cell so that each daughter cell receives one during cell division.

Much more elaborate sets of precisely arranged microtubules occur in **cilia** (sing.: cilium) and **flagella** (sing.: flagellum) (Figs. 3.26 and 3.27), which appear to be identical except that cilia are short (about 2 μm) and occur in groups, whereas flagella tend to be much longer (up to several micrometers) and usually occur either singly or in sets of two or four. Both are present on many types of algae and motile fungi, but in plants only the sperm cells have flagella. In flowering plants and conifers, no cells, not even sperm cells, ever have cilia or flagella.

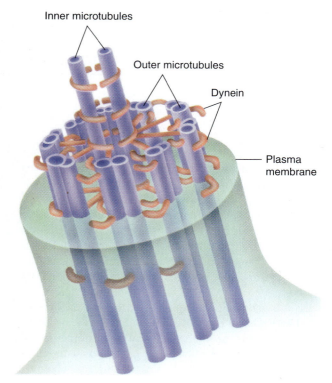

Inner microtubules
Outer microtubules
Dynein
Plasma membrane

FIGURE 3.27
Cilia and flagella are composed of two central single microtubules surrounded by nine sets of two microtubules. The doublets are not merely attached to each other; they actually share tubulin monomers. The outer member of each doublet has two short arms.

BOX 3.1   *Cell Storage Products—Crystals*

Many plants contain numerous crystals; the most common crystalline compound is calcium oxalate, which often forms round masses of angular crystals. We are not certain what role calcium oxalate crystals play in the biology of plants, but two hypotheses have been suggested which could both be correct.

One hypothesis postulates that the crystals are a means of reducing the amount of calcium ion (Ca^{2+}) in the cytoplasm. Calcium regulates the activity of many enzymes, activating some as its concentration increases and inactivating others. Consequently, it is important for the amount of free calcium to be controlled carefully. By combining calcium with oxalic acid and forming a crystal, an atom of calcium is inactivated as far as any enzyme is concerned.

The second hypothesis postulates that the presence of crystals makes plant tissues unpalatable for animals to eat. Consequently, tissues that have crystals are protected from feeding animals. If this hypothesis is a good model of reality, we would expect to find crystals particularly abundant in important tissues.

The accompanying illustrations are micrographs of some tissues that are consistent with the second hypothesis. In each, crystals occur predominantly near tissues that might be the target of insect feeding. The micrographs on the left of each pair were made with ordinary light; those on the right used polarized light to reveal crystals more clearly.

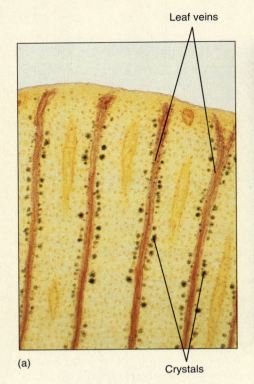

(a)

(b)

Leaf veins

Crystals

(a) A leaf clearing of maidenhair tree (*Ginkgo*), showing several red-stained leaf veins that conduct sugars out of the leaf. Such veins are the targets of aphids and other sucking insects ($\times 15$).

(b) The same tissue, in polarized light ($\times 15$).

In cross-section each cilium or flagellum has a "**9 + 2**" arrangement: Nine pairs of fused microtubules surround two individual microtubules. The outer doublets each have two arms composed of the protein dynein, and each doublet is connected to the central pair of microtubules by protein spokes. The dynein arms convert the chemical energy of ATP into kinetic energy and bend, "walking" along the adjacent microtubule doublet. One set of doublets slides relative to the adjacent set, causing that side of the cilium or flagellum to become shorter and bend the structure. Then, as another set of microtubules slides, the cilium or flagellum bends in a different direction. The result is a powerful beating motion. If the cilia or flagella are on small organisms such as algae, fungi, or protozoans, the organism swims rapidly and gracefully.

Cilia and flagella can polymerize and depolymerize rapidly, but they never form autonomously; each is always associated with a **basal body**. Basal bodies appear to be identical to centrioles by electron microscopy. It was assumed that basal bodies organize the formation of flagellar microtubules, but recent studies have shown that as a flagellum grows, new monomers of tubulin and dynein are added to the tip, not the base where the

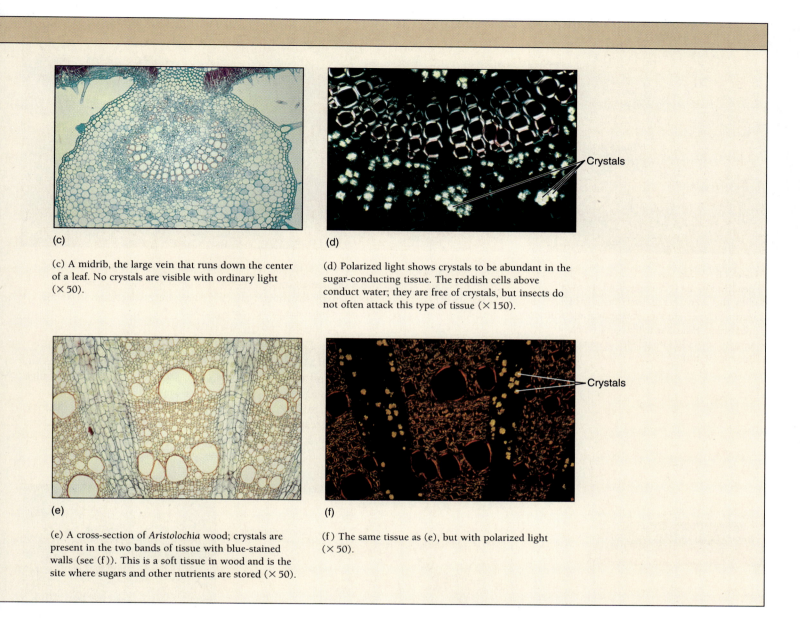

(c) A midrib, the large vein that runs down the center of a leaf. No crystals are visible with ordinary light (× 50).

(d) Polarized light shows crystals to be abundant in the sugar-conducting tissue. The reddish cells above conduct water; they are free of crystals, but insects do not often attack this type of tissue (× 150).

(e) A cross-section of *Aristolochia* wood; crystals are present in the two bands of tissue with blue-stained walls (see (f)). This is a soft tissue in wood and is the site where sugars and other nutrients are stored (× 50).

(f) The same tissue as (e), but with polarized light (× 50).

basal body is located. The exact relationship between flagella and basal bodies is still not known.

MICROFILAMENTS

Like microtubules, **microfilaments** are constructed by the assembly of globular proteins —in this case just one type, **actin.** Microfilaments are narrower than microtubules (only 3 to 6 nm in diameter), and they have been implicated tentatively in different types of structure and movement.

STORAGE PRODUCTS

Many cells exist in an environment in which resources alternate between abundance and scarcity. To survive times of scarcity, cells must accumulate and store extra nutrients. Most often, the reserves consist of sugars that have been polymerized into starch in amyloplasts

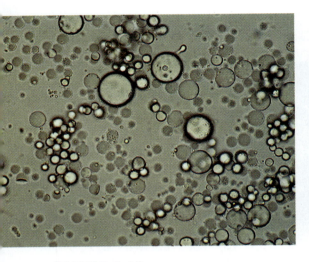

FIGURE 3.28
Avocado contains very large numbers of oil bodies; this view was prepared merely by smearing a small amount of ripe avocado fruit on a microscope slide. Virtually every visible thing is an oil body; other organelles are much less abundant (× 200).

FIGURE 3.29
Calcium oxalate crystals. When calcium is part of a crystal, it is physiologically inactive and cannot affect membranes or the tertiary structure of proteins. These are cubic crystals in wood (× 1000). *(Courtesy of G. Montenegro, Universidad Católica, Santiago, Chile)*

or converted into lipids and stored as large oil droplets (**spherosomes** or **lipid bodies**) in oily material like peanuts and sunflower seeds (Fig. 3.28). But with many other storage products, the function and advantages are not so obvious. Many plants store crystals of calcium oxalate or calcium carbonate (Fig. 3.29); others accumulate large amounts of silica, tannins (Fig. 3.30), or phenols. Because plants have no excretory mechanism, numerous waste products must be stored within the cells.

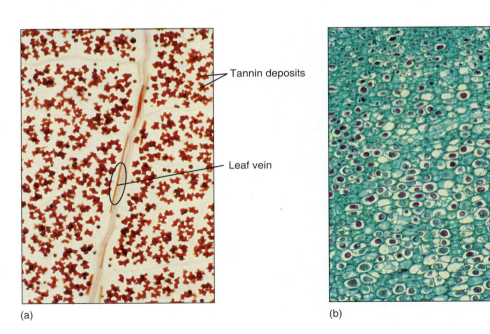

FIGURE 3.30
(a) This type of preparation is called a leaf clearing: A leaf was treated to make its tissues transparent; then it was stained to reveal certain contents. The dark red, irregularly shaped vacuolar contents are tannins, which can denature proteins. They have a bitter taste and damage the mouth and stomach proteins of insects that try to eat them. The band running vertically through the micrograph is a leaf vein (× 200). (b) The bark of a cactus. As its cells aged, they accumulated irregular masses of tannin that have stained red (× 80).

CELL WALL

Almost all plant cells have a **cell wall**; only sperm cells of some seed plants lack one. It is often regarded as simply an inert secretion, providing only strength and protection to the protoplasm inside. However, considerable metabolism occurs in the wall, and it should therefore be considered a dynamic, active organelle.

The cell wall contains a considerable amount of the polysaccharide cellulose (Figs. 3.5b and 3.31a). Adjacent, parallel cellulose molecules crystallize into an extremely strong **microfibril** 10 to 25 nm wide. Numerous microfibrils are wound around the cell, completely covering the plasma membrane. Each cellulose polymer grows only at one end, where a complex of enzymes adds new glucose residues, one molecule at a time (Fig. 3.31b). The enzymes are intrinsic proteins of the plasma membrane, and as they add new sugars to the chain, the enzymes float forward in the membrane (the chain is too heavy to be pushed backward). New cellulose molecules can be added only on the inner side of the wall, adjacent to the plasma membrane.

Cellulose microfibrils are bound together by other polysaccharides called **hemicelluloses,** which are produced in dictyosomes and brought to the wall by dictyosome vesicles. Hemicelluloses are deposited between the cellulose microfibrils and bind chemically to the cellulose, producing a solid structure that resembles reinforced concrete (Fig. 3.32). In multicellular plants, the wall of one cell is glued to the walls of adjacent cells by an adhesive layer, the **middle lamella,** composed of a third class of polysaccharides, **pectic substances.**

All plant cells have a thin wall called the **primary cell wall** (see Fig. 3.5a). In certain cells that must be unusually strong, the protoplast deposits a **secondary cell wall** between the primary wall and the plasma membrane (see Fig. 3.5b). The secondary wall is usually much thicker than the primary wall and is almost always impregnated with the compound **lignin,** which makes the wall even stronger than hemicelluloses alone can make it. Lignin resists chemical, fungal, and bacterial attack. Both primary and secondary cell walls are permanent; once deposited, they are almost never degraded or depolymerized, as can be done with microtubules and microfilaments.

Cells with secondary walls, called sclerenchyma cells, are discussed in Chapter 5.

FIGURE 3.31
(a) Two layers of cellulose microfibrils are visible; other layers are present deeper in the wall. Each layer provides strength in the direction parallel to the microfibrils. Other wall components have been dissolved away to reveal the cellulose (× 5000). *(Courtesy of Dr. Shun Mizuta, Kochi University, Japan)* (b) Cellulose-synthesizing enzymes are large proteins embedded in the plasma membrane; several proteins form a cluster called a rosette. Sugar monomers are absorbed by the enzymes of the membrane inner face, then passed across the membrane and polymerized into cellulose on the outer face. The growing microfibril extends out from the rosette and lies against microfibrils already in place. Because the microfibril cannot push into the wall, the rosette is pushed through the membrane in the direction of the arrows as the microfibril grows.

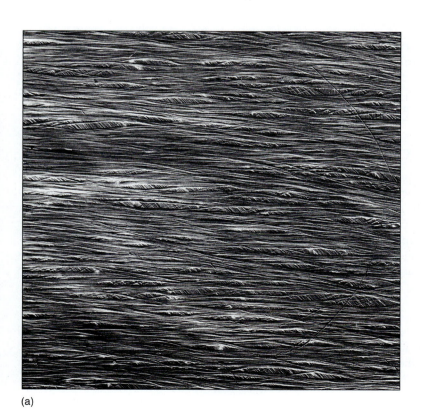

(a)

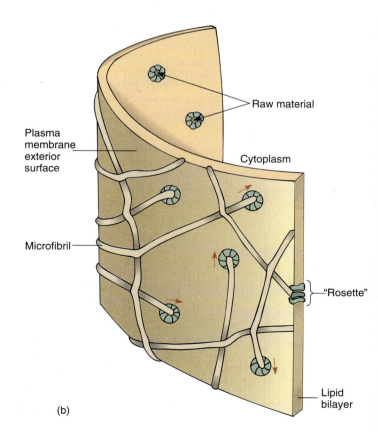

(b)

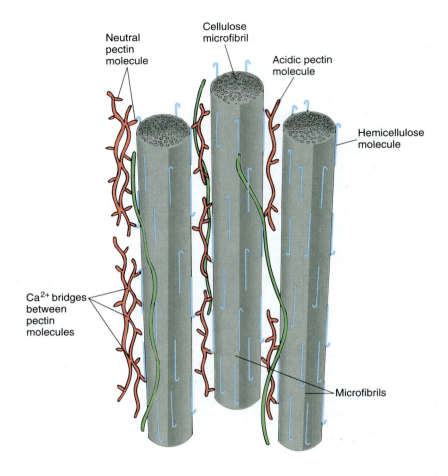

Neutral
pectin
molecule

Cellulose
microfibril

Acidic pectin
molecule

Hemicellulose
molecule

Ca²⁺ bridges
between
pectin
molecules

Microfibrils

FIGURE 3.32

Although cellulose microfibrils lie close together, they do not bond with each other, so a wall of pure cellulose is weak. Hemicelluloses and pectins are short, branched molecules that interact with several adjacent microfibrils, linking them together and forming a solid three-dimensional mesh. The relative amounts of these components can vary, resulting in some walls that are flexible, others that are rigid; some strong, others weak.

Chapter 19 contains details of the cell biology of prokaryotes; Chapter 20 describes fungal cells.

FUNGAL CELLS

Cells of fungi are similar to plant cells, with two important differences: (1) They do not contain plastids of any type, and (2) their walls contain **chitin,** not cellulose, except for one unusual group of fungi, the Oomycetes. Chitin is physically similar to cellulose, being tough, inflexible, and insoluble in water, but it contains nitrogen and is synthesized by a different mechanism than that used for cellulose.

ASSOCIATIONS OF CELLS

In unicellular organisms, such as simple algae, protozoans, and most bacteria, each cell is a complete organism and does not interact directly with other cells. Such cells can communicate indirectly, at least to a small extent, by releasing chemicals that affect the other cells, but such interactions are rare, usually occurring only during the attraction of other cells during sexual reproduction.

In multicellular organisms each cell automatically, unavoidably interacts with its neighboring cells; they must share the same sources of photosynthate, oxygen, carbon dioxide, salts, and water. Whereas a unicellular organism can merely excrete its wastes across its plasma membrane, cells of a multicellular organism are not so free to do so.

Just as important are cellular interactions by which cells not only sense that they are part of a larger organism but also identify which part they are and how they should differentiate. In a developing embryo, the proper cells must be instructed to begin forming an embryonic shoot while other cells are induced to form the embryonic root, vascular tissues, seed leaves, and so on. This requires extensive, sophisticated intercellular communication.

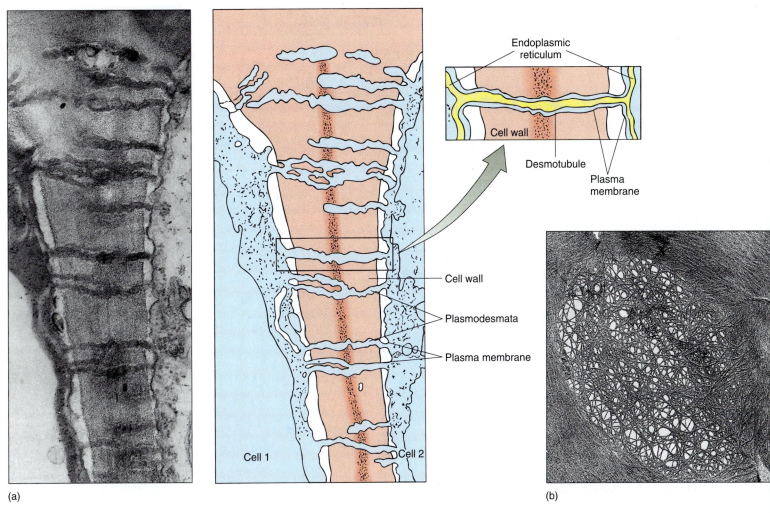

(a)   (b)

FIGURE 3.33
(a) Plasmodesmata are complexes that consist of fine holes in the primary walls; they also contain plasma membrane, liquid, and a tubule (called a desmostubule) attached to the endoplasmic reticulum (× 68,000). *(Courtesy of W. W. Thomson and Mark Lazzard, University of California, Riverside)*
(b) Face view of a cell wall, showing an oval-shaped thin area, a primary pit field. The small open areas outlined by single cellulose microfibrils are holes where plasmodesmata were located. Protoplasm was removed during specimen preparation (× 45,000). *(Courtesy of Dr. Alan L. Koller, University of Georgia, and Dr. Thomas L. Rost, University of California, Davis)*

One communication method is like that used by unicellular organisms: A cell secretes specific compounds that "inform" the surrounding cells of what it is doing metabolically and developmentally. A second method is connections between the cells. Direct physical contact between cells, as exists in animals, cannot occur in plants because the two primary walls and middle lamella are located between any two adjacent protoplasts. However, plant cells are interconnected by fine holes (**plasmodesmata**; sing.: plasmodesma) in the walls (Fig. 3.33). A plasmodesma is only about 40 nm in diameter, but the plasma membrane of one cell passes through it and is continuous with the plasma membrane of the adjacent cell. A small channel of cytosol also passes through, as does a short section of specialized ER.

The abundance of plasmodesmata in a wall is quite variable; they can occur singly or in clusters of 10 to 20 or more. In regions of clustered plasmodesmata, the two primary walls are often particularly thin, and the area is called a **primary pit field** (Fig. 3.33b). In tissues in which considerable transport occurs (into glandular or conducting cells), many primary pit fields and plasmodesmata are present, but in regions where little movement of material happens, few plasmodesmata are found (Table 3.9, p. 82).

BOX 3.2   *The Metric System and Geometric Aspects of Cells*

THE METRIC SYSTEM

In 1791, the French Academy proposed that a new system of measurement be adopted to simplify and regularize weights and measures. This **metric system** is used not only for all scientific measurements but also for engineering, commercial, and ordinary purposes in every country of the world except the United States (see conversion tables on page 81). The fundamental unit is the meter, but for long distances the kilometer is used, and for small objects, centimeters, millimeters, and so on are more convenient. Volumes are also based on the length measurements, and weights are based on the gram, the weight of one cubic centimeter of water. The metric system has two advantages: (1) Measurements and objects made in one country can be understood or used immediately in any other country, except the United States; (2) measurements in one unit can be converted easily to other units: 35,645 meters is 35.645 kilometers, but 35,645 feet is how many yards or miles?

1 kilometer (km) = 1000 meters (m)

1 meter = 1000 millimeters (mm)

1 millimeter = 1000 micrometers (μm)

1 micrometer = 1000 nanometers (nm)

10^9 nm = 10^6 μm = 10^3 mm
$= 10^0$ m = 10^{-3} km

Notice that these are based on multiples of 1000. The centimeter (cm) is only 1/100 meter or 10 mm, but it is such a convenient size for measuring plants and animals that it is used even though it disrupts the otherwise perfect regularity of increments of 1000.

GEOMETRY OF CELLS

Too few botany students bother to learn or apply mathematical analysis to their studies of plants, probably because botany professors and textbooks do not present many mathematical treatments. But most techniques and computations are remarkably simple, and a quantitative understanding of plants provides extremely valuable insight into their biology. The following formulas do not require mathematical knowledge:

area of a rectangle = (length) $\times$ (width)

area of a triangle = $\frac{1}{2}$(length) $\times$ (width)

volume of a cubic space
$\quad$ = (length) $\times$ (width) $\times$ (height)

diameter of a circle = 2 $\times$ (radius)

circumference of a circle = 2π $\times$ (radius)

area of a circle = π $\times$ (radius)^2

surface area of a sphere = 4π $\times$ (radius)^2

volume of a sphere = $4\pi/3$ $\times$ (radius)^3

volume of a cylinder
$\quad$ = (area of base) $\times$ (height)
$\quad$ = π $\times$ (radius)^2 $\times$ (height)

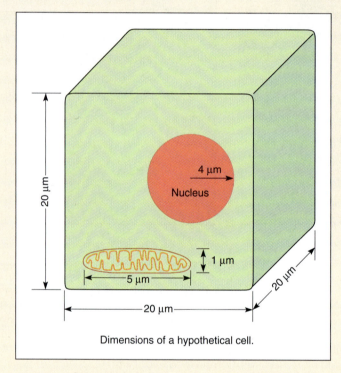

Dimensions of a hypothetical cell.

Dimensions of a hypothetical cell. Notice that the mitochondrion is not drawn to scale.

Because plasmodesmata are actually cytoplasmic channels from one protoplast to another, the individuality of the cells is diminished; all the protoplasm within a single plant is part of one interconnected mass, the **symplast.** In some of the nutrient-conducting tissues (sieve tube members and companion cells—see Chapters 5 and 12) and also in some of the reproductive tissue (microspore mother cells and endosperm—see Chapter 9), plasmodesmata are so large and numerous that the cells act more like one single large cell than like individual cells. In most tissues, however, any two adjacent cells act rather independently; the communication through the plasmodesmata is not enough to prevent individuality.

The fact that walls keep protoplasts physically separated from each other has another consequence. Walls act as a second, nonliving compartment inside plants. Water moves

surface area of a cylinder
= (circumference of circle) × (height)
   + (ends)
= π × (diameter)
   × (height) + (ends)

Many plant cells are almost cube shaped and have edges about 20 μm long; the volume of such a cell is

volume of a cube
= L × W × H
= 20 μm × 20 μm × 20 μm
= 8000 μm^3

Don't forget to multiply the units to get μm^3 (see figure).

Now consider the nucleus of the cell. Many plant nuclei have a diameter of about 8 μm, so their radius is 4 μm. The volume of a nucleus is V = (4) (3.141)/3 × (4 μm)^3 = 4.188 × 64 μm^3 = 268 μm^3. The nucleus, although an important part of a cell, may not be a large part. In our average cell of 8000 μm^3 volume, the proportion that is nucleus is (volume of nucleus)/(volume of cell) × 100% = 268 μm^3/8000 μm^3 × 100% = 3.3%, a very small part of the cell volume.

Mitochondria are about 1 μm in diameter and quite variable in length; let us assume that we have measured many and found the mean length to be 5 μm. Mitochondria are almost cylinders, so we can use that formula to calculate their volume:

volume of a cylinder = (area of a
   circular cross-section) × height
= (3.141) × (0.5 μm)^2 × (5 μm)
= 0.7852 μm^2 × (5 μm)
= 3.93 μm^3

Mitochondria often constitute about 7.5% of a cell's volume; for our typical cell, that would be 8000 μm^3 × 0.075 = 600

μm^3. How many actual mitochondria would there be in each cell? If each mitochondrion has a volume of 3.93 μm^3, the number is (total volume)/(volume of each

Length Conversion

1 in = 2.5 cm	1 mm = 0.04 in
1 ft = 30 cm	1 cm = 0.4 in
1 yd = 0.9 m	1 m = 40 in
1 mi = 1.6 km	1 m = 1.1 yd
	1 km = 0.6 mi

Weight Conversion

1 oz = 28 g	1 g = 0.035 oz
1 lb = 0.45 kg	1 kg = 2.2 lb

Volume Conversion

1 tsp = 5 mL	1 mL = 0.03 fl oz
1 tbsp = 15 mL	1 L = 2.1 pt
1 fl oz = 30 mL	1 L = 1.06 qt
1 cup = 0.24 L	1 L = 0.26 gal
1 pt = 0.47 L	
1 qt = 0.95 L	
1 gal = 3.8 L	

Temperature Conversion / Interval Equivalents

Temperature Conversion	Interval Equivalents	
$°C = \dfrac{(°F - 32) \times 5}{9}$	°C	°F
	1° =	1.8°
	5° =	9°
$°F = \dfrac{°C \times 9}{5} + 32$	10° =	18°

mitochondrion) = 600 μm^3/3.93 μm^3 = 153 mitochondria in each cell.

The surface of a cell or organelle is the space through which material enters and leaves the object by diffusion, facilitated diffusion, or active transport. Objects with large surface areas can absorb or lose material more rapidly than those with smaller surface areas. Also, surface area is a measure of the room available for placing membrane-bound enzymes and other intrinsic membrane proteins. In our hypothetical cubical cell, each surface has an area of 20 μm × 20 μm = 400 μm^2. A cube has six sides, so the total surface area of the plasma membrane is 6 × 400 μm^2 = 2400 μm^2. That is about one thirtieth of the surface area of the period at the end of this sentence. The nucleus of the cell has a surface area of 4π × (radius)^2 = 12.56 × 16 μm^2 = 201 μm^2, less than one tenth that of the cell. The cylindrical mitochondria each have a surface area of their outer membrane of 3.141 × 1.0 μm × 5 μm = 15.7 μm^2 (we are ignoring the ends). But what of their inner membrane? Assume that measurements reveal that each crista is a cylindrical tube 0.2 μm wide and 0.9 μm long and each mitochondrion contains an average of 45 cristae; then the surface area of each is 3.141 × 0.2 μm × 0.9 μm = 0.565 μm^2. The 45 in each mitochondrion have a total surface area of 45 × 0.565 μm^2 = 25.4 μm^2 available for respiratory enzymes. The 153 mitochondria of the cell would have a total inner surface area of 3886 μm^2, greater than that of the plasma membrane area (2400 μm^2). Thus, folding the inner membrane into cristae provides a large amount of extra surface area for enzymes.

through cell walls by capillary action, just as it moves through tissue paper. In addition, many cells do not abut each other tightly but instead have **intercellular spaces** between them, at least at their corners (see Fig. 3.34). In some tissues, especially leaves, cells are so loosely connected that most tissue volume is intercellular space; less than half is actually symplast. These spaces plus the cell wall constitute the **apoplast**; the apoplast and the symplast together make up the whole plant. The apoplast acts as a series of channels and spaces that permit the rapid diffusion of gases, which is necessary because plants do not have lungs. Diffusion through a gas-filled space is approximately 10,000 times faster than through a liquid-filled space. Without an apoplast, large, bulky plant tissues such as tree trunks, tubers, and fruits would be impossible; the interior tissues would suffocate.

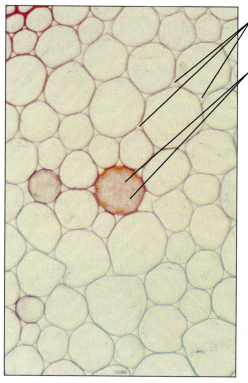

Intercellular spaces

Primary pit
fields
(white areas)

FIGURE 3.34
In this stem tissue of *Ranunculus* (buttercup),
the corners of cells are slightly rounded
and do not fit snugly together, creating
intercellular space. Intercellular spaces here
are rather small, but in other tissues they are
much larger. Most cells appear to be empty
because when this tissue was being prepared,
it was cut into thin slices (called sections),
and both the front and the back of the cell
were cut away. In the red cell in the center,
the section contains the back wall, so the cell
appears to be full even though it is not. The
pale white spots in the back wall are primary
pit fields; plasmodesmata are much too small
to be seen here ($\times 180$).

TABLE 3.9   *Frequency of Plasmodesmata*

Tissue		Plasmodesmata/ μm^2
Area of expected high rates of transport		
Salt bush (*Tamarix*)	Gland cell	17
Russian thistle (*Salsola*)	Sugar-loading cell	14
Pine (*Pinus*)	Sugar-loading cell	20–25
Abutilon	Nectary cell	12.5
Area of expected low rates of transport		
Oat (*Avena*)	Leaf cell	3.6
Onion (*Allium*)	Root cell	1.5
Tobacco (*Nicotiana*)	Outer stem tissue	0.2

SUMMARY

1. All organisms are composed of cells, the fundamental units of life. Most cells contain the same types of organelles, and there are only a few basic types of cells.

2. In multicellular organisms, cells become specialized for specific tasks, making the entire organism more efficient. However, each cell then becomes dependent on the other cells of the organism.

3. Selectively permeable membranes and active transport are critically important for life; they make compartmentalization possible, in which an enclosed compartment has a chemical composition and metabolism distinct from those outside the compartment.

4. The outermost membrane of a cell is the plasma membrane; the protoplasm it surrounds is a protoplast.

5. Plant cells contain the following membrane-bounded organelles:
nucleus—heredity and control of metabolism
vacuole—temporary or long-term storage, growth
mitochondria—respiration
plastids—photosynthesis (chloroplasts)
endoplasmic reticulum—intracellular transport, lipid synthesis
dictyosomes—material processing
microbodies—detoxification or lipid metabolism

6. Plant cells contain these nonmembranous organelles:
ribosomes—protein synthesis
cytosol—the liquid matrix surrounding all other organelles

microtubules and microfilaments—skeletal structure, movement of organelles
cell wall—structure, cell shape
various storage products such as crystals and lipid droplets

7. Animal cells do not have a cell wall, plastids, glyoxysomes, or a central vacuole; fungal cells do have a central vacuole, but they have no plastids and their wall is made of chitin, not cellulose.

8. A plant is composed of both its protoplasm (symplast) and its intercellular spaces and walls (apoplast). The apoplast can be important for diffusion of air and liquids.

IMPORTANT TERMS

apoplast
bilayer
cell wall
centriole
chloroplast
chromoplast
cilium
compartmentalization
cytoplasm
cytosol

dictyosome
endocytosis
endoplasmic reticulum
    (RER and SER)
eukaryotic cell
exocytosis
extrinsic protein
flagellum
fluid mosaic membrane
freely permeable membrane

glycoprotein
Golgi body
impermeable membrane
intercellular space
intrinsic protein
lumen
microbody
microfilament
microtubule
middle lamella

mitochondrion
molecular pump
nucleolus
nucleus
organelle
plasmodesma
plastid
primary pit field
prokaryotic cell
ribosome

selectively permeable
    membrane
symplast
tonoplast
tubulin
vacuole

REVIEW QUESTIONS

1. Why are cells considered the basic units of life? Describe the selective advantage that multicellular organization offers. What are the disadvantages?

2. What are the two basic components of membranes? What are three ways that material can move from one side of a membrane to the other? Which method requires the plant to use energy?

3. Draw a "typical" plant cell and include all the organelles mentioned in this chapter. Most cells have specialized functions and increased proportions of certain organelles. How would you change your drawing if the cell is involved in photosynthesis? Is part of a yellow or orange petal? Secretes protein-rich digestive fluid? Needs a great deal of ATP? Stores starch? Has just grown a great amount?

4. Which of the organelles in Question 3 are composed primarily of membranes? What are some of the ways that the membranes participate in the metabolic activities of those organelles?

5. Describe how material is brought to a dictyosome. What happens to material while it is in a dictyosome and how is it released? What are some of the things that might happen to it after it is released?

6. How do microtubules act as a cytoskeleton? Are there times when they are particularly abundant? What are centrioles, basal bodies, and flagella?

7. What are the most abundant components of a plant cell wall? How do the components interact and how are they arranged with respect to each other? What is the symplast and the apoplast in plants?

BotanyLinks .net Questions

www.jbpub.com/botanylinks

Visit the .net Questions area of BotanyLinks (http://www.jbpub.com/botanylinks/netquestions) to complete this question:

1. Plant cells differ from those of animals. How do they differ? Go to the BotanyLinks home page to investigate.

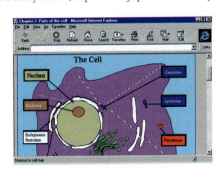

BotanyLinks includes a Directory of Organizations for this chapter.

4

GROWTH AND DIVISION OF THE CELL

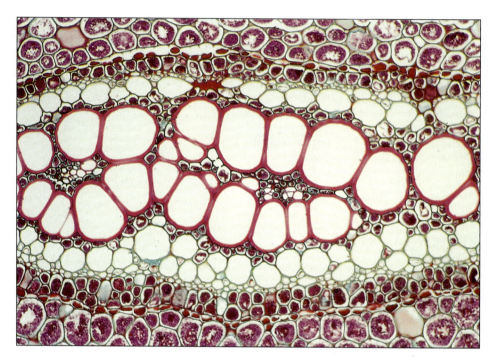

The many cells of a plant's body are produced by nuclear and cell division.

CONCEPTS

The life cycle of individual organisms includes the stages of initiation, growth, and death. Individual cells also have a life cycle, the **cell cycle.** Cells are initiated by division of a mother cell, grow for a period, and usually cease to exist by dividing and producing two daughter cells (Fig. 4.1). This is not a real death because the substance of the mother cell continues to exist in the daughter cells.

As a multicellular organism grows, this type of cell cycle is common (about 25 million cell divisions occur every second in your own body). But as parts of a plant reach their final form, most cells stop dividing **(cell cycle arrest)** and enter an extended period of growth, during which they differentiate and mature. For example, leaf cells stop dividing when the leaf is only a few millimeters long; they continue to grow as the leaf expands and then function for the rest of the leaf's life. Leaf cells cannot normally be stimulated to begin dividing again; if the leaf is damaged, very little regeneration is possible. In contrast, cortex cells in the stem or root also stop dividing before the organ is mature, but if damage occurs the remaining cells resume division and remain active until the damaged portion has been sealed off with a layer of protective cork.

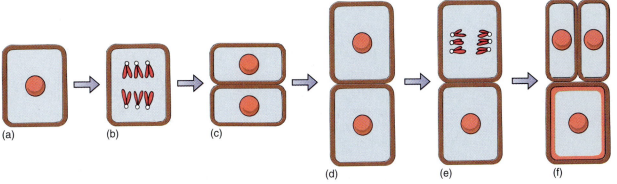

(a)   (b)   (c)   (d)   (e)   (f)

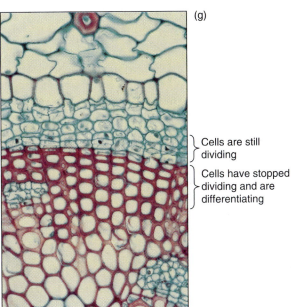

(g)

Cells are still dividing

Cells have stopped dividing and are differentiating

FIGURE 4.1

The cell cycle consists of division and growth phases. (a) The cell has grown and is ready for division; (b) first the nucleus divides, (c) then the cytoplasm is divided by the formation of a new wall. (d) The new cells enter a new cell cycle by growing. (e) Then one divides again, but the other may begin to differentiate; in this example its wall (red) thickens as it matures into a fiber. (f) The upper cell has finished dividing, and both may enter new cell cycles. The lower cell continues differentiation; in some fiber cells the nucleus may divide once or twice, but usually once fiber differentiation begins, the nucleus never divides again. (g) A cross-section of a set of cells similar to those depicted in (a) to (f); the red cells have stopped dividing and are differentiating into fiber cells. The blue cells with thin walls are still capable of growth and cell division (× 250).

Some cells live for many years, even hundreds of years, but others die shortly after they mature. Bark cells are more protective if they are dead (Fig. 4.2); many flower parts die only a few days after the flower first opens; and gland cells often die after a brief period of secretion.

Some cells never stop dividing. Cells in the growing points at the tips of roots and shoots are constantly cycling (Fig. 4.3), as are those that form wood and bark (the cambium). When a cambial cell divides, one of the new daughters becomes a wood or bark cell and the other remains part of the dividing layer. The wood or bark cell divides a few times, then differentiates, matures, and usually does not divide again. The daughter cell that stays in the dividing layer grows back to its original size and divides again.

The cell cycle can be divided into a growth phase and a division phase.

Cells of stem (cortex)

Living bark cells

Dead bark cells

FIGURE 4.2

Cells in this geranium bark divide only a few times, then differentiate by placing antimicrobial compounds in their walls along with waterproofing compounds. The cells then die and degenerate to cell walls without protoplasts. The lack of a protoplast is selectively advantageous because if a fungus or insect does penetrate the wall, it encounters no nutritious protoplasm and starves (× 120).

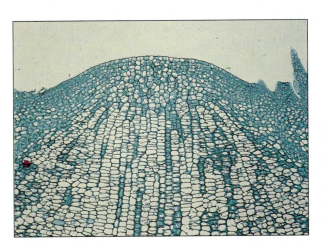

FIGURE 4.3

The growing shoot tip of a cactus (*Cleistocactus*). All its cells are alive and undergo repeated cell division. One of the basic roles of shoot tips is to convert nutrients into new cells that can later differentiate into shoot tissues. At the base, cells are lightly stained because their large vacuoles do not absorb the stain; these cells cycle more slowly than the smaller, less vacuolate cells at the tip (× 80).

Plant Biology Tutor CD-ROM

The length of a cell cycle varies tremendously, depending on the type of cell, the type of plant, its health, age, temperature, and many other factors.

GROWTH PHASE OF THE CELL CYCLE

In the 1800s, when the cell cycle was first being studied intensively, researchers gave greatest attention to the division activities because many events could be identified. They assumed that between divisions cells were "resting," so the growth phase was called the **resting phase,** or **interphase** (Fig. 4.4). We now know that the cell is actually most active during interphase, and it is possible to detect three distinct phases within interphase: G_1, S, and G_2.

G_1 PHASE

In G_1 (or **gap 1**), the first stage after division, the cell is recovering from division and conducting most of its normal metabolism (Fig. 4.4). One important process is the synthesis of nucleotides that are used for the next round of DNA replication. In single-celled organisms such as some algae, the cell cycle may be as brief as several hours; short cycle times also occur in rapidly growing embryos and roots. On the other hand, cell cycle times of 2 to 3 days or even weeks or months are not unusual in tissues or plants that grow slowly. During winter dormancy, the cell cycle may last from autumn until spring; when dormancy is broken and metabolism accelerates, the cell cycle may last only a few hours. The actual length of time that any cell spends in G_1 is similarly variable. Although there are exceptions, in general G_1 is the longest part of the cell cycle (Table 4.1). Once a cell undergoes cell cycle arrest, stops dividing, and begins to differentiate and mature, it may enter a state similar to G_1 and remain in it for the rest of its life.

S PHASE

During the **S** (or **synthesis**) phase, the genes in the nucleus are replicated (see Chapter 16). A gene is a polymer of nucleotides, and each gene has a unique sequence of nucleotides. Although many genes may have similar sizes, different genes never have exactly the same sequence of nucleotides. It has been estimated that many higher plants and animals need about 100,000 types of genes to store all the information required to make the proper enzymes, structural proteins, and hormones necessary for the organism's life. The whole complex of genes for an organism is its **genome.**

Obviously, if each individual gene floated around the cell or nucleus as an independent piece of DNA, then during division, it would be extraordinarily difficult to find each of

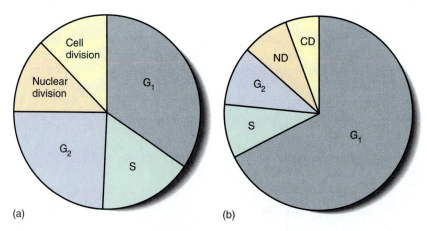

(a)   (b)

FIGURE 4.4

Comparison of cell cycles of different tissues or species. In a shoot tip (a), cell division may occur as soon as the cell grows to its proper size, and G_1 is only a small part of the cell cycle. Below the shoot tip (b), cells still divide but spend more of their time growing and carrying out nondivision metabolism; G_1 is a greater proportion of the cell cycle. Note that both (a) and (b) give only the relative amount of time spent in each phase, not the absolute time in hours or days. The cell cycle in cells like (b) is often much longer than in shoot tips (a).

TABLE 4.1 *Cell Cycle Phase Durations in Root Tip Cells and Cell Cultures*

Species	Cycle Phase Duration (Hours)				
	Total	G_1	S	G_2	M
Daucus carota (carrot)					
Root tip	7.5	1.3	2.7	2.9	0.6
Percent of cell cycle	100%	17%	36%	39%	8%
Daucus carota					
Cell culture	51.2	39.6	3.0	6.2	2.4
Percent of cell cycle	100%	77%	6%	12%	5%
Haplopappus gracilis					
Root tip	10.5	3.5	4.0	1.4	1.6
Zea mays (corn)					
Root tip	9.9	1.7	5.0	2.1	1.1
Allium cepa (onion)	17.5	1.5	10	4.0	2.0

Note that when carrot cells are grown in cell culture, the cell cycle becomes much longer, mostly because of G_1, which goes from being only 17% of the cell cycle in roots to 77% in cell culture.

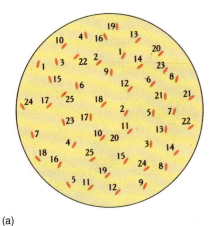

(a)

(b)

those 200,000 genes (100,000 originals and 100,000 copies) and make certain that each daughter cell received one of each (Fig. 4.5). This is not a problem, however, because all genes are attached to other genes by short pieces of linking DNA. Thousands of genes can be attached together in a linear sequence; the whole structure is a **chromosome** (Fig. 4.6). Although it might seem reasonable for all genes to link together in a single chromosome, such a long chromosome would probably become hopelessly tangled. Only a few plants have as few as two chromosomes for their genome; most have between 5 and 30 chromosomes (Table 4.2).

FIGURE 4.5
(a) If each gene were a distinct piece of DNA, ensuring replication of each would be difficult, and ensuring that each daughter nucleus received one copy of each would be even more difficult. (b) Grouping the genes into chromosomes makes them easier to manipulate. Here only two types of chromosomes are shown —long ones with genes 1 to 15 and short ones with genes 16 to 25.

TABLE 4.2 *Number of Chromosomes in a Haploid Set*

Species	Common Name	Number of Chromosomes
Machaeranthera gracilis	None: plant of US deserts	2
Vicia faba	bean	6
Pisum sativum	pea	7
Beta vulgaris	beet	8
Allium cepa	onion	8
Zea mays	corn	10
Lilium pyrenaiecum	lily	12
Oryza sativa	rice	12
Triticum aestivum	wheat	21
Homo sapiens	humans	23
Solanum tuberosum	potato	24

A haploid set is a single set of chromosomes, as carried by sperm cells and egg cells. A fertilized egg has two sets and is diploid.

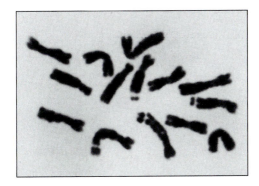

FIGURE 4.6
The chromosomes of peas (*Pisum sativum*), showing 7 types of chromosomes. The nucleus of each cell has two of each type, one inherited from the paternal and one from the maternal parent. (*J. Van't Hoff*)

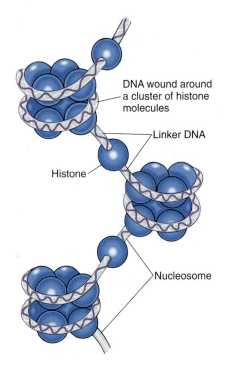

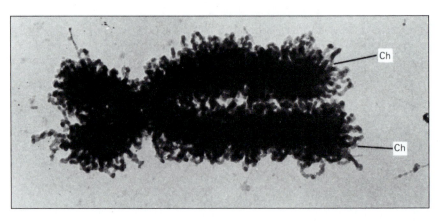

FIGURE 4.8

This chromosome was taken from a cell during division. It had gone through S phase of the cell cycle, so it has two chromatids (Ch), not just one. The chromatin has coiled tightly (condensed) and is visible as loops. Condensation has made the chromosome short and thick, and it can be moved around the cell easily. In its uncondensed condition, it is too long to be pulled to the ends of the cell during division. The pinched region in the center is the centromere, which is holding the two chromatids together. *(Courtesy of E. J. DuPraw)*

FIGURE 4.7

Histone proteins associate into short cylinders; then DNA winds around each cylinder. The DNA-histone particles then associate into a compact arrangement.

Even with 20 different chromosomes, an organism with 100,000 genes would have an average of 5000 genes on each chromosome, resulting in long pieces of DNA that might break if unprotected. In onion, the DNA in *each* nucleus is 10.5 meters long when all DNA molecules are measured; in lilies it is 21.8 meters long. In eukaryotes a special class of proteins called **histones** complexes with the DNA and gives it both protection and structure (Fig. 4.7). Chromosomes also have another structural feature, the **centromere,** which is usually located near the center of the chromosome (Fig. 4.8).

During the S phase, linking pieces of DNA as well as genes are replicated and new histone molecules complex with the new DNA. Thus entire chromosomes, not just DNA, are replicated (Fig. 4.9). Once replication is finished, the duplicate DNA molecules remain attached at the centromere. Although it would be justifiable to call this a "double chromosome" or a "pair of chromosomes," it is also called a chromosome. Now that we know how a chromosome changes during S phase, we call each half of the doubled chromosome a **chromatid.** After S phase each chromosome has two identical chromatids; before S phase each chromosome has just one chromatid. It is important to remember that a chromosome after S phase is twice as large as it was before S phase.

Although many cells stop in G_1 when they cease dividing and begin to mature, most plant cells enter S phase and replicate their DNA before they begin to differentiate. This may involve just a single cycle of DNA replication, resulting in a nucleus that is twice as large as would be expected, or it may continue for many rounds of DNA synthesis and the nucleus becomes gigantic (Fig. 4.10). This process is **endoreduplication,** estimated to occur in 80% of all maturing plant cells (Table 4.3). The resulting nucleus has many copies

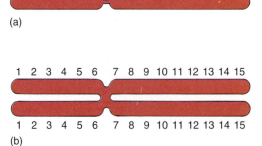

FIGURE 4.9

(a) Before S phase, each chromosome has one chromatid and one copy of each gene. (b) After replication in S phase, each chromosome has two chromatids and two copies of each gene. The constriction represents the centromere.

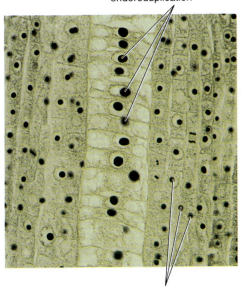

Nuclei that have undergone endoreduplication

Nuclei that have not undergone endoreduplication

FIGURE 4.10

In this root of *Scirpus* (bulrush, sedge) cells in the central column have enormous nuclei that have undergone endoreduplication. Each nucleus now has many copies of each gene so they can produce messenger RNA very rapidly. Consequently, the cytoplasmic ribosomes produce protein quickly as well (× 200).

of each gene—as many as 8192 copies in some cells of kidney beans (*Phaseolus vulgaris*). Endoreduplication occurs most often in hairs, glandular cells, and other cells that must have an extremely rapid, intense metabolism. The normal complement of two copies of each gene does not seem to make messenger RNA rapidly enough for such active cells. Such cells are also rich in ribosomes and produce large amounts of protein.

Gene amplification is similar to endoreduplication but involves only some genes that are repeatedly replicated. The amplified genes are those needed for the specialized metabolism of the mature cell. For example, as a protein-rich seed develops, its cells need large

TABLE 4.3 *Examples of Endoreduplication and Gene Amplification*

Species	Tissue	Degree*
Carex hirta	Tapetum (anther)	8 C
Viola declinata	Elaiosome	16 C
Cucumis sativus	Anther hairs	64 C
Cymbidium hybridum	Protocorm	128 C
Papaver rhoeas	Antipodal cells	128 C
Triticum aestivum	Antipodal cells	196 C
Urtica caudata	Stinging hairs	256 C
Corydalis cava	Elaiosome	512 C
Geranium phaeum	Integument	512 C
Phaseolus vulgaris	Suspensor (embryo)	2,048 C
Echinocystis lobata	Endosperm	2,072 C
Scilla bifolia	Elaiosome	4,096 C
Phaseolus coccineus	Suspensor (embryo)	8,192 C
Arum maculatum	Endosperm haustorium	24,576 C

* "C" is a symbol that represents the amount of DNA present in a nucleus; a normal body cell has 2 C before S phase and 4 C after S phase. Values that are a power of 2 represent endoreduplication; those that are not represent gene amplification.

quantities of the messenger RNA that codes for the storage protein, and those genes are amplified. But it would be a waste of energy and resources and deleterious to the plant to replicate all the other genes as well. The most extreme case of gene amplification known is in *Arum maculatum* (a relative of *Philodendron*): Each nucleus has enough DNA for 24,576 copies of every gene. We do not know how many genes are being amplified and how many are present as only two copies per nucleus, so we cannot compute the actual number of copies of each amplified gene, but it is certainly more than 24,576.

G_2 PHASE

Following S phase, the cell progresses into **G_2 (gap 2)** phase, during which cells prepare for division. This phase usually lasts only about 3 to 5 hours. The alpha and beta tubulin necessary for the spindle microtubules is synthesized, and the cell is believed to produce proteins necessary for processing chromosomes and breaking down the nuclear envelope. In cultured animal tissue, if a cell whose nucleus is dividing is forced to fuse with one in G_1, the second cell's nucleus also begins the first steps of nuclear division. This is evidence that during G_2, the first cell produced factors necessary to start nuclear division and that these factors are located in the cytoplasm.

G_1, S, and G_2 constitute the interphase portion of the cell cycle. After G_2, division can occur.

DIVISION PHASE OF THE CELL CYCLE

The actual division involves two processes: (1) division of the nucleus, called **karyokinesis,** and (2) division of the protoplasm, called **cytokinesis.** There are two types of karyokinesis: mitosis and meiosis.

MITOSIS

Mitosis is **duplication division,** and it is the more common type of karyokinesis. It is the method that any multicellular organism uses as its body is growing and the number of its cells is increasing. It is also used by eukaryotic unicellular organisms when they are not undergoing sexual reproduction. Mitosis is called duplication division because the nuclear genes are first copied; then one set of genes is separated from the other and each is packed into its own nucleus (Fig. 4.11). Each daughter nucleus is basically a duplicate of the original mother nucleus and a twin of the other. Mitosis produces nuclei that are more or less exact copies of the original nucleus, except for occasional errors.

As mentioned above, thousands of genes are linked together into just a few chromosomes, making the separation of gene sets much easier: It is necessary to transport only one of each type of chromosome to each daughter nucleus. This is made even easier because the two new chromatids remain together as the chromosome is replicated in S phase. It is necessary only to make certain that one half of the doubled, large chromosome goes to each end of the cell and the other half to the other end. If that happens with each chromosome, each end of the cell automatically receives one full set of genes. The mechanism that ensures this orderly separation of chromatids is quite straightforward and logical. It consists of the following four phases:

Prophase. During interphase, the DNA of a chromosome exists as a long, extended double helix associated with histone protein (see Fig. 4.7). This open configuration allows enzyme complexes to find specific genes that must be read for the information they contain. But in this condition, the chromosome may be several centimeters long. It could be wrapped around the cell hundreds of times, making it impossible to pull one chromatid from the other. But during prophase the chromosomes **condense**—that is, they undergo a change in the way the histones associate. The chromosomes begin to coil repeatedly (see Fig. 4.8), becoming shorter and thicker each time. In addition, a protein framework develops to which the DNA apparently binds. Condensation continues until chromosomes are only 2 to 5 μm long; in this form they can be moved around the cell much more easily.

In a few types of cells, G_2 may be quite long and during much of it the cell physiology is similar to that in G_1.

In this condensed form we can actually see the chromosomes. The uncoiled molecules of interphase are too narrow to be resolved by light microscopy.

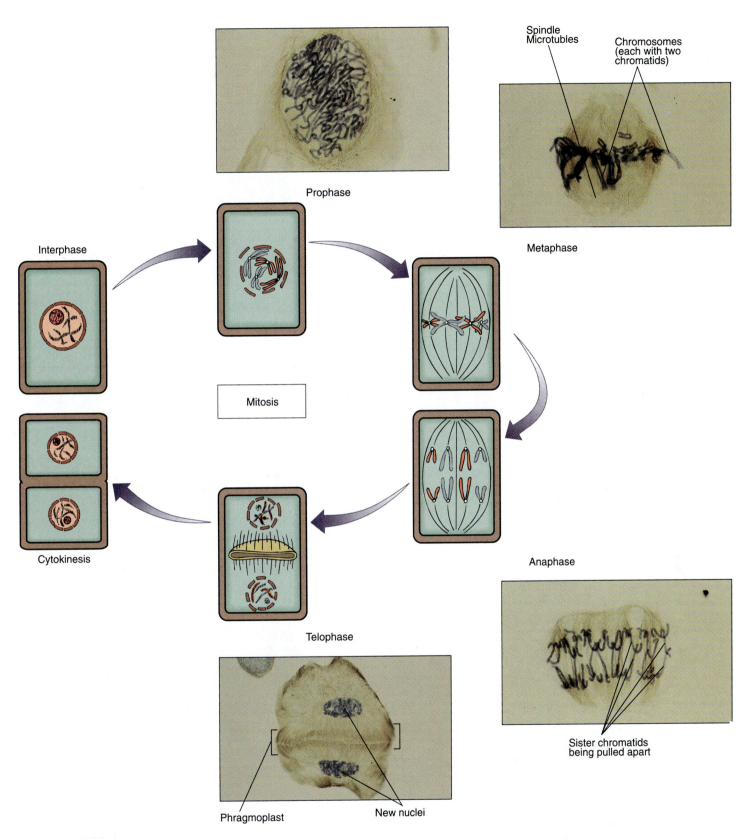

Spindle Microtubles

Chromosomes (each with two chromatids)

Prophase

Metaphase

Interphase

Mitosis

Cytokinesis

Telophase

Anaphase

Sister chromatids being pulled apart

Phragmoplast   New nuclei

FIGURE 4.11

During mitotic nuclear division, one chromatid from each chromosome is pulled to one end of the cell, and the other chromatid of each chromosome goes to the other end. Details are given in text.

(Andrew Bajer)

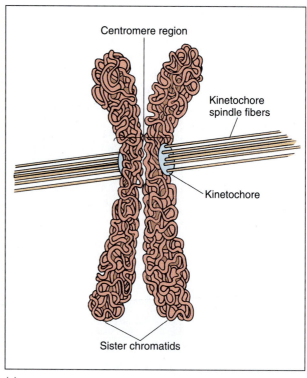

(a)

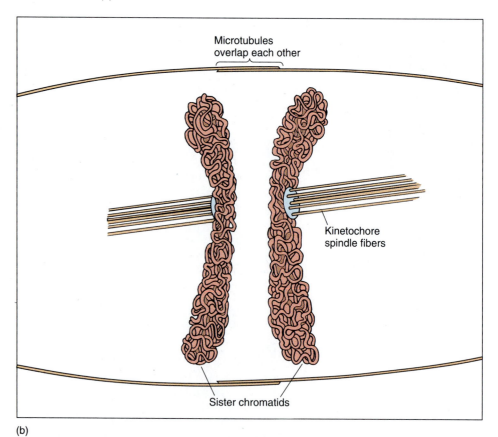

(b)

FIGURE 4.12

(a) Early in mitosis, spindle microtubules attach to the chromosomes at the kinetochore of the centromere. Notice that each chromatid is attached only to the microtubules coming from one end of the cell. (b) Other microtubules pass from pole to pole or overlap the ends of other microtubules at the center of the spindle. As the kinetochore-attached microtubules shorten, the two chromatids of each chromosome are pulled in opposite directions. The long microtubules that do not attach to chromatids act as a framework such that chromatids are pulled apart rather than having the spindle poles pulled into the center of the cell.

As chromosomes condense and become visible during prophase, other events occur as well (see Fig. 4.11). The nucleolus becomes less distinct and usually completely disappears by the end of prophase. The nuclear envelope also breaks down, apparently owing to the action of enzymes synthesized late in G_2. In those organisms that have a set of centrioles (mostly algae, fungi, and animals), the daughter sets that were duplicated during the previous interphase now migrate to opposite **poles** (sides) of the cell. It had been thought that they simultaneously acted as microtubule-organizing centers and produced a long set of microtubules, the **spindle**, between themselves. But recent experiments in animals have shown that even if the centrioles are destroyed by laser microbeams, the spindle still forms. It could be that the centrioles themselves are separated by the formation of the spindle, and the association is a mechanism that ensures that each daughter cell receives a centriole.

In the spindle, some microtubules extend from one pole to the center of the cell, where their ends overlap the ends of other microtubules that extend from the opposite end of the spindle. The two sets together, overlapping in the center, form a large framework (Figs. 4.12 and 4.13). Other microtubules run from a pole to a centromere. Each end of the spindle is attached to one of the two faces of the centromere on each chromosome. The point of attachment is a **kinetochore**; in electron micrographs it appears to be a multilayered structure, but little is known of its actual composition. About 15 to 35 microtubules attach to each kinetochore.

Metaphase.   Once the spindle microtubules attach to the centromeres, they push and pull on the chromosomes and gradually move them to the cell center; their arrangement there is called a **metaphase plate** (see Fig. 4.11). Viewed from the side, the chromosomes appear to be aligned in the very center, but viewed from the end, they are seen to be distributed throughout the central plane of the cell.

No distinct boundary occurs between prophase and metaphase; the chromosomes gradually become visible and gradually move to the metaphase plate. Fragments of nucleolus and nuclear envelope may persist well into metaphase. However, a distinct boundary is evident between metaphase and the following anaphase. At the end of metaphase, the centromeres are duplicated and the two chromatids of each chromosome are suddenly free of each other. In this step, **the number of chromosomes is doubled, but the size of each chromosome is halved.** Each chromosome is like it was in G_1, before S phase. Although the number of chromosomes in the cell is suddenly doubled, the doubling is transitory because the cell soon divides. The exact cause of the separation of the chromatids is not known, but it is thought that the centromere may be a segment of DNA that was not replicated during S phase. Instead, its DNA may be replicated during metaphase and, once it is doubled, the two chromatids are no longer joined.

Anaphase.   Anaphase begins just after the centromere divides; the spindle microtubules that run to the centromeres shorten, depolymerizing at the end near the spindle pole. This is believed to be the primary force responsible for pulling each daughter chromosome away from its twin. Agents that inhibit microtubule depolymerization, such as deuterium oxide, also inhibit chromosome movement, whereas those that speed depolymerization, such as low levels of colchicine, speed movement. The amount of energy necessary to move a chromosome from the metaphase plate to the end of the spindle is small: Just 20 ATP molecules are sufficient. Long chromosomes may tangle somewhat, but microtubules exert sufficient pull to untangle them and drag them to the ends of the spindle. Because the spindle is shaped like a football, as chromosomes on each side get closer to the end, they are pulled together into a compact space.

Telophase.   As chromosomes approach the ends of the spindle, fragments of nuclear envelope appear near them, connect with each other, and form complete nuclear envelopes at each end of the cell. The total surface area of the two new nuclei is larger than that of the envelope of the original mother nucleus; the extra membrane may be derived from endoplasmic reticulum (ER). It is not known how new nuclear pores are formed. Chromosomes become less distinct because they start to uncoil. Gradually, new nucleoli appear as the ribosomal genes become active and produce ribosome subunits. The spindle depolymerizes completely and disappears. Most of the events in telophase are reversals of those in prophase.

To summarize mitosis: After G_2 is completed at the end of interphase, each chromosome has been replicated and consists of twin chromatids. Spindle microtubules from opposite poles attach to the centromeres, then pull the twin chromatids of each chromosome away from each other. Two new nuclei form, each containing a full set of chromosomes and each chromosome having one chromatid. The new nuclei are identical to each other and to the original nucleus that began mitosis. The nuclei can then enter G_1 of

Spindle microtubules

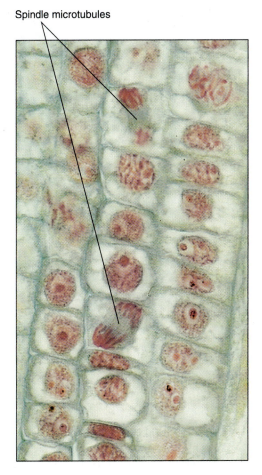

FIGURE 4.13

Anaphase in hyacinth root tips; the gray material between the red chromosomes consists of masses of spindle microtubules. These roots grow quickly and cell cycle times are short, so at any particular time many cells are in mitosis rather than just in interphase ($\times$ 250).

PLANTS & PEOPLE

CONTROLLED GROWTH VERSUS CANCEROUS GROWTH

The actual steps of karyokinesis and cytokinesis must be controlled if cells are to divide properly, but other aspects must be regulated as well. First, the rate and frequency of cell division are important in determining how rapidly or slowly one cell produces a mass of progeny cells. Second, the orientation of both cell division and cell growth affect the shape of the growing mass of cells: If all cells divide with their new walls parallel to each other, the mass grows as a column, but if the new cell walls occur in any plane, the mass grows in all three dimensions. The mass grows as a sheet if the new cell walls can occur in two planes but not in the third. Finally, it is important to control which cells divide: If only some cells undergo cell division, they may produce a lump or outgrowth while the rest of the mass of cells remains unchanged.

All these factors are controlled accurately in plants. In a young embryonic plant, all cells divide but later the cells at the tips of roots and shoots become the centers of cell division and growth while the rest of the stem and root tissues mature and carry out their functions. By controlling the rate, orientation, and location of cell division, plants produce cylindrical stems and roots, thin flat leaves and petals, and massive, three-dimensional fruits.

Plants are able to impose quiescence on certain cells and then reactivate them to growth and division later: Plants produce buds that are forced to remain quiescent for a long time, even years, but can then be stimulated by the plant to grow out as a branch or a flower. In young stems, epidermal cells grow rapidly enough to keep the stem covered, but then they mature and remain mitotically quiescent as they protect the plant. In many species, these can be reactivated years later and be directed to undergo cell division, thus producing the bark cells.

Cell division and growth must be controlled in animals as well. During the early stages of fetal development, all cells of human embryos divide and grow. Later, some cells divide more rapidly than others, but basically most of our cells are mitotically active during much of our growth before birth. Then, cells in certain tissues and organs undergo cell cycle arrest; they mature and never divide again, such as the cells of our eyes and our brains. Other cells never stop dividing and are active until we die, such as the layer of cells that produces our skin and hair and the bone marrow that generates most of our blood cells. Just as in plants, some of our cells can enter a prolonged state of quiescence and then be activated to division later. For example, surgical removal of part of the liver causes the cells of the remaining portion to divide and restore the organ to an adequate size. Of course, this is not true of most of our organs.

For reasons that we still do not understand, some of our cells may release themselves from cell cycle arrest and begin growing uncontrolled by the rest of the body. This is cancerous growth, and its severity depends on which types of cells and organs are involved, how rapidly the cells divide, and whether the cells can migrate from their original site and invade surrounding tissues. It is well known that certain environmental factors act as carcinogens—agents that cause cancer by interfering with cell cycle arrest. Cigarette smoke is known to cause cancer of the lung and throat, and ultraviolet light triggers skin cancer.

Whereas uncontrolled cancerous growth in humans may be fatal, it does not seem to be a problem in plants. Irregular lumps and growths, called galls, may occur, but these are often caused by insects or microbes, not by the plant's own cells undergoing a spontaneous, self-induced release from cell cycle arrest. It may be that plants do form cancerous growths but that they are not a serious problem for several reasons. First, cells cannot migrate through a plant body the way that our cells migrate through our bodies; consequently any uncontrolled growth is localized, not invasive. Second, whereas we have many organs that are each critical to our life and which occur singly (heart, brain) or in pairs (kidneys, lungs), plants have many leaves, roots, and flowers, and no single one is indispensable. Damage to one part of the plant may have little effect on the rest of the plant. By lacking the highly differentiated, complex, and tightly integrated body of humans and many animals, plants are not so threatened by diseases involving control of nuclear and cellular division.

---

interphase. The steps of mitosis (prophase, metaphase, anaphase, and telophase) are part of a continuous process and intergrade with each other. Similarly, prophase and telophase intergrade with interphase.

CYTOKINESIS

The division of the protoplast is much simpler than the division of the nucleus. Although it is necessary for each daughter cell to receive some of each type of organelle, random distribution of the organelles in the mother cell usually ensures this. No matter how the cell is divided, each half typically contains some mitochondria, some plastids, some ER, some vacuoles, and so on. It is not necessary for each daughter cell to get exactly half of each type of organelle. A single mitochondrion can divide, or a fragment of ER can grow until the cell has an adequate amount. The same is not true for genes: If one daughter cell is missing a

Plant Biology Tutor CD-ROM

Karyokinesis must be elaborate because the division must be exact; cytokinesis can be simple because almost any division is adequate (see Fig. 16.23).

Phragmoplast

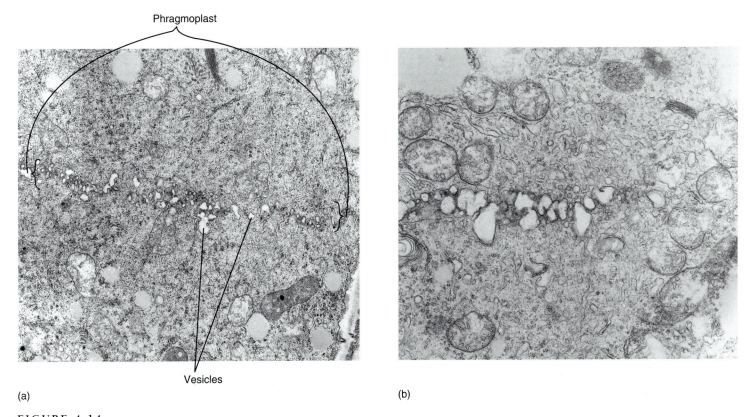

Vesicles

(a)   (b)

FIGURE 4.14
(a) Late telophase in onion. The row of small vesicles is the beginning of a cell plate. At the edges are short segments of microtubules, the phragmoplast. (b) Cell division has been interrupted by the herbicide DCPA (dimethyl tetrachloroterephalate) (both × 20,000). *(Photographs courtesy of Dr. Kevin C. Vaughn, Southern Weed Science Laboratory)*

gene or chromosome, the other genes cannot regenerate the information of the missing gene.

In plants, cytokinesis involves the formation of a **phragmoplast** made up of short microtubules aligned parallel to the spindle microtubules. The phragmoplast forms in the center of the cell where the metaphase plate had been (Figs. 4.14 and 4.15). Phragmoplast microtubules trap dictyosome vesicles that then fuse into a large, flat, platelike vesicle in which two new primary walls and a middle lamella begin to form. The phragmoplast then grows outward toward the walls of the original cell. It is not known if phragmoplast microtubules actually migrate toward the walls. New microtubules may polymerize near the outer edge of the phragmoplast while those on the inner side depolymerize. New dictyosome vesicles are trapped and added to the edge of the large vesicle; it too grows outward, following the phragmoplast. Similarly, the new walls extend outward along their edges. The phragmoplast, vesicle, and walls are called the **cell plate.** This process continues until the large vesicle meets the mother cell's plasma membrane, the two fuse, and the vesicle membrane becomes a part of the plasma membranes of the two daughter cells. Simultaneously, the new walls meet and fuse with the wall of the mother cell, completing the division of the mother cell into two daughter cells.

MEIOSIS

In mitosis, daughter nuclei are replicates of the original mother nucleus. This is necessary for the growth of an organism but creates a problem when sexual reproduction occurs. Two **sex cells (gametes)** fuse together, forming a **zygote,** which then grows into a new adult. Each gamete contains one complete set of chromosomes (Fig. 4.16). Nuclei, cells, and

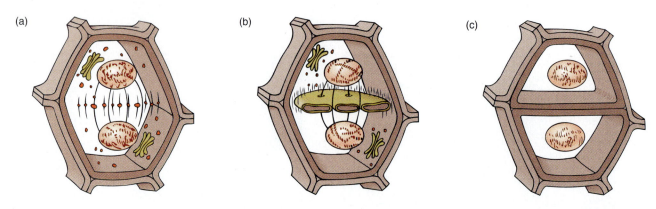

(a)   (b)   (c)

FIGURE 4.15
(a) Cytokinesis begins as dictyosome vesicles are trapped by phragmoplast microtubules in the space between the two new nuclei. (b) The small vesicles fuse into one large vesicle in which the new middle lamella and two primary walls will form. Plasmodesmata form at this time. (c) The cell plate enlarges toward the existing cell wall as more dictyosome vesicles fuse with the edges of the cell plate vesicle. When the cell plate reaches the existing cell walls, the vesicle membrane fuses with the plasma membrane and thus becomes plasma membrane itself. The new cell plate abuts the old cell wall, and the two become glued together with hemicelluloses and pectins.

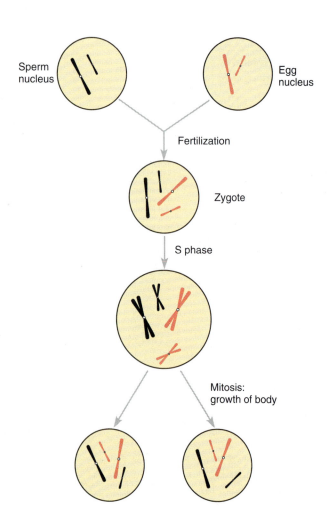

Sperm nucleus

Egg nucleus

Fertilization

Zygote

S phase

Mitosis: growth of body

FIGURE 4.16
Sperm cells and egg cells are haploid, each having just one set of chromosomes; in this case the set contains a long chromosome and a short one. After fertilization, the zygote is diploid with two complete sets: two long chromosomes and two short ones. The zygote grows into a mature plant by mitosis, so all nuclei are duplicates of the original zygote nucleus. Each is diploid, with one set each of paternal and maternal chromosomes. See Chapter 9 for details of sexual reproduction.

BOX 4.1 *Rates of Growth*

An organism, or a part of an organism, can produce more cells in a variety of ways. The two ways that are most important for you to understand are called arithmetic and geometric increase.

In arithmetic increase, only one cell is allowed to divide. Of the two resulting progeny cells, one continues to be able to divide but the other undergoes cell cycle arrest and begins to develop, differentiate, and mature. After each round of cell division, only a single cell remains capable of division and one new body cell exists. For example, starting with a single cell, after round 1 of cell division there is one dividing cell and one body cell, after round 2 there are two body cells, after round 3 there are three, and so on. To obtain a plant body containing one million cells (which would be a very small body), the plant's single dividing cell would undergo one million rounds of nuclear and cellular division. If each round requires one day, this type of arithmetic increase would require one million days, or 2739.7 years. As you can imagine, this arithmetic rate of increase is too slow to ever produce an entire plant or animal. However, it is capable of producing the small number of cells present in very small parts of plants. For example, the hairs on many leaves and stems consist of just a single row of cells produced by the division of the basal cell, the cell at the bottom of the hair next to the other epidermal cells. The hair may contain five to ten cells, so all its cells could be produced in just five to ten days by divisions of the basal cell; this would be quick compared with the rest of leaf development, which might take weeks or months.

An alternative pattern of growth by cell division, geometric increase, results if all cells of the organism or tissue are active mitotically. Again starting with a single cell, after round 1 of cell division, there are two cells capable of cell division, after round 2 there are four, after round 3 there are eight, and so on. The number of cells increases extremely rapidly and can be calculated as a power of 2. After round 3 there are $2^3 = 8$ cells, and after round 10 there are $2^{10} = 1024$ cells (these calculations should be easy for you on any pocket calculator). Notice that after round 20 there are already $2^{20} = 1,048,576$ cells. If, as in our previous example, each round of cell division requires one day, it now takes the plant only 20 days, not thousands of years, to reach a size of one million cells. With geometric increase, a large plant or animal body can be produced quickly.

In fact, geometric growth is common in animals but rarely occurs in plants except when they are extremely young and small. The problem is that in plants, dividing cells are typically simple and do not carry out many of the specialized metabolic processes necessary for plant survival. They cannot be the tough, hard fiber cells that make wood strong enough to hold up a tree; they cannot be the waxy, waterproofing cells of the plant's skin that protect it from fungi, bacteria, and loss of water.

Plants grow by a combination of arithmetic and geometric increase. A young, embryonic plant grows geometrically and rapidly, with all its cells dividing. Then cell division becomes restricted to certain cells at the tips of roots and shoots. After this point, growth is of the slower arithmetic type, but some of the new cells that are produced can develop into their mature condition and begin carrying out specialized types of metabolism. Plants are thus a mixture of older, mature cells and young, dividing cells.

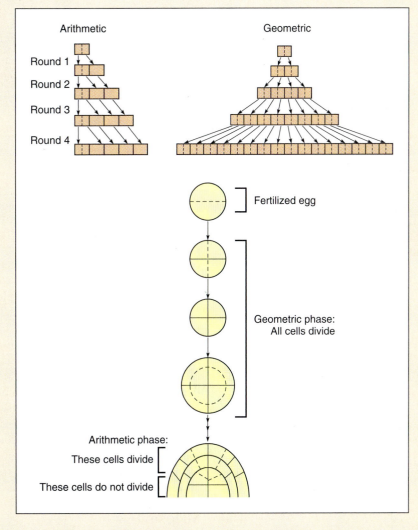

Possible types of cell division.

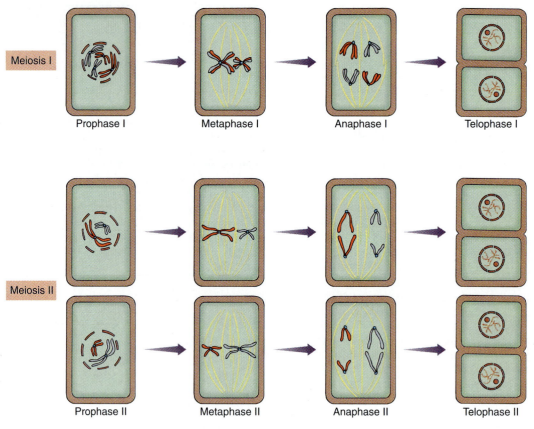

FIGURE 4.17
Meiotic nuclear division consists of two divisions without an intervening S phase. Details are given in text.

organisms with one set of chromosomes in each nucleus are said to be **haploid.** The zygote has two complete sets (one from each gamete), so it is **diploid.** In most species, the zygote grows into an adult by cell divisions in which the nucleus divides by mitosis. As a result, all cells of the adult are diploid because all nuclei are replicates of the mother nucleus.

If the adult were to produce gametes by mitosis, the gametes would be diploid as well, and the next zygote would be **tetraploid** with four sets. It is not possible biologically to double the number of sets per nucleus with each new generation. It is necessary for a **reduction division,** called **meiosis,** to occur somewhere. During reduction division, the two sets of chromosomes present in a diploid nucleus are separated into the daughter nuclei. But a diploid cell is actually faced with separating out four sets of genes because each of the two sets is replicated during the S-phase replication. The cell goes about this in this fashion: Meiosis involves two rounds of division without allowing the S phase to occur after the first division. The two divisions are called **meiosis I** and **meiosis II,** and each contains four phases similar to those of mitosis (Fig. 4.17).

Meiosis occurs only in the production of reproductive cells: gametes in animals and some algae and fungi, and spores in plants and other algae and fungi. In seed plants, meiosis occurs only in a few cells in the stamens and ovaries. Meiosis is never used in the growth of the body of any organism.

MEIOSIS I

Prophase I.   All the events that characterize prophase of mitosis also occur in prophase I: Nucleolus and nuclear membrane break down; centrioles (if present) separate; a spindle forms; microtubules attach to centromeres; and chromosomes condense and become visi-

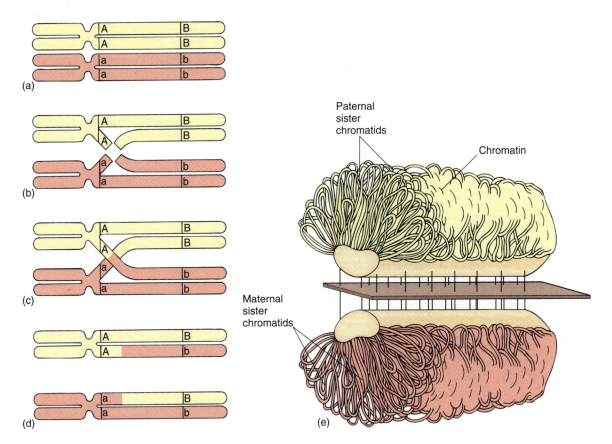

FIGURE 4.18

(a) During prophase I of meiosis, each paternal chromosome somehow finds and pairs with the equivalent maternal chromosome. They lie parallel to each other. (b) and (c) Breaks occur in similar sites on equivalent chromatids, and repair enzymes attach maternal pieces to paternal pieces, resulting in new chromatids (d). (e) The synaptonemal complex between paired homologous chromosomes. It is not known how it facilitates crossing-over.

ble (Fig. 4.17). In addition to these processes, special interactions of the chromosomes occur which are unique to prophase I as opposed to prophase of mitosis. Because of this, prophase I is divided into five stages:

Be careful with terminology: This is prophase I, not prophase.

1. In **leptotene**, chromosomes begin to condense and become distinguishable, although they appear indistinct.

2. During **zygotene**, a remarkable pairing of chromosomes occurs (Fig. 4.18a). Remember that there are two sets of chromosomes, one from the paternal and one from the maternal gamete. Because these have gone through S phase in the preceding interphase, each chromosome has two chromatids; altogether there are four sets of genes on two sets of chromosomes. During zygotene, each chromosome of one set pairs with the equivalent chromosome (its **homologous chromosome**, or **homolog**) of the other set; this pairing is **synapsis**. With remarkable accuracy the two homologous chromosomes in each pair become almost perfectly aligned from end to end. A structure, the **synaptonemal complex**, is present between the paired homologous chromosomes (Fig. 4.18e). This complex is composed of a linear central protein element connected by fine transverse fibers to two lateral elements bound to the DNA of the homologous chromosomes.

3. As chromosomes continue to condense, they become shorter and thicker; this stage is **pachytene.** The synaptonemal complex seems to be involved in the **crossing-over** that occurs now: In several places in each chromosome, the DNA of each homolog breaks (Fig.

4.18). Breaks occur in almost identical places in the paired homologs, and the enzymes that repair the breaks hook the "wrong" pieces together. A piece of the maternal homolog is attached to the paternal homolog, and the equivalent piece of the paternal homolog is attached to the maternal one. The full consequences of this are explained in Chapters 16 and 17. If the maternal and paternal chromosomes are absolutely identical, nothing significant has happened. But if the genes on the paternal and maternal chromosomes are slightly different, the new chromosomes that result from synapsis and crossing-over are slightly different from the original chromosomes. There are no new genes, but rather **new combinations of genes on each chromatid.**

The shuffling of genes during synapsis of prophase I is an important event in sexual reproduction and evolution.

4. After pachytene is **diplotene.** The homologous chromosomes begin to move away from each other but do not separate completely because they are held together at their paired centromeres and at points (**chiasmata;** sing.: chiasma) where they appear to be tangled together. Some biologists believe that chiasmata are the points at which crossing-over occurred, but evidence is accumulating that chiasmata are only tangles and are not related to crossing-over at all. Under good conditions it is possible to see all four chromatids of the paired homologous chromosomes; they are called **tetrads** at this stage.

5. In the final stage, **diakinesis,** homologs continue to separate, and chiasmata are pushed to the ends of the chromosome. The homologous chromosomes become untangled and are paired only at the centromeres.

Prophase I is the most complicated stage of meiosis; the remaining stages are simple and quite similar to the stages of mitosis.

Metaphase I.   Spindle microtubules move the tetrads to the center of the cell, forming a metaphase plate (see Fig. 4.17).

Anaphase I.   The homologous chromosomes separate completely from each other, moving to opposite ends of the spindle. **The centromeres do not divide, and each chromosome continues to consist of two chromatids.** Notice how this is different from the metaphase-anaphase transition of mitosis. In mitosis the centromeres replicate and each chromosome divides into two chromosomes, each with just one chromatid. But in the metaphase I–anaphase I transition, homologous chromosomes separate from each other, and each still has two chromatids. One set of chromosomes is pulled away from the other set, and two new nuclei are formed. **These nuclei are now haploid,** because each has only one set of chromosomes; the homolog of each chromosome in one nucleus is now in the other nucleus.

Each chromosome in anaphase I has two chromatids, so each nucleus has two sets of genes, but this does not make them diploid. Any cell in G_2 has at least two sets of genes; the critical factor is to have two sets of chromosomes.

Telophase I.   Because the chromosomes are still doubled and are in a G_2 state, they do not need to undergo an interphase with its G_1, S, and G_2. Also, because telophase I is basically the opposite of prophase II, some organisms go directly from anaphase I to metaphase II, skipping telophase I and prophase II. In most organisms, however, there is at least a partial telophase I in which chromosomes start to uncoil, and the nucleolus and nuclear envelope start to reappear. If the cells actually progress fully to interphase, no replication of the DNA occurs—the S phase is completely missing. This interval is called **interkinesis.** Cytokinesis may occur, but it is not unusual for this to be absent also and for both daughter nuclei (i.e., both masses of chromatin) to stay in the original, undivided mother cell.

MEIOSIS II

If a telophase I occurs, then **prophase II** is necessary to prepare the nucleus for division. Prophase II is not subdivided into stages like prophase I. **Metaphase II** is short, and at the end of it, **the centromeres divide, thereby separating each chromosome into two chromosomes,** just as in metaphase of mitosis, but different from metaphase I. **Anaphase II** then separates each new chromosome from its replicate, and in **telophase II,** new nuclei are formed. Each nucleus contains just one set of chromosomes, each with a single chromatid.

To summarize, during meiosis I, each chromosome of the paternal chromosomes undergoes synapsis with the homologous chromosome of the maternal set. Crossing-over results in new combinations of genes on the chromosomes. Spindle microtubules pull the paired homologs to opposite poles, so at each pole there is only one set of chromosomes, not two. The new nuclei that form temporarily after meiosis I are thus haploid. When these nuclei undergo meiosis II, the two chromatids of each chromosome are separated, as in mitosis, and the resulting nuclei each have a haploid set of chromosomes, each chromosome with only one chromatid.

LESS COMMON TYPES OF DIVISION

Cytokinesis and karyokinesis are often so closely associated that we tend to think of them as two aspects of one process. In numerous tissues, however, karyokinesis occurs without cytokinesis, and multinucleate cells are formed. The nutritive tissue of many seeds goes through a phase in which its cells are multinucleate (Fig. 4.19), and in many algae all the cells are like this. If the cell becomes very large and has hundreds or thousands of nuclei, it is called a **coenocyte.**

On the other hand, it is possible for cell division to occur without nuclear division; this is most common in algae, fungi, and the nutritive tissues of seeds. The cells become multinucleate and persist in that condition for some time; then cell division begins as new walls are organized around each nucleus and its accompanying cytoplasm (see Fig. 20.13). At present the selective advantage of this behavior is not known.

In meiosis, the processes of cytokinesis and karyokinesis are often not directly linked. In some species, cytokinesis happens after both meiosis I and meiosis II, and four haploid

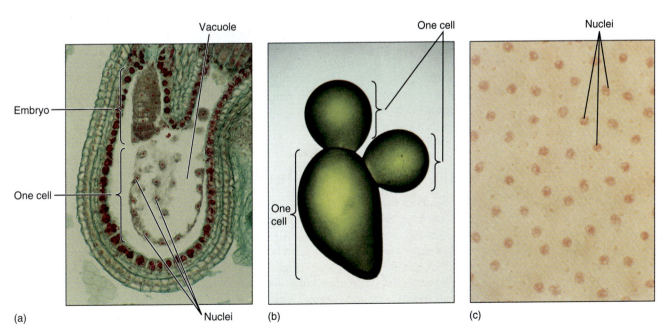

(a)   (b)   (c)

FIGURE 4.19

(a) A multinucleate cell in the developing seeds of *Capsella* (shepherd's-purse). Although the numerous nuclei have undergone mitotic divisions, the cell has not divided but has enlarged and now has a giant vacuole. The small ball of cells is the embryo, which grows into the multinucleate coenocyte and draws nutrients from it (× 80). (b) Multinucleate cells are typical of many algae. This is *Valonia*; the entire alga pictured here consists of just three giant cells (this photo is only four times life size), but each has thousands of nuclei. (c) A magnification of (b), showing some of the many nuclei, which have been stained pink (× 320). *(b and c, Courtesy of J. W. LaClaire, III, University of Texas).*

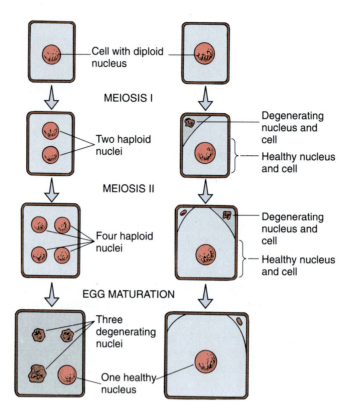

FIGURE 4.20

In animals and some plants, the meiosis that is part of egg production produces only one large haploid cell. (a) In some species, meiosis results in four haploid nuclei; no cytokinesis occurs and all four nuclei are temporarily in one cell until degeneration eliminates three of them. (b) In other species, after meiosis I occurs, cytokinesis places one nucleus in a small cell, and both nucleus and cell degenerate before they can undergo meiosis II. The nucleus of the larger cell does undergo meiosis II, and again one daughter nucleus is placed in a small cell that degenerates. Only three nuclei are ever formed, and only one survives.

cells result from each original diploid mother cell. In other species, no cytokinesis occurs after meiosis I but a double cytokinesis occurs after meiosis II, again resulting in four haploid cells. For example, pollen grains can form either way, and meiosis results in four haploid cells. But in many organisms, no cytokinesis occurs at all during the meiosis that leads to the formation of eggs. The final cell is tetranucleate and may remain that way, or three of the nuclei may degenerate and produce only a single uninucleate haploid egg cell by meiosis (Fig. 4.20). It seems to be selectively advantageous to produce one large egg rather than four small ones.

CELL DIVISION OF PROKARYOTES

The events of mitosis and meiosis occur only in eukaryotes. In prokaryotes—bacteria, cyanobacteria, and archaebacteria—mitosis and meiosis do not occur and cytokinesis is much simpler. All the genetic material occurs as one ring of DNA, which does not have any protective histones. The loop of DNA is attached to the cell's plasma membrane, and when the DNA is replicated, the two daughter loops are both attached to the membrane (Fig. 4.21). Because no centromere holds them together, the circles of DNA are pulled apart as the cell and its membranes grow. A new round of replication may begin even before cell division occurs, so each cell may have as many as 20 identical circles.

Cytokinesis occurs by a process of infurrowing: The plasma membrane pulls inward and finally pinches in two, although the contractile mechanism that pulls it inward has not been found yet. As the plasma membrane furrows inward, a new cross wall grows inward, starting from the existing wall. Once complete, the cross wall splits, becoming two walls and releasing the two daughter cells; multicellular aggregates are rarely formed in bacteria. In cyanobacteria the cross walls do not break apart, and multicellular bodies are formed (see Chapter 19). The cell cycle of many bacteria can be short—only 20 minutes under ideal conditions. For many species, however, the cycle may last several days or weeks even under optimal conditions.

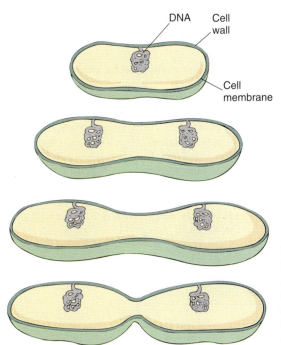

FIGURE 4.21
In prokaryotes, the circles of DNA are attached directly to the plasma membrane. As the cell and its membrane grow, the attached circles of DNA are separated. Cell division then occurs between the circles.

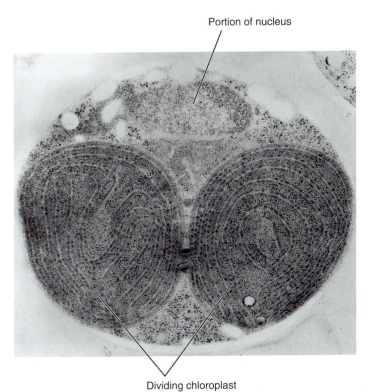

FIGURE 4.22
Micrograph of an algal cell; algae are especially convenient for studying chloroplast division. Because most have only one chloroplast per cell, chloroplast division must be coordinated with cell division. As the chloroplast divides, its central region is pinched inward by a contracting ring of proteins, which constricts the chloroplast into a dumbbell shape (× 1200). (*Courtesy of M. Kuroiwa, University of Tokyo*)

DIVISION OF CHLOROPLASTS AND MITOCHONDRIA

Mitochondria and plastids are constructed similarly to prokaryotes; they also contain circles of naked DNA that become separated by membrane growth. Division of the organelles is accomplished either by infurrowing or by being pulled in two (Fig. 4.22). Because they contain DNA necessary for their growth and functioning, each daughter cell must receive at

BOX 4.2   *Chloroplast Division During Leaf Growth*

Elegant studies have been done of the growth and division of plastids in relation to the growth and development of leaves in spinach. In very small leaves, 1 mm long or less, plastid DNA constitutes 7% of the total cell DNA, and an average of 76 DNA circles are present in each plastid (Table 4.4). As the leaf doubles in size to 2 mm long, plastid DNA is replicated at about the same rate as nuclear DNA, so it remains low, about 8% of total cellular DNA. As the leaf continues to expand to 20 mm long, plastid DNA is replicated much more rapidly than nuclear DNA and increases to 23% of total cellular DNA. At the same time the number of plastids per cell triples from 10 to 29, so neither plastid division nor plastid DNA replication is controlled by the mechanisms that govern cell or nuclear DNA replication. At this point, each plastid has 190 DNA circles, and each cell has a total of 5510. In the next stage of leaf growth, to 100 mm long, no synthesis of DNA occurs and no new cells form. Instead, those already present expand. However, plastids continue to divide even though they are not making any more DNA; consequently, the number of DNA circles per plastid drops from 190 to 32 while the number of plastids per cell increases from 29 to 171. From these data, it is reasonable to form the hypothesis that plastid growth, DNA replication, division, and development are correlated predominantly with tissue or organ development rather than with the cell cycle.

TABLE 4.4   *Development of the Plastid Genome in Spinach Leaves*

Leaf Size	1 mm	2 mm	20 mm	100 mm
Genome copies per plastid	76	150	190	32
Plastids per cell	10	10	29	171
Genome copies per cell	760	1500	5510	5470
Plastid DNA as percentage of total	7%	8%	23%	23%

Data from Scott, N. S., and J. V. Possingham, 1983. Changes in chloroplast DNA levels during growth of spinach leaves. *J. Experimental Bot.* 34:1756–67.

least one mitochondrion and one plastid during cytokinesis; if not, the cell lacks that organellar portion of its genome and cannot produce the organelle. This happens occasionally with plastids and is often not a serious problem; the cell survives by importing sugar from neighboring cells and can grow along with the tissue. All daughter cells also lack plastids, so if this occurs in a young leaf, a white spot forms (see Fig. 16.23a).

The same phenomenon probably occurs with mitochondria but is more difficult to detect. The first daughter cell to lack mitochondria is unable to respire and form adequate amounts of ATP. Because it cannot import ATP from adjacent cells, the cell grows slowly or even dies.

The metabolic stimulus that triggers the replication of nuclear DNA is not the stimulus that controls replication of organellar DNA. During the cell cycle, replication of nuclear DNA is episodic, occurring as discrete episodes that occupy a small portion of the total cell cycle, S phase. But the "duplication" of the rest of the cell seems to be continuous: The volumes of plastids, mitochondria, cytosol, endoplasmic reticulum, and other organelles appear to increase gradually and steadily throughout interphase rather than in discrete episodes.

SUMMARY

1. Cells have a "life cycle," the cell cycle, consisting of alternating phases of growth and division. In organs that live just briefly, cells cycle only while the tissue is young, then enlarge and differentiate.

2. The interphase portion of the cell cycle consists of G_1, S, and G_2. During G_1 the nucleus controls cell metabolism; during S the DNA is replicated; in G_2 the cell prepares for division.

3. Each DNA molecule contains thousands of regions—genes—that contain the information for proteins. A chromosome is a complex in which histone proteins bind to DNA and stabilize it. Each chromosome consists of a centromere and either one chromatid before S phase or two chromatids after S phase.

4. Cell division is cytokinesis, and nuclear division is karyokinesis. The words "mitosis" and "meiosis" technically refer only to karyokinesis but are frequently used to describe cytokinesis as well.

5. During mitosis, spindle microtubules pull one chromatid of the replicated chromosome to each end of the cell. Because the two chromatids are virtually identical, the two new nuclei are also almost identical to each other.

6. Nuclei of any ploidy level—haploid, diploid, and higher ploidies—can undergo mitosis. During growth of the plant body, all nuclear divisions are mitotic divisions.

7. Meiosis is reduction division: During prophase I, homologous chromosomes pair (synapse) and exchange pieces of chromatid (crossing-over). At the end of prophase I, the nucleus no longer has purely paternal and purely maternal chromosomes. During anaphase I, homologous chromosomes are pulled away from each other, reducing the number of sets of chromosomes to half the original number. Haploid, triploid, and all other odd-ploid nuclei cannot undergo meiosis.

8. Meiosis occurs only in the production of reproductive cells, either gametes or spores. Meiosis is never used in the growth of the body of any organism.

9. Cytokinesis in plants and algae occurs by the formation of a large vesicle between the forming daughter nuclei. Within the vesicle, the middle lamella and two new walls form and expand until they reach existing walls.

10. Cytokinesis and karyokinesis are usually closely coordinated, but nuclear division can occur without cell division, resulting in multinucleate cells.

IMPORTANT TERMS

anaphase	crossing-over	genome	meiosis	resting phase
cell cycle	cytokinesis	haploid	metaphase	S phase
cell plate	diploid	histone	metaphase plate	spindle
centromere	duplication division	homologous chromosomes	mitosis	spindle microtubules
chromatid	G_1 phase	interphase	phragmoplast	synapsis
chromosome	G_2 phase	karyokinesis	prophase	telophase
chromosome condensation	gamete	kinetochore	reduction division	zygote

REVIEW QUESTIONS

1. In a woody plant such as a tree, which parts have cells that live only briefly and die quickly? Which parts have cells that live for several years?

2. What are the four phases of the cell cycle? What is the principal activity in the cell during each phase? Can any phase be eliminated or bypassed?

3. Why is mitosis called duplication division and meiosis called reduction division? What is reduced and what is duplicated: the chromosomes, the number of chromosomes, or the number of sets of chromosomes?

4. What are the four phases of mitosis, and what is the principal activity in the nucleus during each phase?

5. Draw a single, imaginary chromosome as it would appear just as mitosis is ending. Now describe what happens to it during interphase and then during mitosis. Be especially careful to consider how many chromatids and how many copies of each gene it has at each stage.

6. How does cytokinesis occur in plants? Which organelle produces vesicles that fuse to form the phragmoplast? What membrane is transformed into new plasma membrane?

7. What are the five stages of prophase I, and what is the principal activity of the nucleus and chromosomes during each stage?

8. Draw all stages in the cell cycle and meiosis for a nucleus that has just one pair of homologous chromosomes. Then do the same for a nucleus that has three different types of chromosomes (six chromosomes in three sets of homologs). Draw all stages (this is not easy).

BotanyLinks .net Questions

www.jbpub.com/botanylinks

Visit the .net Questions area of BotanyLinks (http://www.jbpub.com/botanylinks/netquestions) to complete this question:

1. How many chromosomes do plants have? Go to the BotanyLinks home page to investigate.

TISSUES AND THE PRIMARY GROWTH OF STEMS

Stems of roses produce thorns, an adaptation that deters animals from eating the plants. This increases the likelihood that the plants will survive.

CONCEPTS

The body of an herb contains just three basic parts: leaves, stems, and roots (Fig. 5.1). When the first land plants evolved about 420 million years ago, they were basically just algae that either washed up onto a shore or were left there as lakes and streams evaporated. They had no roots, stems, or leaves, and they just lay on the mud. As the shores gradually became crowded with such plants, some grew over others, shading them. Any plant which had a mutation that allowed it to grow upright into the sunshine above the others had a selective advantage. However, being upright is not easy: Elevated cells are out of contact with the moisture of the mud, so water must be transported up to them. Elevated tissues act as a sail and tend to blow down, so supportive tissue is necessary. Absorptive cells in the mud are shaded and cannot photosynthesize, so sugars must be transported down to them. Shortly after plants began living on land, distinct, specialized tissues and organs began to evolve.

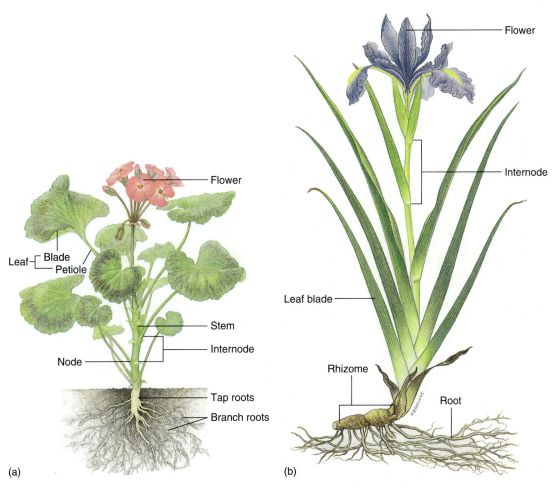

FIGURE 5.1

(a) The primary body of an herb like this geranium consists of roots, stems, and leaves; buds are located in the axil of each leaf and may grow to be either a vegetative branch or a set of flowers. (b) This *Iris* is also an herb and never produces wood or bark. All flowering plants are either broadleaf plants (dicots) like the geranium or plants with grasslike leaves (monocots) like the *Iris*.

As these early populations of land plants continued to evolve and became taller, their stems functioned primarily as transport and support structures, as they do still. But stems of modern flowering plants have additional roles. They produce leaves and hold them in the sunlight, and during winter they store sugars and other nutrients, such as the sugary sap of maples. Stems may also be a means of survival: Underground bulbs and corms remain alive when above-ground leaves die. Stems of many species are a means of dispersal. They spread as runners or vines, or pieces break off and are carried by animals or water to new areas where they sprout roots and grow into new plants.

Although all flowering plants possess the basic structures of leaves, stems, and roots, these parts have been modified so extensively in some species they may not be recognizable without careful study. For example, cacti are often described as leafless, but they actually have small green leaves between 100 and 1000 μm long (Fig. 5.2). A large, broad leaf would be selectively disadvantageous for these desert plants because the extensive surface area of such leaves gives up so much water to the dry air that the plant would desiccate.

Similarly, all flowering plants have stems, but in some they are only temporary, reduced structures. Orchids such as *Campylocentrum pachyrrhizum* and *Harrisella porrecta* consist of a mass of green photosynthetic roots connected to a tiny portion of stem; roots

Leaves

FIGURE 5.2

This prickly pear (*Opuntia*) shows that one plant can have two types of shoot: the "pad" is the main shoot, and the spine clusters are highly modified branch shoots. One of the spine-bearing branches has been stimulated to develop into the first type of shoot and become the earlike branches. The plant also has two types of leaves, small fleshy leaves on the young buds and spines on the axillary branches.

constitute almost the entire plant body (Fig. 5.3). The shoot becomes active only when flowers are to be produced. It has been hypothesized that this unusual body evolved because the ancestors of these species had roots that were more resistant to water stress than their stems were. Such plants could occupy drier habitats if either of two things happened: (1) Mutations were selected that caused stems and leaves to be more water conserving—this happened in most orchid species (Fig. 5.4), or (2) mutations occurred that enhanced the root's ability to absorb carbon dioxide and carry on photosynthesis. Although thousands of plant species are capable of withstanding harsh conditions, only a handful do so by being "shootless" and having photosynthetic roots. Mutations that permit this type of body may be either rare or generally detrimental.

Some plants in the bromeliad family are nearly rootless. In the coastal deserts south of Lima, Peru, fog is frequent but rain never falls. Because the soil is always dry, roots would be of little use. The *Tillandsia straminea* of that region is a small herbaceous vine that lies on top of the soil (Fig. 5.5). Plants absorb moisture through leaves made wet by fog, and they derive minerals from wind-blown dust that dissolves on the moist leaf surface. The plants are not anchored to the soil but roll over the coastal dunes as the wind blows them. Such a life style would be impossible for a tree or bush because large plants could not absorb enough water or minerals without roots.

In each plant described, leaves, stems, or roots have become highly modified by natural selection, permitting survival in unusual habitats. But in no species have any of these organs been completely lost evolutionarily. We must assume that the organ carries out some essential function. In the "shootless" orchids, the residual shoot is necessary for flowering and sexual reproduction. Cactus leaves are involved in the formation of the buds that produce the defensive spines. Although the roots of *T. straminea* neither absorb water and minerals nor anchor the plant, they may be essential for the production of critical hormones, as are the roots of other plants. In all cases, a careful analysis of many aspects of

Leaves

Pseudobulbs
(swollen stems)

FIGURE 5.4

Most epiphytic orchids resist the stresses of temporary drying because their shoots are fibrous and have a thick cuticle composed of cutin and wax.

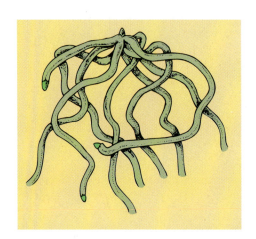

FIGURE 5.3

This orchid plant of the genus *Polyradicion lindenii* is composed almost completely of photosynthetic roots; only a small portion of shoot remains. Unlike most roots, these occur above ground and are green, being rich in chloroplasts.

FIGURE 5.5
These bromeliads are not rooted into the coastal sand dunes—a strong wind can blow them around. All water is absorbed from fog condensing on the leaves.

the plant's biology reveals how the structure and metabolism of a particular organ are adaptive, how modifications affect other plant parts, and how the organs have subtle functions not always obvious in more "typical" plants (Box 5.1).

The flowering plants discussed in this and the next several chapters are formally classified as division Magnoliophyta (also frequently called division Anthophyta), but they are known informally as **angiosperms.** This group consists of about 235,000 species and is the largest division in the plant kingdom. Angiosperms have branched into two main lines of evolution: the **dicots** or **broadleaf plants** such as roses, asters, maples, and others, and the **monocots** such as grasses, lilies, cattails, palms, philodendrons, and bromeliads.

This chapter began by stating that the body of an herb contains three parts; it is necessary now to explain more precisely just what an herb is. Plant bodies are of two fundamental types: an herbaceous body, also called a **primary plant body,** and a woody body, known as a **secondary plant body.** An herb is a plant that never becomes woody and covered with bark; it often lives for less than a single year. Its tissues are primary tissues. In woody plants like trees and shrubs, the wood and bark are secondary tissues (Table 5.1).

BASIC TYPES OF CELLS AND TISSUES

Despite the diversity of types of stems that have originated by natural selection, all share a basic, rather simple organization. The same is true for leaves and roots. Although we might suspect that numerous types of cells are present within a plant, actually the various kinds of plant cells are customarily grouped into three classes based on the nature of their walls: parenchyma, collenchyma, and sclerenchyma.

TABLE 5.1 *Types of Plant Body*

Primary plant body
    Derived from shoot and root apical meristems
    Composed of primary tissues
    Constitutes the herbaceous parts of a plant
    An herb consists only of a primary plant body.

Secondary plant body
    Derived from meristems other than apical meristems (see Chapter 8)
    Composed of secondary tissues: wood and bark
    Constitutes the woody, bark-covered parts of a plant
    A woody plant has primary tissues at its shoot and root tips, and a seedling consists only
        of primary tissues. But after a few months, wood and bark arise inside the primary
        tissues of stems and roots.

BOX 5.1 *How Much Can a Plant Lose and Still Be a Plant?*

The first part of this chapter stated that all flowering plants have roots, stems, and leaves, but some plants come close to being exceptions. The examples given in the chapter are interesting because they are almost completely leafless or stemless or rootless, but other flowering plants are more dramatic because each plant lacks all three basic organs for most of its life: They do not have any roots or stems or leaves, but they are angiosperms. In addition to being dramatic, these plants can teach us a great deal if we consider their own particular and peculiar biology.

Tristerix aphyllus is a mistletoe that infects some of the cacti in central Chile (Fig. 1). When a bird deposits a seed of *T. aphyllus* onto a cactus, the seed germinates and the seedling root presses against the host epidermis. Narrow hairlike filaments pass through the stomata or small tears in the host epidermis. Once the filaments reach the highly cytoplasmic cells of the host cortex, they branch and invade all parts of the cactus (Fig. 2). The rest of the seedling on the outside of the host dies.

For much of its life *T. aphyllus* consists of nothing more than a network of slender

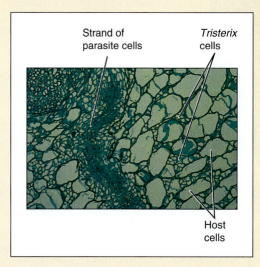

The large cells are those of the host cortex, and the small cells between them are those of *Tristerix aphyllus.* In the center is a wide strand of cells of the parasite (× 150).

filaments that grow between the cells of the host. It has no organs and usually not even much vascular tissue; the entire body is just an extended web of parenchyma. At some point, some filaments grow close to the surface of the host and form a compact, dense nodule. Shoot buds form in the nodule. By some unknown mechanism, the nodules are able to break open the tough epidermis of the host, allowing buds to protrude and form flowers. At this point, *T. aphyllus* does have stems and leaves. These may persist, but often they die after the fruit is ripe, and the plant again consists only of parenchyma inside the host.

Other species of *Tristerix* grow in the same area, attacking trees and shrubs; these other species look more like typical mistletoes, having green leaves and stems at all times. What selective pressures cause one species to become so highly modified while others are less drastically altered? The nature of the host is important in understanding this unusual parasite. The host cacti are very well-adapted to water stress, having a thick, resistant cuticle and greatly reduced

leaves. If *T. aphyllus* had large leaves and stems like other species of *Tristerix,* it would probably lose so much water that the host would die, which would also kill the parasite. By growing totally inside the host, the parasite does not cause water loss. The mode of entry is also especially gentle and causes little damage to the host, whereas most other parasites cause significant breaks in their host's epidermis. Such damage could permit fungi to invade the wound and kill the cactus, which is especially susceptible to fungal attack. After a *T. aphyllus* penetrates, there is virtually no sign of harm. If the seeds of other mistletoe species were to germinate on a cactus and successfully invade it, that particular host plant would almost certainly die from a fungal attack through the entry site or from water loss through the leaves of the mistletoe. Natural selection favors only those parasites that have greatly reduced leaves and stems. Although *T. aphyllus* appears at first glance to be bizarre, an understanding of its special habitat reveals that its anatomy and development are quite logical and adaptive.

The cactus host *Trichocereus chilensis* with the parasite *Tristerix aphyllus.* Part of the parasite (brown) has died, but the rest (red) is alive.

PARENCHYMA

Parenchyma cells have only primary walls that remain thin (Table 5.2). **Parenchyma tissue** is a mass of parenchyma cells. This is the most common type of cell and tissue, constituting all soft parts of a plant. Parenchyma cells are active metabolically and usually remain alive once they mature. Numerous subtypes are specialized for particular tasks (Fig. 5.6).

(Text continues on page 112.)

TABLE 5.2	*Three Basic Types of Plant Cells and Tissues, Based on Cell Wall*
Parenchyma:	Thin primary walls. Typically alive at maturity. Many functions.
Collenchyma:	Unevenly thickened primary walls. Typically alive at maturity. Provide plastic support.
Sclerenchyma:	Primary walls plus secondary walls. Many dead at maturity. Provide elastic support and some (tracheary elements) are involved in water transport.

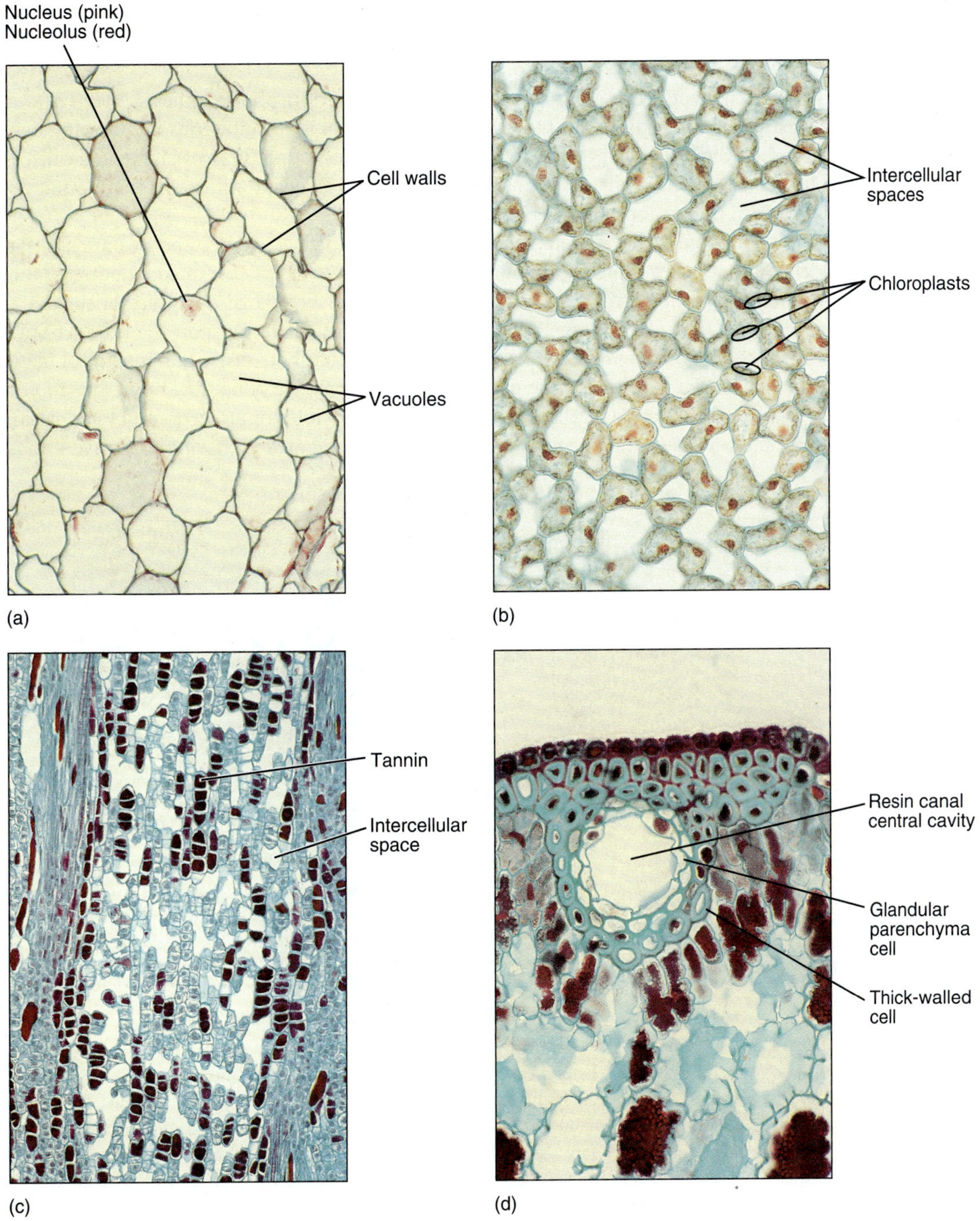

(a)

(b)

(c)

(d)

FIGURE 5.6

(a) Parenchyma cells of geranium; their walls (green) are thin, and their vacuoles are large and full of watery contents that did not stain. Nuclei were present in all cells, but because these cells were so large and the section was cut so thin, most nuclei were cut away during the preparation of this slide. One nucleus is still present (× 160). (b) Chlorenchyma cells from a leaf of privet. Because these cells are small and the section is thicker than that in (a), most of these cells still have nuclei (red). The green structures close to the wall (blue) are chloroplasts. The large white areas are intracellular spaces where the cells have pulled away from each other. The spaces permit carbon dioxide to diffuse rapidly throughout the leaf (× 160). (c) Material taken from the center of a pine (*Pinus*) stem. The cells that have stained dark purple are filled with chemicals called tannins; these are bitter and deter insects from eating the tissue (× 50). (d) A resinal canal in a pine leaf. The white area is the central cavity where the resin is stored, and the cells that line the cavity are glandular parenchyma cells that synthesize and secrete the resin. The innermost cells have thin walls, which permit movement of resin from the cells to the cavity. The outer cells have thick walls, which provide strength; cells with thick walls are not parenchyma cells (× 160).

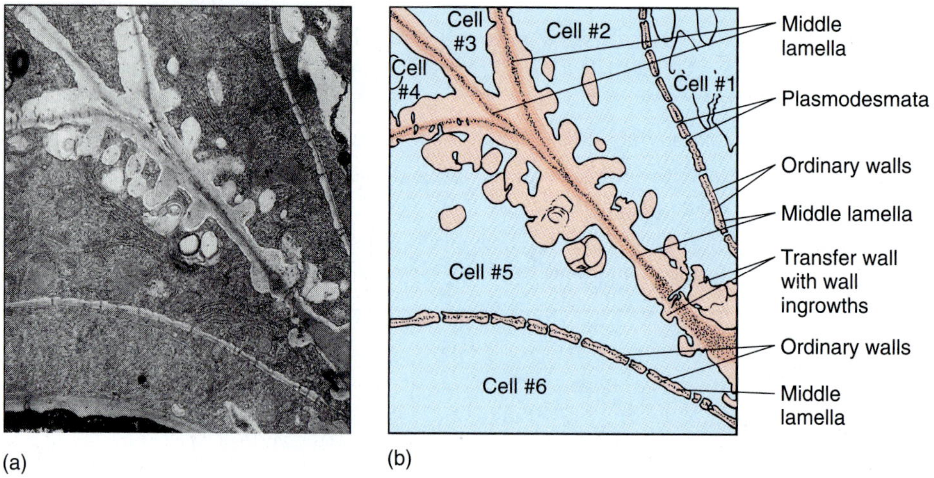

FIGURE 5.7

Transfer cells in the salt gland of *Frankenia grandifolia.* The wall ingrowths increase the surface area of the cell membrane, providing more room for salt-pumping proteins in the membrane ($\times$ 20,000). *(Courtesy of W. W. Thomson and R. Balsamo, University of California, Riverside)*

(a)

(b)

Chlorenchyma cells are parenchyma cells involved in photosynthesis; they have an abundance of chloroplasts, and the thinness of the wall is advantageous for allowing light and carbon dioxide to pass through to the chloroplasts. Other types of pigmented cells, as in flower petals and fruits, also must be parenchyma cells with thin walls that permit the pigments in the protoplasm to be seen.

Glandular cells that secrete nectar, fragrances, mucilage, resins, and oils are also parenchyma cells; they typically contain few chloroplasts but have elevated amounts of dictyosomes and endoplasmic reticulum. They must transport large quantities of sugar and minerals into themselves, transform them metabolically, then transport the product out.

Transfer cells are parenchyma cells that mediate the short-distance transport of material by means of a large, extensive plasma membrane capable of holding numerous molecular pumps. Unlike animal cells, plant cells cannot form folds or projections of their plasma membranes; instead, transfer cells increase their surface area by having extensive knobs,

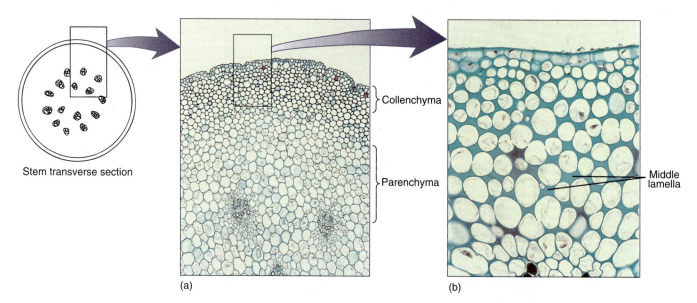

Stem transverse section

(a)

(b)

FIGURE 5.8

(a) Masses of collenchyma cells often occur in the outer parts of stems and leaf stalks; this is part of a *Peperomia* stem. The collenchyma forms a band about 8 to 12 cells thick. The inner part of the stem is mostly parenchyma ($\times$ 50). (b) Look closely between the cells and notice the fine dark lines—the original, thin primary walls and middle lamellas. In collenchyma cells, the primary wall gradually becomes thicker at the corners so the protoplast becomes rounded. No intercellular spaces are present ($\times$ 150).

ridges, and other ingrowths on the inner surface of their walls (Fig. 5.7). Because the plasma membrane follows the contour of all these, it is extensive and capable of large-scale molecular pumping.

Some parenchyma cells function by dying at maturity. Structures such as stamens and some fruits must open and release pollen or seeds; the opening may be formed by parenchyma cells that die and break down or are torn apart. Large spaces may be necessary inside the plant body; some of these are formed when the middle lamella decomposes and cells are released from their neighbors. In other cases, the space is formed by the degeneration of parenchyma cells. In a few species, such as milkweeds, as parenchyma cells die, their protoplasm is converted metabolically into mucilage or a milky latex.

Parenchyma tissue that conducts nutrients over long distances is phloem; it is discussed later in this chapter.

Parenchyma cells are relatively inexpensive to build because little glucose is expended in constructing the wall's cellulose and hemicellulose. Each molecule incorporated into a wall polymer cannot be used for other purposes such as the generation of ATP or the synthesis of proteins. Consequently, it is disadvantageous to use a cell with thick walls any time one with thin walls would be just as functional. Most leaves are soft, composed almost entirely of parenchyma, and are therefore not very expensive metabolically. After several weeks of photosynthesis they replace the sugar used in their construction, and all photosynthesis after that point is net gain for the plant.

COLLENCHYMA

Collenchyma cells have a primary wall that remains thin in some areas but becomes thickened in other areas, most often in the corners (Fig. 5.8). The nature of this wall is important in understanding why it exists and how it functions in the plant. Like clay, the wall of collenchyma exhibits plasticity, the ability to be deformed by pressure or tension and to retain the new shape even if the pressure or tension ceases. Collenchyma is present in elongating shoot tips that must be long and flexible, such as those of vining plants like grapes, as a layer just under the epidermis or as bands located next to vascular bundles, making the tips stronger and more resistant to breaking (Fig. 5.9). But the tips are still capable of elongating because collenchyma can be stretched. In species whose shoot tips are composed only of weak parenchyma, the tips are flexible and delicate and often can be damaged by wind; the elongating portion must be very short or it simply buckles under its own weight.

It is important to think about the method by which collenchyma provides support. If a vine or other collenchyma-rich tissue is cut off from its water supply, it wilts and droops; the collenchyma is unable to hold up the stem. Parenchyma cells are needed in the inner tissues for support. Collenchyma and turgid parenchyma work together like air pressure and a tire: The tire or inner tube is extremely strong but is useless for support without air pressure. Similarly, air pressure is useless unless it is confined by a container. In stems, the tendency for parenchyma to expand is counterbalanced by the resistance of the collenchyma, and the stem becomes rigid.

Because the walls of collenchyma cells are thick, they require more glucose for their production. Collenchyma is usually produced only in shoot tips and young petioles, where the need for extra strength justifies the metabolic cost. Subterranean shoots and roots do not need collenchyma because soil provides support, but the aerial roots of epiphytes such as as orchids and philodendrons have a thick layer of collenchyma.

SCLERENCHYMA

The third basic type of cell and tissue, **sclerenchyma**, has both a primary wall and a thick secondary wall that is almost always lignified (Fig. 5.10). These walls have the property of elasticity: They can be deformed, but they snap back to their original size and shape when the pressure or tension is released. Sclerenchyma cells develop mainly in mature organs that have stopped growing and have achieved their proper size and shape. Deforming

FIGURE 5.9
The shoot tips of long vines need the plastic support of collenchyma while their stems are elongating.

"Turgid" means that the cells are so full of water that the protoplasts press firmly against their cell walls.

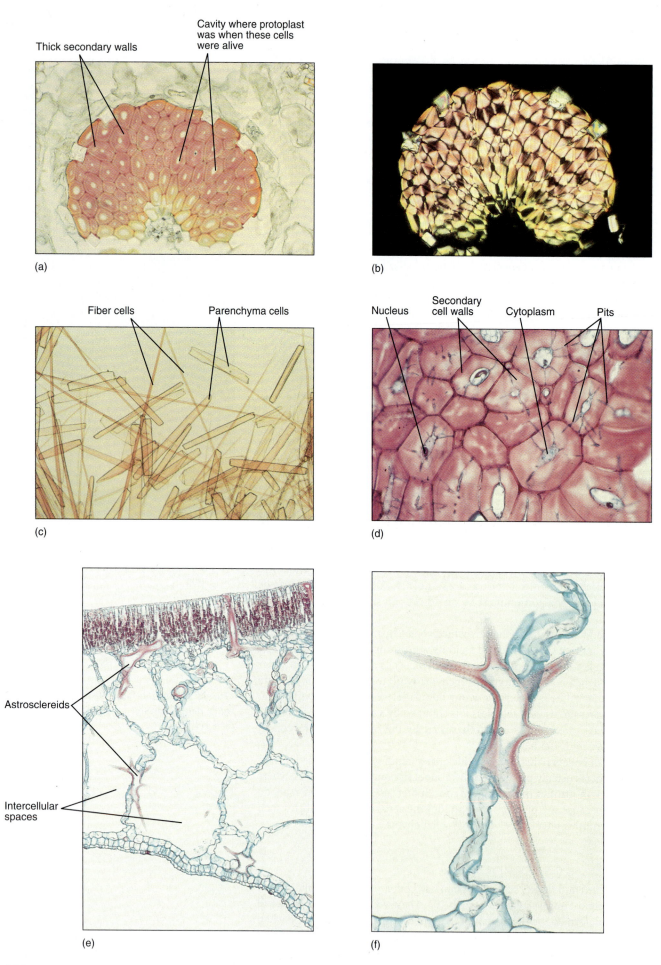

(a) Thick secondary walls — Cavity where protoplast was when these cells were alive

(b)

(c) Fiber cells — Parenchyma cells

(d) Nucleus — Secondary cell walls — Cytoplasm — Pits

(e) Astrosclereids — Intercellular spaces

(f)

◄ **FIGURE 5.10**

(a) A mass of fiber cells in the leaf of *Agave*. These are large, heavy, tough leaves, and the fiber masses give the leaves strength. Notice that each cell consists mostly of thick secondary cell wall; the small white space in each is an area where the protoplast had been before it died ($\times$ 150). (b) This is the same mass of fibers as in (a) but viewed with polarized light. The thick secondary wall shines brightly because its cellulose molecules are packed in a tight, crystalline form, giving the wall extra strength ($\times$ 150). (c) A stem of bamboo was treated with a mixture of nitric acid and chromic acid to dissolve the middle lamellas and allow the cells to separate from each other. In this preparation you can see that the fibers are long and narrow. The shorter, wider cells are parenchyma ($\times$ 80). (d) These are sclereids; they are more or less cuboidal, definitely not long like fibers. These have remained alive at maturity, and nuclei and cytoplasm are visible in several. The blue-stained channels that cross the walls are pits with cytoplasm. The pits of each cell connect with those of the surrounding cells so that nutrients can be transferred from cell to cell, keeping them alive ($\times$ 150). (e) This portion of a leaf of water lily contains large, irregularly branched cells that have stained red. These are known as astrosclereids (star-shaped sclereids). The large white spaces are giant intercellular spaces; this is an aerenchyma type of parenchyma ($\times$ 40). They are shown at higher magnification in (f). (f) A star-shaped sclereid, showing only part of its long, arm-like extensions. The ends of most of the extensions were cut off when the material was cut to make this slide. Tiny cubic crystals are present in the wall ($\times$ 150).

forces such as wind, animals, or snow would probably be detrimental. If mature organs had collenchyma for support, they would be reshaped constantly by storms or animals, which of course would not be optimal. For example, while growing and elongating, a young leaf must be supported by collenchyma if it is to continue to grow. But once it has achieved its mature size and shape, some cells of the leaf can mature into sclerenchyma and provide elastic support that maintains the leaf's shape. Unlike collenchyma, sclerenchyma supports the plant by its strength alone; if sclerenchyma-rich stems are allowed to wilt, they remain upright and do not droop.

Parenchyma and collenchyma cells can absorb water so powerfully that they swell and stretch the wall, thereby growing; sclerenchyma cell walls are strong enough to prevent the protoplast from expanding. The rigidity of sclerenchyma makes it unusable for growing shoot tips because it would prevent further shoot elongation.

Sclerenchyma cells are of two types—conducting sclerenchyma and mechanical sclerenchyma. The latter type is subdivided into long **fibers** (Fig. 5.10a to c) and short **sclereids** (Fig. 5.10d to f; Table 5.3), both of which have thick secondary walls. Because fibers are long, they are flexible and are most often found in areas where strength and

TABLE 5.3 *Types of Sclerenchyma*

Mechanical (nonconducting) sclerenchyma

Sclereids:	More or less isodiametric; often dead at maturity.
Fibers:	Long; many types are dead, other types remain alive and are involved in storage.

Conducting sclerenchyma (tracheary elements)

Tracheids:	Long and narrow with tapered ends; contain no perforations. Dead at maturity. Found in all vascular plants.
Vessel elements:	Short and wide with rather perpendicular end walls; must contain one or two perforations. Dead at maturity. Found almost exclusively in flowering plants. Among nonflowering plants, only a few ferns, horsetails, and gymnosperms have vessels.

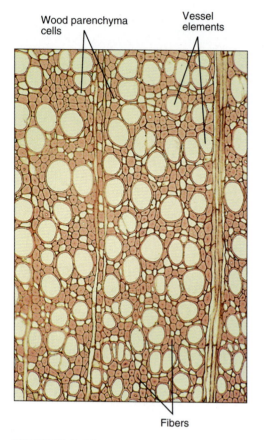

FIGURE 5.11
Wood is composed of several types of cells. The numerous small cells with thick walls and extremely narrow lumens are fibers. These give the wood strength and flexibility. The large round cells that appear to be empty are vessel elements, discussed later in this chapter; the small cells with thin walls and large lumens are wood parenchyma cells (×60).

elasticity are important. The wood of most flowering plants contains abundant fibers, and their strength supports the tree while their elasticity allows the trunk and branches to sway in the wind without breaking (usually) or becoming permanently bent (Fig. 5.11). The fiber-rich bark is important not in holding up the tree but in resisting insects, fungi, and other pests.

Sclereids are short and more or less isodiametric (cuboidal). Because sclereids have strong walls oriented in all three dimensions, sclerenchyma tissue composed of sclereids is brittle and inflexible. Masses of sclereids form hard, impenetrable surfaces such as the shells of walnuts and coconuts or the "pits" or "stones" of cherries and peaches. Flexibility there would be disadvantageous because the soft seed inside might be crushed even though the shell remained unbroken.

When strength or resistance is the only selective advantage of sclerenchyma, the protoplast usually dies once the secondary wall has been deposited. But in some species, certain sclerenchyma cells, especially fibers, remain alive at maturity and carry out an active metabolism (Fig. 5.12). These living sclerenchyma cells most often are involved in storing starch or crystals of calcium oxalate. Some have rather thin secondary walls, but in others the secondary walls are just as thick as those of fibers that die at maturity and provide only support.

Like all cells, sclereids and fibers develop from cells produced by cell division; when newly formed, they are small and have only a primary wall—they are parenchyma cells. If the cell is to differentiate into a sclereid, it may expand only slightly, but if it is to develop into a fiber, it elongates greatly. When immature sclereids and fibers reach their final size, the cellulose-synthesizing rosettes of the plasma membrane begin to deposit the secondary wall. As this wall becomes thicker and is impregnated with lignin, it becomes waterproof,

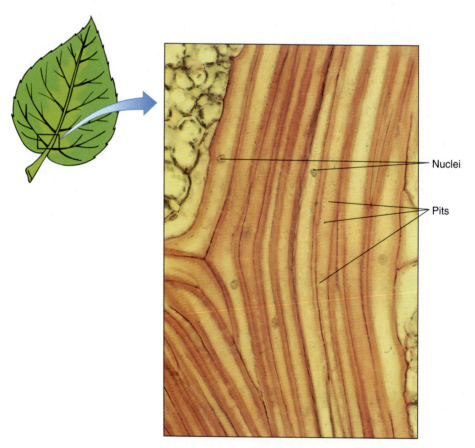

FIGURE 5.12
These fiber cells have nuclei, indicating that they are living cells. The secondary walls are thick, but not so thick as the walls in Figure 5.11. The small dots visible in the walls are pits; these are much narrower than the pits of Figure 5.10d. Leaf of *Smilax* (×150).

so nutrients can enter the cell only through plasmodesmata rather than everywhere, as with parenchyma and collenchyma cells. It is important that the secondary wall not be deposited along the entire inner surface of the primary wall; small, plasmodesmata-rich areas must remain free of the secondary wall. At first these areas are low depressions in the developing secondary wall, but as wall deposition continues these areas become narrow **pits** in the secondary wall (see Figs. 5.10d, 5.31, and 5.32a). The pits of adjacent sclerenchyma cells must meet; if the pits of one cell met areas of secondary wall in the neighboring cells, no water or sugars could be transferred and the protoplasts would starve.

Conducting sclerenchyma transports water and is one of the types of vascular cells; it is discussed later in this chapter.

The secondary wall is located interior to the primary wall; as it becomes thicker, the proto-plast must shrink, usually by removing water from its central vacuole.

EXTERNAL ORGANIZATION OF STEMS

The terms "stem" and "shoot" are sometimes used interchangeably, but technically the stem is an axis, whereas the shoot is the stem plus any leaves, flowers, or buds that may be present. All flowering plants have the same basic stem organization: There are **nodes** where leaves are attached, and **internodes**, the regions between nodes (see Fig. 5.1). The stem area just above the point where the leaf attaches is the **leaf axil.** Within it is an **axillary bud,** a miniature shoot with a dormant apical meristem and several young leaves (Fig.

Terminal bud

(a)   Axillary buds   (b)

FIGURE 5.13

(a) In these buds of pistachio, the leaves are modified into bud scales, just as prickly leaves of prickly pear buds are modified into spines. The bud scales are waterproof and protective, but they are shed when the bud begins to grow in the spring and produces photosynthetic leaves. Notice that there are buds in the axils of the leaves and a bud at the very tip of the shoot. (b) The two axillary buds of this *Viburnum* have already begun to grow into branches, and their first young, expanding leaves are visible. These buds were formed in the spring and started growing immediately; they did not form dormant buds with bud scales.

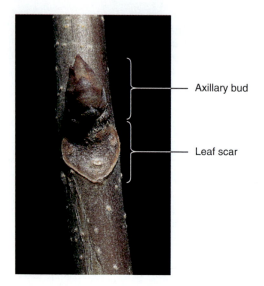

FIGURE 5.14

As a leaf falls from a stem, it tears cells and creates a wound; a leaf scar is a layer of cork that seals the wound, keeping fungi and bacteria out and preventing water loss.

5.13); it is either a vegetative bud if it will grow into a branch, or a floral bud if it will grow into a flower or group of flowers. The bud is covered by small, corky, waxy **bud scales** (modified leaves) that protect the delicate organs inside. At the extreme tip of each stem is a **terminal bud.** In winter, when all leaves have abscised, leaf scars occur where leaves were attached (Fig. 5.14).

The arrangement of leaves on the stem, called **phyllotaxy,** is important in positioning leaves so that they do not shade each other (Fig. 5.15). If only one leaf is present at each node, the stem has alternate phyllotaxy (the leaves alternate up the stem); two leaves per node is opposite phyllotaxy, and three or more per node is whorled phyllotaxy. The orientation of leaves at one node with respect to those at neighboring nodes is also important. In distichous phyllotaxy, the leaves are arranged in only two (di-) rows (-stichies), as in corn and irises. The leaves may be alternate or opposite. In decussate phyllotaxy, the leaves are arranged in four rows; this occurs in only some of the species with opposite leaves. Finally, in spiral phyllotaxy, each leaf is located slightly to the side of the ones immediately above and below it, and the leaves form a spiral up the stem. This is the most common arrangement and may involve alternate, opposite, or whorled leaves (Table 5.4). Different types of phyllotaxy are selectively advantageous, depending on leaf size and shape.

All flowering plant shoots are based on this simple arrangement of nodes and internodes, and diversity and specialization are variations of this arrangement; in vines internodes are especially long, whereas in lettuce, cabbage, and onions, internodes are so short that leaves are packed together (Fig. 5.16). Internodes can be wide (asparagus), intermediate, or narrow (alfalfa sprouts). The diversity is not random—the different types provide particular adaptive advantages in certain situations. For example, plants are physically bound to the site where they happen to germinate, which by accident may be a shady area near a more optimal sunny spot. Vines, with their elongated internodes, are a means by which a plant can "explore" its immediate surroundings, and shoots that happen to grow into a sunnier site may flourish (Fig. 5.17). In some species of climbing vines, support and attachment are provided by **tendrils**—modifed leaves or lateral branches capable of twining around small objects.

(a)

(b)

(c)

FIGURE 5.15

Phyllotaxy. (a) The leaves of this *Erythroxylon coca* (coca, source of cocaine) are arranged with alternate phyllotaxy—one leaf per node occurring in all directions around the stem. (b) The leaves of this melostome show opposite phyllotaxy—two leaves per node, each pair pointing 90 degrees away from the previous and subsequent pairs. (c) Whorled phyllotaxy in *Fuschia*: There are more than two leaves per node. The axillary buds are obviously floral buds.

Leaves

Stem

FIGURE 5.16
The nodes of cabbage are packed closely together, and all leaves are tightly clustered. Such closely packed, large leaves with spiral phyllotaxy are poor at photosynthesis and did not evolve by natural selection; they were produced by plant geneticists.

TABLE 5.4	*Phyllotaxy*
Alternate:	Leaves one per node
Opposite:	Leaves two per node
Decussate:	Leaves located in four rows
Whorled:	Three or more leaves per node
Spiral:	Leaves not aligned with their nearest neighbors
Distichous:	Leaves located in two rows only

The capacity to explore is even more advanced in **stolons,** also called runners (Fig. 5.18). Because their internodes are especially long and thin and their leaves do not expand, stolons extend greatly without using much of the plant's nutrient reserves. Once the stolon encounters a suitable microhabitat, subsequent growth is by shorter, vertical internodes and fully expanded leaves; new roots are established and the end of the stolon resembles a

FIGURE 5.17
These grape vines (*Vitis*) are able to place their leaves in the full sun at the top of the forest canopy, even though they have not invested much sugar in building strong trunks. The pines that support these vines may ultimately die because they are so heavily shaded by the grape leaves.

Stolon   Leaves

FIGURE 5.18
Airplane plant (*Chlorophytum*) is a popular ornamental plant that spreads by runners in nature. Most people are familiar with it as a hanging basket plant, with its runners arching out, then drooping down and forming new plants at the tips. The new plant here has already produced roots.

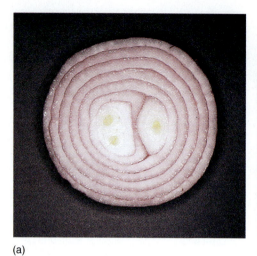

(a)

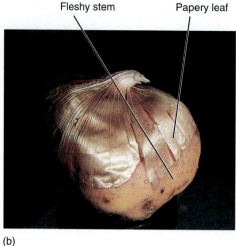

Fleshy stem   Papery leaf

(b)

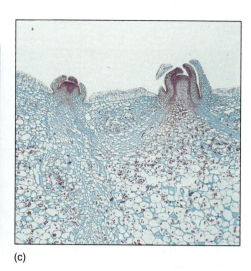

(c)

FIGURE 5.19

(a) Onions are bulbs: They have short vertical stems and fleshy leaves. (b) Gladioluses have corms with fleshy stems and papery leaves. (c) A tuber (potato: *Solanum tuberosum*) is a short horizontal stem that grows only for a limited period of time. Its leaves are microscopic, and its axillary buds are the potato's "eyes," shown here by light microscopy (× 15).

new plant. If older parts of the plant die, these vertical shoots become independent plants. After they have started growing vigorously, these plants send out stolons of their own.

In some shoots, nutrient storage is particularly important for survival; these shoots are often massive and quite fleshy, thus providing room for accumulating starch. The form of the shoot depends on whether starch accumulates in leaves or stems. **Bulbs** are short shoots that have thick, fleshy leaves (onions, daffodils, garlic), whereas **corms** are vertical, thick stems that have thin, papery leaves (crocus, gladiolus; Fig. 5.19). There is no obvious selective advantage of one over the other; the type seems to depend on whether mutations affecting the stem or leaf happen to occur first. **Rhizomes** are fleshy horizontal stems that

FIGURE 5.20

Division of labor: False Solomon's seal (*Smilacena racemosa*) consists of subterranean rhizomes (which survive harsh winters and spread through the soil) and aerial shoots (which carry out photosynthesis and flowering).

allow a plant to spread underground (bamboo, irises, canna lilies). **Tubers** are horizontal like rhizomes, but they grow for only a short period and are mainly a means of storing nutrients (potatoes).

All these storage shoots are subterranean, which is more advantageous for storage than an exposed surface location. A plant needs storage capacity only if it is perennial and goes through a dormant period. Quiescence is the means by which perennial plants of harsh climates survive the stress of winter cold or summer heat and dryness; they often shed their leaves, reducing water loss. In order to produce new twigs, leaves, and roots in the favorable season, they must draw on stored carbohydrates. Protection of their nutrient reserves —the corms, bulbs, rhizomes, and tubers—is most easily accomplished by burying them at depths that do not freeze or dry out. Their subterranean location also hides them from most herbivores.

Each of these specialized tasks is accomplished by characteristic modifications in stem structure and metabolism. As with multicellularity, specialization is accompanied by division of labor: Modifications that increase a stem's ability to survive, spread, or store nutrients decrease its efficiency at other tasks. No plant consists only of rhizomes, which are not exposed to light, because most plants must photosynthesize. Therefore, at least some of the rhizome's axillary buds must grow upward above ground and have green leaves (Fig. 5.20). Similarly, a plant cannot be made up only of stolons: Modifications that allow quick, low-cost exploration are not appropriate for growth, photosynthesis, and sexual reproduction (flowering). It is common for individual plants to have several types of stems and leaves, each of which contributes uniquely to the plant's survival.

Although the axil of every leaf contains a bud, only a few buds ever produce a branch; the others remain dormant or produce flowers. Axillary buds do not become active at random; each species has a particular pattern. In shady environments, it may be advantageous for axillary buds to remain dormant so that all resources are concentrated in the growth of the vertical main shoot, the **trunk,** allowing the plant to reach brighter light in the top of the forest canopy (Fig. 5.21). In an open environment, such dominant upward

(a)

(b)

FIGURE 5.21

(a) In a heavily shaded forest, suppression of branching allows all resources to be concentrated on placing one shoot up into the light. (b) With abundant light, one main trunk is not so advantageous; instead, a high degree of branching allows the rapid production of many leaves.

growth is not particularly advantageous; it may be better for all buds to grow and thereby maximize the rate at which new leaves are formed. On almost all plants, at least a few axillary buds remain dormant and serve as reserve growth centers. As long as the apical meristem is healthy and growing well, some axillary buds are unneeded; if the apical meristem is killed by frost, insects, or pruning, axillary buds become active and replace it, allowing the growth of the shoot to continue.

INTERNAL ORGANIZATION OF STEMS

THE ARRANGEMENT OF PRIMARY TISSUES

In order to function properly, a tissue must contain the right cells in the proper arrangement. The same principle is true on a larger scale: In order to function properly, the tissues of an organ must be arranged correctly.

Epidermis.   The outermost surface of an herbaceous stem is the **epidermis,** a layer of parenchyma cells (Fig. 5.22). Their outer tangential walls are the interface between the plant and the environment and regulate the interchange of material between the plant and its surroundings.

Because air is almost always drier than living cells, even when the relative humidity is high, preventing the loss of water to the air is critical for land plants. Also, the epidermis is a barrier against invasion by bacteria and fungi as well as small insects. The epidermis shields delicate internal cells from abrasion by dust particles, passing animals, or leaves and stems that might rub together, and its reflectivity protects the plant from overheating in bright sunlight.

The outer walls are encrusted with cutin, a fatty substance that makes the wall impermeable to water (Fig. 5.22). In species that occur in mild, moist habitats, such encrustation may be sufficient for water retention, but in most plants cutin builds up as a more or less pure layer called the **cuticle.** Under more severe conditions, even a cuticle is not sufficient, and a layer of wax may be present outside of the cuticle (Fig. 5.23; Table 5.5). Cutin and wax provide defense against pathogens like fungi and bacteria because digestive enzymes secreted by these microbes cannot break down either cutin or wax. In some species, the cuticle is so smooth that fungal spores cannot even stick to it and are washed off by rain or shaken off by wind. Waxes are indigestible and non-nutritious, so a thick layer of wax makes it difficult and unrewarding for an insect to chew into a stem.

Unfortunately, cutin and wax also inhibit the entry of carbon dioxide needed for photosynthesis—a totally impermeable epidermis would lead to the plant's starvation.

Plant Biology Tutor CD-ROM

FIGURE 5.22
(a) The cuticle on this ivy (*Hedera helix*) has been stained dark pink; the lighter layer is wall material impregnated with cutin. The epidermal protoplasts have stained dark blue. Interior cells are photosynthetic collenchyma (× 120).
(b) This epidermis was taken from a desert plant, *Agave.* Its cuticle is much thicker than that of ivy in (a), and the outer walls of the epidermal cells have become extremely thick. The thickness of both the wall and cuticle strengthens the leaf surface against insects and prevents water loss (× 200).

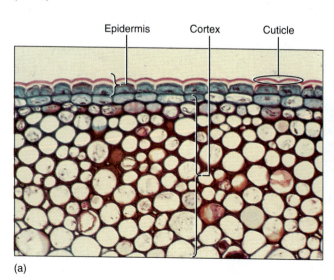

Epidermis   Cortex   Cuticle

(a)

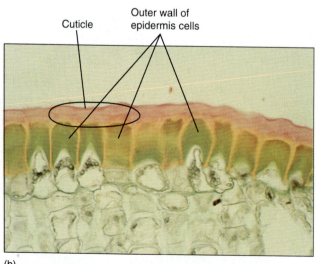

Cuticle   Outer wall of epidermis cells

(b)

TABLE 5.5 *Special Chemicals in Walls*		
Cell Type	**Chemical**	**Special Property**
Collenchyma	Pectins	Plasticity (?)
Sclerenchyma	Lignin	Strength; waterproofing
Epidermis	Cutin/waxes	Waterproofing; indigestible by bacteria, fungi, animals
Endodermis (Chapter 7)	Suberin	Waterproofing
	Lignin	Waterproofing
Cork (in bark) (Chapter 8)	Suberin	Waterproofing

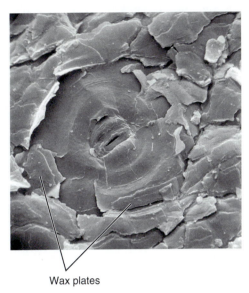

Wax plates

FIGURE 5.23
On this epidermis, wax is present as flat plates, but in other species it can occur as threads, beads, flakes, or a liquid (× 550). *(Courtesy of U. Eggli, Municipal Succulent Collection, Zurich)*

However, the epidermis contains pairs of cells (**guard cells**) with a hole (**stomatal pore**) between them (Fig. 5.24). Guard cells and a stomatal pore together constitute a **stoma** (plural: stomata). Stomatal pores can be opened during the daytime, permitting carbon dioxide to enter the plant.

The opening of the stomatal pore is made possible by the structure of the guard cells. The walls of the guard cells have a special radial arrangement of cellulose microfibrils

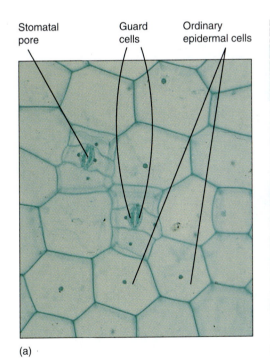

Stomatal pore   Guard cells   Ordinary epidermal cells

(a)

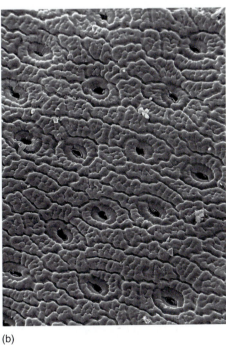

(b)

(c)

FIGURE 5.24
(a) Light microscope view of the epidermis of *Zebrina*; the stomatal pore, guard cells, and adjacent cells are visible, as are ordinary epidermal cells (× 150). (b) Scanning electron micrograph of the epidermis of a cactus showing numerous stomata. Notice how close together the stomata are; no chlorenchyma cell inside the cactus (on the other side of the epidermis) is far from a point where carbon dioxide can enter (× 50). (c) The stomatal pore is surrounded by guard cells, then other epidermal cells. Notice the small particle of debris at one end of the pore; clogging by debris is a problem in dusty areas (× 800). *(b and c, Courtesy of U. Eggli, Municipal Succulent Collection, Zurich)*

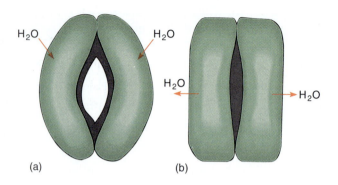

FIGURE 5.25
(a) When guard cells absorb water and swell, they become curved and the inner walls are pulled apart by the expansion of the back walls. (b) If guard cells lose water and shrink, they straighten, closing the stomatal pore.

which causes some parts to be weaker and more extensible than others. Guard cells swell by absorbing water; the walls next to the stomatal pore are rigid and do not extend, but the back walls are weak and stretch (Fig. 5.25), causing the cells to form an arc and push apart, opening the pore and allowing the entry of carbon dioxide and the exit of oxygen by diffusion. This unavoidably allows water vapor to escape. After sunset, when photosynthesis is impossible, it is not advantageous to keep the stomata open; guard cells in most species shrink and collapse toward each other, closing the pore and preventing water loss.

In most plants, some epidermal cells elongate outward and become **trichomes,** also called **hairs** (Fig. 5.26). Trichomes make it difficult for an animal to land on, walk on, or

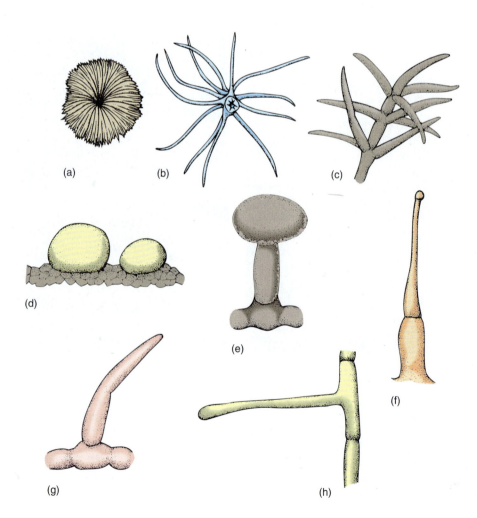

FIGURE 5.26
Trichomes. (a) Flat, scale-shaped trichome. (b) and (c) Branched trichomes. (d) Glandular trichomes—these are two large single cells. (e) Glandular trichome with a stalk cell elevating the secretory head cell. (f) Glandular trichome, capable of injecting poison. (g) and (h) Simple, unbranched, nonglandular trichomes.

chew into a leaf. They can shade the leaf by blocking some of the incoming sunlight, which may be too intense in summer. Trichomes can also create a layer of immobile air next to the leaf surface which allows water molecules that diffuse out of a stoma to bounce back in rather than be swept away by air currents.

Trichomes exist in hundreds of sizes and shapes; many are unicellular, being just long, narrow epidermal cells. Multicellular ones may be a single row of cells or several cells wide, and many branch in elaborate patterns. Most trichomes die shortly after maturity and their cell walls provide protection, but others remain alive and act as small secretory glands. Some secrete excess salt, others produce antiherbivore compounds, and those in the carnivorous plants secrete digestive enzymes onto the trapped insects. The poisonous, irritating compounds of stinging nettle are held in trichomes.

Cortex.   Interior to the epidermis is the **cortex** (Fig. 5.27). In many plants it is quite simple and homogeneous, composed of photosynthetic parenchyma and sometimes collenchyma. In other species it can be very complex, containing many specialized cells that secrete latex, mucilage, or pitch (resin). Some cortex cells contain large crystals of calcium oxalate or deposits of silica.

Cortex cells of most plants fit together compactly, but in fleshy stems such as tubers, corms, and succulents, the cortex parenchyma is aerenchyma, an open tissue with large intercellular air spaces (see Figs. 3.34 and 6.19). A few angiosperms, such as water lilies and the plants grown in aquaria, have become aquatic, living submerged in lakes or oceans. These plants have large cortical air chambers that provide buoyancy. The stems have such a strong tendency to float that no sclerenchyma is necessary for support.

Vascular Tissues.   For very small organisms whose bodies are either unicellular or just thin sheets of cells, diffusion is adequate for the distribution of sugars, minerals, oxygen, and carbon dioxide throughout the body. But if any cells of the organism are separated from environmental nutrients and oxygen by just four or five cell layers, diffusion is too slow and a vascular system is necessary. Two types of vascular tissues occur in plants: **xylem,** which conducts water and minerals, and **phloem,** which distributes sugars and minerals.

The vascular systems of plants are quite different from those of animals. The plant vascular system is not a circulatory system: Water and minerals enter the xylem in the roots and are conducted upward to the leaves and stems. During its passage the xylem sap travels through dead cells, not through tubes composed of living cells like blood vessels. Once in the shoots, water evaporates from the surface of the leaves and is lost; the minerals are absorbed and used by surrounding cells. Phloem cells are living; they pick up sugar from areas where it is abundant, usually leaves during summer and tubers or rhizomes in spring, and transport it to areas where it is needed, especially the growing tips of shoots, roots,

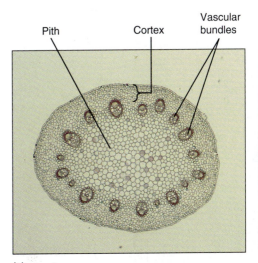

(a)

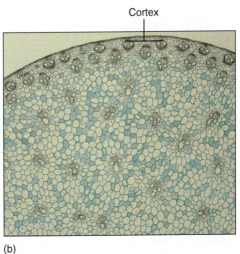

(b)

FIGURE 5.27
(a) The cortex of this stem of buttercup (*Ranunculus*) is the narrow band of cells between the epidermis and the vascular bundles ($\times 20$). (b) The cortex of this corn stem is even narrower, in some places being only two or three cells wide ($\times 20$).

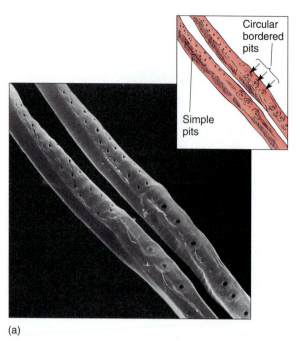

Circular bordered pits

Simple pits

(a)

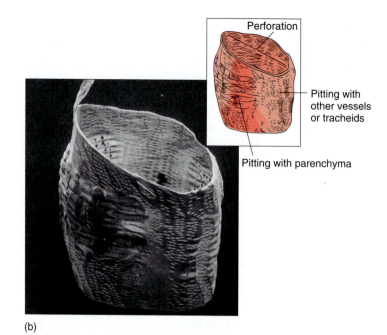

Perforation

Pitting with other vessels or tracheids

Pitting with parenchyma

(b)

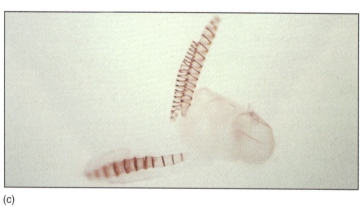

(c)

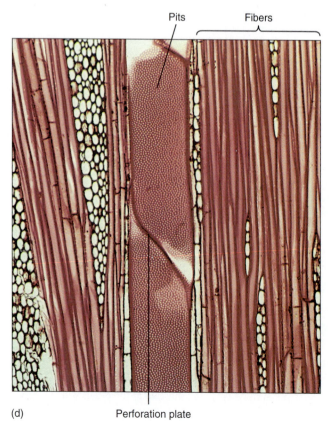

Pits

Fibers

Perforation plate

(d)

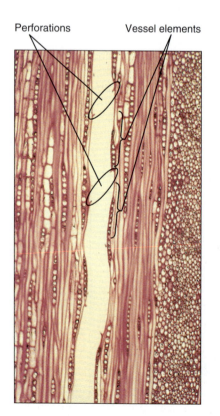

Perforations

Vessel elements

(e)

TABLE 5.6	*Tracheary Elements (Conducting Sclerenchyma)*	
	Tracheids	**Vessel Elements**
Shape	Long/narrow	Short/wide
	Ends pointed	Ends usually flat
Secondary wall	Annular	Annular
	Helical	Helical
	Scalariform	Scalariform
	Reticulate	Reticulate
	Circular bordered pits	Circular border pits
Perforations	None	Usually two: one in each end wall; terminal members with only one

young leaves, and flowers. In the later months of summer, phloem carries sugar into developing fruits and into the storage organs of perennial plants. Because sugar must be dissolved to be conducted, water is transported simultaneously in phloem.

Xylem.   Within xylem are two types of conducting cells: **tracheids** and **vessel elements;** both are types of sclerenchyma (Fig. 5.28; Table 5.6). The term **"tracheary element"** refers to either type of cell. As a young cell matures into a tracheary element, it first must enter cell cycle arrest and stop dividing. It is initially a small parenchyma cell with only a thin primary wall, but the cell becomes long and narrow and then deposits a secondary wall that reinforces the primary wall. The cell then dies and its protoplasm degenerates, leaving a hollow tubular wall (Fig. 5.29).

The secondary wall is impermeable to water, so areas of the permeable primary wall must remain uncovered if water is to enter and leave the cell. In the simplest type of tracheary element, there is only a small amount of secondary wall, organized as a set of rings, called **annular thickenings,** on the interior face of the primary wall (Fig. 5.29e). This arrangement provides a large surface area for water movement into and out of the cell, but it does not provide much strength. The walls must be strong because the movement of the water tends to cause them to collapse inward (Fig. 5.30). With **helical thickening,** the secondary wall exists as one or two helices interior to the primary wall (see Fig. 5.29f). **Scalariform thickening** provides much more strength because the secondary wall covers most of the inner surface of the primary wall and is fairly extensive. Just as in nonconducting sclerenchyma, the area where secondary wall is absent is called a pit (Fig. 5.31a). In tracheary elements with **reticulate thickening,** the secondary wall is deposited in the shape of a net, as the name suggests. The most derived and strongest tracheary elements are those with **circular bordered pits** (Figs. 5.31b and 5.32). In such tracheary elements, virtually all the primary wall is covered by the secondary wall. The pits that allow water

Scalariform and reticulate pits can either have borders or lack them; circular pits in tracheary elements always have borders.

◀ **FIGURE 5.28**
Tracheary elements. (a) and (c) Tracheids are long cells with tapered ends. The primary wall is complete over the whole surface, and in (c) it is so thin that the inner, spiral secondary wall is visible. In (a), the primary wall tore during specimen preparation, permitting views of circular bordered pits and simple pits. (b) and (d) Vessel elements tend to be wider and shorter than tracheids, but the most important feature is the perforation, the large hole at each end. (e) The perforation of one vessel element must be aligned with that of the next vessel element if water is to pass through with little friction; the stack of vessel elements is a vessel. The vessel here has been cut down the center, so the front wall and back wall are missing and the vessel appears to be just an open space. (a) ×600; (b) ×270; (c) ×100; (d) ×200; (e) ×50. *(a and b, Courtesy of W. A. Côté, Jr., N. C. Brown Center for Ultrastructure Studies, State University of New York)*

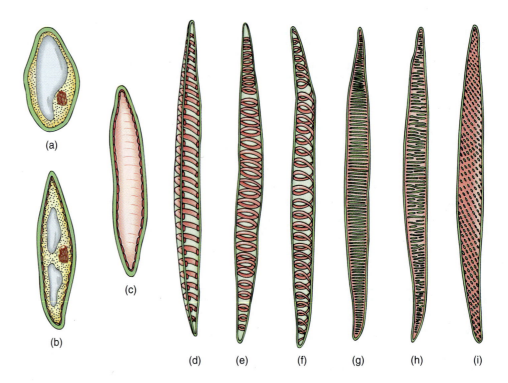

FIGURE 5.29

Tracheids. (a) to (d) The growth of a young cytoplasmic cell into a tracheid. Once the cell has almost reached mature size, the secondary wall is deposited interior to the primary wall (c). After the secondary wall is finished, the protoplasm dies and degenerates, leaving only the primary and secondary walls. (d) A tracheid cut open, showing a secondary wall in the form of helical thickenings. (e) A whole cell in which the primary wall is so thin that the annular secondary walls inside are visible. (f) Whole cell with helical secondary wall. (g) Scalariform secondary wall. (h) Reticulate secondary wall. (i) Pitted secondary wall.

movement are weak points in the wall, but the weakness is reduced by a **border** of extra wall material around the pit—hence the name.

Tracheary elements with annular thickenings are weak, but a large percentage of their primary wall is uncovered and available for water movement. Pitted tracheary elements are just the opposite: They are extremely strong, but so much of the surface is covered by secondary wall that water enters or leaves the cells slowly. Each is selectively advantageous

FIGURE 5.30

(a) As water enters a tracheary element, it passes through the primary wall between regions of secondary wall. (b) Water adheres to cellulose molecules, pulling the walls inward and upward as it rises in xylem. The primary wall tends to collapse under the friction of the passage, but the secondary wall provides strength to hold the primary wall in place. (c) Under severe water stress (dry air and dry soil), annular and helical walls may not be strong enough to prevent collapse.

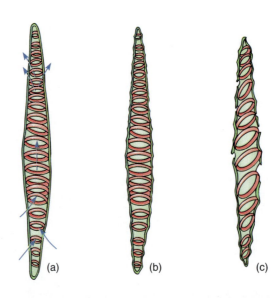

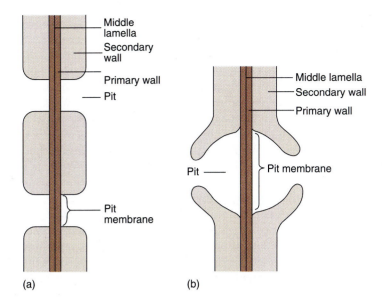

(a)

(b)

FIGURE 5.31
(a) Side view of a pit that has been bisected. In a sclereid or fiber, the secondary wall would be thicker and the pit narrower and longer. The diameter of the pit can vary, being broad as in Figure 5.10d or narrow as in Figure 5.12. (b) A bordered pit has a rim of extra wall material around it, which strengthens it and prevents the pit from being a weak spot in the cell. With borders, pits can be broad enough to allow rapid flow of water from tracheary element to tracheary element, but not so weak as to be detrimental. Only tracheary elements have bordered pits; fibers and sclereids do not.

under certain conditions and disadvantageous under others. If a species grows in perpetually wet soil and water moves easily, the cells are not in danger of collapsing, so tracheary elements with annular or helical secondary walls are adaptive, not only because the large open regions of primary wall allow water to pass through quickly but also because the plant minimizes the amount of sugar it spends building secondary walls. But if the soil dries even slightly, water is pulled with more force and exerts inward traction on the wall of the tracheary elements. The stronger scalariform, reticulate, or circular bordered pitted walls are necessary to keep the tracheary elements open. Although they cannot conduct water as

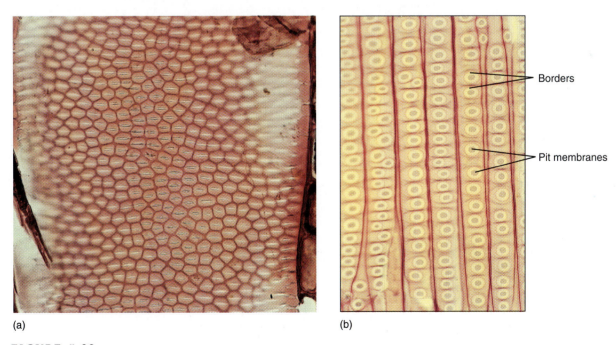

(a)

(b)

FIGURE 5.32
(a) Bordered pits in face view on front wall of a vessel element. The white slits are the actual openings of the pits, the pinkish area around each white region is the border, and the dark red hexagons are the regions where the secondary walls lie against the primary walls ($\times$ 400).
(b) Bordered pits in face view on the front walls of tracheids of pine. This is the same magnification as (a); notice that tracheids are much narrower than vessel elements and that pits in pine are extremely large.

quickly, water movement is automatically slower because the soil is dry. Under dry conditions, using extra glucose to build stronger secondary walls can be a valuable investment.

Tracheids obtain water from other tracheids below them and pass it on to those above; they must occur in groups, lying side by side with some and having their ends overlapping the ends of others (Fig. 5.33). Also, the pits of adjacent tracheids are aligned so that water can pass through. The aligned set of pits is a **pit-pair,** and the set of primary walls and middle lamella between them constitutes a **pit membrane,** just as in fibers and sclereids (see Fig. 5.31). Although very permeable to water, pit membranes do offer slight resistance. If each tracheid is 1 mm long, each water molecule passing from root to leaf in even a short plant of only 1 meter (1000 mm) must pass through 1000 pit membranes. The friction adds up.

Vessel elements provide a way to move water with less friction. Vessel elements, like tracheids, are individual cells that produce both primary and secondary walls before they die at maturity. In vessel elements, however, an entire region of both primary and secondary wall is missing. During the final stages of differentiation, a large hole called a **perforation** is digested through a particular site of the primary wall, often removing the entire end wall (see Figs. 5.28 and 5.33). Because perforations are wide and have no pit membrane,

"Pit membrane" is a very old, misleading term; it should not be confused with the lipid/protein membranes such as the plasma membrane.

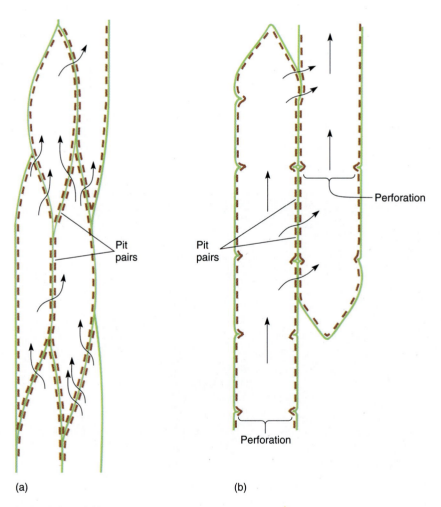

(a)   (b)

FIGURE 5.33

(a) Water can pass between tracheids only through pit-pairs. Other than plasmodesmata (which are too tiny to be important in water conduction), no holes are present in the primary walls of tracheids. (b) Water can pass between vessel elements through perforations, but it can pass from one vessel to another only through pit-pairs. Vessels are long, but not as long as a whole plant, so water must pass from one vessel to another as it moves upward from roots to stems to leaves. Green = primary wall; brown = secondary wall.

TABLE 5.7	Length of Vessels in Trunks of Trees	
	Vessel Length (cm)	
Species	Average	Maximum
Fagus grandifolia (American beech)	302	556
Populus tremuloides (quaking aspen)	122	132
Betula lutea (yellow birch)	119	142
Alnus rugosa (speckled alder)	105	122
Fraxinus americana (white ash)	97.1	1829
Quercus rubra (red oak)	94.8	1524
Ulmus americana (American elm)	94.6	853
Acer saccharum (sugar maple)	15	32
Vaccinium corymbosum (swamp blueberry)	11	120

they provide a pathway with little friction. The perforations of adjacent vessel elements must be aligned, and each element must have at least two perforations, one on each end. An entire stack of vessel elements is a **vessel.** The vessel elements on each end of a vessel have only one perforation. Whereas vessel elements, being individual cells, are only about 50 to 100 μm long, vessels can be many meters long, running all the way from a root tip to a shoot tip, although some are only a few centimeters long (Table 5.7).

Vessels must absorb water from parenchymal cells, tracheids, or other vessels, and they must pass it on. Their side walls must have pits for this lateral transfer, and all the types of wall thickening mentioned above for tracheids also occur in vessel elements. The only constant difference is that vessel elements have perforations—complete holes—in their walls, whereas tracheids only have pits (see Table 5.6). The perforation greatly reduces the friction, and water moves much more easily through a set of vessels than through a set of tracheids.

All plants with vascular tissue have tracheids, which evolved at least 420 million years ago. Vessels evolved more recently and occur almost exclusively in flowering plants, where they perform virtually all long-distance water conduction in roots and stems. In flowering plants, tracheids occur mostly in the fine veins of leaves but can also be found in the wood of some species. Almost all nonangiosperms such as conifers and ferns have xylem composed exclusively of tracheids.

Phloem. Like xylem, phloem has two types of conducting cells, **sieve cells** and **sieve tube members;** the term **"sieve element"** refers to either one (Fig. 5.34). These are very different from tracheary elements: They must remain alive in order to conduct, and because they have only primary walls they are parenchyma cells. As immature sieve elements begin to differentiate, their plasmodesmata enlarge to a diameter of more than 1 μm (1000 nm) and are called **sieve pores.** Plasmodesmata occur in groups (primary pit fields), so sieve pores also occur clustered together in groups called **sieve areas.** Sieve elements remain alive during differentiation, and the plasma membrane that lined the plasmodesma contin-

Tracheids and vessel elements have plasmodesmata in their primary walls, but these are too narrow to be important in long-distance water conduction. Typical diameters—plasmodesmata: 0.04 μm; pits: 1 to 2 μm; perforations: 20 to 50 μm.

BOX 5.2   *Resin-casting: A New Method for Studying Cell Shapes*

It can be difficult for experienced anatomists, as well as beginning biology students, to truly understand the shapes and sizes of cells. It is not easy to visualize complex subcellular features like perforations and pit chambers. Recently, a new method, resin-casting, has been developed that allows us to see cells from a new perspective. The plastics with which we are familiar in all aspects of our lives are produced as liquids, then injected into molds and allowed to harden. Once the mold is removed, the plastic retains the shape of the

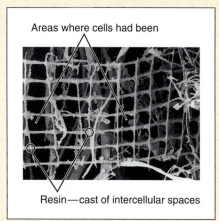

Areas where cells had been

Resin—cast of intercellular spaces

This network is a resin-cast of the intercellular spaces in the wood of an oak, *Quercus accutissima*. The large cubical spaces had been occupied by cells arranged in regular rows and columns, but the cells did not fill with resin and were later dissolved away. The resin flowed between the cells both horizontally and vertically. This cast shows that the intercellular spaces are interconnected and form a network that allows oxygen to diffuse rapidly throughout the wood.

Perforation plate   Vessel element

Pit casts

interior of the mold. Certain types of liquid plastic, called resins, can be injected into plants and the intercellular spaces of the plants act as molds, providing a shape for the resin as it solidifies. Similarly, empty cells such as vessels, tracheids, and fibers also act as molds. After the resin has hardened, the plant material is removed by dissolving it with acids and hydrogen peroxide; many resins are resistant to these chemicals, so the shape of the hardened plastic is not altered by this process. Once the plant material is dissolved and washed away, what remains are resin-casts of cell interiors. These can be examined by scanning electron microscopy for details of pits, perforations, size and shape of the cell interior, the distribution of various types of pits on the walls of vessel elements, and so on. A few examples are shown; as you look at them, remember that the solid resin represents the space of the original cell, and the places where there is no resin represent areas occupied by the cell wall material before it was dissolved with acid.

The resin flowed rapidly into these large vessel elements in the wood of *Castanea* and was able to pass from vessel element to vessel element easily because the perforations are just large holes. The perforation plates are visible as narrow indentations because a small amount of the original vessel element end wall (the perforation plate) remained as a rim when the vessel element was differentiating and digesting out the perforation. If the end wall had been completely removed during xylem maturation, this cast would be smooth and we could not tell one vessel element from another. The small bumps on the surface of the vessel are casts of pit chambers.

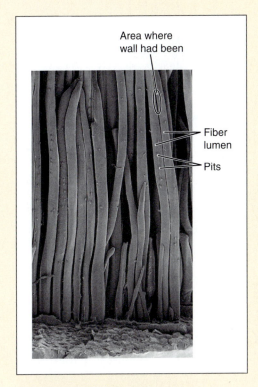

Area where wall had been

Fiber lumen

Pits

Casts of the interiors of wood fibers. The narrow spaces between most casts represent the thickness of the wall that existed before it was dissolved with acid. The smoothness of the surface of the casts indicates that the inside face of the fiber wall was quite smooth. The small bumps are casts of pits that were parts of pit-pairs.

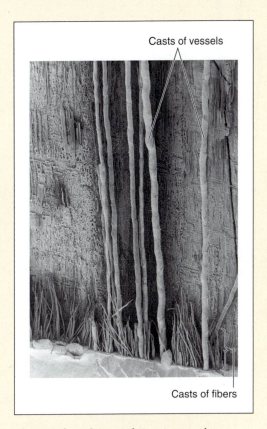

Casts of vessels

Casts of fibers

This cast of vessels in wood may give you a better appreciation of how long vessels are and how numerous vertically aligned vessel elements must coordinate their development if a vessel is to form properly. Actually, these are just tiny fractions of complete vessels, which can be many meters long. *(All micrographs courtesy of Dr. Tomoyuki Fujii, Forest and Forest Research Institute, Tsukuba, Japan)*

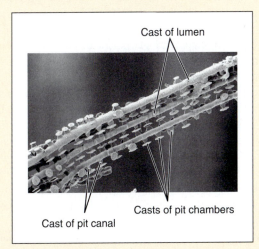

Cast of lumen

Casts of pit chambers

Cast of pit canal

Casts of three fibers that had very narrow lumens (indicated by the narrow columnar portions of the casts) and thick walls (represented by the wide spaces separating the lumen casts). The objects that look like octopus suckers are casts of bordered pits. The flat disks are casts of the broad pit chambers, and the trapezoidal structures connecting the chamber casts to the lumen casts are casts of the pit canal. The liquid resin was not able to flow through the pit membrane from one pit chamber into the facing pit chamber of the pit-pair. Consequently, when the cell material is dissolved with acid, the pit membrane breaks down and the cast of one cell tends to fall away from the casts of the surrounding cells.

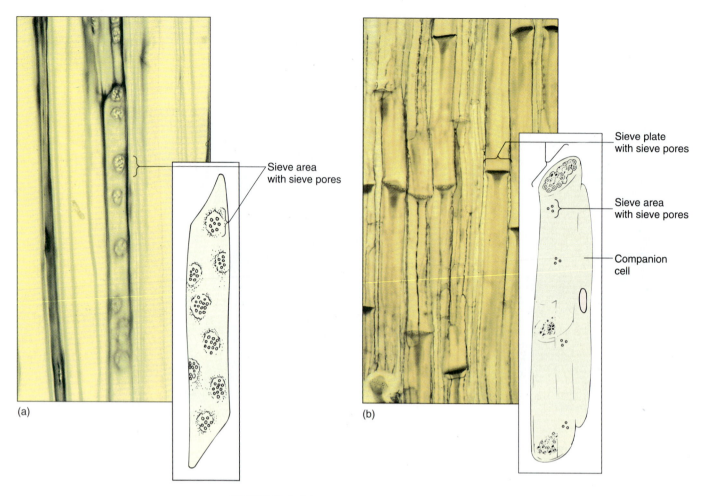

FIGURE 5.34

(a) A sieve cell is shaped like a tracheid and has sieve areas with sieve pores over much of its surface (×250). (b) A sieve tube member has flat end walls (sieve plates) with large sieve pores. On the side walls are only a few small sieve areas with very narrow sieve pores (×60). Note that the micrograph of sieve cells is a high magnification; they are much narrower than the sieve tubes of (b).

ues to line the sieve pore (Fig. 5.35). The amount of cytoplasm within the pore also increases, and rapid, bulk movement from cell to cell becomes possible. Sieve pores of adjacent cells must be aligned.

The two types of sieve elements differ in shape and placement of sieve areas (Fig. 5.34). If the cell is elongate and spindle-shaped (like a tracheid) and has sieve areas distributed over all its surface, it is a sieve cell (Table 5.8). This type evolved first and is found in older fossils and in nonangiosperm vascular plants. Because phloem conducts mostly longitudinally, it is most efficient for the sieve areas on the two ends of the cell to be especially large and to have very wide sieve pores, whereas sieve areas on the sides of the cell can be rather small. Cells like this are sieve tube members; they are stacked end to end with their large sieve areas aligned, forming a **sieve tube.** These end-wall sieve areas with large sieve pores are **sieve plates.** Like vessels, sieve tubes are shorter than the plant, and the small, lateral sieve areas are important for transport between sieve tubes. Sieve tube members must have evolved at about the same time as the flower; all angiosperms have sieve tube members, but none of the nonangiosperms has them.

An unusual feature of sieve elements is that their nuclei degenerate but the cells remain alive. Cytoplasm cannot carry on complex metabolism without a nucleus, and in phloem the necessary nuclear control is provided by cells that are intimately associated with the conducting cells. Sieve cells are associated with **albuminous cells,** and sieve tube members

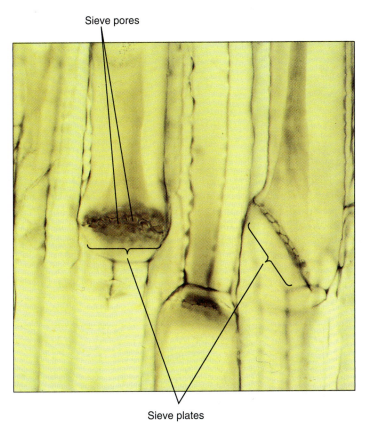

Sieve pores

Sieve plates

FIGURE 5.35
Large sieve areas present on the sieve plates of sieve tube members. Phloem sap flows from one tubelike sieve tube member into the next through the wide sieve pores (× 300).

are controlled by **companion cells** (Fig. 5.36). These cells are often smaller than the accompanying conducting cell and have a prominent nucleus and a cytoplasm that is dense and filled with ribosomes. Walls between conducting cells and controlling cells have many complex passages that are sieve areas on the conducting cell side and large plasmodesmata on the controlling cell side.

Recent evidence shows that the companion cells have another important role, that of loading sugars into and out of the sieve tube members. The extent and importance of this function are unknown.

Vascular Bundles.   Xylem and phloem occur together as **vascular bundles,** located just interior to the cortex (Fig. 5.37). In dicots, vascular bundles are arranged in one ring surrounding the **pith**, a region of parenchyma similar to the cortex. In monocots, vascular bundles are distributed as a complex network throughout the inner part of the stem; between the bundles are parenchyma cells. Monocot bundles are frequently described as

TABLE 5.8 *Sieve Elements*		
	Sieve Cells	**Sieve Tube Members**
Shape	Long/narrow	Short/wide
	Ends pointed	Ends usually flat
Sieve areas	Small, located over all the cell surface	On side walls: small; on end walls: very large end wall is sieve plate.
Associated cells	Albuminous cells	Companion cells
Plant division	All nonangiosperm vascular plants. Some relictual angiosperms	Angiosperms only

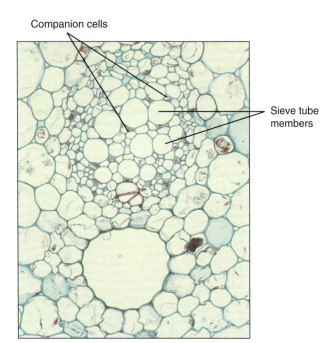

Companion cells

Sieve tube members

(a)

FIGURE 5.36

(a) In a vascular bundle, sieve tube members and companion cells occur together. Companion cells are small and rich in cytoplasm; sieve tube members are large and appear empty because most phloem sap is lost while preparing the tissue for microscopy ($\times 150$). (b) Electron micrograph showing two companion cells and two sieve tube members. (c) The connection between a companion cell and a sieve tube member is a primary pit field with plasmodesmata on the side of the companion cell *(bottom)*, but it is a sieve area with sieve pores on the side of the sieve tube members *(top)* ($\times 45,000$). *(b and c, Courtesy of R. Warmbrodt, Climate Stress Laboratory, USDA)*

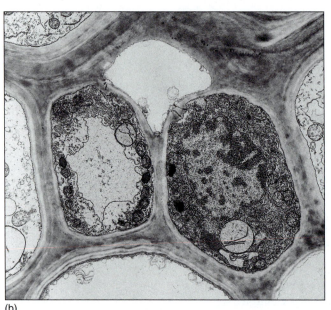

(b)

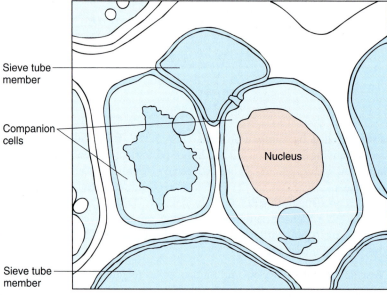

Sieve tube member

Companion cells

Nucleus

Sieve tube member

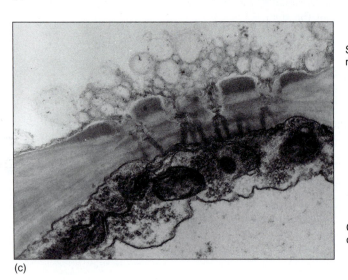

(c)

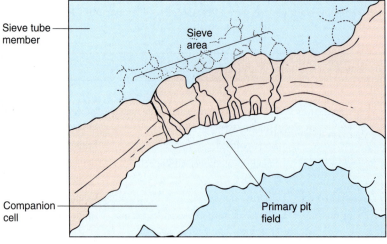

Sieve tube member

Sieve area

Companion cell

Primary pit field

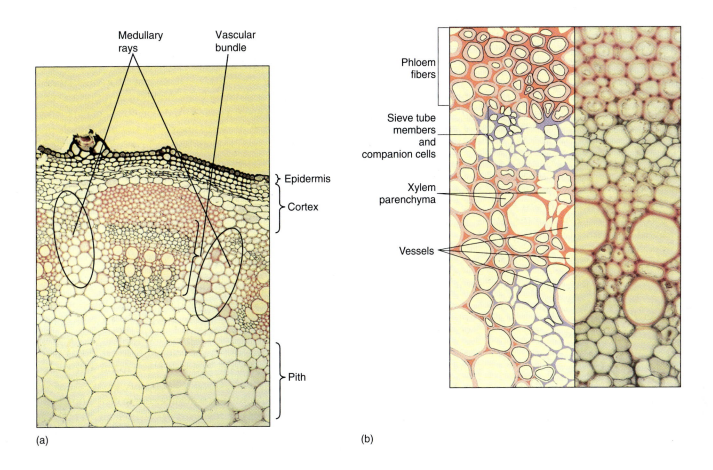

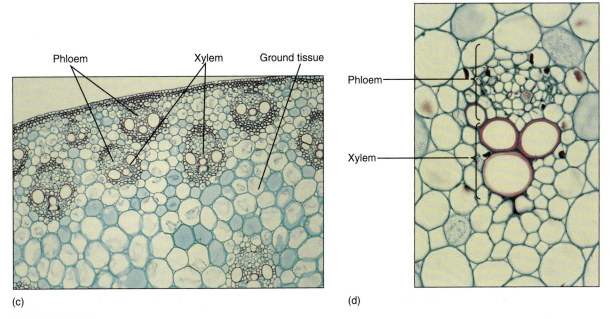

FIGURE 5.37

(a) In this vascular bundle of sunflower *(Helianthus),* the xylem contains rows of vessels alternating with rows of xylem parenchyma. The phloem contains sieve tube members and companion cells. There is a large sclerenchyma cap of phloem fibers (×60). (b) The same bundle, shown at higher magnification (×250). (c) The vascular bundles of monocots have a more complex arrangement. In this corn stem, many large bundles are located near the cortex and fewer, smaller bundles in the ground tissue (×50). (d) High magnification of a corn bundle, showing primary xylem and phloem (×250).

The relative amounts of each cell type within a bundle may vary: Storage stems may have large amounts of parenchyma in the vascular tissues; vines and tendrils have extra fibers.

Plant Biology Tutor CD-ROM

"scattered" in the stem, suggesting that they occur at random, but they actually have precise, specific patterns that are too complex to be recognized easily.

All vascular bundles are **collateral**; that is, each contains both xylem and phloem strands running parallel to each other (Fig. 5.37). The xylem of a vascular bundle is **primary xylem** because it is part of the primary plant body. In addition to conductive tracheids and vessel elements, there is usually a large proportion of xylem parenchyma and even mechanical sclerenchyma in the form of xylem fibers. The vascular bundle phloem is **primary phloem,** and mixed with the sieve elements and companion cells or albuminous cells may be storage parenchyma and mechanical sclerenchyma, usually as phloem fibers, although phloem sclereids also occur. Cells of primary xylem are larger than cells of primary phloem, and the tracheary elements on the inner side of each bundle are much smaller than the outer ones (Fig. 5.37). This is true of the primary xylem in all stems, and the reason for this becomes clear when the shoot growth is examined.

STEM GROWTH AND DIFFERENTIATION

Stems grow longer by creating new cells at their tips, in regions known as **shoot apical meristems;** cells divide by mitosis and cytokinesis, producing progenitor cells for the rest of the stem (Fig. 5.38). When each cell divides, the two daughter cells are half the size of the mother cell, but they quickly grow back to the original size. As they expand, they automatically push the meristem upward; the lower cells are left behind as part of the young stem. This region just below the apical meristem is the **subapical meristem,** and its cells are also dividing and growing, producing cells for the region below.

In the subapical meristem, visible differentiation begins; certain cells stop dividing and start elongating and differentiating into the first tracheids or vessel elements of the vascular bundles (Figs. 5.37 and 5.39). Because they constitute the first xylem to appear, they are called **protoxylem.** The cells around them continue to grow and expand, but soon those just exterior to them also begin to differentiate; because these had expanded for a longer time, they develop into tracheary elements that are even larger than the first. This process continues until the last cells mature and this portion of the stem stops elongating. Because these cells have had the longest time for growth before differentiation, they develop into the largest tracheary elements of all, called **metaxylem.**

Size is not the only difference between tracheary elements of the protoxylem and the metaxylem. Because protoxylem elements differentiate while the surrounding cells are still elongating, they must be extensible and so must have either annular or helical secondary walls. Although protoxylem cells are dead at maturity, they stretch because they are glued by their middle lamellas to the living, elongating cells that surround them. If protoxylem had scalariform, reticulate, or pitted secondary walls, the cells could not be stretched and either they would prevent the surrounding cells from growing or the middle lamellas would break and the cells would tear away from each other. This is not a problem for the metaxylem, however, because its cells differentiate only after the surrounding stem tissues have stopped elongating. Because metaxylem cells do not have to be extensible, any type of secondary wall is feasible.

A similar process occurs in the outer part of each vascular bundle: The exterior cells mature as **protophloem,** and the cells closest to the metaxylem become **metaphloem.** Because the sieve elements do not have secondary walls, the walls of protophloem and metaphloem are identical. However, sieve elements are extremely sensitive to being stretched and die when stressed too much. Consequently, protophloem cells are extremely short-lived, often functioning for only 1 day. Perhaps because they are so temporary, protophloem cells never become well differentiated; their sieve areas are small and rudimentary, and many do not have companion cells. The sieve elements of the metaphloem do not differentiate until later, when all surrounding cells have stopped expanding. Metaphloem cells differentiate fully, having large sieve areas and conspicuous companion cells.

You might expect metaphloem cells to be very large, like those in the metaxylem, but they are much smaller, even though they have undergone long periods of expansion before

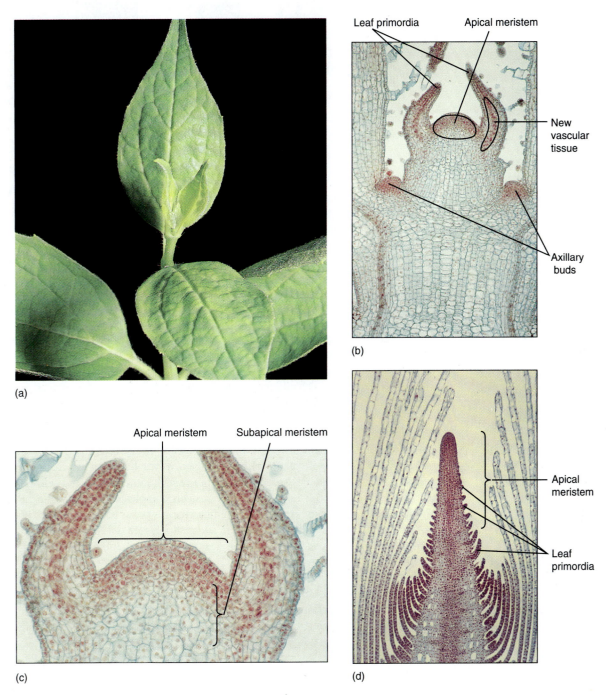

FIGURE 5.38

(a) This shoot tip has several fully expanded leaves, one set of small leaves, and another set of even smaller, younger leaves. Dissection would reveal even more, younger leaves. (b) A longitudinal section through a shoot tip of *Coleus* showing the apical meristem and leaf primordia of various ages. Two axillary buds are just recognizable as small outgrowths (× 100). (c) A higher magnification of the *Coleus* shoot apical meristem. Vascular tissue is differentiating in the center of each leaf primordium (× 200). (d) This apical meristem of *Elodea* is much taller and narrower than that of the *Coleus* in (c), and it has many more leaf primordia (× 50).

(a)

(b)

(c)

(d)

(e)

(f)

FIGURE 5.39

Development of three tracheary elements. (a) All cells are young, small, and cytoplasmic. (b) The first protoxylem element forms while all cells are still small. (c) Cells 2 and 3 continue growing, stretching cell 1, which is dead. (d) and (e) Cell 2 matures while cell 3 continues to grow, stretching cells 1 and 2. Cell 1 has been stretched so much that it is on the verge of being torn open. (f) All elongation has stopped, and cell 3 differentiates as a pitted element. The primary wall of cell 1 has been torn and no longer conducts. Cell 2 is probably still capable of carrying water.

differentiation. All cells of the region are expanding, whether they are dead functioning protoxylem, dead nonfunctioning protophloem, or living undifferentiated metaxylem and metaphloem. Cell size differs because cell division is still occurring in some cells but not in others. In xylem, all cells stop dividing, so each cell becomes larger as the tissue expands —protoxylem cells by being stretched, the soon-to-be metaxylem cells by their own growth. Protophloem cells stop dividing and differentiate, but other phloem cells continue to divide so they remain small.

Other tissues also differentiate in the subapical region. In the epidermis, the first stage of trichome outgrowth may be visible in the youngest internodes, those closest to the apical meristem. In lower, older internodes, trichomes are more mature, and guard cells and stomatal pores may be forming. The cuticle is extremely thin at the apical meristem, but it is thicker in the subapical region and may be complete several internodes below the shoot apex. Differentiation of the pith typically involves few obvious changes: Cells enlarge somewhat, intercellular spaces expand but remain rather small, and cell walls continue to be thin and unmodified. All changes are usually completed quickly, while the cells are in or just below the subapical meristem. Similar changes occur in the cortex, except that plastids develop into chloroplasts. Protoxylem and protophloem develop quickly while the cells are still close to the apical meristem, but metaxylem and metaphloem do not begin to differentiate until the nodes and internodes have stopped elongating.

Because the terms "epidermis," "cortex," "phloem," "xylem," and "pith" typically refer to the mature cells and tissues, it is confusing to use them for the young, developing cells during differentiation. To prevent misunderstandings, a special set of terms has been agreed on: **Protoderm** refers to epidermal cells that are still meristematic and in the early stages of differentiation. Young cells of xylem and phloem are referred to as **provascular tissues;** the equivalent stages of pith and cortex are **ground meristem** (Fig. 5.40). Thus the subapical meristem consists of protoderm, ground meristem, provascular tissue, and more ground meristem. The rate of maturation varies greatly from species to species and also from one type of stem to another on the same plant. In some cases, maturation is slow and immature tissues can be found many internodes below the shoot apical meristem. In others, differentiation may be completed quickly and all tissues are mature close to the stem apex.

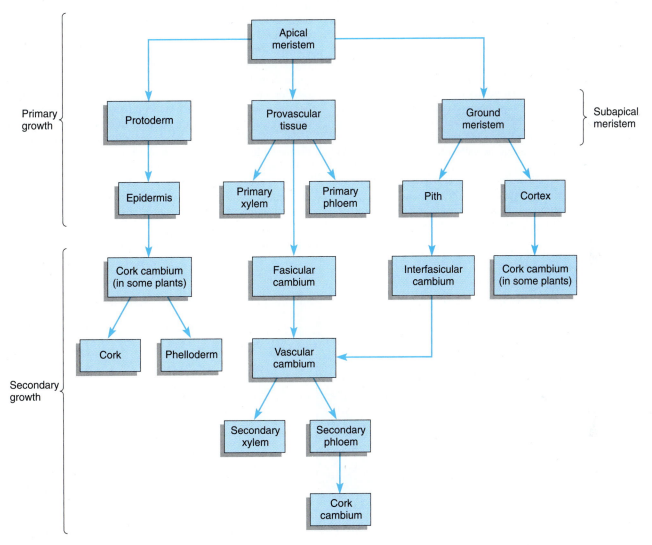

FIGURE 5.40
Cells of the apical meristem give rise to three types of subapical cells, which in turn give rise to particular types of mature primary tissues. Epidermis develops only from protoderm, never from provascular tissue or ground meristem; other cell lineages in the primary body are also quite simple and linear.

PLANTS & PEOPLE

FIBERS

We depend on plants to supply many of the needs of our daily lives, food being particularly obvious, but also lumber, drugs, oils, dyes, and the raw materials for beer, wine, and liquor. Most of these come from leaves, fruits, seeds, or woody trees. Asparagus (*Asparagus officinalis*), potatoes (*Solanum tuberosum*), and sugar and molasses from sugarcane (*Saccharum officinarum*) are examples of the few foods derived from herbaceous stems.

An important product that occurs in herbaceous stems is soft fibers, technically known as **bast fibers.** These fibers are not individual cells, as described earlier in this chapter, but entire masses of phloem fibers that are part of the stem vascular bundles; in some cases, they even include xylem, actually consisting of the whole vascular bundle.

Even though synthetic materials have replaced natural fibers in products like fishing line and nets, sailcloth, and many types of rope, bast fibers are still important. Flax is extracted from *Linum usitatissimum*, the same plant that gives us linseed oil. Its fibers are stronger than those of cotton and are smooth and straight; they are woven into linen, considered one of our most elegant cloths. Flax is also used to make fine writing paper. Unfortunately, all harvesting and extraction must be done by hand, so linen has never become inexpensive. A much coarser fiber, jute, is harvested from stems of *Corchorus capsularis;* the fibers are up to 2 m long and are rougher and stiffer than flax. Jute is probably present in almost every home as the backing on carpets, and many types of canvas, twine, and burlap bags are made of it. World production of jute exceeds 4 million tons per year. Until it was replaced by manmade polymers recently, hemp was the fiber of choice for rope, canvas, sailcloth, and yacht cordage because it has a high lignin content that makes it strong and resistant to water, even seawater. Hemp is harvested from the phloem of *Cannabis sativa,* otherwise known as marijuana; there are several varieties of *Cannabis,* and the fiber-producing type has only low concentrations of the psychoactive drug.

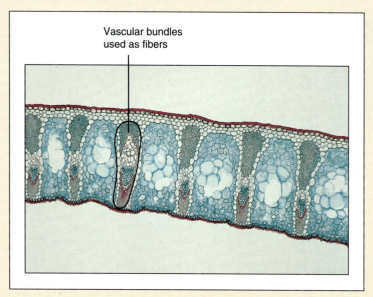

Vascular bundles
used as fibers

Stems and leaves of flax plants have masses of fibers that are used to make fine paper and cloth.

SUMMARY

1. Cells must be arranged in the proper patterns and with the proper interrelationships in order to function efficiently.

2. All angiosperms have roots, stems, and leaves, although each can be highly modified in particular environments.

3. Virtually all stems function in producing and supporting leaves and transporting water, minerals, and sugars between leaves, buds, and roots. Other stems are involved in storage, reproduction, dissemination, and surviving stress.

4. All stems consist of nodes and internodes, and they have leaves and axillary buds. Numerous types of stems exist; most individual plants have two or three types of shoots.

5. The three basic types of plant cells, based on walls, are parenchyma (responsible for most metabolism), collenchyma (plastic support), and sclerenchyma (elastic support).

6. The vascular tissues are xylem and phloem. Tracheary elements consist of vessel elements (with perforations) and tracheids (without perforations). Sieve elements consist of sieve cells (with small sieve pores) and sieve tube members (with large sieve pores on the end walls).

7. Stems of both dicots and monocots have epidermis, cortex, and collateral vascular bundles. In dicots the bundles occur as one ring surrounding pith, whereas in monocots they have a complex distribution in conjunctive tissue.

8. Protoxylem and protophloem form while an organ is elongating and therefore must be extensible. Metaxylem and metaphloem form after elongation ceases.

9. Both primary xylem and primary phloem are complex tissues with a variety of cell types, not just conducting cells.

IMPORTANT TERMS

angiosperm	dicot	perforation	sclereid	stoma
apical meristem	epidermis	phloem	sclerenchyma	tracheid
axillary bud	fiber	phyllotaxy	sieve cell	trichome
bordered pit	guard cell	pit	sieve plate	vascular bundle
collenchyma	internode	pit membrane	sieve pore	vessel
companion cell	monocot	pit-pair	sieve tube	vessel element
cortex	node	pith	sieve tube member	xylem
cuticle	parenchyma			

REVIEW QUESTIONS

1. Describe each of the following types of specialized shoots; be certain to account for modifications of the leaves, internodes, and orientation of growth: stolon, rhizome, tuber, bulb, corm, tendril. Each provides a plant with a selective advantage. What is the adaptive value of each type of specialized shoot?

2. What are the important differences between parenchyma, collenchyma, and sclerenchyma cells?

3. Describe plasmodesmata, pits, perforations, and sieve pores. Which connect living cells? Which connect nonliving cells? Which occur in secondary walls?

4. List the five types of secondary wall deposition that may occur in tracheary elements. Which two types are most characteristic of protoxylem? What is the selective advantage of vessel elements over tracheids?

5. During the differentiation of a young cell into a sieve tube member, what are some of the changes that occur in the cell wall and the cytoplasm? How long do most sieve tube members live after they become mature?

6. Describe the arrangement of tissues seen in a stem cross-section; consider dicots separately from monocots. Is the arrangement different in a stolon than in a rhizome, tuber, or corm?

BotanyLinks .net Questions

www.jbpub.com/botanylinks

Visit the .net Questions area of BotanyLinks (http://www.jbpub.com/botanylinks/netquestions) to complete this question:

1. Many plants, such as cacti, have stems with unusual shapes, but is their anatomy also unusual? Go to the BotanyLinks home page to think about this.

6

LEAVES

One function of foliage leaves is photosynthesis, but another function of leaves is to not endanger the plant. They must not lose water to dry winter air; many are allowed to die and are discarded in the autumn.

CONCEPTS

The term "leaf" usually calls to mind foliage leaves—the large, flat, green structures involved in photosynthesis. However, natural selection has resulted in numerous types of leaves that are selectively advantageous because they provide protection (bud scales, spines), support (tendrils), storage (fleshy leaves of bulbs), and even nitrogen procurement (trapping and digesting insects). Because protecting a bud from freezing or drying is very different from photosynthesizing, leaf structures and metabolisms that are selectively advantageous for a bud scale are different from those that are selectively advantageous for a foliage leaf. This principle is true for all types of leaves: For each function, certain modifications are advantageous whereas others are not. Studying the various leaf types, their modifications, and the roles they play in the plant's biology leads to an understanding of how structure, metabolism, and function are related.

The shoot system, containing both stems and leaves, demonstrates both division of labor and integration of distinct plant organs. Stems and leaves must function together if the plant is to survive and reproduce, but the optimal features for each organ are quite distinct. Leaves should be flat and thin for maximum absorption of light and carbon dioxide, and most of their tissues must be alive and differentiated into chlorophyll-rich chlorenchyma to carry out photosynthesis. Stems elevate leaves and conduct material to and from them, among other activities. Maximum conduction and support with minimum expenditure of construction material require a cylindrical stem; also, much of the stem— all its tracheary elements and most fibers—must die to be functional. In woody species, new cells are added to the stem xylem and phloem each year, creating a massive wood and bark and an increasingly stronger and more conductive stem. Yearly accumulation of new cells onto leaves is not feasible; such leaves would become bulky, heavy, and opaque— features that would not be selectively advantageous. Almost all leaves contain only primary tissues; secondary production of wood and bark in leaves is extremely rare.

The initial discussion in this chapter centers on foliage leaves because they are the most familiar. Then other leaf types are described and analyzed; by considering how these leaves differ functionally from foliage leaves, it is possible to determine the environmental factors important to leaf structure and how natural selection has produced diverse classes of leaves.

EXTERNAL STRUCTURE OF FOLIAGE LEAVES

The most obvious function of foliage leaves is photosynthesis, but other functions often taken for granted are just as important. Leaves must not lose excessive amounts of water; they must not allow entry of fungi, bacteria, or epifoliar algae; they must not be so nutritious and delicious to animals that they are a liability to the plant; they must not be such effective sails that the plant is blown over in a mild wind; and they must be cheap enough that the plant spends less carbohydrate building them than it recovers by their photosynthesis. If a foliage leaf fails in any of these aspects, the plant dies and the leaf's photosynthesis has been useless. Many of the structural and physiological aspects of leaves that make them waterproof, pathogen resistant, and so on in some way interfere with photosynthesis.

During photosynthesis, leaves absorb carbon dioxide and convert it to carbohydrate by using light energy. Because sunlight comes from one direction at a time, it is best for a portion of the leaf to be flat and wide, allowing maximum exposure to light and maximum surface area for carbon dioxide absorption (Fig. 6.1). Because chlorophyll absorbs sunlight efficiently, light penetrates only a short distance, so leaves can be quite thin; in a thick layer of chlorenchyma, the lowest cells would be in almost complete darkness, unable to photosynthesize. This flat, light-harvesting portion is the **leaf blade** (also called the **lamina**).

Although one might assume that foliage leaves are maximally adapted for photosynthesis, these other factors affect leaf structure and metabolism and must be considered as well.

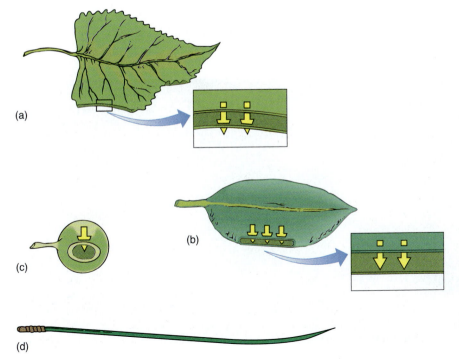

FIGURE 6.1
Several possible leaf shapes, all with equal volumes. (a) The common type with a thin, flat blade. All cells are exposed to light, there is a large surface area for absorption of CO_2, and none of the veins is long. (b) In a thick leaf, light cannot penetrate to the bottom of the leaf and those cells cannot photosynthesize. Relative to the volume, little surface area is available for CO_2 absorption, but this helps conserve water. (c) A spherical leaf has maximum internal self-shading, and only a small fraction of the cells receive enough light for photosynthesis. Surface area for gas exchange is minimal. (d) A cylindrical leaf could be thin with a large surface area and no self-shading, as in (a); however, to have the same volume as (a), it would be extremely long, and conduction through the veins might be difficult.

PLANTS & PEOPLE

LEAVES, FOOD, AND DEATH

Leaves impact our lives every day. Examples that come to mind readily are leafy foods such as artichokes (*Cynara scolymus*), cabbage (*Brassica oleracea* variety *capitata*), celery (petioles; *Apium graveolens*), lettuce (*Lactuca sativa*), onions (*Allium cepa*), and spinach (*Spinacia oleracea*). Also important are numerous herbs and spices: basil (*Ocimum basilicum*), bay leaves (*Laurus nobilis*), marjoram (*Origanum majorana*), oregano (*Origanum vulgare*), parsley (*Petroselinum crispum*), sage (*Salvia officinalis*), tarragon (*Artemisia dracunculus*), thyme (*Thymus vulgaris*), and the flavorings mint (several species of *Mentha*), spearmint (*Mentha spicata*), and peppermint (*Mentha piperita*). The flavors and pungency of these are due to chemicals located within the leaves themselves or in trichomes in the leaf epidermis. Most of these chemicals probably serve as antiherbivore defensive compounds, causing animals other than humans to avoid the plants. Many classes of antiherbivore chemicals have evolved, ranging from only mildly effective ones, such as these flavors, to others that are much more powerful, such as alkaloids, many of which are toxic in small amounts and kill quickly. The alkaloids in poison hemlock and death camas are particularly effective.

An alkaloid of the leaves of one plant in particular is of interest to us—it is quite lethal but acts only slowly: **nicotine** in the leaves of tobacco (*Nicotiana tabacum*). If tobacco leaves are eaten, they cause vomiting, diarrhea, and even death due to respiratory failure. But most tobacco leaves are smoked, of course. Americans smoked an all-time high of 4345 cigarettes per capita in 1963. Since then, consumption has declined for white males, but recently it has been increasing for other groups. Tobacco leaves contain between 0.6% and 9.0% nicotine, and an ordinary filter cigarette has 20 to 30 mg of the alkaloid, about 10% of which is absorbed by the lungs. Nicotine dissolves readily into the mucous membranes and passes quickly into the blood stream. Because it is transferred across the placenta, women who smoke during pregnancy may give birth to babies addicted to nicotine. Blood-borne nicotine also affects the heart, causing coronary problems: People who smoke a pack or more a day are over three times more likely than nonsmokers to die of heart disease. Once nicotine is taken into the cells of the mouth and lungs of a smoker, it can cause cancer—of the lungs especially, but also of the throat, larynx, and mouth. If detected early enough, nicotine-induced lung cancer can be combated with surgery and chemotherapy, but once the cancer has spread to the lymph system, the prognosis is not good. Lung cancer causes more than 100,000 deaths per year in the United States

Most leaves have a **petiole** (stalk) that holds the blade out into the light (Fig. 6.2), which prevents self-shading of leaf blades by those above them; self-shading would defeat the basic usefulness of the leaf.

The presence of a petiole has other consequences: Long, thin, flexible petioles allow the blade to flutter in wind, cooling the leaf and bringing fresh air to the leaf surface. If the leaf and air are still, carbon dioxide is absorbed from the vicinity of the leaf faster than

FIGURE 6.2

The petioles of each leaf of this *Astragalus* are long enough that the leaves do not shade each other and all are fully exposed to light. The leaves are palmately compound.

TABLE 6.1 *Movement of Carbon Dioxide*

Average velocity of a molecule at 30°C:	381 m/s
Mean free path (distance a molecule moves before striking another molecule) at 1 atm pressure:	39 nm
Number of collisions of one molecule at 1 atm pressure:	9.4 billion collisions per second

As any molecule diffuses, it collides with other molecules and changes its direction of motion. Although air is almost entirely empty space, a molecule of carbon dioxide undergoes 9,400,000,000 collisions per second and moves only 0.039 μm each time.

diffusion can bring more to it, depleting the carbon dioxide and decreasing photosynthesis (Table 6.1). Leaf flutter also makes it difficult for insects to land on a leaf and knocks off some that have already alighted.

If leaves are small or very long and narrow, self-shading is not a problem, and there may be no petiole; the leaf is then **sessile** instead of **petiolate.** For example, *Aeonium* grows in arid, sunny regions, and its fleshy leaves are packed close together (Fig. 6.3). Because the sunlight is so intense, even with self-shading the leaves receive enough light for adequate photosynthesis. The close packing of these leaves helps trap water molecules and prevent their escape from the plant.

In many monocots such as grasses, irises, lilies, and yuccas, foliage leaves have a shape quite different from those of dicots. They tend to be very long and tapering, and self-shading occurs only at the base; most of the blade is well-exposed to light (Figs. 6.4 and 6.13). They also typically lack a petiole; instead, the base of the leaf wraps around the stem to form a **leaf sheath.** The blade can flex and flutter even without a petiole. Not all monocots have linear, grasslike leaves. Palms, aroids (such as *Monstera* and *Philodendron*), and bird-of-paradise plants have what appear to be petioles and laminas, but the leaves of these species are believed to have evolved from grasslike leaves, and the "petiole" is actually a modified portion of the lamina (Fig. 6.5).

A leaf blade may be either simple or compound. A **simple leaf** has a blade of just one part (Fig. 6.6), whereas a **compound leaf** has a blade that is divided into several individual

Leaves

FIGURE 6.3
Aeonium tabuliforme and its relatives grow in regions of intense sunlight; self-shading is not a problem, but water conservation is. Close packing minimizes water loss from stomata.

Leaf blades

Stem

Petiole

(a)   (b)   Sheathing leaf bases   (c)   Sheathing leaf base

FIGURE 6.4

(a) Long, narrow leaves are common in many monocots of intensely sunny areas. Shading is no problem, although the leaf bases probably do not photosynthesize at maximum efficiency. (b) These are the leaves of St. Augustine grass; the blades project horizontally and are attached to the stem by sheathing leaf bases. (c) In cattails, the sheathing leaf base can be half a meter long and provide firm but flexible attachment to the stem.

parts (Fig. 6.7). Think of how this might be affected by wind: If the blade is either large or flimsy, it twists and flexes and may tear. Tearing can be prevented by making the leaf small or tough or "pretorn" (compound). A compound leaf has many small blades, each attached by a **petiolule** to an extension of the petiole, the **rachis.** The combination of a petiolule and a small blade is a **leaflet.** Leaves may be palmately compound, with all leaflets attached at the same point, or pinnately compound, with leaflets attached individually along the rachis (Fig. 6.7). Even a large compound leaf has only small leaflets, all of which can flex individually without tearing.

Compound leaves have other advantages. When a mild breeze blows over two leaves of equal size, one simple and the other compound, it tends to flow smoothly over the simple

FIGURE 6.5

Maranta, a common houseplant, is a monocot that has broad leaves.

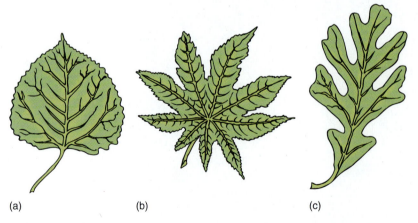

(a)   (b)   (c)

FIGURE 6.6

Simple leaves have numerous shapes, some so deeply lobed as to be almost compound. Poplar (a) and oak (c) have one dominant midrib from which secondary vascular bundles extend. In the castor bean (b), the petiolar bundles immediately diverge into several main veins.

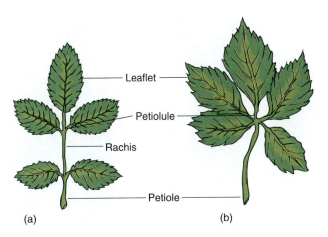

FIGURE 6.7
Leaflets of compound leaves have the same structure and metabolism as simple leaves. In pinnately compound leaves (rose, a) the petiolules of the leaflets are attached to the rachis, but in palmately compound leaves (Virginia creeper, b) they attach to the end of the petiole. In either type, any leaflet can be damaged or can abscise without affecting the remainder of the leaf.

leaf but turbulently over the more complex surface of the compound leaf. Turbulence brings in carbon dioxide and removes excess heat. Also, an insect can crawl or a fungus can spread across the entire blade of a simple leaf rather easily, but the edges and petiolule of a leaflet act as barriers that either prevent or at least slow movement to the rest of the leaf. In some plants, if pathogens severely damage a leaf, it abscises, carrying the pathogens with it and helping to keep pests away from healthy leaves.

Is it reasonable to conclude that plants with compound leaves are better adapted than plants with simple ones? Why are there any simple leaves at all? Most very large leaves are compound, apparently the only feasible architecture. The large simple leaves of banana are formed with a large, single lamina, but they have numerous special lines of weakness and are quickly torn to "compound" condition by wind. The largest simple leaves, those of the Victoria water lily, float on water and thus do not need large amounts of sclerenchyma for support, and adhesion to the water's surface prevents wind damage. Among medium and small leaves, simple ones are common, perhaps because a greater percentage of a simple leaf is composed of photosynthetic cells, whereas a compound leaf has a great deal of nonphotosynthetic rachis, petiolules, and edges.

Although most pinnately compound leaves are easily recognizable as leaves, some can be mistaken for a stem with simple leaves. Close examination reveals that leaflets never bear buds in the axils of their petiolules, and the tip of the rachis never has a terminal bud. Leaflets are always arranged in two rows, never in a spiral, whorled, or decussate phyllotaxy.

Some of the tremendous range of shapes of leaves and leaflets is shown in Figure 6.8. There is speculation, but no total agreement, on how the different leaf shapes may be adaptive; many biologists believe that all function so well that none is clearly superior or strongly disadvantageous selectively. However, a large number of species have several types

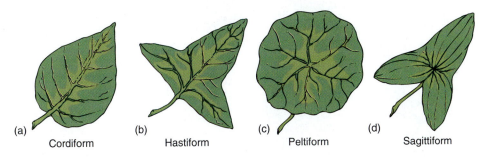

FIGURE 6.8
Four common leaf shapes; numerous other types also exist.

(a)

FIGURE 6.9
(a) Beans have two types of leaves: the very first leaf formed by the seedling is simple *(right)*, but all later leaves are compound *(left)*.
(b) This *Azara lanceolata* shoot has two types of leaves: Some leaves are large, whereas those closest to the axillary buds are small.

(b)

In species such as ivy or citrus, the transition from juvenile foliage to adult leaves does not take place until the plant is old enough to flower, perhaps more than 10 years old.

of leaves (Fig. 6.9a). In the simplest cases, the first few leaves of a seedling are distinctly different from all leaves produced later; if they were merely smaller, it could be argued that the seedling does not have enough stored energy resources to construct the large leaves that a mature plant can afford. But very often the juvenile leaves differ from the adult leaves not only in size but also shape, texture, and even simple versus compound structure. It seems reasonable to hypothesize that juvenile leaves of seedlings are adaptive in the microhabitat close to the soil surface where a seedling is located, whereas adult leaves are adapted to the more aerial microhabitat inhabited by an older, larger plant.

Other species produce two types of leaves simultaneously (Fig. 6.9b), often one type on long shoots and the other on short spur shoots. In cacti, the spines are short-shoot

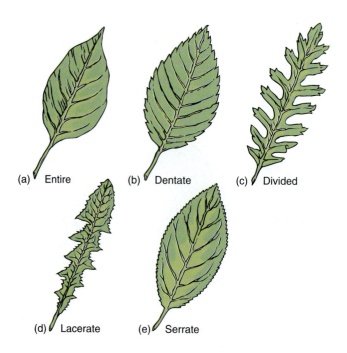

(a) Entire   (b) Dentate   (c) Divided

(d) Lacerate   (e) Serrate

FIGURE 6.10
Several common types of leaf margins; in nature, hundreds of types exist, many being intermediate between two or among several other types.

leaves, whereas the long-shoot leaves are green, fleshy, and in most species almost microscopic (see Fig. 5.2). In a small percentage of species, leaf shape and size are irregularly variable even on one stem, being influenced by environmental conditions that occur at the time of leaf initiation and expansion; amounts of sunlight and minerals are perhaps the most important factors.

Despite the numerous possibilities for variation within a species, leaf shape is a valuable tool for plant identification: The shape of the leaf, including its base and apex, are important, as is the nature of the margin. The margin may be entire (smooth), toothed, lobed, or otherwise modified (Fig. 6.10).

Within a leaf are **veins** or bundles of vascular tissue; in a dicot they occur in a netted pattern called **reticulate venation** (Figs. 6.11 and 6.12). In monocots with long, strap-shaped leaves, the larger veins run side by side with few obvious interconnections: This is **parallel venation** (Figs. 6.13 and 6.14). Veins distribute water from the stem into the leaf and simultaneously collect sugars produced by photosynthesis and carry them to the stem for use or storage elsewhere.

At the leaf base, usually in the petiole, is an **abscission zone** oriented perpendicular to the petiole; its cells are involved in cutting off the leaf when its useful life is over (Fig. 6.15).

FIGURE 6.11

A leaf of *Amaranthus retroflexus* treated to remove all tissues except vascular bundles. The extensive reticulate venation is obvious in this dicot leaf; no part of the leaf is without veins. (*Courtesy of D. Fisher, University of Hawaii*)

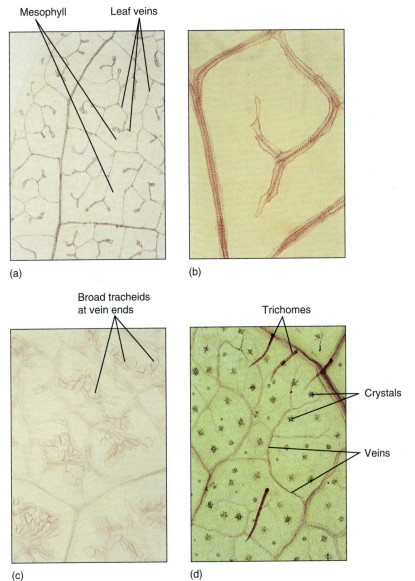

(a)   (b)

(c)   (d)

FIGURE 6.12

These leaves were treated (cleared) to make their cells transparent; then the xylem was stained to make the leaf veins visible. (a) In passionflower (*Passiflora*) leaves, the veins are narrow and highly branched. At this magnification ($\times 40$), the height of the photograph represents 1 mm, so the veins occur close together. (b) Higher magnification, of privet leaf; at any point in the minor veins, only two or three tracheary elements are present ($\times 150$). (c) These minor veins of crown of thorns (*Euphorbia millii*, a species that grows in deserts) are broad and have many wide tracheids at the ends ($\times 50$). (d) This leaf of *Acalypha* has numerous crystals within the leaf and trichomes on the leaf surface ($\times 80$).

FIGURE 6.13
Cattails (*Typha latifolia*) have long, straplike leaves common in monocots. Vascular bundles run parallel to each other from leaf base to tip.

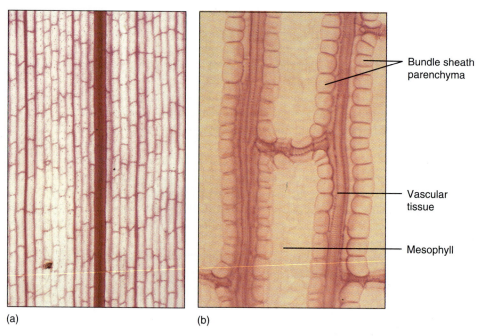

(a)   (b)

FIGURE 6.14
(a) Leaf clearing of corn (*Zea*), a monocot, with several sizes of longitudinal, parallel veins. If a longitudinal vein is broken by insect damage or any other problem, conduction can detour around the site by means of the fine transverse veins (×50). (b) Higher magnification showing the parenchyma cells that surround the leaf bundles in corn (×250).

In autumn, as deciduous leaves begin to die, abscission zone cells release enzymes that weaken their walls. As the leaf twists in the wind, cell walls break and the leaf falls off. Adjacent undamaged cells swell and become corky, forming protective scar tissue, the **leaf scar,** across the wound (see Fig. 5.14). Without an abscission zone, dead leaves might tear off irregularly, leaving an open wound vulnerable to pathogens.

FIGURE 6.15
(a) Longitudinal section through a leaf axil, showing the abscission zone in the petiole and the axillary bud just above (×15).
(b) Magnification of the abscission zone while the leaf is still healthy. The xylem and phloem are not interrupted until the leaf actually falls away (×80).

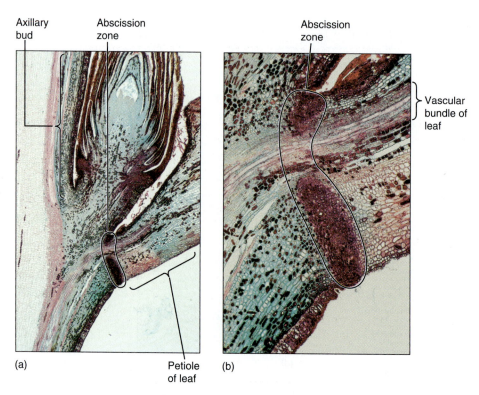

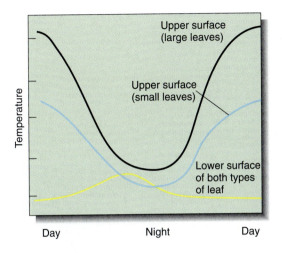

FIGURE 6.16
At night, both the upper and lower surfaces of a leaf are close to air temperature; some loss of water causes evaporative cooling. After the sun rises, the upper surface rapidly becomes warmer than the surrounding air; in large simple leaves this is much more striking than in small compound leaves.

INTERNAL STRUCTURE OF FOLIAGE LEAVES

EPIDERMIS

Flat, thin foliage leaves (optimal for light interception) have a large surface area through which water can be lost. Water loss through the epidermis is called **transpiration,** and it is a serious problem if the soil is so dry that roots cannot replace the water lost from the leaves. The epidermis must be reasonably waterproof but simultaneously translucent, and it must allow entry of carbon dioxide. The structure of the leaf and stem epidermis is basically quite similar, consisting of a large percentage of flat, tabular (shaped like paving stones), ordinary epidermal cells; guard cells and trichomes (either glandular or nonglandular) may be abundant. However, the dorsiventral nature of the leaf causes the upper and lower epidermis to exist in significantly different microclimates (Fig. 6.16). On a clear, sunny day, a leaf is usually warmer than the surrounding air, so it heats the air and convection currents rise from it. If stomata in the upper epidermis are open and losing water, the water molecules are swept away by this convection. However, air tends to be trapped on the underside of a leaf, so water loss from the stomata there is not so great. The molecules are trapped in quiet air and may diffuse back into the stomata. In most leaves, the number of stomata per square centimeter is much greater in the lower than in the upper epidermis; in many species, no stomata at all are found in the upper epidermis (Table 6.2).

Such a unilateral distribution of stomata has other beneficial consequences. Air-borne spores of fungi and bacteria are continually landing on leaves. Rye leaves, for example,

The dorsal surface is the underside, and the large veins protrude like backbones; synonym: abaxial side. The ventral surface is the upper side; synonym: adaxial.

		Stomata Per Cm^2	
Species	**Common Name**	**Upper**	**Lower**
Tilia europea	linden	—	37,000
Allium cepa	onion	17,500	17,500
Helianthus annuus	sunflower	12,000	17,500
Pinus sylvestris	pine	12,000	12,000
Zea mays	corn	9800	10,800
Vicia faba	bean	6500	7500
Sedum spectabilis	stonecrop	2800	3500
Larix decidua	larch	1400	1600

TABLE 6.2 *Frequency of Stomata in Leaf Upper and Lower Epidermis*

BOX 6.1   *Leaf Structure, Layer by Layer*

The internal organization of ordinary leaves is not very complicated but it can be difficult to visualize from looking only at transverse sections. A series of sections through a leaf of privet (*Ligustrum*) is presented here. These are paradermal sections; that is, they are parallel to the epidermis. The first is at the level of the palisade parenchyma, and each successive section was taken from deeper in the leaf. Try to imagine the appearance of the leaf from the perspective of a carbon dioxide molecule.

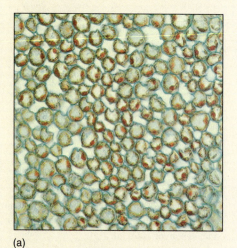

(a)

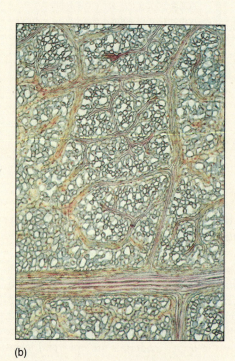

(b)

Perforations

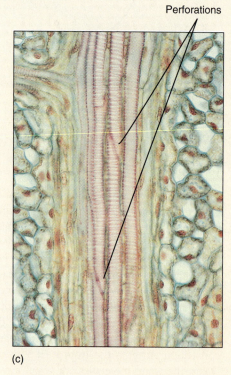

(c)

(a) Palisade parenchyma (×250). The cells are cylindrical and are not closely packed; carbon dioxide can diffuse easily throughout this layer and be absorbed into the cells through their large surface area.

(b) Leaf veins (×50). Just below the palisade parenchyma lies the network of leaf veins. Small lateral veins branch off of larger veins and in turn give off smaller veins.

(c) High magnification of a large lateral vein (×250). This section passes through the xylem of the vein, and perforations are visible, so these are vessel elements, not tracheids. Notice that these have spiral thickenings: They could differentiate while the leaf was still small; then as the leaf expanded, they could be stretched and still conduct water.

typically have 10,000 fungal spores per square centimeter, but most are on the exposed upper surface, where stomata are few, so few fungi are able to penetrate the leaf.

Leaf stomata are frequently sunken into epidermal cavities that create a small region of nonmoving air. In oleander, the lower leaf epidermis contains numerous crypts (areas where the epidermis is depressed into the leaf), and stomata and trichomes are abundant in the epidermis that lines the crypts (Fig. 6.17). Both the recessed nature of the crypt and the presence of trichomes keep air in the crypt extremely quiet; many water molecules that diffuse out of open stomatal pores bounce around the crypt and re-enter the stomata rather than being blown away.

Leaf epidermises are often remarkably hairy, and the trichomes affect the leaf biology in numerous ways. They provide some shade on the upper surface of the leaf, deflecting excessive sunlight, an adaptation common in desert plants (Fig. 6.18). On the lower surface, they prevent rapid air movement and slow water loss from the stomata. In any

Bundle sheath
parenchyma

Intercellular
spaces

Stomata

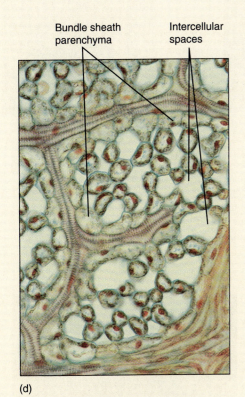

(d)

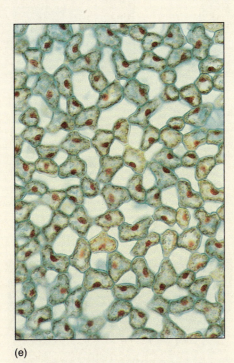

(e)

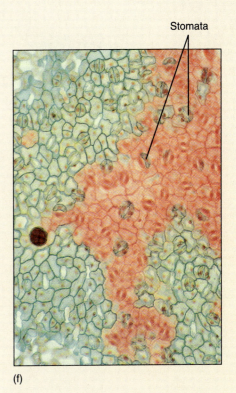

(f)

(d) High magnification of minor veins
(× 250). Notice that all mesophyll cells are
close to one vein or another. Once sugar has
been produced by photosynthesis, it does not
diffuse very far before it is picked up by a
sieve tube member and transported to the
stem.

(e) Spongy mesophyll (× 250). In many
leaves, more than half the spongy mesophyll
consists of intercellular air space. This
allows the carbon dioxide molecules to
diffuse rapidly into the leaf, away from the
stomata. If the spongy mesophyll were more
dense, carbon dioxide molecules might be
trapped near the stomata, then diffuse back
out of the leaf before they could be
absorbed by a chlorenchyma cell.

(f) Lower epidermis (× 150). The epidermis is
slightly undulate, so as it was cut to make this
slide, the knife passed through the center of the
cells in some areas (green) and through the
outer wall and cuticle in other areas (red). The
stomata are abundant, but notice that the actual
stomatal pore is short and narrow; each is only
a small opening through which carbon dioxide
can enter.

position, trichomes make walking or chewing difficult for insects, and many glandular
trichomes secrete powerful stinging compounds that prevent even large animals from
eating the leaf. Insects either do not bother with the leaf or must walk so slowly on it that
they become more vulnerable to their own predators. Conversely, hairs provide excellent
footholds against leaf flutter for insects of the appropriate size. Poisonous glandular trichomes in stinging nettle and other species prevent mammals from eating leaf tissue but
also protect leaf-borne insects small enough to walk, feed, and reproduce safely between
the stinging trichomes.

Like stem epidermis, leaf epidermal cells contain a coating of cutin and usually also
wax on their outer walls. These retain water and make digestion by fungi difficult. Furthermore, their smooth, slippery surface prevents spores from sticking or allows them to be
washed off by rain.

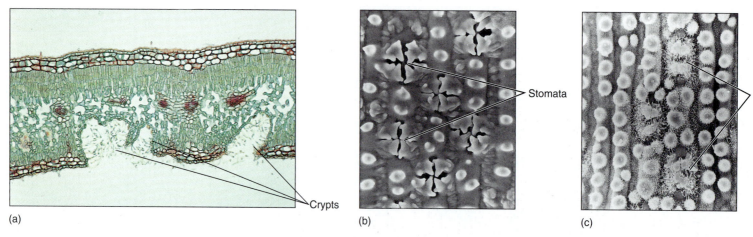

(a)   (b)   (c)

FIGURE 6.17
(a) Oleander (*Nerium oleander*) leaves have stomatal crypts filled with trichomes and stomata. This organization minimizes air movement near the stomata, and water molecules that diffuse out of the stoma may re-enter the stoma by random motion. Crypts occur only in the lower surface (× 80). (b) Each of the stomata of this grass (*Elytrostrachys*) is overarched by four protrusions from surrounding cells. This too reduces air movement near the stoma (× 940). (c) In this bamboo (*Chusqua*), stomata are protected by protrusions and wax rods (× 940). (*b and c, Courtesy of L. G. Clark and X. Londoño, Iowa State University*)

FIGURE 6.18
(a) The leaves of lamb's ears (*Stachys*) are so densely covered with hairs that they often appear white. The hairs greatly reduce the amount of heat that these leaves absorb from sunlight.
(b) These plants of *Sempervivum* are native to high alpine regions where ultraviolet radiation is intense. The hairs of the leaf epidermis reflect part of the ultraviolet radiation away from the plant.

(a)   (b)

MESOPHYLL

The ground tissues interior to the leaf epidermis are collectively called the **mesophyll** (Figs. 6.19 and 6.21). Along the upper surface of most leaves is a layer of cells, the **palisade parenchyma,** which is the main photosynthetic tissue of most plants. Palisade cells are separated slightly so that each cell has most of its surface exposed to the carbon dioxide of the intercellular space. Because carbon dioxide dissolves into cytoplasm slowly, the large surface gives maximum area for dissolution; tightly packed cells could not absorb enough carbon dioxide for efficient photosynthesis (Fig. 6.20). The palisade parenchyma is often only one layer thick, but in regions with intense, penetrating sunlight, it may be three or four layers thick (Fig. 6.21).

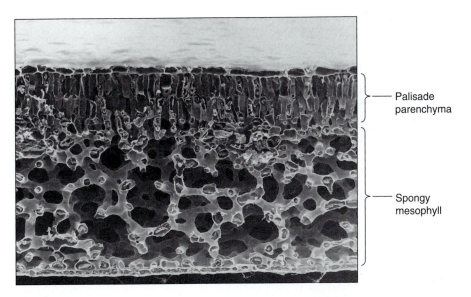

FIGURE 6.19
This leaf of *Laurelia* has a layer of palisade parenchyma that is two cells deep, and the spongy mesophyll is an extensive aerenchyma (× 180). *(Courtesy of G. Montenegro, Universidad Católica, Santiago, Chile)*

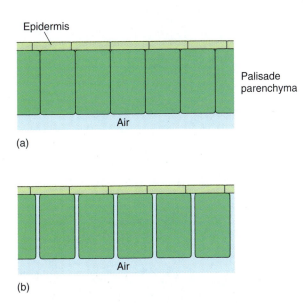

FIGURE 6.20
(a) If palisade cells are closely packed with no space between them, only the bottoms of the cells are exposed to the carbon dioxide in the spongy mesophyll. (b) With only slight separation of the cells, the volume of the palisade is almost unchanged, but the surface area available is increased enormously; carbon dioxide can be absorbed much more rapidly. Assume that the palisade parenchyma cells are rectangular columns 20 μm × 20 μm × 100 μm long; compute the surface area exposed to CO_2 by 100 cells arranged as in (a) versus (b).

In the lower portion of the leaf is the **spongy mesophyll**—open, loose aerenchyma tissue that permits carbon dioxide to diffuse rapidly away from the stomata into all parts of the leaf's interior (see Figs. 6.19 and 6.21). If stomata were surrounded by closely packed cells, a molecule of carbon dioxide might simply bounce off a cell and escape back out of the stomatal pore.

Although this arrangement is the most common, some plants have a layer of palisade parenchyma along both surfaces of the leaf; the spongy mesophyll either occurs in the center or is lacking (Fig. 6.22). The relationship between leaf position and the sun is important: For leaves that are held horizontally (most leaves), the sun is usually overhead, so having palisade parenchyma on the upper surface permits maximum absorption of light and photosynthesis. For plants that hold their leaves vertically (*Iris, Gladiolus, Eucalyptus*), both sides are equally illuminated and palisade parenchyma is equally functional on either side.

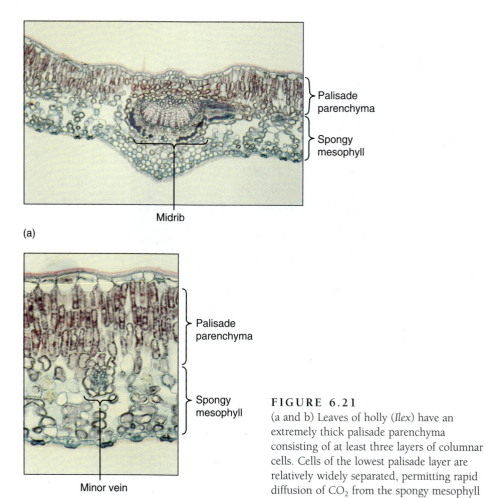

(a)

(b)

FIGURE 6.21
(a and b) Leaves of holly (*Ilex*) have an extremely thick palisade parenchyma consisting of at least three layers of columnar cells. Cells of the lowest palisade layer are relatively widely separated, permitting rapid diffusion of CO_2 from the spongy mesophyll to the upper palisade (a, $\times 50$; b, $\times 150$).

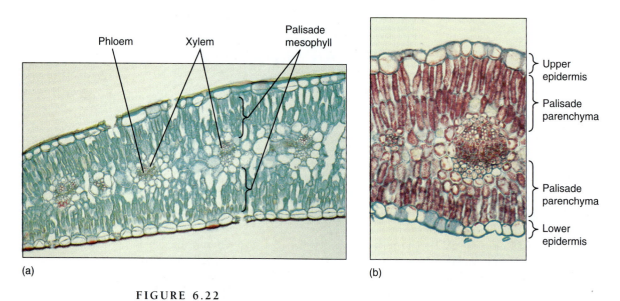

(a)

(b)

FIGURE 6.22
Leaves of both carnations (a) and creosote bush (b) have palisade parenchyma along both surfaces. The little spongy mesophyll present is located in the middle of the leaf (a, $\times 75$; b, $\times 150$).

VASCULAR TISSUES

Between the palisade parenchyma and the spongy mesophyll are the vascular tissues. A dicot leaf usually has a large **midrib,** also called a midvein, from which **lateral veins** emerge that branch into narrow **minor veins** (see Figs. 6.11 and 6.12). Minor veins are most important for releasing water from xylem and loading sugar into phloem, whereas the midrib and lateral veins are involved mostly in conduction. Vein structure changes with size: The midrib and lateral veins always contain both primary xylem on the upper side and primary phloem on the lower side (Fig. 6.23). Because they both conduct and support the leaf blade, they may have many fibers arranged as a sheath, called a **bundle sheath,** around

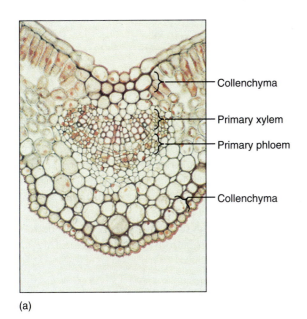

— Collenchyma

— Primary xylem

— Primary phloem

— Collenchyma

(a)

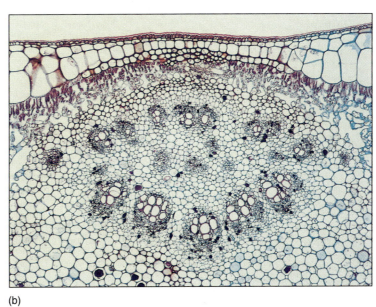

(b)

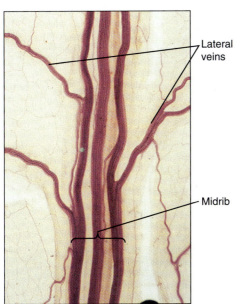

Lateral veins

Midrib

(c)

FIGURE 6.23

Transverse sections through leaf midribs. (a) This midrib of a *Ligustrum* leaf is rather simple, containing a single vascular bundle. Xylem is present along the top, phloem along the bottom. A small amount of collenchyma occurs on the top and bottom of the midrib ($\times$ 150). (b) This midrib of rubber tree has numerous separate bundles embedded in a large mass of mesophyll ($\times$ 50). (c) This is a midrib prepared as a leaf clearing (similar to Figure 6.12); the three distinct bundles have no interconnections in this region. Each transports to and from a particular portion of the blade ($\times$ 15).

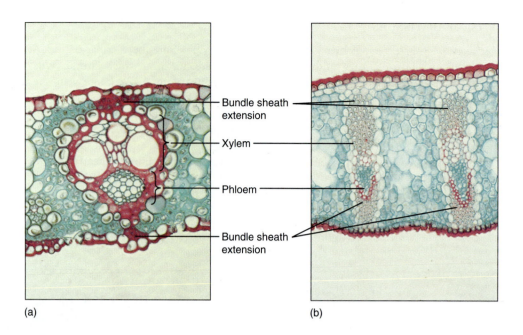

(a)   (b)

FIGURE 6.24
(a) This leaf bundle of sugar cane has bundle sheath extensions contacting both the upper and lower epidermis. Water diffuses from the xylem to the epidermis by capillary action, moving around the surfaces of the fibers (× 100). (b) The leaves of flax have large bundle sheath extensions consisting of many layers of fibers (× 45).

the vascular tissues (Fig. 6.24). The sheath also makes it difficult for insects to chew into the vascular tissues. Many other types of nonconducting cells such as mucilage, tannin, or starch storage cells may be present in the larger veins. Minor veins are the sites of material exchange with the rest of the mesophyll and must have a large surface area in contact with the palisade and spongy mesophyll; they do not contain fibers or other nonconducting cells whose presence would interrupt this contact (Fig. 6.25). The endings of the minor veins in some species consist of only xylem and in others only phloem.

Veins, especially the larger ones, often have a mass of fibers above, below, or both—the **bundle sheath extension** (see Fig. 6.24). The fibers help give rigidity to the leaf and are

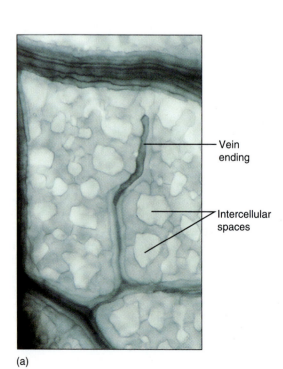

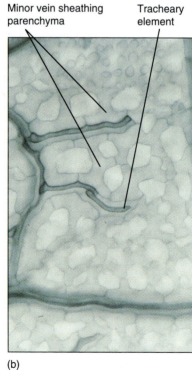

FIGURE 6.25
(a and b) These leaf veins are minor veins, treated to reveal individual cells. The long dark cells are tracheary elements, and the phloem is out of focus, in line with the xylem. Notice how narrow these veins are: The xylem is only one element wide, so there is maximum surface area for releasing water to the leaf mesophyll. In this species, each minor vein is lined with parenchyma cells that extend outward into the spongy mesophyll (× 250). (*Courtesy of N. R. Lersten and J. D. Curtis, Iowa State University*)

(a)   (b)

believed to provide an additional means by which water moves from the bundle out to the mesophyll. Apparently water moves by capillary action around the fibers rather than through them.

THE PETIOLE

Although the petiole is technically part of the leaf, it serves as the transition between the stem and the blade. Consequently, the arrangement of tissues differs at the two ends. The epidermis may be similar to that on the lamina but often contains fewer stomata and trichomes. The mesophyll is rather like cortex—somewhat compact and not especially aerenchymatous—and considerable collenchyma may be present, providing support for the lamina. Vascular tissues are the most variable; one, three, five, or more vascular bundles, called **leaf traces,** exit the stem vascular cylinder and diverge toward the petiole (Fig. 6.26a). They may remain distinct or fuse into a single trace at, near, or in the petiole. In some species, they divide into numerous bundles and 10 or 20 (in large palm leaves, several hundred) may be found in the petiole (Fig. 6.26b). Vascular bundles may either fuse with each other within the petiole or branch further; they can be arranged in a ring, a plate, or a number of other patterns. If the lamina has a strong midrib, most of the petiole bundles fuse together and form the midvein, but other bundles may enter the lamina as small lateral veins.

We do not understand the significance of all the bundle patterns in petioles. Certain patterns may ensure the proper distribution of sugars out of different parts of the lamina, especially of large leaves, into the various bundles of the stem. However, it may be that almost any pattern functions well, so mutations that cause new patterns are not selectively disadvantageous. On the other hand, future studies may show correlations between lamina, petiole, and stem.

In many species, the petiole bears two small flaps of tissue at its base called **stipules,** which serve various functions. They may protect the shoot apical meristem while the leaf is young and small. In other plants, they are large enough to contribute a significant amount of photosynthesis, but usually when the leaf is mature the stipules are still small and they die early.

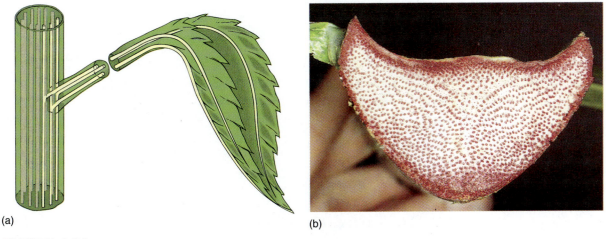

(a)

(b)

FIGURE 6.26

(a) This section of stem has 11 bundles, 3 of which produce leaf traces that enter the petiole. Later, the leaf traces diverge and spread throughout the blade. These traces are drawn without any branching or merging, although both are possible. (b) Cross-section of a petiole of the palm *Attalea* stained to show its numerous vascular bundles. Notice how numerous the bundles are compared with the leaves of Figure 6.23. This leaf, like those of most palms, is very large and must have a massive conduction system.

The entire pattern is completed while the leaf is still very small, often less than one tenth its mature size.

INITIATION AND DEVELOPMENT OF LEAVES

DICOTS

Leaves are produced only through the activity of a shoot apical meristem. At the base of the meristem, cells just interior to the protoderm grow outward, forming a protrusion known as a **leaf primordium** (see Figs. 4.3, 5.38, and 6.27) that extends upward as a narrow cone, growing so rapidly that it becomes taller than the shoot apical meristem. During this stage the primordium consists of leaf protoderm and leaf ground meristem, and all cells are meristematic, with dense cytoplasm and small vacuoles. A strand of cells in the center differentiates into provascular tissue and then into primary xylem and phloem, forming a connection with the young bundles in the stem.

As the leaf primordium grows upward, it increases in thickness, establishing the bulk of the midrib. A row of cells on either edge of the primordium grows outward, initiating the formation of the lamina (Fig. 6.28). As a result of their activity, the young leaf consists of a midrib and two small, thin wings. All cells in the wings are meristematic, and their division and expansion enlarge the lamina rapidly (Fig. 6.29). In a compound leaf, two rows of loci initiate leaflets, which then grow like simple leaves. During lamina expansion, stomata, trichomes, and vascular bundles differentiate, and the petiole becomes distinct from the midrib.

In many perennial plants, leaves are initiated in the summer or autumn before they mature. Once they reach the developmental state just described, they become dormant, part of a resting terminal or axillary bud. During the next growth period, usually the

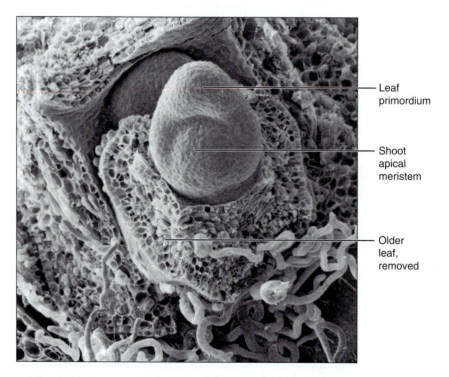

Leaf primordium

Shoot apical meristem

Older leaf, removed

FIGURE 6.27

The shoot apical meristem of grape (*Vitis riparia*) can be seen in three dimensions in this scanning electron micrograph. Notice the large leaf primordium above it; another leaf primordium below has been removed; the phyllotaxy is established at the very apex (× 200). *(Courtesy of Christian R. Lacroix and Usher Posluszny, University of Guelph, Canada)*

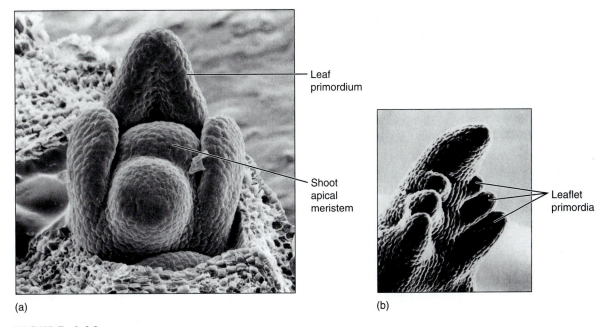

(a)

(b)

FIGURE 6.28

(a) The uppermost leaf primordium is the oldest, and its sides have begun growing out as the lamina of a simple leaf of *Vitis* (× 270). *(Courtesy of C. R. Lacroix and U. Posluszny)* (b) In this compound leaf, rather than forming a lamina, six leaflet primordia have been established (× 150). *(Courtesy of B. Sugiyama and N. Hara, University of Tokyo)*

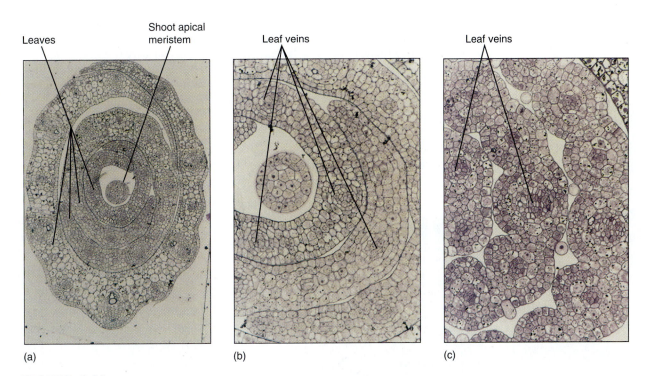

(a)

(b)

(c)

FIGURE 6.29

(a) A low-magnification transverse section through young, developing leaves of the grass *Panicum effusum*. The central round structure is a section of the shoot apical meristem, and the leaves encircle it (grasses have sheathing leaf bases) (× 200). (b) At higher magnification, you can see that all cells in the leaves are still small and cytoplasmic, and the central vacuoles are small. At this stage, all leaf cells are still undergoing mitosis and cytokinesis. The leaf veins are starting to differentiate (× 500). (c) At an older stage, the leaf veins are more distinct. The first protoxylem tracheary elements are mature in some veins, but most cells that will differentiate into tracheary elements are still enlarging (× 500). *(Courtesy of Nancy Dengler, University of Toronto)*

Foliage leaves   Bud scales

FIGURE 6.30
Terminal bud of beech (*Fagus grandifolia*) beginning to expand in the spring. The brownish bud scales are being pushed back and will soon abscise as the new foliage leaves swell. The bud's apical meristem may be initiating new leaf primordia that will expand a little later in the spring, but in some species it may be producing flowers at this time instead.

following spring, the bud opens and the primordial leaves expand rapidly as each cell absorbs water into its vacuole and swells (Fig. 6.30). Little or no mitosis or cell division may occur—only maturation, especially the synthesis of chlorophyll, cutin, and wax. As the leaves expand, the immature, exposed epidermis is very vulnerable, especially to insects (Fig. 6.31). In many species of annual plants, the process is similar, with the embryo in the seed acting like the buds of perennial plants. They initiate leaves before the seed becomes dormant and dry, and those leaves absorb water and expand rapidly during germination.

MONOCOTS

Monocot leaves, like those of dicots, are initiated by the expansion of some shoot apical meristem cells to form a leaf primordium (Fig. 6.32). Apical meristem cells adjacent to the primordium grow upward along with it, becoming part of the primordium and giving it a hoodlike shape. More apical meristem cells become involved until the primordium is a cylinder that completely or almost completely encircles the shoot apical meristem. As the shoot apex elongates and forms new stem and leaf tissue, the tubular portion of the leaf primordium continues to surround it as the sheathing leaf base. The original conical leaf primordium, now located on one side of the top of the tube, gives rise to the lamina.

In some monocots, the lamina becomes broad and expanded like a dicot lamina. But grasses, lilies, and many others have linear, strap-shaped leaves that grow continuously, having no predetermined size. Their lamina grows by a meristem located at its base where it attaches to the sheathing leaf base. The meristematic cells remain active mitotically, producing new cells that extend the leaf. Even if a grazing animal, a range fire, or a lawnmower destroys much of the leaf, the meristem can form a new lamina. This type of regeneration is not possible in dicot leaves.

The constant basal expansion means that the protoxylem and protophloem are constantly being stretched and disrupted in the basal meristem, but new vessel elements and sieve tube members differentiate rapidly enough that conduction is never interrupted. Just

FIGURE 6.31
Aphids and other sucking insects typically attack young leaves and stems that have not yet developed any protective sclerenchyma around the vascular bundles.

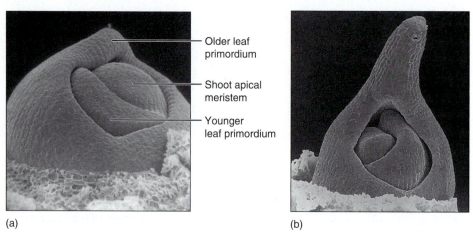

Older leaf primordium

Shoot apical meristem

Younger leaf primordium

(a)   (b)

FIGURE 6.32
In a monocot, after the center of the leaf primordium has been initiated (a), neighboring regions of the shoot apical meristem become active and contribute to the primordium. This results in a leaf base that encircles the stem, forming a sheath around it (b). (a, × 150; b, × 120) (*Courtesy of B. K. Kirchoff, University of North Carolina, and R. Rutishauser, Institute of Systematic Botany, Zurich*)

above the basal meristem is a region where tissues differentiate: More primary xylem and phloem are initiated, as are stomata and other features. Higher above the basal meristem, tissues are mature; differentiation is similar to that for stems but is oriented upside down in monocot leaves.

MORPHOLOGY AND ANATOMY OF OTHER LEAF TYPES

SUCCULENT LEAVES

Numerous adaptations permit plants to survive in desert habitats, one of the most common being the production of succulent leaves (Fig. 6.33). This is characteristic of species in the families Crassulaceae (contains *Kalanchoe* and *Sedum*), Portulacaceae (contains *Portulaca* and *Lewisia*), and Aizoaceae (ice plant), among others. The leaves are thick and fleshy, a shape that reduces the surface-to-volume ratio and favors water conservation. Some leaves are cylindrical or even spherical, the optimal surface-to-volume shape. The reduction in surface area, which is advantageous for water retention, has the automatic consequence of reducing the capacity for carbon dioxide uptake.

Inside the leaf, the mesophyll contains very few air spaces, reducing the internal evaporative surface area and, in turn, water loss through stomata. Lack of air spaces also makes the leaves more transparent, just as pure water is more transparent than soap bubbles or foam, allowing light to penetrate farther into the leaf. Photosynthesis can occur much more deeply than it would in the foliage leaves described earlier. In some members of the genera *Lithops* (stone plants) and *Frithia,* leaves are so translucent that they act as optical fibers; the leaves are located almost completely underground, where it is cool and relatively damp (Fig. 6.34). The exposed leaf tips allow sufficient light to enter and be conducted to the subterranean chlorenchyma. Although the plants live in a harsh desert in Madagascar, photosynthesis actually occurs in a rather mild microclimate.

SCLEROPHYLLOUS FOLIAGE LEAVES

Foliage leaves must produce more sugars by photosynthesis than are used in their own construction and metabolism, or the plant would lose energy every time it produced a leaf. This requirement limits the amount of sclerenchyma in foliage leaves, so most leaves tend to be soft, flexible, and edible.

FIGURE 6.34
Lithops, stone plant, is closely related to *Dinteranthus;* both are in the family Aizoaceae. *Lithops* also has only two leaves; but they are located almost entirely underground. The flat, translucent tips project above the soil and conduct light to the subterranean chlorenchyma.

(a)

(b)

FIGURE 6.33
(a) Leaves of *Senecio rotundifolia* are spherical, giving them an optimal surface-to-volume ratio for conserving water. (b) The leaves of the succulent *Dinteranthus* are hemispherical and attached to the opposite leaf, greatly reducing the exposed leaf surface. Interior tissues are mostly water-storage parenchyma. The low levels of chlorophyll result in low levels of photosynthesis, but lack of water is a more significant danger for these plants. Each plant consists basically of two leaves and a microscopic stem; the root system (not visible) is substantial. *Dinteranthus* and many related species have only two leaves at a time; as two new leaves expand, the existing two collapse and wither away.

Rachis   Leaflet

FIGURE 6.35
The leaves of barberry (*Berberis*) are tough and hard; it is difficult for insects to bite into them or lay eggs in them. However, the plant must invest considerable glucose to make the sclerenchyma cells, so the leaves must photosynthesize longer before they reach the break-even point.

In some species (barberry, holly, *Yucca*), leaves have evolved which are perennial, existing on a plant for two or more years (see Figs. 6.4 and 6.35). With this extended lifetime and increased productivity, sclerenchymatous leaves are feasible, and their hardness makes them more resistant to animals, fungi, freezing temperatures, and ultraviolet light. Such plants are said to be sclerophyllous, and the leaves are sclerophylls. The sclerenchyma is often present as a layer just below the epidermis and in the bundle sheaths, although the epidermis itself can be composed of thick-walled cells (Fig. 6.36). The cuticle is usually very thick, and waxes are abundant on leaves of many sclerophyllous species.

LEAVES OF CONIFERS

In almost all species of conifers, the leaves are sclerophylls; they have an extremely thick cuticle and the cells of their epidermis and hypodermis have thick walls. Most conifer leaves contain abundant chemicals that make them unpalatable. Conifer leaves are always simple and have only a few forms. Needles, either short or long, occur in all pines, firs, and spruces (Fig. 6.37). Needles of longleaf pine can be 40 cm long, although most other species of pine have needles about 10 cm long. Small, flat, scalelike leaves form a shieldlike covering on stems of junipers, cypresses (*Cupressus*), arborvitae (*Thuja*), and others. In *Agathis, Araucaria,* and *Podocarpus*, all genera of the Southern Hemisphere, leaves are rather large, broad scales held away from the stem. Lengths up to 12.5 cm and widths of 3.5 cm have been measured in leaves of *Podocarpus wallichianus*.

Conifer leaves are mostly perennial, remaining on the stem for many years; consequently, the plants are evergreens. Needles of bristlecone pine live for at least 5 years; their vascular bundles can produce new phloem each year, but no new xylem (Fig. 6.38). The small scale leaves of juniper also persist and are photosynthetically active for many years, and leaves of *Agathis* and *Araucaria* remain even on very old trunks. Some conifers have annual leaves that are shed each autumn; larch (*Larix*), bald cypress (*Taxodium*), and dawn redwood (*Metasequoia*) are all deciduous.

(Text continues on page 169.)

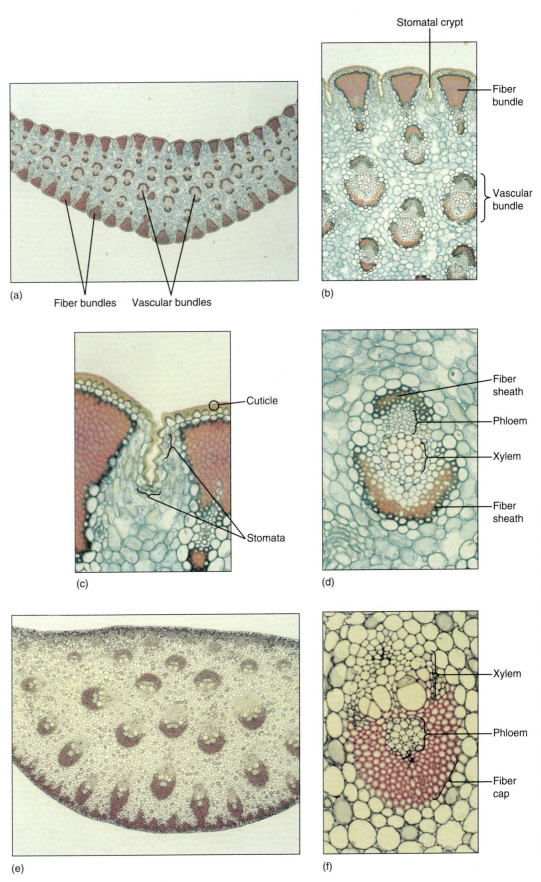

(a)

Fiber bundles   Vascular bundles

(b)

Stomatal crypt

Fiber bundle

Vascular bundle

(c)

Cuticle

Stomata

(d)

Fiber sheath

Phloem

Xylem

Fiber sheath

(e)

(f)

Xylem

Phloem

Fiber cap

FIGURE 6.36
(a) *Yucca* leaves are excellent examples of sclerophylls; they are tough and fiberous, and the cuticle is thick. These features reduce the amount of light reaching the chlorenchyma, but plant growth is probably limited by scarcity of water rather than reduced photosynthesis. These leaves may function for 5 to 10 years or more ($\times$ 15). (b) Bundles of fibers occur just interior to the epidermis, and the thick mesophyll surrounds many vascular bundles ($\times$ 50). (c) Stomata occur in long, shallow grooves that run between the fiber bundles ($\times$ 150). (d) Even the vascular bundles of these sclerophylls have fiber sheaths ($\times$ 150). (e) The sclerophylls of *Dracaena* are very thick and have leaf veins throughout the mesophyll, unlike the single row in ordinary leaves ($\times$ 15). (f) The leaf veins of *Dracaena* have extremely thick fiber sheaths ($\times$ 150).

(a)   (b)

FIGURE 6.37
(a) Many conifers, such as this pine, have needle-shaped leaves. (b) Incense cedar (*Libocedrus*) and many other conifers have scale-shaped leaves.

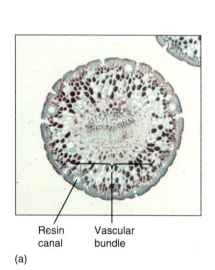

Resin   Vascular
canal   bundle

(a)

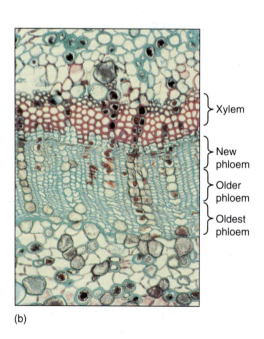

— Xylem

— New phloem

— Older phloem

— Oldest phloem

(b)

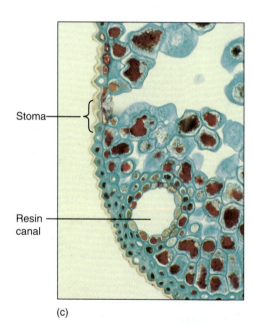

Stoma —

Resin canal —

(c)

FIGURE 6.38
(a) The needle leaves of pine are round in transverse section ($\times 15$). (b) The needles last for many years, producing more phloem each year ($\times 150$). (c) Pine needles are sclerophylls: they have thick-walled epidermal cells with a thick cuticle. They also contain resin canals ($\times 150$). (d and e) Leaves of yew (*Taxus*) are flat, but they also have sclerophyllous features and produce phloem year after year (d, $\times 50$; e, $\times 150$).

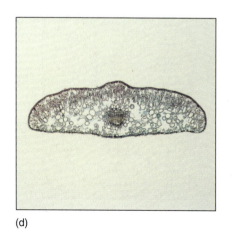

(d)

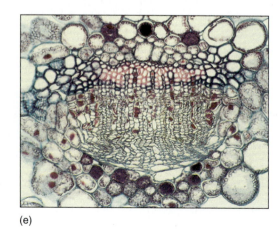

(e)

(a)

(b)

FIGURE 6.39

(a) Bud scales of *Rhododendron* fit together tightly during winter, protecting the enclosed apical meristem and young leaves and flowers. (b) The bud scales of *Magnolia* have a dense covering of hair; this improves the insulation.

BUD SCALES

One of the most common modifications of leaves is their evolutionary conversion into **bud scales** (see Figs. 5.13 and 6.39). In perennial plants, dormant shoot apical meristems must be protected from low temperatures and the drying action of wind during winter. Bud scales provide this protection by forming a tight layer around the stem tip. Because their role is primarily protection, not photosynthesis, they have a different structure from foliage leaves. Bud scales are small and rarely compound, so mechanical wind damage is not a risk for bud scales. Their petiole is either short or absent because they must remain close to the stem and be folded over it. To be protective, they must be tougher and more waxy than regular leaves; bud scales frequently produce a thin layer of corky bark, at least on exposed portions, which provides greater protection than the simple epidermis of foliage leaves.

SPINES

Many spines are modified leaves, especially in the cacti that grow in arid regions. The succulent, moist cactus body would be an excellent source of water for animals were it not for the protective spines (Fig. 6.40). As with bud scales, spines have a distinct structure related to their function. The soft, flexible blade of a photosynthetic leaf is generally useless as a protective device against herbivores, but spines have no blade and are needle-shaped; mutations that inhibit lamina formation have been selectively advantageous. No mesophyll parenchyma or vascular tissue is present; the mesophyll instead consists of closely packed fibers. Once fibers mature, they deposit lignin in their walls, which makes them hard and resistant to decay. The cells then die and dry out, hardening even further. Because cacti carry out photosynthesis in their stem cortex, loss of the leaf lamina was selectively advantageous. But for plants that have no alternative photosynthetic tissues, mutations that cause loss of lamina are extremely disadvantageous.

(a)

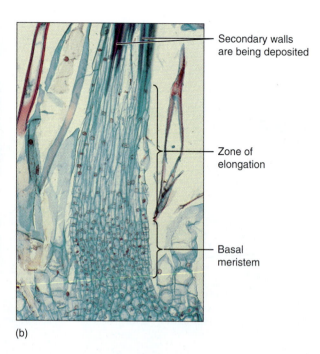

Secondary walls
are being deposited

Zone of
elongation

Basal
meristem

(b)

FIGURE 6.40
(a) The spines of this barrel cactus (*Ferocactus*) are modified leaves. The cluster of spines is actually an entire axillary bud. (b) Cactus spines have a basal meristem; as new cells are formed, older cells are pushed upward. As they move out of the meristem, they fill their central vacuole with water, elongate, then deposit a thick secondary wall and differentiate into fiber cells (× 50).

Remember that in other species (ivy, grape, Virginia creeper; see Chapter 5) tendrils are modified shoots.

The metabolism of this sensation and response is described in Chapter 14.

Kranz is the German word for "wreath." See Chapter 10 for details of C₄ photosynthesis and its relationship to Kranz anatomy.

TENDRILS

The tendrils of many plants (peas, cucumbers, and squash) are another form of modified leaf (Fig. 6.41). Unlike photosynthetic leaves, tendrils grow indefinitely and contain cells capable of sensing contact with another object. When the tendril touches something, the side facing the object stops growing but the other side continues to elongate, causing the tendril to coil around the object and use it for support. A lamina would be detrimental, and none forms. Whereas many foliage leaves are capable of sensing the direction of sunlight and reorienting the lamina for maximum photosynthesis, tendrils respond by sensing solid objects and growing around them.

LEAVES WITH KRANZ ANATOMY

A distinct type of leaf anatomy occurs in plants that have a special metabolism called C_4 photosynthesis (see Fig. 10.27). These plants lack a palisade parenchyma and spongy mesophyll but have prominent bundle sheaths composed of large chlorophyllous cells. Surrounding each sheath is a ring of mesophyll cells that appear to radiate from the vascular bundle. These plants possess a mechanism of carbon dioxide transport that requires this special **Kranz anatomy** and adapts C_4 plants to arid environments.

INSECT TRAPS

The ability to trap and digest insects has evolved in several families. Insectivory has evolved in plants that grow in habitats poor in nitrates and ammonia; by digesting insects, plants

FIGURE 6.41
This tendril is a highly modified leaf: It has no leaf blade, and whereas foliage leaves stop growing after they reach a specific size, this tendril continues to grow.

obtain the nitrogen they need for their amino acids and nucleotides. Trap leaves can be classified as either active traps that move during capture or passive traps incapable of movement. The best-known passive traps are the pitcher-leaves of *Nepenthes, Darlingtonia,* and *Sarracenia* (Fig. 6.42a). Although the leaf appears highly modified, it is actually quite similar to many foliage leaves. It is thin, parenchymatous, and capable of photosynthesis. It has numerous stomata and vascular bundles as well as a mesophyll containing aerenchyma and chlorenchyma. The most significant differences are that the lamina is tubular rather than flat and that it secretes a watery digestive fluid. The epidermis in the digestive region must be absorptive rather than impermeable. Also, the throat of the pitcher contains numerous trichomes that point toward the liquid; it is easier for insects to walk in the direction of the trichomes, and they are thus led to their death.

The leaves of sundew (*Drosera;* Fig. 6.42b) are active traps; they too have many features in common with foliage leaves, but their upper surface is covered with glandular trichomes that secrete a sticky digestive liquid. Once an insect is caught on a single trichome, adjacent trichomes are stimulated to bend toward the victim, placing their digestive drops on it as well. The entire leaf blade curls around the insect so that many trichomes come into contact with it. The modifications in these leaves have been largely metabolic rather than structural: Trichomes and lamina must be able to sense and respond to the presence of an insect, carry on secretion and absorption, and unfold after digestion is complete.

Venus' flytraps (*Dionaea muscipula*) have leaves that are held flat, like most foliage leaves, but in *Dionaea* this position is maintained only because motor cells along the upper side of the midrib are extremely turgid and swollen (Fig. 6.42c). When an insect walks

(a)   (b)   (c)

FIGURE 6.42

(a) Leaves of *Nepenthes* are the most elaborate of the pitcher leaves; they have an ordinary petiole and a broad lamina, but the leaf tip is long, narrow, and pendant. Its extreme tip turns upright and develops into the hollow pitcher that contains a digestive mixture and an epidermis that can absorb nitrogen. The ultimate portion of the leaf tip is a broad flat roof that prevents rain from falling into the pitcher and diluting the digestive juices. (b) Sundew leaves have many shapes; those of *Drosera capensis* are long and narrow. Just after an insect is caught, the leaf curls and places numerous tentacles (trichomes) on it. The drop at the end of each tentacle is both sticky and digestive. Once digestion is complete, the leaf uncurls and is ready for the next meal. (c) Each half of the blade of a Venus' flytrap leaf has three hairs; if an insect touches any two, it stimulates the trap to close. (*c, Ray Coleman/Photo Researchers*)

across the trap, it brushes against trigger hairs. If two of these are stimulated, they cause the midrib motor cells to lose water quickly, and the trap rapidly closes as the two halves of the lamina move upward. On the margins are long interdigitating teeth that trap the insect; short glands begin to secrete digestive liquid. Once digestion and absorption are complete, the midrib motor cells fill with water, swell, and force the trap open, ready for a new victim.

SUMMARY

1. Natural selection has resulted in the evolution of numerous types of leaves involved in photosynthesis, protection, support, water and nutrient storage, and nitrogen absorption.

2. Foliage leaves must be resistant to pathogens and stresses, must produce more sugar than was used in their construction, and must not act as sails. Modifications related to these functions may prevent foliage leaves from being optimally adapted for photosynthesis.

3. Foliage leaves typically consist of a blade with or without a petiole; the blade may be simple or either pinnately or palmately compound.

4. Leaf variability may involve distinct juvenile and adult leaves or distinct leaves on long shoots and short shoots.

5. An abscission zone typically contains a separation layer in which cell walls rupture and a protective layer whose cells form a corky leaf scar.

6. Internally, most foliage leaves have an upper palisade parenchyma rich in chlorophyll and a lower spongy mesophyll that allows gas circulation. Vascular bundles composed of primary xylem and phloem are usually arranged in a netted pattern in dicots and a parallel pattern in monocots. Minor veins load sugars and unload water.

7. Leaves are initiated only at shoot apical meristems, beginning as small conical leaf primordia. Dicot leaves generally have diffuse growth in all parts simultaneously; most monocot leaves grow indefinitely from a basal meristem.

IMPORTANT TERMS

abscission zone
bud scale
bundle sheath
compound leaf
lamina
leaf blade
leaf primordium
leaf scar
leaf trace
leaflet
mesophyll
midrib
minor veins
palisade parenchyma
parallel venation
petiole
sessile leaf
simple leaf
spongy mesophyll
stipule
rachis
reticulate venation

REVIEW QUESTIONS

1. In what ways does the upper epidermis of a leaf differ from the lower epidermis? How are these structural differences adaptive?

2. Draw a cross-section of a foliage leaf and label each part; be certain to show all the types of vascular tissues in the bundles.

3. Which parts of the leaf in Question 2 would be emphasized and which would be reduced in each of the following modified leaves: tendrils, spines, bud scales, scale leaves of a bulb, succulent leaves of a desert plant?

4. How do the growth of dicot and grass leaves differ? Which type would be more capable of recovering from attack by leaf-eating insects or grazing deer?

5. How can you distinguish between a compound leaf and a twig with several simple leaves? Assuming both a simple and a compound leaf have the same texture, which is more easily eaten by an insect larva?

6. Dormant buds, such as the terminal and axillary buds of twigs in winter, consist of bud scales, young leaves, and leaf primordia. Which were formed by the apical meristem first and which were initiated last?

7. Imagine a leaf that is rectangular in shape and has the following dimensions: thickness—0.3 mm; width—10 cm; length—25 cm. What is its surface area? (Do not forget to compute both the upper and lower surfaces.) What is its volume? If fibers are reinforcing its margin, how many centimeters of margin does the leaf have? Now imagine that the leaf is twice as thick; how does that affect its surface area (capacity for carbon dioxide absorption and water loss) and volume (capacity for water storage)?

BotanyLinks .net Questions
www.jbpub.com/botanylinks

Visit the .net Questions area of BotanyLinks (http://www.jbpub.com/botanylinks/netquestions) to complete this question:

1. Leaves must survive insect attack. Go to the BotanyLinks home page for more information on this subject.

BotanyLinks includes a Directory of Organizations for this chapter.

ROOTS

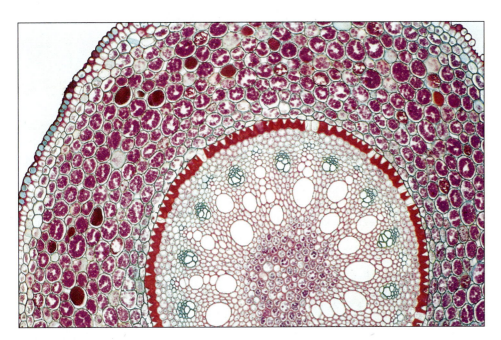

7

The band of cells with thick red walls protects the enclosed conducting tissues of the root; these cells regulate which minerals are transported from the soil to the shoot.

CONCEPTS

Most roots have three functions: (1) anchoring the plant firmly to a substrate, (2) absorbing water and minerals, and (3) producing hormones. Firm anchoring provides stability and is therefore important for virtually all plants. Stems, leaves, flowers, and fruits then can be properly oriented to the sun, to pollinators, or to fruit distributors. Without proper root attachment, trees and shrubs could not remain upright, and epiphytes would be blown from their sites in the tree canopy. A highly branched rhizomatous or stoloniferous plant might resist being blown over even without roots, but the horizontal stems are usually so flexible that roots are necessary to stabilize the aerial structure.

Although roots, like leaves, have an absorptive function, the two organs have totally different shapes. Sunlight always comes from above, but water and minerals are distributed on all sides of a root. Its cylindrical shape allows all sides to have the same absorptive capacity. Consider a leaf and a system of thin roots, both with equal volumes; the root system has a higher surface-to-volume ratio, ideal for absorption. The lower surface-to-volume ratio for leaves reduces the carbon dioxide absorption capacity but not the absorption

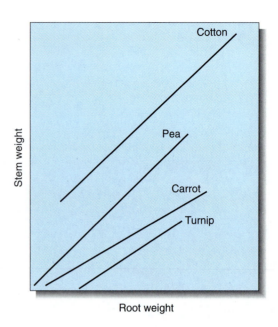

FIGURE 7.1

The ratio of root system to shoot system is critically important; for each species, as the root system becomes heavier, so does the shoot system by a characteristic amount. However, the relationship differs for each species: Obviously for a given amount of turnip root there is relatively little shoot, but there is much more shoot for the same amount of pea root, and even more for cotton.

of light and is actually beneficial for water conservation. Roots do not need to be adapted for light absorption, and they absorb water rather than needing to conserve it. Their cylindrical shape is also undoubtedly related to the growth of the roots through a semisolid, resistant medium. Even a light, porous soil can be most easily and thoroughly penetrated by narrow cylinders rather than thin sheets. Thus, although both leaves and roots have absorptive functions, it is selectively advantageous for them to have different shapes.

Roots are quite active in the production of several hormones; shoot growth and development depend on the hormones cytokinin and gibberellin imported from the roots. This reliance of the shoot on root-produced hormones may be a means of integrating the growth of the two systems. It is selectively advantageous for a plant to control the size of its shoot so that transpiration by its leaves does not exceed absorption by its roots. Nor should a plant waste carbohydrate by constructing a larger root system than its shoot needs; the extra carbohydrate could be used for leaves or reproduction (Fig. 7.1).

In many cases, roots have functions in addition to or instead of anchoring, absorption, and hormone production. Fleshy taproots, such as those of carrots, beets, and radishes, are the plant's main site of carbohydrate storage during winter. As the roots of willows, sorrel, and other plants spread horizontally, they produce shoot buds that grow out and act as new plants. This method of vegetative reproduction is quite similar to that of stoloniferous and rhizomatous plants, except that roots rather than stems are involved. In the palms *Crysophila* and *Mauritia,* roots grow out of the trunk and then harden into sharp spines. Ivy and many other vines have modified roots that act as holdfasts, clinging to rock or brick. Finally, many parasitic flowering plants (mistletoe, dodder) attack other plants and draw water and nutrients out of them through modified roots.

Distinct sets of characteristics are adaptive for the different root functions. As roots specialize and become more efficient for particular tasks, they become poorly adapted for other tasks, so several types of roots may occur in one plant, resulting in division of labor. For example, in addition to the large storage root, carrots and beets also have fine absorptive roots, and ivy has both holdfasts and absorptive roots. The characteristic types of structure and metabolism of each should be analyzed in terms of the function of the particular root.

EXTERNAL STRUCTURE OF ROOTS

ORGANIZATION OF ROOT SYSTEMS

Roots must have an enormous absorptive surface; in order for a single root to have sufficient surface area, it would have to be hundreds of meters long, which would make conduction impossible (Table 7.1). Instead, plants have a highly branched root system (Fig. 7.2). Most dicots have a single prominent **taproot** that is much larger than all the rest and numerous small **lateral roots** or **branch roots** coming out of it (Fig. 7.3). This taproot develops from the embryonic root, called the **radicle,** that was present in the seed; after germination it grows extensively and usually becomes the largest root in the system. Carrots, beets, turnips, and other taproots sold in stores have dozens of fine lateral roots while growing, but these are removed before the products are shipped to market.

(a)

(b)

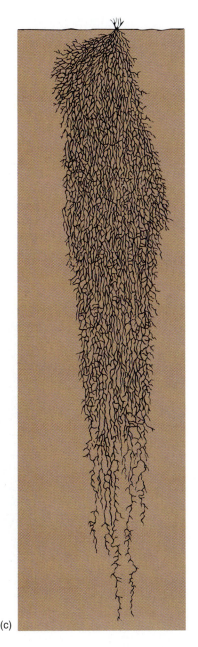

(c)

FIGURE 7.2

Some taproots, such as carrot (a), become extremely swollen and are much larger than the numerous lateral roots, whereas in other species, such as sunflower (b), the taproot is about the same size as the laterals. The important criterion is that the taproot develops from the root of the embryo, the radicle. (c) A fibrous root system, as in this winter wheat, consists of many roots, none of which is the radicle. Instead, all the main roots are adventitious roots that originated in stem tissue.

TABLE 7.1	*Dimensions of the Root System of Barley (Hordeum vulgare), Four Weeks Old*
Number of main roots	
Seminal roots	6.5
Adventitious roots	19.2
Total length of main roots	800 cm
Number of lateral roots	2000
Total length of lateral roots	4900 cm
Diameter	
Main roots	0.6 mm
First-order laterals	0.2 mm
Second-order laterals	<0.1 mm

Lateral roots may also produce more lateral roots, resulting in a highly ramified set of roots analogous to the highly branched shoot system of most plants. Lateral roots can become prominently swollen like a taproot, as in sweet potatoes and the tropical vegetable manioc (cassava). If the plant is perennial and woody, the roots also undergo secondary growth, increasing the amount of wood and bark.

Most monocots and some dicots have a mass of many similarly sized roots constituting a **fibrous root system** (Fig. 7.4). This arises because the radicle dies during or immediately after germination; root primordia at the base of the radicle grow out and form the first stages of the fibrous root system. As the plant ages, more root primordia are initiated in the stem tissue. Because these roots do not arise on pre-existing roots and because they are not radicles, they are known as **adventitious roots.** Adventitious roots increase the absorptive and transport capacities of the root system.

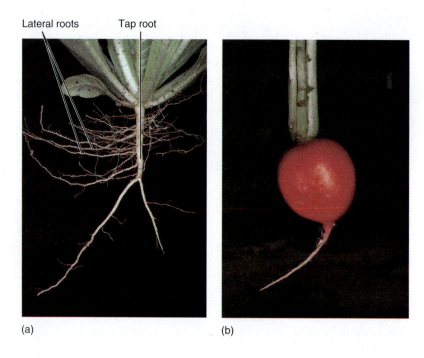

FIGURE 7.3

(a) The taproot of this tobacco seedling (*Nicotiana tabacum*) is definitely larger than the lateral roots, but more importantly, its development can be traced directly to the embryo. Relative size is not the critical factor; in many species lateral roots are the ones that become enlarged. (b) The roots of radishes are obviously taproots.

(a)   (b)

The functional significance of taproots versus fibrous root systems becomes apparent when the general growth forms of dicots and monocots are considered. Many dicots are perennial and undergo secondary growth, resulting in an increased quantity of healthy, functional wood (xylem) in both the trunk and the roots (Fig. 7.5a and b). This enlarging conduction capacity permits an increase in the number of leaves and fine, absorptive roots.

Most monocots cannot undergo secondary growth; once their stem is formed, the number of vascular bundles, tracheary elements, and sieve tubes is set, and their conducting capacity cannot be increased. Extra leaves could not be supplied with water, nor could their sugar be transported (Fig. 7.5c and d). Such a shoot could not supply sugars to an ever-increasing taproot system. However, some monocots do increase their size by means of stolons or rhizomes: Their horizontal shoots branch and then produce adventitious roots (Fig. 7.5e). Because these roots are initiated in the new stem tissues, they transport water directly into the new portions of the shoot, unhindered by the limited capacity of the older portions of the shoot. By this mechanism, monocot shoots can branch and grow larger, as long as they remain close enough to the substrate to produce new adventitious roots (see Fig. 7.18). For them, a fibrous root system is functional, whereas a taproot system is not.

The ability to form adventitious roots is not limited to monocots; many rhizomatous and stoloniferous dicots also grow this way naturally. Furthermore, many dicots that never produce adventitious roots in nature do so if they are cut; this is important in the process of asexual propagation by cuttings.

FIGURE 7.4
Onions (*Allium*), like other bulbs and monocots, have a fibrous root system. The radicle died shortly after germination; no root here has developed from the radicle. (*J. L. Pontier/Earth Scenes*)

(a)

(b)

(c)

(d)

(e)

FIGURE 7.5
(a) A young dicot has a few leaves and a small root system; the narrow trunk with a few vascular bundles can conduct water and nutrients between them. (b) An older dicot has more leaves and a larger root system; the stem has more wood and bark, which increases the capacity to conduct water and sugar. Because most monocots do not undergo secondary growth, the stem of an older plant is not wider (d) than that of a young plant (c), and it has no increased conducting capacity. Consequently, the old plant has no more leaves or roots than the young plant. (e) If a plant can produce adventitious roots, the bottleneck of the monocot stem does not matter. New roots originate near the aerial shoots and conduct directly into them, and little or no long-distance conduction occurs in the rhizome. No part of the monocot stem needs to conduct all the water from all the roots to all the leaves and flowers, as does the trunk of a dicot.

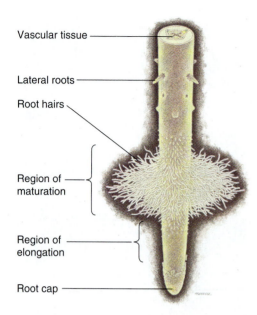

Vascular tissue

Lateral roots

Root hairs

Region of maturation

Region of elongation

Root cap

FIGURE 7.6

The tip of a root consists of a root cap, the root apical meristem, a zone of elongation growth, and a region where root hairs are formed.

STRUCTURE OF INDIVIDUAL ROOTS

An individual root is fairly simple; because it has no leaves or leaf scars, it has neither leaf axils nor axillary buds (Figs. 7.6 and 7.7). The tip of the root, like that of the shoot, is the region where growth in length occurs. In roots, growth by discrete apical meristems is the only feasible type of longitudinal growth. In most animals, all parts of the body grow simultaneously (diffuse growth), whereas the roots and stems of plants elongate only at small meristematic regions (localized growth). Because the root is embedded in a solid matrix, it is impossible for all parts to extend at once; the entire root would have to slide through the soil. With apical growth, only the extreme tip must push through the soil.

Whereas the shoot apical meristem is protected by either bud scales or young, unexpanded foliage leaves, the root apical meristem is protected by a thick layer of cells, the **root cap** (Figs. 7.7 and 7.8). Although some soils may appear soft and easily penetrable, on a microscopic scale, all contain sand grains, crystals, and other components that can easily damage the delicate apical meristem and root cap. Because the cap is forced through the soil ahead of the root body, it is constantly being worn away and must be renewed by cell multiplication.

The dictyosomes of root cap cells secrete a complex polysaccharide called **slime** or **mucigel,** which helps to lubricate the passage of the root through the soil. It causes the soil to release its nutrient ions and permits the ions to diffuse more rapidly toward the root. Mucigel is rich in carbohydrates and amino acids, which foster rapid growth of soil bacteria around the root tip. The metabolism of these microbes is believed to help release nutrients from the soil matrix.

Just behind the root cap and root apical meristem is a **zone of elongation** only a few millimeters long within which the cells undergo division and expansion (see Fig. 7.7). Behind it is the **root hair zone,** a region in which many of the epidermal cells extend out as narrow trichomes. Root hairs can form only in a part of the root that is not elongating or they would be shorn off.

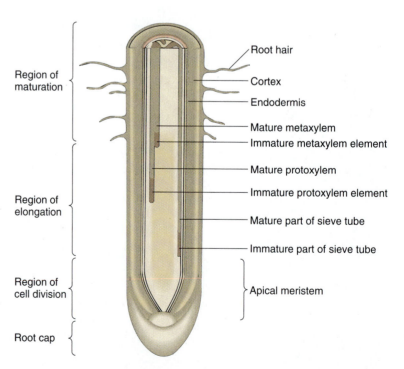

Region of maturation

Region of elongation

Region of cell division

Root cap

Root hair

Cortex

Endodermis

Mature metaxylem

Immature metaxylem element

Mature protoxylem

Immature protoxylem element

Mature part of sieve tube

Immature part of sieve tube

Apical meristem

FIGURE 7.7

Although the exterior of a root appears rather uniform, several distinct zones of differentiation are present internally. Root hairs form only above the elongation zone, and the endodermis and first vascular tissues appear earlier than do root hairs.

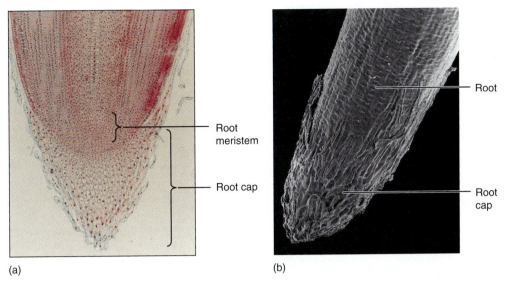

(a)   (b)

Root meristem

Root cap

Root

Root cap

FIGURE 7.8
(a) The root cap of this corn root is distinct from the root proper. The root apical meristem is located at the apex of the root proper but is buried under the root cap. Because no leaves, leaf traces, or branches are present, cells develop in an extremely orderly fashion in regular files. It is easy to see that these cells of the primary body of the root are derived directly from the meristem ($\times$ 150). *(Bruce Iverson)*
(b) Scanning electron micrograph showing the root cap margin. As cells are damaged by being pushed through the soil, they die and break off the cap ($\times$ 1000). *(Photo Rearchers)*

Root hairs greatly increase the root's surface area. In a study of rye, a single plant was found to have 13 million lateral roots with 500 km of root length and a surface area of 200 m^2. Because of the abundant production of root hairs, however, the total surface area was doubled. The presence of root hairs has other effects that should not be overlooked. Most pores in soil are too narrow for a root (usually at least 100 μm in diameter) to penetrate (Fig. 7.9). But root hairs, being only about 10 μm in diameter, can enter any crevice and extract water and minerals from it. Furthermore, carbon dioxide given off by the respiration of root hairs combines with soil water to form carbonic acid, which helps to release ions from the soil matrix. Without the acid, ions would be too firmly bound to soil

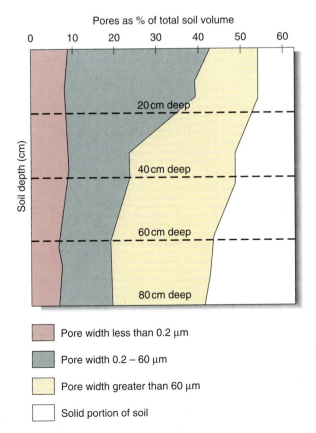

Pores as % of total soil volume

0   10   20   30   40   50   60

Soil depth (cm)

20 cm deep

40 cm deep

60 cm deep

80 cm deep

Pore width less than 0.2 μm

Pore width 0.2 – 60 μm

Pore width greater than 60 μm

Solid portion of soil

FIGURE 7.9
Much of a soil is composed of spaces between soil particles; in the uppermost 20 cm of a soil composed of sandy loam, 54% of the soil is pore space. Of this, 9% consists of extremely fine spaces less than 0.2 μm wide, 33% spaces ranging between 0.2 and 6.0 μm, and 12% large spaces wider than 60 μm. At deeper levels, the soil is more compacted, containing only about 42% pore space; the decrease is due mostly to a compaction of the intermediate-size spaces.

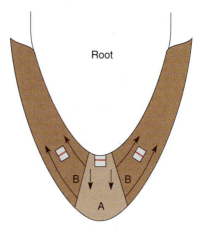

FIGURE 7.10
In the root cap, cells in the central portion (A) divide so that the two daughter cells are aligned with the root and the root cap grows forward. On the edges (B), cells divide and expand in such a way that cells flow radially outward.

particles for the root to absorb them (see Chapter 13 for more details). Root hairs are unicellular, never have thick walls, and are extremely transitory. They die and degenerate within 4 or 5 days after forming.

Behind the root hair zone is a region where new lateral roots emerge. They may occur in rows or may appear to be randomly distributed on the parent root. The outgrowth of lateral roots often depends on the microenvironment of the soil. If the parent root grows into a zone of rich, moist soil, numerous lateral roots form and the pocket is fully exploited. If the soil is poor, hard, or dry, few lateral roots emerge.

INTERNAL STRUCTURE OF ROOTS

ROOT CAP

To remain in place and provide effective protection for the root apical meristem, the root cap must have a specific structure and growth pattern. The cells in the layer closest to the root meristem are also meristematic, undergoing cell division with transverse walls and forming files of cells that are pushed forward (Fig. 7.10). Simultaneously, the cells on the edges of this group grow toward the side and proliferate. Although the cells appear to extend around the sides of the root, the root is actually growing through the edges of the root cap.

Cells are small and meristematic when first formed at the base of the root cap, but as they are pushed forward, they develop dense starch grains and their endoplasmic reticulum becomes displaced to the forward end of the cell (see Fig. 14.10). These cells are capable of detecting gravity because the dense starch grains settle to the lower side of the cell.

As the cells are pushed closer to the edge of the cap, their structure and metabolism change dramatically. Endoplasmic reticulum becomes less conspicuous, starch grains are digested, and the cell's dictyosomes secrete copious amounts of mucigel by exocytosis. Simultaneously, the middle lamella breaks down and releases cells into the mucigel which are usually crushed by the expansion of the root. It has been estimated that only 4 or 5 days pass from cell formation in the root cap to its sloughing off. Consequently, the cap is constantly regenerating itself. A dynamic equilibrium must be maintained between these two processes.

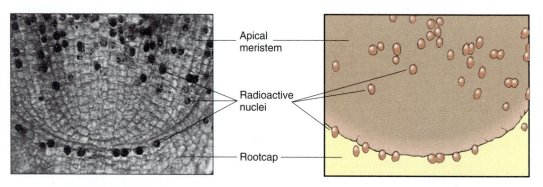

FIGURE 7.11
These roots were grown briefly in a solution of tritiated thymidine, a radioactive precursor of DNA. In cells that were undergoing the cell cycle, during S phase the radioactive thymidine was incorporated into the nuclei. After a few hours, the root was killed, sliced into sections, and placed on photographic film in the dark. The radioactive nuclei caused black spots to form in the film next to the nuclei. The slide was then given a brief exposure to light so that the outlines of the cells and nuclei would be faintly visible (too much light would obscure the radioactivity-induced black spots). The quiescent center is the region where no nuclei became radioactive. Apparently no cell in the region passed through S phase while tritiated thymidine was available—these cells were in cell cycle arrest. (*Courtesy of L. J. Feldman, University of California, Berkley*)

ROOT APICAL MERISTEM

If the root apical meristem is examined in relation to the root tissues it produces, regular files of cells can be seen to originate in the meristem and extend into the regions of mature root tissues (see Fig. 7.8). The root is more orderly than the shoot because it experiences no disruptions due to leaf primordia, leaf traces, or axillary buds. Because the cell files extend almost to the center of the meristem, one might assume that cell divisions are occurring throughout it. However, use of a radioactive precursor of DNA, such as tritiated thymidine, can demonstrate that the central cells are not synthesizing DNA; their nuclei do not take up the thymidine and become radioactive (Fig. 7.11). This mitotically inactive central region is known as the **quiescent center.** These cells are more resistant to various types of harmful agents such as radiation and toxic chemicals, and it is now believed that they act primarily as a reserve of healthy cells. If part of the root apical meristem or root cap is damaged, quiescent center cells become active and form a new apical meristem. Once the new meristem is established, its central cells become inactive, forming a new quiescent center. Such a replacement mechanism is extremely important because the root apex is probably damaged frequently by various agents—sharp objects, burrowing animals, nematodes, and pathogenic fungi.

ZONE OF ELONGATION

Just behind the root apical meristem itself is the region where the cells expand greatly; some meristematic activity continues, but mostly the cells are enlarging. This zone of elongation is similar to the shoot subapical meristem region. Cells begin to differentiate into a visible pattern, although none of the cells is mature. The outermost cells are protoderm and differentiate into the epidermis. In the center is the provascular tissue, cells that develop into primary xylem and primary phloem. As in the stem, protoxylem and protophloem, which form earliest, are closest to the meristem. Farther from the root tip, older, larger cells develop into metaxylem and metaphloem. Between the provascular tissue and the protoderm is a ground tissue, a rather uniform parenchyma that differentiates into the root cortex.

In the zone of elongation, the tissues are all quite permeable; minerals can penetrate deep into the root through the apoplast simply by diffusing along the thin, fully hydrated young walls and intercellular spaces. This zone is so short that little actual absorption occurs there, and much that is absorbed is probably used directly for the root's own growth.

ZONE OF MATURATION/ROOT HAIR ZONE

In the maturation zone, several important processes occur more or less simultaneously. Root hairs grow outward, greatly increasing absorption of water and minerals. In some electron micrographs, a thin cuticle appears to be present on root epidermal cells, but this may be just a layer of fats. The zone of elongation merges gradually with the zone of maturation; no distinct boundary exists because the differences between the two represent the gradual continued differentiation of the cells.

Although cortex cells continue to enlarge, their most significant activity is the transfer of minerals from the epidermis to the vascular tissue. This can be either by diffusion through the walls and intercellular spaces (called apoplastic transport) or by absorption into the cytoplasm of a cortical cell and then transferal from cell to cell, probably through plasmodesmata (symplastic transport; Fig. 7.12a). Cortical intercellular spaces are also important as an aerenchyma, allowing oxygen to diffuse throughout the root from the soil or stem (Fig. 7.12b).

In the zone of maturation, minerals do not have free access to the vascular tissues because the innermost layer of cortical cells differentiates into a cylinder called the **endodermis** (Fig. 7.13). The cells of the endodermis have tangential walls, those closest to the vascular tissue or the cortex, which are ordinary thin primary walls. But their radial walls —the top, bottom, and side walls, that is, all the walls touching other endodermis cells—

A tritiated chemical is made with tritium, the heaviest isotope of hydrogen (see Fig. 2.1); because it is radioactive, it can be located after being incorporated metabolically into a plant.

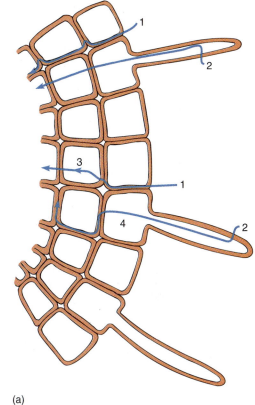

(a)

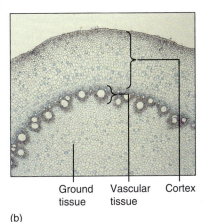

Ground tissue   Vascular tissue   Cortex

(b)

FIGURE 7.12

(a) Many diffusion paths are possible through the root epidermis and outer cortex. 1 is entirely apoplastic: The water or mineral diffuses only through walls in intercellular spaces. 2 is symplastic transport: The material has passed through a plasma membrane and is now in the protoplasm. At 3, a molecule changes from the apoplast to the symplast, and at 4 the opposite occurs. (b) The cortex may be broad in many roots, and its interior boundary is the endodermis ($\times 15$).

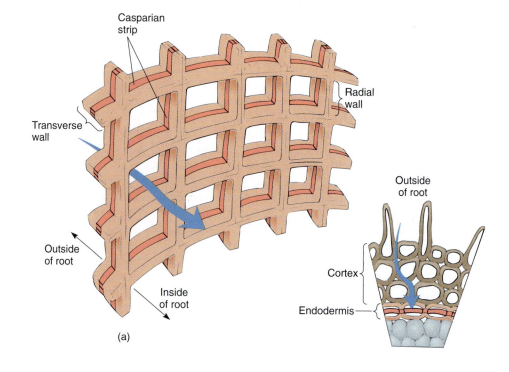

Casparian strip

Radial wall

Transverse wall

Outside of root

Inside of root

(a)

Outside of root

Cortex

Endodermis

Plant Biology Tutor CD-ROM

FIGURE 7.13

(a) The endodermis is a cylinder, one layer of cells thick, each with Casparian strips. The best analogy is a brick chimney: It is a cylinder, one layer thick. The cement is analogous to the Casparian strips. (b) No apoplastic transport occurs across the endodermis. Materials that have been moving apoplastically (1) are stopped and can proceed farther only if the endodermis plasma membrane accepts them (2). Symplastic transport (3) is not affected. Plasma membranes that are impermeable to a particular type of molecule prevent that molecule from crossing the endodermis. (c) The endodermis in this corn is relatively easy to see: In all cells, Casparian strips are visible as red lines on the radial walls. In a few cells, only Casparian strips are present, but in many the radial walls and inner tangential walls have started to thicken; it is unusual for this development to start so early (× 200).

1

2

3

(b)

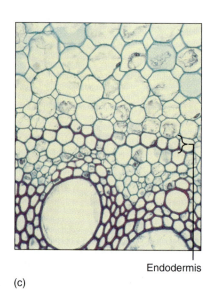

Endodermis

(c)

Suberin is related chemically to cutin and wax and is extremely hydrophobic.

are encrusted with lignin and suberin, both of which cause the wall to be waterproof. The bands of altered walls, called **Casparian strips,** are involved in controlling the types of minerals that enter the xylem water stream. Cortex cells exert no control over the movement of minerals within intercellular spaces; without an endodermis, minerals of any type could move from the soil into the spaces, then into the xylem, and then into all parts of the plant. However, because Casparian strips are impermeable, minerals can cross the endodermis only if the endodermal protoplasts absorb them from the intercellular spaces of the cortex apoplast or from cortical cells and then secrete them into the vascular tissues (Fig. 7.13b). Many harmful minerals can be excluded by the endodermis. It is not a perfect barrier against uncontrolled apoplastic movement, because in the zone of elongation, where the endodermis is not yet mature, minerals do have free access to the protoxylem, but this seems to represent only a low level of uncontrolled movement. Many glands and secretory cavities also have Casparian strips, which prevent the glands' secretory product from seeping into the surrounding tissues.

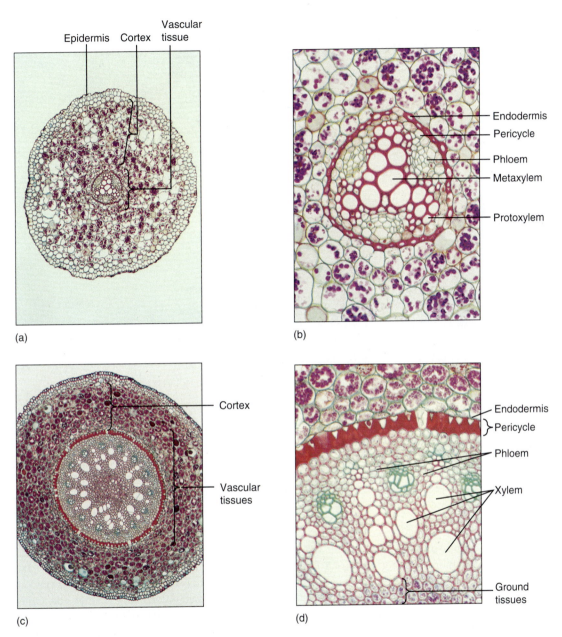

Epidermis   Cortex   Vascular tissue

(a)

Endodermis
Pericycle
Phloem
Metaxylem
Protoxylem

(b)

Cortex

Vascular tissues

(c)

Endodermis
Pericycle
Phloem
Xylem
Ground tissues

(d)

FIGURE 7.14

(a) Low-magnification view and (b), high-magnification view of a transverse section of a root of buttercup, *Ranunculus,* showing the broad cortex and the small set of vascular tissues. The xylem has three sets of protoxylem, the narrow cells at the tips of the arms (b). The central, larger xylem cells are metaxylem. Three masses of phloem are also present; this is a triarch root. The endodermis is visible, but Casparian strips are difficult to see at this magnification. The pericycle is the set of cells between the endodermis and the vascular cells (a, × 50; b, × 250).

(c) Low-magnification view and (d) high-magnification view of a transverse section of a root of *Smilax.* The vascular tissue of many monocot roots (such as this *Smilax*) consists of numerous large vessels and separate bundles of phloem. The endodermis is very easy to see (c, × 15; d, × 150).

Within the vascular tissues, many of the much larger cells of the metaxylem and metaphloem become fully differentiated and functional in the zone of maturation. The arrangement of these tissues differs from that in stems: Instead of forming bundles containing xylem and phloem, the xylem of almost all plants except some monocots forms a solid mass in the center, surrounded by strands of phloem; no pith is present (Fig. 7.14a). In the roots of many monocots, strands of xylem and phloem are distributed in ground tissue (Fig. 7.14c). Within the xylem, the inner wide cells are metaxylem and the outer

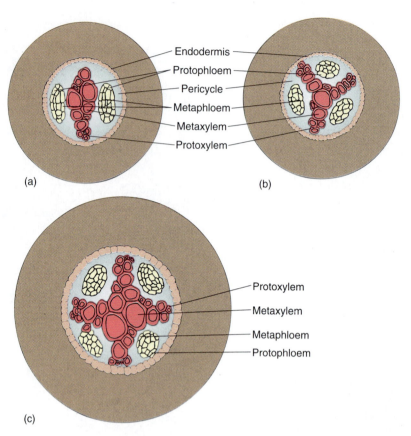

(a)   (b)

(c)

FIGURE 7.15
The number of strands of protoxylem in a root varies from one species to another and among the roots of one plant. Generally, wider, more robust roots have more protoxylem masses. Having two masses of protoxylem (and two of phloem) is diarch (a); three is triarch (b); and four is tetrarch (c).

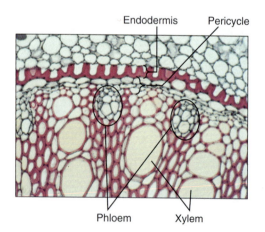

FIGURE 7.16
The endodermis here is in its final stage of differentiation. All walls except the outer tangential ones have thickened so that almost no room remains for the protoplast. This endodermis effectively prevents water leakage from the vascular tissues into the cortex (× 180).

narrow ones are protoxylem. Two to four or more groups of protoxylem may be present, the number being greater in larger roots with a wider mass of xylem (Fig. 7.15); the number of phloem strands equals the number of protoxylem strands. Within the phloem strands, protophloem is found on the outer side, and metaphloem on the inner side. Other than the arrangement, the vascular tissues of the root are similar to those of the stem and leaf. Those formed first are narrowest and most extensible, often finally being torn by the continued elongation and expansion of the cells around them (see Fig. 5.39). Those formed after the adjacent cells have stopped expanding are larger and, in the xylem, have heavier walls, often with bordered pits.

Between the vascular tissue and the endodermis are parenchyma cells that constitute an irregular region called the **pericycle** (see Figs. 7.14 and 7.15). When lateral roots are produced, they are initiated in the pericycle.

MATURE PORTIONS OF THE ROOT

Root hairs function only for several days, after which they die and degenerate. Absorption of water and minerals in this area is then greatly reduced but does not stop entirely. Within the endodermis, the cells may remain unchanged, but usually there is a continued maturation in which a layer of suberin is applied over all radial surfaces, the inner tangential face, and sometimes even the outer tangential face. This can be followed by a layer of lignin and then more suberin (Fig. 7.16). This is an irregular process, and some cells complete it earlier than others; thus in fairly mature parts of the root it is possible to find occasional cells that have only Casparian strips. These cells are called **passage cells** because they were once thought to represent passageways for the absorption of minerals; it is now suspected that they are merely slow to develop.

The result of the continued endodermis maturation is the formation of a watertight sheath around the vascular tissues. In the older parts of the root, it functions to keep water in. The absorption of minerals in the root hair zone causes a powerful absorption of water (see Chapter 12), and a water pressure, called **root pressure,** builds up. If it were not for the mature endodermis, root pressure would force the water to leak out into the cortex of older parts of the root instead of moving up into the shoot. This is presumably also a function of the endodermis in rhizomes and stolons; if water leaked into the stem cortex and filled the intercellular spaces, oxygen diffusion would be prevented and the tissues would suffocate.

Many important events occur at the endodermis. The cortex and epidermis, aside from root hairs, may be superfluous, because shortly after the root hairs die, the cortex and epidermis often die also and are shed from the root. The endodermis becomes the root surface until a bark can form. This happens mostly in perennial roots that persist for several years. In apple trees, the cortex is shed as early as 1 week after the root hairs die, although it may persist for as long as 4 weeks; only the root tips have a cortex. The large fibrous roots of many monocots are strictly annual; they are replaced by new adventitious roots that form on new rhizomes or stolons. In these plants, the entire root dies.

ORIGIN AND DEVELOPMENT OF LATERAL ROOTS

Lateral roots are initiated by cell divisions in the pericycle. Some cells become more densely cytoplasmic with smaller vacuoles and resume mitotic activity. The activity is localized to just a few cells, creating a small root primordium that organizes itself into a root apical meristem and pushes outward. As the root primordium swells into the cortex, the endodermis may be torn or crushed or may undergo cell division and form a thin covering over the primordium. As it pushes outward, the new lateral root destroys the cells of the cortex and epidermis that lie in its path, ultimately breaking the endodermis (Fig. 7.17). By the time the lateral root emerges, it has formed a root cap, and its first protoxylem and protophloem elements have begun to differentiate, establishing a connection to the vascular tissues of the parent root.

OTHER TYPES OF ROOTS AND ROOT MODIFICATIONS

The type of root just described is the most common, generalized type, comprising some or all of the roots of most plants. In other species, some roots are modified and carry out different roles in the plant's survival. Various structures and organizations are selectively advantageous for various functions.

STORAGE ROOTS

Roots frequently provide long-term storage for carbohydrates that accumulate during summer photosynthesis. This occurs only in perennial plants, because the carbohydrates are used in the production of new shoots or flowers in the spring, when photosynthesis is impossible owing to the lack of leaves. Annual plants can survive without such storage capacity.

Deciduous perennials also store significant amounts of nutrients in the stem during winter, but the roots offer certain advantages. Being subterranean, they are less available as food than are swollen, highly nutritious, easily visible stems. Roots also have a much more stable environment, subjected to less extreme changes in temperature and humidity; this may be important for the survival of the storage parenchyma cells. In many perennials, most of the stem is annual, dying back to a few nodes located below ground at the top of the root. It seems less economical in these species to winterproof the shoot than to replace it using nutrients stored in the roots.

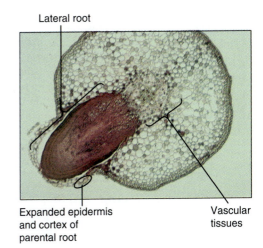

Lateral root

Expanded epidermis and cortex of parental root

Vascular tissues

FIGURE 7.17

This young lateral root of willow (*Salix*) was initiated in the pericycle, next to a mass of protoxylem; its growth caused considerable damage to the parent root cortex, and it broke the surface open. Entry of pathogens is prevented by formation of wound bark around the tear it caused (× 50).

PLANTS & PEOPLE

ROOTS AND WORLD POLITICS

Taproots are important to us as foods, examples being carrots (*Daucus carota*), sweet potatoes (*Ipomoea batatas*), beets (*Beta vulgaris*), turnips (*Brassica campestris*), and radishes (*Raphanus sativus*). None of these is a dominant food; they are eaten in small amounts or only occasionally. One reason is that their food value is rather low. These fleshy taproots store carbohydrates, and the root parenchyma cells tend to be large and filled with amyloplasts—little cytoplasm, proteins, vitamins, or essential fatty acids are present. For example, one sweet potato has only 2 g of protein but 37 g of carbohydrate; for other root crops, the values are: beets, 1 versus 7; carrots, 1 versus 11; turnips, 1 versus 8. In contrast, one cup of peanuts has 37 g of protein versus 27 g of carbohydrates; red kidney beans have 15 versus 42; wheat, 13 versus 71. The roots of horseradish (*Armoracia rusticana*) are also poor in protein, but they are consumed for their pungent allyl isothiocyanate, not for nutrition.

High carbohydrate content is of value in one particular fleshy taproot: sugar beets (*Beta vulgaris*). This is the same species as the red beet we eat as a vegetable; one has been bred for high sugar content, the other for pigments and flavor. The history of this breeding is an interesting example of the interaction of food, botanical science, and global politics. Throughout most of history, honey was the only readily available sweetener until the ancient Arabians developed the process of refining sugar from sugarcane. During the great colonial period of Spain and Portugal in the 1500s and 1600s, sugarcane was spread throughout the tropics. With changing political fortunes, by the early 1800s the British controlled many of the sugar-producing territories directly or the sea lanes by which sugar was shipped to Europe. After Napoleon came to power, Britain was able to blockade France, cutting off supplies of sugar and other materials.

Fortunately for Napoleon, sugar had been discovered in beets in 1590, and by 1747 a German chemist, Andrea Marggraf, had discovered that some beets are sweeter than others. Plant breeders began improving on this by artificially selecting plants with the greatest sugar and eliminating those with the least; sugar content was raised gradually from 2% to 20%. Napoleon realized the potential of these plants, which grow in Europe whereas sugarcane does not, and encouraged their cultivation in France as a means of defying the British. Before this could be of much significance, however, Napoleon was defeated at Waterloo in 1815. Interest in sugar beets shifted to Germany, where national self-sufficiency was considered a worthy cause. German landowners became leaders in the scientific studies of agriculture in the 1830s. They performed carefully controlled experiments in the application of fertilizers and the value of plowing and tilling soil. In 1840, sugar beets had supplied only 5% of the world's sugar, but by 1890 this amount had increased to 50%.

Sugar beets and world politics have also affected the United States. At one time most of our sugar was made from imported Cuban sugarcane. After Fidel Castro came to power in 1959, all imports of sugar from Cuba were banned and sugar beet cultivation in the United States increased tremendously. It also caused a problem for the Union of Soviet Socialist Republics: They have had to buy the entire Cuban sugarcane crop every year at great expense, even though Russia is the world's leading producer of sugar beets.

Sugar beets being harvested.
(*Jeff Greenberg/Photo Researchers*)

PROP ROOTS

The stem of a monocot can become wider, with more vascular bundles, if it can produce adventitious roots that extend to the soil. In some cases, the roots are capable of extensive growth through the air: In many palms, the exposed roots can be 20 to 50 cm long (Fig. 7.18a). In the screwpine (*Pandanus,* a monocot) they are often 3 or 4 meters long, and the roots may grow through the air for months before reaching the ground (Fig. 7.18b). Less

Prop roots

(a)

(b)

(c)

(d)

FIGURE 7.18

(a) Many monocots like this palm (*Ptychosperma*) produce adventitious roots near the base of the stem. These provide both extra absorptive capacity and extra stability. (b and c) Screwpine is able to produce extremely long adventitious prop roots that not only stabilize the large, heavy trunk but also bring water and minerals into the stem. (d) In older plants of screwpine, even the branches produce adventitious roots. This water and mineral supply to the branches bypasses the trunk, where the vascular bundles are already conducting at full capacity.

FIGURE 7.19
Banyan trees produce adventitious roots which, as in the monocots, provide increased support and absorptive capacity. Because the giant branches are supported along their entire length, they can become much larger and more extensive than branches that are supported only at the point of attachment to the trunk.

dramatic examples are found at the bases of corn and many other grasses. Once the prop roots do make contact with the soil, they transport additional nutrients and water to the stem. Just as importantly, they contract slightly and place some tension on the stem, thus acting as stabilizers, much like guy wires on telephone poles. If the roots undergo secondary growth and become woody, they can be extremely strong supports, permitting a branch to extend even farther from the trunk without breaking or sagging (Fig. 7.19). In banyan trees (dicots of the genus *Ficus*), the prop roots and branches can spread and produce massive trees many meters in diameter.

Mangroves also have prop roots, but these seem to be selectively advantageous for other reasons. The plants grow in intertidal marshes and are subjected to powerful water currents during storms and even normal tide changes; the bracelike prop roots provide much more stability than a taproot system would. In addition, the aerial portion of the prop root is covered with numerous air chambers—lenticels—and its cortex is a wide aerenchyma similar to that in the knees of bald cypress. The subterranean portion of the root grows in a stagnant muck that has little or no oxygen; it is able to respire only because the aerenchyma permits the rapid diffusion of oxygen from the aerial lenticels to the submerged root tissues. If the roots were entirely subterranean, respiration would be extremely reduced.

AERIAL ROOTS OF ORCHIDS

Many orchids are epiphytic, living attached to the branches of trees. Their roots spread along the surface of the bark and often dangle freely in the air. Although these plants live in rainforests, the orchids are actually adapted to drought conditions. In the few hours when rain does not fall, the air and bark become dry and could easily pull water out of the orchid's roots if there were no water-conserving mechanism. The root epidermis, called a velamen in these orchids, is composed of several layers of large dead cells that are white in appearance (Fig. 7.20). Apparently the velamen acts as a water-proof barrier, not permitting water to leave the sides of the root.

An intertidal zone is the region along a coast located between the low and high tide lines.

Velamen   Root tips

FIGURE 7.20
Aerial roots of orchids, which grow along the surface of tree bark, are covered by a white velamen that helps retain the water absorbed by the green root tip.

CONTRACTILE ROOTS

In *Oxalis, Gladiolus, Crinum,* and other plants (many with bulbs), roots undergo even more contraction than prop roots do. After extending through the soil and becoming firmly anchored, the uppermost portions begin to contract (Fig. 7.21). Because the root is firmly fixed to the soil, the stem is pulled downward so that the base of the shoot is either kept at soil level or, in the case of bulbs, actually buried deeper. The contraction is caused by changes in the shape of cortical cells: They simultaneously expand radially and shorten, losing as much as one half to two thirds of their height. The vascular tissues buckle and become undulate but are able to continue conducting.

Contractile roots may be more common than is generally appreciated. Many seeds germinate at or near the soil surface; root contraction may be the means by which the shoot becomes anchored in the soil. In bulbs, corms, rhizomes, and other subterranean stems, contractile roots can be important in keeping the stems at the proper depth.

MYCORRHIZAE

The roots of most species of seed plants (at least 80%) have a symbiotic relationship with soil fungi in which both organisms benefit. The associations are known as **mycorrhizae** (sing.: mycorrhiza), and two main types of relationships are known. In nearly all woody forest plants, an **ectomycorrhizal** relationship exists, in which the fungal hyphae (slender threadlike cells) penetrate between the outermost root cortex cells but never invade the cells themselves. Herbaceous plants have an **endomycorrhizal** association, in which the hyphae penetrate the root cortex as far as the endodermis; they are able to pass through the walls of the cortical cells but cannot pass the Casparian strip (Fig. 7.22). They invade the cell but do not break the host plasma membrane or vacuole membrane. Inside the cell they branch repeatedly, forming a small structure called an arbuscule; this fills with granules of phosphorus which later disappear as the phosphorus is absorbed by the plant.

FIGURE 7.21
The roots of this hyacinth have contracted, causing their surfaces to become wrinkled.
(Courtesy of Judith Jernstedt, University of California, Davis)

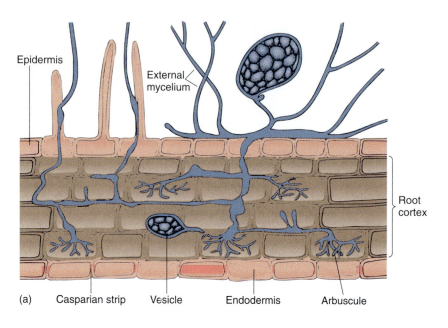

(a)

(c)

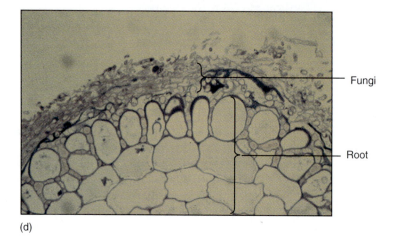

(b)   (d)

FIGURE 7.22

(a) Diagram of endomycorrhizae showing hyphae penetrating the root cortex and forming arbuscules and vesicles. (b) Root cells filled with fungal hyphae in the form of arbuscules. You can tell that this is a symbiotic fungus rather than a pathogenic one because the plant cells are healthy, with large normal nuclei ($\times 220$). (c) Roots of beech encased in ectomycorrhizae. The presence of the fungus stimulates the roots to be short and broad. (d) Transverse section of an ectomycorrhizal root showing that the fungal filaments are restricted to just the outermost part of the root; they do not penetrate the cortex. (*c and d courtesy of R. L. Peterson and D. Hern, University of Guelph*)

Other hyphae fill with membranous vesicles. The plant cells lack starch grains, presumably because the sugars are being transferred to the fungus. The fungus is unable to live without the sugars from the plant, and, in many cases, the plant is severely stunted if the fungus is killed. Apparently, this is a critical means of absorbing phosphorus into the roots, the mycorrhizal fungi being much more effective than root hairs (Fig. 7.23).

All the modified roots mentioned to this point—storage roots, prop roots, contractile roots, aerial roots of orchids, and mycorrhizae-associated roots—have a structure virtually identical to that of the more common anchoring/absorbing roots described above. All grow

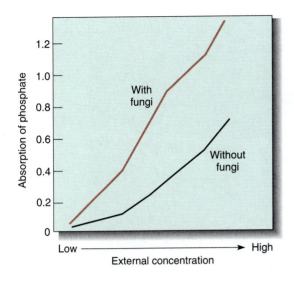

FIGURE 7.23

The ability of barley roots to absorb phosphorus from solutions of different concentrations is shown on this graph. Both sterilized roots (fungi killed) and natural roots (fungi still alive) absorb phosphorus more rapidly from the solutions with the high concentrations of phosphorus, but at every concentration the roots with mycorrhizal fungi absorb much more phosphorus than the sterile roots.

by an apical meristem, and a cross-section through a mature region of each reveals epidermis, cortex, endodermis, pericycle, and vascular tissue. In nitrogen-fixing nodules and haustoria, however, the structure is highly modified.

ROOT NODULES AND NITROGEN FIXATION

For most plants, the scarcity of nitrogenous compounds in the soil is one of the main growth-limiting factors. Although nitrogen is abundant in the air (78% of the atmosphere is N_2), plants have no enzyme systems that can use that nitrogen. Only some prokaryotes can use N_2 by incorporating it into their bodies as amino acids and nucleotides; when they die and decompose, the nitrogenous compounds are available to plants. The chemical process of converting atmospheric nitrogen into usable compounds is nitrogen fixation.

In a small number of plants, especially legumes, a symbiotic relationship has evolved with nitrogen-fixing bacteria of the genus *Rhizobium*. Bacteria free in the soil secrete a substance that causes root hairs to curl sharply; the bacterium then attaches to the convex side of the hair and pushes into the cell by means of a tubelike invagination of the plant cell wall. The tube is an **infection thread,** and the bacterium sits in it. The infection thread extends all the way into the root's inner cortex, where adjacent cortical cells undergo mitosis and form a **root nodule** (Fig. 7.24); the bacterium is released from the infection thread and enters the host cell cytoplasm, where it proliferates rapidly, filling the host cells with bacterial cells (known as bacteroids) capable to converting N_2 into nitrogenous compounds that are released to the host cells. Energy for this process is supplied by sugars from the legume root cells, so both the *Rhizobium* and the legume benefit.

The nodule may remain rather simple or may become complex, with a meristematic region, vascular tissue, and endodermis. Such nodules are able to function for extended lengths of time. Numerous metabolic changes have also evolved: The critical bacterial enzymes are sensitive to oxygen, being immediately poisoned by even traces of free oxygen. But it is the plant, not the bacterium, that produces a special chemical—leghemoglobin—that binds to oxygen and protects the bacterial enzymes from oxygen. These root nodules represent a sophisticated symbiosis; without the bacterium the complex development of nodules does not occur.

This process is often described thus: "The bacterium supplies the plant with nitrogenous compounds and in return the plant gives it sugars." This inaccurately suggests voluntary action, choice, and decision-making on the part of both organisms. Instead, think in terms of natural selection: Because the bacterium receives sugars and other nutrients from the plant, any genetic mutation in the bacterium that makes the plant healthier and more vigorous is beneficial to the bacterium as well, whereas any mutation that is harmful to the plant—such as a bacterial plasma membrane that does not allow nitrogenous compounds

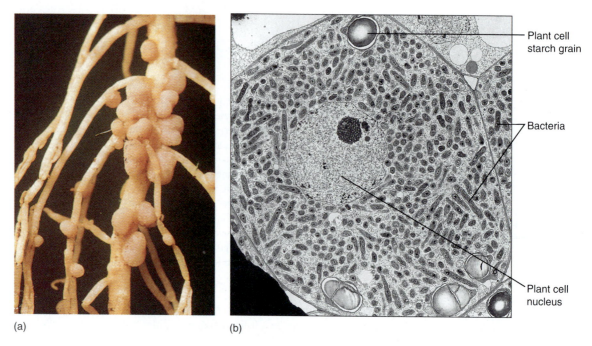

FIGURE 7.24

(a) The nodules on these alfalfa roots contain bacteria capable of absorbing atmospheric nitrogen and converting it to ammonia, which the plant can use to make amino acids. *(Hugh Spencer/Photo Researchers, Inc.)* (b) Bacteria in an infected root nodule cell of cowpea, *Vigna unguiculata* (×5900). *(Courtesy of E. H. Newcomb, University of Wisconsin, Madison)*

to move to the plant—is deleterious to the bacterium. The same principle applies to the plant: Genetic mutations that aid the bacterium ultimately benefit the plant as well. We hypothesize that these organisms will continue to co-evolve and become more fully co-adapted, but this hypothesis will have to be checked by someone a few million years from now.

Working backward, we can imagine less sophisticated legumes and *Rhizobium* bacteria that did not work so well together but evolved to the present state as mutations occurred and survived by natural selection. Certain critical mutations must be rare, or perhaps several unusual mutations must occur almost simultaneously in plant and bacterium; this hypothesis is based on the observation that, although this symbiosis would probably be beneficial to virtually all plant species, only a few have it.

HAUSTORIAL ROOTS OF PARASITIC FLOWERING PLANTS

A number of angiosperms are parasites on other plants; because their substrate is the body of another plant, a normal root system would not penetrate the host or absorb materials from it effectively. Consequently, the roots of parasitic plants have become highly modified and are known as **haustoria** (Fig. 7.25). In most cases, very little rootlike structure remains. Parasitism has evolved several times, so the structures that are termed haustoria are not all related to each other and generalizations are difficult. However, haustoria typically must adhere firmly to their host either by secreting an adhesive or by growing around a small branch or root. Penetration occurs either by forcing a shaft of cells through the host dermal system or by expanding the haustorium radially, cracking the host epidermis. After penetration, cells of the parasite make contact with the host xylem. In many cases both host and parasite cells divide and proliferate into an irregular mass of parenchyma, and then a column of cells differentiates into a series of vessel elements. This results in a continuous vessel from host to parasite constructed of cells of both. Surprisingly, many parasites attack only the xylem; they do not draw sugars from the host but carry out their own photosynthesis. Others do contact both xylem and phloem and perform little or no photosynthesis.

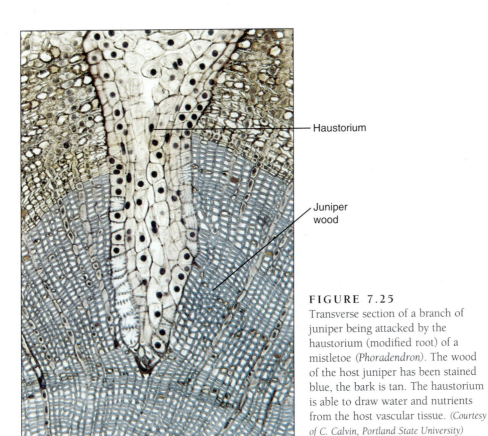

— Haustorium

— Juniper
wood

FIGURE 7.25
Transverse section of a branch of
juniper being attacked by the
haustorium (modified root) of a
mistletoe (*Phoradendron*). The wood
of the host juniper has been stained
blue, the bark is tan. The haustorium
is able to draw water and nutrients
from the host vascular tissue. (*Courtesy
of C. Calvin, Portland State University*)

For such parasites, mutations that result in the loss of leaves or vegetative stems are not
disadvantageous. In the case of *Tristerix,* the parasite spends much of its life as nothing
more than a diffuse "root" system embedded in the host (see Box 5.1).

The typical organization of a root would be nonfunctional in haustoria: The root cap,
root apical meristem, and cortex would all prevent the vascular tissues of the parasite from
contacting the appropriate tissues of the host. Conversely, haustoria are completely inadequate for growth in soil, however rich and moist it might be. For each substrate and each
microhabitat, specific adaptations are selectively advantageous.

SUMMARY

1. Most roots have a variety of functions, including anchorage, absorption, and hormone production. Other roots may be specialized for nutrient storage, vegetative reproduction, or surviving harsh conditions, some
even being modified into spines.

2. Roots have a root cap but nothing equivalent to leaves, nodes,
internodes, or buds.

3. Adventitious roots form in organs other than other roots or the
embryo; they are especially important in stoloniferous and rhizomatous
plants.

4. Roots, like shoots, elongate by localized growth (apical meristems). Only the root tip and zone of elongation must slide between soil
particles.

5. Root hairs greatly increase the absorptive surface area of the root
system, and the carbonic acid that results from their respiration helps release minerals from soil particles.

6. An endodermis with Casparian strips prevents minerals from
moving from the soil solution into the xylem. To enter the xylem, minerals
must at some point cross a plasma membrane.

7. Lateral roots arise in the pericycle, deep inside the root, unlike
axillary buds in stems, which arise in the outermost stem tissues.

8. Prop roots provide additional stabilization and transport for certain plants with narrow stems. Contractile roots aid in burying certain
bulbs, corms, and rhizomes. Haustoria are modified roots that penetrate the
tissues of host plants.

9. Most plants absorb much of their phosphorus from mycorrhizal
fungi that form an extensive network both in the soil and within the root
cortex.

10. A small number of plant species form symbiotic relationships
with nitrogen-fixing prokaryotes. The bacteria or cyanobacteria often live in
root nodules, passing nitrogen compounds to the plant and receiving carbohydrates and minerals.

IMPORTANT TERMS

adventitious root
Casparian strip
ectomycorrhiza
endodermis
endomycorrhiza
fibrous root system
haustorium

lateral root
mycorrhiza
pericycle
prop root
quiescent center
radicle
root apical meristem

root cap
root hair
root nodule
tap root
zone of elongation

REVIEW QUESTIONS

1.   What are the two types of root systems? Give several examples of plants that have each type. Which type is associated with nutrient storage in biennial species like carrots and beets? Which is associated with rhizomes and stolons?

2.   Draw cross-sections of a root showing its structure at three levels: the mature region, the root hair zone, and the zone of elongation. At which level is the endodermis complete with Casparian strips?

3.   Which part of the root produces the primordia for lateral roots? How does the vascular tissue of the lateral root connect with that of the parent root?

4.   What is a mycorrhizal association? What benefit does the plant derive from the association?

5.   Describe the structure of a nitrogen-fixing nodule; consider especially the relationship with the plant's vascular tissue.

6.   Imagine a plant that has ten roots, each 1 cm long. What is the total length of the root system? Imagine that at the tip of each root are ten lateral roots, each 1 cm long. Now what is the total length of the root system? What is the maximum distance a water molecule must travel from the farthest root tip to the base of the shoot? Distance traveled increases as a simple addition but total absorbing capacity increases exponentially. Why is length traveled important? Consider the friction of moving through tracheids and vessel elements.

STRUCTURE OF WOODY PLANTS

8

The trunks, branches, and roots of woody plants become broader every year. Each year the plant has greater conducting capacity and supports increased numbers of leaves, flowers, and fruits.

CONCEPTS

In previous chapters, growth by means of apical meristems was described. From the meristems are derived sets of tissues: epidermis, cortex, vascular bundles, pith, and leaves. These primary tissues together constitute the primary plant body. In plants known technically as herbs, this is the only body that ever develops, but in woody species, additional tissues are produced in the stem and root from other meristems—the vascular cambium and the cork cambium. The new tissues themselves are the wood (secondary xylem) and the bark (secondary phloem and cork); they are **secondary tissues,** and they constitute the plant's secondary body. Examples of woody plants are abundant: Trees such as sycamores, chestnuts, pines, and firs are woody, as are shrubs like roses, oleanders, and azaleas.

The ability to undergo secondary growth and produce a woody body has many important consequences. In an herb, once a portion of stem or root is mature, its conducting capacity is set: All provascular cells have differentiated into either primary xylem or primary phloem. This capacity is correlated with the needs of leaves and roots. If the plant produces so many leaves that they lose water faster than the stem xylem can conduct, some or all of the leaves die of water loss. Similarly, it would not be selectively advantageous for

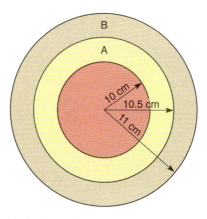

FIGURE 8.1

The cross-sectional area of a ring of wood is given by the formula π times the square of the outer radius minus the square of the inner radius. If each annual ring is 0.5 cm wide, when the wood has a radius of 10.5 cm, its newest ring has a cross-sectional area of 32.2 cm². Next year's ring will be larger, 33.8 cm², an increase of $(33.8 - 32.2)/32.2 = 1.6/32.2 \times 100\% = 5\%$ (notice that this is not drawn to scale).

Gymnosperms consist of three divisions of non-flowering plants; the most familiar division is Coniferophyta, the cone-bearing trees. Gymnosperms are covered more fully in Chapter 24.

the plant to have so many leaves that they could produce sugar faster than the phloem could conduct it to roots, flower buds, or developing fruits.

But many herbs live for several years. How do they respond to the bottleneck of the limited conducting capacity of the first-year stem? In some, the first year's leaves die during the winter, and in the second year, the plant produces only as many leaves as it had during the first year. In other species, adventitious roots are produced which supply conduction capacity directly to the new section of stem being formed, thus bypassing the older portions of the stem (see Fig. 7.5). Most of these plants must remain low enough for adventitious roots to reach the soil, so these are often rhizomatous, such as irises, bamboo, and ferns.

Woody plants not only become taller through growth by their apical meristems but also become wider by the accumulation of wood and bark. Because wood and bark contain conducting tissues, their accumulation gives plants a greater capacity to move water and minerals upward and carbohydrates downward. The number of leaves and roots that the plant can support increases, as does the photosynthetic capacity.

For example, consider a tree with a trunk radius of 10.5 cm (Fig. 8.1); imagine that the plant produced a layer of wood 0.5 cm thick in the previous year. As a consequence, the tree has 32.2 cm² of new wood that conducts water from roots to leaves. Assume that each leaf loses water at a rate equal to the conduction capacity of 0.1 cm² of wood; this plant can conduct enough water through its trunk to support 323 leaves. If the tree produces another 0.5 cm of wood this year, the new ring of wood will have a cross-sectional area of 33.8 cm². It is larger than the previous ring of wood and can support conduction to a greater number of leaves—338. Even if a ring of wood could conduct for only 1 year, the plant could still produce a greater number of leaves every year, so its annual photosynthetic capacity would always increase. The consequence of this ever-increasing capacity is that annual production of seeds and defensive chemicals also increases.

Only those seeds that germinate in a suitable site are able to grow into adults and reproduce. Once a seed of a woody, perennial plant germinates and becomes established, it occupies its favorable site year after year. Other seeds may not be able to germinate because they do not encounter suitable sites. In the springtime, all the sites that had been occupied by herbs are vacant and available to the seeds of both annual herbs and woody perennials, but virtually all the sites occupied by last year's perennials are still occupied by them. Many pines, oaks, and other long-lived trees hold on to the same piece of Earth for as long as 600 years and a few for up to 3000 years; one particular tree has lived for 5000 years. For all that time the trees produce seeds of their own and, by their very presence, prevent the seeds of their competitors from growing at that site. Annual herbs give up their sites when they die; their seeds must compete for new sites every year.

Secondary growth also has disadvantages: A 5000-year-old plant is 10,000 times older than an herb that germinates in April, lives 6 months, then sets seed and dies by September. It has had to battle insects, fungi, and environmental harshness 10,000 times longer, and it is a bigger, more easily discovered target for pathogens. Perennials have a greater need for defenses, both chemical and structural, than annual herbs have, and they must use a portion of their energy and nutrient resources for winterizing their bodies if they live in temperate climates. It is also expensive metabolically to construct wood and bark. The fact that wood burns so readily shows that it is energy-rich. If no secondary growth occurred, this energy could be used immediately for reproduction. In fact, most woody plants do not reproduce until they are several years old; if they are killed by disease or environmental stress before they reproduce, all growth and development have been for nothing.

It must be difficult for secondary growth to arise by evolution; it has evolved only three times in the 420 million years that vascular plants have been in existence, and two of those evolutionary lines later became extinct (see Box 8.2). All woody trees and shrubs alive today have descended from just one group of primitive woody plants that arose about 370 million years ago (see Fig. 24.4). This group has been very successful, evolving into many species and dominating almost all regions of Earth; they include all gymnosperms and many angiosperms. Within the flowering plants, herbaceousness is a new phenomenon; all early angiosperms were woody perennials, but many plants have evolved to be herbs, foregoing the woody life style. At present, true secondary growth occurs in all gymnosperms and many dicots, but not in any ferns or monocots (Table 8.1).

TABLE 8.1	*Presence of Secondary Growth*
Modern ferns	Absent in all species
Gymnosperms	Present in all species
Dicots	Present in many species, but many other species are herbs
Monocots	Ordinary type is absent in all species, but some have anomalous secondary growth

As you study this, remember that a woody plant is a combination of primary and secondary tissues: The tips of the stems and roots, as well as the leaves, flowers, and fruits, are herbaceous and primary; only as portions of stems and roots become older do they begin to undergo secondary growth and become woody.

VASCULAR CAMBIUM

INITIATION OF THE VASCULAR CAMBIUM

The **vascular cambium** (plural: cambia) is one of the meristems that produce the secondary plant body (Fig. 8.2). In an herbaceous species, the cells located between the metaxylem and metaphloem of a vascular bundle ultimately stop dividing and differentiate into

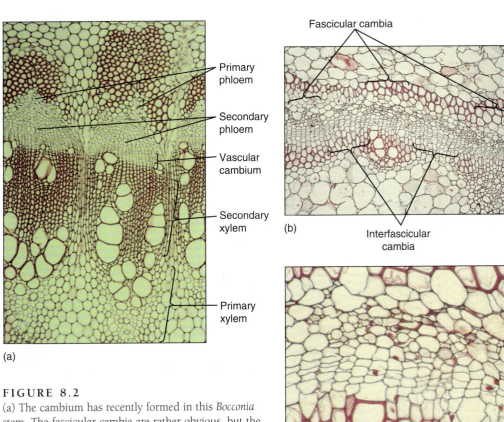

(a)

(b)

(c)

Fascicular cambia

Primary phloem

Secondary phloem

Vascular cambium

Secondary xylem

Primary xylem

Interfascicular cambia

FIGURE 8.2

(a) The cambium has recently formed in this *Bocconia* stem. The fascicular cambia are rather obvious, but the interfascicular cambia are difficult to see. They have produced more secondary xylem than secondary phloem (×180). (b and c) This cambium has only recently formed; it contains two or three layers of secondary xylem and one or two of secondary phloem. Three groups of primary xylem are visible, so you can tell where fascicular and interfascicular cambia arose (b, ×50; c, ×250).

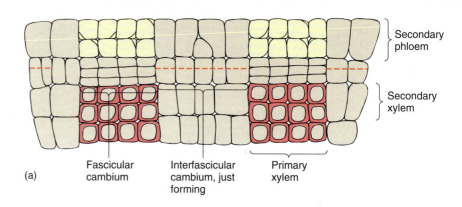

Secondary phloem

Secondary xylem

Fascicular cambium

Interfascicular cambium, just forming

Primary xylem

(a)

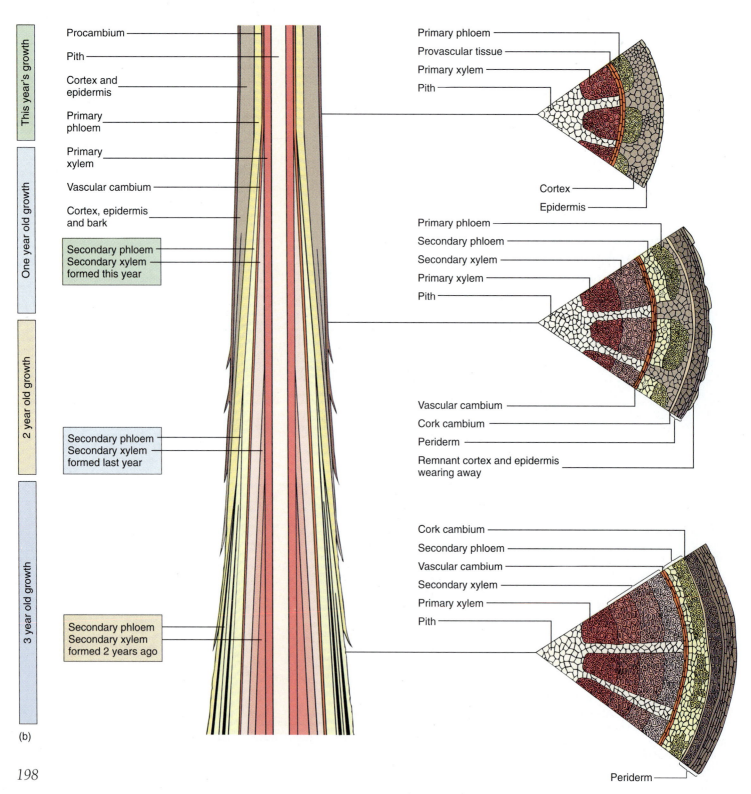

This year's growth

One year old growth

2 year old growth

3 year old growth

Procambium

Pith

Cortex and epidermis

Primary phloem

Primary xylem

Vascular cambium

Cortex, epidermis and bark

Secondary phloem
Secondary xylem formed this year

Secondary phloem
Secondary xylem formed last year

Secondary phloem
Secondary xylem formed 2 years ago

(b)

Primary phloem

Provascular tissue

Primary xylem

Pith

Cortex

Epidermis

Primary phloem

Secondary phloem

Secondary xylem

Primary xylem

Pith

Vascular cambium

Cork cambium

Periderm

Remnant cortex and epidermis wearing away

Cork cambium

Secondary phloem

Vascular cambium

Secondary xylem

Primary xylem

Pith

Periderm

◀ **FIGURE 8.3**
(a) Two vascular bundles and the parenchyma located between them. Some of the parenchyma cells have begun renewed cell division and constitute an interfascicular cambium. This zone of renewed mitotic activity is located between two fascicular cambia, so one full cambial zone will result. (b) The vascular cambium usually does not form until after a portion of shoot or root is several weeks or even many months old, so branch tips have only primary growth. Once a region is old enough, the vascular cambium forms and secondary xylem begins to accumulate to the inside of the cambium; secondary phloem accumulates to the exterior. At first, the vascular bundles are still recognizable. After several years of activity, considerable amounts of secondary xylem and phloem accumulate.

conducting tissues. But in a woody species, the cells located in this position never undergo cell cycle arrest; they continue to divide instead of maturing, and they constitute the **fascicular cambium** (Fig. 8.3a). In addition, some of the mature parenchyma cells between vascular bundles come out of cell cycle arrest and resume mitosis, forming an **interfascicular cambium** that connects on each side with the fascicular cambia. Once this happens, the vascular cambium is a complete cylinder. The terms "fascicular" and "interfascicular" are used only while the cambium is young; after 2 or 3 years the two regions are usually indistinguishable, and then only the term "vascular cambium" is used.

Vascular cambia must be extended each year. The tips of roots and stems initially contain only primary tissues, but at some time in the same season, a vascular cambium arises and that portion of the root or stem then contains both primary and secondary tissues (Fig. 8.3b). During the next growing season, the apical meristem extends the axis beyond this point; a new segment of vascular cambium forms within it and joins at its base to the top of the vascular cambium formed in the previous season. The vascular cambium within a tree consists of segments of distinct ages, those near the ground being oldest and those closer to the tips of the axes being younger.

Very rarely a vascular cambium forms in leaves that stay on a tree for many years, but just a tiny amount of secondary tissues, usually only secondary phloem without any secondary xylem, is formed in the midrib; the other veins contain only primary tissues. Secondary growth never occurs in flowers, fruits (except some fruit peels that are actually bark), or seeds.

Although a vascular cambium shares many features with an apical meristem, it is unique in certain aspects. It is a rather simple meristem in that it has only two types of cells, **fusiform initials** and **ray initials** (Figs. 8.4 and 8.5).

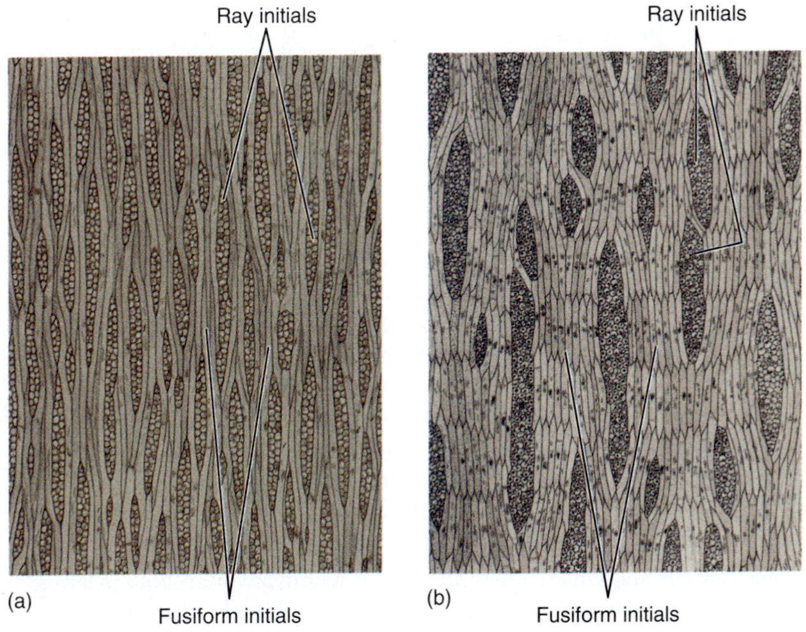

The old term for bundle is "fascicle." Fascicular cambium develops in the bundles, interfascicular cambium between them.

Ray initials

Ray initials

(a)

(b)

Fusiform initials

Fusiform initials

FIGURE 8.4
Tangential sections through a nonstoried vascular cambium of apple (*Malus sylvestris,* a) and a storied vascular cambium of black locust (*Robinia pseudoacacia,* b). Notice that the storied fusiform initials are extremely long and the ends are not aligned. In the storied cambium, the fusiform initials are much shorter and occur in horizontal rows. (*Courtesy of R. Evert, University of Wisconsin*)

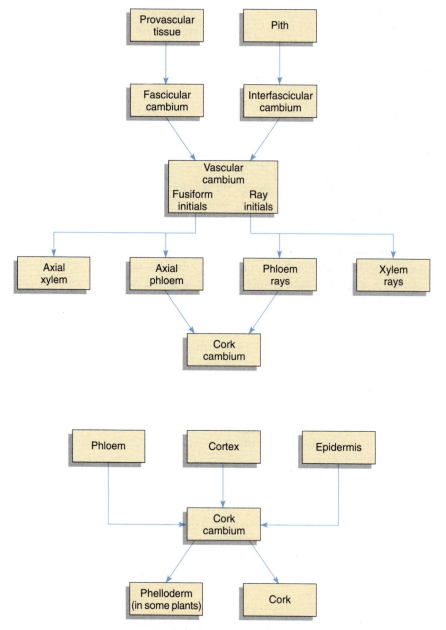

FIGURE 8.5
Cell lineages during secondary growth.

FUSIFORM INITIALS

Fusiform initials are long, tapered cells; typical lengths for fusiform initials are 140 to 462 μm in dicots and 700 to 8700 μm in gymnosperms. When a fusiform initial undergoes longitudinal cell division with a wall parallel to the circumference of the cambium (a **periclinal wall**), it produces two elongate cells (Fig. 8.6). One continues to be a fusiform initial, and the other differentiates into a cell of secondary xylem or secondary phloem. If the outer daughter cell remains a cambium cell, the inner cell develops into secondary xylem. But if the inner one continues as cambium, the outer cell differentiates and matures into secondary phloem. This orientation is constant: Wood never forms to the exterior of the vascular cambium, and bark never forms on the interior side. Regardless of which cell differentiates, one always remains as cambium. It is not known what factors determine which cell remains cambial and which differentiates, but within any year, both xylem and phloem are produced, almost always much more xylem than phloem.

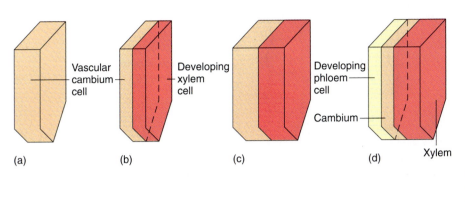

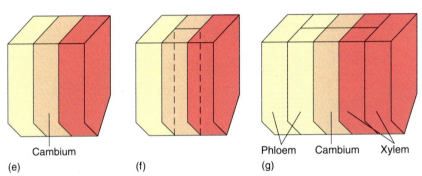

FIGURE 8.6

(a) The lower half of a fusiform initial before division (to simplify the diagram, the top half has not been drawn in). (b) Division by a periclinal wall results in two thin cells; the outer one remains a fusiform initial and the inner cell develops into secondary xylem. (c) Both cells enlarge to the size of the original cell. (d) The fusiform initial divides again; this time the outer cell matures as secondary phloem while the inner one remains a fusiform initial. (e) The cells grow back to the original size. (f) The fusiform initial divides by an anticlinal wall, resulting in two fusiform initials. (g) After the radial division in (f), a new row of cells is initiated in the secondary xylem and phloem.

Cambial cells produce narrow daughter cells, all of which enlarge during differentiation. Daughter cells located on the inner side, which mature into secondary xylem, increase greatly in diameter, causing the cambial cells to be pushed outward (Figs. 8.7 and 8.8). Because the cambium is a cylinder, such outward movement results in a cylinder of larger circumference. Vascular cambium cells must divide longitudinally by **anticlinal walls,** thereby increasing the number of cambial cells (Fig. 8.6f). Without anticlinal divisions, cambial cells would be stretched wider circumferentially and finally could not function (see Box 8.2).

Like apical meristem cells, fusiform initials have thin primary walls, and plastids are present as proplastids. After nuclear division, a phragmoplast forms and elongates toward the ends of the cell. The phragmoplast grows about 50 to 100 μm per hour, and cell division may take as long as 10 days in species with long fusiform initials, whereas the cell cycle may be as short as 19 hours in apical meristem cells of the same plant.

Anticlinal: perpendicular to a surface; here, the surface of the cambium. Periclinal: parallel to the reference surface.

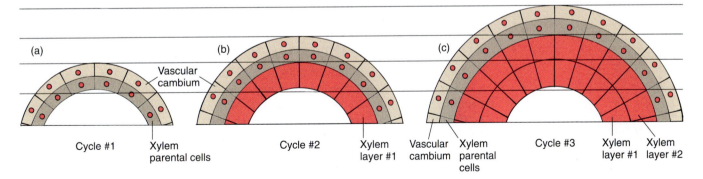

FIGURE 8.7

After vascular cambium cells divide, the progeny cells on the interior side become xylem parental cells. As all cells expand to their mature size, the cambial cells are pushed outward. The cambial cells divide again, depositing a new layer of xylem parental cells exterior to the xylem that just formed (xylem layer 1). The new xylem parental cells expand, pushing the vascular cambium farther outward. Cambial cells divide again. Notice that each xylem cell is formed in place; they are not pushed inward.

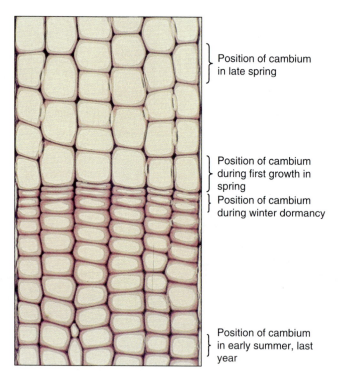

Position of cambium in late spring

Position of cambium during first growth in spring

Position of cambium during winter dormancy

Position of cambium in early summer, last year

FIGURE 8.8

Six rows of wood cells are visible in this transverse section of a pine trunk. All the cells of each row were produced by a single vascular cambium cell, so all the wood visible in this micrograph was produced by just six cambium cells. The cells at the bottom of the picture were produced when the cambium cells were located at that site; similarly, the cells in the middle of the picture were produced when the cambium exited there, having been pushed outward by the production of the cells at the bottom of the picture ($\times$ 150).

RAY INITIALS

Ray initials are similar to fusiform initials except that they are short and more or less cuboidal. They too undergo periclinal cell divisions, with one of the daughters remaining a cambial ray initial and the other differentiating into either xylem parenchyma if it is the inner cell, or phloem parenchyma if it is the outer cell. One of the most significant differences between fusiform and ray initials is that the elongate fusiform initials produce the elongate cells of the wood (tracheids, vessel elements, fibers) and the phloem (sieve cells, sieve tube members, companion cells, fibers). Ray initials produce short cells, mostly just storage parenchyma and, in gymnosperms, albuminous cells.

ARRANGEMENT OF CAMBIAL CELLS

Ray and fusiform initials are organized in specific patterns. Ray initials are typically grouped together in short vertical rows only one cell wide (a uniseriate ray; see Fig. 8.4), two cells wide (a biseriate ray), or many cells wide (multiseriate). Fusiform initials may occur in regular horizontal rows (a **storied cambium**) or irregularly, without any horizontal pattern (a **nonstoried cambium**). Storied cambia have envolved more recently than nonstoried cambia and occur in only a few advanced dicot species, for example, redbud and persimmon. The fusiform initials of storied cambia tend to be short, less than 200 μm long. The selective advantage of storied tissues is unknown.

Typically, a vascular cambium never has large regions of just fusiform initials or just ray initials. If anticlinal divisions result in many fusiform initials side by side, a central one may undergo transverse divisions and be transformed into a set of ray initials (Fig. 8.9). Likewise, if a group of ray initials becomes unusually broad, one or several central ones may elongate and be converted to a fusiform initial. The ratio of fusiform initials to ray initials within a species is quite constant and apparently under precise developmental control.

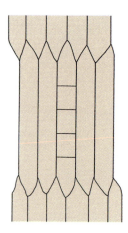

FIGURE 8.9

A fusiform initial can divide transversely and become a row of ray initials, after which all its derivatives differentiate as ray cells.

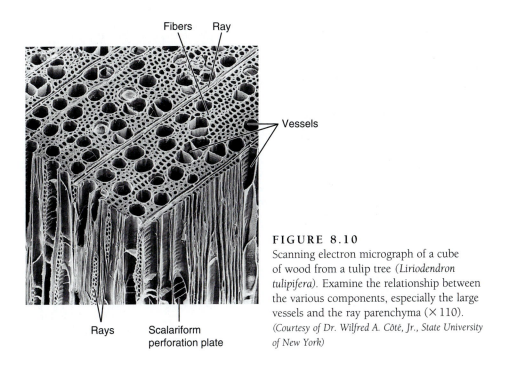

Fibers Ray

Vessels

Rays   Scalariform
        perforation plate

FIGURE 8.10
Scanning electron micrograph of a cube
of wood from a tulip tree (*Liriodendron
tulipifera*). Examine the relationship between
the various components, especially the large
vessels and the ray parenchyma (× 110).
*(Courtesy of Dr. Wilfred A. Côté, Jr., State University
of New York)*

Plant Biology Tutor CD-ROM

SECONDARY XYLEM

TYPES OF WOOD CELLS

All cells formed to the interior of the vascular cambium develop into secondary xylem,
known as **wood.** The secondary xylem contains all the types of cells that occur in primary
xylem but no new ones. Wood may contain tracheids, vessel elements, fibers, sclereids, and
parenchyma. The only real differences between primary and secondary xylem are the origin
and arrangement of cells. The arrangement of secondary xylem cells reflects that of the
fusiform and ray initials: An axial system is derived from the fusiform initials and a radial
system develops from the ray initials (Fig. 8.10; Box 8.1).

The axial system always contains tracheary elements (tracheids or vessels or both),
which carry out longitudinal conduction of water through the wood (Fig. 8.11). In many
species of dicots, the axial system also contains fibers that give the wood strength and
flexibility (Fig. 8.12; Table 8.2). Most commercially important dicot woods contain large

	Gymnosperms	Dicots*
TABLE 8.2 *Cell Types Present in Wood*		
Axial system		
Tracheids	Present	Present
Vessels	Absent (except in three groups)	Present
Fibers	Very rare	Present
Parenchyma	Very rare	Present
Radial system		
Ray parenchyma	Present	Present
Ray tracheids	Present	Absent

* Dicot wood is extremely variable; certain species lack some of these cell types,
other species have them all. The relative amounts of each vary greatly.

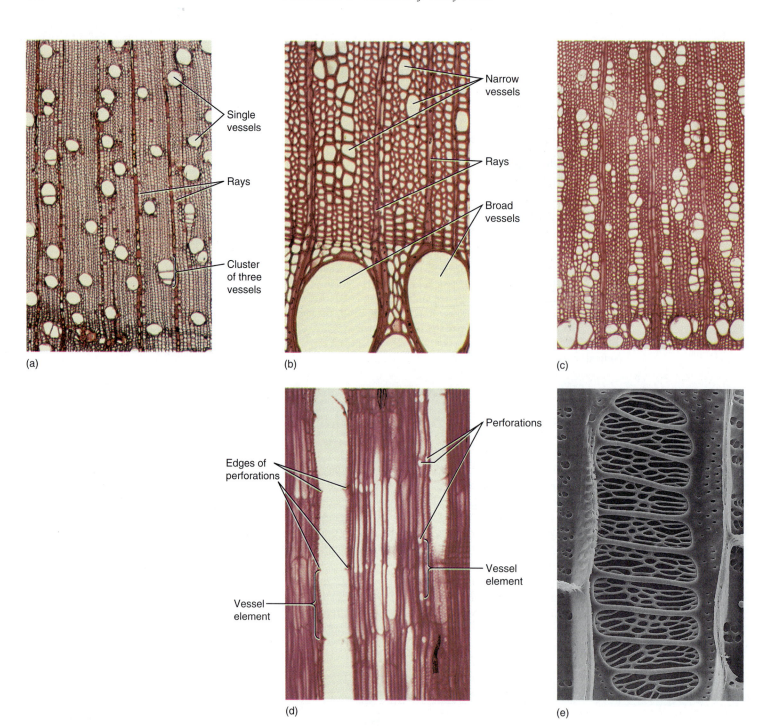

FIGURE 8.11

Vessels in wood. (a) In this wood, rather medium-size vessels occur in a mass of fibers. Each vessel occurs by itself or in a small cluster of two or three vessels (transverse section; × 40). (b) This wood has vessels of two very different sizes: extremely broad (at bottom of the micrograph) and very narrow (near the top). When water is plentiful, it can be transported in large quantities through the broad vessels. The narrow vessels are more effective at carrying water when the soil is dry and only a small amount of water is available to be transported (*Cotinus americana,* transverse section; × 150). (c) The vessels of the *Pistasia mexicana* are arranged in long radial groups. Because most vessels touch two or three other vessels, they can share water through the pits on their side walls (transverse section; × 50). (d) This longitudinal section passes through the lumen of one of the vessels. The front and back walls of the vessel elements were cut away during specimen preparation, so the vessel appears empty. The perforations are visible as short projections into the lumen (radial section; × 150). (e) Perforation between two vessel elements: The nearer cell has a scalariform perforation, and the farther cell has a reticulate perforation (× 700). *(Courtesy of J. Ohtani, Hokkaido University)*

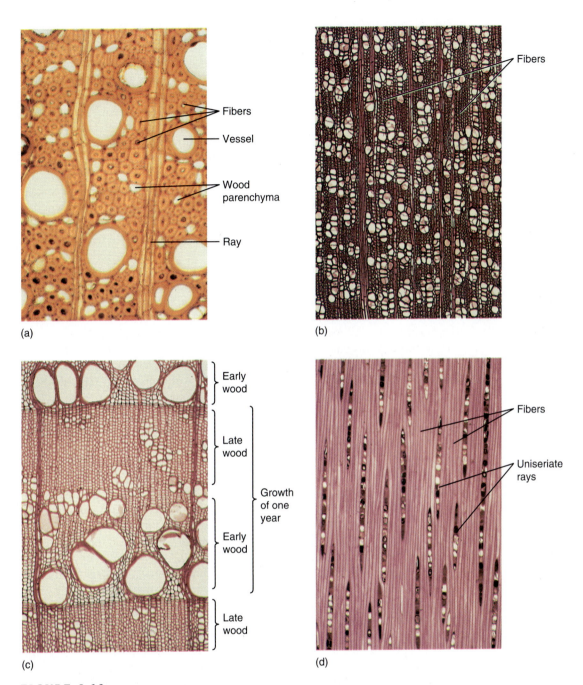

(a)

Fibers
Vessel
Wood parenchyma
Ray

(b)

Fibers

(c)

Early wood
Late wood
Growth of one year
Early wood
Late wood

(d)

Fibers
Uniseriate rays

FIGURE 8.12

Wood fibers. (a) The fibers of this wood have extremely thick walls. Also, the fibers are abundant, so this wood is quite strong (transverse section; × 150). (b) Vessels make up over half the volume of this wood, so fibers are not very abundant. Furthermore, the fiber cell walls are rather thin; this would not be a strong wood (transverse section; × 40). (c) In this wood, regions that have many large vessels and few fibers (called early wood, discussed later in this chapter) alternate with regions that have abundant fibers and only a few narrow vessels (late wood). Early wood provides maximum conduction capacity, whereas late wood provides strength (transverse section; × 50). (d) A longitudinal section through fiber-rich late wood. All fiber cells are long and narrow and have thick walls (tangential section; × 50).

amounts of fibers, making them strong, tough, and useful for construction. They are called **hardwoods,** a term now used for all dicot woods, even those that lack fibers or are very soft, such as balsa. Woods from conifers such as pines and redwoods have few or no fibers and thus have a softer consistency. These are known as the **softwoods,** even though in many instances they are actually much harder than many hardwoods.

Tracheary elements and fibers are elongate, as are the fusiform initials, but in many species, some immature cells undergo transverse divisions and differentiate into columns of xylem parenchyma (Fig. 8.13). Axial xylem parenchyma is important as a temporary reservoir of water; on cloudy or humid days and at night when leaves are losing little water, wood has a temporary surplus of conducting capacity and water is moved from roots into wood parenchyma and held there. When the air is hot and dry, leaves may lose water very rapidly; water can then be drawn from wood parenchyma if the soil is too dry to supply enough water. Conifers such as pine, cedar, juniper, and redwood have little or no axial parenchyma, so they have little reserve water. For them, tough, waxy, water-conserving leaves are selectively advantageous.

The complexity of the axial system of wood varies greatly. Most gymnosperms contain only tracheids in their axial systems; fibers and parenchyma cells are sparse or absent (see Table 8.2). Some dicots, especially the very relictual ones, have mostly just tracheids, but in most dicots all possible types of cells are present, and numerous cell-cell relationships are possible. Water-storing parenchyma may be immediately adjacent to vessels, or it can be arranged such that it never touches vessels. Fibers provide maximum strength if grouped together in masses, which is how they are usually arranged. If fibers are located around a vessel, their secondary walls reinforce the walls of the vessel and help it resist collapse. But the presence of fibers excludes parenchyma cells.

The radial system of xylem is much simpler. In flowering plants, it contains only parenchyma, arranged as uni-, bi-, or multiseriate masses called **rays** (see Figs. 8.10 and 8.14). Ray parenchyma cells store carbohydrates and other nutrients during dormant periods and conduct material over short distances radially within the wood. The two basic types of ray parenchyma cells are **upright cells** and **procumbent cells** (Fig. 8.14). At least in some plants, procumbent ray cells have no direct connection with axial cells, but upright cells do. The ray/axial interface can take many forms. If the upright ray parenchyma cell is adjacent to axial parenchyma, plasmodesmata occur. If the ray parenchyma is adjacent to an axial tracheid or vessel element, the tracheary element has pits in its secondary wall and the ray cell has very thin walls facing the pits. In early springtime, when trees such as maples are drawing on their nutrient reserves, the starch that had been stored in the upright cells is the first to be digested into sugar and passed into the axial tracheary elements for conduction to newly expanding buds, leaves, and flowers. Starch in procumbent cells is not digested until later and presumably must first be routed through upright cells for transfer to axial conducting cells.

In gymnosperms, xylem rays are almost exclusively uniseriate; they are multiseriate only if they contain a resin canal (Fig. 8.15a). In addition to ray parenchyma cells, they may contain **ray tracheids**—horizontal, rectangular cells that look somewhat like parenchyma cells but have secondary walls, circular bordered pits, and protoplasts that degenerate quickly after the secondary wall is completed (Fig. 8.15b).

ANNUAL RINGS

In regions with strongly seasonal climates, the vascular cambium is quiescent during times of stress, either winter cold or summer drought. But when quiescence ceases, the vascular cambium becomes active and cell division begins. At the same time, the new, expanding leaves are thin and delicate, and their cuticle is neither thick nor fully polymerized. Leaves like this lose water at a rapid rate, so trees need a high capacity for conduction at this time. The first wood formed is **early wood**, also called spring wood, and it must have a high proportion of wide vessels (Fig. 8.16a) or, in gymnosperms, wide tracheids (Fig. 8.16d). Later, the cuticle has thickened, transpiration is less, and large numbers of newly formed vessels are conducting rapidly. Wood produced at this time, called **late wood** or summer wood, can have a lower proportion of vessels. But the plant is a year older, it is larger and heavier, and it needs more mechanical strength to hold up the increased number of leaves and the larger branches. Late wood is stronger if it contains numerous fibers or, in gymnosperms, if it contains narrow, thick-walled tracheids. Finally, at the end of the growing season, the cambium becomes dormant again. The last cells often develop only as heavy fibers with especially thick secondary walls. In a tree with wood like that just described, it is

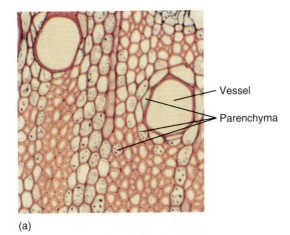

(a)

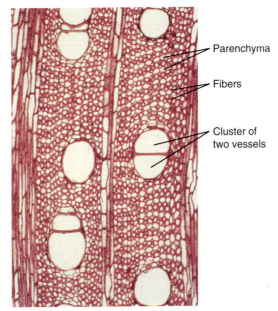

Parenchyma

Fibers

Cluster of
two vessels

(b)

Vessel

Parenchyma

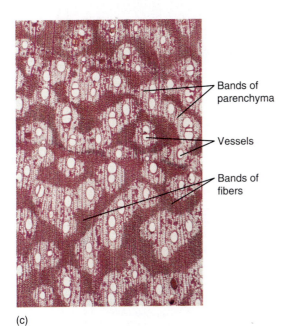

Bands of
parenchyma

Vessels

Bands of
fibers

(c)

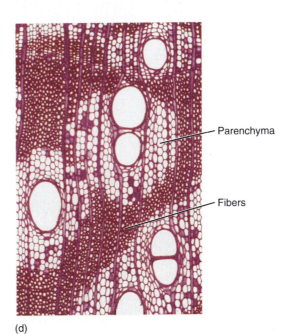

Parenchyma

Fibers

(d)

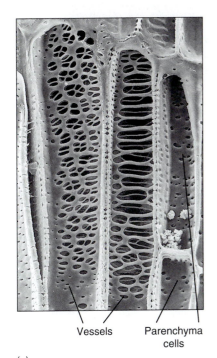

Vessels   Parenchyma
cells

(e)

FIGURE 8.13

Wood axial parenchyma. (a) To prepare wood for microscopy, it is often boiled to make it soft enough to cut into thin specimens; unfortunately this destroys the protoplasm of any parenchyma cells. This wood was not boiled, so the starch-filled parenchyma cells are easily visible. The large cells surrounding the vessels are also parenchyma, and they can release water to the vessels in times of water shortage (temporarily dry soil) (transverse section; ×150). (b) This wood has abundant axial parenchyma—the small cells with thin walls. The small cells with thick walls are fiber cells. Because this wood was boiled, it is difficult to be certain which is parenchyma (transverse section; ×50). (c and d) The axial parenchyma in this wood forms large bands; notice that all the vessels occur in the bands, surrounded by parenchyma. None is in the fiber masses. This parenchyma may act as a "water jacket" around vessels, absorbing excess water when water loss from leaves is low (cool nights) and releasing it when water loss is rapid (hot days) (transverse sections; c, ×15; d, ×50). (e) Axial parenchyma cells adjacent to vessels (×700). (*Courtesy of J. Ohtani, Hokkaido University*)

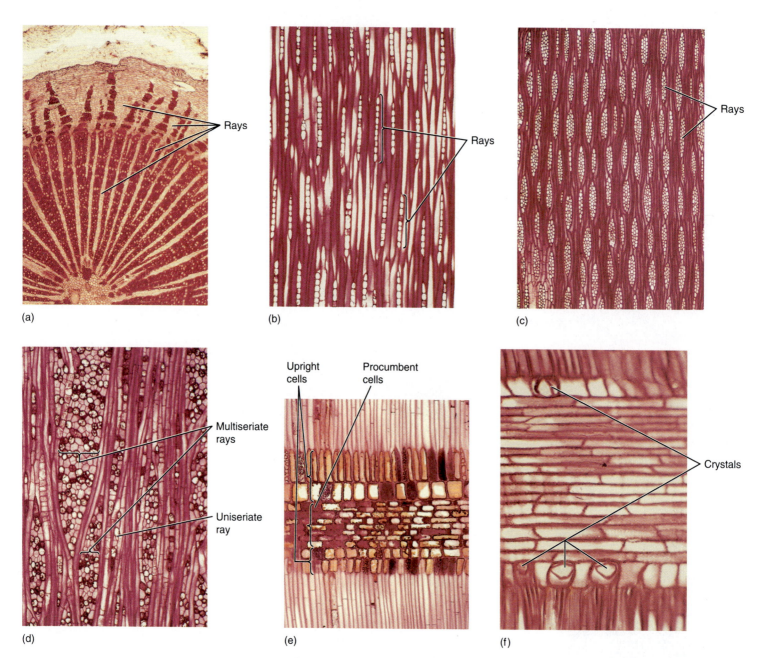

FIGURE 8.14

Wood rays. (a) The rays in this wood are extremely broad, so they are easy to see. Notice that each ray extends from the xylem into the phloem (× 15). (b) These rays are uniseriate, only one cell wide; they are rather short, only about ten cells tall. This is from late wood, so vessels are not present in this section (tangential section; × 50). (c) These rays are multiseriate, several cells wide. Over half the volume of this area of the wood is storage parenchyma, not conductive tracheary elements or strengthening fibers (tangential section; × 40). (d) This wood has giant multiseriate rays and small uniseriate rays (tangential section; × 50). (e) This ray has procumbent cells in the central part and upright cells along the edges (radial section; × 50). (f) Ray cells often contain crystals of stored material; four crystals are visible here in the cells along the margin. None occurs in the procumbent cells (radial section; × 150).

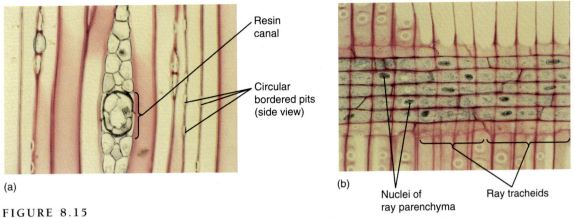

(a)

Resin canal

Circular bordered pits (side view)

(b)

Nuclei of ray parenchyma

Ray tracheids

FIGURE 8.15

(a) Conifer rays often contain secretory canals that produce resin (pitch), important in preventing insects from burrowing through the wood (tangential section; × 150). (b) The central ray cells are living ray parenchyma, with large nuclei that have been stained black. The procumbent cells along the top and bottom of the ray are ray tracheids that have circular bordered pits but no protoplasm when mature (radial section; × 150).

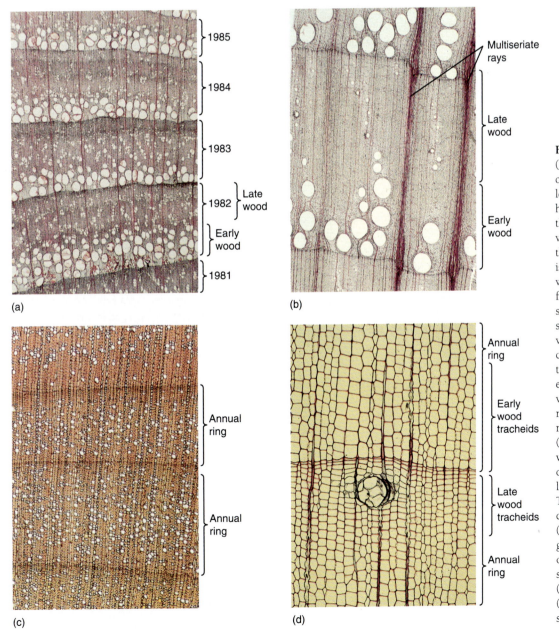

(a)

1985

1984

1983

1982 } Late wood

Early wood

1981

(b)

Multiseriate rays

Late wood

Early wood

(c)

Annual ring

Annual ring

(d)

Annual ring

Early wood tracheids

Late wood tracheids

Annual ring

FIGURE 8.16

(a) In this transverse section, portions of five annual rings are visible. The lowermost ring is oldest, the next higher ring was formed 1 year after the lowest ring, and so forth. The vascular cambium is located beyond the top of the photograph. The wood is ring porous, and vessel-rich early wood is easily distinguished from fiber-rich late wood (transverse section; × 50). (b) This wood is very strongly ring porous, with large vessels formed only when the cambium first becomes active. Later, the cambium produces fibers exclusively, except for a few rare, very narrow vessels. Two large multiseriate rays are visible, but most rays are narrow and uniseriate (transverse section; × 15). (c) This wood is diffuse porous, with vessels occurring rather uniformly in both late wood as well as early wood. The annual rings are not as conspicuous as in ring porous wood (transverse section; × 40). (d) Most gymnosperms have wood composed only of tracheids in their axial system, but they still have early wood (wide tracheids) and late wood (narrow tracheids) (transverse section; × 50).

FIGURE 8.17

The vascular cambium was almost perfectly circular in cross-section in this tree trunk. Note that the innermost annual rings, formed when the tree was a sapling, are wider than the outer rings—the tree grew more vigorously when it was younger. The dark region is heartwood and the narrow light region is sapwood; if the tree had lived a few years longer, several more of the innermost rings of sapwood would have converted to heartwood. Bark—secondary phloem—is also present. *(William E. Ferguson)*

easy to see late wood and early wood, the two together making up 1 year's growth, an **annual ring.**

An alternative arrangement exists for wood cells: In some species, vessels form throughout the growing season. Those produced in spring are neither more abundant nor obviously larger than those produced in summer. Because the wood of an annual ring has vessels located throughout it, it is said to be **diffuse porous** (Fig. 8.16c), whereas species with vessels restricted mainly to early wood are **ring porous.** Examples of trees with diffuse porous wood are yellow birch, aspen, sugar maple, and American holly; trees with ring porous wood include red oak, sassafras, and honey locust. In mild tropical climates, the cambium may remain active almost continuously, and the wood of one year is difficult to distinguish from that of another; annual rings are indistinct.

HEARTWOOD AND SAPWOOD

The center of a log is almost always darker in color than the outer wood, and it is usually drier and more fragrant (Fig. 8.17). The dark wood is **heartwood** and the lighter, moister outer region is **sapwood.** The different regions exist because vessels and tracheids do not function forever in water conduction; water columns break due to freezing, wind vibration, tension, wood-boring insects, and other factors. Once the water column breaks, the upper portion of the water column is drawn upward to the leaves and the lower portion falls toward the roots. After contact with the upper portion is broken, there is no means of pulling the water in the lower portion upward; vessels and tracheids in which this has occurred usually never conduct water again (see Chapter 12). Although only a few water columns break at any time, all water columns in an annual ring eventually snap. The conducting capacity of the wood is decreased, but more tracheary elements are produced by the cambium during the next year. However, a vessel is wide enough that a fungus can easily grow up through it; for tracheary elements that are not conducting, a mechanism that seals them off is selectively advantageous. The adjacent wood parenchyma cells push bubbles of protoplasm through the pits into a vessel, forming a plug, called a **tylosis** (plural: tyloses), completely across it (Fig. 8.18). This occurs repeatedly, and the vessel may become filled. These and other parenchyma cells, both axial and ray parenchyma, undergo numerous metabolic changes and produce large quantities of phenolic compounds, lignin, and aromatic substances that inhibit growth of bacteria and fungi. Ultimately, the parenchyma cells die. This heartwood is aromatic and decay resistant.

The outer, younger region, the sapwood, contains living parenchyma as well as water-filled tracheary elements; it is full of "xylem sap"—hence its name. It is gradually converted to inactive wood as its water columns break; then as the parenchyma cells become resistant to fungi and die, it is converted to heartwood. A new layer of sapwood is formed each year by the vascular cambium, and on average one annual ring is converted to heartwood each

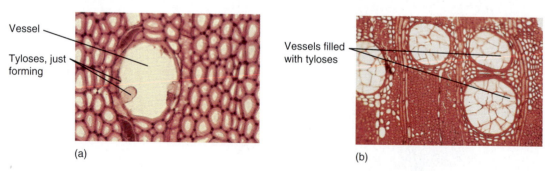

FIGURE 8.18

(a) Tyloses form as protoplasm from surrounding parenchyma cells pushes into the vessel. This was just beginning to form tyloses when the wood was collected for microscopy. Boiling destroyed the protoplasm (transverse section; × 250). (b) The vessels of this wood are completely occluded by tyloses; this sample came from a piece of heartwood (× 50).

TABLE 8.3 *Thickness of Sapwood in Dicot Trees*	
	Number of Xylem Rings in Sapwood
Catalpa speciosa (catalpa)	1–2
Robinia pseudoacacia (black locust)	2–3
Juglans cinerea (butternut)	5–6
Maclura pomifera (Osage orange)	5–10
Sassafras officinale (sassafras)	7–8
Aesculus glabra (Ohio buckeye)	10–12
Juglans nigra (black walnut)	10–20
Prunus serotina (wild black cherry)	10–12
Gleditsia triacanthos (honey locust)	10–12

year. Thus, whereas the heartwood becomes wider with age, the sapwood has a more or less constant thickness (Table 8.3). Of course this is not true of young stems and roots: Black walnut typically has wood that functions for 10 years before converting to heartwood. In a seedling or a branch only 9 years old, no heartwood is present yet.

In branches or trunks that are not vertical, gravity causes a lateral stress; if not counteracted, the branch would droop and become pendant. In response to such stress, most plants produce **reaction wood** (Fig. 8.19). In dicots, this develops mostly on the upper side of the branch and is known as tension wood. In a cross-section of such a branch, the growth rings are eccentric, being much wider on the top of the branch. Tension wood contains many special gelatinous fibers whose walls are rich in cellulose but have little or no lignin. These fibers exert tension on the branch, preventing it from drooping, or the tension wood may even contract, slowly lifting a branch to a more vertical orientation. Conifers form reaction wood that is located on the underside of the branch and is known as compression wood. It is enriched in lignin and has less cellulose; the growth rings are especially wide on the lower side of the limb.

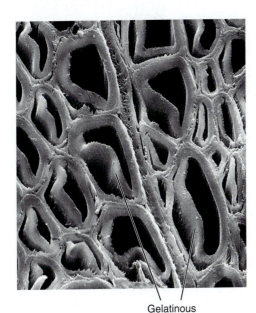

Gelatinous wall

FIGURE 8.19
Reaction wood in poplar *(Populus)*. These are called gelatinous fibers because they have a soft inner layer that is deformed when specimens are cut. Reaction wood is rich in gelatinous fibers (× 600). *(Courtesy of W. A. Côté, State University of New York)*

BOX 8.1 *Wood in Three Dimensions*

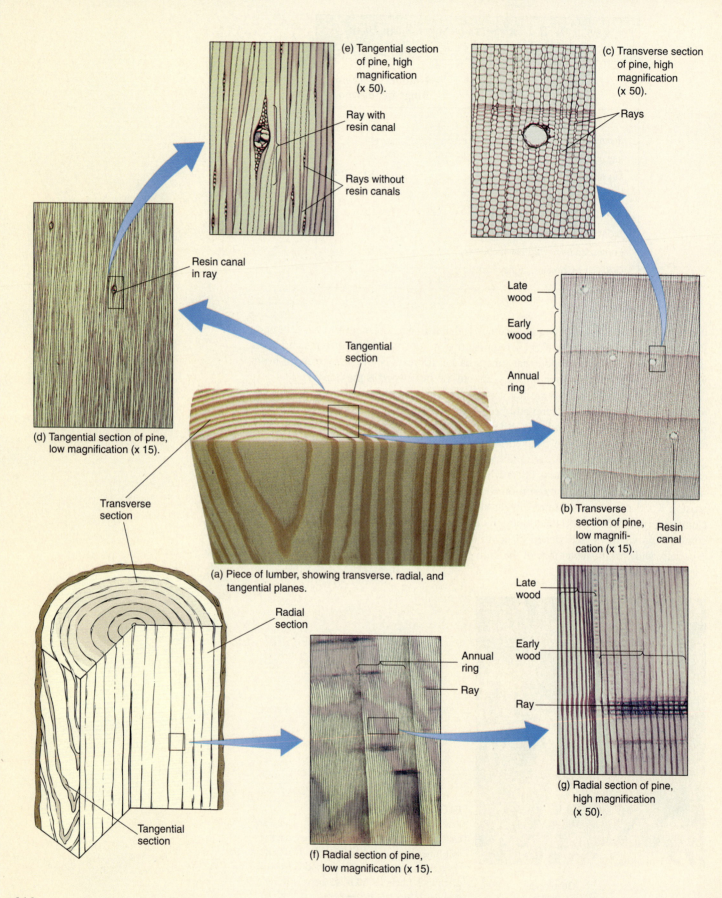

(e) Tangential section of pine, high magnification (x 50).

Ray with resin canal

Rays without resin canals

(c) Transverse section of pine, high magnification (x 50).

Rays

Resin canal in ray

(d) Tangential section of pine, low magnification (x 15).

Tangential section

Late wood

Early wood

Annual ring

Transverse section

(b) Transverse section of pine, low magnification (x 15).

Resin canal

(a) Piece of lumber, showing transverse, radial, and tangential planes.

Radial section

Late wood

Annual ring

Ray

Early wood

Ray

Tangential section

(f) Radial section of pine, low magnification (x 15).

(g) Radial section of pine, high magnification (x 50).

It is important to understand wood in all three dimensions, but that is not particularly easy. As you study these illustrations, think about how the cells are formed by the vascular cambium. Also try to picture how the various cells touch each other in three dimensions, which ones conduct water, which store water and nutrients, and so on. Pine wood is simpler, because its axial system contains only tracheids, so you need to think mostly about tracheids and rays. Oak wood is more complex, with vessels, fibers, and axial parenchyma all touching the rays at some point. For some serious practice of your skill at visualizing three-dimensional objects, imagine a vascular cambium existing on one tangential face: How are the new cells formed each year?

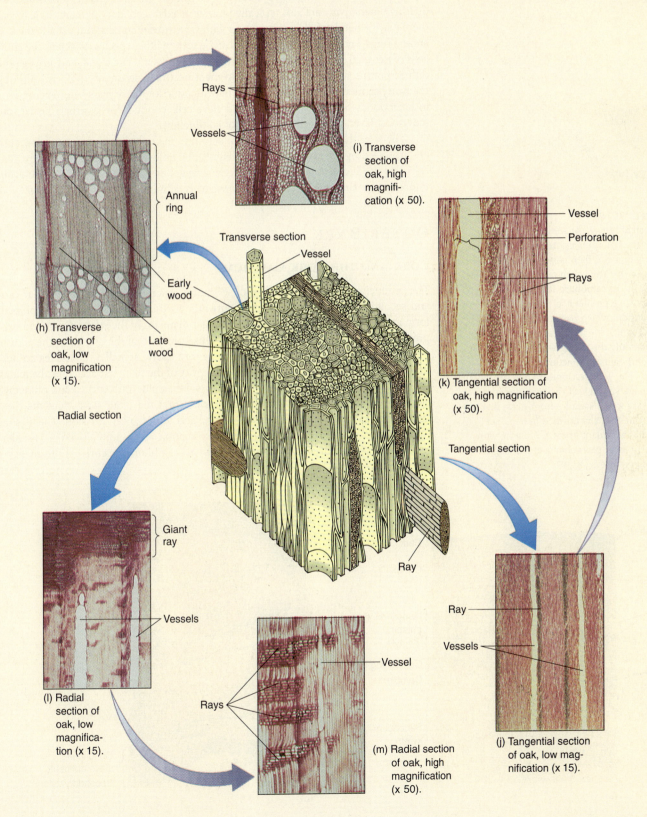

Rays

Vessels

(i) Transverse section of oak, high magnification (x 50).

Annual ring

Transverse section

Vessel

Early wood

Late wood

(h) Transverse section of oak, low magnification (x 15).

Radial section

Giant ray

Vessels

(l) Radial section of oak, low magnification (x 15).

Rays

Vessel

(m) Radial section of oak, high magnification (x 50).

Ray

Vessel

Perforation

Rays

(k) Tangential section of oak, high magnification (x 50).

Tangential section

Ray

Vessels

(j) Tangential section of oak, low magnification (x 15).

213

SECONDARY PHLOEM

Because secondary phloem is formed from the vascular cambium just as secondary xylem is, it too has an axial and a radial system (Fig. 8.20). The axial system is responsible for conduction up and down the stem or root; it contains sieve tube members and companion cells in dicots, or sieve cells in gymnosperms. In both groups of plants, fibers and nonconducting parenchyma are also typically present in axial secondary phloem. In some species, there may be bands of fibers alternating with sieve tube members, but usually these are not annual rings, and many other patterns can be seen in other species. Whereas the equivalent cells of the axial secondary xylem are arranged as early and late wood, and as ring porous or diffuse porous wood, no similar arrangement occurs in secondary phloem. Although the tracheary elements of secondary xylem may function for many years before being converted to heartwood, sieve tube members and sieve cells conduct for less than 1 year; only the innermost layer of phloem is capable of conduction.

The size, shape, and number of phloem rays match those of xylem rays because both are produced by the same ray initials (Fig. 8.21). Phloem rays consist only of parenchyma cells that are used for storage, as are xylem rays, but phloem rays seem to be even more important for this (Fig. 8.22). In gymnosperms, albuminous cells are ray cells.

OUTER BARK

CORK AND THE CORK CAMBIUM

The production and differentiation of secondary xylem cells cause the vascular cambium and secondary phloem to be pushed outward. As the youngest, innermost phloem cells form and mature, they contribute to the larger diameter of the stem or root and increased pressures acting on the outermost tissues. This requires that the tissues on the periphery of the plant either grow in circumference or be torn apart. Actually, the tissues do both, but both must be controlled (Fig. 8.20). The integrity of the plant surfaces must be maintained against invasion by fungi, bacteria, and insects if the epidermis, cortex, and phloem simply rupture. As circumferential stretching increases and the older sieve elements die, some storage parenchyma cells become reactivated and undergo cell division. This is similar to the activation of parenchyma cells during formation of the interfascicular vascular cambium, but in secondary phloem it results in a new cambium, the **cork cambium**, also called the **phellogen** (Fig. 8.23).

Plant Biology Tutor CD-ROM

FIGURE 8.20
(a) This stem transverse section has secondary xylem at the bottom, vascular cambium region, secondary phloem, then cortex and bark at the top. Notice that the xylem rays and the phloem rays meet at the cambium. Parenchyma cells of some rays have proliferated through cell division and expansion, which has prevented tearing of the bark (× 40). (b) In this secondary phloem of *Artabotrys,* only the youngest axial phloem (at the bottom) contains functional sieve tube members. The youngest phloem and the vascular cambium are so soft that they are often damaged while trying to cut samples from trees; that is why they are partially crushed here. Older phloem has abundant fibers alternating with bands of phloem parenchyma and collapsed sieve tube members (× 50).

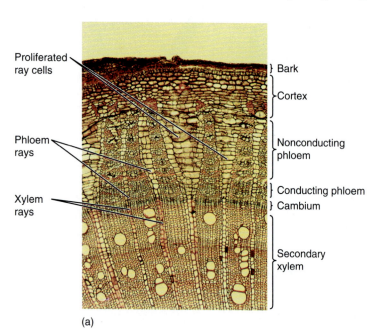

(a)

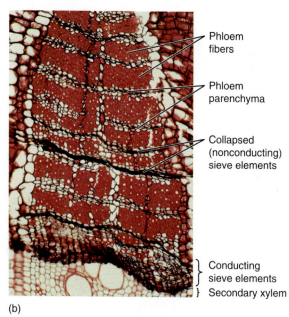

(b)

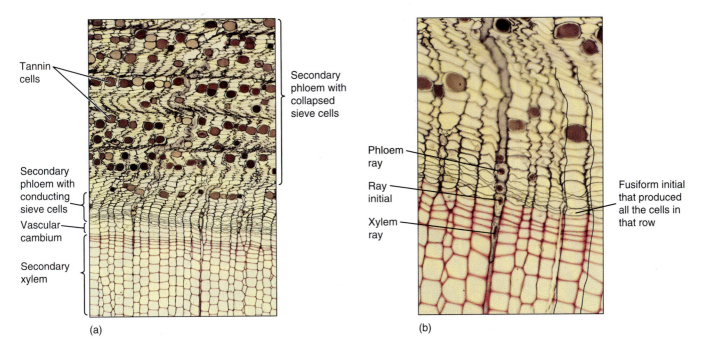

Tannin cells

Secondary phloem with collapsed sieve cells

Secondary phloem with conducting sieve cells

Vascular cambium

Secondary xylem

(a)

Phloem ray

Ray initial

Xylem ray

Fusiform initial that produced all the cells in that row

(b)

FIGURE 8.21

(a and b) The outermost, youngest wood and the bark of pine. As the sieve cells stop functioning and collapse, the phloem shrinks and becomes undulate. This causes the rays to become wavy. Notice that the phloem rays meet the xylem rays at the ray initials. Also, each row of sieve cells was produced by the same fusiform initial that produced the corresponding row of tracheids (transverse sections; a, × 50; b, × 150).

The cork cambium differs greatly from the vascular cambium in both structure and morphogenic activity. All its cells are cuboidal, like ray initials. After each division, the inner cell almost always remains cork cambium while the outer cell differentiates into a **cork cell** (also called a **phellem cell**). In a few species, the cork cambium may produce a cell or two to the inside which matures into a layer of parenchyma called **phelloderm.** The cork cambium, the layers of cork cells, and the phelloderm (if any) are known as **periderm.** Maturing cork cells increase slightly in volume; then the thin primary walls become encrusted with suberin, making them waterproof and chemically inert; then they die. Cell death is probably a critical part of maturation because after death the protoplasm breaks down, leaving nothing digestible or nutritious for an animal to eat. In many species, some of the cells deposit secondary walls and mature into lignified sclereids; these usually occur in layers that alternate with cork, resulting in a periderm that is both impervious and tough (Fig. 8.24). Because periderm is such an impermeable barrier, all plant material exterior to it, such as epidermis, cortex, and older secondary phloem, dies for lack of water and nutrients.

Mature cork provides excellent protection because it cannot be broken down and it is water repellent and relatively fire resistant.

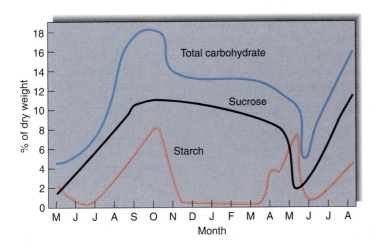

FIGURE 8.22

Carbohydrate accumulates in the bark of stems and roots during early summer while leaves are still present and photosynthesizing; it reaches its peak in September and October, then drops and remains steady through winter (December to April). It is released during the spring growth season, April and May. These data are for black locusts near Ottawa, Canada, where spring comes late. For plants that grow farther south, the spring release occurs earlier. *(Based on data published by D. Simonovitch, C. M. Wilson, and D. R. Briggs)*

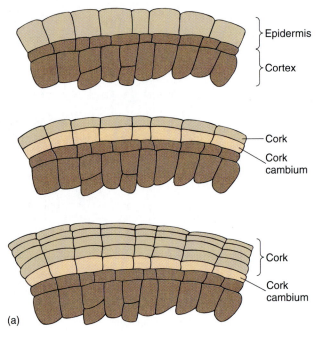

(a)

FIGURE 8.23
(a) The first cork cambium may form as epidermal cells resume mitotic activity. All cell divisions are by periclinal walls, and the inner cell continues as cork cambium while the outer cell differentiates into cork. (b) The hypodermal cells have just started to undergo cell division, resulting in the formation of a cork cambium. This is a young stem of geranium; notice the base of a trichome (×150). (c) Older stem of geranium; many layers of phellem (cork cells) have accumulated owing to cork cambium activity. Notice that the phellem cells are dead and empty. A trichome was present here also, and the formation of bark blocked transfer of nutrients to the trichome and other epidermal cells, killing them (×150).

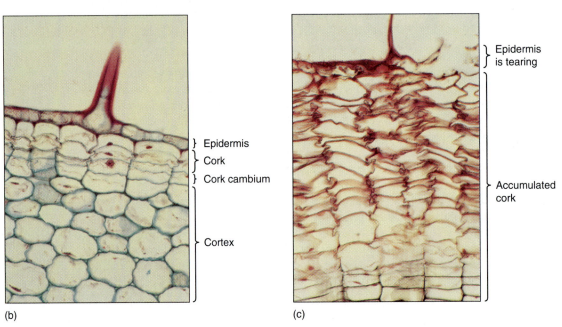

(b)

(c)

Periderm offers only temporary protection, because the root or stem continues to grow interior to it, pushing it outward and stretching it circumferentially. Unlike vascular cambium, cork cambium is typically short-lived; it produces cells for only a few weeks, after which all cells differentiate into cork cells and die. The layer of cork cells cannot expand much circumferentially, and after one or several years a new cork cambium must be formed in younger secondary phloem closer to the vascular cambium. These new cork cells act as a further barrier and also block water and nutrients from reaching any secondary phloem cells located between layers of cork cambium. In this fashion, several layers of cork can build up.

All tissues outside the innermost cork cambium comprise the **outer bark** (Fig. 8.25). All the secondary phloem between the vascular cambium and the innermost cork cambium is the **inner bark.** The amount of bark is quite variable from species to species; in some only a small amount of cork is formed, so the bark is thin and consists mostly of dead secondary phloem; in others, cork is produced in large amounts and becomes 3 or 4 cm

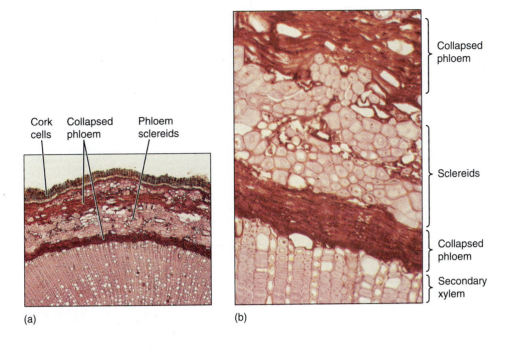

Cork cells   Collapsed phloem   Phloem sclereids

(a)

Collapsed phloem

Sclereids

Collapsed phloem

Secondary xylem

(b)

FIGURE 8.24

(a) A thick layer of uncollapsed cork cells is present on this stem. Many parenchyma cells in the cortex and secondary phloem have converted to sclereids (b), making the bark stronger and more protective. The next cork cambium will form deep in the secondary phloem, where numerous parenchyma cells are capable of becoming mitotically active again; the sclereids cannot resume cell division (transverse sections; a, × 40; b, × 150).

Outer bark

Inner bark

(a)   (b)   (c)   (d)

FIGURE 8.25

(a) As this pecan trunk increases in diameter, the bark is stretched and ultimately cracks. The deepest bark is the youngest, that on the surface the oldest. (b) The bark of maple peels off in large thin sheets because numerous cork cambia form close together and each is sheetlike. (c) The cork cells of sycamore contain many chemicals; as outer patches of bark peel away, fresh patches are exposed. When their pigments oxidize, they turn grey. Each cork cambium forms as a small patch; the size and shape of the cork cambia affect the nature of the bark. (d) The bark of cork oak (*Quercus suber*) becomes extremely thick and is composed mostly of phellem, with few sclereids. Cork oaks grown in Spain and Portugal provide most of our commercial corks for bottles. When the bark has become sufficiently thick, the outer bark is peeled away; after a few years, the bark is once again thick enough to harvest.

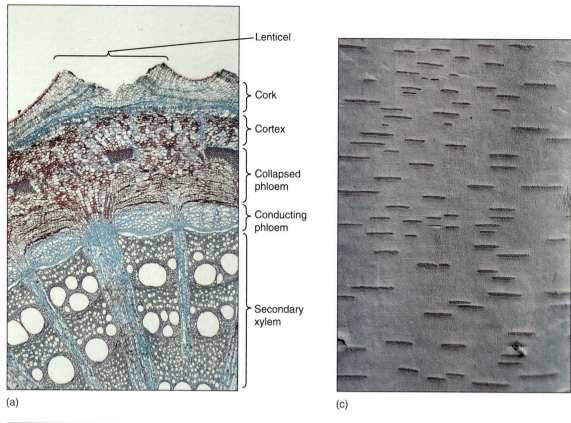

(a)

(c)

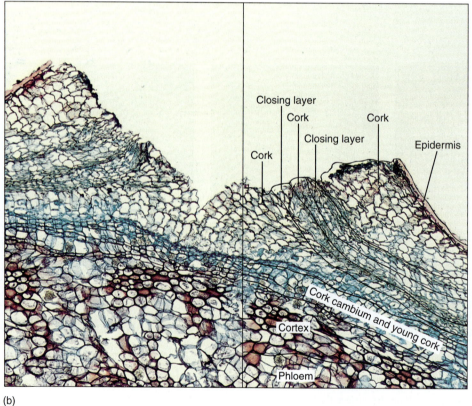

(b)

FIGURE 8.26

(a) Low magnification of stem of *Aristolochia,* with a large lenticel (× 15). (b) The rapid production of cork cells in the lenticel causes the epidermis to rupture. Although they cannot be seen here, small intercellular spaces allow oxygen to diffuse through the lenticel into the trunk or root (× 50). (c) The lenticels of birch (*Betula*) become wider as the trunk increases in circumference. Occasionally they divide into two lenticels when the central cork cambium cells begin producing compact cork instead of aerenchymatous cork.

thick. Although it is usually not obvious, bark is continuously falling off the tree, but it does not accumulate at the base because wind and water carry it away.

LENTICELS AND OXYGEN DIFFUSION

The impermeability of cork has negative as well as advantageous consequences. While it keeps out pathogens and retains water, it also blocks absorption of oxygen, interfering with the respiration of the sapwood, vascular cambium, and inner bark. Bark becomes permeable to oxygen when cork cambium produces cork cells that become rounded as they mature. Because rounded cells cannot fit tightly together, intercellular spaces through which oxygen diffuses penetrate the cork layer. These regions of aerenchymatous cork are **lenticels** (Fig. 8.26). When a new cork cambium arises interior to this one, it too forms a lenticel in the same place; the outer and inner lenticels are aligned, permitting oxygen to penetrate across all layers of the bark. Lenticel-producing regions are more active than adjacent regions that produce only ordinary impermeable cork; consequently lenticels contain more layers of cells and protrude outward. In species that have smooth bark, even small lenticels are easy to identify. On plants that have thick rough bark, lenticels can be almost impossible to see, but generally they are located at the bases of the cracks in the bark. When the bark of cork oak is made into bottle corks, it is necessary to cut them so that the lenticels do not run from the top of the cork to the bottom.

INITIATION OF CORK CAMBIA

The timing of initiation of the first cork cambium is far more variable than that of the vascular cambium. In some species, the first cork cambium arises before a twig or root is even 1 year old. On stems this is often detectable as the surface color changes from green to tan. In other species, the first cork cambium forms only when that region is several years old; until then, the epidermis and cortex are retained. Epidermises more than 40 years old have been reported. The first cork cambium may arise in a number of tissues: epidermis, cortex, primary phloem, or secondary phloem. Subsequent cork cambia may form shortly afterward, sometimes in the same season, but usually a year or two later. If the growth in diameter is slow, new cork cambia may arise at intervals of as much as 10 years. These later cork cambia usually form deep in the secondary phloem.

The first bark on young stems usually differs from bark formed when the stem is older. If the first cork cambium arises by reactivation of epidermal cells, the first outer bark contains only periderm and cuticle and is very smooth. If the first cork cambium arises in the cortex, the first outer bark contains periderm, cortex, and epidermis; this too is smooth and contains any cortical secretory cells that were present. As the first bark is shed and later cork cambia arise in the secondary phloem, they produce an outer bark that contains only cork and phloem. The nature of these later barks depends greatly on the cell types present in the trapped secondary phloem: Fiber cells produce fibrous, stringy bark, sclereid-filled phloem produces hard bark, and so on. It is not unusual for the bark of a young tree to be dramatically different from the bark it will have when it is older (Fig. 8.27).

SECONDARY GROWTH IN ROOTS

The roots of gymnosperms and woody dicots undergo secondary growth, as do the stems. A vascular cambium arises just like the interfascicular cambium, when parenchyma cells located between the primary xylem and primary phloem become active mitotically, as do pericycle cells near the protoxylem (Fig. 8.28). The new vascular cambium has the same star shape as the primary xylem, but it soon becomes round as the cambium in the sinuses of the primary xylem produces more secondary xylem than do the regions of cambium near the arms of protoxylem. Consequently, some portions of the cambium are pushed outward more rapidly than others. When a circular cambium is achieved, the unequal growth stops, and all parts grow at similar rates.

FIGURE 8.27
This trunk of *Casuarina* (red beefwood) still has fragments of its first bark, but the underlying newer bark is also visible. Young branches have only the first type of bark, older parts of the trunk only the second type.

Delayed formation of bark is common in plants that depend on the cortex chlorenchyma for much of their photosynthesis, as cacti do.

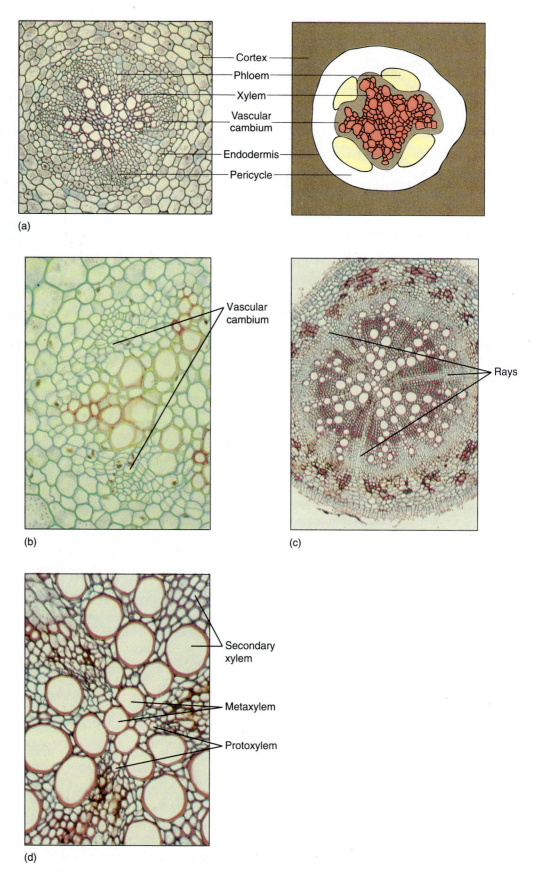

FIGURE 8.28

(a and b) Young roots. The vascular cambium of a root such as this baneberry arises as an undulate cylinder located around the xylem, interior to the phloem (a, × 50, b, × 150).
(c and d) Older roots. The organization of root wood is similar to that of stem wood: an axial system and a set of rays. Notice in (c) that there are three large rays; the high magnification of (d) reveals that these rays are aligned with the protoxylem of the primary xylem (c, × 40; d, × 150).

BOX 8.2   *Geometry, Structure, and Function*

The vascular cambium is an excellent tissue for analyzing some of the relationships between geometry, structure, and function. First consider how geometry affects structure. As the secondary xylem accumulates, the vascular cambium is pushed outward. In a cylindrical axis, for each x μm of added diameter of xylem, the vascular cambium must increase $x \pi$ μm in circumference ($C = \pi d$). Because the cambium has height, its entire area is increased. The automatic geometric consequences of this are that (1) the area of each cambial cell increases while the number of cells remains constant; or (2) the area of each cell remains constant but the number increases by means of anticlinal cell divisions; or (3) the cells increase both in number and in area. In the common type of vascular cambium, the size and area of the cells remain relatively constant, but the number of cells increases. This is obviously a successful, functional mechanism because trees can attain truly amazing circumferences, some being more than 20 m around. Their cambia contain billions of cells, yet secondary growth proceeds without problems.

A different type of vascular cambium existed in some species that are now extinct: *Paralycopodites*, *Sigillaria*, *Stigmaria*, and *Sphenophyllum* (see Figs. 23.19, 23.20, and 23.29) had fusiform initials that were unable to divide by anticlinal walls. As the circumference of the plant increased, the fusiform initials became larger by becoming longer, and as the tips of the cells elongated they intruded farther between the cells beside them. As this continued, the number of cells present at any given cross-section increased, although the total number of cells remained constant (see figure). This mechanism probably was not optimal, because as the fusiform initials gradually extended,

longitudinal cell division must have slowed and perhaps become more difficult. In long cells, the cell plate and the wall-synthesizing organelles are far removed from the nucleus by the time they reach the tips of the cell. There must be a practical limit on the length of a cell whose predominant activity is cell division, and it is reasonable to suspect that this limit adversely affected prolonged secondary growth of these plants.

Now consider the relationship between function and structure in the common type of vascular cambium found in woody dicots. The vascular cambium contains both short cells (ray initials) and long cells (fusiform initials), but no evidence indicates that this is necessary for the working of the vascular cambium itself. More likely the structure of the wood is the key: An axial system composed of elongate cells is obviously necessary for secondary xylem to conduct and to act as a mechanical skeleton for the plant. The rays are also important for storage and lateral conduction, for which

short or horizontal shapes are superior. In a hypothetical vascular cambium in which all the cells are long, rays theoretically could be produced as all the derivatives of one patch of fusiform initials undergo several transverse divisions. But because longitudinal division in elongate cells is probably not easy, all longitudinal divisions would be difficult in a vascular cambium composed only of long cells.

A different type of hypothetical vascular cambium could have all short cells, making all longitudinal divisions easy. But with such a meristem, all the derivatives that must be elongate (tracheary elements and fibers) would have to undergo extensive elongation. Although animal cells can easily intrude between other cells and even migrate through a tissue as they grow, plant cells encased in walls cannot easily do so. Typical structure is therefore optimal, short ray initials producing the short derivatives and long fusiform initials producing the elongate components.

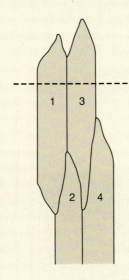

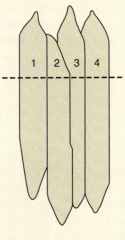

As the fusiform initials of *Sphenophyllum* elongated, the number of cells at any level increased even though the total number of cells remains constant.

The root vascular cambium contains both ray and fusiform initials, and in many cases, wood produced in the root is quite similar to that of the shoot, having sapwood and heartwood and being ring porous or diffuse porous as the stem is. Typically, however, the wood of roots is not identical to that of stems of the same plant, and they may be totally dissimilar when the conductivity requirements of root and stem differ. For example, many cactus roots are extremely long, nonsucculent, cablelike structures. After a brief rain, the numerous root tips absorb water from a large region of surface soil. Because the soil can dry within hours after such a rain, water must be conducted rapidly and in large quantity into the succulent cactus body where it can be stored. Once in the body, however, conduction requirements are totally different; the body may be quite small and consist mostly of parenchyma, with very little wood and just a few small vessels. The water is stored so effectively that most conduction is probably by diffusion through the cytoplasm rather than by the xylem. Furthermore, the narrow roots are strengthened by numerous wood fibers, whereas the small succulent shoots are strengthened only by turgor. In the stem the rays are gigantic, much more voluminous than the axial vessels, providing a large region for water storage. In the roots, which do not store water, rays are small and narrow and consist mostly of sclerenchyma.

Perennial roots also form a bark; the first cork cambium usually arises in the pericycle, causing the endodermis, cortex, and epidermis to be shed. The cork cambium produces cork cells to the outside, forming a protective layer, and in some species a layer or two of phelloderm as well. Lenticels also occur, being especially prominent near lateral roots. The bark on roots may be similar to that on the stem of the same plant, but in some species there are significant differences, again related to the differing metabolisms and microhabitats of the two organs.

Several mechanisms exist by which the storage capacity of a woody root, such as the carrot, can be increased. The ray parenchyma of the secondary xylem offers considerable volume, and storage capacity is increased if the rays become larger. Axial parenchyma in the wood can also be used, and in many storage roots the wood is almost pure parenchyma.

ANOMALOUS FORMS OF GROWTH

ANOMALOUS SECONDARY GROWTH

The development, cellular arrangement, and activity of the vascular cambia in gymnosperms and most woody dicots are remarkably similar. There are alternatives to this type of vascular cambium, and an analysis of the consequences of various types of arrangements or activities can reveal a great deal about secondary growth. Because these cambia produce secondary bodies that differ from the common type, their growth is called **anomalous secondary growth.**

Roots of Sweet Potatoes. In sweet potatoes (*Ipomoea batatas*), the amount of storage parenchyma is increased dramatically by an anomalous method of secondary growth.

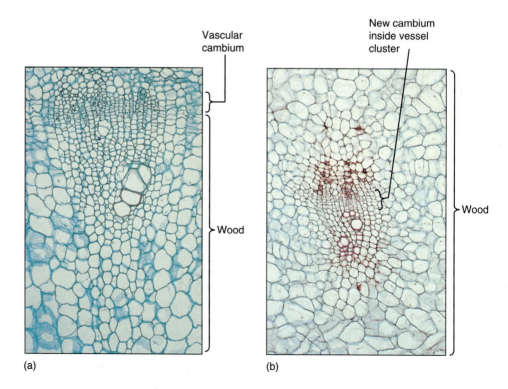

FIGURE 8.29

(a) The part of a radish that we eat is the wood of the storage root. It does not look like wood at first glance. The wood is mostly parenchyma with a few vessels and no fibers. This anomalous wood is well-adapted for the rapid production of, and conduction into, long-term storage tissue (× 40). (b) In the wood of beets (again, the part that we eat), additional cambia arise within clusters of vessels (× 40).

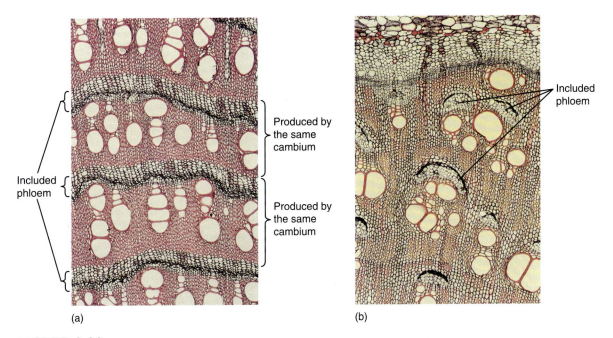

FIGURE 8.30

(a) This stem of *Iresine* shows bands of secondary xylem alternating with bands of included phloem. Each band of xylem and the phloem just external to it were produced by one cambium; the next set was produced by a different cambium (× 40). (b) *Bougainvillea* also has included phloem, but it occurs in patches, not complete bands (× 40).

Numerous vascular cambia arise, not around the entire mass of primary xylem, but around individual vessels or groups of vessels (Fig. 8.29). The cambia act normally, except that the xylem and phloem produced are almost purely parenchyma. New vessels may also be surrounded by another new cambium, and the process is repeated. As the sweet potato becomes quite large, it may contain hundreds of cambia of various ages; the secondary tissues are an irregular matrix of parenchyma, a few sieve tubes, some vessels, and vascular cambia.

What is the selective advantage of producing so many vascular cambia instead of just one? It may be that rate of cell production is important. Because the root must become large very quickly, having just one vascular cambium may be too slow. Multiple vascular cambia all functioning simultaneously speed up the production of storage capacity.

Included Phloem. In several dicots, a vascular cambium of the common type arises and produces ordinary secondary xylem and phloem. But after a short period, the cambium cells stop dividing and differentiate into xylem, and there is no cambium. But cells in the outermost, oldest secondary phloem become reactivated and differentiate into a new vascular cambium that acts just like the first, producing ordinary secondary xylem and phloem, then differentiating and ceasing to exist. Notice in Figure 8.30 that because the second cambium arose in the outermost phloem, the xylem it produces is located exterior to the phloem of the first cambium. From interior to exterior there is first xylem, first phloem, second xylem, second phloem. Then a new vascular cambium arises in the outermost second phloem, and so on. This type of secondary phloem, located between two bands of xylem, is **included phloem.** The selective advantage may be protection of the phloem from insects and other pests by one to several layers of wood. Also, the tissue relationships between xylem and included phloem differ from those between xylem and ordinary phloem, but the advantages of this have not yet been studied.

Included phloem is relatively rare but appears to have evolved several times.

Unequal Activity of the Vascular Cambium. In ordinary growth, all areas of the vascular cambium have about equal activity, so the stem or root is round in cross-section. But in some species of *Bauhinia* and certain other woody vines, two sectors of the cambium are

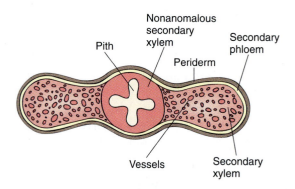

FIGURE 8.31

The cambium on two opposite sides of this *Bauhinia* is rapidly producing cells, many of which differentiate into wide vessel elements. The two alternate portions of cambium produce very few cells, and all of them mature without much expansion.

very active while two are almost completely inactive (Fig. 8.31), so the stem grows outward in two directions but remains thin in the other two and soon becomes a thin, flat, woody ribbon. The selective advantage of this may be related to the relationship between conductivity and flexibility. As an ordinary round stem becomes wider in all directions, its conducting capacity increases but its flexibility decreases. As the *Bauhinia* stem becomes wider, its conducting capacity increases but its flexibility remains about the same. For many vines, flexibility is selectively advantageous, and this type of anomalous secondary growth is adaptive.

Secondary Growth in Monocots. None of the monocots has secondary growth like that in dicots, but some do become treelike and "woody," such as Joshua trees, dragon trees, and palms. The first two groups undergo a process of anomalous secondary growth; the palms have an unusual type of primary growth.

In Joshua trees (some members of the genus *Yucca;* Fig. 8.32) and dragon trees (in the genus *Dracaena*), a type of vascular cambium arises just outside the outermost vascular bundles (Fig. 8.33) which originates from cortex cells in the same manner that the interfascicular vascular cambium arises in dicots. This cambium, however, produces only parenchyma; conducting cells are completely absent. Some of the parenchyma cells undergo rapid division and produce columns of narrow cells that differentiate into **secondary vascular bundles** containing xylem and phloem. The outermost cells of each bundle develop into fibers with thick secondary walls. The parenchyma cells that do not divide like this form a secondary ground tissue, the arrangement of which is almost identical to that of primary tissues. They are "woody" because of the fibers, and they have more conducting capacity and greater strength each year, so branching is feasible.

UNUSUAL PRIMARY GROWTH

Palm trees are unusual in that their trunks do not taper at the tips like those of dicot or gymnosperm trees, and they do not branch. The palm trunk is all primary tissue consisting of vascular bundles distributed throughout ground tissue, as described in Chapter 5; each bundle contains only primary xylem and primary phloem. A vascular cambium never develops, and true wood and secondary phloem do not occur; the trunk does not grow radially. The trunk is hard and "woody" because each vascular bundle is enclosed in a sheath of strong, heavy fibers.

A palm seedling does not have a full set of leaves and a wide trunk. For the first few years of life, the palm trunk becomes wider and the number of leaves increases. This happens without secondary growth because during the seedling years, palms produce numerous adventitious roots from the base of the short stem. Each root adds extra vascular bundles, and the portion of stem above each new root can have that many more bundles

FIGURE 8.32

Joshua tree *(Yucca brevifolia)* is an arborescent monocot that belongs to the lily family. Because it has secondary growth, even though of an unusual type, its ability to conduct increases, and both branching and increased numbers of leaves are feasible without adventitious roots of the type necessary for screwpine (see Fig. 7.18).

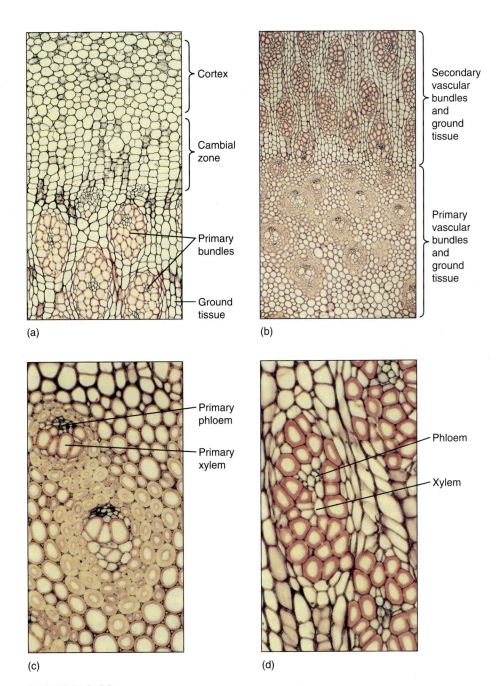

FIGURE 8.33
Secondary growth in arborescent monocots. (a) The broad zone of the vascular cambium region and the immature secondary ground tissues are visible. The bundles are secondary, and a secondary cortex also forms ($\times 50$). (b) The inner part of the stem, with the primary vascular bundles and ground tissue on the bottom and the secondary bundles at the top ($\times 15$). (c) Primary bundles ($\times 50$). (d) Secondary bundles ($\times 50$).

than it does below the root. For example, if the stem has 100 bundles at one point and if just above this it produces five adventitious roots that each have eight bundles, then above these roots the stem can have 140 bundles. This increase in width and addition of adventitious roots in palms is establishment growth, a form of primary growth. At some point this process ceases, no new adventitious roots are established, and the conducting capacity is set for the lifetime of the plant. This same type of primary growth occurs in other monocots with extremely broad stems such as corms and bulbs.

PLANTS & PEOPLE

DENDROCHRONOLOGY—TREE RING ANALYSIS

The amount of wood produced by the vascular cambium is closely correlated with climate. Species that occur in extremely harsh climates produce very little wood, often only a single layer of vessels each year. Slow growth such as this occurs in cold regions at high altitudes or high latitudes, or in hot, dry desert zones. In the tropics, however, the amount of growth each year can be large, with each annual ring being several millimeters wide and consisting of 50 to 100 layers of new cells. Even within a single plant, the vascular cambium produces more wood in an optimal year than it does in a year with poor temperatures or too little rain. This can be seen easily in the varying width of tree rings.

Because annual rings reflect the climate so accurately, they are used for many critical studies. Starting from the most recent, outermost ring, it is possible to count toward the center of the tree, encountering older and older rings. Not only can the age of the tree be determined, but climatic fluctuations can be inferred from the width of the rings. The oldest living trees of a region can be studied for climatic changes that occurred during the lifetime of the tree, which in some individuals is more than 5000 years. However, the analysis of past climates can be extended beyond the age of the oldest living trees. Historical or archaeological sites may contain houses, bridges, or ships constructed with large wooden beams. Some of the outermost rings of those beams may match the patterns of the inner rings of living trees. By counting back and finding the age of that set of rings in the living wood, we can determine how old that portion of the beam is and therefore when the house, ship, or bridge was constructed. This technique has been valuable not only in dating ancient settlements but also in establishing whether people lived in times of good or poor climate.

SUMMARY

1. Cells and tissues produced by apical meristems are primary tissues, whereas those produced by the vascular cambium and cork cambium are secondary tissues.

2. The vascular cambium contains fusiform initials that produce the elongate cells of secondary xylem and phloem and ray initials that produce ray cells.

3. The secondary xylem contains the same types of cells as the primary xylem, but the arrangement differs. Annual rings with early wood and late wood are usually present.

4. In an old trunk, branch, or root, the central xylem is heartwood; it is dry and the parenchyma has died. The outer, younger xylem is sapwood; it is involved in water conduction and its parenchyma is alive.

5. Cork cambium produces resistant cork cells; the cork and all exterior tissues are the protective outer bark. Lenticels permit oxygen diffusion into the organ.

6. Palms appear woody because their numerous vascular bundles contain many fibers. Certain other monocots have anomalous secondary growth. A vascular cambium produces secondary vascular bundles.

IMPORTANT TERMS

annual ring	included phloem	reaction wood
anticlinal wall	late wood	ring porous
bark	lenticel	sapwood
cork	periclinal wall	secondary phloem
cork cambium	periderm	secondary tissues
diffuse porous	phellem	secondary xylem
early wood	phellogen	softwood
fusiform initial	primary tissues	tylosis
hardwood	ray	vascular cambium
heartwood	ray initial	wood

REVIEW QUESTIONS

1. Which groups of plants have secondary growth? Which never do?

2. What are the two types of cells in the vascular cambium? Can one type be converted into the other?

3. What types of cells are derived from fusiform initials? What types of cells are derived from ray initials? Is it theoretically possible to have a vascular cambium without ray initials or without fusiform initials?

4. What changes occur as sapwood is converted to heartwood?

5. In a cross-section of a tree, where are the oldest growth rings—in the outer region or nearer the pith? Where is the oldest secondary phloem —near the outside of the tree or near the cambium?

6. In which tissues does the first cork cambium form? When does it usually arise? In which tissues do later cork cambia form?

7. Why can a monocot like an iris branch and increase its number of leaves? Is the fact that the shoot is a rhizome with adventitious roots important? Is water transported from one end of the shoot to the other?

BotanyLinks .net Questions

www.jbpub.com/botanylinks

Visit the .net Questions area of BotanyLinks (http://www.jbpub.com/botanylinks/netquestions) to complete this question:

1. How is wood related to the environment? Go to the BotanyLinks home page for more information on this subject.

BotanyLinks includes a Directory of Organizations for this chapter.

9

FLOWERS AND REPRODUCTION

During sexual reproduction, one individual combines part of its genetic information with that of another individual. The new offspring may have better combinations of traits than does either parent by itself.

CONCEPTS

Reproduction can serve two very different functions: (1) producing offspring that have identical copies of the parental genes, or (2) generating new individuals that are genetically different from the parents. Under certain environmental conditions, species that are genetically diverse survive better than genetically homogeneous species; under other conditions just the opposite is true. A plant that has been able to survive and grow to reproductive maturity is relatively well adapted to its location, so any progeny that are genetically identical to it are at least as well adapted as it is. Any progeny that are not genetically identical to the parent may or may not be well adapted to the conditions to which the parent is adapted.

If the environment is stable during several lifetimes, it is selectively advantageous for an organism to reproduce asexually by budding or sending out runners, thus producing new similarly adapted individuals. However, if the environment is not stable, such offspring may find themselves in conditions for which they are poorly adapted; if all are identical, all may die. Instability of the environment can result from many factors; for instance, landslides, avalanches, and roadbuilding in forests kill existing vegetation, opening up new, sunny sites that are good for quickly growing, sun-loving weedy plants. Irregular climatic events, such as unusually severe freezes, droughts, floods, or hurricanes, also disrupt plant communities. If all members of a species are equally susceptible to low

TABLE 9.1 *Sexual and Asexual Reproduction*

Sexual reproduction
  Progeny are genetically diverse.
  Some are less adapted than the parent but others are more adapted.
  Offspring cannot colonize a new site as rapidly because not all progeny are adapted for it,
    but some can colonize different sites with characteristics not suitable for parents.
  Changes in habitat may adversely affect some progeny, but others may be adapted to the
    new conditions.
  Isolated individuals cannot reproduce.
Asexual reproduction
  All progeny are identical genetically to parent and to each other.
  All are as adapted as parent is, but none is more adapted.
  Rapid colonization of a new site is possible.
  All may be adversely affected by even minor changes in the habitat.
  Even isolated individuals can reproduce.

temperatures, all may be killed by a rare freeze. But if the species is genetically diverse, some individuals may survive. Even though most die, the few survivors may be sufficient to repopulate the site (Table 9.1).

With asexual reproduction, progeny are never more fit than the parent, but during sexual reproduction sex cells of one plant combine with those of one or several others, resulting in many new gene combinations. Sex cells are so small that many can be produced by a single plant, and many new combinations of genes can be "tested" rather inexpensively. For example, a single large tree can produce thousands of flowers and millions of pollen grains, each genetically unique, yet the tree uses only a few grams of carbohydrate, protein, and minerals (Fig. 9.1). Similarly, thousands of egg cells can be produced using

(a)   (b)

FIGURE 9.1
(a) These apple fruits have developed from flowers, and the seeds inside the fruits developed from egg cells inside the flowers. All the egg cells produced by this tree have similar genes. But sperm cells were necessary to fertilize the eggs, and bees may have brought many types of sperm cells from many different apple trees. Although every seed here has the same maternal parent (the tree in the photo), it is possible that all the seeds have different paternal parents. The seeds are not genetically identical. (b) This bee is emerging from a squash flower, covered with pollen. As it visits the next squash flower, it will accidentally leave some of this pollen there and pick up more pollen. In this process, genes are transferred from plant to plant. *(William E. Ferguson)*

In both stable environments and changing ones, sexual reproduction provides enough diversity of progeny that at least some are well adapted.

only a small amount of resources. The pollen from one plant can be blown or carried by pollinators to the flowers of hundreds or thousands of other plants, and one plant may receive pollen from numerous other individuals. The thousands of seeds produced by a single sexually reproducing plant represent thousands of natural genetic experiments. During seed and fruit maturation, those embryos with severely mismatched genes abort and use no further resources. The tree finally produces hundreds or thousands of fruits and seeds. The total reproductive effort may be a significant drain on the tree's resources, but it produces numerous embryos, many of which are at least as genetically fit as the tree is and perhaps even more fit.

As a further example, think of sexual reproduction in humans: The children produced by a particular couple are variable, not identical to each other or to either parent. Some of the children may have a particularly advantageous combination of genes and be more healthy or intelligent or athletic or creative than either parent. Others may have combinations of genes that result in congenital problems such that the children can survive only through medical help. Most children are more or less the same as the parents. The diversity is important.

Sexual reproduction also has negative aspects. Two individuals are required, and sex cells must move from one plant to another. In seed plants, pollen may be carried by wind, insects, and birds, but each results in the loss of many pollen grains or the need to produce nectar. Furthermore, potential sex partners may be widely scattered. For example, in a population of trees, those few individuals growing at the highest altitudes may have no neighbors, while those growing at lower altitudes have numerous close neighbors. The flowers of the highest individuals may receive no pollen and thus produce no seeds during some years. By contrast, plants that reproduce asexually can do so at any time, even when completely isolated. Some flowering plants are self-fertile and can undergo self-pollination, but they lose the benefit of receiving new genes from another plant.

Some plants reproduce both sexually and asexually. Strawberries have flowers and sexual reproduction involving genetically diverse embryos and seeds, but they also spread rapidly and asexually by runners. Bamboos are perennial grasses that flower and set seed only occasionally (in some species, only once every 80 years), but their rhizomes grow vigorously and establish many new plants asexually. Kalanchoes produce large numbers of seeds each year, but they also produce such large numbers of plantlets along their leaf margins that they can be weeds in both nature and in greenhouses (Fig. 9.2a).

Seeds, which are produced by sexual reproduction, often have a means of long-distance dispersal: Strawberries are eaten and the seeds later defecated; bamboo fruits and seeds are carried by winds. The consequence of this is that seeds may become widely scattered and germinate in numerous diverse sites, each site differing from the others in its microclimate, soil conditions, and exposure to predators and pathogens. Of course, a seed carrying an embryo with a combination of genes selectively advantageous for growth in a dry site may land in a wet site and not survive. But with the large numbers of seeds produced, it is statistically probable that some seeds will land in sites for which they are well suited.

In contrast, new plants that are produced asexually usually are not capable of long-distance dispersal; runners, rhizomes, and plantlets result in new plants that become established in the same microhabitat as their parent. Once a single plant becomes established in a suitable site, by reproducing asexually it can quickly fill the area with replicas of itself, all of which are as fit as it is.

ASEXUAL REPRODUCTION

Within the flowering plants, numerous methods of asexual reproduction have evolved. One of the most common is **fragmentation**: A large spreading or vining plant grows to several meters in length, and individual parts become self-sufficient by establishing adventitious roots. If middle portions of the plant die, the ends become separated and act as individuals. Certain modifications improve the efficiency of fragmentation. In many cacti, branches are poorly attached to the trunk, and the plant breaks apart easily. The parts then

(a)

(b)

(c)

(d)

FIGURE 9.2

(a) Kalanchoe plants are called maternity plants because they produce plantlets complete with stems, leaves, and roots along their leaf margins. Although hundreds of plantlets can be produced, all nuclear divisions are mitotic—duplication division—so all these plantlets are genetically identical to the parent; none is superior. (*James L. Castner*) (b) Chollas, species of *Opuntia*, have branches only weakly attached to the trunk. If an animal brushes against a branch, the spines stick to the animal and the branch is pulled from the plant. After the animal dislodges the branch, it roots and grows into a new plant. Long-distance distribution almost as extensive as that of seeds can occur. (c) All the trees in this photograph are a part of the same plant, each a sprout from a single root system. The plant actually covers several acres: All the trees in the dense stands around the valley (d) are a single individual. (*c and d, Courtesy of Michael Grant, University of Colorado*)

form roots and become independent (Fig. 9.2b). In some members of the saxifrage, grass, and pineapple families, plantlets are formed where flowers would be expected; these look like small bulbs and are called bulbils.

In willows and many thistles, adventitious shoot buds form on roots, then grow into plants. Adventitious buds may grow out even while the parent plant is still alive, and a small cluster of trees may in fact consist of just a single individual. A grove of aspens that covers several acres in Utah has just been discovered to be a single plant (Fig. 9.2c and d).

SEXUAL REPRODUCTION

Sexual reproduction in flowering plants involves the flower itself, which produces the necessary cells and structures. To understand flower structure, one must first understand the plant life cycle.

THE PLANT LIFE CYCLE

The life cycle of mammals such as humans is simple: Diploid adults have sex organs that produce haploid sex cells called **gametes**, either **sperms** or **eggs**, by meiosis. Individuals that produce sperms are called males, of course, and individuals that produce eggs are females. One sperm and one egg are brought together forming a new single diploid cell, the

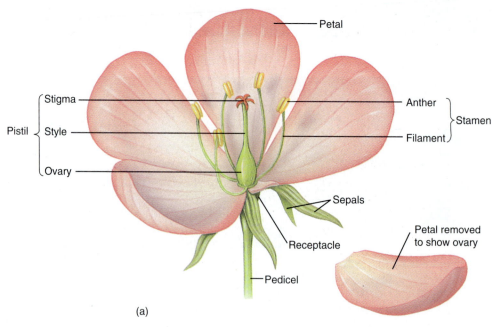

(a)

FIGURE 9.3

(a) Most flowers have four types of structures: (1) sepals that typically protect the young flower, (2) petals that usually attract pollinators, (3) stamens that produce pollen, which produce sperm cells, and (4) a pistil whose ovary is involved in producing egg cells (the pistil shown here is actually composed of five parts called carpels which have merged into one structure). Whereas no one mammal ever has both sperm-producing organs and egg-producing organs, almost all plants do have two types of spore-producing organs, the stamens and the carpels. (b) In the anthers of stamens, the central cells undergo meiosis and each produces four daughter cells called microspores or pollen grains. (c) In the flower's central organs, the carpels, are ovules, each containing only one cell that undergoes meiosis; often three of the daughter cells die and the one survivor becomes the megaspore.

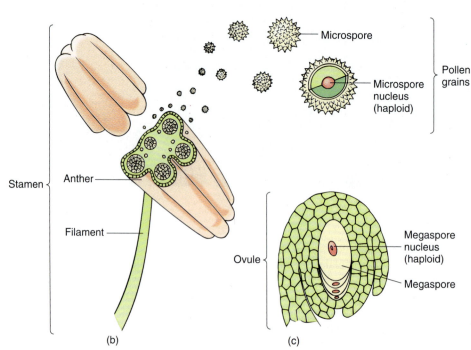

(b)   (c)

fertilized egg or **zygote,** which then grows to become a new individual that is diploid and resembles its parents.

In plants, the life cycle is more complex. The plants you are familiar with—trees, shrubs, and herbs—are all just one phase of the plant life cycle, called the **sporophyte phase** or **sporophyte generation.** A critical factor is that sporophytes are always diploid, like most adult animals, and they have sex organs (located in the flowers in angiosperms) with cells that are capable of undergoing meiosis. In animals, meiosis results in haploid gametes, but in plants it results in haploid **spores** (Fig. 9.3). The difference between gametes and spores is great: Gametes can fuse with other gametes in a process called **syngamy** or **fertilization,** thereby producing the diploid zygote. A gamete that does not undergo syngamy dies because it cannot live by itself and usually cannot grow into a new, haploid individual.

Plant spores are just the opposite: They cannot undergo syngamy but they do undergo mitosis and grow into an entire new, haploid plant called a **gametophyte.** It is called a gametophyte because it is the plant (-phyte) that produces the gametes (gameto-). During sexual reproduction, when a sporophyte reproduces, **it does not produce a new diploid plant like itself** but rather a haploid plant. Furthermore, in all vascular plants, the haploid gametophyte does not even remotely resemble the diploid sporophyte. It is a tiny mass of cells with no roots, stems, leaves, or vascular tissues, but it is an entire plant (Fig. 9.4). Gametes are formed by the haploid plants by mitosis, not meiosis. The gametes then undergo syngamy, forming a zygote that grows into a new, diploid sporophyte, and the life cycle is complete.

The unfertilized eggs of some insects such as bees are exceptional and develop into sterile workers.

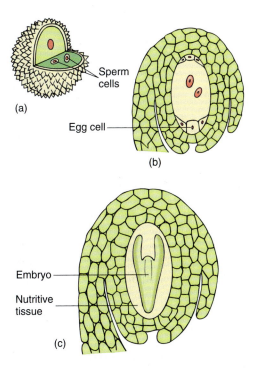

(a)

Sperm cells

Egg cell

(b)

Embryo

Nutritive tissue

(c)

FIGURE 9.4
All seed plants produce two types of gametophytes. (a) Microspores (pollen grains) develop into microgametophytes (also pollen grains). The microgametophyte body is so small that it has just three cells and fits inside the pollen cell wall. From the outside, it is not possible to tell if a pollen grain is a microspore or a microgametophyte. The microgametophyte produces two sperm cells located within its own protoplasm. (b) The megaspore develops into a megagametophyte. It is slightly larger than a microgametophyte and has seven cells, one of which has two nuclei. One of the cells is the egg. The megaspore and megagametophyte are never released from the flower of the parent sporophyte, so they appear to be just sporophyte tissues, not one plant growing inside another. (c) After one of the sperm cells fertilizes the egg cell, the new egg cell nucleus is diploid and the cell is a zygote. It develops, by mitosis, into a new sporophyte, shown here as the embryo in an immature seed. The entire ovule develops into a seed; the ovary develops into a fruit.

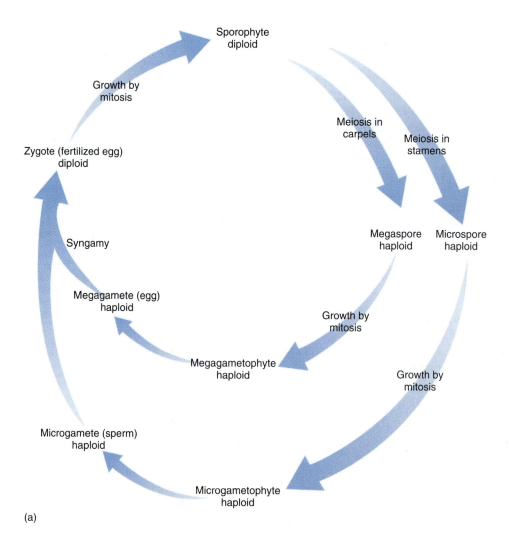

(a)

FIGURE 9.5

(a) Diagram of the life cycle of a flowering plant: A sporophyte produces micro- and megaspores and each develops into micro- and megagametophytes, respectively. These produce micro- and megagametes that undergo syngamy (the sperm fertilizes the egg), producing a zygote, the fertilized egg. The zygote grows into a new sporophyte. The sporophyte generation alternates with the gametophyte generation. (b) A flowering plant life cycle showing the actual structures involved.

Mammalian gametes are of two types: small sperm cells (**microgametes**) that swim, and large eggs (**megagametes**) that do not. This is also true of many plants and is known as oogamy. In oogamous plants, just as in oogamous mammals, sperms are produced by one type of individual and eggs by a different type of individual; hence, there are "male" or **microgametophytes** and "female" or **megagametophytes** (Fig. 9.4). The two types of gametophytes have grown from two types of spores: microgametophytes from **microspores** and megagametophytes from **megaspores.** Having two types of spores is known as heterospory. Typically just one kind of sporophyte occurs in a life cycle, and it produces both microspores and megaspores.

A life cycle like this, with two generations—sporophyte and gametophyte—is said to be an **alternation of generations** (Fig. 9.5). Because gametophytes do not resemble the sporophytes at all, this is an alternation of heteromorphic generations. This is a complex life cycle, with at least three distinct plants (one sporophyte and two gametophytes). The human life cycle, like that of all other animals, does not have anything equivalent to the haploid generation.

FLOWER STRUCTURE

A flower is basically a stem with leaflike structures, so almost everything discussed for vegetative shoots in Chapter 5 also applies to flowers. Flowers never become woody; secondary growth does not occur in flowers.

The flower stalk is a **pedicel,** and the very end of the axis, where the other flower parts are attached, is the **receptacle** (Fig. 9.6). There are four types of floral appendages: sepals,

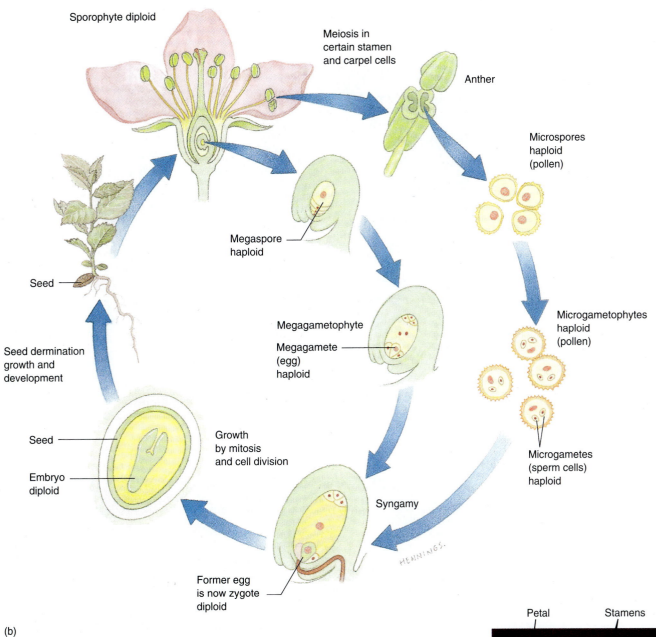

Sporophyte diploid

Meiosis in certain stamen and carpel cells

Anther

Microspores haploid (pollen)

Megaspore haploid

Microgametophytes haploid (pollen)

Megagametophyte

Megagamete (egg) haploid

Seed

Seed dermination growth and development

Microgametes (sperm cells) haploid

Seed

Growth by mitosis and cell division

Embryo diploid

Syngamy

HENNINGS.

Former egg is now zygote diploid

(b)

petals, stamens, and carpels. Most flowers have all four types and are **complete flowers**. They typically have three, four, five or more appendages of each type; for example, lilies have three sepals, three petals, three stamens, and three carpels. It is not uncommon for flowers of certain species to lack one or two of the four basic floral appendages, thus being **incomplete**. Flowers of willow (Fig. 9.7) are incomplete, lacking both sepals and petals but having either stamens or carpels; flowers of pigweed have all parts except petals.

Sepals. **Sepals** are the lowermost and outermost of the four floral appendages. They are modified leaves that surround and enclose the other flower parts as they mature. Sepals are typically the thickest, toughest, and most waxy of the flower parts. They protect the flower bud as it develops, keeping bacterial and fungal spores away, maintaining a high humidity inside the bud, and deterring insect feeding (Fig. 9.8a and b). Sepals also protect the flower from nectar-robbing insects and birds. If flower buds develop in a protected position such as beneath a spiny cover or surrounded by regular leaves and branches, sepals are less important for protection and may be quite reduced or absent. It is not uncommon for sepals to be colorful (petalloid) and help attract pollinators. All the sepals together are referred to as the **calyx.**

Petal   Stamens

Sepal   Carpels

FIGURE 9.6
All flower parts are present and easily visible in this pear blossom. The anthers of the outermost, tallest stamens have opened, releasing their pollen. *(Runk/Schoenberger from Grant Heilman)*

FIGURE 9.7

In most species of flowering plants, the stamens and carpels are produced together in the same plant, but in a few species like these willows, stamens occur in flowers without carpels on one plant (a), while carpels occur in stamenless flowers on different plants (b). In contrast to mammals, the two types of plants look so much alike that they usually cannot be identified as staminate (having stamens) or carpellate (having carpels) without looking at the flowers. *(a, Dwight R. Kuhn; b, John D. Cunningham/Visuals Unlimited)*

(a)   (b)

(a)   (b)

(c)   (d)

FIGURE 9.8

(a) The sepals of this rose form a tight covering over the rest of the flower as it develops, protecting the inner parts. Notice that the sepals and the pedicel have glands on them, helping to deter insects. When the microspores (pollen) and megaspores (in the ovules) are ready, the sepals bend outward and the flower opens. (b) Sunflowers and daisies are actually clusters of many small flowers that together have the appearance of a single flower. In this case, the whole cluster of flowers is protected by a set of bracts that look and function like sepals. It is not necessary for the individual flowers to have protective sepals. (c) Petals, like these of the mallow, are leaflike in being thin and flat, but in most species, they are more highly modified and less leaflike than sepals. (d) Leaves and petals of *Arbutus* can be compared here; these petals are highly modified, and the five petals of each flower have fused together, forming a tubular corolla.

FIGURE 9.9
Some flowers do not produce nectar; the pollinator eats part of the stamens and pollen instead. This is more expensive metabolically for the plant because pollen is rich in protein, whereas nectar is composed of carbohydrate. In these flowers of cannonball tree *(Couropita)*, the lower stamens are modified for edibility; they are large and showy and produce little or no pollen. The pollen-producing stamens are small and inconspicuous, easily overlooked by pollinators.

Edible decoy
stamens

Petals.   Above the sepals on the receptacle are **petals,** which together make up the **corolla.** Sepals and petals together constitute the flower's **perianth.** Petals are also "leaflike," being broad, flat, and thin, but they differ from leaves in that they contain pigments other than chlorophyll, have fewer or no fibers, and tend to be thinner and more delicately constructed (Fig. 9.8c and d). Petals are important not merely in attracting pollinators, but rather the correct pollinators. Each plant species has flowers of distinctive size, shape, color, and arrangement of petals, allowing pollinators to recognize specific species. Sexual reproduction cannot occur efficiently if pollen is carried to other plants indiscriminately. Reproduction results if pollen is carried only to other flowers of the same species. If a flower has a distinctive pattern and offers a good reward such as nectar or pollen (Fig. 9.9), the pollinator is likely to search for and fly to other flowers with the same pattern, thereby enhancing cross-pollination; mutations are advantageous selectively if they cause flowers to have characters that are easily recognized by their pollinators. In addition to visible colors, many flowers have pigments that absorb ultraviolet light, creating patterns only insects can see (Fig. 9.10).

Pollen is carried by wind in many species and by water in a few. Petals cannot attract wind or water, so loss of petals is not disadvantageous for these species. Mutations that inhibit their differentiation prevent plants from wasting carbon and energy in the construction of nonfunctional petals. Typically, petals do not develop in wind-pollinated species.

Stamens.   Above the petals are the **stamens,** known collectively as the **androecium.** Stamens are frequently referred to as the "male" part of the flower because they produce the **pollen,** but technically they are not male because the flower, being part of the sporophyte, produces spores, not gametes. Only gametes and gametophytes have sex.

Without light, colors cannot be seen, and night-blooming species have white flowers lacking pigments. Their petals produce volatile fragrances, and insects and bats follow the aroma gradient to the flower.

(a)

(b)

FIGURE 9.10
(a) A flower as seen by the human eye. (b) The same flower photographed through filters that reveal the pattern of its pigments that absorb ultraviolet light. Although we cannot see this pattern, most insects can. Distinctive patterns like this, in either ultraviolet or visible light pigments, direct the pollinator to the nectar and position the animal to pick up pollen; the markings are known as nectar guides. *(Thomas Eisner)*

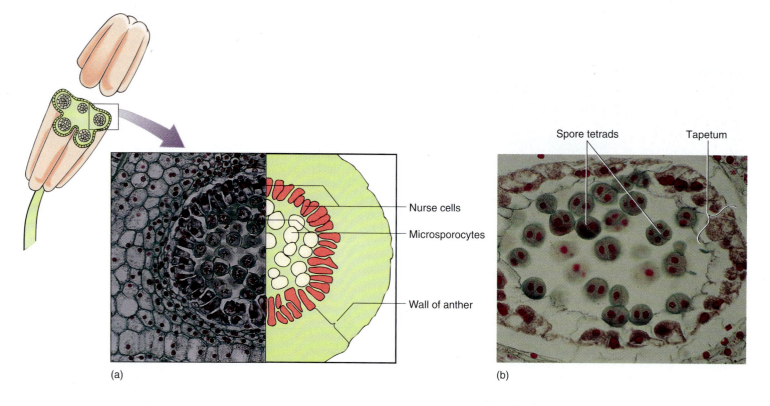

Spore tetrads   Tapetum

Nurse cells

Microsporocytes

Wall of anther

(a)

(b)

FIGURE 9.11

(a) Transverse section through an anther of a lily, showing the microsporocytes in prophase of the first meiotic division that will produce haploid microspores. The surrounding cells constitute the tapetum and aid in the maturation of the pollen grains (×200). *(G. I. Bernard/Earth Scenes)* (b) After meiosis, the four microspores remain together temporarily, but as they develop their specialized pollen cell wall, they usually separate from each other (×500).

Stamens have two parts, the **filament** (its stalk) and the **anther**, where pollen is actually produced. As part of the sporophyte, the anther is composed of diploid cells, and in each anther four long columns of tissue become distinct as the cells enlarge and prepare for meiosis (Fig. 9.11a). These **microspore mother cells** or **microsporocytes** continue to enlarge and then undergo meiosis, each producing four microspores (Fig. 9.11b; Table 9.2). Neighboring anther cells, in a layer called the **tapetum**, act as nurse cells, contributing to microspore development and maturation. Microspores initially remain together in a tetrad, but later separate, expand to a characteristic shape, and form an especially resistant wall. They are then called **pollen**. The anthers open (**dehisce**) along a line of weakness and release the pollen.

The pollen wall is a cell wall; however, it is quite complex structurally. It has an inner layer called the intine, composed of cellulose, and an outer layer called the exine that

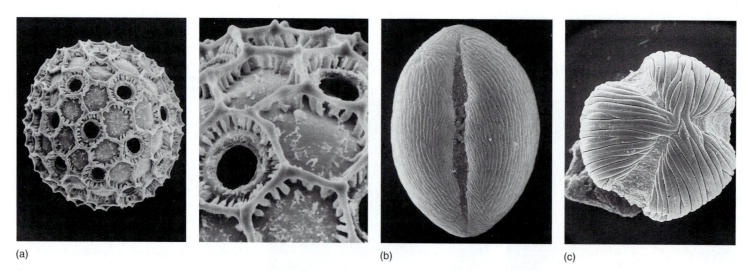

(a)   (b)   (c)

FIGURE 9.12

(a) Pollen grains of *Cobaea*, with the pollen wall forming hexagonal ridges. The numerous holes are germination pores where pollen tubes emerge after the pollen lands on a stigma (×2000). *(Courtesy of Alan Prather, University of Texas)* (b) Pollen of *Lycium* has a single long groove from which the pollen tube emerges (×4000). (c) Pollen of *Macrolobium* has three germination grooves (×4000). *(b and c, Courtesy of Beryl Simpson, University of Texas)*

consists of the polymer sporopollenin. It has one or several weak spots, germination pores, where the pollen opens after it has been carried to the stigma of another flower. Sporopollenin is remarkably waterproof and resistant to almost all chemicals; it protects the pollen grain and keeps it from drying out as it is being carried by the wind or animals. The exine can have ridges, bumps, spines, and numerous other features so characteristic that each species has its own particular pattern (Fig. 9.12). In many cases, it is possible to examine a single pollen grain and know exactly which species of plant produced it. Because the sporopollenin is so resistant, pollen grains and their characteristic patterns fossilize well. By examining samples of old soil, botanists can determine exactly which plants grew in an area at a particular time in the ancient past.

Carpels. **Carpels** constitute the **gynoecium,** located at the highest level on the receptacle (Fig. 9.13a). Carpels have three main parts: (1) a **stigma** that catches pollen grains, (2) a **style** that elevates the stigma to a useful position, and (3) an **ovary** where megaspores are produced. A flower can have zero (some imperfect flowers) to many carpels; usually they

TABLE 9.2	*Stamen and Carpel Structure*

Stamen
  Filament
  Anther
    Microsporocytes

Carpel
  Stigma
  Style
  Ovary
    Ovary wall
    Placenta
    Ovule(s)
      Funiculus
      Integuments and micropyle
     Nucellus
      Megasporocyte

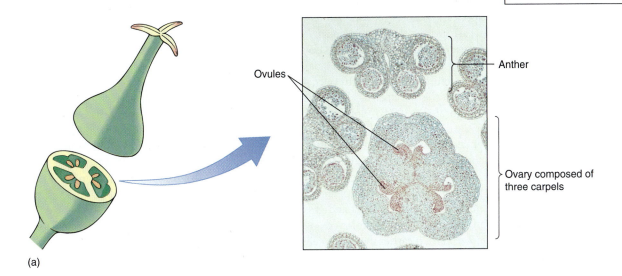

(a)

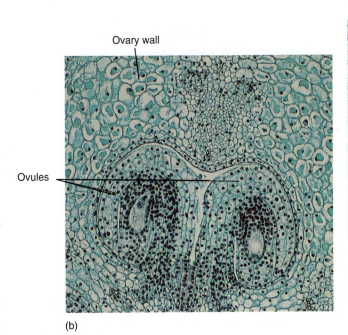

(b)

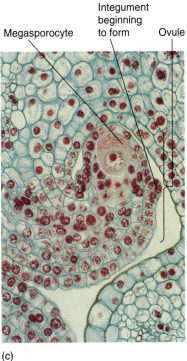

(c)

FIGURE 9.13
(a) In lily, the pistil consists of three carpels fused together. This lily ovary has been cut across to reveal the three ovary locules or chambers. In each chamber two ovules are visible (× 15). (b) Ovules are attached to the placenta by means of their funiculus, which contains both xylem and phloem. The central part of the ovule is the nucellus with (usually) just one megasporocyte, unlike the anthers which have numerous microsporocytes (× 50). (c) Magnification of a single ovule. Surrounding the nucellus are the integuments; here they are extremely short and do not yet cover the tip of the nucellus. Growth of the ovule has been by mitosis, but now one cell, the megasporocyte—megaspore mother cell—is preparing for meiosis (× 120). *(b, M. I. Walker, Science Source/Photo Researchers, Inc.)*

239

are fused together into a single compound structure, frequently called a pistil. Inside the ovary are placentae (sing.: placenta), regions of tissue that bear small structures called **ovules.** Ovules have a short stalk, called a funiculus, that carries water and nutrients from the placenta to the ovule by means of a small vascular bundle. The ovule has a central mass of parenchyma called a **nucellus.** Around the nucellus are two thin sheets of cells (integuments) that cover almost the entire nucellus surface, leaving only a small hole (micropyle) at the top (9.13c). As in anthers, some nucellus cells, usually only one in each ovule, enlarge in preparation for meiosis; these are **megaspore mother cells** or **megasporocytes.** After meiosis, usually three of the four megaspores degenerate, and only one survives, becoming very large by absorbing the protoplasm of the other three. Megaspores differ from microspores (pollen) because the ovule and the carpel do not dehisce and the megaspore remains enclosed inside the carpel.

THE GAMETOPHYTES

Microgametophyte.   Microspores develop into microgametophytes. In flowering plant species, the microgametophyte is very small and simple, consisting of at most three cells located within the original pollen cell wall (Fig. 9.14). The microspore nucleus migrates to the side of the pollen grain and lies next to the wall. There it divides mitotically, producing a large **vegetative cell** and a small lens-shaped **generative cell**, which subsequently divides and forms two sperm cells. The entire microgametophyte consists of the vegetative cell and the two sperm cells (the microgametes); this is a full-fledged plant.

In about 30% of flowering plant species, the formation of sperm cells occurs even while the pollen is still located within the anther. In the majority of angiosperm species, the pollen is released from the anther at about the time the generative cell has formed, and sperm cells are not produced until after the pollen has been carried to a stigma.

After the pollen lands on a stigma, it germinates by producing a **pollen tube** that penetrates into the loose, open tissues of the stigma (Fig. 9.15). The pollen tube absorbs nutrients from the stigma and grows downward through the style toward the ovary. The microgametophyte is protected and nourished by the style tissue. As the pollen tube grows downward, it carries the sperm cells to the ovule. Almost all the pollen cytoplasm is located at the tip of the pollen tubes; the rest of the tube and the pollen grain are filled with a giant vacuole. Pollen grains are too small to store much starch or protein; if they could not absorb nutrients from the style, the pollen tubes could not grow long enough to reach the ovules.

FIGURE 9.14

Only with transmission electron microscopy can the details of a microgametophyte be studied well. The large cell is the vegetative cell, and the two smaller cells (only one visible here) are the sperm cells that arose by division of the generative cell. In this species, the sperm cells have no cell walls; they are separated from the vegetative cell only by plasma membranes (× 5000). *(Courtesy of S. Nakamura and H. Miki-Hirosige, Kanagawa Dental College)*

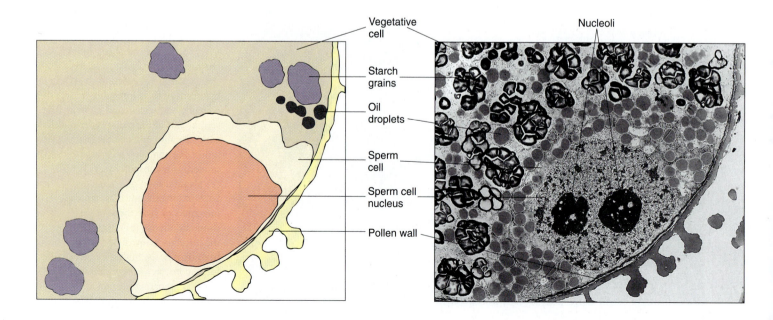

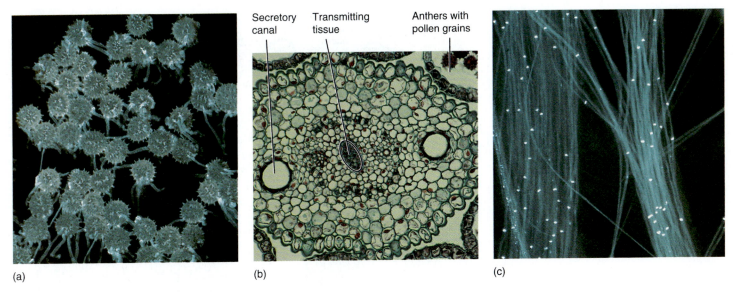

(a)

Secretory canal   Transmitting tissue   Anthers with pollen grains

(b)

(c)

FIGURE 9.15
(a) Shortly after landing on a stigma of the correct species, a pollen grain may germinate by pushing out part of the wall as a pollen tube. This penetrates the loose tissues of the stigma and style. Photographed with ultraviolet light which causes the pollen wall and callose to fluoresce and shine (× 50). (b) Tissues of the style allow the pollen tubes to grow through them easily. Some styles are hollow and lined with a rich, nutritious transmitting tissue; other styles are solid but also have transmitting tissue. Sunflower (*Helianthus*) style (× 80). (c) These pollen tubes have penetrated deep into the style. The bright spots are callose plugs which seal off the protoplasm at the tip from the empty parts of the pollen tube (× 200). *(a and c, Courtesy of Allison Snow, Ohio State University)*

Megagametophyte.   Within the ovule the surviving megaspore develops into a megagametophyte. In one type of development, the nucleus undergoes three mitotic divisions, producing two, four, and then eight haploid nuclei (Fig. 9.16). The nuclei migrate through the cytoplasm, presumably pulled by microtubules, until three nuclei lie at each end and two in the center. Walls form around the nuclei, and the large, eight-nucleate megaspore becomes a megagametophyte with seven cells, one of which is binucleate. The seven cells are one large **central cell** with two **polar nuclei**, three small **antipodal cells**, and an **egg apparatus** consisting of two **synergids** and an **egg** (the megagamete) (Table 9.3). Like the microgametophyte, the megagametophyte is a distinct plant. As with the pollen, the megagametophyte obtains all its nourishment from the parent sporophyte.

TABLE 9.3   *Gametophyte Development*

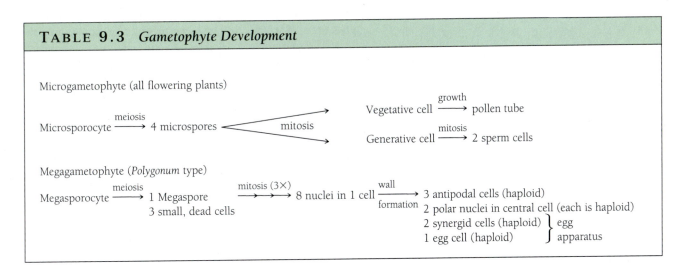

Microgametophyte (all flowering plants)

Microsporocyte —meiosis→ 4 microspores —mitosis→ Vegetative cell —growth→ pollen tube
Generative cell —mitosis→ 2 sperm cells

Megagametophyte (*Polygonum* type)

Megasporocyte —meiosis→ 1 Megaspore —mitosis (3×)→ 8 nuclei in 1 cell —wall formation→ 3 antipodal cells (haploid)
3 small, dead cells
2 polar nuclei in central cell (each is haploid)
2 synergid cells (haploid) } egg
1 egg cell (haploid) } apparatus

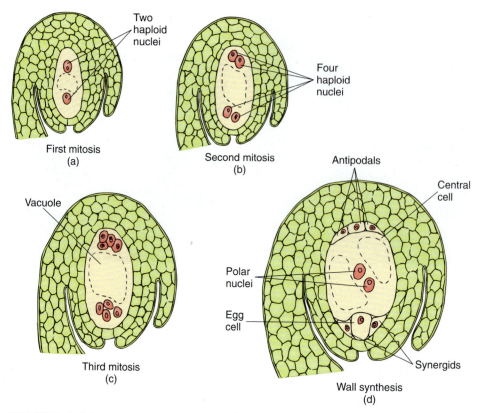

FIGURE 9.16

As a megaspore develops into a megagametophyte, three sets of mitosis occur (a to c) without any cytokinesis, so the megagametophyte is temporarily an eight-nucleate coenocyte (c). Then nuclei migrate, cytoplasm accumulates around each, and walls are established (d). Several studies have shown that much of the egg is not covered by wall. The central cell is binucleate and is mostly vacuole.

FERTILIZATION

Syngamy of sperm and egg involves both **plasmogamy,** the fusion of the protoplasms of the gametes, and **karyogamy,** the fusion of the nuclei. As a pollen tube grows downward through the style toward the ovule, it is guided to the ovule's micropyle by some means. It penetrates the nucellus and reaches the egg apparatus, then enters one synergid. The pollen tube tip bursts and releases both sperm cells. One of these migrates toward the egg, and as it does so, its plasma membrane breaks down. The sperm cell loses most of its protoplasm before it reaches the egg. As the sperm fuses with the egg plasma membrane, only the sperm nucleus enters the egg. The sperm nucleus is drawn to and then fuses with the egg nucleus, establishing a diploid zygote nucleus.

Because the sperm sheds its protoplasm as it passes through the synergid, it contributes only its nucleus with the set of nuclear genes during karyogamy. The sperm does not carry mitochondria or plastids into the egg, so organellar genes from the pollen parent are rarely inherited by the zygote. Instead, the mitochondrial and plastid genes of the embryo are inherited only from the ovule parent.

In flowering plants only, the second sperm nucleus released from the pollen tube migrates from the synergid into the central cell. It undergoes karyogamy with both polar nuclei, establishing a large **endosperm nucleus** that is triploid, containing three full sets of genes. Because both sperm nuclei undergo fusions—one with the egg nucleus, the other with the polar nuclei—the process is called **double fertilization.** The endosperm nucleus is extraordinarily active and begins to divide very rapidly by mitosis, with cell cycles lasting only a few hours.

The endosperm nucleus initiates a dynamic cytoplasm and the central cell enlarges enormously, usually without cell division, into a huge cell with hundreds or thousands of nuclei (see Figs. 4.19 and 9.17). Finally, nuclear division stops, and dense cytoplasm gathers around the nuclei. Walls are constructed, thus forming cells. An example of this is a coconut full of "milk." The hollow center of the coconut is one single cell, and the milk is its protoplasm. The white coconut "meat" is the region where nuclei form cells. A green, immature coconut is full of milk but has almost no meat; as it ripens, the coenocytic milk is converted to a thick layer of cellular meat. All this tissue, both coenocytic and cellular, is called **endosperm,** and it nourishes the development of the zygote. No other megagametophyte forms as much endosperm as a coconut does. A more typical example is corn, in which most of the kernel is endosperm. Corn endosperm also passes through a milk stage, but most people are not familiar with it.

EMBRYO AND SEED DEVELOPMENT

As the endosperm nucleus proliferates, the zygote also begins to grow, but always by both nuclear and cellular divisions; a coenocytic stage never occurs in the embryo. The zygote grows into a small cluster of cells, part of which later becomes the embryo proper, and the other part becomes a short stalklike structure, the **suspensor,** which pushes the embryo deep into the endosperm (Fig. 9.18). The suspensor is usually delicate and ephemeral in flowering plants; it is crushed by the later growth of the embryo and is not easily detectable in a mature seed.

The cells at one end of the suspensor continue to divide mitotically, developing into an embryo. The cells at first are arranged as a small sphere, the globular stage. The end of the embryo farther from the suspensor initiates two primordia that grow into two **cotyledons** in dicots (see Figs. 9.19 and 9.20), such as beans and peanuts. While young, the cotyledon primordia give the embryo a heart shape; this is the heart stage. In monocots such as corn, only one cotyledon primordium grows out. Later, in the torpedo stage, the embryo is an elongate cylinder: A short axis is established, consisting of **radicle** (primordial root), **epicotyl** (primordial stem), and **hypocotyl** (the root/shoot junction). Finally, vascular tissue differentiates within the embryo. The epicotyl may bear a few small leaves, and the radicle

Endosperm   Embryo

FIGURE 9.17
In many species, endosperm development is accompanied by some cytoplasmic division, so the endosperm is a mass of multinucleate "cells," each with variable amounts of protoplasm and nuclei and with irregular shapes. This is the only plant tissue in which nuclear and cytoplasmic divisions are so poorly coordinated, possibly because the endosperm is just a temporary tissue that is consumed by the embryo before or soon after seed germination ($\times 180$).

Embryo   Suspensor   Embryo   Suspensor

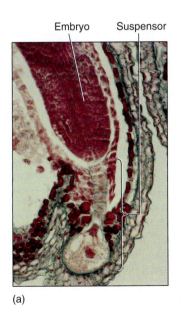   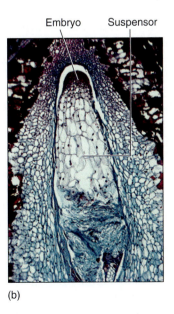

(a)   (b)

FIGURE 9.18
(a) The suspensor of shepherd's purse (*Capsella*) has one large bulbous cell and a stalk of smaller cells. The young embryo is being pushed deep into the endosperm ($\times 600$). (b) The suspensor of *Tristerix* is much more massive than that of *Capsella*, being larger than the embryo until the last stages of seed maturation. The embryo is still just a small ball of densely cytoplasmic cells ($\times 300$).

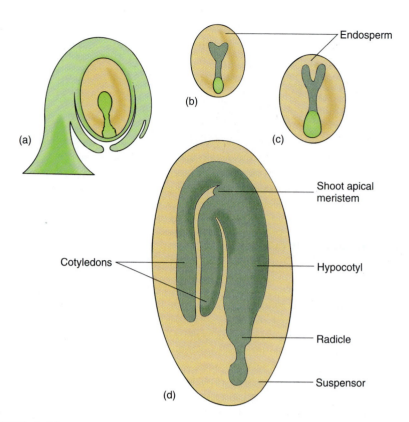

FIGURE 9.19

Embryo development. (a) Globular stage. (b) Heart stage. The two mounds that cause the heart shape are the first stage of the outgrowth of cotyledons. (c) Torpedo stage. Embryo root (radicle), cotyledons, and hypocotyl are present. At this stage, the first xylem and phloem may become distinguishable in the hypocotyl. (d) Mature embryo. A shoot apical meristem is shown; in some species, even some leaf primordia and a small stem are present.

"Dicot" is an abbreviation of "dicotyledon," those plants whose embryos have two cotyledons. Monocots are monocotyledons, plants with only a single cotyledon on their embryos.

often contains several primordia for lateral roots in its pericycle. Once mature, the embryo becomes quiescent and partially dehydrates, and the funiculus may break, leaving a small scar, the hilum.

In most dicots, cotyledons store nutrients that are used during and after germination; during embryo development, the cotyledons become thick and filled with starch, oil, or protein while the endosperm, which is supplying the nutrients, shrinks. When the seed is mature, the cotyledons are large and the endosperm may be completely used up.

In monocots, the one cotyledon generally does not become thick and full; instead, the endosperm remains and is present in the mature seed. During germination, the cotyledon acts as digestive/absorptive tissue, transferring endosperm nutrients to the embryo. Some dicots are intermediate: The cotyledons store some starch and protein, but a considerable amount of endosperm remains in the seed and both methods of nutrition are used during germination. A mature seed in which endosperm is rather abundant, such as corn, is called an **albuminous seed** (Fig. 9.20). If endosperm is sparse or absent at maturity (peas, beans), the seed is **exalbuminous.**

The amount of embryo growth and development that occurs before dormancy sets in is extremely variable. Orchids, bromeliads, and a few other species have small, dustlike seeds in which the embryo is only a small ball of cells with no cotyledons, radicle, or vascular tissue. In seeds of most angiosperm species, all the parts are present and the epicotyl contains two or three young leaves in addition to cotyledons; these leaves can begin photosynthesis immediately after germination (Fig. 9.20). In corn, the embryo is even more advanced: It contains as many as six young leaves while in the seed. A fully mature corn plant often has only 10 or 12 leaves. Over half of the leaf production of the new sporophyte occurs while it is embedded in the parental gametophyte, which is itself embedded in the previous sporophyte.

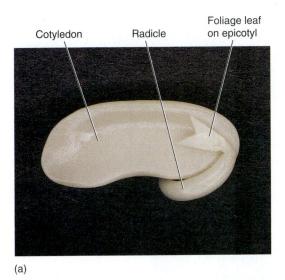

(a)

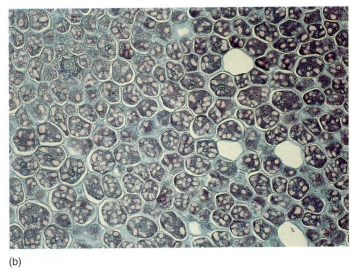

(b)

(c)

(d)

(e)

FIGURE 9.20

(a) This bean seed has begun germinating; the radicle has extended and will develop into a taproot. The two cotyledons (one was removed) were digesting their carbohydrates and proteins and transporting nutrients to the root and shoot apical meristems. Beans have a well-developed epicotyl with several partially expanded leaves; one is visible here. (b) These cells of bean cotyledon are full of starch (stained pink) and reserve protein (stained blue) ($\times$ 100). (c) The small leaves present on the epicotyl are expanding during germination; because they were rather completely formed before the embryo became dormant, the leaves can now mature rapidly and photosynthesis begins almost as soon as germination is complete. Notice how the hypocotyl is curved: This allows the leaves to be dragged gently up through the soil, protected by the cotyledons. (d) Most of a corn seed is the endosperm; the embryo is less than half the volume of the seed. The one cotyledon is large and contacts the endosperm. During germination it secretes digestive enzymes into the endosperm and absorbs the resulting monomers. *(Barry L. Runk from Grant Heilman)* (e) Corn seeds also produce many small leaves before becoming dormant, and they can begin photosynthesis immediately after germination. The leaves are protected from the soil by a tubular sheath called a coleoptile.

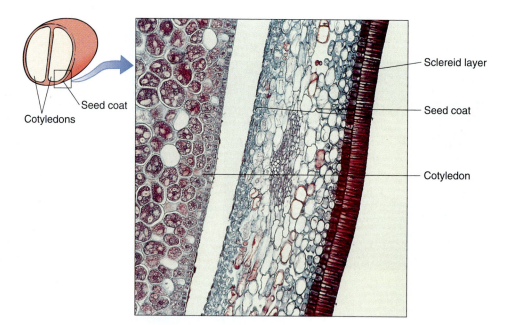

FIGURE 9.21

This seed coat of bean has an outer layer of sclereids and inner parenchymatous layers. The cotyledon is part of the embryo, and you might expect to see endosperm or nucellus between the cotyledon and seed coat, but those have been crushed during the later stages of seed development (× 200).

The embryo and endosperm develop from the zygote and megagametophyte central cell, respectively, both located in the nucellus of the ovule. Soon after fertilization, or even before, synergids and antipodals break down in most species. The nucellus expands somewhat but later is crushed by the expansion of the embryo and endosperm and is usually not detectable in mature seeds.

The integuments that surround the nucellus expand and mature into the **seed coat** (also called the **testa**) as the rest of the ovule grows. In their last stages of maturation, they may become quite sclerenchymatous and tough (Fig. 9.21).

FRUIT DEVELOPMENT

As the ovule develops into a seed, the ovary matures into a **fruit.** Development varies with the nature of the carpels as well as the nature of the mature fruit. The stigma and style usually wither away, as do sepals, petals, and stamens, although they may persist at least temporarily (Fig. 9.22). Often three layers become distinct during growth: The **exocarp** is the outer layer—the skin or peel; the middle layer is the **mesocarp,** or flesh; and the innermost layer, **endocarp,** may be tough like the stones or pit of a cherry or it may be thin

FIGURE 9.22

In apples, the petals die and fall off after pollination. Stamens and sepals die and dry out, but they remain present while carpels and receptacle develop into a fruit. In most species, sepals, petals, and stamens all abscise after pollination.

(Fig. 9.23). The relative thickness and fleshiness of these layers vary with fruit type, and often one or two layers are absent. The entire fruit wall, whether composed of one, two, or all three layers, is the **pericarp.**

FLOWER STRUCTURE AND CROSS-POLLINATION

The production and development of spores, gametes, zygotes, and seeds are complex, elaborate processes, but they are not the only functions of flowers. Flowers are also involved in the effective dispersal of pollen and seeds. Because numerous mechanisms carry out these processes, numerous types of flowers and fruits exist.

CROSS-POLLINATION

Cross-pollination is the pollination of a carpel by pollen from a different individual; **self-pollination** is pollination of a carpel by pollen from the same flower or another flower on the same plant. In any plant population, there is genetic diversity. Random mutations will have produced some new genes that offer improved fitness and some that are deleterious. With cross-pollination, sperm cells and egg cells from different plants unite, resulting in new combinations of genes, at least a few of which may be better adapted than either parent. But self-pollination has about the same result as asexual reproduction because all genes come from the same parent. No possibility exists of bringing in new genes that might provide more fitness than those inherited from the parent. Many new plants may result, but none can be better adapted than the parent. However, if a plant is isolated by distance or lack of pollinators from potential cross-pollination partners, self-pollination allows it to set seed and propagate its genes rather than lose them when the plant dies.

Many mechanisms have evolved that decrease the probability of self-pollination and increase the chances of cross-pollination with its accompanying genetic diversity.

STAMEN AND STYLE MATURATION TIMES

Self-fertilization in flowers that have both stamens and carpels is prevented if anthers and styles mature at different times (Fig. 9.24). In many species, anthers release pollen while the stigma tissues are immature and unreceptive; the style may not have elongated yet, so the stigmas may be near the base of the flower while the anthers are at the top, elevated by elongated filaments. When the stigma and style become mature there may be no living pollen left in the flower, so all pollination is effected by younger flowers just opening their anthers.

FIGURE 9.23
Peach fruits have all three fruit layers: Their endocarp is a solid mass of sclerenchyma around the seed, the exocarp is the skin, and the mesocarp is the flesh. Fruits with endocarps like this—a stone or pit—are drupes. (*Jonas Kahn*)

Exposed pollen lives only briefly, being susceptible to dessication in dry air and to damage to its DNA by ultraviolet light.

FIGURE 9.24
The stamens of this cactus flower are mature and shedding pollen, but the stigmas (green) are pressed together and are unreceptive. Later, the stamens wither and the stigmas spread open, ready to receive pollen from a different flower.

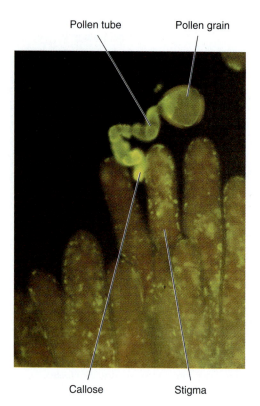

Pollen tube   Pollen grain

Callose   Stigma

FIGURE 9.25
This pollen grain has started to grow on an incompatible stigma, but the stigma has blocked it, causing callose to form. The callose fluoresces in ultraviolet light, so it shines brightly in this micrograph (× 200). *(Courtesy of J. B. and M. E. Nasrallah, Cornell University)*

Dioecy means "two houses"—each type of individual is a "house." Monoecy is "one house"; one plant bears both types of flowers.

This is not a very effective means of ensuring cross-pollination. On plants with many flowers that do not open simultaneously, older flowers could be self-pollinated by freshly opened flowers of the same plant. Even if pollen does come from other plants, many of the near neighbors are probably closely related, because many of the seeds of a plant fall and germinate near the plant, producing a cluster of plants that are at least as closely related as first cousins.

STIGMA AND POLLEN INCOMPATIBILITY

In many species, especially many important crop species, self-pollination is inhibited by **compatibility barriers,** chemical reactions between pollen and carpels that prevent pollen growth. In one incompatibility system, as a pollen tube grows, the stigma and style test proteins on the tube's surface; if one of these proteins is produced by a gene that matches an incompatibility gene in the carpel-bearing plant, the stigma and style block any further growth of the pollen tube. In self-pollination, all pollen tube genes match those of the stigma and style, and blocking occurs (Fig. 9.25).

In another common system, the critical proteins are deposited on the outer surface of the pollen grain by the anther. Any match of proteins produced by the incompatibility genes blocks germination of the pollen grain. The pollen tube–style interaction involves a haploid genome (pollen tube) and a diploid one (carpel), whereas the second system involves two diploid genomes (anther and carpel). The diploid/diploid system involves twice as many genes, so it has a much greater probability of pollen rejection. With the diploid/diploid system, not only is self-pollination prevented, but inbreeding between close relatives is blocked as well.

MONOECIOUS AND DIOECIOUS SPECIES

Among incomplete flowers there is a significant difference between flowers that lack sepals or petals and those that lack stamens or carpels. The latter two organs are **essential organs** because they produce the critically important spores; if either organ is absent, sexual reproduction is dramatically affected. Flowers that lack either or both essential organs are not only incomplete but also **imperfect flowers** (see Fig. 9.7). If a flower has both, it is **perfect** even though it may lack either sepals or petals or both. Sepals and petals do not produce spores and are considered **nonessential organs.**

It is necessary to consider the whole plant and the whole species as well as individual flowers. Stamens produce pollen that results in sperm production, and carpels are involved in egg production, so a *species* must have both types of organs. Plants that have perfect flowers satisfy this requirement. But if some flowers of the species are imperfect, having no stamens for instance, then other flowers must have stamens by being either perfect or by being imperfect due to lack of carpels. A large number of combinations is possible: A species may have individuals that produce only staminate flowers and others that produce only carpellate flowers—this is **dioecy,** and the species (not the flower or the plant) is said to be **dioecious.** Examples of dioecious species are marijuana, dates, willows, and papaya. In dioecious species, the life cycle actually consists of four types of plants: (1) microgametophytes, (2) megagametophytes, (3) staminate sporophytes, and (4) carpellate sporophytes.

Monoecy is the condition of having staminate flowers located on the same plant as carpellate flowers; **monoecious species** include cattails and corn—ears are clusters of fertilized carpellate flowers, and tassels bear numerous staminate flowers (Fig. 9.26). In some members of the cucumber family, the type of flower produced varies. Young plants and those growing in a poor environment produce staminate flowers, whereas older plants and those growing in a good environment produce perfect or carpellate flowers. If fertilization occurs, the carpels develop into large fruits, and only a healthy, robust plant can afford to do this. A young or poorly growing plant cannot supply enough carbohydrate and protein for fruit development, but it can supply enough to produce pollen.

Dioecy is an extreme adaptation that ensures cross-pollination; a plant that produces only one type of spore cannot pollinate itself. It is similar to the condition of separate sexes

(a)

(b)

FIGURE 9.26
(a) Ears of corn are really large groups (inflorescences) of carpellate flowers surrounded by protective leaflike bracts (the husks). The corn "silks" are long stigmas. (b) Corn tassels are inflorescences of staminate flowers.

in mammals—no individual can fertilize itself. Both conditions ensure that fertilization is by sex cells that are not identical, thus increasing the genetic diversity of the offspring.

ANIMAL-POLLINATED FLOWERS

The evolution of animal-mediated pollination had a dramatic impact on the evolution of flowering plants. In wind-pollinated gymnosperms, things such as brightly colored petals, fragrances, and nectar are a waste of material and energy. But once insects began visiting early angiosperms, mutations that resulted in pigments, fragrances, or sugar-rich secretions became adaptive. For pollen that is carried by wind, the probability that any particular grain will actually land on a stigma is very low, whereas if it is carried by an insect that flies from flower to flower, the probability of a pollen grain reaching a stigma is much improved.

When this insect-flower association began around 120 million years ago, neither insects nor flowers were particularly sophisticated. Insects probably visited all kinds of flowers, not recognizing the different species. As a result, pollen often landed on the stigma of the wrong species, where it was useless. Mutations that increased a plant's distinctiveness, its recognizability by an insect—flower color, size, shape, fragrances, and so on—became selectively advantageous. Rewards for the insect such as sugary nectar or protein-rich pollen and stamens were advantageous. Mutations were adaptive if they increased the capacity of insects to recognize flowers that offered abundant nectar or pollen. Because it is expensive energetically for an insect to fly, the ability to recognize the most nutritious flowers from a distance while flying is advantageous. As a result, many lines of insects and flowers underwent **coevolution,** a flower becoming adapted for visitation by a particular insect and the insect for efficient exploitation of the flower. Coevolution has also occurred between flowers and birds and between flowers and bats.

The shape of the flower is particularly important, as a pollinator actually makes contact with it. Most flowers are radially symmetrical; that is, any longitudinal cut through the middle produces two halves that are mirror images of each other. These flowers are **actinomorphic** or **regular** (Fig. 9.27). But all insects, birds, and bats are bilaterally symmetrical—only one longitudinal plane produces two halves that are mirror images. In many species, flowers and pollinators have coevolved in such a way that the flowers are now also

Plant Biology Tutor CD-ROM

(a)

(b)

(c)

FIGURE 9.27

(a) Any median longitudinal section of this poppy results in two halves that are mirror images of each other; it is radially symmetrical, that is, actinomorphic. (b) *Dendromecon* is also radially symmetrical. (c) This orchid flower is bilaterally symmetrical, zygomorphic; only if cut from top to bottom are the two halves mirror images. A pollinator, also bilaterally symmetrical, fits well only if it approaches the flower properly. (*Courtesy of Todd Barkman, University of Texas*)

bilaterally symmetrical—**zygomorphic** (Fig. 9.27c). When a pollinator approaches a zygomorphic flower, only one orientation is comfortable for it; any misalignment prevents the pollinator's head or body from fitting the flower's distinctive shape. As a result, as the pollinator feeds at the flower, pollen is placed on a predictable part of its body. When it visits the next flower, pollen is rubbed directly onto the stigma. Not only is pollen carried to the appropriate flower, but it is carried directly to the stigmas. By contrast, a pollinator can approach an actinomorphic flower from any direction and pollen may be brushed onto any part of the body; when it visits the next flower, the pollen-carrying part may not come into contact with the stigmas and effect cross-pollination.

WIND-POLLINATED FLOWERS

In species that are wind-pollinated, a totally distinct set of modifications is adaptive. Attracting pollinators is unnecessary, so mutations that prevent the formation of petals are selectively advantageous, and the energy saved can be used elsewhere in the plant. Sepals are also often reduced or absent and the ovaries need no special protection, so the whole flower may be tiny. Zygomorphy provides no selective advantage. Once pollen is released to the wind, the chance of any particular pollen grain landing on a compatible stigma is small, so huge numbers of grains must be produced. Large, feathery stigmas are adaptive by increasing the area that can catch pollen grains (Fig. 9.28). In general, wind-pollinated individuals produce up to several thousand small flowers; although each flower is tiny, the entire plant has a large total stigmatic surface area.

Pollination is aided by the growth pattern of the plant population. Wind-pollinated species—like grasses, oaks, hickories, and all conifers—grow as dense populations in rangelands or forests. Within 1 km² may be found thousands of plants and, more importantly, millions of stigmas. Species that occur as widely scattered, rare individuals must rely on animal pollination.

Stigma

Stamens

FIGURE 9.28

This is an inflorescence of St. Augustine lawn grass (*Stenotaphrum secundatum*); if lawns are mowed frequently, inflorescences and flowers are not seen. Like many wind-pollinated species, this has feathery, highly branched stigmas that catch wind-borne pollen. (*James L. Castner*)

OVARY POSITION

The ovary and ovules must be well protected from pollinators. Paradoxically, a flower must bring a hungry animal to within millimeters of protein-rich ovules in order to effect pollination. Adaptations that maximize the separation are long styles and stamen filaments.

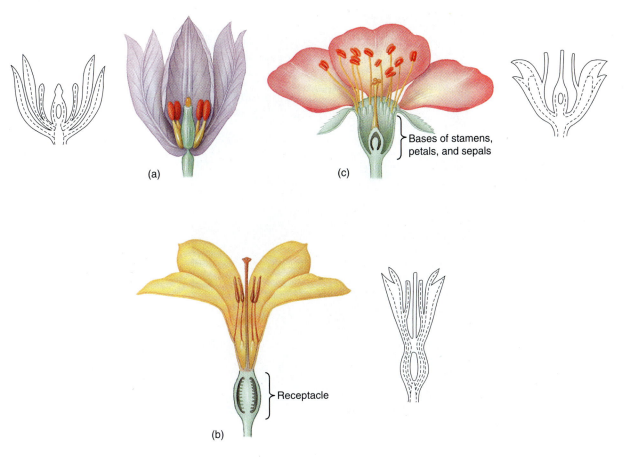

FIGURE 9.29
(a) In flowers with superior ovaries, carpels sit above the other organs, as would be expected because they are initiated last by the flower apical meristem. (b) This is an inferior ovary resulting from the fusion of the bases of sepals, petals, and stamens to the base of the carpel. Four appendages are involved whose vascular bundles (*dashed lines*) are still present and distinct. (c) In some species, after all the appendage primordia are initiated, the receptacle tissue grows upward and surrounds the ovaries. This can be detected because the vascular bundles around the ovary are receptacle bundles, not appendage bundles.

Burying the ovaries deep within the flower provides further protection. In some species, after all flower organ primordia are initiated at the receptacle apex, the primordia crowd together and the bases of the stamens, petals, and sepals fuse, creating a thick layer of protective tissues around the ovaries, which *appear* to be located below the other organs (see Figs. 9.22 and 9.29a and b). Two terms describe this: We can either say that the ovary is an **inferior ovary** or that the other parts are **epigynous.** Inferior ovaries also can result if receptacle tissue grows upward around the ovary (Fig. 9.29c). The more common arrangement, in which no fusion to the ovary occurs and the ovary is obviously above the other flower parts, is a **superior ovary** or **hypogynous** parts. Intermediate, partially buried ovaries are **half-inferior** with **perigynous** flower parts.

Epi—above; gynous—the gynoecium, the carpels; hypo—below; peri—around.

INFLORESCENCES AND POLLINATION

The positioning of flowers on an individual plant is important; few species have plants that produce only a single flower. Instead, many flowers are produced either within a single year or over a period of many years. A large mountain ash or cherry tree produces thousands of flowers every year for well over 100 years. Many important factors determine whether a flower is seen and visited by pollinators—for example, the positions of flowers relative to other flowers, leaves, and trunk; height from the ground; and distance from an open, uncluttered flight path for pollinators (Fig. 9.30).

FIGURE 9.30
Flowers of sausage tree *(Kigelia)* are pollinated by bats, which do not like to fly among the clutter of leaves because their sonar does not work well there. Long stalks allow flowers to hang in open air where bats have free access to them.

Reproductive success is measured in terms of the number of healthy, viable seedlings that become established. One important factor for this is pollination. It might seem optimal to have large flowers that can be easily seen by pollinators; unfortunately, they can also be seen by herbivores, and because ovules and pollen are rich in protein, they are good food sources for pests. Production of the pedicel, receptacle, sepals, and petals can be thought of as packaging and advertising cost; the larger the flower, the larger the cost. These costs can be made more acceptable (less disadvantageous selectively) by increasing the number of ovules (potential seeds) per flower. Large flowers tend to have numerous ovules and small flowers fewer. Small flowers exist because small plants cannot afford large flowers and their accompanying large seeds, which are very expensive. Also, a large flower with many ovules is a big risk—if it is found and eaten, a major reproductive investment has been lost. But damage to a small flower with only a few ovules is less significant. Also, if many flowers are grouped together, an **inflorescence,** they give a collective visual signal to pollinators: One small flower may be overlooked, but not a hundred close together (Fig. 9.31). Further-more, in an inflorescence, the plant is able to accurately control the timing of the initiation, maturation, and opening of the flowers. Consequently, the plant can be in bloom and available to pollinators for several weeks even though each flower lasts only a day or two.

Large flowers can produce more nectar than small flowers, but this makes them a more tempting target for nectar robbers, animals that take nectar without carrying pollen. Fur-thermore, a flower should not produce enough nectar to completely satisfy even a legiti-mate pollinator, because no incentive would remain for the pollinators to go to another flower; it is better if the pollinators get only enough to make them interested in searching for more.

Many other factors affect the relative selective advantage or disadvantage of flower number and size as well as the number of ovules per flower. Considering the numerous types of plants, pollinators, and environments, the diversity of flower and inflorescence types is not surprising. In the simplest arrangement, flowers occur individually in leaf axils or as a transformation of the shoot apex. When grouped into inflorescences, two basic arrangements occur: (1) **determinate inflorescences** and (2) **indeterminate inflores-cences.**

(a)

(b)

FIGURE 9.31
(a) Flowers of this *Clerodendrum* are small and inconspicuous, but when grouped together their inflorescence has even more impact than a single large flower. (b) The inflorescence of *Combretum* is even more striking than the individual flowers. Notice how the flowers mature and open sequentially.

A determinate inflorescence has only a limited potential for growth because the inflorescence apex is converted to a flower, ending its possibilities for continued growth (Fig. 9.32a and b). Typically, but not always, the terminal flower opens first and then lower ones open successively. In the simplest type, below the terminal flower is a bract with an axillary flower, which also may have a bract and axillary flower, and so on.

In an indeterminate inflorescence, the lowest or outermost flowers open first, and even while these flowers are open, new flowers are still being initiated at the apex. A raceme has a major inflorescence axis, and the flowers are borne on pedicels that are all approximately the same length (Fig. 9.32c). Catkins are similar to racemes, differing in that the flowers are

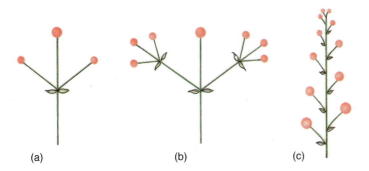

(a)   (b)   (c)

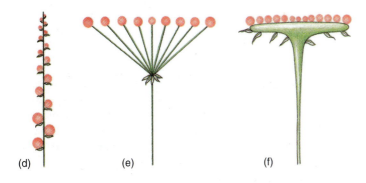

(d)   (e)   (f)

FIGURE 9.32
Inflorescence types: (a) Simple determinate inflorescence. (b) Compound determinate inflorescence. (c) Raceme. (d) Spike. (e) Umbel. (f) Head. Larger circles represent flowers that open earlier than those depicted as smaller circles.

PLANTS & PEOPLE

FLOWERS, FRUITS, SEEDS, AND CIVILIZATION

Flowers and the fruits and seeds that result from them have always been important to us. One of the first hominids—animals on our line of evolution—was *Australopithecus africanus;* fossils of its teeth, 3 million years old, show that it was adapted to eating plants. *Homo habilis* lived only 1.8 million years ago and likewise had a diet strongly dependent on plants. Our species, *Homo sapiens,* appeared evolutionarily about 500,000 years ago and survived by hunting game and collecting fruits, seeds, and edible roots.

About 11,000 years ago, a momentous change occurred. Small groups of humans began cultivating plants rather than just gathering them. This happened in the Middle East, an area that is now Iran, Iraq, and Syria, also known as the Fertile Crescent. The plants were common, wild species such as wheat, barley, peas, and lentils. Farming required major changes in human society: People had to stay in one place to tend and protect crops rather than follow herds of game animals. Permanent villages were necessary, as were the rules and regulations needed when people live together in close proximity. Land, huts, and harvests are tempting targets for thieves, so defense, both individual and collective, became a necessity—government had to be created.

Civilization advanced rapidly: Neolithic (New Stone Age) agricultural societies were widespread by 6000 BC, and the ox-drawn plow was in use as early as 4500 BC. New plants were cultivated: olives, date palms, and grapes for eating and for wine. Independently, the peoples of Southeast Asia domesticated rice and soybeans; in the New World the Incas, Mayans, and Aztecs cultivated potatoes, corn, tomatoes, beans, squashes, cocoa, pineapples, and peanuts.

The importance of fruits and seeds to the survival of both individual humans and societies is reflected in the prominent position they were given in early art. Ancient Egyptians carved likenesses of date palms, barley, and wheat as long ago as 3000 BC, and on Crete in 1800 BC, Minoan artists depicted, in addition to the plants already mentioned, figs, saffron crocus, pomegranate, and lupin. They also showed several plants that must have been grown purely as ornamentals: lilies, myrtle, narcissus, and roses.

As societies become wealthier, time and resources become available for leisure and pleasurable pursuits. Gardens dedicated to beautiful, fragrant flowers, not food crops or medicinal plants, arose early. Expeditions in search of exotic ornamental plants are described in the oldest epic poem in the world, the story of Gilgamesh of Sumer, and an ancient Egyptian monument, the Palermo Stone, records a plant-gathering expedition by King Snefru in 2900 BC. Inventories list the extensive plant-collecting trips and gardens of Ramses III.

Growing beautiful flowers and enjoying exotic fruits have never lost their popularity; after the Spanish discovered America, the initial exploitation was for gold and silver, but there soon followed expeditions dedicated solely to gathering plants, especially ornamental ones. Two of the many famous plant explorers sent to the United States are David Douglas, who explored the Pacific Northwest for the Horticultural Society of London in the 1820s, and the father and son team, André and François-André Michaux, sent by the government of France in 1785.

Even today, gardening and growing flowers for pleasure are considered by many an essential part of our civilization. As technology increases, cultivating flowers, or at least having a potted plant on a window sill, seems to maintain contact with our past.

---

imperfect, either staminate or carpellate, and all flowers of a single inflorescence are the same, so each species must have both staminate catkins and carpellate ones. Catkins almost always contain very small wind-pollinated flowers. A spike is similar to a raceme except that the flowers are sessile, lacking a pedicel (Fig. 9.32d). A spadix (plural: spadices) is a spikelike inflorescence with imperfect flowers, but both types occur in the same inflorescence, most often with staminate flowers located in the upper portion of the inflorescence and carpellate flowers in the lower portion, although they can intermingle. The main inflorescence axis is thick and fleshy with minute flowers embedded in it; the entire inflorescence is subtended or enclosed by a petal-like bract called a spathe (see Fig. 18.7). A panicle is a branched raceme with several flowers per branch.

Several types of indeterminate inflorescences have no dominant main axis. In umbels, the inflorescence stalk ends in a small rounded portion from which arise numerous flowers (Fig. 9.32e). Their pedicels are long and arranged so that all flowers sit at the same height, forming a flat disk. A head is similar to an umbel except that the flowers are sessile and attached to a broad expansion of the inflorescence stalk (Fig. 9.32f); numerous bracts may surround the inflorescence during development. Heads are almost synonymous with the aster family, sunflowers and dandelions being easily recognizable examples. In this group, the inflorescences are so compact and highly organized that they mimic single flowers; what appear to be the petals are really entire flowers, ray flowers, in which the petals are very large and fused together. The center of the inflorescence is composed of a different type of flower, disk flowers, in which the corollas are short and inconspicuous.

FRUIT TYPES AND SEED DISPERSAL

Fruits are adaptations that result in the protection and distribution of seeds. Many different agents disperse fruits and the seeds they contain: Gravity, wind, water, and animals are the most common (Table 9.4). The principles involved in fruit function are somewhat opposed to each other: Fruits that are tough and full of fibers or sclereids, such as pecans, walnuts, brazilnuts, and coconuts, offer maximum protection but are heavy and expensive metabolically. Also, the protected seed must be able to break out to make germination possible; a more fragile fruit is better for that. If animals are to disperse the seeds, part of the fruit must be edible or otherwise attractive while the seed and embryo must be protected from consumption. A division of labor often occurs—some parts being protective, others attractive, and still others allowing germination.

TRUE FRUITS AND ACCESSORY FRUITS

The term "pericarp" refers to the tissues of the fruit regardless of their origin. In most cases, this is the ovary wall, but in many species, especially those with inferior ovaries, the receptacle tissues or septal, petal, and stamen tissues may also become involved in the fruit. The terms "pericarp" and "fruit" have been applied to both types of fruit, so now the term **true fruit** is used to refer to fruits containing only ovarian tissue, and **accessory fruit** (or false fruit) is used if any nonovarian tissue is present (Fig. 9.33). Apples develop from inferior ovaries and the bulk of the fruit is enlarged bases of sepals and petals; only the innermost part is true fruit derived from carpels (see Fig. 9.22).

Fusion of carpels also affects the nature of fruits. If the fruit develops from a single ovary or the fused ovaries of one flower, it is a **simple fruit,** the most common kind. If the separate carpels of one gynoecium fuse during development, an **aggregate fruit** results, such as raspberries. If during development all the individual fruits of an inflorescence fuse into one fruit, it is a **multiple fruit,** as in figs, mulberries, and pineapple (Fig. 9.33b). These are also largely accessory fruits because in addition to the ovary tissue, the inflorescence axis, bracts, and various flower parts contribute to the mature fruit.

TABLE 9.4	*Agents of Dispersal*
Agent	**Descriptive Term**
Animals	Zoochory
Attached to animal	Epizoochory
Eaten by animal	Endozoochory
Birds	Ornithochory
Mammals	Mammaliochory
Bats	Chiropterochory
Ants	Myrmecochory
Wind	Anemochory
Water	Hydrochory
Dispersed by the plant itself	Autochory

Seeds, fruits, and asexual propagules can be dispersed by many means. These are a few of the most common types.

Each type of dispersal agent favors certain adaptations and eliminates others; fruits become specialized for particular types of dispersal.

(a)

(b)

FIGURE 9.33

(a) The red, edible flesh of strawberry is really the receptacle, not carpel tissue, so it is an accessory fruit. The "seeds" are true fruits, each derived from one carpel of the flower.
(b) Pineapples develop from the coalescence of all the many true fruits of one inflorescence, so it is a multiple fruit. In addition, many noncarpellary tissues become involved, so it is an accessory fruit as well. Bracts are visible here. *(Courtesy of Dole, Inc.)*

CLASSIFICATION OF FRUIT TYPES

There are several ways of grouping or classifying fruits. In one method, emphasis is placed on whether the fruit is **dry** or **fleshy**. A dry fruit is one that is not typically eaten by the natural seed-distributing animals; fleshy fruits are those that are eaten during the natural seed distribution process.

A further classification of dry fruits emphasizes fruit opening: **Dehiscent fruits** break open and release the seeds, whereas **indehiscent fruits** do not (Table 9.5). For the most part, fleshy fruits are indehiscent. Animals chew or digest the fruit, opening it; if uneaten, the fruit rots and liberates the seeds. Although many dry fruits are dehiscent, others must rot or the seed must break the fruit open itself.

Perhaps the simplest fruits are those of grasses: Each fruit develops from one carpel containing a single ovule. During maturation, the seed fills the fruit and fuses with the fruit wall, which remains thin. It has little protection and no attraction for animals. The fruit/seed falls and germinates close to the parent sporophyte. The fruits are indehiscent, and during germination the embryo absorbs water, swells, and bursts the weak fruit. These fruits are caryopses.

This refers to natural frugivores (fruit eaters), not to humans who cook food and add sugar to make dry fruits edible.

TABLE 9.5   *Fruit Types*

1. Fleshy fruits
   Berry: a fleshy fruit in which all three layers—endocarp, mesocarp, exocarp—are soft (grape, tomato).
   Pome: similar to a berry except that the endocarp is papery or leathery (apple).
   Drupe: similar to a berry except that the endocarp is hard, sclerenchymatous (stone fruits: peach, cherry, plum, apricot).
   Pepo: a fleshy fruit in which the exocarp is a tough, hard rind; the inner soft tissues may not be differentiated into two distinct layers (pumpkin, squash, cantelope).
   Hesperidium: exocarp is leathery (*Citrus*).

2. Dry fruits
   Indehiscent fruits
   Developing from a single carpel
      Caryopsis: simple and small, containing only one seed, and the testa (seed coat) becomes fused to the fruit wall during maturation (grasses: wheat, corn, oats).
      Achene: like a caryopsis, but the seed and fruit remain distinct. Fruit wall is thin and papery (sunflowers).
      Samara: a one-seeded fruit with winglike outgrowths of the ovary wall (maples, alder, ash).
   Developing from a compound gynoecium (a compound pistil)
      Nut: although the gynoecium originally consists of several carpels and ovules, all but one ovule degenerate during development. Pericarp is hard at maturity (walnut).
   Dehiscent fruits
   Developing from a single carpel
      Legume: fruit breaks open along both sides (beans, peas)
      Follicle: fruit breaks open on only one side (columbine, milkweeds)
   Developing from a compound gynoecium
      Capsule: opens many ways:
         Splitting along lines of fusion (*Hyperium*)
         Splitting between lines of fusion (*Iris*)
         Splitting into a top and bottom half (primrose)
         Opening by small pores (poppy)
      Schizocarp: Compound ovary breaks into individual carpels called mericarps.

3. Compound fruits
   Aggregate fruits: carpels of flower not fused, but grow together during fruit maturation (raspberry).
   Multiple fruit: all the fruits of an inflorescence grow together during fruit maturation (pineapple).

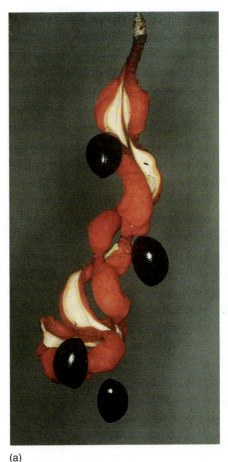

(a)

(b)

FIGURE 9.34

(a) At maturity different layers in the fruit contract in opposite directions, causing the pod to twist and break open. (b) As these maple fruits drop, the wings cause them to rotate and float, thus remaining aloft and traveling much farther on the wind.

Beans and peas form from a single carpel that contains several seeds. The fruits, known as legumes, are dry and inedible (except to humans and insects) like caryopses but do not fuse to the seeds. At maturity, the two halves of the fruit twist and break open (dehisce) along the two specialized lines of weakness (Fig. 9.34). After liberation from the legume, the seeds are protected from small animals and fungi only by the seed coat.

Fruits or seeds carried by wind must be light; they often have wings (maples) or parachutes (dandelions) that keep them aloft as long as possible. Such fruits are dry and weight much less than 1 g. Fruits and seeds that are transported by water (ocean currents, streams, floods) can be larger and heavier, but they must be buoyant and resist mildew and rot. An excellent example is coconuts that float from one island to another.

Animals carry fruit in a variety of ways. Dry fruits with hooks or stickers catch onto animal fur or feathers; sticky, tacky fruits glue themselves onto animals; and fleshy, sweet, colorful fruits are eaten. Edible fruits require particular specialization because the fruit must be consumed but the seeds must not be damaged; they must not be crushed by teeth or gizzard or be digested. Immature fruits deter frugivores by being hard, bitter, or sour, such as unripe apples or bananas. At maturity, the fruit is soft, sweet, flavorful, and typically strikingly different in color. The enclosed seeds often have hard seed coats or endocarps (peach, cherry) that allow them to pass through an animal unharmed (see Fig. 9.23).

Grapes and tomatoes are examples of simple fleshy fruits: They are berries because all parts of the fruit are soft and edible (Table 9.5). The seed coat provides some protection against being crushed during chewing, but perhaps more important are the size of the seeds and their slippery seed coat, which allows the seeds to slip out from between our molars. Pomes (apples and pears) differ from berries in two ways. First, they develop from inferior ovaries, so the inner tissue is the true fruit and the outer tissue is accessory fruit. Second, the seeds are protected because the innermost fruit tissues, the cores, are bitter and tough. The most elaborate fleshy fruits are drupes, in which the innermost tissues, the endocarp, are extremely hard and totally inedible, full of sclerenchyma. The mesocarp becomes thick, fleshy, juicy, and sweet and the exocarp forms the peel, whose color informs the frugivore of the fruit's ripeness. Drupes provide maximum attraction to animals with minimum danger to the seed.

The endocarp of a drupe is known as the pit and is often mistaken for the seed, but the true seed is located inside.

Because frugivores are a major danger to seeds, it is necessary to think about how edible fruits evolved: They must offer some advantage that outweighs the danger. Animals have predictable habits and migration patterns. Ants, birds, and mammals carefully select the sites where they feed, nest, and sleep; therefore, seeds are not distributed at random but are moved through the environment to specific sites. One example is mistletoe: Its seeds are sticky and adhere to a bird's beak as it eats the fruit. The bird probably feeds and cleans itself in the same species of tree, so the seeds are rubbed off the beak directly onto a proper host tree. Few seeds are ever lost by dropping to the ground or being deposited in the wrong tree, as would happen with wind or water dispersal. Just as efficient is dispersal of fruits and seeds of marsh plants. The birds and animals that live in marshes typically spend little or no time in other habitats, so seeds carried by these animals are almost certainly distributed to favorable sites.

A special benefit of seed distribution by animals is that the "deposited" seed may find itself in a small (or large) mound of "organic fertilizer" (Fig. 9.35). Seeds that are adapted to pass through an animal's digestive tract are resistant to digestive enzymes and can tolerate the high level of ammonium in fecal matter. Because many seeds of plants cannot tolerate these two factors, the adapted seeds find themselves in a microenvironment that is not only nutrient-rich but also excludes some competitors.

FIGURE 9.35
This seedling is as happy as a clam at high tide: It has enough fertilizer to last for months. Many of its competitors have been digested; others are being suffocated.

SUMMARY

1. Asexual reproduction results in new individuals genetically identical to the parent. The parent and all progeny are equally adapted to the habitat.

2. Sexual reproduction results in progeny that differ from each other genetically. There is a range of fitness, and some progeny may have a combination of maternal and paternal genes that causes them to be even better adapted than the parents.

3. All seed plants have a life cycle that consists of an alternation of heteromorphic generations. Micro- and megagametes are produced by micro- and megagametophytes that grow from micro- and megaspores. Most species have only one type of sporophyte, capable of producing both types of spores. Dioecious species have two types of sporophytes, one of which produces microspores and the other megaspores.

4. Flowers contain all or some of the following: nonessential organs —pedicel, receptacle, sepals, and petals; and essential organs—stamens and carpels.

5. In flowering plants, the microgametophyte consists initially of a vegetative cell and a generative cell, but the latter divides into two sperm cells. The megagametophyte consists of the egg cell, two synergid cells, a binucleate central cell, and three antipodal cells.

6. After fertilization, the zygote (fertilized egg) grows into an embryo; polar nuclei fuse with the second sperm nucleus, establishing the endosperm nucleus and endosperm; and the integuments develop into a seed coat. The entire structure is a seed. The ovary develops into a fruit.

7. Inflorescences are groups of many flowers. They may be determinate or indeterminate, and may contain several types of specialized flowers.

8. Fruits protect seeds during development and may aid in their dispersal and maturation. Fruits are either dry, inedible, dehiscent, or indehiscent, or they are fleshy, edible, and indehiscent.

IMPORTANT TERMS

agamospermy	double fertilization	imperfect flower	petal	sepal
alternation of generations	endosperm	incomplete flower	plasmogamy	spore
anther	epicotyl	inflorescence	pollen	sporophyte
carpel	fruit	karyogamy	pollen tube	stamen
coevolution	gamete	ovary	radicle	stigma
compatibility barrier	gametophyte	ovule	receptacle	style
complete flower	heteromorphic generations	pedicel	seed	syngamy
cotyledon	hypocotyl	perfect flower	seed coat	zygote
cross-pollination				

REVIEW QUESTIONS

1. Under what conditions is it selectively advantageous for a plant to produce offspring identical to itself?

2. Draw and label a microgametophyte of a flowering plant. What type of gamete does it produce? Draw and label a megagametophyte of a flowering plant. What type of gamete does it produce?

3. What is the difference between a perfect and an imperfect flower? Between a complete and an incomplete flower?

4. Explain the following terms: inferior ovary, superior ovary, actinomorphic flower, zygomorphic flower. How is each of these modifications selectively advantageous?

5. After pollination and then fertilization, what usually happens to each of the following: stigma, style, carpel, ovule, integuments, zygote?

6. The pericarp often consists of three layers. What are the layers called? In a cherry, which layers correspond to the skin, flesh, and stone?

BotanyLinks .net Questions

www.jbpub.com/botanylinks

Visit the .net Questions area of BotanyLinks (http://www.jbpub.com/botanylinks/netquestions) to complete these questions:

1. How does living underground affect flowering, pollination and seed dispersal? Go to the BotanyLinks home page for more information on this subject.

BotanyLinks includes a Directory of Organizations for this chapter.

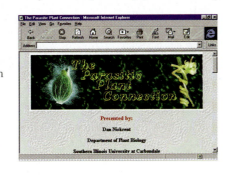

2. What kinds of flower modifications would be selectively advantageous for desert plants? Go to the BotanyLinks home page for more information on this subject.

These apples (which contain the seeds necessary to initiate the next generation of apple trees), are the result of the plants' successfully obtaining energy, water, and nutrients as well as the plants' controlling their growth and development.

II

PLANT PHYSIOLOGY AND DEVELOPMENT

When you are studying the subject of Part I—plant structure—you can often see the material directly with the naked eye, although light or electron microscopy may be necessary to see some structures. But in plant physiology and development, the objects of our study—chemical reactions and metabolic pathways—cannot be seen at all. Instead, the results of experiments, measurements, and analyses are studied and hypothetical reactions and pathways are set down on paper. From these, predictions are made and new observations planned to test the hypotheses. It is easy for us to study and memorize metabolic pathways, chemical formulas, and diagrams of physiological control mechanisms without appreciating that these have never been, and cannot ever be, seen directly. Every one is a theoretical model that is consistent with the majority of the available data and that is logically and internally consistent.

When experienced anatomists see an unusual structure, they may be able to recognize instantly that it is new to science or is at least a significant variation. For physiologists, it is not so easy: If an experiment on photosynthesis does not come out as expected, it may be that a new type of photosynthesis is being discovered; it may be that the current theories of photosynthesis are not completely accurate and have made an erroneous prediction; or it may be that an experimental error has occurred. It can be difficult for students to appreciate the tremendous amount of careful, ingenious work that must be done just to establish that a particular theoretical metabolic pathway truly represents the reactions that occur in certain plants. Whereas it is relatively easy to determine that natural selection has resulted in the evolution of many types of leaves, stems, roots, flowers, xylem, phloem, and so on, it is much more difficult in physiology. Numerous differences in microhabitats, water availability, heat, cold, soils, pests, and plant diseases have resulted in diverse types of structures that are selectively advantageous under various conditions. It is logical to expect the same to be true of metabolism; we do know that there are several varieties of photosynthesis and respiration, and there may be others that have not yet been discovered.

As you study the material in this section, keep in mind that, just as is true for structure, organization is of fundamental importance for metabolism. The chemical and physical reactions that constitute plant metabolism are highly ordered and not at all random. This orderliness is maintained by the input of energy (see Chapters 10 and 11) acting on materials brought into plants from the environment (see Chapters 12 and 13). There are many types of order, and the information necessary to establish the proper reactions acting on the proper material is stored in the genes, both in the nucleus and in the plastids and mitochondria. The mechanisms by which the genes control the interaction of energy and matter, such that a plant of a particular species results, are discussed in Chapters 14 and 15.

10

ENERGY METABOLISM: PHOTOSYNTHESIS

Plants gather energy by photosynthesis, a process carried out by pigments and enzymes in the chloroplasts of green tissues.

CONCEPTS

Probably the most important concept concerning cells and all of life itself is that living organisms are highly ordered, highly structured systems. The universe as a whole is constantly becoming less and less orderly; its disorder (**entropy**) is increasing. Bacteria, protists, fungi, plants, and animals, however, represent phenomena in which particles become more orderly. A plant absorbs diffusely scattered molecules of carbon dioxide, water, and minerals and organizes them into organic molecules, cells, tissues, and organs. Each plant carries this out with such precision that each species of plant is easily distinguishable from others. After death, decay is the process by which an organism's molecules become more disordered and scattered—their entropy increases.

    Because living organisms are part of the natural world that is described by the laws of physics and chemistry, the decrease in the entropy of living organisms must obey physical laws. This is accomplished by putting energy into the living system, the source of energy being sunlight. To be accurate, we must consider the sun and life together: The atomic reactions that generate sunlight cause greater disorder in the sun than sunlight causes order in living organisms. The whole system (sun + life) becomes more disordered. Because there is no means of putting energy into an organism's body after death, an increase in entropy cannot be prevented.

Sunlight maintains and increases the orderliness of life by two methods: (1) directly, in the process of photosynthesis, which produces complex organic compounds, and (2) indirectly, in the respiration of those organic compounds, either by the organism itself or by another organism that eats it. These two methods of supplying energy and maintaining orderliness—photosynthesis and respiration—are the basis for a major, fundamental distinction in the types of organisms. **Photoautotrophs** are organisms that gather energy directly from light and use it to assimilate small inorganic molecules into their own tissues. Photoautotrophs include all green plants, all cyanobacteria, and the few bacteria capable of photosynthesis. **Heterotrophs** are organisms that cannot do this but instead take in organic molecules and respire them, obtaining the energy available in them. Heterotrophs include all animals, all completely parasitic plants (Fig. 10.1), all fungi, and the nonphotosynthetic bacteria. Gathering energy by taking in organic material has the advantage that part of the material can be used as construction material instead of fuel. At least some of the amino acids, fatty acids, and sugars in food can be built into the organism's own polymers and the rest respired for energy. Photoautotrophs must build all their own molecules using just carbon dioxide, water, and various nitrates, sulfates, and other minerals.

Tremendously important consequences follow from the fact that photoautotrophs and heterotrophs differ in their sources of energy and building material (Table 10.1). Sunlight and carbon dioxide do not need to be stalked, hunted, and captured, so sensory organs, muscles, and central nervous systems like those of animals are unnecessary. Conversely, the ocean is full of microscopic bits of food, and animals such as sponges and corals can gather it the way plants gather carbon dioxide. The mode of nutrition has had overriding influence on the bodies and metabolisms of plants and animals.

Tissues and organs are also either photoautotrophic or heterotrophic. Chlorophyllous leaves and stems are photoautotrophic, whereas roots, wood, and flowers are heterotrophic and survive on carbohydrates imported through phloem. During winter, if all leaves have abscised, the entire plant may be composed of heterotrophic tissues, and it maintains its metabolism by respiring stored starch.

Tissues often change their type of metabolism; young seedlings are white and heterotrophic while germinating underground (see Fig. 9.20a); they survive on nutrients stored in cotyledons or endosperm. Seedlings become photoautotrophic only after they emerge into sunlight. Immature fruits may be green and photosynthetic, but in the last stages of maturation, chloroplasts are converted to chromoplasts and metabolism then depends on imported or stored nutrients (chromoplasts are plastids that contain large amounts of pigments other than chlorophyll). Young leaf primordia are green, but they grow more rapidly than their own photosynthesis would permit; they have a mixed metabolism of photosynthesis and carbohydrate import.

Photosynthesis is a complex process by which carbon dioxide is converted to carbohydrate. This involves endergonic reactions driven by ATP and requiring new bonding orbitals filled by electrons carried to the reaction by NADPH (see Chapter 2). Before this can happen, ATP and NADPH themselves must be formed in highly endergonic reactions driven by light energy. In order to understand this, you must first understand the nature of light and pigments along with the concept of reducing power.

FIGURE 10.1
Total parasites such as this broomrape (*Orobanche*) are heterotrophs like animals. Like parasitic tapeworms or blood flukes, they absorb monomers such as monosaccharides and amino acids.

TABLE 10.1	*Differences Between Photoautotrophs and Heterotrophs*	
	Photoautotrophs	Heterotrophs
Source of energy	Sunlight	Food: carbohydrates, proteins, fats
Source of building material	Carbon dioxide	Food
Organisms	Photosynthetic plants	Animals, protozoa
	Algae	Nonphotosynthetic parts of ordinary plants: wood, roots, bark
	Cyanobacteria	Completely parasitic plants
	Photosynthetic bacteria	Fungi
		Most bacteria

ENERGY AND REDUCING POWER

ENERGY CARRIERS

Energy enters the biological world through photosynthesis, a process that converts light energy to chemical energy. The sun's light is captured by certain plant pigments that use the energy in chemical reactions. Unfortunately, the energized pigments can enter into only two chemical reactions, although plants have thousands of different reactions in their metabolism. Several theoretical ways exist of transporting energy from energized pigments into endergonic reactions:

1. Allow the pigments to enter into every reaction necessary. A problem is that the energized pigments are large molecules, so they are not very mobile and never move across membranes (Fig. 10.2). Furthermore, they are too energetic; they can react with almost anything and would be difficult to control.

2. Allow the energized pigments to make one or several smaller, less energetic intermediates that can be moved and controlled easily. Such a method has evolved: Photosynthetic reactions produce adenosine triphosphate (ATP), an extraordinary molecule (see Fig. 2.24). Its high-energy phosphate bonds carry enough energy to force almost any reaction to proceed, and it can enter into almost every reaction for which energy is needed. In those that it does not enter, other energy carriers, often relatives of ATP, are involved; the most frequent is **guanosine triphosphate (GTP),** which also carries high-energy phosphate bonds.

Although ATP is an essential molecule, it constitutes only a tiny fraction of the plant body. Each molecule is recycled and reused repeatedly, thousands of times per second. ATP is converted to ADP and phosphate by metabolic reactions, but the phosphate can be reattached with a high-energy bond by the reactions of either photosynthesis or respiration. Each molecule is an energy carrier, shuttling between reactions that release energy and those that consume it.

There are three methods by which ADP can be phosphorylated to ATP (Table 10.2). The first, **photophosphorylation,** involves light energy in photosynthesis; animals, fungi, and nonchlorophyllous plant tissues cannot perform photophosphorylation because they lack the necessary pigments and organelles. Instead, they respire some of the high-energy compounds that they have consumed as food or imported by phloem. Compounds with high-energy phosphate groups are produced, and these compounds force their phosphate onto ADP, making ATP. This is **substrate-level phosphorylation.** In the last stages of respiration, ADP is phosphorylated to ATP by **oxidative phosphorylation.** Each process occurs in a distinct site within the cell, and each captures energy from distinct types of exergonic reactions. Substrate-level and oxidative phosphorylation occur in all parts of the plant at all times, but photophosphorylation occurs only in chloroplasts in light.

Substrate-level phosphorylation and oxidative phosphorylation are discussed in Chapter 11.

REDUCING POWER

Earth's atmosphere is about 21% oxygen, so many compounds are found in their oxidized form: carbon as carbon dioxide (CO_2), sulfur as sulfate (SO_4^{-2}), nitrogen as nitrate (NO_3^-), and so on. "**Oxidized**" means that an atom does not carry as many electrons as it

TABLE 10.2 *Methods of Synthesizing ATP*		
	Energy Source	**Site**
Photophosphorylation	Sunlight	Chloroplasts
Substrate level phosphorylation	Reactions not involving oxygen	Cytosol
Oxidative phosphorylation	Oxidations with oxygen	Mitochondria

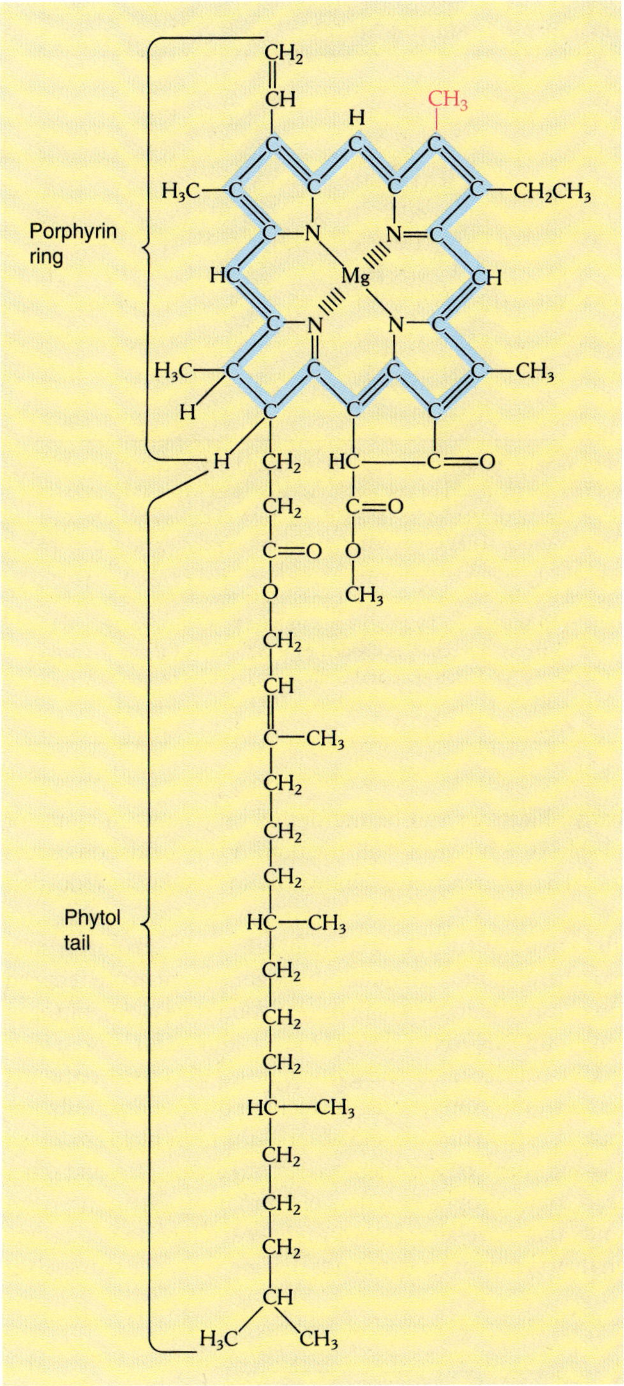

FIGURE 10.2
The tail of chlorophyll *a* contains only hydrogen and methyl functional groups, so it is hydrophobic and dissolves into the chloroplast's membrane lipids, immobilizing it. The porphyrin ring system of alternating double and single bonds acts as an antenna that captures light energy. The magnesium atom carries the electrons involved in photosynthesis. If the colored methyl group were an aldehyde group (—CHO), the pigment would be chlorophyll *b*.

could. In carbon dioxide, each oxygen can be considered to have pulled two electrons almost completely away from the carbon, and the carbon is said to be at a +4 **oxidation state.** This is speaking figuratively; electrons spend more time near the oxygen than they do near the carbon, sulfur, or nitrogen, but they are not torn completely away; these bonds are covalent, not ionic (Table 10.3).

When electrons are added to an atom, it becomes **reduced.** Think of a **reduction reaction** as one that reduces the positive charge on an atom and an **oxidation reaction** as one that increases the positive charge (Table 10.4). Oxygen has an electronegativity of 3.5 (see Table 2.3), a strong tendency to pull electrons away from an atom and raise that atom's partial positive charge. But hydrogen becomes more stable by giving up electrons, reducing

A preliminary rule of thumb is that oxidized compounds often (but not always) contain a great deal of oxygen, whereas reduced compounds contain hydrogen.

TABLE 10.3   Calculation of Oxidation States

CO_2 (carbon dioxide)	O -2	O -2	C $+4$	
CH_2O (carbohydrate)	O -2	H $+1$	H $+1$	C 0

$$\begin{array}{l} \text{COOH} \\ | \\ \text{H—C—H} \quad \text{(malic acid)} \\ | \\ \text{H—C—OH} \\ | \\ \text{COOH} \end{array}$$

$5 \times O$	$6 \times H$	$4 \times C$
$5 \times 2 = -10$	$6 \times +1 = +6$	$4 \times +1 = +4$

Calculating the oxidation state is often simple. Always assume that H is $+1$ and O is -2; carbon, sulfur, and nitrogen are variable, because they can add or lose electrons rather easily (see Table 2.3). Carbon dioxide and carbohydrates are neutral molecules, so the oxidation state of carbon must be opposite and equal to that of all other oxidation states combined. C is $+4$ in carbon dioxide and $+0$ in carbohydrate. During photosynthesis, four electrons must be added to each carbon in carbon dioxide so that the bonding orbitals of carbohydrate can exist; this lowers its oxidation state from $+4$ to 0. In malic acid the five oxygens contribute -10 and the six hydrogens add up to $+6$, so there are still 4 + charges to be accounted for by the four carbons: 4 + charges divided by 4 carbons equal one + charge per carbon. Carbon in malic acid is at the $+1$ oxidation state and is not as reduced as carbon in carbohydrate.

its partner's partial positive charge. Electrons can be transferred only between atoms or molecules, so each reaction in Table 10.4 is only a "half reaction." Every oxidation occurs simultaneously with a reduction. The full reaction is known as an "oxidation-reduction reaction," or "redox reaction."

Whereas compounds in the environment are predominantly in the oxidized state because of our oxygen-rich atmosphere, most compounds in organisms are in the reduced state. Carbon is often in the form of carbohydrate, where its oxidation state is $+0$; nitrogen is frequently present as an amino group, NH_3 ($N^{-3} H^{+1} H^{+1} H^{+1}$); and sulfur is present as SH_2 ($S^{-2} H^{+1} H^{+1}$). Thus in addition to needing energy, organisms also need **reducing power**, the ability to force electrons onto compounds. Reducing power is especially important to plants because they take in carbon dioxide and water—the most highly oxidized forms of carbon and hydrogen—and then convert them to carbohydrates, fats, and other compounds that are very reduced. Heterotrophs have less of a problem: When they consume plants, they get compounds that have already been reduced. They do need some reducing power, however, when they synthesize compounds that are very reduced, such as fatty acids.

TABLE 10.4   Reductions and Oxidations

	Oxidation State
Reductions	
$X + e^- \rightarrow X^-$	$0 \rightarrow -1$
$X^{+2} + e^- \rightarrow X^+$	$+2 \rightarrow +1$
$X^- + e^- \rightarrow X^{-2}$	$-1 \rightarrow -2$
Oxidations	
$X \rightarrow X^+ + e^-$	$0 \rightarrow +1$
$X^+ \rightarrow X^{+2} + e^-$	$+1 \rightarrow +2$
$X^{-2} \rightarrow X^- + e^-$	$-2 \rightarrow -1$

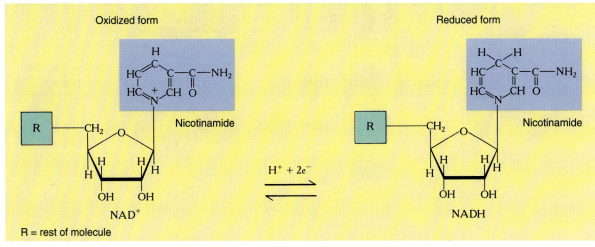

(a)

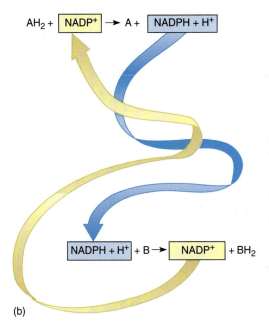

(b)

FIGURE 10.3

(a) The full structure of NAD^+ is presented in Figure 2.27; here, only the portion involved in carrying electrons is shown. The oxidized state (NAD^+ and $NADP^+$) carries a partial positive charge on the nitrogen and three double bonds in the ring. When it is reduced by two electrons, the positive charge disappears from the nitrogen, numerous bonding orbitals within the ring are changed, and the top carbon picks up a proton. No bonding orbitals are formed between NAD^+ and the electron donor, so NADH is free to diffuse away after picking up electrons. Similarly, as it donates electrons to some substrate, the ring bonding orbitals revert to the NAD^+ form, none of which binds it to the substrate, so the NAD^+ diffuses away. The same is true for $NADP^+$ (b) In the upper reaction, two electrons are passed from AH_2 to $NADP^+$: AH_2 has been oxidized to A, and $NADP^+$ has been reduced to $NADPH + H^+$. The NADPH is free to diffuse to another site, where it passes the two electrons onto another molecule, B, reducing it to BH_2 and becoming oxidized back to $NADP^+$.

Just as with energy, an optimum solution for moving and handling reducing power—electrons—is to use small molecules that are semistable and mobile. The two molecules used most often are nicotinamide adenine dinucleotide (NAD^+; see Figs. 2.27 and 10.3) and the closely related nicotinamide adenine dinucleotide phosphate ($NADP^+$). Both can pick up a pair of electrons and a proton, thereby becoming reduced to NADH and NADPH. When they reduce a compound by transferring their electrons to it, the proton is released and NAD^+ or $NADP^+$ is regenerated. Rather than having a large number of carrier molecules, the cell recycles each molecule, using it thousands of times a second as it shuttles between electron-producing and electron-consuming reactions.

Because NAD^+ and $NADP^+$ take electrons away from other molecules, they are **oxidizing agents**—they oxidize the material they react with. It is also possible to say that the material has reduced the NAD^+ or $NADP^+$. During the process, NADH and NADPH, two strong **reducing agents,** are formed. They have a powerful tendency to place electrons onto other molecules, reducing those molecules and becoming oxidized themselves (Fig. 10.3b). The tendency to accept or donate electrons varies greatly and is known as a molecule's **redox potential** (Table 10.5). Cells contain a variety of electron carriers that differ in their tendency to accept or donate electrons.

TABLE 10.5   *Redox Potentials of Electron Carriers*

Oxidized	Reduced	Number of Electrons Transferred	Redox Potential (volts)
X (Fe_4S_4)	X^-	1	−0.73
Ferredoxin (ox)	Ferredoxin (red)	1	−0.43
Q	Q^-	1	−0.35
$NADP^+$	$NADPH + H^+$	2	−0.32
NAD^+	$NADH + H^+$	2	−0.32
FAD	$FADH_2$	2	−0.22
FMN	$FMNH_2$	2	−0.22
Cyt b^{+3}	Cyt b^{+2}	1	−0.00
Ubiquinone	Ubiquinone-H_2	2	+0.10
Cyt c_1^{+3}	Cyt c_1^{+2}	1	+0.22
Cyt c^{+3}	Cyt c^{+2}	1	+0.25
Cyt a^{+3}	Cyt a^{+2}	1	+0.25
Cyt f^{+3}	Cyt f^{+2}	1	+0.36
Plastocyanin (ox)	Plastocyanin (red)	1	+0.37
Cyt a_3^{+3}	Cyt a_3^{+2}	1	+0.38
$1/2 O_2 + 2H^+$	H_2O	2	+0.82

Compounds with a large negative redox potential (at the top of the list) tend to donate electrons and exist in the oxidized state: X, ferredoxin, $NADP^+$, FAD. Compounds with a large positive redox potential (at the bottom of the list) tend to accept electrons and exist in the reduced form: Cyt a_3^{+2}, H_2O.

OTHER ELECTRON CARRIERS

Cytochromes.   **Cytochromes** are small proteins that contain a cofactor, heme, which holds an iron atom (Fig. 10.4); the iron carries electrons and cycles between the +2 and +3 oxidation states. Cytochromes are intrinsic membrane proteins; they are an integral part of the chloroplast's thylakoid membranes and cannot be removed without destroying the membrane. Consequently, they carry electrons only between sites that are extremely close together within a membrane rather than diffusing throughout the stroma as NADPH does.

Plastoquinones.   Plastoquinones, like cytochromes, transport electrons over short distances within a membrane (Fig. 10.4b). After they pick up two electrons, they also bind two protons. Their long hydrocarbon tail causes them to be hydrophobic, so they dissolve easily into the lipid component of chloroplast membranes.

Plastocyanin.   Like cytochromes, plastocyanin is a small protein that carries electrons on a metal atom, in this case copper. When oxidized, the copper ion is in the +2 oxidation state, but as it picks up the electron, it goes to the +1 oxidation state—it is reduced one level. Plastocyanin is loosely associated with chloroplast membranes; it can move a short distance along the surface, but it does not travel far.

PHOTOSYNTHESIS

As the name implies, photosynthesis is a process that uses light energy to synthesize something. The term is so general that it could be applied to many different types of reactions, but whenever a botanist uses the term "photosynthesis," the reaction being discussed is the combination of carbon dioxide with water to form carbohydrate (Fig.

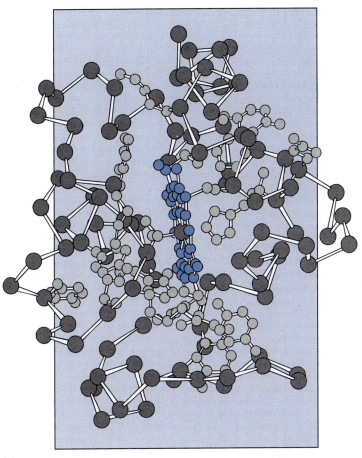

(a)

(b)

Plastoquinone A$_{45}$

Plastoquinone C

FIGURE 10.4

(a) In cytochromes, iron is not bonded directly to any amino acid but is held by heme, a boxlike porphyrin ring similar to the portion of chlorophyll that holds the magnesium ion. (b) The distinguishing feature of the quinone class of electron carriers is that each has two ketone groups (the double-bonded oxygen) whose carbon atoms are part of a ring structure.

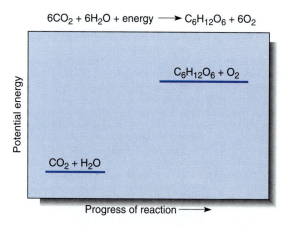

$$6CO_2 + 6H_2O + energy \longrightarrow C_6H_{12}O_6 + 6O_2$$

FIGURE 10.5

Although this chemical equation succinctly summarizes photosynthesis, it reveals virtually nothing of the reaction mechanism or the many carriers and enzymes that participate. We cannot draw a reaction diagram because photosynthesis does not occur by the direct interaction of six molecules of carbon dioxide with six of water; however, the relative potential energies can be shown, indicating that this is an endergonic process.

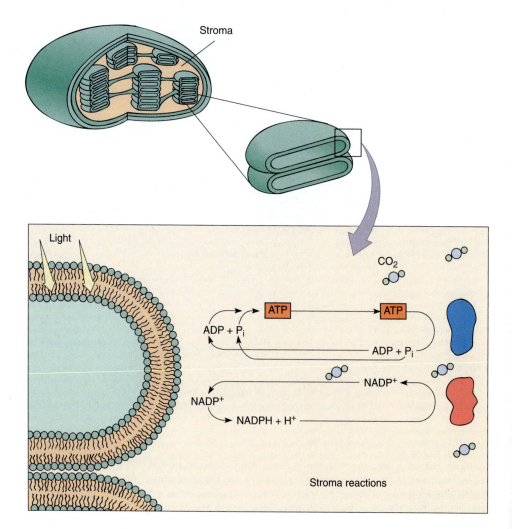

FIGURE 10.6

Light-dependent reactions of photosynthesis occur by means of membrane-bound carriers, but the actual formation of carbohydrate occurs in the chloroplast liquid (stroma). ATP-ADP and $NADP^+$-NADPH diffuse between the two regions. No region of the chloroplast is far from a membrane, so the distances traveled are only a few hundred times the diameter of a molecule.

10.5). Think about why these particular compounds are part of photosynthesis. First, both carbon dioxide and water are abundant and cheap, occurring almost everywhere in large quantities. The exception is the lack of water in severe deserts like the Sahara, where very little life exists simply because water is scarce. It is important to have a metabolism based on abundant compounds. It is necessary also for raw materials to be cheap; that is, the plant must be able to obtain them without expending much energy. Water and carbon dioxide are excellent because they diffuse into the plants automatically from soil, air, or water.

Another important quality of carbon dioxide and water is that they are very stable and contain little chemical energy, so it is possible to deposit a large amount of energy into them. The carbohydrates they form are a good means of storing energy because all reactions leading to carbohydrate breakdown have high energy-of-activation barriers. Despite being energy-rich, carbohydrates are stable and chemically unreactive.

Finally, both the reactants and the products of photosynthesis are nontoxic; it is safe to absorb large quantities of carbon dioxide and water and to store high concentrations of carbohydrates. Many substances that are critical for life are extraordinarily toxic if they become even slightly concentrated; chlorine, sodium, ammonium, and many vitamins are just a few examples.

During photosynthesis, the carbon of carbon dioxide is reduced and energy is supplied to it, converting it to carbohydrate (Fig. 10.5). The carbon atom in carbon dioxide is at the +4 oxidation state, while the carbon atoms in carbohydrate are, in general, at the +0 oxidation state. Four electrons must be found and placed into new bonding orbitals around the carbon atom to reduce it. This is not easy, because carbon is more stable in the oxidized state than in the reduced state: Carbohydrates such as wood and sugar can burn, releasing energy, but carbon dioxide does not. Therefore, a source of electrons and a source of energy are necessary for photosynthesis: The electron source is water and the energy source is light. But water and light do not act on carbon dioxide directly; instead they create the intermediates ATP and NADPH by a process called the **light-dependent reactions.** In a separate set of reactions, the **stroma reactions** (formerly known as the **dark reactions**), ATP and NADPH interact with carbon dioxide and actually produce carbohydrate (Fig. 10.6).

THE LIGHT-DEPENDENT REACTIONS

The Nature of Light. Light is one small segment of the **electromagnetic radiation spectrum,** which encompasses gamma rays, X-rays, ultraviolet light, infrared light, microwaves, and radio waves in addition to visible light (Fig. 10.7). Radiation can be thought of and treated physically either as a set of particles called **quanta** (sing. quantum), also called **photons,** or as a set of waves. The various types of radiation differ from each other only in their wavelengths and the amounts of energy each individual quantum contains. Those radiations with short wavelengths (cosmic rays, gamma rays, ultraviolet light) have relatively large amounts of energy in each quantum, whereas those with long wavelengths (infrared, microwaves, radar, and radio waves) have relatively little. Because humans use visible light for vision (that is why it is visible), we are more sensitive to and familiar with this region of the spectrum. We cannot see other wavelengths, but we can feel infrared radiation as heat, and our bodies react to ultraviolet light by becoming tanned or burned. We distinguish differences in quantum energy (wavelength) of visible light as differences in **color.** Most of us see all wavelengths from red (760 nm) through orange, yellow, green, blue, indigo, to violet (390 nm). Certain insects see some near ultraviolet (see Fig. 9.10), but in general all animals see in the range from 350 to 760 nm, which is also the radiation that plants use for photosynthesis.

The Nature of Pigments. Most materials absorb certain wavelengths of light much more than other wavelengths. If a substance absorbs all wavelengths except red, then red light either bounces off or passes through it, and the substance appears red to us. Any material that absorbs certain wavelengths specifically and therefore has distinctive color is a **pigment,** but we more often think of pigments as substances that absorb light as part of their

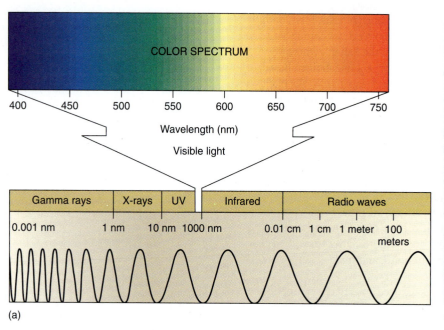

(a)

(b)

FIGURE 10.7

(a) On the left end of this electromagnetic spectrum are gamma rays. If treated as waves, they have extremely short wavelengths, less than 1 nm. If treated as particles, each quantum is highly energetic. Moving to the right, wavelengths become longer and quanta less energetic. The region to which our eyes respond—the visible region—is enlarged. (b) These apples have an abundance of pigments that absorb all colors except red. Earlier, while still developing, they had the pigment chlorophyll which absorbs red and blue strongly, but reflects green.

biological function. Some pigments, such as melanin (the pigment of suntanned skin), absorb light and thereby protect other light-sensitive substances. Other pigments, such as those in flowers, fruits, or the skins of animals, are important for the light they do not absorb, which gives the pigments their color and allows them to be useful to the organisms in attracting mates, pollinators, or frugivores or in hiding from predators (Fig. 10.7b).

Photosynthetic pigments transfer absorbed light energy to electrons that then enter chemical reactions. Because only absorbed energy can be used, a theoretically ideal photosynthetic pigment would be black: It would absorb and use all light, not letting any escape. The pigment should at least absorb high-energy radiation (ultraviolet light and gamma rays) instead of the fairly weak visible light. Rather surprisingly, the critical pigment, **chlorophyll *a,*** is not like this at all (see Fig. 10.2). It absorbs only some red and some blue light, letting most of the rest pass through, especially high-energy radiation. It is important to consider why chlorophyll *a* does not have what seem like ideal characteristics.

First, chlorophyll, like all other biological pigments, does not utilize high-energy quanta because they have too much energy. Each is so powerful that it would knock the electrons completely away from the pigment, disrupting bonding orbitals and causing the molecule to break apart. Notice in Figure 10.2 that all bonds in the chlorophyll ring system are double bonds that alternate with single bonds (conjugated double bonds); this bond system is excellent for absorbing quanta, but if even a single electron is knocked out, the whole structure becomes useless. It is selectively disadvantageous for a plant to produce a photosynthetic pigment that is destroyed by the light it absorbs; the molecule would break down and all the ATP that had been expended earlier in its construction would be wasted.

Long wavelength radiations, such as infrared and microwaves, have so little energy per quantum that they cannot appreciably boost an electron's energy. Instead, they make the pigment molecule warmer, as in a microwave oven, but this is not especially useful for chemical synthesis. Visible light contains just the right amount of energy per quantum. When one of these quanta is absorbed by the pigment, an electron is **activated**—raised to an orbital of a higher energy level. We say that the electron and the molecule go from the **ground state** to an **excited state** (Fig. 10.8). Under the right conditions, this high-energy,

Fortunately for our own molecules, the atmosphere's ozone layer protects us by absorbing high-energy radiation.

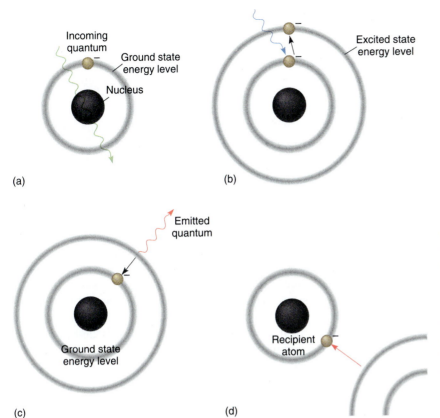

(a) (b)

(c) (d)

FIGURE 10.8

(a) If a quantum has the wrong wavelength for a pigment, it passes through the pigment's bonding orbitals without being absorbed. Chlorophyll looks green because most green light passes through it. (b) If the quantum has the correct wavelength, the correct amount of energy, it is absorbed and the electron must move to a new orbital whose energy level corresponds to the electron's new energy load. (c) The excited state is unstable; it may stabilize itself by having the electron emit enough energy (fluoresce) to drop back to its original ground state energy level. (d) The electron can also be stabilized by moving to a more stable orbital on an entirely different atom. This is the critical process in photosynthesis; without this step, life would not exist.

excited electron can be used in chemical reactions. If it is not used, it returns to its original, stable ground orbital by emitting a new quantum of light, one with less energy and a longer wavelength than the one that it absorbed. The release of light by a pigment is called **fluorescence** (Fig. 10.8c).

Two of the most useful pieces of information about a photochemical process are its action spectrum and the absorption spectrum of its pigment (Fig. 10.9). An **absorption spectrum** is a graph that shows which wavelengths are most strongly absorbed by a pigment, whereas an **action spectrum** shows which wavelengths are most effective at

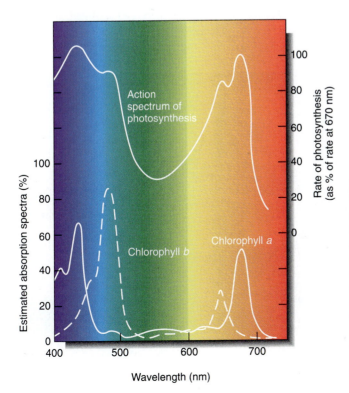

FIGURE 10.9

The absorption spectra of chlorophyll *a* and chlorophyll *b*, and the action spectrum of photosynthesis. On the bottom axis is the wavelength of light, with short (blue) wavelengths to the left and long (red) ones to the right. The vertical axis of the absorption spectra is the amount of light absorbed by the pigment; for the action spectrum, it is the amount of photosynthesis carried out. Chlorophylls absorb little of the very short wavelength light at 400 nm, and little photosynthesis occurs. But light at slightly longer wavelengths, about 425 nm, is absorbed well by chlorophyll *a* and photosynthesis proceeds. Quanta with intermediate wavelengths pass right through the pigment and photosynthesis is low, but in the 650 to 680 nm range (red) considerable absorption occurs. Because the absorption spectra of chlorophyll *a* and *b* are different, more wavelengths can be efficiently harvested. If the two matched perfectly, chlorophyll *b* would be useless.

powering a photochemical process. To initiate a photochemical process, light must first be absorbed, so the action spectrum of a process must match the absorption spectrum of the pigments responsible. The absorption spectrum of chlorophyll *a* shows that it absorbs red light (especially 660 nm) and blue light (440 nm) very well and the other wavelengths only slightly. It would be better if it could absorb a greater number of wavelengths, but it simply does not do so. Chlorophyll *a* is the essential photosynthetic pigment in all plants, algae, and cyanobacteria and has existed unaltered by evolution for about 3 billion years. To put this in perspective, the entire Milky Way Galaxy takes 250 million years to make one rotation about its center, so in 12 full rotations of our galaxy, no alteration in the structure of chlorophyll *a* has been selectively advantageous.

Accessory pigments are molecules that strongly absorb wavelengths not absorbed by chlorophyll *a*. The absorbed energy is passed on to chlorophyll *a*. In effect, the accessory pigments overcome the narrow absorption of chlorophyll *a* and broaden the action spectrum of photosynthesis. We know that accessory pigments are involved because the action spectrum of photosynthesis does not perfectly match the absorption spectrum of chlorophyll *a*. The most common accessory pigments in land plants are **chlorophyll *b*** and the **carotenoids** (Fig. 10.10); algae have other types. Chlorophyll *a* and *b* are large, flat molecules with almost identical porphyrin ring structures. Their phytol tails are hydrophobic and dissolve into the lipid portion of the thylakoid membrane. When packed tightly in a membrane by their phytol tails, their porphyrin rings lie parallel to each other, which causes the electron orbitals of one molecule to interact with those of the two adjacent

Lutein

β-Carotene

(a)

FIGURE 10.10

(a) There are two types of carotenoids: Carotenes lack oxygen but xanthophylls (such as lutein) have it. Both are accessory pigments that protect the chlorophyll from excess sunlight. (b) Carotenoids are always present in leaves, but usually the abundant chlorophyll masks their presence. In autumn, chlorophyll breaks down and we can see the carotenoids.

(b)

PLANTS & PEOPLE

PHOTOSYNTHESIS, AIR, AND LIFE

The living creatures we are familiar with could not survive if the composition of Earth's atmosphere were very different. In addition to supplying the oxygen that all animals need for respiration, the atmosphere protects us from harmful radiation from space by absorbing high-energy cosmic rays and x-rays so that little of it reaches ground level. Ultraviolet light can induce mutations in our DNA and cause skin cancer, but fortunately most of the ultraviolet light is absorbed by a special form of oxygen, **ozone** (O_3), in the upper atmosphere. All oxygen involved in respiration and in the formation of ozone is supplied by photosynthesis based on chlorophyll *a*: Water-splitting, proton-pumping, and oxygen liberation have caused the gradual accumulation of oxygen in the atmosphere over billions of years—there is no other source. Oxygen itself is effective at blocking harmful radiation, but a by-product of absorbing high-energy quanta is the conversion of oxygen to ozone ($3O_2 \rightarrow 2O_3$), which offers even more effective protection.

Man-made chlorofluorocarbons from cans of hairspray and deodorant, among other products, have been escaping into the atmosphere and destroying the ozone, thus increasing the amount of dangerous radiation that reaches us. Some effort has been made to eliminate the production of chlorofluorocarbons worldwide, but so far only a gradual phase-out has been accepted by most governments.

Another atmospheric component, carbon dioxide, traps long wavelength radiation and is involved in the **greenhouse effect**. Earth receives much of its energy in the visible wavelengths that pass readily through the atmosphere. A little of this is reflected directly back to space, but much is absorbed by soil, water, and plants, causing them to become warmer. As their temperature rises, they give off the heat as infrared radiation; because the atmospheric carbon dioxide traps some of the heat, Earth is warm enough to support life. If there were no carbon dioxide, much more of earth would be frozen. But the concentration of carbon dioxide is rising rapidly as a result of burning of gas, oil, firewood, and rainforests. As the carbon dioxide concentration increases, the amount of heat trapped increases and Earth's temperature rises.

The level of atmospheric carbon dioxide is controlled by a complex interaction of many processes, which appear to prevent **rapid** changes. As carbon dioxide increases, not only do plants photosynthesize more rapidly, removing carbon dioxide faster, but algae grow more abundantly also, which in turn supports increased populations of coral, clams, and other shellfish. All of these build their shells of calcium carbonate, and they remove carbon dioxide from seawater in the process, causing more carbon dioxide to dissolve in the seawater from the atmosphere. As these animals die and their shells sink to the ocean bottom, carbon dioxide goes with them and remains there for millions of years. However, these biological processes can only slow the greenhouse effect, not halt it altogether.

Many people seem undisturbed by the idea of global warming, perhaps because a rise of only 3 to 5°C is predicted. The real threat is not so much a generalized warming of the entire Earth as a drastic alteration in patterns of air circulation and ocean currents as continents heat more than oceans. It is possible that normally moist areas will become deserts and vice versa. Ecological disasters may occur as species are unable to adapt as rapidly as their environments change.

(a) Clear-cutting destroys all plants in an area; trees are no longer able to remove carbon dioxide from the atmosphere. (*Thomas Kitchin/Tom Stack & Associates*)

(b) Tropical rainforests are not the only forests being damaged. This area is in the northwestern United States, in Oregon. Note the clear-cut patches in all parts of the region. (*David J. Cross/Peter Arnold, Inc.*)

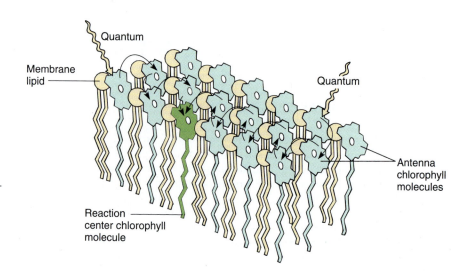

FIGURE 10.11

Only special chlorophyll *a* molecules—reaction centers—can actually undergo the initial photochemical reaction of photosynthesis, but they are surrounded by other chlorophyll *a* molecules as well as accessory pigments; regardless of which pigment absorbs light, the energy is transferred to the reaction center.

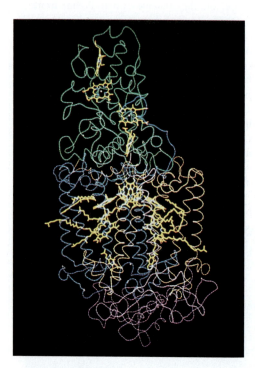

FIGURE 10.12

Computer model of the reaction center of a photosynthetic bacterium. At the top in green are cytochromes, whose heme groups are yellow. In the middle in blue and brown are proteins with hydrophobic alpha helix secondary structure that fits into the chloroplast membranes. The proteins surround bacteriochlorophylls, carotenoids, and bacteriophaeophytin (yellow). The bottom purple is another protein. Reaction centers in plants have many analogous components that probably have similar positions. (*Courtesy of Johann Deisenhofer and Hartmut Michel, University of Texas, Southwestern Medical Center*)

molecules. Hundreds of chlorophylls act somewhat like one molecule, and the energy absorbed by one can be rapidly transferred to another in a different part of the complex. This transfer, called **resonance**, allows chlorophyll *b* to absorb wavelengths that chlorophyll *a* would miss and then transfer the energy to chlorophyll *a* for use in chemical reactions. Carotenoids are poor at this type of resonance and transfer only about 10% of their energy; they seem to be more important in absorbing excessive light and thus protecting chlorophylls.

Photosystem I.   When light excites an electron in chlorophyll *a*, the electron is so unstable that it either reacts with almost anything or fluoresces, wasting its energy. Chlorophyll molecules and the electron carriers involved in photosynthesis must therefore be tightly bound to the thylakoid membranes of the chloroplast. Chlorophyll's electron must react only with the proper molecule, which should be close enough for the reaction to take place instantly, before fluorescence can occur. All pigments and carriers that work together are packed into a granule called a **photosynthetic unit,** and the thylakoid membranes are filled with millions of these granular arrays (Fig. 10.11). Each photosynethetic unit contains about 300 molecules of chlorophylls *a* and *b* and carotenoids. In some photosynthetic units, chlorophyll *b* is plentiful and in others it is less abundant. Those with little chlorophyll *b* have been named **photosystem I;** those in which chlorophyll *b* is present at levels almost equal to *a* are **photosystem II.** The photosystem I units are involved in the following reactions.

When light strikes any pigment of a photosystem I array, either chlorophyll *a* or an accessory pigment, the energy is transferred to the **reaction center,** a structure that contains a pair of special molecules of chlorophyll *a* whose properties differ from all other molecules of chlorophyll *a* in the unit (Fig. 10.12). This pair of chlorophylls of the photosystem I reaction center is given the special name **P700** because they absorb red light of 700 nm most efficiently. The energy excites an electron of P700, which is then absorbed by a membrane-bound electron acceptor known as "X." This is a transfer of an electron; no bonding orbital is formed. The exact chemical nature of X is not known, but it contains iron and sulfur and is sometimes designated Fe_4S_4. When X absorbs an electron from P700, it becomes a powerful reducing agent, with a redox potential of -0.73 volts (Table 10.5). The transferred electron is still extremely unstable, and the reduced X immediately passes it onto **ferredoxin,** which is also located in the thylakoid membrane (Fig. 10.13). Ferredoxin is a small protein (10,500 to 11,000 daltons; a dalton is the weight of one hydrogen atom) with an active site consisting of two iron atoms bound to two sulfur atoms. Reduced ferredoxin is also a strong reducing agent, with a redox potential of -0.43 volts. Electrons are passed from ferredoxin to an enzyme, **ferredoxin-NADP$^+$ reductase,** which then reduces NADP$^+$, converting it to NADPH, as its name indicates. Ferredoxin carries only one electron, but two are needed simultaneously to reduce NADP$^+$. Ferredoxin-NADP

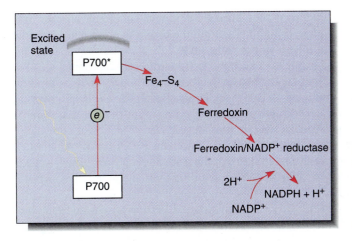

FIGURE 10.13

In photosystem I, energy is absorbed by a pair of P700 chlorophyll *a* molecules, raising two electrons to an excited energy level; from here, they pass onto $Fe_4\text{-}S_4$ ("X"), then onto ferredoxin, and finally onto ferredoxin-NADP reductase. After two electrons have reduced ferredoxin-NADP reductase, they are transferred simultaneously to $NADP^+$, reducing it to $NADPH + H^+$.

reductase carries two electrons, but it can be reduced one electron at a time; then it transfers those two electrons together to $NADP^+$. Although NADPH is also a strong reducing agent, it is stable enough to move away from the membrane safely without the risk of reducing things indiscriminately, as the previous electron carriers might.

Photosystem II. Photosystem I efficiently produces NADPH, but the reaction center P700 chlorophyll *a* loses electrons during the process. In this oxidized state, bonding orbitals could easily rearrange, causing the molecule to break down and be destroyed. There must be a mechanism that adds electrons back to the P700, reducing it so that it can work repeatedly.

The mechanism that reduces P700 is located in the photosynthetic unit rich in chlorophyll *b*, photosystem II (Fig. 10.14). Photosystem II can be best described by working backward from photosystem I: A molecule of **plastocyanin,** which contains copper, donates an electron to the chlorophyll *a* of the photosystem I reaction center. The plastocyanin is now oxidized, lacking an electron; it must reacquire one because it also is too expensive a molecule to donate just one electron and then never work again. It receives its new electron from a complex of cytochrome molecules, called the **cytochrome b_6/f complex,** which in turn gets an electron from a molecule of **plastoquinone.** This receives electrons from another carrier, **Q,** a molecule of quinone, which in turn receives electrons from **phaeophytin.** Phaeophytin is actually a chlorophyll *a* molecule that does not contain a magnesium atom. Phaeophytin becomes oxidized as it donates an electron to Q, so it

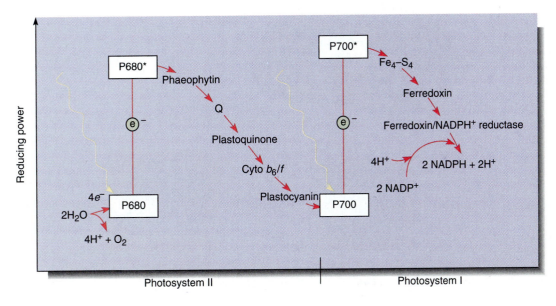

FIGURE 10.14

The two photosystems work together to transfer electrons from water to NADPH. The scale on the left indicates reducing power, the redox potential (see Table 10.5). The higher a molecule is in the chart, the greater its capacity to force electrons onto another molecule. Because of its shape, this diagram is called a **Z scheme.**

must obtain another electron, which it does when a chlorophyll *a* molecule absorbs light and is activated. This is a different chlorophyll *a* from the one in photosystem I; it is the reaction center of photosystem II and has the name **P680.**

We may seem to be going in circles: taking an electron from one chlorophyll *a*, P680, to pass it onto another chlorophyll *a*, P700, which then sends it to NADP$^+$. But the physical differences between the two molecules of chlorophyll *a* are crucial. The one in photosystem II gets new electrons from water, not plastocyanin. The important thing is that water is cheap enough to just throw away after the electrons are removed. Water breaks down into protons (H$^+$), which the plant uses, and oxygen (O$_2$), which it discards. Whereas all electron carriers are large, expensive molecules that the plant must construct itself, water is simply brought in. The electrons are stripped off, the protons are used, and the oxygen is discarded through stomata.

Photosystems I and II together are an efficient system. Electrons are passed from water to P680 in photosystem II, their energy is boosted by light, then they move through an **electron transport chain**—the various electron carriers—to P700 in photosystem I. Their energy is boosted by light again, and they pass through a short second electron transport chain to NADP$^+$, reducing it to NADPH. This last step requires that protons be added to NADP$^+$; these protons are present in the water surrounding the membrane (water is always a mixture of H$_2$O, H$^+$, and OH$^-$). It would be simpler if photosystem I could receive electrons directly from water, but that does not happen. Besides, the electron transport chain between P680 and P700 is necessary for the production of ATP.

The Synthesis of ATP.   The light-dependent reactions produce the reducing agent NADPH that actually places electrons onto the carbon of carbon dioxide in the stroma reactions. But the stroma reactions are highly endergonic and must be driven by being coupled to the exergonic splitting of ATP. The necessary ATP is also generated by the light reactions, but the process is indirect. It is photophosphorylation because light is involved, but a more specific name is often used: **chemiosmotic phosphorylation.** To understand it, we must take a closer look at the structure of chloroplasts. As mentioned in Chapter 3, the inner membrane of chloroplasts folds inward, forming flattened sacs called thylakoids (Fig. 10.15). In certain regions, these swell slightly and form rounded vesicles. All thylakoids in one region form vesicles at the same spot, so they occur in sets called **grana** (singular: granum). Thylakoids that lie between grana are **frets.** The liquid surrounding the thylakoid system is the stroma, but notice especially that there is another compartment, the thylakoid lumen.

FIGURE 10.15
Grana are stacks of small thylakoid vesicles compressed together; frets are regions of thylakoid that connect one granum to another. The lumen of the thylakoid region is continuous with that of the fret region. The liquid surrounding all the thylakoids is stroma.

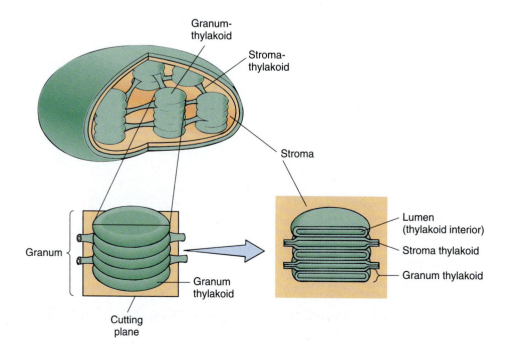

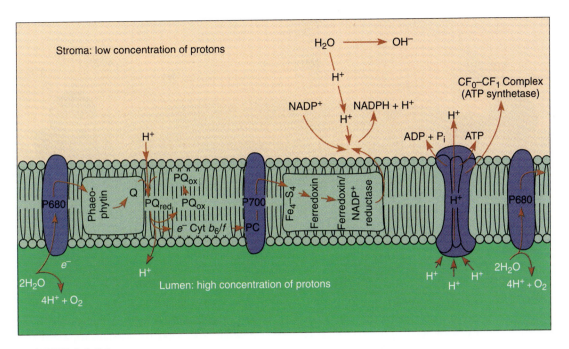

FIGURE 10.16

The water-splitting, proton-producing reactions of photosystem II take place on the lumen side of the thylakoid membrane. Plastoquinone is like $NADP^+$ in that when it picks up electrons, it also picks up a proton. This occurs on the stroma side of the membrane, but the reduced plastoquinone must diffuse to the other side of the membrane to pass electrons on to the cytochrome b_6/f complex. The proton then dissociates and is deposited in the lumen, adding to the growing pool of protons. When NADPH is formed, it picks up protons from the stroma, so this and the plastoquinone pumping result in a deficiency of protons in the stroma. Protons return to the stroma by passing through ATP synthetases; their passage is exergonic and powers the phosphorylation of ADP to ATP.

The thylakoid lumen is a critically important compartment because some of the enzymes and electron carriers of the photosystems are embedded in the membrane layer facing the lumen, whereas other enzymes are in the membrane layer that faces the stroma (Fig. 10.16). The reactions that break down water and produce oxygen and protons are located on the lumen side of the thylakoid membrane, in the granum areas. This membrane is not permeable to protons, so as the light reactions run, the thylakoid interior accumulates protons and their concentration increases. The molecules of ferredoxin-NADP reductase that generate NADPH are located on the other side of the membrane, facing the stroma. The protons they attach to $NADP^+$ are those present as a result of the natural breakdown of water: $H_2O \rightarrow H^+ + OH^-$. As the protons are absorbed, their concentration in the stroma decreases. Furthermore, during electron transport between P680 and P700, the electron carrier plastoquinone moves a proton from the stroma to the thylakoid lumen every time it carries an electron between phaeophytin and the cytochrome b_6/f complex. This also contributes to the increased concentration of protons in the thylakoid lumen and to the decreased concentration of protons in the stroma.

The strong difference between the concentrations of protons inside the thylakoid lumen and exterior to it in the stroma quickly becomes so powerful that protons begin to flow out of the lumen through special channels in the membrane. These channels are complex sets of enzymes that can synthesize ATP from ADP and phosphate; the entire complex is called **ATP synthetase** (Fig. 10.16). The ATP synthetase of chloroplasts is known specifically as the **CF_0-CF_1 complex.** CF_0 is the portion of the enzyme spanning the membrane where the actual proton channel is located. CF_1 is the portion of the enzyme that phosphorylates ADP to ATP. The power required to force phosphate onto ADP and establish the high-energy bonding orbitals of ATP comes from the flow of protons through the ATP synthetase channels.

In a car battery, the flow of electrons through wires has enough power to turn the starter; in chloroplasts, the flow of protons through ATP synthetase channels has enough power to phosphorylate ADP to ATP.

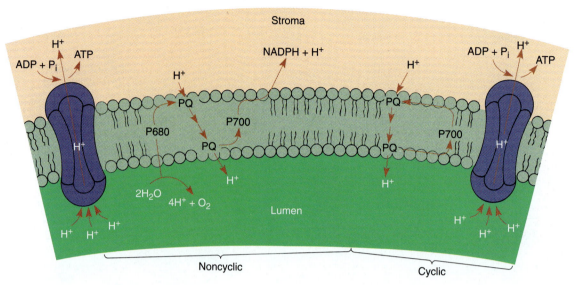

FIGURE 10.17
In noncyclic electron transport, electrons flow through the Z scheme from water to NADPH. Cyclic electron transport is much simpler: Electrons flow from P700 to plastoquinone, which carries a proton to the lumen and returns the electron to P700.

When electrons flow smoothly from water to NADPH, the process is called **noncyclic electron transport** (see Figs. 10.14 and 10.17). The chemiosmotic potential that builds up does not produce quite enough ATP for the stroma reactions: There is too little ATP relative to the amount of NADPH produced. This problem is overcome by an alternate route for electrons. Once they reach ferredoxin in photosystem I, they can be transferred to the plastoquinones of photosystem II instead of being used to make NADPH. The plastoquinones carry the electrons along just as though they had gotten them from Q and use their energy to pump more protons into the thylakoid lumen. This is **cyclic electron transport,** and with it chloroplasts make extra ATP without making extra NADPH, thus producing ATP and NADPH in the proper ratios for the stroma reactions. Cyclic electron transport is a simple light-powered proton pump, and similar types occur in bacteria. This may have been the original power system that evolved first, billions of years ago.

THE STROMA REACTIONS

The conversion of carbon dioxide to carbohydrate occurs in the stroma reactions, which are also called the **Calvin/Benson cycle,** or the **C_3 cycle** (Fig. 10.18). These reactions take place in the stroma, mediated by enzymes that are not bound to thylakoid membranes (see Fig. 10.6). In the first step, an **acceptor molecule (ribulose-1,5-bisphosphate; RuBP)** reacts with a molecule of carbon dioxide. Because RuBP contains five carbons and one more is added from carbon dioxide, you might expect a product that contains six carbons. However, the new molecule breaks apart immediately, while still on the enzyme; stable bonding orbitals cannot be formed between all six carbon atoms while so many oxygen atoms are present and pulling electrons to themselves. Instead, orbitals rearrange and two identical molecules are formed that each contains three carbons: **3-phosphoglycerate,** hence the name C_3 cycle. The abbreviation PGA is often used for 3-phosphoglycerate.

The enzyme that carries out this reaction has many names; the most common is **RuBP carboxylase (RUBISCO).** This is one of the largest and most complex enzymes known— a giant complex of two kinds of protein subunits. There are eight copies of a small protein, each with a molecular weight of 14,000 to 15,000 daltons, and eight copies of a large protein, each with a molecular weight of 53,000 to 55,000 daltons (Fig. 10.19). The entire enzyme has a molecular weight of about 480,000 daltons. Not only is the tertiary structure of each protein subunit important, but their quaternary structure as a complex is critical.

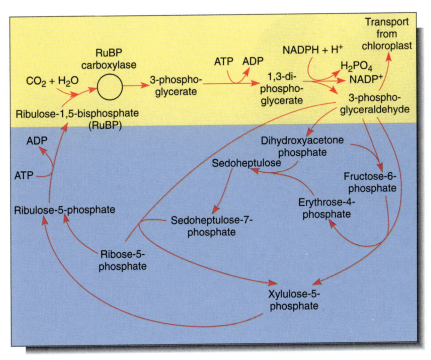

(a)

(b)

FIGURE 10.18

(a) In the yellow area are the first steps of the stroma reactions, also known as the C_3 cycle; the product is two molecules of 3-phosphoglyceraldehyde. Some of this is transported out of the chloroplast, and the rest undergoes reactions (blue area) that form a new molecule of the acceptor, RuBP. (b) At various times acids such as phosphoglycerate and malate are written as phosphoglyceric acid and malic acid; the "-ic acid" ending refers to the whole acid. The "-ate" ending refers to the acid's anion, the negatively charged portion left after the proton dissociates. In protoplasm, most of the acids occur as free anions, not intact neutral acids still holding their protons.

RuBP carboxylase can constitute up to 30% of the protein in a leaf, making it the most abundant protein on Earth. Without it, there would be almost no life at all; all photosynthesis that produces oxygen is mediated by this enzyme. Only a few rare photosynthetic bacteria use a different enzyme. RuBP carboxylase is crucial to the production of food; without it heterotrophs would starve.

Like chlorophyll *a*, RuBP carboxylase is by no means ideal. Its active site recognizes and binds to carbon dioxide only poorly, and it has low substrate specificity, frequently putting oxygen rather than carbon dioxide onto RuBP. Yet this enzyme is highly conserved evolutionarily: The amino acid sequences of RuBP carboxylase from all plants are virtually identical. Apparently all mutations that cause any change in structure, however slight, disturb the active sites and are selectively disadvantageous.

It is important to realize that the first step of the stroma reactions is a carboxylation only. Electrons and energy are added in the next two steps: ATP donates a high-energy phosphate group to the 3-phosphoglycerate, converting it to **1,3-diphosphoglycerate**,

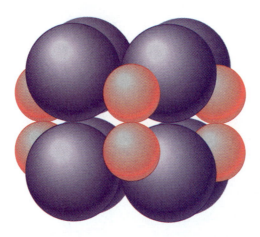

FIGURE 10.19
RuBP carboxylase. This is a large enzyme in which quaternary structure is critical; the individual proteins have no useful activity until combined in the full complex.

which then is reduced by NADPH to **3-phosphoglyceraldehyde** (abbreviated PGAL); a phosphate comes off in this step also. The carbon is now both reduced and energized.

The rest of the stroma reactions are complex, but the important point is that as they operate, some 3-phosphoglyceraldehyde can be taken out of the chloroplast and used by the cell to build sugars, fats, amino acids, nucleic acids—basically anything the plant needs. The rest of the 3-phosphoglyceraldehyde remains in the chloroplast and undergoes several more stroma reactions which convert it to RuBP, the original acceptor molecule. The principle involved is important: To incorporate carbon dioxide, the plant needs the acceptor RuBP, and the two react on a one-to-one basis. To assimilate large amounts of carbon dioxide, the plant either needs large amounts of RuBP or needs to use a few RuBP molecules repeatedly. It is the second strategy that the plant uses. As the 3-phosphoglyceraldehyde is formed, some of it is reconverted to RuBP by the rest of the stroma reactions and some is exported to the cytoplasm. The chloroplast does not need to import quantities of RuBP from the rest of the cell; it just recycles the small amount that it has. Imagine a chloroplast that has 1,000,000 carbon atoms inside it as the various intermediates of the stroma reactions; after three carbon dioxides have been assimilated there are 1,000,003 carbon atoms. When one molecule of 3-phosphoglyceraldehyde is exported, the carbon pool returns to 1,000,000, and a steady state is maintained. In very young leaves with growing chloroplasts, little or no 3-phosphoglyceraldehyde is exported; it is retained and the pools of C_3 metabolites increase in numbers of molecules.

ANABOLIC METABOLISM

3-Phosphoglyceraldehyde is an amazingly versatile molecule: Using it plus water, nitrates, and minerals, plants construct everything inside themselves. The entire fabric of the organism can be synthesized. This is also the basis of all animal metabolism, because animals either eat plants or eat other animals that eat plants.

Most biological molecules are larger than 3-phosphoglyceraldehyde, so it must be rearranged and altered in the cytoplasm to build up larger, more complex molecules. This constructive metabolism is called **anabolism** and it consists of **anabolic reactions.**

Anabolic pathways are numerous, but two are especially important with regard to energy metabolism: the synthetic pathways of polysaccharides and fats, which are storage forms of energy and carbon. The NADPH and ATP produced by photosynthesis are excellent sources of energy, but they cannot be stored for even a short time. They are so reactive and unstable that they would break down. A plant cannot stockpile them to survive times when photosynthesis is impossible. Nor can they be transported over long distances, so even if leaves had an abundant supply, roots would starve.

Several types of storage compounds have evolved which solve these problems.

1. **Short-term storage:** ATP and NADPH can be used within the cell and last only briefly.

2. **Intermediate-term storage:** The simple sugar glucose and the disaccharide sucrose are stable enough to be moved from cell to cell, either in the vascular tissue of a plant or in a blood stream. They are also sufficiently stable to last for weeks or months. A problem with storing large quantities of mono- or disaccharide is that they cause cells to absorb water by osmosis.

3. **Long-term storage:** Starch is a large, high-molecular-weight polymer of glucose, too large to be transported. It is even more stable than glucose, lasts for years, and does not cause the cell to absorb water. Lipids are an even more concentrated storage form of energy which can be synthesized rapidly and stored in large quantities.

The Synthesis of Polysaccharides. The anabolic synthesis of glucose is **gluconeogenesis** (Fig. 10.20). In reactions similar to those of C_3 metabolism, part of the 3-phosphoglyceraldehyde exported to the cytoplasm is converted to **dihydroxyacetone phosphate;** one molecule of this condenses with one molecule of unconverted 3-phosphoglyceraldehyde to form the sugar **fructose-1,6-bisphosphate.** This loses a phosphate to become **fructose-6-phosphate,** and part of this is rearranged, converting it to **glucose-6-phosphate.** Both fructose-6-phosphate and glucose-6-phosphate are versatile, useful molecules that enter many metabolic pathways. In plants, the glucose-6-phosphate is polymerized into polysaccharides: amylose, amylopectin, or cellulose. It is similarly essential in animals.

ENVIRONMENTAL AND INTERNAL FACTORS

A plant's photosynthesis is affected by its environment in many ways.

LIGHT

From a plant's viewpoint, light has three important properties: (1) quality, (2) quantity, and (3) duration.

Quality of sunlight refers to the colors or wavelengths that it contains. Sunlight is pure white because it contains the entire visible spectrum. During sunset and sunrise, sunlight passes tangentially through the atmosphere and a large percentage of the blue light is deflected upward; consequently, light at ground level is enriched in red, which is easily visible. This period of red-enriched light lasts only a few minutes and probably has little effect on photosynthesis. At noon, sunlight passes nearly vertically through the atmosphere, more blue light is transmitted, and even though the blueness of the sky suggests that all of the reds, greens, and yellows have been blocked, in fact enough of all of these wavelengths penetrate to the Earth's surface to allow efficient photosynthesis. This is true of plants in deserts, grasslands, and the top layer, the canopy, of a forest. But herbs and shrubs that grow near soil level in a forest are understory plants, and the light they receive has already passed through the leaves of the canopy (Fig. 10.21). As light penetrates those leaves, red and blue are absorbed by chlorophyll, so the dim light received by understory plants is especially depleted in these critical wavelengths. It is selectively advantageous for them to have extra amounts of accessory pigments so that they can gather the wavelengths available and pass the energy on to chlorophyll *a*. Similarly, algae that grow near the surface of lakes or oceans receive complete light, but water absorbs red and violet. Algae at deep regions receive mostly green and blue light and must have special accessory pigments capable of absorbing these wavelengths efficiently.

Quantity of light, which refers to light intensity or brightness, is affected by several factors. More light is available for photosynthesis on a clear than on a cloudy day; understory plants receive dim light; lower branches and branches on the shaded side of a plant receive less light (Fig. 10.21b). Plants growing in the shadow of mountains or in deep canyons receive much less light than plants that grow on slopes that face the sun. Plants growing near the equator receive intense light because the sun is always more or less directly overhead at noon, whereas plants near the poles receive very little light. Even during the summer the sun is low at noon and light is scattered by the atmosphere.

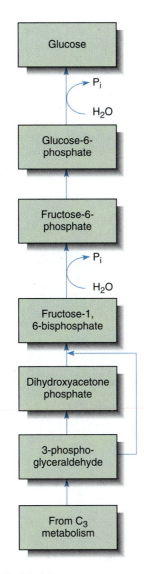

FIGURE 10.20
Gluconeogenesis is an anabolic pathway in which large molecules are built up from small ones. This process may occur in chloroplasts, amyloplasts, or cytosol.

Plant Biology Tutor CD-ROM

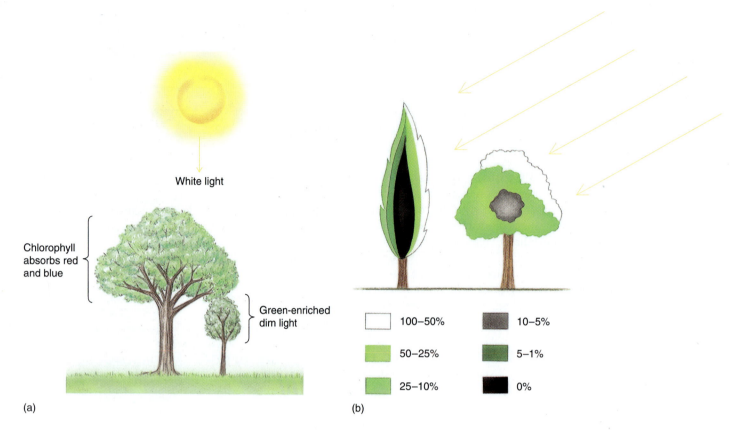

White light

Chlorophyll
absorbs red
and blue

Green-enriched
dim light

☐ 100–50%		■ 10–5%	
☐ 50–25%		■ 5–1%	
☐ 25–10%		■ 0%	

(a)   (b)

FIGURE 10.21

(a) Plants growing in the shade of other plants receive not only dim light, but light depleted of red and blue. (b) Not all leaves of a tree receive equal amounts of light; this self-shading is more severe in latitudes farther away from the equator. Colors indicate the percentage of full sunlight that reaches the various parts of a tree.

Think about how intensity of sunlight varies during the day and affects photosynthesis. Examine the solid line labeled "300 ppm CO_2" in Figure 10.22. This was derived from many experiments in which the rate of photosynthesis was measured for plants grown under different intensities of light, but all with 300 ppm carbon dioxide in the air. Near the left side at point **a**, those plants that received dim light absorbed little carbon dioxide, whereas those grown in brighter light absorbed more carbon dioxide. Under these normal levels of carbon dioxide, light is the limiting factor. Photosynthesis is slow on dull, overcast days but faster on brighter days. At point **b** in the graph, plants that received more light did not photosynthesize faster than those that received slightly less. Where the curve turns flat, there was enough light to saturate the process. In these conditions, the limitation was lack of carbon dioxide; in the experiments in which more carbon dioxide was available (blue line, 1000 ppm CO_2), photosynthesis went faster. Thus at point **b**, carbon dioxide was the rate-limiting factor. At point **c**, light was so intense that it damaged the plant by overheating it and bleaching the pigments.

At point **a**, if the lack of light prevents photosynthesis from proceeding faster, there must be adequate amounts of carbon dioxide. As soon as ATP and NADPH are produced, they move to the stroma and are used by the waiting enzymes and carbon dioxide; then ADP and $NADP^+$ diffuse back to the thylakoid membranes and wait for another quantum. Conversely, at point **b**, where light is bright and the low concentration of carbon dioxide is rate limiting, there is so much light that as quickly as ADP and $NADP^+$ come to the thylakoids, they are reprocessed immediately into ATP and NADPH. These then move to the stroma, where they must wait for a carbon dioxide molecule.

To the left of the **light compensation point**, it appears that there was no photosynthesis, even though some dim light was provided. The problem actually lies with the

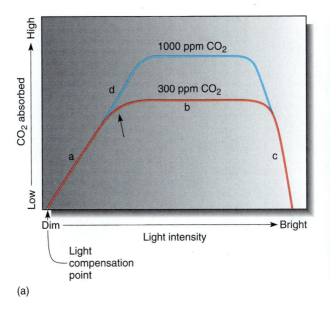

(a)

(b)

FIGURE 10.22

(a) Light and photosynthesis; details are given in text. The unit ppm is parts per million; for 300 ppm, in 1 million liters of air, 300 liters are carbon dioxide. (b) These plants have abundant sunlight. They could probably photosynthesize more rapidly if more carbon dioxide were present.

technology of measuring photosynthesis. Either the amount of carbon dioxide absorbed or the amount of oxygen released must be measured. But both of these gases are involved in respiration as well as photosynthesis. To the left of the light compensation point, photosynthesis was absorbing carbon dioxide more slowly than respiration was releasing it, so it *appears* as though no photosynthesis occurred. The same problem arises when we try to measure respiration: Photosynthesis distorts the measurement, but then we can turn off the lights and stop photosynthesis. The light compensation point is the level of light at which photosynthesis matches respiration. Plants that are grown for a long time in conditions below the light compensation point respire faster than they photosynthesize; they gradually consume their reserve carbohydrates and fats and starve to death. For plants grown in light brighter than the light compensation point, photosynthesis exceeds respiration and the extra sugar can be used for growth and reproduction.

In the brightest of environments, the air is so clear that sunlight is frequently too intense during summer months. Protective adaptations are necessary, and in many species mechanisms have evolved that provide shade. A common method is the production of a thick layer of dead trichomes, plant hairs (Fig. 10.23). A heavy coating of wax can also reflect light, and cutin is especially good at absorbing the more harmful short wavelengths. Part of the value of carotenoids and other accessory pigments is that they shade the chlorophyll, absorbing some of the most damaging wavelengths.

The intensity of light, the actual number of quanta that strike a given area per unit time (e.g., that strike 1 cm^2/s) is greatest at noon in midsummer when the sun is most directly overhead and is less whenever the sun is lower—morning, evening, and winter. Light may be too intense at midday but optimal when the sun is lower; some species (iris, eucalyptus) have adapted to this by means of vertical leaves. The lamina face is exposed fully only in mornings and afternoons, but at noon only the leaf edge is exposed. In other plants, leaves orient vertically automatically when stressed—they wilt and hang down.

Understory plants of forests are adapted to low light. If a roadway is cut into a forest and plants adjacent to the cut are suddenly exposed to full sunlight, the shock may cause them to wilt and die. The same phenomenon occurs when trees are blown down during storms, floods, or avalanches and the surrounding plants are exposed.

(a)

(b)

FIGURE 10.23
(a) While young, the leaves of dusty miller are completely obscured by trichomes, protecting the leaf from strong sunlight and insects.
(b) These cacti (*Epithelantha*) live in environments where sunlight is extremely intense; their spines are so abundant and closely spaced that they shade the stem and prevent the chlorophyll from being damaged. (Big Bend National Park)

Duration of light refers to the number of hours per day that sunlight is available. At the equator, days are 12 hours long throughout the year. Farther north or south, days become longer in summer; maximum length occurs near the poles, where the day is 24 hours long in midsummer and only night occurs in midwinter. In middle latitudes, winter days are short and sunlight is weak because the sun is so low in the sky. Under these conditions, even evergreen plants are unable to undergo very much photosynthesis. However, because temperatures are low, the plants are growing little and have a low rate of respiration. Even deciduous trees and biennials can survive by means of stored nutrients.

During summer, days are longer and light is brighter because the sun is higher in the sky. The amount of energy obtained by photosynthesis easily exceeds the amount consumed by respiration and growth. In many plants, longer days cause greater amounts of photosynthesis, but in others, chloroplasts become so full of starch that photosynthesis stops, even though light is present. At night, starch is converted to sugar, which is then transported out of chloroplasts and can be used for growth or stored in amyloplasts in tubers, corms, or other such organs. By morning, leaf chloroplasts can resume photosynthesis.

LEAF STRUCTURE

Leaf structure of most temperate and tropical plants is quite standard: palisade parenchyma above and spongy mesophyll below. This structure is excellent for absorbing carbon dioxide but inefficient for conserving water. If plants of hot, dry habitats had this leaf architecture, they would have to keep their stomata closed so much of the time that they would starve. Instead, their leaf cells are frequently packed together without intercellular spaces. Water loss is reduced because the small internal surface area retards water evaporation. But with so little surface area, it is difficult to dissolve carbon dioxide from the air into the cytoplasm. This slows photosynthesis, but apparently this trade-off, slow growth versus water conservation, is selectively advantageous.

Another method of minimizing water loss while maintaining photosynthesis is to reduce external surface by means of cylindrical leaves. Water movement from interior air spaces to exterior air is minimized because so few stomata are present (Fig. 10.24). Photosynthesis is reduced because absorption of carbon dioxide is slowed.

FIGURE 10.24
These living stone plants (*Lithops*) of African deserts conserve water in several ways. They have only two leaves at a time; when two new ones form, the old two die. Not enough water is available for four leaves. The leaves are fleshy and pressed together, such that they form a cylinder with minimal surface area through which water can be lost.

WATER

The amount of water available greatly affects photosynthesis. Most plants keep their stomata open during the day, permitting the entry of carbon dioxide, but water is inevitably lost. At night, carbon dioxide cannot be used, and stomata are closed, retaining water

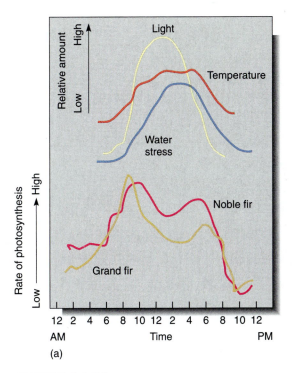

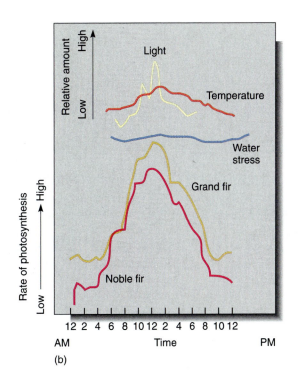

FIGURE 10.25

(a) Light intensity exceeded the light compensation point for these trees just after 6 AM, and photosynthesis increased rapidly. However, by early afternoon, photosynthesis dropped even though light and temperature were adequate; the problem was lack of water (water stress), and probably stomata had begun to close around noon. (b) On a cloudy day, water stress never became a problem and stomata remained open. Even though there was less total light than in (a), there was more photosynthesis for the day.

within the plant. If the soil becomes dry and water is not readily available, a plant keeps its stomata closed even during the day, and carbon dioxide cannot enter (Fig. 10.25). The small amount of carbon dioxide produced by the respiration of the leaves can be reused photosynthetically, but this is a minor amount. In addition to numerous structural modifications that conserve water, metabolic adaptations also exist; two of the most important are C_4 metabolism and crassulacean acid metabolism.

C_4 METABOLISM

An important factor for plants is the amount of water lost for each molecule of carbon dioxide absorbed. Ideally, this ratio is low. Carbon dioxide diffuses into a leaf faster if its concentration in air is higher or if its concentration inside leaf protoplasm is lower. Plants can do nothing about the carbon dioxide level in the external air, and many also have poor control over the protoplasmic concentration of carbon dioxide. RuBP carboxylase has a low affinity for carbon dioxide; as carbon dioxide concentration drops, enzyme-substrate binding slows. Even while carbon dioxide is still rather abundant in protoplasm, the enzyme is only rarely picking it up. With this relatively high concentration, carbon dioxide diffusion into the leaf is slow whereas water loss may be high.

RuBP carboxylase occasionally binds to oxygen instead of carbon dioxide, acting as an oxygenase and producing one molecule of 3-phosphoglycerate and one of phosphoglycolate. This latter molecule is transported from the chloroplast to peroxisomes and mitochondria, where some of it is converted to the useful amino acids glycine and serine. But much of the phosphoglycolate is broken down to two molecules of carbon dioxide (Fig. 10.26). The breakdown is **photorespiration,** an energy-wasting process. The energy and reducing power used to produce the two reduced carbons of phosphoglycolate are completely lost. Photorespiration is extremely exergonic but is not used to provide power to

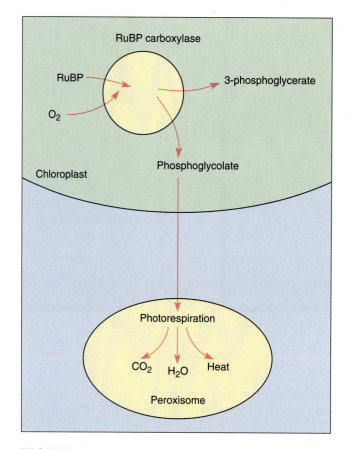

FIGURE 10.26

If RuBP carboxylase puts oxygen onto RuBP, one of the products is phosphoglycolate, which is transported to a peroxisome and broken down during photorespiration. All the energy that was present in the phosphoglycolate is wasted. Mitochondria are also involved.

The term "C_4 metabolism" seems more appropriate because this process is only a supplement to the C_3 stroma reactions. C_4 metabolism by itself is not full photosynthesis.

endergonic reactions. Because phosphoglycolate may be toxic, photorespiration protects the plant whenever RuBP carboxylase picks up oxygen. However, it is an expensive defense because up to 30% of all ATP and NADPH produced by the chloroplast can be immediately lost by photorespiration.

Apparently RuBP carboxylase cannot be significantly improved. At the time when RuBP carboxylase originated by evolution, virtually no free oxygen was present in the atmosphere, and carbon dioxide levels were high (see Chapter 17). RuBP carboxylase has existed for billions of years, and virtually no structural mutations have survived, so they must not have produced superior versions of the enzyme. An alternative is to improve its working conditions. RuBP carboxylase should be compartmentalized in a site where carbon dioxide concentration is high and oxygen concentration is low. This has evolved in some plant groups and is known as **C_4 metabolism** or **C_4 photosynthesis.** Basically, C_4 metabolism is a mechanism by which carbon dioxide is absorbed, transported through, and concentrated in a leaf, whereas oxygen is kept away from RuBP carboxylase.

C_4 metabolism occurs in leaves with Kranz anatomy (Fig. 10.27). In such leaves, mesophyll is not distributed as palisade and spongy parenchyma, but rather each vascular bundle has a prominent chlorophyllous sheath of cells and around the sheath are mesophyll cells. Mesophyll cells contain the enzyme **PEP** (phosphoenolpyruvate) **carboxylase,** which has a very high affinity for carbon dioxide. Unlike RuBP carboxylase, as the carbon dioxide concentration drops lower and lower, PEP carboxylase continues binding to it rapidly and firmly. Carbon dioxide concentrations inside the leaf are kept very low and carbon dioxide diffuses inward rapidly whenever stomata are open. The ratio of water lost to carbon dioxide absorbed is favorably low. Also, PEP carboxylase has a high specificity for carbon dioxide; it never picks up oxygen. PEP carboxylase is an ideal enzyme except that **it**

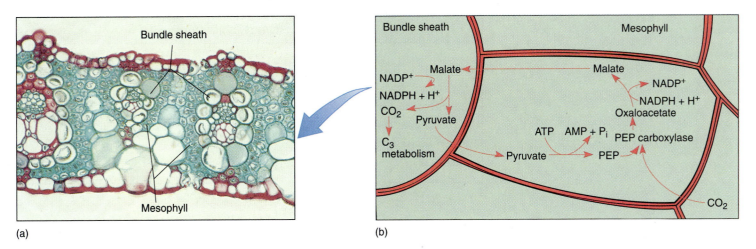

(a)   (b)

FIGURE 10.27

(a) In leaves with Kranz anatomy, such as this sugarcane, the bundle sheath around all veins is prominent and rich in chloroplasts. The mesophyll cells are arranged in a sheath around the bundle sheath ($\times 150$). (b) Knowing the reactions of C_4 metabolism is not sufficient for understanding it; only by realizing that some reactions occur in separate compartments does it become logical.

does not perform the critical reaction that results in 3-phosphoglycerate. Despite its shortcomings, RuBP carboxylase is still the only enzyme that carries out the necessary reaction.

PEP carboxylase adds carbon dioxide to PEP, producing oxaloacetate, which has four carbons; hence the name C_4 metabolism. The oxaloacetate is reduced to malate by a molecule of NADPH, and further reactions may occur, depending on the species. Malate from throughout the mesophyll moves into the bundle sheath and breaks down into pyruvate by releasing carbon dioxide (Fig. 10.27b). This reaction is powerful enough to drive the formation of a new molecule of NADPH, so the process results in the transport by malate of both carbon dioxide and reducing power. Because all the malate from a large volume of mesophyll decarboxylates in the small volume of the bundle sheath, carbon dioxide concentration in the sheath is very high (Fig. 10.28). Also, because NADPH is synthesized by this unusual method in bundle sheath cells, the bundle sheath chloroplasts primarily carry out cyclic electron transport, pumping protons and making ATP. Without noncyclic electron transport, there is no breakdown of water or production of oxygen.

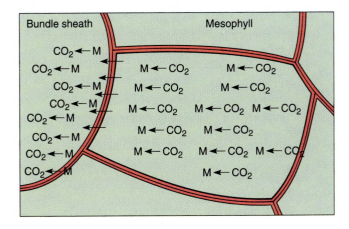

FIGURE 10.28

Mesophyll cells are much larger than bundle sheath cells, so transport of carbon dioxide from mesophyll to sheath automatically results in a higher concentration in the sheath.

TABLE 10.6	*Types of Carbon Dioxide Processing*		
	C$_3$ Plants	**C$_4$ Plants**	**CAM Plants**
Ultimate carboxylase	RuBP carboxylase	RuBP carboxylase	RuBP carboxylase
Adjunct metabolism	None	CO$_2$ transfer	CO$_2$ storage
Adjunct carboxylase	None	PEP carboxylase	PEP carboxylase
Photorespiration	High	Low	Moderate
Stomata open	Day	Day	Night

RuBP carboxylase, located exclusively in the bundle sheath chloroplasts, is in ideal conditions—high carbon dioxide concentration with low or no oxygen—and carries out efficient C$_3$ stroma reactions with low production of phosphoglycolate. C$_4$ plants have little photorespiration (Table 10.6).

Like any other carrier, malate must be shuttled back to its recharge site, the mesophyll. In the bundle sheath it is converted to pyruvate by the release of carbon dioxide; pyruvate moves back to the mesophyll and receives a phosphate group from ATP, which converts it to PEP.

The selective advantage of C$_4$ metabolism depends on the environment. Photorespiration increases with temperature, so it is more of a problem in hot climates. Under warm, dry conditions, C$_4$ metabolism has a strong selective advantage over C$_3$ metabolism: Much less water is lost during carbon dioxide absorption. Also, abundant light is available to generate the extra ATP needed to convert pyruvate to PEP. The ATP used to make PEP means that C$_4$ metabolism is not free; under cool conditions, photorespiration may be slow enough that it loses less energy than C$_4$ metabolism. Also, many cool habitats are also moist, so water conservation by stomatal closure is not as critical. The critical temperature above which C$_4$ metabolism is more advantageous selectively than C$_3$ metabolism varies among species but averages around 25°C.

C$_4$ metabolism and Kranz anatomy have evolved several times; most C$_4$ species are monocots of hot climates such as corn, sugarcane, sorghum, and several other grasses, but a considerable number of dicots, also from warm, dry regions, are also C$_4$ species (Table 10.7).

TABLE 10.7	*Plant Families Having C$_4$ Species*
Family	**Familiar Example**
Aizoaceae	Ice plants
Amaranthaceae	Amaranths
Asteraceae	Asters, daisies
Chenopodiaceae	Pigweeds, beets
Cyperaceae	Sedges
Euphorbiaceae	Spurges
Poaceae	Grasses
Nyctanginaceae	Four o'clocks, bougainvillea
Portulacaceae	Purslanes
Zygophyllaceae	(no familiar example)

BOX 10.1 *Global Warming—Will 2 or 3°C Really Matter?*

The carbon dioxide in our atmosphere acts like a light filter, allowing certain wavelengths (those of visible light) to pass through but absorbing others (especially infrared light). As visible light passes from the sun through the atmosphere and strikes Earth's surface, some is reflected immediately back out into space. A small amount is absorbed by certain biological pigments such as chlorophyll in leaves or rhodopsin in eyes, where it powers photosynthesis or vision. But most is absorbed by other molecules and has no effect other than to warm the molecules, causing them to radiate the extra energy away as long-wavelength infrared light. Many of these low-energy infrared quanta can pass directly through the atmosphere without hitting a carbon dioxide molecule, because the concentration of carbon dioxide is so low (0.03% of the air). But many quanta do strike atmospheric carbon dioxide molecules and are absorbed, causing them to become warmer. Rather than being lost to space, this energy is trapped in the Earth/atmosphere system and warms our world. This is called the greenhouse effect because the glass in greenhouses works the same way, as does the glass in a parked car.

An important balance exists between atmospheric concentration of carbon dioxide and life: With less carbon dioxide, more heat would be lost and Earth would be frozen, like Mars. With more, more heat would be trapped and our world would be as hot as Venus, at 800°C, with lakes of molten lead.

During the industrial age, we have been adding to the atmosphere's carbon dioxide by burning oil, gas, and coal, and we have destroyed the forest trees that can remove the carbon dioxide by photosynthesis. The concentration of carbon dioxide is increasing in the atmosphere, and the average temperature is also increasing. This global warming could cause mean temperatures to be 2 or 3°C (3 or 4°F) warmer in the next century.

This seems like too little to worry about; certainly in Texas a summer high of 98°F instead of 95°F would not be news. And direct temperature effects on crops or on the date of the last freeze in spring planting time might not really be serious.

But imagine what would happen if the surfaces of the oceans were slightly warmer. Much of our weather comes as winds blow eastward across the North Pacific. The water is cold and the air picks up only enough moisture to keep the Pacific Northwest wet; by the time it moves to the Central Plains states, it has so little moisture left that only grass, not forests, thrive. But if the surface waters of the Pacific were slightly warmer, vastly more moisture would evaporate into the wind and be carried to the Mississippi drainage basin. The floods of 1993 show what can happen when extra rain hits an area once.

What would happen if this occurred year after year? At what point would we decide that the climate had truly changed and whole cities should be abandoned due to annual flooding? Where would they be rebuilt? New areas might be optimal for farming, but it would take a great deal of confidence (or foolishness) to be certain that the new weather patterns would be stable enough for people to risk starting again in an unknown area. Millions of lives would be disrupted, some suddenly and catastrophically in floods, others slowly and inexorably as conditions imperceptibly but steadily declined.

CRASSULACEAN ACID METABOLISM

Crassulacean acid metabolism (CAM) is a second metabolic adaptation that improves conservation of water while permitting photosynthesis. It is so named because it was first discovered in those members of the family Crassulaceae that have succulent leaves (Fig. 10.29; Table 10.8). The metabolism is almost identical to that in C_4 plants: PEP is carboxylated, forming oxaloacetate, which is then reduced to malate or other acids. These acids are not transported but simply accumulate, in effect storing carbon dioxide. This occurs at night. These plants differ from C_3 and C_4 plants in that their stomata are closed during the hottest periods and open only at night when it is cool. Coolness reduces transpiration, and usually air is calmer at night so water molecules near stomata are not blown away immediately and may diffuse back into the plant. Opening stomata at night is effective for conserving water, but the lack of light energy creates a problem for photosynthesis: A plant cannot store ATP and NADPH during the day for use at night; they are not stable enough and each cell has too little. Hence, carbon dioxide is stored on acids until daytime, when stomata close and the malate or other acids break down, releasing carbon dioxide for C_3 metabolism. The released carbon dioxide cannot escape because stomata close at dawn.

Crassulacean acid metabolism is not particularly efficient; the total amount of carbon dioxide is so small that it may be entirely used in C_3 metabolism after just a few hours of sunlight. Furthermore, RuBP carboxylase is protected from photorespiration only in the morning when internal carbon dioxide levels are high. Later in the day, photorespiration may be high, but the carbon dioxide released by the breakdown of phosphoglycolate is trapped and refixed.

(a)

FIGURE 10.29

(a) *Sempervivum arachnoideum,* a member of the Crassulaceae. Like many desert succulents, it has CAM metabolism as well as other adaptations that conserve water. (b) This large barrel cactus (*Echinocactus*) and short pincushion cactus (*Mammillaria*) are CAM plants. (c) Photosynthetic cells of CAM plants have large vacuoles, a feature that permits the accumulation of C_4 acids.

(b)   (c)

Crassulacean acid metabolism is selectively advantageous in a hot, very dry climate where survival rather than rapid growth is most important. In these habitats, unaided C_3 metabolism is so wasteful of water that C_3 plants cannot survive. In the most arid regions, C_4 plants barely get by, growing in slightly less stressful microhabitats such as near temporary streams and ponds, or growing and flowering quickly during the cool, moist months of winter and spring, then dying in the summer and surviving only as seeds. Under such conditions, CAM plants have a selective advantage; they may not grow luxuriantly, but they do grow and reproduce more succesfully than C_3 and C_4 plants.

TABLE 10.8 *Plant Families Having CAM Species*

Family	Familiar Example
Agavaceae	Agaves, yuccas
Aizoaceae	Ice plants
Asclepiadaceae	Milkweeds
Asteraceae	Asters, daisies
Bromeliaceae	Bromeliads
Cactaceae	Cacti
Crassulaceae	Stone crops, sedums
Cucurbitaceae	Cucurbits, squash
Didiereaceae	(no common, familiar example)
Euphorbiaceae	Spurges
Geraniaceae	Geraniums
Labiatae	Mints
Liliaceae	Lilies
Orchidaceae	Orchids
Oxalidaceae	*Oxalis*
Piperaceae	Peppers, *Peperomia*
Polypodiaceae	(a fern)
Portulaceae	Purslanes
Vitaceae	Grapes
Welwitschiaceae	(a gymnosperm)

Under milder, moister conditions, crassulacean acid metabolism is not selectively advantageous. Water conservation is less of a benefit, and the limited capacity to absorb and store carbon dioxide is a distinct disadvantage. C_3 and C_4 plants photosynthesize all day, whereas CAM plants may stop before noon.

Crassulacean acid metabolism has evolved several times and is also present in the cactus family, many orchids, bromeliads, lilies, and euphorbias. All have other metabolic and structural adaptations for water conservation, such as succulent bodies filled with water-storing parenchyma and covered by a tough epidermis with a thick cuticle and wax layer.

SUMMARY

1. All physical systems have a tendency to become disordered, to increase in entropy. Living organisms have a high degree of order and regularity, maintained by the input of energy, the ultimate source of which is the sun.

2. All photosynthetic plants are autotrophs, but some parasitic plants and many plant tissues are heterotrophic.

3. When electrons are passed from one atom to another, the recipient becomes reduced and the donor becomes oxidized. Most of a plant's raw materials are highly oxidized and must be reduced as they are assimilated.

4. Photosynthetic pigments respond to light quanta that have just enough energy to boost electrons one or two energy levels.

5. Accessory pigments have absorption spectra different from that of chlorophyll *a*, so they absorb different wavelengths and transfer energy to chlorophyll *a*.

6. Photosystems I and II work together, transferring electrons from water to NADPH. The electron's energy is boosted twice, once at P680 and again at P700.

7. The light-dependent reactions result in a chemiosmotic gradient. Driven by concentration differences, protons flow from the thylakoid lumen to the stroma through ATP synthetase channels, powering the phosphorylation of ADP to ATP.

8. Cyclic electron transport permits the production of extra ATP without the synthesis of NADPH or the production of free oxygen. Noncyclic electron transport results in the production of both ATP and NADPH, but the amount of ATP is not sufficient for the stroma reactions.

9. All photosynthetic plants use the C_3 stroma reactions mediated by RuBP carboxylase: A molecule of RuBP is carboxylated, energized, and reduced, resulting in two molecules of 3-phosphoglyceraldehyde.

10. C_4 metabolism is an adjunct to C_3 metabolism, not a replacement. PEP carboxylase, which acts as the initial carboxylating enzyme, has a great affinity for carbon dioxide.

11. Crassulacean acid metabolism is similar to C_4 metabolism, except that it accumulates and stores carbon dioxide at night while stomata are open and releases it during the day while stomata are closed.

IMPORTANT TERMS

absorption spectrum
action spectrum
anabolic reactions
ATP synthetase
C_3 (Calvin/Benson) cycle
C_4 metabolism
chemiosmotic phosphorylation
chlorophyll
Crassulacean acid metabolism (CAM)
cyclic electron transport
cytochromes
electron transport chain
gluconeogenesis
greenhouse effect

heterotroph
light compensation point
light-dependent reactions
noncyclic electron transport
oxidation state
oxidative phosphorylation
oxidized compound
ozone
3-phosphoglyceraldehyde
photoautotroph
photon
photophosphorylation
photorespiration
photosystem I

photosystem II
pigment
quality of light
quantum
reaction center
reduced compound
reducing power
ribulose-1, 5-bisphosphate (carboxylase)
RUBISCO
stroma reactions
substrate-level phosphorylation
Z scheme

REVIEW QUESTIONS

1. What is a reduction reaction? Why does a reduction reaction always occur simultaneously with an oxidation reaction?

2. In photosynthesis, what is the ultimate source of electrons? What are the benefits of this molecule in terms of its toxicity and the cost of the plant to obtain it?

3. Describe the absorption spectrum of chlorophyll. Why does it match the action spectrum of photosynthesis?

4. Name the electron carriers that transport electrons from photosystem II to photosystem I. Which ones contain metal atoms and which do not?

5. When electrons are removed from water, protons are liberated. Does this occur in the stroma or inside the thylakoid lumen? Can protons move directly across the membrane? Describe the chemiosmotic mechanism of ATP synthesis in chloroplasts.

6. What chemical is the acceptor of carbon dioxide in the C_3 cycle? What enzyme catalyzes the reaction, and what is the product?

7. Imagine a leaf in bright light but an atmosphere with no carbon dioxide. Would RuBP carboxylase be functioning? Would the NADP be in the reduced or oxidized form?

8. In a C_4 plant, where is PEP carboxylase located? Where is RuBP carboxylase located?

BotanyLinks includes a Directory of Organizations for this chapter.

ENERGY METABOLISM: RESPIRATION

11

After sundown, when plants are in darkness, chloroplasts cannot carry out photosynthesis. Instead, plants obtain ATP and other energy carriers by respiration, just as animals do.

CONCEPTS

The light-dependent reactions of photosynthesis produce an excess of energy, which is stored as glucose and starch; an important corollary must be the ability to recover that energy and reduced carbon. Recovery may occur later, when photosynthesis is impossible, such as at night or in winter when the plant is leafless. Also, energy recovery may occur at a different site from photosynthetic capture: Glucose may be converted to sucrose, transported to apical meristems, vascular cambia, or any other heterotrophic tissue, and broken down to recover the energy (Fig. 11.1).

Respiration is the process that breaks down complex carbon compounds into simpler molecules and simultaneously generates the ATP used to power other metabolic processes (Fig. 11.2). During respiration, carbon is oxidized. Its oxidation state goes from +0 to +4 as electrons are removed by NAD^+, which is converted to NADH in the process. This is basically the opposite of photosynthesis, in which NADPH carries electrons to carbon, reducing it. The NADH generated by respiration is a good reducing agent, but it is produced in much larger quantities than needed for constructive reduction reactions. Plants use mostly NADPH from photosynthesis for their reductions, not NADH, and most of the

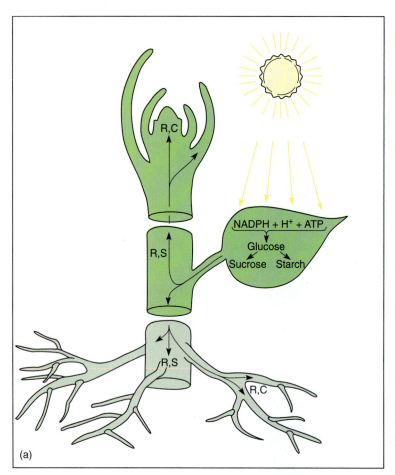

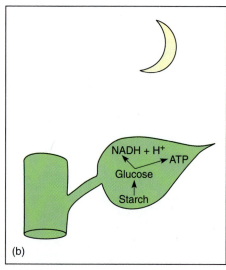

FIGURE 11.1

(a) When light is available, autotrophic tissues produce more ATP and NADPH than needed for cell metabolism. Glucose is produced, part of which is stored as starch in the leaf and part converted to sucrose and transported to areas that are either completely heterotrophic or are growing more rapidly than their own photosynthesis can support. Sucrose is converted back to glucose and is either respired (R) or used in the construction (C) of cell structures such as cellulose, lignin, amino acids, and nucleic acids. In roots, trunks, fruits, and seeds, some glucose may be polymerized to starch and stored (S) for months or years. (b) Even autotrophic tissues are heterotrophic in the dark, surviving on the respiration of stored starch.

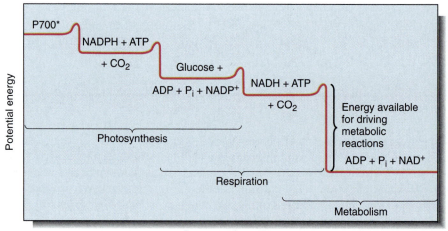

FIGURE 11.2

This hypothetical reaction diagram shows the relative levels of potential energy of the major compounds of photosynthesis and respiration. Energized P700 has the greatest potential energy, most of which is trapped in the formation of ATP and NADPH. When these reduce carbon dioxide and make glucose, much of the energy is conserved. Respiration transfers the energy to ATP and NADH. During other metabolic reactions, the breakdown of ATP and NADH yields large amounts of energy. Depending on the metabolic pathway, some of this energy is retained and some lost.

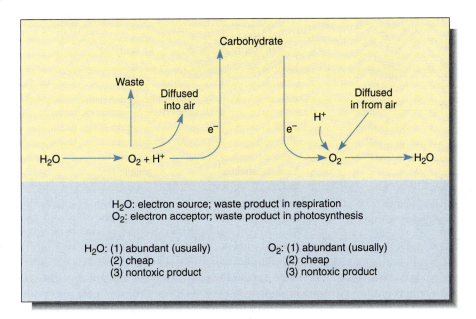

FIGURE 11.3

It is selectively advantageous for an organism's metabolism to be based on raw materials that are abundant and cheap and produce nontoxic waste products. Mutations that cause an individual to require rare or expensive compounds or those that break down into toxic products are selectively disadvantageous.

compounds that animals and fungi consume are already reduced. Because NADH contains a great amount of energy, it is selectively advantageous for an organism to be able to oxidize it so as to generate even more ATP.

The oxidation of NADH to NAD^+ requires transferring electrons from it onto something else. The ideal recipient would be abundant, cheap, and nontoxic after it is used (Fig. 11.3). These are also the three ideal characteristics of the source of electrons in photosynthesis, and the chemical involved in both processes in water. As it turns out, the result of respiration is the reverse of photosynthesis: Electrons are transferred from carbon in carbohydrate by means of reduced NADH, which carries them to an electron transport chain, which in turn deposits them onto oxygen, reducing it. As electrons are added, protons are attracted and incorporated, converting oxygen to water. If this reverse analogy were carried one step further, the electron transport chain would have to give off light. But that would be a complete waste of energy; instead it conserves the energy as high-energy phosphate-bonding orbitals of ATP.

Respiration is also useful to the plant because the numerous intermediate compounds of its many steps are useful as the starting materials for several anabolic pathways. A molecule of glucose, after entering the respiratory pathway but before being broken down to carbon dioxide and water, may be picked up in the form of an intermediate by a nonrespiratory enzyme that diverts it into a pathway that produces amino acids, fats, nucleic acids, lignin, and other molecules.

TYPES OF RESPIRATION

Cellular respiration falls into two categories: aerobic and anaerobic. Respiration that requires oxygen as the terminal electron acceptor is **aerobic respiration.** Under certain conditions, oxygen is not available and an alternative electron acceptor must be used. This is **anaerobic respiration,** respiration without oxygen, often called **fermentation.** Because animals and plants must have oxygen for their respiration, they are known as **obligate** or **strict aerobes** (Table 11.1). At the opposite extreme are certain bacteria called **obligate anaerobes,** which carry out anaerobic respiration exclusively; such bacteria are actually

TABLE 11.1	*Relationships Between Types of Organisms and Presence or Absence of Oxygen*

	Oxygen Present	Oxygen Absent
Obligate aerobes	Aerobic respiration; able to live	No respiration; death
Obligate anaerobes	Oxygen destroys certain vital metabolites; death	Fermentation; able to live
Facultative organisms	Aerobic respiration; able to live	Fermentation; able to live

killed by oxygen. Many fungi and certain types of tissues in animals and some plants are **facultatively aerobic** (or **facultatively anaerobic**): If oxygen is present, they carry out aerobic respiration, but when oxygen is absent or insufficient, they switch to anaerobic respiration. Although many fungi, especially the yeasts, can live indefinitely anaerobically, plant and animal tissues can survive this way for only a short time. They must eventually obtain oxygen and switch back to aerobic respiration or they die.

ANAEROBIC RESPIRATION

Because glucose is broken down during anaerobic respiration, the metabolic pathway is given the name **glycolysis** (from Greek: *glykys*—sweet, and *lysis*—to break down). It is also called the **Embden-Meyerhoff pathway** in honor of the physiologists who first elucidated its steps. A comparison of Figures 10.20 and 11.4 shows that glycolysis and gluconeogenesis are essentially the same pathway, with the reactions running in opposite directions. Although all intermediates are the same, as are many of the enzymes, the two processes use different enzymes at certain key steps. This allows a cell to regulate the two processes so that one is stopped while the other runs; it would be useless for both pathways to operate simultaneously within a single cell.

In glycolysis, ATP phosphorylates glucose to glucose-6-phosphate, which is then converted to fructose-6-phosphate. A second molecule of ATP then phosphorylates this to fructose-1,6-bisphosphate, which breaks down into 3-phosphoglyceraldehyde and dihydroxyacetone phosphate. The latter can be converted to 3-phosphoglyceraldehyde, of course, which can be oxidized to 1,3-diphosphoglycerate. During this oxidation step electrons are transferred from a carbon of 3-phosphoglyceraldehyde to NAD^+, converting it to NADH. The 1,3-diphosphoglycerate is energetic enough that an enzyme can transfer one of its phosphate groups onto an ADP, converting it to ATP and changing the 1,3-diphosphoglycerate into 3-phosphoglycerate, a process called **substrate level phosphorylation** (see Table 10.2). The enzyme is phosphoglycerate kinase; the kinases constitute a large group of enzymes that remove phosphate groups from substrates. Phosphorylases are just the opposite, adding phosphates to substrates.

Although this is basically a reversal of the stroma reactions, ribulose-1,5-bisphosphate does not occur next as one might expect. Instead, 3-phosphoglycerate is converted first to 2-phosphoglycerate and then to phosphoenolpyruvate (PEP), the same metabolite that is the carbon dioxide acceptor in C_4 metabolism. PEP is also energetic enough that an enzyme can transfer its phosphate group onto ADP to make ATP; dephosphorylation causes PEP to become pyruvate.

As with NADPH, to be strictly correct when balancing equations, NADH should be considered to be associated with a proton: NADH + H^+.

FIGURE 11.4   ▶

Glycolysis, also called the Embden-Meyerhoff pathway, constitutes the major portion of anaerobic respiration and is also the first part of aerobic respiration.

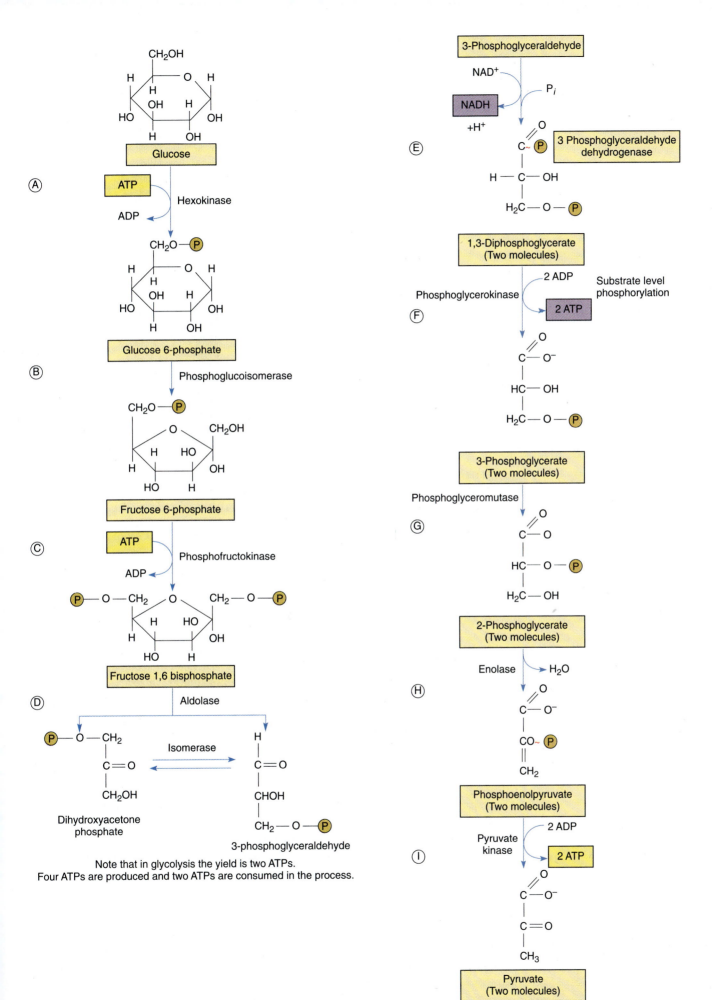

Note that in glycolysis the yield is two ATPs.
Four ATPs are produced and two ATPs are consumed in the process.

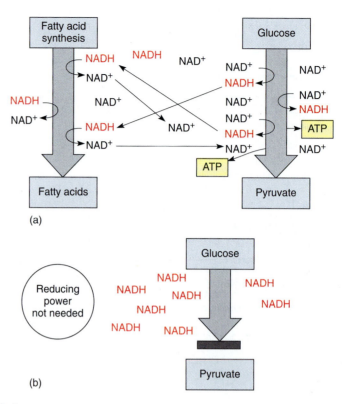

FIGURE 11.5

(a) If a cell needs reducing power, as for the synthesis of fatty acids, those reactions are actually electron-accepting reactions and they regenerate NAD$^+$, allowing the continued production of ATP by glycolysis. (b) If reducing power is not needed, NAD$^+$/NADH accumulates as NADH. Without NAD$^+$, glycolysis stops: 3-phosphoglyceraldehyde cannot be oxidized to 1,3-diphosphoglycerate, and no ATP can be formed.

If no oxygen is present, anaerobic respiration has now removed all the energy possible. From each molecule of glucose, four ATPs were generated and two were consumed, so there is a net production of two ATPs. Occasionally the cell can start with glucose-6-phosphate and use one less ATP to get started. The ATP generated in anaerobic respiration is used for other metabolic reactions such as protein synthesis, nucleic acid replication, microtubule assembly, and ion transport. Indeed, these metabolic pathways are the reasons for respiration, and the ADP they generate migrates back to the sites of glycolysis and is rephosphorylated to ATP.

The reduction of NAD$^+$ to NADH during glycolysis is a problem. If the cell needs reducing power, the NADH can be used and it regenerates the NAD$^+$ necessary to keep glycolysis running. For example, roots absorb nitrates and sulfates that must be reduced, and during the synthesis of fatty acids, large amounts of reducing power are needed. Many of these reductions can be carried out with this extra NADH (Fig. 11.5). But usually a cell does not need as much reducing power as is produced during respiratory ATP production; consequently, NADH accumulates, all the NAD$^+$ is consumed, and glycolysis stops for lack of NAD$^+$. Without further glycolysis, no ATP would form and death would result. The big problem is converting NADH back to NAD$^+$ by dumping its electrons. In animal tissues under anaerobic conditions, the electron acceptor is pyruvate: NADH reacts with it to form **lactate**, the anion of lactic acid (Fig. 11.6). In plants and fungi, pyruvate is first converted to **acetaldehyde** and then NADH reacts with that, forming **ethanol** (**ethyl alcohol**; Fig. 11.7a). Anaerobic conditions occur in plants and fungi growing in mud beneath stagnant water, especially in swamps and marshes (Fig. 11.7b). Rice seeds germinate and grow anaerobically until the shoots reach oxygenated water (Fig. 11.8).

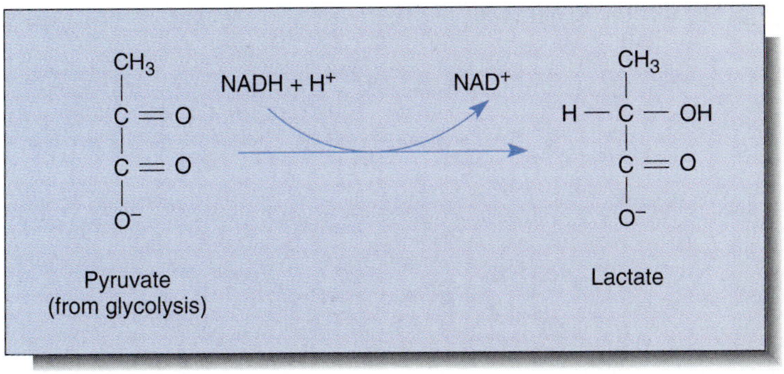

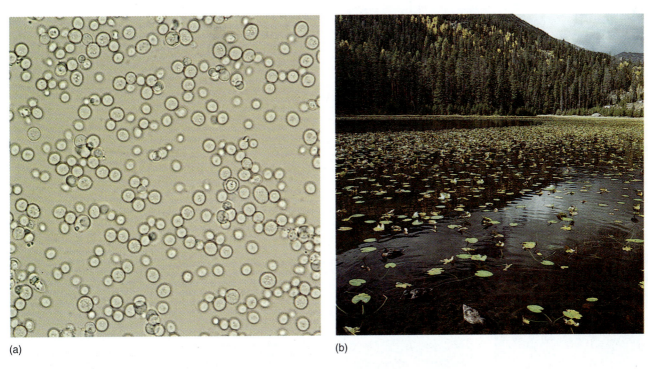

(a)

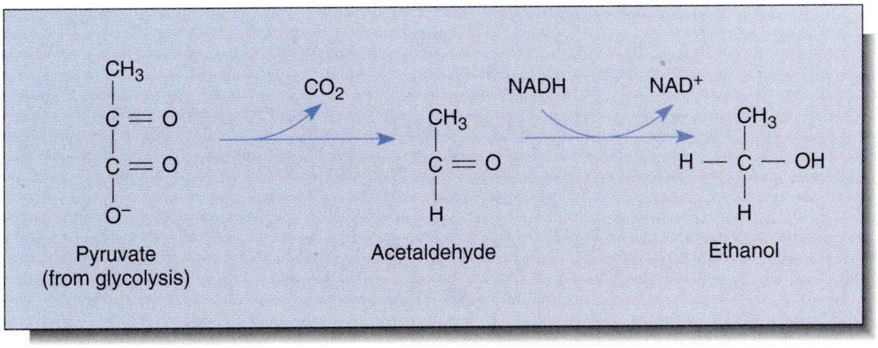

(b)

FIGURE 11.6

(a) When we exercise slowly enough for blood circulation to keep up, all our muscular activity is aerobic. But with rapid, intense, and prolonged activity, blood does not carry oxygen to the muscles rapidly enough and lactic acid fermentation begins. Lactate accumulation causes cramps and muscle pain. (b) Alcoholic fermentation involves the conversion of pyruvate to acetaldehyde before reduction by NADH + H$^+$.

(a)

(b)

FIGURE 11.7

(a) In wine-making, yeast cells (*Saccharomyces*) ferment the glucose present in grapes and excrete the waste product, ethanol. Naturally fermented wine has a maximum alcohol content of 14%; at that point, the waste product kills the yeast. Beer has a lower alcohol content, 5.5%, not because its yeast is more sensitive to the ethanol but because it is the legal limit. Fermentation must be stopped artificially, usually by heating the beer to kill the yeast. Alcoholic beverages with alcohol contents higher than 14% have extra alcohol added or part of their water removed ($\times$ 430). (b) The leaves and upright stems of water lily (*Nuphar*) exist in a highly oxygenated aerobic environment, but the rhizomes and roots exist in an anaerobic muck. Aerenchyma tissues in the stems permit the diffusion of some oxygen from leaves to rhizomes and roots, but some anaerobic respiration may be occurring. (*b, © Butch Gemin*)

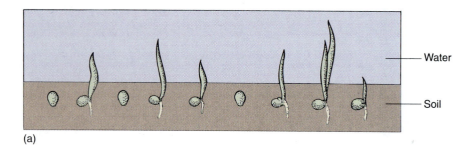

(a)

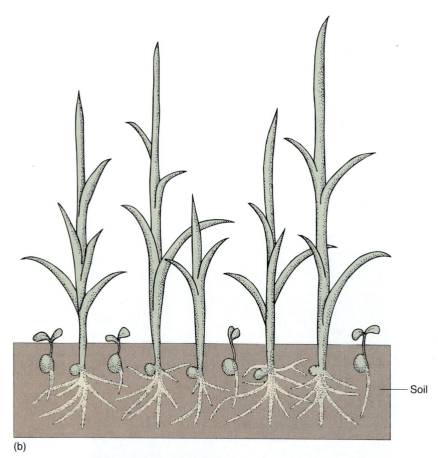

(b)

(c)

(d)

FIGURE 11.8

(a) Even under flooded, anaerobic conditions, rice seeds germinate and begin to grow because their embryos carry out facultative anaerobic respiration. Other seeds (round) remain dormant because they are incapable of fermentation; being dormant, they have a low rate of metabolism and the small amount of oxygen present may keep them from dying immediately. (b) When flooding subsides, oxygen is available and non-rice seeds germinate, but they are shaded by rice plants, and the soil is already filled with rice roots. The rice plants have a tremendous advantage. (c) Rice seedlings, still flooded. (d) Fields of rice. The lines are the dikes used to flood the fields for irrigation. *(c and d, Courtesy of M. Woods/The Rice Council)*

Ethanol and lactate are not especially good solutions to the problem of NADH accumulation. The pyruvate consumed is always present in adequate amounts, of course, because glycolysis itself produces it, but it is not really "cheap" because many of its bonding orbitals have high-energy electrons. Furthermore, pyruvate could be used as a monomer for many types of synthesis. Lack of oxygen forces a cell to use a very valuable molecule as an electron dumping ground. Even worse, the products, either lactate or ethanol, are toxic. If they accumulate in the tissues or environment, they damage or even kill the cells producing them (Table 11.2).

Considering the negative aspects of anaerobic respiration, how could natural selection have produced something so inefficient? At the time life arose and respiratory pathways were evolving, Earth's atmosphere contained reduced, hydrogen-rich compounds but no oxygen (see Chapter 17). Consequently, pyruvate and acetaldehyde had to be used as electron acceptors. Millions of years later, photosynthesis based on chlorophyll *a* evolved, and oxygen was released to the environment as a result of the water-splitting involved. After free oxygen became relatively abundant, the mutations leading to aerobic respiration began to be selectively advantageous.

Even in the presence of an oxygen-rich atmosphere and aerobic respiration, retaining the capacity for anaerobic respiration still has some selective advantage. Although the method is far from ideal, it does allow certain organisms to survive in particular environments. For example, because rice seedlings are capable of anaerobic respiration, they can germinate and establish themselves during times of floods when other plants or seedlings are suffocating (Fig. 11.8). In the absence of oxygen, the alternative to anaerobic respiration is death (see Table 11.1). Although this metabolism is expensive for the rice, by the time the flooding subsides, rice plants are already well-rooted, have several leaves, and outcompete other species whose seeds are just beginning to germinate.

AEROBIC RESPIRATION

With oxygen present, the problems of using pyruvate or acetaldehyde as an electron acceptor are eliminated; oxygen is absorbed and acts as the terminal electron acceptor. Oxygen is inexpensive because it is absorbed and distributed by molecular diffusion, which requires neither active transport nor ATP consumption. Oxygen is abundant in most situations, and the product of reduction is water, which is not only nontoxic but actually beneficial. Aerobic respiration consists of three parts: (1) glycolysis, (2) the citric acid cycle, and (3) oxidative phosphorylation in an electron transport chain.

Glycolysis. The initial steps of aerobic and anaerobic respiration are identical: glycolysis by the Embden-Meyerhoff pathway to pyruvate (see Fig. 11.4). This produces ATP and NADH just as before, but with oxygen present the NADH migrates to electron carriers that oxidize it back to NAD^+, permitting glycolysis to continue. Glycolysis occurs in the cytosol.

The Citric Acid Cycle. Because pyruvate is not needed as an electron acceptor, it can be used in a number of metabolic pathways. One of the main pathways takes advantage of the large amount of energy remaining in pyruvate and, rather than using it for its structure, breaks it down and generates more ATP. This pathway is called the **citric acid cycle**, the **Krebs cycle**, or the **tricarboxylic acid cycle**. These names reflect different facts about the cycle. One of the intermediates is citrate, the anion of citric acid. Much of the pioneering work on this metabolism was carried out by Hans Krebs. Finally, several of the intermediates are tricarboxylic acids—that is, each has three carboxyl (—COOH) groups.

In the citric acid cycle, pyruvate is transported from the cytosol, where glycolysis occurs, across the mitochondrial membranes to the mitochondrial matrix. There it is oxidized and decarboxylated: As the electrons are transferred to NAD^+, the bonding orbitals holding the last COO^- rearrange and CO_2 is liberated. Carbon dioxide and NADH are produced, along with a two-carbon fragment called **acetyl** (Fig. 11.9). The carbon dioxide and NADH remain free in the matrix solution, but the acetyl becomes attached to a carrier molecule, **coenzyme A** (CoA), the resulting combination being **acetyl CoA**. Like pyruvate, acetyl CoA can be used in many synthetic pathways, but here we are interested in its entry

Remember that the ideal electron acceptor is abundant and cheap and does not produce a toxic product.

TABLE 11.2 *Electron Acceptors*
Pyruvate
1. Abundant when needed
2. Expensive
3. Toxic product (lactate)
Acetaldehyde
1. Abundant when needed
2. Expensive
3. Toxic product (ethanol)
Oxygen
1. Usually abundant when needed
2. Cheap
3. Nontoxic product (water)

BOX 11.1   *Fungal Respiration—The Prehistoric Industrial Revolution*

Most college students know at least a little about fermentation—you probably already know that beer and wine are fermented even though you never much worried about the Embden-Meyerhoff pathway. Perhaps you know already that aerobic respiration of the fungi known as yeast causes bread to rise and holes to form in Swiss cheese. But you may not have realized that the use of these fermentations started out as lifesaving processes several thousand years ago.

When we think about the origins of agriculture between 2000 and 3000 B.C., we often think of the domestication of wheat, rice, and corn, but the use of fruits and animal products was necessary not only for diversity in the diet but also for diversity of vitamins and essential nutrients. Think about life in 3000 B.C.—no supermarkets, of course, and no canned food, frozen food, or refrigerated food. People grew their own food, or collected it, or traded for it.

Now imagine the first cold winds and rains of autumn—how do you preserve the abundant food from summer to allow your survival through winter? It will be 7 long months before grains can be planted, before

the dormant buds on grape vines open and put out their new leaves and flowers. Even after spring, the first crops cannot be harvested until May or June at the earliest. Hopefully, cows will give birth and there will be milk (cows lactate only while they have calves too young to eat grass). How do you preserve food for the winter?

Grains like wheat dry naturally just before harvest and can be stored rather easily; they are so dry that microbes cannot grow. One problem is that they must be kept dry —protected not only from winter's dampness but also from the water produced by their own slow respiration.

Fruits also can be dried, but these cannot be kept dry reliably without cellophane or plastic bags. Under ancient storage methods, dried fruit often drew enough moisture from wet winter air to allow fungi and bacteria to grow on them. One way to prevent microbial growth was discovered early: fermentation. If the sugary fruits were partially respired in a sealed jar (anaerobic conditions), the production of ethanol would finally kill all microbes, sterilizing the fruit. As long as the jar remained sealed, the fruit (or fruit juice) was safe—

including all its minerals, vitamins, amino acids, and so on. Beer is most often fermented barley, and barley grows best in cool, damp, northern climates where it is most difficult to keep grains dry.

Natural, unpasteurized milk becomes sour in just 2 or 3 days at room temperature (and of course in 3000 B.C. just about all temperatures were room temperature) as bacteria grow rapidly. But if the right fungi are added, the milk is only partially degraded and the resulting cheese is dry and stable enough to stop bacterial growth. Covered with a layer of wax to protect it from air-borne bacteria, it can last for years.

Using yeast to leaven bread (to make it rise and be spongy) has basically the opposite effect. If wheat is ground to flour, mixed with water and then baked, the resulting bread can be so hard and dry that it will last forever, even in your mouth— mastication (chewing) is an ordeal. But the carbon dioxide from respiring baker's yeast becomes trapped by the bread dough, causing the bread's texture to be lighter and more open. This bread does not preserve as well, but it is easier to eat.

In animal mitochondria, breakdown of succinyl CoA to succinate and CoA drives the phosphorylation of guanosine diphosphate (GDP) to GTP, which immediately reacts with ADP, resulting in GDP and ATP.

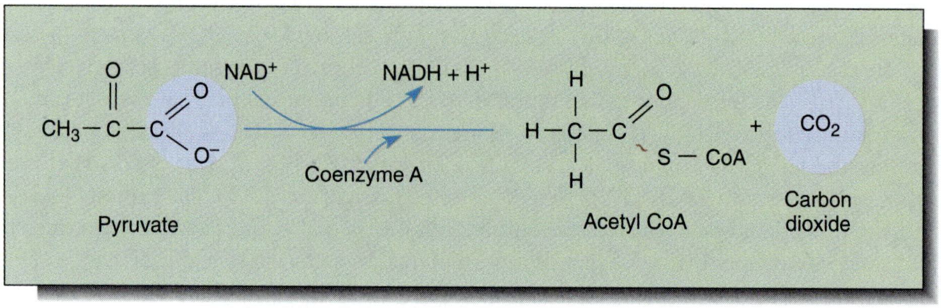

FIGURE 11.9

The process by which pyruvate is attached to CoA releases a carbon dioxide molecule and forms NADH. This step does not occur during anaerobic respiration; not only would it use up pyruvate needed as an electron acceptor, but it would generate even more NADH. Both of those results would be detrimental under anaerobic conditions, but both are beneficial when oxygen is present.

into the citric acid cycle (Fig. 11.10) by transfer of the acetyl group to an acceptor molecule, **oxaloacetate,** a compound with four carbons. The oxaloacetate is converted to a six-carbon compound, **citrate,** which is then rearranged to *cis-aconitate*, which in turn is transformed to **isocitrate** (Fig. 11.10, steps A, B, C, D). In the next step, one of the carbons of isocitrate is oxidized by passing electrons onto NAD^+, creating NADH. The oxidized carbon is liberated as carbon dioxide, leaving **alpha-ketoglutarate,** which has only five carbons (step E). This too is oxidized by NAD^+, liberating another carbon dioxide, and the four-carbon remnant becomes attached to a new molecule of CoA in the process, forming **succinyl CoA** (step F). The energy released by the breakdown into free CoA and free succinate can power phosphorylation of ADP to ATP (step G). The succinate still contains considerable energy and is oxidized to **fumarate** as electrons and protons are passed to **flavin adenine dinucleotide** (FAD), reducing it to $FADH_2$ (Fig. 11.11). Although the molecule becomes oxidized by this step, no carbon dioxide is lost; the fumarate is also a

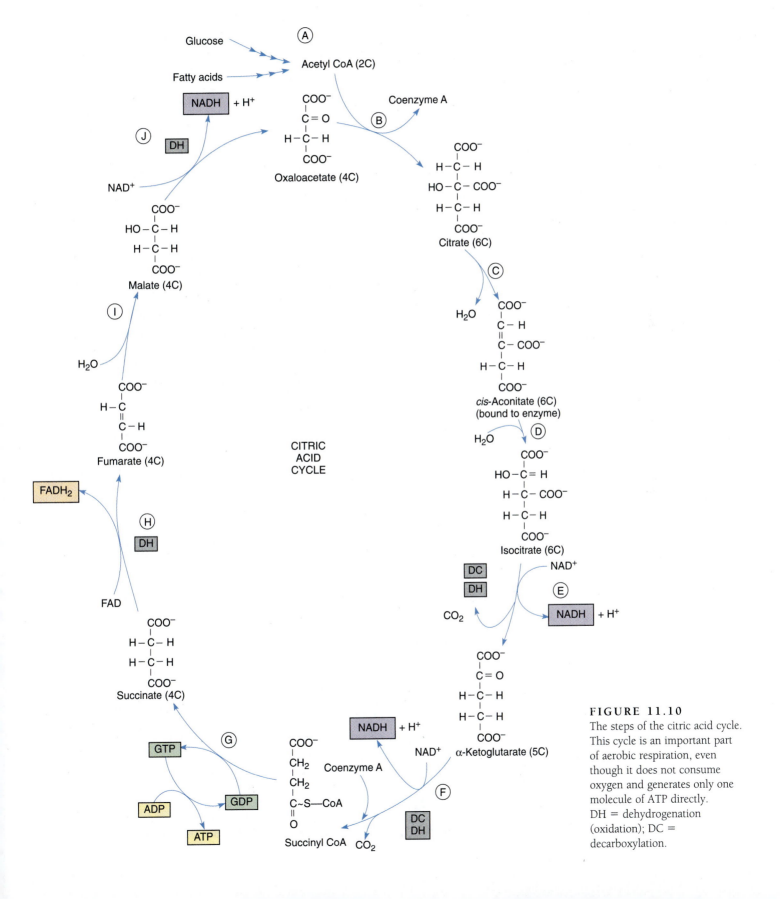

FIGURE 11.10
The steps of the citric acid cycle. This cycle is an important part of aerobic respiration, even though it does not consume oxygen and generates only one molecule of ATP directly. DH = dehydrogenation (oxidation); DC = decarboxylation.

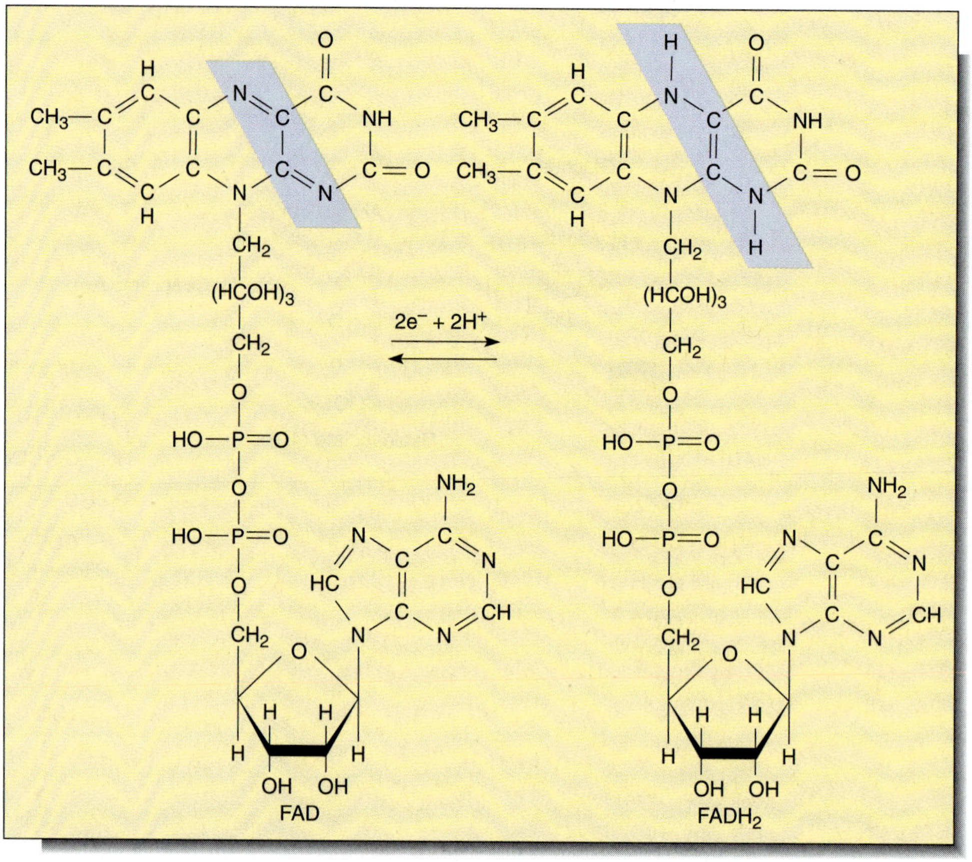

FIGURE 11.11

Flavin adenine dinucleotide (FAD) is a large organic electron carrier similar to NAD$^+$. When it is reduced by two electrons, it picks up two protons, becoming FADH$_2$. Unlike NADH + H$^+$, both protons are covalently attached to FADH$_2$ (shown in blue box).

four-carbon compound. It reacts with a water molecule and becomes **malate,** which passes a final set of electrons onto NAD$^+$ and is transformed into the original acceptor molecule, oxaloacetate (steps H, I, J).

It was mentioned above that the benefit of the citric acid cycle is the generation of more ATP, yet at only one step has ATP been produced. Instead, there are four steps in which more NAD$^+$ is reduced to NADH and one in which FAD is reduced to FADH$_2$. Excess NADH and the related deficiency of NAD$^+$ were seen to be problems in anaerobic respiration, so the citric acid cycle at first seems to be a real contradiction. But in the next step, the electron transport chain, the energy in NADH and FADH$_2$ drives the synthesis of ATP, and NADH is simultaneously oxidized back to NAD$^+$.

The Mitochondrial Electron Transport Chain: Chemiosmotic Phosphorylation.   The mitochondrial inner membrane, like the chloroplast inner membrane, contains sets of compounds capable of carrying electrons (Fig. 11.12; see Table 10.5). Although many of the actual carriers differ in the two organelles, the principles of electron transport are the same. The carriers that react with NADH or oxygen are placed precisely and asymmetrically in the cristae membranes, both on the matrix side only (Fig. 11.13). At present, the exact order of carriers in plant mitochondria is not known, and numerous types exist. Only the most well-understood are presented below.

NADH diffuses to the membrane and passes electrons to a protein that has **flavin mononucleotide (FMN)** bound to it as a cofactor. The FMN is reduced and the NADH simultaneously oxidized to NAD$^+$, which can migrate back to the site of the citric acid cycle. The reduced FMN (FMNH$_2$) passes the electrons to a set of iron- and sulfur-containing, proteinaceous electron carriers, which transfer the electrons to one or several

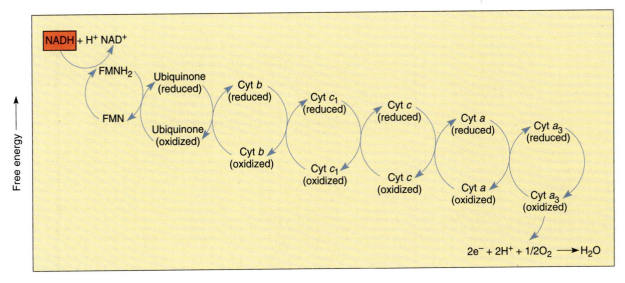

FIGURE 11.12

Membrane-bound electron carriers of the mitochondrial electron transport chain. As in chloroplasts, their positions and movements are important (see Fig. 11.13). Electrons are brought to membrane-bound carriers by mobile carriers such as NADH and $FADH_2$, which are produced in the matrix.

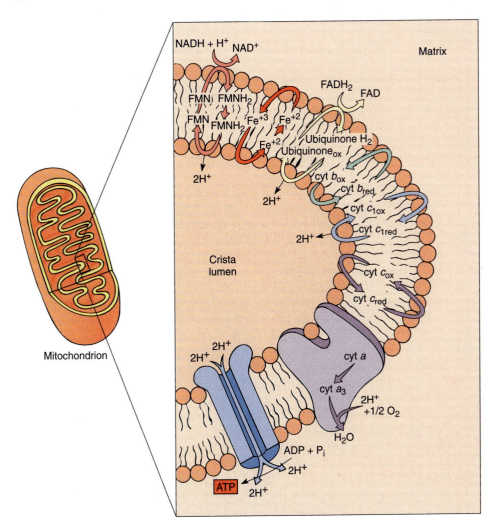

FIGURE 11.13

Only the important components of the mitochondrial electron transport chain are shown in this diagram—the steps that transport protons from the matrix to the crista lumen, establishing a proton/hydroxyl chemiosmotic gradient, just as in chloroplasts. The formation of water contributes to the gradient because protons are absorbed from the matrix but not from the cristae. FMN is flavin mononucleotide, an electron carrier similar in structure to FAD.

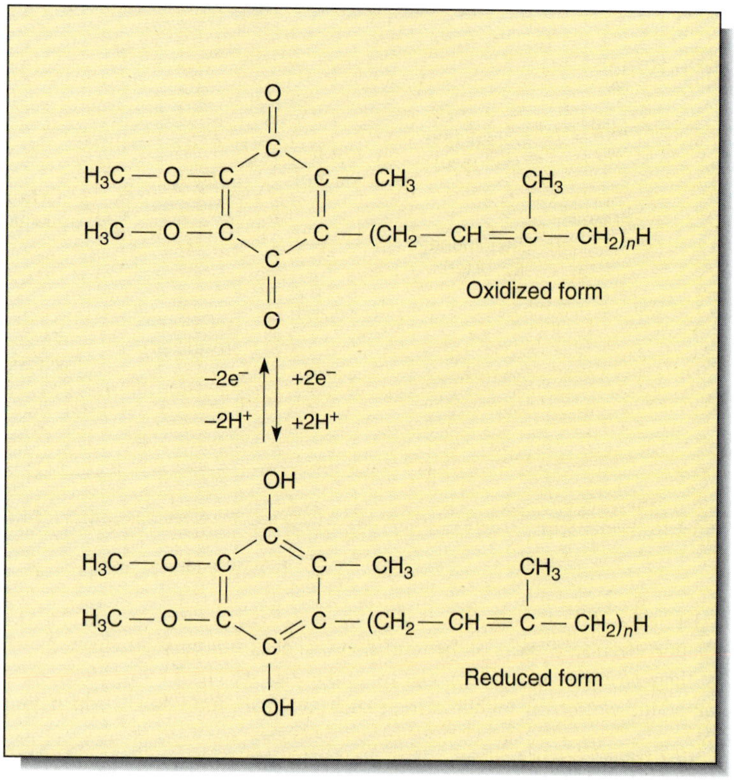

FIGURE 11.14
Ubiquinone (coenzyme Q) is a quinone that can carry two electrons and two protons simultaneously.

quinones, one of which is **ubiquinone** (also called coenzyme Q; Fig. 11.14). From the pool of quinones, electrons are transferred to **cytochrome b.** The next carriers in sequence are more quinones, then **cytochrome** c_1, **cytochrome c, cytochrome a,** and **cytochrome** a_3. Cytochromes a and a_3 are part of a large enzyme complex known as **cytochrome oxidase,** which contains several proteins and two copper ions that mediate the transfer of electrons from the iron in cytochrome to oxygen. As oxygen is reduced, it picks up two protons and becomes water.

If this electron transfer were the only thing to happen in the electron transport chain, it would be better than anaerobic respiration because the cell could avoid the waste of pyruvate and the synthesis of the toxic lactate or ethanol, but much more is accomplished. Mitochondria also use chemiosmotic phosphorylation. As with chloroplasts, the fate of the protons released and absorbed during electron transport is important. In mitochondria, when NADH reacts with FMN, two protons are transferred as well as electrons. One proton comes from the NADH and the other from the water, both on the matrix side of the membrane. When the $FMNH_2$ reduces any quinone, the two protons are released on the other side of the membrane into the lumen of the crista. This step of electron transport acts as a proton pump; a large concentration of protons begins to build up in the lumen while there is a lack of protons in the matrix. Ubiquinone and the other quinones just before cytochrome c_1 act in a similar manner, pumping protons out every time they carry electrons. Also, as oxygen receives electrons from the electron transport chain, it forms water by picking up two protons from the matrix, decreasing the proton concentration. In a short time, a proton concentration gradient develops which is strong enough to cause the protons to migrate back into the matrix.

The flow of protons from the crista lumen to the mitochondrial matrix can be used to synthesize ATP. As in chloroplasts, ATP synthetase channels in the membrane allow the flow of protons to force a phosphate group onto ADP, creating ATP. This is a chemiosmotic phosphorylation, and because of it, the NADH is an excellent if indirect source of ATP rather than a problem. The electrons that each molecule of NADH contributes to the mitochondrial electron transport chain provide enough power to create three ATPS.

A few water molecules are always splitting spontaneously ($H_2O \longrightarrow H^+ + OH^-$); water is not split specifically as in photosystem II of photosynthesis.

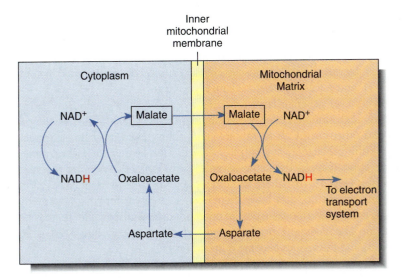

FIGURE 11.15
The malate-aspartate shuttle allows reducing power to be transferred into mitochondria when NADH itself cannot cross the mitochondrial membranes.

The FADH$_2$ also passes its electrons to the mitochondrial electron transport chain, but it reacts with ubiquinone instead of FMN, so the first step in proton pumping does not use these electrons. Therefore the FADH$_2$ contributes less to the proton gradient than does NADH, providing enough power for the production of two ATPs rather than three.

The NADH Shuttle. NADH produced by glycolysis cannot cross the mitochondrial inner membrane and donate electrons directly to the electron transport chain because the inner membrane is impermeable to such large molecules. Instead, a series of chemical reactions carries (shuttles) reducing power across the membrane (Fig. 11.15). Several shuttle mechanisms occur. In the malate-aspartate shuttle, NADH in the cytosol powers the conversion of oxaloacetate to malate, which crosses to the mitochondrial matrix and powers the formation of a new molecule of NADH. Malate is converted to aspartate and transported back out of the mitochondrion, where it is converted to oxaloacetate again and can repeat the cycle. In this shuttle, each cytosolic NADH drives the formation of a matrix NADH and the consequent oxidative phosphorylation of three ADPs to three ATPS.

In a second type of shuttle, the glycerol phosphate shuttle, cytosolic NADH reduces dihydroxyacetone phosphate to glycerol phosphate, which is transported across the inner membrane to the matrix. There it converts back to dihydroxyacetone phosphate, reducing FAD to FADH$_2$ in the process. Each cytosolic NADH results in the formation of one matrix FADH$_2$, which can drive the formation of only two ATPs; not as much energy is conserved in this shuttle.

In at least some plants, NADH can cross the outer mitochondrial membrane and react directly with ubiquinone at the outer surface of the inner membrane. A shuttle mechanism is not necessary, but the step of proton pumping by FMN is bypassed, decreasing the amount of ATP that can be generated.

HEAT-GENERATING RESPIRATION

During glycolysis, the citric acid cycle, and mitochondrial electron transport, small amounts of energy are lost in each step even though a great deal of energy is conserved by the synthesis of ATP (see Fig. 11.2). The total chemical energy of all products (ATP, carbon dioxide, and water) is less than that of all reactants (glucose-6-phosphate and oxygen); the difference is "lost" as heat and increased entropy (disorder). For example, compost piles become warm because of the heat loss during the respiration of the fungi and bacteria that decompose the compost. Heat loss is usually an inefficient aspect of respiration; in most

FIGURE 11.16
By generating heat internally, skunk cabbage
(*Symplocarpus foetidus*) maintains a high enough
temperature to carry out active metabolism even
when covered by snow. When it is ready to
emerge, it produces even more heat and melts
the snow, revealing the inflorescence to
pollinators. (*L. L. Rue III/Earth Sciences*)

cases, it would be selectively advantageous for a plant to produce more ATP from each molecule of glucose respired and lose less heat to the environment.

The heat "lost" during respiration by warm-blooded mammals is vitally necessary to maintain body temperature, and when we are chilled we must generate even greater amounts of heat by shivering: Our muscles contract and relax rapidly, breaking down large amounts of ATP and releasing the stored energy as heat.

Some plants also generate large amounts of heat. In the voodoo lily (*Sauromatum guttatum*), parts of the inflorescence become much warmer than the surrounding air, causing amines and other chemicals to vaporize and diffuse away as chemical attractants for pollinators. Skunk cabbage (*Symplocarpus foetidus*) often begins floral development while covered with snow; it melts the snow cover and exposes its flowers by generating large quantities of heat (Fig. 11.16). These and other plants are able to produce heat much more efficiently than humans do; they have alternative electron carriers that apparently do not pump protons during electron transport in mitochondria. Consequently there is no proton gradient and no chemiosmotic production of ATP. The energy in NADH is converted entirely to heat, and the tissues become quite warm (Fig. 11.17).

In ordinary mitochondria, cyanide (CN^-), azide (N_3^-), and carbon monoxide (CO) interfere with the last electron carrier, cytochrome oxidase. When they are present, electrons cannot pass from cytochrome *c* to oxygen; without electron flow and consequent ATP generation, both animals and plants die unless they are capable of anaerobic respiration. In plants that generate heat, the alternative electron carriers do not interact with cyanide, azide, or carbon monoxide, and heat is generated even if these chemicals are present. Heat-generating respiration can thus be studied by poisoning normal aerobic respiration with cyanide; consequently, heat-generating respiration is usually called **cyanide-resistant respiration.** A better name is **thermogenic respiration.**

The term "cyanide-resistant respiration" is somewhat misleading because it suggests that plants can be immune to cyanide poisoning. The analogy that because anaerobic respiration allows cells to survive in anaerobic conditions, cyanide-resistant respiration must allow cells to survive in the presence of cyanide is incorrect. Plants never encounter high concentrations of cyanide in natural conditions; if they did they would be killed because aerobic ATP generation is blocked. "Thermogenic respiration" and "heat-generating respiration" are less confusing terms.

Many aspects of thermogenic respiration are still completely unknown. We do not know if it occurs only in specialized mitochondria or if a single mitochondrion can have

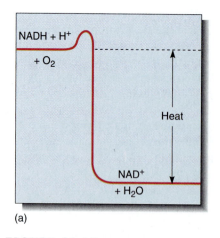

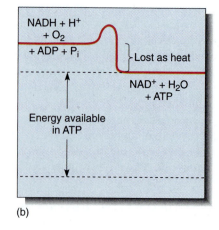

(a)   (b)

FIGURE 11.17
If NADH breaks down to NAD^+, a large amount of energy is liberated. (a) If this powerful exergonic reaction is not coupled to any endergonic reaction, all the energy is converted to heat. (b) When coupled to the endergonic synthesis of ATP from ADP, most of the energy is conserved and little is converted to heat.

PLANTS & PEOPLE

PLANTS, BABIES, AND HEAT

Most plants resemble cold-blooded animals in having a temperature close to that of the immediate environment and no thermogenic mechanisms. The few plant tissues and organs that heat up do so in response to their own developmental state, such as being ready to open their flowers; plants are not known to have any ability to sense their own temperature.

In warm-blooded animals like humans, internal body temperature—core temperature as opposed to skin temperature—is monitored by the hypothalamus portion of the brain, which detects changes in the core temperature of as little as 0.01°C. When the core temperature drops, the hypothalamus causes muscle tone to increase and shivering begins. Because there is no overall movement of the entire body, all the energy of ATP-driven muscle contraction is converted to heat.

Newborn human babies are unable to shiver; they have instead a thermogenic mechanism more like that of plants and hibernating mammals. Fat cells of adult humans can only convert stored lipids to fatty acids, which are then circulated by the blood stream to cells that can respire them. But newborns, like mammals that hibernate, have a special type of fat, called brown fat, whose cells have many mitochondria and can fully respire lipids and fatty acids, similar to the fat storage cells in plants. Brown fat cell respiration results in NADH, which contributes electrons to the mitochondrial electron transport chain, but little or no ATP is produced. Mitochondria of brown fat cells in newborns have membranes that are permeable to protons; proton pumping occurs, but no chemiosmotic gradient builds up, so all energy is converted to heat.

Brown fat cell and plant thermogenic respiration have similar results—the generation of heat. It is not known whether one is more suited to animals and the other to plants. Perhaps in animals the mutation of proton-permeable membranes occurred earlier than mutations of the genes for alternative carriers, whereas in plants the opposite occurred.

both types of electron carriers. Although we know that plants turn this respiration on and off at specific times, we do not know what the controlling mechanism is. We are not certain how widespread thermogenic respiration is; it has been found in many plant tissues as well as several algae, fungi, and bacteria.

PENTOSE PHOSPHATE PATHWAY

The **pentose phosphate pathway** is so named because it involves several intermediates that are phosphorylated five-carbon sugars (pentoses). It is usually included in discussions of respiration because it begins with glucose-6-phosphate, gives off carbon dioxide, and involves oxidations that produce NADPH (Fig. 11.18). However, its importance as a source of respiratory energy is much less significant than its role as a synthetic pathway. The pentose phosphate pathway transforms glucose into four-carbon sugars (erythrose) and five-carbon sugars (ribose) that are essential monomers in many metabolic pathways. The ribose-5-phosphate produced can be shunted into nucleic acid metabolism, forming the basis of RNA (*ribo*nucleic acid) and DNA (deoxy*ribo*nucleic acid) monomers, the nucleotides. In meristematic cells, large amounts of DNA must be synthesized during the S-phase of a short cell cycle; the pentose phosphate pathway is an extremely important part of the metabolism of these cells.

The four-carbon sugar erythrose-4-phosphate is the starting material in the synthesis of many compounds. Two important types are lignin and anthocyanin pigments. Tissues such as wood, fibers, and sclereids deposit large amounts of lignin into their secondary walls during development, and erythrose-4-phosphate is in great demand. During differentiation, these cells use the pentose phosphate pathway but pull out erythrose-4-phosphate, not ribose-5-phosphate. Flower petals and brightly colored fruits divert erythrose-4-phosphate from the pentose phosphate pathway to anthocyanin production while synthesizing their pigments. The pentose phosphate pathway also occurs in plastids, where it supplies erythrose-4-phosphate for synthesis of amino acids such as tyrosine, phenylalanine, and tryptophan.

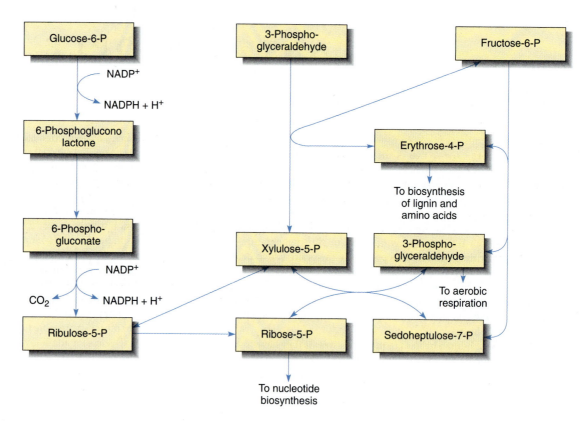

FIGURE 11.18

The pentose phosphate pathway, showing all intermediates. If ribose-5-phosphate is drawn off into nucleic acid metabolism, the pentose phosphate pathway is shifted in favor of ribose production. But if erythrose-4-phosphate is diverted to lignin metabolism, the pentose phosphate pathway reaction equilibria are shifted toward erythrose production.

In addition to these four-carbon and five-carbon sugars, the pentose phosphate pathway also produces NADPH. Although many anabolic reductions in the cytoplasm use NADH rather than NADPH, the reduction of nitrate (NO_3^-) to amino acids (NH_3^+) can be accomplished only by NADPH, which can be generated by the pentose phosphate pathway.

Many of the reactants in the pentose phosphate pathway are the same as those in glycolysis, and both pathways occur in the cytosol. They are best understood as interconnected and simultaneous pathways. In meristematic cells the pentose phosphate pathway is active and shifted in favor of ribose production. In wood cells it is also active but produces erythrose. In other cells it may be much less active and glycolysis may dominate, producing NADH that then powers ATP production in mitochondria. Imagine a cell as it is produced in the vascular cambium (meristematic, needs nucleic acids), then differentiates into lignified wood parenchyma (needs lignin), which, after the wall is mature, takes on the role of storing starch during the summer and releasing it in the spring (Fig. 11.19). Energy metabolism is adjusted at each stage in a major way, and smaller changes in the rates of reactions may occur on a day-to-day basis.

RESPIRATION OF LIPIDS

Some tissues, especially oily seeds and dormant apical meristems, store large amounts of lipid, usually as triglycerides or phospholipids. During germination or release from dormancy, lipids undergo catabolic metabolism in which they are broken down into glycerol and three fatty acids (triglycerides) or glycerol phosphate and two fatty acids (phospholipids). Fatty acids are then further broken down into two-carbon units—acetyl CoA—by a process called **beta-oxidation** in either cytosol or microbodies called glyoxisomes (see

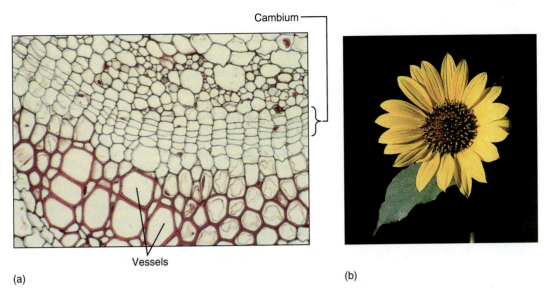

Cambium

(a)

Vessels

(b)

FIGURE 11.19

(a) At the cambium, the pentose phosphate pathway may be producing ribose-5-phosphate, but in the differentiating vessels it is producing erythrose-4-phosphate. In all cells, glycolysis and the rest of aerobic respiration are also occurring simultaneously with the pentose phosphate pathway ($\times$ 400). (b) In petals, erythrose-4-phosphate, used in the production of pigments, is produced by the pentose phosphate pathway.

Chapter 3). For example, an 18-carbon fatty acid would be converted into nine acetyl CoA units. As each acetyl CoA is formed, one FAD is reduced to $FADH_2$ and one NAD^+ is reduced to NADH, both of which can carry electrons to mitochondria and drive production of ATP by means of the electron transport chain (Fig. 11.20). Acetyl CoA may be used for synthesis of carbohydrates and other compounds or may enter the citric acid cycle and be further respired.

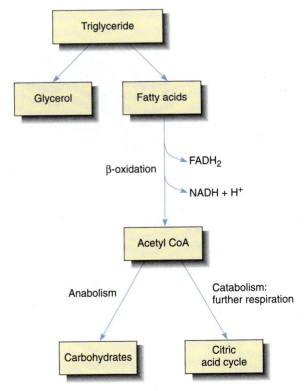

Triglyceride

Glycerol

Fatty acids

β-oxidation

$FADH_2$

$NADH + H^+$

Acetyl CoA

Anabolism

Catabolism: further respiration

Carbohydrates

Citric acid cycle

FIGURE 11.20

During the respiration of lipids, fatty acids are separated from glycerol and undergo beta-oxidation to acetyl CoA units, producing $FADH_2$ and NADH. In germinating seeds, much of the acetyl CoA is converted to glucose and fructose by gluconeogenesis and then is polymerized into sucrose. The sucrose is transported by phloem from cotyledons to embryo meristems, where it is used in construction or respired for energy.

PHOTORESPIRATION

Photorespiration occurs only when RuBP carboxylase adds oxygen rather than carbon dioxide to ribulose-1,5-bisphosphate, resulting in one molecule of 3-phosphoglycerate and one of phosphoglycolate (see Chapter 10). Phosphoglycolate is dephosphorylated to glycolate, which is then transported to microbodies called peroxisomes. Glycolate can be converted to glycine in the peroxisomes, and the glycine may be transferred to mitochondria, where it is respired to carbon dioxide and water with no conservation of energy in either ATP or NADH; all energy is wasted. In some cases, some of the glycine can be converted to serine; both are useful amino acids. Other mechanisms also produce these amino acids, however, and direct measurements show that for many C_3 species, photorespiration wastes as much as 30% of the energy trapped by photosynthesis.

ENVIRONMENTAL AND INTERNAL FACTORS

As with photosynthesis, numerous environmental factors influence the rate of respiration. It is necessary to consider the integration of respiration into the total biology of the plant as well as the metabolic pathways involved.

TEMPERATURE

Temperature greatly influences respiration in a plant growing under natural conditions. In all environments shoots located in air are at a different temperature than roots in soil. Shoots are subjected to great changes of temperature from day to night, often more than 20°C.

In most tissues, an increase in temperature of 10°C, in the range between 5 and 25°C, doubles the respiration rate, a magnitude of increase seen in many enzyme-mediated reactions. Below 5°C, respiration decreases greatly (Fig. 11.21). Above 30°C, respiration still increases, but not so rapidly; at such high temperatures, oxygen probably cannot diffuse into tissues as rapidly as the tissues use it. Above 40°C, respiration, like many other processes, slows greatly, probably owing to enzyme damage or disruption of organelle membranes.

FIGURE 11.21
Many fruits and vegetables are stored in refrigeration to reduce respiration and thus preserve starch and sugar content. (© *Tom Myers*)

LACK OF OXYGEN

Because plants are not as active as animals, much lower oxygen concentrations—as little as 1% to 2%—are sufficient to maintain full rates of plant respiration. Oxygen concentration in the atmosphere is so stable that it does not cause variations in respiration, but variations occur in a cell's access to oxygen. During daylight hours, chlorophyllous tissues and organs produce oxygen, some of which is used in respiration. At night oxygen is not produced, but it can diffuse into the large intercellular spaces of the plant if it can penetrate the closed stomata. Because the cuticle is not absolutely impermeable to oxygen and because the atmosphere contains such a high concentration, oxygen diffuses in rather well, even through closed stomatal pores. Less oxygen is available, but the temperature is lower and so the demand for oxygen is reduced.

Oxygen availability is much more variable for roots; in well-drained soil, the large quantity of gas located between soil particles is depleted in oxygen owing to respiration by roots, fungi, bacteria, protists, and soil animals. During and after rain, soil air is displaced by water, and roots either have lesser amounts of oxygen (hypoxia) or none (anoxia) available. Anaerobic respiration allows roots to survive for a short period, but it is not sufficient for root growth and healthy metabolism. If continually flooded, the roots of almost any species die. Those that survive do so primarily by oxygen diffusion from the stems through aerenchyma channels in the cortex or pith, such as in petioles of water lilies and cattails. Rice seedlings are the only certain example of plants that can grow normally without oxygen for a prolonged period.

In thick tubers (potatoes) or bulky roots (beets, carrots), it is not well established whether respiration is completely aerobic or at least partially anaerobic. Such organs usually have a significant amount of intercellular space through which oxygen might diffuse, but the concentration of oxygen is very low. Similarly, the sapwood and cambium of trees with thick bark may be hypoxic; even though lenticels are present, the diffusion path is long and the dense inner bark, cambium, and sapwood lack intercellular spaces. Developing embryos inside large seeds and fruits are reported to respire anaerobically.

INTERNAL REGULATION

Like virtually all other processes, respiration is subject to specific metabolic controls. Cells that have an active metabolism, such as glands that secrete protein or epidermal cells that secrete waxes, have a high level of aerobic respiration, whereas meristematic cells produce ribose by means of the pentose phosphate pathway. Nearby cells, perhaps those of the spongy mesophyll or collenchyma, may have a much lower respiration rate (Fig. 11.22). During fruit maturation, respiration usually remains steady or increases gradually until just before the fruit is mature, at which point a sudden burst of respiration is triggered by endogenous hormones. Conversely, within seeds, once an embryo is mature, its respiration

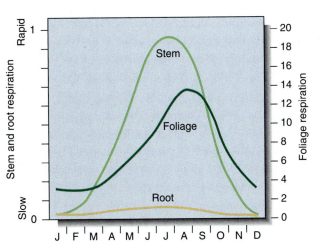

FIGURE 11.22

Respiration rates for various parts of loblolly pine trees in a 14-year-old plantation. Respiration is much higher in summer than in winter for all organs, and leaves have the highest rate—20 times the scale for the other organs. The rate is given as "grams of carbon dioxide released per square meter of soil surface per hour." This type of measurement is more useful to a forest manager than to a cell physiologist. Different scientists or studies need different types of measurements of the same function. A cell physiologist might express respiration as "grams of carbon dioxide released per cell per hour," "per gram of protein per hour," or "per gram of tissue per hour." Each gives a different result: Leaves have small cells with thin walls, so a gram of leaf tissue has much protein and many cells, whereas a gram of woody stem or branch is mostly inert, nonrespiring cellulose walls with few cells and little proteinaceous protoplasm. It can be difficult to select an appropriate basis for stating rates of respiration, and comparing studies that use different bases is not easy.

TABLE 11.3 *Production of ATP by Various Types of Respiration*

Anaerobic respiration	2 ATP ⟶ 2 ATP
Aerobic respiration	
Glycolysis	2 ATP ⟶ 2 ATP
	2 NADH ⟶ 4 or 6 ATP ⟶ 4 or 6 ATP
Pyruvate/acetyl CoA	1 NADH ⟶ 3 ATP (×2) ⟶ 6 ATP
Citric acid cycle	1 ATP (×2) ⟶ 2 ATP
	3 NADH ⟶ 9 ATP (×2) ⟶ 18 ATP
	1 $FADH_2$ ⟶ 2 ATP (×2) ⟶ 4 ATP
Pentose phosphate pathway	2 to 6 NADPH ⟶ 0 to 12 ATP
Heat-generating respiration	None
Lipid respiration	Variable
Photorespiration	None

decreases so dramatically that it is difficult to measure, and the seed becomes dormant. In seeds with a true dormant period, virtually no respiration occurs even if the embryo is surgically removed and given water, warmth, and oxygen, because metabolic inhibitors suppress respiration.

TOTAL ENERGY YIELD OF RESPIRATION

During anaerobic glycolysis, four molecules of ATP are synthesized, while either one or two ATPs must be used to initiate the process, depending on whether glucose or glucose-6-phosphate is the initial substrate. The NADH + H$^+$ generated cannot be used for energy, so the net result is two molecules of ATP for every molecule of glucose fermented (Table 11.3).

During aerobic respiration, glycolysis again yields two ATPs directly; in addition, the two NADHs can be transported to mitochondria, where their electrons power the formation of two or three more ATPs. The conversion of each pyruvate to acetyl CoA yields another NADH. Because two pyruvates are produced from each initial glucose, six more ATP per glucose are produced. Within the citric acid cycle, each original molecule of glucose yields two molecules of ATP, six of NADH, and two of $FADH_2$; the total is 24 ATPs. Aerobic respiration can produce as many as 38 molecules of ATP, making it significantly more efficient than anaerobic respiration.

The pentose phosphate pathway yields only two NADPH per glucose-6-phosphate if either ribulose-5-phosphate or erythrose-4-phosphate is drawn off for anabolic metabolism. If neither of these is removed, the various intermediates continue to cycle until all carbon of the glucose is completely oxidized to carbon dioxide and six NADPHs are produced. These may be used for cellular reductions of nitrate to amino acids, sulfates to sulfhydryls, or carbohydrates to fats. NADPH not consumed in anabolic reductions may contribute protons to the mitochondrial electron transport chain and indirectly produce two molecules of ATP.

The amount of ATP produced by fatty acid respiration depends on the length of the fatty acid and whether acetyl CoA enters the citric acid cycle; with maximum respiration, up to 40% of the total energy in a fatty acid is conserved in ATP. Thermogenic respiration and photorespiration produce no ATP. Also, keep in mind that intermediates may be diverted from all respiratory pathways and be used in synthetic reactions, so complete respiration may not occur; ATP production thus is less.

RESPIRATORY QUOTIENT

An action spectrum is a valuable tool for studying light-mediated phenomena such as photosynthesis. For respiration, a similar type of information is useful. A theoretical calculation can be made of the amount of oxygen consumed by each type of respiratory substrate. For example, the complete aerobic respiration of glucose should consume six molecules of oxygen and produce six molecules of carbon dioxide (Table 11.4). This ratio of carbon dioxide liberated to oxygen consumed is known as the **respiratory quotient (RQ)**; for glucose RQ = 1.0. Acids can enter the citric acid cycle and be oxidized to carbon dioxide, producing NADH for the mitochondrial electron transport chain. Because acids contain relatively large amounts of oxygen in their molecular structure, less is needed to convert them to carbon dioxide and water, so their RQ is high, greater than 1.0. On the other hand, fatty acids contain virtually no oxygen, so when they are respired, enough oxygen must be consumed to oxidize not only every carbon but every hydrogen as well, so the RQ is very low, often about 0.7. Of course, anaerobic respiration consumes no oxygen, although its carbon dioxide production is very high.

Once these RQ values are calculated, it is relatively easy to measure the amounts of gases exchanged during respiration and thus gain information about the respiratory metabolism. For example, peanut seeds are rich in lipids and oils. As they germinate, the seedling's RQ is initially low, indicating that the lipids are being respired. If they are kept in the dark, the RQ remains low until the seedling exhausts its nutrient reserves and dies. But if they are allowed to germinate in the light, the RQ gradually increases toward 1.0 as photosynthesis begins to produce glucose needed for respiration (Fig. 11.23). The increase is not immediate, because much of the first products of photosynthesis are used for constructive anabolic reactions in the leaf cells, so a rise in the RQ indicates that leaf photosynthesis is meeting anabolic needs and producing extra for respiration.

TABLE 11.4	*Respiratory Quotients of Various Compounds*

$$RQ = \frac{CO_2 \text{ liberated}}{O_2 \text{ consumed}}$$

Glucose

$$C_6H_{12}O_6 + 6O_2 \longrightarrow 6CO_2 + 6H_2O$$

$$\frac{6CO_2}{6O_2} = 1.0 = RQ$$

Lipid

$$C_{18}H_{34}O_2 + 25.5O_2 \longrightarrow 18CO_2 + 17H_2O$$

$$\frac{18CO_2}{25.5O_2} = 0.71 = RQ$$

Fermentation

$$C_6H_{12}O_6 \longrightarrow \text{alcohol} + CO_2$$

No O_2 consumption

Citric acid

$$C_6H_8O_7 + 4.5O_2 \longrightarrow 6CO_2 + 4H_2O$$

$$\frac{6CO_2}{4.5O_2} = 1.33 = RQ$$

FIGURE 11.23

Sunflower seeds store oils in their cotyledons; when they germinate, the oils are respired, giving the seed a low RQ value. The cotyledons of this seed, although still enclosed by the seed coat, have turned green and are carrying out photosynthesis, so both oils and sugars are available for respiration. After a few days, all the oil will be consumed, then only photosynthetically produced sugars will be available, and the seedling will have a high RQ value, near 1.0.

SUMMARY

1. Respiration provides a cell with both ATP and small carbon compounds that are important in various metabolic pathways.

2. The type of respiration carried out by a cell depends on environmental factors and the state of differentiation of the cell.

3. Anaerobic respiration—fermentation—is inefficient because only two ATP molecules are produced for each glucose respired; however, under anaerobic conditions, it is selectively more advantageous than its alternative, death.

4. Some plant and animal tissues are facultatively anaerobic, but no large plant or animal can live for prolonged periods without aerobic respiration. They are all obligate aerobes.

5. Three common electron acceptors are oxygen, acetaldehyde (in plants), and pyruvate (in animals); only oxygen is abundant and cheap and results in a harmless waste product, water.

6. During the complete aerobic respiration of glucose, electrons are removed and transported, ultimately, to oxygen. As oxygen accepts electrons, it picks up protons and water is formed.

7. Aerobic respiration consists of glycolysis, the citric acid cycle, and oxidative phosphorylation via the mitochondrial electron transport chain. Passage of electrons through the electron transport chain creates a chemiosmotic gradient of protons and hydroxyl ions.

8. Some tissues of some plants generate heat by carrying electrons on an alternative set of electron carriers that do not pump protons; no ATP is formed, so all energy of NADH is released as heat.

9. The pentose phosphate pathway is a complex interaction of metabolites that can produce erythrose (for lignin), ribose (for nucleotides), or NADPH (for reducing power).

10. Respiration increases as temperature increases, within the range of about 5 to 25°C. During warm days and nights, respiration is much more rapid than during cool periods.

IMPORTANT TERMS

acetyl CoA	cytochrome oxidase	flavin adenine dinucleotide	lactate (lactic acid)	respiration
aerobic respiration	Embden-Meyerhoff pathway	(FAD)	nicotinamide dinucleotide	respiratory quotient
anaerobic respiration	ethanol (ethyl alcohol)	flavin mononucleotide (FMN)	(NAD$^+$; NADH)	strict aerobe
citric acid cycle	facultative aerobe	glycolysis	obligate aerobe	thermogenic respiration
cyanide-resistant respiration	fermentation	Krebs cycle	pentose phosphate pathway	tricarboxylic acid cycle

REVIEW QUESTIONS

1. When a storage organ becomes active after dormancy and mobilizes its reserves, the starch is usually converted to sucrose for transport. Why isn't starch transported?

2. Under what conditions does plant tissue experience lack of oxygen? How is ATP generated from glucose without oxygen?

3. Anaerobic and aerobic forms of respiration differ in the ultimate electron acceptors for each process. What is the electron acceptor in each? What are the advantages and disadvantages of each?

4. What are the three basic parts of aerobic respiration? Which steps occur in mitochondria and which in cytosol?

5. Describe the generation of the chemiosmotic potential in mitochondria and its use in the phosphorylation of ADP to ATP.

6. Consider a meristematic cell preparing for mitosis and a young xylem cell differentiating into a fiber or a tracheary element; each uses the pentose phosphate pathway, but for different products. Explain why.

BotanyLinks .net Questions

www.jbpub.com/botanylinks

Visit the .net Questions area of BotanyLinks (http://www.jbpub.com/botanylinks/netquestions) to complete this question:

1. How does fermentation affect your life (other than through alcoholic beverages)? Maybe more than you think. . . . Go to the BotanyLinks home page for more information on this subject.

BotanyLinks includes a Directory of Organizations for this chapter.

TRANSPORT PROCESSES

12

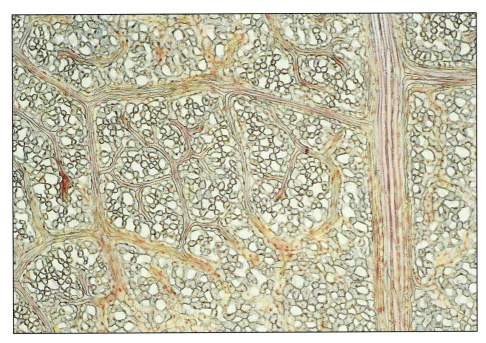

Water and minerals are obtained from soil, then transported throughout the plant body. The presence of sufficient water is often the most important factor determining whether plants live or die.

CONCEPTS

One fundamental aspect of life itself is the ability to transport specific substances to particular sites, moving molecules against the direction in which they would diffuse if left alone. After death occurs, atoms, ions, and molecules diffuse, moving from regions of higher to lower concentration, and the organization of protoplasm decays; the disorder of the components increases. Diffusion also occurs during life but proceeds more slowly than the controlled and oriented transport processes that tend to increase the order within the plant or animal body. Transport processes consume energy, and many are driven by the exergonic breaking of ATP's high-energy phosphate-bonding orbitals.

Specific transport occurs at virtually every level of biological organization: Enzymes transport electrons, protons, and acetyl groups; membranes transport material across themselves; cells transport material into and out of themselves as well as circulate it within the protoplasm; whole organisms transport water, carbohydrates, minerals, and other nutrients from one organ to another—between roots, leaves, flowers, and fruits.

319

Plants have only a few basic types of transport processes, and the fundamental principles are easy to understand. They are grouped here into **short distance transport,** which involves distances of a few cell diameters or less, and **long distance transport** between cells that are not close neighbors.

Many types of short distance transport involve transfer of basic nutrients from cells with access to the nutrients to cells that need them but are not in direct contact with them. Such transport requirements arose when early organisms evolved such that they had interior cells that were not in contact with the environment. Short distance transport became necessary to the survival of internal cells.

Long distance transport is not absolutely essential in the construction of a large plant. Many large algae have no long distance transport, nor do sponges, corals, or similar animals. However, the ability to conduct over long distances is definitely adaptive, especially for land plants. Before xylem and phloem evolved, a plant's absorbing cells could not penetrate deep into soil, because they would starve so far from photosynthetic cells. Nor could they have transported their absorbed nutrients very far upward. Being limited to the uppermost millimeter or two of soil meant that the absorptive cells could not reach the more permanently moist, deep soil where there are more minerals; the uppermost layers dry quickly and free minerals are leached away by rain. With xylem and phloem, roots that penetrate deeply can be kept alive and their gathered nutrients can be carried up to the shoot.

Vascular tissues make it selectively advantageous for shoots to grow upright, elevating leaves into the sunlight above competing plants. This elevation is feasible because photosynthetically produced sugars can be transported downward to other plant parts. Such elevation of photosynthetic tissues resulted in tall plants that could also place their reproductive tissues at a high elevation, enabling spores or pollen to be distributed more widely and effectively by the wind. After insect-mediated pollination evolved, it was adaptive to have flowers located high, in an easily visible position. The evolution of transport processes affected all aspects of plant biology and permitted later evolution of many new types of plant organization.

Vascular tissues also act as a mechanism by which nutrients are channeled to specific sites, resulting in rapid growth and development of those sites (Fig. 12.1). If meristematic zones relied exclusively on their own photosynthesis, growth and leaf primordium initiation would be extremely slow. At other times of the year, nutrients can be directed to flower buds or young fruits, promoting their growth while inhibiting the formation of new leaves.

FIGURE 12.1

(a) Imagine an angiosperm that has no vascular tissue. The leaves would produce large amounts of glucose, far in excess of the metabolic needs of the tissue itself. But because diffusion is slow over long distances, the sugar would diffuse out of the leaf only slowly and in small quantities. The shoot apex has almost no chlorophyll; if it had to depend solely on its own photosynthesis, growth, leaf initiation, and leaf expansion would be extremely slow, even though they would be occurring only a few centimeters away from large quantities of glucose in leaves. (b) With vasculature, glucose can be transported from regions of excess to regions of need; apical meristems and leaf primordia can thereby grow very rapidly. Vascular tissues also make the minerals gathered by an extensive root system available to regions that need them.

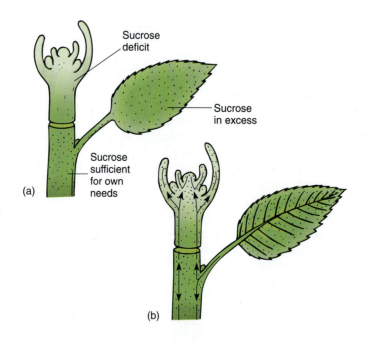

Cuticle

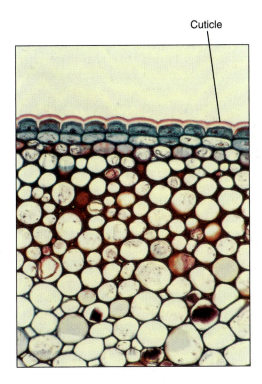

FIGURE 12.2

The cuticle (pink), composed of cutin, is a waterproof epidermal layer that acts as an isolation mechanism, retaining water within the plant and keeping pathogens out ($\times$ 80).

The combination of short and long distance transport has resulted in the ability of some plants to become large and complex enough to survive temporary adverse conditions, such as drought, heat, or attack by pathogens.

Because almost everything transported by a plant is dissolved in water, the ability of water to move from cell to cell or throughout a plant is an important factor. Water is an unusual liquid: It is rather heavy and viscous, and it tends to adhere to cell components as well as to soil, factors that affect transport processes.

Related to transport processes are **isolation mechanisms** that inhibit the movement of substances. Plants are adept at synthesizing organic polymers impermeable to a variety of substances. The epidermis with its cutin-lined walls is an excellent means of keeping water in the shoot after it has been transported there by the xylem (Fig. 12.2). The Casparian strips of the endodermis prevent the diffusion of minerals from one part of a root to another. Isolation mechanisms are essential if transport is to be useful.

DIFFUSION, OSMOSIS, AND ACTIVE TRANSPORT

The first thing to consider in transport is the mechanism by which material moves through a solution and crosses a membrane. The simplest method is **diffusion,** in which the random movement of particles in solution causes them to move from areas where they are in relatively high concentration to areas where they are in relatively low concentration. Diffusion through a membrane is technically known as **osmosis.**

Membranes are of three types: **Freely permeable** membranes allow all solutes to diffuse through them and have little biological significance. **Completely impermeable** membranes do not allow anything to pass through and occur as isolation barriers. **Differentially** or **selectively permeable** membranes allow only certain substances to pass through; all lipid/protein cell membranes are differentially permeable. Hydrophobic molecules diffuse easily through any cell membrane, whereas many polar, hydrophilic molecules can cross differentially permeable membranes only if the membranes have special protein channels through which the molecules can diffuse. Water molecules, even though highly polar, pass through all membranes.

Most membranes also have membrane-bound **molecular pumps** that use the energy of ATP to force molecules across the membrane, even if that type of molecule is extremely

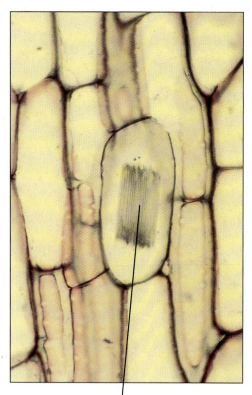

Crystals in vacuole

FIGURE 12.3

These crystals are located in a cell vacuole. Calcium and oxalic acid are transported across the vacuole membrane by molecular pumps, then combine into these needle-shaped crystals ($\times$ 80).

concentrated on the receiving side; this process is **active transport.** The molecular pump, which is a protein molecule, binds to both the molecule and ATP; when ATP splits into ADP and phosphate, the energy is transferred to the pump, forcing it to change shape, carry the molecule across the membrane, and release it. The membrane must otherwise be extremely impermeable to the molecule or it leaks back as rapidly as it is transported. Proton pumping in photosynthesis and respiration are examples of active transport (see Chapters 10 and 11).

All cell membranes are important in transport processes; the plasma membrane governs movement of material into and out of the cell. Substances can move across the vacuolar membrane by either osmosis or active transport; the vacuole acts as an accumulation space for sugars, pigments, crystals, and many other compounds (Fig. 12.3). The endoplasmic reticulum and dictyosome membranes transport material that then accumulates in vesicles. These vesicles may be relatively permanent, remaining in the cell for long periods of time, or they may be a means of **intracellular transport,** in which the vesicles migrate through the cytoplasm and fuse with another organelle. During fusion, the membranes merge and the vesicle contents are transferred into the organelle (see Fig. 3.9). This is a common means of moving material from the endoplasmic reticulum to dictyosomes or from either of these organelles to the cell exterior by fusion with the plasma membrane. During cell division, the new cell plate (the two primary walls and middle lamella) is formed by the coalescence of vesicles from both endoplasmic reticulum and dictyosomes.

WATER POTENTIAL

Like any other chemical, water has free energy, a capacity to do work. For most chemicals this energy is called its *chemical potential.* Because water is so important in botany, its chemical potential is usually referred to as **water potential** and has the symbol ψ (pronounced "sigh"). Water potential, the free energy of water, can be increased several ways: Water can be heated, put under pressure, or elevated. The energy of water can be decreased by cooling it, reducing pressure on it, or lowering it.

These examples of change in water's capacity to do work are easy to understand, but it can be changed in other ways as well. When water adheres to a substance, these water molecules form hydrogen bonds to the material and are not as free to diffuse as are other water molecules; their capacity to do work has decreased. Consider a small beaker of water; if it is pure water, it can flow, move, dissolve material, and hydrate substances. But if a sponge is added, water molecules adhere to the sponge material and can no longer flow or easily dissolve things. If a large amount of sugar is added instead of a sponge, the results are the same. Syrup is just a sugar solution, but the water molecules in syrup have less capacity to do work than do the molecules of pure water.

Water potential has three components:

$$\psi = \psi_\pi + \psi_p + \psi_m$$

In this equation, ψ_p is the **pressure potential,** the effect that pressure has on water potential. If water is under pressure, the pressure potential increases and so does water potential. If pressure decreases, so do the pressure potential and water potential. Pressure can be positive (when something is compressed) or negative (when something is stretched). Most liquids cannot be stretched very much, but because water is cohesive, it actually can resist considerable tension. When water is under tension, pressure potential is a negative number. Potential is measured in units of pressure, usually in **megapascals** (MPa) or **bars.** One megapascal is approximately equal to 10 bars or 10 atmospheres of pressure. Pure water at one atmosphere of pressure is defined by convention as having a water potential of zero.

ψ_π is the **osmotic potential,** the effect that solutes have on water potential. In pure water, no solutes are present and osmotic potential is given the value of 0.0 MPa. Adding solutes can only decrease water's free energy because water molecules interact with solute molecules and cannot diffuse easily. Therefore, osmotic potential is always negative. If water molecules do not interact with the added molecules, the substance does not dissolve.

It is important to be careful here: Adding acid to water only seems to make water more active. The solution may have more free energy than the water, but it does not have more free energy than both pure water and the original concentrated acid.

Osmotic potential is related to the *number of particles* present in solution; that is, a solution composed of 2 g of glucose in 100 mL of water has an osmotic potential twice as negative as a solution of only 1 g of glucose per 100 mL of water. This has some unexpected results: If a molecule of starch containing 1000 molecules of glucose is hydrolyzed to 1000 free glucose molecules, the osmotic potential of the solution becomes much more negative because there are now 999 more particles in solution than previously. Using terms such as "increase," "decrease," "larger," and "smaller" can be confusing when dealing with negative numbers. If the osmotic potential goes from -0.01 to -0.1 MPa, is it increasing or decreasing? It is least confusing to use the terms "more negative," "less positive," and so on.

ψ_m is the **matric potential,** water's adhesion to nondissolved structures such as cell walls, membranes, and soil particles. Adhesion can only decrease water's free energy, so matric potential is always negative. In soils, matric potential is important because so much of the soil water is tightly bound to soil particles. But in living cells, matric potential usually is much less important than osmotic potential or pressure potential and usually is ignored entirely (Table 12.1).

The movement of water is related to water potential; substances diffuse from regions where they are more concentrated to regions where they are more dilute. This can be stated more precisely for water: **Water moves from regions where water potential is relatively positive to regions where water potential is relatively more negative** (Fig. 12.4). This statement contains several important points:

1. **Water moves whenever there is a difference in water potential** within the mass of water. All protoplasts are interconnected and most cell walls are fully hydrated, so basically all of a plant body is one mass of water; water can move between regions in

TABLE 12.1	*Possible Values of Water Potential Components*
$\psi =$	$\psi_\pi + \psi_p + \psi_m$
ψ_π	is 0.0 for pure water, negative for all solutions
ψ_p	can be either positive or negative.
ψ_m	is always negative and is usually insignificant except for relatively dry materials.
ψ	for living cells is always 0.0 or negative.

Because matric potential is typically so small, the water potential equation for living cells is usually considered to be $\psi = \psi_\pi + \psi_p$.

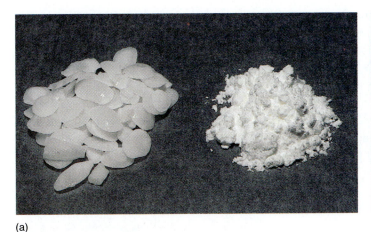
(a)

(b)

(c)

FIGURE 12.4
(a) The material on the left is potassium hydroxide; that on the right is starch. They were photographed immediately after being exposed to air, while they were dry. (b) Photographed after 1 hour in humid air. Water has moved from where it was more concentrated (the air) to less concentrated. The potassium hydroxide holds water by forming a solution with a very negative osmotic potential. Water is held to the starch by adhering to the long polysaccharide molecules. Water is not obvious in the starch, but think of saltine crackers left unwrapped. (c) By adding salt to eggplant, water can be drawn from the tissues, making them easier to cook.

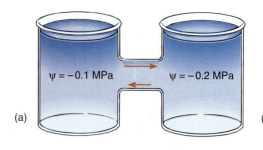
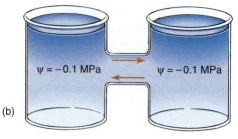
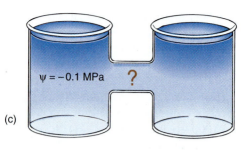

(a)   (b)   (c)

FIGURE 12.5

(a) If a solution whose water potential is -0.1 MPa is connected to a solution whose water potential is -0.2 MPa, the two are not in equilibrium. Water moves from a region of relatively positive water potential to a region of more negative potential, as indicated. Water molecules move in both directions, but more move to the right in any particular instant. Both beakers must have solute dissolved in them (otherwise each would have $\psi = 0.0$ MPa), and the -0.2 MPa solution must have twice as much solute. Because the concentration of solute molecules in the left beaker is lower, they are less able to restrict the movement of water molecules than those in the right beaker, where the greater concentration of solute molecules is more able to restrict water movement.

(b) If the water potential of two solutions is identical, they are in equilibrium and no net movement of water occurs; in each second, equal numbers of water molecules move to the right and left.

(c) A ψ of -1.0 MPa is very negative and has a strong tendency to absorb water, but we cannot be certain that water will move to it in this case; the right beaker may contain a solution with a ψ of -1.1 MPa.

the plant if the water potentials of the regions are not equal (Fig. 12.5a). As a consequence, the water potential of any particular cell may change many times a day as various parts of a plant lose or gain moisture (Fig. 12.6).

2. **If the water potentials of two regions are equal, the regions are in equilibrium and there is no net movement of water.** Water still diffuses back and forth, but, on average, equal numbers of water molecules diffuse into and out of a site (Fig. 12.5b).

3. **Water potentials must always be considered in pairs or groups.** Because water moves from one site to another, the water potentials of the two sites are important. Knowing one single water potential does not allow us to predict whether water will move (Fig. 12.5c).

Temperature is not a factor, because the solutions being compared are assumed to be at the same temperature. This is not strictly correct if the water potentials of leaves and roots are compared, but the difference is not significant.

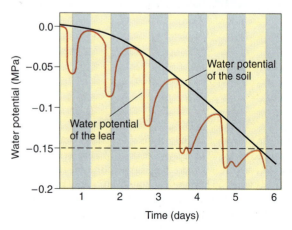

FIGURE 12.6

If soil is watered well and then allowed to dry over a period of days, the water potential of the soil solution gradually and smoothly becomes more negative. Every day, leaf transpiration loses water more rapidly than xylem transport replaces it, so leaves dry slightly and leaf water potential becomes more negative. At night, stomata close and xylem transport rehydrates the leaf tissue, but in each cycle, the leaves dry more than the previous night. This is not very serious until daytime leaf water potential becomes more negative than the wilting point (*dashed line*). When leaves wilt, many metabolic processes are adversely affected. Wilting point varies from species to species and is much more negative for xerophytes than for mesic plants. Once soil becomes extremely dry, even nighttime rehydration does not bring leaf water potential above the wilting point; the plant is at its permanent wilting point and severe stress damage may occur. All growth stops, and leaves and developing flower buds or fruits may die and be abscised.

CELLS AND WATER MOVEMENT

Some examples may help to explain the importance of water potential. Imagine a cell with a water potential of -0.1 MPa. It contains solutes that cause the osmotic potential to be some unknown negative number. The cell is turgid and presses against the cell wall, but the cell wall presses back equally, causing pressure on the cell, and the pressure potential is some unknown positive number. We can ignore matric potential because it is usually such a small number. The osmotic potential, whatever it is, plus the pressure potential, whatever it is, equals -0.1 MPa (Fig. 12.7a). Now imagine the cell being placed in a beaker of solution that also has a water potential of -0.1 MPa. The two water potentials are equal, the cell and solution are in equilibrium, and no net water movement occurs—the cell neither shrinks nor swells. Water molecules do move between the cell and the solution, but approximately equal numbers move in each direction every second (Fig. 12.7b).

Now imagine the same cell placed in a solution with a water potential of -0.3 MPa (Fig. 12.7c). The water potentials are not equal, that of the solution being more negative (has more solutes) than that of the cell, so water moves from the cell into the solution. How much water moves? As water leaves the cell, the solutes that remain in the protoplasm become more concentrated. Because more solutes are present per unit water, osmotic potential becomes more negative. As water moves out, the protoplasm volume becomes smaller, the protoplast presses against the wall with less force and the wall presses back less, so pressure potential becomes less positive. Because both osmotic potential and pressure potential are decreasing (becoming more negative), so is the water potential of the cell. At some point the cell's water potential (ψ_{cell}) reaches -0.3 MPa and is in equilibrium with the solution; then net water movement ceases (Fig. 12.7d). Of course, as water moves from

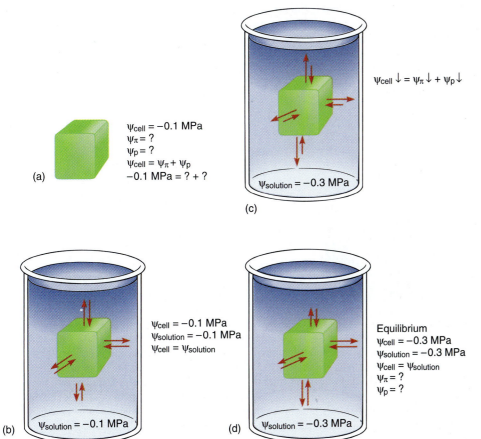

(a)

$\psi_{cell} = -0.1$ MPa
$\psi_\pi = ?$
$\psi_p = ?$
$\psi_{cell} = \psi_\pi + \psi_p$
-0.1 MPa $= ? + ?$

(c)

$\psi_{cell} \downarrow = \psi_\pi \downarrow + \psi_p \downarrow$

$\psi_{solution} = -0.3$ MPa

(b)

$\psi_{cell} = -0.1$ MPa
$\psi_{solution} = -0.1$ MPa
$\psi_{cell} = \psi_{solution}$

$\psi_{solution} = -0.1$ MPa

(d)

Equilibrium
$\psi_{cell} = -0.3$ MPa
$\psi_{solution} = -0.3$ MPa
$\psi_{cell} = \psi_{solution}$
$\psi_\pi = ?$
$\psi_p = ?$

$\psi_{solution} = -0.3$ MPa

FIGURE 12.7
(a) A cell whose ψ is -0.1 MPa but whose values of ψ_π or ψ_p are unknown. (b) If the cell is placed in a solution with a ψ of -0.1 MPa, the cell is in equilibrium with the solution and no net movement of water occurs. (c) If the cell is placed in a solution with a ψ of -0.3 MPa, the cell loses water to the solution until the cell's water potential is also -0.3 MPa (d), even if the cell is killed by the loss of water. The fact that the cell's water potential becomes more negative means that the osmotic potential or the pressure potential or both also become more negative.

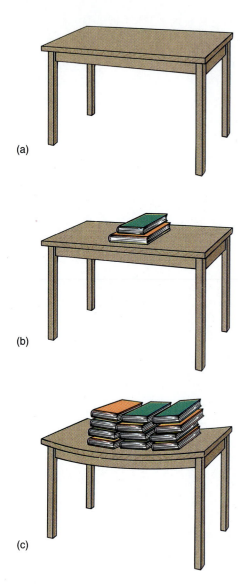

the cell into the experimental solution in the beaker, the solution becomes more dilute and its osmotic potential and water potential become less negative; its pressure potential does not change because pressure cannot build up in an open beaker. Therefore, equilibrium actually occurs slightly above − 0.3 MPa; but because most beakers are much larger than most cells, the amount of water that moves is much more significant to the cell than to the beaker solution.

Consider the relative importance of osmotic potential and pressure potential in this example. In order for osmotic potential to become twice as negative (for example, from − 0.15 MPa to − 0.3 MPa, the cell has to lose half its water or double the number of solute particles. Either action is drastic; almost any cell dies if it loses half its water, so osmotic potential does not usually increase or decrease more than a few megapascals. Pressure potential can change enormously, however, usually with movements of only small amounts of water. Consider the top of a table: Its molecules are pressing upward with exactly the same force that gravity is pulling them downward. Placing a book on the table causes the table's molecules to exert more pressure upward as their bonding orbitals are stretched (Fig. 12.8). The table changes the amount of upward pressure it can exert with very little change in shape.

A similar process occurs in cell walls; imagine placing the cell, now with a water potential of − 0.3 MPa, in a beaker of pure water. Some water moves inward, diluting the solutes and causing the osmotic potential to become less negative, but the change is not significant. However, the extra volume of the water causes the protoplast to swell and press against the wall with more force. The wall presses back with equal force and pressure potential rises rapidly, even though only a small amount of water moves in (Fig. 12.9). How high will it rise? We cannot predict its value, but it will go high enough that osmotic potential plus pressure potential will equal zero, and the cell will be in equilibrium with pure water.

Can a cell ever absorb so much water that it bursts? Animal cells often burst if placed in pure water, a process called **lysis,** but plant cells can never burst (Fig. 12.10). Walls, either primary or secondary, are always strong enough to resist breakage by water absorption. Even the thinnest, most delicate walls of mature parenchyma cells can exert enough pressure on the protoplast to raise the pressure potential high enough to counterbalance osmotic potential, however negative it might be.

FIGURE 12.8
(a) The molecules in this table top are exerting just enough pressure upward to counteract the gravitational force. We know this because there is no movement upward or downward of the table top, so forces must be in equilibrium. (b) Adding several books increases the gravitational force, but no net movement occurs. The table top is slightly bent, which stretches its molecules, but they resist just enough to counter the new force. (c) Even more force is perfectly balanced by the table top. If the books were removed, the stretching on the table would stop and the molecules would go back to exerting only enough pressure to counteract its own weight—the table would not fly upward.

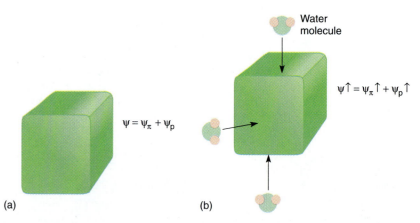

FIGURE 12.9
(a) A healthy cell, turgid and full of water and protoplasm; it is swollen and firm, just like the cells of unwilted leaves. Its walls are stretched and are pressing back against the protoplasm that is pressing on them. (b) If even a small amount of water enters the cell, the slight increase in volume causes the walls to stretch even more, but they push back, and pressure builds inside the cell. Consequently, the ψ_p becomes much more positive and so does the water potential of the cell. Because of the slight increase in the amount of water in the cell, the salts, sugars, amino acids, and all other solutes are now slightly more dilute, so ψ_π becomes slightly less negative also, but this change is insignificant compared with the change in ψ_p.

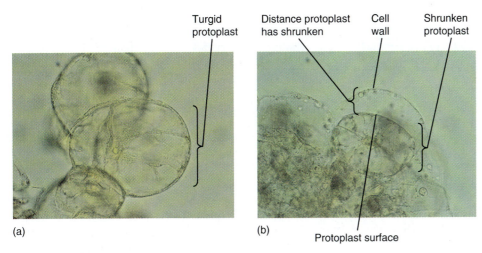

FIGURE 12.10
Some aspects of biotechnology processes require that the cell wall be digested away with cellulase enzymes, leaving behind a naked protoplast. The digestion mixture must contain just enough solute—usually sucrose or the sugar-alcohol mannitol—that the cells lose water and shrink away from the wall. If not, the protoplasts burst when the wall is removed. The solute concentration must be adjusted carefully; if it is too strong, too much water is pulled out of the cells and they die. (a) Cells in suspension culture before treatment. (b) Cells in 6% mannitol with the protoplasts just pulling back from the wall (both × 500).

Immature, growing cells have weak, deformable walls and cannot generate enough pressure to stop water absorption. The cell grows rather than bursts. Under these conditions, the cell may increase greatly in size. With such a large influx of water, solutes in the cell may become significantly diluted, osmotic potential and water potential may become less negative, and the cell reaches hydraulic equilibrium with surrounding cells and growth stops. However, in growing regions, such as the tips of stems and roots, and in expanding leaves, cells can keep their osmotic potential and water potential very negative despite the influx of water, either by actively pumping in solutes through the plasma membrane or by hydrolyzing giant starch molecules into thousands of glucose molecules. Once the proper size is reached, growth can be stopped either by strengthening the wall so that it exerts more pressure and raises the pressure potential or by stopping the import of solutes or the hydrolysis of starch, allowing the osmotic potential to rise. Either way, rising pressure potential or osmotic potential causes the cell's water potential to rise and reach equilibrium with the surrounding cells, stopping the net inflow of water.

Although plant cells cannot absorb so much water that they burst, water loss can be a serious problem. Imagine that our demonstration cell, now in pure water and with a water potential of 0.0 MPa, is placed in a strong sugar solution with a water potential of − 2.0 MPa. Water moves out of the cell, osmotic potential becomes slightly more negative, and the pressure potential drops rapidly. In such a strong solution, long before the cell reaches equilibrium it loses so much water that the protoplast shrinks in volume and no longer presses against the wall. The wall is not stretched and does not exert any pressure back, so the pressure potential drops to 0.0 MPa. The point at which the protoplast has lost just enough water to pull slightly away from the wall is called **incipient plasmolysis** and is quite important (Figs. 12.11 and 12.12). Up to that point the cell has lost very little water, so its volume change and osmotic potential change have not been great. But because the pressure potential is now zero, the water potential equation is

$$\psi = \psi_\pi + 0$$

If the cell has not reached equilibrium at the point of incipient plasmolysis, it continues to lose water and the protoplast pulls completely away from the wall and shrinks. The cell has become **plasmolyzed.** Water potential continues to become more negative entirely

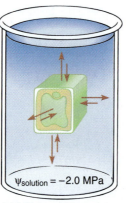
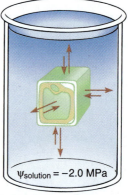

(a) Initial conditions

$\psi_{cell} = 0.0$ MPa
$\psi_\pi = ?$
$\psi_p = ?$
$0.0 = \psi_\pi + \psi_p$
$\psi\downarrow = \psi_\pi\downarrow = \psi_p\downarrow$

(b) Incipient plasmolysis

$\psi_{cell} = ?$
$\psi_p = 0.0$ MPa
$\psi_\pi = ?$
$\psi_{cell} = \psi_\pi + 0.0$
$\psi\downarrow = \psi_\pi\downarrow + 0.0$

(c) Equilibrium

$\psi_{cell} = \psi_{solution} = -2.0$ MPa
$\psi_p = 0.0$ MPa
$\psi_{cell} = \psi_\pi + 0.0$
-2.0 MPa $= \psi_\pi$

FIGURE 12.11

(a) When placed in a solution with a strongly negative water potential, the cell loses water rapidly and cell volume drops. ψ_p becomes less positive and ψ_{cell} becomes more negative; ψ_π changes only a small amount. (b) Incipient plasmolysis is the point at which the protoplast has shrunk just enough to pull away from the wall, so ψ_p is zero and ψ_{cell} equals ψ_π. (c) If the cell does not reach equilibrium at incipient plasmolysis, it continues to lose water, and ψ_{cell} continues to become more negative until it reaches -2.0 MPa. The pressure potential here cannot become a negative number, so the changing water potential is due to a changing osmotic potential. During plasmolysis, the cell loses enough water to change the concentration of solutes significantly.

Shrunken protoplasts

Fragments of dead protoplast

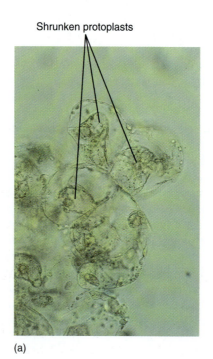
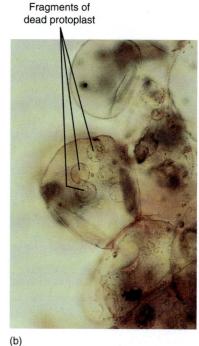

(a)

(b)

FIGURE 12.12

(a) These cultured cells have been placed in 12% mannitol for several hours and are severely plasmolyzed (compare with Fig. 12.10). (b) After a few days of severe plasmolysis, the cells have died (both $\times$ 500).

TABLE 12.2	*Water Potentials of Various Tissues Under Certain Conditions (MPa)*

Megapascals	
0.0	Leaves at full turgor
− 0.05	Fertilizer solution
− 0.2	Most roots in dry soil
− 0.5	Leaves of plants in well-watered soil; leaf growth good
− 1.0	Leaves of plants rooted in dry soils; leaf growth slow
− 2.5	Seawater
− 2.7	1 molal sucrose or mannitol
− 3.0	Leaves of plants rooted in very dry soil; for many plants, growth is slow or has stopped. Some plants have died.
− 6.0	Leaves and twigs of desert shrubs in very dry soil
− 20	Dry, viable seeds capable of germination

owing to the osmotic potential as the solutes become more concentrated. Most plants at the equilibrium point of − 2.0 MPa would die of severe water loss.

Although such severe desiccation kills most cells, some can survive it easily. The embryos in most seeds are much drier, having water potentials as low as − 20 MPa. Less dramatically, the leaves of desert shrubs in dry soil have water potentials as low as − 2.0 to − 6.0 MPa (Table 12.2). For most plants of temperate climates, a leaf water potential below − 1.0 MPa stops leaf growth, although leaves can survive such desiccation for many days or weeks.

A pressure of − 20.0 MPa equals − 3000 lb/in²; bicycle tires have a pressure of only 90 lb/in².

SHORT DISTANCE INTERCELLULAR TRANSPORT

Most plant cells communicate with their neighboring cells, transferring water, sugars, minerals, and hormones at least. This movement occurs by a variety of mechanisms. First, all living cells are interconnected by plasmodesmata, the fine cytoplasmic channels that pass through primary cell walls (see Chapter 3 and Table 3.9). All the protoplasm of one plant can be considered one continuous mass, referred to as the **symplast.**

Material also is transferred from one cell to another by transport across the plasma membrane. The methods involve osmosis, molecular pumps in the plasma membrane itself, or fusion between transport vesicles and the plasma membrane. Once across a plasma membrane, a molecule initially resides in the cell wall. The wall is probably thin and permeable, and the molecule can penetrate it easily, diffusing across it to an intercellular space or laterally through it, spreading along the cell surface (Fig. 12.13). Most small molecules can move easily through both the wall and the intercellular spaces; the two together are called the **apoplast** of the plant.

In glands, the apoplast is mostly intercellular space through which molecules move easily, usually toward the surface of the gland. In nonglandular regions, the apoplast is mostly cell wall. The secreted molecule is probably absorbed by a cell neighboring the one that secreted it. In most parenchymatous tissues, primary walls are thin (less than 1 μm thick), and the contact faces between two cells are so extensive (10 to 20 μm²) that the probability is much greater that a molecule will either diffuse more or less directly into the next cell or re-enter the cell from which it came (Fig. 12.14). If the molecule was originally secreted by active transport, the original cell membrane is probably impermeable to it, at least in that area, so return to the original cell is usually not possible. This is probably the most common mechanism for the movement of water, sugar, and other nutrients between parenchyma cells within the cortex, pith, or leaf mesophyll.

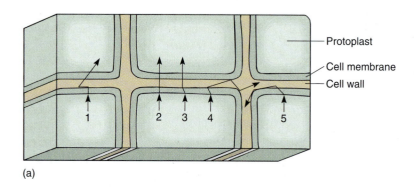

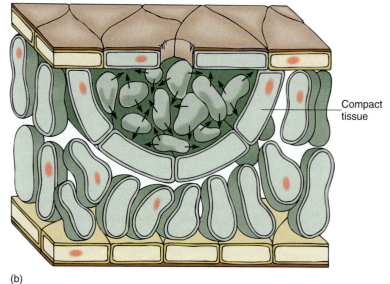

(a)

(b)

FIGURE 12.13

(a) Once released from a cell, a molecule diffuses in a series of random short paths as it collides with and bounces off other molecules. Probability favors its entering a neighboring cell (molecules 1, 2, and 3), but it can diffuse laterally along the wall as well (molecules 4 and 5). (b) In many glands, the apoplast is large, so movement between cells may be faster and easier than movement within cells. Such glands often have a lining of compact tissue that isolates the gland, preventing the secreted material from permeating the whole region.

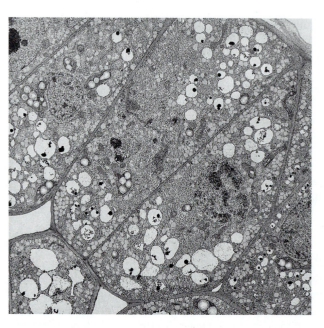

FIGURE 12.14

Primary walls are usually so thin that few molecules can diffuse sideways; most move rather quickly into the next cell. *Washingtonia filifera* seed epithelium (×4,700). *(Courtesy of Darleen A. DeMason, University of California, Riverside)*

GUARD CELLS

The opening and closing of stomatal pores are based on short distance intercellular transport. At night, when stomata are closed (except for CAM plants), guard cells are somewhat shrunken and have little internal pressure. Water can enter or leave them easily, because water crosses any biological membrane rather freely. But the guard cells are in hydraulic equilibrium with surrounding cells: Water enters and leaves guard cells at approximately the same rate, and no net change occurs in the amount of water. When guard cells must open, such as just after sunrise, potassium ions (K^+) are actively transported from surrounding cells into guard cells (Fig. 12.15). Once inside the guard cells, the potassium cannot leave because the plasma membrane is impermeable to it: Potassium pumping is possible but diffusion is not. The loss of potassium causes the water potential in adjacent cells to become less negative, whereas absorption of potassium causes the water potential in guard cells to become more negative. Adjacent cells and guard cells are thrown out of hydraulic equilibrium by the pumping of potassium ions, and water diffuses out of surrounding cells across their plasma membrane, across the two primary walls and middle lamella, and into guard cells across their plasma membrane. The extra water and potassium cause guard cells to swell, bend, and push apart, opening the stomatal pore. Once they are open, potassium pumping stops, water movement brings guard cells and adjacent cells into water potential equilibrium again, and net water movement stops. When guard cells must close, the process is reversed: Potassium is pumped from guard cells into surrounding cells and water follows. Notice that guard cells and adjacent cells are in equilibrium when stomata are fully open and fully closed.

Notice also that guard cells of fully opened and fully closed stomata are both in equilibrium with surrounding cells (and thus also with each other), even though they all have different internal conditions: When open, the guard cells' abundant potassium gives them a very negative osmotic potential which is countered by turgor pressure, giving a large pressure potential. This results in only a small negative water potential. When closed, the guard cells have less potassium, so only a small negative osmotic potential; this is coun-

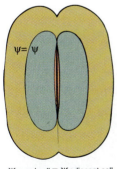

Ψ guard cell = Ψ adjacent cell

(a) Night

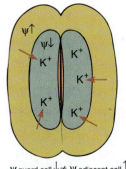

Ψ guard cell ↓ ≠ Ψ adjacent cell ↑

(b) Sunrise

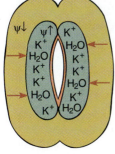

Ψ guard cell ≠ Ψ adjacent cell

(c) Morning

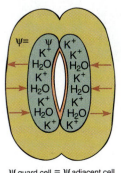

Ψ guard cell = Ψ adjacent cell

(d) Morning

FIGURE 12.15

At night, guard cells and adjacent cells are in hydraulic equilibrium (a), but at sunrise, potassium is pumped into the guard cells, increasing solute concentration (b). Osmotic potential and water potential become more negative and water flows in (c), causing the guard cells to swell and open the pore. As pressure builds, pressure potential rises, counteracting the falling osmotic potential, so the guard cell's water potential rises and moves back into equilibrium with the adjacent cell, but by the time that happens, the stomatal pore is open (d). At night, when the stomata must close, all steps are reversed.

tered by less turgor, and therefore a less positive pressure potential. These cells too have only a small negative water potential. Even though two cells might have very different internal conditions, they can have the same water potential and be in hydraulic equilibrium.

MOTOR CELLS

The leaves of sensitive plant *(Mimosa pudica),* prayer plant *(Oxalis),* and many other species are able to move slowly and reorient themselves by flexing and folding in response to a variety of stimuli (Fig. 12.16). The location of flexure is either the entire midrib or the point at which the petiole attaches to the lamina or stem. The cells at these "joints," called **motor cells,** are similar to guard cells: They can either accumulate or expel potassium and thus adjust their water potential and turgidity.

In Venus' flytrap, the leaf can close rapidly, in less than a second, but it requires several hours to reopen. Motor cells are located along the midrib, and when they are shrunken, pressure in other midrib cells causes the two halves of the blade to be appressed, and the trap is closed. Trap opening occurs as potassium is slowly accumulated by motor cells, water diffuses in, and the motor cells become turgid. Closure is not caused by pumping potassium out of the motor cells; that would be too slow to be effective. Instead, the membrane suddenly becomes freely permeable to potassium and it rushes out instantly. The water balance is rapidly changed and water too floods out, allowing the motor cells to virtually collapse, and the trap shuts quickly enough to catch insects.

TRANSFER CELLS

The rate at which material can be actively transported depends on the number of molecular pumps present, which in turn depends on the surface area of the plasma membrane: The larger the membrane, the more molecular pumps it can hold. In certain specialized **transfer cells,** the walls are smooth on the outer surface but have numerous finger-like and ridgelike outgrowths on the inner surface. The plasma membrane is pressed firmly against

(a)   (b)

FIGURE 12.16

(a) Each leaflet of the compound leaf of *Oxalis* is joined to the petiole by motor cells; here they were photographed in the morning when it was cool and the sunlight was not intense. Motor cells are turgid and leaflets are held into the sunlight. (b) *Oxalis* plants in full, intense sunlight. The motor cells have lost potassium and water; thus they are not turgid. As a result, the leaflets hang down, minimizing their exposure to light. Later in the afternoon, when sunlight is not so intense, or if the shadow of a tree moves across the plant, the motor cells will absorb potassium, then water, and raise the leaflets again.

all the convolutions and thus has a much larger surface area than it would if the wall were flat. Consequently, room is available for many molecular pumps, and high-volume transport can occur across these transfer walls. Transfer cells are found in areas where rapid short distance transport is expected to occur: in glands that secrete salt, in areas that pass nutrients to embryos, and in regions where sugar is loaded into or out of phloem.

LONG DISTANCE TRANSPORT: PHLOEM

Although the exact mechanism by which water and nutrients are moved through phloem is not known, most evidence supports the **pressure flow hypothesis.** Membrane-bound molecular pumps and active transport are postulated to be the important driving forces.

The sites from which water and nutrients are transported are **sources.** During spring and summer, leaves are dominant sources as their photosynthetically produced sugars are exported to the rest of the plant. At other times, such as early spring before new leaves have been produced by deciduous trees, the sources are storage sites such as tubers, corms, wood and bark parenchyma, and fleshy taproots. Cotyledons and endosperm are sources for embryos during germination. Within sources, sugars are actively transported into sieve elements—sieve tube members in angiosperms, sieve cells in nonflowering plants. The location of the molecular pumps is not known for certain; it may be the sieve element plasma membrane. But in many species, cells surrounding sieve elements, especially companion cells, appear to be transfer cells of great importance in loading phloem.

As sugars such as sucrose are pumped into sieve elements, the sieve element protoplasm tends to become more concentrated (Fig. 12.17a). Consequently, both its osmotic potential and its water potential become more negative. This causes hydraulic disequilibrium between the sieve elements and surrounding cells, and water diffuses into the sieve

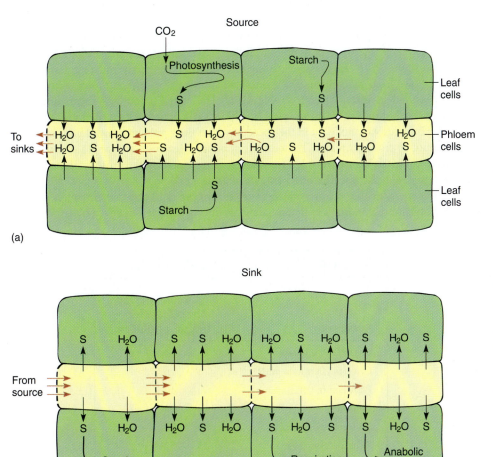

FIGURE 12.17

(a) As sucrose (s) is actively transported into sieve elements, ψ_π and ψ_{cell} become more negative, moving away from equilibrium with companion cells and other neighboring cells. Water moves into the sieve elements, squeezing phloem sap out through the sieve pores. Because the sugary water escapes through sieve pores, high pressure does not build up, so the pressure potential does not rise and stop the influx of water. The water potentials of sieve elements and surrounding cells never reach equilibrium as long as sucrose is being pumped. The source of the glucose may be photosynthesis or the breakdown of starch. (b) In sinks, sucrose is actively transported out of sieve elements, and all processes work in reverse compared with sources. In some sinks, most of the sucrose is converted to glucose and then polymerized to starch (storage organs), is respired (most tissues), or is used to construct new compounds (growing tissues).

TABLE 12.3	*Speed of Phloem Sap Translocation*
	Maximum Speed (cm/hr)
Picea (spruce) stem	13.2
Pinus (pine) stem	48
Fraxinus (ash) stem	48
Ipomoea (morning glory) stem	72
Ulmus (elm) stem	120
Triticum (wheat) leaf	168
Heracleum (cow parsnip) stem	210
Helianthus (sunflower) stem	240
Zea (corn) leaf	660

elements. In any other cell, the increased volume of sugars and water would cause the protoplast to expand and press against the cell wall, but sieve elements are unique, being living cells with relatively large holes in the walls, up to 14 μm wide in cucumbers and pumpkins (see Fig. 5.34). When pressure starts to build in these cells, "protoplasm" is squeezed through the sieve pores into the next cell. Sieve element protoplasm is not like that of most cells: The vacuolar membrane disintegrates, allowing vacuolar water to mix with part of the cytoplasm, creating an extremely watery, nonviscous substance—the phloem sap. The majority of the protoplasm is held firmly to the walls, probably by microtubules or microfilaments, and is not carried away with the watery central phloem sap.

In sources, phloem loading occurs along numerous vascular bundles such as the fine veins in leaves, the network of bundles in tubers and corms, and the inner bark in storage roots and stems. With this massive loading, pressure builds quickly and a large volume of material flows from the source. The rate of transport can be high; up to 660 cm/hr has been measured in leaves of corn (Table 12.3). The actual amount of sugars and other nutrients (excluding water) transported by phloem per hour is called the **mass transfer.** Vascular bundles vary not only in the speed at which their phloem translocates but also in the amount of phloem present. The number of bundles leaving a source is also important. To make comparisons easier, mass transfer can be divided by the cross-sectional area of phloem to obtain the **specific mass transfer** (Table 12.4).

Sinks are sites that receive transported phloem sap, and they are extremely diverse. Storage organs are important in perennial plants during summer, but even more important

TABLE 12.4	*Specific Mass Transfer*
Leaves of	**Maximum Specific Mass Transfer (g/hr/cm² of phloem)**
Digitaria (crab grass)	4.7
Cynodon (Bermuda grass)	6.6
Paspalum (a grass)	12.7
Cocos (coconut)	32
Triticum (wheat [roots])	180
Ricinus (castor bean)	252

are meristems, root tips, leaf primordia, growing flowers, fruits, and seeds. On even a small tree there may be thousands of sinks, each receiving nutrients. Not all sinks are active simultaneously; most plants do not produce flowers and leaves at the same time, and fruits can develop only after flowers. Within sinks, sugars are actively transported out of sieve elements into surrounding cells. The loss of sugar causes phloem sap to be more dilute, and its osmotic potential and water potential tend to become less negative, so water diffuses outward into the surrounding cells (Fig. 12.17b). As a result, even though phloem sap flows rapidly into a sink, the end cells of the phloem do not swell. As rapidly as nutrients are loaded at sources, they are unloaded at sinks.

Because phloem sap is under pressure, the danger exists of uncontrolled "bleeding" if the phloem is cut. Vascular bundles are broken open frequently, especially by chewing insects and larger animals. Two mechanisms seal broken sieve elements. The first is **P-protein** (P for phloem), found as a fine network adjacent to the plasma membrane inner surface of uninjured sieve elements. When phloem is ruptured, the phloem sap initially surges toward the break; this rapid movement sweeps the P-protein into the cell center, where it becomes a tangled mass. When it is carried to a sieve area or sieve plate, the P-protein mass is too large to pass through and forms a **P-protein plug** (Fig. 12.18).

Within uninjured phloem there is another polymer as well, **callose**. Apparently, it can stay in solution only if it is under pressure; without pressure it precipitates and forms a flocculent mass. When injury causes a pressure drop, the callose precipitates and is carried along with the P-protein to the nearest sieve areas; there the callose contributes to the plug and leaking is prevented.

In monocots with long-lived stems, such as palms and Joshua trees, sieve tube members live and function for many years, even hundreds of years. But in all other plants, individual sieve elements have a lifetime of only months or even weeks. They stop transporting and are replaced by new phloem cells from the provascular tissues or vascular cambium. Once they cease to function, callose deposits seal them permanently.

A further aspect of phloem transport is important to consider. As sugar is actively transported into phloem in sources, what happens to the water potential of the cells losing the sugar? Shouldn't it become less negative? No, it remains unchanged, because the sugars are being exported at the same rate they are being synthesized in leaves. Chlorenchyma cells absorb carbon dioxide molecules, but this does not cause the water potential to become more negative because the carbon dioxide is synthesized into sugar and exported. Millions of molecules of carbon dioxide may pass through a chlorenchyma cell without any long-term impact on its osmotic or water potentials. In sources such as tubers, the export of sugar has no impact on the storage cell water potential because as rapidly as it is exported, new sugar appears by the depolymerization of starch. The same is true at sinks: Storage cells do not accumulate sugar as sucrose but polymerize it into starch. Thousands of molecules may be absorbed, but because they are polymerized into one molecule, no change in water potential occurs. In growing cells of sinks such as meristems and buds, the imported sugar is polymerized to cellulose, hemicellulose, and other carbohydrates. It can also be metabolized into amino acids, fatty acids, and nucleotides; these are then polymerized into proteins, fats, and nucleic acids, so again little or no change in osmotic potential and water potential occurs. In all cells, part of the imported sugar is respired, but this converts it to carbon dioxide that is expelled, so again there is no osmotic effect.

Plants control the direction and rate of flow of phloem sap. While dormant in late winter and early spring, buds receive very little phloem sap, but once they become active, phloem transport increases greatly. Not all buds are equally affected; some grow rapidly whereas others, even though located quite close by, receive virtually nothing. While flowers are open, phloem transport is low, but after fertilization, when the ovary begins to develop into a fruit, transport increases.

The direction of transport can also change. Leaf primordia and young expanding leaves are sinks; imported sugar allows them to develop much more rapidly than they would if they had to be completely autotrophic (see Fig. 12.1). In addition, they also need large amounts of nitrogen, sulfur, potassium, and other minerals. Once leaves reach a critical size, they become self-supporting, able to photosynthesize rapidly enough to meet

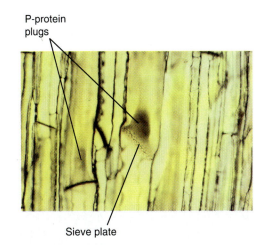

FIGURE 12.18

When this material of squash (*Cucurbita*) was being prepared for microscopy, it was cut open, causing the phloem sap to surge toward the cut, sweeping P-protein along and forming P-protein plugs, visible as dark brown masses. Sieve pores in the sieve plates are also visible (× 500).

Young, expanded leaves

Shoot tip with meristem and leaf primordia

Cotyledons

FIGURE 12.19

This bean contains two prominent sources, the cotyledons, which are supplying sugars, amino acids, and minerals to the rest of the seedling. The shoot tip with its meristem and leaf primordia are sinks, as are the roots. The first two leaves are expanded and are probably sources now, but they were sinks while they were developing.

all their own needs (Fig. 12.19). Shortly afterward they become sources, exporting material. Phloem transport is reversed in the leaf and petiole; molecular pumps must now load the phloem, not unload it. Plasma membranes may be altered, or one set of sieve elements may cease to function and may be replaced by a whole new set of cells. The early primary phloem often lives for less than a few weeks.

Leaves near a stem tip export upward to the shoot apical meristem, while leaves farther back export toward the trunk and roots. As the shoot apex grows, leaves that had been near the apex are left behind and their transport shifts from upward to downward. If the apex produces flowers and then fruit, the direction of transport may shift again so that all leaves send sugars upward.

LONG DISTANCE TRANSPORT: XYLEM

PROPERTIES OF WATER

The movement of water through xylem is based on a few simple properties of water and solutions. One property is that water molecules interact strongly with other water molecules, behaving as if weakly bound together; when frozen, the molecules become strongly bound to each other. Because of this, liquid water is said to be **cohesive,** and any force acting on one molecule acts on all neighboring ones as well.

Another property of water is that its molecules interact with many other substances— it is **adhesive.** Almost all substances in plants, except lipids, interact with water; water molecules adhere to these substances and cannot be easily removed from them. Cellulose molecules, enzymes, DNA, sugars, and so forth, have a shell of water molecules rather firmly attached to them. Occasionally a water molecule vibrates out of one of these shells and is replaced by another water molecule, but in general, adhesion makes these water molecules less free to move around than other water molecules.

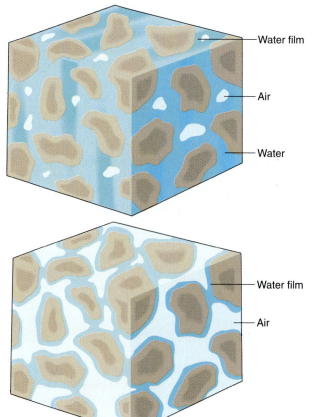

Water film

Air

Water

Water film

Air

FIGURE 12.20

Wet soil contains water both as a film covering all surfaces of soil particles and as small masses held in the capillary spaces formed where two soil particles touch each other (a). The latter is held weakly and can be easily absorbed by roots. (b) Dry soil contains only the tightly bound films of water. This adhesion is measured by ψ_m, which can be so negative that the soil's water potential is also very negative. This water cannot be absorbed by the roots of most plants.

Water also adheres firmly to soil particles. When soil is quite moist, roots can absorb the free liquid water between soil particles. But as the soil dries, the remaining water adheres firmly to the soil and cannot be absorbed easily, if at all (Fig. 12.20). Even though the soil may contain considerable water, it is unavailable to the plant. The same is true of seawater; the water molecules interact so strongly with the salt molecules that land plants cannot pull the water away.

Another property of water is that it is heavy, and lifting it to the top of a tree requires a great deal of energy. If water were lighter, less energy would be involved.

WATER TRANSPORT THROUGH XYLEM

Water movement through xylem and plants as a whole is governed by the principles of water relations just described. The **cohesion-tension hypothesis** is the most widely accepted model of the process. When stomatal pores are open, they unavoidably allow water loss. The apoplastic space of spongy mesophyll and palisade parenchyma is filled with moisture-saturated air, so water molecules have a strong tendency to diffuse from intercellular spaces (the point of less negative water potential) to the atmosphere (the point of more negative water potential). This water loss is called **transstomatal transpiration.** The cuticle and waxes on the epidermal surfaces are fairly efficient isolation mechanisms, being so hydrophobic that very little water passes through them. However, some water is lost directly through the cuticle by **transcuticular transpiration** (Table 12.5).

Consider a leaf in early morning: Stomata are closed, air is cool, and relative humidity is high. The air may have cooled enough during the night to allow dew to form. Cells within the leaf are turgid and probably in equilibrium with each other, all having a water potential between 0.0 and − 1.0 MPa (see Fig. 12.6). As the sun rises, stomata open and begin losing water; the air warms and its relative humidity decreases. As transpiration causes epidermal cells and mesophyll cells near stomata to lose water, their water potentials become more negative, going out of equilibrium with surrounding cells. The disequilibrium does not become major because water diffuses into these cells from other cells and apoplastic spaces deeper within the leaf. But this water movement out of the deeper mesophyll cells causes their water potentials to become more negative, away from equilibrium with even deeper cells (Fig. 12.21).

Finally, this gradient of water potentials reaches a tracheid or vessel member. As water molecules move out of tracheary elements into mesophyll parenchyma cells, the water potential within the xylem water column becomes more negative. The loss of water from tracheary elements does not really affect the xylem osmotic potential because solutes are very dilute to begin with. Here water's cohesive properties are more important: As a water

Even relatively humid air has a tremendous capacity to absorb water; at 50% relative humidity, warm air can have a water potential as negative as − 50.0 MPa.

TABLE 12.5   *Transstomatal and Transcuticular Transpiration**

	Transstomatal	Transcuticular
Herbaceous plants		
Coronilla varia (crown vetch)	1810	190
Stachys recta (mint family)	1620	180
Oxytropis pilosa (locoweed)	1600	100
Woody plants		
Betula pendula (European white birch)	685	95
Rhododendron ferrugineum (rhododendron)	540	60
Pinus sylvestris (Scots pine)	527	13
Picea abies (Norway spruce)	465	15
Fagus sylvatica (European beech)	330	90

* Rates are mg H_2O/dm²/hour (dm is decimeters; 1 dm = 10 cm). Surface area includes both sides of the leaf.

FIGURE 12.21
As water moves out of the leaf into the air, the tissues dry and a water potential gradient becomes established. Water flows from the xylem, where water potential is least negative, toward air, where water potential is most negative.

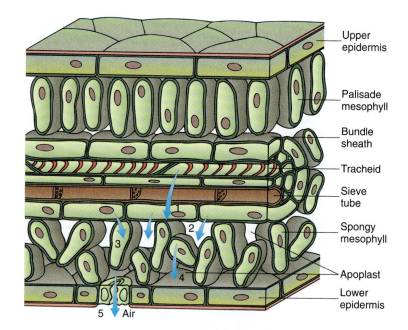

1. ψ tracheids > ψ sheath

2. ψ sheath > apoplast

3. ψ sheath > spongy mesophyll

4. ψ spongy mesophyll > apoplast

5. ψ apoplast > air
   ψ air = −50 MPa

molecule leaves the xylem, it does not leave a hole behind but instead drags other water molecules along with it. All water molecules of the plant are hydrogen bonded together, but the water molecules in the xylem can move upward most easily. That water is purest, is not bound to proteins and cellulose, is not locked into hydration shells around solutes, and so on. As water molecules diffuse out of xylem in the leaves, cohesive forces pull water upward through the xylem, all the way from the roots (Tables 12.6 and 12.7). Imagine the water being frozen solid; if the top molecules are pulled upward, the entire mass of ice is lifted.

Water is heavy, and water molecules in the uppermost tracheary elements must lift the weight of the entire water column. There is tension on these molecules, and consequently the **pressure potential is a negative number;** as water moves into the leaf mesophyll, the xylem water potential becomes more negative owing to an increasingly negative pressure potential. In vertical stems, water must move directly upward in the xylem, and the water's weight is a significant consideration. In the examples discussed earlier, water could move laterally between a cell and a beaker, so no lifting was involved. In vertical xylem, the weight of water counteracts its tendency to rise into areas of more negative water potential.

TABLE 12.6	Speed of Xylem Sap Translocation
	Maximum Speed (cm/hr)
Evergreen conifers	120
Sclerophyllous plants	150
Diffuse porous trees	600
Ring porous trees	4,400
Herbs	6,000
Vines	15,000 (0.6 mi/hr)

TABLE 12.7	Transpiration Rate
	Average Transpiration (g/dm²/day)
Pinus taeda (loblolly pine)	4.6
Acer negundo (box elder)	6.4
Liriodendron tulipifera (tulip tree)	9.8
Myrica cerifera (wax myrtle)	10.8
Quercus rubra (red oak)	12.0
Ilex glabra (inkberry)	16.1

Values are given in the most common form, based on transpiration per unit of epidermis surface. Transpiration per unit xylem cross-section, equivalent to the specific mass transfer of Table 12.4, is rarely calculated because xylem provides capacity but little control over water movement.

Consequently, if leaf xylem water potential is only slightly more negative than root xylem water potential, the water does not move. For every 10 meters of height, leaf water potential must be at least 0.1 MPa more negative than root water potential (Fig. 12.22). In trees such as elms and sycamores that are typically more than 30 meters tall, leaf water potential must be at least 0.3 MPa more negative than root water potential simply to overcome the weight of water. This is accomplished automatically: When stomata open in the morning, leaf cells lose water and their water potentials become more negative. However, water does not begin moving upward in the xylem until drying causes the water potential of leaf cells to become sufficiently more negative. Stolons, rhizomes, and horizontal vines have no such problem; long distance transport is horizontal and no lifting is involved, so gravity is not a factor. A

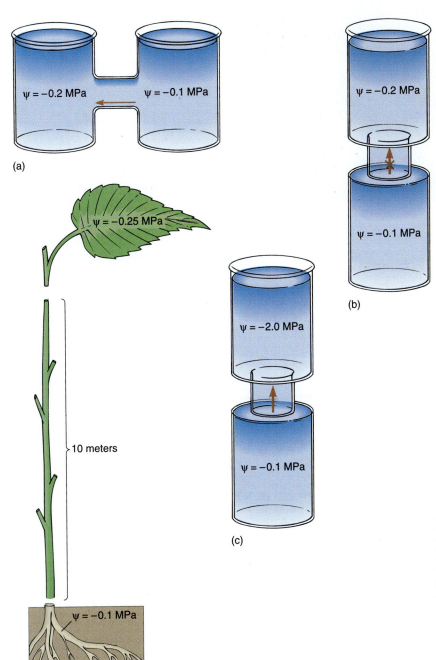

(a)

$\psi = -0.2$ MPa $\psi = -0.1$ MPa

$\psi = -0.25$ MPa

10 meters

$\psi = -0.1$ MPa

(d)

(b)

$\psi = -0.2$ MPa

$\psi = -0.1$ MPa

(c)

$\psi = -2.0$ MPa

$\psi = -0.1$ MPa

FIGURE 12.22

(a) If water can move laterally between two solutions, no lifting is involved. The slightest difference in water potential results in water movement. (b) If water must move upward against gravity, work must be done. A slight difference in water potential may not be enough to cause water movement; only a large difference will (c). (d) To overcome gravity and friction, the water potential of plant tissues receiving water must be at least 0.2 MPa more negative than that of roots for every 10 meters of height separating them. In the case illustrated here, with a difference of 0.15 MPa, water would not move up the stem. The water potential of the leaf would have to become 0.5 bars more negative.

FIGURE 12.23
Because the transpiration surfaces of this epiphytic cactus (*Rhipsalis*) are lower than its roots, water flows downward from the roots and gravity actually assists xylem conduction rather than hindering it, as in most plants.

In narrow tracheary elements, almost all water molecules are held in place; in wide tracheary elements such as vessels, water in the center is relatively unaffected and moves easily.

few plants grow as pendant epiphytes, their stems dangling down from the branches of the host plant. Their stems and leaves are lower than their roots, and gravity assists water movement (Fig. 12.23).

Water is extremely adhesive, and its molecules interact strongly with the polymers of the cell walls of tracheids and vessel elements. Water molecules adjacent to the walls tend to remain fixed to the walls and also tend to prevent neighboring water molecules from being drawn upward by transpiration/cohesion. This results in a layer of relatively immobile water that does not move easily. In narrow tracheids and vessel elements, this immobile water is a significant fraction of the water column. The resulting friction hinders water's movement and contributes to the tendency of water to not move upward even when leaves have a more negative water potential than roots. As a rough approximation, to overcome friction, leaf water potential must be at least 0.1 MPa more negative than root water potential for every 10 meters of height. Therefore, considering both friction and gravity, a difference of 0.2 MPa is needed for every 10 meters of height. In plants with numerous wide vessels, friction is less, and less than 0.2 MPa is needed, but in plants with narrow tracheary elements, even more than 0.2 MPa is necessary. Also, in plants that have only tracheids, the water molecules must be pulled through pit membranes when entering and leaving each tracheid, further contributing to friction.

Returning to the plant in our example, transpiration causes leaf cells to lose water and their water potentials become more negative; water moves into them from tracheary elements and tension builds on water molecules in the xylem. When the water potential in the uppermost tracheary elements has become sufficiently more negative than that in the lower elements, friction and gravity are overcome and water moves upward. This causes the lowermost xylem cells to pull water inward from the root cortex. The cortex water potential drops lower than that of the root epidermis and root hairs, and water flows from them to the cortex. In the final step, the water potential of the root epidermis and root hairs becomes more negative than that of the soil, and water moves automatically into the root from the soil.

Long distance water transport occurs in this manner as long as the soil is sufficiently moist. A sandy soil that has 30% moisture has a water potential of about -0.001 MPa, almost equal to that of pure water (Table 12.8). Water is held in soils by cohesion and

TABLE 12.8	*Water Potentials of Soils*	
	Water Potential (Mpa)	
Soil Type	10% Moisture	30% Moisture
Sand	− 0.05	− 0.001
Loam	− 0.5	− 0.005
Clay	− 10.0	− 0.1

adhesion as wedges and droplets between soil particles, and in sandy soils, root hairs can easily draw water from moist soils. Gravity and evaporation also pull water away from the droplets in sandy soils, so these soils dry quickly after a rain. As the soil dries, the most mobile molecules are removed and those tightly bound to soil particles remain (see Fig. 12.20). In a dry soil, not only is less water present, but it is also relatively immobile. Clay soils are composed of thin flakes with a high surface-to-volume ratio. When wet, they hold a great deal of water, but it is firmly bound as a hydration layer. No root can pull water away from clay that is even slightly dry. Loam soils consist of sand, silt, and clay and have a diversity of pore sizes. During a rainfall, loam soils absorb large amounts of water, then hold it for weeks after the rain has stopped. The diversity of pore sizes allows root hairs to absorb water but prevents gravity from pulling the water so deep into the soil that roots cannot reach it.

Many roots remain healthy with their water potentials as low as − 2.0 MPa; they still absorb water from soils that are quite dry (Fig. 12.24; see Table 12.2). But if the plant is 10 meters tall, the leaves would have to have a water potential at least slightly more negative than − 2.2 MPa to overcome friction and gravity. This is also possible in some species, but typically the leaves would either be dormant or preparing for abscission.

When both soil and air are dry, plants are greatly stressed. Even if stomata close, transpiration continues, at least through the cuticle. Leaf water potentials become more negative, but water cannot move upward easily because the soil is so dry. Tension on the

(a)

(b)

FIGURE 12.24

(a) Many desert shrubs withstand severe desiccation; even though their water potentials become extremely negative, the cells survive, although they may become inactive. (b) Plants such as this marsh plant (*Nymphoides peltata*) do not tolerate water stress at all; if their water potential falls very far below − 0.2 MPa, they die quickly.

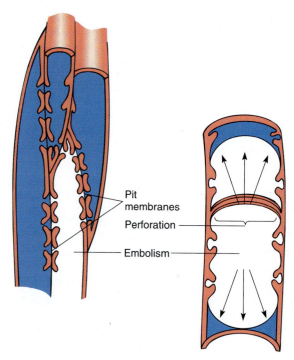

FIGURE 12.25
Severe tension can overcome the cohesion of water molecules and cause an embolism to form and expand rapidly. The embolism can pass through holes such as perforations but is stopped by pit membranes. If an embolism occurs in a tracheid, it cannot spread beyond the tracheid; but if a vessel element cavitates, the embolism spreads throughout the entire vessel.

An embolism is often called an air bubble, but it is a partial vacuum with some water vapor; it does not contain air.

water columns increases and at some point cohesion is overcome: Hydrogen bonding is broken over a large region and the water column breaks. This breaking is called **cavitation**, and the water column acts just like a broken cable. Molecules above the cavitation point are drawn rapidly upward because they are now free of the weight of the water below them; those below the cavitation point rush downward because nothing supports their weight. Between the two portions is space called an **embolism**, which expands until its surface encounters a solid barrier such as a pit membrane. The water/embolism surface cannot pass through pit membranes, but it can pass through perforations because they are open holes (Fig. 12.25). When an embolism forms in a tracheid, only that tracheid loses its water, but when an embolism forms in a vessel element, it may expand through perforation after perforation until the entire vessel has been emptied.

Cavitation usually means that that tracheid or vessel can never conduct water again. The cohesive bonding that permits leaf transpiration to draw water upward has been disrupted. Under unusual conditions, embolisms are occasionally "healed." If all the surrounding cells are full of water and if the night is so cool and humid that transpiration stops, enough water may seep into the embolism to fill it and re-establish a continuous water column. More typically, any water that seeps in simply flows down the side of the tracheary element but is not able to fill it.

Because cavitation destroys the usefulness of an entire vessel or tracheid and because a plant invests considerable energy and reduced carbon in making tracheary elements, features that minimize cavitation are selectively advantageous. Adhesion between water and the cell wall is just such a feature, giving the water column extra strength so that it does not cavitate easily. This is most effective in narrow elements, where the reinforcement affects all water molecules, even those in the center. In wide elements, the central molecules are freely mobile and cavitate almost as easily as pure water.

This wall-induced reinforcement of water columns is believed to be the feature that allows plants to reach the heights they do. *Eucalyptus* trees in Australia are the tallest plants known; they grow to 100 meters in height, and water is pulled upward the entire distance through their tracheary elements. This cannot be duplicated with glass or metal capillary tubes; the water columns are too fragile to support their own weight without the reinforcing that cell walls provide. One hundred meters appears to be the limit for xylem; even with reinforcement, the cohesive forces at the top of the water column cannot support the weight of 100 meters of water hanging from them.

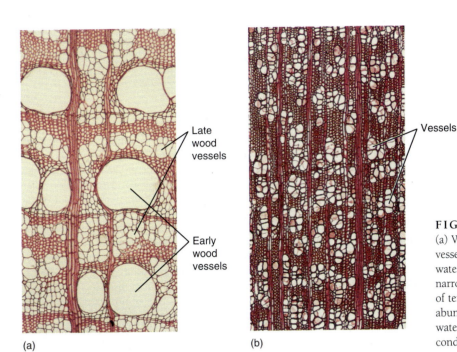

FIGURE 12.26

(a) Wood from a tropical tree. There are many broad vessels in the early wood, each of which can conduct water rapidly from the moist soil. Late wood has narrower vessels ($\times$ 100). (b) This wood is from a tree of temperate climates; the vessels are narrow and abundant. No single vessel can conduct very much water, but if one cavitates, only a small fraction of the conducting capacity of the wood is lost ($\times$ 100).

Some remarkably exquisite types of wood have evolved that are elegantly adapted for the various conditions in which plants live. In the moist tropics, water is always abundant, the soil is never dry, and water always moves easily; reinforcement is not necessary and the wood is full of wide vessels (Fig. 12.26; see also Fig. 8.16a). In drier temperate areas, especially rocky slopes, water is frequently scarce and water stress common. It is selectively advantageous for plants in such an environment to have narrow vessels or even wood with tracheids only, as conifers have. In temperate areas with good rainfall, the plants usually have a moist spring and produce early wood (also called spring wood) with large vessels; the summer is drier and they then produce late wood (summer wood) with narrower vessels or only tracheids. During the summer, the wide vessels of the early wood cavitate, and conduction occurs primarily or entirely in the late wood (Fig. 12.27).

Eventually all vessels and tracheids cavitate. Dry conditions in summer cause many cavitations, as do freezing in winter, vibration in wind, and damage by burrowing insects. After tracheary elements cavitate, surrounding parenchyma cells may block them off by

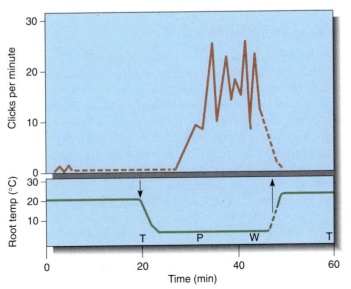

FIGURE 12.27

Cavitation of vessels and tracheids causes audible clicks that can be heard by sensitive microphones and counted. This graph shows the induction of cavitation in castor bean xylem; plants were grown in water solutions that permitted water uptake to match transpiration loss—there was little tension on xylem water columns and no cavitations occurred. Then the roots were cooled (*arrow at left*) to 5°C to inhibit root absorption of water, but shoots were kept warm to encourage transpiration. Water columns were stretched and immediately began breaking. After 20 minutes, embolisms were forming at almost 30 per minute and the plants were wilting. At 45 minutes the root solution was warmed to permit water absorption, and cavitation quickly stopped.

PLANTS & PEOPLE

FARMING "WASTELANDS"

It has been proposed that we develop plants that tolerate high levels of salt and use them as crop plants in arid, marginal "wastelands" where fresh water is scarce. This might be possible in two ways: We could examine desert-adapted plants to see if any have useful properties such as nutritious seeds, medicinal drugs, or useful fibers. Alternatively, desert-adapted plants could be studied to identify which of their features make them drought-resistant; then the corresponding genes could be identified, cloned, and transferred to some of our crops that are not now drought-adapted. Once either of these types of plants is available, they could be grown in semi-desert regions, or perhaps even in deserts using a little irrigation with sea-water or brackish water.

This concept has many problems. The first is that it is one thing to find plants with enough drought- and salt-tolerance to grow in sea-water during a 1-year-long experiment, but it is another thing to irrigate an area with sea-water year after year. Because they have little rain, most arid regions have only small, temporary rivers that have not been able to carve effective drainage channels: They often end in dry lakes with no outlet. After a rain, water is present for a few weeks, but it evaporates, leaving its minerals behind. The rivers do not carry minerals out to sea continuously as do rivers in the eastern United States and other moist areas. If sea-water is used for irrigation in an arid area, tremendous amounts of salt would be deposited as the water evaporates. This would continue year after year until the salt concentration becomes so high that a salt desert is created, such as occurs in the Great Salt Lake in Utah, the Dead Sea in the Middle East, or the Devil's Golf Course in Death Valley, where the accumulated salt is so abundant it forms crystals 2 or 3 feet tall. Nothing could grow—not the new crop plants, not the genetically engineered plant, and certainly not the original plants, which would all be destroyed when the region was plowed to start the project. Even using fresh water rather than sea-water causes a gradual accumulation of salt, and California's Central Valley is already facing a serious problem of salt accumulation in the soil.

Another problem with this type of project is the concept that deserts and semi-deserts are wasteland, that they must be "developed" and "improved." Although this type of thinking may have been popular at one time, many people now disagree with it. These areas, in their natural state, have an intrinsic worth. They are home to a great diversity of plants, animals, fungi, and other species. Even if none of these "wasteland" species is ever discovered to contain a medicinal drug or other useful feature, does that give us the right to exterminate them? Is it really necessary to bring more land into cultivation when so many Americans are overweight or eat foods that have been processed specifically to remove calories from them?

---

secreting gums and resins. If the tracheary element has wide pits, the adjacent parenchyma cells can soften the wall and swell into the vessel, blocking it with a mass of protoplasm called a tylosis (see Fig. 8.18). As more tracheary elements cavitate in a region of wood, the adjacent parenchyma synthesizes phenolic compounds that are resistant to attack by bacteria and fungi. These chemicals are often dark and aromatic, causing the wood to become brown and fragrant. The parenchyma cells then also die, and the region becomes part of the heartwood (see Fig. 8.17).

CONTROL OF WATER TRANSPORT BY GUARD CELLS

Bulk water movement through xylem is influenced and powered primarily by water loss to the atmosphere. Although water loss through the cuticle is important, transstomatal transpiration is usually more significant whenever stomatal pores are open. Open stomata always represent a trade-off between carbon dioxide absorption and water loss. Whenever water supply in the soil is adequate, water loss is actually advantageous—the water movement is the primary means of carrying minerals upward from roots to shoots, and the evaporative cooling that results from transpiration can prevent heat stress in leaves and young stems. However, if the soil is too dry to supply water, transpiration represents an immediate, potentially lethal threat due to desiccation. Numerous mechanisms have evolved that control stomatal opening and closing. Each mechanism is keyed to a particular environmental factor, and their interaction results in great sensitivity to potential stresses in the habitat.

If the leaf has an adequate moisture content, light and carbon dioxide are the normal controlling factors. For most healthy, turgid plants, light most often controls guard cell water relations. Blue light is the important, triggering wavelength, and the action spectrum of opening closely matches the absorption spectrum of a flavin or flavoprotein pigment. It is not yet known how absorption of light by the pigment leads to potassium pumping.

The presence of light also leads to the photosynthetic fixing of carbon dioxide; the decrease in internal carbon dioxide concentration may also lead to stomatal opening. Artificial manipulation of the amount of carbon dioxide available can stimulate guard cells to open or close in light or dark. At night, with no photosynthesis, carbon dioxide levels are high and presumably contribute to stomatal closing.

All these mechanisms in healthy plants are completely overridden by a much more powerful mechanism that is triggered by water stress. As leaves begin to dehydrate, they release the hormone abscisic acid. This hormone immediately causes guard cells to lose potassium and close the stomatal pore. Closure occurs even in blue light and low concentrations of carbon dioxide, factors that would otherwise favor opening. Water stress—induced closure often occurs in the early afternoon on a warm, dry day if root uptake and xylem conduction cannot keep up with transpiration. Under these circumstances, lack of carbon dioxide stops photosynthesis even though light is available.

In plants with crassulacean acid metabolism, stomatal opening and closing are reversed when the plant is water-stressed. Stomata open at night and close in the morning. Temperature is particularly important for these plants; if night temperatures are too high, stomata may remain closed for days or weeks at a time. The low night temperatures typical of the desert are essential for stomatal opening and carbon dioxide absorption. Under conditions of mild temperatures and abundant moisture in the tissues, such as after a spring rainfall, CAM plants convert to C_3 metabolism, opening their stomata in the morning and picking up carbon dioxide with RuBP carboxylase directly. When the soil dries after several days, they revert to CAM and night opening of stomata.

SUMMARY

1. Living organisms transport materials over short distances (within organelles and cells) or over long distances (between nonadjacent cells).

2. Active transport is the forced pumping of material from regions where it is relatively unconcentrated to regions where it is more highly concentrated; active transport is an energy-consuming process.

3. Water potential (ψ) measures the capacity of the water to do work; in cells, it has two important components, osmotic potential (ψ_π) and pressure potential (ψ_p). In soils and rather dry materials, a third component, matric potential (ψ_m), becomes important.

4. Water moves from regions where it is relatively concentrated (similar to pure water, ψ near zero) to regions where it is less concentrated (dissimilar to pure water, ψ more negative).

5. A cell's water potential can be made more negative by pumping solutes such as K^+ and sucrose into it or by depolymerizing polymers to monomers, especially starch to glucose. Reversing these processes increases a cell's tendency to lose water.

6. A cell grows if its wall is too weak to counteract the tendency of water to enter the cell. Pressure potential cannot rise high enough to raise the cell's water potential and bring the cell into equilibrium with its environment.

7. Incipient plasmolysis is the point at which the protoplast has lost just enough water that it no longer presses against the wall, and ψ_p equals zero. If the cell continues to lose water, it becomes plasmolyzed.

8. The water potentials of guard cells, motor cells, and sieve elements become more negative as these cells are forcibly loaded with solutes by active transport. Water flows into the cells, causing guard cells and motor cells to swell but the phloem sap to be squeezed out of sieve elements through sieve pores.

9. Water begins to flow from roots to shoots when the shoots lose water to the air and their water potential becomes negative enough to draw water from the roots and overcome the effects of gravity and friction.

10. As both the air and the soil become dry, tension on the water columns in xylem increases; cohesion may be overcome and the water column cavitates, forming an embolism.

11. Water's adhesion to the sides of tracheary elements helps prevent embolisms; the presence of tracheids as opposed to vessels limits the amount of damage done if an embolism does form.

12. For non-CAM plants, light and low carbon dioxide stimulate stomata to open if moisture is adequate. Water stress overrides all other controls and causes stomata to close.

IMPORTANT TERMS

active transport	cohesion-tension hypothesis	incipient plasmolysis
adhesive	cohesive	long distance transport
callose	diffusion	matric potential
cavitation	embolism	molecular pump

osmosis
osmotic potential
P-protein
P-protein plug

plasmolysis
pressure flow hypothesis
pressure potential
selectively permeable membrane

short distance transport
sink
source
water potential

REVIEW QUESTIONS

1.   What are the three components of water potential? Which of these potentials measures water's interaction with dissolved material?

2.   In a beaker of pure water, what is the water potential? Does water potential become more positive or more negative as you add solute to it? Put pressure on it? Add it to dry clay? Add acid to it? In each case, which water potential component is changing?

3.   What can you say about the water potentials of two solutions (or of a solution and a cell, or of two cells) when they are in equilibrium? At equilibrium, is there any *net movement* of water?

4.   In each of the following pairs, circle the one that would probably have the more negative water potential: cell of a wilted leaf—cell of a turgid leaf; guard cells of an opening stoma—regular epidermal cells; guard cells of a closing stoma—regular epidermal cells; root cortex cell—moist soil; phloem cell being loaded with sucrose—leaf chlorenchyma cell; phloem cell unloading sucrose—tuber cell storing starch; clay with 10% moisture —silt with 10% moisture.

5.   Describe the cohesion-tension model of water movement through xylem. Would the weight of water be more of a problem in an upright tree or in a stolon? Why?

6.   Consider the pressure flow model of phloem transport. How do sugars and water enter the phloem from the source? How do sugars and water move from one phloem cell to another?

SOILS AND MINERAL NUTRITION

13

As rocks weather into soil, mineral nutrients are released and become available to plants. Remarkably small amounts of nutrients are sufficient to permit plant survival.

CONCEPTS

In addition to the carbon, hydrogen, and oxygen discussed with regard to photosynthesis and respiration, all organisms need elements such as nitrogen, phosphorus, calcium, magnesium, and sulfur. Plants must absorb these from soil and then use them and the glyceraldehyde-3-phosphate from chloroplasts to build all their chemical components, however complex. This is an important concept: Plant metabolism is based on sunlight and chemicals present in water, air, and soil. No animal is able to survive on just minerals and one simple carbohydrate; they must obtain minerals and complex organic compounds in their food.

Most of the elements that are essential for plant growth and development are present in the crystal matrix of minerals. The elements become available to roots as rocks weather and break down, creating soil. During soil formation, rocks are converted gradually into dissolved ions and inorganic compounds. Because they are derived from the rock minerals, their role in plant nutrition is called **mineral nutrition.**

The term "mineral nutrition" covers a variety of types of plant metabolism. For some elements, once the mineral is absorbed from the soil, it can be used immediately as it is. An example is potassium, which is used by cells such as guard cells to adjust their turgor and water relations. Simple potassium ions are sufficient. Mineral elements such as iron and magnesium are more complex because they must be incorporated into compounds such as cytochromes or chlorophyll molecules before they are useful. Nitrogen is even more complicated: Like carbon, its oxidation state is important. Consequently, it must be reduced, and elaborate electron transport chains are necessary to convert it to useful forms.

The term "soil" covers a wide variety of substances. The various soils are important to plants not only in supplying minerals and harboring nitrogen-fixing bacteria, but also in holding water, supplying air to roots, and acting as a matrix that stabilizes plants, preventing them from blowing over. Critical aspects of soil are its chemical nature, which determines which mineral elements are present; its physical nature, which reflects is porosity, texture, and density; and its microflora and microfauna—the small animals, fungi, protists, and bacteria that live, respire, and gather food within the soil.

The microflora and microfauna of the soil deserve special mention. It is easy to think of soil in terms of its chemical and physical properties only, but that would be an incomplete concept of soil. Most soils contain large amounts of microbes and tiny animals that are extremely important to plants (Table 13.1). Although they are microscopic to us, they are about the same size as roots and root hairs, and they interact extensively with root systems. For example, many soil microbes supply plants with nitrogen: Nitrogen is not found in rock matrixes, so soil formation does not make nitrogen available to plants. Instead, the primary source of nitrogen is the molecular nitrogen (N_2) of the atmosphere, but only certain bacteria and cyanobacteria have enzymes that convert molecular nitrogen into forms useful for metabolism. When these microbes die and decay, their organic nitrogen compounds are released to the soil and become available to plants.

Just as the foods of animals vary, soils also vary in the quantities of minerals present, the texture of the soil, and the organisms present. Also, plants vary with regard to the amounts of minerals they require and their capacity for absorbing and processing minerals. All these factors affect a plant's health and, to a large extent, determine the types of plants that exist in a particular area.

TABLE 13.1	*Number of Organisms in Soil*
Insects	670,000,000 per hectare*
Arthropods	1,880,000,000 per hectare
Bacteria	1,000,000,000 per gram of soil
Algae	100,000–800,000 per gram of soil
Earthworms	1,800,000 per hectare
Weight of worms	990 kg
Worm casts	396,800 kg

* 1 hectare = 10,000 m^2 = 2.45 acres.

ESSENTIAL ELEMENTS.

Much research in mineral nutrition involves experiments in which a single element is supplied to a plant in excessive quantities or withheld from it. This is accomplished by growing the plant in a **hydroponic solution** whose chemical composition is carefully controlled. Hydroponic experiments were formalized and used extensively by Julius von Sachs in 1860. In such experiments, a solution is developed that supports plant growth. At first, this must be by trial and error, with numerous chemicals being added to the solution, and plants are tested to see if they survive. Many tests are necessary because some of the chemicals may be toxic, even in low concentration. Other chemicals needed by plants are toxic if present at too high a concentration. Also, the form in which a chemical is present makes a difference: Nitrogen is important, but in addition to nitrate and ammonia, which plants use, other nitrogen compounds exist which plants do not use. Also, when numerous compounds are added to the same solution, unsuspected reactions may occur that create toxic compounds or convert useful compounds to useless ones.

After a solution of known composition is found which supports plant growth, an identical solution can be prepared except that one component is left out (Fig. 13.1). If that component is not necessary for plant growth—that is, if it is not **essential**—plants grow normally. But if the excluded element is essential, the plant cannot grow correctly. For example, an experimental solution might contain nickel; if nickel is left out of the second solution, plants would grow well, perhaps even better than in the first solution; therefore, nickel is not an essential element. The second solution then becomes the main test solution. If a third solution is prepared similar to the second but lacking potassium, for example, no plant could survive; therefore potassium is an essential element and must always be included (Fig. 13.2). Using these hydroponic techniques, Sachs was able to establish a minimal nutrient solution that would support plant growth; it contained calcium nitrate, potassium nitrate, potassium phosphate, and magnesium sulfate. At present, several solutions are known to support reasonable growth of most plants; two are Hoagland's solution and Evan's modified Shive's solution. Many experiments begin with one of these.

It may seem simpler just to grind up a plant and then extract and measure all the chemical elements present. But plants actually absorb many elements that they do not need—the endodermis simply cannot exclude them completely. A living plant usually contains at least trace quantities of every element present in the soil, whether essential to the plant or not.

FIGURE 13.1

In a hydroponics experiment such as this one at the University of California at Davis, the test solution is often just a water solution in a bottle of boron-free glass. Air must be bubbled through the liquid to permit root respiration (clear plastic tubes). The bottles are wrapped in foil or painted black to exclude light, more closely resemble a soil environment, and prevent the growth of algae. (*L. Migdale, Science Source/Photo Researchers, Inc.*)

FIGURE 13.2
These plants are being grown in a hydroponic solution that contains all known essential elements except one—magnesium. Even though nitrogen, sulfur, and the other elements are abundant, they are of little use to these tomato plants if one essential element is missing. Growth and reproduction are governed by the least abundant factor, not the most abundant. *(John Colwell from Grant Heilman)*

The essential elements discovered by Sachs are called the **major** or **macro essential elements** because they are needed in large quantities by plants (Table 13.2). If dry plant material is analyzed, calcium, nitrogen, potassium, phosphorus, magnesium, and sulfur are present at concentrations of between 0.1% and 3.0% of the plant's dry weight. Even as the first hydroponic experiments were being performed, botanists realized that the available chemicals were quite impure and that trace quantities of other elements were present. Despite their best efforts, they could not prepare solutions that had absolutely no copper, zinc, or many other elements. Thus, it was not possible to conclude that any particular element was completely nonessential; it was possible to determine only that relatively large amounts of most chemical elements were not essential.

As manufacturing methods improved in the chemical industry, purer compounds were produced, and mineral nutrition experiments could be repeated with more certainty that the element tested had been almost completely excluded and was present in much lower amounts than previously. With such improved chemicals, it was soon discovered that there exists a group of **minor** or **micro essential elements**, also called **trace elements.** Iron, boron, chlorine, copper, manganese, molybdenum, and zinc are required in extremely low concentrations by plants. Iron is the exception, being needed in amounts intermediate between those of the major and minor elements.

Although our reagents today are much purer than those of 100 years ago, it is still impossible for us to create a solution that contains absolutely only the nine major and seven minor elements. Our purest water has traces of many contaminants, the very best glass containers release silicon and boron into the solution, and all forms of chemical reagents have some trace contaminants. Chlorine was only recently discovered to be essential, at least for some plants; it remained undetected until it was realized that despite the purity of the test solutions, experimental plants receive adequate chlorine if scientists touch the plant because the skin of one fingertip has sufficient chlorine for an entire plant. Typical hydroponic experiments had shown that plants could grow in chlorine-free solutions, but when the studies were repeated with all experimenters wearing plastic gloves to prevent transfer of chlorine, the plants did not survive. Because such small quantities of chlorine are sufficient, we must suspect that other elements may also be necessary in similarly minute quantities; testing will continue whenever purer reagents become available.

TABLE 13.2	*Elements Essential to Most Plants*

Macronutrients

Carbon	Organic compounds
Oxygen	Organic compounds
Hydrogen	Organic compounds
Nitrogen	Amino acids; nucleic acids; chlorophyll
Potassium	Amino acids; osmotic balance; enzyme activator; movement of guard cells and motor cells
Calcium	Controls activity of many enzymes; component of middle lamella; affects membrane properties
Phosphorus	ATP; phosphorylated sugars in metabolism; nucleic acids; phospholipids; coenzymes
Magnesium	Chlorophyll; activates many enzymes
Sulfur	CoA; some amino acids

Micronutrients

Iron	Cytochromes; nitrogenase; chlorophyll synthesis
Chlorine	Unknown; possibly involved in photosynthetic reactions that liberate oxygen
Copper	Plastocyanin
Manganese	Chlorophyll synthesis; necessary for the activity of many enzymes
Zinc	Activates many enzymes
Molybdenum	Nitrogen reduction
Boron	Unknown

CRITERIA FOR ESSENTIALITY

An element must meet three basic criteria to be considered essential.

1. **The element must be necessary for complete, normal plant development through a full life cycle.** If a particular element is required for any aspect of a plant's growth, differentiation, reproduction, or survival, that element is essential. This logical and necessary criterion can be difficult to test. In a hydroponic experiment, the solution and plant must be carefully protected from contamination by dust and insects, both of which are mineral-rich. The experiments must occur in laboratories, growth chambers, or specially controlled greenhouses. Laboratory conditions are not the same as the natural environment, however, and part of a plant's ability to survive depends on its response to stress: cold, heat, drought, pathogens. It may be that certain elements needed only under unusual conditions are not anticipated in a growth chamber experiment.

   It is not feasible to study large trees in greenhouses or growth chambers, both because of their size and because it takes so many years to complete a life cycle. Because it is virtually impossible to test such species, the practical assumption is made that the major elements and most minor ones are essential for all plants, having been tested and found essential for many small herbs. This is a safe assumption if the role of the element is known. For example, nitrogen is present in amino acids and nucleic acids; because nothing can live without these, nitrogen is essential. The same is true for iron (cytochromes), calcium (middle lamella, enzyme control), and others. But the roles of boron and chlorine are still not known for certain, so we cannot assume with confidence that all plants require them. Furthermore, some desert plants have large amounts of silica in their epidermal cell walls; silicon is an essential element for them, but most plants appear not to need it at all.

2. **The element itself must be necessary, and no substitute can be effective.** This criterion is relatively straightforward; in most instances, if one essential element is absent, the presence of a chemically related element does not keep the plant alive. In

some cases, elements can be substituted in specific enzymatic reactions or transport processes when studied in vitro, but even so, the whole plant cannot live with only the substitute. Some plants that require chlorine can survive if given large amounts of bromine, but such elevated levels of bromine do not occur in nature, so the substitution works only in the laboratory.

3. **The element must be acting within the plant, not outside it.** The complexity of test solutions makes it difficult to analyze results in many cases. Iron is an essential element, but it is soluble only in a very limited range of acidity. Although iron may be added in adequate quantities, it often reacts with other chemical components and forms an insoluble precipitate. A test chemical can cause the precipitated iron compound to break down and form a soluble iron compound. With the new availability of iron, the plant grows well, even if the test element is not needed or absorbed. Without careful analysis, the test element would appear to be essential even though it is not used inside the plant.

   This criterion has interesting ramifications. Most plants absorb phosphorus only poorly and depend on an interaction with soil fungi that absorb phosphorus and transfer it to the plant. If soil fungi are killed, plants grow poorly. If an element is essential to the fungus, is it then also essential to the plant even if the plant does not use it directly? This is one of those points that can be debated endlessly without being resolved; it is better to understand the biology than to dispute definitions.

MINERAL DEFICIENCY DISEASES

CAUSES OF DEFICIENCY DISEASES

Virtually all types of soil contain at least small amounts of all essential elements; under natural conditions it is rare to encounter plants whose growth and development are seriously disrupted by a scarcity or an excess of mineral elements. It is especially uncommon to find plants suffering from an overabundance of a particular mineral. In many cases, the unnecessary ions are not even absorbed by the roots. Certain types of excess minerals, if absorbed, can be precipitated in vacuoles as crystals; although a cell may contain large amounts of the mineral, only the ions actually in solution have any significant effect on metabolism.

FIGURE 13.3
As water evaporates from the soil, more water moves upward, carrying dissolved minerals that crystallize as the water evaporates. When soil water is present, its osmotic potential is extremely negative, as is its water potential (its pressure potential is zero). Roots cannot pull water out of such osmotically dry soil; they would die before their water potential became sufficiently negative.

(a)   (b)

FIGURE 13.4

(a) Certain mangroves *(Avicennia)* of tidal marshes secrete salt through special glands. Manipulation and transport of such large amounts of salt require tremendous energy use, but salt excretion not only permits growth in saline habitats but also provides protection against herbivores. *(Robert & Linda Mitchell)* (b) Most plants neutralize excess salts by precipitating them as crystals such as these. *(Courtesy of H. J. Arnott, University of Texas, Arlington)*

Desert soils often have excessive amounts of all available minerals because ground water moves upward, carrying dissolved minerals with it. These can reach such strong concentrations that the water potential of the soil solution is extremely negative and roots are unable to extract water from it. Plants are unable to grow, not because of mineral toxicity but because of osmotic drought (Fig. 13.3). Some species are adapted to less severely salty regions; salt bush *(Atriplex)* absorbs both water and salt but passes salt directly through the body and secretes it from salt glands located on leaves. This produces a coating of salt crystals that is thought to be selectively advantageous (Fig. 13.4). The salt reflects away some of the excessive sunlight. It also makes the bush an unacceptable food because desert-dwelling animals must avoid salty foods in order to balance their salt/water intake.

Mineral deficiency disease does not seem common in natural populations. Some soils may have such low concentrations of certain essential elements that some species are unable to thrive on them. For example, a type of soil called serpentine soil is extremely deficient in calcium and few plants grow on it. Some species are more sensitive than others to low concentrations of essential elements (Fig. 13.5). Owing to competition with other plants and pathogens, the sensitive plants probably weaken and die early and do not reproduce successfully. Consequently, species especially sensitive to a particular deficiency typically do not occur in the plant community growing on soil deficient in that element.

Deficiency diseases are most commonly encountered in non-native crop plants or ornamentals. Crop plants especially have undergone human artificial selection (not natural selection) for traits such as rapid growth and high fruit/seed yield that require large amounts of nitrogen and mineral nutrients. Without fertilization, these plants often have poor growth and symptoms of deficiency diseases. As shown in Table 13.3, entire forests can be fertilized. The data show the results of fertilizing forests in Tennessee with 335 kg/ hectare of fixed nitrogen, either with or without phosphorus. Addition of nitrogen was always beneficial, as virtually all soils are deficient in fixed nitrogen. But addition of phosphorus often did not improve tree growth more than nitrogen alone. Soils naturally contain adequate phosphorus for many forest species.

FIGURE 13.5

Beans are especially sensitive to deficiency of zinc. On soils low in zinc, many plants grow well and are healthy but leaves of beans develop white necrotic spots where the cells die. *(Ken Wagner/Phototake)*

TABLE 13.3 *Fertilizing Forests*		
	Increase in Growth (%)	
	Nitrogen Alone	**Nitrogen + Phosphorus**
Hickory	261	172
Northern red oak	167	100
Cucumber tree	107	207
Chestnut oak	61	70
Black cherry	52	71
Yellow poplar	48	69
Dogwood	36	173

The very act of harvesting crops leads to soil depletion: Fruits, seeds, tubers, and storage roots often have the greatest concentration of minerals in a plant. Harvesting them removes those minerals from the area; the rest of the plant body is relatively mineral-poor and does not contribute much to re-enriching the soil even if it is plowed back in. With human populations, consumption of food may occur thousands of miles from the site of plant growth; even worse, waste products are dumped into rivers and carried away, never being returned to the soils. Under natural conditions, minerals are returned in the form of manure, which is typically deposited in the general region where feeding occurs. The most extensive crop removal is harvesting a forest for timber (Table 13.4). But several methods of harvesting are possible: removal of every part of the tree versus trimming and debarking, which permit the small branches, leaves, and bark to remain in the ecosystem.

SYMPTOMS OF DEFICIENCY DISEASES

The particular symptoms of a mineral deficiency are more closely related to the particular element that is lacking than to the plant species; usually all plants that suffer from a scarcity of a particular essential element show the same symptoms. One common symptom is **chlorosis**; leaves lack chlorophyll, tend to be yellowish, and are often brittle and papery. Deficiencies of either nitrogen or phosphorus cause another common symptom, the accumulation of anthocyanin pigments that give the leaves either a dark color or a purple hue. A lack of certain elements causes **necrosis,** the death of patches of tissue. The location of the necrotic spots depends on the particular element: Potassium deficiency causes leaf tips and margins to die, whereas manganese deficiency causes the leaf tissues between veins to die even though all the veins themselves remain alive and green.

"Necrosis" is a general term for localized death of tissues; it can also be caused by bacterial, viral, and fungal infections.

TABLE 13.4 *Mineral Loss Through Crop Removal*					
	Minerals Removed from Forest (kg/hectare/year)				
Harvest Method	**N**	**P**	**K**	**Ca**	**Mg**
Complete removal, with roots	17.6	2.3	12.6	12.8	3.6
Complete removal of shoot, leaving roots	16.1	1.9	10.3	11.7	2.9
Wood only, leaving bark, narrow branches, and roots	4.6	0.8	3.8	4.3	1.3

Box 13.1 *Limiting Factors*

Many plants waste cellulose. Many tree trunks are larger and stronger than is really necessary to hold up the plant, and throughout the year the plant may freely shed petals, fruit, bark, leaves, even entire branches. Yet no mechanism exists to depolymerize the cellulose back to glucose and conserve the energy present in the cell walls of the tissues about to be abscised.

Ecologists often evaluate organisms and processes in terms of their energy budget: If a new type of growth evolves that requires less energy than the current type of growth, then the extra energy can be used for reproduction and might be selectively advantageous. Or if a new type of energy-saving nest-building process evolves in birds, the birds should have more energy to lay more eggs and raise more young. The concept of energy budget seems to work well for animals, but it often seems to have little relevance to plants. Many plants deposit multilayered bundle sheaths composed of fiber cells with thick, cellulose-rich walls. Many fruits and nuts are almost rock-like with the amount of sclereids they contain. In some trees, the wood consists of a few water-conducting vessels in an extensive matrix of thick-walled fibers.

These plants appear to have energy to spare, and this may in fact be true. Imagine desert plants. They certainly do not need more sunlight—if given even more, they would not grow faster. The factor that restricts their growth is the lack of water. We say that water is a "limiting factor" for them. If all other conditions were held the same but the plants were given more water, they would grow and reproduce more. Now imagine cool, moist temperate forests: Their trees usually do not grow better if given more light or more water. Instead, lack of nitrogen is the limiting factor for them. Although plants need many factors (light, carbon dioxide, essential elements, water), at any given time for any given plant, only one is the limiting factor. If it is supplied, the plant grows more rapidly until another factor becomes limiting. If you supply an abundance of everything, the plant's own metabolic capacity becomes limiting—no plant can grow infinitely rapidly.

Consider the temperate forest again. With their leaves in full sunlight and the ground often moist, but with nitrogen the limiting factor, the trees can use their photosynthetically produced glucose to make polymers that do not contain nitrogen: cellulose, carbohydrates, cutin, lignin, and so on. It does not really matter much if they squander these components; they could not use them to make proteins or nucleic acids anyway because they do not have any extra nitrogen. But the herbs and small shrubs on the forest floor live in the shade of the large trees. They have much less light available and for them light is the limiting factor. They have no more nitrogen available than the large trees do, but with their low rate of photosynthesis, the nitrogen is not limiting. It may be better for these plants to make as little cellulose as possible, and instead divert a greater percentage of their glucose into building proteins, nucleic acids, and protoplasm in general rather than building sclerenchyma. These plants are often soft and very leafy, with a great part of their body as protoplasm-rich leaves and meristems and less as energy-rich woody trunks and branches.

MOBILE AND IMMOBILE ELEMENTS

An important diagnostic aspect is whether symptoms appear in young leaves or older leaves. This is related to the **mobility** of the essential element. Boron, calcium, and iron are **immobile elements;** once they have been incorporated into plant tissue, they remain in place. They do not return to the phloem and cannot be moved to younger parts of the plant. A plant that grows in a soil deficient in boron, calcium, or iron is probably able to grow relatively well until the few available ions have been absorbed. Growth is normal until the soil is exhausted; further growth suffers mineral deficiency, and the newly formed tissues are affected (Fig. 13.6a).

The elements chlorine, magnesium, nitrogen, phosphorus, potassium, and sulfur are **mobile elements;** even after they have been incorporated into a tissue, they can be translocated to younger tissue. Once the soil becomes exhausted of one of these elements, older leaves are sacrificed by the plant. The mobile elements are salvaged and moved to growing regions (Fig. 13.6b). The adaptive value of this is easy to understand: A leaf photosynthesizes most efficiently right after it has first expanded and less efficiently as it ages. The plant can increase its overall photosynthetic rate by sacrificing the old, inefficient leaves and using the minerals to construct new, efficient leaves.

The immobility of certain ions is not understood; boron, calcium, and iron are initially moved upward from roots into shoots, flowers, and fruits, so transport mechanisms do exist for them. Mutations that would result in the degradation of cytochromes in old leaves and the recovery of iron should be selectively advantageous. Animals have trouble with mineral recovery as well. For example, humans do not recycle the large quantities of iron in dead red blood cells; it is simply discarded, even if this results in anemia.

(a)   (b)

(c)

FIGURE 13.6

(a) This rose leaf is suffering from iron deficiency. Because the element is immobile, the little iron present in the plant cannot be transferred from older leaves to younger ones, so young leaves show the disease symptoms. Cells near the veins have chlorophyll; those farther away are chlorotic. *(Earth Scenes; Holt Studios)* (b) These mature grape leaves are suffering from a deficiency of phosphorus. Because the element is mobile, the plant can transfer atoms of phosphorus from older leaves to newer ones. *(Nigel Cattlin/Earth Scenes; Holt Studios)* (c) These potato leaves, suffering from manganese deficiency, have black necrotic spots. *(Earth Scenes; Holt Studios)*

SOILS AND MINERAL AVAILABILITY

Soils are derived from rock by processes of **weathering.** The initial rock may be volcanic (granite, basalt), metamorphosed (marble, slate), sedimentary (sandstone, limestone), or other types, but two things are important: Rock has a crystalline structure, and trapped within the structure are numerous types of contaminating ions and elements (Fig. 13.7). Its crystal matrix prevents rock from being a suitable substrate for plant growth; it may contain most essential elements, but as long as they are part of the matrix or trapped by it, they cannot be absorbed and used by plants. Also, any water held by the rock as part of the crystal structure is unavailable to plants.

Two fundamental processes of weathering convert rock to soil: **physical weathering** and **chemical weathering.** As its name implies, physical weathering is the breakdown of rock by physical forces such as wind, water movement, and temperature changes. Ice is an important agent. During winter days, water from rain or melted snow seeps into capillary spaces within rock; at night, when temperatures fall below freezing, the expansion of water as it becomes ice causes cracks to widen. Portions of rock, ranging from small flakes to large pieces, can be broken off. This is a slow process, but gradually the average particle size of the rock is reduced.

Runoff from rainstorms, avalanches, and similar forces wash rock fragments and pebbles into streams and rivers, where physical weathering accelerates. In rapidly moving streams the rocks are scoured by suspended sand grains; with high flow rates, rocks and

(a)

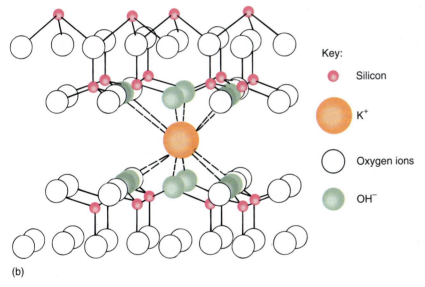

Key:

● Silicon

● K⁺

○ Oxygen ions

● OH⁻

(b)

FIGURE 13.7

(a) Rock is a complex, highly contaminated crystal; as it weathers, it gradually breaks down into soil. (b) Within the crystal matrix of rock, numerous atoms of many different elements occur. As rock weathers, it breaks apart and atoms at the surface are liberated into the soil solution. This is potassium trapped in vermiculite, a type of clay.

boulders are carried downstream, grinding against each other and the stream bed. Wind-blown sand is a powerful erosive force, as are glaciers, which are extensive during the periodic ice ages.

Physical weathering produces a variety of sizes of soil particles; the largest ones that are technically important to soil are grains of **coarse sand** with a size range of 2.0 mm to 0.2 mm. Particles only one tenth this large (0.2 to 0.02 mm) are **fine sand**, and those one tenth of this (0.02 to 0.002 mm) are **silt.** The finest particles, smaller than 0.002 mm in diameter, are **clay particles,** technically known as **micelles** (Fig. 13.8).

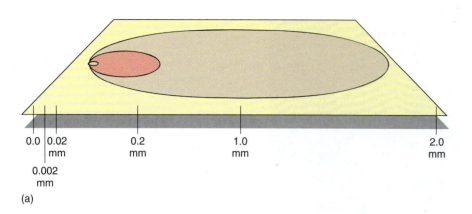

(a)

(b)

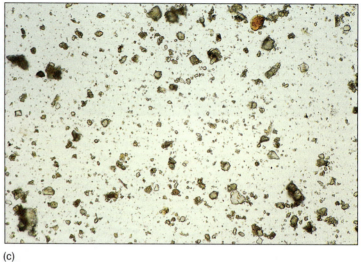

(c)

FIGURE 13.8

(a) Relative sizes of soil particles: From 0.2 to 2.0 mm is coarse sand, the next smaller is fine sand, and so on. It is easy to calculate the amount of surface area, the volume of rock, and the volume of air/water space in a soil composed of only one type of particle. Assume you have 1 cm³ of pure coarse sand; the number of particles present is calculated by dividing 1 cm³ by the volume of a cube measuring 2.0 mm on each side (the volume of the particle plus its surrounding air/water space). Next, multiply the number of particles by the volume of a sphere with a radius of 1.0 mm (each sphere has a diameter of 2.0 mm). This gives the total volume of rock in 1 cm³ of soil. Subtract this from 1 cm³ to obtain the volume of air or water that can be held by the soil. Finally, calculate the surface area of each sphere and multiply that by the number of particles to obtain the total surface area that is releasing minerals into the 1 cm³ of soil. Now do this for fine sand, silt, and clay. Which soil has the most surface area? The greatest amount of air and water? (b) Particles of fine sand, viewed by light microscopy (× 60). (c) Particles of silt and micelles of clay (× 60).

The various sizes produced by physical weathering are important factors in soil texture and porosity. In sands, particles are large and fit together poorly, so a great deal of space remains between particles (see Fig. 7.9). The spaces permit rapid gas diffusion, and roots in sandy soils typically are never starved for oxygen. During rain, the spaces fill with water, but typically they cannot hold it against gravity because the spaces are too broad to act as capillary tubes. Once rain stops, most of the water percolates downward and enters aquifers, flowing underground to wells, springs, and streams. Water that remains in the soil is held by capillary adhesion/cohesion (Fig. 13.9; see Fig. 12.20) and is said to be the **field capacity** of the soil. Much of this water is available to roots.

Chemical weathering involves chemical reactions, and the most important agents are acids produced by decaying bodies, especially those of plants and fungi. In addition, many organisms secrete acids while alive, and the carbon dioxide produced during respiration

(a)

(b)

FIGURE 13.9

(a) This sandy soil has such large spaces between its particles that it cannot hold water well; it dries quickly after a rain. It has a low field capacity. (b) As water is pulled away from soil by gravity, it percolates downslope into valleys. Soil at the bottom of ravines has water available longer than soil on the sides of a slope; consequently, more vegetation grows at the bottom of a valley than on the sides.

can combine with water, forming carbonic acid. When an acid dissolves in water, it dissociates into a proton and an anion, both of which interact chemically with rock's crystal matrix and with embedded contaminant elements. In regions with a great deal of warmth, moisture, and abundant decaying vegetation, such as tropical regions, chemical weathering can be extremely rapid, and thick soils accumulate in a short time. In drier regions with long, cold winters and less vegetation, such as temperate mountains and the prairies of the plains states and Canada, fewer acids are available and chemical weathering is slower. It may take as long as 10 million years for just 1 cm of soil to form. Chemical weathering is greatly increased if rock has already been reduced to sands and silts by physical weathering; these have a large surface area for the acids to attack.

Chemical weathering decreases soil particle size, but more importantly, it alters soil chemistry. As the crystal matrix dissolves, matrix elements become available to the plant and trapped elements are liberated. As the matrix breaks down, positively charged cations are freed; thus, the residual undissolved matrix has a negative charge (Fig. 13.10). With coarse particles such as sand and silt, the surface-to-volume ratio is so small that the charge is not important. But with clay micelles, the total amount of surface per unit of soil volume is great. Because the particles have a negative charge, cations such as K^+, Ca^{2+}, Mg^{2+}, and Cu^{2+} are held near the particles' surfaces. The bonding is much weaker than that of the crystal matrix, so roots can absorb the cations. This attraction to the micelle surface is beneficial; without it, many important cations would be washed deep into the soil by passing rainwater.

CATION EXCHANGE

Because cations are loosely bound to micelle surfaces owing to their charge, roots cannot absorb them directly. Instead, the cations must first be freely dissolved in the soil solution; this is done by **cation exchange.** Roots and root hairs respire, giving off carbon dioxide. As this dissolves in the soil solution, some reacts chemically with water, forming carbonic acid, H_2CO_3. This breaks down into a proton and a bicarbonate ion (Fig. 13.11), which can further dissociate into a second proton and carbonate ion. The presence of protons acidifies

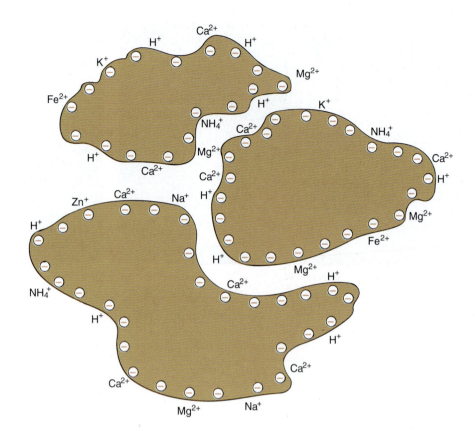

FIGURE 13.10

As rock weathers, the negatively charged components are most resistant, so the rock fragments (micelles) have a negative surface charge. Because of this, the cations that are released by weathering do not completely leave the rock, but are held by very weak electrical attraction.

the soil solution adjacent to roots and root hairs; as the protons diffuse, they bounce close to a bound cation at a micelle surface. The presence of the proton's positive charge disrupts the electrical attraction of the cation, liberating it and trapping the proton. Because the proton was derived from waste carbon dioxide, its loss does not hurt the plant. The liberated cation may diffuse in the direction of the root and be absorbed and transported upward by xylem, or it may diffuse away from the root or strike another micelle, liberating either a proton or another cation. Over time, however, large quantities of cations are absorbed. Acidity caused by secretion of acids by bacteria and fungi, by the decomposition of humus, and by acid rain also result in the liberation of cations.

$$CO_2 + H_2O \longrightarrow H_2CO_3 \longrightarrow H^+ + HCO_3^- \longrightarrow H^+ + H^+ + CO_3^{2-}$$

(a)

FIGURE 13.11

(a) The reaction of water and carbon dioxide results in carbonic acid (H_2CO_3), most of which dissociates into a proton and the bicarbonate anion. Some of the bicarbonate dissociates further, releasing another proton and the carbonate anion. (b) Protons from carbonic acid may diffuse close enough to a cation to disrupt its attraction to a soil micelle, liberating it (c). As the cation then diffuses through the soil, it may encounter a root (arrow b) or it may not (arrow c).

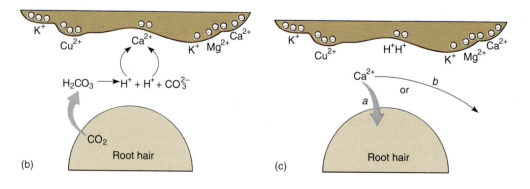

Soil particles are not the only structures that hold cations. Much of the decaying organic matter also forms negatively charged matrixes. The cellulose crystals of cell walls in mulch and humus are especially valuable not only for holding cations liberated from rock weathering but also for retaining the essential elements released by decaying protoplasm. Such organic matter also holds water and greatly improves the quality of any type of soil.

SOIL ACIDITY

Soil pH, the concentration of free protons in the soil solution, is important for cation exchange and the retention of cations in the soil during heavy rain. As acidity increases (pH becomes lower), the greater concentration of protons causes more cations to be released from soil micelles; these may be absorbed by roots or washed away in ground water. An extremely acid soil (pH of 4.0 to 5.0) tends to lose cations too rapidly and becomes a relatively poor soil. On the other hand, highly alkaline soils (pH of 9.0 to 10.0), which are frequent in dry climates, have too few protons to allow cation release, and concentrations of minerals can become excessively high.

Soil pH affects the chemical form of certain elements, causing them to change solubility. In acidic soils, aluminum and manganese can become so soluble as to reach toxic levels. In alkaline soils, iron and zinc become quite insoluble and unavailable to plants, but molybdenum is more soluble at a high pH. In general, a pH between 6.5 and 7.0 is best for many elements, especially iron, zinc, and phosphorus.

Many factors affect soil acidity, such as the chemical nature of the original rock, but probably the most important factor is rainfall. With high rainfall, there tends to be an abundance of vegetation that produces acids by means of respiration, excretion, and decay. With low rainfall, not only is there little vegetation, but there may not be any washing out of the soil. Cations build up, increasing soil alkalinity by increasing the concentration of hydroxyl ions. Just as dissociation of an acid produces a proton and an anion, dissociation of a base produces a hydroxyl and a cation:

$$NaOH \longrightarrow Na^+ + OH^-$$

Because some soils are more acidic than others, plants have adapted to differences in the availability of essential elements (Table 13.5). Natural selection favors mutations that allow desert plants to cope with alkaline soils, whereas plants of wet areas must become adapted to acid soils. For example, azaleas, camellias, gardenias, and rhododendrons absolutely must have acidic soil; if planted into alkaline soils, they show stress symptoms immediately, often suffering from lack of iron as well as general poor health (Fig. 13.12).

FIGURE 13.12
This azalea leaf is from a plant growing in alkaline soil. Azaleas require acid soils and suffer iron deficiency in alkaline soils.

TABLE 13.5	Species Adapted to Acidic, Neutral, and Alkaline Soils	
Acidic (pH 4.5–5.5)	**Neutral (pH 5.5–6.5)**	**Alkaline (pH 6.5–7.5)**
azalea	carrot	apple
blueberry	*Chrysanthemum*	*Asparagus*
Camellia	corn	beet
cranberry	cucumber	cabbage
fennel	pea	cauliflower
Gardenia	*Poinsettia*	lettuce
potato	radish	onion
Rhododendron	strawberry	soybean
sweet potato	tomato	spinach

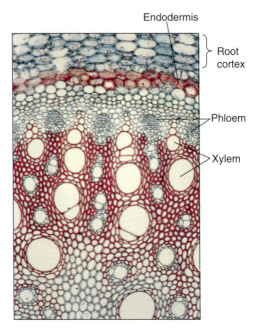

FIGURE 13.13
The endodermis in roots prevents uncontrolled diffusion of minerals into the xylem ($\times 50$).

Alfalfa, apples, broccoli, and hydrangeas require an alkaline soil, whereas most plants do best if the soil pH is near 6.0 to 6.5. Some plants are able to tolerate a wide range of soil acidity, but typically plants show best health and vigor only in soil with the optimal pH.

THE ENDODERMIS AND SELECTIVE ABSORPTION OF SUBSTANCES

Elements in the soil solution, whether essential or not, can enter roots either by crossing a plasma membrane and entering the symplastic protoplasm phase of the plant or by diffusing along cell walls and intercellular spaces in the apoplastic phase. In the first method, the selective permeability of the plasma membrane and the presence or absence of molecular pumps control the entry of ions and molecules; certain substances can be excluded and others can be actively transported in. However, a substance can penetrate the root epidermis and cortex simply by moving through the water in cell walls and intercellular spaces. No metabolic control exists, and even harmful substances can enter. If all substances could enter the xylem transpiration stream, they would have access to all cells of the plant body.

The endodermis prevents uncontrolled, apoplastic diffusion in roots. The Casparian strips on all radial walls are impermeable to water and water-borne solutes; nothing can cross it simply by diffusion (Fig. 13.13). For a substance to penetrate beyond the root cortex, it must first enter the endodermal cell protoplasm by being accepted across the plasma membrane of a cell in the root epidermis, cortex, or endodermis. Its highly selective permeability allows the endodermis to control which elements enter the transpiration stream.

MYCORRHIZAE AND THE ABSORPTION OF PHOSPHORUS

The roots of most plants form a symbiotic association with soil fungi, and this relationship is called a **mycorrhiza**; the symbiosis permits plants to absorb phosphorus efficiently (see Chapter 7). In the most common type, vesicular/arbuscular mycorrhizae, some of the fungal filaments penetrate root cortex cells, then branch profusely, forming a small tree-shaped arbuscule inside the cell; other filaments swell into balloon-like vesicles (see Fig. 7.22). The fungus collects phosphorus from the soil and transports it into arbuscules, where it accumulates as granules. After the arbuscules fill with phosphorus, the granules gradually disappear as phosphorus is transported into the root cell protoplasm. Once the transfer is complete, the arbuscule collapses and the root cell returns to normal. This mycorrhizal symbiosis is essential to most plants; plants in sterilized soil grow poorly and show signs of phosphorus deficiency even if the soil contains adequate amounts of phosphorus (Fig. 13.14). In soils with very high levels of available phosphorus, mycorrhizae may be less important.

FIGURE 13.14
In the pot on the left, soil fungi were killed, and the plant suffers from phosphorus deficiency. In the two pots on the right, mycorrhizal fungi are present. (*R. Roncadori/Visuals Unlimited*)

BOX 13.2 *Acid Rain*

Acid rain is a silent killer, destroying forests, streams, and lakes throughout the world. As we burn fuels rich in sulfur, such as much of the coal used to generate electricity, the sulfur burns to sulfur dioxide and is emitted through the smokestack of the generating plant. In the air, sulfur dioxide reacts with water to form sulfuric acid, which then dissolves into the water droplets of clouds. As the drops fall as rain, snow, or sleet, they carry the sulfuric acid with them as acid rain, also called acid precipitation.

Acid precipitation damages plants in many ways. Because the cuticle on the epidermis is not absolutely impermeable, some acid slowly moves directly into the plant tissues and damages leaves, flowers, fruits, and cones. Perhaps more significantly, most of the acid enters the soil and accelerates cation exchange, causing positively charged ions to be released from the soil particles and to be washed away in the rain. The soil is left depleted of nutrients, and plants suffer from mineral deficiency. Downwind of the most heavily polluting industrial centers of Germany, entire forests are dying or dead. Pollution from the United States and Canada is causing extensive damage to North American forests.

As acid rain accelerates cation exchange, minerals are washed from the soil and enter streams, in effect fertilizing them and causing rapid growth of algae. In small quantities, this provides more food for fish, turtles, and other aquatic animals, but in many cases the algal growth is so abundant it forms a massive, impenetrable layer across the top of a lake or slowly moving river. As the algae die, their bodies sink and are attacked by decomposers, mostly bacteria. Bacterial decomposition consumes oxygen, and before long the bottom of a lake or quiet river is an anaerobic dead zone with too little oxygen to support any animal life. This process is called eutrophication, and the result is a lake or river that is basically dead, having little life other than a mat of algae at its surface.

These conifers have been damaged by acid rain. Notice that the broadleaf trees in the foreground appear healthy. The acid rain does not always destroy a forest but instead may alter the species composition and diversity. *(Terraphotographics/BPS)*

NITROGEN METABOLISM

Nitrogen occurs neither as a component of rock matrixes nor as a contaminant in rock; the most abundant source of nitrogen in the environment is the atmospheric gas N_2. In this form, nitrogen is relatively inert chemically and is useless to almost all organisms. To be a useful substrate for most enzymes, it must first be converted to chemically active forms. This process, called **nitrogen metabolism,** consists of (1) nitrogen fixation, (2) nitrogen reduction, and (3) nitrogen assimilation.

NITROGEN FIXATION

Nitrogen fixation is the conversion of N_2 gas into nitrate, nitrite, or ammonium, all forms of nitrogen that are substrates for a variety of enzymes. One means of nitrogen fixation is human manufacturing; the fertilizer industry can synthesize either nitrate or ammonium from atmospheric nitrogen, but this is an extremely expensive and energy-intensive process. About 25 million tons of nitrogen fertilizer are produced annually.

TABLE 13.6	*Oxidation States of Nitrogen Compounds*	
Nitrate	NO_3^-	$N = +5$
Nitrite	NO_2^-	$N = +3$
Ammonia	NH_3	$N = -3$
Ammonium	NH_4^+	$N = -3$
Amino group in amino acid	$-NH_2$	$N = -3$

Like carbon, nitrogen can hold various numbers of electrons in more or less stable bonding orbitals. The partial charge on nitrogen can be calculated by adding the partial charges on protons ($+1$) and oxygens (-2). Reducing nitrate to nitrite requires two electrons; reducing nitrite to ammonia requires six.

Natural processes fix over 150 million tons of nitrogen annually. Lightning is important; the energy of a lightning strike passing through air converts elemental nitrogen to a useful form that dissolves in atmospheric water and falls to the earth in raindrops. However, nitrogen-fixing bacteria and cyanobacteria are by far the most important means of fixing and utilizing atmospheric nitrogen, annually converting 130 million tons of nitrogen to forms that plants and animals can use. These organisms have **nitrogenase**, an enzyme that can use N_2 as a substrate. In a complex reaction, it forces electrons and protons onto nitrogen, reducing it from the $+0$ oxidation state to the -3 oxidation state (Table 13.6). Ammonia, NH_3, is the product; it immediately dissolves in the cell's water and picks up a proton, becoming ammonium, NH_4^+. Nitrogenase is a giant enzyme or enzyme complex; it has a molecular weight of 100,000 to 300,000 daltons and contains 15 to 20 iron atoms and 1 or 2 molybdenum atoms. It is extremely sensitive to oxygen and functions only if oxygen is completely excluded from it.

Some nitrogen-fixing microorganisms are free-living in the soil; examples are *Azotobacter, Clostridium, Klebsiella,* and some cyanobacteria like *Nostoc* (Fig. 13.15; see Chapter 19). The nitrogen that they fix is used in their own metabolism and becomes available to plants and fungi only when they die and their bodies decay. Other nitrogen-fixing organisms live symbiotically, growing inside the tissue of host ferns and seed plants (Table 13.7). The best-known examples are the root nodules on legumes such as alfalfa; the nodules are growths of root tissue whose cells contain bacteria of the genus *Rhizobium* (see

Note that ammonia is the non-ionized form and ammonium is the dissolved, ionized form; in both, nitrogen is in the -3 oxidation state.

FIGURE 13.15
Cyanobacteria are common components of most soils, although usually they are quite inconspicuous. This species of *Nostoc* forms large masses (a) easily visible to the naked eye and covering large areas of the soil. They swell and fix nitrogen rapidly when wet (a), but they become dormant and crisp when dry (b). Even though extremely desiccated with an extraordinarily negative water potential, the cells are alive and revive within seconds of receiving water.

(a)

Nostoc

(b)

Nostoc in dry condition

TABLE 13.7	*Plants that Form Associations with Nitrogen-fixing Prokaryotes*

Prokaryote	Plant
Cyanobacteria	
Anabaena	Ferns: *Azolla*
Nostoc	Cycads: all genera examined
Actinomycetes	
Frankia	Angiosperms
	Betulaceae (birch and alder family)
	Casuarinaceae (beefwood family)
	Eleagnaceae (oleaster, Russian olive family)
	Myricaceae (bayberry family)
	Rhamnaceae (buckthorn family)
	Rosaceae (rose family)
Eubacteria	
Rhizobium	Angiosperms
	Fabaceae (legume family)
	Ulmaceae (elm, hackberry family)

Chapter 7 and Fig. 7.24). Plants such as alders (*Alnus*), bog myrtle (*Myrica gale*), and *Casuarina equisetifolia* are pioneer plants that are the first to grow in poor, nitrogen-deficient soils such as bogs, sand dunes, and glacial rubble. These species obtain their nitrogen from a symbiosis with the prokaryote *Frankia*. The symbiotic bacterial cells use part of the fixed nitrogen for their own growth and reproduction, but they also permit large amounts to leak out into the protoplasm of the surrounding root cells. Symbiotic nitrogen-fixers usually produce fixed nitrogen at a much greater rate than free-living microorganisms, perhaps because they have the energy resources of the plant at their disposal.

The rate at which plant/prokaryote symbioses fix nitrogen is strongly influenced by the stage of development of the plant: When soybeans begin to produce their protein-rich seeds (40% protein, the richest seeds known), nitrogen fixation in the roots increases greatly. As much as 90% of all nitrogen fixation occurs during the phase of seed development, whereas only 10% occurs during all the vegetative growth that precedes it. Furthermore, if legume crop plants are given high levels of nitrogen fertilizer, plants that have not yet formed nodules do not produce them, and those that already have bacteroid-filled nodules decrease the amount of nitrogen fixed and even allow the nodule to senesce. With adequate nitrogen available in the environment, it is selectively advantageous not to pass glucose on to bacteria.

NITROGEN REDUCTION

Nitrogen reduction is the process of reducing nitrogen in the nitrate ion, NO_3^-, from an oxidation state of $+5$ to the -3 oxidation state of ammonium, which is also the oxidation state of nitrogen in amino acids, nucleic acids, and many other biological compounds (see Table 13.6). Nitrogenase automatically reduces nitrogen during the fixation process, and if a plant can absorb that form of nitrogen, no further reduction is necessary. Also, as organic matter decays in the soil, ammonium is released and becomes available. Unfortunately for plants, ammonium is an extremely energetic compound that numerous species of soil bacteria can use as "food," oxidizing it to produce ATP (see Chapter 19). In the process, ammonium is converted to nitrate. Such soil bacteria are so common that ammonium lasts only a short time in the soil, and the predominant form of nitrogen available to roots is the oxidized form, nitrate.

PLANTS & PEOPLE

FROM FERTILITY GODS TO FERTILIZERS

The field of mineral nutrition is excellent for studying the development of biological thought. It is easy to assume that people have always understood the importance of nitrogen, potassium, phosphorus, and other elements, even if they did not study it explicitly. After all, humans have been farming for thousands of years. But the way we think of things now is different from other approaches to understanding the world. The first agricultural societies, those of Sumer, Egypt, India, and China, thought in terms of fertility gods; plants were believed to grow well or poorly according to the whims of divine intervention.

The earliest attempts to understand plant growth in a nonmystical way were formulated by ancient Greeks. Their thoughts are usually summarized by the phrase "Plants eat dirt." This is not as simplistic as it sounds. The Greeks believed that the universe contained only four "elements": earth, water, fire, and air. All things, plants and animals included, were constituted of various combinations of those elements. In such a world, the idea of plants as transmuted earth makes sense. Plants do not grow unless their roots are in earth, they must have water, and of course they can burn, so the element fire must be contained as well.

This concept remained basically unchanged until the 1600s. By then it was known that there are many elements, although the concept of "element" was not perfectly clear. The periodic chart had not been developed yet by Mendeleev, and the belief that lead could be changed into gold was still held. In 1644 J. B. van Helmont published the results of an important experiment. He had made a cutting of a willow and allowed it to form roots. When it weighed 5 pounds, he planted it in a container holding exactly 200 pounds of soil. He watered the willow and allowed it to grow for 5 years, then removed it and cleaned off the roots, being careful not to lose any soil. The plant had increased in weight to 169 pounds, but the soil had lost almost nothing—it still weighed 199 pounds, 14 ounces. Clearly the plant was not composed primarily of transformed earth; van Helmont concluded that water was the important transformed element (carbon dioxide was not yet known to exist).

This was a tremendous advance for two reasons. If water could be transmuted to vegetable matter, then it must be a compound and not an element. The second reason lay in the concept of experimentation. Greek science had been based on observations combined with thought, reasoning, and logic, but without experimentation. Van Helmont's results showed not only that the Greek conclusion was incorrect, but that their method of study without experimentation was inadequate.

Van Helmont's work was followed by that of John Woodward, who tested various types of water without soil: rainwater, stream water, and water from mud puddles. He found that rainwater was least effective in promoting plant growth, even though it was known to be the purest type of water. Chemical methods still were not advanced enough for him to be able to analyze why water from puddles was best. We now know that it is richest in the minerals necessary for plant growth, whereas rainwater is almost completely lacking in them.

A problem that impeded study of plants and animals was the belief that living creatures contain a vital force, something that was not chemical or physical and was assumed to be beyond study. It was believed that when an organism died and decayed, its vital force passed into the soil and made it fertile. Try to think like someone in 1700, only 50 years after the Pilgrims landed at Plymouth Rock. You would know that soil could be fertilized with bonemeal, manure, fish scraps, or compost (Table 13.8). All these things had been living and could presumably add vital force to the soil.

At this time two scientists, Nehemiah Grew and Marcello Malpighi, were making the first studies of the microscopic structure of plants and animals. Unfortunately, some of the early observations were misinterpreted and reinforced the concept of the existence of vital force. For example, the presence of microscopic holes in various types of cells caused people to think in terms of filtration: It was postulated that roots contained fine pores that allowed water and vital humours (liquids) to pass into the plant while non-nutritive soil sap was excluded. Once partially purified by root filtration,

Reducing nitrate back to ammonium requires eight electrons for each nitrogen atom and a great deal of energy. In the first step, two electrons are added, reducing nitrogen from +5 to +3 and forming **nitrite**, NO_2^-. The enzyme is nitrate reductase, and it carries electrons by means of a molybdenum atom (Fig. 13.16). Just like electron carriers in photosynthesis or respiration, when nitrate reductase reduces nitrate, it becomes oxidized and must pick up more electrons. It gets these from $FADH_2$ and NADH; the ultimate source of energy and electrons is respiration.

In the second step, nitrite reductase adds six electrons to nitrite, reducing it to ammonium. The process is not well understood, but it is known that extremely strong reducing agents are needed. Apparently, nitrite reduction in leaves is powered by reduced ferredoxin from the light reactions of chloroplasts, but roots are a more important site of nitrite reduction, and of course they have no light reactions. Instead they use NADPH, which is

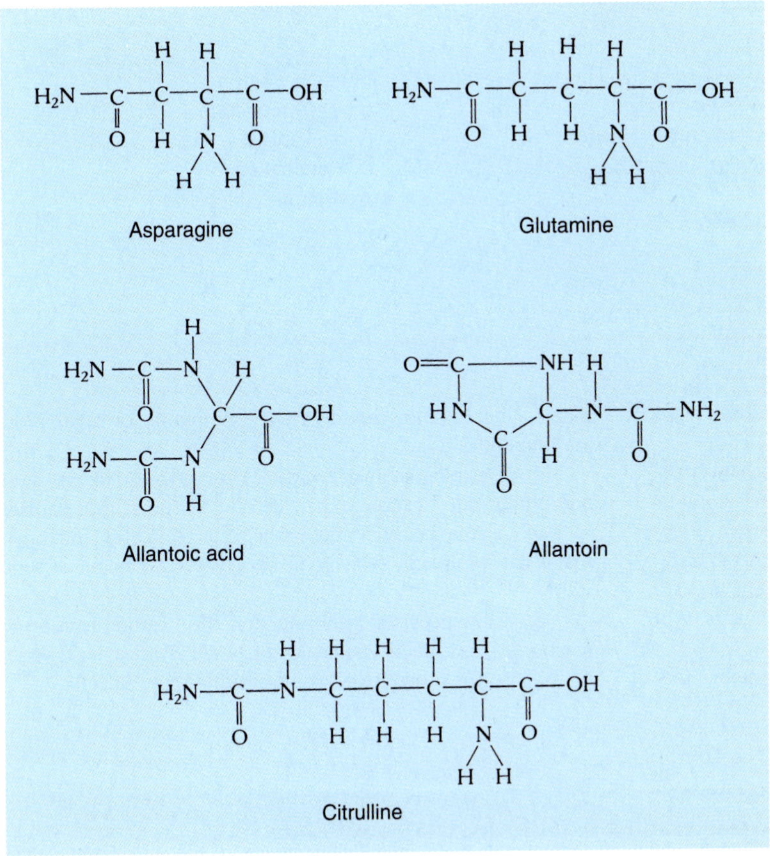

FIGURE 13.18
These chemicals each have a high nitrogen-to-carbon ratio and are transport forms of reduced nitrogen. Once moved through the phloem to a nitrogen sink, the compounds are catabolized and the amino group is used in the synthesis of amino acids, nucleic acids, and other compounds.

SUMMARY

1. Plants synthesize all of their body's materials using only carbohydrate and a few essential elements.

2. Three criteria must be met for an element to be considered essential: (1) it is necessary for the completion of a full life cycle; (2) it cannot be replaced by a chemically similar element; (3) it must act inside the plant.

3. Macro or major essential elements are needed in relatively large amounts, whereas micro or minor or trace elements are required in only extremely small amounts.

4. Deficiency diseases are rare in nature, probably because susceptible plants are outcompeted by deficiency-tolerant plants. However, deficiency symptoms are frequent in crops and horticultural plants if proper fertilizing is not maintained.

5. A deficiency of an immobile element produces symptoms in young leaves and buds, whereas a deficiency of a mobile element results in symptoms in older organs.

6. Cations are released from the surface of soil particles if protons, which result from the respiration of roots and soil organisms, disrupt their electrical attraction.

7. Mycorrhizae are the principal means by which most plants absorb phosphorus.

8. Nitrogen metabolism consists of fixation, reduction, and assimilation.

9. Nitrogen must be obtained from the air, but only certain prokaryotes—some free-living, others symbiotic—have the necessary enzyme, nitrogenase.

10. Most plants absorb nitrate, which must be reduced by the plants, using large quantities of NADH, NADPH, and ferredoxin.

11. Once formed within plant cells, ammonium is assimilated by transamination, being passed from glutamate to various amino acids.

IMPORTANT TERMS

amino group
cation exchange
chlorosis
essential element
hydroponic solution
immobile elements

major (macro) essential elements
micelle
minor (micro) essential elements
mobile elements
necrosis
nitrogen assimilation

nitrogen fixation
nitrogen reduction
trace element
transamination
weathering

REVIEW QUESTIONS

1. What are the three criteria an element must meet to be considered essential?

2. Name the essential element involved in each process: changes the osmotic potential in guard cells as they swell or shrink; present in all nucleotides and amino acids; carries electrons in chlorophyll; present in cytochromes; present in plastocyanins; involved in nitrogen metabolism; present in the water-splitting enzyme of photosystem II; present in ATP; present in the middle lamella and important in regulating the activity of many enzymes.

3. What is the difference between a mobile and an immobile element? Where are the first deficiency symptoms for each? Which elements are immobile?

4. Describe cation exchange. How would decomposition of mulch and humus affect cation exchange? Why is acid rain so damaging?

5. Name the three steps in the conversion of N_2 into organic nitrogen that is part of a plant. Which type of organism is capable of performing each step?

6. How does nitrogen reduction affect carbon fixation in the stroma reactions? How many molecules of NADPH must be diverted from one pathway to the other? How would respiration be affected?

BotanyLinks .net Questions
www.jbpub.com/botanylinks

Visit the .net Questions area of BotanyLinks (http://www.jbpub.com/botanylinks/netquestions) to complete this question:

1. How important is acid rain? Is it an extensive problem? Go to the BotanyLinks home page to begin your search for more information on this subject.

BotanyLinks includes a Directory of Organizations for this chapter.

DEVELOPMENT AND MORPHOGENESIS

14

Plants respond to environmental factors. A change of season may cause flowering. This flower is oriented with respect to gravity, as are the insects that will visit it.

CONCEPTS

If all plants were extremely small and lived in completely uniform, nonvarying environments, most would probably be simple and would experience little selective pressure for the evolution of complex shapes, tissues, organs, and metabolism. The most uniform, constant conditions occur in oceans and large lakes, where water buffers rapid changes in temperature, acidity, oxygen concentration, and other factors. Under such stable conditions, small organisms such as algae, protozoans, and sponges are simple. But most organisms exist in a heterogeneous environment: Gravity comes from only one direction; the sun is either to the side or overhead but never below; temperatures are lower on the shady side of a plant; moisture depends on depth below or height above the soil surface. This mosaic of conditions is not stable; it changes over minutes, days, seasons, or longer periods of time. It is selectively advantageous for organisms—plants included—to be able to sense these differences and changes and respond to them.

371

Most plants are so large that their bodies exist in several different microenvironments. Consider a small tree: Its roots are in soil, which is usually moister, cooler, and darker than air; the highest branches are in open air, exposed to full sunlight and the full force of wind, storms, rain, and snow (Fig. 14.1). The trunk base and lower branches are in an intermediate environment—less stable than soil and less variable and severe than open air. A vertical tree trunk is oriented to best resist gravitational attraction (its own weight), whereas horizontal branches are highly stressed unilaterally by gravity.

In springtime, the shoot can become warm enough for active metabolism even though the soil remains cold or frozen. The plant parts must communicate with each other, or shoot buds would become active and expand before roots were capable of transporting water to new leaves. In autumn, decreasing day length and declining air temperatures signal impending winter and the need for dormancy; roots are informed about changing seasons by chemical signals from the shoot. Otherwise, the plant would waste energy constructing a root system larger than necessary.

The need for intercommunication and coordination also exists within a limited region of the body. For example, it is reasonable to hypothesize that leaf parts act in a coordinated fashion during development such that the petiole has enough xylem and phloem to facilitate the transport needs of the blade. Too little conductive tissue causes the blade to suffer water stress or an inability to export sugars to the rest of the plant; too much conductive tissue is a waste of energy and material. Even on the intracellular level, organelles must communicate with each other because their metabolisms are interrelated.

All levels of communication have in common a basic mechanism. Information about the environment or the metabolic status of the organ must be **perceived**. The plant must

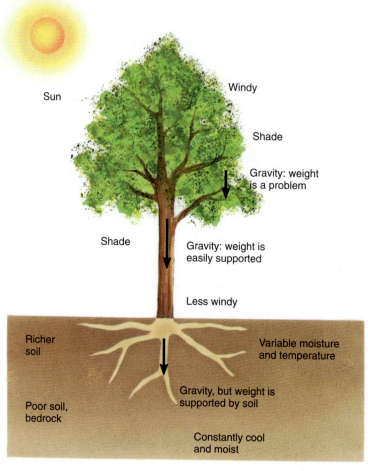

FIGURE 14.1

All plants, especially large ones, live in a complex mosaic of microenvironments that differ from each other, often dramatically. Each factor also changes throughout a day or a season.

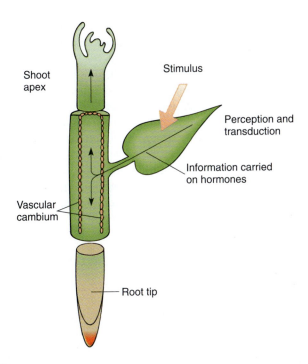

FIGURE 14.2

The signal for dormancy and preparation for winter is short days in autumn, perceived by leaves. The chemical messenger that is transported from the leaves causes the shoot tip to produce bud scales instead of leaves; the vascular cambium fills with many small vacuoles and stops mitosis and cytokinesis; young xylem mother cells differentiate to a predetermined stopping point, then become dormant; roots slow their growth greatly but usually do not stop completely. Roots and the cambial region cannot perceive the approach of winter themselves but depend on leaves as sites of perception. Although the shoot apices could perceive it themselves, the entire plant is integrated as a whole by being cued by the leaves.

sense environmental cues such as changes in temperature, moisture, or day length, or the nucleus must receive chemical signals if conditions in the surrounding cytoplasm change. Next, information must be **transduced**, or changed to a form that can be either acted upon or transported. Finally, there must be a **response:** The plant must enter dormancy, produce flowers, change the type of leaf production, and so on (Fig. 14.2). If any of these steps is missing, the plant cannot respond to the environment. These principles apply to all organisms, and of course higher animals have the most elaborate mechanisms. They have sophisticated sense organs for sight, taste, touch, hearing, and smell that monitor external conditions. Less familiar sensing mechanisms monitor internal, metabolic conditions such as heartbeat, tissue oxygenation, blood pH, and insulin levels. These sensory mechanisms transduce the perceived information to a transportable form such as nerve impulses or hormones that are secreted into the blood stream, making rapid, long-distance integration of the animal possible. Animal responses are also typically highly elaborate, involving precisely controlled movement of muscles and increased activity of organs such as adrenal glands and pancreas and the immune system.

Sensory systems and response mechanisms of plants tend to be simpler than those of animals, and the signal transport is usually slow, involving movement through cortex parenchyma or phloem. This does not mean that animal systems are superior to those of plants. "Superior" has no meaning: The proper question is, "Which is more adaptive, which is more advantageous selectively?" In terms of evolution and natural selection, animal systems are successful adaptations for organisms that must detect food or predators and must move in order to capture that food or avoid being captured. Any animal with a plant-type mechanism of perception, transduction, and response would soon starve or be eaten. Perception of and response to mates for sexual reproduction also require sensitive, rapid response mechanisms.

On the other hand, rapid response mechanisms would not be adaptive in plants: They are extremely expensive to build and maintain and are unnecessary because most conditions important to plants change only slowly. Eyes, nerves, muscles, or adrenal glands are not needed by plants for the absorption of carbon dioxide, water, minerals, or light or for perception of autumn and preparation for winter dormancy. For sexual reproduction, animal-pollinated plants do need sophisticated perception and response mechanisms, but basically, they simply "rent" those of their pollinators, paying with nectar or other rewards. Finally, the very sophistication of animal sensory/response systems makes animals more vulnerable, more easily injured, and more dependent on avoiding dangers such as fires, floods, freezes, and predators. Plants are typically much more resilient than animals, being able to survive burning because of thick bark or resprouting from rhizomes and bulbs; many withstand flood by being tough or flexible. Predators can consume most of a plant's leaves, wood, or roots without actually killing the plant.

SENSING ENVIRONMENTAL STIMULI

Numerous factors in the environment affect plant growth, development, and morphogenesis, but probably only a few are sensed by plants and initiate a response. It is important to think about this carefully: Most factors in the environment that impinge directly on a plant affect its growth but do not initiate a specific adaptive response. For most perennial plants, cool temperatures *in autumn* initiate responses that prepare the plant for cold, stressful winter conditions. This response is adaptive, increasing the plant's chances of survival. But cool temperatures *in midsummer* do not cause this response; instead, the plant merely grows more slowly, as would be expected for any enzyme-mediated process. This response can be predicted easily from ordinary chemical/physical considerations and probably is not adaptive; no new or specialized response is initiated. Mineral deficiency may result in severe stress symptoms such as chlorosis or necrosis, but these are unavoidable disease symptoms, not adaptive responses—they do not increase the plant's ability to find the deficient element or conserve what it already has. In contrast, when the root of a legume perceives the presence of nitrogen-fixing *Rhizobium* bacteria, the root begins an entirely new type of development, initiating root nodules. This is an adaptive response, many aspects of which cannot be accounted for simply; it probably involves activation of certain specific genes.

Among the environmental factors that are perceived and transduced and initiate adaptive responses are the following.

LIGHT

Besides energy for photosynthesis, light also provides two important types of information about the environment: (1) the **direction** or, more precisely, the gradient of light. It is advantageous for a plant to be able to grow or to orient its leaves toward a region of bright light in order to gather the most light available for photosynthesis. (2) The **duration** of light or, more specifically, the length of the day, provides information about time of year. Air temperature is unsuitable because cool autumn temperatures may be followed so quickly by severe cold that plants do not have enough time to become dormant. But day length is an infallible indicator of season.

GRAVITY

It is selectively advantageous for many plants to orient themselves or their parts with respect to the direction of gravity. In some cases, gravity itself is important as a cause of weight stress. Whenever a plant is bent or tilted because of flooding or the slipping of a hillside, the plant must change its growth back to an upright direction. In some situations, the direction of gravity is a guide to other important factors. Roots that grow downward are more likely to encounter water and minerals. Shoots that grow upward grow above other plants and encounter better conditions for photosynthesis, pollination, and seed distribution. Normally shoots do this by growing toward the brightest light, the open sky, but

FIGURE 14.3
Weight is a source of information to the plant about the amount of collenchyma or sclerenchyma needed to counteract the gravitational attraction on the plant. Initially, the flower stalk was strong enough only to support the weight of a flower; by sensing and responding to gravity, it is now strong enough to support the weight of an apple.

shoots of seeds that germinate deep in the soil must determine which way is up while in the dark. Direction of gravity is their only reliable guide. Most bilaterally symmetrical flowers must be aligned with the body symmetry and flight pattern of their pollinator. Such flowers must be oriented vertically because bees and moths do not fly upside down or sideways. The flower must orient itself along the same environmental gradient that the pollinator uses—the gravitational gradient of up versus down.

Although gravity does not change with time, the amount of force it exerts on a particular organ does change as the weight supported by the organ changes. The pedicel of an apple flower supports almost no weight, but the same pedicel must later support the weight of a fully grown fruit (Fig. 14.3). The extra fibers are not produced until needed. Similarly, a young branch must be strong enough to support a small amount of weight, whereas a larger branch must support more.

TOUCH

Although plants do not move around like animals, their parts frequently grow against objects and respond to this contact. Certain types of contact are detrimental, for example, when a root grows against a stone or a branch rubs against another branch. In these cases, a thick bark is adaptive as a protective layer. Other types of contact are beneficial: After a tendril touches an object, it grows around the object and uses it as a support. When a fly touches sensitive trigger hairs on a Venus' flytrap, the trap closes, catching and holding the insect for digestion. In some cases, the contact is between two growing primordia and is a normal developmental feature. Many flowers that have fused petals or carpels start with separate primordia that grow together and fuse, acting as a single unit during development (Fig. 14.4). In each case, the physical action of touching is similar, but each organ responds in a distinct way that is adaptive for the plant, and the response of each would be inappropriate if it occurred in the others.

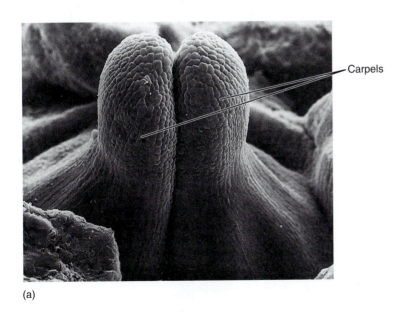

(a)

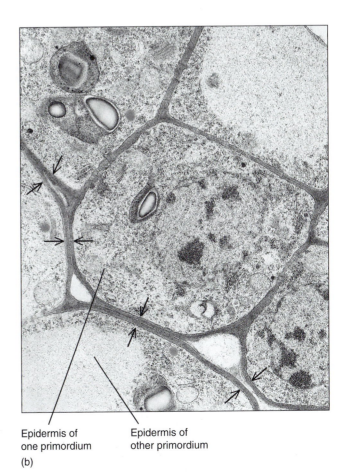

Epidermis of one primordium

Epidermis of other primordium

(b)

FIGURE 14.4

(a) The carpels of *Catharanthus roseus* are initiated separately and consist of protoderm and ground meristem. But they soon crowd into each other and fuse into one syncarpelous gynoecium ($\times 70$). (b) At the points of contact, the protoderm changes into mesophyll rather than epidermis ($\times 6000$).

(Courtesy of Judith A. Verbeke, University of Arizona)

Carpels

TEMPERATURE

Temperature fluctuates in a predictable pattern on both a daily and a yearly basis. Changing temperatures can signal many specific types of plant development. Cold temperatures are required for the normal flowering of biennial and many perennial plants. This effect, called **vernalization,** causes the plant to switch to a state in which it is capable of sensing and responding to a subsequent stimulus that induces flower formation (Fig. 14.5). Other species, such as apples, require very cold temperatures to break the dormancy of flower buds formed in the previous summer; if grown in areas with warm winters, they do not bloom. Although most plants appear to be quiescent and virtually lifeless in winter, a considerable amount of critically important metabolism is occurring. This metabolism usually does not proceed at temperatures above 1 to 7°C.

Low temperatures are required to induce the deep dormancy of temperate trees and shrubs. Short days induce plants to prepare for winter, but the most resistant stages of dormancy are not entered until the plant actually experiences cool temperatures.

WATER

Although water is an absolute prerequisite for all forms of life, its presence probably does not act like a signal in the way that other factors do. If enough water is available, plants grow; if not, plants wilt and perhaps even die. Although roots often appear to grow toward water, they actually grow in all directions, and those which, by chance, grow toward water grow more rapidly because they are in a favorable environment. Roots that grow away from water grow slowly, but only because they enter an environment too dry to permit growth. Roots do not turn and grow toward water in the way that they turn and grow toward gravity.

The beginning of water scarcity triggers specific adaptive responses. One of the first is the production of the hormone abscisic acid, which causes guard cells to lose potassium and close stomatal pores. This mechanism is adaptive and occurs in most plants even while cells have enough water to carry on basic metabolism. If water stress continues or becomes more severe, new responses are triggered that may inhibit the production of new leaves, increase the cuticle on existing leaves, or even initiate the abscission of leaves.

If a biennial plant is never vernalized but is always maintained in warm conditions, it never flowers, and plants such as carrots, beets, and lettuce become old and shrubby.

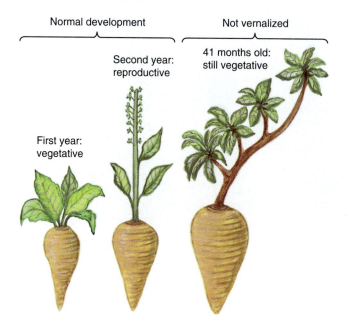

FIGURE 14.5

If a beet is protected from cold temperatures, it remains in a juvenile condition and is never able to flower. Cold is the stimulus that induces biennial plants to become competent to respond to a later floral stimulus. If such a plant is exposed to flower-inducing conditions, but without vernalization, it never flowers. We do not know whether the incompetence is an inability to perceive, transduce, or respond.

TABLE 14.1	*Types of Plant Responses to Stimuli*	
Response	**Mechanism**	**Oriented**
Tropic	Growth	Yes
Nastic	Turgor changes	No
Morphogenic	Changes in basic metabolism	No
Taxic	Swimming	Yes

RESPONDING TO ENVIRONMENTAL STIMULI

Plant responses to the diverse types of information present in the environment can be grouped into four simple classes (Table 14.1):

TROPIC RESPONSES

A **tropic response**, also called a tropistic response, is a growth response oriented with regard to the stimulus. For example, when a plant grows toward a bright light, it is showing a phototropic response because the growth response is not random but oriented toward the light (Fig. 14.6; Table 14.2). A **positive** tropism is growth toward the stimulus, a **negative** one is growth away from the stimulus, and a **diatropic** tropism is growth at an angle. Most tap roots are positively gravitropic, growing downward in response to gravity, whereas shoots are negatively gravitropic, growing upward in response to gravity. Branches and secondary roots grow horizontally or at an angle, diagravitropically. Experiments in which plants are temporarily grown in darkness have shown that the upward growth of shoots is negative gravitropism, not positive phototropism.

When touch is the stimulus, the response is thigmotropism. Positive thigmotropism occurs when a tendril touches an object and, by growing toward it, wraps around it. We are not yet certain if touch is the true stimulus in all cases, because an object that touches a plant may also shade it, so growth toward touch may sometimes actually be a response to a lack of light. Contact also blocks air movement, so the contact side may be more humid because the transpired water vapor is trapped; thus the true stimulus may be increased humidity. At least in pea plants, touch is known to be a stimulus: After a brief rubbing, tendrils coil toward the stimulated side, even if a physical object is no longer present to provide shade or trap humidity.

Pollen tubes of flowering plants display positive chemotropism, growing along the style to the ovary by following a gradient of chemical released from the ovule, probably from the synergids. Although the chemical responsible is not known for certain, the pollen tube is apparently sensitive to extremely slight variations in its concentration.

Thigmo *is derived from Greek: "to touch."*

FIGURE 14.6
Construction of porch steps trapped these plants, but by growing toward light (positive phototropism), their shoot tips found spaces between boards and emerged from a suboptimal dark environment into a more suitable sunny environment.

TABLE 14.2	*Prefixes for Stimuli*
Stimulus	**Prefix**
light	photo-
gravity	gravi- (formerly geo-)
touch	thigmo-
chemical	chemo-

NASTIC RESPONSES

A **nastic response** is a stereotyped nongrowth response that is not oriented with regard to the stimulus. For example, the trap-leaf of a Venus' flytrap has six large, sensitive trichomes. If a fly or other insect touches any two of these, the trap closes. It does not matter if the fly was moving north or south, up or down; the trap always closes in the same manner in this thigmonastic response. Furthermore, the trap does not grow shut; it closes as motor cells on the midrib upper side suddenly lose turgor.

Many pollinators are active only at night or during the day, and the flowers they pollinate are open only at the appropriate time. Sepals and petals spread open when the sun rises in **diurnal** species (active during daylight) and as it sets in **nocturnal** ones (active at night). Although presence of the pollinator is the critical factor for pollination, the cue that stimulates flower opening is the presence or absence of light, not the presence or absence of pollinators. The opening and closing always happen in the same manner, even if light is given artificially from the west, north, or south. This response is photonastic.

Because the response is not oriented with regard to the stimulus, positive and negative nastic responses are not possible.

MORPHOGENIC RESPONSES

A **morphogenic response**, sometimes called a **morphogenetic response**, causes a change in the "quality" of the plant; that is, a fundamental change occurs in the metabolism of a tissue or even the whole plant. Photomorphogenic responses are numerous: the induction to form flowers (which later open photonastically; Fig. 14.7), the induction of dormant seeds to germinate (Fig. 14.8), the induction of buds to become dormant. Many more occur because day length is such an excellent indicator of season. An example of a gravimorphogenic response is the formation of fibrous wood when a stem or branch is tilted and becomes stressed by gravity (Fig. 14.9). Thigmomorphogenic responses include formation of extra bark where branches rub against an object, and formation of a suture when petal or carpel primordia grow against each other (see Fig. 14.4).

FIGURE 14.7
Conversion from the vegetative to the floral condition is a common photomorphogenic response. Day length is controlled in these commercial greenhouses to ensure that all the poinsettias bloom simultaneously at Christmas. The growers could just as easily make them bloom on the 4th of July by controlling day and night length. *(Matt Meadows/Peter Arnold)*

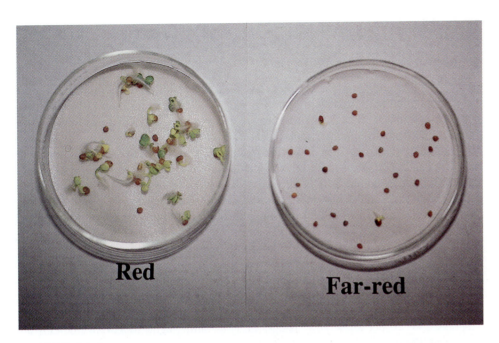

FIGURE 14.8

In these seeds and many others, a two-part mechanism ends dormancy: Cold winter temperatures and rain destroy or wash out an inhibitory chemical, and then light triggers germination. The light-dependent mechanism ensures that the seeds do not germinate while deeply buried under leaf litter and soil. Red light induces germination, but far-red light (infrared) blocks germination.

TAXIS

Taxis is a response in which a cell swims toward (positive taxis) or away from (negative taxis) a stimulus. Even in plants like mosses, ferns, cycads, and maidenhair tree *(Ginkgo)*, sperm cells swim to the egg cell by following a chemical gradient (chemotaxis). In algae, chemotaxis is similarly important for reproduction, and in many species, phototaxis allows them to swim toward light for photosynthesis or away from light that is too intense.

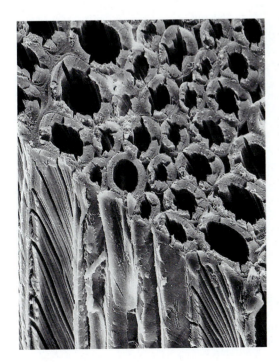

FIGURE 14.9

"Reaction wood" in a branch of Douglas fir *(Pseudotsuga)*. In a vertical trunk, the orientation of normal wood cells is ideal for resisting gravitational attraction (weight), but normal wood cannot hold up large horizontal branches. A gravimorphogenic response is the formation of reaction wood. In dicot trees, reaction wood is known as tension wood and is rich in gelatinous fibers whose walls are almost exclusively cellulose with little lignin or hemicellulose ($\times 640$). *(Courtesy of W. A. Côté, N. C. Brown Center for Ultrastructure Studies, SUNY College of Environmental Science and Forestry, Syracuse, NY)*

COMMUNICATION WITHIN THE PLANT

PERCEPTION AND TRANSDUCTION

Many, possibly most, responses occur in tissues or organs different from those that sense the stimuli. The site of perception is not the site of response, so a form of communication must exist. In plants, most sites of perception and response are not specialized for those functions but seem to be rather ordinary cells. Day length is probably perceived by all living leaf cells; no specialized region of cells has been discovered. Low temperatures for vernalization appear to be detected by buds, which do not contain a particular group of cells specialized just for temperature perception. In root caps, certain cells called **statocytes** do have large starch granules, statoliths, that sink in response to gravity; they are too dense to float in cytoplasm and always settle to the bottom of the cell, thereby distinguishing "down" from "up" (Fig. 14.10). This is our best example of a set of specialized perceptive cells. The trigger hairs on Venus' flytrap leaves are also a discrete perceptive mechanism, but it is not known which cells within the hairs are responsible.

The site of perception is tentatively assumed to be the site of transduction, where the stimulus is converted into a form that can be transmitted and can trigger a reaction at a response site (see Fig. 14.2). Transduction is still a complete mystery in almost all plant responses; we do not know how changes in temperature, light, weight, or humidity are converted into chemical signals.

Two factors are important in perception and transduction: presentation time and threshold. **Presentation time** is the length of time the stimulus must be present for the perceptive cells to react and complete transduction. Presentation time for root gravitropism is easy to understand: A root must lie on its side long enough for statoliths to sink to the new bottom of the cell. If the root is returned to vertical before they can settle, no perception occurs. In many tropic responses, only a brief touch or unilateral lighting is sufficient to cause curvature; presentation times are often only a few seconds.

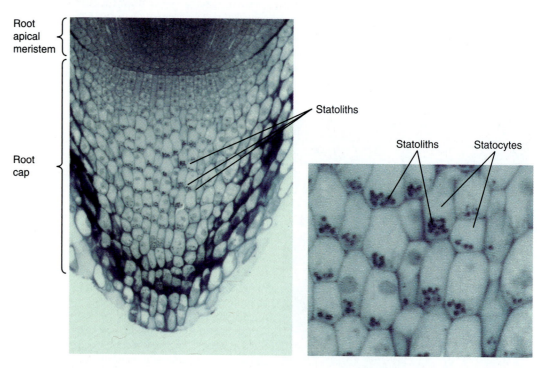

FIGURE 14.10

Cells located centrally in this root cap are statocytes, and their starch grains are statoliths. Regardless of vertical or horizontal position, the statoliths are located at the gravitationally lower side. It is necessary to distinguish between the gravitational and the morphological bottom in gravity-sensing systems. *(Courtesy of R. Moore and E. McClellan, Baylor University)*

Once the stimulus has acted long enough to fulfill the presentation time, a response occurs even if the stimulus is removed. For example, the vernalization of many biennial plants has a presentation time of only one or a few days; after this, the plants still flower at the proper time even if kept in warm, nonvernalizing conditions. Tendrils of peas do not bend thigmotropically in the dark, but if they are rubbed for several seconds—their presentation time—in the dark, they bend when placed in light even though they are no longer being touched.

Threshold refers to the level of stimulus that must be present during the presentation time to cause perception and transduction. In phototropism, plants are extremely sensitive to very dim unilateral light if they are in an otherwise dark environment; the threshold for curvature is low. However, in bright conditions the threshold is higher, and the unilateral light must be much stronger to trigger curvature. In Venus' flytraps, the threshold for stimulating leaf closure is moderate: The trigger hairs must be firmly bent. This is advantageous in preventing wind or rain from triggering trap closure; the moderate threshold almost guarantees that the trap contains an insect every time it closes (Fig. 14.11).

Related to threshold is the level of response relative to the level of stimulation; the alternatives are all-or-none responses and dosage-dependent responses. In an **all-or-none response,** once the threshold and presentation time requirements are met, the stimulus is no longer important; the response is now completely internal. Individuals respond identically whether they received strong or weak stimuli, regardless of whether the stimulus was brief or long lasting. For example, many species are induced to flower by environmental conditions; once the minimal threshold and presentation time requirements are met, the plants flower fully, limited only by their general health, vigor, and nutrient reserves. Until they receive the proper stimulus, they produce no flowers; their flowering is all or none. Examples are poinsettia, chrysanthemum, *Hybiscus syriacus,* and oats.

In **dosage-dependent responses,** the amount or duration of the stimulus affects the amount or duration of the response. In species of this type, individuals that receive only minimum stimulation flower poorly, even if the plant is quite healthy. Those that receive longer or stronger stimulation produce more flowers. Examples are turnip, marijuana (*Cannabis sativa),* and some varieties of cotton and potato.

CHEMICAL MESSENGERS

Almost all plant communication is by a slow mechanism: transport of hormones through the plant. **Hormones** are organic chemicals that are produced in one part of a plant and then transported to other parts, where they initiate a response. A critical aspect is that hormones act at very low concentrations. Hormones are synthesized or stored in regions of transduction and are released for transport through either phloem or mesophyll and cortical cells when the appropriate stimulus occurs. At the site of response, hormones bind to

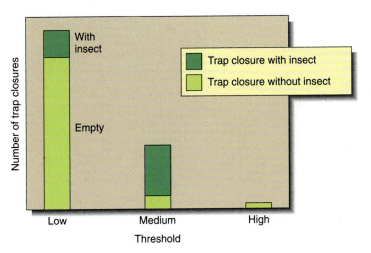

FIGURE 14.11

Threshold must be appropriate to the amount of change a stimulus can cause; Venus' flytrap hairs with an extremely low (sensitive) threshold would be capable of detecting and catching every insect, but wind and rain would cause so many useless closings that the leaf would be inefficient because it would miss insects whenever it was closed unnecessarily. With a medium threshold, it captures more insects because it is open and ready much of the time. Small insects may escape because they cannot bend the trigger hairs enough to meet the threshold. If the threshold were too high, no insects would be caught because none could bend the hairs. The trap would close only when larger animals brushed against it, but these animals are too big to be enclosed in the trap, so all closures would be unproductive. We can hypothesize that natural selection results in a threshold appropriate for the most abundant size of insects.

receptor molecules, probably located in the plasma membrane or some other membrane, and thereby trigger a response. Hormones appear to be released into general circulation and are not carried specifically to the target. Many regions that are not target regions are exposed to the hormone but do not respond because they do not have the proper receptor molecules. In some instances, a plant hormone acts directly on the cells that produce it.

At one time, plant hormones were believed to carry in their structure much of the information necessary for the response. We now know that plant hormones are quite simple in structure. The receptor cell and its nucleus contain almost all the information necessary for proper response, and hormones serve only to activate the response. An analogy is a computer, its programs, and commands. The computer is capable of carrying out numerous functions and processes but only if properly controlled; the same is true of cells, tissues, organs, and whole plants. Computer programs contain the information needed to run the computer, just as the nuclear, plastid, and mitochondrial genes contain the information needed to run cells. Both computer programs and genomes contain numerous sets of information. On a computer, commands such as PRINT, STORE, COPY, and EDIT select subroutines or programs. Hormones are thought to act as commands that activate programs within the target cells (Fig. 14.12a and b). In the early 1980s it was finally proven that the activation of genes and the initiation of the production of new proteins needed for the new type of metabolism that is part of the response is one of the effects of hormone action.

In higher animals, because so many systems and responses must be activated, many distinct hormones are necessary. Because plants are much simpler, their responses can be controlled by fewer hormones; even so, the handful of known plant hormones seems

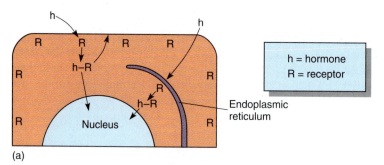

(a)

FIGURE 14.12

A cell responds to a hormone only if it has receptors for that hormone. (a) Evidence suggests that some receptors (R) are in the plasma membrane and others are in membranes such as the endoplasmic reticulum. Once bound, the enzyme-receptor complex (h-R) may cause a metabolic change immediately, or the complex may migrate to another site, such as the nucleus. (b) In many responses, some nuclear genes are activated and others are repressed. Cells may have receptors for several hormones (R_A and R_B); if hormone A is present, it binds and activates (or represses) program A. Other programs are unchanged. One of the results of program A might be to withdraw the receptors of either A or B from the membrane or to add receptors for C or D, thus changing the sensitivity and type of response possible. A second cell (the cell on the right) may have a different program (A program 2) activated by hormone A; the response is cell specific, not just hormone specific. (c) The effects of a hormone are often quite different when the hormone is applied alone, or with or after a second hormone. In this hypothetical cell, development occurs only if hormones A and B are applied simultaneously or if hormone D is applied after hormone C.

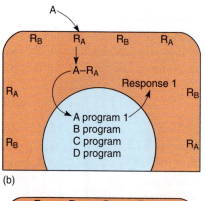

(b)

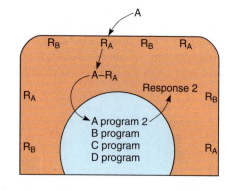

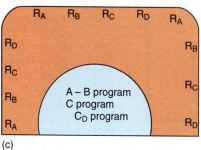

(c)

inadequate. It is likely that many plant hormones are still unknown to us. Also, many responses are activated not by one hormone but by a combination or a sequence of several hormones (Fig. 14.12c), or a particular hormone elicits different responses when present at different concentrations. The following are examples of the most well-studied hormones.

Auxins. The first plant hormone discovered was **auxin**. In 1926 it was identified as the chemical messenger involved in positive phototropism in oat seedlings. Identifying it chemically at that time was impossible because it is present in such low concentrations. Experiments had to be performed by allowing auxin to diffuse out of a seedling leaf tip into a small block of agar, which was then used as if it were a small dose of auxin (see Fig. 14.24g and h). The auxin was later identified as being **indoleacetic acid (IAA)**, which could be synthesized artificially and applied to plants under various conditions to find other responses that IAA might either mediate or inhibit (Fig. 14.13). The search was successful—dozens of responses were found (Table 14.3). Many compounds chemically related to IAA were also found to be as effective or even more so. It was hypothesized that IAA was only one of many natural auxins, each with its own effect and role, but further findings were not consistent with that hypothesis. Analysis of IAA metabolism showed that the compounds were converted to IAA by the plant's enzymes. At present some evidence suggests that phenylacetic acid and chlorinated IAA act directly as natural auxins without being converted to IAA.

Because we are not certain how many natural auxins there are, any compound that imitates IAA in controlling aspects of plant differentiation is called an auxin. For example, any chemical that gives a response similar to IAA in oat seedling phototropism is considered to be an auxin. Experiments are often performed by applying IAA to a plant and examining the responses. When such extra IAA causes a response, we suspect that in normal development, auxin is involved. But we cannot be certain that it is actually IAA that participates in natural conditions. It may be that another, undiscovered auxin is normally involved and in the experiment we have overwhelmed it with IAA. Because of this uncertainty, the word "auxin" is often used when we have not yet identified the chemical messenger involved.

Many synthetic compounds mimic the effect of auxin or the other hormones; for clarity, only natural products are called hormones. The term "**plant growth substance**" is used for any hormone-like compound, natural or artificial (Fig. 14.13). Naphthaleneacetic acid, an artificial compound, produces effects in plants that are for the most part indistinguishable from those of IAA. 2,4-Dichlorophenoxyacetic acid (2,4-D) is auxin-like but so

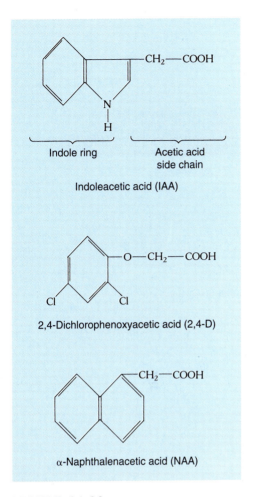

FIGURE 14.13
Indoleacetic acid is the natural hormone auxin; the others are synthetic so they are called plant growth substances.

TABLE 14.3	*Examples of Responses Involving Hormones*
Hormone	**Response**
Auxin	Abscission suppression; apical dominance; cell elongation; formation of roots in cuttings; fruit maturation; tropisms; xylem differentiation
Cytokinin	Bud activation; cell division; fruit and embryo development; mimics the effects of phytochrome and red light in several cases; prevents leaf senescence
Gibberellin	Converts some plants from juvenile to adult condition; converts other species from adult back to juvenile condition; involved in flowering; releases some seeds and buds from dormancy; stem elongation; stimulates pollen tube growth
Abscisic acid	Initiation of dormancy; resistance to stress conditions; stimulation of growth at very low doses; stomatal closure; probably not involved in abscission
Ethylene	Formation of aerenchyma in submerged roots and stems; fruit ripening and abscission; initiation of root hairs; latex production

TABLE 14.4	Conjugation of Indoleacetic Acid to Various Substances
Species	**Conjugate***
corn (*Zea mays*)	High-molecular-weight polysaccharides *myo*-inositol *myo*-inositol-arabinose *myo*-inositol-galactose
rice (*Oryza sativa*)	*myo*-inositol
oat (*Avena*)	glycoprotein
soybean (*Glycine max*)	aspartate

* Results of being conjugated: (1) protected from destruction; (2) storage form; (3) can release IAA rapidly—conjugation controls the level of auxin without affecting synthesis or destruction; (4) transport form (the low-molecular-weight conjugates such as IAA-aspartate).

powerful that it disrupts most normal growth and development even in low concentrations, making it valuable as an herbicide.

Indoleacetic acid is related to the amino acid tryptophan; so far, at least four separate metabolic pathways are known that convert tryptophan to IAA. Some plants, corn for example, synthesize IAA by different pathways at different stages of development. The significance of this is not clear, but each pathway has its own characteristic set of controls. The most active centers of auxin synthesis are shoot apical meristems, young leaves, and fruits. It is present in root tips, but is believed to be transported there from the shoot rather than being synthesized there.

The concentration of substances as powerful as hormones can be controlled not only by synthesis but also by destruction and by conversion to an inert storage form. Two pathways for IAA destruction are known: removal of the side group and oxidation of the five-member ring. IAA is converted to an inactive form by **conjugating** (attaching) it to various compounds (Table 14.4). In the conjugated form, IAA is safe from destruction, it can be stored indefinitely in seeds, and it can be transported from cotyledons to the epicotyl during germination. Conjugation allows rapid regulation of the level of free IAA. Up to 80% of the IAA in oat seeds is conjugated, but this can be deconjugated, releasing free IAA during germination more quickly than synthesis could.

In addition to hormone transport by phloem, a second mechanism exists for auxin only: **polar transport.** In shoots and leaves, auxin moves basipetally—from the apex to the base of the plant, and in roots it moves acropetally toward the root apex. Movement is about 11 mm/hr regardless of whether the tissue is in a vertical, horizontal, or upside-down orientation. By means of the polar transport system, auxin movement through the

FIGURE 14.14
Cells of most dicots, such as this tobacco, can be grown in culture if provided with auxin, cytokinin, some vitamins, minerals, and sugar. The ratio of auxin to cytokinin is important, as are the absolute concentrations. (a) At one ratio, the cells proliferate as a callus composed of parenchyma. (*Biological Photo Service*) (b) At another ratio of auxin to cytokinin, buds form in the callus and then grow into shoots. (*Courtesy of Dennis Gray, University of Florida*)

(a)

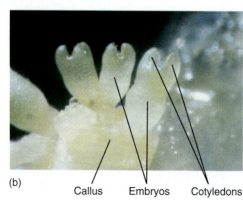

(b)   Callus   Embryos   Cotyledons

plant can be maintained independently of the variation in phloem transport caused by changing sinks and sources for carbohydrates and minerals.

Cytokinins. **Cytokinins** were named for the fact that their addition to a tissue culture medium containing auxin and sugar stimulates cell division—cytokinesis (Fig. 14.14). The first one discovered, kinetin, is an artificial cytokinin, but two natural ones, zeatin and isopentenyl adenine, have been found, and more are suspected to exist (Fig. 14.15). Cytokinins are purines, related to adenine; extensive testing of adenine analogs has been done to determine which aspects of its chemical structure are critical to the molecule's ability to act as a cytokinin. The most active compounds have a side group containing four to six carbon atoms attached to C_6. If this side group is longer or if complex groups occur at other areas, the molecule does not act like a cytokinin, apparently lacking the proper shape and charge to bind with the cytokinin receptor molecule.

Several enzymes carry out the destruction or conjugation of cytokinins, but their significance is not obvious. The enzymes work rapidly on cytokinins added to tissues experimentally, but natural, endogenous cytokinins often appear to be immune to them. Corn kernels have high levels of both cytokinin oxidase and zeatin. Perhaps the two are compartmentalized in separate organelles within cells.

Like auxin, cytokinins are involved in dozens of responses in all parts of the plant (see Table 14.3). One important response is root-shoot coordination. As roots begin to grow actively in the spring, they produce large amounts of cytokinins that are transported to the shoot, where they cause the dormant buds to become active and expand. The richest concentrations of cytokinins generally occur in endosperm and are apparently involved in controlling the development and morphogenesis of the embryo and seed.

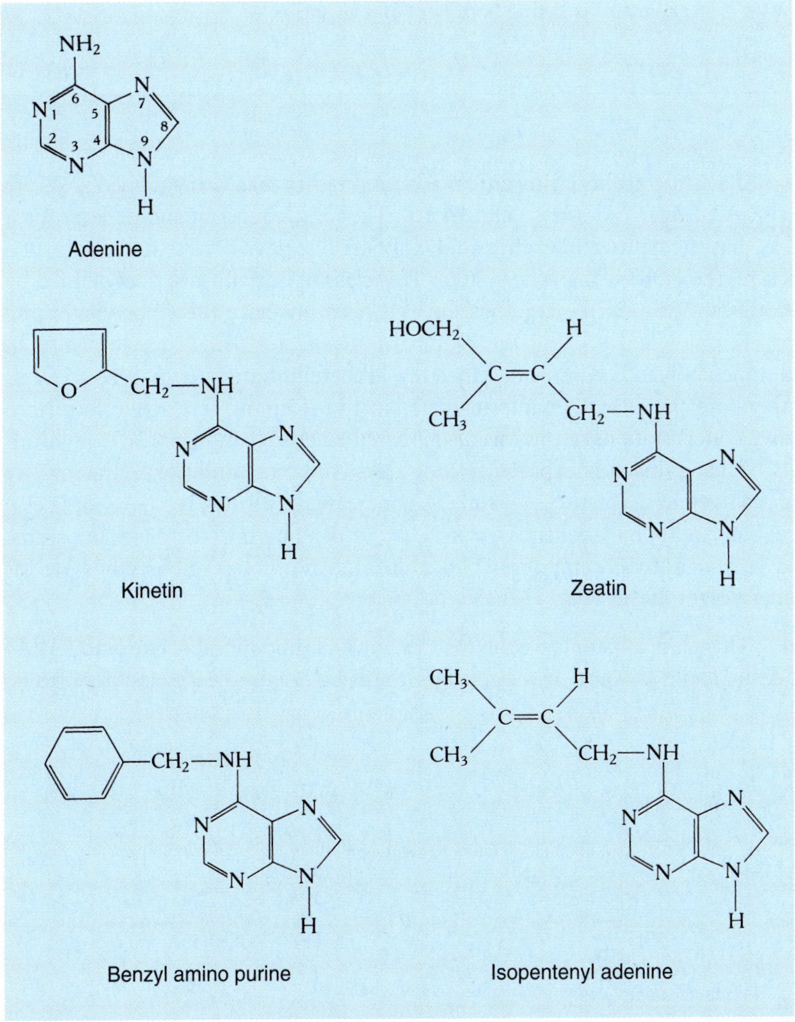

FIGURE 14.15
Both natural and artificial cytokinins are related chemically to adenine. The size and chemical nature of the group on C_6 is critical.

PLANTS & PEOPLE

PLANT TISSUE CULTURE AND MEDICINE

Despite our advanced chemical industry, it is still easier and cheaper to extract many important drugs, dyes, perfumes, and chemicals from plants rather than attempt to synthesize the compounds artificially. For many compounds, it has been proposed that rather than grow the plants in fields, we grow the plant cells in cell cultures and extract the compounds from the cultures. To do this, the cells would be grown in giant vats, called fermenters, supplied with water, hormones, sugars, and minerals. This process would be similar to that used to obtain beer, wine, and similar alcoholic products, as well as many antibiotics and other drugs that are extracted from cultured bacteria and fungi. Some problems exist with this high-tech approach: It is expensive and in many cases the rapidly-growing cultured cells do not produce the desired compound— only mature differentiated cells do. In these cases, botanists must search for culture medium components that might induce the cells to activate the desired synthetic pathway: Usually we try altering the hormones that are added to the culture medium, or we adjust the photoperiod or temperature and so on.

This research is expensive and not always successful, so it is usually limited only to those compounds that are particularly valuable. One example is taxol, a drug that is extremely effective against ovarian and breast cancer. It can be found in small quantities in the bark of Pacific yew trees (*Taxus brevifolia*) and it is such a complex molecule that it cannot be synthesized in the laboratory. As soon as taxol's anticancer properties were discovered, people started harvesting yew trees for their bark. Unfortunately, the tree grows slowly and is rather rare, and it soon became evident that to provide treatment for even a small number of women, the species would be harvested to the extinction—it simply could not be grown on forest plantations rapidly enough.

This presented an ethical dilemma: Save some human lives or protect an endangered species? In Chapter 1, we discussed scientific questions as opposed to ethical ones: Science can tell us how to extract the taxol and administer it to cancer patients, but it cannot tell us if we should. Fortunately, botany seems to be providing a solution. Many botanists are searching for conditions that will induce cultured *Taxus* cells to produce taxol. So far, that has not been successful, but some cultures do produce a compound so similar to taxol that it can be harvested and then chemically converted into taxol. Thus, a combination of high-tech botany and industrial chemistry is saving both lives and species.

Auxins and cytokinins have strictly functional definitions: Any newly discovered compound that mimics indoleacetic acid is an auxin; anything that elicits the same response as zeatin or isopentenyl adenosine is a cytokinin.

Gibberellins.   At least 62 **gibberellins** are known, and rather than being named, they are just numbered: GA_1, GA_2, . . . GA_{62} (Fig. 14.16). The most abundant and perhaps most important, GA_3, has the name **gibberellic acid.** Gibberellins have diverse functions but a unifying structure, the gibberellane ring system. This class of hormones is thus defined by structure: A compound cannot be a gibberellin if it does not have the gibberellane ring system.

Gibberellin metabolism is complex. Only a few gibberellins are known to be active as hormones; others are precursors or intermediates in transforming one active form into another. Relative concentrations of the various gibberellins change in response to environmental signals. When spinach is exposed to long days (summer conditions), the level of GA_{19} undergoes a fivefold decrease, GA_{20} and GA_{29} increase drastically, but GA_{17} and GA_{44} do not change. Gibberellin metabolism appears to occur in all parts of the plant, but seeds, roots, and leaves are especially important. Like all other hormones, gibberellins are involved in numerous responses (Fig. 14.17).

Abscisic Acid.   This class apparently contains the single compound **abscisic acid (ABA)** (Fig. 14.18). As its name suggests, it was thought to play a role in the abscission of fruits,

FIGURE 14.16
All gibberellins are based on gibberellane; the most common forms are GA_3 and GA_7. Not all of the 62 gibberellins occur in plants; at least 15 have been found only in fungi.

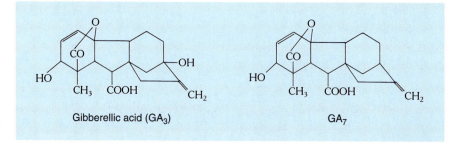

Gibberellic acid (GA_3) GA_7

(a)

(b)

FIGURE 14.17

(a) Many biennial plants grow as a rosette in their first year, then elongate rapidly (bolt) in their second year, producing a tall shoot that bears numerous flowers. The bolting is controlled by gibberellic acid; if it is absent, the plants remain short. *(Robert E. Lyons)* (b) In crop plants, having tall stalks often makes harvesting difficult, and it is a waste of energy in fields where there is no competition for light. Plant breeders often artificially select mutant plants that are dwarfed or short. Many of these dwarf forms either lack the particular gibberellin needed for stem elongation or are unresponsive to it, perhaps because they lack the proper receptors. *(Courtesy of B. O. Phinney, University of California, Los Angeles)*

leaves, and flowers, but at present we do not know whether that is true of many species or only sycamore, the species in which it was discovered. ABA is widely regarded as a growth inhibitor, possibly involved in inducing dormancy in buds and seeds. However, dormancy is not just cessation of growth, but a complicated set of changes that prepare the plant or seed for adverse conditions. It is incorrect to equate induction of dormancy with mere growth inhibition.

The role of ABA in all types of stress resistance is receiving great attention; heating of leaves, waterlogging of roots, chilling, and high salinity have all been found to cause sudden increases in ABA. If healthy plants are pretreated with ABA, they become much more resistant to stressful conditions. When plants begin to wilt, the concentration of ABA in leaf cells increases dramatically from about 20 μg/kg fresh weight to 500 μg/kg, and guard cells close stomatal pores. Wilt-induced production of ABA overrides all other stomatal controls; other mechanisms that normally cause stomata to open are ineffective when ABA is present. The stimulus for ABA production appears to be loss of turgor. Large amount of ABA are exported from wilted leaves to the rest of the plant, moving through phloem.

ABA can be removed by being converted to phaseic acid, which has no known hormonal activity. ABA can be conjugated to glucose, but the conjugation may not be a useful, recoverable storage form. As plants wilt, the concentration of free ABA rises suddenly but that of conjugated ABA does not drop. Beets (*Beta vulgaris*) labeled with radioactive ABA conjugate did not release any of the radioactive ABA when allowed to wilt.

Ethylene.　**Ethylene** is the only gaseous plant hormone, and it has the simplest structure (Fig. 14.19). Its most commonly studied effects occur during fruit development. Fruits

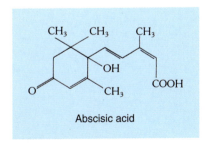

Abscisic acid

FIGURE 14.18

Abscisic acid is transported rapidly between cells and through phloem, so its presence in a tissue is not proof that it was produced there. ABA can be synthesized from mevalonic acid in roots, stems, leaves, fruits, and seeds of various species.

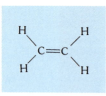

FIGURE 14.19
Ethylene is a simple, small molecule. Many of its effects are blocked by carbon dioxide, whose size and shape are similar enough to those of ethylene that it can bind to ethylene's receptor and block normal response.

such as apple, avocado, banana, mango, and tomato are **climacteric fruits:** They ripen slowly as they mature, but in the final stages, numerous developmental changes occur rapidly. Starches are converted to sugars, cell walls break down and soften, flavors and aromas develop, and color changes. Ethylene stimulates these changes, but at first so little ethylene is present that the changes occur slowly. However, one effect of ethylene in these fruits is the production of more ethylene, which constitutes a positive feedback system: The concentration increases exponentially and rapidly. The sudden burst of ethylene and the rapid completion of maturation of the fruit are known as a **climacteric;** at this time ethylene production can be as high as 320 nl/g/hr (one nl = one nanoliter = one billionth of a liter). In **nonclimacteric fruits,** such as cherry, lemon, and orange, ethylene does not stimulate its own production, so ethylene levels remain stable and no sudden change occurs just before maturity.

Because ethylene controls the ripening of most of our important food fruits, it is important commercially. It is used in harvesting cherries, cotton, and walnuts by causing their uniform abscission; it also synchronizes flowering and fruiting in pineapple, making harvesting easier. Ethylene can be drawn out of unripe fruits by storing or transporting them in a partial vacuum. When they reach market, air pressure is returned to normal, ethylene accumulates, and ripening occurs. Fruits may also be treated with 2-chloro-ethylphosphonic acid (commercial trade name Ethrel), which breaks down and releases ethylene.

Being a gas, ethylene moves rapidly through tissues by diffusion rather than by specific transport mechanisms. In many cases, it acts as a final effector for auxin. Arrival of auxin at a target site often causes that site to produce ethylene, which can diffuse rapidly and trigger responses in the adjacent area more quickly than the auxin itself could. Ethylene can be broken down to either carbon dioxide or ethylene oxide, and when radioactively labeled ethylene is supplied experimentally to bean tissues, it becomes bound. It is not known whether binding is a form of storage or if the hormone is attaching to its receptor.

ACTIVATION AND INHIBITION OF SHOOTS BY AUXIN

Auxin is often described as a growth hormone, whereas abscisic acid is considered an inhibitor; unfortunately, such characterizations are confusing. Hormones simply carry information about the status of a particular region, nothing more; whether the elicited response is inhibition or activation depends on the site of response. An example of the complexity is provided by shoot tips.

As shoot apical meristems grow and initiate the new cells of shoots and leaf primordia, they also produce the auxin, IAA. Young leaves are also a rich source of this hormone. No external signal must be perceived to initiate auxin production; instead, this is a means of integrating the plant during ordinary growth. Large quantities of auxin indicate to cells that shoots are elongating and producing new leaves. Although neither signal perception nor transduction occurs, transport takes place. In stems auxin undergoes basipetal, polar transport through the cortex, perhaps by means of molecular pumps in plasma membranes. This downward flow of auxin surrounds all stem cells, and at least three cell types are set to respond to it, each response unique to the particular cell type.

CELL ELONGATION

In cells of the young internodes just below the apical meristem, auxin triggers cell elongation. When IAA contacts these responsive cells, which are prepared for growth, the cells begin to transport protons actively out across the plasma membrane (Fig. 14.20). This has the effect of acidifying the cell wall. The protons break some of the chemical bonds that hold one cellulose microfibril to another and activate enzymes that weaken other bonds so that the wall becomes weaker. If the protoplast is turgid and pressing against the wall, it exerts enough pressure to stretch the weakened wall and growth results. Immature cells neither excrete protons nor grow if auxin is lacking. At lower internodes, fully grown, mature cells apparently lack the proper auxin receptors, because auxin does not cause them to extrude protons or grow.

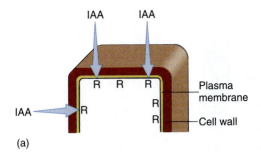

(a)

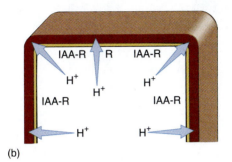

(b)

(c)

FIGURE 14.20

Cells in the subapical region have auxin receptors. If the apex is growing and producing auxin, IAA is present to bind to the receptors (a). After stimulation by auxin binding, the plasma membrane pumps protons from the cytoplasm into the wall (b), weakening it and allowing turgor pressure to stretch it. (c) Cell elongation stops once the maximum cell size is reached, and adding more auxin does not cause any more elongation; perhaps the receptors have been removed from the membrane.

APICAL DOMINANCE

The second site of response to apically produced auxin is the buds located in leaf axils; their response is not cell elongation but rather inhibition of growth. Apically produced auxin induces dormancy in these axillary buds, the result being that each shoot tip has only one active apical meristem, a phenomenon called **apical dominance** (Fig. 14.21). This is a threshold response: As the terminal shoot apical meristem grows away, the concentration of auxin around an axillary bud gradually decreases until at some point it drops below the threshold. Inhibition cannot be maintained, and the axillary bud becomes active and grows out as a branch or flower. As the axillary bud grows, it produces auxin but does not inhibit itself, although it does inhibit all its own newly formed axillary buds.

DIFFERENTIATION OF VASCULAR TISSUES

The third site of response to auxin produced in shoot tips is the vascular cambium; the response is cell division and morphogenesis. In springtime, as air temperatures rise and buds become active, their auxin moves basipetally, activating the dormant vascular cambium. Auxin not only stimulates cambial cells to begin mitosis and cytokinesis but also causes new daughter cells to differentiate into xylem cells. If an apical meristem is destroyed, by insects or a late frost for example, basipetal flow of auxin stops, vascular differentiation is interrupted, internode elongation ceases, and apical dominance is broken. Some axillary bud, now free of apical dominance, becomes active and re-establishes a flow of auxin that maintains the vascular cambium and any other cells that depend on it.

(a)

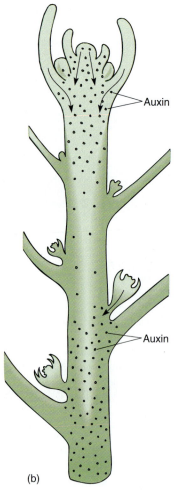

Auxin

Auxin

(b)

FIGURE 14.21

(a) Many cacti show almost 100% apical dominance; as long as the apex is undamaged, all axillary buds (spine clusters) remain dormant. Here, one bud has become active but is now producing enough auxin to keep its own buds dormant and suppress nearby buds on the parent stem. (b) The concentration of auxin is greatest at the shoot apex and less at lower levels; at sites where it drops below the threshold level, it can no longer suppress axillary buds, and one or several become active. Other factors must be involved, because it is not always the lowest bud that becomes active first, even though it should be the first to encounter sufficiently low auxin concentrations.

Three separate target tissues give three distinct responses, not because there are three separate chemical messengers—there is only auxin—but because part of their previous differentiation was preparation to respond to auxin in a particular way. It is important to realize that the auxin carries no information except that the shoot apex is healthy and active. Each target site must have receptor molecules that interact with IAA and therefore detect the presence or absence of auxin. It is not known whether all three targets have the same receptors or if each has a unique type, nor is it known how the interaction of IAA with the receptor triggers the response. That topic is discussed in Chapter 15.

INTERACTIONS OF HORMONES IN SHOOTS

In some species, apical dominance may involve only the presence or absence of auxin; in others, there is an interplay of two or three hormones. Active roots synthesize cytokinins that are transported to the shoot and stimulate axillary buds. Whether buds become active or remain dormant depends on the relative amounts of the two hormones. If a plant is growing vigorously, its roots are active and cytokinin levels are high; many buds at a distance from a shoot apical meristem have a low auxin/high cytokinin ratio and become active. Such a mechanism is adaptive because if a plant is growing well, activating more dormant buds increases the rate of new leaf production. The role of ABA in apical dominance is uncertain. It is present in quiescent buds but does not decrease either just before or as buds are becoming active and growing out.

Apical dominance in prickly pear cacti (*Opuntia polyacantha*) is more elaborate. The spine clusters are short shoots and the spines are highly modified leaves. If the spine cluster is excised and placed in tissue culture with cytokinin, the dormant short shoot apical meristem grows out as a long shoot—a new "pad" similar to a normal branch. If the culture medium contains gibberellins instead of cytokinin, the short shoot apical meristem produces more spines—it acts as a rejuvenated short shoot.

A second interaction of hormones occurs in the vascular cambium. Auxin alone activates the cambium and elicits differentiation of xylem. But gibberellin is also present in a healthy stem and causes some of the new cells to differentiate as phloem. Without the interaction of both auxin and gibberellin, a normal, functional vascular system would not develop.

HORMONES AS SIGNALS OF ENVIRONMENTAL FACTORS

LEAF ABSCISSION

Whereas normal growth of shoots and roots results in large flows of auxin and cytokinin, respectively, environmental factors also influence hormone concentrations. Hormones communicate to various parts of a plant the information that a particular part has encountered an environmental change. Export of ABA by wilted leaves has been mentioned, and another example involves abscission of leaves and fruits. A young leaf produces large amounts of auxin, but production falls to a low but steady level in a mature leaf. As long as auxin flows out through the petiole, activity in the abscission zone is inhibited (Fig. 14.22). If the leaf is damaged by animal feeding or water stress, auxin production drops to such a low level that its flow through the petiole does not keep the abscission zone quiescent. Perception and transduction in this case may be simply that insect or wilt damage makes it impossible for the impaired cells to produce enough auxin to inhibit the abscission. Old age of the leaf may also result in the lack of sufficient auxin, but evidence suggests that autumn conditions stimulate production of ethylene, which then suppresses auxin production and transport in time for abscission before winter.

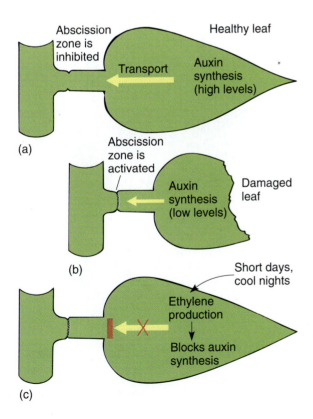

FIGURE 14.22

(a) A healthy leaf produces and transports enough auxin to suppress activity in the abscission zone. (b) A damaged leaf produces less auxin, insufficient to prevent abscission. (c) Autumn stimuli may cause the production of ethylene, which then suppresses auxin synthesis and transport. An alternative hypothesis postulates that autumn conditions directly suppress auxin production and ethylene is not necessary.

FIGURE 14.23

Auxin's effects on fruit ripening are often studied in the strawberry because the tiny true fruits (achenes) are located on the outside of the large red false fruit. The achenes are sources of auxin necessary for development of the false fruit. If they are removed, no development occurs unless auxin is applied experimentally.

Normal development

Seeds removed

Seeds removed; auxin applied

Fruits are prevented from abscising prematurely by the presence and export of sufficient amounts of auxin through the pedicel. Fruit ripening is under the control of both auxin and ethylene, at least in edible, fleshy fruits. Initial transformation of the ovary wall into a fruit is a response to auxin synthesized in developing embryos and transported to the ovary wall (Fig. 14.23). Auxin stimulates many changes, including cell enlargement and differentiation; there is usually surprisingly little cell division during the formation of fruits, even large ones. One effect of auxin is release of ethylene by the developing fruits, which leads to other aspects of ripening in both climacteric and nonclimacteric fruits. At maturity the high concentration of ethylene stimulates the pedicel abscission zone, overriding the presence of auxin.

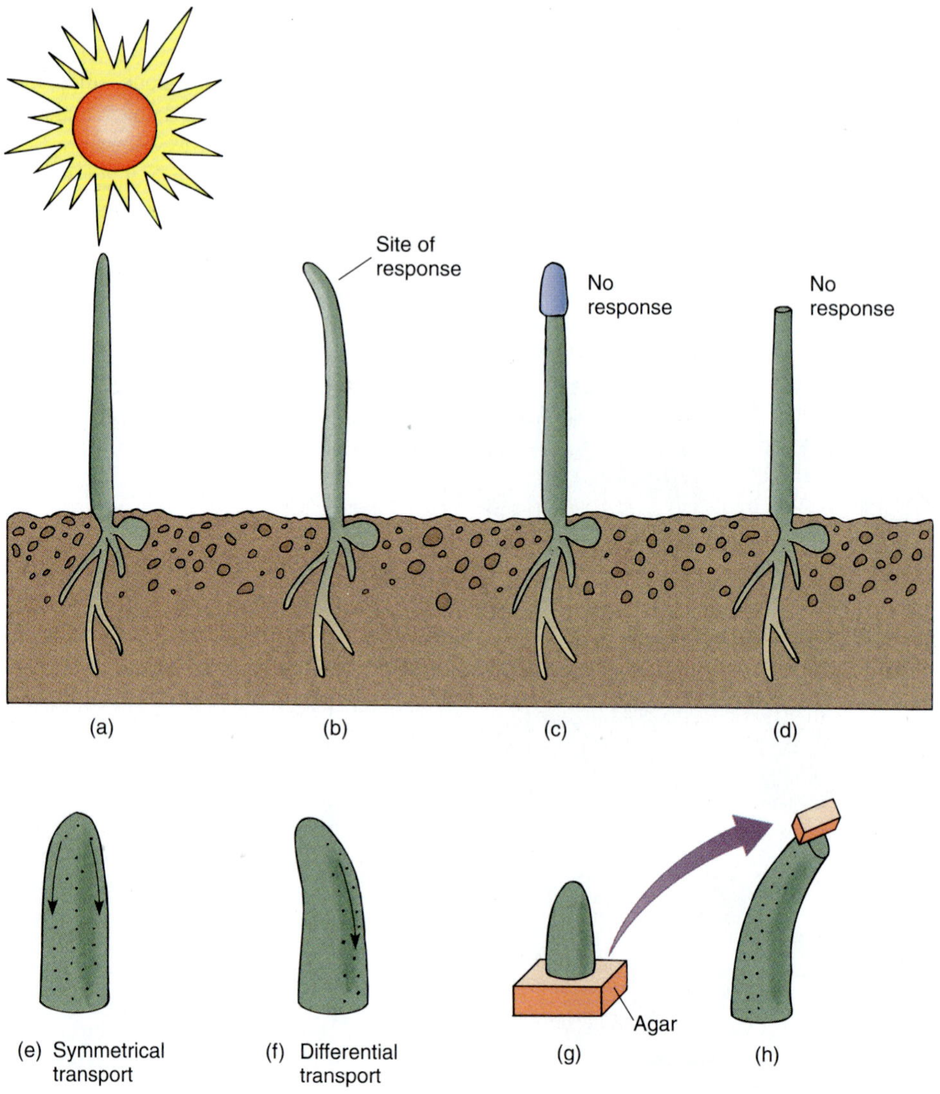

FIGURE 14.24

(a) When illuminated from directly above, oat seedlings grow upward. (b) When a young oat seedling is exposed to light from one side, its outermost sheathing leaf, the coleoptile, bends and grows toward the light If the coleoptile apex is covered (c) or cut away (d), no response occurs to unilateral illumination, so the tip is the site of perception. If the site of response is covered, bending occurs, so the site of response is not involved in perception at all. (e) In dark conditions or with overhead lighting, auxin is transported symmetrically down the coleoptile, causing equal amounts of growth everywhere. With unilateral illumination (f), auxin is redistributed, with the darker side transporting more auxin than the lighted side, so the darker side grows faster, resulting in curvature. (g) and (h) Auxin can be collected in small blocks of agar or other absorptive material and then placed asymmetrically on a decapitated coleoptile; the side receiving auxin grows, but the other side does not.

Site of response

No response

No response

(a)   (b)   (c)   (d)

(e) Symmetrical transport

(f) Differential transport

(g)

Agar

(h)

TROPISMS

Bending of plant parts toward or away from stimuli requires **differential growth;** that is, one side of the responding organ grows more rapidly than the other. In phototropism, the mechanism responsible is asymmetrical distribution of auxin. In positive phototropism of oat seedlings, the tip of the outermost, protective leaf, the **coleoptile,** is the site of perception; if it is covered with an opaque hood, light direction is not detected. The site of response, where differential growth causes bending, is about 5 mm below the coleoptile tip (Fig. 14.24). It has not been possible to determine conclusively which pigment is the photoreceptor for phototropism, but the result of unilateral illumination is a redistribution of auxin to the darker side of the coleoptile apex (Fig. 14.25). Careful measurements have shown that auxin synthesis is not affected; the gradient is not established by destruction of auxin on the lighted side or extra synthesis on the darker side. Once the differential distribution of auxin is established, the darkened side of a coleoptile, or the stem in other plants, receives extra auxin, so it grows more rapidly and the stem bends toward the light. When the stem points directly at the light, neither side is brighter nor darker, so differential auxin transport stops and the stem grows straight ahead.

In positive gravitropism in roots, the root cap acts as the organ of perception (see Fig. 14.10). Once the lower side of the root is detected, a growth inhibitor is transported to that lower side of the root cap and then into the root, where it slows growth on the lower side.

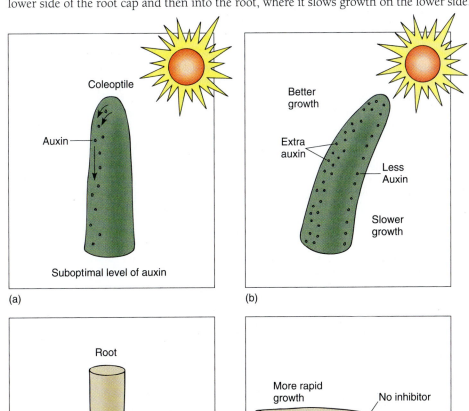

(a)

(b)

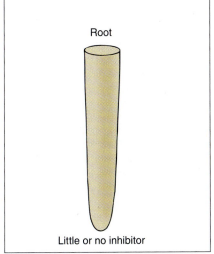

(c)

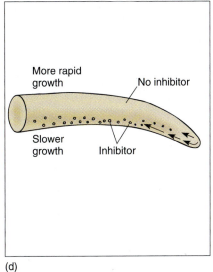

(d)

FIGURE 14.25

(a and b) Shoots may contain slightly too little auxin for fastest growth. Thus, redistribution in a coleoptile apex due to unilateral light causes the side with more auxin to have a level that is more nearly optimal, whereas the other side, which receives less, has even poorer growth than before. (c and d) In root positive gravitropism, the horizontal position of a root cap causes statoliths to fall to the side of statocytes, causing a downward redistribution of growth inhibitor. The side that receives inhibitor becomes inhibited, whereas the side that receives none has normal growth.

Unfortunately, we know almost nothing about how the position of the statoliths is transduced, nor do we know the identity of the growth inhibitor. Much circumstantial evidence indicates that it may be abscisic acid, but claims have been made for a different compound that has not yet been chemically identified.

FLOWERING

RIPENESS TO FLOWER

Almost all plants must reach a certain age before they can be induced to flower; only a few species can be induced as seedlings. Annual plants need to be only several weeks old before they become competent to respond to a floral stimulus, but many perennials must be 5, 10, or even 20 years old. Before this time, conditions that should induce flowering have no effect. Virtually nothing is known about the metabolic difference between the **juvenile** stage, when plants are incapable of being induced to flower, and the **adult** stage, when they are sensitive to floral stimuli.

Cold temperatures are the stimulus responsible for converting biennial plants to the adult condition, in the process of vernalization (see Figs. 14.5 and 14.17a). The site of perception is the shoot apex itself; if it is cooled while the rest of the plant remains warm, vernalization occurs. But if the rest of the plant is cooled while a small heater keeps the apex warm, no vernalization occurs. Presentation time is as short as 1 day in some plants. The transduction process is known to require oxygen, carbohydrates as an energy source, and an optimal temperature just above freezing, between 1 and 7°C. Vernalization results in a stable change. If the plants are returned to warm conditions but are not given the floral stimulus of short days and long nights, they continue to grow vegetatively, without flowering, year after year. But they retain their vernalization and flower whenever the floral stimulus is finally given.

PHOTOPERIODIC INDUCTION TO FLOWER

The conversion of an adult plant from the vegetative to the flowering condition may be the most complex of all morphogenic processes. This is not one process—different mechanisms exist in different species. In certain annual species, size appears to be the only

TABLE 14.5 *Photoperiodic Species**

Long-night Plants (Short-day Plants)	Short-night Plants (Long-day Plants)	Day-neutral Plants
chrysanthemum (*Chrysanthemum*)	barley (*Hordeum vulgare*)	balsam (*Impatiens balsamina*)
cockleburr (*Xanthium strumarium*)	cabbage (*Brassica* sp.)	bean (*Phaseolus*)
Japanese morning glory (*Pharbitis nil*)	clover (*Trifolium pratense*)	corn (*Zea mays*)
kalanchoe (*Kalanchoe blossfeldiana*)	hibiscus (*Hibiscus syriacus*)	cotton (*Gossypium hirsutum*)
poinsettia (*Euphorbia pulcherrima*)	petunia (*Petunia* sp.)	holly (*Illex aquifolium*)
strawberry (*Fragaria chiloensis*)	radish (*Raphanus sativus*)	rhododendron (*Rhododendron* spp.)
violet (*Viola papilionaceae*)	wheat (*Triticum aestivum*)	tomato (*Lycopersicon esculentum*)

* All plants fall into one of these three photoperiod categories; this table lists only a few familiar examples.

TABLE 14.6 *Photomorphogenic Responses for Which Phytochrome is the Photoreceptor*
Inhibition of internode elongation
Development of proper leaf shape
Increase in number of stomata per leaf
Increase in amount of chlorophyll
Decrease in apical dominance; formation of a more highly branched, bushy plant
Increased accumulation of carotenoid pigments in tomato fruits

important factor: Peas and corn initiate flowers automatically after a particular number of leaves has been produced, regardless of environmental conditions; flowering is controlled by internal mechanisms. In many species, perhaps most, transition to the flowering condition is triggered by **photoperiod**—day length—which acts as a season indicator (Table 14.5). One subclass of these plants blooms when days are short (spring or fall) and are **short-day plants**. Another subclass, **long-day plants**, are induced to bloom when days are long, in summer. Plants that do not respond to day length are **day-neutral plants**.

We know much about how plants measure day length. First, the pigment **phytochrome** detects the presence or absence of light (Table 14.6; Fig. 14.26). Phytochrome has a light-absorbing portion attached to a small protein of about 124,000 daltons. When phytochrome absorbs red light with a wavelength of about 660 nm, the protein changes its folding. This affects many of its properties, one of which is its hydrophobicity; in the refolded state, it is more hydrophobic and binds to membranes. A second altered property is its absorption spectrum; it now absorbs not at 660 nm but in the far-red (almost infrared) region of 730 nm. But when this form absorbs far-red light, it refolds back to the red-absorbing form and releases from the membrane. The two forms are called P_r (red-absorbing) and P_{fr} (far red-absorbing). Also, P_{fr} reverts to P_r in darkness. Apparently P_r is inactive metabolically but becomes morphogenically active and exerts its effect when it absorbs red light and is converted to P_{fr}: P_{fr} is the active form and may bring about metabolic responses. After a plant is given red light, converting phytochrome to the active P_{fr} form, exposure to far-red light converts the phytochrome back to the inactive P_r form. If far-red light is given quickly enough after red light, phytochrome does not have enough time to affect cell

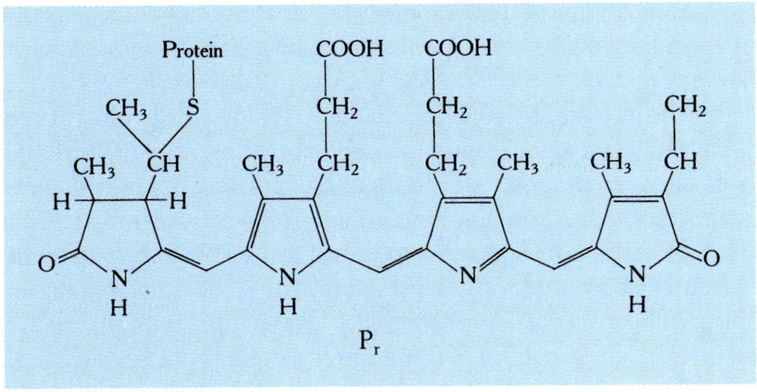

FIGURE 14.26
The chromophore or light-detecting portion of phytochrome. The rest of the molecule is a protein. Notice that the two portions are joined by a sulfur atom.

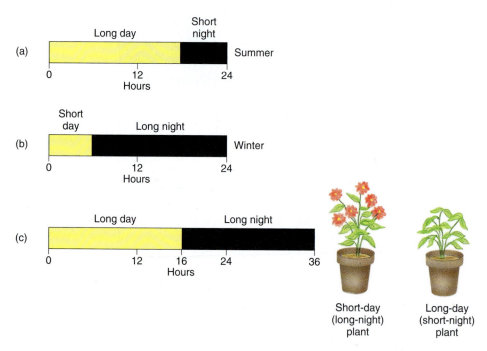

FIGURE 14.27

(a) and (b) Natural 24-hour light/dark cycles must have short nights any time the days are long (summer) and long nights when days are short (winter), so plants could detect season by measuring either day or night. When photoperiodism was first discovered, it was assumed that day length was important, so all our terminology was based on that. (c) With electric lights, it is possible to create a long photoperiod—36 hours long here—with both long days and long nights. Under these conditions, short-day (long-night) plants bloom, but long-day (short-night) plants do not.

metabolism significantly, and no effect is seen. But if the far-red comes long enough after the red for the presentation time to be fulfilled, the P_{fr} is able to complete the transduction process and far-red light can no longer cancel the red light stimulation.

A plant experiencing a short day in nature automatically receives a long night (Fig. 14.27). Similarly, long days are always accompanied by short nights. Night length is actually the critical factor. A long-day plant is in reality a short-night plant. It can be placed in a growth chamber and artificially given both long days and long nights—for example, 16 hours of light and 16 hours of dark in a 32-hour "day." If day length is the important factor, the plant should flower because it has long days; if night is the critical factor, then it should not flower because it does not have short nights. When the experiment is done, the long-day plant does not flower, indicating that night length is critical, not day length. Similar experiments have shown that short-day plants really are long-night plants: If given a 16-hour cycle (8 hours light/8 hours dark—both day and night are short), they do not flower. But if given long nights, even accompanied by artificially long days, the plants are induced to flower.

Each species has its own particular requirements for long or short nights; that is, not all "long nights" have to be the same length. Instead each species has a **critical night length**; if a short-night plant receives nights shorter than this critical length, it flowers, whereas a long-night plant must receive nights longer than its own critical night length. Because critical night length varies from species to species, it is possible for a long-night plant and a short-night plant to bloom under the same conditions if the critical night length for the short-night plant happens to be longer than that for the long-night plant (Table 14.7).

Day length/night length control of flowering and other processes such as the initiation or breaking of dormancy is more common at locations farther from the equator. Away from the equator, nights become progressively shorter from winter to summer, then progressively longer from summer to winter. The greater the distance from the equator, the greater

TABLE 14.7	*Concurrent Flowering of Long-day and Short-day Plants*	

Species	Critical Night Length	Response
Xanthium strumarium	8.4 hr	Flowers if nights are longer than 8.4 hr
Hyoscyamus niger	13.6 hr	Flowers if nights are shorter than 13.6 hr

Although *Xanthium strumarium* is a short-day plant and *Hyoscyamus niger* is a long-day plant, both flower if given days 14 to 12 hr long; those days have nights 10 to 12 hr long, longer than the 8.4-hr critical night length of *Xanthium* but shorter than the 13.6-hr critical night length of *Hyoscyamus*.

the length of the longest winter night and the shorter the length of the shortest summer night. Thus, if two species are to bloom just after the beginning of May, a species in the southern United States or Mexico must have a critical night length shorter than that of a species in the northern United States or Canada.

Presentation time varies considerably; in some morning glories, one photoperiod of the proper length induces flowering, whereas at least 1 or 2 weeks of proper photoperiods are necessary for other species. The accuracy with which night lengths can be measured varies, but the most accurate species known is the long-day (short-night) plant henbane (*Hyoscyamus niger*); it must have nights shorter than 13 hours, 40 minutes. If the nights are even 20 minutes too long, 14 hours long for example, it does not flower.

When phytochrome was discovered to be responsible for measuring night length, it was hypothesized that most phytochrome was converted to P_{fr} by the end of a day and then reverted slowly back to P_r during the night. It was postulated that if it could completely revert in the dark, before the next sunrise, metabolic changes would be triggered. However, it is now known that virtually all P_{fr} converts back to P_r within 3 or 4 hours, a time far too short to be a night-measuring clock by itself, although it may be the initial part of a longer-acting clock.

The sites of perception for night length are young leaves. It is possible to stimulate one leaf with a spotlight and induce the plant to flower, even if the apical meristem, the site of response, is not illuminated (Fig. 14.28). Because the site of perception is not the site of response, a chemical messenger must be transmitted between the two. If a leaf is photoinduced and then immediately cut off the plant, no flowering occurs; if it is allowed to remain attached for several hours, the flowering stimulus is synthesized and transported out of the leaf. If the leaf is then removed, the plant still flowers. An obvious experiment is to induce a leaf by giving it the proper night length, then collect the sap that is transported through the

(a)

(b)

FIGURE 14.28
If a short-night plant is given long nights, it does not flower. But it is possible to cause flowering by illumination with 15 minutes of dim red light; the plant acts as though it has received two short nights separated by a 15-minute day. (a) The red light "night break" does not have to be given to the whole plant; if a narrow beam of red light shines on a single leaf while all the rest of the plant remains in darkness, the plant flowers (b).

petiole and assay it for the hormone that acts on the apical meristem. This has been done hundreds of times by many people, without any repeatable success; the process is not as simple as we had at first thought.

An extremely interesting set of results has been obtained by grafting together plants with different photoperiod requirements. Individuals of the tobacco species *Nicotiana silvestris* are long-day plants. The species *N. tabacum* has two types of individuals: Those of the cultivar *N. tabacum* cv. Trabezond are day-neutral plants, and those of *N. tabacum* cv. Maryland Mammoth are short-day plants. When long-day and day-neutral plants are grafted together and given long days, the day-neutral plant is also induced to flower, presumably by a floral stimulus that passes through the graft union. Similarly, day-neutral plants are induced to flower if grafted to short-day plants and given short days (Fig. 14.29). Grafting one day-neutral plant to another does not increase flowering, so the grafting by itself has no effect.

When the combination of short-day plants grafted to day-neutral plants was given long days, the short-day plants did not flower, as expected, and the day-neutral partners flowered about the time they would have if not grafted to anything. But when long-day plants

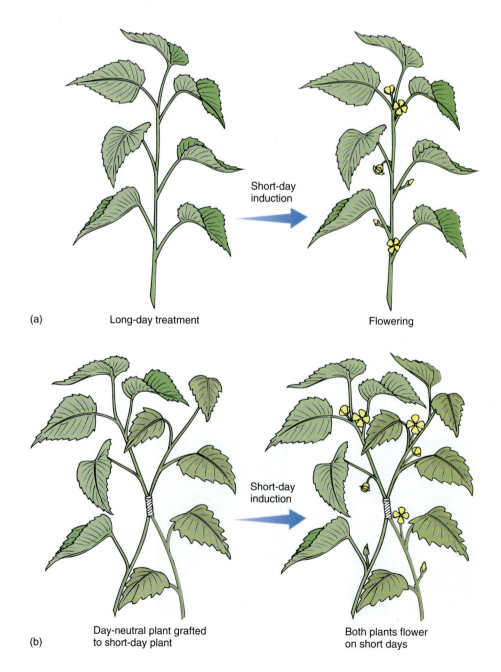

(a)   Long-day treatment   Flowering

Short-day induction

Day-neutral plant grafted   Both plants flower
(b) to short-day plant   on short days

Short-day induction

FIGURE 14.29
(a) Under short days, short-day plants are induced to flower but day-neutral ones are not. (b) If grafted together and given short days, both short-day plants and day-neutral plants flower. Apparently the floral stimulus that was induced in the short-day plant has been transported to, and is active in, the day-neutral plant.

grafted to day-neutral plants were exposed to noninductive short days, not only did the long-day plants not flower, but the day-neutral partners were also prevented from flowering. Apparently, under noninductive conditions, long-day plants actually produce an inhibitor of flowering. Flowering in these long-day tobacco plants seems to be controlled by a switch from inhibitor production to promoter production, whereas in the short-day tobacco plants, flowering is controlled only by the presence or absence of a promoter. Neither the promoter nor the inhibitor has been isolated, but gibberellins are suspected to be involved in the synthesis or activation of the promoter.

ENDOGENOUS RHYTHMS AND FLOWERING

Plants contain **endogenous rhythms**; that is, certain aspects of their metabolism cycle repeatedly between two states, and the cycle is controlled by internal factors. The most obvious example of this is in the "sleep movements" of the leaves of plants like prayer plant (*Oxalis*). In the evening, leaflets drop down, and in the morning, they raise themselves to the horizontal position as motor cells increase their turgor. It is easy to assume that this is a photonastic response, but if the plants are placed in continuous darkness, the leaflet position continues to change, returning to the up position about every 24 hours (see Fig. 12.16). In many flowers, the production of nectar and fragrance is also controlled by an endogenous rhythm and occurs periodically even in uniform, extended dark conditions.

Endogenous rhythms are involved in numerous aspects of plant metabolism that are not easily observed, such as many aspects of photosynthesis, respiration, growth rate, and exudation of mucilage from roots (Table 14.8). The underlying mechanism that constitutes the clock is poorly understood but is known to be independent of temperature and general health of the plants. Endogenous rhythms are not affected by cooling or warming. Also, extensive studies have shown that the rhythm is truly endogenous and not controlled by an unsuspected exogenous rhythm related to Earth's rotation. Plants have been taken to the South Pole, where planetary rotation would have no effect, and have been taken into orbit (Fig. 14.30). If the rhythms were actually exogenous, plants at the South Pole should lose their rhythmicity, whereas those in orbit should have a more rapid rhythm that matches the orbital period. In both cases, normal rhythm was maintained.

Many types of endogenous rhythms have a period that is not 24 hours long. Cytoplasmic streaming and the spiralling motion of elongating stem tips have periods of only a few minutes to a few hours; these are ultradian rhythms. If a period is approximately 24 hours long, it is a **circadian rhythm**, the most common kind. The release of gametes in brown algae is controlled by a 28-day lunar rhythm. Some seeds have an annual rhythm of germinability: If stored in uniform conditions and periodically provided with moisture and warmth, they germinate only at times of the rhythm that correspond to springtime.

Endo-: *from the Greek word for "internal";*
exo-: *from the Greek for "external."*

TABLE 14.8 *Metabolic Processes that Undergo Endogenous Rhythms in Plants*
Accumulation of metabolites for CAM
Changes in leaf position
Growth rate of stems and roots
Mitosis
Opening and closing of stomata
Rates of enzyme activity
Respiration and other metabolisms
Root absorption of minerals

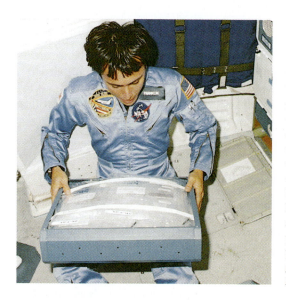

FIGURE 14.30

Ruling out exogenous factors such as light and temperature fluctuations does not prove that rhythms are endogenous; the plants could be responding to an exogenous rhythm we have not yet thought to test. But if it is exogenous, then in orbit the rhythm should match the period of the orbit. Experiments have shown that even in orbit, the 24-hour cycle is maintained. The space shuttle has made such experiments possible. This is F. Chang-Diaz performing an experiment on root growth for Dr. Randy Moore. (*Courtesy of NASA, R. Moore, and E. McClellan*)

"Circadian" means that the rhythm is only approximately, not exactly, 24 hours long. When placed in uniform conditions, the true cycle typically differs slightly from 24 hours, being either somewhat longer or shorter (Fig. 14.31). However, in nature, the rhythm is exactly 24 hours long because light is able to **entrain** (reset) the rhythm. The pigment responsible for detecting the light for entrainment is phytochrome. Each morning, sunrise resets the rhythm so it can never get out of synchronization with exogenous light/dark cycles.

The involvement of endogenous circadian rhythms in flowering was discovered during dark interruption experiments: A short-day (long-night) plant can be prevented from flowering by interrupting long nights with a brief (15 minutes or less) exposure of red light.

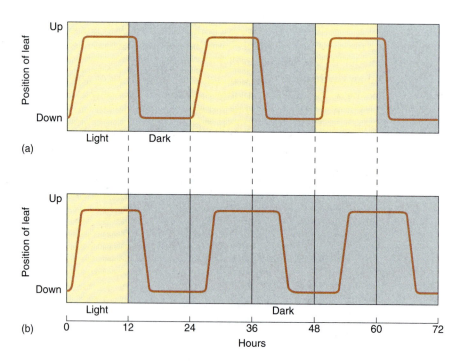

FIGURE 14.31

(a) Under natural conditions, a circadian rhythm matches the cycle of light and dark, being exactly 24 hours long. (b) In continuous darkness, most circadian rhythms have periods slightly longer than 24 hours. In nature, sunrise resets the clock by acting on phytochrome every morning.

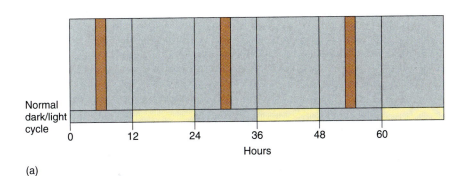

(a)

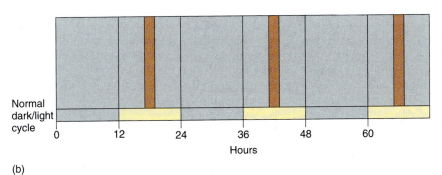

(b)

FIGURE 14.32
If a long-night plant is placed in continuous darkness, it can be prevented from flowering by giving it red light night breaks, but these are effective only if given at those times of the endogenous rhythm when the plant is "expecting" darkness (a). If given when the rhythm is "expecting" light, the light breaks are ineffective (b). Whatever metabolism red light interrupts does not occur continuously in prolonged darkness, but rather periodically, controlled by the internal clock.

This is detected by phytochrome, and the plant acts as though it has received two short nights separated by a 15-minute day. Short-day plants given a very long night—continuous darkness—have an endogenous rhythm of sensitivity to light breaks. If the light break is given at 6 hours into the dark period, or at 30 hours (24 + 6), 54 hours (24 + 24 + 6), and so on, the light break prevents flowering (Fig. 14.32). These times correspond to darkness in a normal environment. But if the light break is given at a time when the endogenous rhythm would be "expecting" normal daylight conditions, such as at 16, 40 (24 + 16), or 64 (24 + 24 + 16) hours after the beginning of the dark treatment, the light break does not stop flowering. Plants kept in uniform, dark conditions undergo an endogenous cyclic sensitivity and insensitivity to red light interruption of the critical night length. Just how the endogenous rhythm and the critical night length work together to stimulate flowering is not known.

SUMMARY

   1. It is selectively advantageous for organisms to be able to sense and respond to significant aspects and changes in their environments.

   2. Communication between the various body parts of an organism is essential to the integration and coordination of the organism's metabolism and development. Certain parts may need to respond to environmental or metabolic changes that they cannot sense themselves.

   3. Plants must perceive important environmental information, transduce it to a communicable form, and respond to the transduced information.

   4. Four ways in which plants respond to stimuli are tropic responses (oriented growth), nastic responses (stereotyped turgor changes), morphogenic responses (changes in quality), and taxis (oriented swimming).

   5. In general, sites of perception are different from sites of response and must be linked by a means of communication. Presentation time and threshold are important elements in perception. Most communication appears to be by hormones.

   6. At present, the known classes of plant hormones are auxins, cytokinins, gibberellins, abscisic acid, and ethylene.

   7. The response to a hormone depends on which hormone is acting, the preparation of the responding cell, and the simultaneous or sequential presence of other hormones.

   8. Flowering may involve the following steps: competence ("ripeness") to be induced, occurrence of inductive conditions, sufficient health to produce flowers, and later stimuli to induce flowers to open.

   9. Flowering and many other seasonal responses are controlled by night length in many species. Phytochrome is the pigment involved in measuring night length.'

   10. Plants contain endogenous rhythms, cyclic changes in their metabolism. The rhythms most frequently are circadian, having a period of approximately 24 hours. These rhythms affect numerous aspects of plant metabolism.

IMPORTANT TERMS

abscisic acid (ABA)
all-or-none response
apical dominance
auxin
chemo-
circadian rhythm
critical night length
cytokinin
day-neutral plants
differential growth
dosage-dependent response

endogenous rhythm
ethylene
gibberellin
gravi-
hormone
indoleacetic acid
long-day plants
morphogenic response
nastic response
perception of a stimulus
photo-

photoperiod
phytochrome
presentation time
response to a stimulus
short-day plants
taxis
thigmo-
threshold for a stimulus
transduction of a stimulus
tropic response (negative and positive)
vernalization

REVIEW QUESTIONS

1. Name five environmental factors that plants are able to detect as stimuli. In each case, what types of information are provided to the plant?

2. In the perception of a stimulus, what are presentation time and threshold?

3. Name the five known classes of plant hormones. What are the characteristics of plant hormones?

4. What are the four ways that a plant can respond to a stimulus?

Define and give examples of each. Why are some responses classified as "positive," "negative," or "dia-," whereas others are not?

5. What are long-day plants? Short-day plants? Day-neutral plants? What is the critical night length?

6. What is the difference between endogenous and exogenous rhythms? Give several examples of each type, and be certain to include several that have either short periods or long ones.

BotanyLinks .net Questions

www.jbpub.com/botanylinks

Visit the .net Questions area of BotanyLinks (http://www.jbpub.com/botanylinks/netquestions) to complete this question:

1. How can tissue culture be used to study plant growth and development? Go to the BotanyLinks home page for more information on this subject.

BotanyLinks includes a Directory of Organizations for this chapter.

GENES AND THE GENETIC BASIS OF METABOLISM AND DEVELOPMENT

15

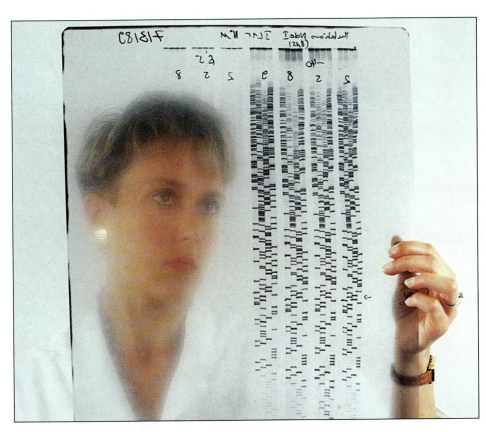

Techniques of DNA analysis and manipulation permit botanists to study how plants activate or suppress the genes that control metabolism and development. (*Jean Claude Revy/Phototake*)

CONCEPTS

Plants are composed of numerous types of cells. Each cell type is unique because it has a distinct metabolism, based largely on proteins such as enzymes, microtubules, and membrane proteins. Although all cells carry out a fundamental metabolism involving respiration, amino acid synthesis, and so on, some of the reactions in one cell type differ from those in other cell types. Each cell type may have one or more characteristic types of enzymes or other proteins. For example, enzymes involved in synthesis of flower color pigments are present in petal cells but not in cells of roots, wood, and bark (Fig. 15.1a).

(a)

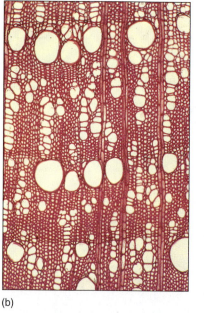

(b)

Intercellular spaces   Chlorenchyma cells

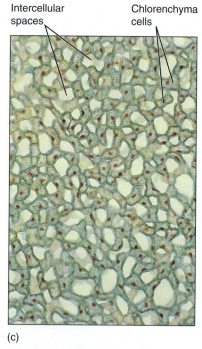

(c)

FIGURE 15.1

Cell differentiation is controlled by regulating particular genes in each type of cell. (a) These petals have enzymes necessary for the synthesis of pigments. These enzymes are not produced in most of the other cells of the plant, although all cells contain the necessary genes. (b) These wood cells had many enzymes not found in the petal cells of the same plant. During differentiation, each cell type became unique as different genes were activated, different proteins were produced, and their metabolisms diverged into unique pathways ($\times 200$). (c) Chlorenchyma cells differ from other types by having well-developed chloroplasts ($\times 200$). (d) These cells have differentiated such that starch storage and release are the dominant aspects of metabolism. They probably have no unique enzymes: All cells can metabolize starch. Degree of activity of a particular set of proteins is important in differentiation ($\times 180$). (e) These vessels and fibers have similar if not identical metabolisms for synthesis and lignification of walls; they differ primarily in cell shape and pattern of secondary wall deposition. Precise positioning of cellular elements is also critical during differentiation ($\times 150$).

Starch grains

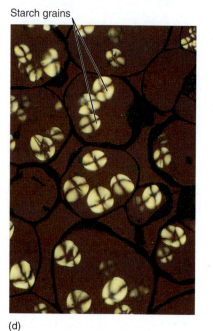

(d)

Fibers   Vessel

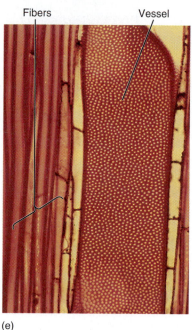

(e)

Also, sclerenchyma cells contain all the types of enzymes necessary for producing and lignifying secondary walls, but these enzymes and metabolic pathways are not present in parenchyma cells (Fig. 15.1b, c, and d).

Cells also differ in shape, again largely owing to differences in their proteins. All tracheids and vessel elements probably have the same enzymes and metabolism for secondary wall deposition, but the wall can vary (Fig. 15.1e). We believe that wall deposition is guided by a pattern of proteins in the protoplasm, probably a set of microtubules. Similarly, many cell divisions occur in precise patterns, so there must be an underlying pattern in the cell that causes the mitotic spindle to have the proper alignment.

The information needed to construct each type of protein is stored in genes, but because an organism grows by mitosis—duplication division—all its cells have identical sets of genes. As each cell differentiates and develops a unique suite of proteins, the underlying developmental process is the **differential activation of genes**. In the maturing epidermis, genes that code for cutin-synthesizing enzymes are turned on, whereas in other cells they remain turned off (Fig. 15.2). On the other hand, genes for P-protein remain quiescent in all except phloem cells. Studies of development and morphogenesis examine the mechanisms by which some genes are activated and others are repressed.

During protein synthesis, the correct amino acids must be incorporated in the proper sequence, because this determines both the structure and all other properties of the protein. The cell must contain a source of information that holds the sequence information for all its proteins; this information archive is **DNA, deoxyribonucleic acid** (see Figs. 2.19 and 2.20). DNA is a linear, unbranched polymer composed of four types of deoxynucleotide monomers, usually abbreviated A, T, G, and C (Table 15.1). Once actually polymerized into DNA, the base portion of each nucleotide monomer protrudes as a side group. It is the sequence of nucleotide side groups that is the information needed to synthesize proteins correctly. A **gene** is each region of DNA that is responsible for coding the amino acid sequence in a particular protein. Each type of protein has its own gene.

Both environment and protoplasm also contain information vitally important for plant growth, morphogenesis, and survival. As described in Chapter 14, the environment provides informative cues about season, moisture availability, time for seed germination, time for flowering, and direction of gravity. These environmental and metabolic signals must be converted into chemical messengers that enter the nucleus and interact with genes. If the signals indicate that the cell is to differentiate into a vessel element, all genes that produce enzymes necessary for the synthesis and lignification of a secondary wall must be located and activated. Genes that code for the proteins that guide a particular pattern of wall deposition also must be turned on. Conversely, the cell must be inhibited from undergoing any further cell division; genes involved in mitosis and cytokinesis must be repressed.

Recently, techniques have been developed that permit botanists to locate the genes for many proteins; the genes can then be isolated in vitro and duplicated and their nucleotide sequences revealed. At present our knowledge is still limited, but these techniques of **DNA sequence analysis** are so powerful that progress is extremely rapid. Similar techniques make it possible to alter the DNA sequence and then insert the gene back into a plant cell. As the cell grows, divides, and differentiates, the altered DNA either produces an altered protein if the coding region was changed or produces the protein at an unusual time or place if its control site was changed. These **recombinant DNA techniques,** sometimes called **genetic engineering,** are helping us understand the processes that occur between the perception of a stimulus and the plant's response to that stimulus. In addition, recombinant DNA techniques permit us to change features of plants—for example, making them more resistant to insects or having seeds and fruits that are more nutritious for us.

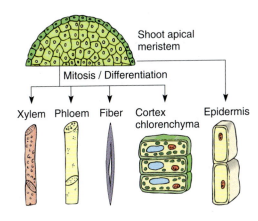

FIGURE 15.2

Because all body cells are produced by mitosis, they all have the same genes. Certain basic metabolism genes are probably active in all cells, but specific pathways become active during differentiation, probably because the genes that code for the enzymes of those pathways become active. Once fully mature, the cells may go back to basal metabolism. Epidermal cells often produce cutin only when differentiating, not after maturity. Sieve elements lose their nuclei as part of maturation, and tracheary elements digest away all of their protoplasm; they have no metabolism at all when mature.

TABLE 15.1 *Nucleotides*

Symbol	Base	Nucleoside	Ribonucleotide	Deoxyribonucleotide
T	thymine	thymidine	—	dTTP
A	adenine	adenosine	ATP	dATP
G	guanine	guanosine	GTP	dGTP
C	cytosine	cytidine	CTP	dCTP
U	uracil	uridine	UTP	—

Nucleic acids contain five bases, abbreviated T, A, G, C, and U. One base plus a five-carbon sugar is a nucleoside, and with a phosphate attached, a nucleoside becomes a nucleotide:

base + sugar = nucleoside

base + sugar + phosphate = nucleotide

When DNA is synthesized, deoxyribonucleotides—those that contain the sugar deoxyribose—are used; when RNA is made, ribonucleotides—containing ribose—are used. Uracil does not occur in DNA, and thymine is absent from RNA.

STORING GENETIC INFORMATION

PROTECTING THE GENES

It is critically important that the information in DNA be stored accurately for a long time; if storage is not safe, the information produced by the DNA will be inaccurate and probably useless or even harmful. There are several ways in which DNA is kept relatively inert and safely stored.

1. *DNA does not participate directly in protein synthesis.* Instead, DNA produces a messenger molecule, **messenger RNA (mRNA),** which carries information from DNA to the site of protein synthesis. The mRNA, not DNA, is exposed to the numerous enzymes, substrates, activators, and controlling factors of protein synthesis (Fig. 15.3). If mRNA is damaged, it can be replaced with more copies of mRNA. Within a single cell, thousands of individual molecules of a particular enzyme may be needed; if the DNA itself had to direct the synthesis of each protein molecule, it would probably be damaged long before enough protein had been synthesized. But instead, DNA directs the production of several copies of mRNA, each of which directs the production of hundreds of protein molecules.

2. *Most DNA is stored in the nucleus, protected from the cytoplasm by the nuclear envelope.* During interphase, the nuclear envelope forms the outer boundary of the nucleus, keeping most cytoplasmic components out and the proper nuclear substances in. The DNA of plastids and mitochondria are protected from cytosol enzymes by being located within plastids and mitochondria themselves.

3. *Histone proteins hold most nuclear DNA in an inert, resistant form.* **Histones** are a special class of proteins found in all organisms that have nuclei (plants, animals, fungi, algae, and protozoans). There are five types—H1, H2A, H2B, H3, and H4. The last four are among the most highly conserved proteins known; that is, the sequence of amino acids in the histones of one organism is virtually identical to the sequence in any other organism. For example, the H4 histone contains 103 amino acids, and its sequence in higher animals, such as cows, differs from its sequence in higher plants, such as peas, at only two sites (Fig. 15.4). Histone proteins are so

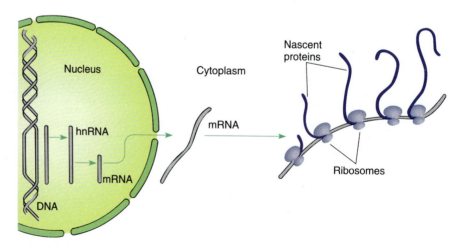

FIGURE 15.3

When a gene is active, its sequence of nucleotides guides the synthesis of "heterogeneous nuclear RNA," whose nucleotide sequence is complementary to that of the gene. The hnRNA is modified into messenger RNA, which is transported to the cytoplasm; it binds to ribosomes that translate (read) the nucleotide sequence in the mRNA and polymerize amino acids in the proper order, thus creating a protein. Ribosomes on the left have just started, so their proteins are still short; ribosomes on the right have read almost all the RNA, so their proteins are longer, almost complete.

```
                                              −200
                                               .
AGCACCCTCCCACCTCATCCCACCCTTCTGATCTCAATCCAACGTCGCATCTCCACCGTCTCG
GCCACAGCTCTGACCCAGCCCGCCAAACCACCGGTCCAATCTCTCGGCCACGTCACCGATCCG

                                              −150
                                               .
CGGATCGACCCAGCGAAGTCCCTCCCGCCCCCAAAGTCCCCCAAATCTTGCAGTTCCCTCCTA
CGGCATCTCTCCCCCGGATCGCCGTCTCGACCGTCCACTCCATCCGCATCCAACGGCAGCCAC

                                              −100
                                               .
AATCCTCCCCATATAAACCAACCCCCCGCCCTCAGATCCCTAATCCCATCGCAAGCATCAGAC
ACGCCTCCTCCAACCTCTCGACCCCTTTAAGACGCCCTTCGCCCCACCCAGCAAATCACAGCA

             −50
              .
TCCCTCCAAAGCAGGCAGCAGCTCCTCTTCTTCCTAATCACACTATCTCGGAGAGGAGCGGCC
CCAGACGCCACCCACCACCGTTCCTCCCATCCCACACTCGCTCGCAGCTCGAGATCGTCGGCC
```

```
+1
 .
ATG  TCT  GGG  CGC  GAC  AAG  GGC  GGC  AAG  GGG  CTG  GGC  AAG  GGC  GGC  GCC
---  --C  ---  ---  -G-  ---  --A  ---  ---  --C  --A  ---  ---  ---  ---  ---
     Ser  Gly  Arg  Asp  Lys  Gly  Gly  Lys  Gly  Leu  Gly  Lys  Gly  Gly  Ala
                    Gly
```

```
+50
 .
AAG  CGG  CAC  CGG  AAG  GTC  CTC  CGC  GAC  AAC  ATC  CAG  GGC  ATC  ACC  AAG
---  --C  ---  ---  ---  ---  ---  ---  --T  ---  ---  ---  ---  ---  ---  ---
Lys  Arg  His  Arg  Lys  Val  Leu  Arg  Asp  Asn  Ile  Gln  Gly  Ile  Thr  Lys
```

```
 +100
  .
CCG  GCG  ATC  CGG  AGG  CTG  GCC  AGG  AGG  GGC  GGC  GTG  AAG  CGC  ATC  TCC
---  ---  ---  C--  ---  --G  C--  C--  ---  ---  ---  ---  ---  ---  ---  --G
Pro  Ala  Ile  Arg  Arg  Leu  Ala  Arg  Arg  Gly  Gly  Val  Lys  Arg  Ile  Ser
```

```
     +150
      .
GGC  CTC  ATC  TAC  GAG  GAG  ACC  CGC  GGC  GTC  CTC  AAG  ATC  TTC  CTC  GAG
--G  ---  ---  ---  ---  ---  ---  ---  --G  ---  ---  ---  ---  ---  ---  ---
Gly  Leu  Ile  Tyr  Glu  Glu  Thr  Arg  Gly  Val  Leu  Lys  Ile  Phe  Leu  Glu
```

```
          +200
           .
AAC  GTC  ATC  CGC  GAC  GCC  GTC  ACC  TAC  ACC  GAG  CAC  GCC  CGC  CGC  AAA
---  ---  ---  ---  --T  ---  ---  ---  ---  ---  ---  ---  ---  ---  ---  --G
Asn  Val  Ile  Arg  Asp  Ala  Val  Thr  Tyr  Thr  Glu  His  Ala  Arg  Arg  Lys
```

```
          +250
           .
ACC  GTC  ACC  GCC  ATG  GAC  GTC  GTC  TAC  GCG  CTC  AAG  CGC  CAG  GGC  CGC
---  ---  ---  ---  ---  ---  ---  ---  ---  ---  ---  ---  ---  ---  ---  ---
Thr  Val  Thr  Ala  Met  Asp  Val  Val  Tyr  Ala  Leu  Lys  Arg  Gln  Gly  Arg
```

```
          +300
           .
ACC  CTC  TAC  GGC  TTC  GGA  GGC  TAG  ATTTGTGTGGTGAAGCAACTTCCTCGTTTGC
---  ---  ---  ---  ---  --C  ---  --A  GGGCCGGCCGGCCGACGGGAGTCACTCTTTG
Thr  Leu  Tyr  Gly  Phe  Gly  Gly
```

```
       +350                                             +400
        .                                                .
TCTGTGATCTGTGCTGTCGTAGATGAGATTTACTGATTTGGCGTGCGCCGGTTGTATTCTGTC
TCGCCGCCTGCAGATTCCAGAAGCCTGATGAAGCCCCGACTTGTTTAGTTCGCTATTTCCTCT
```

FIGURE 15.4

The nucleotide sequence for the gene for histone H4 in wheat is presented in the top row of each set of lines. In the second row is the sequence for the same gene for a different type of wheat. Where the two genes are identical, only a dash appears for the second gene. The portion of the gene that codes for protein begins in the fifth line (labeled +1), and the amino acid sequence—the primary structure of the protein—is given in the third row of each set of lines. Wherever a mutation has caused the second gene to code for an amino acid different from the first gene, that amino acid is given in the fourth row. Although 18 mutations have occurred, the resulting amino acid sequence is unchanged except at one site (most of these mutations have no effect because the genetic code is redundant—see later in chapter). The noncoding regions of the gene—from the beginning to +1, and from +309 to the end—are not highly conserved, and the two genes differ greatly in these sites. (*Sequence data obtained by T. Tabata and M. Iwabuchi*)

essential that virtually any change in their amino acid sequence causes the organism to die or at least not reproduce.

Histones form aggregates and DNA wraps around them, forming a spherical structure called a **nucleosome.** Histone H1 then binds nucleosomes into a tightly coiled configuration. In this mode the DNA/protein structure—chromatin—is so dense that enzymes cannot penetrate it and DNA is relatively inert. Even if it is exposed directly to DNA-digesting enzymes, called **DNases** (also written as DNAases), histone-bound DNA is not extensively damaged. However, chromatin is still sensitive to regulatory molecules and can be unpacked in preparation for synthesis of mRNA or for replication of DNA during the S phase of the cell cycle.

THE GENETIC CODE

Twenty types of amino acids are used in synthesizing proteins, but only four different nucleotides are present in DNA or mRNA; consequently, it is not possible for one nucleotide alone to specify one amino acid, because 16 amino acids would be left without nucleotides to code for them. Similarly, nucleotides cannot be used simply in pairs of two, such as AU for isoleucine or CC for proline, because there are only 16 possible pairs. It is necessary for nucleotides to be read and used in groups of three; 64 possible triplets, known as **codons,** can be made using four nucleotides. Table 15.2 shows which amino acid is coded by each codon.

With 64 possible triplets, a surplus of 44 codons remains after each amino acid is paired with a codon. Three of the codons—UAA, UAG, and UGA—are **stop codons;** they signal that the ribosome should stop protein synthesis. The codon AUG is the **start codon** that signals the point in mRNA where protein synthesis should begin. The extra 40 codons also code for amino acids, so most amino acids have two or more codons. For example, both UUU and UUC code for phenylalanine, and CAU and CAC code for histidine. Because multiple codons exist for most amino acids, the genetic code is said to be **degenerate.** Degeneracy further protects DNA: A mutation in DNA might change a codon in mRNA from UUU to UUC for example, but because both code for phenylalanine, the same protein is produced before and after this particular mutation.

The genetic code is almost perfectly universal; all organisms and genetic systems but one share the genetic code shown in Table 15.2. Viruses, bacteria, fungi, animals, and plants all use the same codons to specify particular amino acids, and the same is true for plastid DNA. Only in mitochondria are several codons changed. This almost universal commonality of the genetic code is one of the strongest pieces of evidence that life arose only once on Earth and that all living organisms have evolved from one ancestral organism.

Notice that the name "codon" is applied to triplets that occur in mRNA, not in DNA.

THE STRUCTURE OF GENES

Most genes, up to 90% in any cell, are quiescent most of the time and are activated and read only when the cell needs the particular enzymes they code for. Each gene must have a structure that allows controlling substances to recognize the gene, bind to it, and activate it at the proper time.

Genes are composed of a **structural region** that actually codes for the amino acid sequence, and a **promoter,** a controlling region involved in regulating the synthesis of mRNA from the structural region (Fig. 15.5). The promoter is located "upstream" from the structural region, that is, to the 5′ side. It varies in length from gene to gene but can be

The carbons in ribose and deoxyribose are numbered from 1′ to 5′, with phosphate attached to the 5′ carbon and the hydroxyl used in polymerization attached to the 3′ carbon. Any nucleic acid has a 5′ end and a 3′ end.

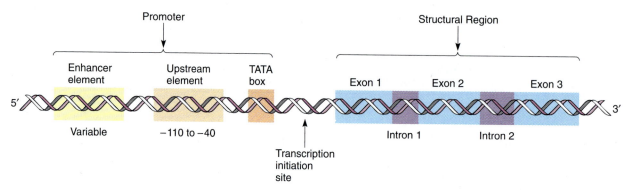

FIGURE 15.5

A gene is always written with the 5′ end, and thus the promoter, on the left. Nucleotides are numbered beginning at the left boundary of the structural region; nucleotides to the left of this are given negative numbers and are said to be upstream. Because DNA is double-stranded, its length is measured as the number of nucleotide pairs or base pairs, whereas RNA, being single-stranded, is measured as the number of nucleotides or bases. The decision to speak of nucleotides versus bases is just personal choice.

First Base	Second Base				Third Base
	U	**C**	**A**	**G**	
U	UUU Phenylalanine	UCU Serine	UAU Tyrosine	UGU Cysteine	U
	UUC Phenylalanine	UCC Serine	UAC Tyrosine	UGC Cysteine	C
	UUA Leucine	UCA Serine	UAA *STOP*	UGA *STOP*	A
	UUG Leucine	UCG Serine	UAG *STOP*	UGG Tryptophan	G
C	CUU Leucine	CCU Proline	CAU Histidine	CGU Arginine	U
	CUC Leucine	CCC Proline	CAC Histidine	CGC Arginine	C
	CUA Leucine	CCA Proline	CAA Glutamine	CGA Arginine	A
	CUG Leucine	CCG Proline	CAG Glutamine	CGG Arginine	G
A	AUU Isoleucine	ACU Threonine	AAU Asparagine	AGU Serine	U
	AUC Isoleucine	ACC Threonine	AAC Asparagine	AGC Serine	C
	AUA Isoleucine	ACA Threonine	AAA Lysine	AGA Arginine	A
	AUG *(START)* Methionine	ACG Threonine	AAG Lysine	AGG Arginine	G
G	GUU Valine	GCU Alanine	GAU Aspartic acid	GGU Glycine	U
	GUC Valine	GCC Alanine	GAC Aspartic acid	GGC Glycine	C
	GUA Valine	GCA Alanine	GAA Glutamic acid	GGA Glycine	A
	GUG Valine	GCG Alanine	GAG Glutamic acid	GGG Glycine	G

TABLE 15.2 *The Codons of mRNA*

The triplets in mRNA are codons; the DNA triplets in the gene itself are their complements. The first base of the codon is listed on the left and the second base at the top. Within each large box are the four possible third bases and the amino acid specified by each.

several hundred nucleotides long. Certain regions are particularly important; one, called the **TATA box,** is a short sequence about six to eight base pairs long rich in A and T. If the TATA box is damaged by either mutation or experimental treatment, the RNA-synthesizing enzyme **RNA polymerase II** does not bind well. Most eukaryotic genes have other promoter sequences called **enhancer elements** located even farther upstream, as many as several hundred base pairs away from the structural region of the gene. It is hypothesized that when a hormone alters cell metabolism, it does so by producing intracellular chemical messengers that activate genes either by binding directly with the promoter region or by binding with proteins that then interact with the promoter. After activating agents have bound to the promoter, RNA polymerase II can attach.

Once RNA polymerase II binds to the promoter, it migrates downstream (toward the 3' end of the DNA strand) toward the structural region. However, it does not create any RNA until it is about 20 to 30 nucleotides below the TATA box. The RNA polymerase might be expected to search for the DNA equivalent of the AUG start codon of mRNA, but that is not the case. If some of the DNA is artificially removed between the TATA box and the normal start site, the RNA polymerase begins synthesizing RNA farther downstream than normal.

The structural portion of genes contains two distinct types of regions: exons and introns. **Exons** are sequences of nucleotides whose codons are eventually expressed (**ex**on, **ex**pressed) as sequences of amino acids in proteins, and **introns** are sequences of nucleotides that are not expressed, but instead **inter**vene between exons (Figs. 15.5 and 15.6). Several plant genes have just two or three introns: the gene for RuBP carboxylase and the genes for the storage proteins glycinin and phaseolin of legume cotyledons. The gene that codes for the protein portion of phytochrome has five introns, one of which is 1500 base pairs long. Genes with 25 introns occur, as do genes with no introns.

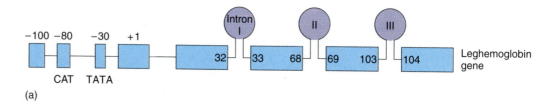

(a)

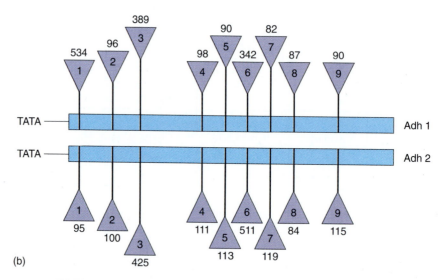

(b)

FIGURE 15.6

Two ways of illustrating introns and exons in maps of genes. (a) The gene for leghemoglobin in legumes (see Chapters 7 and 13) has three introns: Intron I occurs between bases 32 and 33 of the finished mRNA; intron II between bases 68 and 69, and so on. (*After Brown et al.*, J. Molecular Evolution *21:19–32, 1984*) (b) Two genes in corn produce two similar enzymes, both called alcohol dehydrogenase and distinguished as Adh1 and Adh2. The numbers above and below the triangles indicate the number of bases present in each intron. The two genes are similar, having nine introns located at the same positions; corresponding exons in Adh1 and Adh2 are the same length, but corresponding introns may be quite different. Intron 1 is 534 bases long in Adh1 but only 95 bases long in Adh2. (*After Llewellyn et al.*, Molecular Form and Function of the Plant Genome. *Edited by Vloten-Doting et al. New York, Plenum Press, 1985*)

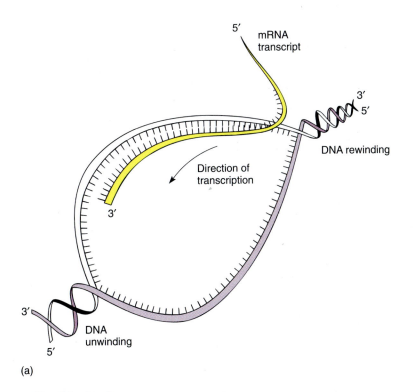

(a)

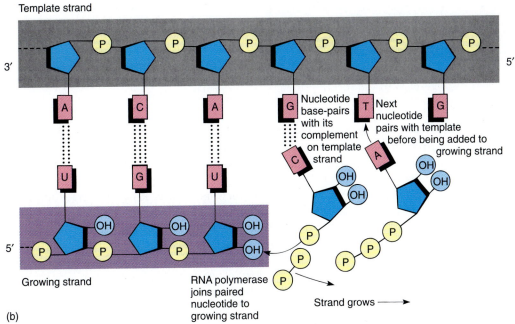

(b)

FIGURE 15.7
When a gene is turned on, the DNA double helix separates over a short region (a); free ribonucleotides can now diffuse in and pair with the region of temporarily single-stranded DNA (b). The formation of two or three hydrogen-bonds between the DNA deoxyribonucleotides and the free ribonucleotides allows the DNA sequence to control the sequence of the RNA being formed.

TRANSCRIPTION OF GENES

After RNA polymerase binds and encounters the start signal, it begins actually creating RNA, a process called **transcription.** The two strands of DNA separate from each other over a short distance, and free ribonucleotides diffuse to the region (Fig. 15.7). If a ribonucleotide containing cytosine approaches a DNA nucleotide that contains guanine, the two form three hydrogen-bonds and remain together, at least temporarily. Wherever DNA contains T, a free A can form two hydrogen-bonds to it; similarly, free U forms two hydrogen-bonds with A in DNA, and so on with G and C. RNA polymerase binds the free ribonucleotide, holds it, and catalyzes the formation of a covalent bond, forming RNA. As

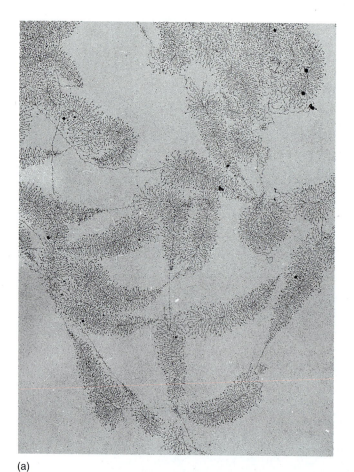

(a)

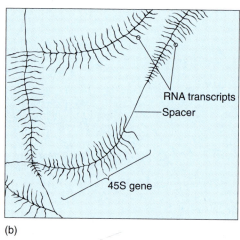

(b)

FIGURE 15.8

DNA can be carefully extracted from a nucleus, allowed to spread out, and then prepared for examination in an electron microscope (a). These are ribosomal genes being transcribed into long RNAs that will later be cut into three separate rRNAs. *(Courtesy of O. L. Miller, University of Virginia)* The diagram (b) explains each type of line.

Free ribonucleotides diffuse to and pair with both strands of DNA, but RNA polymerase is located only on the strand that has the gene; the complementary strand is not read.

each covalent bond is formed, two high-energy phosphate-bonding orbitals are broken, so polymerization is a highly exergonic process that cannot be easily reversed.

Transcription proceeds rapidly, incorporating about 30 ribonucleotides per second, with the DNA double helix unwinding ahead of the moving enzyme. Once RNA polymerase moves off the promoter/initiation site, a new molecule of RNA polymerase binds and begins synthesizing another molecule of RNA (Fig. 15.8). Whereas two molecules of DNA wrap around each other into a double helix, RNA/DNA duplexes do not; the RNA polymer that emerges from RNA polymerase releases from the DNA.

RNA polymerase continues to act until it encounters a transcription stop signal in the DNA. The stop signal of several genes consists of two parts. The first is a short series of DNA nucleotides that are self-complementary; that is, the RNA transcribed from them can double back on itself and hydrogen bond to another part of itself. This results in a small kink, a **hairpin loop** that is believed to affect RNA polymerase. Just downstream of this region of DNA is a long series of adenines. Various protein factors also are involved; in their absence RNA polymerase sometimes continues transcribing, reading the next region of DNA as if it were also part of the gene.

RNA polymerase transcribes both introns and exons into a large molecule of **hnNRA** (heterogeneous nuclear RNA) that is rapidly modified by nuclear enzymes. Introns are recognized, cut out, and degraded back to free ribonucleotides. Exons are spliced together, resulting in an RNA molecule, all of which codes for amino acids. All RNA destined to become mRNA is somehow recognized by an enzyme that binds to it and attaches a series of adenosine ribonucleotides on its 3′ end, forming a **poly(A) tail** about 200 bases long, the only exception being the mRNAs for histone proteins (Fig. 15.9). Another step in message processing involves changing the first nucleotide into 7-methyl guanosine. Ultimately a completed mRNA is produced and transported from nucleus to cytoplasm.

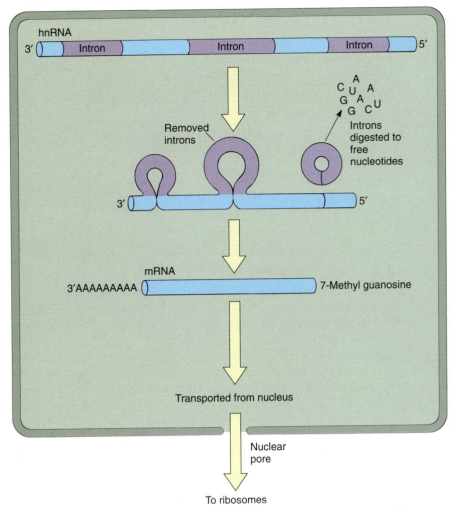

3' AAAAAAAAA 7-Methyl guanosine

Transported from nucleus

Nuclear pore

To ribosomes

FIGURE 15.9

The primary transcript, hnRNA, has its introns cleaved out and the exons spliced together; the introns are depolymerized back to free nucleotides. An enzyme adds up to 200 adenosine ribonucleotides to the end of the RNA; these are not coded by thymidines in the DNA. The first nucleotide at the 5' end is converted to 7-methyl guanosine.

PROTEIN SYNTHESIS

In the process of protein synthesis, ribosomes bind to mRNA and "read" its codons. Guided by the information in the nucleotide sequence of the mRNA, the ribosomes catalyze the polymerization of amino acids in the order specified by the gene from which the mRNA was transcribed.

RIBOSOMES

Ribosomes are small, cytoplasmic particles that "read" the genetic message in mRNA and construct proteins guided by that information. Each is composed of two subunits, one larger than the other, and each is made up of both proteins and **ribosomal RNA, rRNA** (Fig. 15.10; Table 15.3). The small subunit contains one molecule of rRNA, the large subunit one molecule each of three types of rRNA. The number of proteins present in eukaryotic ribosomes is known to be greater than 80, but the exact number is still uncer-

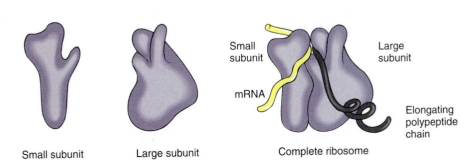

Small subunit   Large subunit   Complete ribosome

FIGURE 15.10

Ribosomes have two subunits, one large and one small. When they fit together there is a groove for mRNA to pass through, a channel through which the growing protein emerges, and a channel into which amino acid carriers enter. This diagram is based on ribosomes of the bacterium *Escherichia coli*; eukaryotic cytoplasmic ribosomes are similar, but we are not as certain about details of their shape.

TABLE 15.3	*Components of 80S Cytoplasmic Ribosomes*	
	Size of RNA Molecules	Number of Proteins
Small subunit	18S	approx. 33 ?
Large subunit	28S	approx. 49 ?
	5.8S	
	5S	
	4 molecules	approx. 84

tain. Ribosomes found in the cytoplasm of eukaryotes are designated **80S,** meaning that they are relatively large and dense; ribosomes of plastids, mitochondria, and prokaryotes are smaller, lighter **70S ribosomes.** (S is a Svedberg unit, used to measure the rate at which a particle sediments in a centrifuge.)

The genes for the three largest rRNAs are unusual because they occur tightly grouped together and act as a single gene with just one promoter (Fig. 15.11). Transcription of this cluster, which is located in the nucleolus, produces a long RNA molecule that is then cut into three pieces (see Fig. 15.8). Short regions are digested off the ends of these pieces, resulting in three rRNA molecules. The gene for the smallest rRNA molecule (5S RNA) is not located in the nucleolus but out in the chromatin along with protein-coding genes.

Once transcribed, rRNA molecules combine with ribosome proteins in the nucleolus, forming one large particle that then is cleaved into the large and small subunits of a ribosome. These subunits are then transported out of the nucleolus through the nucleus to the cytoplasm.

All cells need large numbers of ribosomes. Most genes are present in a diploid nucleus as only two copies. But just two copies would be inadequate for producing the hundreds of thousands of rRNAs needed. Instead, rRNA genes are highly amplified; that is, many copies of each are present. In flax plants, each nucleus may have up to 120,000 copies of the gene for the small rRNA and 2700 copies of the gene cluster for the three large rRNA molecules.

tRNA

During protein synthesis, amino acids are carried to ribosomes by ribonucleic acids called **transfer RNA, tRNA.** tRNAs are necessary because a codon cannot interact directly with an amino acid; the genetic code can be read only by a ribonucleic acid that has a three-nucleotide sequence, called an **anticodon,** that is complementary to and hydrogen bonds to

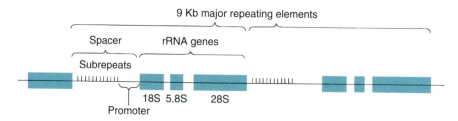

FIGURE 15.11

The genes for the 18S, 5.8S, and 28S ribosomal RNAs occur together as a cluster, separated by short regions of spacer DNA. All three genes are transcribed as just one molecule of RNA, spacers included; spacer RNA is cut out, resulting in three individual rRNA molecules. Upstream from the 18S gene is the promoter and then a region made up of 10 to 15 copies of a short sequence about 135 base pairs long, called subrepeats; they are not transcribed. On either side of the rRNA gene cluster are more rRNA clusters, as many as 2700, all occurring end to end as a gene "family." Kb is kilobases; 1 Kb = 1000 bases.

the codon. For example, UUU and UUC are both codons for phenylalanine; the tRNAs that carry phenylalanine have the complementary anticodons of either AAA or GAA.

There are as many types of tRNA as there are codons that specify amino acids; stop codons do not have tRNAs. All tRNAs have the same parts, an anticodon and an **amino acid attachment site** at its 3′ end consisting of the sequence CCA (Fig. 15.12). A special class of enzymes recognizes each tRNA and attaches the correct amino acid to it. This step, called amino acid activation, must be precise. If the wrong amino acid is placed onto a tRNA, it becomes incorporated into the protein as if it were the correct amino acid, causing the protein to have an erroneous structure.

GAA looks backward; by convention nucleotide sequences are written 5′ to 3′: GAA. An alternative is to write 3′-AAG-5′, the numbers showing that it is written backward, allowing easier comparison with the codon.

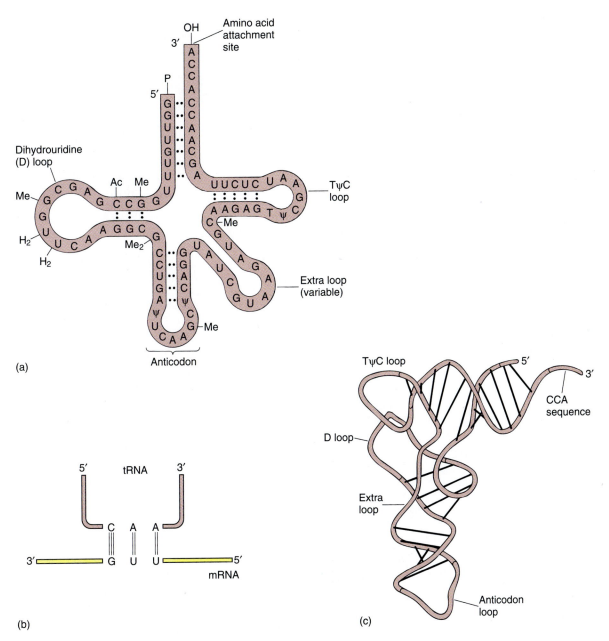

FIGURE 15.12
(a) and (b) All tRNAs have this general shape, with at least three loops caused by self-complementary base pairing. The small fourth loop is present only in some. The anticodon is at the middle loop and the amino acid is attached to the CCA at the 3′ end. This is one of the six tRNAs that carry leucine; it has the anticodon CAA, which recognizes the codon 3′-GUU-5′, normally written UUG. The unusual symbols (Ac, Me, ψ) indicate bases that are chemically modified into unusual forms. (c) tRNA folds into an **L** shape rather than lying flat.

Transfer RNAs contain bases not found in other nucleotides. tRNAs are transcribed from tRNA genes, and like all ribonucleic acids, they contain the common bases A, U, G, and C. But after these are polymerized into tRNA, enzymes modify some bases to unusual forms (Fig. 15.12a).

Each tRNA carries an amino acid only briefly; after it is activated, it rapidly encounters a ribosome that is "reading" the codon complementary to its anticodon. It gives up its amino acid and shuttles back to the cytosol and is reactivated. Even though each tRNA can cycle like this many times per second, millions of tRNAs are needed in every cell, and hundreds or even thousands of copies of each tRNA gene may be present in each nucleus.

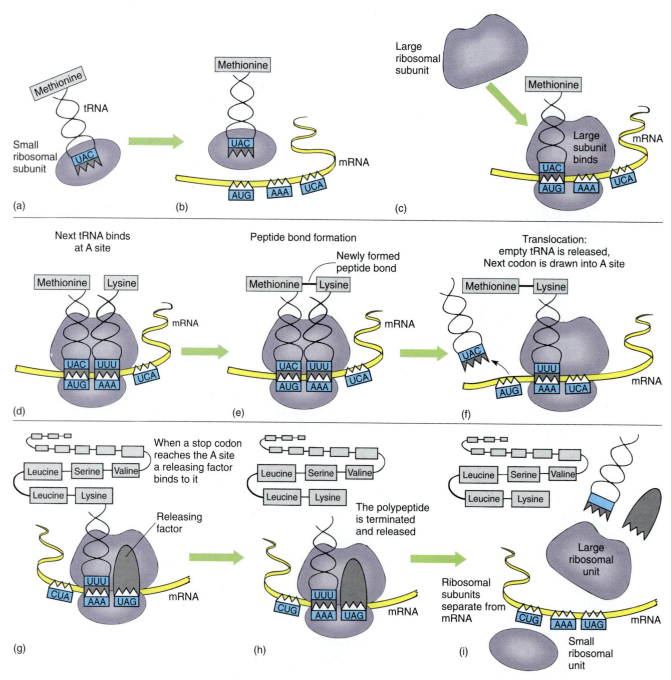

FIGURE 15.13
Protein synthesis; the steps are explained in the text.

mRNA Translation

Initiation of Translation.   The synthesis of a protein molecule by ribosomes under the guidance of mRNA is called **translation**. Protein synthesis begins with a complex **initiation** process involving the start codon AUG. This codes for the amino acid methionine, but two types of tRNA actually carry methionine, and they have different properties. One, called initiator tRNA, binds to the ribosome small subunit even though no mRNA is present (Fig. 15.13). Several initiation factors also bind to the small subunit. In this condition, the complex is competent to bind to mRNA, after which the large subunit of the ribosome binds to the complex and the initiation factors are released. Just how mRNA is recognized is not known, but the initiator tRNA is important for finding the AUG start codon and positioning the small subunit.

It is critically important that the ribosome be properly aligned on mRNA, because the sequence of nucleotides can be read in any set of three. The sequence CUUGCACAG can be read as CUU GCA CAG and would code for leucine alanine glutamine (Fig. 15.14a). But if the ribosome binds incorrectly, shifted downstream by one nucleotide, mRNA is read as _ _ C UUG CAC AG _, the codons for phenylalanine histidine and either serine or arginine, depending on the next nucleotide. Reading nucleotides in the wrong sets of three is a **frameshift error;** because virtually all codons are misread, frameshift errors typically result in completely useless proteins.

Elongation of the Protein Chain.   Messenger mRNA lies in a channel between the two ribosome subunits. Extending outward from the mRNA channel are two grooves, each wide enough for a tRNA to fit into it such that the tRNA anticodon can touch an mRNA codon. When both channels contain activated tRNAs, adjacent codons are being read (Fig. 15.13).

At the time of initial binding, the small subunit already contains initiator tRNA in the **P channel** (P for protein). The adjacent **A channel** (A for amino acid) is empty, but numerous molecules enter it at random. Some are activated tRNAs, but if their anticodons do not complement the exposed codon at the bottom of the A channel, they diffuse out. If a tRNA with the proper, complementary anticodon enters, hydrogen-bonds form between the codon and the anticodon. This holds the tRNA in place long enough for an even more stable binding to occur. Enzymes located in the large ribosomal subunit break the bond between the methionine and its tRNA in the P channel, simultaneously attaching the methionine to the amino group of the amino acid on the tRNA in the A channel. This reaction needs no outside source of power: The broken bond is a high-energy bond, whereas the newly formed peptide bond is only a low-energy bond.

The empty tRNA, freed of its methionine, is released from the P channel, and the ribosome pulls itself along the mRNA for a distance of three nucleotides. This movement is powered by GTP and seems to be performed by proteins on the large subunit. As the ribosome slides along the mRNA, the A channel slides to the next codon and the tRNA with the two amino acids attached becomes surrounded by the P channel, not the A channel.

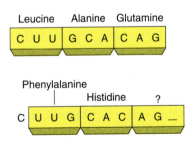

FIGURE 15.14
Because there are no spacers between codons, the ribonucleotides can be read in three possible ways. Each results in completely different proteins, only one of which has the proper primary structure.

When the proper activated tRNA diffuses into the A channel and hydrogen bonding between codon and anticodon occurs, the large subunit enzymes again release the short protein chain (now two amino acids long) from the tRNA in the P channel and attach it to the amino acid on the tRNA in the A channel, creating a protein three amino acids long. This process repeats until the ribosome reaches a stop codon.

Termination of Translation.   When a stop codon is pulled into the A channel, normal elongation cannot occur; no tRNA is present with an anticodon complementary to a stop codon. Instead, a release factor enters the channel and stimulates the large subunit enzymes to initiate the normal reactions. The high-energy bond holding the protein to the P-channel tRNA is broken, and a new bond is formed with water, releasing the protein from both the tRNA and the ribosome. The tRNA is released, and the small subunit disassociates from the large subunit and releases the mRNA. All components diffuse away, but as soon as the small subunit encounters an initiator tRNA and other initiating factors, it binds to another mRNA and the process begins again.

In summary, protein synthesis involves the following steps: RNA polymerase transcribes hnRNA, being guided by the sequence of DNA nucleotides in a gene. The hnRNA is processed into mRNA as introns are cut out and exons are spliced together. After moving from nucleus to cytoplasm, mRNA binds with ribosomes, and tRNAs carry amino acids to the ribosome-mRNA complex. The ribosome translates the mRNA codon by codon, and tRNAs fit into the ribosome only when their anticodon is complementary to the codon in the A channel. Once the protein is completed, it is released and the ribosome subunits detach from the mRNA. All components diffuse away from each other, but the ribosome subunits can translate the next mRNA they happen to meet. The mRNA can be translated again by other ribosomes.

CONTROL OF PROTEIN LEVELS

As cells undergo differentiation and morphogenesis, their metabolism and structure become different from those of other cells, owing to the presence of proteins, especially enzymes, unique to that cell type. A central question in developmental biology is the mechanism by which distinct cell types control the activities of genes so that they undergo the proper differentiation and obtain the proper set of proteins. There are several points at which protein synthesis and activity theoretically could be controlled: making a gene physically available for transcription; nature of the promoter region; processing of hnRNA into mRNA; transport of mRNA from nucleus to cytoplasm; binding of mRNA to the ribosome small subunit; rate of translation; processing of protein; activation or inactivation of the protein.

Biochemical studies have shown that many enzymes and structural proteins may be present in a cell in an inactive form. Tubulin is an excellent example: A cell may contain a large pool of tubulin monomers, then aggregate them into microtubules at a specific time. Microtubules can appear rapidly without the need for gene activation or protein synthesis. Similarly, enzymes are often completely inactive until phosphate groups are added by a class of enzymes, called kinases, that are themselves activated by the arrival of hormones. Hormones arrive at the plasma membrane and bind to receptors, which then synthesize second messengers that enter the cell. These activate the phosphorylating enzymes that in turn activate the dormant enzymes, leading to a significant change in the cell's metabolism. This is mostly an activating rather than a differentiating mechanism; these cells are already prepared for highly specific responses.

It would be extremely inefficient if a cell transcribed all its genes and synthesized all of its possible proteins when only a few are needed. For maximum efficiency, a cell should not even synthesize the mRNA for proteins it does not need. We expect that the most fundamental level of control of morphogenesis would occur at the level of transcription.

Unfortunately, we know very little about the control of transcription in eukaryotes, especially plants. Our lack of specific knowledge in this area is not due to lack of effort but to the tremendous technical problems involved. Except for the genes that code for rRNA and tRNA, most genes are present as only two copies in a diploid nucleus. When hormones cause a cell to produce a new protein, it may be that only one or two activator molecules

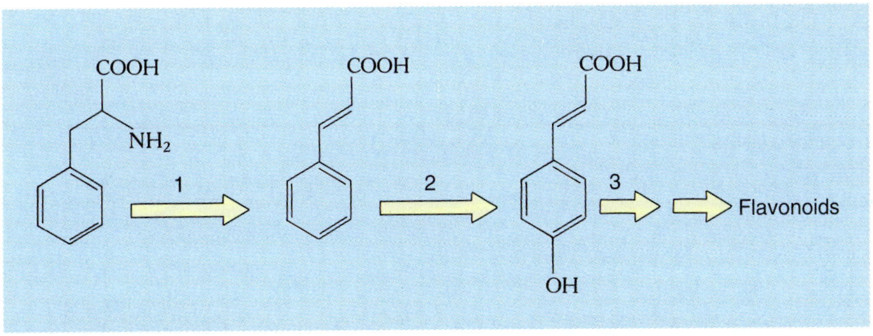

FIGURE 15.15
The genes for the first three enzymes in this pathway for the synthesis of flavonoids and lignin are turned on simultaneously. Perhaps they have identical promoters and are activated by the same type of intracellular chemical messenger. The enzymes are (1) phenylalanine ammonia lyase, (2) cinnamate 4-hydroxylase, and (3) 4-coumarate: CoA ligase.

bind to the promoter region of one or both copies of the gene that codes for that protein. Consequently, when we study the activation of a gene, we study a process in which a change occurs in only one or two molecules out of the billions in the nucleus. Because genes can also be turned off, the chemical messengers can also release from DNA, making it extremely difficult to find the actual molecules responsible for activating a particular gene.

Theoretically, chemical signals could activate genes in several ways. The hormone/receptor complex might actually bind to DNA, or presence of the hormone might cause the receptor to be liberated from a membrane so that the receptor alone binds to DNA. Or the hormone/receptor complex might cause synthesis or release of a second messenger that then enters the nucleus. Although past research in this area has been slow, DNA analysis techniques offer great promise. The genes for many proteins have now been identified, and their promoter regions are being studied. It is hoped that genes turned on simultaneously have aspects of their promoters that will indicate how they are activated in unison. For example, phenolic compounds are important in many aspects of a plant's metabolism, and their biosynthetic pathway is well understood. When cell cultures of parsley, which have few phenolics, are transferred from darkness to light, phenolic pigments called flavonoids appear. This is preceded by the sudden synthesis of the first three enzymes of their metabolic pathway, followed a few hours later by the appearance of the next two enzymes (Fig. 15.15). The first three enzymes may be coded by genes whose promoters are identical or are activated by similar chemical messengers.

Control can be exerted in the processing of hnRNA into mRNA. Rather surprisingly, the great majority of RNA transcribed in nuclei is immediately degraded back to free ribonucleotides. Even accounting for the large number of introns that must be removed, many exons are probably degraded also. Even though a gene is transcribed, its information may not be converted into mRNA. Some of the chemical messengers from the cytoplasm may bind to hnRNA rather than to DNA; they would regulate hnRNA processing rather than DNA transcription.

ANALYSIS OF GENES AND RECOMBINANT DNA TECHNIQUES

NUCLEIC ACID HYBRIDIZATION

The two halves of a DNA double helix can be separated by heating them just enough to break the hydrogen-bonds between complementary bases. This separation, which produces a solution of single-stranded DNA molecules, is called both **DNA melting** and **DNA denaturation**. If the solution of single-stranded DNAs is cooled slowly to a temperature low enough for hydrogen-bonds to reform, two molecules with complementary sequences form hydrogen bonds and stick together whenever they collide. If the two halves of one piece encounter each other, all of their bases pair and adhere firmly even at a relatively high temperature (Fig. 15.16a). But just by chance, many pieces have short sequences that are complementary, whereas most of their sequences are not. If these encounter each other, they form only a small number of hydrogen-bonds and probably fall apart again if the mixture is cooling only slowly. As the degree of complementarity decreases, the stability of pairing decreases.

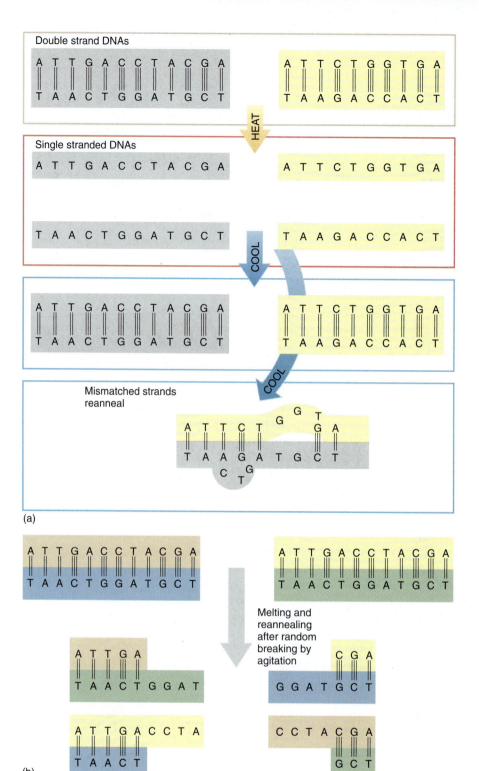

FIGURE 15.16

(a) Mild heating of DNA disrupts hydrogen bonding, allowing the two halves to separate into single-stranded molecules; cooling allows them to reanneal. In the millions of nucleotides in a single DNA molecule, short sequences occur many times simply by chance, so almost any two pieces have a few short complementary sequences. However, these are not usually enough to hold the two molecules of DNA together if the solution is still warm. The poorly bonded DNA duplexes fall apart, giving each piece another chance to encounter its true complementary partner. (b) If two identical pieces of DNA are broken at random, each produces pieces perfectly complementary to those produced by the other, but because the pieces are unequal in length, a large amount of single-stranded DNA remains.

The reformation of double-stranded DNA by cooling a solution of single-stranded DNAs, called both **DNA hybridization** and **reannealing,** is used to determine the relatedness of two types of DNA. For example, DNA can be extracted from two organisms, cut into small pieces, and melted. The sample from one is attached to a filter, and the other sample, which has been made radioactive, is poured over it at a temperature that permits hybridization. If the two organisms are related, a large amount of the radioactive DNA hydrogen bonds firmly to the DNA on the filter, so the filter becomes very radioactive. If the two species are not closely related, they have so few sequences in common that few hydrogen-bonds form and most of the radioactive DNA pours through, so the filter does not become radioactive.

This method is also used to measure the number of copies of a gene that occurs in a nucleus. It had always been assumed that any diploid nucleus contains two copies of every gene, one from the paternal parent and one on the homologous chromosome inherited from the maternal parent. If such DNA is broken into small pieces, denatured, and allowed to cool, the reannealing is extremely slow. Of the hundreds of thousands of pieces, each has only two that can pair with it—its own that had melted away from it and one from the homologous chromosome. Thus each piece undergoes thousands of collisions before its complement is encountered and stable reannealing occurs. Although most genes do act this way and reannealing is extremely slow, a portion does reanneal very rapidly, seeming to easily encounter complementary pieces, as if thousands of the same gene occur. These are genes for rRNA and tRNA; this method has revealed that up to 120,000 copies of certain ribosomal genes exist in each nucleus.

RESTRICTION ENDONUCLEASES

Natural DNA is such a long molecule that it cannot be worked with easily. It can be broken into smaller pieces by chemical treatment or simply by agitating a solution violently (Fig. 15.16b). When it is broken into fragments of a more manageable size, it is often critical to cut it at specific known sites so that repeat experiments can yield the same pieces as the first experiment. Prior to the 1970s, this was impossible; both chemical treatment and enzymatic cleavage cut DNA at random, and no two experiments ever yielded the same pieces of DNA. Then a class of bacterial enzymes, **restriction endonucleases,** was discovered. Each restriction endonuclease recognizes and binds to a specific sequence of nucleotides in DNA and then cleaves the DNA (Table 15.4). Owing to these properties, we always know exactly where DNA will be cut by a particular restriction endonuclease, and when two identical batches of DNA are treated with the same restriction endonuclease, the resulting fragments are always the same.

The sequence recognized by a restriction endonuclease is present in both strands, running in opposite directions; such sequences are **palindromes.** The sequence can be read "forward" or "backward," depending on the strand. Also, because most restriction endonucleases cut each DNA strand near the ends of the palindrome, the two cuts are not aligned. Each end is complementary to and capable of pairing with any other end made by that type of restriction endonuclease; the ends are said to be "sticky" (Fig. 15.17). At

TABLE 15.4	Sites Recognized by Restriction Endonucleases
Endonuclease	**Recognition Site**
EcoR1	↓ 5'-GAATTC-3' 3'-CTTAAG-5' ↑
BamH1	↓ 5'-GGATCC-3' 3'-CCTAGG-5' ↑
HindIII	↓ 5'-AAGCTT-3' 3'-TTCGAA-5' ↑
HpaI	↓ 5'-GTTAAC-3' 3'-CAATTG-5' ↑
PstI	↓ 5'-CTGCAG-3' 3'-GACGTC-5' ↑

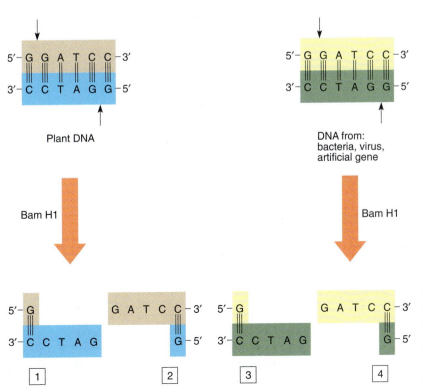

FIGURE 15.17

Most restriction endonucleases cut near the ends of palindromes, resulting in pieces with sticky ends. DNA pieces from different sources can be mixed together and tend to adhere, making it much simpler to work with them. Notice that both pieces 1 and 3, even though from different organisms, have the same sequence at their ends because Bam H1, like all restriction endonucleases, binds to and cuts the DNA only if it finds a particular sequence. The same is true of pieces 2 and 4.

present, more than 100 restriction endonucleases have been discovered, giving us a large choice of cleavage sites.

All fragments produced by a particular class of restriction endonucleases have exactly the same sequence in their single-stranded ends. Thus, if fragments made from the DNA of one organism are mixed with those of another organism, they adhere to each other. A DNA repair enzyme, **DNA ligase,** can be added to the mixture to repair the cuts so that the two fragments join together. DNA prepared by this method is **recombinant DNA.**

IDENTIFYING DNA FRAGMENTS

Evolutionary Studies.   Once restriction endonucleases have acted, the DNA fragments can be identified and used. Most simply, fragments are used directly to study the evolution of the DNA. For example, plastid DNA is a small molecule containing only 60,000 to 80,000 base pairs. Treatment with the Pst I restriction endonuclease produces about 10 to 20 fragments. These can be separated by gel electrophoresis and then made visible by staining, producing a **restriction map** of the plastid DNA (Fig. 15.18). The number of fragments reveals the number of Pst I sites and the number of base pairs between them in the plastid DNA. If plastid DNA is extracted from the chloroplasts of a second, closely related species, it should have the same number of Pst I sites and the same spacing between them as in the first species. If two species are not closely related, their fragment profiles may be quite different. A mutation may have altered one of the Pst I sites so that Pst I neither recognizes nor binds to it. If so, one less fragment is present, and one of the remaining fragments is extra long. Conversely, a mutation may have altered an ordinary sequence, converting it to a new Pst I site and thus producing an extra fragment. In addition, two species may have equal numbers of fragments, but with a fragment that is longer in one species than the corresponding fragment in the other: A mutation has added extra base pairs to one species or removed some from the other (see Fig. 18.9).

This sort of analysis can be done with various restriction endonucleases, each providing its own information about how much change has occurred in the DNA. For further analysis, some or all fragments can be sequenced; that is, the identity of every nucleotide can be revealed in order by techniques described below. Similar analyses can be performed with mitochondrial DNA. Nuclear DNA cannot be used easily because there is too much of it; among thousands of fragments, the addition or loss of a few cannot be detected.

FIGURE 15.18

(a) Plastid DNA from species A can be cut with a particular restriction endonuclease, resulting in several pieces that can be identified by gel electrophoresis. The enzyme digest is placed on a gel slab and a voltage is applied. Phosphate groups on the nucleotides cause the pieces to move in the electrical field, and the shortest DNA pieces slip quickly through the gel matrix, so they move farthest. The largest DNA pieces can barely move through the gel, so they stay at the top. After staining, the gel reveals the number and length of pieces. Here, we have indicated which band on the gel corresponds to each position in the plastid DNA circle, but you would not know that from this type of experiment. (b) In a closely related hypothetical species B, a mutation has changed a nucleotide between a and b so that that region does not have the proper sequence for the enzyme, so no cutting occurs. Electrophoresis shows that the two species have three bands in common and that species B has one band as long as the combined length of the missing two.

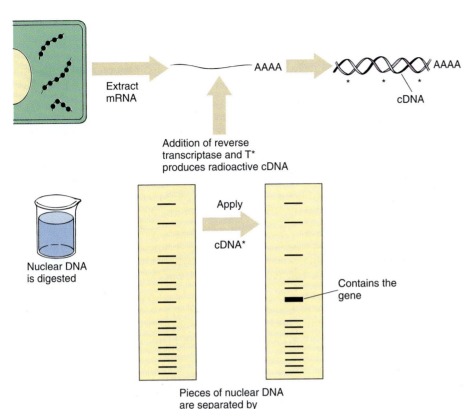

Extract mRNA

AAAA

cDNA

AAAA

Addition of reverse transcriptase and T* produces radioactive cDNA

Nuclear DNA is digested

Apply

cDNA*

Contains the gene

Pieces of nuclear DNA are separated by electrophoresis

FIGURE 15.19

The method for isolating a particular piece of DNA that contains a gene of interest. T* indicates radioactive thymidine; the gel slab contains fragments made from restriction endonuclease treatment of nuclear DNA. The method is described in the text.

Physiological Studies. In many cases, the objective of the experiment is to locate and isolate one particular fragment, such as the one that contains a specific gene. The following method can be used if any cell forms large quantities of the protein; for example, developing cotyledons of legumes such as beans and soybeans produce large amounts of the storage proteins phaseolin and glycinin, respectively. While the cells are actively synthesizing the protein, ribosomes can be extracted from them and mRNA obtained. The mRNA can be mixed with **reverse transcriptase**, a virus enzyme that synthesizes DNA using RNA as a template (Fig. 15.19). This **complementary DNA**, or **cDNA**, is complementary to the exons of the gene. It is synthesized using radioactive nucleotides; it can then be placed on the various DNA fragments produced by the restriction endonuclease treatment of nuclear chromatin. The cDNA forms hydrogen-bonds to whichever fragment contains the gene, and this bonding can be detected by assaying for the radioactivity. The gene itself may contain a restriction site, so that it is cut in two and separated into two fragments; the radioactive cDNA probe hybridizes with both fragments. If this happens, the experiment may have to be repeated with other restriction enzymes that do not cleave inside the gene itself. The best enzyme can be found only by trial and error. Few cells in plants, other than those in cotyledons, produce large amounts of just one protein, so most cells contain hundreds of types of mRNAs, and it is not possible to know which one codes for the protein of interest. However, reserve proteins of seeds are some of our most significant, important foods, and hundreds of billions of dollars are spent every year buying storage proteins in the form of beans, wheat, corn, and soybeans. These significant proteins are the objects of intensive study and genetic manipulation (see Fig. 16.1).

A second method can be used if mRNA is difficult to obtain but something is known about the amino acid sequence in the protein, even just a little of it. If the sequence of 5 to 10 amino acids is known, the genetic code in Table 15.3 can be used to guess the probable sequence of nucleotides in the corresponding exon. For example, if the known protein sequence contains phenylalanine, the exon may contain either UUU or UUC. Once two or three probable nucleotide sequences have been chosen, they can be synthesized by commercially available "gene machines." These are automated chemical apparatuses that contain the reagents necessary to synthesize any sequence of nucleotides; the experimenter types in the sequence on a keyboard and a few hours later the DNA is ready in the form of a radioactive probe that is used to detect which fragment contains the complementary sequence.

If neither of these approaches is possible, all the restriction endonuclease fragments from the plant are modified and mixed with bacteria. The bacteria take up the fragments; some take up many, others absorb one, and some take up none. In many cases the plant DNA is digested by the bacterium's own restriction endonucleases, but often it is incorporated into the bacterium's DNA and replicated. The bacteria are spread over dozens of Petri dishes, and every bacterium grows into a small colony. Each colony is tested to see if it is making the protein of interest. Any colony that is producing the protein must be replicating the gene as well as transcribing and translating it. That colony can be transferred to a new Petri dish and grown. Then DNA is isolated from some of its members and is treated with the same restriction endonuclease that was used to obtain the original plant fragments. At least one fragment of the bacterial digest should match in size one fragment of the plant digest; that fragment contains the gene.

It is not possible to compare each of the dozens of bacterial fragments with the tens of thousands of fragments from the original plant material, but that is no longer a problem. Genetically modified bacteria are used, and we know every type of fragment that the bacterial DNA will be digested into when treated with any restriction endonuclease if it has not picked up foreign DNA from our experiment. Any fragment that does not match the maps of the bacterial DNA must correspond to the plant fragment that was introduced.

DNA Cloning

The method of placing DNA fragments into bacteria, as just described, is an extremely useful technique of **DNA cloning**. The colony that contains the important fragment can be subcultured and grown easily, and each time a bacterium divides, a new copy of the experimental DNA fragment is made. The bacteria can be cooled, induced to become dormant, and stored for years. When the gene is needed again, the bacteria are revived and cultured. As long as the proper restriction endonuclease is used, the gene can be obtained easily.

Placing the original plant DNA fragments into bacteria is much easier than it may seem. Fragments are not simply mixed with bacteria, but are typically combined with plasmids or virus DNA. A **plasmid** is a short, circular piece of DNA that occurs in bacteria and acts like a tiny bacterial chromosome. Plasmids can move from one bacterium to another and are readily taken up from solution by bacteria. Recall that many restriction endonucleases cut DNA near the ends of a palindrome, resulting in sticky ends. If the plasmid or virus DNA is cut with the same enzyme used to isolate the DNA fragment from the plant, then when the two batches of DNA are mixed, some of the plant DNA adheres to plasmid DNA by means of their complementary ends. DNA ligase is then added to bond the pieces covalently into one continuous double helix (Fig. 15.20).

Several plasmids, such as pBR322, have been genetically engineered to be ideal DNA fragment **vectors** (carriers). In their present, highly modified form, they have factors that allow them to enter bacteria readily and be replicated, they have antibiotic resistance genes, and they have single sites for several restriction endonucleases so that fragments can be incorporated in precisely known positions. Once the fragment is added, the plasmid is mixed with bacteria and absorbed. After a few hours, all bacteria are treated with the appropriate antibiotics; those that did not absorb a plasmid die. The survivors have the plasmid and therefore DNA fragments from the original plant digest.

Several viruses are being used like plasmids; they contain a short fragment of DNA and can infect bacteria and then replicate. As with plasmids, certain viruses are now being modified into highly efficient vectors for genetic engineering.

The Polymerase Chain Reaction.   The newest technique for DNA cloning is the **polymerase chain reaction (PCR)**, in which only enzymes, not living bacteria, are used. The sequence to be amplified is heated to separate the two strands of the helix; after cooling, DNA polymerase is added and it replicates both strands of DNA. This enzyme can add

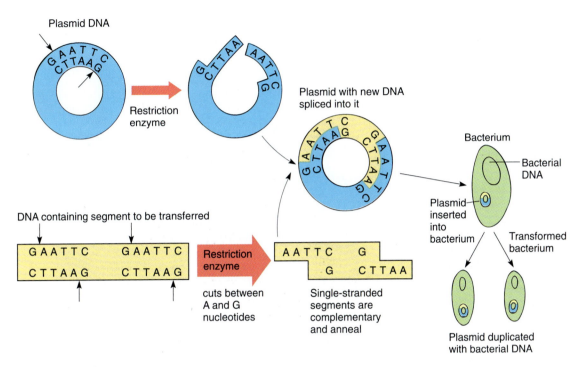

FIGURE 15.20

The method by which a plasmid is used to insert plant DNA into a bacterium for cloning. A virus could be used instead of a plasmid. The method is described in the text.

nucleotides only to a pre-existing nucleic acid, so before the replication can begin, it is necessary to add two types of primer DNA, each complementary to a short region at either end of the sequence to be cloned. End sequences are known if closely related genes are being studied, but often no part of the nucleotide order is known. In that case, artificial DNA—usually a short sequence such as AAAAAA—is added chemically to both ends; then the primer TTTTTT will be effective.

After the primer has hydrogen bonded to the sequence, a molecule of DNA polymerase attaches and begins working toward the other end. Once replication is completed, the mixture is heated temporarily to separate the two strands; then the process is repeated. In the past (late 1980s), this heating denatured the tertiary structure of the DNA polymerase and inactivated it; new enzymes had to be added at each step. Now, a heat-stable enzyme, Taq polymerase extracted from hot springs bacteria, is used.

After each heating/replication cycle, there are twice as many copies of the sequence being amplified. Consequently, extremely small amounts of DNA can be cloned very rapidly, and the PCR is used for rare copies of DNA. Examples are small amounts of cDNA made from the mRNA of just a few cells, as well as the DNA present in the early stages of viral infection (HIV, which causes AIDS, is detected very early with PCR) or the DNA present in a hair, a drop of blood, or a bit of skin at the scene of a crime.

DNA SEQUENCING

Until about 1980, most biologists did not think that discovering the sequence of nucleotides in a gene would be feasible; chemical and biochemical methods were too cumbersome and error-prone. Then, two simple, effective methods were developed that made DNA sequencing a routine, ordinary technique. In one method, DNA to be sequenced is first cloned to obtain a large sample, then divided into four batches. To each batch are added all the enzymes and free nucleotides necessary to carry out DNA duplication. To one tube a small amount of dideoxyadenosine is also added. A dideoxynucleotide can be added to a growing DNA, but it cannot react any further, and the growth of the DNA stops; nucleo-

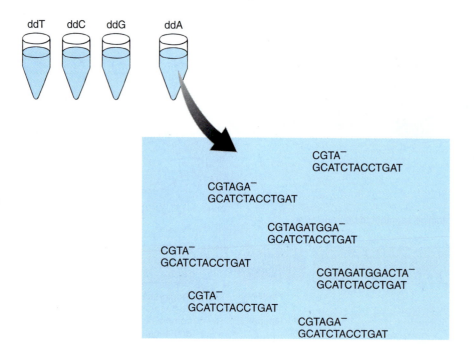

FIGURE 15.21

Into four test tubes, identical pieces of DNA have been placed for sequencing; all received everything necessary, but each received a small amount of chain terminator; the one depicted here received dideoxyadenosine. If too much of the chain terminator is added, almost all stop at the first T; if too little is added, almost no chains stop at any T and most finish the whole template. The correct amount produces a good number of copies of every possible length.

tides cannot be added to it. In this tube with dideoxyadenosine, the DNA acts as a template and replication begins; when a T is reached in the template molecules, a few incorporate a dideoxyadenosine and stop. Most growing DNA strands incorporate a normal A and keep on growing; then a few stop at the next T, a few more stop at the next, and so on. When the reaction is complete, the test tube contains thousands of DNA molecules of hundreds of sizes, but each size corresponds to the point where a T occurred in the template DNA being analyzed. Similarly, the second test tube contains a small amount of dideoxy T, the third dideoxy C, and the fourth dideoxy G (Fig. 15.21).

When all four batches are finished, they are loaded into separate lanes on a gel and allowed to separate, as in Figure 15.18. Each lane contains bands corresponding to the sizes of the DNA molecule. For example, if the DNA contained T as the third, seventh, eighth, and fifteenth base, the lane containing dideoxyadenosine has bands corresponding to DNAs that are 3, 7, 8, and 15 nucleotides long. The DNA sequence can be read immediately (Fig. 15.22).

FIGURE 15.22

(a) Diagram of a gel for a DNA fragment. You should be able to read the nucleotide sequence, starting at the top with the shortest, fastest nucleotide. A short column like this can analyze only short pieces of DNA. (b) An actual gel is typically long with many columns to do several samples at once. (© *Matt Meadows/Peter Arnold, Inc.*)

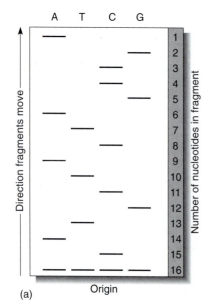

GENETIC ENGINEERING OF PLANTS

Recombinant DNA techniques have made it possible for botanists to identify, isolate, and study the structure and activity of many genes. They also allow botanists to insert genes into plants that do not normally contain those genes. The US Department of Agriculture has approved field trials of many genetically engineered plants. For example, cotton plants have received a gene from the bacterium *Bacillus thuringiensis* which codes for a protein toxic to caterpillars but not to other insects or mammals (including humans). If the gene protects the plant, it could prevent as much as $100 million in crop losses annually. Botanists have transferred genes from desert petunias into normal petunias, resulting in plants that require 40% less water. If such drought-resistance genes can be transferred to crop plants, the amount of water needed for irrigation will be greatly reduced. Tomatoes have been engineered such that the enzymes which cause them to become mushy are inhibited without inhibiting the other enzymes involved in developing flavor and aroma. The tomatoes can be allowed to ripen fully on the vine but are still firm enough to ship to market.

Obtaining some genes is not very difficult, but at present inserting them into nuclear DNA properly so that they can be transcribed and translated is not easy. A gene may code for an improved, more nutritious storage protein and should be expressed in cotyledons during seed formation. But if the gene inserts into a region of nuclear DNA that codes for root, stem, or wood characteristics, the engineered gene either may not be activated or may become active in the wrong place. Research on the nature of promoter sites and their interaction with chemical messengers is especially intensive. Studies are attempting to locate genes expressed only in particular tissues such as cotyledons, wood, chlorenchyma, or epidermis. Once found, the promoters of these genes will be valuable, because a "cotyledon" promoter could be attached to our foreign gene before insertion. It is assumed that with the correct promoter attached, the gene will respond to the appropriate chemical messenger, regardless of where it inserts into the DNA (Fig. 15.23).

Once a gene and promoter have been prepared, they are attached to an insertion vector, usually the **ti plasmid** from the bacterium *Agrobacterium tumefasciens* (Fig. 15.24). The plasmid has been modified by botanists by adding genes for herbicide resistance. The completed plasmid-promoter-gene is mixed with plant cells in tissue culture; they are allowed to grow and then are treated with the herbicide. Those cells that did not take up the plasmid, and those that took it up but are not expressing it, are killed by the herbicide. Those cells that have taken up the plasmid and are transcribing the gene and translating the mRNA into protein are resistant to the herbicide and survive. They can be cultured further. The surviving cells are induced to form new plants, and in many cases these plants not only carry the gene and express it but pass it on to their progeny when they undergo sexual reproduction.

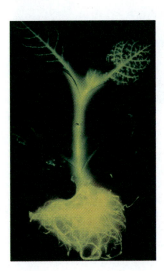

FIGURE 15.23

The gene for a luminescent protein from fireflies was inserted into a plant cell in tissue culture; then the cell was induced to grow into a full plant. Apparently, this gene had a promoter that is normally activated in every cell, because all cells are glowing. By attaching different promoters to this gene before the genetic engineering, it is possible to study when and where the promoter is normally active. Biologists hope to find promoters that, when attached to this gene, cause only specific cells or tissues to luminesce; then those promoters can be used whenever we want to affect specifically those types of cell or tissue. *(Courtesy of Dr. Marlene DeLuca, University of California, San Diego. From Science 234:856–859, 14 November 1986. © 1986 by The American Association for the Advancement of Science)*

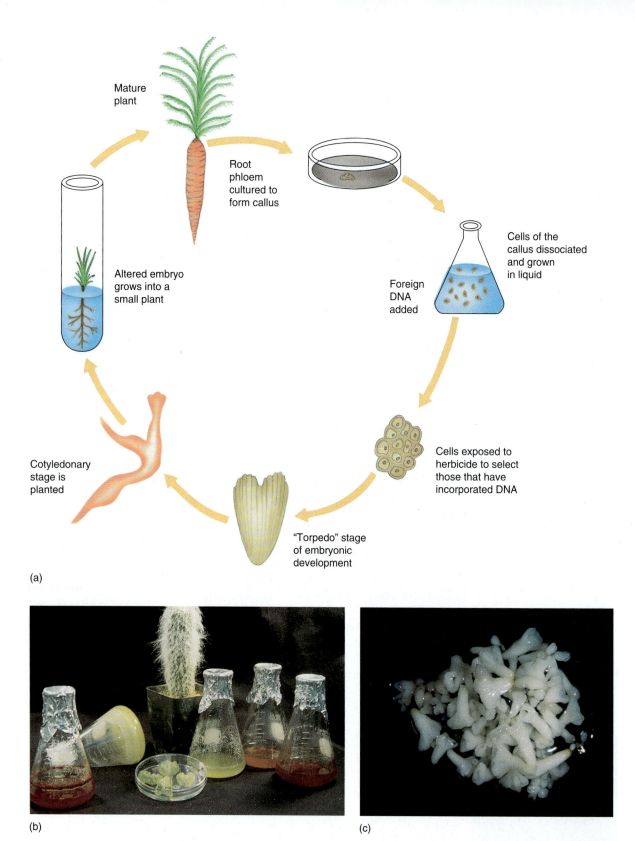

(a)

(b)

(c)

FIGURE 15.24

(a) This diagram illustrates the method involved in using recombinant DNA techniques to genetically alter a plant. The method is explained in the text. (b) Several components of a tissue culture experiment: the original plant (the cactus *Cephalocereus senilis*), callus culture on agar in foreground, liquid cultures in flasks. The red liquid cultures have been experimentally induced to form pigments. The flask tipped on its side shows that these cells can grow so extensively that they consume all the liquid medium and form a solid mass of cells. (*Courtesy of M. Bonness and T. J. Mabry, University of Texas, Austin*) (c) Tissue culture experiments can also produce embryos. If they have been genetically engineered, they grow into altered plants. (*Courtesy of Dennis Gray, University of Florida*)

PLANTS & PEOPLE

GENETIC ENGINEERING—BENEFITS AND RISKS

Genetic engineering of plants, animals, and microbes has become a reality. We already have available several new drugs, such as insulin and human growth hormone, that are produced by genetically engineered bacteria, and several species of modified plants are on the market. The benefits and risks of genetic engineering are being debated. Some questions are quite philosophical: For example, is it ethical for humans to alter a species or does that species have the "right" to evolve naturally? But as genetic engineering has already begun, let's look at some practical examples.

1. *Herbicide resistance.* It is now possible to give plants a gene that codes for an enzyme that can destroy an herbicide, thus making the plant resistant to the herbicide. Fields of altered crop plants could then be treated with the herbicide and all nonengineered plants—the weeds—would be killed. Benefits: Without the weeds, the crops would grow better and yield more, thus decreasing the amount of land that must be used for agriculture. Risk: This might allow even heavier uses of herbicide, thus increasing the amount of chemical residues in our food, possibly with unknown, serious health risks. A possible solution is to design new types of herbicide that are either completely harmless to animals or that break down quickly. Such compounds may exist now, but they kill both crop plants and weeds; by genetically engineering the crop plants, these safe compounds would become useful herbicides.

2. *Drought resistance.* Plants have already been genetically engineered to require less water for their survival. These plants require less irrigation, so less water could be diverted from rivers and lakes, which would then retain more of their natural condition. The temptation will be to divert the same amount of water, as even more arid land is plowed up and converted from its natural condition into cropland. But is this increased production necessary? If we ended government subsidies and price support systems that distort the true need for foods, food would not be grown just to be warehoused or discarded. The United States overproduces many products that are stored until they rot or decay.

3. *Bruise resistance in fruits.* As fruits like apples and tomatoes ripen, the fruit's cell walls are altered enzymatically such that the hard walls of the immature "green" fruits are converted to the soft edible walls of the ripe fruits. Simultaneously, flavors, aromas, and color develop. Because ripe, good-tasting fruits are too soft to ship, the fruits must be picked while still green and hard. At the store they can be treated with ethylene, which causes them to soften, but often this artificially induced ripening is not complete and good flavor and color do not develop. Several fruits, particularly tomatoes, have been genetically engineered such that the enzyme for wall softening is greatly inhibited. Thus fruits ripen naturally on the plant, but remain firm enough to ship. By the time they are picked and transported to market, the enzyme has finally started to soften the fruit, making it both edible and flavorful. This is one demonstration of genetic engineering that seems to have no drawbacks.

As is so often the case, advances in technology bring with them both benefits and risks. By analyzing each opportunity for its merits and dangers, we should be able to increase our quality of life without harming the environment.

VIRUSES

In the past, much of the emphasis in studying viruses in plants was on preventing the spread of virus diseases in crops. Now many viruses are especially interesting as possible vectors for plant genetic engineering. Also, because most are RNA viruses (described below), they contain many unusual enzymes that are useful for the experimental manipulation of nucleic acids.

VIRUS STRUCTURE

Viruses are extremely small particles that usually contain only protein and nucleic acid. They were originally discovered in 1892 as factors that cause disease but were so small that they could not be seen with a microscope and could pass easily through filters with fine enough pores to trap even bacteria. They were never actually seen until the development of the electron microscope (Fig. 15.25). They were originally thought to be very small cells, but now we know that is not true: They have no protoplasm, no organelles, and no membranes. They cannot carry out their own metabolism independently—they must always parasitize cells of some sort. The great majority of viruses, especially those that attack plants, consist of only one or a few types of proteins and a small amount of nucleic acid (Fig. 15.26).

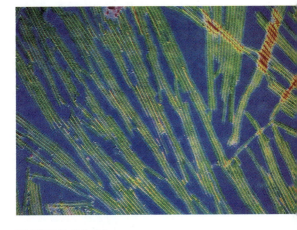

FIGURE 15.25
Tobacco mosaic virus particles are long, narrow filaments. In this preparation, viruses have been isolated from plant cells, washed, and treated with heavy metals to make them visible in an electron microscope (×180,000). (*Dennis Kunkel/ Phototake NYC*)

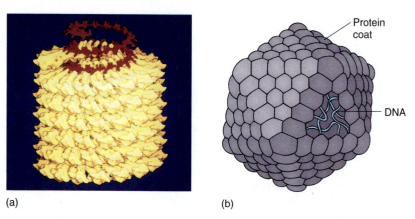

(a)   (b)

FIGURE 15.26

Most viruses, but especially plant viruses, are extremely simple. (a) Tobacco mosaic virus contains only one type of protein, which binds to the DNA, forming the long rod. In this form the DNA is well protected. *(Courtesy of G. Stubbs, K. Namba, Vanderbilt University, and D. Caspar, Brandeis University)* (b) In the virus shown here, the protein does not bind so closely to the nucleic acid but rather forms a coat, often called a capsid, around the nucleic acid.

Plant viruses always have a simple morphology, either long or short rods or even round particles. Tobacco mosaic viruses are rods 15 × 300 nm, whereas citrus tristeza viruses are rods up to 2000 nm long. Spherical or polyhedral virus are common, often having a diameter of about 60 nm.

The diversity of nucleic acids in plant viruses is great. Of the known viruses, the greatest number (400) are **retroviruses,** which contain single-stranded RNA. Retroviruses are our source of reverse transcriptases, needed to make cDNA. Many of the viruses that cause cancer in humans are animal retroviruses. Twelve known plant viruses have double-stranded DNA, and there are two groups of unusual types: 10 contain double-stranded RNA and 15 contain single-stranded DNA. Furthermore, many plant viruses are **split genome viruses**—not all of their nucleic acid is packaged as one particle. Instead the virus consists of at least two different particles, and both must be transmitted to a new host cell for viral metabolism to occur. Alfalfa mosaic virus consists of four distinct particles.

Most plant viruses have enough nucleic acid to code for only a very small number of proteins. Tobacco mosaic virus contains only 6400 nucleotides; its one and only coat protein contains 158 amino acids, so a minimum of 158 × 3 = 474 nucleotides are needed just for that. Three other genes exist in this virus; two large ones code for replicase enzymes and a smaller one codes for a protein thought to mediate spread of the virus from cell to cell. Most other plant viruses are similarly small.

VIRUS METABOLISM

Viruses must always invade a living cell in order to reproduce, and all known types of organisms are attacked by viruses: plants, animals, fungi, protozoans, algae, and prokaryotes. Viruses by themselves never have metabolism. They carry out no reactions, no energy transductions, no material exchange, no growth, no reproduction. Viruses that attack bacteria are called **bacteriophages** or phages, but these are viruses just like the others. Viruses that attack bacteria and animals are usually extremely specific: A particular type of virus can attack only a certain species of host. The same is true of many plant viruses, but others are able to attack many related plant species. Fungal and algal viruses are less well known.

To invade a plant, viruses rely on damage to living cells, such as through the action of aphids, chewing insects, or open wounds left by pruning or breaking. If an insect has chewed or sucked on an infected plant, the virus particles adhere to its mouth parts and feet. If these penetrate a host cell, some virus particles are transferred, allowing a virus particle to enter the protoplasm directly, completely bypassing the cell wall. During experimental studies of virus activity, healthy plants are infected by first abrading the epidermis with sandpaper to create fine breaks in the cells and then rubbing into the wound a solution of virus particles extracted from an infected plant; alternatively, freshly crushed, infected leaves are rubbed into the abrasion.

Once the virus is inside a suitable host cell, the protein molecules of the coat fall away from each other, releasing virus nucleic acid into the cell (Fig. 15.27). Coat dissociation is caused by normal conditions in the cell, such as pH or the concentration of magnesium or manganese ions. If the virus is a DNA virus, the liberated DNA is recognized by the plant's own RNA polymerases, and transcription of viral genes into mRNA begins. These mRNAs are translated by the plant's ribosomes, and within minutes of infection, new viral protein appears in the cell. Details vary greatly with the type of virus. Transcription and translation of some viruses are rather slow, and the virus does little harm to the host cell. In other types, viral mRNAs dominate the ribosomes, and the plant's protein-synthesizing apparatus is completely taken over. At the same time that RNA polymerases are transcribing viral DNA, DNA-replicating enzymes bind to and copy the DNA, making new viral double helixes. As the number of viral DNA molecules increases, the rate of viral transcription and translation also increases. Finally, all plant cell metabolism is redirected to viral metabolism.

Retroviruses, which have RNA but no DNA, may act as an mRNA and be picked up by a plant ribosome and translated. Usually one of the first proteins to result is reverse transcriptase, which binds to the RNA and synthesizes a complementary molecule of DNA. Another enzyme then synthesizes the other half of the DNA duplex, and the rest of the steps are the same as for a DNA virus. Some retroviruses bring molecules of reverse transcriptase with them as part of their coat, so reverse transcription is the first step.

As virus components become more abundant, viral nucleic acid moves into surrounding cells by means of plasmodesmata, and perhaps by passing through the wall as well (Fig. 15.28). Viral nucleic acid that enters the phloem can spread rapidly throughout the plant; the only tissues that are somewhat safe are root and shoot apical meristems, which do not contain phloem. Even in a heavily infected plant, shoot apical meristems may be able to produce cells as rapidly as the virus can spread acropetally, so the apical meristem itself remains virus-free.

The rate at which viruses take over host cell metabolism is variable; in many plants, it is slow and the viruses never become abundant in the protoplasm. Such plants show few

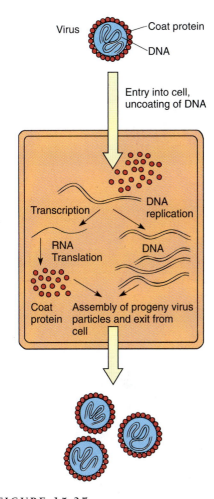

FIGURE 15.27
The method by which viruses reproduce. Details are given in the text.

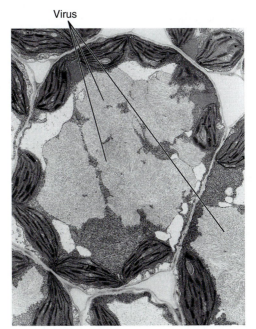

FIGURE 15.28
These cells of tobacco are infected with tobacco mosaic virus. Numerous new particles have formed, almost killing the cell, and the infection has spread to many cells. Whole particles are probably not necessary for the spread of a virus within a plant; the nucleic acid alone may be sufficient (× 9000). *(Courtesy of K. Esau, University of California, Davis)*

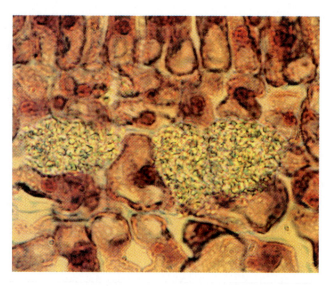

FIGURE 15.29
These cells of tobacco leaf are infected with tobacco mosaic virus. Although the cytoplasm is almost filled with virus particles, the surrounding cells appear healthy, and the vacuoles are virus-free (× 400). *(Runk/Schoenberger from Grant Heilman)*

disease symptoms, and the presence of virus is detected only by electron microscopy. This type of virus is common in many crop plants. Its abundance in wild plants is not known because even if native plants are examined by electron microscopy, small particles or rods can be assumed to be storage materials, normal structures, or artifacts. Proving that a plant is infected with a slow-acting virus is difficult. If the virus multiplies quickly and dominates the cell, symptoms such as chlorosis, necrosis, and leaf curling may appear (Figs. 15.29 and 15.30).

The rate at which a virus reproduces is affected by many factors; slowly reproducing viruses often are limited by promoter regions that RNA polymerases do not recognize well. In addition, the plant's DNA replicating enzymes may not recognize and bind well to replication initiation sites on viral DNA. Also, virus mRNA may contain codons that specify rare types of tRNA, thus slowing the progress of ribosomes.

Some viruses, once freed of their coat protein, can insert viral DNA into the host DNA. Little is known about this process, but apparently viral DNA lies next to host DNA and then breaks are made in both molecules. The ends join and the cuts are healed by DNA ligase. Such viral DNA usually lies dormant, perhaps simply because so little of the eukaryotic DNA is expressed in any particular cell. However, if this insertion happens in young plant tissues, viral DNA is replicated along with plant DNA during the S phase of every cell cycle, and both daughter cells receive infected DNA during mitosis. Such viruses are termed **temperate** and may produce no symptoms whatsoever; they may remain hidden forever and die when the plant dies. But in some they become active and reproduce just like virulent viruses. The factors that induce conversion from temperate to virulent are not known. Perhaps the viral DNA inserted itself downstream from the promoter region for a gene that is now being activated by a chemical messenger as part of the cell's differentiation. Once the promoter region allows RNA polymerase to bind and transcribe, the viral DNA is transcribed as well. This hypothesis has not yet been proven, and other factors may well be important.

This ability of virus DNA to insert itself into host DNA makes it useful as a vector for genetic engineering. All cells derived from the original cell contain viral DNA as part of one of their chromosomes.

FORMATION OF NEW VIRUS PARTICLES

As viral nucleic acids and proteins are synthesized within a cell, their concentrations increase. At some point, the components assemble into new viral particles (see Fig. 15.26). For many viruses, such as tobacco mosaic virus, this is a self-assembly process. Viral coat protein has a tertiary structure that causes it to bind to viral DNA. This binding then

(a)

(b)

FIGURE 15.30
(a) This tobacco is infected with TMV. *(Runk/Schoenberger from Grant Heilman)* (b) Oat leaves infected with barley yellow dwarf virus. *(Holt Studios, Ltd./Earth Scenes)*

permits it to attract and adhere to more viral protein, and a new protein/DNA viral particle is quickly assembled. The protein monomers by themselves do not adhere to each other; viral DNA must be present. No other components are necessary, and even in a test tube the two self-assemble into infectious virus particles.

In many types of animal viruses and bacteriophages, one of the last proteins made is an enzyme that destroys the host cell, causing it to burst (lyse) and release virus particles into the environment. This does not happen with plant viruses; perhaps the cell wall is too massive to be digested. Instead, virus particles remain in the cell until it is broken open by insects or larger animals. If an infected leaf is never eaten but instead is abscised in autumn, virus particles are released as the leaf decomposes; these too can be spread by animals.

ORIGIN OF VIRUSES

Evidence is growing that many if not most viruses are actually portions of genes of the host species or a species closely related to the host. As living organisms are damaged or die and decay, their nuclei break down along with the rest of the cell material. It is possible that occasionally a fragment of a chromosome codes for proteins that can self-assemble into a crude coat and have some infectious potential. Such a fragment has the possibility of acting as the forerunner of a virus; the fragment does not have to be as efficient as a full-fledged virus because it can evolve just like any other genetic system. With a long span of time, natural selection favors mutations in this DNA fragment that improve its survival rate.

Some of the first observations on which this hypothesis is based concerned the specificity of viruses to their hosts. Every virus attacks only those organisms that have a metabolism similar to its own. Many aspects of plant virus metabolism are also aspects of the growth and developmental metabolism of healthy plants. The same is true of animals and animal viruses: Viral diseases like AIDS are difficult to treat because virtually any treatment that damages a virus also damages some aspect of human metabolism. The DNAs of several cancer-causing viruses in animals have been sequenced and found to be very similar to host DNA sequences. The sequence homology is too long and too close to perfect to have evolved twice. Initially it was suspected that the host sequences may have been temperate viral DNA that had infected the organism or its ancestors years ago, but then many of these sequences were found to code for normal host proteins.

There are two prominent, alternative hypotheses that do not postulate that viruses evolved from the host's own DNA. The first proposes that they are the result of the evolution of extremely efficient parasites. Whenever an organism can obtain required compounds from its environment, it is selectively advantageous not to waste energy synthesizing them. But even the most reduced, simplified parasites are vastly more complex than viruses. Could viruses be reduced parasitic bacteria instead? This seems doubtful for most viruses, because the metabolism of plant and animal viruses should then resemble bacterial metabolism, but instead it matches plant or animal metabolism.

The other hypothesis postulates that viruses were an early stage in the evolution of life; this is based on the theory that life in general has evolved from simpler forms (prokaryotes) to more complex forms (algae and protozoans) to even more complex forms (seed plants and mammals). Extending this sequence backward, forms simpler than bacteria are viruses. A big problem with this hypothesis is that viruses must invade living cells for their metabolism to occur, so if they arose before bacteria, what did they invade?

Until recently, the problem was mostly philosophical; a subject cannot be studied scientifically until a hypothesis is formulated which makes predictions that can be tested. DNA hybridization and sequencing make testing and verification possible, so much more scientific interest is now being shown in the origin of viruses.

PLANT DISEASES CAUSED BY VIRUSES

Plants suffer from at least a thousand different virus-caused diseases. Because symptoms are similar to those of mineral deficiency or other environmentally caused problems, detection can be difficult (Fig. 15.30). Few effective treatments exist for plants infected with viruses. We do not try to cure whole crops of viral disease; rather we try to maintain healthy, uninfected breeding stock for virus-free seeds or cuttings. Antiviral drugs such as ribavirin appear to have cured some plants of virus infection, and gibberellic acid at least counteracts virus disease symptoms in some cases. Heat treatment inactivates some viruses; whole plants can be kept in hot (35 to 40°C) growth chambers or greenhouses for several weeks to several months, after which they may be free of virus. Even these techniques are ineffective in most plants and against most virus diseases; the best policy is to use virus-free plants and protect them from infection. Virus-free plants are obtained by shoot meristem propagation or sometimes by normal sexual production of seeds, if stamens and carpels can grow and set seed more rapidly than virus particles can infect them.

SUMMARY

1. All information required to specify protein primary structure—the sequence of amino acids—is stored as the sequence of deoxyribonucleotides in DNA.

2. Cell differentiation is based largely on differential activation of genes and control of the processing of heterogeneous nuclear RNA into messenger RNA.

3. The exact details of the mechanism by which a plant hormone induces differential activation of either nuclear or organellar genes are not known. The hormone receptor may either produce or act as an intracellular messenger that binds to DNA promoter regions.

4. The genetic code consists of triplets of nucleotides, each triplet coding for only one amino acid, or for STOP or START. The code is degenerate, each amino acid being coded by several codons.

5. Genes consist of a promoter region that contains enhancer elements and a structural region that usually contains both exons and introns.

6. In transcription, RNA polymerase attaches to the promoter region, moves to a start site, then polymerizes RNA, being guided by base pairing in a short region of single-stranded DNA. Both introns and exons are transcribed.

7. Heterogeneous nuclear RNA is processed to mRNA, then transported to the cytoplasm where it binds to ribosomes. Each ribosome has a large and a small subunit, four molecules of rRNA, and about 80 proteins.

8. Amino acids are carried to ribosomes as part of an activated tRNA, each of which has an anticodon complementary to the codon for the amino acid it carries. All tRNAs have similar structures, consisting of three loops caused by four pairs of self-complementary regions.

9. Restriction endonucleases cut DNA at specific sequences; the resulting pieces can be melted to the single-stranded state. Single-stranded nucleic acids from different sources can be mixed and allowed to hybridize, either as a measure of their relatedness or as part of the construction of a new molecule of DNA.

10. Specific sequences of DNA can be synthesized artificially in gene machines, or longer sequences can be synthesized by incorporating one copy into a vector and then inserting the vector into a bacterium. As the bacterium reproduces, the sequence of DNA is reproduced as well.

11. Most viruses appear to be short pieces of DNA or RNA that contain a few genes closely related to normal host genes. Most plant viruses have RNA, not DNA, and a coat of just one type of protein.

12. Viruses infect plants through wounds, then divert the plant's nucleic acid and protein-synthesizing metabolism to the synthesis of more virus molecules, which then self-assemble into complete virus particles.

IMPORTANT TERMS

anticodon
bacteriophage
codon
complementary DNA (cDNA)
differential activation of genes
DNA cloning
DNA denaturation
DNA hybridization
DNA ligase
exon

gene
intron
messenger RNA (mRNA)
palindrome
polymerase chain reaction
promoter region of a gene
recombinant DNA
restriction endonuclease
restriction map
retroviruses

reverse transcriptase
ribosomal RNA (rRNA)
ribosome
start codon
stop codon
structural region of a gene
transcription
transfer RNA (tRNA)

REVIEW QUESTIONS

1. Cutin, lignin, and chlorophyll are not proteins, so how is it possible for genes to control the synthesis of these polymers?

2. What is a codon? How many exist? Why is the genetic code described as being degenerate?

3. Describe the promoter and structural region of a nuclear gene. What is a TATA box? What are exons and introns?

4. What is a frameshift error? Does the binding of methionine tRNA to the small subunit help reduce frameshift errors? If so, how?

5. If a molecule of mRNA had the following sequence, what would be the sequence of the gene that coded the mRNA? Which amino acids would be incorporated into the protein?

gene: 3'-_____-5'

mRNA: 5'-UCGAACAAUUGUCCCGGCCUC-3'

protein:

6. Describe the steps of polypeptide elongation by ribosomes.

7. What are palindromes, and how are they related to restriction endonucleases? Why are they useful for inserting one piece of DNA double helix into another?

8. Briefly describe how a piece of DNA can be cloned using bacteria. Why would it be important to use the same type of restriction endonuclease in the harvest phase as in the preparation phase? Once the DNA had been cloned, how would you go about sequencing it?

9. What are ti plasmids and *Agrobacterium tumefasciens*? What are viruses? How are they used by botanists in the genetic engineering of plants?

BotanyLinks .net Questions

www.jbpub.com/botanylinks

Visit the .net Questions area of BotanyLinks (http://www.jbpub.com/botanylinks/netquestions) to complete these questions:

1. How many viruses attack plants? Do they attack organisms like algae, fungi, or other viruses? Go to the BotanyLinks home page to begin researching this subject.

2. Does being a scientist have moral and ethical implications? Now that we really can genetically engineer our food plants and clone animals (even humans?), what ethical responsibilities do scientists have? Go to the BotanyLinks home page for discussion on this subject.

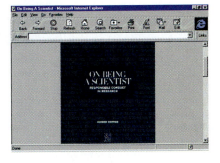

BotanyLinks includes a Directory of Organizations for this chapter.

The brown alga *Ectocarpus*. (*Cabisco/Visuals Unlimited*)

III

GENETICS AND EVOLUTION

The focus of this section is the mechanism by which DNA and its information are passed from parent to progeny; how that DNA changes over time, resulting in new types of organisms; and how the environment and natural selection have interacted to produce hundreds of thousands of species of prokaryotes, algae, fungi, animals, and plants.

DNA serves as an archive of information that, interacting with environmental information and resources, produces an organism's body—its structure, metabolism, and biology. DNA also carries the information needed to make more DNA, so that during reproduction —either of the cell or of the whole organism—each progeny receives copies of the DNA and the information it contains. Errors occur occasionally, and molecules of DNA are not copied perfectly; each version may code for slightly different enzymes or in some other way produce an organism slightly different from parents or siblings. If the world were absolutely uniform, we might expect that whenever a mutation resulted in a new type of individual, either the new type or the original would be favored by natural selection until it was the only type left; a more advantageous metabolism or structure would be more advantageous everywhere.

But our world is complex, containing aquatic and terrestrial environments, areas with mild temperatures and others with temperature extremes, and a diversity of rainfall, soil types, pathogens, pollinators, and so on. Consequently, alternative biologies may each be selectively more advantageous in different conditions. The diversity and richness of the environment result in diversity and richness of organisms.

GENETICS

As this bee carries pollen from flower to flower, it carries genes, contributing to "gene flow." (*William E. Ferguson*)

CONCEPTS

Genetics is the science of inheritance. The chemical basis of genetic inheritance is the gene, the sequence of DNA nucleotides that guides the construction of RNA and proteins and also controls the construction of more copies of the genes themselves. If all plants of a species had exactly identical nucleotide sequences in their DNA, then all those plants would be identical physically. But virtually all genes occur in multiple forms known as **alleles,** the alleles of a particular gene differing from each other in their sequence of nucleotides (Fig. 16.1).

Alleles arise by mutation; if the nucleotide sequence is changed (mutated) in any way, the new sequence is a new allele. Mutations can occur in any gene in any individual, so gradually a population of separate plants comes to have a variety of alleles. The types of alleles that a single individual has are called its **genotype,** and the expression of those alleles in the individual's size, shape, or metabolism is its **phenotype** (Fig. 16.2). As a result of mutations, the population of individuals has varying genotypes and phenotypes. They are not identical, as is apparent from considering humans.

An important concept of inheritance is the selective advantage of reproduction, which may be either sexual or asexual (see Chapter 9). In asexual reproduction, each offspring is identical to its parent and its siblings, having exactly the same DNA and thus the same alleles. Although this might seem like the safest, most efficient mechanism for producing

```
TCTTACTATT  GTGAGCGGTA  TGGGCAATTA  TTTGGGCCAT  CTAAACTCTC  AAAGAAGGAT
AGATCGTGAG  TTCGAGTTGT  CAGGATGAAC  GTTCCCAAGT  CGTTCTTATA  AGATCTCCGG
AGTATGCTGT  GGTTTAAGCT  CCTCTATTTG  TTCCAAGACA  AACCATCTCT  CCTCCCCGTC
GTTGTTCCCC  TCCTCTCCGA  CGTTCTCTCA  CACTAACACC  TTTAGAGTTT  CTTTGTTTAA
GCCCTTGACT  CGTTTGTACG  GTTTAGATCA  AGTTCCTTTT  GGTAAAGAAG  ACTTCTATTT
GGAAAGTTGA  ACCCTTCGGC  GCTGGGGTAG  ATAAGGTTGT  TCGAACCGTT  CAACAAACTC
TAATGGGTCT  CTTTGGGAGT  CGAAGCCCTG  AACCTACAGA  AGGAGTCACA  ACACCTATAC
TTGCTCCCTC  GAGAAAAAGA  CGGTGTGAAG  TTAAGTTTCC  GGTATCACCA  TGATCACTAA
TTACTTCCTC  TTCGTTTGTA  ACTTGAACAA  CCGTAATTTC  TTGTTGTTGT  CTCCGTCGTT
GTCCTTCTCG  TTGGAAACCT  TCATGCCTTT  ATATCTCGAC  TTAACAGACT  TGTTCTATAT
AAACATTAGG  ATCGTCCAAT  AGGTCAACAC  CAGTTGCGAT  GGAGTCTAGA  CTTAAAGAAA
CGAAAACCAT  AGTTACGGCT  CTTGTTGGTC  TCCTTGAAGG  AACGTCCAAG  CTTTCTGTTA
CACTATTCGG  TCTATGGATC  AGTTCACGTC  CTCGAACGCA  AGGGATCCAG  ACGTTTCTA
TAACTCTTGG  ATTATTTCTC  GGTTTGACTC  AGGATGAAAC  ACCTACGAGT  CGGAGTCGTC
TTTCTCCTCC  CCTTGTTCCC  TTCTTTCCCA  AGAAACAGAA  GTTAAAACTC  CCGAAAAATG
ACTTATTCAT  ACATCATGAT  TTTACATACT  ACATTATCGA  GTATCACTCG  CTCCTTTCAT
AGCCCGATAA  ATTGATACTG  AACTCGAGGT  AGATACTTAT  TTATTTAGTC  GTATACTACG
AAAACAAAAC  ACATG
```

FIGURE 16.1

This is one of numerous alleles of the gene that codes for the storage protein glycinin in soybeans. Each allele varies slightly in the sequence of nucleotides, and, as a result, produces a slightly different form of protein. Although each allele has a unique sequence, they are considered variations of the glycinin gene rather than distinct genes, because they all produce more or less the same protein at the same time, in the same place. See Figure 15.4 for alleles of the histone H4 gene.

large numbers of offspring to carry the parent's genes into future generations, it is effective only for the genotype and phenotype that have the greatest survival value. During a drought, plants with poor capacity to withstand water stress are outcompeted by plants whose alleles confer superior water stress resistance. During an infestation of pathogenic fungi, individuals without alleles that confer immunity are destroyed. There may be no individual resistant to both stresses.

Sexual reproduction is a mechanism by which an organism combines its alleles with those of other, possibly better-adapted individuals, thereby increasing the probability that copies of its own alleles survive. If a plant is susceptible to drought, sexual reproduction is advantageous because some of its sex cells may fuse with those of a drought-resistant plant.

FIGURE 16.2

Flower color in four o'clocks (*Mirabilis jalapa*) is controlled in part by a gene that has two alleles. The DNA sequence of one results in a protein whose primary and tertiary structures cause it to synthesize red pigment. The DNA sequence of the other allele codes for a protein whose tertiary structure is misformed; it has no enzymatic activity and no pigment is produced, so flowers are white. If only a small amount of pigment is produced, the flower is pink. Flower color is the plant's phenotype; the type of alleles present is its genotype. (*R. Calentine*)

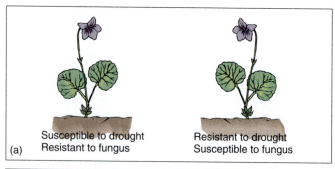

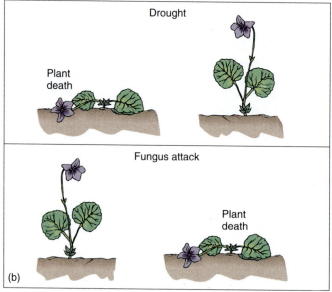

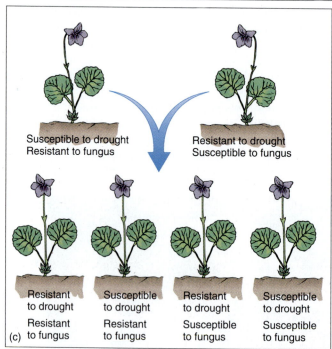

FIGURE 16.3

(a) A population probably contains a diversity of genotypes and phenotypes. (b) Without sexual reproduction, selectively advantageous alleles of one individual cannot be combined with those of another, and multiple stresses might kill all individuals. (c) With sexual reproduction, traits of one individual are combined with those of another, often producing individuals more fit than either parent.

If so, the new zygote should grow into an individual resistant to water stress, thus adding some protection to the alleles derived from the susceptible parent. Although the original plant may die during a drought, copies of its alleles survive in its progeny, protected by alleles from the resistant parent (Fig. 16.3). The second parent should benefit as well, as it is almost certainly not superior in all attributes. The drought-sensitive parent may carry alleles for resistance to fungal attack which, during times of fungal attack, protect the alleles of the drought-resistant parent in the offspring, making the progeny of this sexual reproduction resistant to both fungi and water stress.

Sexual reproduction involves a large degree of chance and risk. Fitness and survival capacity are not governed by just one or two genes, but by almost all genes, including those responsible for the proper construction of membranes, the functioning of organelles, production and transport of hormones, and so on. One cannot say that any aspect is trivial and does not matter. Therefore, the presence of one or two particular alleles in a sex cell is not the key feature, but rather the combination of all the alleles. Because of synapsis and crossing-over in meiosis, each plant produces thousands of types of sex cells, each with a unique genotype, which then fuse with the sex cells of many other individuals. For example, wind-pollinated trees produce millions of pollen grains that blow away and fertilize ovules on hundreds of trees; simultaneously, its own ovules are receiving wind-borne pollen from numerous trees. As a result, thousands of fertilizations may occur that involve this one plant's alleles. Many may produce poor combinations of alleles that have little survival value, but at least a small percentage should have the best attributes of both parents and should produce healthy, genetically sound plants with a high capacity to survive all stresses. It is not necessary or even advantageous for all potential progeny to survive, only the most fit. The plants and animals around us today are the successful survivors of evolutionary experimentation.

REPLICATION OF DNA

Before a cell can undergo nuclear division, either mitosis or meiosis, DNA must be replicated during S phase of the cell cycle. Replication doubles the amount of DNA, and each gene exists in at least two copies, one on each of the two chromatids, one of which goes to each daughter nucleus during anaphase.

As DNA replication begins, chromatin first becomes less compact, opening sufficiently to allow entry of the necessary replicating factors. DNA does not release from the histones; instead the nucleosome structure remains intact. Next, one strand of the DNA double helix is cut, and the two strands separate from each other in a short region, forming a small "bubble" called a **replicon** (Figs 16.4 and 16.5). With the double helix open, free nucleotides diffuse to the regions of single-stranded DNA and pair with its bases along both strands. These are ribonucleotides, not deoxyribonucleotides, and they are polymerized into short pieces of **primer RNA** about 10 nucleotides long. The primer RNAs then act as substrates for the DNA-synthesizing enzyme, **DNA polymerase**. It now enters and adds deoxyribonucleotides onto the end of the primer RNA using the open DNA as a guide. This method of replication, in which each strand of DNA acts as the template for making the complementary strand, is **semiconservative replication,** because each resulting double helix contains one new molecule and has conserved one old one.

DNA polymerase can add deoxyribonucleotides only to the 3′ end of the growing nucleic acid, so one strand of the open DNA is copied in one direction and the other strand in the opposite direction (Fig. 16.5: top strand toward the left, bottom strand toward the right; both are actually 5′ → 3′). As DNA continues to unwind and open in both directions, new pieces of primer RNA form and new pieces of DNA grow from them toward the existing fragments (called Okazaki fragments). The original primer RNAs are depolymerized, and the new Okazaki fragments are joined to the first ones. On the other end of each strand, replication is continuous, one long chain forming from the original primer RNA.

Each chromosome is believed to contain only a single DNA double helix, and in each chromosome hundreds or thousands of sites occur where replication can be initiated. As

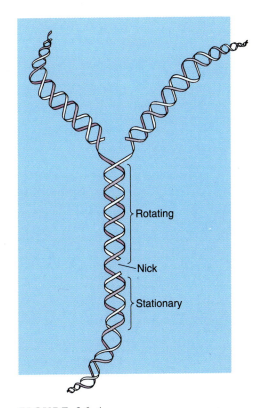

FIGURE 16.4

In order for DNA to unwind, one strand must be cut; otherwise, the entire unreplicated part would have to spin rapidly, which is impossible. With a single cut, only the region between the replicating enzymes and the cut must rotate.

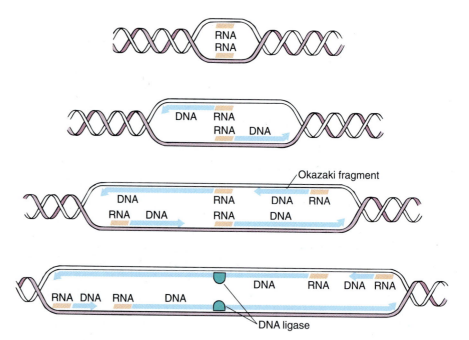

FIGURE 16.5
Once a replicon opens, two pieces of primer RNA are formed; then DNA polymerase adds new DNA to the 3′ end of the primers, and both strands elongate as the replicon unwinds at both ends. As the replicon continues to open, Okazaki fragments are initiated periodically; they grow to the downstream fragments, and the enzyme DNA ligase joins them into single molecules.

DNA uncoils and then separates, it has a forked appearance; this region is called a **replication fork.** As DNA continues to uncoil and open, replication forks at each end of each replicon travel along the DNA. Finally, each replication fork runs into one of the adjacent replicon; then all new pieces of DNA are **ligated** (attached to each other with covalent bonds) into two new, complete molecules. The large number of initiation sites allows numerous DNA polymerases to work simultaneously. Each is able to incorporate about 1000 nucleotides every second into growing DNA fragments, and replicons extend at the rate of about 0.5 μm/min. Each replicon is about 45,000 to 180,000 base pairs long (45 to 180 kbp) and can be replicated in about 1 to 3 hours. Because S phase of the cell cycle is 3 to 10 hours long, only about one third to one tenth of the replicons are active at any particular moment. If only one DNA polymerase could act on each chromosome, S phase would require 40 to 60 days (see Tables 4.1 and 16.1).

As replication forks advance, DNA partially dissociates from the histone octamers; but as all enzymes migrate forward and the two new DNA double helixes are complete, they immediately reassociate with histones into complete nucleosomes. Some of the old histone

TABLE 16.1	Amount of DNA per Haploid Set of Chromosomes		
		Nucleotide Pairs (in billions)	Meters of DNA
Beta vulgaris	beet	1.2	0.41
Zea mays	corn	2.3	0.78
Pisum sativum	pea	4.7	1.62
Vicia faba	bean	14	4.8
Allium cepa	onion	16.3	5.62
Lilium	lily	31.7	10.9

The values are known for only a few species at present, but more are being analyzed. Diploid nuclei have twice as much as shown in this table.

octamers go to one double helix, some to the other, and new histone octamers are added to both. Any particular segment of DNA remains unpackaged for only a few moments.

MUTATIONS

A **mutation** is any change, however large or small, in DNA. The smallest mutation, affecting the least amount of DNA, is a **point mutation** in which a single base is converted to another base by any of various methods (Fig. 16.6). If a piece of DNA is lost, the mutation is a **deletion**; the addition of extra DNA is an **insertion**. Under some conditions, a piece of DNA becomes tangled and breaks, and during repair it is put in backward as an **inversion**.

CAUSES OF MUTATIONS

A **mutagen** is something that causes mutations. Several that are important are certain chemicals, ultraviolet light, X-rays, and radiation from radioactive substances. Many chemical agents that are mutagens are manmade and are increasing in our environment. One chemical mutagen, nitrous acid, reacts with cytosine and converts it to uracil, giving the DNA a G-U base pair in place of a G-C base pair (Fig. 16.7). Numerous classes of DNA repair enzymes exist, one of which recognizes this G-U pair as abnormal and changes it back to G-C. But in meristematic cells that are replicating their DNA, DNA polymerase may arrive before the repair enzyme; if so, one new strand is formed with A complementing the U. In later replication, the A guides the incorporation of thymine, and the original G-C base pair becomes A-T.

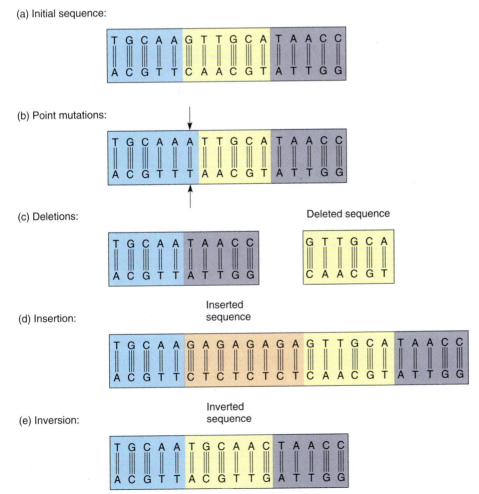

(a) Initial sequence:

(b) Point mutations:

(c) Deletions: Deleted sequence

(d) Insertion: Inserted sequence

(e) Inversion: Inverted sequence

FIGURE 16.6

(a) Consider this initial sequence of DNA; the box is not important. (b) A point mutation consists of the change in just a single base pair—here G-C is converted to A-T. If this sequence is part of a gene rather than just spacer DNA, this change in sequence has resulted in a new form of the gene, a new allele. (c) The six base pairs in the yellow box of the initial sequence have been deleted. Deletions often remove hundreds of base pairs. On the upper strand, the right boundary of the deletion has removed a C and an A and left a T, changing CAT to AAT. The AUG mRNA start codon is coded by the DNA triplet CAT, so this may be an important mutation, destroying a start codon. But this may not be a triplet; the real triplets may have been GCA TAA or TGC ATA. The diagram provides too little information to predict the significance of the mutation. (d) An insertion is the addition of one or more base pairs. (e) In this inversion, the DNA has broken at two points, then flipped over and been reinserted; the original top strand of the insert is now the bottom strand.

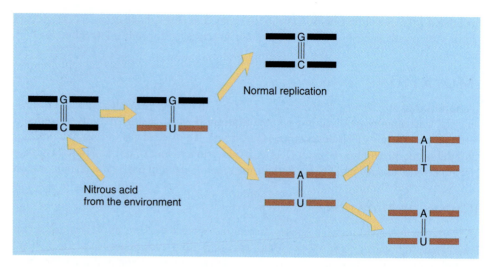

FIGURE 16.7

Nitrous acid converts cytosine to uracil. This strand is mutated, and all strands derived from it will be mutated, having A-T base pairs. The guanine is not affected and produces the original allele. One new allele has come into existence.

Deletions can be caused in several ways; one method is due to short regions of a self-complementary sequence. As DNA unwinds ahead of the DNA polymerase, one strand with a self-complementary sequence may form a small loop. DNA polymerase may pass by this loop without reading any of the bases in it; the bases complementary to those in the loop are left out of the new DNA molecule, so it is shorter than it should be.

Insertion mutations can be caused by many methods, because a variety of enzymes cut and rejoin DNA as part of repair processes. If a small piece of foreign DNA is present after cutting, it may accidentally be incorporated into the chromosomal DNA by DNA ligase.

One of the most interesting and potentially most useful causes of insertion and deletion mutations is the action of **transposable elements.** These are pieces of DNA that readily change their positions from one chromosome to another. Transposable elements have two basic forms—insertion sequences and transposons. **Insertion sequences** are only a few thousand base pairs long and contain the genes that code for the enzymes actually involved in cutting the insertion sequence out and splicing it into DNA somewhere else. A **transposon** is like an insertion sequence except that it may be much longer and carries genes that code for proteins not associated with transposition (Fig. 16.8). The deletion and insertion mutations caused by transposable elements can vary in severity. If the element inserts into spacer DNA, the effect is not very important, but if it inserts into a gene, it totally disrupts either the promoter or the structural region.

Mutations caused by transposable elements may be one of our most powerful tools for genetic analysis and engineering. Often, a metabolic pathway is studied by exposing thousands of plants to mutagens such as X-rays or nitrous acid, then examining the plants to find any that have mutations disrupting the metabolic pathway of interest. These are then studied further to determine how the pathway was affected, but we never actually know which gene was disrupted. The gene is given a name (Table 16.2), but we do not know its DNA sequence, its location on a chromosome, or the nature of its promoter or structural regions. However, transposons are now being sequenced and engineered to contain markers such that the transposons can be applied to plants or cultured cells, where they cause insertion mutations. The mutants of interest are located (Fig. 16.9), and DNA from the mutants is cut into pieces by restriction endonucleases, denatured, and combined with radioactive DNA complementary to the transposon. Once the piece is found, it can be sequenced. Because we know the transposon's sequence already, the DNA on either side of the transposon must be the gene of interest. Transposon mutagenesis offers the possibility of quickly correlating metabolic pathways, proteins, genes, and even promotors for virtually any aspect of an organism.

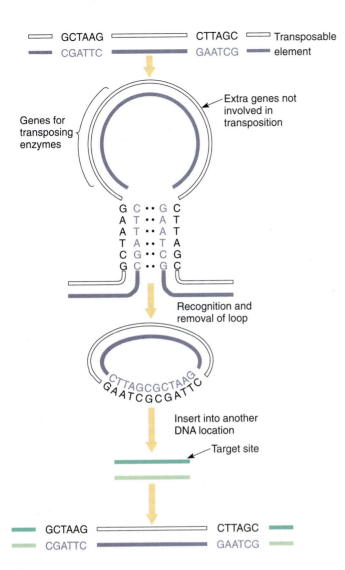

FIGURE 16.8

How transposable elements move is not known for certain, but the sequences at their ends are thought to allow self-pairing, which defines the ends of the transposon during its release from a chromosome.

TABLE 16.2 *Names of Genes in Pea Plants*

Gene	Symbol	Phenotype
Chlorophyll synthesis	*alb, alt, au, auv, ch₁, ch₂, cov, cvit, lum, pa, py, vac, xa₁ xat, yg*	Each affects chlorophyll synthesis, changing the color from green to dark bluish green
Leaf structure		
Aeromaculata	*f₁*	Few airspaces between epidermis and palisade parenchyma
Stomata	*sa₁, sa₂, sa₃*	Leaves have fewer stomata than in the wild type
Leaf development		
Afila	*af*	Leaflets develop as tendrils
Clavicula	*t₁*	Tendrils develop as leaflets
Unipetiole	*up*	Leaves have only one pair of leaflets
Unifoliata	*uni*	Leaflets are converted to just a single leaf
	cri, crif, cris	Folded, crinkled leaves
	cont, lat, st, x	Leaves of unusual size

Although there are thousands of alleles in plants, only a few ever become obvious and only a small number of those are ever studied and given names. These are just a few of the named alleles in pea (*Pisum*).

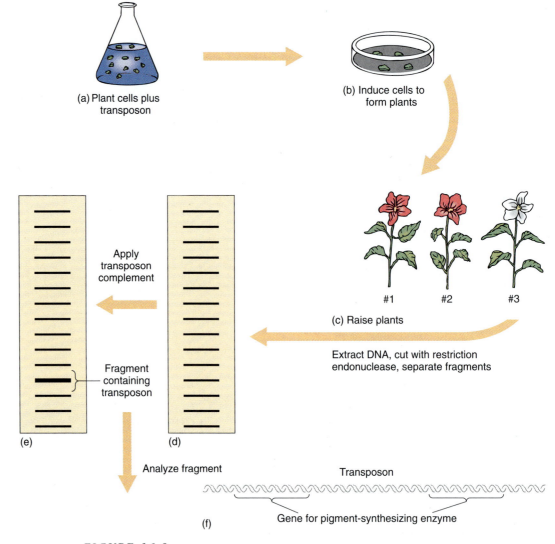

(a) Plant cells plus transposon

(b) Induce cells to form plants

#1   #2   #3

(c) Raise plants

Extract DNA, cut with restriction endonuclease, separate fragments

Apply transposon complement

Fragment containing transposon

(e)   (d)

Analyze fragment

Transposon

(f)

Gene for pigment-synthesizing enzyme

FIGURE 16.9

Transposon mutagenesis; the method is described in the text. In this case, flower pigment synthesis is of interest; plant 1 is a normal plant; plant 2 has a mutation for leaf shape, which is not our objective; plant 3 has white flowers, so we have induced a mutation in some gene that is necessary for pigment synthesis.

EFFECTS OF MUTATIONS

The effect and significance of a mutation depend on its nature, its position, and its extent. If it occurs in spacer DNA between two genes, it may have no effect whatsoever. Also, point mutations and small insertions and deletions in introns appear to be unimportant generally—they change a portion of hnRNA that will not be incorporated into mRNA. Within exons, a small mutation may not be important if it only changes a codon into another codon that specifies the same or a similar amino acid. Changing codons to ones that specify very different amino acids may not matter if they are located in a part that is not critical to the protein's functioning.

On the other hand, a mutation in an exon may cause the gene to code for a protein whose active site is disrupted and the protein cannot function (see Fig. 16.6). An insertion mutation may cause the gene to code for protein so long that it cannot fold properly. Also, mutations in the promoter regions can completely inactivate a gene or cause it to be active at the wrong time or place. Even point mutations can have profound effects; for example, the nitrous acid conversion of a G-C base pair to A-T might cause the formation of a new start codon. Other simple changes can eliminate start codons, the recognition site for distinguishing the boundaries of exons and introns, TATA boxes, and so on. The larger the mutation, the greater the probability that a critical part of the DNA is affected.

Statistically, mutations are almost always harmful. Enzymes tend to be about 300 to 400 amino acids long, and hundreds of trillions of proteins could possibly exist. Yet only a small fraction would be useful in living organisms. Any mutation that changes the structure of proteins, rRNA, or tRNA is more likely to produce a less useful than a more useful form. The majority are deleterious, the minority beneficial. Natural selection eliminates the deleterious mutations and preserves the beneficial ones.

SOMATIC MUTATIONS

Mutations can occur at any time in any cell, but if they happen in cells that never lead to sex cells, they are called **somatic mutations.** For example, a gene in a leaf primordium cell may undergo a mutation, but because leaves are not involved in sexual reproduction, the mutation is somatic and is not passed on to the plant's offspring, regardless of whether the mutation is advantageous or disadvantageous. The same is true of any mutation in roots, wood, or bark. In general, somatic mutations are not very important for most plants because they affect such a small portion of the plant and are not passed on to the offspring. A somatic mutation might not ever result in an altered phenotype. The mutation may occur in a leaf cell nucleus, but in a gene that is inactive in leaves, such as a gene that affects root hair growth or bark formation. In species that undergo extensive vegetative reproduction, such as blackberries, alder, and prickly pear cactus, somatic mutations can be important if they affect a part of a plant that gives rise to a vegetative offshoot. As the offshoot grows and reproduces vegetatively, the patch of mutated cells may finally include an axillary bud and then affect a flower and its sex cells. The mutation then becomes a sexual one.

DNA REPAIR PROCESSES

Because most mutations are deleterious and occur frequently enough to be a significant problem, it is selectively advantageous for organisms to have DNA repair mechanisms that recognize and remove mutations. Certain mechanisms recognize base mismatches, loops, or other problems; other enzymes minimize the number of errors that occur in the first place.

The DNA repair rate must be neither too efficient nor too ineffective. In organisms that have very short genomes, such as bacteria and perhaps algae and protozoans, DNA polymerases can replicate a full genome without errors. Perfect replication does not occur every time, but nonmutated replication can sometimes occur. With the larger genomes of higher animals and higher plants, one set of chromosomes can virtually never be replicated without mutations if no repair mechanisms are present. DNA polymerase may make an error only once in every 1 million nucleotides, but if the genome is several billion base pairs long, every replication produces numerous errors. Most angiosperms have about 10 to 100 billion base pairs per haploid set of chromosomes. The smallest number known is in *Arabidopsis,* with about 200 million base pairs; every round of replication before a mitotic or meiotic division results in about 200 errors (see Table 16.1).

As a zygote grows into an adult, every cell cycle introduces new mutations, and in the adult, no two nuclei are exactly alike; after several cell cycles, DNA would be useless if there were no repair mechanisms. It would be impossible to produce sperms or eggs that were not extensively mutated, and under such circumstances, complicated organisms could not exist. It is estimated that a minimum of 15,000 genes are needed to code for all the information required for a flowering plant. This is about 20 million base pairs, so DNA polymerase is not accurate enough to provide error-free replication for even the simplest angiosperm. But DNA proofreading and repair systems bring the error rate down to an acceptable level. Actual rates as low as one mutation per 500,000 genes have been measured in corn.

Mutations occur in the genes that code for repair enzymes, resulting in serious problems. In humans, the disease xeroderma pigmentosum is caused by an inability to repair mutations caused by ultraviolet light. People with this disease are sensitive to sunlight and develop skin cancers easily. Sunlight has the same effect on all of us, but most of us can repair the damage. Our bodies have mechanisms that repair other types of mutations as well, but our modern chemical society may be contaminating our environment with mutagens for which we have no repair mechanisms.

DNA repair mechanisms must not be perfect or even extremely efficient. If every replication were perfect, if no mutations ever arose, all cells of an individual would have absolutely identical nuclei, as would all sperm and egg cells. Sexual reproduction would be useless, because all eggs and sperms of a species would carry identical genes. Mutations must occur at a low enough rate that they do not endanger every individual but rapidly enough that a species evolves as its environment changes. With no variation, there would be no differences for natural selection to act on; if all are identical, none has a selective advantage.

MONOHYBRID CROSSES

Sexual reproduction between two individuals is called a **cross.** The meiotic divisions that precede a cross reduce the number of sets of chromosomes per cell from the diploid number to the haploid number. Consequently, each sex cell—that is, each sperm cell and egg cell—contains one complete set of genes. Furthermore, each sperm cell contains all the genes necessary to construct a new plant; the same is true of each egg cell. The zygote (the fertilized egg) has two complete sets of genes.

Within a population, mutations produce new alleles, and the genotypes of individuals within the population differ. Of the plants that grow in an area and that can interact sexually, many may have the same allele of a particular gene but other individuals may have other alleles, other versions of the gene. Consequently, the alleles carried by a particular sperm cell may or may not be identical to the homologous alleles of the egg it fertilizes.

MONOHYBRID CROSSES WITH INCOMPLETE DOMINANCE

In a **monohybrid cross** only a single character is analyzed and studied; the inheritance of other traits is not considered. For instance, a plant with red flowers might be crossed with one that produces white flowers, and only the inheritance of that flower color trait is studied. Characters involving flower shape, leaf structure, and photosynthetic efficiency are also being inherited simultaneously, but in a monohybrid cross, only one is studied. This makes the analysis and understanding of the results much simpler. Once basic principles of a particular trait are known, then its interaction with another factor (dihybrid cross) or two other factors (trihybrid cross) can be studied.

Consider the cross just mentioned: A plant with red flowers is bred to one with white flowers. It does not matter which flower produces pollen and which produces ovules. Once the cross is made, seeds and fruit develop; the seeds are planted, and when mature, the new plants are allowed to flower so that their flower color can be examined. The flowers of all plants in this new generation are pink, resembling each parent somewhat, but not exactly like either (Fig. 16.10). The parents are called the **parental generation,** the offspring of their crossbreeding are the F_1 or **first filial generation,** and if these interbreed, their offspring are the F_2 generation.

The molecular biology of this monohybrid, flower color cross is easy to understand. Each parent is diploid and thus has two copies of the gene involved. In the red-flowered parent, both alleles produce mRNA that is translated into functional enzymes involved in the synthesis of red pigment. In a white-flowered parent, both alleles are defective. It may be that each produces an mRNA that when translated results in a protein unable to perform the necessary reaction. Or the promoter region may be mutated and can no longer interact with a chemical messenger. Whatever the cause, there is no pigment, and the flower is white.

Using genetic symbols, the red-flowered parent is *RR* and the white-flowered one is *rr* (Fig. 16.10). Each parent is said to be **homozygous,** because each has two identical alleles for this gene. The pink-flowered F_1 has received an *R* allele from one parent and an *r* allele

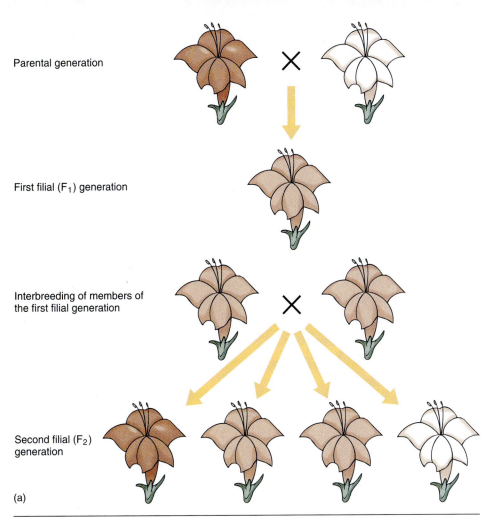

Parental generation

First filial (F₁) generation

Interbreeding of members of the first filial generation

Second filial (F₂) generation

(a)

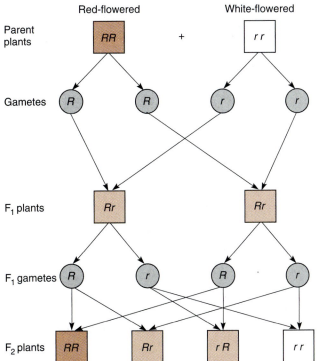

Red-flowered   White-flowered

Parent
plants   RR   +   r r

Gametes   R   R   r   r

F₁ plants   Rr   Rr

F₁ gametes   R   r   R   r

F₂ plants   RR   Rr   r R   r r

Genotype ratio: 1*RR* : 2 *Rr* : 1 *rr*
Phenotype ratio: 1 Red : 2 pink : 1 white

Key: *R* = red flower, *r* = white flower

(b)

FIGURE 16.10

(a) A monohybrid cross analyzing the trait of flower color. Details are explained in the text. (b) The phenotypes of parents, gametes, and progeny.

from the other, so its genotype is *Rr*; it is **heterozygous** because it has two different alleles for this gene. We use these symbols even though we have neither isolated the gene nor analyzed its nucleotide sequence; *R* and *r* are simply labels. Most genes are known only by their phenotypes and the labels given to them by geneticists (Table 16.2). With a genotype of *Rr*, the plant produces mRNA, half of which carries the defect; thus only half the normal amount of enzyme is produced. This results in less pigment being formed, only enough to make the flower look pink, not red. Neither parental trait dominates the other, so this pair of alleles shows **incomplete dominance:** The heterozygous phenotype differs from both homozygous phenotypes.

When analyzing the possible outcomes of crosses and breeding, one must understand the types and quantities of gametes involved. All the plants we are considering are diploid and form haploid spores by meiosis. The *RR* parent has chromosomes as shown in Figure 16.10b, and regardless of how chromosomes separate during meiosis, all spores receive an *R* allele. In the *rr* parent, all spores receive an *r* allele. Because the *RR* parent produces only

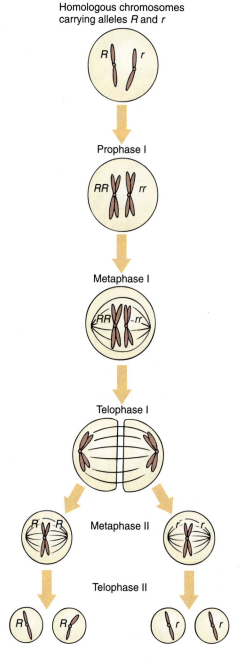

FIGURE 16.11

Production of haploid cells by meiosis in a heterozygote. During anaphase I the paternal chromosome is separated from the maternal chromosome, so one type of allele is separated from the other. As sister chromatids separate during anaphase II, each daughter nucleus receives identical alleles.

R spores, it also produces only *R* sex cells, both sperms and eggs in typical bisexual flowers. Similarly, the *rr* parent produces, indirectly, only *r* gametes. When the two plants are interbred, an *R* gamete unites with an *r* gamete, establishing a heterozygous (*Rr*) zygote that grows into a heterozygous adult by means of mitotic cell divisions. No other outcome is possible; it does not matter which gamete is which, an *R* sperm and an *r* egg result in the same type of zygote as an *r* sperm and an *R* egg.

When the heterozygote matures and flowers, each spore mother cell produces two types of spores (Fig. 16.11), not just one as is true of a homozygote. Because each cell has one *R* allele and one *r* allele, during the first meiotic division, one daughter cell receives *R* and the other receives *r*. The second division of meiosis results in two *R* spores and two *r* spores. If this is occurring in the anther, half the pollen grains and hence half the sperms have *R*, and the other half have *r*. In the nucellus of many plants, only one megaspore survives. If this gene has no effect on spore metabolism (and a gene for flower color is probably inactive in spores), then half the time *R* cells survive and half the time *r* cells live. In heterozygote parents, the two types of sperms and eggs are produced in equal numbers.

Spores are not sex cells, but as they develop into gametophytes, they divide by mitosis— duplication division—so all cells of the micro- gametophyte (pollen) have the same allele, as do all cells of the megagametophyte (inside the ovule).

CROSSING HETEROZYGOTES WITH THEMSELVES

When a plant's own pollen is used to fertilize its own eggs, the cross is a **selfing**. A plant can also be selfed by being crossed with another plant with exactly the same genotype. Selfing heterozygotes has interesting, instructive consequences; 50% (on average) of all sperms and eggs contain the *R* allele and 50% have *r* (Fig. 16.11). Not all zygotes are identical genotypically: Some are *RR*, having resulted from an *R* sperm and an *R* egg, others are *rr* (*r* sperm and *r* egg), and some are *Rr* (either *R* sperm and *r* egg or *r* sperm and *R* egg). Selfing a heterozygote produces three types of F₁s, some of which (*Rr*) resemble the parents and others (*RR* and *rr*) the grandparents. Again, to analyze the results, we must wait for the zygote to develop into an embryo, then plant the seeds and wait until the new plants are old enough to flower. Because each genotype produces a distinct phenotye, the genotype of each plant is known simply by looking at the flowers. If a large number of heterozygotes (pink-flowered plants) are selfed and large numbers of F₁ plants grown, about one fourth are red (*RR*), one half pink (*Rr*), and one fourth white (*rr*). This is an important ratio, typically represented as 1 : 2 : 1, and should be memorized immediately.

The reason for the proportion of these genotypes is explained in Figure 16.12. A **Punnett square** can be set up in which all types of one gamete, say, the egg, are arranged along the top of the square, and all types of the other gamete are arranged on the left side. The boxes are then filled in with the allele symbol above it and to the left. Because the gametes are produced in a ratio of 1 *R* : 1 *r*, listing them as in Figure 16.12 automatically represents their relative numbers in nature.

The Punnett square does not represent the outcome of any one cross; if a single heterozygous flower is selfed, it may produce only one or two seeds. If you plant just one seed, there is one chance in four that it will have red flowers, two chances in four pink, and one chance in four white.

The 1 : 2 : 1 ratio was one of the great discoveries of Gregor Mendel (1822–1884), an Austrian monk who performed experiments with pea plants. His cross-breeding experiments became the basis for modern genetics. Mendel discovered that in a selfing of this type, the recovery of the parental types means that genetic material must be composed of particles, such that the *R* genetic material can be separated from the *r* genetic material in the *Rr* heterozygote. Prior to Mendel's work, it was thought that the genetic material was a fluid. Two fluids cannot mix and then separate again perfectly. The constancy of the 1 : 2 : 1 ratio made it more logical to think in terms of discrete particles, genes, that never lost their identity regardless of the crosses in which they participated.

Furthermore, the 1 : 2 : 1 ratio can be realistically interpreted only in terms of each individual plant having two copies—being diploid—and each sex cell having one copy —being haploid. If each plant had only one copy, pink-flowered heterozygotes would be impossible, whereas if each had three, phenotypes such as dark pink (*RRr*) and light pink (*Rrr*) should also be present. Keep in mind the state of scientific knowledge in 1865 when

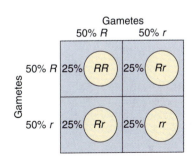

FIGURE 16.12
A Punnett square makes it easy to analyze and understand the results of a cross. Gametes from one parent are listed across the top; those of the other parent are listed on the left side.

Mendel was working. The concept that all organisms are composed of cells with nuclei had only recently been proposed; mitosis and cell division were very poorly understood, and meiosis would not be discovered for another 23 years. The concepts of chromosomes and homologous pairs would not be well established until the 20th century, and the existence of mRNA was not confirmed until the mid-1960s.

Mendel's work was well known but was years ahead of its time. Advances in microscopy and chromosome staining were necessary to recognize homologous chromosomes and the diploid nature of most plant and animal cells. But as soon as these were discovered, scientists immediately realized that chromosomes must be the carriers of Mendel's hypothetical genes.

Evolution by natural selection had just been discovered and was revolutionizing not only biology but also all aspects of philosophy, theology, anthropology, and cosmology—the very ways in which we think about our world. Natural selection required diversity and change, whereas Mendel's discoveries appeared to show that differences in phenotype were due only to the mixing and separating of unchanged, constant genes; early proponents of evolution were certain that Mendel's theory of genes as the basis of inheritance had to be incomplete. In 1927, mutations were proven to occur; genes are just changeable enough to be the basis of both evolution by natural selection and Mendelian genetics. With the discovery of mutations, the two monumental discoveries of the 1800s were reconciled.

MONOHYBRID CROSSES WITH COMPLETE DOMINANCE

The situation in which only half as much product of an enzyme, such as the red pigment discussed above, is produced in a heterozygote is not a universal situation (see Chapter 15). In certain species or with other traits, cytoplasmic control mechanisms may cause the enzyme to function until a specific amount of product is synthesized. The enzyme may have to work faster or longer, but the final amount of product is the same whether the plant has two functional alleles or only one. In other situations, the amount of enzyme might be monitored such that the nonmutant mRNA of the heterozygote is translated more frequently or rapidly.

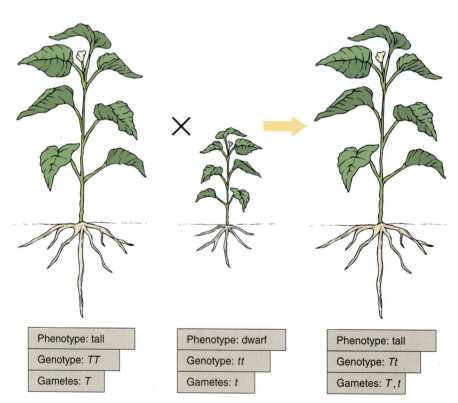

FIGURE 16.13
In this monohybrid cross, the trait "tall" shows complete dominance over the trait "dwarf." In a *Tt* nucleus, the *T* allele may be transcribed twice as much as each *T* allele in a *TT* nucleus, or the resulting mRNA may be translated twice as much, or the protein may work twice as long or twice as fast. Or a *TT* plant may contain a level of *T* protein far above the threshold for responsiveness.

Phenotype: tall
Genotype: *TT*
Gametes: *T*

Phenotype: dwarf
Genotype: *tt*
Gametes: *t*

Phenotype: tall
Genotype: *Tt*
Gametes: *T*, *t*

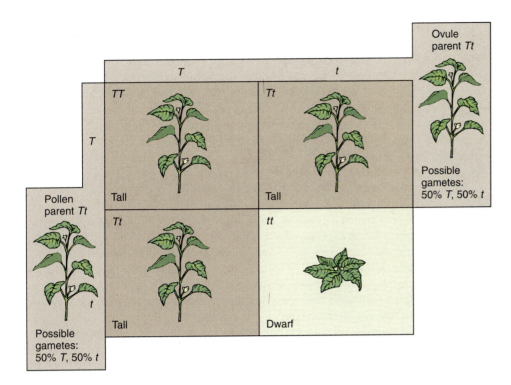

FIGURE 16.14
In setting up the Punnett square for a selfing of *Tt* plants, first establish the genotypes of the two parents. Then determine what types of gametes are produced and in what proportions, and fill in the squares with the genotypes. From the genotypes, the phenotypes in each square can be determined.

In either case, the phenotype of the heterozygote is like that of the parent with two effective alleles. That trait is said to be **dominant** over the other version of the trait, which is **recessive**. An example is height. A tall plant with a *TT* genotype produces sex cells that are all *T*. A short plant, genotype *tt*, produces only *t* sex cells (Fig. 16.13). When the tall plant is crossed with the short one, all F_1 progeny have the *Tt* genotype, but they all have the phenotype of being tall, indistinguishable from their *TT*, homozygous dominant parent. We do not know what protein is produced by the *T* allele, but even with only one functional allele, enough product is made to permit normal growth. *Tt* plants are tall, and the tall character completely dominates the "dwarf" character.

Knowing the molecular biology of genetic systems, you can predict that when heterozygotes are selfed, two types of sperm cells (*T* and *t*) and two types of eggs (*T* and *t*) are produced. The Punnett square for the cross is as in Figure 16.14, and a genotype ratio of 1 *TT* : 2 *Tt* : 1 *tt* is expected. The prediction is correct, but what will the phenotype ratio be? One out of four plants will be tall due to a *TT* genotype, and two out of four will be tall due to a *Tt* genotype. Thus, three fourths have the tall phenotype and one fourth have the short phenotype. Whenever a cross is made and a phenotype ratio of 3 : 1 is seen, we should suspect that the parents are heterozygous and the character shows complete dominance.

TEST CROSSES

When traits with incomplete dominance are studied, the genotype of any plant is easy to determine from its phenotype. If the trait has complete dominance, it is difficult to know what the genotype of any particular plant is unless the plant shows the recessive trait. Imagine you are studying the inheritance of tallness; you have selfed some heterozygotes and planted the resulting seeds. In your greenhouse or garden, there are now hundreds of plants, approximately 75% of which are tall and 25% short. You know the short ones are *tt*, and if you need a plant with the *tt* genotype for experimentation, you know automatically to chose short plants for the experiment. But if you need to experiment on plants with the *TT* genotype, how can you tell which they are? A tall plant picked at random is more likely to be a *Tt* plant, because there are twice as many of them as *TT* tall plants.

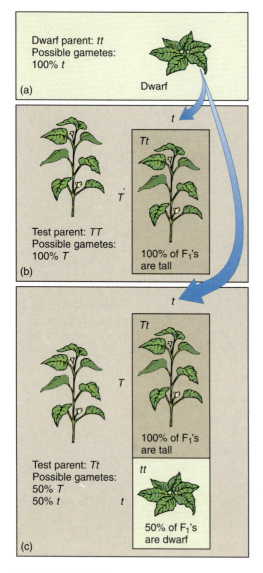

(a) Dwarf parent: *tt*
Possible gametes: 100% *t*

Dwarf

(b) Test parent: *TT*
Possible gametes: 100% *T*

t

Tt

T

100% of F₁'s are tall

(c) Test parent: *Tt*
Possible gametes: 50% *T* 50% *t*

t

Tt

T

100% of F₁'s are tall

tt

t

50% of F₁'s are dwarf

FIGURE 16.15
Test crosses to determine if a plant with the dominant phenotype is heterozygous or homozygous. Of the data in this diagram, you would not know the genotypes of the test parents before the test cross; those data are what you are trying to discover. If you want to find only a single homozygote or heterozygote, you would need to do test crosses on only one or a few plants. But imagine that you had done an experimental cross for a trait that you assumed would give you a 3 : 1 ratio; it would be necessary to do test crosses on a large number of progeny plants with the dominant phenotype just to confirm your experiment.

The genotype can be revealed by a **test cross,** a cross involving the plant in question and one that is homozygous recessive for the trait being studied. All gametes produced by a homozygous recessive parent carry the recessive allele, which is unable to mask the homologous allele in the resulting F₁ zygote. If the plant being tested is actually homozygous dominant *(TT),* 100% of the progeny will be heterozygous *(Tt)* and tall (Fig. 16.15). If the tall parent is heterozygous, half its progeny in the test cross will be tall *(Tt)* and half short *(tt).* If only a few seeds resulting from the test cross are grown, it is possible statistically to choose all of them, by accident, from the *TT* group; but if a large number of seeds are grown, the chance of accidentally not choosing any of the *Tt* seeds is very low, and if even one of the F₁ plants is short, the test parent must have been *Tt.*

Once the actual genotypes of plants are known, those that are homozygous dominant can be gathered and planted in special areas, kept free of all natural pollinators, and allowed to breed only among themselves. All their progeny will be homozygous dominant and can be used in breeding experiments. The homozygous recessives can also be kept as a special line, being selfed and kept pure. Such groups are **pure-bred lines** and are both useful and valuable. It is not possible to maintain the heterozygotes like this, because they do not breed true; that is, their progeny are not exactly like them. One fourth are homozygous recessive; one fourth are homozygous dominant and cannot be distinguished visually from the one half that are heterozygotes.

So far, we have considered only one trait, height, but for plants that are important crop or horticultural species, dozens or even hundreds of traits are cataloged. Seed companies maintain hundreds of different lines of corn, for instance, in which the genotype of many characters is known for each line (Fig. 16.16). Many times, plants are collected from the wild, so nothing is known about their genotype unless it is immediately obvious from the phenotype. However, from looking at a few collected plants it is not possible to tell which characters are dominant and which are recessive; carefully controlled and recorded crosses must be made.

When test crosses must be made on annual plants, the results are usually not known until after the plants have died. Their genotypes are then known, but the plants cannot be used for experimentation or breeding. In such cases, it usually is necessary to do both test crosses and experimental crosses simultaneously, not knowing which plants have the correct genotype for the experiment being performed. After the test cross results are complete, the parents that had the proper genotype can be identified and the experimental crosses that involved them can be analyzed.

MULTIPLE ALLELES

Each gene may have many alleles, not just two as in the examples discussed so far (*T* and *t*, *R* and *r*). A protein of average size consists of about 300 amino acids, so the coding portion of its mRNA must have about 300 codons, each containing three nucleotides. The gene is

FIGURE 16.16
Maintaining pure-bred lines is relatively easy for perennial plants; once the plant is growing, it can produce pollen or ovules for experimentation for years. Annual plants are much more difficult because pollination, seed gathering, planting, and record keeping must be done annually. Here, wheat seeds are being frozen in liquid nitrogen; after several years, some will be thawed and planted. Their seeds will be frozen for further storage.
(Courtesy of the USDA Agricultural Research Service, National Seed Storage Laboratory, Fort Collins, Colorado)

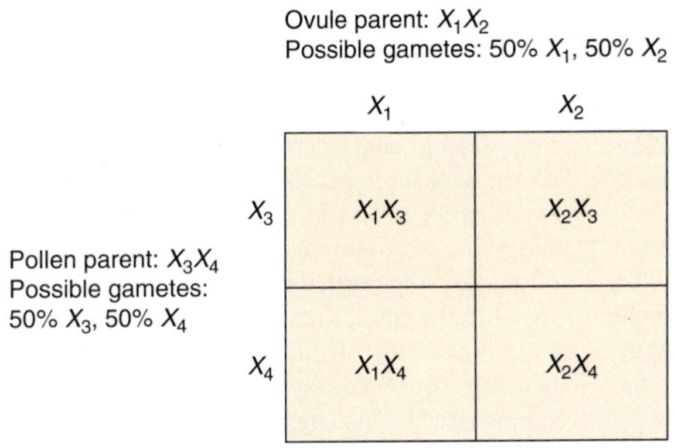

Ovule parent: X_1X_2
Possible gametes: 50% X_1, 50% X_2

	X_1	X_2
X_3	X_1X_3	X_2X_3
X_4	X_1X_4	X_2X_4

Pollen parent: X_3X_4
Possible gametes:
50% X_3, 50% X_4

FIGURE 16.17
With multiple alleles for a single character, numerous types of crosses become possible. However, we still determine, from the parental genotype, all possible gametes, then construct a Punnett square in the usual fashion.

therefore at least 900 nucleotides long, not counting introns and promoters. At least 900 sites exist at which point mutations can occur, and of course any mutation may involve several nucleotides. Consequently, the gene may exist in many forms, called **multiple alleles** (see Fig. 16.1). When genes are polymorphic, having multiple alleles, numbers, such as X_1, X_2, X_3, X_4, and so on, are used rather than capital and lower-case letters. Certain mutations still result in the production of a protein with the normal sequence, but most lead to altered protein structure. Some of these proteins are quite similar to the original protein, perhaps having similar or even identical enzymatic activity. However, many carry out the proper reaction more slowly or are not accurately controlled by regulatory mechanisms in the cytoplasm; a normal, wild-type phenotype may not be produced. With multiple alleles, the concept of dominance is more complex; one allele may produce the proper quantities of the functional protein, whereas a different allele produces more, another produces less, and a fourth produces one with altered activity. Many different phenotypes are possible.

Within a population of plants, as many types of gametes can be produced as there are different types of alleles. A heterozygous X_1X_2 plant can be crossed with a heterozygous X_3X_4 one, resulting in progeny such as in Figure 16.17: X_1X_3, X_1X_4, X_2X_3, and X_2X_4. Four distinct types of F_1 plant are produced, none of which has the genotype of either parent.

DIHYBRID CROSSES

A **dihybrid cross** is one in which two genes are studied and analyzed simultaneously, rather than just one, as in a monohybrid cross. Every cross involves all the genes in the organism, but the terms "monohybrid" and "dihybrid" refer only to the number being analyzed.

When two genes are studied, the results of the crosses depend upon the positions of the genes on the chromosomes. If they are on different chromosomes, the alleles for one gene move independently of the alleles for the other gene, but when two genes are close together on the same chromosome, the alleles for one gene are chemically bound to the alleles for the other gene and move together. The situation in which the genes are on separate chromosomes is easier to understand and is explained first.

GENES ON SEPARATE CHROMOSOMES: INDEPENDENT ASSORTMENT

Consider a plant heterozygous for two traits, for instance seed coat texture and color, with a smooth seed coat (*S*) showing complete dominance over a wrinkled seed coat (*s*), and a yellow seed coat (*Y*) being dominant over a green seed coat (*y*). The plant's genotype is *SsYy*, and its phenotype is smooth yellow seeds. We know that if we consider only the gene for color, the plant produces two types of gametes in approximately equal numbers, some carrying *Y* and some carrying *y*. Also, if we consider only texture, half carry *S* and half *s*. How do the alleles of the two genes relate to each other? As in monohybrid crosses, knowing the types of gametes that can be formed is the key to understanding the patterns of inheritance.

BOX 16.1 *Botanical Philosophy and Popular Culture*

When thinking about genetics and the contributions of Gregor Mendel, it is important to keep in mind the state of scientific knowledge at that time. Some of the greatest scientific geniuses of all time were discovering and documenting the fundamental concepts of biology. Only 27 years earlier than Mendel's work, M. J. Schleiden and T. Schwann had proposed the cell theory, the concept that all organisms are composed of cells with nuclei, and that all existing cells come from pre-existing cells. We take this for granted today, but it means that all living organisms are fundamentally the same, that life may have arisen only once, and that all cells are descended from those first cells. In 1849, Wilhelm Hofmeister proved that during plant reproduction, the new embryo develops primarily from the egg after receiving some "influence" from the pollen tube. This was revolutionary because until that time, everyone firmly believed that in both plants and animals, the new embryo developed from the sperm cell—the female parent might provide nourishment and protection, but the new generation was really a continuation of the paternal parent (could this belief have been based not on observation of nature but on observation of society and history dominated by males?). Hofmeister's conclusion that the female parent also contributes to the new offspring was both

shocking and also very important for Mendel's later observation that genes are passed equally from both parents.

In 1859, Charles Darwin published *Origin of Species by Natural Selection,* which postulated that organisms change through time, that they are not constant (A. R. Wallace came to the same conclusion at the same time). The discovery of evolution by natural selection caused a great deal of trouble for Mendel: Natural selection requires diversity and change, whereas Mendel's theories of genetic inheritance showed that differences in phenotype were due only to the mixing and separating of unchanged, constant genes. The idea of evolution had aroused great controversy and serious philosophical battles between proponents of evolution and of biblical creation. This battle had been raging for 6 years when the paper by Mendel (a monk) appeared. Had he published 10 years earlier, he might have found instant acceptance and Darwin would have had a more difficult time.

But a further difficulty for genetics was its postulation of genes as particles, not as fluids. This was at a time when fluids were considered the basis for most of biology, with sieve pores in the phloem and pits and perforations in the xylem considered to be effective by acting as filters that separated the various fluids that were thought to exist. Fortunately, in 1879, H. Fol showed

that during angiosperm reproduction, a sperm nucleus (a particle) entered the egg, and in 1888, E. Strasburger described meiosis, giving a firm foundation to the concept of genes being particulate. If genes are on the chromosomes, then suddenly it is possible to see how Mendel's ratios and independent assortment could occur. This was summarized in 1915 by T. H. Morgan in *The Mechanism of Mendelian Heredity.* Finally, in 1927 and 1928, H. J. Muller and L. J. Stadler showed that genes could be artificially changed (mutated) by use of X-rays. Thus, genes could be stable enough to permit mendelian genetics to operate and yet variable enough to allow natural selection to operate as well. It is important to keep in mind that these were not the only biologists working at the time, nor were these the only discoveries. A huge volume of information was being generated, much of which concerned unusual, exceptional, or aberrant cases and which thus made the fundamental discoveries more difficult to recognize as being important. Also, we must always try to appreciate the profound influence of the general philosophy of the time—a male-dominated society produced male-dominated theories of biology. It is likely that our own sincerely held beliefs are causing us to improperly evaluate some of our own biological observations.

If the two genes are on separate chromosomes, the alleles of one gene move independently of the alleles of the other gene during meiosis I; this is called **independent assortment.** All chromosomes align on the metaphase plate during metaphase I; the chromosomes have duplicated and each has two copies of each allele (Fig. 16.18a). The homologous chromosomes have also paired, so there are two *Y* alleles, two *y* alleles, two *S* alleles, and two *s* alleles at the metaphase plate. During anaphase I, homologous chromosomes separate from each other, and both *Y* alleles move to one pole because they are on the two chromatids of one chromosome, still held together by its centromere. Both *y* alleles, located on the two chromatids of the homologous chromosome, move to the other pole. There is no way to predict which pole will receive which type of allele. Similarly, the *S* alleles separate from the two *s* alleles and move randomly to the poles. In some cells, the pole that receives the *Y* alleles at telophase I also receives the *S* alleles, but in other cells the *Y* and *s* alleles end up together. During meiosis II, the two chromatids of each chromosome separate from each other, resulting in four types of haploid cells in equal numbers: *SY, sY, Sy* and *sy.* Any single microspore or megaspore mother cell produces only two types of haploid cell; the set *SY* and *sy* or the set *sY* and *Sy.* All four types are produced by a single plant because some mother cells produce one set and some produce the other set.

Once the possible types of gametes are known, the Punnett square can be set up, as in Figure 16.18b. Any single fertilization results from the syngamy of one sperm cell and one egg cell and produces only one of the 16 possible zygote genotypes shown in Figure

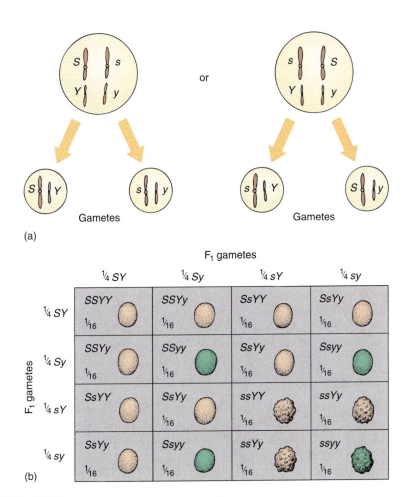

FIGURE 16.18

(a) During anaphase I, chromosomes move independently of each other, so one pole receives *S* and the other receives *s*; likewise, one gets *Y* and the other *y*. But in about half the cells both *S* and *Y* move to the same spindle pole by chance, whereas in the other dividing cells, one spindle pole receives *S* and *y*, again by chance. (b) A Punnett square for a dihybrid cross is set up just like one for a monohybrid cross; establish the types and relative abundance of gametes, then fill in the squares. The table looks a little formidable, but it really consists of two 3 : 1 ratios intermingled.

16.18b. All 16 types of zygote occur only if we study many fertilizations; to have the ratios come out accurately, we have to analyze hundreds of progeny. In plants, it is usually easy to obtain large numbers of fertilizations because pollen and ovules are produced in large amounts. Once pollinated, most plants produce dozens or even thousands of seeds, enough progeny that all 16 zygote genotypes occur in about the expected ratios. But in large animals such as mammals, reproduction may be infrequent and only one or two offspring are produced each year; a great deal of work is necessary to get enough progeny to verify the results of a dihybrid cross.

In a dihybrid cross involving independent assortment of two heterozygous genes, each gene showing complete dominance, a characteristic phenotype ratio occurs, just as is true of the 3 : 1 ratio in a monohybrid cross. The ratio is 9 : 3 : 3 : 1, with 9/16 of the plants having the dominant phenotype for both traits (in our example smooth yellow seed coats), 3/16 with the dominant phenotype of the first trait and the recessive phenotype of the second (smooth, green), 3/16 with the first trait recessive and the second dominant (wrinkled, yellow), and 1/16 in which the plants have the recessive phenotype of both traits (wrinkled, green) (Table 16.3). The 9 : 3 : 3 : 1 ratio results only if all four types of gametes are produced in equal numbers and have equal opportunity to participate in reproduction. The alleles *Y* and *y* must be independent of *S* and *s* during meiosis I; this automatically happens if they are on different chromosomes.

TABLE 16.3   *Genotype and Phenotype Ratios of a Dihybrid Cross**

Phenotype	Ratio	Genotype	
Smooth, yellow	$\frac{9}{16}$	$\frac{1}{16}SSYY : \frac{2}{16}SSYy : \frac{2}{16}SsYY : \frac{4}{16}SsYy$	or $S\_\_Y\_\_$†
Smooth, green	$\frac{3}{16}$	$\frac{1}{16}SSyy : \frac{2}{16}Ssyy$	or $S\_\_yy$
Wrinkled, yellow	$\frac{3}{16}$	$\frac{2}{16}ssYy : \frac{1}{16}ssYY$	or $ssY\_\_$
Wrinkled, green	$\frac{1}{16}$	$\frac{1}{16}ssyy$	must be $ssyy$

* The two individuals are heterozygous for two characters showing complete dominance and independent assortment.

† The notation ___ indicates that either allele may be present; e.g., $S\_\_yy$ represents either $SSyy$ or $Ssyy$, both of which produce a smooth green phenotype.

Notice that if only one trait is considered, it behaves as in a monohybrid cross: Plants with smooth seeds outnumber those with wrinkled seeds by 3 : 1 and those with yellow seeds are three times more abundant than those with green seeds. Similarly, the monohybrid genotype ratios are also present—1 SS : 2 Ss : 1 ss and 1 YY : 2 Yy : 1 yy. Considering two genes simultaneously does not affect their inheritance at all.

CROSSING-OVER

Independent assortment can also occur if two genes are located far apart on the same chromosome such that crossing-over occurs between them during prophase I, after homologous chromosomes have paired and a synaptonemal complex is formed (see Chapter 4). Because no preferential sites for crossing-over seem to exist, the farther apart two genes are, the greater the possibility that crossing-over will occur between them. Most plant chromosomes are so long that crossing-over occurs several times within each chromosome during each prophase I. Consequently, the two ends act like separate entities; if the gene for seed coat color were at one end of the chromosome and the gene for seed coat texture were at the other, they would still undergo independent assortment. However, if the two genes are close together on a chromosome, crossing-over may not occur and the two alleles may move together during meiosis I, as described below.

GENES ON THE SAME CHROMOSOME: LINKAGE

If two genes occur close together on a chromosome, they usually do not undergo independent assortment during meiosis I; instead, the two genes are said to be **linked.** Consider a plant heterozygous for the two traits of the seed coat again, but now imagine the genes occurring close together on one chromosome. During meiosis, haploid cells that are either S or s, Y or y are formed, but a new complexity arises. Because the genes are linked, several types of heterozygote are possible: an $SsYy$ individual may have the alleles S and Y on one chromosome and the alleles s and y on the homologous chromosome. But an $SsYy$ individual may have s and Y linked together and S and y linked (Fig. 16.19a). We must consider the types of gametes that can be formed. If crossing-over is ignored, the first individual produces haploid cells with the genotypes SY and sy only, whereas the second plant would produce sY and Sy gametes only. The result of selfing is not a 9 : 3 : 3 : 1 ratio, but something drastically different.

The most instructive cross is a test cross using a double homozygous recessive parent: $ssyy$. If we continue to exclude crossing-over (imagine that the two genes are extremely close together), the gametes from the first type of heterozygote will be SY and sy and the gametes from the homozygous recessive parent will be all sy. Two types of F_1 will be produced, as shown in Figure 16.19b, smooth yellow and wrinkled green, and they occur in a ratio of 1 to 1. The two phenotypes are like those of the parents.

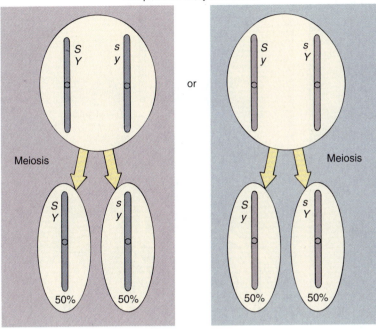

Types of gametes
if no crossing over occurs

(a)

Double homozygous recessive parent: *ssyy*
Phenotype: wrinkled, green
Possible gametes: 100% *sy*

Test parent: *SsYy*, *S* linked to *Y*, *s* linked to *y*
Possible gametes (no crossing over):
50% *SY*, 50% *sy*

(b)

	sy 100%
SY 50%	*SsYy* 50% smooth yellow
sy 50%	*ssyy* 50% wrinkled green

FIGURE 16.19

(a) Two types of *SsYy* individuals are possible, but in most instances their phenotypes are identical. Only genetic tests can distinguish which is which, based on the unique gametes produced by each. (b) A test cross, with the plant on the left in (a) being the test parent; try setting up the Punnett square for a test cross with the plant on the right in (a). A test cross of two closely linked genes produces results very different from those of a test cross involving nonlinked genes that assort independently. What would the Punnett square be like if *S* and *Y*, *s* and *y* were not linked?

If the genes are not extremely close together, crossing-over may occur; for example, in 10% of the cells undergoing meiosis, a crossing-over might happen, so the plant would produce four types of gametes, **but not in equal numbers.** Of the gametes for the first type of heterozygote, 47.5% would be *SY*, 47.5% *sy*, 2.5% *sY*, and 2.5% *Sy* (Fig. 16.20a). The last two are **recombinant chromosomes** formed from a crossing-over of the homologous chromosomes and recombination of alleles. The first two are **parental type** chromosomes. In the majority of the cells, *Y* is still linked to *S* and *y* is linked to *s*, but in a minority, *Y* is linked to *s* and *y* to *S*. A Punnett square alone is not sufficient to show the relative proportions of progeny, so the percentages of gamete must be added. A test cross now would result in 47.5% of the F₁ progeny being smooth and yellow, 47.5% wrinkled and green, 2.5% wrinkled yellow, and 2.5% smooth green (Fig. 16.20b). The first two types are like the parents, but the last two are recombinant types produced at the same percentage as crossing-over occurs. We cannot actually "see" crossing-over with a microscope, but we can infer that because 5% of the F₁s have a recombinant phenotype, then 5% of the chromatids must have undergone crossing-over.

The rate of crossing-over is directly proportional to the physical spacing between the genes on a chromosome. Two genes that produce 6% recombinant F₁s are closer together

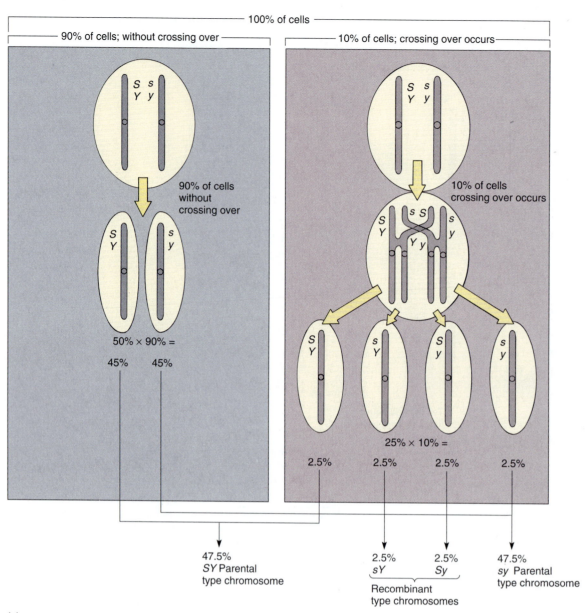

(a)

Double homozygous recessive parent: *ssyy*
Phenotype: wrinkled, green
Possible gametes: 100% *sy*

Test parent *SsYy*, *S* linked to *Y*; *s* linked to *y*
Phenotype: smooth, yellow
Possible gametes with 5% crossover:
47.5% *SY*, 47.5% *sy*
2.5% *sY*, 2.5% *Sy*

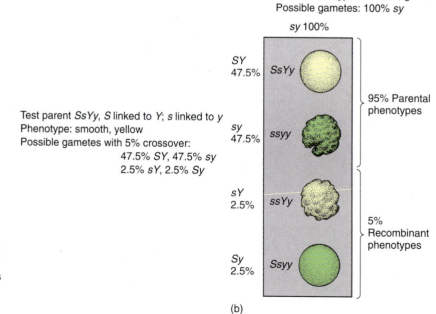

(b)

FIGURE 16.20

(a) With linkage, four types of haploid genotypes might be produced, but not in equal numbers. Recombinant types are less abundant than parental types—in this case 5% recombinant chromosomes and 95% parental type. These two genes are 5 map units apart. (b) This test cross, unlike that of Figure 16.19b, gives four types of progeny: two parental types and two recombinant types. The alleles *S* and *Y* show linkage rather than independent assortment.

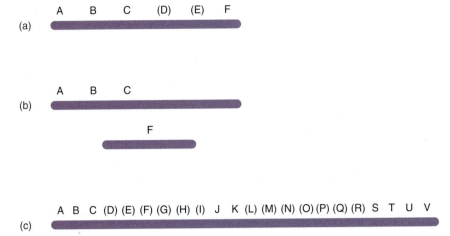

FIGURE 16.21

(a) A and C might be far enough apart for crossing-over to occur in virtually 100% of the nuclei, so they undergo independent assortment. If each can be shown to be linked to B, they must be on the same chromosome. If A, B, and C are not known to be linked to F, it might be because they are on separate chromosomes (b) or because they are on the same chromosome but intervening genes have not yet been mapped, as shown in (a). (c) Because of incomplete mapping, ABC, JK, and STUV are three linkage groups even though there is only one chromosome.

than two genes that produce 10% recombinant F_1s. By analyzing as many mutant alleles as possible, we can measure "space" between them in recombination percentages, each 1% being called one **map unit** or one **centimorgan**; one map unit, on average, is about one million base pairs. By this means, a genetic map can be constructed. If enough genes can be mapped, we may find that A is linked to B which is linked to C; even if A and C are so far apart that they assort independently, by knowing that both are linked to a common gene, B, we know that all three are on the same chromosome. On the other hand, if A, B, and C cannot be shown to be linked to another gene F, that does not mean that F is on a different chromosome; it might be that intervening genes have not yet been mapped (Fig. 16.21). A set of genes known to be linked is called a **linkage group;** when all genes are mapped, there will be exactly as many linkage groups as there are chromosomes. But if only a few genes are mapped, there may be more linkage groups than chromosomes.

MULTIPLE GENES FOR ONE CHARACTER

Individual phenotypic traits are the result of complex metabolic processes involving numerous enzymes; therefore, many separate genes may affect any single trait. The gene *R* was described as affecting flower color by producing an enzyme that synthesized a red pigment. But that enzyme requires the proper substrate, which is present only if it is synthesized by a different enzyme controlled by a distinct gene. If this gene is present as an ineffective allele, there will be no substrate and thus no pigment, regardless of whether the first flower color gene is present as allele *R* or allele *r* (see Fig. 15.15). In *Arabidopsis thaliana*, abscisic acid (ABA) causes developing seeds to become dormant at maturity. In several mutants (Table 16.4), mature embryos continue to grow as if germinating and do not enter dormancy. One mutant, *aba*, is unable to produce enough ABA. A second mutant, *abi*-3, produces plenty of ABA but does not respond to it—perhaps its ABA receptor is the affected protein. Although *aba* and *abi*-3 involve different genes, they both result in the same phenotype. They are multiple genes for one trait even though they are not part of one metabolic pathway.

Most synthetic pathways involve at least four or five intermediates, and their four or five genes all affect the same trait. Having multiple genes for each trait is referred to as **epistasis.** Some intermediates may be produced by several pathways, each with its own enzymes; a mutation in one is not necessarily particularly severe, because the alternate pathway may increase its activity and produce sufficient amounts of the intermediate. If this happens, the mutation is masked and is not reflected in the phenotype, even if it is present in the homozygous condition. Conversely, when an intermediate is part of several metabolic pathways and is produced by only one enzyme, a mutation in that enzyme's gene affects all the pathways and alters several different traits. Multiple phenotype effects of one mutation are called **pleiotropic effects.** For example, any mutation that affects the protein portion of phytochrome affects all developmental processes controlled by phytochrome, and mutations that alter the level of pyruvate affect the citric acid cycle, amino acid synthesis, and C_4 metabolism.

TABLE 16.4	Multiple Genes for One Character
Allele	**ABA***
wild type	169
aba	10
abi-3	279

* ng/g fresh weight of seed tissue

OTHER ASPECTS OF INHERITANCE

MATERNAL INHERITANCE

In the crosses described above, the alleles of both parents are transmitted equally to the progeny, a situation called biparental inheritance. All genes in the nucleus undergo biparental inheritance. An unexpected feature of both plants and animals is that during fertilization, the sperm cell loses most of its cytoplasm and only the sperm nucleus enters the egg. Consequently, the zygote obtains all its plastid and mitochondrion genomes from the maternal parent; this is **uniparental inheritance**, more specifically **maternal inheritance.**

Mitochondrion genetics are difficult to study for several reasons. First, each cell has many mitochondria, up to hundreds, and each mitochondrion has several circles of DNA. Each cell has multiple copies of each mitochondrion gene and may have received diverse types of alleles from its maternal parent. Simple mendelian ratios are not produced. Second, the presence of many of the mitochondrion DNA mutants does not result in easily recognizable phenotypes. The plants usually simply grow somewhat more slowly, or, if the mutation has a severe effect, they may die as embryos. Consequently, the presence of mitochondrion mutants may not be detectable by examination.

Plastid genetics are somewhat easier if alleles affecting chlorophyll synthesis are studied. Consider a plastid whose DNA has mutated so that the plastid can no longer produce any chlorophyll; it may be red or orange because the carotenoids are no longer masked by chlorophyll, or it may be white if no pigment at all is produced. If a zygote receives only plastids like these, the embryo and seedling will be red, orange, or white. Such seedlings are rare, but in large commercial nurseries that grow millions of seedlings, they appear from time to time. The plants can be kept alive artificially if they are grafted onto a normal plant that supplies them with carbohydrates (Fig. 16.22). Grafted plants can then be used for genetic experiments.

When an achlorophyllous plant is crossed with a green, chlorophyll-bearing plant, the outcome depends on which plant is the **pollen parent** and which is the **ovule parent.** If the achlorophyllous parent provides the pollen, 100% of the zygotes will be green, because they receive all their plastids from the normal parent by way of the egg; all mutant plastids are destroyed during sperm differentiation and syngamy. But if the achlorophyllous plant is the ovule parent, all progeny also have mutant plastids and are achlorophyllous.

Plastid inheritance is also responsible for certain types of **variegation** in plants, the presence of spots or sectors that are white, red, or orange on a plant that is otherwise green (Fig. 16.23). Because plant cells have many plastids, only rarely do all the plastids of an egg have the same alleles unless the parent has only one type, as in the cactus of Figure 16.22. More often, each cell has a mixture of plastid types. During cell division, some plastids move into each end of the dividing cell; usually random chance alone results in each daughter cell getting some of both types of plastid (Fig. 16.23b). However, occasionally one daughter cell by chance receives only mutant plastids. It is white, red, or orange, and is able to survive because surrounding normal cells supply it with sugar. If this occurs in a leaf primordium or a young internode, the cell continues to grow and divide along with the surrounding cells, resulting in a patch of cells that lack chlorophyll (Fig. 16.23a). If it occurs in a very young leaf, the cells grow quite a lot and produce a large spot; if it happens in an older leaf that has reached almost full size, the patch is small.

When variegation affects stems, the achlorophyllous patch may include an axillary bud. If so, the entire bud and the branch that grows from it are achlorophyllous. Any flowers it produces can be used for genetics experiments, providing either pollen or ovules. Experiments on both corn and morning glory have shown the achlorophyllous character to be maternally inherited; thus it is a feature of the plastid DNA as just described. Recombinant DNA technology is now greatly accelerating our efforts to understand plastid and mitochondrion genetics. In both organelles the DNA is a circular double helix, one of its ends being attached to the other, and no histones are present. Plastids contain many circles, 200 in each spinach plastid and 1000 in each of those of wheat, so rather than being haploid or diploid, wheat plastids are 1000-ploid. Plastid DNA may constitute up to 21% of the total DNA content of a cell (see Table 4.4). In angiosperms, plastid DNA tends to be about 150 kbp long. In plants, mitochondrion DNA circles are very large, well over

FIGURE 16.22

Plastid mutations in the cactus genus *Gymnocalycium* seem to be common; quite often totally red or orange seedlings appear in cactus nurseries. These lack chlorophyll but can be kept alive by grafting them onto a green cactus that supplies them with sugar. Many plants like *Coleus* and maples have bright red leaves; they do have chlorophyll, but it is masked by large amounts of other pigments.

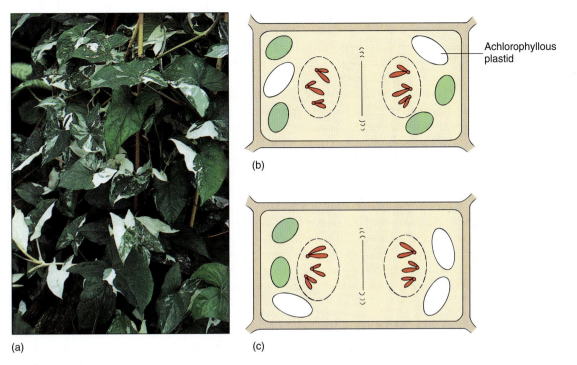

FIGURE 16.23

(a) Variegated plants of *Syngonium podophyllum*. Each green cell probably has a mixture of normal chloroplasts and plastids with a mutation that prevents chlorophyll synthesis. Cells in the white area have only mutant plastids. (b and c) Cells usually have many plastids and mitochondria, so random distribution of the organelles is usually sufficient to ensure that both cells receive some (b). If mutant plastids are present, then occasionally when cytokinesis occurs, one end of the cell may by chance have only mutant plastids (c). Most cells have only two sets of (nuclear) chromosomes, so all the elaborate mechanisms of mitosis are necessary if each daughter cell is to receive one complete set.

200 kbp (up to 2400 kbp), more than ten times larger than mitochondrion DNA in animals.

Genomes of plastids contain genes for numerous enzymes as well as a complete set of plastid ribosomal RNAs and transfer RNAs. At present it is estimated that 80 to 100 enzymes and membrane proteins are coded by plastid DNA, and many of these are essential: the large subunit of RuBP carboxylase, components of the proton channel/ATP synthetase complex, cytochromes, and proteins that bind chlorophyll *a* (Fig. 16.24). Mitochondrion DNA also contains genes for rRNA and tRNA as well as those that code for proteins involved in ATP synthetase complexes, electron transport, and NADH dehydrogenase complex. In many cases, organelle genes must work in coordination with nuclear genes. The interaction of these three genomes within one cell is now an area of very active research.

LETHAL ALLELES

The phenotypic result of a mutation can vary in severity from almost undetectable to **lethal**; that is, its presence can kill the plant. Genetically inherited lethal mutations are most often recessive and are fatal only if present in the homozygous condition; a heterozygous plant has enough normal protein to survive. The reason for the recessive nature of most lethals is simple: A dominant lethal would kill the plant, even in the heterozygous condition, so it would rarely be passed on to any offspring.

The presence of a lethal allele can be difficult to detect if its effect occurs early. If it affects basic metabolic functions such as respiration, DNA replication, cellulose synthesis, or the structure of histones, plants that are homozygous for it probably die while still very young; in some instances the gametes die even before fertilization. The only seeds that

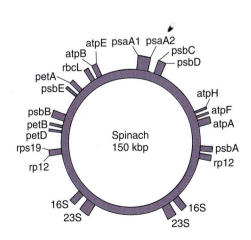

FIGURE 16.24

A map of the plastid DNA circle from spinach; the letters are the abbreviations for the names of the genes. For several species, the entire plastid circle has been sequenced so that we know the exact position of every nucleotide. *(Based on data of J. D. Palmer)*

complete development are heterozygotes or dominant homozygotes. This results in unusual genotype and phenotype ratios. A selfing of a heterozygote for a lethal condition, *Ll*, produces zygotes in the typical 1 *LL* : 2 *Ll* : 1 *ll* genotype ratio, but the *ll* zygotes die. The seeds and plants that live are 1 *LL* : 2 *Ll* genotypically; such a ratio should always lead one to suspect a recessive lethal allele.

Plants have an alternation of heteromorphic generations, the diploid sporophytes alternating with the haploid gametophytes (see Chapter 9). In gametophytes, each gene is present only as a single allele. No homologous chromosomes occur, and recessive and lethal alleles are not masked by the presence of a dominant homologous allele. In animals, this haploid phase is restricted to the sperm and egg, which are simple and have many inactive genes. As a result, numerous genes with deleterious mutations are not active during the brief haploid condition; they are turned on only during growth and differentiation from a zygote to an adult, when their effect is mitigated by their diploid condition. In plants, however, the haploid phase consists of at least a pollen grain with three cells and a megagametophyte with seven cells. This phase is still quite simple and most genes are inactive, but in both gametophytes many types of central metabolism are occurring, and any highly deleterious allele causes that gametophyte to develop poorly or not at all. If the gametophyte dies without producing a sperm or an egg, the allele is eliminated.

MULTIPLE SETS OF CHROMOSOMES

Whereas most animals have diploid body cells and haploid gametes, plants are much more diverse. Rarely, a spore is formed without undergoing meiosis, resulting in diploid gametophytes and diploid gametes. If these fertilize a normal haploid gamete, a **triploid** zygote results. Although this would be instantly lethal in almost all animals, plants tolerate it well and a triploid sporophyte develops from the zygote. However, triploids are sterile because they cannot undergo pairing of homologous chromosomes during prophase I; chromosomes come together in threes and the rest of meiosis is aberrant.

Occasionally, cells fail to undergo mitosis after S phase DNA replication, and the cell becomes tetraploid (or diploid if it happens in a gametophyte). Tetraploid cells are usually perfectly healthy; if they produce a part of the plant that initiates flowers, diploid pollen and eggs are produced, again raising the possibility of triploid zygotes if fertilized by haploid gametes. These processes can occur in virtually every conceivable combination, and plants that are 3n (three sets of chromosomes, triploid), 4n, 5n, 6n, . . . up to several hundred n are common. All plants with more than two sets of chromosomes are **polyploid**. All plants with an even ploidy level can undergo meiosis and are fertile; all odd ploidy levels are sterile like triploids.

Change in chromosome number per nucleus in plants can also come about by **nondisjunction:** During the second division of meiosis, the two chromatids may remain together (fail to disjoin), so one daughter cell receives both copies of the chromosome (the two chromatids) whereas the other receives none. The cell with the extra chromosome probably survives quite well and forms a functional sperm or egg cell. If this gamete, which is diploid for one chromosome, is involved in fertilization, the new zygote will be triploid for that chromosome. It received a normal, haploid chromosome set from one parent and a haploid set plus the extra chromosome from the parent in which nondisjunction occurred. In plants, this condition is tolerated often, and only by studying the karyotype can it be discovered. In animals, this is almost invariably fatal. The human congenital disease Down syndrome is caused by nondisjunction of the very small chromosome 21. Down syndrome individuals are triploid for just a few genes yet have severely disrupted metabolism. Fetuses triploid for longer chromosomes abort spontaneously.

Plants with even ploidy levels grow and reproduce successfully, often more vigorously than diploid individuals of their species. Initially, these have complex genetics. For example, a tetraploid can have any of the following genotypes for red/white flower color: *RRRR*, *RRRr*, *RRrr*, *Rrrr*, or *rrrr*. Its gametes can be *RR*, *Rr*, or *rr*. Notice that polyploids have more copies than needed for every gene. In all diploid plants, even heterozygotes with only one functional allele are usually healthy. For nonessential genes, a complete absence of functional alleles may not be harmful, such as when *rr* produces white flower color. Conse-

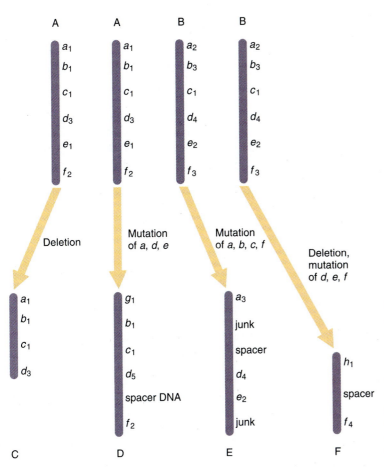

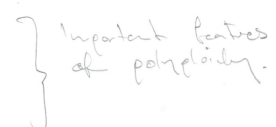

Important features of polyploidy.

FIGURE 16.25

Once a plant becomes tetraploid, mutations in the extra copies of a gene are tolerable as long as (1) at least one allele is left functional, and (2) the extra copies do not mutate into a lethal allele. In this example, two chromosomes have changed morphology, becoming shorter; they could have become longer by picking up pieces of DNA by unequal crossing-over. Several copies have mutated into spacers and some into junk that still has much sequence homology with functional alleles. Others have evolved into new alleles of the original genes (d_3 to d_5; a_2 to a_3; f_3 to f_4), and some have evolved into entirely new genes (a_1 to g_1; d_4 to h_1).

quently, the two or three extra copies of each gene in a tetraploid plant are surplus DNA. Mutations that completely incapacitate one of the four alleles of a tetraploid usually are not deleterious, as one, two, or three functional alleles remain in every nucleus. For polyploid plants, deleterious mutations tend to have little effect on phenotype, so they do not affect survival as much as they would in a haploid cell or a diploid plant. As a result, deleterious mutations are not eliminated quickly by natural selection, and the extra copies of a gene may rapidly become nonfunctional; before many generations have passed, polyploids act like diploids, their "extra" copies having mutated. Almost half of all flowering plant species are actually polyploids; those that now appear to be perfectly normal diploids probably underwent conversion to the polyploid condition so long ago that all their extra alleles have mutated extensively and can no longer be considered alleles of those that are still functional. Nondisjunction, insertions, and deletions may have changed the numbers and sizes of the chromosomes so much that the four originally homologous chromosomes now appear as just two different pairs. Conversion back to the diploid state is an uneven process, and some species act like diploids for certain genes but tetraploids for others.

What do the extra alleles mutate into? At first they may simply be alleles that produce defective proteins. Mutations in the promoter regions may prevent them from ever being activated, so they turn into "junk," acting as nothing more than spacer DNA. Deletions, inversions, and movements by transposons may break them up. DNA sequencing has now revealed that there are large families of genes in which one gene codes for a useful protein while the other genes have nucleotide sequences that are obviously related to the useful gene but are inactive and apparently code for nothing.

This extra DNA is actually extremely valuable—it is the raw material for the evolution of new genes. Some may mutate into genes that produce enzymes almost identical to those being coded by the original form of the allele. The original enzyme may work best at low temperature, whereas the new form may have a higher optimum temperature. The plant can now produce two types of enzymes and function well in both warm and cool days, or the species may be able to extend its range, the new enzyme allowing it to survive in hot deserts and the original enzyme allowing it to live in its original habitat (Fig. 16.25).

Mutations in the promoter region instead may allow the structural portion to produce the same protein as the original gene but at a different time or place in response to a different chemical messenger.

In other cases, mutations in the extra DNA may result in totally new genes that produce proteins whose function is not at all related to that of the original gene. Mosses, ferns, conifers, and flowering plants have evolved from green algae. Because algae have no genes for flower color, lignin synthesis, and so on, the evolution of these species has involved the evolution of whole new metabolic pathways. All of these genes had to arise by the gradual mutation of surplus alleles into nucleotide sequences that code for useful proteins.

SUMMARY

1. Most genes occur as slightly different forms, alleles, in which the DNA sequence varies. Alleles arise by mutations.

2. An organism's physical and physiological features constitute its phenotype, whereas its suite of alleles is its genotype.

3. During sexual reproduction, two sets of alleles, maternal and paternal, are brought together into one cell. During prophase I of meiosis, the alleles of homologous chromosomes are rearranged into new chromosomes.

4. Mutations are any changes in the sequence of nucleotides in DNA. Larger mutations are often more significant than small ones, but any type can potentially affect the phenotype. The great majority of mutations are selectively disadvantageous.

5. DNA proofreading and repair enzymes keep the mutation rate low enough to allow complex organisms to live, but they are imperfect enough to allow a low rate of mutation, permitting evolution by natural selection.

6. If a diploid organism has two identical alleles for a gene, the organism is homozygous for the gene; if it has two distinct alleles, the organism is heterozygous for that gene. If the phenotype of heterozygotes is indistinguishable from that of one of the homozygotes, the trait shows complete dominance. If the heterozygote's phenotype differs from that of both homozygotes, the trait shows incomplete dominance.

7. When heterozygotes with complete dominance are selfed, the F_1 generation should have a phenotype ratio of three dominant to one recessive, and a genotype ratio of one homozygous dominant to two heterozygotes to one homozygous recessive.

8. A test cross is one in which one parent is known to be homozygous recessive for the trait of interest; this allows the phenotype of the other parent to be expressed in every progeny, thus revealing the genotype of the parent.

9. Almost any gene has multiple alleles; that is, there are many mutant forms of the gene. Any particular phenotype trait is probably the result of the action of many enzymes or other proteins; the alleles for all these genes may affect the expression of the phenotype.

10. A dihybrid cross is one in which two traits are studied simultaneously; if the responsible genes are on separate chromosomes or if they are widely separated on the same chromosome, they show independent assortment.

11. Two genes that occur close together on a chromosome are linked. Crossing-over during prophase I breaks up linkage groups.

12. In sperm cells of both plants and animals, organelles are destroyed, so sperms do not contribute mitochondrion or plastid genes. All organelle genes are provided in the egg and show maternal rather than biparental inheritance.

13. Lethal genes cause such a severe disruption of metabolism that the individual dies. Many kill the organism before it reaches reproductive maturity, so they can be passed on only rarely, masked in the heterozygous condition.

IMPORTANT TERMS

alleles
biparental inheritance
complete dominance
cross fertilization
dihybrid cross
DNA polymerase
dominant allele
F_1 (first filial) generation
genotype
heterozygous
homozygous

incomplete dominance
independent assortment
lethal allele
linked genes
map units
maternal (uniparental) inheritance
monohybrid cross
mutagen
mutation
parental generation
phenotype

polyploid
Punnett square
pure-bred line
recessive allele
recombinant chromosomes
self-fertilization
somatic mutation
test cross
transposon

REVIEW QUESTIONS

1. What is a gene? What is an allele? Why do some genes have only a few alleles? What happens to their alleles as they arise by mutation?

2. What is a plant's phenotype? Is the shape of a leaf an aspect of the phenotype? The shape of a nuclear pore?

3. Describe DNA replication. What are Okazaki fragments? Why does each chromosome have thousands of replication start sites instead of just one?

4. Name the types of mutations that may occur and describe how some of them happen. With regard to UV-induced mutations, think about the fact that most leaves last for just a few months and then are replaced by new leaves the next spring. The next time you are getting a suntan, remember that your dermis must last as long as you live.

5. What are monohybrid, dihybrid, and trihybrid crosses? Is it possible to make a cross in which only one single character is actually involved?

6. In a field you find three types of the same plant: some with long leaves, some with short leaves, some of intermediate length. What would you suspect to be the genotypes of each? You measure hundreds of plants and find that about half have leaves of intermediate length, whereas one fourth have long leaves and one fourth have short leaves. Does this support your estimate of the genotypes? What kinds of crosses would you do to test it further? In each cross, what types of gametes would each parent produce and in what ratios?

7. What is independent assortment? What are linkage and crossing-over? Can two genes act independently if they are on the same chromosome?

8. Why is it easier to study plastid inheritance than mitochondrial inheritance? What is unusual about both? Do plastids and mitochondria undergo sexual reproduction? Syngamy?

BotanyLinks .net Questions

www.jbpub.com/botanylinks

Visit the .net Questions area of BotanyLinks (http://www.jbpub.com/botanylinks/netquestions) to complete these questions:

1. To study the genetics of inheritance, we need to know which genes are on which chromosomes. How are genes mapped? Go to the BotanyLinks home page for information on this subject.

2. Genetics and molecular genetics are becoming increasingly important, but how do they affect your life? Go to the BotanyLinks home page to begin researching this subject.

BotanyLinks includes a Directory of Organizations for this chapter.

17

POPULATION GENETICS AND EVOLUTION

All the yellow flowers are individuals of the species *Coreopsis bigelowii*; they constitute a population that can interbreed. *(William E. Ferguson)*

CONCEPTS

Evolution is the gradual conversion of one species into one or, in some cases, several new species. It occurs for the most part by natural selection: Mutations cause new alleles or new genes to arise which affect the fitness of the individual, making it less or more adapted to the environment than other individuals without the new allele or gene. If the mutation is deleterious, the individual may grow or reproduce slowly or even die early without reproducing. Either way, the new allele has less chance of increasing in the population and is more likely to be eliminated as the individuals that carry it die out. If the mutation is beneficial, the individual should grow and reproduce better than other individuals, producing a greater number of gametes and ultimately a greater number of seeds than the other individuals. Abundance of the new allele increases relative to the original alleles (Fig. 17.1). As evolution by natural selection continues, the types and abundances of alleles present in the species change, and consequently the phenotype also changes.

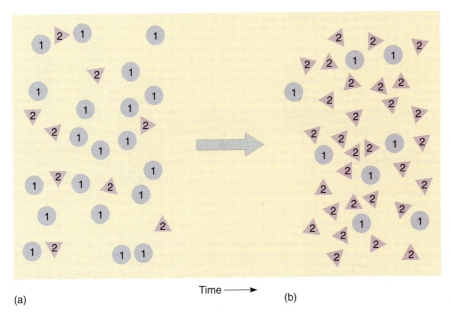

(a) Time ⟶ (b)

FIGURE 17.1

(a) The population in this diagram originally consisted of 29 individuals, 20 of which carried allele 1 and 9 allele 2. Allele 2 produces individuals that are more vigorous, so allele 2 has increased in the next generation (b) from 9/29 = 31% to 30/38 = 78.9%. Allele 2 produces a phenotype (triangle) distinct from allele 1 (round), so as the allele frequency of the population changes, so does the phenotype. In this diagram, the extra vigor of allele 2 is allowing the population as a whole to enlarge, from 29 to 38 individuals, perhaps by crowding out other species or by entering new habitats.

Evolution is an extremely slow process that may require thousands of generations and million of years to produce obvious changes in a species. Because it is so slow compared with the length of a human lifetime, it is easy to understand that it went undetected until recently, just as continental drift, mountain building, and valley formation by erosion went unnoticed. A human lifetime is too short to perceive changes in these processes; even the full history of accurate observation and detailed record keeping is not long enough to detect the fact that, since the times of the ancient Chinese, Sumerians, and Egyptians, continents have moved closer together, mountains have risen higher, and species of plants and animals have changed. It is not surprising that people thought in terms of a constant, unchanging world composed of specific mountain ranges, rivers, lakes, continents, plants, and animals.

During the explorations of Africa, the Americas, and Australia near the time of Columbus, so many plants and animals were discovered that people could not help but notice the remarkable similarities between many species. The various types of roses resembled each other more than they resembled other plants, just as did the many new types of lilies, orchids, and so on. People began to realize that giant basins of sediment, such as the deltas at the mouth of the Nile, the Ganges, and the Mississippi, held enough eroded material to account for whole mountain ranges; the concept of a changing landscape—geological evolution—began to develop (Fig. 17.2). An important corollary was the realization that in

FIGURE 17.2

As the science of geology developed, it was soon realized that river deltas, such as this delta of the Mississippi, contained so much sediment that they must represent the transformation of landscapes, that mountain ranges must have eroded and no longer exist. But this means that the world is not constant, as God had supposedly created it, but changing and evolving. *(NASA)*

order for so much sediment to have formed, Earth must be millions (actually billions) of years old, not merely 6000 years old as had been calculated from the genealogy of the Book of Genesis in the Old Testament. This is one of those discoveries that is easy for us to underappreciate; the quantity of sediments proved that either Earth had a very long history or God had created sediments. This gave theologians a difficult dilemma and caused many people to be even more skeptical of priestly interpretations of natural phenomena. One of the areas of doubt was the idea that all species had been created at once. Everyone was vividly aware that the Church already had made a grave error by censuring Galileo, insisting that Earth was the center of the universe and that the sun, planets, and stars all revolved around it (Fig. 17.3).

As careful observations of nature became common after the Renaissance and as the scientific method proved to be an accurate method for analysis, the concept of the evolution of species became widespread. Scientists searched for an understanding of the mechanism by which it occurred. Finally, Alfred Russel Wallace and Charles Darwin independently discovered the basis—natural selection—in the mid-1800s, the critical explanation being given in Darwin's *Origin of Species* published in 1859. The concept of natural selection—survival of the fittest—had an electrifying effect on all biologists and led to a revolution in all aspects of thinking. Suddenly, many observations had a rational explanation; every character could be interpreted in terms of whether it made a species less or more adapted to its environment. And because so many environments exist, it is logical to expect that many types of plants and animals should also exist, each having genotypes and phenotypes that make it particularly adapted to its own environment. Prior to the discovery of natural selection, flowers were thought to have been created by God for the delight of the human eye and nose; that philosophy did little to explain bilaterally symmetrical flowers, inferior ovaries, or wind pollination. Science had been greatly hampered by trying to interpret the world in terms of the mind of God and what had been created to feed, clothe, and house the descendants of Adam and Eve.

Fortunately, the discovery of evolution by natural selection happened at about the same time as other critical discoveries—the discovery of genes and chromosomes, cell theory, proving that spontaneous generation does not occur, sophisticated hydroponic cultures of plants, the discovery of enzymes and the carrying out of some metabolic steps in a test tube without living cells, and the artificial synthesis of biological compounds. All these combined to move biology firmly out of the realm of metaphysics/theology and into that of scientific analysis and interpretation.

FIGURE 17.3

Galileo Before the Inquisition. Painting by Robert Fleury, 1847. Kepler, Newton, and Galileo proved that the sun is the center of the solar system, but church doctrine considered this heresy; as the home of the one organism created in the image of God, the Earth should be the center of the Universe. A board of inquiry forced Galileo to recant his belief in a sun-centered solar system and placed him under house arrest for the rest of his life. Attempts were made to confiscate and burn all copies of his book; only a few survived. Less famous scientists were frequently punished more severely. Misinterpretation of the physical world by religious doctrine is not a fundamental aspect of religion, but rather the rejection of observation and experimentation in favor of speculation without confirmation. Ancient Greeks also came to many erroneous conclusions for the same reason. *(The Bettmann Archive)*

For those of you who believe in divine creation rather than evolution by natural selection, keep in mind that all of our experience has shown the scientific method to be infinitely superior to theological interpretation of scriptures for understanding and predicting the nature of the physical/biological world. Equally important, only ethical/philosophical systems allow us to understand and solve moral problems; science cannot do that. It is not logical or consistent for a person to accept certain discoveries of science (photosynthesis, DNA, respiration, vessels, sieve tubes) while rejecting others (evolution) if one type of discovery is just as well-documented as the other.

POPULATION GENETICS

In the last several chapters, the genetics and reproduction of individual organisms were discussed; the genetics of groups of individuals of one species is now considered. **Population genetics** deals with the abundance of different alleles within a population and the manner in which the abundance of a particular allele increases, decreases, or remains the same with time. The genetic recombination that occurs during sexual reproduction is important only if the two sexual partners have differing genotypes. A cross between two plants that are $A_1A_1B_1B_1 \times A_2A_2B_2B_2$ produces offspring that have the genotype $A_1A_2B_1B_2$, which is different from that of either parent. Crossing-over also increases genetic diversity in populations.

The total number of alleles in all the sex cells of all individuals of a population constitutes the **gene pool** of the population. Imagine that gene A has four alleles—A_1, A_2, A_3, and A_4. If the population consists of 1 million individual plants, each with 100 flowers and each flower producing on average 100 sex cells (sperm cells and egg cells), the gene pool contains 10 billion haploid sex cells. The alleles are probably not present in equal numbers; for instance, 60% of all gametes may be A_1, 20% A_2, 15% A_3, and 5% A_4. Will this ratio be the same for the population next year or in the next generation? Early in the 20th century, two biologists, G. H. Hardy and G. Weinberg, demonstrated mathematically that *if only sexual reproduction is considered,* the ratio remains constant over time; even if an increase or decrease occurs in the total number of individuals or gametes, the ratios do not change. Sexual reproduction alone does not change the gene pool of a population; if no other factors were involved, the gene pool would remain constant forever.

FACTORS THAT CAUSE THE GENE POOL TO CHANGE

Although it is possible theoretically for a gene pool to remain constant, in reality, changing allele frequencies are the rule, because populations are always affected by factors other than sexual reproduction.

Mutation. All genomes are subjected to mutagenic factors, and mutations occur continually. Because of mutation, existing alleles decrease in frequency and new alleles increase. Whether or not mutation is significant depends in part on the population's size. In the population described earlier, there were 10 billion copies of gene A, and 60% were the allele A_1. Therefore, there were $10,000,000,000 \times 0.6 = 6,000,000,000$ copies of A_1; if one mutates to become a new allele, A_5, then the frequency of A_1 drops to 5,999,999,999, which is still pretty close to 60%. A_5 exists as just one copy out of 10 billion, so its frequency is extremely small; however, the presence of one copy is an infinite increase over zero copies, and the existence of even a single copy should not be ignored.

Accidents. Accidents are events that an organism cannot adapt to, such as the collision of a large meteorite with Earth. When a meteorite of sufficient size strikes Earth, a large region of Earth's surface is destroyed, killing all life in the area. All organisms, along with all their alleles, are eliminated. If by chance the area of impact had included the single individual that had just obtained A_5 by mutation, then A_5 would no longer exist. If the population in the impact area has the same gene frequencies as the general population, the alleles are eliminated in the same proportions as they exist generally, and no change in allele frequency occurs. If the impact area had an unusually high number of a particular allele, for

FIGURE 17.4

This large forest fire in Yellowstone Park in 1988 was an accident for certain plants, a selective force for others. Many small plants cannot possibly become adapted to such fires; the heat not only kills the plants but also destroys the humus layer and kills buried seeds. All their alleles are destroyed indiscriminately. However, certain large trees can survive most fires; their bark is thick enough to insulate the vascular cambium. Plants that have alleles for extra thick bark survive; any plants that have alleles for thin bark die. For this species, the allele frequency after the fire differs from that before the fire. (© *Jeff Henry/Peter Arnold, Inc.*)

FIGURE 17.5

In earlier times, Antarctica was located in temperate latitudes (see Fig. 27.8) and supported many plants, such as this *Glossopteris*. Continental drift moved Antarctica to the South Pole, into conditions for which no alleles are able to produce an adapted phenotype. All plants perished. (*Courtesy of T. Taylor, Ohio State University*)

instance A_3, then A_3 is affected more than the other alleles, so the allele frequency of the gene pool of the survivors is altered.

Many phenomena qualify as accidents. A volcanic eruption produces poisonous gases and molten rock that destroy everything within a limited area. Infrequent floods, hailstorms, or droughts can act as accidents for plants too small and delicate to become adapted to those events; all individuals in the affected area are killed. For other species that consist of plants with larger bodies, such natural phenomena act as selective forces, removing the weaker, less well-adapted individuals but not affecting the more well-adapted members (Fig. 17.4). The continental drift of Antarctica southward from a temperate region to the South Pole was an accident for all the plants living on it. While located in the temperate latitudes, Antarctica had a rich flora with abundant plant life; as it drifted southward, it entered a region too cold and severe for any plant life to survive. All individuals and their alleles were eliminated (Fig. 17.5).

Accidents can be small events as well as large ones. Once an allele is formed, A_5 for instance, its numbers may increase as it is used in sexual reproduction; after a few years there may be ten individuals with the A_5 allele. The individuals are closely related and probably located close together because most seeds do not travel very far. It is possible for a local disturbance to eliminate all ten: an avalanche, a herd of grazing animals, or the construction of a highway.

Artificial Selection.   **Artificial selection** is the process in which humans purposefully change the allele frequency of a gene pool. The most obvious examples are the selective breeding of crop plants and domestic animals (Fig. 17.6). Plant breeders continually examine both wild populations and fields of cultivated plants, searching for individuals that have desirable qualities such as resistance to disease, increased protein content in seeds, and ability to survive with less water or fertilizer. When plants with beneficial qualities are found, they are collected and used in breeding programs to produce seed for future crops. Consider just wheat, rice, and corn: Almost the entire world populations of these three species consist of cultivated plants; very few of the natural ancestors still exist in the wild. Consequently, the gene pool for each is made up almost entirely of alleles that have been artificially selected for thousands of years.

Artificial selection is also used to produce ornamental plants that flower more abundantly or for a longer time. Artificial selection has also been used to alter flower color and size and to make the plants hardy in regions where they otherwise could not grow. The trees that are cultivated for lumber and paper are also subjected to artificial selection.

(a)

(b)

(c)

(d)

FIGURE 17.6
Artificial selection by crop breeders has resulted in increased frequency of certain alleles and the elimination of others. The allele frequency of the population changes dramatically from year to year, depending on whether farmers decide to grow one variety or the other. Cabbage (a), broccoli (b), and cauliflower (c) are all the same species (as are brussel sprouts as well) and all evolved from the same ancestor, but their evolution was controlled by artificial selection for certain traits in each variety. Recently, artificial selection has produced the variety Violet Queen that stores large amounts of pigment (d).
(Courtesy of Stokes Seeds, Fredonia, NY)

Artificial selection is often carried out in conjunction with artificial mutation. Plants are exposed to mutagens such as acridine dyes or irradiation with ultraviolet light or gamma rays to increase the number of new alleles that come into existence. The plants are allowed to grow to see how the new alleles affect the phenotype, and those plants with the desired phenotypic traits are used in selective breeding programs.

Natural Selection. **Natural selection,** which is the most significant factor causing gene pool changes, is usually described as survival of the fittest: Those individuals most adapted to an environment survive whereas those less adapted do not. However, natural selection is such an important factor in evolution that it must be given careful attention. Two conditions must be met before natural selection can occur:

1. The population must produce more offspring than can possibly grow and survive to maturity in that habitat. This condition is almost always valid for plants anywhere on Earth. Most plants produce hundreds of seeds, which often germinate near the parent plant (Fig. 17.7). Even in species with wind-dispersed seeds, such as maples or milkweeds, most seeds do not travel far. Consequently, the ground can be covered with hundreds of seedlings crowded closely together, and there simply is not enough room to accommodate the physical bulk of so many plants as they grow.

   Besides limited resources, the number of individuals that can survive in a particular habitat is affected by predators, pathogens, and competitors. All plants are faced with attack by herbivorous animals, ranging from almost microscopic mites and nematodes to much larger beetles, reptiles, birds, and mammals. Animals not only eat

FIGURE 17.7
Even if animals carry away many of the fruits of this red elderberry *(Sambucus racemosa),* most fall and germinate in its immediate neighborhood, and there are too many seedlings to fit into the space physically. Also, the tree will produce seed abundantly for many years. *(Zig Leszczynski/Earth Scenes)*

plants but may also lay eggs in them, bore into tree trunks for nesting sites, walk on them, and rob nectar without carrying out pollination. Pathogenic fungi and bacteria are similarly harmful. Competitors are other organisms that use the same resources the plant needs to survive. When root systems grow together, the two plants compete for the same water and nutrients. If two species are pollinated by the same species of insect or bird, they must compete for the attention of the pollinators. In a forest, plants compete for light: Those that can grow tallest receive the most light; those that have shorter trunks receive only dim light. All these activities adversely affect the plant's ability to reproduce and may cause the plant's death.

2. The second condition necessary for natural selection is that the progeny must differ from each other in their types of alleles (Fig. 17.8). If they are all identical, all are affected by adversity in the same way and to the same degree. Under crowded conditions, probably all are stunted similarly, all grow poorly, and finally none reaches reproductive maturity. If all individuals of a species are equally susceptible to a pathogenic fungus, no increase in survivability and fitness occurs as the result of a fungal attack. Even if some survive and reproduce, they are identical genetically to those that died, so no change occurs in allele frequency; natural selection has not occurred.

When genetic diversity exists among individuals, differential survival can occur. If some members of the population have an allele that gives them increased resistance to fungi, those plants should fare much better during an outbreak of fungus in the population than those lacking the allele. If the fungus is so virulent that it often kills the plants it attacks, the allele frequency of the population is changed radically after infection—the resistance allele constitutes a much greater percentage of the gene pool. Natural selection operates even if the fungus only weakens plants but does not kill them outright; the weakened plants should produce fewer seeds than do the resistant, healthy plants.

Competition for water and nutrients among crowded root systems also acts as a selective force; alleles that allow roots to absorb water and nutrients more efficiently have an advantage. If most roots can no longer absorb water when the soil has a water potential of − 0.8 MPa, an allele that alters root metabolism such that it can extract water even at − 1.0 MPa enables the plant to grow when others cannot. The mutation has a selective advantage—the plant is more fit or adapted to conditions of crowding.

Natural selection refers to the differential survival among organisms that have different phenotypes. Natural selection can act only on pre-existing alleles; it does not cause the mutations. The presence of a fungus does not cause plants to become resistant; if none had been resistant before, all individuals would be adversely affected. The advantageous allele must exist first. But if an allele for resistance does exist, natural selection can cause the

FIGURE 17.8
The summer of 1988 produced one of the most severe droughts on record in the American midwest, our primary corn-growing area. Under natural conditions, such a drought would have resulted in the natural selection of drought-resistant plants; all corn plants of 1989 would have been the progeny of the resistant plants that survived, and all would have carried alleles for drought resistance. But natural selection did not occur because there was no genetic variability among the corn plants; corn is one of our most highly inbred crops, and all seed corn is produced by careful crosses of absolutely uniform parents. Consequently, of the trillions of plants affected by the drought, none was more or less resistant than any other. Any that survived did so not by genetics, but by accident: receiving a little extra rain or being in a valley where extra moisture collected. (*Grant Heilman/Grant Heilman Photography*)

population to become resistant by the preferential survival of resistant individuals, even though it cannot cause an *individual* to become resistant. A population or species evolves, but an individual does not. Similarly, a population may become more adapted but a single individual does not change; it cannot become more adapted during its lifetime.

Natural selection does not always result from the action of an agent outside the organism. Although many selection pressures are external—pathogenic fungi and dry, hot climate—any factor that causes one plant to produce more progeny than other plants is a selective factor (Fig. 17.9). If an allele causes chloroplasts to photosynthesize more efficiently, plants with that allele can produce carbohydrates more rapidly than plants that lack the allele; the former plants grow faster and produce more seeds, at least half of which carry the advantageous allele.

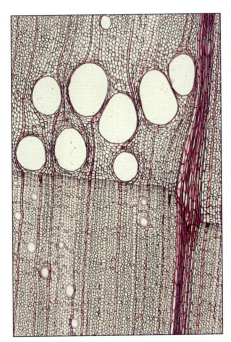

(a)   (b)

FIGURE 17.9
(a) This oak wood contains wide vessels that can conduct water with relatively little friction ($\times 80$).
(b) This *Casuarina* wood has vessels so narrow that the water in them is strongly reinforced and cavitation is almost impossible under natural conditions. Even though vessels are completely internal, they still affect the plant's survival relative to its environment ($\times 80$).

Factors that are Not Part of Natural Selection.   Natural selection can be understood more clearly and accurately if you realize that certain factors are not part of natural selection at all: Natural selection does not include **purpose, intention, planning,** or **voluntary decision-making.** Whenever we say that "plants do something in order to . . . ," we are suggesting that the plants can plan their activities and have purpose, which is not true. It is not correct to say that plants produce leaves in order to photosynthesize more efficiently than those that do not have leaves. The resistance of a population of plants to a fungus is nothing more than the result of the preferential survival of the plants' ancestors because they had the allele for resistance whereas their competitors did not. Although plants that have this allele in the presence of the fungus have a selective advantage, **the plants do not have the allele in order to protect themselves.** Similarly, plants do not produce nectar *in order to* attract pollinators; those plants in the past that did produce nectar happened to be pollinated more often than others that did not produce nectar, so the alleles for nectar production were increased in the population. At present those plants that secrete nectar are visited by pollinators, but there is no purpose, intent, or planning by the plant. Only humans and other primates act with intent and purpose, and this applies only to our voluntary actions. Even humans do not digest food in order to have an energy supply; rather, our autonomic nervous system and cell metabolism have automatic responses to the presence of food in the small intestine that cause the secretion of digestive juices, the absorption of monomers, their distribution through the hepatic portal system, and their respiration by cytoplasm and mitochondria. We have no control over this result of our evolutionary history. This may seem to be just a trivial problem of semantics, because everyone knows what we mean, but statements should be accurate, not sloppy. If we are not meticulous in how we express our ideas, we will not be meticulous in how we think, and important details will be lost.

SITUATIONS IN WHICH NATURAL SELECTION DOES NOT OPERATE

Further understanding of natural selection can be gained by considering several cases in which it does not operate. As already mentioned, it cannot operate if all individuals of a population are identical genetically or if it is impossible to become adapted to a certain condition. Competition does not occur in a habitat that can support the growth and reproduction of all individuals; if survival is universal, natural selection does not occur. Situations like this occur in newly opened habitats such as a plowed field. All seeds present may germinate and grow vigorously, even the ones not well adapted for competition. Because all survive, no natural selection has occurred. Other examples are the sides of a road cut, a recently burned area, or a recently flooded plain covered with rich sediments. We must be careful here; if the road cut passes through a heavy, dark forest, the newly exposed sides may be too bright and exposed for seedlings from plants adapted to the forest shade. In this case, the environment favors those plants that can tolerate full sunlight and suppresses those that require shade; many selection pressures, but not all, have been eliminated, and some natural selection can still occur.

MULTIPLE SELECTION PRESSURES

In many cases, the loss of individuals and reduced reproduction are not caused by a single factor, such as a pathogenic fungus. Instead, the plants are also affected by insect attack, drought, cold, need for pollinators, and need for a mechanism to disperse their seeds, as well as the efficiency of their own metabolism, such as the ability of their membranes to pump ions, the capacity to reduce nitrogen, or the efficiency of producing just enough P-protein in the phloem without a wasteful, nonuseful excess.

A mutation that produces an allele that would result in improved fitness is **potentially** advantageous selectively, but it may never have the opportunity to improve the fitness of the plant or the species. A mutation that results in improved cold hardiness may be eliminated from the gene pool if the plant carrying the new allele is killed by fungus or drought or cannot reproduce because of poor competition for pollinators. Such a loss of this allele is simply an accident.

However, if the new allele for cold hardiness does survive, it may be able to improve the species. If cold winters are common, this allele greatly improves fitness, and its frequency may increase rapidly. If cold winters are infrequent, they do not exert a strong selection pressure, the allele does not improve fitness very much, and its frequency may remain low for years until a harsh winter does occur. Until that time, the allele's frequency is determined by several factors. It may be tightly linked to an allele that is strongly advantageous for an important condition. For instance, the cold hardiness allele may be on the same chromosome as the allele for resistance to fungi. If the two genes are so close on the same chromosome that crossing-over virtually never occurs between them in prophase I, the presence of fungi selects not only for the antifungal allele but also for the cold hardiness allele, just by coincidence.

The cold hardiness allele may affect the plant in various ways besides the ability to withstand cold; that is, pleiotropic effects may operate. If these are also advantageous under common conditions, the allele increases in frequency. But often many of the "side effects" are disadvantageous. Many improvements to phenotype have some negative aspects, at least in terms of cost. Increased cold hardiness may be due to thicker, more sclerified bud scales with a thick layer of wax. These require the input of increased amounts of nutrients and energy which could have been used to produce more seeds. Thus, whereas this allele may be strongly advantageous in an environment with frequent cold winters, it may be disadvantageous in an environment where winters are always mild. Whether a particular allele is beneficial or not depends entirely on the habitat, which may change with time. As the environment changes, the selection pressures change and certain features become more or less advantageous.

(a)

(b)

RATES OF EVOLUTION

From the examples given, it seems that the allelic composition of a population could change rapidly, within a few generations, but that is not typically the case. Most populations are relatively well adapted to their habitat, or they would not exist. Very few mutations produce a new phenotype so superior that it immediately outcompetes all other members of the population.

It is difficult to identify the presence of particular alleles in a population unless they result in an easily identifiable effect on the phenotype. Consequently, most studies of evolution concern the changes in gross structures such as flowers, leaves, fruits, shoots, and trichomes. But these complex structures are the product of the developmental interaction of many genes. Any new mutation results in a more adaptive structure only if the effects of the new allele fit into the already existing highly integrated mechanism of morphogenesis without causing serious disruptive effects. As systems become more intricate, the probability decreases that any random change is beneficial.

Evolutionary changes that result in the loss of a structure or metabolism can come about quickly, however, and for the same reason: complexity. If a feature becomes selectively disadvantageous, many of the mutations that disrupt its development become selectively advantageous. Because disruptive mutations outnumber constructive mutations, loss can occur relatively rapidly. For example, the ancestors of cacti lived in a habitat that became progressively drier; large thin leaves were advantageous because they carried out photosynthesis but disadvantageous because too much water was lost by transpiration (Fig. 17.10). Mutations that disrupted formation of the lamina were advantageous, and cacti lost their leaves in perhaps as little as 10 million years, whereas the evolutionary formation of leaves in seed plants had required over 200 million years. Leaves could not be lost too quickly, however, because the plants would be left with virtually no photosynthetic surface area. Mutations that caused the complete absence of leaves could not be selectively advantageous until other mutations had occurred that permitted the stem to remain green and photosynthetic, that prevented the early formation of an opaque bark, and that slowed the metabolism of the plant to a level compatible with reduced photosynthesis. The loss of leaves could occur only simultaneously with or after these modifications of the stem.

FIGURE 17.10

(a) The ancestors of the cactus family were large woody trees with rather ordinary dicot leaves. The cactus genus *Pereskia* still contains members quite similar to the ancestors, as shown here. Apparently few genes had been modified by the time *Pereskia* appeared. (b) This *Gymnocalycium* is also a cactus, but its phenotype is significantly different from the ancestral condition; apparently all critical genes involved in leaf production have mutated so much that they are nonfunctional or absent. Genes involved in stem elongation now produce short stems, and this species contains genes for succulence that were not present in the ancestral *Pereskia*-like species; these new genes probably are highly mutated forms of "extra" genes from a tetraploid ancestor or arose by other methods of gene duplication.

SPECIATION

As natural selection operates on a population for many generations, the frequencies of various alleles and consequently the phenotype of the population change. At some point, so much change has occurred that the current population must be considered a new species, distinct from the species that existed at the beginning. Natural selection has caused a new species to evolve, a process called **speciation.**

At what point can we conclude that a new species exists? It is not possible to give an exact definition of species that is always valid, but generally, two organisms are considered to be members of distinct species if they do not produce fertile offspring when crossed. Many exceptions exist, however. The individuals of maples in the western United States cannot naturally reproduce sexually with those in the eastern United States, but that is only because the pollen does not travel that far; if an eastern plant is brought close to a western one artificially, they can cross-fertilize and are therefore considered the same species. On the other hand, many orchid species grow together in nature without producing hybrids, but artificial cross-breeding may produce healthy, fertile hybrids. Nevertheless, they are considered distinct species because they look different from each other. So, if two plants freely interbreed in nature, they are members of the same species; if they do not interbreed even when manually cross-pollinated they are separate species; and if they do not inter-breed naturally but they do artificially, it is a judgment call on the part of the scientists who work with the plants and are most familiar with them.

Speciation can occur in two fundamental ways: (1) **phyletic speciation,** in which one species gradually becomes so changed that it must be considered a new species (Fig. 17.11a

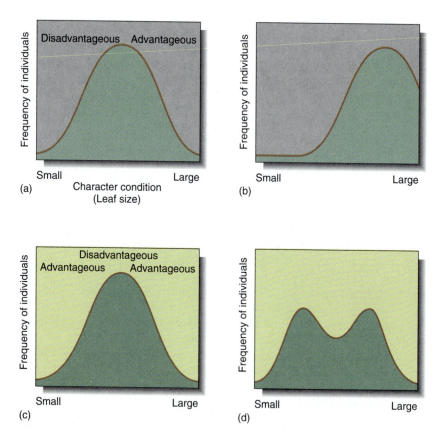

FIGURE 17.11

(a) and (b) In phyletic speciation, all of a species gradually changes because one particular aspect of a character is advantageous for all individuals. Here, all leaves become larger; perhaps the climate is becoming more humid or herbivorous insects are less of a problem. (c) and (d) In this scenario of divergent evolution, both extremes of the condition are more advantageous than intermediate values; the climate may become drier in some areas, favoring the smaller leaves, but moister in other areas, favoring plants with larger leaves.

and b), and (2) **divergent speciation**, in which some populations of a species evolve into a new, second species while other populations either continue relatively unchanged as the original, parental species or evolve into a new, third species (Fig. 17.11c and d).

PHYLETIC SPECIATION

Millions of years are often required for a species to evolve into a new one. The critical feature is that as new beneficial alleles arise and are selected for, they become spread throughout the entire population. This movement of alleles physically through space, called **gene flow**, occurs in many ways, such as by pollen transfer, seed dispersal, and vegetative propagation.

Pollen Transfer. Pollen grains each carry one full haploid genome, and all alleles of a plant are present in its pollen grains. Wind-distributed pollen, such as that of ragweed, grasses, and conifers, can travel great distances (Fig. 17.12). If a new allele is carried by some of the pollen grains, it can move to very distant plants; if the pollen grain's sperm cells fertilize an egg, a new seed is formed whose embryo contains the new allele.

Animal-mediated pollination also contributes to gene flow; both birds and insects tend to spend most of their time in a small area, so although birds can fly long distances, they usually spread pollen through a smaller area. Nevertheless, allele movement can be rapid.

Seed Dispersal. The fruits and seeds of some plants fall close to the parent, but many species have long-distance dispersal mechanisms. Seeds and fruits can be carried by wind, floods, and stream flow. They can be carried to islands by rafting, in which they are trapped above water on a tangled mat of floating debris. Seeds or fruits that are spiny or gummy stick to the fur or feathers of animals; migratory animals can be especially important in dispersing seeds. Most birds reduce their weight before migration by preening themselves of all adhesive seeds. But if just one seed is overlooked, its alleles are transferred; as the seed germinates, grows into a new plant, and reproduces, the new alleles can be spread throughout the new site.

Vegetative Propagation. If a species produces small, mobile pieces that reproduce vegetatively, these too contribute to gene flow.

If these various mechanisms are sufficient to enable alleles that arise in one part of the species' range to travel to all other parts, the species remains relatively homogeneous, even as the entire species evolves into a new species. Alleles that arise at various geographic sites ultimately come together by gene flow; then meiosis, crossing-over, and genetic recombination rearrange them into thousands of combinations (Fig. 17.13).

FIGURE 17.12
Strong winds can blow pollen great distances, resulting in gene flow between widely separated points. Much of the pollen actually settles near the parent plant, but large numbers of pollen grains may travel a long distance and then successfully fertilize eggs. The new sporophytes pass on the pollen's genes to their progeny. *Pinus canariensis. (Robert and Linda Mitchell)*

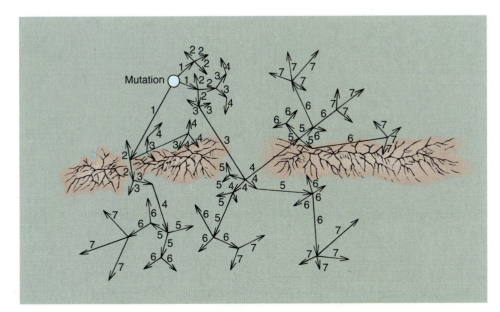

FIGURE 17.13
Gene flow. This range is almost divided in two by a series of mountains over which pollen and seeds cannot travel; this slows down gene flow and homogenization of the population. The numbers indicate the generation in which the allele flows: 1—gene flow in the first generation; 2—gene flow from plants that grew from seeds produced by the gene flow of the first generation, and so on. Here, three generations were necessary for the new allele to become distributed to the lower side of the mountains, seven generations to become widespread throughout the range. In a species of annual plants, each generation is 1 year long; for biennials, a generation is 2 years long and gene flow is slower. In trees that do not produce their first flowers, pollen, or seeds until they are many years old, gene flow can be very slow.

DIVERGENT SPECIATION

If gene flow does not keep the species homogeneous throughout its entire range, divergent speciation may occur; if alleles that arise in one part of the range do not reach individuals in another part, the two regions are **reproductively isolated.**

Reproductive isolation can occur in many ways, but the two fundamental causes are abiological and biological reproductive barriers.

Abiological Reproductive Barriers.   Any physical, nonliving feature that prevents two populations from exchanging genes is an **abiological reproductive barrier.** The original species is physically divided into two or more populations that cannot interbreed; if speciation results, it is called **allopatric speciation.** Mountain ranges are frequently reproductive barriers because pollinators do not fly across entire mountain ranges while feeding or gathering pollen. Seeds may occasionally be carried across mountains by birds or mammals, but probably too rarely to be significant. Rivers are often good barriers for small animals, but they rarely prevent plant gene flow by means of seed dispersal. Deserts and oceans are effective barriers. Plants that are adapted to the harsh conditions of mountain tops are reproductively isolated from plants on adjacent mountain tops by the intervening valleys because pollinators and seed dispersers do not often travel from mountain to mountain. Any seeds dropped into the valley would not be able to compete with lowland plants and would die. Ultraviolet (UV) light and dry air are barriers to very long-distance wind dispersal of pollen; during the ride on the wind, pollen is damaged by the UV light and dry air and often dies before it can travel far.

Biological Reproductive Barriers.   Any biological phenomenon that prevents successful gene flow is a **biological reproductive barrier.** Differences in flower color, shape, or fragrance can be effective barriers if the species is pollinated by a discriminating pollinator. A mutation that inhibits pigment synthesis and causes a normally colored plant to have white flowers might prevent the pollinator from recognizing the flower. The flower is not visited and gene flow no longer occurs between the mutant plant and the rest of the species even though the individuals grow together. Timing of flowering can be important: If some flowers open in the evening, they probably do not interbreed with those that open in the morning. Differences in flowering date are critical because of the brief viability of pollen once it leaves the anthers. When two groups become reproductively isolated even though they grow together, the result is **sympatric speciation.**

Evolutionary changes in pollinators can also act as reproductive barriers for plants. If a plant population covers a large area, some parts of the range probably have characteristics different from those of other parts, such as elevation, temperature, and humidity. These variations may be important to pollinators and seed distributors if not to the plant. If so, the animals may diverge evolutionarily into distinct species, each limited to a small part of the plant's range. Little or no flow of the plant's genes occurs owing to the restricted movements of pollinators and seed dispersers. These examples of biological reproductive barriers prevent pollen from moving from one plant to another, so neither pollination nor fertilization occurs. Consequently, these are called **prezygotic** isolation mechanisms: They act even before a zygote can be formed.

The environmental diversity of a large geographic range can lead directly to divergence of the plants themselves. Although a plant species may be able to occupy an extensive, heterogeneous range, with time mutations arise that are particularly adaptive for certain aspects of specific regions of the range. When these new alleles arrive at that part of the range by gene flow, their frequency there is increased by natural selection. When they arrive at other parts of the range, their frequency remains low or they are eliminated there if they are neutral or selectively disadvantageous for conditions in these sites (Fig. 17.14). Even with active gene flow and interbreeding, different subpopulations of the plant species emerge, each adapted to its own particular portion of the total range. As this process continues, each subpopulation becomes progressively more distinct and may be recognized and named as a subspecies. Finally, because of the large number of unique, characteristic alleles, each may become sufficiently distinct phenotypically that the two subpopu-

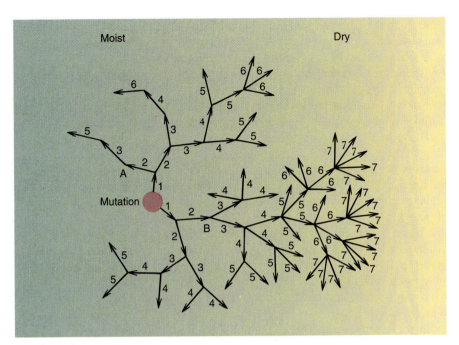

FIGURE 17.14
A mutation has produced an allele highly adaptive in dry parts of the range but neutral or deleterious in moist parts. Because it originated in a less-than-optimal environment, in generation 1 only two offspring survive. In generation 2, each of these produces two more that survive, one in a moister habitat (A), one in a drier region (B). The allele is deleterious at A, and only a single progeny survives in generation 3; at B, two survive, both in drier habitats, and in generation 4 each produces three surviving progeny. As the alleles flow, by chance, into regions that are more optimal (drier), reproductive success is greater and the allele frequency increases. As they flow, by chance, into unsuitable habitats, reproductive success is low and the allele never becomes abundant.

lations can no longer interbreed: Their genomes are too different. At this point, **postzygotic internal isolation barriers** are in place, and the two subpopulations must be considered separate species.

One of the earliest postzygotic barriers to arise is **hybrid sterility:** The two populations occasionally interbreed or are artificially cross-pollinated and produce viable seed, but the seed grows into a sterile plant. Spore mother cells in anthers and ovules in the sterile hybrid are unable to complete meiosis because of a failure of synapsis: "Homologous" chromosomes are no longer truly homologous, and they fail to form pairs and synaptonemal complexes during prophase I. Without fertile pollen or ovules, no seed is formed and the mixture of the two genomes ends when the hybrid plant dies. The two populations continue to diverge and become even more distinctly separate species. If cross-pollination occurs, alleles from one parent may code for proteins incompatible with those coded by alleles from the other, and not even a sterile hybrid can result; instead, the zygote or embryo dies early in development.

Once internal isolation barriers are established, evolutionary divergence should be even more rapid because new alleles cannot be shared; gene flow between the two populations comes to a stop. Each species becomes more easily recognizable as new characters evolve in each which are not present in its sibling species. The two species are initially so similar that they are obviously related; such closely related species are grouped together into genera by taxonomists.

Divergent evolution may result in numerous types of new species; in some cases, one subpopulation changes into a new species while the remaining part of the population continues relatively unchanged as the original species. In other cases, both subpopulations change so much that two new species emerge and the original species no longer exists. The original species has become extinct, although it has numerous progeny that form the members of two new species.

Adaptive Radiation.   **Adaptive radiation** is a special case of divergent evolution in which a species rapidly diverges into many new species over an extremely short time, just a few million years. This usually occurs when the species enters a new habitat where little or no competition or environmental stress exists. The best examples are the colonization of newly formed oceanic islands such as the Hawaiian or Galapagos Islands. After being formed by volcanic activity, they are initially devoid of all plant and animal life, but eventually a seed arrives, carried by either a bird, the wind, or ocean currents. Once the seed germinates and begins to grow, it is free of danger from herbivores, fungi, bacteria, or competition from

FIGURE 17.15

One of the pioneer plants that arrived and colonized the Hawaiian Islands was a composite, related to daisies. Because of the low levels of competition and because so few individuals existed initially, each mutation became a significant part of the gene pool even while only one or a few copies existed, so genetic drift was rapid. *Argyroxyphium,* silver swords, grow only in Haleakala National Park in Hawaii. Goats ate them almost to extinction before the park was established. *(Stephen J. Krasemann/© Peter Arnold, Inc.)*

other plants. It must be relatively adapted to the soil, rainfall pattern, and heat/cold fluctuations, but otherwise its life is remarkably free of dangers. If this plant is self-fertile or if it reproduces well vegetatively, it successfully colonizes the area.

All offspring greatly resemble the first, **founder** individual(s), because the initial gene pool is extremely small; if just one seed is the founder, the original gene pool consists of its two sets of alleles. This homogeneity may last only briefly because, with the lack of competition, pathogens, and predators, fewer forces act as selective agents. Consequently, new alleles build up in the population much more rapidly than can occur in the parental population on the mainland (Fig. 17.15). While the population is small, it is more subject to accidents, so the gene pool can change rapidly and erratically; this is **genetic drift,** and the island population soon becomes heterogeneous. Natural selection for adaptation to soil types, drainage, climate, and metabolism is still operative, so divergence is based largely on physical factors in the environment.

If the island is very far from any seed source (the mainland or other islands), like Hawaii is, hundreds of years may pass before the arrival of more seeds of either the same or another species. Even if it is relatively close to the mainland, like the Galapagos, seeds arrive only rarely. For thousands of years the only animals might be birds or airborne insects, typically spiders; land reptiles (lizards) and mammals (mice) arrive by rafting, but this can be extremely rare.

Adaptive radiation can also occur in mainland populations if the environment changes suddenly and eliminates the dominant species of a region. With the absence of these species, competition changes and other species that had not been able to compete well before can now occupy the new areas; while few in number, they undergo genetic drift, rapidly producing many unusual genotypes. Within a short time, many new species are recognizable, and adaptive radiation has occurred.

CONVERGENT EVOLUTION

If two distinct, unrelated species occupy the same or similar habitats, natural selection may favor the same phenotypes in each. As a consequence, the two may evolve to the point that they resemble each other strongly and are said to have undergone **convergent evolution.** The most striking example is the evolutionary convergence of cacti and euphorbias (Figs. 17.10b, 17.16 and 17.17). Cacti evolved from leafy trees in the Americas; as deserts formed, mutations that prevented leaf formation were advantageous because they reduced transpiration. Other selectively advantageous mutations increased water storage capacity (succulent trunk) and defenses against water-seeking animals (spines). In Africa, the formation of deserts also favored a similar phenotype, and the euphorbia family became adapted by these means. As a result, the succulent euphorbias are remarkably similar to the

FIGURE 17.16

Euphorbia gymnocalycioides, often mistaken for a cactus. Careful examination shows that many anatomical features are also different. *(Courtesy of Ron LaFon)*

FIGURE 17.17

Cacti and euphorbias have undergone extensive convergent evolution, and each is parasitized by distinct types of mistletoe that have also undergone convergent evolution. *Viscum minimum* infects *Euphorbia lactea;* compare it with *Tristerix aphyllus,* which infects the cactus *Trichocereus chilensis.* The selective advantages associated with *T. aphyllus* and its succulent, desert-adapted host were discussed in Box 5.1. Those same considerations are also valid for *Viscum minimum* and *Euphorbia lactea.*

Euphorbia
(host)

Viscum
(parasite)

PLANTS & PEOPLE

ZOOS, BOTANICAL GARDENS, AND GENETIC DRIFT

Genetic drift, adaptive radiation in small populations, and artificial selection are important factors in conserving endangered species. With only a small number of individuals maintained in zoos, botanical gardens, and national parks, the gene pools of the species of plants and animals are so small that rapid fluctuation and drift can cause the species to evolve rapidly. An endangered species protected in captivity may be lost because it evolves into one or several species adapted to highly artificial, humanly maintained conditions. Even if it still looks like the ancestral species that was placed into the zoo, it differs in its resistance to infections, parasites, and predators and its ability to understand and perform mating interactions. Most zoologists now believe that habitat preservation is the only true means of maintaining an endangered animal species. Zoos cannot accommodate the numbers of individuals necessary and cannot provide realistic habitats that act as agents of natural rather than artificial selection. Many of the progeny born in zoos would be poorly adapted for survival in the wild; if they are kept alive by antibiotics and special diets, the presence of nonbeneficial alleles is maintained in the population. Natural populations have a tremendous death rate among juvenile animals, a factor necessary for natural selection; with this eliminated in zoos, the gene pool of the captive population rapidly diverges from that of the natural population, being greatly enriched in deleterious mutations. The zoo population may become incapable of surviving without human care. The same is true of plant species maintained in botanical gardens, where often the most unusual varieties are given special attention, increasing the presence of exotic alleles that would be selectively disadvantageous in natural conditions. Artificial and natural selection result in very different gene pools. This is not a condemnation of zoos or botanical gardens: They have provided an essential service in saving many species from extinction, which at present would be the only alternative for them. But it should be remembered that a species in the wild consists not only of its individuals but of its genetic diversity interacting with its environment.

In order to provide truly safe sanctuary for endangered species, each park or habitat preserve may have to be expanded to encompass a population large enough to be stable against genetic drift. Even Yellowstone National Park is not large enough to maintain a genetically stable population of grizzly bears, although the buffalo seem to be doing well. Many individuals of a plant species may survive in a small area, such that 1 km² may hold a large enough number for their genetic stability on a theoretical basis. An important aspect, however, is to preserve enough habitat to maintain the pollinators and seed dispersers as well. The habitat must also be diverse enough to contain realistic selection pressures, including pathogens, herbivores, and so on. If these are eliminated owing to insufficient habitat preservation, the plant species experiences unnatural selection. The forest fires of the late 1980s are a good example. Fire is a natural part of many ecosystems and is an agent of natural selection for many features; if fires are suppressed, nonresistant individuals survive and the allele frequency of the population is altered.

Although artificial selection improves our cultivated plants, it can have disastrous effects on natural populations. For many species, individual specimen plants are valuable commercially, such as orchids, bromeliads, and cacti with particularly beautiful shapes or flowers. In many cases, it is easier and cheaper to collect these plants from natural populations than to cultivate them in nurseries. Extensive plant collecting actually threatens some species with extinction, but more often it has the very serious impact of removing most of the healthy plants from a population. Because the plants that are left are those that are weak, unhealthy, or misshappen, the remaining population has an increased frequency of alleles that produce unhealthy, poorly adapted plants. Even though the collector does not wipe out an entire population, he may do significant damage to the population's genetic resources. We inflict the same condition on parrots, toucans, and tropical fish.

---

succulent cacti, even though the ancestral euphorbias are quite distinct from the ancestral cacti. Two groups cannot converge to the point of producing the same species; only the phenotypes converge, not the genotypes. For example, cactus spines are modified leaves, whereas euphorbia spines are modified shoots.

EVOLUTION AND THE ORIGIN OF LIFE

The species present today have evolved from those that existed in the past, which evolved from those that existed before them. It is appropriate now to consider how life arose and what it was like initially.

The most seriously considered hypothesis about the origin of life on Earth is that of **chemosynthesis.** Basically, the chemosynthetic hypothesis attempts to model the origin of life using only known chemical and physical processes, rejecting all traces of divine intervention. It was first proposed by the Russian scientist A. Oparin in 1924 and then by J. B. S.

Haldane in England. They postulated that prior to the origin of life, the surface of Earth was different from the way it is now, and the chemicals present then could react spontaneously, producing more complex chemicals that could in turn continue to react. Over millions of years, reactions might produce all the molecules necessary for life, and these might aggregate into primitive protocells. From the protocells, natural selection would guide the evolution of true, living cells. Four conditions would have been necessary for the chemosynthetic origin of life: The primitive Earth would have to have had (1) the right inorganic chemicals, (2) appropriate energy sources, (3) a great deal of time, and (4) an absence of oxygen in its destructive molecular form, O_2.

CONDITIONS ON EARTH PRIOR TO THE ORIGIN OF LIFE

Chemicals Present in the Atmosphere.   Earth condensed from gases and dust about 4.6 billion years ago; it was initially hot and rocky and had an atmosphere composed mostly of hydrogen. Because hydrogen is such a light gas, most of this first atmosphere was lost into space. It was replaced by a **second atmosphere** produced by release of gases from the rock matrix composing Earth and from heavy bombardment by meteorites. Both sources would have provided gases such as hydrogen sulfide (H_2S), ammonia (NH_3), methane (CH_4), and water. All these are found in volcanic gases and in meteorites that still strike Earth. Molecular oxygen was absent; it had already combined with other elements, resulting in compounds such as water and silicates. The early second atmosphere was a **reducing atmosphere** due to the lack of molecular oxygen and the presence of powerful reducing agents (Fig. 17.18).

Energy Sources.   There must have been a complex chemistry in the early second atmosphere because it was exposed to powerful sources of energy. Foremost was intense UV and gamma radiation from the sun. These radiations have energetic quanta that knock electrons from atoms, creating highly reactive free radicals. Part of the ammonia would have decomposed to hydrogen and nitrogen, and some of the methane would have converted to carbon monoxide and carbon dioxide, increasing the complexity of the atmosphere.

Heat was another source of energy available to power reactions. One heat source was the coalescence of gas and dust to form Earth: As the particles fell toward the center of gravity, they accelerated and then collided, converting kinetic energy to heat. A second heat source was the radioactive decay of heavy elements like uranium and radium; this decay was extremely intense 4.5 billion years ago. Even today, enough radioactive decay remains to keep Earth's core molten. The chemicals present in the early second atmosphere were also dissolved in the ocean's water, and whenever they came into contact with hot or

FIGURE 17.18
The outer planets, except for Pluto, are so massive that their gravity has retained their original hydrogen atmosphere. In addition, meteor bombardment and other activities have added ammonia, methane, and other components that make these atmospheres similar to Earth's early second atmosphere. Reactions in the atmosphere may be producing organic compounds. *(NASA/Peter Arnold, Inc.)*

FIGURE 17.19
Rapid movement of gases during a volcanic eruption generates the electrical potential necessary for lightning. Electrical discharges through the volcanic gases produce organic compounds. Mt. Kilauea, Hawaii. *(E. R. Degginger)*

molten rock, endergonic reactions could proceed, resulting in the formation of more complex chemical compounds.

Electricity was abundant on a gigantic scale; much of Earth's water was initially suspended in the atmosphere because of the high temperature of the air and of the planet's surface. When sufficient planetary cooling occurred, rains fell, but the rainwater evaporated upon hitting the hot surface. The rainstorms must have been immense, lasting for thousands of years and generating tremendous amounts of lightning. As each lightning strike flashed through the atmosphere, it triggered more chemical reactions, and the resulting products were washed downward by the rain: On the hot surface, they would have boiled, producing even more chemical reactions. After further planetary cooling, the surface temperature dropped to less than 100°C, and liquid water began to accumulate as streams, lakes, and oceans.

Volcanoes also produce lightning around their throats as they erupt (Fig. 17.19). Considering the immense amount of vulcanism that occurred, this lightning would have supplied a significant amount of energy. Also, volcanic lightning occurs through the clouds of venting gases where hydrogen sulfide, methane, ammonia, and water are most concentrated.

Time Available for the Origin of Life. The time available for the chemosynthetic origin of life basically had no limits, simply because of the lack of free molecular oxygen. Without oxygen no agent was present to cause the breakdown and decomposition of the chemicals being created. If molecular oxygen had been present, the chemicals either would have not formed or would have oxidized soon after formation; without oxygen, they could accumulate for millions of years. The ocean of that time has been called a "dilute soup" or a "primordial soup" containing water, salts, and numerous organic compounds that become increasingly complex as time went on. As much as 1.1 billion years may have elapsed between the time Earth solidified and life arose.

CHEMICALS PRODUCED CHEMOSYNTHETICALLY

After the writings of Oparin and Haldane, the first experimental tests of the chemosynthetic hypothesis were performed in 1953 by a graduate student, S. Miller, at the University of Chicago. He constructed a container that had boiling water in the bottom and a reducing atmosphere in the top; electrodes discharged sparks into the gases, simulating lightning: As the water boiled, steam rose, mixed with the atmosphere and was reacted upon by the

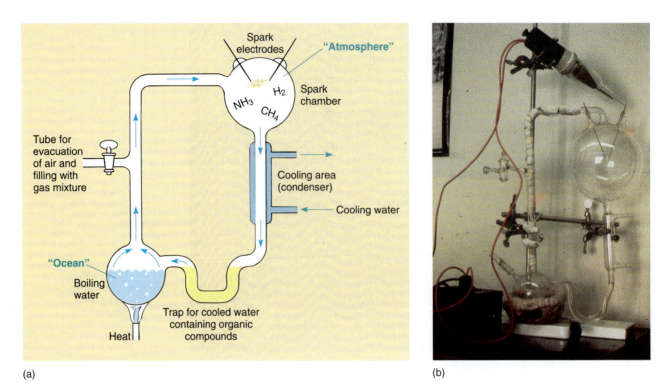

(a)   (b)

FIGURE 17.20

(a) Diagram and (b) photograph of apparatus used by Miller to show that the first steps in the chemosynthetic origin of life were possible. *(Courtesy of Dr. Stanley Miller)*

electrical sparks, then condensed and fell back into the water to be cycled again (Fig. 17.20). Miller let his first experiment cycle for a week and noticed that the solution had become dark from the accumulation of complex organic compounds that had formed. When he analyzed their composition, he found that many different substances were present, including amino acids. Since then, this type of experiment has been performed numerous times, testing the effects of varying atmospheric compositions, using several types of energy sources, or including metal ions in the water. Virtually all the small molecules essential for life can be formed this way: amino acids, sugars, lipids, nitrogen bases, and so on.

These experiments tell us what is theoretically possible; direct analysis of meteorites and lunar samples reveals what has actually happened in nonliving environments. Rock samples brought back from the moon by the Apollo astronauts contain various organic compounds, including amino acids. The interiors of meteorites, uncontaminated by the fall through the atmosphere or contact with soils, have contained alcohols, sugars, amino acids, and the nitrogenous bases that occur in nucleic acids. With regard to the formation of monomers, the chemosynthetic hypothesis represents a plausible model.

THE FORMATION OF POLYMERS

Monomers present in the early ocean had to polymerize if life were to arise, but polymerization required high concentrations of monomers. Given enough time, the oceans would have changed from a dilute soup to a concentrated one, but that probably was not necessary. Numerous mechanisms would have produced pools of highly concentrated reactants.

An obvious method of concentration is the formation of seaside pools at high tide that evaporate after the tide goes out. With intense sunlight the pools would have been warm, perhaps even hot, and polymerization reactions could occur. With the return of high tide, the polymers would be washed into the sea and accumulate. Monomers could also have accumulated when ponds and seaside pools froze; ice is relatively pure water and the monomers become increasingly concentrated in whatever water has not yet frozen. This might produce a class of polymers distinct from those formed by evaporation at high temperature.

Absorption by clay particles could have concentrated monomers, and clays are receiving great attention now. Because clay particles are tiny fragments of rock, they have a regular, crystalline surface; organic molecules adhere to them in a particular orientation, not simply at random. Thus, binding to a clay particle is similar to binding to an enzyme. Furthermore, the crystalline matrix of clay contains contaminating ions of iron, magnesium, calcium, phosphate, and other charged groups that are typically present at the active sites of enzymes. Considerable experimentation is being done to determine whether clays might have both concentrated monomers and acted as the first primitive catalysts.

AGGREGATION AND ORGANIZATION

The next step in the possible chemical evolution of life would have been the aggregation of chemical components into masses that had some organization and metabolism. Fatty, hydrophobic material would have accumulated automatically as oil slicks in quiet water or as droplets in agitated water (Fig. 17.21). Fatty acids would have occupied the outermost layer, accompanied by other molecules such as proteins that had a hydrophobic portion and a hydrophilic one. The interior may have been mostly hydrophobic, but proteins that had a hydrophobic exterior and hydrophilic interior would have added complexity. Aggregation of certain types of proteins would have resulted in large regions of hydrophilic sites.

These first aggregates would have formed basically at random, controlled only by relative solubility. If some of the proteins had some enzymatic activity by chance, the aggregate would have had some simple "metabolism"—perhaps the conversion of some molecule, absorbed from the sea, into another molecule; the aggregates would have been heterotrophs completely.

These aggregates are not postulated to have been alive or even to have been early stages of life, because at that point no means of storing genetic information existed. Without genetics, natural selection cannot occur. Countless numbers of these aggregates might have formed and disassociated throughout all the oceans over millions of years. Their existence may have had a significant effect on the chemistry of the oceans. Some may have provided appropriate conditions for certain proteins to be very active enzymes that might have produced quite complex products. However, others may have been able to degrade such products, so an equilibrium may have been established between the formation and destruction of ever more elaborate molecules and polymers.

At some point, presumably an aggregate formed that did have a heritable information molecule able to direct the synthesis of products useful to the aggregate. By "useful" we mean that the product helped the aggregate persist longer, without being broken up by wave action or dissipating by diffusion. It may have helped the aggregate grow or even helped the information molecule replicate. With the presence of heredity, everything changed; any mutation that caused the production of a more efficient, more advantageous enzyme or structural protein provided a strong selective advantage and could be passed on to progeny aggregates. At present, attention is focusing on RNA as the first heritable information molecule. As the aggregate increased in size, perhaps largely by absorbing material from the ocean, its information molecule would replicate. If the aggregate divided into two by either wave action or surface tension and if each half contained an information molecule, reproduction occurred.

EARLY METABOLISM

The aggregates would have been complete heterotrophs, absorbing all material from the ocean and modifying only a few molecules. However, as aggregates continued to consume certain nutrients, scarcity occurred; any aggregate that could produce an enzyme capable of synthesizing the scarce molecule from an abundant one still available in the ocean would have had a strong selective advantage. With this, there would have been a metabolic pathway two steps long involving two enzymes (Fig. 17.22). Being able to synthesize a valuable, scarce molecule from an abundant, free one would have given that aggregate great

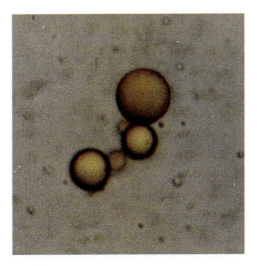

FIGURE 17.21

If dry proteins are heated mildly, they form a substance called proteinoid; if water is added, small round bodies called proteinoid microspheres are formed. They are round because of surface tension, just like plant protoplasts. They have an outer membrane that is differentially permeable; and some inner regions are aqueous, others hydrophobic. These cannot be considered living at all, but their differential permeability and their internal heterogeneity make them good sites for a primitive metabolism. *(SEM by Steven Brooke, color copy arranged by Richard Le Duc)*

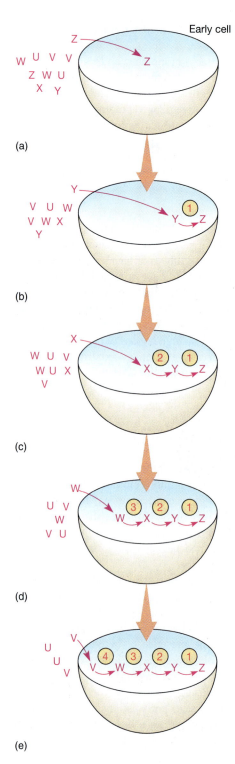

FIGURE 17.22
We believe that many metabolic pathways evolved backward as raw materials in the environment became scarce. At present, the pathway for photosynthesis has gone as far as it can, requiring only water and carbon dioxide.

advantage over the others; it would have had a more rapid metabolism and would have grown and reproduced more rapidly.

As its abundance increased, it and its descendants would finally have become so numerous as to use the second chemical faster than it could be formed chemosynthetically, and scarcity would have occurred again. Natural selection would again favor any genetic system that could extend the metabolic pathway another step to include an abundant precursor. Recall that in photosynthesis in modern plants, the only two precursor molecules that need to be drawn from the environment are water and carbon dioxide—both are abundant and cheap. From them, metabolic pathways extend and ramify such that all carbon-containing compounds are derived from them. No preformed organic molecules are necessary.

Energy metabolism must have been important also, and glycolysis must have evolved early because it is present in virtually all organisms. The aggregates and first cells may have absorbed some ATP and generated more by fermentation with glycolysis for millions of years. At some point, electron transport systems and hydrogen ion pumping systems would have evolved. The first would have been powered by light, and an early form of photosynthesis would have evolved. Higher plant photosynthesis is elaborate, but much simpler types occur even today in certain photosynthetic bacteria. The evolution of photosynthesis would not have been as complex or as difficult as chloroplasts would lead us to believe. Oxidative electron transport, as occurs in mitochondria, could not have evolved because free molecular oxygen still did not exist.

OXYGEN

The evolution of chlorophyll *a* and photosynthesis that liberates oxygen had two profound consequences: (1) it allowed the world to rust, and (2) it created conditions that selected for the evolution of aerobic respiration. Until that time, photosynthesis had involved the bacterial pigment bacteriochlorophyll, which liberates sulfur from hydrogen sulfide rather than oxygen from water (see Chapter 19). Earth retained its reducing atmosphere, and chemosynthesis of complex precursor molecules could continue. But once chlorophyll *a* evolved, the raw material for photosynthesis became water, and free molecular oxygen, O_2, was released as a waste product. We know that this evolutionary step occurred 2.8 billion years ago because the oxygen rapidly combined with iron and formed ferric oxide—rust. Sedimentary rocks of this age contain a thick red layer of rust, indicating when oxygen-liberating photosynthesis arrived and how long the oxygen/iron reactions continued. Only after all the free iron in Earth's oceans had oxidized did oxygen finally begin to accumulate in the atmosphere. The atmosphere present today was derived from the early second atmosphere by this addition of oxygen from photosynthesis. It is an **oxidizing atmosphere**.

The period of rusting was critically important for all life because it kept the concentration of free oxygen very low. It can be difficult for us humans—obligate aerobic animals—to appreciate just how toxic free oxygen is; we need it to live, but our bodies have numerous mechanisms to keep it under control and to prevent it from reacting at random in our bodies. The same is true of plants, fungi, algae, and many bacteria. Had iron not been present, the concentration of free oxygen in water and air might have increased rapidly and killed everything; possibly there would have been no survivors. But with iron, oxygen concentration remained low for millions of years; it was an environmental danger but was not instantly, universally lethal. Any mutation that produced a mechanism that could detoxify oxygen had great selective advantage. Of course, the best way is to add two electrons to it and let it pick up two protons and turn into water, which is exactly what cytochrome oxidase does. This is the last electron carrier in the mitochondrion transport chain, and with its evolution, aerobic respiration became possible. Oxygen was transformed from a dangerous pollutant to a valuable resource. The evolution of aerobic respiration did not occur in mitochondria but rather in bacteria, and some species seem to retain very relictual, early versions.

The build-up of atmospheric oxygen had other important effects; under the influence of UV light, oxygen is transformed into ozone, O_3. Ozone is extremely opaque to UV light and thus prevents most of it from penetrating deep into the atmosphere and reaching Earth's surface. This immediately removed UV light as an energy source and must have greatly slowed the chemosynthetic formation of organic molecules. Biological photosynthesis became the main means of bringing energy into the living world. The lack of UV radiation made the surface a safer place to live; UV radiation damages nucleic acids, and while the atmosphere lacked ozone, probably nothing lived near the top of the oceans or on land. With the ozone shield, organisms could move higher in the oceans and onto the mud of seashores. The transition to terrestrial life became possible.

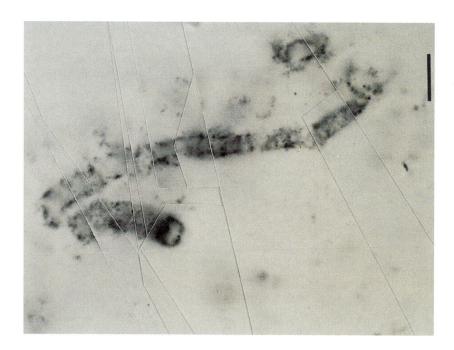

FIGURE 17.23
The fossil remains of an early prokaryote that lived about 3.5 billion years ago. *(Biological Photo Service)*

THE PRESENCE OF LIFE

The chemosynthetic theory postulates a long series of slow, gradual transitions from completely inorganic compounds to living bacteria. At which stage can we say that life came into being? Can the aggregates be considered alive? This is a difficult question, because the theory delineates no absolute demarcation between living and nonliving objects (Fig. 17.23). The chemistry of living creatures is more complex than that of nonliving objects, but it does not possess any unique properties. The physics of living and nonliving systems is identical. As so often in the living world, we are dealing with a continuum, not simply two mutually exclusive alternatives. To look for a dividing line would be simplistic; it is more important for us to understand the processes in their complexity.

SUMMARY

1. When considering the reproductive effort of an entire population, it is important to consider all possible alleles for each gene, their relative frequency in the gene pool, and their effects on reproductive success.

2. Sexual reproduction alone does not cause the relative abundance of alleles in the gene pool to change. Mutations bring about new alleles but do not greatly affect allele frequency.

3. Both artificial and natural selection increase the fitness of a population to its habitat by preventing certain individuals from breeding and passing their alleles on to the next generation. Artificial selection often increases the plant's fitness for artificial habitats.

4. In order for natural selection to operate, a population must produce more offspring than can possibly survive in the environment, and the offspring must be genetically diverse. These two conditions occur almost everywhere in nature at all times.

5. Natural selection operates only on the phenotypes of pre-existing alleles.

6. Plants and most other organisms never do anything with purpose or intent. Only humans and higher primates have purpose, intent, and will, and then only with regard to voluntary actions.

7. Organisms face multiple selection pressures and must be adapted to all aspects of their environment. Certain aspects are more important than others, and it may be less dangerous to be poorly adapted to one factor than to another.

8. Most adaptations have negative consequences as well as advantageous ones; retention or loss of these alleles from the gene pool is related to the relative benefit versus cost of the adaptation. As an environment changes, the selective value of the phenotype may also change.

9. In phyletic speciation, all of a species evolves into one new species. In divergent speciation, part of a species evolves into a new species while the rest of the population continues as the original species or evolves into a different new species.

10. Gene flow is inhibited or completely prevented by reproductive barriers, either abiological aspects of the environment or biological aspects of the organisms themselves or of their pollinators and seed dispersers.

11. Life is believed to have originated on Earth by the process of chemosynthesis. This hypothesis postulates that reactions of inorganic compounds in Earth's early second atmosphere resulted in the formation of organic compounds that could have coalesced into simple aggregates with a rudimentary metabolism. Once a system of heredity developed, evolution by natural selection made it possible for truly living cells to come into existence.

IMPORTANT TERMS

abiological reproductive barrier
adaptive radiation
allopatric speciation
artificial selection
biological reproductive barrier
chemosynthesis

convergent evolution
divergent speciation
founder
gene flow
gene pool
genetic drift

hybrid sterility
natural selection
phyletic speciation
population
reducing atmosphere
sympatric speciation

REVIEW QUESTIONS

1. What is the gene pool of a species? Why are some alleles more common than others?

2. Many mutations can be corrected by DNA proofreading/repair enzymes. What would happen to the further evolution of a species if a perfect repair mechanism evolved such that no mutation ever went uncorrected?

3. How does an accident differ from natural selection? Would an unusually severe drought with no rainfall at all for 50 years be a selective or a nonselective destructive force?

4. How does artificial selection differ from natural selection? Do you think the ancestors of lettuce had such soft, nonbitter leaves? Could today's lettuce plants survive in natural conditions?

5. Change each of the following sentences into ones that are true: (a) Plants produce root hairs in order to absorb more water. (b) Plants close their stomata at night to conserve water. (c) DNA produces mRNA in order to keep ribosomes and potentially harmful high-energy enzymes away from itself. (d) Stems and roots produce bark to keep pathogens out. (e) Cacti and euphorbias have succulent bodies in order to store water. (f) Certain plants have C_4 metabolism in order to reduce photorespiration. (g) Some plants time their flowering by night length in order to avoid damage caused by blooming too early in an unusually mild spring. (h) It is best for plants of harsh climates to become dormant in winter.

6. What is gene flow? How does it occur?

7. Describe the conditions present on Earth at the time of its early second atmosphere. What chemicals and energy sources were present? What chemical in particular was absent? Can you speculate about what conditions would be like on Earth today if meiosis, crossing-over, and sexual reproduction had never evolved?

18

CLASSIFICATION AND SYSTEMATICS

This is an alpine willow, about 3 inches tall. Which characters allow plant taxonomists to recognize it as being closely related to other willows?

CONCEPTS

Nomenclature is the science of giving things names. One of the critical goals of biological nomenclature is to provide each species with a unique name, thereby permitting easy and effective communication about organisms. The confusion and misunderstandings that would arise if several organisms shared the same name must be avoided. We humans and our prehuman ancestors have always been concerned with plant identification: Herbivores must recognize which plants are suitable for food and which are poisonous. Medicinal plants had to be named and categorized as well. As civilization developed in Asia, the Middle East, and the Americas, each society devised its own system of nomenclature, many of which were quite elaborate and grew to encompass thousands of cultivated or gathered plants (Fig. 18.1).

   With the discovery of evolution by natural selection, the basis of naming plants suddenly changed: Natural selection showed that all organisms are related to each other genetically, some closely, others distantly. Taxonomists, scientists who specialize in classification and naming, immediately realized that the most scientifically valid system of assigning names to species would be one that reflected evolutionary relationships. At the end of the last century, taxonomists adopted the goals of (1) developing a **natural system of classification,** a system in which closely related organisms are classified together, and (2)

FIGURE 18.1

This wall painting shows an Egyptian garden about 1400 B.C., but extensive collecting, cultivation, and classification occurred long before this. Grave inscriptions from 3000 B.C. show that this basic garden pattern had been developed 5000 years ago. Ancient records list cultivated and collected plants, some used for food or medicine, others for beauty and fragrances. Shown here are date palms, sycamore figs, and papyrus stalks. *(Copyright © by the Metropolitan Museum of Art)*

assigning plant names on the basis of evolutionary relationships. The nomenclature would reflect the natural system of classification. Because organisms range from closely to distantly related, taxonomists devised a classification and a nomenclature with numerous levels. In this system, closely related species are placed into a genus, closely related genera are grouped together into a family, and so on.

The development of a natural system of classification had profound, far-reaching effects on taxonomy. Prior to this, the task of taxonomists was to discover and identify new species and give them unique names. Now, with a natural system of classification, any new species that is identified must be studied to determine which other species it is related to, so that the correct genus and family names can be given to it. In many cases this is relatively simple. Explorations of Mexico frequently result in the discovery of new species of composites, members of the sunflower family. These often have distinctive features shared by other composites in the region, so it is not difficult to determine the genus to which a new species belongs. Other cases can be much more difficult because the new species has features common to several genera, or it has some features that seem to indicate that it definitely belongs in one genus but other features that seem to definitely indicate that it should not be a member of that genus. In such cases, taxonomists must consider carefully the significance of the conflicting features. Part of the classification system may be incorrect and may not really be reflective of evolutionary relationships. The problem of inconsistent or conflicting features occurs at all levels of the classification system, and in many cases we are not certain which genera should be classified in which families or which families should be considered closely related.

Starting about 400 million years ago, a long series of mutations and natural selection gave rise to the vascular land plants. During the intervening eons since then, that evolutionary line has progressed and diversified, branching into more and more lines of evolution. Thousands of these have become extinct and are known only by fossils; thousands of other evolutionary lines are represented by the approximately 400,000 species of living vascular plants. The goal of modern taxonomy is to understand each of these evolutionary lines and to have the classification reflect their relationships accurately. Although the evolutionary diversification pattern is a reality, the classification system is a hypothesis, a model that attempts to map that evolution. Our knowledge is incomplete and imperfect, and the current classification system is only an approximation. It is probably a close approximation; we do not expect any *major* changes in it, but we do expect numerous smaller changes to be made periodically for many years.

TYPES OF CLASSIFICATION SYSTEMS

The natural system of classification, which attempts to follow the evolutionary history—the **phylogeny**—of the organisms being categorized, is only one of several types of classification system. Another fundamental type is an **artificial classification system,** in which several key characters, often very easy to observe, are chosen as the basis of classification. Good examples are roadside floras and picture guides to plants, birds, and mammals of national and state parks. The botanical classifications in these are often based primarily on flower color: All plants with white flowers are grouped together, as are all those with red flowers, and so on. Within each category, the next classification category might be the plant's habit: Trees are grouped together, as are shrubs, herbs, vines, and so on. These systems are described as artificial because many plants in a category are not closely related to each other by descent from a common ancestor; furthermore, they are separated from their close natural relatives in other categories simply because they differ in flower color or habit.

Both artificial and natural classification systems have distinct advantages. Natural systems facilitate our study of evolution—not just the evolution of species, but also the evolution of various aspects of a species, such as its structure, metabolism, and reproductive biology (Fig. 18.2). Artificial systems typically have the goal of easy plant identification by means of characters such as flower color and plant habit. Alternatively, the artificial system may be designed to group together plants with economically or scientifically important features. From a practical standpoint, carpenters and woodworkers are more interested in color, texture, grain, and hardness of a wood than its phylogeny. Gardeners might classify plants according to their ability to tolerate shade, full sun, frost, or alkaline soils.

Although artificial classifications are extremely useful, they can only be adjuncts to natural systems of classification which take evolution into account. As physiologists examine the phylogeny of all species with C_4 metabolism as revealed by a natural classification, they find that C_4 metabolism has evolved several times, so all C_4 metabolisms are not expected to be identical. Similarly, someone studying the metabolism of petal pigmentation might classify flowers artificially according to color, deciding to investigate the synthesis of red pigment first. It would be necessary to check the phylogeny of the experimental plants because some plants produce red anthocyanin pigments, but a different evolutionary line produces red betalains that are metabolically very different (see Fig. 25.26). If this phylogenetic difference were not known, the scientist would get inexplicably conflicting results.

Because we do not yet know all details of plant phylogeny, we cannot be certain how to construct the definitive natural system that is correct in every detail. In many cases, botanists must make educated guesses, and of course they do not all agree. Because investigators are free to publish their own views and opinions, numerous "natural" systems have been proposed. At present, most agree with each other on most major points, and the areas of disagreement are the groups of plants that need more study.

A third type of classification, used for fossil organisms, combines features of both artificial and natural systems. The goal is to understand the evolution of the fossil and to identify both its ancestors and its relatives that later evolved into another species. This requires a natural system. Typically, we do not know much about the fossil, so superficially similar ones are grouped together for convenience—a basically artificial system. The groupings are **form genera:** All fossils with the same basic form or structure are classified together. For example, a piece of fossil wood similar to the wood of modern pines, spruces,

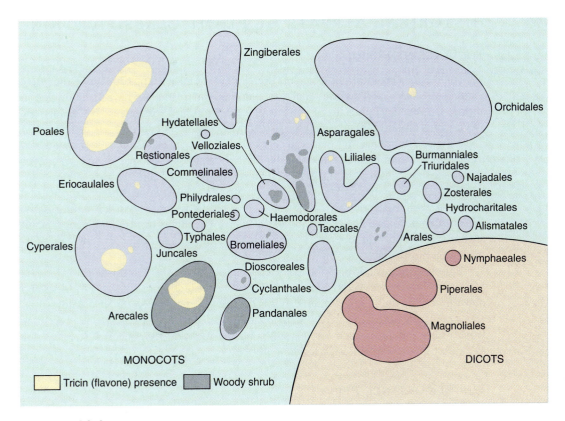

FIGURE 18.2

This diagram represents one interpretation of the evolutionary relationships of the monocots; each enclosed space represents an order (a group of related families). The size of the space represents the number of species in the order: The small order Typhales contains only one family, Typhaceae, one genus, *Typha,* and three species (cattails), whereas the large order Poales contains the grass family Poaceae with several thousand species. Orders drawn in the lower right, such as Arales and Alismatales, are thought to be relatively unchanged from the ancestors of the monocots, whereas all orders farther to the left or top have undergone a great deal of modification. Chapter 25 gives common names for many of these orders. A map of the phenotype of treelike or shrublike body is given in green. The entire order Arecales (palms) and most of the order Pandanales ("screw pines") are treelike, whereas a few members of Asparagales are treelike (Joshua trees, Australian grass trees). In Poales, bamboos are treelike grasses. The woody, treelike growth of palms, screw pines, and Cyclanthales is similar, so probably these three orders are closely related and their close placement in this diagram may be correct. The growth of bamboos is not at all like that of palms; their treelike, woody growth has evolved separately as a type of convergent evolution. Although bamboos and palms have a somewhat superficial resemblance, they are not really closely related, so the separation of the two orders in this diagram is probably justified. Chemical features are also analyzed in this way; the presence of tricin, a chemical known as flavone, is given in brown. It is almost universally present in the grasses (Poales) and is very common in sedges (Cyperales) and palms (Arecales). Could this also be a case of convergent evolution? It could, because tricin also occurs sporadically in the Asparagales, Liliales, Orchidales, and even a few dicots; these species are so distantly related that the genes for the tricin-synthesizing metabolic pathway must have arisen several times as independent mutations in each group. The possibility of such convergent evolution makes it difficult to determine whether the presence of tricin in Poales, Cyperales, and Arecales is evidence of natural relatedness or convergence.

and larches is classified in the form genus *Pityoxylon.* If the piece of fossil wood was part of a branch with leaves and cones attached, there would probably be enough characters present to allow us to assign the wood to a natural genus. But if the fossil contains only wood, not enough characters are present to determine whether it came from an ancient pine, spruce, larch, or some other group that has since become extinct.

HISTORICAL ASPECTS OF PLANT CLASSIFICATION

THE ANCIENT PERIOD

Although early classifications of plants were prepared in ancient China and the Aztec Empire of Mexico and Central America, our modern treatments have their origins in ancient Greece. Theophrastus, a student of both Plato and Aristotle, wrote extensively about plants around 300 B.C. and established many important concepts. He distinguished flowering and nonflowering plants, recognized sexuality in plants, and understood that fruits develop from carpels. Theophrastus described almost 500 species, and our genus names *Asparagus, Narcissus,* and *Daucus* (carrot) can be traced directly back to him.

Pliny the Elder (Caius Plinius Secundus; A.D. 23–79), a Roman lawyer and natural historian, wrote voluminously on almost every subject. His largest work, *Natural History,* was an attempt to describe everything in the world. Despite its inclusion of many fanciful creatures based only on folktales, it served as the definitive, authoritative source of information on most subjects for more than 1000 years.

The most important book on plant classification from the ancient world is *Materia Medica* by Dioscorides. Written in the first century A.D., it describes 600 plant species and how they can be used to treat disease. It was the best, often the only, source of information about preparing herbal medicines. *Materia Medica* was at last superseded by more accurate work during the Renaissance, but it is still published today for its immense historical value as a direct link to early Greek science.

THE RENAISSANCE PERIOD

Between the time of Dioscorides and the Renaissance, almost no works of great value in natural history were written. However, changes began to occur in the 15th century, and Europe entered an age of exploration. Prince Henry the Navigator sent ships on expeditions down the west coast of Africa, and the explorers returned with new plants and animals and knowledge of new lands and peoples. Simultaneously, exploration of physics, chemistry, astronomy, and geology began. An important result of these explorations was the discovery that Pliny and other ancient authorities had been wrong in many subjects. Until then people accepted the idea that the ancient Greeks and Romans had represented a Golden Age and were basically infallible; any observation that contradicted Pliny was assumed to be an error on the part of the observer. The realization that the ancients were indeed fallible meant that the answers to questions about the world had to be sought in the world itself, not in ancient books. Exploration—not only geographical but also scientific, philosophical, and religious exploration—became an obsession.

In botany, this new, independent thinking first became apparent in the publication of medically oriented plant books, called **herbals,** which were published in large numbers. By the middle of the 16th century, they began to contain extremely careful, precise descrip-

FIGURE 18.3
Herbals prepared during the Renaissance relied on direct observation of plants. Illustrations such as this drawing of purslane (*Portulaca oleracea*) were rendered more carefully and accurately then they had been previously. (*Culver Pictures, Inc.*)

FIGURE 18.4
Carolus Linnaeus (1707–1778)
developed the binomial system of
nomenclature in which each species has
both a genus name and a species epithet.
*(Biophoto Associates, National Portrait Gallery,
London)*

tions of plants based on first-hand observation of the plants, not by paraphrasing *Materia Medica* (Fig. 18.3). Most herbals contained fewer than 1000 plants, so a reasonably good botanist could become familiar with all of them.

However, as exploration continued, especially after the discovery of the Americas, the number of plant species became too large for this type of familiar treatment. Rather than listing or describing all the plants of the known world, it was necessary to develop a classification system, so that a person would have some means of identifying an unfamiliar specimen and finding it in the ever-larger herbals.

Several important ideas developed at this time. One was the concept of a genus as a group of similar species, established by Gaspard Bauhin (1560–1624). Although this may seem like an obvious notion, it was a profound breakthrough at the time. The concept of species had been easy: God was believed to have created all the types of organisms—the species. But this concept did not explain why there should be groupings of species, as the concept of genus implies. Why would God create several types of roses, several types of mints? This was not a trivial question then. Until the theory of evolution by natural selection, the world was viewed as a reflection of the mind of God, who presumably had created it. If there were thousands of species of grasses but only three of cattails (see Fig. 18.2), this would reveal something about how God thinks. The ancient Greeks were of no help. Their philosophy was based on the concept of idealism: All reality—plants, animals, humans— are merely imperfect physical manifestations of ideal, theoretical types. They did not group the types into anything equivalent to genera. Yet it was obvious to anyone familiar with the increasing number of plant and animal species that some species resembled each other very closely.

The classification system, particularly scientific names, can be traced directly to Carolus Linnaeus, a professor of natural history at the University of Uppsala in Sweden during the middle and latter part of the 18th century (Fig. 18.4). Of major significance was the fact that he was the primary professor for 180 students, many of whom became excellent botanists and large numbers of whom traveled widely. Not only did they send specimens to Linnaeus from all over the world, but they also transmitted his ideas to the people they worked with, and vice versa.

The number of species known to exist by the time of Linnaeus was too great for any single person to be familiar with even half of them. An efficient system of classification and nomenclature was desperately needed, and Linnaeus supplied it. He adopted the genus system of Bauhin and standardized it. He created a large numbers of genera and placed every species into one genus or another. Every species had both a genus name and a species name, the basis of our present **binomial system** of nomenclature. Linnaeus's system was easy to use; he decided early that the numbers of stamens and carpels were the most important features of a species, so his classification was based on that. He created class Monandria for species with one stamen, and it was divided into two subclasses, Monogynia (one carpel) and Digynia (two carpels). Other classes were Diandria, Triandria, and so on. The system was expandable, so that as new species were discovered, they could easily be fitted in. This system was entirely artificial: The fact that plants share the same number of stamens does not mean that they resemble each other in more fundamental, important ways. Few botanists shared Linnaeus's conviction that the numbers of stamens and carpels were the most significant aspects of species. But there was no concept of a natural classification system at the time, more than a century and a half before the theory of evolution by natural selection was developed.

In 1753, Linnaeus published *Species Plantarum,* a treatment of all plant species known in the world at that time. This book made botanical studies of all types much easier, because specimens could be identified quickly. However, the system often placed obviously dissimilar species into the same genus.

THE MODERN PERIOD: EVOLUTION AND CLASSIFICATION

The idea of the evolution of species developed slowly over many years; it did not begin with Darwin and Wallace by any means. Bauhin recognized resemblance early, and an expansion of the idea of similarity occurred at the Jardin des Plantes in Paris. The administrators of the garden wanted to organize it in a logical way, so that plants resembling each other were planted together. Antoine-Laurent de Jussieu was given the task of organization; the result was the book *Genera Plantarum* (1789). The first major attempt at a natural system of classification had no theoretical or philosophical basis; science was still dominated by the idea of divinely created types, but it was impossible to ignore the various levels of similarity.

Unfortunately, J. B. P. de Lamarck at this time presented his theory of **evolution by inheritance of acquired characteristics.** Lamarck's theory was based on the incorrect idea that all cells of the body produced fluids that diffused to the genitalia, where the fluids were concentrated and formed into sperm cells or egg cells. As an individual changed, the secreted fluids and therefore the characteristics carried by the gametes would change. The evolution of giraffes was explained as follows: Ancestors to giraffes had been born with short necks, but as they stretched their necks to reach leaves in high trees, their necks lengthened—long necks in the adults would be the acquired characteristic—and subsequently the genetic fluid passed to their gametes carried information about long necks. Consequently, their offspring had longer necks. Mendel showed that this could not be right because genes are nonvarying particles, not variable fluids. Evolution by natural selection explains giraffes by postulating that variability of neck size occurred among the earliest ancestors of giraffes; alleles in some animals produced short necks and in others slightly longer necks. Those with longer necks could reach more leaves, so they had better survival and reproductive success. Over millions of years, various mutations produced new alleles that resulted in even longer necks, and these new alleles were increased by natural selection. The theory of inheritance of acquired characteristics was never widely accepted, and it gave evolution a bad reputation for many years.

In 1859 Charles Darwin and Alfred Wallace each propounded the theory of evolution by natural selection, which gave natural systems of classification an immediate validity. Taxonomists quickly understood why some species resemble each other so strongly that they must be grouped into a genus: They are closely related by evolution. Similarly, closely related genera constitute a family. From that point on, every effort has been made to construct natural classification systems.

The theory of evolution by natural selection became accepted while two German botanists, A. Engler and K. Prantl, were working on a monumental classification of all the world's plants, *Die Natürlichen Pflanzenfamilien,* published in 1915. It organized the species on a phylogenetic, natural basis as understood by Engler and Prantl. In those early days of evolutionary studies, organisms were assumed to have evolved from the simple to the complex (amoebae to humans). Thus in flowering plants, the wind-pollinated species that have no sepals and petals were placed first, and the sunflowers, with their complex floral structures, were considered more advanced. It was quickly discovered that simplicity may also be the result of evolutionary simplification. Because no automatic correlation exists between simple and primitive or between complex and advanced, the Engler and Prantl classification has many errors, although as a compendium of information it is still an unmatched vital resource. C. E. Bessey (early 1900s) of the University of Nebraska was the first to reclassify plant species in a more phylogenetic system, paying closer attention to which features are relictual and which are derived in each group. The most recent works are those of A. Takhtajan in Russia and Arthur Cronquist at the New York Botanical Garden (see Chapter 25).

LEVELS OF TAXONOMIC CATEGORIES

Plants have varying degrees of relatedness, and a natural classification system reflects this in its numerous levels.

The most fundamental level of classification is the **species,** which ideally and theoretically is a set of individuals closely related by descent from a common ancestor. Members of a species can interbreed with each other successfully but cannot interbreed with individuals of any other species (see Chapter 17; Table 18.1). Most species are not so predictable; they may not interbreed well with closely related species, but an occasional cross-pollination results in a viable seed that grows into a fertile adult. If this occurs frequently, the two plant groups may best be considered **subspecies** of a single species. The two subspecies must be so similar genetically that the chromosomes of the two parents can function in the same nucleus and, when the plant is mature, spore mother cells can undergo meiosis successfully. For all this to happen, the organisms must be extremely closely related, having undergone divergent evolution from a common ancestor only recently, so that few mutations have accumulated since they diverged. But as mutations continue to accumulate in each subspecies, finally either no viable seed results when the two are crossed or the seed grows into a sterile plant. The two then must be considered separate species.

Closely related species are grouped together into **genera** (singular: genus; Table 18.1). Deciding whether several species are closely related enough to be placed together in the

The word "species" is both singular and plural; "specie" refers to money.

TABLE 18.1	*The Taxonomic Categories*	
	Examples	
	Dicot	**Monocot**
Kingdom	Plantae	Plantae
Division	Anthophyta	Anthophyta
Class	Magnoliopsida	Liliopsida
Order	Fabales	Liliales
Family	Fabaceae (Leguminosae)	Liliaceae
Genus	*Lupinus*	*Hymenocaulis*
Species	*Lupinus texensis*	*Hymenocaulis caribaea*
Common name	Texas bluebonnet	Spider lily

The scientific name of a species is always a binomial, consisting of the genus and species. The species epithet is often descriptive: *Lupinus texensis* occurs in Texas; other common species epithets are "longifolia," "grandiflora," "vulgare" (common), "acaulis" (rosette, without a stem), and so on.

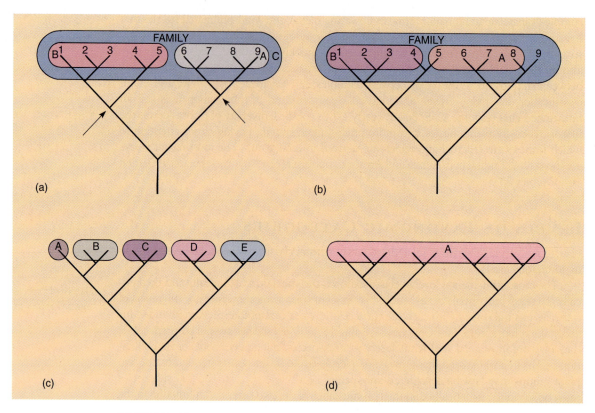

FIGURE 18.5

(a) In this diagram two natural genera, A and B, have been designated. Both are natural because all members of each can be traced back to common ancestors *(arrows)*. The two genera would constitute a family, C, which would also be natural. (b) Genus A is not natural because it is not monophyletic: species 5 does not share an appropriate common ancestor with the others; there is a common ancestor, but it is in the remote past, at the family level. Technically, it is acceptable to leave species 9 out, making it a monospecific genus (a genus with only one species), but because it is even more closely related to species 8 than are 6 and 7, leaving it out would be inconsistent. Genus B is monophyletic, so it is a natural group, but species 5 should be included for consistency. Even though the genera are either incomplete or polyphyletic, still the family, C, is perfectly monophyletic and complete. (c) The family classified by a "splitter." The concept of a monospecific genus is valid: If genera B and C were extinct, then A should be in a genus by itself because it is not closely related to any other living species. (d) The family classified by an extreme "lumper"; this is also a valuable concept. If rapid adaptive radiation is occurring, then all species may be so similar that no strong distinctions exist to use as guidelines for creating smaller genera.

same genus is difficult. No objective criteria exist; the decision is entirely subjective and often the cause of great dispute. Some taxonomists, generally referred to as "lumpers," believe that even relatively distantly related species should be grouped together in large genera. Other taxonomists, called "splitters," prefer to have many small genera, each containing only a few species that are extremely closely related. For example, some taxonomists believe that cranberries and blueberries are so similar that they should go into the same genus, *Vaccinium;* others think that cranberries are distinct enough that segregating them into their own genus, *Oxycoccus,* more accurately reflects evolutionary reality. Both groups of taxonomists agree that the two sets of species are closely related, but they have different opinions as to how much evolution has occurred since the time of the most recent common ancestor. The critical concern is that the genera are natural, that all the species included in the genus are related to each other by a common ancestor, and that all descendants of that common ancestor are in the same genus. Such a group is **monophyletic** (Fig. 18.5). In an unnatural, **polyphyletic** group, members have evolved from different ancestors and may resemble each other only as a result of convergent evolution.

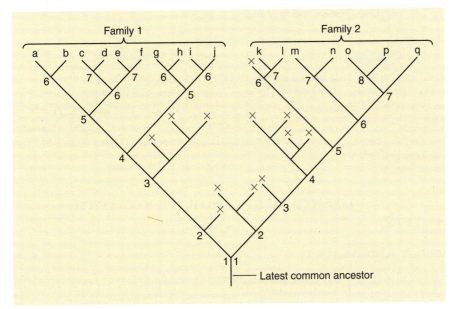

FIGURE 18.6

In this hypothetical evolutionary history, two lines diverged long ago, then many of the early species became extinct (lines marked with x), and only some continued on, speciating into two separate families. In family 1 all extant species have at least six speciation events in their history since the latest common ancestor shared with family 2. Family 2 has undergone slightly more rapid evolution: The minimum number of branch points is seven, and two species have eight. Thus, since these two lines of evolution diverged, many changes have occurred in each, so they are quite distinct. In reality, there are about 400,000 species of vascular plants with unknown hundreds of thousands or millions of extinct species, so the true phylogeny is gigantic. The latest common ancestor for vascular plants probably occurred about 420 million years ago.

The level above genus is **family,** each family being composed of one, several, or often many genera. Most families are well defined, with widespread agreement as to which species and genera belong in a particular family. As examples, consider how easy it is to recognize the following families: cacti, orchids, daisies, palms, and grasses. The reason for this probably is the age of most families. A line of evolution must be very old—usually at least several tens of millions of years—to diversify into several genera and many species. Even closely related families have been separate evolutionary lines for at least 20 or 30 million and often 50 million years. The common ancestor and many of the earliest species probably have become extinct. Therefore, although two families have many characters in common, enough features are unique to each family to make them recognizable as distinct from each other (Fig. 18.6). Problems arise when the earliest species and genera of a family persist; these have characteristics intermediate between the families, so it can be argued reasonably that all should be considered a single family. The term "subfamily" is used for the groups of closely related genera within such a single large family.

The levels above family are **order, class, division,** and **kingdom.** Although you might expect universal agreement at the level of kingdom, even the boundaries of the plant kingdom, kingdom Plantae, are disputed. Some believe that the green algae should be included because they are almost identical biochemically to vascular plants. Others believe that, despite the biochemical similarity, the green algae should be excluded because they are so different morphologically and anatomically.

The classification used in this book is based on that proposed by Drs. H. Bold, C. Alexopoulos, and T. Delevoryas in *Morphology of Plants and Fungi* (1987), which recognizes a large number of divisions. Other botanists have proposed systems with many fewer divisions, with many of the divisions of Bold, Alexopoulos, and Delevoryas considered subdivisions. There is no "correct" view; it is most important to understand the reasons used to support each view.

Except for kingdom, genus, and species, the names must have a certain ending to indicate the classification level. Division names end in *-phyta,* for example Chlorophyta (green algae), Coniferophyta (conifers), and Anthophyta (flowering plants) (Table 18.1). Class names end in *-opsida,* subclass names in *-idae,* order names in *-ales,* and family names in *-aceae* (pronounced as if you were spelling the word "ace"). Genera and species names do not have standard endings. The **scientific name** of a species is its genus and species designations used together and either underlined or italicized; for example, tomato is *Lycopersicon esculentum.* Note that the species name is not "esculentum" but *Lycopersicon esculentum.* Esculentum is the species **epithet,** the word that distinguishes this species only from the other species of the genus *Lycopersicon.* We cannot refer to tomatoes as "esculentum" because that species epithet is used for many different species in different genera: for example, buckwheat—*Fagopyrum esculentum,* and taro—*Colocasia esculenta.*

Zoologists use the term "phylum" instead of division, and "phylum" was used in botany until recently.

501

Some families have two names; the above rules were adopted to regularize family names so that each family is named after one of its genera, using the *-aceae* suffix. The rose family was named Rosaceae, based on the genus *Rosa.* This worked well for most families, but for some, the old name was so well-known and so familiar that it was kept as well. Examples are Asteraceae (Compositae), Fabaceae (Leguminosae, legumes), Aracaceae (Palmae, palms), Poaceae (Gramineae, grasses), Brassicaceae (Cruciferae, mustards), and Apiaceae (Umbelliferae, umbels).

TYPES OF EVIDENCE USED FOR TAXONOMIC ANALYSIS

HOMOLOGY AND ANALOGY

The study of plant phylogeny centers on examining the similarity of one species to others. A species evolves into two species as different populations accumulate distinct alleles; even when enough divergence has occurred to create separate species, the two still resemble each other strongly. As they continue to evolve each acquires its own mutations, and because they cannot interbreed they cannot share the new alleles (Fig. 18.6). Thus, they differ from each other more as time passes. Distantly related plants have been on separate lines of evolution for millions of years, time enough for so many mutations to accumulate that they resemble each other only slightly.

Taxonomy and systematics consist of studying various species and examining their degrees of similarity, but this study is complicated by the fact that plants can resemble each other for two distinct reasons: (1) they have descended from a common ancestor, or (2) they have undergone convergent evolution. Features similar to each other because they have descended from a common ancestral feature are **homologous features.** For instance, almost all members of the anthurium family are easily recognizable because of their spathe and spadix inflorescence (Fig. 18.7). All members have these structures because all have inherited their inflorescence genes from a common ancestor. The spathe and spadix are

FIGURE 18.7

Members of the family Araceae (the aroids, Jack-in-the-pulpit) all have a spathe (the collar-like sheath) and spadix (the central column), as in this *Amorphophallus* (a) and *Anthurium* (b). In all extant species, the genes that control the formation of these structures are descendents of the genes of the original, ancestral species. The genes that exist today have resulted from thousands of rounds of DNA replication during plant growth, crossing-over, syngamy, and then more rounds of DNA replication as the zygote grows. This has been repeated millions of times since spathes and spadices first evolved millions of years ago. During all this, mutations and natural selection have led to obvious varieties of spathes and spadices, but the basic relatedness, the homology, is still obvious.

(a)   (b)

homologous in all *Anthurium* species. Homologous features are the ones critically important for making taxonomic comparisons and the only ones that can be used to conclude that species are related.

The second cause of resemblance, convergent evolution, results when two distinct evolutionary lines of plants respond to similar environments and selection pressures. Under these conditions, natural selection may favor mutations in each line that result in similar phenotypes. Features like this are **analogous features** and should never be used to conclude that plants are closely related. A striking example is the convergent evolution of cacti and the succulent euphorbias discussed in Chapter 17. Both occur in deserts where a succulent water-storing body is advantageous and spines are selectively advantageous by deterring animals from eating the plants to get the water. The two families are not considered closely related on the basis of spines and succulence, however, because these features do not share a common ancestry; they converged evolutionarily because of similar selection pressure.

Determining whether a similarity is due to homology (common ancestry) or analogy (convergent evolution) can be extremely difficult. In the case of spines, it is easy: Cactus spines are modified leaves that are always smooth and never branched, occurring in clusters on an extremely short shoot. Euphorbia spines are modified shoots that branch and have small, scalelike leaves. Being modified shoots, they occur singly in the axil of a leaf, never in clusters. If we look beyond the analogous similarities, the dissimilar features are dramatically different; flowers of cacti are large and have many parts, whereas those of euphorbias are small and unisexual, either staminate or carpellate. The wood of cacti is parenchymatous and has few fibers whereas that of euphorbias is very fibrous and hard. Most important, the plants that most closely resemble the succulent euphorbias in important homologous features are the nonsucculent euphorbias of Africa. Succulent cacti resemble a completely different set of plants, nonsucculent cacti of the tropical regions of the Americas. Thus, the evolution of spines and succulence occurred independently in each group.

A corollary to the assumption that similar plants are closely related is the assumption that dissimilar plants must not be closely related. Studying lack of similarity can also be difficult because in some cases, a small genetic change results in dramatic phenotypic changes (Fig. 18.8). Mutations that affect production, distribution, and sensitivity to hormones result in large changes of the phenotype between two closely related species. Also, mutations that affect early stages of development such as the embryo or bud meristems can

(a)

(b)

FIGURE 18.8

This small, leafless shrub (a, *Acacia aphylla*) may not appear at first glance to have much in common with this larger, leafy tree (b, *Acacia drepanolobium*), but careful examination reveals homologous similarities in many critical features of the flowers and stems which allow assignment to the same genus, *Acacia*. In this case, genes controlling flower and stem morphogenesis have changed little during evolution, whereas genes controlling elongation growth of stems, formation of leaf primordia, and formation of bark have all changed dramatically.

cause closely related species to look deceptively dissimilar. In many evolutionary lines, species of large woody trees have evolved into herbs; the main transition involved is the ability to flower early and then not survive a harsh winter. It must be a simple evolutionary process involving few genes because it has occurred many times. For example, nearly half the species of the Violaceae are herbs in the genus *Viola*—violets and pansies—but the early members of the family were woody trees, as are the members of the relictual, tropical genus *Rinorea*. Artificial selection by humans has resulted in one species that has dramatically different forms: Broccoli, brussels sprouts, cabbage, and cauliflower are all subspecies of a single species, *Brassica oleracea*.

For taxonomic studies, a mosaic of evidence is available—some valid and useful, some misleading. Taxonomists attempt to look at as many features as possible on the assumption that misleading evidence will be outweighed by valid characters. Also, certain features are considered more significant; those that result from numerous metabolic interactions and the influence of many genes tend to change more slowly than those with a simple metabolism controlled by just one or two genes. Loss of features, such as sepals and petals in wind-pollinated flowers, is often rapid because deleterious mutations occur commonly; without the natural selection of insect pollination to eliminate these mutations, they are retained and cause a loss of perianth. Once those genes have become so mutated that a flower lacks sepals and petals, the evolution of new ones would be slow and difficult.

At present, taxonomists study virtually every aspect of plants using a wide variety of tools. Simple observation of major parts is still important, but scanning electron microscopy to study hairs, stomata, cuticle, and waxes is also used. Internal structure is studied with both light and transmission electron microscopy; the nature of plastids in the phloem has been found to be an extremely valuable character (see Fig. 25.28). Various aspects of metabolism are important, ranging from the types of pigments in flowers to the presence of CAM and C_4 metabolism and specialized defensive, antipredator compounds.

DNA sequencing is a new tool for analyzing evolutionary relationships (Fig. 18.9). Two plants are considered separate species only if they differ in significant, heritable ways; therefore, the sequence of nucleotides in the DNA of each species must differ from that of all other species. When plant phenotypes are studied, the actual objective is to determine differences in genotype. In the past, this could be done only by examining phenotypic features and inferring genotype differences if they could not be related to direct environmental effects. Now DNA can be examined directly and mutations can be identified, even if they do not cause any detectable change in phenotype.

Like traditional taxonomic data, DNA sequencing data have their own ambiguities and limitations. Any two species, even those closely related, must differ at many points

FIGURE 18.9
Chloroplast DNA was extracted from plants of three species of *Machaeranthera* (in the sunflower family), then cut into fragments with the restriction endonuclease ClaI. The restriction map of *M. canescens* plastid DNA is almost identical to that of *M. bigelovii*, which has every one of the 56 restriction sites present in *M. canescens* plus one more (blue to left of middle). *M. riparia* is rather distinct: It lacks only one of the sites common to the other two species (near its right end), but it has five that do not occur in the other species (red). Black—present in all three species; blue—present in two species; red—present in only one species. (*Sequence data provided by David Morgan, University of Texas, Austin*)

(nucleotide pairs) of their DNA. Simply counting the number of sequence differences is not enough, because even individuals of a very uniform, homogeneous species may have hidden mutations that do not affect the phenotype. Furthermore, any two species in the same family probably share the great majority of their DNA sequences in identical form, with only a few genes mutated. How many differences are needed to conclude that two genomes represent two distinct species? There is no simple answer. If it were necessary to examine the entire genome to find the few points of difference, the problem would be insurmountable. A quick glance instantly reveals that two anthuriums are more closely related to each other than to a geranium, cactus, or euphorbia, but demonstrating the same thing using DNA techniques would take many weeks and a great deal of money.

TAXONOMIC STUDIES

EXPLORATION AND DISCOVERY

People have been hunting for and discovering new plant types since before recorded history to find not only new food plants but also ornamental plants and beautiful woods (See Plants and People in Chapter 9). Exploration continues today as biologists enter new areas that either have never been explored or have been visited only once or twice (Fig. 18.10). Tropical rainforests of Brazil, Central America, Africa, and Southeast Asia are so poorly explored that any serious expedition returns with previously unknown species of plants and animals. New species of birds are routinely discovered in the mountainous regions of Peru. Most North Americans have difficulty understanding how inaccessible many of these regions are; exploration must often be done on foot, carrying all supplies, and simply walking 3 or 4 miles in one day can be a major accomplishment. Of course, the pace is slower when one stops to examine, photograph, collect, and preserve plants. Field notes must be made as to flower color, plant height, ecology, pollinators, and neighboring plants. At a different season, other plants are in flower or fruit, some plants have died, and the seeds of different species may have germinated; a new exploration of the same area would encounter many new species but fail to find some that had been present during the first exploration. Some plants are less than 1 cm tall when fully grown and have an extremely limited range; it is easy to walk past them without noticing them (Fig. 18.11). On the other hand, the giant old man cacti are too big to overlook, but they grow in only three small valleys in Mexico and thus can easily be missed by explorers (Fig. 18.12). Finally, a species may resemble others so closely that a scientist may ignore it, thinking it is already well known; the fact that it is a new species might be revealed only by careful study of certain details.

Botanical exploration is also concerned with discovery of new facts about already recognized species. No species is made up of absolutely identical individuals, so numerous

FIGURE 18.10
Collecting plants in southern Chile. Doing field work requires great sacrifice and courage and readily exemplifies the heroic nature of botanists. We risk our own lives to bring new knowledge to students, who are such an important part of the lives of professors.

FIGURE 18.12
These old man cacti (*Cephalocereus senilis*) are easily visible, but they occur in so few areas that it was easy for explorers to miss them for years.

FIGURE 18.11
This *Lipanthes* orchid is fully grown; even in full flower, it could be overlooked easily. Photo is twice life size.

FIGURE 18.13

The geographical distribution of the genus of beans (*Phaseolus*) is already well-known in many regards, but certain species are relatively unstudied, especially their ecology and distribution in some parts of Mexico and Central and South America. This and other familiar genera still must be investigated, both to discover new species and to refine our concepts about which species are most closely related. (*Based on data of A. Delgado, Universidad Nacional Autónoma de México*)

FIGURE 18.14

This is part of the Missouri Botanical Garden in St. Louis, one of the most active botanical gardens in the world, sponsoring research by its own scientists and helping both graduate and undergraduate students to do field work. In addition, it organizes many international conferences and projects. The New York Botanical Garden is equally important and has been responsible for fundamental discoveries of plant classification for more than 100 years. A few other particularly large gardens in the United States are the Fairchild Tropical Gardens in Miami, the Huntington Botanical Garden in Los Angeles, and the USDA Plant Introduction Gardens in Miami.

samples must be collected to gather information about the variability of its features as well as its geographical and ecological range (Fig. 18.13). Exploration may be dedicated to gathering either seeds or live material so that the plant can be propagated for research, horticulture, or food. Now that recombinant DNA techniques make it possible to transfer genes from one species to another, live preservation of as many species as possible is particularly important. Although much botanical exploration is carried out by botanists based at universities, a large share, perhaps even more than half, is done by members of botanical gardens and commercial nurseries (Fig. 18.14).

PRELIMINARY STUDIES OF NEW PLANTS

Preliminary studies of newly collected plants include many diverse activities. Specimens must first be identified using diagnostic keys and personal knowledge. An experienced taxonomist often recognizes the family of an unfamiliar species almost immediately; even the genus may be obvious. If so, further identification is usually simple. In an unfamiliar region in which many species are unknown, identification can be more difficult, and numerous characters must be studied and compared to those in published descriptions of the plants of the area.

If the plant can be identified, it may be studied to see if it provides new information about the species. Even if it turns out to be absolutely typical in every way, that too is important. If the specimen cannot be identified (Box 18.1) or if it does not match any description, the taxonomist may suspect that it is a new species. Typically, the specimen is then sent to a specialist experienced with the species most closely related to the specimen.

BOX 18.1 *Identifying Unknown Plants*

When a plant must be identified, it can either be sent to an expert who will recognize it immediately or it can be identified using a **key**, like the one below. Most keys are constructed of pairs of choices (couplets); start at the first choice and determine which description matches your plant; at the end of each line, there is either the name of the species or a number indicating the next set of couplets to use. The first choices are usually quite broad and easy, such as whether the plant is woody or herbaceous; notice that this divides the plants into two completely artificial categories. This is acceptable because the key is used only for identification, not classification. As you proceed through the choices, you finally come to a name, meaning that you have tentatively identified your plant. The key should be accompanied by descriptions of all the species and, if possible, drawings or photos. After arriving at the name in the key, you should turn to the description to see if it matches.

In this key, I have incorporated only the most common plants found in grocery stores, so you may try to identify a plant that is not in this key. If the plant does not resemble any plant in this key, you will soon run into a couplet in which neither description seems appropriate. In real life, this could mean that you have discovered a new species or a known species growing in an area where it had not been known to exist. Unfortunately, it most often means that you have made a mistake somewhere in the keying process.

Keys in floras or monographs usually are constructed in stages; a preliminary, small key identifies the plant to family or genus, and subsequent keys identify it further to species. As you go to each new key, you can check the description of the family or genus to see if your keying has been correct to that point. With some experience and knowledge of the plants of an area, most people can recognize the family immediately and perhaps even the genus, so they can skip the first keys and go directly to the appropriate level. The key below is really three keys: The first distinguishes between fruits and vegetables, and the other two are separate keys for fruits and vegetables; most of you should be able to identify your specimen as a fruit or a vegetable without having to go through a key. On the other hand, mushrooms present a problem. You do not know how I would treat a mushroom—it is reproductive like a fruit but has not developed from an ovary or associated structure; you have to use the first key for peculiar specimens like this.

As you use the key, keep in mind that you may have selected a plant not available in my store, so it will not be in the key.

Key to Fruits and Vegetables Commonly Found in Grocery Stores

1. Item is neither a fruit nor a false fruit; instead it may be a stem, flower, petiole, root, or fungus; a vegetable 2
1. Item is a fruit or false fruit, having developed from ovaries or associated structures 18

Vegetables

2. Vegetable is a fungus . . . . mushrooms
2. Vegetable is not a fungus 3
   3. Vegetable is a root 4
   3. Vegetable is not a root 7
      4. Vegetable is much longer than wide, gradually tapered, orange in color . . . . carrot
                                                                 Daucus carota
      4. Vegetable is about as wide as long, not orange 5
         5. Vegetable is small, about 2 cm in diameter, definitely less than 4 cm in diameter; red . . . . radish
                                                          Raphanus sativas
         5. Vegetable is not red; much larger than 2 cm in diameter 6
            6. Vegetable is dark purple or purple-red; about 6 cm in diameter . . . . beet
                                                               Beta vulgaris
            6. Vegetable is white or off-white, but not purple or red; about 6 cm in diameter . . . . turnip
                                                              Beta rapa
   7. Vegetable is a cluster of tightly compacted flowers, either green or white 8
   7. Vegetable is not a cluster of flowers; may be a shoot or petiole 9
      8. Vegetable is a compact cluster of white flowers and stalks; complete inflorescence . . . . cauliflower
                                                 Brassica oleracea variety *botrytis*
      8. Vegetable as above, but green . . . . broccoli
                                                        Brassica oleracea var. *italica*
   9. Vegetable is a shoot 10
   9. Vegetable is a petiole or cluster of petioles 17
      10. Shoot is a bulb, white, off-white, purple or yellow-brown; outermost leaves are dry and papery 11
      10. Shoot is not a bulb and not white; may be composed mostly of leaves or mostly of petioles or mostly of stems 12
         11. Vegetable is a cluster of small white bulbs; outermost leaves are dry, papery, and flakey, surrounding fleshy, pungent leaves . . . . garlic
                                                               Allium sativum
         11. Vegetable is a single bulb, larger than above, most leaves fleshy, only one or two papery . . . . onion
                                                               Allium cepa

BOX 18.1   *Identifying Unknown Plants (continued)*

12. Vegetable is composed almost entirely of leaf blades, both the stem and petioles very short 13
12. Vegetable is composed mostly of stem or mostly of petioles 15
   13. Leaves are tubular, base of plant is white with numerous roots . . . . onion
                                                                                            Allium cepa
   13. Leaves are flat, not tubular; without roots when sold 14
      14. Leaves are thin and flat, forming either a loose or a semicompact mass . . . . lettuce
                                                                     Lactuca sativa
      14. Leaves are thicker, fleshy, forming a very compact mass, never loose . . . . cabbage
                                                    Brassica oleracea var. *capitata*
15. Vegetable is composed mostly of stem, with or without small scales 16
15. Vegetable is composed mostly of petioles 17
      16. Stem is green, with small scales, grows above ground . . . . asparagas
                                                              Asparagus officinalis
      16. Stem is brown or tan or reddish, subterranean . . . . potato
                                (if orange with orange flesh, this is a root, yam)
                                                            Solanum tuberosum
   17. Vegetable is almost entirely thick green petioles with just a small amount of highly dissected
       lamina . . . . celery
                                                                 Apium graveolens
   17. Vegetable is mostly white petioles with some green; lamina entire . . . . bok choy
                                                          Brassica chinensis

Fruits, including false fruits

18. Fruit is considered fruit even by most lay people, mostly sweet and edible raw; some sour, acidic 19
18. Fruit is not sweet, commonly considered vegetables by lay people 34
   19. Fruit is sour, very tart 20
   19. Fruit is edible even when fresh, but also used in cooking; many are very sweet 22
      20. Fruit is a compound false fruit, yellow-brown with many scales; usually sold with leaves attached to top . . . . pineapple
                                                                 Ananas comosus
      20. Fruit is simple, not compound, not false fruit 21
         21. Fruit is yellow . . . . lemon
                                                                  Citrus limon
         21. Fruit is green . . . . lime
                                                            Citrus aurantifolia
22. Fruit has an inedible rind or peel that must be removed before the endocarp is eaten 23
22. Entire fruit is often eaten, although some people may peel some of these 29
   23. Fruit is small, each one is usually consumed by one person 24
   23. Fruit is larger, not customarily eaten by one person 26
      24. Fruit is longer than wide, a curved yellow cylinder . . . . banana
                                                                 Musa paradisiaca
      24. Fruit is spherical or slightly flattened 25
         25. Fruit is orange . . . . orange
                                                                 Citrus sinensis
         25. Fruit has sort brown hairs . . . . kiwi fruit
                                                            Actinidia chinensis
26. Rind is extremely hard, like rock . . . . coconut
                                                                 Cocos nucifera
26. Rind is not so hard, can be cut by knife 27
27. Fruit is very large, longer than wide; flesh is red with black seeds . . . . watermelon
                                                                 Citrullus vulgaris

27. Fruit is smaller, more or less spherical, flesh never red 28

  28. Rind is smooth, flesh green . . . . honeydew melon
  Cucumis melo

  28. Rind is netted, flesh orange . . . . cantaloupe
  Cucumis melo var. *cantalupensis*

29. Fruits are sold as clusters of fruits attached to infructescence stalks . . . . grapes
  Vitis vinifera

29. Fruits are not attached to each other at time of sale 30
30. Fruits are blue, small, round . . . . blueberries
  Vaccinium corymbosum

30. Fruits are not as above 31
  31. Fruits are red false fruits with "seeds" on outer surface . . . . strawberry
  Fragaria chiloensis

  31. Fruits are not as above 32
  32. Fruit is red, soft, juicy, not sweet, with many seeds . . . . tomato
  Lycopersicon esculentum

  32. Fruit is not as above 33
    33. Fruit is variable, typically yellow/brown and pear-shaped, but may be greenish and round; always with
    small nests of stone cells that give the flesh a gritty texture . . . . pear
    Pyrus communis

    33. Fruit is variable, often similar to above, but not pear-shaped and never with nests of stone cells; apex of
    fruit (opposite stalk) typically retains five sepal tips and often has withered stamens also . . . . apple
    Malus sylvestris

34. Fruit is not much longer than wide, spherical or oval 35
34. Fruit is more than three times longer than wide 38
35. Fruit is hollow . . . . bell pepper
  Capsicum annuum

35. Fruit is not hollow 36
  36. Fruit is green, solid, oily with one large brown seed . . . . avocado
  Persea americana

  36. Fruit is not as above, is red or purple 37
    37. Fruit is red . . . . tomato
    Lycopersicon esculentum

    37. Fruit is purple . . . . eggplant
    Solanum melangena var. *esculentum*

38. Fruit has many prominent seeds and a cob . . . . corn
  Zea mays

38. Fruit is without a cob 39
  39. Fruit is narrow, flattened, less than 1 cm wide . . . . string bean
  Phaseolus vulgaris

  39. Fruit is much wider than 1 cm 40
    40. Fruit is yellow . . . . summer squash
    Cucurbita pepo var. *melopepo*

    40. Fruit is not yellow 41
      41. Fruit has small white, raised points; stalk less than 0.5 cm broad . . . . cucumber
      Cucumis sativus

      41. Fruit has smooth green and yellow skin, without raised points, stalk at least 1 cm broad . . . . zucchini
      Cucurbita pepo var. *medullosa*

Brittonia, 30(2), 1978, pp. 233–238.
© 1978, by the New York Botanical Garden, Bronx, NY 10458

TWO NEW TAXA OF SEDUM SECTION GORMANIA (CRASSULACEAE) ENDEMIC TO THE TRINITY MOUNTAINS IN CALIFORNIA

MELINDA F. DENTON

Denton, Melinda F. (Department of Botany, University of Washington, Seattle, WA 98195). Two new taxa of *Sedum* section *Gormania* (Crassulaceae) endemic to the Trinity Mountains in California. Brittonia 30: 233–238. 1978.—Two taxa, narrowly endemic to the Trinity Mountains in northwestern California, are described. **Sedum laxum** subsp. **flavidum** is a tetraploid (*n* = 30) and is found on metabasaltic outcroppings, and **Sedum obtusatum** subsp. **paradisum** is a diploid (*n* = 15) and occurs on granite outcroppings.

The taxa included in *Sedum* section *Gormania* are geographically restricted to western North America. The plants occur on exposed rock outcroppings at elevations of 180 to nearly 11,000 feet from the central part of the Sierra Nevada and the North Coast mountains in California northward to the Siskiyou Mountains in southwestern Oregon and along the Oregon Cascades to Mt. Hood, Oregon. The taxa are characterized by horizontal rhizomes, prominent sterile shoots with leaf rosettes, thick and usually glaucous leaves, and petals erect at the base for at least one-tenth of their length. In a recent taxonomic treatment of section *Gormania*, Clausen (1975) recognized 11 taxa. Based on current field, greenhouse, and laboratory studies, I accept ten of Clausen's taxa (*S. laxum* subsp. *latifolium* is synonymized with *S. laxum* subsp. *laxum*) and describe in this paper two new ones endemic to the Trinity Mountains in northwestern California. In an effort to be consistent with Clausen's treatment of *Sedum* (1975) and to recognize only one infraspecific category, the subspecific rank, rather than the varietal one, has been utilized.

Sedum laxum (Britton) Berger subspecies **flavidum** Denton, subsp. nov. (Fig. 1)

Herba perennis glabra parce glauca; caulis erectus 10–24 cm altus; folia basalia rosulata obovata 2–3.5-plo longiora quam lata, 15–25 mm longa, 4.5–10 mm lata; folia caulina obovata tam longa vel 1.2-plo usque longiora quam lata, 7–8 mm longa, circa 7 mm lata; sepala 2.5–4 mm longa, 1.4–2 mm lata, apice obtusa; petala 7.5–10 mm longa, 1.7–2.6 (–3.0) mm lata, flavida vel raro alba nervo medio rosro notata, basi per 1–2 mm connata; antherae flavae vel rubro-brunneae, circa 1 mm longae; *n* = 30.

Perennial, glabrous and slightly glaucous; stem erect, 10–24 cm tall; leaves of basal rosette obovate, 2–3.5 times longer than wide, 15–25 mm long, 4.5–10 mm wide; cauline leaves obovate, as long as wide or up to 1.2 times longer than wide, 7–8 mm long, ca 7 mm wide; sepals 2.5–4 mm long, 1.4–2 mm wide, the apices obtuse; petals 7.5–10 mm long, 1.7–2.6 (3.0) mm wide, pale yellow or seldom white with pink midveins, connate 1–2 mm at the base; anthers yellow to reddish brown, ca 1 mm long; *n* = 30.

TYPE: UNITED STATES. CALIFORNIA: Trinity Co., 4.5 mi NE of Forest Glen along Post Creek, or 1.5 mi NE of intersection of highway 36 and Post Creek Road, narrow canyon with small creek and metabasaltic slopes, elev. 3000 ft, zone of *Pinus sabiniana*, *Quercus kelloggii*, and *Quercus chrysolepis*, plants in bud or flowering or starting to fruit, locally abundant in rock crevices or on ledges, 23 Jun 1976, *Denton 3953* (HOLOTYPE: WTU).

BRITTONIA 30: 233–238. April–June, 1978.

233

FIGURE 18.15
The description of *Sedum laxum* subspecies *flavidum*. (*Reprinted with permission from* Brittonia *30(2), 1978, published by the New York Botanical Garden*)

Declaring a plant to be a new species is easy; proving it and describing it properly are difficult. Anyone can declare a plant to be a new species, name it, and write about it, even if it is obviously not a new species; this has caused countless problems. To overcome this, taxonomists from all over the world have established an **International Code of Botanical Nomenclature** that describes precisely the steps necessary for naming a new species. A valid name, one never previously used, must be declared and must be accompanied by a detailed description of the species in Latin and usually also in English, French, German, Spanish, or Russian. The name and description must be published in a widely circulated journal (Fig. 18.15), a step that prevents many problems. The journal's editors send the description to at least two independent specialists to verify that it is a previously unknown species and that the name has never been used before. The description must also include the designation of a **type specimen;** this is a single preserved plant that truly carries the name (Fig. 18.16).

When new species are named, very little is known about them; as more research is carried out in the following years, enough variation may be discovered to warrant the recognition of a second species. For example, the range of leaf sizes, types of trichomes, and types of nectaries may warrant classification into two or three species. The original type specimen determines which type of leaf size, trichome, and nectary goes with the first name. When doubt arises as to which type of plant goes with which name, the type specimen must be checked. Herbaria tend to be extremely generous and willingly ship thousands of specimens to various botanists whenever a new project is started to study a genus or family. Type specimens, however, are often kept in special fireproof cabinets and are not allowed out of the herbarium. During World War II, Allied bombing destroyed the herbarium in the Berlin Botanic Garden, and thousands of type specimens burned to ashes. Germans had been very active in plant exploration during the first part of this century and much of the 1800s, and the loss of those type specimens has caused tremendous problems. To prevent a recurrence of this disaster, other specimens, as similar as possible to the type specimen, are sent to many herbaria around the world; these are **isotypes.**

(a)

(b)

FIGURE 18.16
The type specimen of *Sedum laxum* subsp. *flavidum* (a) and a plant growing in the field (b). (*Both courtesy of M. Denton, University of Washington*)

BIOSYSTEMATIC AND EXPERIMENTAL STUDIES

As species are discovered and named, they are assigned to families and genera according to the information then known about the specimen and the concept of the family or genus to which it is assigned. But the discovery of a new species must change our concept of the genus, family, and perhaps even the order to which it is assigned, because the species represents new information.

Similarly, new studies on recognized species can change our concept of them and of how they are classified. They may be found to have a particular wood structure or a class of defensive compounds which is unknown in the family in which they are classified but does occur in other groups. There is then the problem of deciding whether the newly discovered features are the result of convergent evolution or the species has been misclassified. Over a period of years, problems such as this accumulate in many groups, and periodically some-one decides the group needs a detailed study of all its members and of other related groups. Specimens are requested from many herbaria, and perhaps field trips are undertaken to gather either new specimens or more accurate ecological, geographical, and pollination information. Journals are searched for studies of anatomy, physiology, and anything else that may be useful. New studies may be undertaken, especially using newer techniques of scanning electron microscopy, phytochemistry, plastid DNA sequencing, rRNA sequencing, and so on, that were not available when the original descriptions were written. Often, data about species and characters are so abundant that computer techniques are used to evaluate and sort them and suggest possible phylogenies.

Such studies often take many years because of the large number of factors being analyzed. In the end, large genera or families may be divided into several smaller ones if the scientist is convinced that there are several major lines of evolution rather than one. On the other hand, he or she may decide to combine genera into a more broadly defined genus, or a few species may be transferred from one genus to another, perhaps even to a different family. Frequently, these studies discover that certain names are not valid: The name had been used previously for a different species or had not been validly published because no type specimen had been designated or Latin description written. Surprisingly often, a particular commonly used scientific name is found never to have been published at all (Fig. 18.17).

(a)

(b)

FIGURE 18.17

(a) This cactus species is currently named *Ariocarpus retusus*, but its name has changed through the years. It was named *Anhalonium prismaticum* in 1839, but later studies (1898) of the literature revealed that the species had actually been named *Ariocarpus retusus* before 1839; as the first name published, it had precedence and the name *Anhalonium* had to be abandoned. The complete list of synonyms used since 1839 includes 17 names. (b) *Ariocarpus furfuraceus*. Some taxonomists believe that plants with this phenotype constitute a subspecies of *Ariocarpus retusus*. Others believe they are closely related but separate species. *(Photographs courtesy of C. Glass and R. Foster, Abbey Garden Press)*

The results of the study are published as a major report, usually called a **monograph.** Naturally these vary in quality, but a good one has extensive summaries of all pertinent data, maps of geographical distribution, names and numbers of the herbarium specimens examined, and carefully prepared documentation of nomenclature. Any change in the scientific epithet of a species, either genus or species name, must be declared and explained, and a complete list of the previous names must be presented. For most species, few studies have resulted in name changes, so there may be only one or two previous names. But for other species, dozens of previous names are not unusual, reflecting the history of uncertainty about the phylogeny of the species or its group.

Periodically an individual publishes a monograph on all flowering plants or even all vascular plants. The most recent comprehensive monograph of the flowering plants is *An Integrated System of Classification of Flowering Plants* by Dr. Arthur Cronquist of the New York Botanical Garden. It presents both classes of flowering plants (see Chapter 25) and all the subclasses, orders, and families, giving diagnostic features of each, comparing each with its relatives, and discussing the major events of evolution in the flowering plants. The books *Paleobotany: An Introduction to Fossil Plant Biology* by Drs. Thomas N. Taylor and Edith L. Taylor and *Paleobotany and the Evolution of Plants* by Drs. Wilson N. Stewart and Gar W. Rothwell describe the major groups of extinct plants and discuss all the known main lines of plant evolution since plants become distinct from algae about 400 million years ago.

THE MAJOR LINES OF EVOLUTION

All organisms are grouped into five kingdoms: kingdom Monera (bacteria and cyanobacteria), kingdom Protista (algae and protozoans), kingdom Myceteae (fungi), kingdom Animalia, and kingdom Plantae. The most significant event in evolution was the origin of life itself, probably about 3.5 billion years ago. The first organisms were simple, consisting of a cell membrane, protoplasm, and DNA but no distinctive nucleus or other membrane-bounded organelles. Such organisms, either living or extinct, are prokaryotes and are classified in kingdom Monera. This line of evolution diversified rapidly into numerous lines, as evidenced by the presence of thousands of living species. A significant step in prokaryote evolution was the development of a type of photosynthesis that liberates oxygen and is based on chlorophyll *a*. We do not know exactly which species of bacterium this occurred in, but their descendents are known as cyanobacteria rather than bacteria, and one or several were the ancestors of chloroplasts.

The next major evolutionary event was the conversion of a prokaryote into a eukaryote, having a membrane-bounded nucleus. This must have been an extremely gradual procedure with many intermediates, because many living species still have characteristics intermediate between prokaryotes and eukaryotes (see Fig. 21.5). A significant aspect was the formation of mitochondria, perhaps by endosymbiosis, with a bacterium living inside an early eukaryote (see Fig. 21.3).

Organisms with an organization similar to that of early eukaryotes are classified in kingdom Protista (see Chapter 21). This classification is not satisfactory because the included members are extremely diverse; clearly they are not closely related and did not evolve from a common ancestor. The criterion for placing an organism in kingdom Protista is simply that it is a eukaryote but not a plant, animal, or fungus, although the immediate ancestors of plants, animals, and fungi are classified as protists.

Once eukaryotes evolved to the level of having mitochondria, endoplasmic reticulum, and a true nucleus, numerous evolutionary lines emerged. In one, mutations and natural selection resulted in the metabolism that produces a cell wall of chitin; these became the fungi. A second line of evolution, resulting in the animals, could be characterized as having a developmental system that permits the formation of complex multicellular bodies with distinctive tissues and numerous types of highly differentiated cells. In fungi, the bodies may be multicellular, but they are rarely complex and have only rudimentary tissues, if any.

A third group of eukaryotes apparently established a symbiosis with cyanobacteria, which then evolved into chloroplasts, producing the first algal cells. These continued to diversify, and about 400 million years ago, some became adapted to living on land, establishing the plant line of evolution, kingdom Plantae. From these early pioneers, the major lines that developed were the simple vascular plants that do not produce seeds (ferns and similar plants; see Chapters 22 and 23) and the seed-bearing vascular plants (gymnosperms and angiosperms; see Chapters 24 and 25). The plants around us today represent thousands of evolutionary lines descended from early land plants, which in turn were derived from algae that evolved from advanced bacteria that descended from primitive bacteria. Our classification system does not yet have every step identified and not all relationships are clearly understood, but the major features appear to be consistent with the data, and we expect that only relatively minor changes will be made as our knowledge increases.

SUMMARY

1. One of the goals of classification and nomenclature is to give each species a single unique name.

2. The classification system being developed at present is a natural one, attempting to reflect the actual evolutionary relationships—phylogeny—of all species.

3. Artificial classification systems are now used only to identify plants or to categorize useful features but are never the basis for naming species.

4. Closely related species are classified in one genus, related genera are grouped into families, and so on through orders, classes, divisions, and kingdoms.

5. Each taxonomic category should be monophyletic, all the organisms in it having evolved from the same ancestral group. Groups discovered to have had two or more distinct origins are the result of convergent evolution; they are polyphyletic and the category is not a natural one.

6. Homologous characters are those that have evolved from the same ancestral character, whether they now are similar to each other or not; these can be used as guides to phylogeny and the construction of a natural classification system.

7. Analogous characters are those that resemble each other but have arisen independently; they are a part of convergent evolution and often cause confusion and misclassification because they appear to be homologous characters.

8. When a new species is declared, it must be given a name never before used, and it must be accompanied by a type specimen and a complete Latin description published in a widely circulated journal. The internationally recognized rules are occasionally not followed, leading to an invalid name.

IMPORTANT TERMS

analogous features
artificial system of classification
binomial system of nomenclature
family
form genus
genus

homologous features
inheritance of acquired characteristics
key
monophyletic group
natural system of classification
phylogeny

polyphyletic group
scientific name
species
type specimen

REVIEW QUESTIONS

1. Of all taxonomic categories, only species has a rather objective definition. What is it? How does a taxonomist decide whether two plants are members of the same species or two closely related species?

2. Give definitions for homology and analogy. Assume that you encounter two species that have many features in common, but you suspect that they are not closely related. How would you investigate whether the similarities are analogous or homologous?

3. What are type specimens? Why are they important?

4. Imagine that two completely unrelated species in very different families were accidentally given the same name. What kind of confusion would this cause? What if both were separately the subject of several scientific studies, and results were published in papers that did not mention the family name. Would it be possible to know which study concerned which species?

5. Imagine that you have discovered a plant with a chemical that is quite effective against cancer. If you wanted to find a species that produced even more of the compound or perhaps an even more effective compound, would you test just any species at random or a closely related species, such as the other species of the same genus in a natural classification system?

BotanyLinks
www.jbpub.com/botanylinks

BotanyLinks includes a Directory of Organizations for this chapter.

19

KINGDOM MONERA: PROKARYOTES

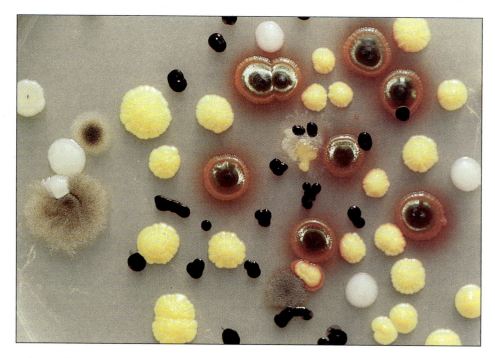

Colonies of bacteria can be grown in laboratory culture for study and analysis. (*William E. Ferguson*)

CONCEPTS

Most prokaryotes are familiar to us as the well-known **bacteria;** a second group, **cyanobacteria,** formerly known as blue-green algae, is less familiar, although they are common in sea water, soils, and fresh water polluted with sewage. Many of the "bacteria" that live in unusual environments, such as hot pools and acidic habitats without oxygen, are unique in numerous fundamental aspects; they are now classified as **archaebacteria.** Prokaryotes are so different from eukaryotes in aspects of structure, metabolism, reproduction, and ecology that they are placed in their own kingdom, kingdom Monera (Tables 19.1 and 19.2).

Since prokaryotes are not plants, why are they included in a botany textbook? One reason is tradition: Until recently, cyanobacteria were thought to be at least somewhat related to algae and were, therefore, studied by botanists. In the last several years, studies of their nucleic acids, metabolism, and ultrastructure have shown that cyanobacteria are not at all closely related to algae but rather are true prokaryotes. Even today, however, they are as frequently studied by botanists as by bacteriologists.

A much more important reason for botanists to be familiar with prokaryotes is that they allow us to perform comparative studies with plants. By comparing and contrasting plant biology and prokaryote biology, we can learn much about plants themselves. In many

TABLE 19.1	*Traditional Classification of Prokaryotes*

I. Prokaryotes
   A. Kingdom Monera
      1. Division Bacteria
      2. Division Blue-green algae (Cyanophyta)
II. Eukaryotes
   A. Kingdom Protista
   B. Kingdom Myceteae
   C. Kingdom Plantae
   D. Kingdom Animalia

Cyanobacteria were thought to be related to true algae, but studies of their DNA, ribosomes, and other features have shown that they are more similar to bacteria. These subdivisions of kingdom Monera are no longer believed to reflect accurately the evolutionary lines of prokaryotes.

TABLE 19.2	*One of Several Current Classifications of Prokaryotes*

I. Prokaryotes
   A. Kingdom Monera
      1. Division Eubacteria
         a. Section Cyanobacteria (plus approximately 32 other sections)
      2. Division Archaebacteria

ways, all plants are so similar to each other that it can be impossible to know whether certain features are alike because they are related evolutionarily or because no alternatives are possible for that feature.

Consider photosynthesis. If the photosynthetic mechanisms of mosses, ferns, conifers, and flowering plants are compared, virtually no differences are found. All plants contain the same chlorophylls, accessory pigments, and electron carriers, and the photosynthesis occurs in chloroplasts and liberates oxygen. Even C_4 metabolism and CAM are merely adjuncts to the regular light-dependent reactions and stroma reactions. It is both logical and reasonable to suspect that this particular photosynthetic mechanism is the only possible one; alternative types either must be theoretically impossible or selectively disadvantageous.

However, in prokaryotes we find not just one type of photosynthesis, but actually five. Two types are similar to that which occurs in plants, but three are remarkably different, having different types of chlorophyll and different electron carriers, and some produce sulfur, not oxygen, as a by-product (Fig. 19.1). Knowledge of these alternative types of

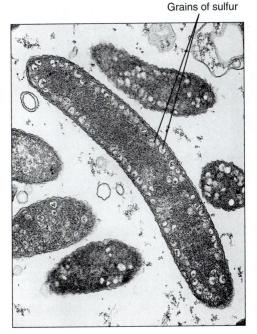

Grains of sulfur

FIGURE 19.1

Photosynthetic bacteria differ from plants in having completely different mechanisms of photosynthesis—none has chlorophyll *a* and none produces oxygen. Instead, hydrogen sulfide is the electron donor and sulfur is the waste product. *Rhodospirillum rubrum* ($\times$ 30,000). (*H. S. Pankratz and R. L. Uffen/ Biological Photo Service*)

prokaryotic photosynthesis immediately places plant photosynthesis in a new perspective. Is one type most advantageous selectively under certain conditions? Could plant photosynthesis be improved if it used bacterial chlorophylls or bacterial electron carriers? If so, could the genes for the synthesis of those bacterial molecules be inserted into plants by genetic engineering?

Respiration provides another example: Many prokaryotes thrive under conditions of low oxygen or even no oxygen. Only a few plants can do this, although many plant parts find themselves in oxygen-deficient conditions. It seems that the prokaryotes' ability to carry on active metabolism without oxygen should be strongly advantageous. Why then do plants lack it? Do theoretical limits prevent large multicellular organisms from surviving without oxygen? Or is it just a matter of bad luck—the necessary mutations either never occurred in plants or were never selected?

Prokaryotes must also be studied because they are valuable economically. They cause numerous diseases of plants and animals—cucurbit wilt, potato ring rot, diphtheria, and tuberculosis being just a few examples. On the other hand, prokaryotes, especially actinomycetes, provide many antibiotics, our most important medicines against bacterial disease (Fig. 19.2). Prokaryotes are everywhere in the environment and play many critical ecological roles. Some convert sulfur and atmospheric nitrogen (which plants cannot use) into sulfates and nitrates, compounds essential for plants. Prokaryotes, along with fungi, are agents of decay: When plants and animals die, microorganisms attack their bodies enzymatically. Molecules containing nitrogen, sulfur, and other minerals are released and become available to living plants (Fig. 19.3).

Finally, prokaryotes, especially bacteria, are increasingly important in genetic engineering and analysis. When a botanist isolates a particular gene for study, too little is available for chemical analysis and sequencing. But the gene can be inserted into a bacterium, which is then allowed to grow and multiply (see Chapter 15). When the bacterium has reproduced to the level of billions, each with an inherited copy of the plant gene, the botanist can collect the bacteria, extract the gene, and have billions of copies. Alternatively, the botanist may be interested in the protein that the gene codes for; he or she inserts the gene into the bacterium in such a way that the gene is not only duplicated and passed on to progeny bacteria but is also expressed. The protein can then be harvested.

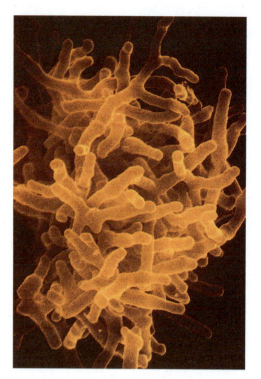

FIGURE 19.2

Although many of the worst human diseases are caused by bacteria, many of our antibacterial drugs are extracted from a particular group of bacteria, the actinomycetes. Many of the diseases do not seem so terrible now, because they last only a few days until we get an injection of antibiotics; but in the days before penicillin, many of these were invariably fatal. The bacteria pictured are in the genus *Actinomyces* (× 14,000). *(David M. Phillips/Visuals Unlimited)*

FIGURE 19.3

After death, organisms can no longer counteract microorganisms and decay begins. The process is complex, involving not only bacteria and fungi, but also small animals that chew dead material into smaller pieces, increasing the surface area available for decay by microbial enzymes. Without microbe-mediated decomposition, the nitrogen, phosphorus, sulfur, calcium, and other minerals of the bodies would not be released back into the environment, and growth of new plants would be severely restricted. The mushrooms present (*Polyporus sanguineus*) are fungi; the bacteria involved in the decay are too small to be seen. *(© Charles Palek/Earth Scenes)*

STRUCTURE OF THE PROKARYOTIC CELL

PROTOPLASM

The cells of prokaryotes tend to be quite small, usually only 1.0 to 5.0 μm in diameter, much smaller than any plant cell. Bacteria known as mycoplasmas are even smaller, some being spheres only 0.2 μm across, whereas certain cyanobacterial cells are up to 60 μm long.

Bacteria that lack walls are **pleomorphic**; that is, they have no permanent morphology but change shape as conditions vary. Wall-less bacteria are rare, however, and almost all prokaryotes have walls that hold them into one of three basic forms: rods **(bacilli)**, spheres **(cocci)**, or long coils **(spirilla)** (Fig. 19.4). Bacteria and archaebacteria are invariably unicellular; small clusters of cells may stay together tenuously after cell division if the medium is not agitated, but within the cluster, each cell is an individual and no intercellular connections occur. Consequently, bacteria never have differentiated tissues and organs.

Cyanobacteria have a mucilage sheath strong enough to hold long filaments together, and the entire filament acts as a single individual (Fig. 19.5). Within it, specialized cells, **heterocysts**, fix nitrogen and pass it to surrounding cells through **microplasmodesmata**, fine holes in their walls. When environmental conditions trigger reproduction, certain cells in the filament form large holes in their walls and then die; the walls and mucilage sheath tear at these spots, so the filaments are broken into multicellular fragments, each capable of growing into a new filament (Fig. 19.6). If environmental cues stimulate dormancy, some cells form a thick wall and become **akinetes**, resistant spores. However, this seems to be the maximum amount of body-level differentiation in prokaryotes.

The singular forms of these are bacillus, coccus, and spirillum.

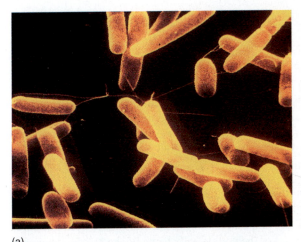

(a)

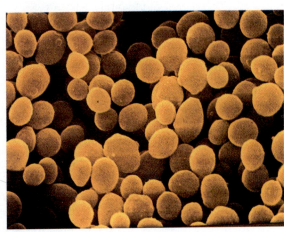

(b)

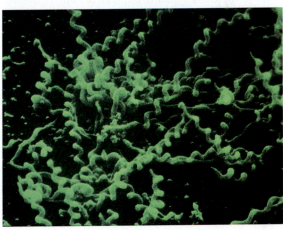

(c)

FIGURE 19.4

All bacteria and archaebacteria with walls have a constant, simple shape; large numbers of species are short rods (a), usually less than 5 μm long; many are spheres (b); and some have a spiral shape (c). In quiet habitats with little motion, cells may form short chains, but they are usually so weakly attached to each other that microscopic currents break the chains into individual cells. (*David M. Phillips/ Visuals Unlimited*)

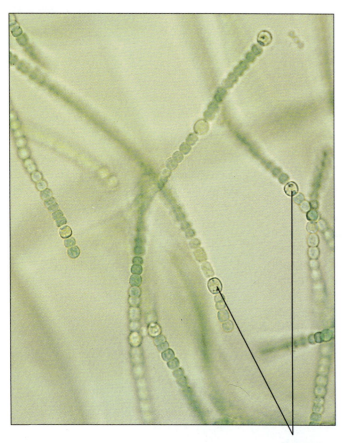

Heterocysts

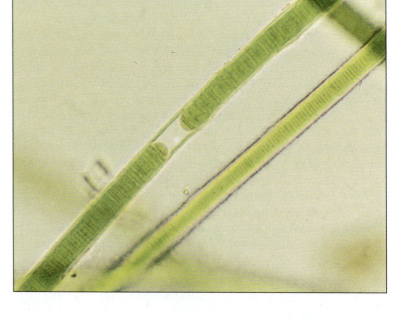

FIGURE 19.5

Cyanobacteria, such as this *Nostoc*, are unique among prokaryotes in that some have truly multicellular bodies that may contain two types of cells. Here, the large round cells are heterocysts, specialized cells that fix nitrogen (N_2) into organic compounds. Once fixed, the nitrogen compounds are transported to surrounding cells through microplasmodesmata ($\times 350$).

FIGURE 19.6

Long filaments of cyanobacteria may be broken into smaller pieces by disturbances in their medium, but some also have a self-controlled means of breaking up: Certain cells, "hormongonia," die, causing the filament to tear at this spot. This is not simply accidental death—the breaks occur with a regular spacing, producing fragments of a characteristic size. *Lyngbya* ($\times 400$). (*Bruce Iverson*)

No nuclear envelope separates DNA from the rest of the prokaryotic cell; instead DNA occurs as one or several masses within the cytoplasm (Fig. 19.7). These are often referred to as "nuclei," but the term **"nucleoid"** or "nuclear region" is less confusing. DNA in nucleoids is in the form of closed circles, one end of the double helix being hooked to the other end. No histones or nucleosomes are present, so the DNA is said to be "naked." The numerous negative charges on the nucleotide bases are neutralized by Mg^{2+} or Ca^{2+} ions.

In nonphotosynthetic prokaryotes, the rest of the cell is filled with cytoplasm, and in electron micrographs it appears quite homogeneous, lacking mitochondria, plastids, dictyosomes, endoplasmic reticulum, and a central vacuole. Photosynthetic prokaryotes have an extensive invagination and folding of the plasma membrane, and this folded region is the site of photosynthesis, much like a chloroplast. However, this set of membranes is embedded in the cytoplasm rather than enclosed by an outer envelope, as a chloroplast is.

Ribosomes are present and have the same appearance as eukaryotic 80S ribosomes; however, when extracted from cells and subjected to centrifugation, they settle at a different rate and therefore are called **70S ribosomes.** They have the same function as eukaryotic ribosomes, but the actual rRNA molecules differ somewhat, as do the proteins. Many antibacterial drugs exert their effect by interfering with ribosome action. The most important ones interrupt only 70S ribosomes and do not affect 80S ribosomes; examples are streptomycin, neomycin, and tetracycline. Other drugs affect only 80S ribosomes (cycloheximide), and some affect both (puromycin).

Photosynthetic prokaryotes such as cyanobacteria as well as the purple and the green photosynthetic bacteria must be buoyant to stay in the upper lighted layers of ponds and lakes, the **photic zone** (Fig. 19.8). These organisms contain a **gas vesicle** that provides buoyancy (Fig. 19.9). Whereas vesicles in eukaryotes are surrounded by a membrane of lipids and proteins, the gas vesicle is surrounded instead by a sheath of pure protein in a semicrystalline arrangement. The gas vesicle has parallel sides and pointed ends; it is more similar to an enlarged microtubule than to a eukaryotic vesicle.

Inclusions of various types are deposited in many prokaryotic cells, and they provide some information about the metabolism. Bacteria that metabolize sulfur often contain particles of sulfur (see Fig. 19.20). Nitrogen-fixing cyanobacteria develop granules of polymerized aspartic acid and arginine as a storage form for reduced nitrogen. Other bacteria contain granules of glycogen, a glucose polymer similar to starch and useful for storing energy. Particles of an unusual compound, poly-beta-hydroxybutyric acid, are a means of storing energy and reduced carbon in certain species.

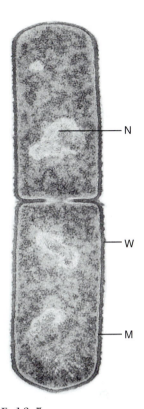

FIGURE 19.7
The central light region is the mass of DNA, the nucleoid (N). In many bacteria, each nucleoid contains many circles of DNA, not just two; note that there is no nuclear envelope. The number of circles and nucleoids may vary with age and vigor. M = cell membrane; W = wall. (*Dr. A. Ryter*)

FIGURE 19.8
The uppermost regions of ponds, rivers, and oceans are the photic zone: Enough light is present to permit photosynthesis. Photosynthetic organisms, bacteria and algae, must be able to float in this layer; when conditions of temperature and mineral nutrients are optimal, the growth can be so great that a thick mat of organisms forms. (*Courtesy of J. Robert Waaland, University of Washington*)

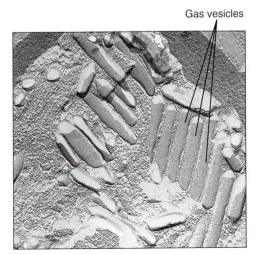

Gas vesicles

FIGURE 19.9
Prokaryotes adjust their density with the formation of gas vesicles; unlike vesicles of eurkaryotes, these are not surrounded by a lipid-protein membrane; instead, the surface is just a layer of protein monomers. The gas vesicles are basically a crystalline shell of proteins surrounding a gas bubble. *Nostoc muscorum* (× 60,000). (*Courtesy of J. Robert Waaland, University of Washington*)

CELL WALL

Eubacteria (or "true bacteria," comprising both bacteria and cyanobacteria) are divided into two groups, gram-positive and gram-negative, depending on their cell's reaction to the Gram's stain, developed by the Danish physician Christian Gram. The difference is related to wall structure, and we now know that in virtually all prokaryotes except archaebacteria, only two types of wall exist.

Cell Wall of Gram-positive Prokaryotes.   In the gram-positive cell, the wall consists of a thick, 15 to 80 nm, layer of a polymer called **peptidoglycan** (Fig. 19.10). This is an unusual polymer, but it basically has a physical structure rather similar to that of a plant cell wall. One component is a polysaccharide made up of two sugars, *N*-acetylglucosamine

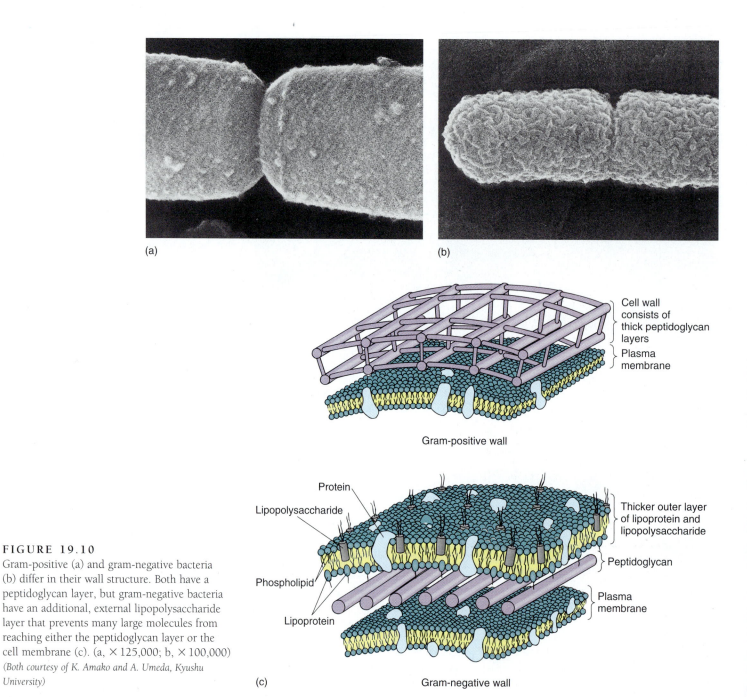

(a)   (b)

Cell wall consists of thick peptidoglycan layers

Plasma membrane

Gram-positive wall

Protein

Lipopolysaccharide

Thicker outer layer of lipoprotein and lipopolysaccharide

Peptidoglycan

Phospholipid

Plasma membrane

Lipoprotein

(c)   Gram-negative wall

FIGURE 19.10

Gram-positive (a) and gram-negative bacteria (b) differ in their wall structure. Both have a peptidoglycan layer, but gram-negative bacteria have an additional, external lipopolysaccharide layer that prevents many large molecules from reaching either the peptidoglycan layer or the cell membrane (c). (a, × 125,000; b, × 100,000)

(Both courtesy of K. Amako and A. Umeda, Kyushu University)

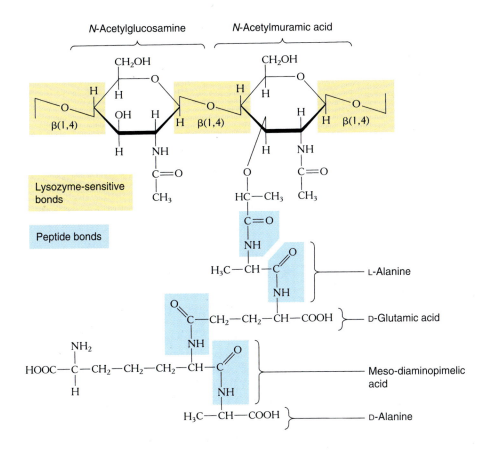

Lysozyme-sensitive bonds

Peptide bonds

FIGURE 19.11

Unlike plant cell walls, which are made up of glucose polymerized into cellulose, the peptidoglycan of bacteria and cyanobacteria is composed of two sugars that carry amino groups. Complex additional groups are added, such as the amino acids shown here. Whereas plant cells cross-link cellulose molecules with hydrogen bonds and sugars, bacteria cross-link theirs with peptide bonds, basically forming a giant carbohydrate-protein molecule.

and *N*-acetylmuramic acid (Fig. 19.11). The disaccharide polymer forms long, strong molecules, but just like cellulose, if molecules cannot be bound together laterally, they cannot form a strong wall. In the walls of plants and algae, cross-linking of cellulose is done by hemicelluloses, but in prokaryotes cross-linking of the acetylglucosamine/acetylmuramic acid polysaccharide is done by small peptide bridges, each containing only a few amino acids. The peptides react chemically with the sugars and are covalently bound to them; every molecule of *N*-acetylmuramic acid can attach to a peptide bridge, so tremendous cross-linking can occur, and the wall becomes one giant molecule. Such a wall is even stronger than the lignin-reinforced walls of plants, strong enough to prevent growth of the protoplast inside. The wall contains enzymes that can break the peptide bonds, keeping the wall just plastic enough to allow growth. Peptidoglycan occurs in almost all prokaryotes but never in eukaryotes; they never make *N*-acetylmuramic acid, nor do they have the unusual amino acid—diaminopimelic acid—that is part of the peptide cross-link (Fig. 19.11). Consequently, prokaryote wall metabolism has numerous steps that are exclusively prokaryotic; certain antibiotics—such as penicillin, which prevents cross-link formation —interfere only with those steps and so have no effect on eukaryotes.

Cell Wall of Gram-negative Prokaryotes. In gram-negative prokaryotes, the peptidoglycan layer is present but is quite thin, only about 10 nm, and a layer of **lipopolysaccharide** (LPS) lies exterior to it (see Fig. 19.10). The LPS layer greatly resembles an ordinary lipid/protein membrane in electron micrographs and is believed to be a lipid bilayer, although the lipids are unusual chemically. Those located on the outer half of the bilayer have polysaccharides attached to them, and these sugars themselves are quite unusual, some having seven carbons and others being dideoxy sugars. Because it is only a thin membrane, the LPS layer cannot contribute strength to the wall. It is now thought that it acts instead as a selectively permeable barrier that keeps many large foreign molecules away from the plasma membrane. The LPS layer also acts to retain many of the cell's own large molecules such as enzymes and molecular pumps that transport nutrients. These pumps can be on the exterior of the protoplast, but they are unlikely to be lost or damaged, as might occur if the LPS were not present. The LPS layer is extremely toxic to animals; it is the reason the food-poisoning bacteria *Salmonella* and *Shigella* are so deadly to us.

Most prokaryotes also secrete a loose, mucilaginous, slimy material around them. This was formerly called a "capsule" or "slime layer" but is now known as a **glycocalyx.** This may form either a light sheath or simply a loose gel; in the cyanobacteria, it is sufficiently strong and copious to hold large colonies together.

FLAGELLA

Many bacteria, but no cyanobacteria, have flagella (Fig. 19.12). These are extremely narrow tubes, about 20 nm across, made of just one type of protein, **flagellin.** The flagellum is hollow, and flagellin molecules pass through it from the protoplast until they reach the tip; there they are attracted by electrostatic forces and "crystallize" into the structure of the flagellum. The flagellum is anchored to the cell by a set of rings, one attached in or near the cell membrane and another in the peptidoglycan layer; in gram-negative bacteria, a third set of rings lies near the LPS layer. Flagella are located at one or both ends of a cell (polar flagellation) or over the entire surface (peritrichous flagellation). They rotate continuously and permit the bacterium to swim rapidly, up to 20 times its length per second.

Motile bacteria usually are able to sense certain stimuli and then swim toward or away from them; for example, light (phototaxis) in photosynthetic bacteria, nutrients (chemotaxis) in others, and even magnetic fields in *Aquaspirillum magnetotacticum* (Fig. 19.13a). It is instructive to consider the mechanism of guidance. In eukaryotes, both plants and animals, organs or tissues of sensation are usually sophisticated enough to detect the actual direction in which a stimulus is strongest, our eyes being excellent examples. Bacteria cannot sense a stimulus gradient; that is, they cannot detect where it is stronger or weaker. Instead, flagellar motion causes the bacterium to swim smoothly (called a **run**), then stop and tumble (a **twiddle**), followed by another period of smooth swimming, usually in a

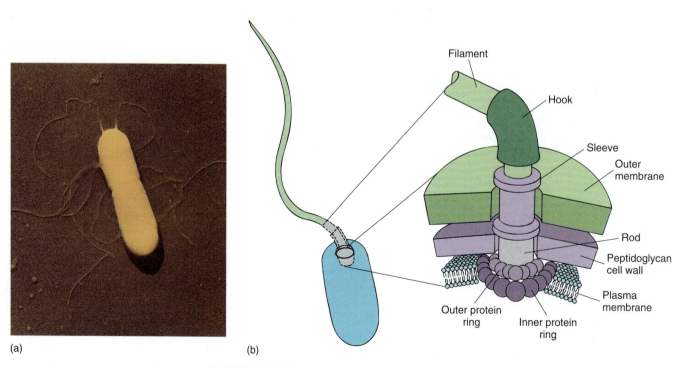

(a)   (b)

FIGURE 19.12

(a) Unlike eukaryote flagella, those of prokaryotes are too narrow to be resolved by light microscopy. Each prokaryote flagellum consists of just one type of protein, flagellin, and the gene for this has numerous alleles. Consequently, the flagella of one species may differ slightly from those of other species ($\times$ 9400). (*Moredun Animal Health, Ltd./Science Source Library/Photo Researchers, Inc.*) (b) Diagram of a flagellum.

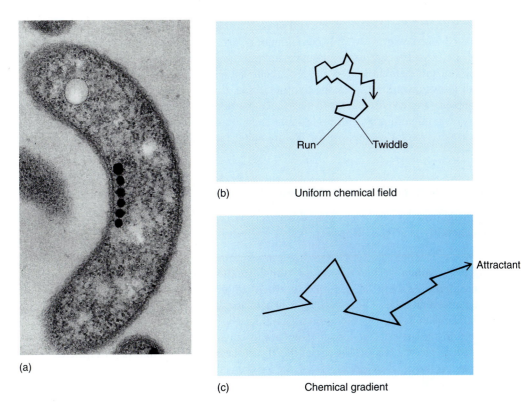

FIGURE 19.13

(a) Magnetotactic bacteria such as this *Aquaspirillum magnetotacticum* contain particles of magnetite (Fe_3O_4) that allow them to detect and orient themselves with regard to Earth's magnetic lines of force. Just as with gravitropism in plants, the detected environmental stimulus by itself is not important but serves as a guide to other factors. Magnetotactic bacteria do not swim to the North or South Poles; rather the magnetic field seems to be disrupted by certain conditions in the environment that are important. These bacteria are microaerophilic, needing a very small amount of oxygen but being killed by large amounts; microaerobic habitats may be associated with magnetic field perturbations ($\times 64,000$). (*T. J. Beveridge, University of Guelph/BPS*) (b) When a chemotactic bacterium is in a uniform chemical field, all runs and twiddles are of equal duration and the bacterium moves randomly in one area. (c) In a chemical gradient, runs that lead away from the attractant are short whereas runs that lead into the attractant are long; each twiddle is still random, but the result is well-oriented movement. If the chemical were a repellent, the relative lengths of runs would be reversed.

different, random direction. When the level of stimulant is low, runs last about 1 second and twiddles last only 0.1 second; the bacterium moves randomly (Fig. 19.13b, c). But if the swimming brings the cell into a region of an attracting stimulant, the stimulant molecules interact with protein chemosensors in the cell's surface, inhibiting tumbling and prolonging swimming. If after the next, delayed twiddle the bacterium is still swimming toward the stimulus, the sensors continue to be activated and twiddling is further suppressed. However, if the cell swims into an area of low stimulant, the stimulant molecules release from the chemosensors and twiddling is increased, increasing the chances that the cell will turn around.

CELL DIVISION AND REPRODUCTION

CELL DIVISION

Cell division in prokaryotes occurs by **binary fission,** in which the cell simply pinches in two (see Fig. 4.21). Nothing equivalent to mitosis, meiosis, or cell plate formation occurs. Instead, as the cell grows, its circles of DNA are duplicated; this is quite similar to the process in eukaryotes, with multiple replication start sites, bidirectional replication, and discontinuous replication on one end of each strand. DNA replication is not closely coordinated with cell growth, and cells commonly have 10 to 20 circles of DNA. All copies of the circles are attached to the plasma membrane and are pulled away from each other as the cell wall and membrane grow between them. When the critical size has been reached, the wall grows inward between the two nucleoids, dividing the cell. Once the new wall is complete, the cells separate. In flagellated bacteria, new flagella arise by an unknown process until both progeny cells have the same number as the original cell. Many bacteria, *Escherichia coli* for example, grow rapidly under ideal conditions, dividing every 20 minutes.

EXCHANGE OF GENETIC MATERIAL

Genetic material can be exchanged between individual bacteria, similar to sexual reproduction in eukaryotes. However, only a few genes are exchanged at a time, and the exchange occurs only rarely.

Transformation.   The first process, **transformation,** occurs when some bacteria die and decompose. As their DNA is broken down, it first breaks into rather large pieces, each with one or several genes. Other bacteria may occasionally absorb one of these pieces, although such large molecules usually cannot pass through the LPS layer or the peptidoglycan wall. Once inside the cell, the foreign genes are usually digested, but they may be accidentally inserted into the cell's DNA circle, thereby automatically becoming part of the cell's genome. As the cell duplicates its DNA, the foreign genes are also duplicated and passed on to the progeny.

Transduction.   **Transduction** is similar to transformation but occurs when a virus invades a cell and reproduces. As new viral particles are formed, small pieces of bacterial DNA can be accidentally incorporated into the virus; when the virus attacks a new bacterium, the DNA from the first bacterium is released into the second. This is not a particularly efficient method because the second bacterium may be killed by the virus; if it survives, its

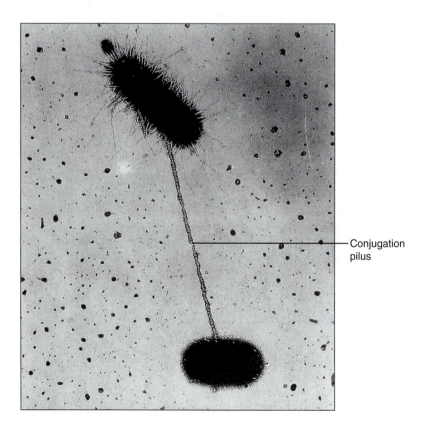

Conjugation pilus

FIGURE 19.14
A conjugation pilus is an extremely narrow, proteinaceous tubule that grows from donor to recipient. It and its attachment are so fragile that the connection is broken before much DNA passes from donor to recipient, even in gentle laboratory conditions. *(Courtesy of Charles C. Brinton, Jr., and Judith Carnahan)*

enzymes will probably depolymerize the foreign DNA rather than incorporate it. However, most ordinary soils have more than 50,000 bacteria per gram, and sludge and sewage have much more, so even if the success rate is less than one in a billion, tremendous amounts of genetic exchange are occurring.

Conjugation.   Whereas both transformation and transduction are basically accidental processes, **conjugation** involves specific structures and metabolisms that are the result of natural selection favoring genetic exchange. If two compatible bacteria come close to one another, the donor extends a narrow proteinaceous tube, a **conjugation pilus,** to the recipient (Fig. 19.14). The donor then duplicates one of its DNA circles, one copy of which moves through the conjugation pilus into the recipient. The conjugation pilus is fragile and usually breaks before all the DNA double helix is transferred, so the recipient becomes only "partially diploid". The foreign DNA may be immediately depolymerized or may be inserted into the recipient's own DNA circle. If so, it first pairs with the homologous genes of the recipient and then replaces them as they are cut out and degraded. The recipient is no longer partially diploid, and normal cell replication results in progeny that carry the new foreign genes.

Whereas genetic exchange—sexual reproduction—is easy to observe in most eukaryotes, it can be studied in prokaryotes only with rather laborious experiments. Cells must be grown in culture, and each type must have some observable trait that the other lacks. The two strains are then cultured together and later examined to see if any progeny have both traits. If so, genetic exchange has occurred. However, it is almost impossible to culture many types of prokaryotes—those from hot pools, deep ocean sediments, and so on—and for other species we do not have any strains with the right kinds of observable traits. Consequently, genetic exchange is known positively to occur in only a few bacteria; despite extensive searches, none has ever been found in cyanobacteria.

METABOLISM

Many aspects of plant metabolism were described in Chapters 10 to 15; these centered on sources of energy, carbon, and the flow of electrons that carry energy. Prokaryotes have numerous alternate methods for each of these processes: several types of photosynthesis, multiple types of respiration, and numerous sources of energy other than sunlight or sugars. These metabolisms are sufficiently out of the ordinary to amaze us, but more importantly, they teach us a great deal about the principles involved in the metabolism of life, not just of prokaryotes but of plants, fungi, and animals as well.

ACQUISITION OF ENERGY

All organisms must acquire energy. Two sources of energy are available to organisms—sunlight and chemical energy. The first is harnessed through photosynthesis, the second by respiration.

Photosynthesis.   Photosynthesis is carried out by plants, algae, cyanobacteria, and the green and purple bacteria, all of which are known as **phototrophs.** As described in Chapter 10, plants carry out photosynthesis by means of chlorophyll *a* and *b* as well as accessory carotenoid pigments, all located in chloroplasts.

In *cyanobacteria* the light reactions of photosynthesis are almost identical to those in chloroplasts, having chlorophyll *a* but lacking chlorophyll *b*. The accessory pigments in cyanobacteria are **phycobilins,** open-chain tetrapyrrole rings (Fig. 19.15). Like carotenoids, phycobilins absorb wavelengths that chlorophyll cannot; this energy is then transferred to chlorophyll and used to activate an electron. One class of phycobilins, **phycocyanin,** absorbs most strongly at about 620 to 640 nm and is blue in color. The other class of phycobilins, **phycoerythrin,** is red because it absorbs maximally at 550 nm. Both types of pigments occur bound to proteins, forming a **biliprotein.** The biliproteins aggregate in small nodules (**phycobilisomes**) visible by electron microscopy.

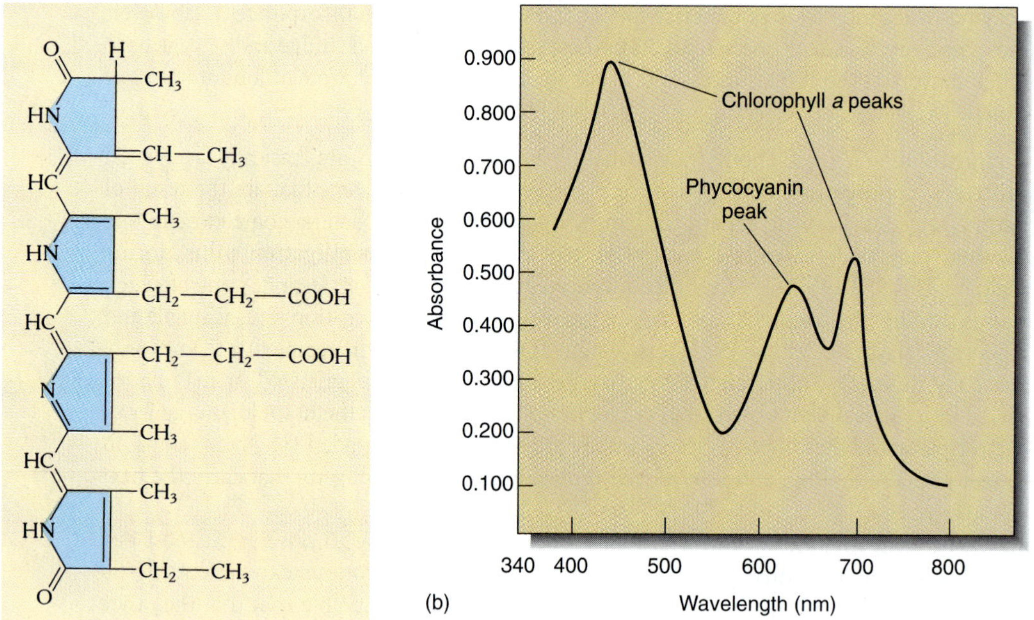

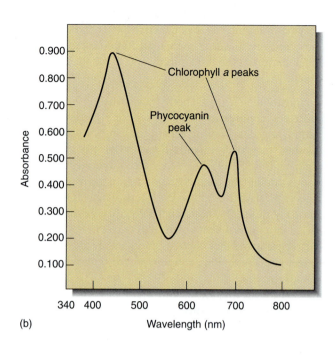

FIGURE 19.15

(a) A typical phycobilin, containing four pyrrole groups—a tetrapyrrole. The two ends of the pyrrole are not attached to each other as they are in chlorophyll, so this is an open-chain tetrapyrrole. These do not occur free in the cytoplasm, but rather are attached to proteins. (b) When the open-chain tetrapyrrole of (a) is attached to its protein, it is known as phycocyanin and has an absorption maximum of about 620 nm. Like carotenoids, phycobilins absorb light that chlorophyll *a* would miss, thus expanding the range of the electromagnetic spectrum that is used.

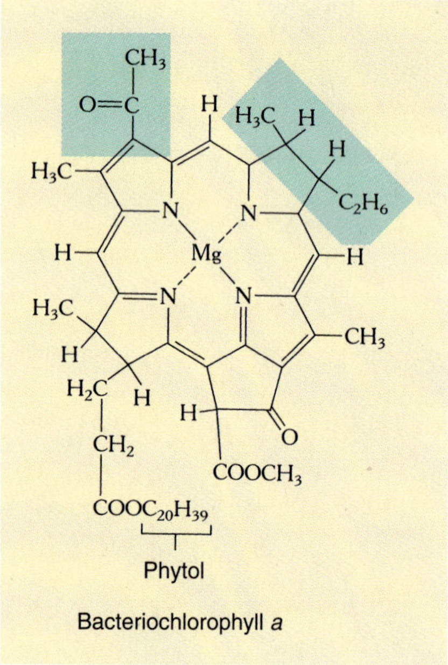

FIGURE 19.16

Bacteriochlorophyll *a* is similar to chlorophyll *a*, differing only at the sites indicated by color. Is this similarity analogous, the product of two independent lines of evolution, or homologous, the genes for one pigment having evolved from those for the other pigment? These two molecules are not sufficient evidence to judge, but we are quite certain that all photosynthetic systems are closely related evolutionarily.

Most other aspects of photosynthesis in cyanobacteria are identical to those of plants; electrons are taken from water, passed through photosystem (PS) II to PS I, and then used to reduce NADP⁺ to NADPH. Oxygen is liberated as a waste product. Cyanobacteria do not have chloroplasts, but they do have extensive sheets of infolded plasma membrane which contain pigments and electron carriers; the folded membrane forms accumulation spaces for protons and the generation of a chemiosmotic gradient.

The *purple bacteria* and the *green bacteria* (see Fig. 19.1) do not contain chlorophyll, either *a* or *b*, but instead have **bacteriochlorophylls** (Figs. 19.16 and 19.17). Like chlorophyll, these are closed tetrapyrroles with a long tail, but they have certain side groups that the chlorophylls lack. Carotenoid accessory pigments are present, as in chloroplasts. PS I operates in these bacteria, but no PS II is present to put electrons back onto the bacteriochlorophyll. Instead, the original electron comes back through a series of carriers, so electron flow is cyclic, similar to that in plant photosynthesis (Fig. 19.18; compare with Fig. 10.17). One of these carriers is quinone, so protons are pumped across the photosynthetic membranes, creating a strong chemiosmotic gradient that causes ATP to be generated, just as in eukaryotic chloroplasts. Because there is no PS II, oxygen is not formed. This is **anoxygenic photosynthesis.**

Whereas plants can have both cyclic and noncyclic electron flow, flow in the purple and the green bacteria is always cyclic. When chlorophyll *a* in plants absorbs light, an electron is activated to a high enough energy state to force the electron onto NADP⁺. However, because of the structure and the absorption of bacteriochlorophyll, its electrons receive less energy and, when fully activated, are not energetic enough to convert NADP⁺ to NADPH. Instead, the bacterium must generate NADPH by chemical reactions (see below under Respiration) or by forcing electrons onto it by using energy available in ATP.

The photosynthetic apparatus of the purple bacteria consists of an extensive array of membranes connected to the plasma membrane (Fig. 19.19). In the green bacteria, photosynthetic membranes are cylindrical vesicles that occur in clusters surrounded by another "membrane." The enclosing membrane does not look like a typical bilayered membrane in electron micrographs and is probably different physiologically as well.

Respiration.   As an energy source, the alternative to photosynthesis is respiration, the use of the chemical energy present in various compounds. In all plants, animals, cyanobacteria, and most bacteria, the compounds used for respiration are organic—sugars, fats, amino acids, and so on—and the organisms are **conventional heterotrophs.** If oxygen is present, they carry out aerobic respiration; if not, they carry out anaerobic respiration—

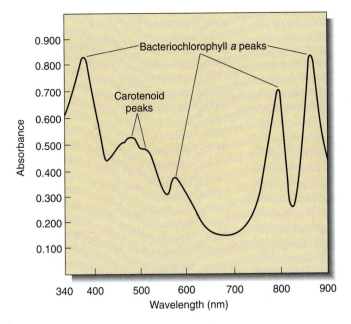

FIGURE 19.17
Bacteriochlorophyll *a* has several peaks of absorption—870, 805, 600, and 360 nm. In addition, several carotenoids are present and absorb between 450 and 550 nm.

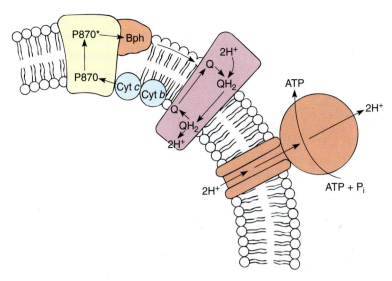

FIGURE 19.18
Activated bacteriochlorophyll *a* is not strong enough to reduce $NADP^+$; instead electrons flow cyclically, pumping protons with quinone carriers. This establishes a chemiosmotic gradient that can generate ATP. Bph = bacteriophaeophytin (see Chapter 10).

fermentation. However, certain bacteria (**lithotrophs**) metabolize inorganic compounds and extract energy from them, examples being hydrogen and compounds of sulfur, iron, and nitrogen.

The principles of hydrogen respiration are rather easy to understand. The reaction $2H_2 + O_2 \rightarrow 2H_2O$ liberates two electrons and releases so much energy that it can easily force the electrons onto NAD^+, producing NADH. This then donates the electrons to an ordinary electron transport chain that produces ATP, just as in mitochondria.

Hydrogen sulfide (H_2S) and thiosulfate ($S_2O_3^{2-}$) can both be converted to sulfur (S^0) (Fig. 19.20) and then to sulfate (SO_4^{2-}). These reactions release energy and electrons that either go into an electron transport chain or react directly to produce ATP. The end product

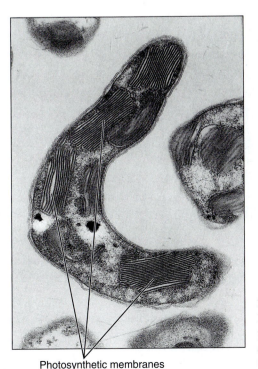

Photosynthetic membranes

FIGURE 19.19
In purple photosynthetic bacteria, the photosynthetic membranes may be either flat sheets or long tubes. *Ectothiorhodospira mobilis* ($\times$ 51,000). (*Journal of Bacteriology*)

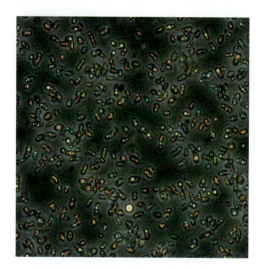

FIGURE 19.20
Sulfur bacteria, *Chromatium*. When hydrogen sulfide is present, it is oxidized to water and sulfur. The sulfur is stored in the cell as an orange granule of nutrient reserve; if hydrogen sulfide becomes depleted, the sulfur particles are oxidized to sulfate ($\times 250$). *(Paul W. Johnson/ Biological Photo Service)*

of these steps is sulfuric acid, which is toxic like the ethanol produced in plant and yeast fermentation. As long as the environment absorbs the acid and keeps it dilute, the bacteria are unharmed. *Thiobacillus thiooxidans* can withstand extremely high concentrations of acid and grows well at pH 2.

The oxidation of ferrous iron (Fe^{2+}) to ferric iron (Fe^{3+}) releases an electron that can be used to form a single molecule of ATP (Fig. 19.21). Likewise, ammonium (NH_4^+) can be oxidized to nitrite (NO_2^-) and then to nitrate (NO_3^-), with the release of six energetic electrons that can be used for ATP synthesis.

Electron Donors, Electron Acceptors, and Oxygen. Electrons carry energy for an organism's metabolism; to do this, the electrons must flow from electron donors to electron acceptors. The most common sources of electrons are water and organic compounds, but hydrogen, sulfur, iron, and nitrogen compounds donate electrons in lithotrophic bacteria.

Prokaryotes can be obligate aerobes, obligate anaerobes, or facultative anaerobes. The most common electron acceptor is oxygen; it gains two electrons and then reacts with two protons and forms water during aerobic respiration. Other molecules can be used as electron acceptors if oxygen is not available. This is anaerobic respiration, and many highly oxidized compounds can be used. The most common is nitrate (NO_3^-), which is converted to either NO_2^- or N_2. Sulfate (SO_4^{2-}) can also act as an electron acceptor and be converted to hydrogen sulfide (H_2S). These processes are the opposite of those described above, in which these same compounds were used as substrates for respiration because they could be oxidized. How can they act as a source of energy and electrons one time but as electron acceptors another time? The key is the presence or absence of oxygen: Oxygen has such a strong tendency to absorb electrons that in its presence the balance between H_2S and SO_4^{2-} is shifted in favor of SO_4^{2-}. Without oxygen, the balance is shifted the other way.

In addition to inorganic compounds, certain organic compounds can also act as electron acceptors during times when oxygen is not present; the process is **fermentation**, similar to the fermentation described in Chapter 11. Bacteria are diverse in the types of substrates that can be fermented, such as most sugars and organic acids, resulting in products like ethanol, lactate, propionate, acetate, acetone, methanol, and carbon dioxide. The fermentations of amino acids and other compounds that contain nitrogen or sulfur are often called putrefactions and produce odiferous compounds: hydrogen sulfide (aroma of rotten eggs), isobutyric acid, cadaverine (from lysine), putrescine (from ornithine), and ammonia.

FIGURE 19.21
The run-off from iron mines is usually both acidic and rich in iron compounds. Iron-oxidizing bacteria use the ferrous iron as a substrate, respiring it to ferric iron and generating ATP. Virtually no other life occurs in such polluted conditions. *(Don Duckson/Visuals Unlimited)*

SOURCES OF CARBON AND REDUCING POWER

Among eukaryotes, all animals are heterotrophs. The food they eat is digested and some is then used to construct more tissue for the animal: protein, nerve cells, and other structures. The majority of the food is used only as a source of electrons in aerobic respiration. Electrons are transferred to oxygen, and their energy drives the formation of ATP. Autotrophs obtain the carbon of their metabolism entirely from carbon dioxide, whereas heterotrophs derive their carbon from organic compounds such as sugars, fats, amino acids, organic acids, and many other compounds. All photosynthetic plants are autotrophs: They take in carbon dioxide and reduce it to carbohydrate by means of the NADPH and ATP produced in the light-dependent reactions of oxygenic photosynthesis. It is then used like the food of animals: Some is used to build more structural elements, and some is transported into mitochondria where it is respired and ATP is formed. Animals lack the Calvin cycle, so they cannot use their food purely for energy and then use some of that energy to reduce carbon dioxide. Such a scheme is theoretically possible, but it does not seem selectively advantageous for animals to reduce carbon while they obtain it in their food automatically.

In prokaryotes, similar relationships hold. Green bacteria cyanobacteria, and some of the purple bacteria are photosynthetic (phototrophic) and have the Calvin cycle (autotrophic), so they are photoautotrophs like plants. Most other bacteria, like animals, take in organic substances and use them both for ATP generation and for polymer construction from monomers; these are conventional heterotrophs. However, prokaryotes also have some different types. Whereas plants rarely have organic food available to them, photosynthetic bacteria commonly do, and most purple bacteria and green bacteria are photoheterotrophs. They absorb and use organic carbon for construction, and use photosynthesis almost exclusively to generate ATP. Little or no photosynthetic energy is used to reduce $NADP^+$ to NADPH because NADPH is not in great demand, since the Calvin cycle is unnecessary if carbohydrate is present in food.

Lithotrophs use carbon dioxide as their primary carbon source; they are lithotrophic autotrophs. Examples are the colorless (nonphotosynthetic) sulfur bacteria, the hydrogen bacteria, nitrifying bacteria, and iron bacteria. As they oxidize hydrogen, sulfur, iron, and so forth, the energy goes to make ATP. Part of the ATP is used for regular metabolism and part to pump protons, creating a pH gradient that can be used to reduce $NADP^+$ to NADPH. Their energy metabolism is somewhat like bacterial photosynthesis. Neither metabolism can produce a compound as energetic as activated chlorophyll, which is powerful enough to reduce $NADP^+$ by itself.

A very few bacteria (*Beggiatoa, Thiobacillus*, and some others) are lithotrophic heterotrophs: They get their energy by oxidizing sulfur compounds and take in organic compounds for structural uses. They do not use carbon dioxide, and apparently they lack the Calvin cycle.

The ability to live as a heterotroph offers great advantages. By taking in carbon in its reduced form, heterotrophs can use all their respiratory or fermentative energy for growth and reproduction. Autotrophs, which must use energy to reduce carbon dioxide, seem to be at a disadvantage. Their one advantage is that an autotroph has probably never starved to death; carbon dioxide is always abundant enough to sustain the life of the autotrophs. The same is not true for the food supply of heterotrophs; famine is a common occurrence not only for animals and fungi, the eukaryotic heterotrophs, but also for bacteria.

CLASSIFICATION OF PROKARYOTES

In plants and animals, the most important and most widely used characters for classification are rather easy to study. Features of anatomy and morphology, such as flower structure or the shapes of bones, teeth, and internal organs, offer abundant information for the analysis of evolutionary relationships. Unfortunately, prokaryotes are almost all unicellular, so they have very little anatomy and virtually no morphology. Consequently, their systematics and classification have been based almost entirely on features of their metabolism, wall chemistry, ability to carry out photosynthesis and use various substrates, and sensitivity to oxygen.

When a character is based on a complex metabolism that is the result of proper integration of numerous components, we are probably safe in assuming that the character is stable and conservative. There is only a slight possibility that it has arisen independently in two or more distinct lines. An example of this is oxygenic photosynthesis of cyanobacteria: chlorophyll *a,* all the electron carriers of PS II, and the metabolic mechanism that ensures their proper placement in the correct membranes. We are confident that this suite of characters arose only once. Cyanobacteria are a natural, monophyletic group.

Some features are more difficult to assess. Mycoplasmas, the wall-less bacteria, share many unusual features: (1) They are always parasitic within plant or animal cells, (2) they have no wall, (3) they are very small and have a simple metabolism, and (4) they have little DNA, only enough for about 650 average-sized genes (Fig. 19.22). This striking suite of characters is uniformly present in all mycoplasmas. However, many other parasites also have much simpler structures and metabolisms than their free-living relatives. Whereas it may take millions of years to build up a pathway evolutionarily, the pathway can be lost in a few generations in a parasite. Although they share many dramatic features, the mycoplasmas actually could be an artificial group, composed of the advanced members of several different lines of convergent evolution, each of which began with a more complex ancestor.

Nucleotide sequencing is becoming a powerful and reliable technique for analyzing prokaryote evolution. The most commonly sequenced nucleic acid is the 16S ribosomal RNA, the prokaryote equivalent of 18S rRNA in eukaryotes. This type of RNA is especially valuable in studying the phylogeny of major groups for two reasons. First, it is always present in all organisms because ribosomes are invariably present. Second, 16S rRNA is extremely conservative; almost any mutation appears to be deleterious, so it evolves extremely slowly. Even if two groups diverged from a common ancestor hundreds of millions of years ago, large portions of their 16S rRNA are still identical. The 16S rRNA of many mycoplasmas has been studied, and the sequences in five of the six genera have been found to be so similar they must have evolved recently from one ancestral rRNA. The 16S rRNA of the sixth genus, *Thermoplasma,* however, is quite different and likely evolved separately from the others; thus the mycoplasmas are an artificial group resulting from convergent evolution (Fig. 19.23).

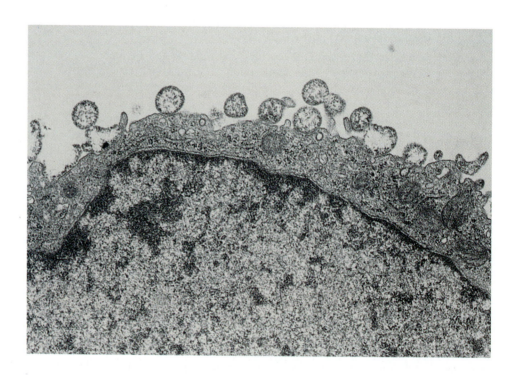

FIGURE 19.22
Mycoplasmas have no cell walls and are always found as parasites inside plants or animals (× 5500). *(CNRI/Phototake)*

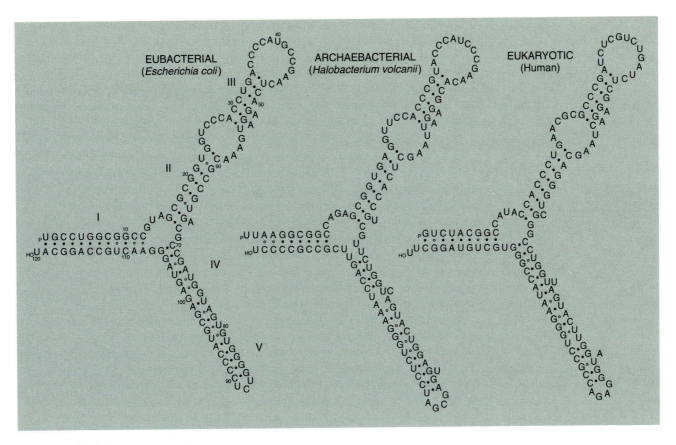

FIGURE 19.23

5S rRNA can be compared in all living groups, even those as diverse as archaebacteria, eubacteria, and humans. Several nucleic acids may share the same nucleotide base at a particular position, whereas others have a different base; this may represent a homologous sequence descended from the same mutant DNA molecule, or they could be analogous, having descended from two identical but independent mutations. Although this sounds like the same problem discussed in Chapter 18, it is actually easier to resolve; we can calculate the probability that any particular base will mutate and the probability that any two nucleic acids will show the same base-sequence changes purely by chance. Comparative statistics allow us to determine the degree to which two nucleic acids are related; that is, it gives an estimation of how long ago the two shared a common ancestor. *(Courtesy of Dr. Carl Woese, University of Illinois)*

At present, few species have been adequately sequenced, and most classification of prokaryotes is still based on metabolism. We know that the current system contains many errors, but we do not know what changes to make to create a more natural system of classification. Research is progressing rapidly, with great confidence that a more accurate, natural classification will soon be developed. The most widely accepted classification is that given in *Bergey's Manual of Systematic Bacteriology*.

For many years, the prokaryotes have been classified as one kingdom, kingdom Monera, with two divisions, division Cyanophyta ("blue-green algae") and division Bacteria (see Table 19.1). Cyanobacteria (note the name change to "blue-green bacteria") are no longer considered unusual, as when they were thought to be related to algae, so at present they are classified as just another group of eubacteria. Recently, microbiologists have begun to recognize many unusual features in several of the archaebacteria. Some have suggested that these organisms be separated out of kingdom Monera completely, whereas others believe that evidence is insufficient to warrant such a fundamental reclassification and recognize them as a second division, division Archaebacteria. The remaining members of the old division Bacteria are now called division Eubacteria (See Table 19.2). The taxonomic categories of classes and orders discussed in Chapter 18 are not used for prokaryotes because we do not understand the phylogenetic relatedness of the species sufficiently. Instead, prokaryotes are grouped in 33 sections, each with one or several families. Many of the families are thought to be natural, monophyletic groups, but even at this level, nucleic acid sequencing is indicating that many families based on a distinctive type of metabolism are actually polyphyletic. We expect that the entire kingdom will be reclassified within the next few years; this monumental task will give us a phylogenetic classification much more useful and satisfying than the present artificial one.

DIVISION ARCHAEBACTERIA

Archaebacteria are some of the most unusual and intriguing of all organisms. They differ from other prokaryotes so strongly and in such fundamental ways that it has been suggested several times that they be set aside in their own groups, such that all living and fossil organisms would be divided into Archaebacteria, Prokaryotes (Eubacteria), and Eukaryotes (Fig. 19.24).

Archaebacteria are distinct metabolically, but they resemble other prokaryotes structurally. They lack a true nucleus, mitochondria, plastids, and all other membrane-bounded organelles (Fig. 19.25). They fall within the size range of other prokaryotes, and they have walls and 70S ribosomes. In electron micrographs, they resemble cells of eubacteria.

Metabolically, however, they are very different from eubacteria: Their walls contain protein, glycoprotein, or polysaccharide, but they lack a true peptidoglycan component; therefore they are immune to most antibiotics that interfere with wall synthesis. Their membranes have unusual lipids in which the fatty acids are attached to glycerol by ether linkages, not the esters that occur in all other organisms. Furthermore, some lipids are diglycerols; that is, each end of the fatty acid is linked to a glycerol. This affects the structure of membranes, making it difficult to modify existing components of the membrane (Fig. 19.26). Many drugs that inhibit ribosomes and protein synthesis in eubacteria

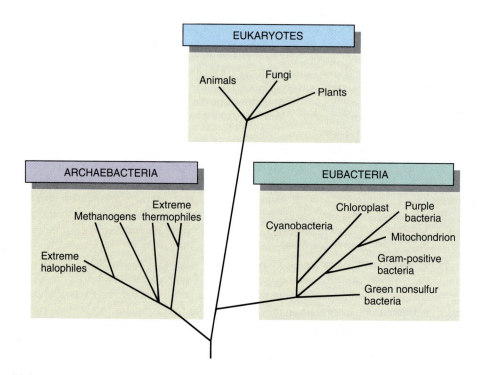

FIGURE 19.24

One current classification proposed for the evolution of all major groups of living organisms, based on rRNA analysis. Further analysis may suggest other changes, so this is not to be regarded as absolute truth. Halophiles are "salt-loving" organisms, thermophiles are "heat-loving," and methanogens produce methane.

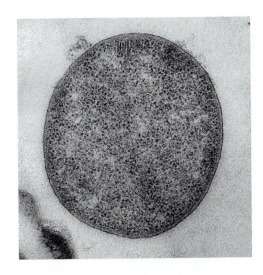

FIGURE 19.25
Archaebacteria such as this *Methanococcus voltal* have the same structure as eubacteria when viewed with an electron microscope; they have no membrane-bounded organelles (× 115,000). *(T. J. Beveridge, University of Guelph and K. Jarrel, Queens University, Ontario/BPS)*

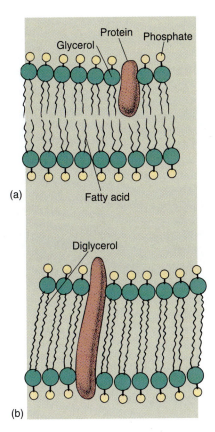

FIGURE 19.26
(a) In typical diglycerides and phospholipids, the fatty acids extend to one side of the glycerol, resulting in a bipolar molecule, and two layers are needed to form a stable membrane that is completely immersed in water. (b) A diglycerol molecule is polar at both ends and hydrophobic along the interior; a single layer can form a stable membrane.

have no effect on archaebacteria. The overall metabolism of archaebacteria is exotic in that they thrive in certain rare environments: Some archaebacteria are methane-producing bacteria, others are extremely halophilic (salt-loving), and some are thermoacidophilic. These three highly specialized types of metabolism are neither interrelated nor obviously related to the metabolism of eubacteria.

Because of their unusual metabolic requirements, archaebacteria are now found in exotic environments: salt pools (*Halobacterium*) (Fig. 19.27), sulfur-rich hot acid pools (*Sulfolobus*) (Fig. 19.28), and anaerobic, H_2S-rich environments such as deep sea sediments, bogs, and sewage treatment plants (methanogenic bacteria). Archaebacterial evolutionary origins may be extremely ancient, having begun when Earth was quite different, with a reducing atmosphere, acidic pools, and extensive areas of hot volcanic habitats. If so, as Earth cooled and the atmosphere became enriched in oxygen, archaebacteria were able to survive only because a few isolated sites continued to have the characteristics of early Earth.

FIGURE 19.27
Salt ponds such as this evaporation pond of the Leslie Salt Company are habitats for halophilic archaebacteria; they must have an environment in which the salt level is almost at the saturation/crystallization point. Although the water potential is extremely negative, the halophiles can still absorb water for their metabolism. *(William E. Ferguson)*

(a)

(b)

FIGURE 19.28

(a) Hot pools such as this one at Yellowstone National Park are required for thermophilic archaebacteria. Some archaebacteria are the most thermophilic organisms known, living in water that is 100°C, the boiling point. These bacteria are not just thermotolerant organisms—tolerating the heat; they are thermophiles—they require the heat and die without such hot conditions. (b) In pools like these at Mammoth Hot Springs, the top pool has the hottest water and the most thermophilic archaebacteria, which are brown here. In the outflow streams where the water is cooler, other species that are less thermophilic (green here) can be found. The highest temperatures tolerated by a few plants and insects is 50°C, and by vertebrates 38°C.

An alternative hypothesis postulates that archaebacteria are more modern and have evolved from eubacteria. This hypothesis is based on the concept that acid pools, brine pools, and H_2S-rich sites acted as selection pressures, and those eubacteria with the proper mutations were able to adapt and survive in them. This is plausible, because many eubacteria have evolved to survive in other types of extreme habitats. However, it is not obvious why these adaptations should have included modification of the 16S rRNA, a major weakness of this hypothesis.

DIVISION EUBACTERIA

Eubacteria are the "true bacteria," those that are not archaebacteria. This is the largest group of prokaryotes, with the greatest metabolic diversity (see Fig. 19.24). The groups described below were chosen because they make nitrogen and sulfur available, decompose cellulose, or are plant pathogens. Most are probably not natural, monophyletic groups.

GLIDING BACTERIA

Gliding bacteria are capable of gliding motility only when in contact with a solid surface; they have no flagella. Many resemble cyanobacteria in consisting of long filaments of short disklike cells. An example is *Beggiatoa*, a eubacterium that inhabits areas rich in H_2S, such as water polluted with sewage, decaying beds of seaweed, and deep mud layers of lakes. *Beggiatoa* oxidizes H_2S first to sulfur (S^0) and then to SO_4^{2-}. This can be important in detoxifying H_2S, which is poisonous to most organisms, and *Beggiatoa* has recently been discovered to form associations with the roots of cattails, rice, and other plants that live in stagnant water. The *Beggiatoa* appears to benefit in that it receives either nutrients or perhaps the enzyme catalase from the roots while protecting them from H_2S.

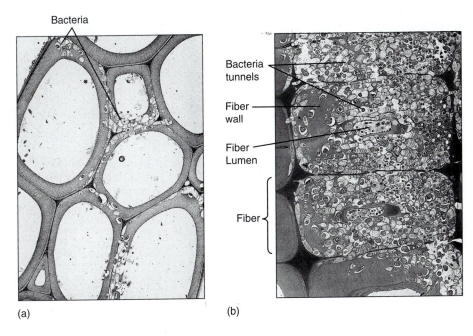

(a)

(b)

FIGURE 19.29

(a) Wood-digesting bacteria are tunneling through all parts of the vessels in this wood (× 4000). (b) Tunneling bacteria have enzymes that permit them to consume virtually all of these fibers, even the most heavily lignified portions (× 5000). *(Courtesy of A. P. Singh, Forest Research Institute, Rotorua, New Zealand; T. Nilsson and G. F. Daniel, Swedish University of Agricultural Sciences, Uppsala, Sweden)*

Cytophaga is a genus of gliding eubacteria that digest cellulose and chitin. The cells must actually bind to the cellulose because the cellulases (cellulose-degrading enzymes) are part of the cell envelope and are not released. *Cytophaga* is easily cultured by putting soil onto moist filter paper: *Cytophaga* cells form yellow or orange colonies that are transparent because the underlying filter paper cellulose has been digested. Cellulose-digesting, gliding bacteria are important in humus formation and soil ecology. Many other genera of bacteria attack wood, some by tunneling through both the primary and secondary walls (Fig. 19.29).

NITROGEN-FIXING BACTERIA

The bacteria capable of fixing nitrogen are divided into free-living (*Azotobacter, Derxia*) and symbiotic (*Agrobacterium, Rhizobium*; see Chapter 13).

Members of *Azotobacter* are large gram-negative rods and are obligately aerobic. Like *Cytophaga, Azotobacter* live in soil where they use a variety of carbohydrates, alcohols, or organic acids as carbon and energy sources. In the absence of nitrogen compounds, they absorb N_2 and fix it into organic compounds; when the cells die and break down, these nitrogen compounds can be absorbed by plant roots.

Rhizobium is a genus of small flagellate gram-negative rods. As described in Chapter 13, they establish a symbiotic relationship with legumes, living inside root nodules and fixing large amounts of nitrogen, much of which is made available to the plant. *Azospirillum* is an unrelated genus of bacteria that are capable of fixing nitrogen and form a loose symbiosis with certain grain crops and some tropical grasses.

NITRIFYING BACTERIA

When organisms die and decay, their nitrogen groups become available to the roots of living plants. However, because most of these nitrogen groups are highly reduced, they can be used as an energy source in an oxygen-rich atmosphere.

Certain soil bacteria (*Nitrosomonas, Nitrosococcus*) oxidize ammonium to nitrite, and others (*Nitrobacter, Nitrococcus*) further oxidize the nitrite to nitrate. The entire process is called **nitrification** (Table 19.3). Surprisingly, no single organism is able to carry out both steps, but even so, both types of bacteria are so common that nitrite never builds up in the soil (Fig. 19.30). Nitrification has important ecological consequences; although plants can absorb and use nitrate, it is more readily washed from soil because it is anionic (negatively

TABLE 19.3 *Processes in Nitrogen Availability*

Organism	Process	Effect
Nitrogen fixation		
Free living and symbiotic eubacteria including cyanobacteria	Conversion of N_2 to organic nitrogen	Very beneficial; it makes nitrogen available to all organisms
Nitrification		
Nitrifying bacteria	Conversion of reduced nitrogen to nitrate or nitrite	Slightly detrimental; plants must reduce it back to amino group oxidation level. Also, it is easily washed from soil.
Denitrification		
Denitrifying bacteria	Conversion of nitrogen compounds to N_2	Very detrimental; nitrogen is no longer available for use by plants or other organisms

charged) and remains in solution, whereas the ammonium (NH_4^+) is cationic and is bound to soil particles. Also, when ammonium is absorbed by a plant, no energy must be expended because the nitrogen is reduced already; however, when nitrate is absorbed, large amounts of ATP must be used to reduce it before it can be incorporated into amino acids, nucleotides, and other compounds.

DENITRIFYING BACTERIA

Denitrification is the process in which certain bacteria (*Hyphomicrobium, Pseudomonas*) reduce nitrate, not to ammonium, but to gaseous nitrogen, N_2. Nitrate or nitrite is used as an electron acceptor during energy production. Whereas nitrification results in nitrate that plants cannot use as easily as ammonium, denitrification results in nitrogen gas, which plants cannot use at all.

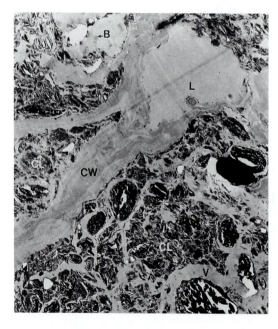

FIGURE 19.30
Soil bacteria are so abundant that most reduced nitrogen compounds are converted to nitrates (nitrification) or nitrogen gas (denitrification) before roots can absorb them. B = bacteria; CW = cell wall of collapsed epidermal cell; L = lumen of collapsed cell; CL = clay particles; V = void, a large space between clay micelles. *(Reprinted by permission from R. C. Foster, A. D. Rovira, and T. W. Cock, 1983,* Ultrastructure of the Root-Soil Interface, *The American Phytopathological Society, St. Paul, Minnesota)*

FIGURE 19.31
Mycoplasmas cause numerous diseases in plants, such as this yellow blight in palm. Other diseases are aster yellows, elm phloem necrosis, pear decline, and corn stunt disease. (© *Peter B. Kaplan, The National Audubon Society Collection/Photo Researchers, Inc.*)

MYCOPLASMAS

The mycoplasmas (*Mycoplasma, Acholeplasma*) are bacteria that completely lack a cell wall, as proven by both chemical analysis and electron microscopy (see Fig. 19.22). They are variable in shape, being filamentous, branched, coccoid, and rod-shaped. They are the smallest living cells known, coccoid forms being only 0.2 to 0.3 μm in diameter, and they pass through extremely fine filters capable of stopping all other cells. This presents an easy method for isolating mycoplasmas: Pass the material through a fine filter, then incubate it on medium that contains an antiwall antibiotic such as penicillin.

Many mycoplasmas are pathogenic in animals or plants (Fig. 19.31). They may cause severe disease, such as citrus stubborn disease caused by *Spiroplasma,* or they may live in plants without producing any obvious disease symptoms. Electron microscopy has shown that many types of diseased plants contain mycoplasma-like inclusions. Very few have actually been isolated, cultured, and identified, so they usually are referred to as **mycoplasma-like organisms** rather than as mycoplasms. Mycoplasmas may be a much larger and more damaging group of bacteria than is appreciated.

SECTION CYANOBACTERIA

Cyanobacteria are now classified as eubacteria, although formerly when they were thought to be closely related to true algae and were called blue-green algae they were placed in their own division (see Fig. 19.24; Box 19.1). Although some have cells as small as those typical of other eubacteria (0.5 to 1.0 μm in diameter), most are larger, and the very largest cells of all prokaryotes are those of the cyanobacterium *Oscillatoria princeps,* up to 60 μm in diameter. Whereas some are unicellular (*Chroococcus*), most occur in large groups. These can be simple colonies held together by a mucilaginous matrix (*Microcystis*), or the cells may be more firmly attached and form long filaments that are either unbranched (*Oscillatoria, Lyngbya*) or branched (*Hapalosiphon*). The walls of cyanobacteria are like those of other gram-negative eubacteria, but they also produce an extra, external layer of mucilage that binds cells and filaments together into a small, loose aggregation. In some genera, such as *Nostoc,* the filaments are bound firmly and result in large spherical structures several centimeters across. In other cases, filaments adhere in large flat sheets (see Fig. 13.15).

Plant Biology Tutor CD-ROM

BOX 19.1   *The "Misclassification" of the Blue-green Algae*

Until very recently (as late as 1971) cyanobacteria were considered to be either primitive eukaryotic (blue-green) algae or perhaps a transition between prokaryotes and eukaryotes. At present there is no doubt that they are completely prokaryotic: They do not have a true nucleus; they have 70S ribosomes; they lack membrane-bounded organelles; and their walls are similar to those of gram-negative bacteria, completely lacking cellulose. It is easy to regard the misclassification of the cyanobacteria as a remarkable mistake and to wonder how scientists could have misunderstood their nature. It is important to understand how this happened.

In comparing an electron micrograph of a eubacterial or cyanobacterial cell with one of a plant or animal cell, it is impossible to confuse either of the prokaryotic cells with the eurkaryotic ones. However, comparison of the cyanobacterial cell with one from certain types of algae or fungi might cause trouble: The cytological differences between eukaryotes and prokaryotes are not always easy to see (see Fig. 21.4). Furthermore, whereas chloroplasts and mitochondria are relatively easy to see in the cells of angiosperms even using a light microscope, they are much more difficult to see in the small cells of many true algae, especially if the algae have thick walls or refractile grains of starch or oil. Many aspects of cell structure that we take for granted were not well understood for many algae even into the mid 1970s.

Furthermore, only late in the 1960s did evidence begin to accumulate in support of the endosymbiont theory of eukaryotic cell origin (see Chapters 1 and 21). Prior to that time, it was generally believed that the prokaryotes had gradually evolved into eukaryotes as their nucleoids and membrane systems became more complex and their cells became larger. Under that hypothesis, cyanobacteria appeared to have many of the characters of an intermediate stage between prokaryotes and eukaryotes. Of course, the oxygen-producing, chlorophyll a–mediated photosynthesis appeared to be a strong piece of evidence. Another supporting factor is that many cyanobacteria are considered to have true bodies. Some are composed of long, wide filaments much more massive than those of other prokaryotes and analogous to the bodies of many eukaryotic algae, especially the green algae (see Fig. 21.18). These bodies can also undergo some differentiation: Many species are able to form heterocysts capable of fixing nitrogen. Thus the bodies consist of at least two distinct types of cells, often arranged in a predictable pattern. Under certain conditions, the cells that do not fix nitrogen can be converted to akinetes. When this happens, differentiation begins in those cells located farthest from heterocysts; this requires a level of whole-body integration that is certainly quite simple but is just as certainly far more sophisticated than anything that occurs in other eubacteria.

We can learn much about science and ourselves as scientists as we look at why cyanobacteria were misclassified for so long. Although they were studied by many very intelligent and reasonable scientists, three factors led to the error: (1) Critically important data, such as the prokaryotic ultrastructure and prokaryotic DNA organization, were not known. (2) Many of the characters that were known are analogous to similar structures in eukaryotic algae and were misinterpreted as homologous. (3) Those scientists were working with a theory no longer considered to be valid, and this affected their evaluation of the evidence.

It is sobering to realize that these three factors may be operating in any given sector of our knowledge at present. In very few areas do we know absolutely everything about a subject; some data are always missing. How do we know if they are trivial data or something that could completely outweigh everything else? For any structure or metabolism or behavior we study, we can usually find similar traits in some other organism. Are the two homologous (related) or analogous (convergent evolution)? How can we be certain? Regardless of what we study, we are working within the framework of numerous large, multifaceted theories. It is always possible that people working in a rather unrelated field will make a basic discovery that will change the theories and theoretical frameworks, thus requiring us to re-evaluate our findings. When we look at scientists of the past and are amused by their mistakes, we should consider what the scientists of the future will think about us.

Cyanobacteria are of special interest because they seem to be most closely related to the organisms that might have given rise to the chloroplast: They all have chlorophyll a and PS II and produce oxygen. One group in particular, **prochlorophytes** (Fig. 19.32), appears especially interesting because they have both chlorophyll a and b and completely lack phycobilin pigments.

A small number of species (some strains of *Nostoc commune* and *N. muscorum*) are able to grow completely heterotrophically, in the dark, if supplied with sugars and inorganic salts. These metabolize the sugars as both a carbon source and an energy source. A few other species can be grown heterotrophically in dim light; they use sugar as a source of carbon but they cannot oxidize it to make ATP. They still must use photosynthesis for that, but the ATP made in the light does not have to be used for carbon reduction.

The ability of many cyanobacteria to fix nitrogen is important ecologically, because these are widespread and often more active than other nitrogen-fixing eubacteria. Typically, nitrogen fixation occurs in heterocysts that develop from vegetative cells (see Fig.

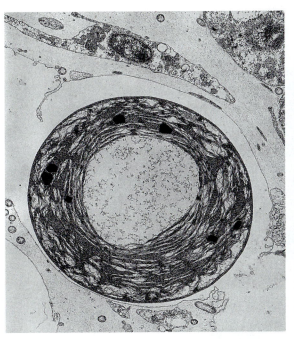

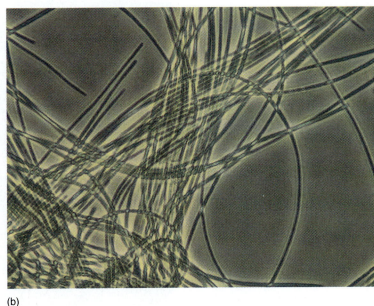

(a)   (b)

FIGURE 19.32

(a) This is *Prochloron*, an organism that closely resembles chloroplasts. It lacks the phycobilisomes characteristic of cyanobacteria (× 15,900). *(Courtesy of E. H. Newcomb, University of Wisconsin-Madison/ BPS)* (b) This is *Prochlorothrix hollandica*, a prochlorophyte like *Prochloron*, also related to the ancestors of chloroplasts (× 200). *(Courtesy of H. Swift, University of Chicago)*

19.5). If ammonium is present in the environment, heterocysts do not form; but when available nitrogen is depleted, heterocysts develop, fix nitrogen, and pass it to surrounding vegetative cells. Most species are free-living, but others form symbiotic associations with plants: *Anabaena* grows with the water fern *Azolla* and the roots of many cycads (Fig. 19.33); *Nostoc* is part of the body of liverworts such as *Blasia* and *Anthoceros*.

Cyanobacteria often store extra nitrogen compounds as **cyanophycin granules** in the cytoplasm. These are composed of a simple polymer of aspartic acid, each molecule of which has an attached molecule of arginine. When the cell needs large amounts of nitrogen, cyanophycin is depolymerized.

FIGURE 19.33

The water fern *Azolla* forms a symbiotic relationship with the cyanobacterium *Anabaena*. Nitrogen compounds are passed from *Anabaena* to *Azolla*, and carbohydrates pass in the other direction. *(P. Gates, University of Durham/Biological Photo Service)*

FIGURE 19.34
Fossil stromatolites from the Bitter Springs rock formation in central Australia are known to be about 1 billion years old. Their cells strongly resemble those of cyanobacteria in living stromatolites still occurring at Shark Bay in Australia. *(William E. Ferguson)*

Cyanobacteria inhabit numerous extremely harsh environments, withstanding intense insolation and periods of almost complete desiccation. *Synechococcus* withstands temperatures of 73°C and lives in hot pools such as those in Yellowstone National Park. In such environments, cyanobacteria often extract lime from the water and secrete it around their own filaments, creating large rocklike masses. In waters that are not quite so heated, such as those in shallow pools in hot, dry regions, cyanobacterial growths can become gigantic and so heavily encrusted with calcium carbonate that they appear to be stones. These deposits are called **stromatolites** (Fig. 19.34) and are currently being formed in only a few localities, but thousands of them are known as fossils. The most ancient stromatolites, 2.7 billion years old, were formed just as oxygen was beginning to accumulate in the atmosphere; they may have been built by the organisms that began the era of rusting and conversion of Earth's second atmosphere from a reducing atmosphere to our present oxidizing one.

Cyanobacteria are abundant in both fresh and salt water. They are normally a rather inconspicuous component of marine ecosystems, but when conditions are optimal, they grow rapidly, forming **blooms**—thick mats of filaments. They become especially conspicuous then because many form gas vacuoles, float, and accumulate on the water's surface, accentuating the bloom. Such blooms have deleterious consequences because they are followed by the death of many of the cells that then sink and decompose. Decomposition consumes oxygen, leaving the water almost completely anaerobic and thus killing many fish.

SUMMARY

1. The kingdom Monera contains two divisions: Eubacteria (including Cyanobacteria) and Archaebacteria. Although cyanobacteria were formerly known as blue-green algae, none of the prokaryotes are closely related to plants.

2. Cells of prokaryotes differ significantly from those of eukaryotes, lacking membrane-bounded organelles and microtubules; walls contain polymers of amino sugars cross-linked by peptides.

3. Multicellular bodies are rare: the largest occur in cyanobacteria and contain only two or three cell types. Differentiated tissues and body-level integration have not evolved.

4. Prokaryotes have no histones; their DNA occurs as naked, closed circles in the cytoplasm. Regular, self-controlled exchange of DNA—conjugation—is known in only a few species.

5. Bacterial flagella are composed of flagellin, not tubulin, and are constructed as just a single tubule, not as a set of microtubules.

6. Prokaryotes carry out the same fundamental types of metabolism that occur in plants, but in many the processes are simpler and help us understand how the more complex eukaryotic processes may have begun and evolved.

7. Photosynthesis in cyanobacteria is almost identical to that in eukaryotic chloroplasts. In purple and green photosynthetic bacteria, light acts only to pump electrons in a cyclic electron transport chain; bacteriochlorophyll is never strong enough to reduce $NADP^+$ to $NADPH + H^+$.

8. Prokaryotes have a tremendous range of sources of energy and carbon, depending on the species. A few are lithotrophic heterotrophs that take in organic compounds but are not able to respire them and trap the energy as ATP. Lithotrophs are important, affecting the availability of nitrogen to plants.

9. Almost all features used for the identification of eubacteria and archaebacteria are chemical or metabolic. DNA and RNA sequence analyses are rapidly becoming the most important tools for analyzing the phylogeny of prokaryotes; their genomes are small enough that complete restriction maps are relatively easy to prepare, and complete sequencing is feasible.

10. Archaebacteria may be members of ancient lines of evolution, having changed little since the beginning of life, more than 3 billion years ago. They share little nucleotide sequence homology with eubacteria.

IMPORTANT TERMS

anoxygenic photosynthesis
archaebacteria
bacillus
bacteria
bacteriochlorophyll
binary fission
bloom
coccus

conjugation
conventional heterotroph
cyanobacteria
heterocyst
lithotroph
mycoplasma
naked DNA
nucleoid

photic zone
phototroph
prochlorophytes
70S ribosome
spirillum
stromatolite
transduction
transformation

REVIEW QUESTIONS

1. What does the name "prokaryote" mean? What are some differences in protoplasm organization between prokaryotes and eukaryotes?

2. What are gram-negative and gram-positive bacteria? How does the structure of the wall affect the cell's sensitivity to drugs?

3. Describe each of the following: binary fission, transformation, transduction, conjugation. Why is it difficult to do genetic studies with many species of prokaryotes?

4. In both bacterial photosynthesis and PS I cyclic electron transport, quinones play an important part in establishing the proton gradient. Do you think these two metabolisms might be homologous?

5. What is a conventional heterotroph as opposed to a lithotroph? What are some of the compounds that can be oxidized as part of lithotrophic respiration? What are some of the organic compounds required during fermentation? During putrefaction?

6. What are some of the habitats inhabited by archaebacteria? Why do we think that some of these resemble conditions on Earth at the time of the chemical evolution of life (see Chapter 17)? Considering some of the chemicals that archaebacteria use in their metabolism, do you think these organisms might ultimately provide genes for genetically engineering new types of bacteria that can clean up toxic wastes?

7. Many eubacteria are important components of soil and play an important part in the cycling of minerals. What effects do the following eubacteria have on the plant life in an area: *Beggiatoa, Cytophaga, Azotobacter, Rhizobium, Nitrosomonas, Nitrobacter, Hyphomicrobium?*

BotanyLinks .net Questions

www.jbpub.com/botanylinks

Visit the .net Questions area of BotanyLinks (http://www.jbpub.com/botanylinks/netquestions) to complete this question:

1. New diseases of plants and animals appear from time to time (AIDS and Ebola virus in animals). How do plant scientists investigate new diseases? Go to the BotanyLinks home page for information on this subject.

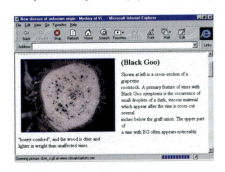

BotanyLinks includes a Directory of Organizations for this chapter.

KINGDOM MYCETEAE: FUNGI

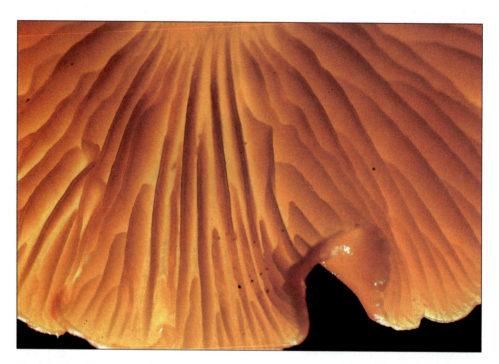

The ridgelike gills on the underside of a mushroom cap produce millions of microscopic spores. *(Ken Kleene/PHOTO/NATS)*

CONCEPTS

At one time, fungi and bacteria were considered parts of the plant kingdom, primarily because they produce spores, have cell walls, and obviously are not animals. With recognition of the differences between prokaryotes and eukaryotes, bacteria were reclassified out of the plant kingdom, but fungi were left in. More recently it has become clear that the organisms grouped together as fungi are definitely not plants. The conclusion that they had a separate evolutionary origin is based on many observations: They lack plastids; their walls do not contain cellulose (with one exception); their bodies are filamentous, not parenchymatous; the organization of large structures such as mushrooms and morels is completely different from that of plants with massive structures; and many biochemical pathways differ significantly.

Furthermore fungi are not a natural, monophyletic group; they are classified together only provisionally until we have enough information to understand their true evolutionary relationships. Many fungi are not closely related to others, and some apparently represent separate lines of evolution from very early eukaryotes. This conclusion is based on the discovery that both mitosis and meiosis are unusual in these fungi, differing from these processes in plants and animals. In contrast, class Oomycetes (Table 20.1; Fig. 20.1)

TABLE 20.1   *Classification of Fungi and Related Organisms*

Kingdom Myceteae
   Division Myxomycota*   Slime molds
   Division Eumycota   True fungi
      Subdivision Mastigomycotina†   Fungi with motile cells
         Class Chytridiomycetes
         Class Hyphochytridiomycetes
         Class Oomycetes
      Subdivision Zygomycotina   Fungi with zygospores
      Subdivision Ascomycotina   Fungi with asci
      Subdivision Basidiomycotina   Fungi with basidia
      Subdivision Deuteromycotina   Fungi without known sexual stages

*Most frequently not considered true fungi; perhaps not even closely related to the ancestors of true fungi.
†Considered true fungi by some, not by others.

---

probably has evolved from algal ancestors fairly recently. Its members have cellulose in their walls, and many of their metabolic pathways and modes of reproduction are similar to those in algae.

Several characters were originally used to unite and define fungi: complete heterotrophy with no photosynthetic stages, formation of spores, presence of chitin in their walls, and lack of complex bodies with organs. But these features may result from convergent evolution as well as common ancestry. Unicellular fungi and those with flagella are grouped together in subdivision Mastigomycotina, but these are now frequently included in kingdom Protista and are not considered true fungi (Table 20.1). Other mycologists do consider some of them either true fungi or at least related to the ancestors of true fungi and on this basis retain them in kingdom Myceteae. Subdivisions Zygomycotina, Ascomycotina, Basidiomycotina, and Deuteromycotina are composed of filamentous individuals without motile stages; these groups may represent a natural line, with subdivision Zygomycotina having originated from unicellular forms and then later given rise to subdivision Ascomycotina, which in turn produced subdivision Basidiomycotina (Fig. 20.1). This hypothesis is still tentative; many fungi are poorly known, and the gaps in our knowledge are extensive.

*The word **fungi** is pronounced as if spelled either funjee or funjai; the hard g (fun guy) is not used. Many names end in -mycetes, pronounced to rhyme either with "my treaties" (a more formal pronunciation) or with "my treats" (more informal).*

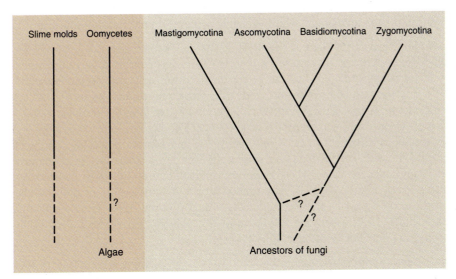

FIGURE 20.1

Most mycologists are reasonably certain that the basidiomycetes have evolved rather recently from ascomycete-like ancestors, which in turn had arisen from the early zygomycetes. However, it is not known if these three groups were derived from the ancestors or early members of the mastigomycetes: if so, all four groups share the same ancestor, form a natural group, and should all be in the same kingdom. If mastigomycetes arose independently, then probably their ancestors and those of the zygomycetes could be traced ultimately to a very distant common ancestor. But if the origin is pushed too far back, we may start to include the ancestors of algae and protozoans. The Oomycetes are almost certainly derived from algae and in some classifications are moved there.

As in other groups, nucleic acid sequencing holds great promise of providing definitive information about phylogeny. The classification here is based on that of Alexopoulos (Bold, Alexopoulos, and Delevoryas: *Morphology of Plants and Fungi,* 1987). However, if division Myxomycota and most of subdivision Mastigomycotina were to be placed in kingdom Protista, then subdivisions Zygomycotina, Ascomycotina, Basidiomycotina, and Deuteromycotina would all be raised in status to be divisions Zygomycota, Ascomycota, Basidiomycota, and Deuteromycota. This would result in a much more natural kingdom Myceteae.

In order to identify and classify a fungus, it is necessary to have the structures involved in sexual reproduction. A specimen cannot be classified definitively without these and is placed in the artificial subdivision Deuteromycotina. This taxonomic group serves basically as a "holding tank" until enough information can be obtained to classify its members correctly. All deuteromycetes are in reality members of some other fungal subdivision.

GENERAL CHARACTERISTICS OF FUNGI

Fungi constitute a large group of organisms; although about 100,000 species have been named, 200,000 more species are estimated to remain undiscovered. Like insects and orchids, fungi are quite possibly speciating more rapidly than they are being discovered; in the future we may know more fungi, but we will know a smaller percentage of the total.

NUTRITION

Modes of Nutrition.   A universal characteristic of fungi is that they are completely heterotrophic; no trace of photosynthesis is found in any stage of any group. However, because they have walls, fungi cannot engulf food as animals do; instead, fungi must obtain nutrients from the environment, from living, dying, or dead organisms. On this basis, fungi are subdivided into three types: (1) **biotrophs (parasites),** which draw nutrients slowly from living hosts, often without killing them; (2) **necrotrophs,** which attack living hosts so virulently that they kill the hosts and then absorb released nutrients; (3) **saprotrophs,** fungi that attack organisms after they have died from other causes.

Many biotrophs secrete chemicals that cause the host cell membrane to become unusually permeable to sugars or amino acids; as they leak from the host cell, the fungus absorbs them. In sophisticated parasites, damage to the host cell is so slight that the plant responds as though the fungus were a normal sink for metabolites, and extra sugars and amino acids are actually transported by the plant from other leaves to the site of infection, just as though the fungus were a developing fruit or other plant part. In biotrophic attacks, fungal cells may remain confined solely to intercellular spaces; in other species, the fungus creates a small hole in the plant cell wall, then inserts a small portion, the **haustorium,** of its filamentous cell through the hole (Fig. 20.2). The haustorium is in close contact with the plant cell plasma membrane, which probably makes it easier to absorb nutrients.

Extracellular Digestion.   Natural release of sugars may be sufficient for the fungi, but typically they secrete digestive enzymes that attack host polymers, converting them to sugars, amino acids, and lipids that can be absorbed. Saprotrophs depend predominantly on this form of extracellular digestion, and many depolymerize and consume cellulose (brown rot fungi), hemicelluloses, and even lignin (white rot fungi). The ability either to absorb host-produced monomers or to secrete extracellular digestive enzymes is often both species specific and tissue specific. Many biotrophic and saprotrophic fungi can attack successfully only a few or just one host species, or even just a single variety of one plant species. Many fungi are even tissue specific: Wilt-inducing fusariums must invade xylem and attack xylary middle lamellas; they cannot survive in cortex, pith, or phloem, even though those cells are rich in free monomers. Fungi transmitted by aphids and other phloem-sucking insects usually are able to attack only phloem; their toxins and extracellular enzymes are not effective against other tissues.

Toxins.   Necrotrophs secrete toxins that kill host cells; an example is the release of alternaric acid by *Alternaria solani* (Fig. 20.3). Many toxins kill host cells by damaging their

Plant Biology Tutor CD-ROM

Plant cell

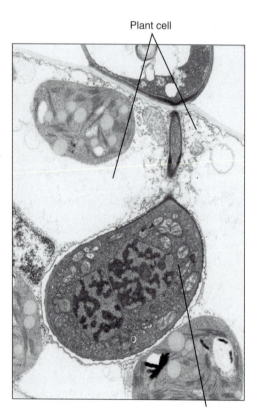

Haustorium

FIGURE 20.2

Some biotrophic fungi insert specialized portions, haustoria, through the walls of host cells. These never break the host cell's plasma membrane, so technically they never enter the cell, and nutrients must still pass across the plasma membranes of host and fungus as well as the fungal cell wall (× 5000). *(Courtesy of C. W. Mims, J. Taylor, and E. A. Richardson, University of Georgia)*

(a)

Alternaric acid

(b)

FIGURE 20.3
(a) Alternaric acid is a toxin that diffuses extensively through a leaf, killing plant cells. If dissolved in water and applied to leaves, alternaric acid is nonspecific, damaging many plant species not normally attacked by the fungus that produces it. Other fungal toxins damage only those species attacked by the fungus. (b) Dark, necrotic spots such as these on peanut leaves are a common symptom of diseases caused by fungi. (*Grant Heilman*)

plasma membranes, causing nutrients to leak out rapidly and be readily available to the fungus; loss of nutrient equilibrium kills the host cells. In many wilt diseases, the total mechanism by which turgor loss in host plants is caused is not known. In some *Fusarium* species, the fungus invades the xylem and then produces pectolytic enzymes that depolymerize the middle lamella, producing a slimy mucilage. This is drawn into tracheids and vessels by the water tension of transpiration, but the mucilage quickly blocks pits and perforations, halting water conduction. "Soft rots" of fruits are also caused by fungi that attack the middle lamella.

Excessive levels of plant hormones are involved in many fungus-induced diseases. Gibberellic acid was discovered because rice plants in Japan suffer from lethal stem elongation induced by gibberellins secreted by the fungus *Gibberella fujikuroi*. Plant hormone may be produced and secreted by the fungus, or the fungus somehow induces the plant to produce increased levels of its own hormones. The increased hormone levels and abnormal structural and metabolic changes they cause may be essential to survival of the fungus; in other cases they may be nonessential side effects.

Some plants have elaborate antifungal defense mechanisms. The simplest defense against an obligate biotroph is a **hypersensitive reaction:** Plant cells die immediately upon contact with fungal cells. Because biotrophs can survive only in living hosts, the fungus cannot draw nutrients from these dead cells and must extend farther into the host. But as it does so, more host cells die before the fungus can parasitize them. The fungus starves to death and the plant has suffered the loss of just a few cells.

Some plants resist biotrophic and necrotrophic fungi by rapidly forming wound cork around any injury caused by the fungus. Cells adjacent to the site of invasion undergo cell division, then differentiate into cork with walls saturated with suberin and lignin; these are highly resistant to enzyme attack, and the fungus is trapped.

High levels of phenolic compounds in a plant vacuole may confer resistance; if the fungus' extracellular enzymes damage the cell membrane and vacuolar membrane, phenols flood out and denature all proteins, digestive enzymes included. With these enzymes inactivated, the fungus obtains no nutrition.

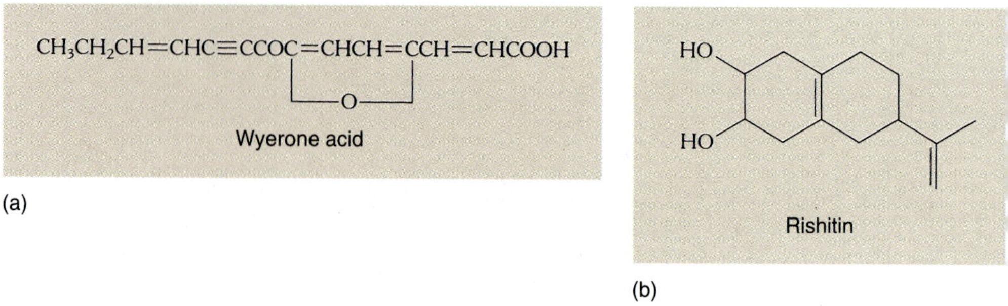

CH₃CH₂CH=CHC≡CCOC=CHCH=CCH=CHCOOH

Wyerone acid

(a)

Rishitin

(b)

FIGURE 20.4
(a) Wyerone acid, a phytoalexin produced by broad beans when attacked by *Botrytis fabae*.
(b) Rishitin, a phytoalexin produced by potato.

Many plants produce **phytoalexins**, lipid-like or phenolic compounds, in response to attack by fungi, bacteria, and even nematodes (Fig. 20.4). During invasion, **elicitors** are released; these are usually large constituents of the fungal wall which interact with the plant cell membrane, stimulating the cell to produce phytoalexins. In some cases, elicitors are extremely specific for either host or pathogen; a given fungus elicits a strong phytoalexin response in resistant plants but little or no response in susceptible ones. On the other hand, certain plants produce phytoalexins in response to some fungi but not others; the latter are pathogenic.

Phytoalexins damage fungi, bacteria, and nematodes in various ways, depending on the type of phytoalexin. Some pathogens are highly sensitive to phytoalexins, whereas others are resistant to the phytoalexins the pathogen induces because the pathogen inactivates them enzymatically. Certain virulent strains of fungi are even able to suppress the host's ability to produce phytoalexins in response to that fungus' elicitor. Very little is known about phytoalexins, but this is a fascinating area of research in co-evolution, because it is essentially a gene-for-gene battle between two organisms.

Essential Nutrients.   Fungi require essential mineral elements, and their macroelement needs are almost identical to those of plants: carbon, hydrogen, oxygen, nitrogen, potassium, phosphorus, magnesium, and sulfur. Calcium, an essential macroelement for plants, either is a microelement for some fungi or is not needed at all, being replaced by strontium. Nitrogen metabolism is similar to that of plants; virtually all fungi absorb and assimilate ammonium and many absorb and reduce either nitrate or nitrite if ammonium is not available. Some fungi cannot assimilate any form of inorganic nitrogen and must have amino acids to survive.

The common essential microelements are iron, zinc, copper, manganese, molybdenum, and either calcium or strontium. Boron and cobalt are not required by many fungi, whereas other species need gallium or scandium.

Many fungi synthesize all their own compounds if supplied with minerals and a carbon/energy source such as sugar; other fungi lack the ability to synthesize certain vitamins and must obtain them from the environment. The most commonly required vitamin is thiamine, which many yeasts and filamentous fungi need. Biotin is required by many species, as is pyridoxine (B_6). Many yeasts are unable to synthesize their own pantothenic acid.

Fungi, along with most bacteria, are agents of decay, rot, spoilage, and decomposition. As plants incorporate minerals and carbon into their molecules, these elements are locked away from other plants and are available only to the animals and biotrophs that consume the plants. Without microbial decay, dead plant and animal bodies would last almost forever, decomposing slowly, geochemically the way rocks decompose. Within a few years all available minerals in the soil would be incorporated and unavailable, and new plant growth would cease. Decomposition mediated by fungi and bacteria is rapid, and within several months, the minerals are made available for new plant growth.

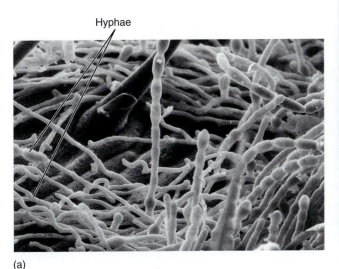

Hyphae

(a)

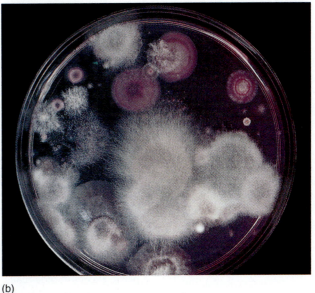

(b)

FIGURE 20.5

(a) All filamentous fungi are composed of hyphae, which are narrow, delicate, and usually transparent. Because hyphae are so long and slender (1.0 to 15 μm in diameter), they have a tremendous surface-to-volume ratio, ideal for absorbing nutrients ($\times$ 1000). *(G. T. Cole, University of Texas/BPS)* (b) All the hyphae of an individual constitute a mycelium; entire mycelia are virtually impossible to study in nature because they are so extensive, and the hyphae are so delicate that they are broken in the attempt to free one from the substrate. On agar in culture dishes, mycelia tend to be very symmetrical, but they probably are more irregular in nature where small differences in the host or substrate affect growth, branching, and sporulation. Each circular colony grew from a single spore. *(Richard Homola, PHOTO/NATS)*

BODY

Cellular Organization. The bodies of all fungi, except unicellular ones, are filamentous. Individual filaments are **hyphae** (sing.: hypha), and they branch profusely, forming a network called a **mycelium** (pl.: mycelia) (Fig. 20.5). In the "lower" filamentous fungi (subdivision Zygomycotina and class Oomycetes in subdivision Mastigomycotina), hyphae are long multinucleate cells known as **coenocytes**, but in the "higher" fungi (subdivisions Ascomycotina and Basidiomycotina), hyphae are septate and cellular.

Fungal cells are eukaryotic, having a membrane-bounded nucleus, a nucleolus, and a cytoplasm that contains membranous organelles (Fig. 20.6). Mitochondria are abundant

(a)

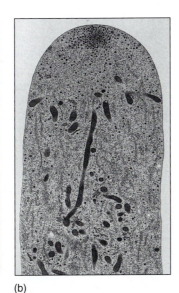

(b)

FIGURE 20.6

(a) Fungal cells can usually be distinguished immediately from plant cells owing to the lack of plastids and because vacuoles are often not particularly abundant. Mitochondria of fungi and plants usually are remarkably similar, as are ribosomes and endoplasmic reticulum. (b) Cytoplasm at the extreme tip of a hypha is dense, having only a few small vacuoles, none in this area. Dictyosomes produce vesicles that carry material to the growing wall (black dots at tip). Nuclei are generally near the tip but not located at the extreme apex. Both are *Sclerotium rolfsii* ($\times$ 4000). *(Courtesy of R. W. Roberson and M. S. Fuller, University of Arizona)*

FIGURE 20.7

The monomers of chitin are amino sugars known either as 2-acetamido-2-deoxy-D-glucopyranose or more simply as *N*-acetylglucosamine (also a component of peptidoglycan in prokaryotes). The linkage is beta-1,4 as in cellulose, and the chitin molecules hydrogen bond into microfibrils, just as cellulose molecules do. Chitin constitutes about 60% of the hyphal wall in most species but is absent in many oomycetes.

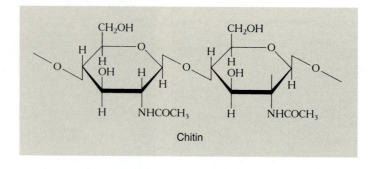

Chitin

and similar to those in plants, but endoplasmic reticulum is sparse and smooth, rarely having ribosomes attached. Vacuoles are present but usually do not dominate the protoplasm except in older cells. Dictyosomes are rare throughout most of a hypha, and plastids of all types are absent. Vesicles are common and are involved in transporting digestive enzymes across the plasma membrane to the exterior. Ribosomes are 80S, the eukaryotic type.

A hypha grows only at its tip, not throughout its length, just like roots, root hairs, rhizomes, and anything else that grows through a dense substrate. Most of the protoplasm and nuclei are aggregated in the most apical several millimeters of the hypha, and new wall materials are added at the apex (Fig. 20.6b). Walls of absorptive, vegetative hyphae are thin, less than 0.2 μm thick. The innermost wall layer is rich in **chitin**, a polymer arranged in microfibrils similar to cellulose, and providing much of the strength of the wall (Fig. 20.7). Exterior to the chitin layer is a predominantly proteinaceous layer, and outermost is a thick layer of polymerized sugars of complex structure.

The cell wall of oomycetes has no chitin at all, but rather cellulose, up to 25% of the wall's dry weight. The cellulose is often not in well-defined, crystalline microfibrils but rather is more amorphous, a feature also found in many algae. The embedding matrix is polysaccharide, similar to hemicelluloses. The oomycete wall contains large amounts of a protein rich in proline, like the wall proteins in algae and plants; it does not occur in wall proteins of other fungi.

Septa.   An important aspect of hyphal walls is the nature of the cross walls or **septa** (sing.: septum). Lower fungi have no septa. Myxomycetes are gigantic coenocytes with no walls of any type, and mastigomycetes generally are small, unicellular, nonhyphal individuals. Zygomycetes and oomycetes are filamentous, composed of hyphae, but these are usually nonseptate, each hypha being a long tubular cell with hundreds of nuclei.

Ascomycetes and basidiomycetes regularly form septa as part of their normal development. In ascomycetes, the septum has a small, simple pore in its center that allows cytoplasmic continuity from cell to cell, and organelles migrate through it (Fig. 20.8a and c). Basidiomycetes also have perforated septa, but the opening has a flange or collar and typically a hemispherical cap on each side (Fig. 20.8b and d). When hyphae are cut or chewed open, the outward rush of cytoplasm causes septal pores to become plugged, so the primary selective advantage of septa appears to be damage control rather than compartmentalization of hyphae into distinct cells.

Nuclei and Mitosis.   Nuclei and nuclear events have been extremely difficult to study in fungi for several reasons. Fungal nuclei are extremely small, less than 2 μm in diameter, whereas those of plants and animals are at least 5 to 8 μm in diameter. Chromatin in most species of fungi is very uniform, showing little differentiation into euchromatin and heterochromatin, and in some of the most important types such as yeast and *Neurospora*, chromosomes do not condense and become visible even during mitosis. In species that do have mitotic condensation, chromosomes are too small to be seen clearly or counted reliably by light microscopy. Consequently, we do not know how many chromosomes most fungi have; for those whose genetics have been studied extensively, we do know that they have fewer than 15 linkage groups.

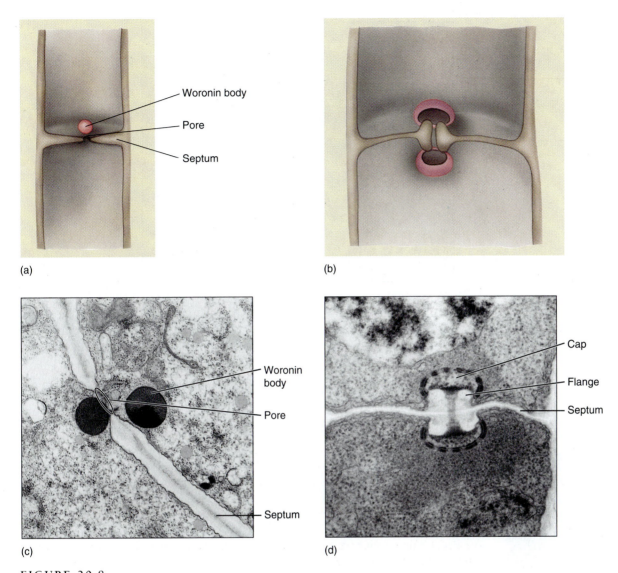

(a)

(b)

(c)

(d)

FIGURE 20.8

Hyphal septa in ascomycetes (a and c) have a characteristic structure, a pore that is closely associated with one or several "Woronin bodies." The Woronin bodies have a granular content and are surrounded by two unit membranes; in healthy hyphae, they are near but not in the septal pore. Upon injury, as protoplasm surges toward the break in the hypha, Woronin bodies are swept to the septal pore and plug it, much the way P protein and callose plug sieve plate pores in phloem. This septal structure is diagnostic for ascomycetes—all fungi with septa like this are assumed to be ascomycetes. The septum of a basidiomycete (b and d) is called a dolipore septum; it has a flange and a cap that is bounded by a layer of endoplasmic reticulum. All organelles, nuclei included, may pass through a dolipore septum, so the cap is no hindrance in normal, healthy hyphae, but the septum can be plugged if damage occurs. Virtually all basidiomycetes have dolipore septa, except for the rusts and smuts ($\times$ 20,000) *(c, Courtesy of J. W. Kimbrough, University of Florida; d, Courtesy of C. W. Mims, University of Georgia)*

The details of mitosis differ greatly, depending on the species; lower subdivisions of slime molds and nonfilamentous fungi have the greatest range of variation. During mitosis, the spindle forms inside the nucleus, which remains quite distinct because the nuclear envelope does not break down as it does in plants and animals (Fig. 20.9). The spindle always contains just a few microtubules, and some studies have shown that only one microtubule attaches to each centromere face. Only species with flagella have centrioles; the rest lack these organelles completely. In the lower forms with centrioles, a second spindle may form outside the nuclear envelope, but this may be just a centriole movement mechanism. Some reports indicate that after the centrioles are separated, the external

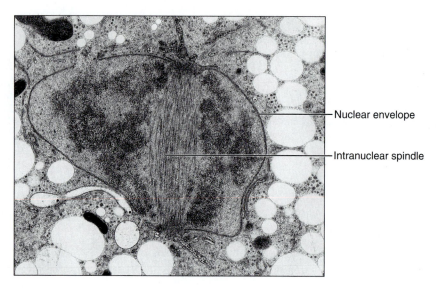

Nuclear envelope

Intranuclear spindle

FIGURE 20.9
During fungal mitosis, a spindle forms inside the nuclear envelope, which remains intact during all or most of the division process. A second spindle may form outside as well (× 19,000). *(Courtesy of E. C. Swann and C. W. Mims, University of Georgia)*

FIGURE 20.10
A morel is the fruiting body of the ascomycete genus *Morchella;* most of the surface consists of hyphae specialized for the production of spores. The fruiting body is the only portion of the mycelium that emerges from the substrate; its exposed position allows spores to be blown away and widely dispersed. *(William E. Ferguson)*

spindle may depolymerize or punch holes in the nuclear envelope of some species, its microtubules mingling with those of the internal spindle.

Chromosomes, if they condense, do not move to the center of the nucleus; there is no metaphase plate. Apparently microtubules encounter the centromeres wherever they are, and during anaphase one chromatid moves only a short distance, whereas the sister chromatid moves very far. In some species, microtubules form a dense column in the center of the nucleus, with chromosomes attached around the periphery.

During anaphase, the nucleus elongates and the nuclear envelope stretches. In some, especially yeasts, the nucleus becomes dumbbell shaped and then pinches in two. In others, the nucleus tears in the center as it stretches, and the two ends aggregate around the two sets of chromosomes. Parts of the original nuclear membrane may remain as fragments in the cytoplasm for quite some time. In other species, the nuclear envelope breaks down at the end of anaphase, and two new sets of nuclear envelope form from endoplasmic reticulum and fragments of the original envelope.

The fate of the nucleolus is similarly variable; in some it is ejected from the nucleus during prophase, whereas in others it remains intact and passes to one of the new nuclei. In others it breaks down within the nucleus during division.

Details of nuclear structure and mitosis are almost identical in plants and animals, and one may forget that the evolution of mitosis must have required millions of years and many intermediate forms. Because fungal mitosis differs from that of plants and animals, the ancestors of fungi may have diverged very early from the ancestors of plants and animals. Those fungi with the most unusual types of mitosis may retain some of the earliest steps in its evolution.

Fruiting Body Organization.   In ascomycetes and basidiomycetes, but not in the other fungi, some mycelia form a large, compact, highly organized structure called a **fruiting body,** such as a morel, truffle, mushroom, bracket, or puffball, which is the principal means of producing spores sexually (Figs. 20.10 and 20.11). Although these become quite large (mushrooms of *Termitomyces titanicus* can have a cap 1 meter in diameter), they are always composed of hyphae compacted together. Fungi never produce true three-dimensional parenchymatous tissues from an apical meristem, and perhaps as a consequence, they do not produce numerous distinct types of highly differentiated cells and tissues. However, some significant hyphal specialization can occur. In large structures, the outermost surface hyphae are often slender and have thick walls impregnated with pig-

FIGURE 20.11

(a) Brackets are persistent, perennial fruiting bodies of the basidiomycete family Polyporaceae. The underside of the bracket consists of hundreds of tubular pores lined with spore-producing hyphae. Because the pores are long and narrow, they must be perfectly vertical for the spores to fall out after they are released; brackets and other polypores are responsive gravitropically. (b) Puffballs are basidiomycete fruiting bodies in which the white outer covering breaks open, releasing millions of spores. *(Muriel V. Williams/PHOTO/NATS)* (c) The fruiting body of a stinkhorn fungus *(Claphrus)* releases spores by decomposing. Its odor of rotting flesh attracts flies, which then disperse the spores. (d) Earthstars *(Geastrum)* are related to puffballs but have an outer protective layer that peels back, forming the star-like shape. The central spherical mass contains millions of spores that are released through the pore. *(c and d, James L. Castner)*

ments called melanins. Melanins are similar to lignins, being composed of phenolic compounds; they absorb ultraviolet light, protecting nuclei inside the body, and they also deter insects from eating the fungus. Melanin may either be deposited on the surface of hyphae or spores in a cuticle-like layer or permeate the wall, binding to chitin microfibrils similarly to the way that lignin binds to cellulose in plant sclerenchyma.

Within large fruiting bodies of some fungi, especially bracket fungi, are three types of hyphae: (1) generative hyphae, which are thin walled and produce spores; (2) skeletal hyphae, which are thick walled and unbranched; and (3) binding hyphae, which are also thick walled but are highly and irregularly branched (Fig. 20.12). Binding hyphae adhere

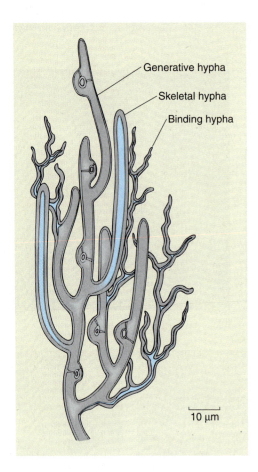

Generative hypha

Skeletal hypha

Binding hypha

10 µm

FIGURE 20.12
In a bracket fruiting body, three types of specialized hyphae occur.

to the others and cement them into a solid structure. However, this complexity is exceptional; most fruiting bodies are much simpler. The equivalents of xylem, phloem, fibers, sclereids, collenchyma, and glandular cells are never formed in even the most complex fungus.

SPORES

A universal character of fungi is their formation of spores, resistant resting stages that are the primary means of reproduction, dispersal, and survival. Spores are produced either asexually or sexually. The method of sexual spore formation differs in each subdivision and is discussed later. Asexual spores are described here.

In subdivisions Mastigomycotina and Zygomycotina, asexual spores are typically **sporangiospores**—that is, spores that form inside the large swollen tip of a hypha. A large amount of cytoplasm and many nuclei aggregate at the tip, a septum seals the region off from the rest of the hypha, then each nucleus organizes the cytoplasm around it into a spore (Fig. 20.13). When the original hyphal wall breaks down, the spores are released (Fig. 20.14). When sporangiospores germinate, they are nonmotile in all but the Mastigomycotina.

In subdivisions Ascomycotina, Basidiomycotina, and Deuteromycotina, asexual spores are more often produced as **conidia** (sing.: conidium), spores that do not form inside a sporangium. In the simplest type, the tip of a hypha forms septa, cutting off many uninucleate cells, each of which acts as an individual spore (Fig. 20.15). In a more elaborate type of development, special flask-shaped cells push out a large bud of material, which forms a cross wall and becomes a conidium (Fig. 20.16). Later, more material is pushed out from the basal cell and a new conidium forms beneath the first, a process that is repeated until long chains of conidia have been formed. The conidium-producing structures, long, upright hyphae called conidiophores, elevate conidia somewhat, improving wind dispersal. When conidia germinate, they are never motile; they always grow out as a hypha that establishes a new mycelium.

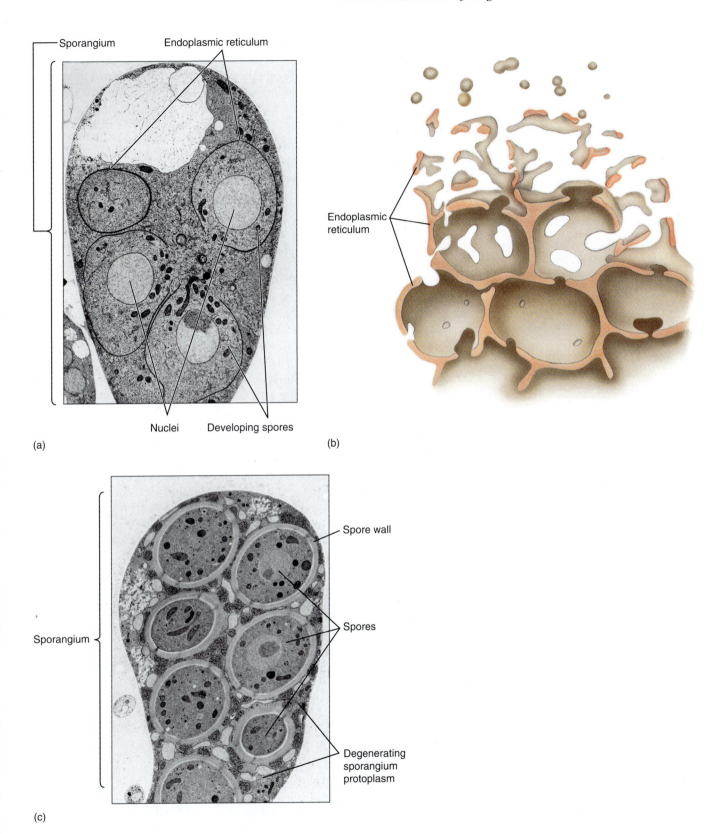

Sporangium

Endoplasmic reticulum

Nuclei   Developing spores

(a)

Endoplasmic reticulum

(b)

Sporangium

Spore wall

Spores

Degenerating sporangium protoplasm

(c)

FIGURE 20.13

(a) and (b) In the sporangium, regions of endoplasmic reticulum located between nuclei swell and merge, gradually forming sheets. As this continues, cytoplasm is partitioned and endoplasmic reticulum is modified into plasma membrane. (a) shows a complete cell; (b) shows just the endoplasmic reticulum. (c) The endoplasmic reticulum contents, which at first are amorphous, are converted to spore walls. (a and c, × 1400) *(a, Courtesy of C. W. Mims, E. A. Richardson, and J. W. Kimbrough. c, Courtesy of C. W. Mims, University of Georgia)*

FIGURE 20.14
The protoplasm of this sporangium of *Saprolegnia* (Oomycetes) has been converted to spores. *(Carolina Biological Supply)*

Fungal spores are so small and light that even mild air currents lift them easily, and strong winds carry them great distances. Spores have been collected in air samples taken at 100 km altitude. The size and number of spores are inversely related: A given mycelium can either make a few large spores or many small ones (a puffball can release 1 billion spores). Larger spores have greater amounts of nutrients and upon germination establish a somewhat extensive mycelium, at least part of which might encounter a suitable substrate or host. A small spore has few reserves; if it does not land in a favorable site, it has few nutrients it can use to grow and explore for a better site (Fig. 20.17). Of the multitudes of spores formed, at least some almost certainly land in good areas and establish a new individual.

Fungi are sensitive to fluctuating environmental conditions—winter cold and summer drought and heat. Many fungi are biotrophic on ephemeral plants and animals that live only briefly, existing for several months merely as seeds, fertilized eggs, or larvae. Because these food sources are present episodically, it is selectively advantageous for fungi to have long-lived spores that survive times of environmental stress or food absence. The simplest type of super-resistant spore is a chlamydospore. It is formed when a mass of protoplasm, rich in reserves of oil or glycogen, accumulates within a short length of hypha, then deposits very thick melanized walls. The rest of the hypha dies as conditions become inhospitable, but the chlamydospore survives, usually at least a year, until favorable conditions or hosts return. Chlamydospores are especially common in soil fungi.

Sclerotia (sing.: sclerotium) are more elaborate and even more resistant; they develop as a section of mycelium branches profusely and the hyphae attract each other, forming a compact aggregate. The outermost hyphae are swollen and globose and have thick, melanized walls. The inner mass consists of large hyphae (filled with nutrients such as oil, glycogen, mannitol, and trehalose) and small, thin-walled hyphae (rich in cytoplasm and organelles). Rather than remaining distinct, the hyphae undergo numerous fusions with each other, forming a highly interconnected mass. Large amounts of mucilage are secreted around the hyphae, which seems to act as a water-holding substance. Sclerotia are often formed by biotrophic fungi, and their germination frequently depends not merely on good conditions but on conditions favorable for the host as well. The sclerotia of *Sclerotium cepivorum* germinate only when a host root happens to grow next to it, and those of *Claviceps* germinate in the spring, just when the host grasses are starting to produce flowers.

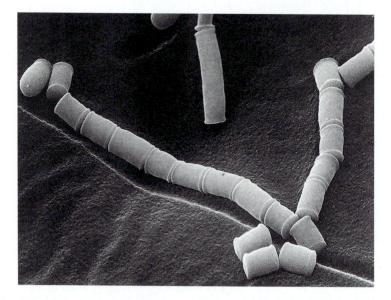

FIGURE 20.15
In the simplest method of conidium formation, nuclei space themselves at uniform intervals along the hypha; then septa form and partition this region into uninucleate spores, conidia. *Geotrichum candidum* (× 3250). *(G. T. Cole, University of Texas/BPS)*

FIGURE 20.16
Much more abundant conidia are produced by conidiophores, as in *Aspergillus.* The tip of the hypha branches into flask-shaped "sterigmata," and these push out a bubble of wall, protoplasm, and a nucleus, which then mature into a conidium. The process is repeated indefinitely, producing new conidia under existing ones. They are only lightly attached to each other, and the oldest ones may blow away while new ones are being formed. *(David M. Phillips/Visuals Unlimited)*

FIGURE 20.17
Even though this spore has landed close to a stoma, the distance it must grow in order to enter the stoma and encounter a vulnerable cell is great compared to its limited nutrient reserves. Hyphal elongation is often much greater than one would expect, however, because all the protoplasm remains in the hyphal tip, leaving the spore and original hyphal portions empty (× 400). *(Dr. Jeremy Burgess/Science Photo Library/Photo Researchers)*

HETEROKARYOSIS AND PARASEXUALITY

In most fungi, sexual reproduction does not involve production of discrete unicellular gametes such as sperms and eggs. Instead, hyphae of one mycelium fuse with hyphae of a different mycelium if the two are compatible. The two hyphae are virtually identical morphologically, with no differentiation into male and female. However, they are distinct biochemically; we say that they are of two different **mating types,** and the designations + and − are used. In order for hyphae to fuse, they must be of different mating types: A + cannot fuse with another +. This requirement ensures that when a fusion does occur, the nuclei brought together are not identical, creating some genetic diversity (Fig. 20.18).

Plasmogamy, the fusion of two hyphae, is usually not followed immediately by karyogamy; nuclear fusion is delayed slightly in zygomycetes and for a long time in ascomycetes and basidiomycetes. In these latter two groups, hyphae grow out that have a mixture of the two types of nuclei, + and −. This condition is termed **heterokaryosis;** a hypha before fusion, in which all its nuclei are identical, is **homokaryotic** (Fig. 20.19a and b). Being

FIGURE 20.18

In plants, flowers are basically functioning in plasmogamy: They are a means by which a cell (sperm) of one plant can move to and undergo syngamy with a cell (the egg) of another plant. But in fungi, this role is carried out by the mycelium itself. Hyphae fuse, undergoing plasmogamy; the fruiting bodies such as mushrooms are only formed afterward. In this Petri dish, two compatible mycelia have been established by inoculating with + hyphae on one side and − hyphae on the other. As they grew together in the center of the dish, the two individuals merged at hundreds of points, each representing a distinct act of sexual reproduction. The dark structures are melanized spores formed after plasmogamy and karyogamy were completed. (Dennis Drenner)

heterokaryotic is similar to being diploid: + nuclei may carry different alleles than − nuclei, so the hypha is heterozygous and the effects of impaired alleles are masked. Heterokaryosis can be superior to diploidy, however, because the various hyphae of a single mycelium can each undergo fusion with different hyphae from numerous other mycelia; one mycelium then has several distinct heterokaryoses simultaneously.

Karyogamy does occur ultimately, in a special reproductive structure characteristic of each subdivision. Two nuclei of compatible mating types pair and fuse into a diploid nucleus (Fig. 20.19c and d), followed immediately by meiosis, during which synapsis and crossing-over occur. Before karyogamy, the heterokaryotic mycelia of ascomycetes and basidiomycetes grow extensively, forming fruiting bodies; karyogamy occurs only in cells at the tip of the hyphae, but in an average-sized fruiting body there can be millions or even billions of these (Fig. 20.20). Each act of plasmogamy results in millions of karyogamies, resulting in millions of types of + and − spores, having undergone recombination of alleles (see Chapters 16 and 17). The spores blow away; each one that germinates grows into a new + or − mycelium.

In at least a few fungi, for example *Aspergillus nidulans* and wheat rust fungus (*Puccinia graminis*), a **parasexual cycle** occurs in which compatible nuclei fuse prematurely, even though they are not part of a fruiting body. This happens only rarely, less than two nuclei in a thousand, but once they are diploid, meiosis may occur accompanied by crossing-over. The hypha returns to a heterokaryotic state with only haploid nuclei. If these nuclei become involved in the production of conidia or sporangiospores, they produce spores with a genetic make-up different from that of the original nuclei. For the many fungi that rarely or never undergo sexual reproduction (the deuteromycetes), the parasexual cycle is the only means of genetic recombination.

FIGURE 20.19

(a) Hyphae of two compatible individuals have encountered each other. Until this point, all nuclei in one hypha have been identical (either + or −), having been produced by mitosis of the original spore nucleus of each. Each mycelium is homokaryotic. (b) Plasmogamy has occurred; a hole is digested between two cells and the cytoplasm fuses, resulting in a binucleate cell. This heterokaryotic cell grows out and establishes a new hypha, each cell of which has two nuclei, one of each type. Each of the original homokaryotic hyphae continues to grow and may fuse again in the future, either with each other or with different mycelia. (c) Ultimately, in the tip cell of a heterokaryotic hypha, the nuclei undergo karyogamy to the diploid condition as shown here. This is immediately followed by meiosis with crossing-over. (d) Four meiospores are formed, two + and two −.

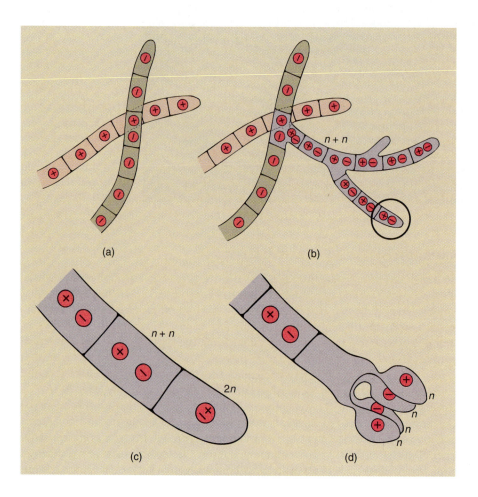

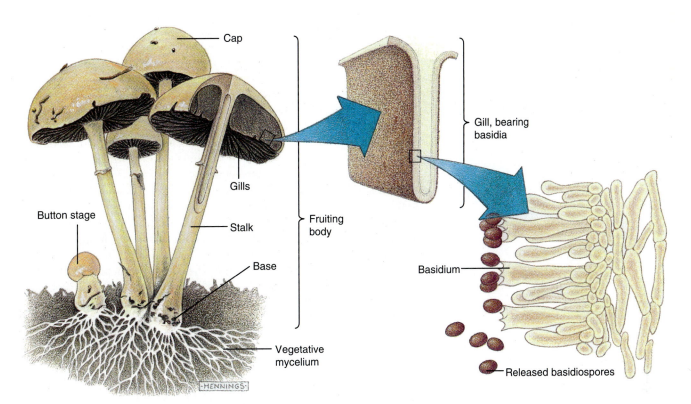

FIGURE 20.20
All hyphae that compose a mushroom are heterokaryotic, but only nuclei in cells at the tips of the hyphae that end in the gills undergo meiosis. These cells are the only ones that ever become truly diploid and then perform meiosis. All other cells remain heterokaryotic and die as the mushroom decomposes. In plants, too, only a few cells in anthers and ovules undergo meiosis.

METABOLISM

Although fungi have remained rather simple morphologically, they have become widely diverse in their metabolism, ecology, and life cycles. Many are capable of either aerobic or anaerobic growth, switching to a fermentative respiration when oxygen is not available. This is the basis of the brewing and vinting (wine making) industries: Yeasts used for making beer and wine are ascomycetes. Similarly, yeast is used to leaven bread, that is, to make it rise by giving off carbon dioxide and thereby forming bubbles in dough. The characteristic flavors of cheeses come from the fungi used in making them: Camembert cheese contains *Penicillium camemberti,* Roquefort cheese is made with *P. roqueforti,* and so on.

Fungi can attack many different substances or hosts; although each species may be very specific in its nutritional requirements, so many fungi exist that almost any plant, animal or prokaryote can be attacked, and fungi even attack other fungi. Common diseases are rusts and smuts, but damping off and heart rot are also fungal diseases. Fortunately for humans, most pathogenic fungi require warm, humid conditions, so we are not often attacked. The two most common fungal diseases for us, at least in temperate zones, are athlete's foot and ringworm, which is not a worm at all. Unfortunately, a common fungus that normally lives quietly within us, a species of *Candida,* has started to become a significant problem. When people are given drugs that impair the immune system in order to allow organ transplants or to treat cancer, or if their immune system is damaged by AIDS, *Candida* becomes uncontrolled by the body and grows rapidly, becoming a serious disease instead of a relatively harmless biotroph.

The physiological diversity of fungi is also shown by the ability of some not only to survive but actually to thrive in harsh environments. As with bacteria, some are extreme thermophiles, growing best at temperatures up to 50°C but poorly at or below 20°C. *Mucor pusillus* and *Thermomyces lanuginosus* are two examples of thermophilic fungi; like most of the others, they tend to inhabit compost piles, wood chip piles, and garbage dumps, all special environments in which decomposition and bacterial respiration produce high temperatures in rich organic matter.

Psychrophilous fungi are those that grow best in cold conditions, in the range of −10 to −15°C. *Sclerotinia borealis* grows most rapidly at 0°C in culture and shows some growth even at −7°C. Such fungi are inhabitants of extreme altitudes and latitudes; Arctic tundra contains many species of all classes of fungi. *Aureobasidium pullulans* and *Sporotrichum carnis* are just two of several fungi that grow on food and meat even while it is frozen; these can be stopped only by keeping the temperature below −15°C. Psychrophilous fungi called snow molds attack turf grasses during winter while they are covered with snow; the plants are healthy in autumn but diseased when snow melts in spring.

Xerophilous fungi grow on "dry" substrates, dryness being due not to lack of water but to a high concentration of solutes or other materials that bind water firmly, causing the water potential to be extremely negative. Examples of dry substrates are jams, jellies, dry grains, leather, dried fruit, and salted fish and meat. These fungi do not merely tolerate high solute content but actually require it: If the moisture content is raised, they do not grow as well. Xerophilous fungi draw water from their substrate by producing a high concentration of solutes, sugar alcohols such as mannitol, in their own cytoplasm. These reach such high concentrations that enzymes and membranes of the fungus are hydrated not just by water but also by the hydroxyl groups of the alcohols.

DIVISION MYXOMYCOTA (SLIME MOLDS)

This division contains slime molds, organisms quite distinct from true fungi. They are heterotrophic and form spores, but they lack walls and have a unique body organization. In true slime molds, the body is a large mass of protoplasm with a volume of several cubic centimeters containing thousands or millions of nuclei, all in the same cytoplasm and covered only by a plasma membrane (Fig. 20.21). This mass of protoplasm, called a **plasmodium** (pl.: plasmodia), is capable of migrating over a substrate, much like an amoeba, but is so large that it is easily visible to the naked eye. The plasmodium digests material from the substrate as it moves along; bacteria, yeasts, and decayed plant material are the most common nutrients and are engulfed just as bacteria are engulfed by an amoeba. Such consumption of particulate matter is not possible in true fungi because of their rigid cell walls.

FIGURE 20.21

Many slime molds are easy to grow in culture; this *Physarum polycephalum* can be raised on oatmeal. This is one large mass of protoplasm with no walls; the cytoplasm streams vigorously and surges forward—on a microscopic scale—causing the entire mass to roll over and engulf leaf litter, dead insects, and bacteria, parts of which are digested and absorbed. Slime molds are often inconspicuous on the decomposing leaves of rich soil, but under warm humid conditions, they can reach large sizes like this one. *(Patrick W. Grace/Science Source/Photo Researchers, Inc.)*

(a)

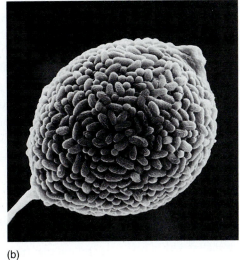

(b)

FIGURE 20.22

(a) The sporangia of *Stremonitis splendens:* The protoplasm aggregated into mounds, then formed a rigid structure which most of the protoplasm climbed. At the top, cell walls formed, spores differentiated, and nuclei underwent meiosis when the spores were about 18 to 30 hours old (× 30). *(Runk/Schoenberger from Grant Heilman)* (b) Scanning electron micrograph of a sporangium of *Dictyostelium discoideum;* the protoplasm has aggregated into spores (× 550). *(Courtesy of M. Sameshima, Tokyo Metropolitan Institute of Medical Science)*

In response to environmental cues, a plasmodium aggregates its protoplasm, forming one or several mounds; it then extends upward, producing sporangia on stalks (Fig. 20.22a and b). Spores with true walls are formed and released; the spores are extremely resistant, surviving for many years even under adverse conditions. Meiosis occurs after the spores form; then three nuclei degenerate. Upon germination, the spores may release either an amoeboid or a flagellated cell, and the two forms are interconvertible. The cell grows and may undergo both nuclear and cellular division, proliferating into a population of haploid cells. Each cell contains mating type factors, and when two compatible cells meet, they fuse into one mass, mixing the two types of nuclei. Fusion is not limited to two cells; many cells may fuse, resulting in a multinucleate plasmodium. Karyogamy occurs shortly thereafter, and nuclei are then diploid. The plasmodium continues to migrate and feed until induced to undergo another round of spore formation.

The phylogenetic—evolutionary—relationships of slime molds to other organisms are not known. They may represent a surviving line of evolution which originated not long after eukaryotes arose and which has changed little since. It is also possible that some are extremely reduced forms that evolved from more advanced organisms. Uncertainty about their evolutionary position is compounded because other groups—cellular slime molds (Acrasiomycetes), protostelids (Protostelidiomycetes), and net slime molds (Labrinthulales—have other combinations of fungal and nonfungal characters. Many of these have received very little study, but some may represent lines of evolution that arose before the genes controlling multicellular organization evolved.

DIVISION EUMYCOTA (TRUE FUNGI)

SUBDIVISION MASTIGOMYCOTINA

Most mastigomycetes (Table 20.2) have chitinous walls like other members of Eumycota, but they are distinct in having flagellated motile cells. Their flagella have the standard 9 + 2 arrangement of microtubules found in flagella of other eukaryotes. The class Chytridiomycetes contains those members that have a single posterior **whiplash flagellum**, one with a

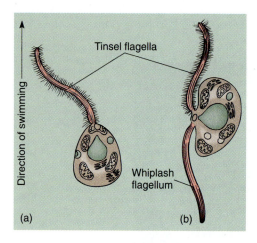

FIGURE 20.23

A whiplash flagellum has a smooth surface, whereas a tinsel flagellum has numerous fine projections. The projections increase the power of the flagellum's movement and can vary in type and arrangement, occurring in one or two rows or on all sides. Tinsel flagella occur only in oomycetes, hyphochytridiomycetes, and algae. The flagella would be much longer and narrower relative to the body size than shown here.

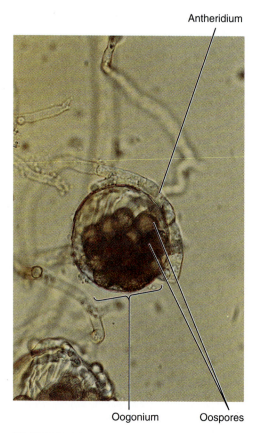

FIGURE 20.24

Saprolegnia: An antheridium has already contacted the oogonium and transferred microgametes. Oospores have formed. This situation is a form of oogamy, which occurs in plants and some algae but in few other fungi (×500). (*J. W. Richardson/CBR Images*)

TABLE 20.2	*Classification of Subdivision Mastigomycotina*
Class Chytridiomycetes	*Allomyces, Chytriomyces, Coelomomyces*
Class Hyphochytridiomycetes	*Rhizidiomyces*
Class Plasmodiophoromycetes	*Plasmodiophora*
Class Oomycetes	*Achyla, Peronospora, Phytophthora, Plasmopara, Pythium, Saprolegnia*

smooth surface (Fig. 20.23), whereas members of class Hyphochytridiomycetes have a single anterior **tinsel flagellum**, one with numerous small projections on its surface. Members of class Oomycetes have two flagella, either subapical or lateral, one a whiplash flagellum that trails backward as the cell swims, the other a tinsel flagellum that projects forward from the cell as it swims. All groups are primarily water molds, with most species living in streams, ponds, or lakes.

Class Chytridiomycetes. Chytridiomycetes (usually called just chytrids) are either unicellular throughout their lives or form a small nonseptate mycelium, as in *Allomyces*. Chytrids live on various substrates, some being parasitic on other water molds and algae and others saprotrophic on dead insects and plant parts. Chytrids have chitin in their walls.

Class Oomycetes. Many botanists have concluded that oomycetes evolved from algae that lost the ability to synthesize chlorophyll. They are the only fungi with cellulose in their walls; they have tinsel flagella as do algae; and their chromosomes are relatively large. At present, class Oomycetes is diverse in structure and nutrition. Most are water molds, but others live in dry habitats such as on and in leaves. They may consist of single cells that live as parasites inside rotifers, minute aquatic animals, whereas other oomycetes form extensive nonseptate mycelia that live as biotrophs, necrotrophs, or saprotrophs.

Sexual reproduction occurs in most oomycetes: A hypha swells into a bulbous *oogonium,* which contains one or several diploid nuclei that undergo meiosis, each in its own mass of protoplasm (Fig. 20.24). An adjacent hypha elongates as an *antheridium* and contacts the oogonium; a hole forms and microgametes pass directly into the oogonium, fertilizing the megagametes. The resulting *oospores* are diploid and are actually zygotes; each matures into a resting spore with a thick cellulose wall. These remain viable for many months even under dry, adverse conditions. Under favorable conditions, they germinate into diploid hyphae.

Whereas most oomycetes such as *Saprolegnia* and *Achlya* are aquatic, others are terrestrial, such as the two severe plant pathogens *Plasmopara viticola* and *Phytophthora infestans.* The first attacks grapes and was introduced to France in the late 1870s on grafting stock imported from the United States. *P. viticola* caused tremendous damage throughout vineyards, except along roadsides where a mixture of lime and copper sulfate had been applied to the plants to keep passersby from stealing grapes. Once it was recognized that this inhibits the fungus without damaging the grape plants, it was used for treating entire vineyards.

Phytophthora infestans, an oomycete disease of potatoes, caused the great potato blight and famine in Ireland in 1846–47. At that time Ireland was overpopulated, and the potato was almost the only food for most people. The infection with *P. infestans* spread so rapidly and virulently that virtually the entire food crop for the country was destroyed; as many as 800,000 people starved to death the following winter, and more than 1 million emigrated, mostly to the United States. *P. infestans* is still a serious disease, and in cool, moist climates, it is controlled only by regular spraying with fungicides such as zineb and chlorothalonil (Daconil).

SUBDIVISION ZYGOMYCOTINA

This subdivision contains about 600 described species (Table 20.3) that are mostly terrestrial and live in decaying plant and animal matter in soil or forest litter. They are familiar as the mold on stale bread, *Rhizopus stolonifer*. Zygomycetes have simple mycelia composed of branched coenocytic hyphae; complex fruiting bodies are not formed. Haploid spores germinate by sending out a hypha that soon becomes multinucleate and branches profusely (Fig. 20.25). Where hyphae touch the substrate, small projections, **rhizoids,** are sent down into the material, both anchoring the hyphae and absorbing nutrients by acting like roots. Above the rhizoids, more horizontal hyphae extend outward, continuing the expansion of the mycelium; other hyphae, sporangiophores, grow directly upward and form spores. Some spores are blown far, whereas others drop close to the sporangiophore. This is not a waste of spores, because the long-distance spores may initiate new, distant colonies, and those that drop nearby reinfect the same substrate that the original mycelium is growing on, thus attacking and digesting it more quickly, before spores of other fungi land on it.

Sexual reproduction in zygomycetes occurs if hyphae of one individual come close to those of another of compatible mating strain. Each mycelium produces short, multinucleate branches that grow toward equivalent branches on the other mycelium (Fig. 20.26). When the branches meet, the contacting walls break down, plasmogamy occurs, and the nuclei pair and fuse. The result is a large **zygosporangium** that has many diploid nuclei; it

TABLE 20.3	*Classification of Subdivision Zygomycotina*
Zygomycetes	*Entomophthora, Mucor, Phycomyces, Pilobolus, Rhizopus*

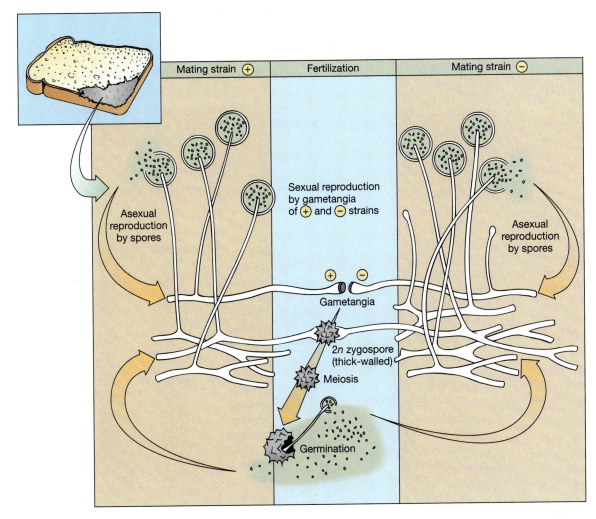

FIGURE 20.25
Life cycle of bread mold, *Rhizopus stolonifer*. Details are given in the text.

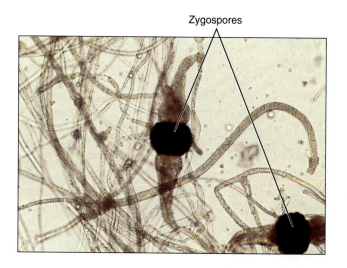

Zygospores

FIGURE 20.26
Zygomycetes are characterized by their method of sexual reproduction, the zygosporangium. After the two specialized hyphae meet, plasmogamy and the mingling of nuclei and cytoplasm occur, and nuclei pair and fuse in a common cytoplasm. This develops into a resting structure, often called a zygospore but more accurately considered a resistant sporangium. As it germinates, meiosis occurs, and a stalked sporangium is formed (× 600). (*J. W. Richardson/ CBR Images*)

becomes dormant and inactive, often for months. When it germinates, nuclei undergo meiosis, a sporangiophore is formed (but not a mycelium), and the new, haploid spores blow away and continue the life cycle. No flagellated cells are produced at any point in zygomycetes.

SUBDIVISION ASCOMYCOTINA

When taxonomists study and classify a group of organisms, they usually hope to find a character that perfectly defines a natural group; that is, all members have the character whereas all nonmembers lack it. Ascomycetes (Table 20.4) have such a character, the **ascus** (pl.: asci), a large saclike cell in which karyogamy and meiosis occur and in which the resulting meiospores (**ascospores**) form (Fig. 20.27). All ascomycetes have asci, and all fungi with asci are ascomycetes. With at least 30,000 described species, ascomycetes are familiar to you: morels, truffles, yeasts, Dutch elm disease, and powdery mildew on fruits (Fig. 20.28). *Neurospora,* an important research organism, is an ascomycete.

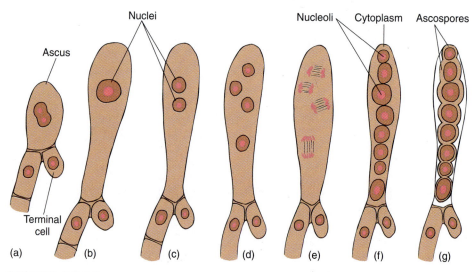

FIGURE 20.27
An ascus is the subterminal cell of a reproductive hypha of an ascomycete; in it, the two compatible nuclei fuse, then immediately undergo meiosis and crossing-over. Subsequent mitotic division may or may not occur, so either four or eight ascospores are formed.

TABLE 20.4	*Classification of Subdivision Ascomycotina*
Class Ascomycetes	
Subclass Hemiascomycetidae	*Saccharomyces*
Subclass Plectomycetidae	*Eupenicillium*
Subclass Hymenoascomycetidae	*Claviceps, Humaria, Monilinia, Morchella, Neurospora, Peziza, Sarcoscypha*
Subclass Loculoascomycetidae	*Guignardia, Venturia*

Ascomycetes are thought to have evolved from a zygomycete-like ancestor. The majority of their life and the largest portion of their bodies are a mycelium of hyphae. The hyphae are septate, but the cross walls are not complete: They each contain a pore large enough for cytoplasm and nuclei to move from one "cell" to the next rather easily (see Fig. 20.8). Ascomycetes are more derived than zygomycetes because most form a rather organized, pseudoparenchymatous fruiting body, the **ascocarp** (see Fig. 20.10). Little differentiation of tissues occurs in an ascocarp—perhaps a firm covering layer and a region of asci (called the **hymenium**; Fig. 20.29) but no specialized tissues for conduction, support, or nutrient storage. Many ascocarps are ephemeral, decomposing a few days after formation.

The initial parts of the life cycle are like those in zygomycetes; germination of a haploid spore results in a mycelium of branched septate hyphae (Fig. 20.30). Cells are either uni- or multinucleate, depending on the movement of nuclei through septa. Asexual reproduction occurs by formation of conidia. Sexual reproduction initially resembles that of zygomycetes, because two compatible hyphae that happen to approach each other each produce special short, multinucleate branches. One is called an **ascogonium** and the other an **antheridium**; the ascogonium in a number of ascomycetes sends a small tube **(trichogyne)** to the antheridium and receives nuclei through the trichogyne (Fig. 20.30). **The nuclei pair but do not fuse;** because the sets of genes are not in one nucleus, the cells are not considered diploid; instead we say they are dikaryotic. The ascogonium does not become dormant like a zygosporangium; instead, new hyphae sprout from it, forming a new myce-

FIGURE 20.28

The ascomycete *Ceratocystis fagacearum* causes oak wilt. This is a relatively new disease that has been spreading southward through the Mississippi Valley area, killing various species of oaks. *(Courtesy of Jieh-Juan Yu and Garry Cole, University of Texas)*

Spores

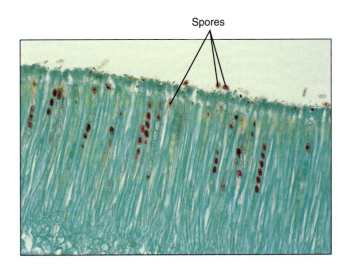

FIGURE 20.29

The hymenium of a *Peziza*, an ascomycete, is composed of thousands, even millions, of asci. Because each ascus produces four or eight ascospores, the reproductive potential of each fruiting body is prodigious (×100).

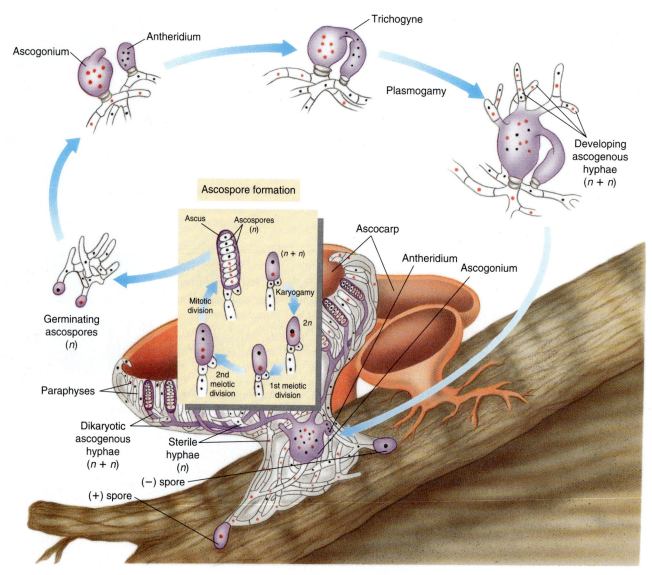

FIGURE 20.30
Life cycle of an ascomycete; details are given in the text.

lium. The nuclei migrate into these **ascogenous hyphae** in pairs, and because they undergo simultaneous mitosis, new pairs of nuclei are produced. Each cell therefore is heterokaryotic and contains two sets of genes.

The ascogenous hyphae, along with new branches of the original haploid hyphae, grow in an organized manner and produce the ascocarp. Ascogenous hyphae form the fertile layer, the hymenium, and the subterminal cell of each hypha forms an ascus. The ascus initial must contain one nucleus from each parent, and this is accomplished in the following manner: The tip bends over strongly, forming a hook (called a crozier) (see Fig. 20.27), and two complementary nuclei undergo simultaneous mitoses with parallel spindles. Two cell walls seal off one nucleus in a small, lateral tip cell and isolate two complementary nuclei in an ascus initial. The nuclei fuse and then undergo meiosis. Meiosis may be followed by mitosis, so the ascus contains either four or eight ascospores, or rarely 16, 32, 64—the record is about 7000 in each ascus of *Trichobolus*. Immediately after nuclear division, all nuclei lie in a single mass of cytoplasm; this is organized into uninucleate spores, as described earlier for sporangiospores (see Fig. 20.13). Depending on the group, the ascus may develop a pore or a lid or may just disintegrate, liberating the ascospores, which blow away and initiate new haploid mycelia.

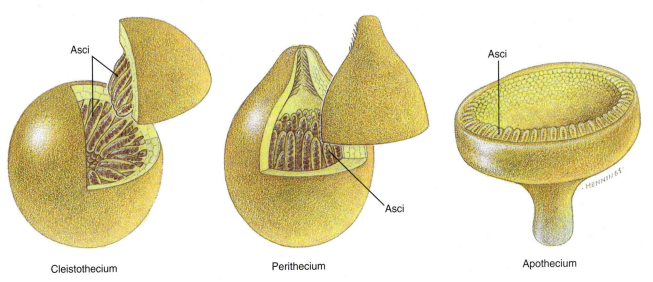

FIGURE 20.31

Types of ascocarps: cleistothecium, perithecium, and apothecium.

Cleistothecium | Perithecium | Apothecium

Ascocarps are of various types, all rather simple. The *cleistothecium* is spherical with no opening; sterile hyphae form the outermost layers, and asci and ascospores are located in the center. Spores are released as the cleistothecium decays (Fig. 20.31). A *perithecium* is a flask-shaped ascocarp that releases ascospores through a narrow opening. *Apothecia* are disk- or saucer-shaped, with the hymenium on the upper surface and sterile hyphae on the underside. Apothecia may be small, simple disks, as in *Humaria*, *Peziza*, and *Sarcoscypha*, or they may be large, as in morels (*Morchella*).

It is important to compare this life cycle to that in plants and animals: There are no sperms, eggs, or zygotes, but syngamy, meiosis, and crossing-over do occur. All the essentials of sexual reproduction occur, but in a different manner. The same is true of basidiomycetes.

Although ascomycetes are generally more complicated than zygomycetes, some have become quite reduced. **Yeasts** are unicellular, without a mycelium, but individual cells still perform most of the functions of a large mycelium. They bud off new cells like conidia, or they fuse with other cells, acting like ascogonia and antheridia, developing into asci with karyogamy and meiosis taking place; ascospores are formed inside the original wall.

SUBDIVISION BASIDIOMYCOTINA

Basidiomycetes (mushrooms, puffballs, bracket fungi; Table 20.5) are the most familiar of all fungi and, like ascomycetes, are clearly delimited by a characteristic feature, the **basidium** (pl.: basidia), a club-shaped terminal cell within which karyogamy and meiosis occur and which produces **basidiospores** externally (see Figs. 20.19 and 20.32). Much of the

Basidium | Basidiospores

FIGURE 20.32

A basidium is the terminal cell of a reproductive hypha of a basidiomycete; in it, the two compatible nuclei fuse, then immediately undergo meiosis and crossing-over. Each basidium has four or eight projections (sterigmata); cytoplasm and a nucleus are pushed out through each, forming a basidiospore. The sterigmata are extremely narrow, causing all organelles, the nucleus included, to be squeezed and distorted severely as they pass through.

(Biophoto Associates)

TABLE 20.5 *Classification of Subdivision Basidiomycotina*	
Class Basidiomycetes	
Subclass Holobasidiomycetidae	*Agaricus, Amanita, Boletus, Coprinus, Geastrum, Nidularia, Phallus, Polyporus, Psilocybe*
Subclass Phragmobasidiomycetidae	*Auricularia, Tremella*
Subclass Teliomycetidae	
Order Uredinales	*Puccinia, Phragmidium*
Order Ustilaginales	*Ustilago*

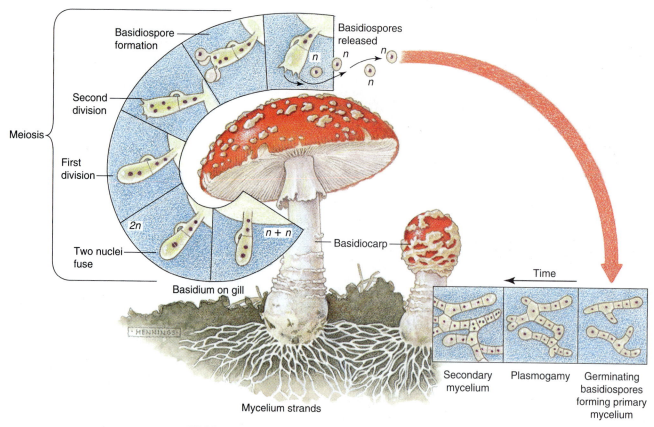

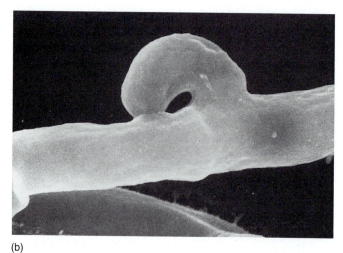

FIGURE 20.33
Life cycle of basidiomycetes; details are given in the text.

(a)

(b)

FIGURE 20.34
(a) Clamp connection in basidiomycetes is somewhat like ascus formation; backward growth occurs from the most apical portion of a hypha. This fuses with a lower portion of the same hypha and simultaneously two mitotic divisions of two nuclei occur, one of each parental type. Two new septa are formed, establishing a heterokaryotic tip cell. (b) Scanning electron micrograph (×20,000). *(Courtesy of S. L. Flegler, Michigan State University)*

FIGURE 20.35
This bracket is a perennial basidiocarp; it persists from year to year, forming a fresh region of hymenium each year.

basidiomycete life cycle is like that of the ascomycetes in that it begins with a haploid mycelium, also containing cross walls with pores (see Fig. 20.8). When compatible hyphae encounter each other, no formation of ascogonia or antheridia occurs. Instead, with the exception of the rust fungi, regular hyphae contact each other and fuse (Fig. 20.33). From this cell a new dikaryotic mycelium is established, with each cell containing one of each type of nucleus.

Ascomycetes maintain a dikaryotic condition in each ascogenous hypha cell by mere nuclear migration, but in basidiomycetes the process is more carefully controlled. Each nuclear division is accompanied by the formation of a **clamp connection** (Fig. 20.34), and each cell always has one nucleus from each parent. The dikaryotic mycelium grows in a diffuse manner, as do the initial haploid mycelia. When they receive the proper stimulus, certain regions undergo an organized growth to form the fruiting body, the **basidiocarp,** which may have generative, skeletal, and binding hyphae. Basidiocarps such as brackets are tough and persistent, growing larger each year (Fig. 20.35). Basidiocarps always contain an extensive surface area, most having numerous gills or pores, and the ends of the hyphae that terminate on these surfaces become basidia (see Fig. 20.20). Karyogamy occurs in the terminal cell, followed by meiosis. Basidiospores do not form inside the basidium, however; instead four tiny projections form (**sterigmata;** sing.: sterigma), and both cytoplasm and one nucleus are squeezed out through each sterigma. This protoplasm expands, forming a basidiospore on the tip of each sterigma. At maturity, the spores are released and blow away, establishing new mycelia.

Basidiomycetes are important as food; the mushrooms most commonly used in cooking are *Agaricus bisporus,* and in the United States over 60,000 tons are eaten annually. Many mushrooms are dangerous. In many cases, mushroom identification is difficult, and poisonous species can greatly resemble edible ones.

SUBDIVISION DEUTEROMYCOTINA (FUNGI IMPERFECTI)

For organisms that reproduce asexually, mutations that cause the loss of sexual reproduction are not immediately disadvantageous selectively. Certainly the new mutant line cannot evolve as rapidly without sexual reproduction, but it may be able to persist indefinitely using only asexual means of propagation. When this occurs in fungi, we say that the organism is **imperfect,** and it is assigned to the deuteromycetes (also called the fungi imperfecti). At present, the group contains about 25,000 named species. In the past, these simple masses of hyphae were virtually impossible to classify because they have so little morphology to examine. However, electron microscopy has made it possible to examine the structure of their septa, so they can be assigned at least to zygomycetes, ascomycetes, or basidiomycetes; almost all are ascomycetes.

Incomplete flowers that lack either stamens or carpels or both are called imperfect flowers, for the same reason (see Chapter 9).

Many, perhaps most, species actually have sexual stages, but we have not found them. Think about how fungi are studied: A sample of soil or decomposing material is examined microscopically for spores and mycelia. Mycelia often look pretty much alike, but spores are typically quite distinctive. If a completely new spore is found, it is named as a new species. But one spore does not tell us much about the whole organism. It would be best to culture the spore along with another of compatible mating type, grow the mycelia, then watch reproduction, both asexual and sexual. Many of the "imperfect" fungi probably are really the vegetative states of normal ascomycetes and basidiomycetes whose sexual stages have not yet been encountered. Whenever these stages are found, the species is reclassified out of the deuteromycetes.

ASSOCIATIONS OF FUNGI WITH OTHER ORGANISMS

LICHENS

Lichens are an association of a fungus with an alga or a cyanobacterium (Fig. 20.36). There are about 13,500 lichen "species"; because they are an association rather than a single organism, they are not considered true species. The fungi in most lichens are ascomycetes (only about a dozen are basidiomycetes), and the algae are most often green algae (especially the genera *Trebouxia* and *Trentopholia*), but cyanobacteria are also frequent. Lichens are commonly assumed to be a symbiotic relationship in which each organism benefits: The fungus (often referred to as the mycobiont) receives sugars, thiamine, and biotin from

(a)

(b)

(c)

FIGURE 20.36
(a) *Psora decipiens,* a crustose lichen growing as a thin mat that adheres firmly to its substrate. (b) A foliose lichen (*Parmotrema* sp.) grows as a "leafy," open network of thin, flat sheets that project away from the substrate. (c) This is a fruticose or "shrubby" lichen (*Cladonia cristatella); * like a foliose lichen, it projects out from the substrate, but its body is composed of cylindrical elements. (R. Calentine)

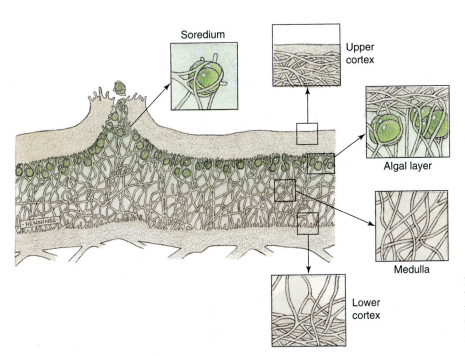

Soredium

Upper cortex

Algal layer

Medulla

Lower cortex

FIGURE 20.37
Diagram of the three-layered structure found in many lichens; different structures are also possible. This "body" has more complex differentiation and more tissue-like organization than most fungi or algae.

the alga (the phycobiont) or cyanobacterium, and the alga is protected from desiccation by the fungus. The body of many lichens is stratified with a thin, tough fungal upper layer that shields the autotroph, a middle layer in which most of the algal or cyanobacterial cells occur, and the lowest and thickest layer, composed of loosely packed hyphae (Fig. 20.37). This layer may be 0.5 mm thick and is involved in storing water and keeping the algae or cyanobacteria moist.

Lichens may actually represent specialized forms of parasitism: The autotrophs might not benefit from the association at all. In culture, they grow much more rapidly when free of the fungus than when combined with it in lichen form. Circumstantial evidence exists that the fungus interferes with the wall metabolism of the algae or cyanobacteria, such that sugars are secreted but polymerization is blocked. The fungus absorbs the unpolymerized sugars; presumably the autotroph secretes more sugar as a response to the continued thinness and weakness of the wall. In some, the fungus actually penetrates the alga, inserting haustoria into it.

Lichens are extraordinarily hardy, possibly because they can dry rapidly and enter a resistant state of suspended animation. With just a little moisture, as during a morning dew, lichens rapidly rehydrate and begin photosynthesis and growth. Many are metabolically active for only an hour or less each day, but this is enough, and lichens grow on bare rock, fence posts, and deserts, and in Antarctica (where about 350 species occur). Because their environments are typically extremely barren, lichens must be especially efficient at absorbing nutrients. This efficiency has become a problem for some, because in many areas they absorb toxic pollutants from the air and die.

Although lichens are associations, in many ways they act remarkably like individuals. They undergo reproduction either by fragmenting or by producing **soredia** (sing.: soredium), small masses of hyphae and autotrophic cells. Soredia are light enough to be distributed by wind, water, and small animals. The fungi produce spores, and the algae in some also produce reproductive cells. Whereas most lichen autotrophs can be grown in culture and their free-living form can be studied, only a few of the fungi can be grown successfully. Consequently, we do not know what the morphology of the fungi would be without the autotroph. Many aspects of the form of the fungus have been found to be controlled by the autotroph. Some lichens, such as *Peltigera*, involve two autotrophs, one a green alga (*Coccomyxa*) and one a cyanobacterium (*Nostoc*). The two occur in separate areas of the lichen, and the fungal structure differs, depending on which alga is surrounded.

FUNGUS-PLANT ASSOCIATIONS

Mycorrhizae, the symbiotic association of roots and soil fungi, were described in Chapters 7 and 13. A different fungus-plant association occurs in orchids; orchid seeds are quite underdeveloped when mature, often having as few as ten cells, no root, no leaves, and no chlorophyll. They remain partly or entirely inactive until invaded by hyphae of soil fungi, usually the basidiomycete *Rhizoctonia*. Once hyphae penetrate the base, the seed becomes active and starts to develop. The seedling enlarges and forms a bulblike structure, then roots and leaves. In *Dactylorchis* the first photosynthetic leaf is not produced until the second year after fungal invasion; in *Spiranthes spiralis* it is not until the 11th year. Until then, the seedling is subterranean and heterotrophic. It actually appears as if the orchid is parasitizing the fungus, because the fungi are capable of living freely without the orchid. In the association, the fungus degrades cellulose in leaf litter and converts it to sugar, which is then transported to the orchid. It is not known if the fungus receives anything in return.

FUNGI AS DISEASE AGENTS OF PLANTS

Mycorrhizal associations are beneficial to plants, as is the release of nitrogen by means of fungus-mediated decomposition; however, fungi mostly interact with living plants as disease organisms. Fungi parasitize plants, either weakening them and slowing their growth or killing them outright. Hundreds of fungus-caused plant diseases are known and are a central concern of plant pathologists. Two particularly important diseases are discussed below.

BROWN ROT OF STONE FRUITS

Stone fruits such as peaches, plums, and cherries are attacked by the ascomycete *Monilinia fructicola,* which causes brown rot (Fig. 20.38). Probably the most severe pathogen of peaches, brown rot causes large crop losses. Ascospores are released from apothecia in early spring, just as peach trees are beginning to flower. Peaches produce flowers before their new leaves expand, so the first wave of infection involves only flowers; these wither and the fungus produces masses of conidia. Conidia spread the infection to other flowers of the orchard, then attack leaves as they appear, causing leaf and twig blight. Any flowers that survive are able to produce fruit, which is resistant while young. However, as a fruit approaches maturity, it becomes susceptible to conidia from infected leaves. After conidia germinate, hyphae invade the fruit, dissolve the middle lamella, and cause the tissues to become soft and brown. The fruit finally rots and shrivels, becoming a **mummy** that may remain on the tree. If the mummified fruit falls, conditions near the soil permit the fungus

FIGURE 20.38
An advanced case of brown rot of peach, caused by *Monilinia fructicola. (© Kathy Merrifield, The National Audubon Society Collection/ Photo Researchers, Inc.)*

to form ascocarps and ascospores that initiate new infections the following spring. If the mummy remains on the tree, it does not undergo sexual reproduction, but it can produce copious amounts of conidia.

No completely effective control of brown rot is available, but several combined measures greatly reduce the damage. Soil between trees is plowed or disked to disturb the fallen mummified fruits to inhibit spore formation. Systemic fungicides such as benomyl and captan can be applied during the blooming season and the time of fruit maturation. Wounding of fruits by handling and insects is minimized because the fungus can invade only through the wound sites.

RUSTS AND SMUTS

Many basidiomycete species cause serious plant diseases; the most important are rusts (order Uredinales) and smuts (order Ustilaginales). These are both in subclass Teliomycetidae, which is characterized by members that do not form basidiocarps. Rusts and smuts can attack hosts so virulently that they kill the plant; even in less severe infections so much damage is done that host growth and reproduction are adversely affected. In economic terms, rusts that attack cereals such as barley, corn, oats, rye, and wheat are most important, causing losses of millions of tons of grain annually (Fig. 20.39). The most notorious species is the stem rust fungus *Puccinia graminis*, which occurs in many forms (each called a forma specialis), each adapted to a particular host: *P. graminis* forma specialis *tritici* attacks only wheat (*Triticum*), *P. graminis* f. sp. *secali* attacks only rye (*Secale*), and so on. More than 250 different special forms and races of *P. graminis* have been identified at present; more are discovered every year.

The life cycle of *P. graminis* is complex but excellent for understanding many aspects of fungal biology. First, it is a **heteroecious** species; that is, it requires two different living hosts to complete its life cycle (Fig. 20.40). Fungi that develop completely to maturity on a single host are **autoecious.** In the spring, basidiospores of *P. graminis* are released into the wind; in order to survive, they must land on young, tender leaves of barberry (*Berberis*). If the leaf is moist, the basidiospores germinate and produce a penetration peg that punctures the leaf cuticle; the hypha then grows into the leaf and parasitizes it. Once the mycelium is large enough, it produces basket-shaped masses of hyphae, called spermogonia, on the upper side of the leaf. Within spermogonia are long receptive hyphae and small cells called spermatia. For development to progress, spermatia of one mating type must contact recep-

FIGURE 20.39
Wheat heavily infected by rust, *Puccinia recondita.* The patches on the wheat leaf are fungal sporangia, in this case, uredinia. (×4).
(Holt Studios, Ltd./Earth Scenes)

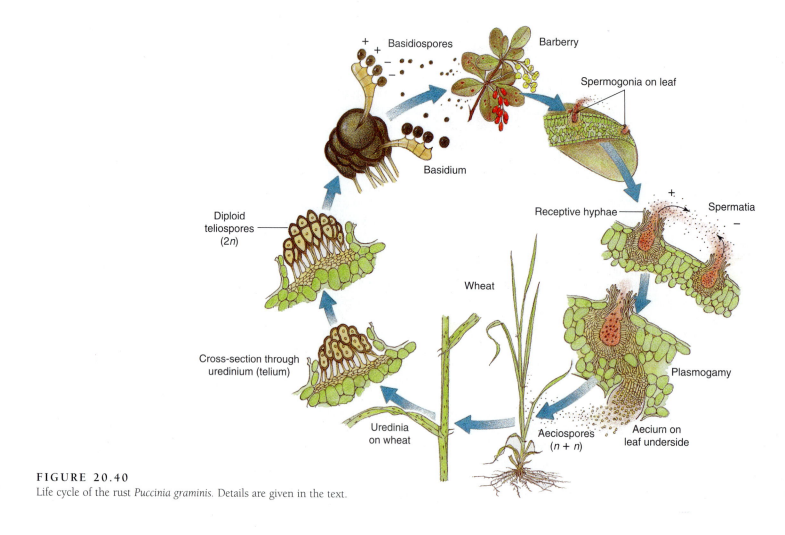

FIGURE 20.40

Life cycle of the rust *Puccinia graminis*. Details are given in the text.

tive hyphae of a compatible mating type; the two fuse, establishing a dikaryotic phase. Once initiated, the dikaryotic condition spreads as the nucleus from the spermatium divides repeatedly, and the daughter nuclei enter adjacent cells of the hyphae by passing through the septal pores.

While this occurs, the haploid mycelium has begun producing another reproductive structure, the aecium, on the lower side of the leaf. This is sterile until nuclei from the spermatium finally arrive through the hyphae and convert its cells to the dikaryotic condition. It then produces dikaryotic, binucleate aeciospores, which cannot germinate on barberry at all, even though they are produced on it. Instead, they must land on leaves of wheat.

Unlike basidiospores, aeciospores cannot penetrate the cuticle when they germinate; instead they must find an open stoma. Once inside the wheat, hyphae spread extensively, then form a new type of spore, the binucleate, dikaryotic urediniospores in a third reproductive structure, the uredinium. Urediniospores are a means of rapid infection, either of the same plant or others. They are extremely effective and are known to blow hundreds of kilometers. Urediniospores are produced continually in massive numbers throughout summer.

In late summer or early autumn, the rust color changes to black as the uredinia stop producing urediniospores and begin producing yet another type of spore, dikaryotic teliospores. The uredinia are then called telia. In spring, the two nuclei in each cell of each teliospore fuse and then undergo meiosis; as the spore germinates, it produces four cells, each of which receives a nucleus and then produces a sterigma and a basidiospore, completing the life cycle.

This life cycle has the basic features of basidiomycetes, plus some additional ones. Spermatia are produced like other spores, but they do not germinate unless they are in contact with receptive hyphae, thus acting more like sex cells than spores. Aecia can be formed by haploid hyphae, but they are sterile unless they become dikaryotic. Their spores can germinate only on wheat, not on barberry. Obviously aeciospores have all the genetic programming to attack barberry because they are formed on this host, yet they themselves cannot live on it. Urediniospores, on the other hand, are formed on wheat and are specifically able to reinfect it.

Control of this disease in North America is almost entirely by means of breeding resistant strains. In eastern Europe, the disease is controlled by eliminating all barberry plants, but in North America the fungus can spread solely by means of urediniospores. The mild winters in Mexico and the southwestern United States allow the fungus to survive by growing on winter wheat and then infecting more northerly fields during summer by means of wind-carried urediniospores. Winter wheat in Mexico is in turn infected in autumn by urediniospores blowing from Canadian fields. To fight the disease, infected fields are searched for resistant wheat plants; fortunately, alleles for resistance are dominant to those for susceptibility, so resistance can be bred quickly into new lines of wheat. However, a resistant strain is good for only a few years before a mutation in the gigantic gene pool of *P. graminis* produces a new virulent pathological race.

SUMMARY

1. Fungi are not considered plants because they lack plastids, they lack cellulose (with one exception), their body construction is filamentous, and many of their metabolic pathways are not homologous with corresponding pathways in plants.

2. The kingdom Myceteae is not a natural group that evolved from a common ancestor. Too little information is available to create an alternative classification that is definitely less erroneous.

3. Fungi have three heterotrophic metabolisms: biotrophs, necrotrophs, and saprotrophs.

4. Fungi and bacteria break down plant and animal bodies, releasing their minerals and permitting nutrient recycling.

5. The fungal body is either a single cell or a mycelium of septate or nonseptate hyphae. Except for oomycetes, which contain cellulose, fungal walls are composed of chitin. In ascomycetes and basidiomycetes, large complex fruiting bodies are formed.

6. Conidia and sporangiospores are spores that are produced asexually.

7. Exchange of genetic material occurs by the fusion of compatible hyphae. Nuclear fusion is delayed, resulting in heterokaryotic hyphae. Shortly before formation of sexual spores, the nuclei fuse and then immediately undergo meiosis, with synapsis and crossing-over.

8. Slime molds consist of a plasmodium containing hundreds of thousands of nuclei but no individual cells.

9. Mastigomycetes produce flagellated cells but are probably not a natural group. The oomycetes appear to have evolved from algae, whereas the chytridiomycetes do not.

10. Zygomycetes have mycelia of branched coenocytic hyphae; complex fruiting bodies are never formed. After specialized multinucleate hyphae grow together, plasmogamy and karyogamy result in a zygosporangium with many diploid nuclei.

11. Ascomycetes have saclike asci where karyogamy and meiosis occur. Asci may be produced in ascocarps such as morels or truffles.

12. Basidiomycetes have clublike basidia where karyogamy and meiosis occur. Basidiomycetes produce basidiocarps such as mushrooms, puffballs, and brackets.

13. Subdivision Deuteromycetes was established for species whose sexual reproduction is unknown. Most have incomplete septa of the ascomycete type.

14. Lichens are associations of fungi and either algae or cyanobacteria. Most of the fungi are ascomycetes; a few are basidiomycetes.

IMPORTANT TERMS

ascocarp
ascogenous hyphae
ascus
basidiocarp
basidium
biotroph
chitin
coenocyte
conidium

fruiting body
haustorium
heterokaryosis
homokaryosis
hymenium
hypha
imperfect fungus
lichen
mating types

mycelium
necrotroph
plasmodium
saprotroph
septum
tinsel flagellum
whiplash flagellum
yeasts
zygosporangium

REVIEW QUESTIONS

1. Why are fungi no longer considered plants? Why were they originally classified as plants?

2. Why are we certain that the fungi are not a monophyletic, natural group? Which fungi seem most certainly unrelated to the ascomycetes and basidiomycetes?

3. How do biotrophs differ from necrotrophs and saprotrophs? Are any fungi autotrophs? Are any fungi lithotrophic like certain bacteria?

4. What are sporangiospores and conidia? What are meiospores? Name several types.

5. Can you think of any selective advantage of heterokaryosis as opposed to diploidy? Is a dikaryotic cell significantly different from a diploid one with respect to control of transcription, translation, and gene activation?

6. Describe the life cycle of an ascomycete; compare it with a life cycle of a flowering plant (see Chapter 9). Would you consider the ascogenous hyphae and ascocarp to be one generation and the homokaryotic mycelia the other? Or would you consider the two just distinct phases or tissues of the same body? Is the megagametophyte of angiosperms an alternate generation or just a tissue in the parent plant's body?

7. Basidiocarps must be perfectly vertical if basidiospores are to fall out; basidiocarps detect gravity and respond gravitropically. Is this metabolism homologous or analogous to the corresponding metabolism in plants (see Chapter 14)?

8. Several fungi such as *Thermomyces, Sclerotinia borealis,* and *Aspergillus glaucus* grow in harsh habitats; archaebacteria also live in extreme environments, and it is thought that they might represent some of the earliest forms of life. Do you think it is likely that fungi of extreme environments are also directly related to early life forms?

BotanyLinks .net Questions

www.jbpub.com/botanylinks

Visit the .net Questions area of BotanyLinks (http://www.jbpub.com/botanylinks/netquestions) to complete these questions:

1. Many mushrooms are delicious, and deadly. Go to the BotanyLinks home page to learn how to cook with them and stay alive.

2. Foods provide nutrients for fungi and bacteria, and keeping our food safe from microbes is a constant battle. Go to the BotanyLinks home page for resources for investigating how this is done.

BotanyLinks includes a Directory of Organizations for this chapter.

ALGAE AND THE ORIGIN OF EUKARYOTIC CELLS

21

Kelps are members of the group known as brown algae. Many have bodies almost as complex as those in flowering plants. *(W. E. Ferguson)*

CONCEPTS

Biology is currently dominated by the five-kingdom concept, the theory that there are five main groups of organisms: plants, animals, fungi, protista, and prokaryotes. Among eukaryotes, plants, animals, and fungi are quite distinct and easy to recognize when derived members are considered. But when the simplest eukaryotes are studied, deciding which should be classified as plants, which should be considered animals, and which are fungi can be difficult. Kingdom Protista was established to hold these problematic organisms. It was thought that as our knowledge increased, we would be able to assign each organism to one or another of the other three eukaryotic kingdoms until nothing was left in Protista.

We no longer think so. Green algae are obviously related to land plants, and certain protozoans are clearly related to multicellular animals (Table 21.1; Fig. 21.1), but many protistans are quite different from plants, animals, or fungi. Among algae, organisms called dinoflagellates (Fig. 21.1b) have so many relictual features that they may not be truly eukaryotes. Each algal group must have diverged from the others quite early judging by the numerous differences in biochemistry, pigmentation, nutrient reserves, and other features.

FIGURE 21.1

(a) Euglenoid algae (division Euglenophyta). *(Peter Parks Animals Animals/Oxford Scientific Films)* (b) A dinoflagellate (division Pyrrhophyta). *(Biophoto Associates)* (c) Diatoms (division Chrysophyta). *(Courtesy of Greta Fryxell, Texas A and M University)* (d) Golden brown algae (division Chrysophyta). (e) Yellow-green algae (division Chrysophyta). *(d and e, E. R. Degginger)* (f) Green algae (division Chlorophyta). *(William E. Ferguson)* (g) Brown algae (division Phaeophyta). *(E. R. Degginger)* (h) Red algae (division Rhodophyta). *(William E. Ferguson)*

We are not certain yet how they are interrelated, and placing them in kingdom Protista emphasizes that uncertainty. Placing the green algae with the Protista is difficult because they are so obviously related to land plants, but this placement seems necessary because green algae constitute a large division with thousands of species, most of which are not closely related to the evolutionary line that gave rise to land plants. Classifying division Chlorophyta in kingdom Plantae would put many unicellular and colonial algae into the plant kingdom, which would be just as artificial as taking the ancestors of true plants out of the plant kingdom. Taxonomy is a human endeavor, and when we assign groups to cate-

TABLE 21.1   *Divisions of Algae*

Division Euglenophyta—euglenoids

There are many nonpigmented euglenoids, and this group is classified by zoologists as a member of their phylum Protozoa. The photosynthetic species have pigments similar to those of green algae. Euglenoids seem to be extremely ancient, having a cell organization similar to that which evolved hundreds of millions of years ago (see Fig. 21.1a).

Division Pyrrhophyta—dinoflagellates

These lack histones and have intranuclear mitosis with chromosomes that never decondense during interphase. Many botanists now consider them to be an isolated line that originated during the very first stages of the evolution of eukaryotic cells (see Fig. 21.1b).

Division Chrysophyta—diatoms, golden brown algae, yellow-green algae

This is a diverse group, sometimes divided into several separate divisions, sometimes included with the brown algae. They and the brown algae have similar biochemistry, especially photosynthetic pigments (chlorophylls *a* and *c*) and unusual storage products (see Fig. 21.1c, d, and e).

Division Chlorophyta—green algae

Green algae are extremely diverse structurally but very homogeneous and well-defined biochemically, being almost identical to true plants in terms of basic metabolism. They are universally considered to be the ancestors of true plants (see Fig. 21.1f).

Division Phaeophyta—brown algae

A large group of species that often have large, complex bodies and that are common along rocky coasts. Some kelps have well-defined tissues, one of which strongly resembles phloem and is involved in long-distance transport of organic molecules (see Fig. 21.1g).

Division Rhodophyta—red algae

Many species of this large group have simple filamentous bodies, but the filaments may be aggregated and resemble parenchyma. Unlike all other eukaryotic groups, the red algae never have flagella at any time. They are also unique in having a strongly prokaryotic organization to their chloroplasts: There are phycobilisomes as in cyanobacteria. However, nuclear structure and mitosis do not appear to be unusual (see Fig. 21.1h).

gories, we must separate some species from rather close relatives. It is necessary to understand why organisms are classified as they are and not merely memorize the classification.

Reproductive structures are the critical factors in distinguishing algae from the plants of kingdom Plantae. Algae are defined as photosynthetic individuals whose reproductive structures are completely converted to spores or gametes that, when released, leave nothing but empty walls (Fig. 21.2). In true plants (technically known as **embryophytes**) the reproductive structures are always complex and multicellular, and only some of the inner

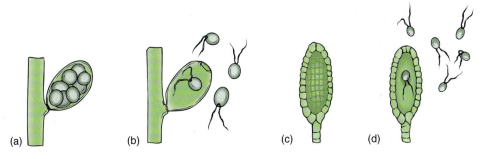

(a)   (b)   (c)   (d)

FIGURE 21.2

Algae are distinguished from plants (embryophytes) primarily by the nature of the reproductive structures. (a and b) In algae, the sporangium or gametangium may be either one large cell, as shown here, or multicellular, but every cell is converted into a reproductive cell. When reproductive cells are released (b), only the wall of the original cell is left. (c and d) Sporangia and gametangia of plants are always multicellular and only the inner cells differentiate into spores or gametes; after they are liberated (d), a residual layer of sterile cells remains.

cells become reproductive. Outer cells constitute a protective layer and persist after the reproductive cells are released. This may seem trivial, but the algal method of reproduction is a cellular process, reflecting the simple, cellular level of organization most species have throughout their bodies. Reproductive structures of true plants involve two tissues—reproductive and sterile—functioning together as an organ; this reflects a more complex organ level of integration in the whole body.

ORIGIN OF EUKARYOTIC CELLS

The differences between prokaryotes and eukaryotes are deep-seated, involving many of the most basic metabolic processes and cellular organization (Table 21.2).

DNA STRUCTURE

In prokaryotes, DNA is "naked," not complexed with any proteins, its numerous negative charges being neutralized by calcium ions instead. Each genome is just a short circle of DNA, often containing only about 3000 genes. The DNA of prokaryotes contains few introns, and processing of mRNA is relatively simple.

In eukaryotes, DNA is more elaborately packaged, being complexed with the nucleosome histones and forming a chromosome (see Chapter 15). Each molecule of DNA is long, capable of carrying thousands of genes, yet no genome ever consists of just a single DNA molecule; at least two occur per haploid set, and 10 to 20 are more common. Many eukaryotic genes contain introns, and large segments of DNA do not code for any type of RNA.

TABLE 21.2   *Differences Between Prokaryotes and Eukaryotes*

Character	Prokaryotes	Eukaryotes
DNA		
form	small circles	chromosomes
introns	rare	common
histones	absent	present
Nucleus		
nuclear envelope	none	present
nucleolus	none	present
true nucleus	none	present
Mitosis		
spindle	none	present but variable
centromeres	none	present
Meiosis	none	present in most
Gene exchange	conjugation or none	sex in most
Membranous organelles	none or lamellae in some	plastids, mitochondria, endoplasmic reticulum, dictyosomes, vacuoles
Ribosomes	70S	80S cytoplasmic 70S in organelles
Flagella	not 9 + 2	9 + 2

NUCLEAR STRUCTURE

The DNA of prokaryotes lies directly in the cytoplasm; no nucleus is present. Most of the DNA of eukaryotes is located within a nucleus, separated from the cytoplasm by two nuclear membranes with nuclear pores; this is the source of their name: *eu-* (true) karyotes. A nucleolus is present as well.

NUCLEAR DIVISION

As a prokaryotic cell grows, its plasma membrane expands; the DNA circles attached to it are separated from each other by cell growth. The cell then pinches in two.

Eukaryotic cells are typically haploid or diploid, having only one or two copies of each of their various chromosomes, so the rather haphazard prokaryotic method of simple membrane expansion would not work reliably. Instead, mitosis has evolved in eukaryotes: Spindle microtubules, cell plate formation, and centromeres ensure that each progeny cell receives one of each type of chromosome (see Chapter 4). In almost all eukaryotes, meiotic nuclear division occurs as part of sexual reproduction. The critical phenomena are pairing of maternal and paternal homologous chromosomes followed by crossing-over and genetic recombination. The complex procedures result in an interchange of alleles, such that a great diversity of spores or gametes is produced.

ORGANELLES

Prokaryotes lack membrane-bounded organelles: Their cytoplasm is rather homogeneous, containing only ribosomes and perhaps gas vacuoles, storage granules, or photosynthetic membranes, depending on the species. Eukaryotes all have nuclei and mitochondria, and plants and algae also have plastids. In addition, dictyosomes, endoplasmic reticulum, vacuoles, and vesicles are all parts of the eukaryotic endomembrane system. These membranes result in the division of the cell into numerous compartments, permitting several types of highly specialized metabolism to occur within the same cell. The ribosomes of prokaryotes are 70S, being smaller and denser than the 80S ribosomes in the cytoplasm of eukaryotes (see Chapter 15). Plastids and mitochondria have 70S ribosomes instead of 80S ribosomes.

Flagella and cilia are remarkably uniform in eukaryotes, consisting of a 9 + 2 arrangement of microtubules. Only a few prokaryotes have flagella, and these have a totally different type of construction, never a 9 + 2 arrangement. They are not composed of microtubules or tubulin.

Numerous other distinctions could be mentioned, especially in gene regulation and biochemical pathways. The differences are numerous and basic, much more so than the differences that distinguish plants from animals.

ORIGIN OF EUKARYOTES: THE ENDOSYMBIONT THEORY

Prokaryotes represent the types of organisms that arose first. Until the early 1970s, it was assumed that some had given rise to eukaryotes by gradually becoming more complex and developing an endomembrane system with mitochondria and plastids. This is the **autogenous theory.** Cyanobacteria, then known as blue-green algae, seemed to be logical intermediates.

However, in the 1960s the **endosymbiont theory** was revived. It had been proposed in 1905 by K. C. Mereschkowsky, who had speculated that plastids and mitochondria might be prokaryotes living inside eukaryotic cells. At that time there was no way to test the theory, but by the late 1960s, plastids and mitochondria were discovered to have their own DNA and ribosomes that are 70S, not 80S. Furthermore, these organelles divide by a type of cleavage similar to that of prokaryotes, and they also lack microtubules. Before long,

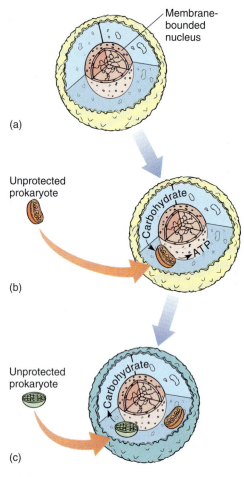

FIGURE 21.3

The endosymbiont theory postulates that a
prokaryote (a) evolved to the level of having a
membrane-bounded nucleus; this must have
been an early step, because all eukaryotes have
a nuclear envelope, and all are rather uniform
in structure. (b) The early eukaryote engulfed
but did not digest a prokaryote, which then
evolved into a mitochondrion. (c) Finally, the
descendants of that early eukaryote would engulf
but not digest a prochlorophyte type of
cyanobacterium, which would evolve into a
plastid; this would have been the point of
separation of the animal and plant lines of
evolution. Not shown in the diagram is the
evolution of mitosis, meiosis, sexual
reproduction, and cell walls.

other prokaryotic traits were discovered in them: Their DNA is a small closed circle of
naked DNA, and organellar genes are organized like prokaryotic genes. Their ribosomes are
sensitive to the same antibiotics that interfere with prokaryotic ribosomes.

The endosymbiont theory postulates that a prokaryotic cell had evolved to the point of
having some eukaryotic features such as 80S ribosomes and a nuclear envelope (Fig. 21.3).
It was postulated to be heterotrophic, living by engulfing and digesting other cells. It may
have engulfed one but then not digested it, perhaps because the larger cell had inefficient
digestive enzymes or because the engulfed cell had a more resistant membrane. Such an
association might be advantageous: The engulfed cell could be extremely efficient at aerobic respiration but poor at obtaining nutrients. If it could get sugars from the larger cell and
"leak" ATP back to it, both cells might be better off. An endosymbiotic relationship might
be more advantageous selectively than the condition in which each partner is living independently. Such an endosymbiosis would still have many obstacles to overcome—
increasing the compatibility between the two cells and coordinating growth, development,
and reproduction—but the descendants of the engulfed organism could possibly evolve
into a mitochondrion.

Origin of Plastids.   Plastids could arise similarly if the engulfed partner were photosynthetic (Fig. 21.3). Because all chloroplasts contain chlorophyll *a* and produce oxygen but
none have bacteriochlorophyll, we suspect cyanobacteria as possible ancestors and do not
believe that the green and purple bacteria are very likely candidates. Prochlorophytes
arouse great interest because they have both chlorophyll *a* and *b* and some lack phycobilin
pigments (see Fig. 19.32). The first prochlorophyte discovered, *Prochloron,* not only resembles chloroplasts biochemically but also exists as an obligate symbiont inside marine
invertebrates, called didemnid ascidians, supplying them with carbohydrates whenever the
ascidians lie in the sunny photic zone. *Prochloron* has never been isolated and cultured
outside the tissues of the ascidian, so research with it has been difficult; however, a second
prochlorophyte, *Prochlorothrix,* has been discovered growing freely in lakes in Holland. 16S
rRNA sequencing has now shown that chloroplasts, *Prochlorothrix,* and *Prochloron* are all
definitely closely related, but neither bacterium is the immediate ancestor of chloroplasts.
Instead, they all share a common ancestor.

How many steps of endosymbiosis might have occurred? At least two, one for mitochondria and a second for plastids. Because all eukaryotes have mitochondria and only
some have plastids, we suspect that mitochondria arose first.

Chloroplasts of red algae may have arisen by an act of endosymbiosis distinct from the
one that produced the plastids of green algae. Red algal chloroplasts are remarkably cyanobacterial in organization and biochemistry, having phycobiliproteins and phycobilisomes
arranged on membranes that do not form grana stacks (Fig. 21.4a). They store a polysaccharide, floridean starch, that resembles glycogen.

Although it is possible to construct a hypothetical pathway by which a red algal
chloroplast could evolve into a green algal chloroplast, such a hypothesis has many difficult
and highly unlikely steps. Also, it would mean that red algae would have been ancestral to
green algae, and many cell characteristics are not compatible with such a conclusion. It is
more plausible that red algae represent a distinct line of evolution from the early eukaryotes, its chloroplasts being evolutionarily distinct from the chloroplasts of green algae and
land plants.

Brown algae may represent a third line of chloroplast evolution (Fig. 21.4b). Brown
algae, along with golden brown algae, diatoms, and yellow-green algae (all three in division
Chrysophyta) and dinoflagellates (division Pyrrhophyta) all contain chlorophyll *c* in addition to chlorophyll *a,* and they all lack chlorophyll *b.* Their nonchlorophyll accessory
pigments are also unique, being types of carotenoids not found in green algae or plants.
Membranes of brown algal chloroplasts associate into grana-like stacks, but they are always
small, each consisting of just two or three membranes.

If there really were three or four separate prokaryotic ancestors to three or four distinct
lines of chloroplasts, the plastid DNA nucleotide sequences of each line should show little
homology to the others. But if one type evolved from another—green algae evolving from
brown algae for instance, considerable sequence homology is expected. Studies are
currently underway, and results should be known soon.

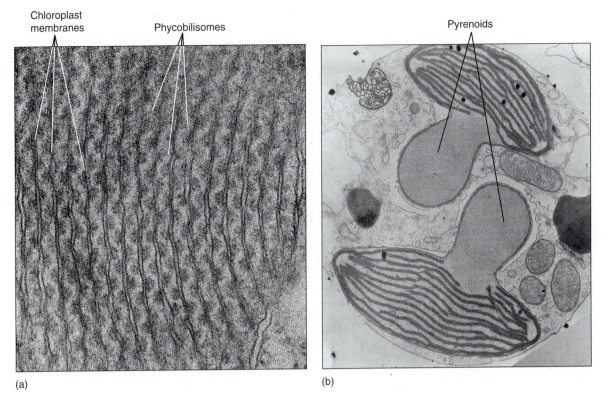

FIGURE 21.4

(a) Internal membranes of red algal chloroplasts do not form grana stacks like those of green plants. Their phycobilisomes are visible, attached to the membrane. *Porphyridium* (×115,000). *(Courtesy of E. Gantt and S. F. Conti, University of Maryland)* (b) Brown algal chloroplast membranes associate into stacks of three or, less often, stacks of two. The large clear structures are pyrenoids, involved in polymerization of sugars into polysaccharides. *(Courtesy of L. V. Evans, University of Leeds, UK)*

Origin of Nuclei.   The nuclei of plants are virtually identical to those of animals in terms of structure, metabolism, mitosis, and meiosis. Apparently the plant and animal lines of evolution diverged only after the eukaryotic nucleus had become quite sophisticated. But nuclei are complex organelles, and mitosis and meiosis are intricate processes with many steps. Certainly plant and animal nuclei could not have arisen from prokaryotic nucleoids quickly; many intermediate steps spanning hundreds of millions of years must have been involved.

Several groups of organisms with unusual nuclear characteristics may represent lines of evolution that originated earlier in the history of eukaryotes than did plants and animals. In Chapter 20, the nuclei of many fungi were shown to have an unusual mitosis, some having both an intranuclear and an extranuclear spindle. In many fungal species the nuclear envelope and nucleolus either do not break down at all or do so only late in mitosis. Similar behavior occurs in many groups of algae, including dinoflagellates, brown algae, euglenoids, and some green algae (Fig. 21.5a). Gaps form in the nuclear envelope, and bundles of microtubules pass completely through the nucleus.

Even more significantly, nuclei of dinoflagellates (see Fig. 21.1b) have no histones. They do have what appear to be chromosomes, but these remain condensed at all times, undergoing only a slight uncoiling during interphase (Fig. 21.5b and c). Because histone and nucleosome structure is so constant in all other eukaryotes, this unique situation in dinoflagellates is considered highly significant. It has been proposed that dinoflagellates be removed from eukaryotes and classified as **mesokaryotes** (*meso*, intermediate). Dinoflagellates are receiving a great deal of attention.

Spindle   Nuclear envelope

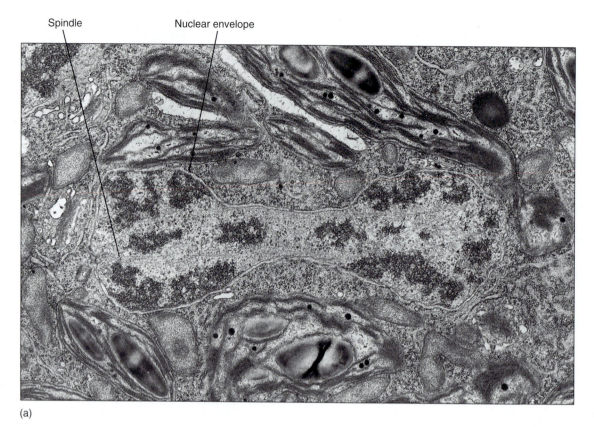

(a)

Nucleus   Chromosomes

Nuclear envelope   Chromosomes

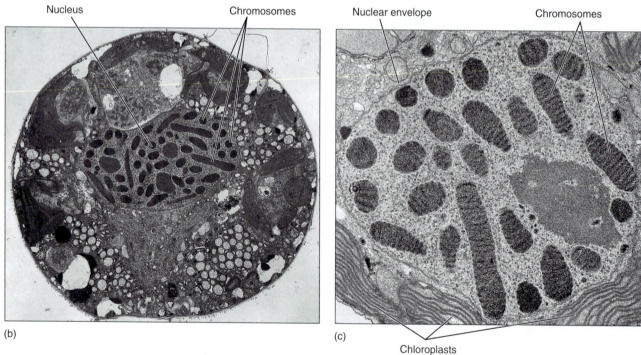

(b)

(c)

Chloroplasts

FIGURE 21.5

(a) Late telophase in the green alga *Cladophora*; not only is the nuclear membrane completely intact, but a portion of the intranuclear spindle is visible (× 26,656). *(Courtesy of J. L. Scott and K. W. Bullock, College of William and Mary)* (b) An interphase nucleus of *Peridinium*, a dinoflagellate; chromosomes never completely decondense. It is not known how chemical messengers, DNA replicases, and RNA synthetases gain access to the DNA molecule (× 14,000). *(Courtesy of J. M. Chesnick, University of Washington)* (c) Magnification of dinoflagellate chromosomes (× 21,000). *(Courtesy of J. Dodge, University of London)*

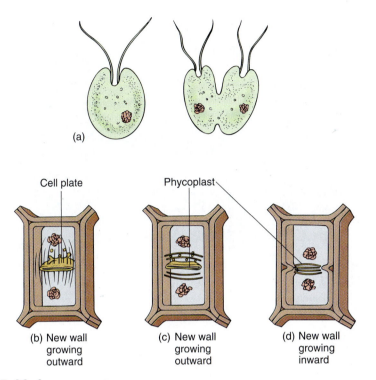

FIGURE 21.6
(a) In unicellular, wall-less forms, cell division occurs by infurrowing of the plasma membrane. (b) Most algae with walls divide by a phragmoplast and cell plate, just as plant cells do. (c) and (d) Some green algae have a phycoplast; the mitotic spindle disorganizes quickly, after which a new set of microtubules polymerizes perpendicular to the spindle axis. Division may then occur by cell plate growth (c) or by infurrowing (d).

Types of Cytokinesis.   Several types of cytokinesis occur in algae. Many groups of unicellular algae have no wall, their plasma membrane being their outermost surface. During cell division, the plasma membrane pinches in two, being pulled inward as a **cleavage furrow**, a process remarkably similar to cytokinesis in animals (Fig. 21.6a).

In almost all algae with walls, cell division is similar to that of plants: After telophase, the spindle microtubules persist temporarily, holding the nuclei apart. Between them a phragmoplast forms, consisting of short microtubules oriented parallel to the spindle microtubules (see Figs. 4.14, 4.15 and 21.6b). The coalescence of dictyosome vesicles establishes a cell plate, which then grows radially outward—**centrifugal growth**—until it meets and fuses with the wall of the parental cell.

In some green algae, cytokinesis occurs by a different method; the mitotic spindle depolymerizes quickly, and the two daughter nuclei lie close together. A new set of microtubules appears between them, oriented parallel to the plane where the new wall will form, which is perpendicular to the orientation of the spindle. This group of microtubules is a **phycoplast** and may be associated with division either by furrowing or by cell plate formation (see Fig. 21.6c and d). Phycoplasts seem to be associated with those green algae whose nuclear envelope does not break down during mitosis, whereas phragmoplasts are associated with the loss of the nuclear envelope. Plants probably arose from green algae that divide with a phragmoplast rather than a phycoplast.

In red algae, cell division occurs by a phragmoplast, but the new wall grows inward from the pre-existing walls. It stops growing while still incomplete, having a large hole called a **pit connection**, which is filled with material (a **pit plug**; Fig. 21.7). Otherwise, most aspects of wall formation are similar to those of other walled algae.

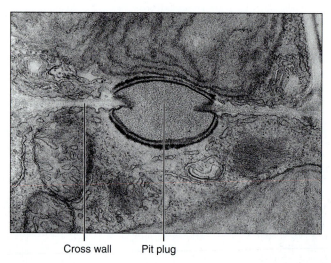

Cross wall   Pit plug

FIGURE 21.7
Cross walls of red algae are unusual in being incomplete, having a hole in their center. This is closed by a pit plug. *Scinaia confusa* (×15,000). *(Courtesy of Curt Pueschell, State University of New York, Binghamton)*

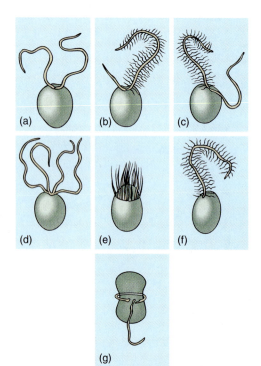

FIGURE 21.8
The most common arrangements of flagella in algae. (a) Two equal whiplash. (b) A short whiplash and a long tinsel. (c) Two equally long flagella, one of each type. (d) Four whiplash. (e) Numerous flagella located in a ring (rare). (f) One tinsel flagellum (mostly just in some brown algae). (g) One flagellum trailing, one in a transverse groove (dinoflagellates).

Origin of Flagella.   The origin of flagella is still unknown. Being composed of microtubules, flagella may be related in origin to the mitotic spindle. Flagella must have occurred extremely early because all flagella and cilia of eukaryotes have virtually identical internal structure. The standard composition is an external membrane that is an extension of the plasma membrane and an internal **axoneme** composed of nine outer doublets and two inner singlet microtubules (see Chapter 3). Because the structure is so constant, we are quite confident in excluding the possibility of multiple origins and convergent evolution. Because virtually all eukaryotes have flagella or cilia, this structure must have arisen before the various lines diverged.

Variability exists in the external covering and length of flagella (Fig. 21.8). Many algal and fungal and all plant flagella have a smooth surface and are known as whiplash flagella (see Fig. 20.23). Others have a surface covered with hairs and are referred to as tinsel flagella. Hairs are of multiple types; some appear to be just a single row of protein monomers, whereas others are hollow tubes of protein that end as a single row.

Considerable attention is being given to both the tips and the bases of flagella, although studies of both are difficult. The central microtubules extend beyond the peripheral ones, and the arrangement at the tip varies from group to group. In many species, the central microtubules extend a long distance beyond, forming an extremely fine hair tip that is an adhesive recognition organelle. In gametes, the hair tips of one gamete recognize and bind to those of a compatible gamete. Once the flagellar tips adhere, the gametes are drawn together and fusion of plasma membranes and cytoplasm occurs. In some swimming spores (**zoospores**), flagellar hair tips appear to recognize suitable substrates; the correct substrate triggers adhesion, and the spore then settles and germinates. If the flagellum encounters an unsuitable substrate, adhesion and germination do not occur.

The base of a flagellum is attached to a basal body, which is part of a large, complex root structure. The root structure of eukaryotic flagella is an extensive, elaborate complex of bracing structures that interconnect the basal body and flagellum to arrays of microtubules extending deep into the cell (Fig. 21.9). Analyzing root structure organization is extremely difficult because they are much too small to be seen by light microscopy yet much too large to extract whole. Serial sections must be cut extremely carefully and every section examined by transmission electron microscopy; from the numerous micrographs, the structure can be reconstructed. Despite the difficulty, phycologists have been successful in analyzing many root structures, which are useful in understanding not only the workings of flagella, but also flagellar growth, duplication, and evolution.

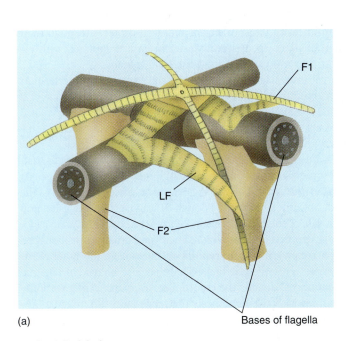

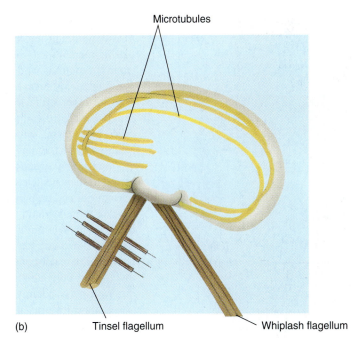

FIGURE 21.9

(a) Model of the root structure of *Urospora,* a green alga, with two massive root fibers (F2) and four narrower ones (F1). All are interconnected and bound to the flagella by lateral fibers (LF). (b) The root structure of brown algae is anchored to the cell by long bands of microtubules. In this generalized diagram, some microtubules extend the entire cell length, and others extend about one third the length.

Many but not all motile algae have **eyespots,** organelles involved in phototaxis (swimming toward or away from light). The structure of eyespots is variable: All contain at least globules of lipid, but some also contain carotenoids dissolved in the lipid, and many are associated with specialized membranes. Whereas eyespots are of special importance to photosynthetic organisms, motile gametes such as sperms swim guided by chemotaxis. The chemoreceptors are not at all well understood, although in many species, the attractant sex hormones have been discovered.

Eukaryotes evolved gradually in many steps, and we have numerous organisms that perhaps retain some intermediate, transitional features. Although we do not know how all the organisms are interrelated, we are certain that no single line of eukaryotes at some point diverged into plants, animals, and fungi. Rather, many protistan groups represent distinct phases in the evolution of eukaryotes. Eventually, the artificial kingdom Protista may be replaced by three, four, or more natural kingdoms.

The divisions of algae that contain predominantly unicellular individuals are discussed below first because they are the simplest and many seem to be relictual, perhaps having originated early. The three groups that contain predominantly multicellular individuals are then considered.

DIVISION EUGLENOPHYTA: EUGLENOIDS

This division comprises a rather large group of organisms: More than 800 species have been discovered, named, and placed into 36 genera (Fig. 21.1a; Table 21.3). Its members, **euglenoids,** illustrate the problem of trying to classify protistans as plants or animals: 25 genera never have chloroplasts at all, but 11 do. Most are unicellular (Fig. 21.10), but in some species a few cells remain together after cell division.

The green euglenoids are capable of photosynthesis, having chlorophylls *a* and *b*, but carbohydrate is stored as the glucose polymer paramylon, not as starch. Euglenoid chloroplasts are extremely sensitive to antibiotics that interfere with 70S ribosomes, and exposure to streptomycin results in the loss of chloroplasts and colorless "bleached" individuals.

TABLE 21.3	*Classification of Division Euglenophyta*
Order Eutreptiales	
Order Euglenales	*Euglena*
Order Heteronematales	

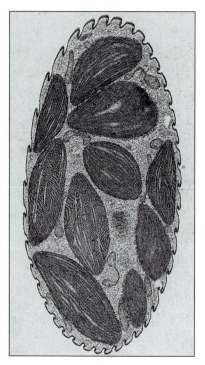

FIGURE 21.10
The pellicle, now often called a periplast, is not a cell wall, but rather a thick layer of elastic proteins that accumulate just below the plasma membrane (× 11,700). *(Paul W. Johnson/Biological Photo Service)*

Many colorless individuals, both natural and bleached, take in food particles by phagocytosis (engulfing their food), and most can absorb sugars and even proteins through their plasma membrane. Even chlorophyllous species grow faster in light if supplied with organic compounds, and all must obtain B vitamins from the environment.

Euglenoids swim actively with two flagella located at the cell's anterior end, one short and the other quite long. An important nonplant character in euglenoids is lack of a cellulose wall; instead their surface contains a layer of elastic proteins called a pellicle (also called a periplast), which is located under the plasma membrane and gives the cell some rigidity and shape, but it is very different from a typical plant wall.

A representative example is *Euglena* (see Fig. 21.1a); individuals are elongate cells that, owing to their pellicle, have a somewhat ovoid shape, although they flex as they swim. At the anterior end is an invagination, the canal. The two flagella are attached at the base of the canal and then extend upward, but only one is long enough to emerge from the canal and be useful for swimming. At the side of the canal, near the base of the flagella, is an eyespot composed of many small, orange droplets of carotenoids. There are numerous disk-shaped chloroplasts, each containing a region called a **pyrenoid**; in most algae the pyrenoid is involved in polymerizing sugars into reserve polymers. That may also be true in *Euglena*, but the reserve paramylon accumulates outside the chloroplast, away from the pyrenoid.

Certain features are consistent with the hypothesis that euglenoids are relictual eukaryotes. During mitosis the nuclear envelope and nucleolus persist, and spindle microtubules form within the nucleus. Chromosomes are aligned lengthwise on the spindle, not crosswise as in other eukaryotes, and each chromatid appears to have its own minispindle consisting of four centromere microtubules and 8 to 12 longer microtubules. Cell division is by longitudinal cleavage, not by phragmoplast and cell plate, and after division chromosomes remain condensed. Neither meiosis nor sexual reproduction has been found in any euglenoid. All this could be interpreted as representing an early stage in the evolution of a true nucleus and mitosis, normal mitosis and meiosis apparently having not arisen until after the origin of euglenoids.

DIVISION PYRRHOPHYTA: DINOFLAGELLATES

Dinoflagellates (see Fig. 21.1b; Table 21.4) show many unusual characters that probably represent a relictual state: The nuclear envelope and nucleolus persist throughout mitosis, and a typical spindle does not form. Instead, channels open in the nucleus and large bundles of microtubules pass through. The chromosomes, which do not have centromeres and are permanently condensed, are attached to the nuclear envelope, whose expansion during the formation of the nuclear channels causes the chromosomes to split. There are no histones.

Dinoflagellates are almost exclusively motile and unicellular; only a few species such as *Gonyaulax monilata* form chains of similar cells with no differentiation. Flagella have a strikingly characteristic arrangement in dinoflagellates: One long flagellum lies in a longitudinal groove with its distal end free and responsible for swimming (Fig. 21.11). The

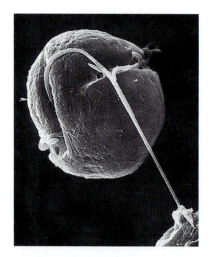

FIGURE 21.11
Dinoflagellates have two flagella, one of which lies in a groove that encircles the cell (× 10,000). *(David Phillips/Photo Researchers)*

TABLE 21.4 *Classification of Division Pyrrhophyta*	
Class Dinophyceae	
Order Gymnodiniales	
Family Gymnodiniaceae	*Gymnodinium*
Order Peridiniales	
Family Gonyaulacaceae	*Gonyaulax*
Family Peridiniaceae	*Peridinium*

FIGURE 21.12
The population density of dinoflagellates can rise so high that they color the water, producing a red tide. A few centimeters away, the density is so low that the water appears normal. (*Carlton Ray/Photo Researchers*)

other flagellum is flat and ribbon-like and lies completely within a transverse groove that encircles the cell. Most species are photosynthetic, but many are completely heterotrophic. Although some dinoflagellate characteristics are similar to those of the euglenoids, others are different: They have chlorophylls *a* and *c*, not *a* and *b*, and their carotenoids are quite unusual. In addition to beta carotene, they also have several unique xanthophylls such as peridinin and dinoxanthin. Their reserve material is starch or oil. Dinoflagellates may lack a wall or have one consisting of cellulose plates whose arrangement and number are useful for identifying the genera and species.

Dinoflagellates have evolved to the level of having sexual reproduction. Vegetative cells release small, naked cells that act as gametes; at the initiation of fusion, they become quiescent but then resume swimming, and the zygote may remain motile for 12 to 13 days. It then becomes thick walled and quiescent for about 2 months. During germination, it undergoes meiosis and a single haploid vegetative cell emerges—a new individual that grows and multiplies by mitosis.

Organisms with many relictual characters are often regarded as poorly adapted, perhaps surviving only in certain specialized, protected environments. But dinoflagellates are abundant throughout the oceans. Under conditions of ideal temperature and nutrients, population growth of some dinoflagellates, especially *Gonyaulax* and *Gymnodinium,* is explosive. Within a few days they become so numerous that their bodies actually color the water reddish brown—a "**red tide**" (Fig. 21.12). The density of dinoflagellates can be as high as 30,000 cells per milliliter of sea water. Red tides are becoming increasingly frequent in the Gulf of Mexico, a phenomenon that is unexplained at present. This fascinating biological spectacle is also dangerous, because these algae produce toxins, and their large concentrations kill fish and make other marine life poisonous to humans. The poisons of many species, *Gonyaulax catenella* for instance, are potent neurotoxins, interfering with the movement of sodium ions across our nerve membranes.

DIVISION CHRYSOPHYTA

Members of division Chrysophyta are grouped into three classes: Bacillariophyceae (**diatoms:** see Fig. 21.1c; Table 21.5), Chrysophyceae (**golden brown algae:** see Fig. 21.1d), and Xanthophyceae (**yellow-green algae:** see Fig. 21.1e). Many phycologists think each of these groups should have its own division, whereas others think they should be classified as

TABLE 21.5	*Classification of Division Chrysophyta*
Class Xanthophyceae	
Order Vaucheriales	
Family Botrydiaceae	*Botrydium*
Family Vaucheriaceae	*Vaucheria*
Class Chrysophyceae	
Order Ochromonadales	
Family Ochromonadaceae	*Ochromonas, Dinobryon*
Family Synuraceae	*Synura*
Class Bacillariophyceae	
Order Centrales	
Order Pennales	

brown algae. Division Chrysophyta all have chlorophylls *a* and *c*, store their reserve carbohydrate as the polymer chrysolaminarin, and have such abundant accessory carotene and xanthophyll pigments that they mask the chlorophylls. Some have cellulose walls that are frequently encrusted with either silica or calcium carbonate; others have walls with no cellulose but only silica embedded in a matrix of pectin. Some have two anterior, unequal flagella.

CLASS BACILLARIOPHYCEAE: DIATOMS

Class Bacillariophyceae contains about 200 genera and 5000 species. Diatoms are easy to recognize because of their distinctive morphology. Each cell has a wall composed of two halves or **frustules** that fit together like a Petri dish and its lid. Each frustule is encrusted with silica, and when the diatom dies and the protoplasm degenerates, the siliceous frus-

FIGURE 21.13

The true immensity of the deposits of microscopic diatom fossil frustules is difficult to imagine. This is a mine of diatomaceous earth or diatomite. The frustules are excellent as polishing material or as filtering agents in water purification plants and swimming pools. *(Courtesy of the Celete Corporation, Lompoc, California)*

tules sink to the ocean floor and accumulate. Such deposits, known as **diatomaceous earth,** can become hundreds of meters thick and cover many square kilometers (Fig. 21.13). Accumulations of this volume are possible because diatoms are the most abundant organisms in the oceans, and a single liter of sea water may easily contain more than 1 million individuals.

It might be thought that organisms consisting of a single cell could not possibly be very complex and certainly would not have enough structure to allow 5000 species to be recognized, but frustules are extremely intricate. Where the two come together, they are held in place by girdle bands, and each species has its own characteristic complicated pattern of ridges, depressions, and pores. The cells are either round in face view (**centric diatoms:** see Fig. 21.1c) or elongate (**pennate diatoms**) (Fig. 21.14).

When a diatom undergoes mitosis and cytokinesis, each progeny cell receives one of the frustules from the parental cell, using it as its outer frustule and synthesizing a new inner frustule. All cells that receive the inner frustule automatically mature into a cell that is smaller than the parent cell, and average cell size decreases with each cell cycle. This obviously cannot go on forever, and when a cell reaches a critical small size, sexual reproduction is triggered. The cells undergo meiosis, some producing 4, 8, or 16 sperm cells, others producing just one or two large egg cells. After fertilization, the zygote grows into a large cell, then becomes dormant, surrounded by a wall different from that of a vegetative cell. The zygote finally "germinates," undergoing a round of nuclear and cell division, and re-initiating the process of decreasing cell size.

CLASS CHRYSOPHYCEAE: GOLDEN BROWN ALGAE

Class Chrysophyceae contains about 70 genera and 325 species of golden brown algae (Table 21.5). These are biochemically similar to diatoms but differ in not having frustules. Instead, the single cells, such as those of *Synura,* are covered with numerous tiny siliceous scales that develop within special vesicles in the endoplasmic reticulum (see Figs. 21.1d and 21.15). Cells may be either uniflagellate or biflagellate; rarely they have no flagella, crawling by ameboid motion instead. Although they are photoautotrophic because of their chloroplasts, they can also ingest bacteria by phagocytosis.

Golden brown algae are particularly important because of their past biology. The living cells float in warm, sunny surface waters, like diatoms, but when they die, the lack of swimming motion allows their dense siliceous scales or frustules to sink and accumulate on the ocean floor. The remains of both diatoms and golden brown algae, especially a group known as coccolithophorids (usually called coccoliths), are relatively inert and do not decompose. Drilling into seafloor sediments makes it possible to analyze the amount of diatom and coccolith remains and to estimate the rate at which they accumulated. It is possible to reconstruct the surface climate: Accumulation was rapid when waters were warm and carbon dioxide was abundant for photosynthesis and slow when the climate was cool or if carbon dioxide was low. Shallow coastal seas tend to be warmer than the open oceans, and this also affects coccolith growth and reproduction. By also analyzing fossil animals, which are affected by temperature but not carbon dioxide, we can reconstruct past climates and ocean basin geology precisely.

CLASS XANTHOPHYCEAE: YELLOW-GREEN ALGAE

Yellow-green algae (see Fig. 21.1e; Table 21.5) occur mostly in fresh water, and formerly many were thought to be green algae until chlorophyll *c* was discovered in them. They are somewhat diverse, some being unicellular, some filamentous, and some forming giant multinucleate cells. Filamentous forms, such as *Tribonema,* have walls like diatom frustules, each cell having two half-walls, but each of these frustule-like walls is firmly attached to the adjacent half-wall of the neighboring cell. Although the cells remain together, the organism can be called multicellular only in the broadest sense, because no differentiation or specialization occurs in any part: The filament is really just a colony of individuals that happen to remain together.

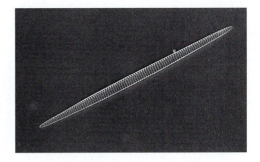

FIGURE 21.14
Like all pennate diatoms, this *Nitzschia pungens* is extremely elongate and bilaterally symmetrical. One of its varieties, *N. pungens* forma *multiseries,* produces a neurotoxin as the blooms senesce. The toxin is domoic acid, which causes the syndrome amnesiac shellfish poisoning (× 1000). *(Courtesy of Greta Fryxell, Texas A and M University)*

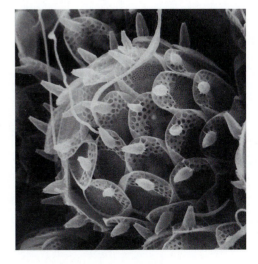

FIGURE 21.15
Each cell of a golden brown alga is covered with many minute scales that are formed within vesicles of the endoplasmic reticulum. The pattern within each scale is constant for each species, but how it is controlled is unknown. *Synura uvella* (× 5000). *(Courtesy of R. E. Andersen, Bigelow Laboratory for Ocean Science)*

Coenocytic cell   Progeny "cells"

Megagametangium   Microgametangium

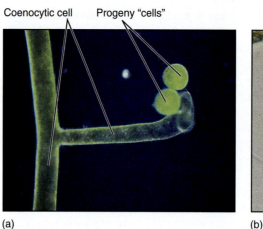

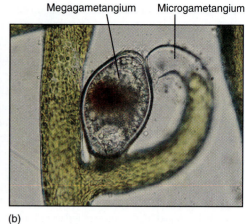

(a)   (b)

FIGURE 21.16
(a) *Vaucheria* occasionally walls off the tip of a coenocyte; then pairs of flagella form near each nucleus. The wall breaks open and the "cell" swims away, settles, and grows into a new individual (× 100). (*T. E. Adams/Visuals Unlimited*) (b) During sexual reproduction, one segment of the coenocyte forms numerous biflagellate sperm cells, and another forms one large uninucleate egg. After the sperm cells swim away, all that is left of the gametangium is wall; unlike gametangia in true plants, no protective layer of sterile cells is present (× 200). (*R. Calentine/Visuals Unlimited*)

Some yellow-green algae, such as *Vaucheria,* have an unusual body that consists of a long tubular coenocyte with a single large central vacuole, a thin peripheral layer of cytoplasm, and thousands of nuclei. During asexual reproduction, the cytoplasm at the tip of a tube is isolated by a transverse wall, and pairs of flagella form near each nucleus; then the old wall breaks open, and the multiflagellate "cell" swims slowly away (Fig. 21.16a). It soon loses its flagella and begins growth at each end, forming a new tube. The greatest complexity is displayed during sexual reproduction. A mass of cytoplasm is walled off and pairs of flagella form, but then each nucleus organizes the protoplasm around it into an individual cell. When the original wall breaks down, numerous biflagellate sperm cells swim away (Fig. 21.16b). An egg cell is formed as a small multinucleate protrusion on a tube, but as the cross wall forms, all nuclei except one migrate out, leaving a single large uninucleate cell. After fertilization the zygote forms a thick wall and is abscised from the parent branch. After a period of dormancy, the zygote undergoes meiosis and germinates, forming a new tube filled with haploid nuclei.

DIVISION CHLOROPHYTA: GREEN ALGAE

From an evolutionary standpoint, the **green algae** (Table 21.6) constitute an extremely important group: Not only could some early green algae organize complex differentiated multicellular bodies, but some moved onto land—the ancestors of true plants, kingdom Plantae. Perhaps because most of the animals and plants with which we are familiar are multicellular and terrestrial, we tend not to appreciate just how complex these phenomena are. Consider that organisms spent as many as 1 billion years at the level of unicellular organization; then among algae true multicellularity evolved only a few times. The transition to living on land was so complex that it occurred no more than two or three times at most, perhaps only once.

Green algae have remarkable developmental and metabolic plasticity: They are resilient and survive many types of disturbances and changes. Mutations that cause cells to adhere after cytokinesis have not been lethal, nor were many mutations that affected swimming, orientation of cell division, or coordination of karyokinesis and cytokinesis. Bodies composed of several types of specialized cells have not evolved in most other algal groups. Although brown algae and red algae do show considerable sophistication in certain types of multicellular bodies, they are metabolically intolerant of ecological changes; few can live in fresh water, soil, air, or inside animals as many green algae do.

The diversity in the green algae is tremendous, but some of the most important evolutionary lines and developments are involved. These give you a good basis for understanding the diversity possible among living organisms and appreciating that the metabolisms and organizations of humans and flowering plants are by no means the only solutions to biological problems.

TABLE 21.6	*Classification of Division Chlorophyta*

Class Chlorophyceae
    Order Volvocales
        Family Chlamydomonadaceae   *Chlamydomonas*
        Family Volvocaceae   *Eudorina, Gonium, Pandorina, Volvox*

    Order Chlorococcales
        Family Hydrodictyaceae   *Hydrodictyon, Pediastrum*
    Order Ulotrichales
        Family Ulotrichaceae   *Ulothrix*
Class Ulvophyceae
    Order Ulvales
        Family Ulvaceae   *Ulva*
    Order Dasycladales
        Family Dasycladaceae   *Acetabularia*
    Order Zygnemetales
        Family Zygnemataceae   *Spirogyra*
        Family Desmidaceae   *Closterium, Micrasterias*
Class Charophyceae
    Order Charales
        Family Characeae   *Chara, Nitella*
    Order Coleochaetales
        Family Coleochaetaceae   *Coleochaete*

Plant Biology Tutor CD-ROM

BODY CONSTRUCTION IN THE GREEN ALGAE

The most relictual and simple body is probably the motile single cell. Numerous green algae are unicellular. From this, many evolutionary possibilities exist:

1. **Motile colonies** (Figs. 21.17a, 21.25, and 21.26): If cells adhere loosely, the resulting structure is a **colony**, not an individual organism. All cells are similar and none is particularly specialized; there may be some differentiation into two or three somewhat distinct cell types.

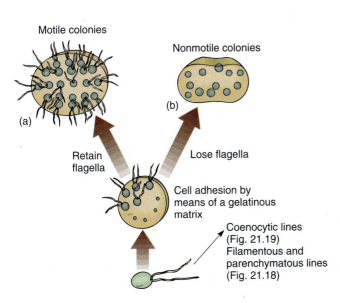

FIGURE 21.17
Evolution of colonial body type in green algae. The association of cells is controlled, not random. Depending on the species, each colony has a specific number of cells held in a particular shape such as a sphere, a flat plate, or a curved plate. No cell is free to divide at random.

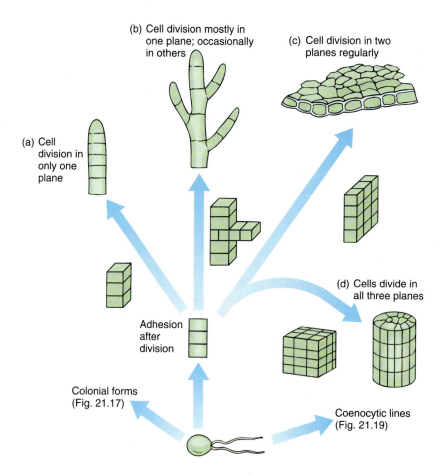

(b) Cell division mostly in
one plane; occasionally
in others

(c) Cell division in two
planes regularly

(a) Cell
division in
only one
plane

(d) Cells divide in
all three planes

Adhesion
after
division

Colonial forms
(Fig. 21.17)

Coenocytic lines
(Fig. 21.19)

FIGURE 21.18

Evolution of filamentous, membranous, and parenchymatous body types in green algae. The formation of a multicellular body requires control over timing, position, and orientation of cell division, along with cell-cell adhesion mediated by a middle lamella. (a) All cells divide in only one plane, resulting in a uniseriate filament. (b) If cells occasionally divide in a different plane, the filament branches. A morphogenic mechanism may signal which cells can divide to branch and which cannot. (c) A two-dimensional sheet of cells results if all cells divide in two dimensions regularly. (d) A three-dimensional body is produced if cells divide regularly in all three planes. All cells share plasmodesmata with all adjacent cells because all walls result from a cell division cell plate. This is a true parenchyma tissue.

2. **Nonmotile colonies** occur if the cells lose their flagella or never develop them (Fig. 21.17b). Although a nonmotile cell is simpler than one with flagella, the flagellated form is believed to be ancestral because all motile algae have the same type of flagella, the 9 + 2 arrangement of microtubules.

3. A **filamentous body** results if cells are held tightly by a middle lamella and if all cells divide transversely (Figs. 21.18a and b and 21.27). If occasional cells undergo longitudinal division, the filament branches. Often some portions of the body differ from others; for example, one end may serve as a means to attach the filament to a rock (a **holdfast**), or other parts may produce spores or gametes.

4. A **membranous body** results if the orientation of cell divisions is controlled precisely such that all new walls occur in only two planes (Figs. 21.18c and 21.28). The result is a sheet of cells that can become more extensive but remains thin. A membranous body is more strongly affected by currents and wave action and more likely to be torn than is a filamentous body.

5. If cell division occurs regularly in all three planes, a bulky, three-dimensional **parenchymatous body** results (Figs. 21.18d and 21.31). All cells are interconnected by plasmodesmata, and a true parenchyma tissue is formed.

6. A **coenocytic** or **siphonous body** results if karyokinesis occurs without cytokinesis, and giant multinucleate cells result (Figs. 21.19 and 21.30). The cells can grow to several centimeters in diameter, but they usually remain fairly simple despite their size. Some coenocytes, such as *Acetabularia,* can be surprisingly complex (Fig. 21.20).

Evolutionary development of control over the orientation of cell division and adhesion must have been difficult. The fossil record indicates that the ability to grow as a siphonous form came into existence 230 million years earlier than the ability to organize even the simplest filament.

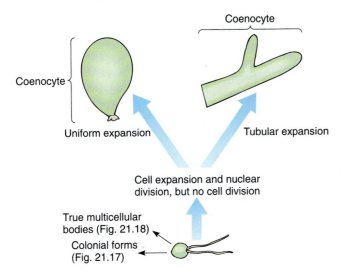

FIGURE 21.19
Coenocytic bodies in green algae may be more or less isodiametric (spherical) or long, branched or unbranched tubes. Tubular forms are called siphonous forms.

LIFE CYCLES OF THE GREEN ALGAE

The angiosperm type of life cycle, an alternation of heteromorphic generations, can be traced to the green algae, and the theory of life cycles can be discussed here. In unicellular algae that have no sexual reproduction, such as euglenoids, the life cycle is really just the cell cycle; mitosis and cytokinesis constitute reproduction. But with the evolution of sex, two processes are critical: *meiosis,* which segregates out sets of chromosomes, and *syngamy,* which brings two sets back together.

After sexual reproduction evolved, while all organisms were still unicellular (no multicellular organism is relictually asexual), the simplest life cycle may have been the following: A diploid cell undergoes meiosis, and each cell then exists as a haploid individual, able to undergo mitosis. Because this is a unicellular organism, cell division is both "growth" and asexual reproduction. Some of these cells can also act as gametes and fuse, producing a diploid zygote that also is able to reproduce asexually—and simultaneously grow—by mitosis. Finally, at some time some of these undergo meiosis again. Little difference exists between gametes, zygotes, and individuals, none of which is very specialized. Both haploid and diploid cells are capable of growth, division, and reproduction. Such species are **dibiontic;** that is, there is an alternation of generations between haploid and diploid.

In **monobiontic** species, specialization occurs in that only one free-living generation exists. In some monobiontic species the haploid phase represents the individual and the

FIGURE 21.20
Individuals of *Acetabularia* are unusual for being giant single cells, but even more remarkable, they are uninucleate. Each cap is composed of hollow tubes of cytoplasm and vacuole, and at times of reproduction, the sole nucleus begins rapid, repeated mitosis, resulting in hundreds or thousands of nuclei. Cytoplasmic streaming carries them upward into the cap segments; when each segment has several hundred nuclei, it either becomes a dormant cyst and is released or produces hundreds of gametes. (*Dr. Dennis Kunkel/Photake*)

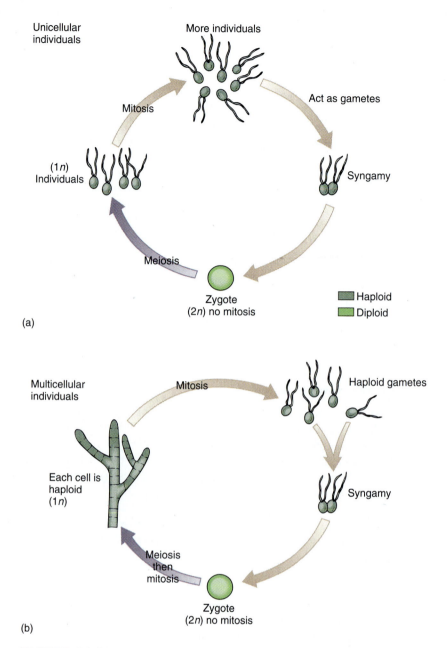

FIGURE 21.21

In monobiontic life cycles, only one generation—one phase—is capable of undergoing mitosis. In most algae (a and b), the haploid phase is dominant, undergoing mitosis that results in either more unicellular individuals or a multicellular individual. Only one diploid cell occurs—the zygote—and it cannot undergo mitosis; it cannot produce more diploid cells. (c and d, facing page) In a few algae and in all animals, the monobiontic life cycle is dominated by the diploid phase, the only haploid cells being the gametes.

only diploid cell is the zygote, which is capable only of meiosis, not growth or mitosis (Fig. 21.21a and b). The haploid body, either uni- or multicellular, carries out photosynthesis and growth. In other monobiontic species, the diploid phase represents the individual, vegetative growth phase; the only haploid cells are the gametes, which can undergo only syngamy (Fig. 21.21c and d).

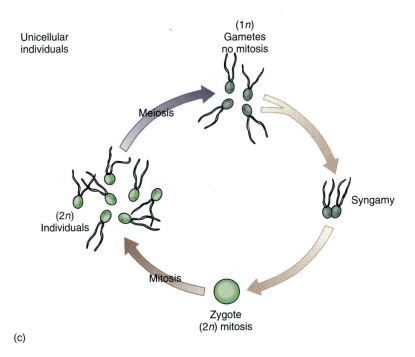

(c)

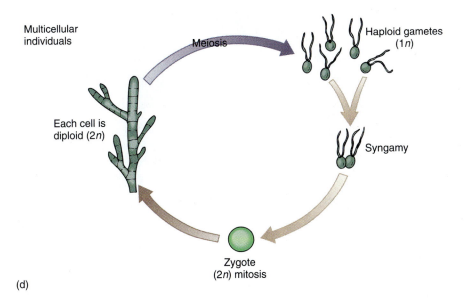

(d)

In a dibiontic species in which both stages are multicellular, the gametophyte (haploid phase) and sporophyte (diploid phase) may resemble each other strongly, and alternation of **isomorphic generations** occurs (Fig. 21.22a). Or the two may be very different in appearance and construction (alternation of **heteromorphic generations**), which allows them to exploit different ecological niches almost as if they were two species, and gametophytes do not compete directly with sporophytes (Fig. 21.22b).

All sporophytes, both in algae and in embryophytes, produce spores by meiosis, but many algal sporophytes also produce spores by mitosis; these are diploid and grow into a new sporophyte in a form of asexual reproduction. Some algal gametophytes produce spores by mitosis; these are haploid and develop into new gametophytes, also a form of asexual reproduction.

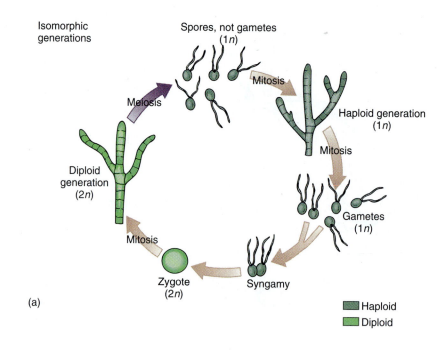

(a)

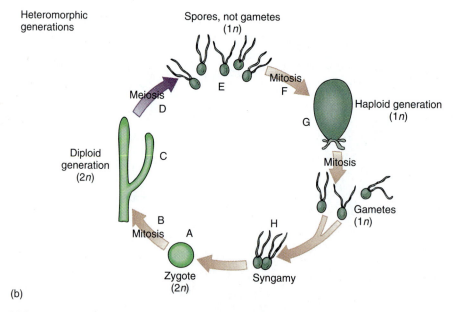

(b)

FIGURE 21.22

Dibiontic life cycles involve an alternation of either isomorphic (a) or heteromorphic (b) generations. With isomorphic generations, the similarity of body types can be so great that it is very difficult to know if an individual is a gametophyte or a sporophyte. (b) In dibiontic algae with an alternation of heteromorphic generations, the two types of individuals could never be confused. In flowering plants, the mitosis at B would be the growth of the embryo into the sporophyte plant body, such as an oak tree or a lily, and the meiosis at D would occur only in the sporogenous cells of anthers and ovules. The mitosis at F would be the formation of the tube nucleus, generative nucleus, and sperm cells in pollen and the growth of the megaspore into the antipodals, central cell, synergids, and egg cell of the megagametophyte. The syngamy at H would be fusion of one sperm nucleus with the egg. In flowering plants, the zygote begins growth immediately without a resting period, and of course flagella are not formed at any stage.

During the earliest stages of the evolution of sex, the gametes were isogamous (identical) but **anisogamy** (slight differences in gametes) and oogamy later evolved (Fig. 21.23). The reproductive tissues and organs of different algal, fungal, and plant groups differ markedly. At one time specialized terms were applied to every group, resulting in an overwhelming nomenclature. It has since been standardized, so that regardless of the group, gametes are produced in **gametangia**. Sperm cells or microgametes form in microgametangia, and egg cells or megagametes form in megagametangia. Spores are formed in **sporangia**, either megasporangia or microsporangia, depending on the size of the spore.

REPRESENTATIVE GENERA OF THE GREEN ALGAE

Unicellular Species.   *Chlamydomonas* is one of the simplest chlorophytes. It is unicellular and, like all green algae, has chlorophyll *a* and *b*, carotenoids, and xanthophylls; its starch is formed in chloroplasts just like that of true plants (Fig. 21.24). It has two anterior flagella, like most motile green algae, and it has normal mitosis, meiosis, and syngamy. Its life cycle is simple: A haploid cell resorbs its flagella, divides mitotically, and forms 2, 4, 8, or 16 new cells that grow new flagella, each of which swims until it encounters a compati-

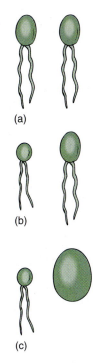

(a)

(b)

(c)

FIGURE 21.23

(a) Isogametes are identical gametes. Theoretically they could be either motile or nonmotile, but in reality, all known isogametes do have flagella. (b) If the gametes have any visible differences but are still similar, they are anisogametes. (c) Oogametes are obviously different in size, and the megagamete is virtually always nonmotile. In most algae and plants the microgamete has flagella. In flowering plants, conifers, and some algae, microgametes are also nonmotile.

Chloroplast   Flagella   Nucleus

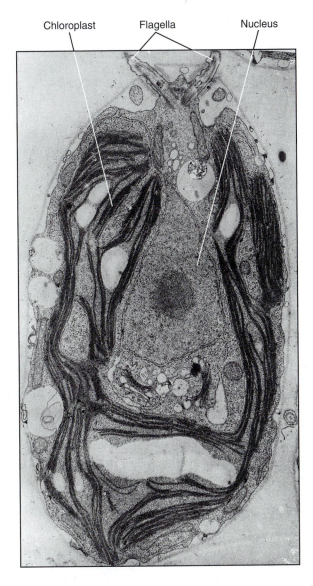

FIGURE 21.24

The unicellular, haploid green alga *Chlamydomonas*. Its nucleus, chloroplast, and pyrenoid are visible, but only the bases of the two flagella can be seen. The flagella are at the anterior end; that is, they pull the cell forward rather than pushing it. Several starch grains are present. (*U. Goodenough, Washington University*)

FIGURE 21.25

Eudorina elegans is a small motile colony of green algae. In each species, the number of cells and their spatial arrangements are constant. The numbers and orientations of cell divisions are controlled. *(Walker/Photo Researchers)*

ble cell. The cells recognize each other by reactions at the tips of their flagella. They undergo plasmogamy and karyogamy, forming a large zygote that sheds the four flagella and sinks to the bottom in a quiescent state. Germination is by meiosis to form four biflagellate haploid individuals; the zygote is the only diploid cell.

Motile Colonial Species. In the motile colonial line of evolution, cells that greatly resemble *Chlamydomonas* are produced when the zygote divides, but the progeny cells are held together by a gelatinous matrix (Fig. 21.25). In *Gonium,* each colony contains only a few cells (4, 8, 16, or 32), and the only sign of organization is that all flagella beat in a coordinated fashion. *Pandorina* is about the same size as *Gonium* but is slightly more derived because it shows a trace of differentiation; the colony swims in one direction, and anterior cells are slightly different from posterior ones. *Volvox* is the stunning conclusion of this line: Its colonies can contain up to 50,000 *Chlamydomonas*-like cells and are easily visible without a microscope (Fig. 21.26). Differentiation exists in that up to 50 cells in the posterior half of a colony are specialized for reproduction only.

Filamentous Species. Members of the genus *Ulothrix* are simple types of filamentous green algae (Fig. 21.27a). Their life cycle is monobiontic; it has only one free-living multicellular generation, and it is haploid. It consists of one row of cells that are all rather similar except for the basal cell, which is modified into a holdfast. Filament cells may produce asexual zoospores mitotically; these have four flagella, swim briefly, then settle and grow into new filaments. During sexual reproduction, some cells produce gametes, which can be identified because they have only two flagella, not four. *Ulothrix* is isogamous, and the gametes, which all look like cells of *Chlamydomonas,* pair and fuse. The zygote germinates by meiosis, producing four haploid zoospores, each of which swims for a period, then loses its flagella, attaches to a substrate, and grows into a new filament.

Spirogyra is an extremely common fresh-water filamentous green alga (Fig. 21.27b) found in streams and ponds throughout North America. Its cells have beautiful spiral band-shaped chloroplasts that wind around the cell just below the plasma membrane. Swimming gametes are not formed; instead filaments undergo conjugation. Each filament is haploid and is either a + or − mating type. If compatible filaments drift against each other, a conjugation tube forms between cells, and the − protoplasts migrate through the tube and fuse with the + protoplasts. Karyogamy follows, although as many as 30 days may pass before nuclei fuse. The cell becomes dormant, thick-walled, and resistant; it later germinates and grows into a new filament. Meiosis occurs immediately after karyogamy, so the spore is haploid.

Laminar Species. *Ulva* is slightly more complex than *Ulothrix,* but many of its stages are almost identical (Fig. 21.28). A quadriflagellate, haploid zoospore settles down and grows into a *Ulothrix*-like filament. The cells divide in two directions and thus form a sheet; then all cells divide once in a third direction so that the sheet becomes two layers thick. *Ulva* has a dibiontic life cycle with an alternation of isomorphic generations (see Figs. 21.22a and

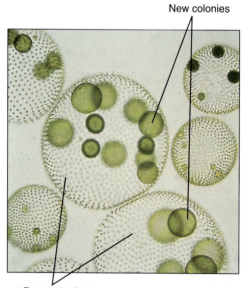

New colonies

Parental colonies

FIGURE 21.26

Individuals of *Volvox* are giant spherical colonies containing up to 50,000 cells. The colonies are easily visible to the naked eye. *(Manfred Kage/Peter Arnold, Inc.)*

(a)

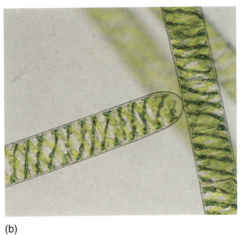

(b)

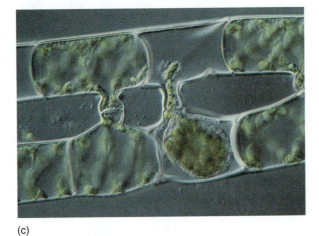

(c)

FIGURE 21.27

(a) Individuals of *Ulothrix* are uniseriate, unbranched filaments of haploid cells. When the spores or gametes are released, only empty cell walls remain. Sporogenous cells do not have a protective layer of sterile cells surrounding them, whereas the sporogenous cells of flowering plants are surrounded by either the anther wall or the nucellus ($\times 80$). *(E. R. Degginger)* (b) Cells of *Spirogyra*, showing the spiral, band-shaped chloroplasts. *(Runk/Schoenberger from Grant Heilman)* (c) Compatible filaments of *Spirogyra* have been brought together in culture and are undergoing conjugation. This type of plasmogamy is unusual in algae, being analogous to the plasmogamy of fungal hyphae. It superficially resembles conjugation in bacteria, but few details are similar. *(M. I. Walker/Science Source/Photo Researchers, Inc.)*

FIGURE 21.28

Individuals of sea lettuce, *Ulva*, are sheets two cells thick. They grow to almost any length or width but do not become thicker. If cut or torn into several pieces, each continues to grow. *(William E. Ferguson)*

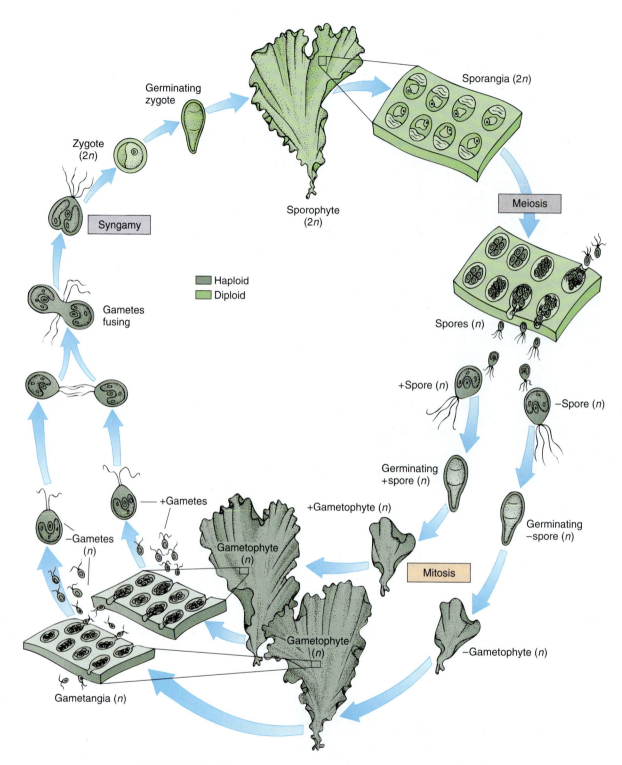

FIGURE 21.29

Life cycle of *Ulva*; details are given in text. Two multicellular, isomorphic generations occur. The types of reproductive structures and the size of the nuclei are the only reliable criteria for distinguishing between sporophytes and gametophytes.

21.29). During sexual reproduction, gametophyte cells produce biflagellate anisogametes, the smaller gametes being produced on one gametophyte and the larger ones on a different gametophyte. Having two different types of individuals in one generation is dioecy, just as in flowering plants. The zygote grows into a filament and then into a double-layered sheet just like the gametophyte.

Coenocytic Species.   A dibiontic life cycle with an alternation of heteromorphic generations is illustrated by *Derbesia,* the organism in which it was discovered (see Figs. 21.22b and 21.30). Before going into the details of the life cycle, think for a second about how organisms are studied. Often a biologist goes on a field trip, collects plants, insects, birds, or whatever, and preserves them by drying or fixing them in formaldehyde. It is difficult to maintain things in a living condition. Now imagine trying to do this 100 years ago when transportation was slow, laboratories did not have good artificial light or temperature control, and few pure chemicals were available for making culture solutions. Growing algae to study their life cycles was and is extremely difficult.

In 1938, P. Kornmann succeeded in obtaining zoospores from *Derbesia marina,* a branching, filamentous alga made up of giant coenocytic cells (Fig. 21.30). The zoospores were carefully maintained, but instead of growing into another *Derbesia*-like individual as

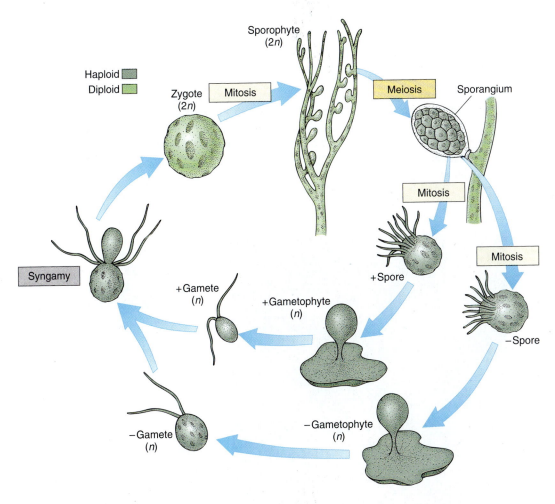

FIGURE 21.30

Derbesia has a dibiontic life cycle with an alternation of heteromorphic generations. The green algae are a large group (approximately 450 genera with 7500 species), and not all life cycles are known, so many more surprises of this type may be awaiting us. Only by growing all stages of a species can we study its life cycle.

expected, they grew into individuals of a totally different genus, *Halicystis ovalis* (Fig. 21.30). A member of *Halicystis* looks nothing at all like *Derbesia;* instead it is composed of a single large, spherical coenocytic cell attached to the substrate by a small holdfast. Almost its entire volume is one giant vacuole, with only a thin layer of protoplasm next to the wall. When mature, *Halicystis* individuals, which are the gametophyte stage, produce either male or female anisogametes that undergo syngamy and establish the zygote. After germination, the zygote grows into a *Derbesia* sporophyte. To classify different life stages as different genera is obviously not correct, so the name *Halicystis ovalis* has been eliminated. Flowering plants also have an alternation of heteromorphic generations, but the gametophytes grow inside the sporophytes, so the question of which goes with which is easily solved.

Parenchymatous Species.   Several groups of green algae undergo a true parenchymatous growth and may be related to the ancestors of true plants. One group, in class Chlorophyceae, divides by means of a phycoplast, which never occurs in plants, so this group is a remote possibility. The other group, in class Charophyceae, undergoes cell division by means of a phragmoplast, just as plant cells do. In Chlorophyceae, the flagellar root apparatus consists of four bands arranged in a cross; no true plant is known to have this type of basal body. In charophytes the flagellar root complex is similar to that of the motile cells of true plants: One major band of microtubules extends down into the cytoplasm from the basal body. Consequently, great attention is being given to charophytes.

An interesting example is *Chara* (Fig. 21.31), which has a stemlike body divided into nodes and internodes, with whorls of branches arising at internodes. The body is several cells thick, composed of true parenchyma tissue derived from cell division in all three planes and originating from an apical meristem that contains a prominent apical cell. Although these features seem to correspond to those of flowering plants, virtually all of the resemblance is spurious, because the earliest vascular land plants had no nodes, internodes, or branches (see Chapter 23). If *Chara's* ancestors were the ancestors of true plants, the only features that may be homologous rather than analogous are the parenchymatous body and growth by an apical meristem. Simple parenchymatous bodies also occur in members of *Coleochaete,* another group of charophytes that are being studied as possible close relatives of land plants.

Reproduction in *Chara* is significant. True plants (embryophytes) have multicellular reproductive structures with sterile cells, and on this basis *Chara* would have to be classi-

FIGURE 21.31
(a) An individual of *Chara* has a strong resemblance to certain aquatic angiosperms, but most of the similarity is analogous (resulting from convergent evolution), not homologous (resulting from descent from a common ancestor). *(Runk/Schoenberger from Grant Heilman)* (b) The microgametangium of *Chara* consists of an outer layer of sterile cells. Some inner cells mature into sperm cells, whereas others remain sterile and act as spacers. (c) The megagametangium is multicellular at maturity but not at initiation. The sterile cells around the egg are not sibling cells of the egg but rather are filaments that grow up from the cell below the egg. This is a multicellular gametangium, but it is filamentous, not parenchymatous.

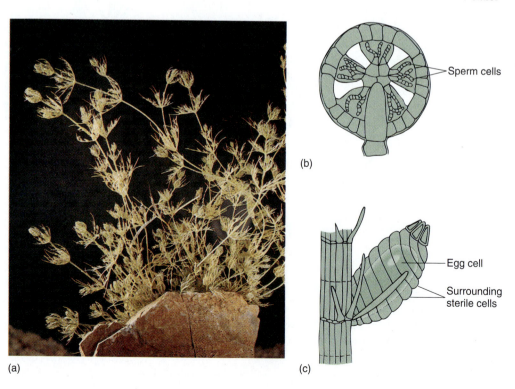

Sperm cells

(b)

Egg cell

Surrounding sterile cells

(a)   (c)

fied as a plant rather than an alga. Its sperm cells are produced in a truly multicellular gametangium whose outer cells are sterile; only the inner cells convert to sperm cells (Fig. 21.31b). At maturity, the outer cells separate slightly and the motile sperm cells swim away. The egg is formed as the terminal cell of a short filament three cells long, but the subterminal cell subdivides, and those cells grow upward and surround the egg (Fig. 21.31c). After fertilization, the sterile cells surrounding the fertilized egg deposit thickenings on their inner walls, those adjacent to the zygote. The resting structure thus consists not only of a thick-walled zygote but also protective sterile cells. When the resting cell germinates, it grows out as haploid filaments that soon establish an apical meristem and parenchymatic growth.

Multicellular gametangia with sterile cells have been used as the primary defining feature of embryophytes, true plants. Morphologists have long debated whether the gametangia of charophytes are ancestral to those of true plants or are merely a fascinating case of convergent evolution. Although they are similar, they are not identical, but *Chara* and *Coleochaete* are groups of living species, none of which is the direct ancestor of true plants which arose 420 million years ago. The question is whether they are members of an evolutionary line that has changed little since its ancestors gave rise to the ancestors of true plants, or whether they are only very distantly related, their parenchyma, apical growth, and multicellular gametangia all being cases of convergent evolution.

DIVISION PHAEOPHYTA: BROWN ALGAE

The **brown algae** (Table 21.7; see Fig. 21.1g) are almost exclusively marine; only a few fresh-water species are known. They prefer cold water that is very agitated and aerated. They can most easily be found on rocky coasts growing in the **littoral zone,** the region between low tide and high tide, also called the intertidal zone, where they are periodically exposed to air and full sunlight. The upper sublittoral zone also contains numerous brown algae if it is rocky and offers stable surfaces for attachment. Over 1500 species are known, grouped into about 250 genera.

The brown algae are the most complex algae anatomically and morphologically, some considerably more complex than mosses and liverworts. Although very distinct from true plants biochemically and ecologically, the two groups have remarkable parallels in the types of bodies and life cycles that have evolved.

The differences between brown algae and green organisms (both green algae and plants) are clear-cut: Brown algae have chlorophyll *a* and *c* and large amounts of a variety of xanthophyll pigments such as fucoxanthin, violaxanthin, and diatoxanthin. Carotenes are also present. Their suite of pigments permits the brown algae to carry out photosynthesis at

TABLE 21.7 *Classification of Division Phaeophyta*	
Class Phaeophyceae	
Order Ectocarpales	
Family Ectocarpaceae	*Ectocarpus*
Order Laminariales	
Family Laminariaceae	*Laminaria*
Family Lessoniaceae	*Macrocystis, Nereocystis, Pelagophycus*
Order Dictyotales	
Family Dictyotaceae	*Padina*
Order Fucales	
Family Fucaceae	*Fucus, Sargassum*

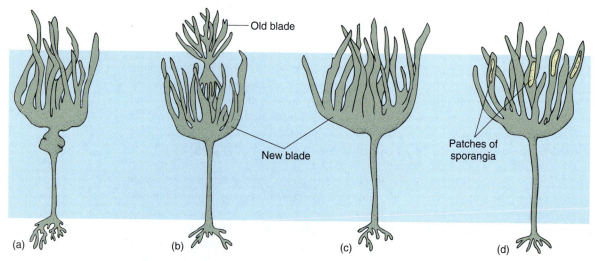

Old blade

New blade

Patches of
sporangia

(a)　(b)　(c)　(d)

FIGURE 21.32

In many kelps, the blade is annual but the stipe and holdfast are perennial. A new blade is formed
each year; if the old blade has not been completely destroyed by wave action, it is sloughed off.

FIGURE 21.33

Many kelps become extremely large, but their
bulk is due to great length, not great width. The
stipe and air bladder are the thickest parts, up
to several centimeters, but the blade is always
thin. Long, narrow, elastically flexible bodies are
well adapted for life in a tidal region. Think of
the damage that wind causes to land plants.
Water is more dense and massive, so it is
virtually impossible to build a large body that
could resist such constant pressure from
currents. The flexible bodies of brown algae, like
those of willows, bend without breaking and so
remain undamaged by the fluid flowing around
them. (*Gregory Ochocki/Photo Researchers*)

numerous levels in the ocean. Sunlight differs not only in intensity but also in quality at
different depths: White light with a full spectrum occurs at the surface, but primarily just
blue-green light reaches depths of 50 meters or more.

The storage product of brown algae is **laminarin** (a polymer of glucose), mannitol, or
fats, but not starch. Some brown algae store large amounts of laminarin, up to 34% of their
body weight. Most algae, especially the marine species, live in such a stable environment
with regard to light, temperature, and nutrients that photosynthesis and growth occur
more or less continuously, so large reserves are unnecessary. But many brown algae,
especially the kelps (order Laminariales) that grow near the surface, have large bodies made
up of three parts: holdfasts and stipes (stalks) that may be perennial, and photosynthetic
blades that are annual (Figs. 21.32 and 21.33). When the blades become moribund and
decompose, the holdfasts and stalks must subsist on stored nutrients until the new blade
can be formed and begin photosynthesis in the spring.

Cell walls of brown algae contain cellulose and alginic acid, an unusual polymer of
D-mannuronic acid and L-guluronic acid not found in other algae. The alginic acid component of the wall is gummy or slimy and causes the filaments of cells to adhere into a
compact body. It may have another, unknown function, however, because as much as 24%
of the body dry weight can be alginic acid in *Ascophyllum*.

A remarkable feature of brown algae is that they are all multicellular; no unicellular
species is known to exist, and individuals of many species of kelps become huge and
complex. Plants of *Nereocystis* that measure up to 45 m—taller than most trees—are not
uncommon, but they are not bulky, being less than 5 cm thick. Any plant that becomes as
long as half a football field must have specialized regions to its body. The holdfasts of a kelp
are located in deep, dark, poorly aerated water, whereas the blades exist in brightly lighted
and well-aerated shallow waters. To keep from being damaged by wave actions, the bodies
must be firm, elastic, and thick—too thick for diffusion alone to mediate the exchange of
gases, nutrients, and wastes. Some species have an epidermis-like outer covering, a parenchymatous middle tissue that resembles cortex, and a cylinder of **trumpet cells** that resemble phloem cells so much that many biologists do call them sieve tube members (Fig.
21.34). Trumpet cells carry out long-distance transport of carbohydrates through the body.
They are elongate, large holes occur in a sievelike arrangement in the end walls, and the
holes are lined with callose. If radioactive carbon dioxide is given to the blade, a short time
later radioactive photosynthates can be detected in the trumpet cells of the stipe. The flow
rate can be as high as 65 to 78 cm/hr.

Are these cells sieve tube members? Although trumpet cells and sieve tube members
are extremely similar structurally and metabolically, they are not at all homologous. They
did not evolve from a common ancestral cell, nor did trumpet cells evolve from sieve tube
members or vice versa. Trumpet cells and sieve tube members are outstanding examples of
convergent evolution.

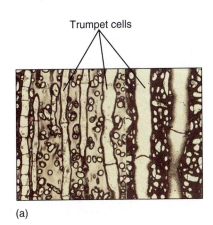

Trumpet cells

(a)

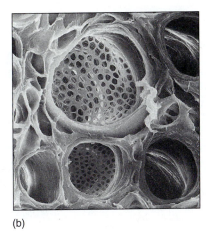

(b)

FIGURE 21.34
(a) A longitudinal section through the stipe of the kelp *Macrocystis.* In the center are numerous trumpet cells (× 290). (b) A young trumpet cell, showing the enlarged end walls where holes have developed (× 500). *(Courtesy of M. L. Shih, J. -Y. Floch, and L. M. Srivastava, Simon Fraser University)*

One of the simplest of the brown algae is *Ectocarpus* (Fig. 21.35). It has an alternation of isomorphic generations, both generations consisting of uniseriate branched filaments that arise from a prostrate branched system attached to rocks, shells, or other, larger algae. In the diploid sporophytes, some of the terminal cells of small lateral branches enlarge greatly and become **unilocular sporangia.** The nuclei divide repeatedly, then 32 or 64 zoospores are released. The first nuclear division is meiotic, so the spores are haploid. After swimming temporarily, zoospores settle down and grow into gametophytes that are almost identical to the sporophytes. The primary difference is that at the ends of the branches are multicellular gametangia, not unicellular sporangia as in sporophytes. Because they are multicellular, they are called **plurilocular gametangia.** The gametes are anisogamous: Some settle and attract others by secreting a sex hormone ectocarpene. After fertilization, the zygote grows into a new sporophyte.

A more complex brown alga is *Fucus,* which is common on rocks in the intertidal zone (see Figs. 21.1g and 21.36). The diploid individuals are exposed at low tide, and their bodies can be seen to be large (up to 2 m), dichotomously branched, and attached to the rock by holdfasts. The bodies are complex histologically with epidermis, cortex, and a central region. The ends of the branches are called **receptacles** and are swollen with large deposits of hydrophilic compounds. Scattered over the surface of the receptacles are mi-

FIGURE 21.35
An individual of the brown alga *Ectocarpus,* a species that has an alternation of isomorphic generations. From this photo, it is not possible to tell if this is a gametophyte or a sporophyte (× 8). *(Visuals Unlimited/Cabisco)*

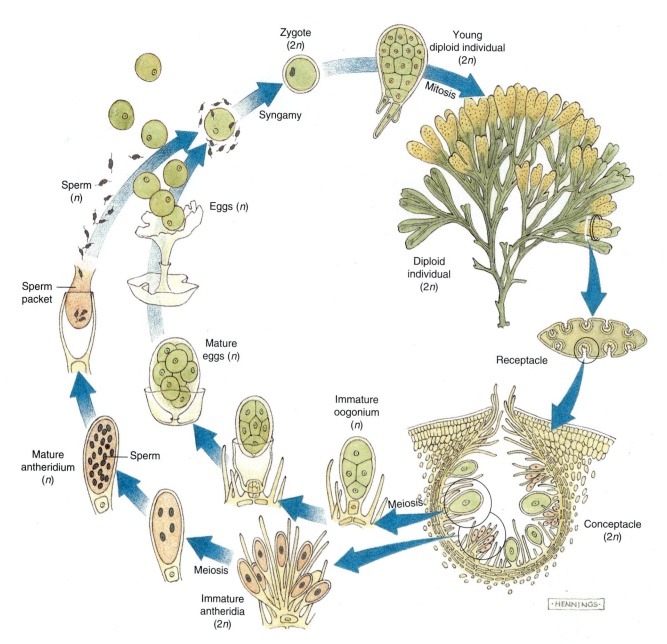

FIGURE 21.36
Life cycle of *Fucus;* details are given in text.

nute openings that lead to small cavities, **conceptacles;** some conceptacle cells undergo meiosis, producing either large eggs or small sperms. At low tide, individuals are exposed to air, and the conceptacles contract, squeezing out gametes. When the tide comes in, gametes are washed free and fertilization occurs in the water. The fertilized eggs settle to the bottom and grow into new diploid individuals. No free-living haploid generation occurs; *Fucus* is monobiontic.

The large kelps, such as *Nereocystis,* have complex bodies and complex life cycles. The large, visible individuals are diploid sporophytes and generally consist of a holdfast, a stipe with an enlarged air bladder (pneumatocyst) that provides flotation, and several leaflike blades on each air bladder. The kelps all have bodies composed of true parenchyma, rather than filaments that adhere to each other (Fig. 21.37). All body parts have an outer meristoderm that is both meristematic and photosynthetic, a cortex of parenchyma, and a central region of elongate cells and trumpet cells.

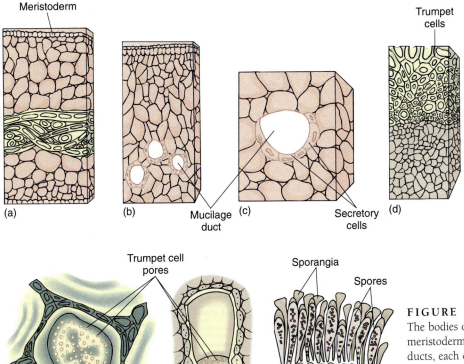

FIGURE 21.37

The bodies of kelps can be quite complex. (a) The outer layer is a meristoderm. (b) Cortex with photosynthetic cells and mucilage ducts, each of which is surrounded by secretory cells (c). (d) The central region of the stipe or the blade midrib is composed of trumpet cells, which have pores in their end walls (e and f). (g) During reproduction, surface cells elongate greatly, becoming sporangia in which meiosis will occur.

Holdfasts, stipes, and air bladders are often perennial, and one specimen of *Pterygophora* is known to be 17 years old. These portions grow in circumference each year through the activity of the meristoderm, which adds new layers to the surface. Trumpet cells, like sieve tube members, function only temporarily, then are replaced by new ones that differentiate from cortex cells.

The junction between an air bladder and a blade is an intercalary meristem capable of prolonged growth, which produces blades several meters long. Growth can be extremely rapid, up to 6 cm/day. Blades are photosynthetic and vegetative for a period after their formation by the intercalary meristem, but then portions become fertile and produce haploid zoospores by meiosis (Fig. 21.37g). Under inductive conditions, sporangia form either in patches or over the blade's whole surface. In *Macrocystis*, some blades remain vegetative and others are specialized and produce sporangia. *Macrocystis* releases as many as 76,000 spores/min/cm² of reproductive blade surface; the mean rate over long periods is 5000 spores/min/cm². Spores grow into tiny, filamentous gametophytes that somewhat resemble small plants of *Ectocarpus*—an alternation of heteromorphic generations. Gametophytes produce oogametes, and after fertilization, a new sporophyte is formed.

DIVISION RHODOPHYTA: RED ALGAE

The **red algae** constitute a large group (about 400 genera and 3900 species) of especially distinct and fascinating algae (see Figs. 21.1h, 21.38, and 21.39, and Table 21.8). Numerous structural, biochemical, and reproductive features set them off from the other algae as well as from true plants. One of the most important biochemical distinctions of red algae is that, like cyanobacteria, they contain phycobilin accessory pigments that are aggregated into phycobilisomes (see Fig. 21.4a). Their red color is due to the presence of phycoerythrin; however, they are often purple, brown, or black owing to the additional presence of

FIGURE 21.38

Many red algae have highly branched, filamentous bodies. *Polysiphonia* (×20). *(Runk/Schoenberger from Grant Heilman)*

FIGURE 21.39
Coralline red algae have walls so heavily impregnated with calcium carbonate that the body is hard and brittle. *(Peter Arnold/Sea Studios, Inc.)*

TABLE 21.8 *Classification of Division Rhodophyta*
Class Rhodophyceae
Subclass Bangiophycidae
Order Bangiales
Family Bangiaceae   *Bangia, Porphyra*
Subclass Florideophycidae
Order Nemalionales
Order Cryptonemiales
Family Corallinaceae   coralline red algae
Order Ceramiales

phycocyanin, just as in cyanobacteria. Carotenoid accessory pigments are also present, as is chlorophyll *a*. The actual quantities of each type of pigment vary with depth. Those algae that grow in bright surface waters have an array of pigments suitable for relatively intense light, whereas algae in deeper waters have a different complex of pigments that is better suited both to the dimmer light and to the altered spectrum present as a result of the water's differential color absorption.

Excess photosynthate is stored as floridean starch, a branched polymer of glucose somewhat similar to glycogen; it occurs as granules in the cytoplasm, never in the chloroplast. Other reserves occur which contain unusual sugars such as floridoside and isofloridoside, indicating that the carbohydrate metabolism of red algae has characteristic features not shared by plants.

In addition to a thin layer of cellulose, the walls of red algae contain a thick layer of slimy mucilages composed of compounds called sulfated galactans. These are important commercially as thickening, suspending, or stabilizing agents in puddings, ice creams, cheeses, and salad dressings. The culture medium **agar** is also extracted from them. The complete wall—cellulose layer plus mucilage—is quite thick, often as thick as the protoplast is wide. Almost all red algae are multicellular, so a large fraction of the individual's volume consists of this apoplastic space. As a result, all cells, even the most internal ones, may have relatively direct contact with the surrounding water. In the largest family of rhodophytes, the Corallinaceae or coralline red algae, such large amounts of calcium carbonate are deposited in the walls that they become rocklike (Fig. 21.39). It was not until 1837 that they were recognized as algae rather than corals.

Walls of red algae lack plasmodesmata of the type that occurs in plants, but they do have distinctive pit connections (see Fig. 21.7). During cell division, the new wall is formed as a ring that grows inward from the original cell wall; the growth of the septum stops before it is complete, leaving a hole in the center. Vesicles then deposit a material that precipitates, forming a lens-shaped plug. Whether these are a means of intercellular transport or communication is not known, but when two separate filaments of cells come into contact, a "secondary pit connection" can form between them.

Red algae are usually multicellular; only a few unicellular species (*Porphyridium, Rhodospora*) have ever been discovered. Most red algae tend to be rather conspicuous, often beautiful individuals and are filamentous, membranous, or foliaceous; true parenchymatous bodies are reported to occur in only a few algae, such as *Bangia* and *Porphyra*. Typically red algae are smaller and less complex than brown algae. Like brown algae, the ancestors of the red algae must have diverged from the ancestors of green algae very early, certainly before multicellularity had evolved.

Despite the rather large size that the bodies of red algae can attain, little differentiation or specialization occurs among the cells. The greatest differentiation typically involves only cell size and pigmentation. Outer cells are smaller and more heavily pigmented; inner cells are larger and have fewer, smaller chloroplasts.

Most species grow attached by rhizoids to rocks, shells, algae, or sea grasses. Numerous red algae (more than 40 genera) are parasitic, usually on other red algae. The basal cells of these species penetrate into the host, forming secondary pit connections with host cells.

The life cycles of most red algae are poorly known, but the few that have been studied well are all extremely complex, almost all involving at least one multicellular stage but none having any motile cells; flagella and centrioles do not occur in any stage of any species. Many variations occur and there is no "typical" red alga life cycle. Figure 21.40 is a generalized life cycle. Gametophytes bear gametangia that produce nonmotile sperm cells called spermatia and egglike cells called carpogonia. The carpogonia are large cells with a long tubular extension that basically acts as a receptor for drifting spermatia. When a spermatium contacts the extension, plasmogamy occurs and the nucleus migrates to the carpogonium base, where karyogamy occurs. In any other group, this cell would be a zygote and would either grow or produce spores. But in many red algae, the fertilized carpogonium puts out another long filament that carries the diploid nucleus out of the carpogonium and deposits it into a totally different cell, an auxiliary cell, where mitosis begins and produces a mass of cells. This mass is a new generation, the carposporophyte,

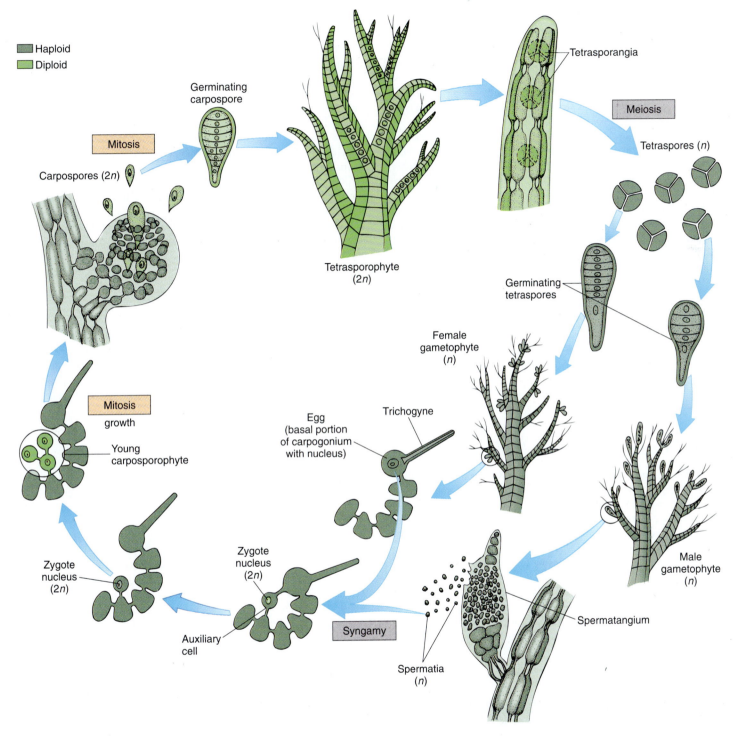

Haploid
Diploid

Germinating
carpospore

Mitosis

Carpospores (2*n*)

Tetrasporangia

Meiosis

Tetraspores (*n*)

Tetrasporophyte
(2*n*)

Germinating
tetraspores

Female
gametophyte
(*n*)

Mitosis
growth

Young
carposporophyte

Egg
(basal portion
of carpogonium
with nucleus)

Trichogyne

Zygote
nucleus
(2*n*)

Zygote
nucleus
(2*n*)

Male
gametophyte
(*n*)

Auxiliary
cell

Syngamy

Spermatangium

Spermatia
(*n*)

FIGURE 21.40
Life cycle of a red alga; details are given in text. Note especially the lack of swimming cells and the presence of three generations, one haploid and two diploid.

and it produces diploid carpospores by mitosis. The carposporophyte is a totally distinct generation from any other mentioned for any other algae or plants. Carpospores are released, float away and settle, and then germinate and grow into diploid plants called tetrasporophytes, equivalent to a regular sporophyte; these have sporangia in which the cells undergo meiosis, producing haploid tetraspores that grow into gametophytes.

SUMMARY

1.   Several significant early evolutionary divergences occurred in the eukaryotes. Dinoflagellates lack histones and normal mitosis; chloroplasts may have arisen at least three times.

2.   Algae are distinguished from true plants by having gametangia and sporangia in which fertile cells are not protected by a surrounding layer of sterile cells.

3.   Eukaryotic plastids and mitochondria are believed to have arisen as ancient prokaryotic endosymbionts in an early, simple eukaryotic cell.

4.   Mitosis is unusual in several algal groups, involving persistence of the nuclear envelope, bundles of microtubules, and persistently condensed chromosomes.

5.   Cell division in unicellular algae is by membrane infurrowing and cell cleavage. In some green algae it is by means of a phycoplast. In most multicellular algae, cell division occurs by phragmoplast and outward growth of the cell plate, but in red algae, walls grow inward from the existing side walls.

6.   In green algae, charophytes have a flagellar root complex of one large band of microtubules, similar to that in plants. Chlorophytes have a root complex of four bands of microtubules, unlike that of plants.

7.   Within the green algae, several body types have evolved from unicellular forms: motile and nonmotile colonies, branched and unbranched filaments, coenocytes, membranous forms, and true parenchyma.

8.   Several types of life cycle exist: dibiontic in which both a diploid and a haploid generation occurs; monobiontic in which only one generation is capable of mitosis.

9.   Dibiontic life cycles may consist of an alternation of either isomorphic or heteromorphic generations. All true plants have an alternation of heteromorphic generations.

10.   Several evolutionary lines are evident in the green algae, but the charophyte line is the one that most strongly resembles true plants.

IMPORTANT TERMS

agar	colony	endosymbiont theory	membranous body	pyrenoid
alternation of heteromorphic generations	diatoms	filamentous body	monobiontic species	red tide
alternation of isomorphic generations	dibiontic species	gametangium	parenchymatous body	sporangium
autogenous theory	dinoflagellates	holdfast	phycoplast	trumpet cells
axoneme	embryophytes	littoral zone	pneumatocyst	zoospores
coenocytic (siphonous) body				

REVIEW QUESTIONS

1.   Which group of algae appears to be most closely related to the ancestors of true plants? Which features appear to be homologous?

2.   What is the main character used to distinguish algae from plants? Why is this character significant?

3.   Describe the endosymbiont theory of plastid origin. How many times might chloroplasts have arisen? How could DNA or rRNA sequencing help us determine if all chloroplasts came from just one endosymbiotic event?

4.   Describe cell division and mitosis in algae. How is the red alga method of cross-wall formation different from that of plants?

5.   Describe and compare the cell walls or cell coverings of dinoflagellates, euglenoids, diatoms, green algae, red algae, and true plants. Which organelles are involved in the formation of each?

6.   Name and describe the six types of body construction that occur in green algae. What role does the middle lamella play in these? What is the importance of controlling the orientation of cell division? Which of these body types occurs in other groups of algae?

7.   What are the two shapes of coenocytic cells? None is ever a flat, two-dimensional sheet. Why? Does a balloon ever have a flat, two-dimensional shape?

8.   Describe monobiontic and dibiontic life cycles. Be careful to mention all possible types. What is the difference between a spore and a gamete? What is the difference between a spore and a zygote?

9.   If it were proven that there were at least three endosymbiotic origins of chloroplasts (red, brown, and green algae), would it be at all possible to include red algae and brown algae in the plant kingdom, even using a very liberal definition of a plant?

BotanyLinks .net Questions

www.jbpub.com/botanylinks

Visit the .net Questions area of BotanyLinks (http://www.jbpub.com/botanylinks/netquestions) to complete this question:

1.   Do you use or eat algae every day? Go to the Botany-Links home page for information on this subject.

BotanyLinks includes a Directory of Organizations for this chapter.

NONVASCULAR PLANTS: MOSSES, LIVERWORTS, AND HORNWORTS

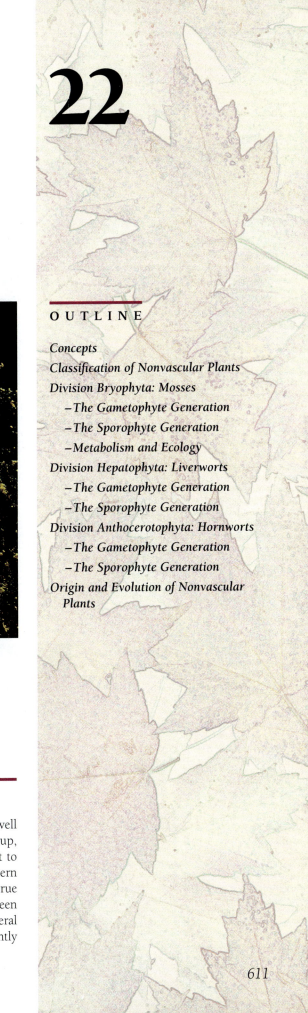

22

The sporophyte of a moss, releasing spores. *(Dwight Kuhn)*

CONCEPTS

Most people have at least a vague concept of nonvascular plants because mosses are well known, and many people have heard of liverworts (Figs. 22.1 and 22.2). The third group, hornworts, is relatively unfamiliar, even to most botanists (Fig. 22.3). It is important to understand clearly what nonvascular plants are *not:* Spanish moss of the southeastern United States is the flowering plant *Tillandsia usneoides* of the pineapple family, not a true moss. Club mosses are lycophytes (see Fig. 23.21), not mosses, and the slimy, bright green "mosses" of ponds and slow-moving streams are green algae, usually *Spirogyra.* Several types of lichens, especially *Alectoria, Bryonia, Usnea,* and "reindeer moss" are frequently mistakenly thought to be mosses or liverworts.

FIGURE 22.1

(a) Although mosses occur in hot, arid areas, they are most abundant in cool, moist regions.
(b) The most abundant green or gray-green moss plants are gametophytes; the small stalk and capsules you may have noticed are sporophytes. *Mnium. (Barry L. Runk from Grant Heilman)*

(a)   (b)

For a more rigorous scientific definition, nonvascular plants are embryophytes (nonalgal plants) that do not have vascular tissue. Being embryophytes, they have multicellular sporangia and gametangia: Reproductive cells are always surrounded by one or several layers of sterile cells (Figs. 22.4 and 22.5). The bodies of nonvascular plants are not composed of filaments as in many algae, but rather of true parenchyma derived by three-dimensional growth from meristems. They are almost exclusively terrestrial and have a cuticle over much of their bodies. Many nonvascular plants have stomata (Fig. 22.6).

These features occur in all other embryophytes as well, but nonvascular plants are distinct in having neither xylem nor phloem. Here we must be careful: A few mosses such as *Polytrichum* do have tissues involved in long-distance transport of water and sugars which are occasionally referred to as "vascular tissues." But the conducting cells in these nonvascular plants do not strongly resemble tracheary elements or sieve elements. The ancestry of nonvascular plants is not known. They may have evolved from the earliest vascular plants (similar to *Rhynia;* see Fig. 23.4) by becoming simplified. If so, then conducting cells of nonvascular plants may actually be reduced tracheids and sieve elements. However, nonvascular plants may be a second, distinct line of evolution out of green algae.

FIGURE 22.2

This is *Marchantia,* the most abundant greenhouse liverwort; unfortunately, it is one of the least typical. As in mosses, the large green plant is the gametophyte, not the sporophyte. The cup-shaped structures contain clumps of cells that can be splashed out by rain and then grow into new plants. *(Robert and Linda Mitchell)*

FIGURE 22.3

Hornworts such as this *Anthoceros* are quite rare, and few people have ever seen them; they are easily confused with liverworts unless the tall "horns"—sporophytes—are present. *(Runk/Schoenberger from Grant Heilman)*

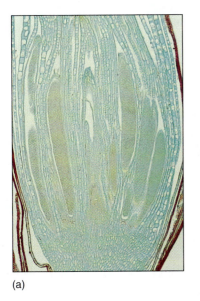

(a)

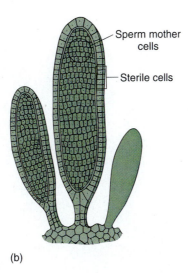

Sperm mother cells

Sterile cells

(b)

FIGURE 22.4
In nonalgal plants, gametangia and sporangia always have an outermost layer of cells that do not become gametes or spores. In these microgametangia of Bryophyta, the sterile layer is only one cell thick, and each microgametangium produces numerous sperm cells (×25). Compare with Figure 21.2, an algal gametangium.

Most other features are less controversial. Nonvascular plants, like vascular plants, have a life cycle with an alternation of heteromorphic generations; as in *Derbesia,* the sporophyte and gametophyte differ from each other structurally (see Chapter 21). Recall that sporophytes in flowering plants are the large plants with leaves and roots, whereas the gametophytes are tiny and occur inside pollen grains and ovules (see Chapter 9). In nonvascular plants, the gametophyte is the larger, more prominent generation and the sporophyte is much smaller, more temporary, and often very inconspicuous. The gametophyte is perennial and photosynthetic; it collects mineral nutrients and typically can spread rapidly by asexual reproduction. Sporophytes are usually very small, and they absorb minerals only from the gametophytes. Sporophytes carry out so little photosynthesis that they could not support even their own respiration, not to mention growth and sporogenesis, without sugar imported from the gametophyte.

Nonvascular plants offer interesting contrasts to vascular plants, particularly flowering plants. The gametophyte, the haploid generation, is always the dominant phase of nonvascular plants, whereas the diploid or polyploid sporophyte is dominant in vascular plants. Is this difference related in any way to the differences in size and complexity of the two groups of organisms? In vascular plants and in animals, the presence of at least two genomes in every nucleus buffers the effects of mutations. A mutation that severely disrupts a

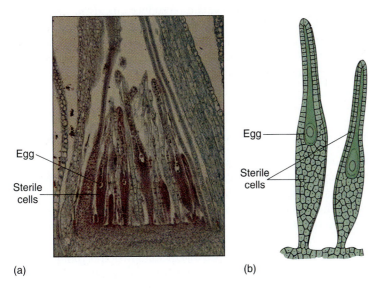

Egg

Sterile cells

(a)

Egg

Sterile cells

(b)

FIGURE 22.5
Megagametangia of nonalgal plants, such as these of a moss, also have a one-layered sterile jacket, but each produces only one egg, not four (×25). *(Robert and Linda Mitchell)*

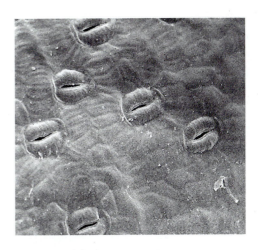

FIGURE 22.6
Most mosses have stomata on their sporophytes. Although the gametophytes are larger and more active in photosynthesis, they never have stomata. *(Courtesy of F. Sack, Ohio State University, and D. Paolillo, Cornell University)*

FIGURE 22.7

Mosses outcompete vascular plants in certain severe habitats such as open rock faces, like this one on Mt. Rainier. However, even here the true champions of adaptation are usually lichens, the association of algae with fungi.

gene for an essential protein or RNA is usually lethal in a haploid organism, but its effects may be masked by the presence of the unmutated, original allele on the homologous chromosome in a diploid organism. The mutant, useless allele is not immediately eliminated by natural selection. Rather it remains in the gene pool and mutates further, perhaps ultimately becoming a new gene that produces a beneficial product. If so, the species now has the original gene and its product plus the new gene and its product: The metabolism or structure of the species has become more complex (see Fig. 16.25). The haploid nature of the dominant generation in nonvascular plants may have made it more difficult for them to retain the extra DNA needed to develop new genes for a more elaborate body and biology. The sporophytes are diploid, but they may be limited in their size and complexity by the limited resources available to them from the gametophyte.

As an alternative, being small and simple may have great selective advantage. Many flowering plants have evolved into species with bodies only 1 or 2 cm tall with simple leaves and stems. These germinate, grow, and reproduce quickly in temporarily adequate environments or under low light. The tiny, nonelaborate bodies of mosses and liverworts are actually superior to those of flowering plants in many microhabitats such as stone walls, fences, and bare rock (Fig. 22.7). Even in these environments, nonvascular plants face stiff competition from very simple vascular plants such as lycophytes with their diploid bodies (see Chapter 23).

CLASSIFICATION OF NONVASCULAR PLANTS

It is not known how closely related mosses, liverworts, and hornworts are. They have many features in common but also differ in significant respects. They are commonly treated as three distinct divisions: mosses, division Bryophyta; liverworts, division Hepatophyta; and hornworts, division Anthocerotophyta (Table 22.1). This classification emphasizes their differences and stresses the need to treat each group individually and study each in its own right.

Alternatively, all three may be grouped together in division Bryophyta as three classes: mosses, Musci; liverworts, Hepaticae; and hornworts, Anthocerotae. This emphasizes their similarities and the possibility that all three are closely related. Inaccurate generalizations may result, obscuring the fact that we actually know very little about many of them, particularly hornworts and liverworts. Numerous studies have been made of moss physiol-

TABLE 22.1	Alternative Methods of Classifying Nonvascular Plants*
A. Kingdom Plantae	
Division Bryophyta	
Division Hepatophyta	
Division Anthocerotophyta	
B. Kingdom Plantae	
Division Bryophyta	
Class Musci	
Class Hepaticae	
Class Anthocerotae	

*A assumes that they are less closely related than B does.

ogy, especially water relations and mineral nutrition, but almost nothing is known of liverwort or hornwort metabolism. Conclusions derived from studies of mosses are sometimes applied to liverworts and hornworts because all are "bryophytes." To discourage such generalizations and to encourage studies of all groups, the three groups are treated here as divisions, and the generic term "bryophyte" is used only for mosses. All groups together are called "nonvascular plants."

DIVISION BRYOPHYTA: MOSSES

THE GAMETOPHYTE GENERATION

Mosses are ubiquitous, occurring in all parts of the world and in almost every environment (Table 22.2). They are perennial and thrive in many places within cities (*Tortula* on walls, *Mnium* on soil).

Morphology.   The leafy stems, technically known as **gametophores,** of many moss plants grow close together, tightly appressed and forming dense mounds (*Grimmia, Pohlia*). In other species, particularly those of cool wet areas, the plants are more open and loose (*Anacolia, Climacium, Platygyrium*; see Fig. 22.1). *Scouleria* gametophores grow as ribbons up to 15 cm long, submerged in rapidly flowing water. Moss plants have stems, usually very short, and almost microscopic leaves (Fig. 22.8). Because the stems and leaves of mosses are parts of a gametophyte, not a sporophyte, they are not homologous with those of vascular plants. That is, the leaves of mosses did not evolve from the same structures as vascular plant stems and leaves did.

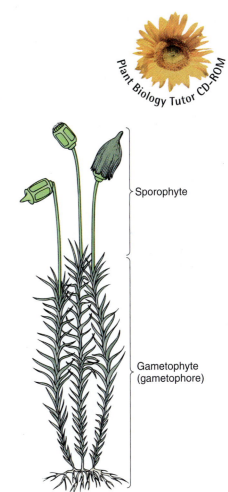

FIGURE 22.8
The individual shoots of a moss superficially resemble those of a flowering plant, having stems, leaves, nodes and internodes, and even buds. None of these structures is homologous to the equivalent organ in flowering plants, but the same set of names is used for both groups. Dark green, haploid tissue; light green, diploid tissue.

TABLE 22.2	Classification of Division Bryophyta
Class Sphagnopsida	*Sphagnum*
Class Andreaeopsida	*Andreaea, Neuroloma*
Class Bryopsida	*Atrichum, Bryum, Buxbaumia, Dicranum, Fissidens, Funaria, Grimmia, Mnium, Physcomitrium, Polytrichum*

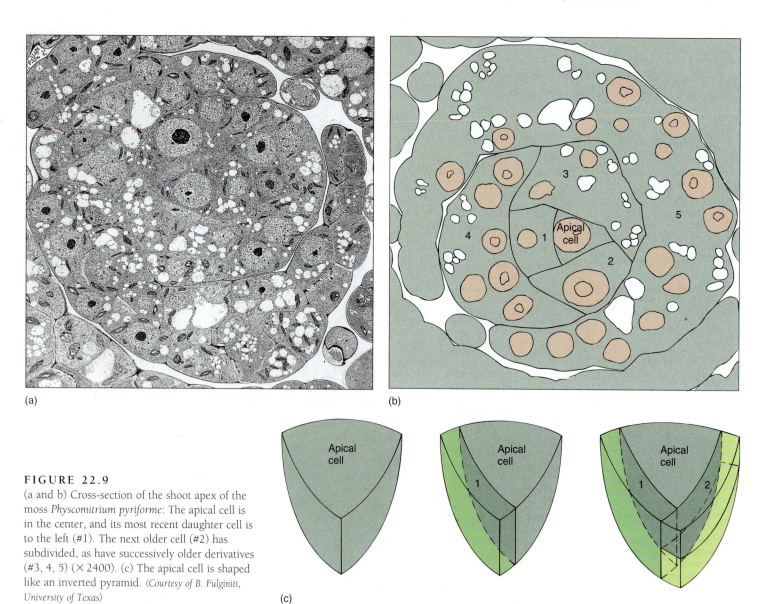

FIGURE 22.9

(a and b) Cross-section of the shoot apex of the moss *Physcomitrium pyriforme*: The apical cell is in the center, and its most recent daughter cell is to the left (#1). The next older cell (#2) has subdivided, as have successively older derivatives (#3, 4, 5) (× 2400). (c) The apical cell is shaped like an inverted pyramid. *(Courtesy of B. Fulginiti, University of Texas)*

Moss gametophores grow from an apical meristem that contains a prominent apical cell (Fig. 22.9a). Derivative cells subdivide, producing the tissues of stem and leaves in rather precise arrangements. Leaves are aligned in three rows on the stem, at least while young; later expansion may obscure this pattern. Moss leaves are broad and most have a midrib (costa). The leaves of almost all mosses are only one cell thick except at the midrib and along the margin (Fig. 22.10a). In the family Polytrichaceae, common along roadsides in forested areas of the northern United States, genera such as *Polytrichum* and *Atrichum* have thin lamellae on the leaf upper surface, greatly increasing the photosynthetic tissue (Fig. 22.10b). Cuticle occurs only on the upper surface of most moss leaves, the underside being uncutinized and capable of absorbing water directly from rain, dew, and fog. Water is usually not transported into leaves from stems, because mosses are nonvascular plants. The lack of a cuticle on the lower or both sides of leaves means that mosses have little protection against desiccation; when the microhabitat dries, so do the leaves. No stomata occur on moss leaves; they would be useless because the leaves are unistratose.

Moss stems are always slender and have little tissue differentiation (Fig. 22.11). The surface layer is at most only slightly different from the underlying layers and is not called epidermis. The stem tissues, all called cortex, may be uniform in all parts, or the outer cells

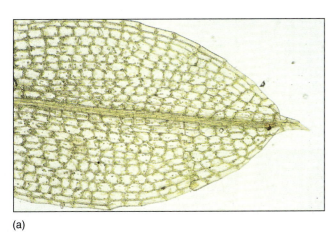
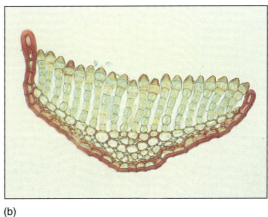

(a)   (b)

FIGURE 22.10

(a) Many moss leaves are only one cell thick, but the leaf "midrib" is thicker and is true parenchyma. (b) In the family Polytrichaceae, leaves bear long sheets of cells on their upper surface; this greatly increases the volume of photosynthetic tissue but does not decrease the high surface-to-volume ratio as it would if this were a solid tissue ($\times$ 200).

may be slightly narrower with walls that are somewhat thickened. Inner cells are larger, more parenchymatous, and chlorophyllous. In a few species of mosses, stems have hairs, but stomata do not occur.

Water Transport.   In some mosses, primarily the family Polytrichaceae, the innermost cortex is composed of conducting cells called **hydroids,** elongated cells that lose their cytoplasm when mature (Fig. 22.12). Their end walls are partially digested and become more permeable to water and dissolved solutes, but they are not removed entirely. Each hydroid is aligned with those above and below it. If a moss is given a soluble form of lead (Pb-EDTA) and then later treated with hydrogen sulfide, a lead precipitate is found in the hydroids. This indicates that hydroids do conduct water and dissolved minerals.

Species that have hydroids typically also have **leptoids,** cells that resemble sieve cells (Fig. 22.12b). They are elongate, have relatively prominent interconnections with adjacent

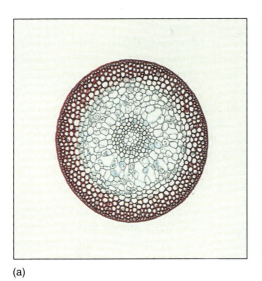
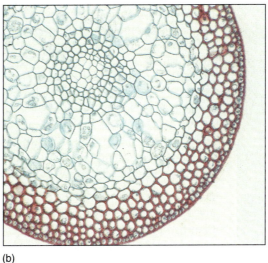

(a)   (b)

FIGURE 22.11

Although most moss stems do not have vascular tissues, they do support the shoot, and most have a layer or two of thick-walled cells. Sugars and minerals must be transported from leaves to the shoot apex, gametangia, and sporophytes, so living parenchyma cells are necessary (a, $\times$ 16; b, $\times$ 40).

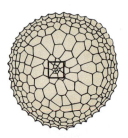

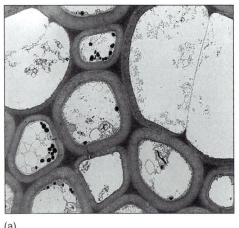

(a)

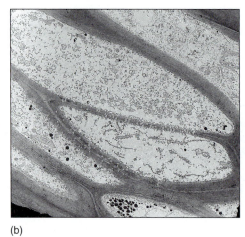

(b)

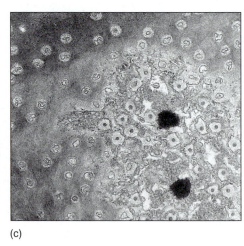

(c)

FIGURE 22.12

(a) Hydroids lose their cytoplasm at maturity, just as do tracheids and vessel elements. Hydroids, however, do not have specialized secondary walls. (b) Leptoids are analogous to sieve cells, phloem cells that are simpler than sieve tube members. All pores are small, and the cells remain alive at maturity. (c) Face view of a leptoid end wall with plasmodesmata. *(Courtesy of Dan Scheirer, Northeastern University)*

FIGURE 22.13

The leaves of many mosses curl and press against the stem as they dry, forming capillary channels that conduct water *(Doug Sokell/Visuals Unlimited)*

cells, and lack nuclei at maturity, although they do retain their cytoplasm. Adjacent parenchyma cells are unusually cytoplasmic and rich in enzymes, just as are companion cells. If radioactive bicarbonate is placed on the leaf, radioactive sucrose can be found later in the leptoids, moving as rapidly as 32 cm/hr. In the mosses that do not have leptoids, sugar is moved simply between parenchyma cells by slow transport.

In the majority of mosses, which lack hydroids and leptoids, water is conducted along the *exterior* of the plant by capillary action. Leaves and stems are so small that they form spaces narrow enough to act as capillary channels and transport water. In some species, such as *Funaria*, the leaves curl and become more closely appressed to the stem as they dry (Fig. 22.13); when rain or dew returns, the dried plant has even more capillary spaces than in the hydrated condition when the leaves are spread away from the stem.

At the base of the stem or along the length of horizontal stems are **rhizoids.** These small, multicellular trichome-like structures penetrate the surface of the substrate. Rhizoids only anchor the stem; they do not appear to be involved in absorbing either water or minerals. They lack chloroplasts and have reddish walls.

Development.   Growth of the gametophore begins when a spore germinates and sends out a long, slender chlorophyllous cell. This cell undergoes mitosis and produces a branched system of similar cells; the entire network is a **protonema** (pl.: protonemata) (Fig. 22.14). A protonema superficially resembles a filamentous green alga but can be distinguished by its numerous small chloroplasts in each cell. Algal cells have only one or two large chloroplasts. Also, the cross walls of the protonema are placed at an angle to the side walls, not perpendicularly as in a green alga. Rhizoids grow from the underside of the protonema. Eventually nodules of small cytoplasmic cells form on the protonema. These buds organize an apical cell, then grow upright as a stem with leaves—the gametophore.

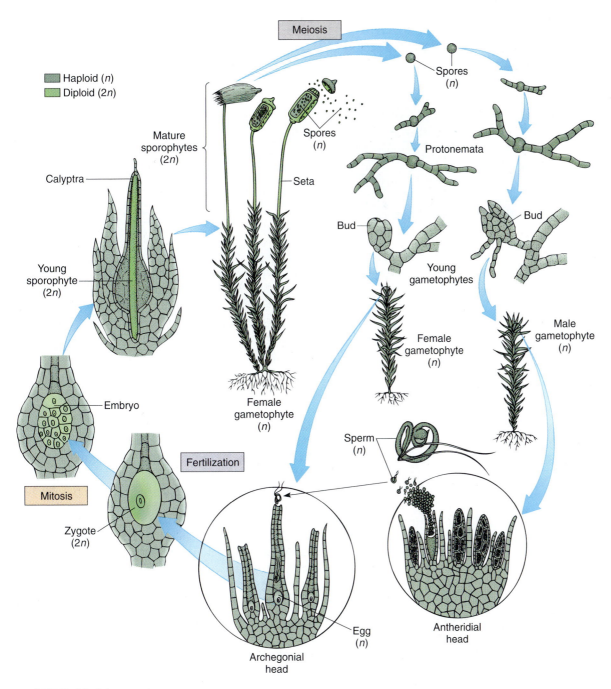

Haploid (*n*)
Diploid (2*n*)

Meiosis

Spores
(*n*)

Mature
sporophytes
(2*n*)

Spores
(*n*)

Protonemata

Calyptra

Seta

Bud

Bud

Young
sporophyte
(2*n*)

Young
gametophytes

Female
gametophyte
(*n*)

Male
gametophyte
(*n*)

Embryo

Female
gametophyte
(*n*)

Sperm
(*n*)

Fertilization

Mitosis

Zygote
(2*n*)

Egg
(*n*)

Antheridial
head

Archegonial
head

FIGURE 22.14
Life cycle of a moss. Details are given in the text.

Protonemata are perennial and can grow extensively, producing many buds. The filamentous cells usually break when mosses are collected, so a tuft of gametophores may appear to be independent plants when in fact they have all arisen from a single protonema.

Reproduction. The gametophore at some point produces gametangia. All mosses are oogamous; that is, every species has small biflagellate sperm cells and large nonmotile egg cells. Sperms are produced in microgametangia called **antheridia,** which consist of a short stalk, an outermost layer of sterile cells, and an inner mass of cells that differentiate into sperm cells (see Fig. 22.4). Eggs occur in megagametangia called **archegonia:** Each is shaped like a vase with a long neck (see Fig. 22.5). The neck is hollow at maturity, and the single egg is located at the base.

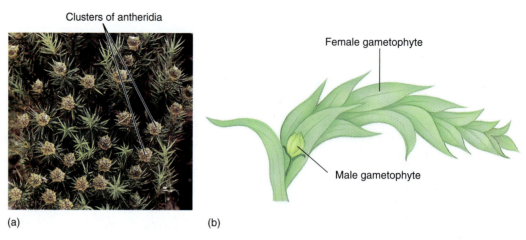

(a)   (b)

FIGURE 22.15

(a) In this moss, antheridia are clustered together at the gametophore apex; in other species, they might be located along the stem. *(E. R. Degginger)* (b) The phenomenon of dwarf males is common in many plants, animals, and algae. Each sperm cell is much smaller and less expensive to produce than each egg, but usually many sperm cells are not carried to eggs, so a large number of sperm cells must be produced. If sperm delivery can be ensured, the number produced can be decreased, and even a small, dwarf male can produce adequate numbers. Delivery can be assured if the male gametophyte grows on the female as an epiphyte.

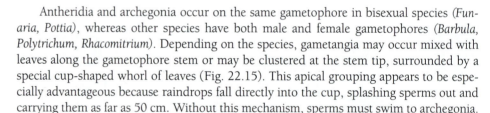

Antheridia and archegonia occur on the same gametophore in bisexual species (*Funaria, Pottia*), whereas other species have both male and female gametophores (*Barbula, Polytrichum, Rhacomitrium*). Depending on the species, gametangia may occur mixed with leaves along the gametophore stem or may be clustered at the stem tip, surrounded by a special cup-shaped whorl of leaves (Fig. 22.15). This apical grouping appears to be especially advantageous because raindrops fall directly into the cup, splashing sperms out and carrying them as far as 50 cm. Without this mechanism, sperms must swim to archegonia.

When sperm cells are mature, the antheridium breaks open and liberates sperm cells either by contracting the sterile outer cells or by accumulating liquid below the sperms and pushing them out. Antheridia open when the gametophore is wet and a film of water is present through which sperms can swim (Fig. 22.16). The microgametes move to the archegonia either by swimming alone or by being splashed and then swimming. Secretion of sucrose from the archegonia guides sperms toward the archegonia and then down the neck to the egg, where one sperm cell effects fertilization.

THE SPOROPHYTE GENERATION

In all embryophytes, the megagamete and subsequently the zygote are retained by the gametophyte. However, unlike those of mosses, megagametophytes of vascular plants are small and offer little nourishment or protection to the embryonic sporophytes. Megagametophytes of flowering plants typically have only six cells other than the egg, and both synergids and antipodals often degenerate just after the egg is fertilized (undergoes syngamy). All nutrition for the zygote is supplied by the grandparent generation, the sporophyte, by way of endosperm formation. In contrast, moss gametophytes are both large and very efficient at photosynthesis, and they support the sporophyte throughout its entire life. The moss sporophyte is never an independent, free-living plant (see Fig. 22.14).

The zygote of a moss undergoes a transverse division, and the basal cell, located at the bottom of the archegonium, develops into a small, bulbous tissue called the foot. The foot is the interface with the gametophore, from which it absorbs sugars, minerals, and water. Its cells are transfer cells in many species. The upper cell grows by cell division and expansion into a simple apical sporangium called the **capsule,** consisting of an outer layer of sterile cells and an inner column of sterile cells (the columella). A ring of sporogenous cells undergoes meiosis, producing haploid spores (Fig. 22.17). Between the foot and the

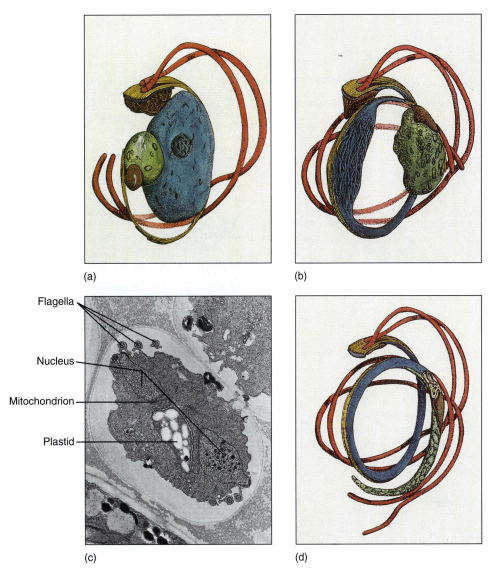

FIGURE 22.16

(a) The organelles of a young sperm cell have rather typical shapes. (Blue, nucleus; green, chloroplast; red, flagella and basal bodies; brown, mitochondrion; yellow, microtubules that anchor the flagella). (b and c) During differentiation, the organelles become modified, resulting in an elaborate spiral-shaped sperm cell at maturity (d). Hyaloplasm and cell membrane are not shown. *(a, b, and d, Courtesy of Karen Renzaglia and Fred Alsop III; c, courtesy of Douglas Bernhard and K. Renzaglia, East Tennessee State University)*

sporangium is a narrow stalk, the **seta** (pl.: setae). All moss sporophytes have this basic, simple structure; none is ever branched or has leaves, bracts, or buds of any kind.

Although morphologically simple, the sporophyte is relatively complex structurally. It is smaller than the gametophyte but more dense, being many cells in diameter, even in the seta and foot regions. The sporophyte has a true epidermis with stomata at least on the base of the sporangium (see Fig. 22.6). Below the epidermis the seta contains long, thick-walled cells and often leptoids and hydroids. Dehiscence of the sporangium is more elaborate than the opening of the gametangia: The apex of the sporangium differentiates as a caplike lid, the **operculum,** which separates from the rest of the sporangium as cells are torn apart. Cell breakage is elaborate and precise, resulting in one or two rows of beautiful, exquisitely complex teeth, called **peristome teeth** (Fig. 22.17). The teeth respond to humidity, bending outward and opening the sporangium when the air is dry and bending inward and trapping the spores when the air is humid. Spores are released when they are light, dry, nonsticky, and easily carried by air currents (Fig. 22.18).

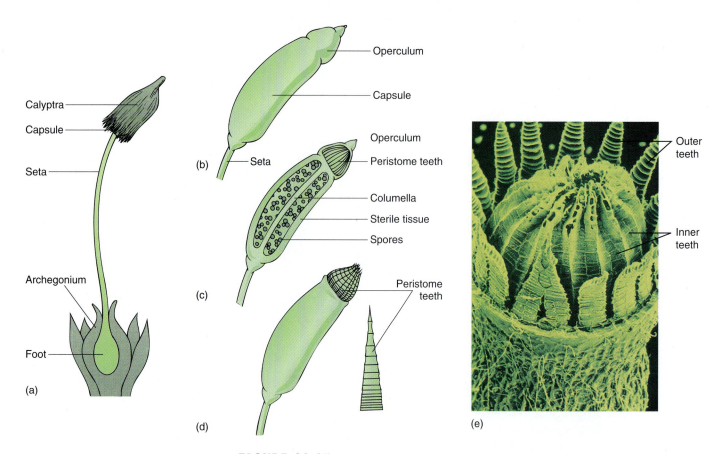

FIGURE 22.17
Aspects of a moss sporophyte. (a) The foot is embedded in the remnants of the archegonium.
(b) External view of the capsule (sporangium) with the operculum still in place covering the teeth.
(c) Median section view showing the columella and spores. (d) External view after the operculum
has fallen off. Spores can be shed when the peristome teeth bend back. (e) Scanning electron
micrograph view of peristome teeth. *(G. Shih and R. Kessel/Visuals Unlimited)*

In addition to the operculum, the apex of the sporangium in many species is covered
by a **calyptra**, a layer of cells derived from the neck of the archegonium. As the embryo
begins to grow, neck cells also proliferate. They keep pace with sporophyte growth at first
but later grow more slowly and are torn away from the gametophore. The calyptra is
important for proper development; if it is prematurely cut or removed surgically, the seta
elongates but a sporangium does not form.

FIGURE 22.18
The transition to land required the evolution of spores
that could resist desiccation. Little is known about the
chemicals that provide waterproofing. Like pollen, moss
spores show many types of surface pattern when studied
by scanning electron microscopy. (a) *Acaulon rufescens*
(×600). (b) *Bruchia drummondi* (×600). *(Courtesy of
D. M. J. Mueller, Texas A and M University)*

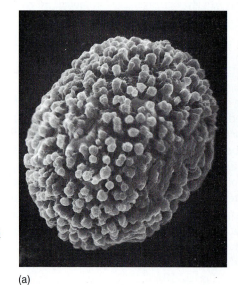

(a)

(b)

Virtually all mosses are homosporous: All spores are the same size and appear to be identical. A few species of *Macromitrium* and *Schlotheimia* produce two types of spores in each capsule. The larger spores develop into large gametophytes with archegonia, and the smaller spores grow into dwarf males that live epiphytically on the females.

METABOLISM AND ECOLOGY

Small size and lack of conducting tissues are two critical factors in the metabolism and ecology of mosses. Vascular plants tend to be larger than nonvascular plants, and their bulk protects them from short-term fluctuations in air humidity and moisture availability. Should the air become dry for several hours, most leaves and stems would not die, even if they were not receiving water from the roots by means of xylem. Very large plants, such as trees and succulents, can withstand dry conditions for weeks, months, or even years before losing a fatal amount of water. Mosses, however, have no such mechanisms to absorb and store water. Leaves on the gametophore have only a thin, incomplete cuticle, and typically at least the undersides are not cutinized. If exposed to dry air for even a few minutes, the plants dry out. Without vascular tissue, stems and leaves can become desiccated, even while the rhizoids are in contact with moist soil or tree bark.

Several mechanisms compensate for the inability of mosses to retain water. Many species grow in permanently moist microhabitats such as rainforests, cloud forests, and the spray zones near waterfalls (Fig. 22.19a). In such situations, the plants can become quite large because water is available to and absorbed by the entire plant surface, so water-conducting tissues are completely unnecessary. Even habitats that are not constantly humid may have microhabitats that are continually moist. The foot of a rock cliff is usually damp because dew and mist concentrate there as runoff from the cliff. Shallow depressions in rock or soil retain moisture and are protected from drying winds. These sites are very small, but so are the mosses that live within them.

Other mosses compensate for their inability to retain water by being tolerant of desiccation. That is, drying does not damage them as it does most vascular plants and algae. Like lichens, many mosses can lose much of their water rather rapidly without dying or even

(a)

(b)

FIGURE 22.19

(a) Most mosses live in habitats that are permanently wet, never drying or subjecting the mosses to water stress. *(Sydney Karp, PHOTO/NATS)* (b) *Sphagnum* moss grows at the edges of streams or ponds where there is little water movement. The plants become large and tangled, creating a mat that is solid enough to support the weight of an animal. These are known as quaking bogs, because they quake and undulate with every step. Given enough time, seeds of vascular plants germinate, and leaves and debris blow onto the bog and decompose, slowly converting it to a rich soil. *(G. Bamgarter, PHOTO/NATS)*

being injured. As long as about 30% of their weight is water, they remain dormant but alive. If rain falls or dew forms, water is absorbed rapidly, and within a few minutes respiration and photosynthesis are occurring at normal levels.

The alternation between dry and turgid states varies with the environment. In mild temperate regions, mosses may remain moist and metabolically active except for certain times during summer months. In deserts and dry range lands, on the other hand, mosses may be turgid primarily during winter. From spring to autumn, the mosses are dry and inactive every day, perhaps receiving enough dew on certain nights to be metabolically active for a few hours in the early morning. In harsh conditions, gametophores grow crowded together, forming an extremely dense clump. The close packing prevents air movement and helps retain transpired water molecules, and dew or rain that touches the top of the tuft is conducted downward by capillary action.

Desiccated mosses are remarkably resistant to high or low temperature and to intense ultraviolet (UV) light. A common moss, *Tortula ruralis,* can be frozen safely in liquid nitrogen (−196°C) when dry, and can also withstand brief periods at +100°C. Resistance to UV light is important for species that grow in full sun on the surface of rocks or at high altitudes on mountains, where the atmosphere is too thin to block UV radiation.

Many moss species thrive at low temperatures near or even below 0°C. This is adaptive in habitats where the larger vascular plants die or abscise their leaves in winter. Leaf fall permits much more sunlight to reach the soil surface or tree trunks where the mosses are growing. Snow or winter rains provide abundant moisture. Mosses that can carry out metabolism at cold winter temperatures can take advantage of the plentiful light and moisture.

Certain mosses can grow on hard, impervious surfaces because they have no roots that must penetrate the substrate. Consequently, mosses can be found on bare rock (*Andreaea, Hedwigia*), brick or mortar (*Tortula muralis*), and bark as well as soil. At the other extreme, *Sphagnum* mosses live on the surface of quiet water at the edges of lakes and ponds (Fig. 22.19b). Their leaves and stems are buoyant, and such thick mats can be built up that a person can walk across the floating colony if enough care is taken.

In almost all environments, mosses are important in the later establishment of other species, particularly vascular plants. Like lichens, they can colonize a bare surface and dissolve the rock with acids that leak from their cells. Leaves and stems of gametophores catch and hold dust particles, creating small pockets of soil in which spores and seeds of vascular plants can become established. A similar process occurs even on the ground; in forests or grasslands, the moss layer on the soil may be thick enough to act as a moist, airy seed bed. It may hold water better than the soil and thus improve the microhabitat for seedlings.

DIVISION HEPATOPHYTA: LIVERWORTS

Like mosses, liverworts are small plants that have an alternation of heteromorphic generations (Table 22.3). The smallest individuals of *Cephaloziella* are only 150 μm by 2000 μm. Few species have plants that ever become large, the maximum being approximately 5 by 20 cm in *Monoclea*. Some species are leafy and greatly resemble mosses (Fig. 22.20a); others form small, solid, ribbon-like gametophytes similar to fern gametophytes (see Figs. 22.2 and 22.20b). The sporophyte is even less conspicuous than in mosses and is also completely dependent on the gametophyte.

THE GAMETOPHYTE GENERATION

Hepatics are divided into two basic groups according to the nature of their gametophytes: **leafy liverworts** (orders Jungermanniales, Haplomitriales, and Takakiales) and **thallose liverworts** (Figs. 22.20 and 22.21). In both, the gametophyte phase is initiated when spores germinate and establish a small, temporary protonematal phase. Liverwort protonemata are never as extensive or ramified or as long-lived as those of mosses. Instead, after only a few cells are produced, an apical cell is established and growth of the gametophore begins.

(a)

(b)

FIGURE 22.20

(a) Leafy liverworts such as this *Lophocolea* can be very easily confused with mosses. Their distinguishing features (not visible at this low magnification) are their typically lobed leaves, which grow from two apical points, and their basal pouch. Moss leaves are never like this. (*L. West/Photo Researchers*) (b) Thallose liverworts such as this *Marchantia* have rather thick bodies and are not easily confused with mosses. (*G. K. Scott/Photo Researchers*)

TABLE 22.3 *Classification of Division Hepatophyta*

Class Hepatopsida	
Order Marchantiales	*Conocephalum, Marchantia Reboulia, Riccia, Ricciocarpus*
Order Sphaerocarpales	*Riella, Sphaerocarpos*
Order Monocleales	*Monoclea*
Order Metzgeriales	*Fossombronia, Pallavicinia, Pellia*
Order Jungermanniales	*Cephaloziella, Frullania, Jungermannia, Porella*
Order Haplomitriales	*Haplomitrium*
Order Takakiales	*Takakia*

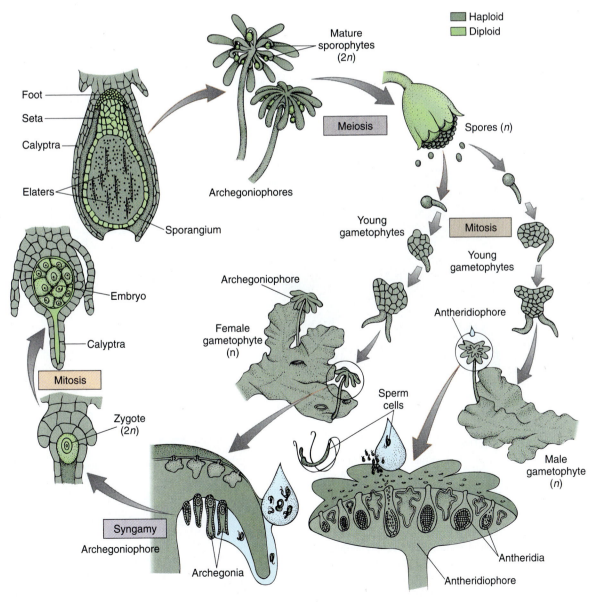

FIGURE 22.21
Life cycle of *Marchantia*. Details are given in text.

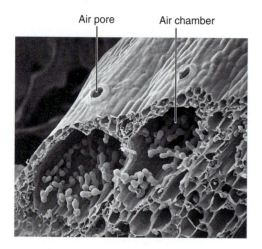

FIGURE 22.22

Although gametophytes of thallose liverworts may become very thick (a few millimeters), they are never solid; rather, they have numerous chambers, and both carbon dioxide and water vapor can diffuse through them easily (× 425). *(Biophoto Associates/Science Source/Photo Researchers, Inc.)*

The gametophore of leafy liverworts greatly resembles that of a moss—thin leaves on a slender stem (see Fig. 22.20a). But liverwort leaves typically have two rounded lobes with no midrib and no conducting tissue, whereas moss leaves are pointed and usually have a midrib. Liverwort leaves are arranged in three clearly defined rows, the leaves of two rows being much larger than those of the third. In prostrate liverworts, in which the shoot grows flat against the substrate, the underside row of leaves is reduced and may even be completely suppressed or replaced by rhizoids.

The gametophore stem grows by an apical cell with three, four, or five sides, one uppermost and the others embedded in the stem. Stem tissue is quite simple: All of it is parenchyma, and no thick-walled cells or conducting cells occur in the majority of species. Two genera, *Takakia* and *Haplomitrium*, have elongate cells with large plasmodesmata; these are suspected to be involved in long-distance transport. Virtually all cells of liverworts contain characteristic oil bodies.

Thallose liverworts show even less resemblance to mosses. They are not leafy at all but rather flat and ribbon-like or heart-shaped and bilaterally symmetrical. The body is sometimes referred to as a thallus (pl.: thalli), a body without roots, stems, and leaves. At present, "body" is much more commonly used than "thallus." Bodies of thallose liverworts are stratified and tend to be much thicker than those of leafy liverworts and mosses. The side next to the substrate bears unicellular rhizoids, and many cells contain large oil drops. Cells in the side away from the substrate have no oil but are rich in chlorophyll. The cells are loosely arranged as an aerenchyma with large air chambers that open to the exterior by means of large pores (Fig. 22.22).

Some species are even simpler. *Sphaerocarpos texanus* is a small, thin ribbon a few cells thick at the center, but the rest of the body is only one cell thick. It has simple rhizoids on the bottom, but no internal air chambers and no scales or other types of vegetative differ-

(a)

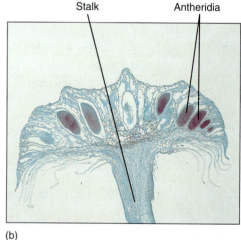

(b)

FIGURE 22.23

(a) These are antheridiophores of *Marchantia*; as their name indicates, these structures bear (*-phore*) antheridia. These are gametophytic tissues, and the antheridia are hidden within the upper surface. (Four archegoniophores are also present.) *(Runk/Schoenberger from Grant Heilman)* (b) Low-magnification view of antheridiophore with antheridia (× 8). (c) Like antheridia in mosses and all other embryophytes, those in liverworts consist of a jacket of sterile cells surrounding fertile cells that differentiate into sperms (× 32).

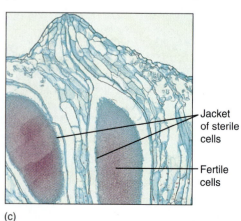

(c)

(a)

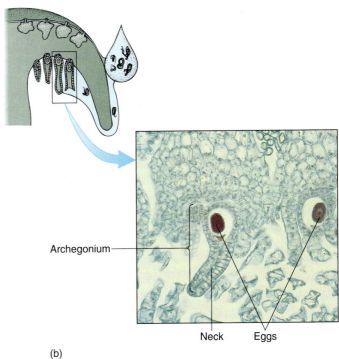

Archegonium

Neck   Eggs

(b)

FIGURE 22.24

(a) Archegoniophores of *Marchantia* are easily recognizable because they have finger-like segments radiating from the stalk. Archegonia are located on the underside (× 3). *(Runk/Schoenberger from Grant Heilman)* (b) Each archegonium consists of a long, tubular neck and a slightly swollen base, all only one cell thick. Each archegonium contains only one egg (× 160).

entiation. These plants are basically as simple as the sea lettuce alga, *Ulva*, or the charophyte, *Coleochaete*. The liverworts *Pallavicinia* and *Pellia* are also this simple.

As in mosses, liverwort gametophores may be either bisexual, producing both antheridia and archegonia, or unisexual, depending on the species. Leafy liverworts may bear their gametangia either mixed with regular leaves or positioned on specialized side branches and surrounded by modified leaves. In thallose species, antheridia and archegonia may be grouped together and surrounded by a tube of chlorophyllous cells. *Marchantia* is probably the most familiar thallose liverwort because it grows easily on moist soil in greenhouses. Its gametangial production is particularly elaborate. Male gametophores of *Marchantia* produce an umbrella-shaped outgrowth called an **antheridiophore** (Fig. 22.23). It has a stalk several millimeters tall, and dozens of antheridia grow from its upper surface, each surrounded by a rim of sterile cells. **Archegoniophores** also are stalked, but their apex is a set of radiating fingers that project outward and droop downward; the underside has numerous archegonia (Fig. 22.24). Archegoniophores occur on separate plants; antheridia and archegonia are not produced by the same gametophyte.

If sperm cells are carried to the archegoniophore by raindrop splashing, they swim into the archegonium neck and fertilize the egg. The zygote is retained and grows into a small sporophyte (Fig. 22.25). Surrounding gametophore tissue expands with it temporarily, forming a protective sheath, and the archegonial neck expands into a calyptra. In *Marchantia*, sporophytes are so small that each archegoniophore can support many of them.

THE SPOROPHYTE GENERATION

Little variability exists in the sporophytes of most liverworts, and their basic morphology is like that of mosses. In fundamental structure, however, the two groups are quite distinct. Most liverwort sporophytes have a foot, seta, and calyptra-covered sporangium, but the seta is extremely delicate, composed of clear, thin-walled cells that collapse quickly (Fig. 22.25). Many members of Marchantiales have no seta at all. The sporangium is globose and while young is bright green and chlorophyllous, even though it has no stomata. The outer layer of sterile cells is much thinner than in moss sporangia, often only one cell thick. These

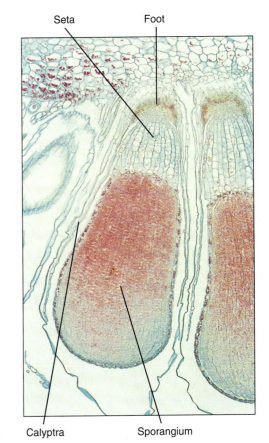

Seta   Foot

Calyptra   Sporangium

FIGURE 22.25

Liverwort sporophytes consist of foot, seta, and capsule (sporangium), but no elaborate set of teeth as in mosses. Instead, the apex breaks into several segments, all of which curl back and release the spores (× 20).

Elaters   Spores

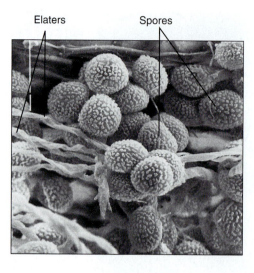

FIGURE 22.26
Spores and elaters of the liverwort
Haplometrium hookeri. (*Courtesy of Sharon Bartholomew-Began, Georgia Southwestern University*)

cells may have nodulose or bandlike thickenings on their walls, a feature mosses never have. The liverwort sporangium lacks a columella, the central mass of sterile cells found in mosses, and dehiscence occurs by means of four longitudinal slits, not by an operculum. Within the sporangium, some cells do not undergo meiosis but rather differentiate into **elaters**—single, elongate cells with spring-shaped walls. When the sporangium opens, the elaters uncoil, pushing the spores out (Figs. 22.26 and 22.27). All liverworts are homosporous.

Strikingly simple sporophytes are produced in *Riccia* and *Ricciocarpus:* The zygote grows into a spherical mass within the archegonium of the gametophore body. No foot or seta is formed; instead the inner cells of this mass undergo meiosis and produce spores. As the surrounding gametophore and sporophyte tissues age, die, and decay, the spores are finally liberated.

DIVISION ANTHOCEROTOPHYTA: HORNWORTS

Hornworts are a group of small, inconspicuous thalloid plants that grow on moist soil, hidden by grasses and other herbs (see Fig. 22.3; Table 22.4). They rarely inhabit tree trunks or bare rock, so one does not often encounter them unless one is looking for them specifically. Roadside cuts and stable eroded soil are good sites for hornworts. Hornworts number about 400 named species in five or six genera, with *Anthoceros* and *Phaeoceros* being common examples. Many of the species names are probably synonyms, that is, several names applied to the same species, because many hornworts have such variable growth forms that individuals of the same species can differ greatly under varied conditions.

Hornworts superficially resemble thalloid liverworts, the gametophores of many species being ribbon-shaped and thin, without a distinct stem or leaves (Fig. 22.28). However, hornworts never contain oil bodies, whereas liverworts almost always do. As in all other embryophytes, an alternation of heteromorphic generations occurs, with the sporophyte depending on the larger, photosynthetically active gametophyte. But hornworts are quite

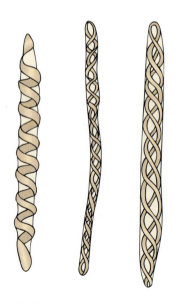

FIGURE 22.27
Thickenings on the walls of elaters are deposited on the interior surface of the wall. As the air dries, the spiral thickenings cause elaters to twist, agitating the surrounding spore mass. Spores are knocked free gradually rather than all falling out simultaneously.

TABLE 22.4 *Classification of Division Anthocerotophyta*	
Class Anthocerotopsida	
Order Anthocerotales	*Anthoceros, Dendroceros, Megaceros, Notothylos, Phaeoceros*

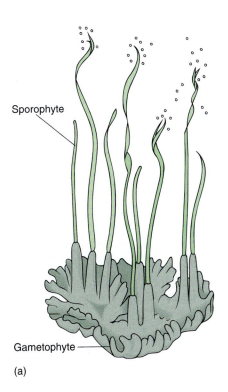

(a)

(b)

FIGURE 22.28
(a) The "horns" of a hornwort are sporophytes that grow continuously from a basal meristem. The lower part of the sporophyte is surrounded by gametophyte tissue. (b) *Anthoceros* species. *(Robert and Linda Mitchell)*

distinct from all other embryophytes, including the other nonvascular plants. One of the most striking features is the presence of a single large chloroplast in each cell as opposed to the numerous small plastids present in all other nonalgal plants (Fig. 22.29). The single chloroplast per cell is characteristic of many algae and is surprising in the hornworts. Furthermore, hornworst chloroplasts have a pyrenoid, an algal feature absent in all other embryophytes. Hornwort chloroplasts may be representative of the condition of the ancestors of all embryophytes, or they may indicate that hornworts had a separate algal origin, distinct from all other embryophytes. Unfortunately, we do not know what the chloroplasts of the earliest land plants were like because they are not preserved in any known fossil.

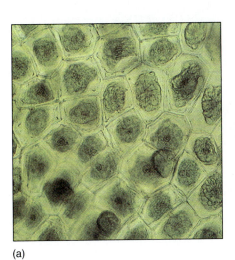

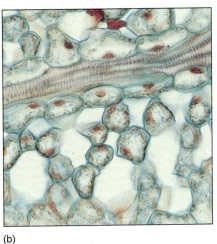

(a)

(b)

FIGURE 22.29
(a) Hornwort cells are unique among embryophytes: Each has just one plastid. Outside of hornworts, this feature is known only in algae. Although it may seem trivial, it must represent a significantly distinct cell/plastid relationship. The plastid's reproduction must be carefully coordinated with that of the cell, and a special mechanism is necessary to ensure that both progeny cells receive one each ($\times$ 60). (b) Cells of privet leaf with many chloroplasts each ($\times$ 60).

THE GAMETOPHYTE GENERATION

The protonema phase in hornworts is even more reduced than in liverworts. As few as three or four protonema cells are produced before the gametophore phase is established in most species. Gametophores are always thin, at least along the edges. Only in the center do they become more than four or five cells thick. They may be shaped like a ribbon or a heart, or they may grow outward irregularly, forming a disk. The upper surface in many species is smooth, but in others leaflike lamellae grow upward. The gametophyte is parenchymatous, rather succulent but brittle. It does not tolerate drying; gametophores of hornworts typically live for less than 1 year in temperate climates. They act as winter annuals, appearing in the cool, moist autumn months, growing during winter, producing sporophytes in the spring, and dying before summer. In some species, they form oil-rich "tubers" as inner cells fill with oil and outer cells die; one thallus can produce several tubers.

Internally, hornwort gametophytes have numerous chambers. Young plants have mucilage chambers formed as cells break down and their contents are altered chemically into mucilage. With age, these dry out and become air chambers and are then invaded by *Nostoc* cyanobacteria. All hornworts form this symbiosis and presumably benefit by receiving nitrogen compounds from the *Nostoc*.

Gametangium development in hornworts is distinctive (Fig. 22.30). Unlike all other embryophytes, the antheridial initials of hornworts are not surface cells on the gametophyte. Instead, a special mucilage chamber forms near the upper surface; then cells lining the chamber grow into it and become antheridia. The chamber expands as the antheridia grow, and finally its roof breaks and only then are antheridia exposed to the environment.

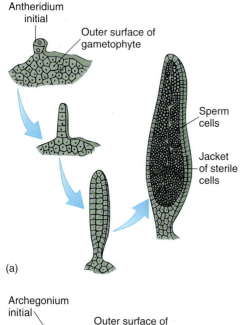

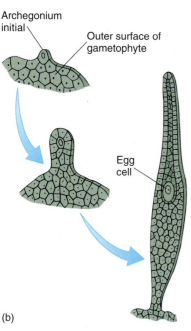

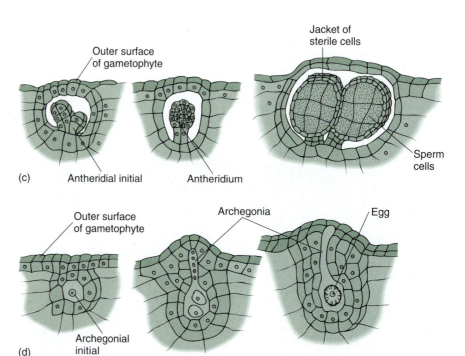

FIGURE 22.30

(a and b) In mosses, the initial cell in the formation of either an antheridium or an archegonium lies on the surface of the gametophyte. Then cell divisions change this single cell into a mass that has sterile jacket cells and reproductive central cells. This pattern occurs in all mosses, liverworts, and vascular plants. The gametophytes of flowering plants (see Chapter 9) are so reduced that the pattern is not obvious. It is discussed in Chapters 24 and 25. (c and d) The gametangia of hornworts are unique; the antheridia (c) do not form on the true surface of the gametophyte, and the egg (d) is not surrounded by discrete archegonial cells.

As sperm cells mature, the sterile outer cells of the antheridia transform their chloroplasts into chromoplasts, becoming orange or yellow.

Archegonia are formed from superficial cells, but the archegonia do not completely surround the egg as do the flask-shaped archegonia of other embryophytes. Rather, the egg lies below a short neck and neck canal but is surrounded by vegetative thallus cells. Even the neck cells are not particularly distinct but rather are difficult to distinguish from ordinary thallus parenchyma cells.

Once fertilization occurs, the zygote divides longitudinally; in mosses and liverworts, it divides transversely.

THE SPOROPHYTE GENERATION

Similarities between sporophytes of hornworts and those of mosses or liverworts are not easy to find. Hornworts have a foot embedded in gametophore tissue, but there is no seta or discrete sporangium (see Figs. 22.28, 22.31, and 22.32). Instead, just above the foot is a meristem that continuously produces new sporangium tissues. As the newly formed cells are pushed upward, they grow, differentiate, mature, and die. They are simultaneously being replaced by more cells from the basal meristem. Consequently, the sporangium is a long, hornlike cylinder, typically 1 or 2 cm long in *Anthoceros* and *Phaeoceros,* but up to 12 cm in some species. At the tip, the sporangium is mature and open as a result of dehiscence along two linear apertures. The outer layer of sterile cells is thick, up to six cells deep, and chlorophyllous. It has stomata in *Anthoceros* and *Phaeoceros* but lacks stomata in *Notothylos, Dendroceros,* and *Megaceros.* Unlike liverworts, hornworts have no nodules or bands on their cell walls. Spores are green, golden yellow, brown, or black and in some species are multicellular when ready to be released. Hornworts have a columella as in mosses, but unlike the elaters of liverworts, those of hornworts (often called pseudoelaters) are multicellular and do not have spirally thickened walls. The basal meristem is active over a long period, depending on moisture availability and temperature, and large numbers of spores can be produced by each sporophyte. Several attempts have been made to remove the

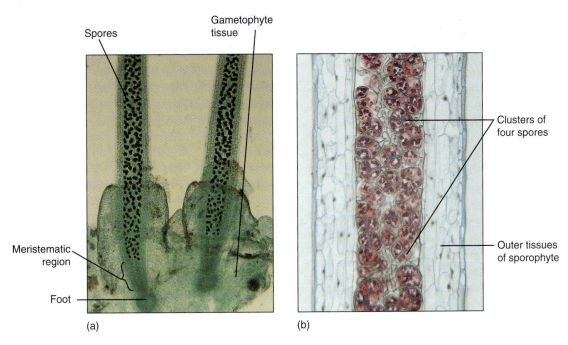

(a)   (b)

FIGURE 22.31

(a) The base of the sporophyte resembles a foot embedded in the gametophyte, and recently transfer cells have been discovered, so active nutrient transport into the sporophyte must be occurring. Just above the foot is a meristematic region. (b) At a higher level, equivalent to Figure 22.32c, spores are mature. (*Robert and Linda Mitchell*)

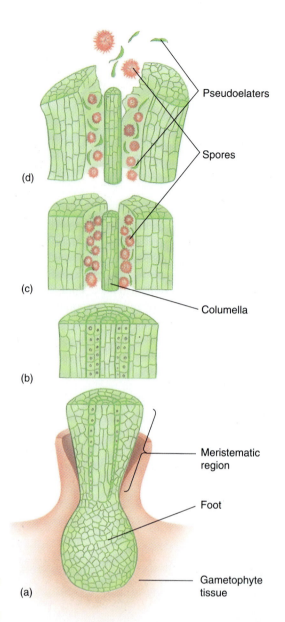

Pseudoelaters

Spores

(d)

Columella

(c)

(b)

Meristematic region

Foot

Gametophyte tissue

(a)

FIGURE 22.32
A longitudinal section through the sporophyte. At the base is a meristematic region (a); higher, above the basal meristem, sporocytes (spore mother cells) undergo meiosis (b). At higher levels the spores become mature (c) and are then released (d). This type of continuous meiosis is unknown in plants other than hornworts.

Rhyniophytes are discussed in detail in Chapter 23.

sporophyte surgically from the gametophyte and grow it in laboratory conditions, but even though it is chlorophyllous, it dies. It is not known if death is caused by a lack of minerals, insufficient photosynthesis, or absence of growth factors, vitamins, or hormones from the gametophyte.

ORIGIN AND EVOLUTION OF NONVASCULAR PLANTS

No consensus exists about the origin of nonvascular plants; the sparse fossil evidence can be interpreted in various ways. A variety of hypotheses have been offered, but many botanists believe that insufficient information is available to form any firm opinion. Two frequently considered hypotheses are (1) that nonvascular plants evolved from early vascular plants, the rhyniophytes, by becoming simpler and reduced and by losing their vascular tissues, and (2) that nonvascular plants are not related to vascular plants at all, but rather evolved from green algae separately and at a different time than the rhyniophytes. A third emerging hypothesis suggests that at least some nonvascular plants evolved from vascular ancestors by simplification, but the divergence between the two groups started even earlier than the rhyniophytes; that is, some nonvascular plants evolved from the same group that later gave rise to the rhyniophytes.

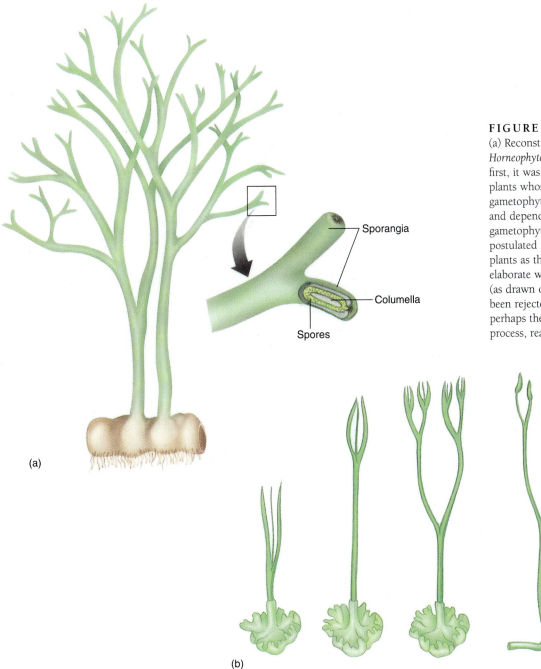

(a)

(b)

FIGURE 22.33

(a) Reconstruction of the extinct plant *Horneophyton* and its terminal sporangia. (b) At first, it was postulated that algae gave rise to land plants whose life cycle was dominated by the gametophyte and whose sporophyte was small and dependent, hemiparasitic on the gametophyte (as drawn on the left). It was postulated that this evolved into the vascular plants as the sporophyte became larger and more elaborate while the gametophyte became reduced (as drawn on the right). This hypothesis has been rejected and it is now suspected that perhaps the hornworts evolved by the reverse process, reading this series from right to left.

Sporangia

Columella

Spores

The last hypothesis centers on living hornworts like *Anthoceros* and fossil plants called *Horneophyton.* The sporophytes of *Horneophyton* were naked axes that branched dichotomously and were up to 20 cm tall (Fig. 22.33a). They had stomata and terminal sporangia. Each sporangium had a short columella, and just as in hornworts, the sporogenous tissue surrounded the columella on all sides and the top. The stem base was swollen, similar to a hornwort foot, except that it bore rhizoids.

The similarities between *Anthoceros* and *Horneophyton* were recognized early in this century, and it was suggested then that vascular plants may have arisen by a process in which (1) algae gave rise to hornworts similar to *Anthoceros,* with small sporophytes that depended on thalloid gametophytes; (2) these species in turn evolved into *Horneophyton*-like plants as the gametophytes became simpler and shorter-lived while the sporophytes became taller and more active photosynthetically (Fig. 22.33b, from left to right); (3) finally, the gametophyte was very small and temporary, and the sporophyte was an independent, free-living plant.

This idea is not accepted by many botanists today for several reasons, but it is possible to think about these transitions in reverse—that hornworts evolved from species like

Horneophyton by means of an elaboration of the gametophyte and a reduction of the sporophyte. In the earliest true land plants, gametophytes were not thalloid like those of *Anthoceros,* but rather had upright stems; there appears to have been an alternation of isomorphic generations. Early true plants appear to have been a group of species that resembled each other but had already diverged to the point that some had sporangia without a columella and others had sporangia with a short, incomplete hornwort-type of columella (discussed further in Chapter 23).

The first group apparently gave rise to the vascular plants, with sporophytes becoming larger and gametophytes becoming smaller. The second group might have been the ancestors to the hornworts, with the gametophytes becoming the dominant phase of the life cycle. If so, vascular plants and hornworts are somewhat related, but the divergence between the two groups would have started very early, even before the full land plant syndrome had evolved. The presence of just one chloroplast per cell and the unusual formation of gametangia can be interpreted as supporting the theory of a completely separate origin in the green algae. Any common, nonalgal ancestor of hornworts and vascular plants must have existed very early, before many features were established definitively.

Only the origin of *Anthoceros* has been discussed, not that of nonvascular plants in general. The few available fossils of nonvascular plants do not support the idea that mosses and liverworts are closely related to hornworts. Mosslike plants do not appear in the fossil record until the Carboniferous Period, about 300 million years ago, about 100 million years after the hypothesized origin of hornworts. By the Permian Period, fossil mosses were as complex morphologically as living ones today; fossils with the cell arrangement pattern of *Sphagnum,* called *Protosphagnum,* occurred about 260 million years ago. Most fossils found in rocks 60 million years old are assigned to modern, extant genera.

The earliest liverwort fossils are much older than those of mosses; *Pallavicinites devonicus,* as its species name indicates, lived in the Devonian Period (400 million years ago), not long after *Horneophyton.* Full liverwort features existed by this early time. A few million years later, in the Carboniferous Period, leafy liverworts in the form of *Hepaticites kidstoni* had evolved; this species is very much like a living liverwort, *Treubia.* Indeed, the earliest fossils of both groups are well developed as either mosses or liverworts; no known fossil is being interpreted as an ancestor to both. Consequently, it seems prudent to favor the hypothesis that the nonvascular plants are polyphyletic, having had separate origins and having been distinct, well-defined groups before the end of the Paleozoic Era.

SUMMARY

1. Nonvascular plants are classified here in three divisions, emphasizing their differences and possible separate evolutionary origins. The divisions are Bryophyta (mosses), Hepatophyta (liverworts), and Anthocerotophyta (hornworts).

2. All are embryophytes: They have gametangia and sporangia in which only the internal cells differentiate into gametes or spores; a jacket of sterile cells is always present.

3. In all three, the life cycle consists of an alternation of heteromorphic generations. The gametophyte is the larger and more persistent and photosynthetically active phase; the sporophyte depends almost entirely on the gametophyte for carbon, energy, and minerals.

4. The setae, stems, and leaf midribs of some mosses have cells that facilitate long-distance conduction: Leptoids transport sugars, and hydroids transport water. It is not known if they are homologous or merely analogous to xylem and phloem.

5. Archegonia and antheridia may be borne on the same gametophore or on distinct, unisexual gametophores.

6. Neither the egg nor the zygote is released. The new sporophyte develops initially within the archegonium. When fully mature, it has a foot, seta, and capsule.

7. A moss spore grows into a filamentous protonema which produces buds that develop into thick parenchymatous gametophores with stems, leaves, and rhizoids.

8. Liverworts may be either thallose or leafy. Liverworts contain oil droplets, have bilobed leaves, and have no thick leaf midrib, no columella in the sporangium, and no operculum or peristome teeth.

9. Hornwort sporophytes differ from those of all other plants: They have a foot and a basal meristem, so sporogenous tissue is formed continuously.

10. Hornwort antheridia do not arise from surface cells, and the egg is not completely surrounded by distinct archegonial tissue.

11. No consensus exists concerning the origin of nonvascular plants. The hornworts are similar to some of the earliest vascular plants, but mosses and liverworts are not.

IMPORTANT TERMS

antheridiophore
antheridium
archegoniophore
archegonium
calyptra

capsule
elater
gametophore
hornwort
liverwort

moss
operculum
peristome
protonema
rhizoids
seta

REVIEW QUESTIONS

1. What are the three groups of nonvascular plants? How would you determine whether an unknown specimen is a vascular plant?

2. What are some characters that distinguish nonvascular plants from algae? What types of genes would be involved? What types of metabolisms and structures would have to be altered?

3. If the leptoids of mosses were found to contain a protein whose gene had the same nucleotide sequence as the gene that codes for P-protein, would that be significant evidence for either the homology or analogy of leptoids and phloem?

4. Draw and label the life cycle of a moss; be certain to show gametangia and sporangia. Which parts are haploid and which are diploid? Where and when does meiosis occur? Plasmogamy? Karyogamy?

5. What are some of the ways in which liverworts differ from mosses? How do hornworts differ from both? Do the three have similar life cycles?

6. An important consideration in the evolution of any organism is gene flow. What are some of the mechanisms by which genes move through the habitat in nonvascular plants? In a dense, cool forest, how strong are wind currents? Could they carry spores very far? What would you guess might be the maximum distance sperms can swim? How far can a raindrop splash a sperm or a spore?

BotanyLinks .net Questions

www.jbpub.com/botanylinks

Visit the .net Questions area of BotanyLinks (http://www.jbpub.com/botanylinks/netquestions) to complete this question:

1. Do bryophytes affect your life in ways you do not realize, as is true for algae? Go to the BotanyLinks home page to begin researching this subject.

BotanyLinks includes a Directory of Organizations for this chapter.

23

VASCULAR PLANTS WITHOUT SEEDS

Ferns and related plants have vascular tissue but reproduce without seeds. *(Peter Arnold/BIOS)*

CONCEPTS

In this and the next two chapters, the origin and evolutionary diversification of vascular plants are discussed. Several critical events occurred in the history of this group: their origin from green algae, the development of the vascular tissues xylem and phloem, and the origin of seeds, leaves, and woody growth. Vascular plants are traditionally divided into those that do not produce seeds—**vascular cryptogams**—and those that do—**spermatophytes**. The former group arose first, and some of its later members were the ancestors of the seed plants. Many of the fundamental aspects of plant biology described in Chapters 5 through 15 were established in the early, simple vascular cryptogams, the biology of which is the subject of this chapter.

Beginning about 420 million years ago, certain green algae began to adapt to living on land. Green algae have both marine and fresh water species, and some of the latter must have adapted to the occasional drying of their streams, smaller lakes, and oceanside mud flats. In environments where long dry periods alternated with long moist periods, the optimal survival strategy probably was formation of dormant, drought-resistant spores. But spore production requires major metabolic conversions; if the dry periods were frequent and temporary, the ability to continue active metabolism by conserving water and avoiding desiccation would have great selective advantage (Table 23.1). A large, compact, multicel-

T ABLE 23.1	*Means of Coping with Dry Periods*
Not coping	This is the least expensive—the organism simply dies. Rapid extinction occurs unless the organism lives in a permanently moist environment such as an ocean, large river or lake, or rain forest.
Desiccation tolerance	Many mosses, liverworts, and lichens and some terrestrial algae have the ability to survive even when their body loses large amounts of water. The mechanism seems to be expensive because these organisms all grow slowly even in continually moist environments. Perhaps desiccation-tolerant membranes, proteins, and organelles have characteristics that do not permit rapid metabolism.
Formation of spores	Many organisms have a strategy in which most of the body is allowed to die during times of stress, while a few cells—spores—are made stress-tolerant. Fewer energy and nutrient resources are required to modify a few cells than the entire body; and the body, while alive, is capable of rapid metabolism. A disadvantage is that once the spores germinate, an entire new plant must be formed; if wet periods are brief, a new period of dryness may kill the sporlings before they are large enough to form new spores.
Desiccation avoidance	The organism avoids dry conditions by either retaining water within itself or tapping a safe, relatively permanent source of water. This method allows some metabolism at all times, and growth and reproduction can be somewhat independent of dry cycles. Such a survival strategy led to the vascular plants; a cuticle that retains water, a large body that has a low surface-to-volume ratio, and later, roots that penetrate deeper regions of soil that tend to be more permanently moist.

lular body would have had a low surface-to-volume ratio and would have automatically retained water better than a small unicellular or filamentous body. In addition, a waterproofing cuticle would have been selectively advantageous. Evolution of a cuticle may not have been difficult, because fatty acids polymerize automatically in the presence of oxygen, producing a cuticle-like layer. All plants produce fatty acids for their membranes and other metabolic functions, so mutations that permitted some fatty acids to leak to the surface would have resulted in a somewhat waterproof coating and would have had adaptive value. Reproduction would have had to change slightly to coordinate gamete production with periods of moisture, because the sperms had to swim. Also, gamete and spore mother cells would have had to be protected from dryness by one or several layers of cells that formed a jacket around them. Such reproductive modifications resulted in the grouping and protection of spore and gamete mother cells into the sporangia and gametangia characteristic of all embryophytes—relatively massive and protected and having an outer layer of sterile cells.

These simple modifications probably allowed the algae not only to survive but also to be metabolically active during short dry periods. But an automatic consequence of much greater advantage was that the algae were safe from predators while they were out of the water. Tremendous selective benefit resulted from mutations that enabled the algae to be active for even longer periods out of water. These mutations probably involved increasing the size of the body and the impermeability of the cuticle, but there had to be the simultaneous evolution of stomatal pores and guard cells because a more protective cuticle also prevents the entry of carbon dioxide.

As a truly terrestrial existence became more successful, the mud flats and stream banks would have become crowded, and some plants must have grown over others and shaded them. With such shading, the environment became selective for mutations that produced an upright body that could grow into brighter light. This simple change had profound consequences: The strengthening material that evolved, xylem, was also good at conducting water. As phloem evolved, the basal part of the plant that remained in the shade could be nourished. Phloem permitted roots to extend deep into the dark soil, and xylem permitted transport upward of the water and nutrients those roots encountered. The presence of xylem, phloem, and roots freed the plants from their muddy habitats and allowed them to grow anywhere moisture was available near the surface. The presence of the two vascular tissues, necessary for individuals even a few centimeters tall, suddenly gave plants the means to become 100 meters tall, to branch, to form leaves, and to put sporangia high into the air, allowing the spores to be blown great distances, carrying the plant's genes over huge areas, and permitting the colonization of distant habitats.

Vascular tissue, especially phloem, made feasible the evolution of truly heterotrophic tissues—roots, meristems, and organ primordia. Without phloem, each part of a plant can grow and develop only as rapidly as its photosynthesis permits. But the presence of phloem allows mobilization of sugars, minerals, and hormones throughout the entire body and transport to a shoot apical meristem or a group of sporangia, thus permitting a more vigorous, robust growth than could otherwise occur.

One obstacle to the total invasion of land was that plants were still reproductively amphibious. They had terrestrial bodies but aquatic reproduction with swimming sperms. At some point in the life cycle, the environment had to be sufficiently wet that sperms could swim from one gametophyte to another. This requirement is a handicap, but not an insurmountable one: All plants without seeds—nonvascular plants such as mosses and vascular cryptogams such as ferns—still reproduce in this fashion.

The production of pollen and seeds eliminates the need for environmental water for reproduction. In the line of evolution that led to seed plants, gametophytes became so reduced that they could form completely within the walls of the spores (Fig. 23.1). Once this happened, retention of the megaspore and megagametophyte inside the parental sporophyte was feasible. The gametophytes functioned basically as tissue of the sporophyte and benefitted from the land adaptations that had evolved. The microgametophyte, inside the microspore wall, could be transferred by wind to the vicinity of the megasporangium, where it released the sperms into fluids secreted by the megagametophyte. The sperms would swim in this tiny artificial pond and effect fertilization; only the last step of reproduction was still wholly aquatic. In the most advanced seed plants a pollen tube carries nonmotile sperms even further, to the egg itself.

Some gametophytes—for example, those of all nonvascular plants—do not need any protection by the parental sporophytes. In most mosses, liverworts, and hornworts, gametophytes remain rather small and delicate, limiting the entire plant to environments mild enough for the gametophytes (a plant can live and reproduce only in environments suitable to all essential aspects of its full life cycle). Land plants should have been able to develop a life cycle in which gametophytes were just as large, tough, and woody as sporophytes, even having bark, leaves, roots, and so on, but this never happened. As discussed in Chapter 22, the very first land plants did have rather large gametophytes, but in all lines of evolution, gametophytes became simpler and smaller. In the three divisions of nonvascular plants, they are the dominant phase of the life cycle, but nowhere have they achieved the complexity of even the simplest vascular sporophyte. Perhaps in the rigorous terrestrial environment, complex haploid plants and animals cannot compete with diploid plants and animals.

It is common to refer to the features of early species as "primitive" and those of more recently evolved species as "advanced." Unfortunately, in nonscientific English these terms carry judgmental value. "Primitive" usually suggests inferior, inefficient, or poorly func-

The evolution of seeds is discussed more fully in Chapter 24, but the first steps, endosporial development of gametophytes, occurred in the vascular cryptogams.

FIGURE 23.1

The megagametophyte of *Selaginella,* which has developed almost completely within the original wall of the megaspore. This is not part of a seed plant but is a necessary first step in seed evolution. Small megagametophytes can be protected and nurtured by their parent sporophyte. By doing so, the gametophyte generation benefits from all those mutations that make the sporophyte adapted for life on land.

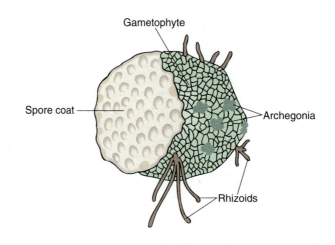

tioning, and "advanced" is often inferred to mean superior, efficient, well-adapted, and so on. Used with regard to evolution, these terms may be taken to mean that any plant with advanced features is "better" than one with primitive features. But "better" has no meaning in biology; the proper concept is whether one feature provides more reproductive success, survival value, or selective advantage than another feature. All this depends on the environment and the total biology of the plant. In mild tropical environments, being evergreen is more advantageous than being deciduous because the plants can photosynthesize more, but in environments with stressful winters or summers, being deciduous is more advantageous selectively.

Biologists continually search for neutral words to replace "primitive" and "advanced"; two words that are accurate with regard to evolution and neutral with regard to adaptive value are "relictual" and "derived." Features that were present in the early species of an evolutionary line are described as "original," but the term can really apply only to fossil organisms. Therefore, in a living group that retains many features similar to those of early species, those features are said to be relictual, not original. Features that were not present in the early species but instead evolved later are "derived"; that is, they were derived evolutionarily from the original features. For example, the very first vascular plants evolved from algae, and the bodies of those original species contained only primary growth, only tracheids in the xylem, and little sclerenchyma. Reproduction was by spores and free-living gametophytes. Some plants living today have similar features, which are described as relictual. However, many flowering plants have secondary growth, vessels in the xylem, and abundant collenchyma and sclerenchyma, and all produce flowers with dependent gametophytes; these are derived characters. It is tempting to call one set of features primitive and the other advanced, but consider this: Ferns contain all the relictual features just mentioned. Could they be called primitive? They outcompete flowering plants in many environments, perhaps because of their relictual features: By producing no secondary growth and no flowers and by having simple bodies, more of their energy is directed to spore production.

EARLY VASCULAR PLANTS

RHYNIOPHYTES

The earliest fossils that definitely were vascular land plants belong to *Cooksonia*, a genus of extinct plants (Fig. 23.2; Table 23.2). These had upright stems that were simple, short cylinders (several centimeters long) with no leaves (they had "naked stems"). They branched **dichotomously,** both branches being of equal size and vigor (Fig. 23.3). Plants of *Cooksonia* had an epidermis with a cuticle, a cortex of parenchyma, and a simple bundle of xylem composed of tracheids with annular secondary walls. The ends of the branches were swollen and contained the sporogenous tissue. Unlike the sporangia of algae, these were large and multicellular. Only the central cells were sporogenous and were surrounded by several layers of sterile cells; the sporangium wall had to open to release the spores. The plants were homosporous; there were no separate microspores and megaspores. Fossils that have these general characters are called **rhyniophytes.**

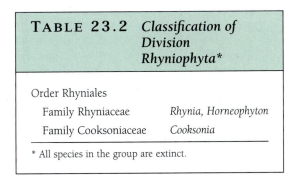

TABLE 23.2	*Classification of Division Rhyniophyta**
Order Rhyniales	
Family Rhyniaceae	*Rhynia, Horneophyton*
Family Cooksoniaceae	*Cooksonia*
* All species in the group are extinct.	

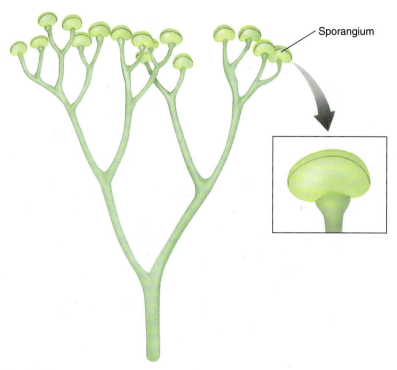

FIGURE 23.2

A reconstruction of *Cooksonia caledonica,* the earliest known plant that had xylem—tracheids with annular secondary walls. Important features are its sporangia at the ends of branches, the lack of leaves, and the dichotomous branching.

(a)   (b)   (c)

FIGURE 23.3

Branching patterns. (a) If a stem forks, resulting in two equal stems, it is a dichotomy or dichotomous branching. (b) It is pseudomonopodial branching if one stem is definitely larger and tends to form a trunk. (c) If one stem dominates the system absolutely, it is monopodial branching, as in most seed plants.

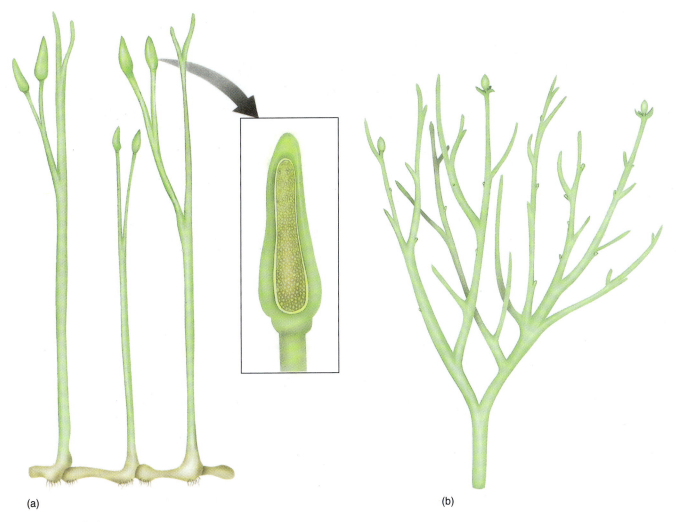

FIGURE 23.4
Reconstruction of *Rhynia major* (a) and *Rhynia gwynne-vaughanii* (b). They strongly resemble *Cooksonia,* and we know that they definitely had rhizomes, upright stems, and rhizoids. Inset shows the sporangium cut away, revealing spores.

Rhynia was another rhyniophyte, an early vascular plant similar to *Cooksonia* (Fig. 23.4). It had a prostrate rhizome, upright naked stems, and terminal sporangia. It had stomata and guard cells in the epidermis and a layer that appears to have been a cuticle.

Xylem Structure of Early Vascular Plants. Early vascular plants had two types of organization of xylem. In both the center is a solid mass of xylem with no pith; this is a **protostele.** In an **endarch** protostele, protoxylem is located in the center and metaxylem differentiates on the outer edge of the xylem mass. Recall from Chapter 5 that protoxylem is the xylem that differentiates while cells are small and narrow; metaxylem differentiates after the cells have expanded for a few more hours or days and are larger. The other type of stele present in early vascular plants is an **exarch** protostele, with metaxylem located in the center of the xylem mass and protoxylem on the edges as several groups next to the phloem. Another type of stele, which did not evolve until later, is the **siphonostele,** one in which pith is present in the center, as occurs in the stems of flowering plants. Because xylem is often preserved well in fossils, its characteristics of exarch/endarch and protostele/siphonostele are usually available for study.

The xylem in many specimens of *Rhynia* is well preserved: It was a round cylinder without pith but with protoxylem in the center and metaxylem on the exterior—that is, an

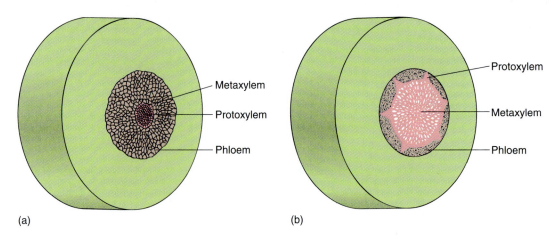

(a)   (b)

FIGURE 23.5

Early vascular plants had two types of organization of xylem. In both, the center is a solid mass of xylem with no pith. This is a protostele, common in roots of almost all plants. (a) In an endarch protostele, protoxylem is located in the center, and metaxylem is on the periphery. Endarch protosteles occurred in the rhyniophytes. (b) In an exarch protostele, metaxylem is located interior to the xylem mass and protoxylem on the edges, as several groups next to the phloem. Exarch protosteles are found in fossils of zosterophyllophytes.

endarch protostele (Fig. 23.5). All xylem cells were tracheids with annular thickenings (Fig. 23.6). Around the xylem was a layer of phloem-like cells, then a parenchymatous cortex and epidermis.

Life Cycles of Early Vascular Plants. In the same rocks with *Rhynia* are other fossils of similar plants. Two of these, *Lyonophyton* and *Sciadophyton,* were gametophytes, not sporophytes. The ends of the stems bore flattened cup-shaped areas that contained gametangia, both antheridia and archegonia, but not sporangia. These plants had upright, dichotomously branched stems, vascular tissue with tracheids, stomata, and a cuticle.

The nature of these early gametophytes is important in understanding the early evolution of vascular plants. The discovery of an alternation of heteromorphic generations in algae was possible because the individuals could be grown in culture and their life cycle could be observed. This is impossible with fossils, but it is possible to suspect that because *Rhynia*-type sporophytes occur together with *Sciadophyton*-type gametophytes, they might have been alternate phases of the same species. If this is true, these plants had an alternation of isomorphic generations, and later evolution into the seed plants involved reduction of the gametophyte to just a few cells and elaboration of the sporophyte into a more complex plant. This hypothesis is the **transformation hypothesis** (Fig. 23.7). An alternative hypothesis, the **interpolation hypothesis,** postulates that the earliest plants were monobion-

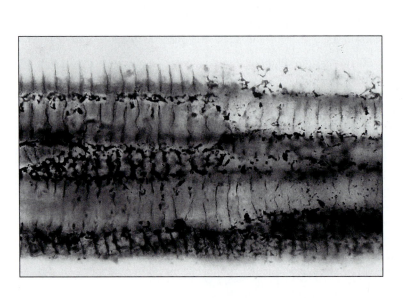

FIGURE 23.6

A tracheid with annular secondary walls found in *Rhynia gwynne-vaughanii* (× 200). *(Courtesy of K. J. Niklas, Cornell University)*

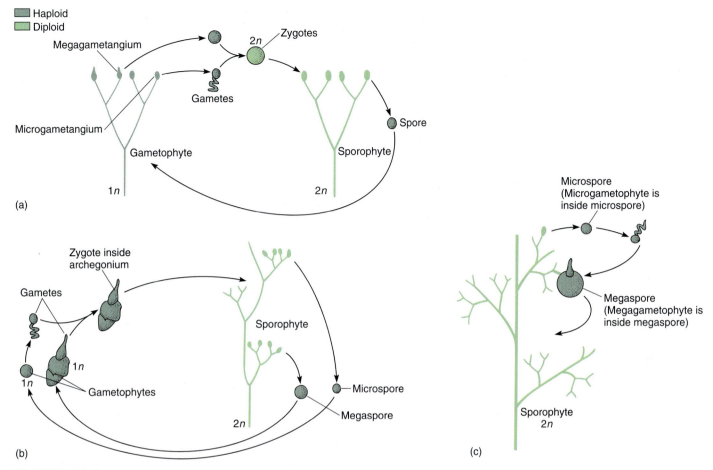

FIGURE 23.7

The transformation theory of the origin of the vascular plant life cycle postulates that in early land plants (a), gametophytes were upright and dichotomously branched, with epidermis, cuticle, and vascular tissue, just like sporophytes. (b) With time, sporophytes became larger and more complex and gametophytes became simpler. Even though these are two plants of the same species, different phenotypes are adaptive for each, because each contributes to the survival of the plant in different ways. In the species illustrated here, the gametophytes have become so small that the microgametophyte develops within the spore wall and the megagametophyte protrudes from the spore only slightly. (c) With continued reduction, it is possible—but neither necessary nor inevitable—for the megaspore and its megagametophyte to be retained inside the megasporangium and remain on the parental sporophyte, an important step in the process of seed evolution.

tic without a multicellular sporophyte (Fig. 23.8). A small sporophyte was presumed to come into existence when a zygote germinated mitotically instead of meiotically. The sporophyte generation would have gradually evolved in complexity while the gametophyte generation remained small. A sporophyte generation would be inserted (interpolated) into the monobiontic life cycle. In this hypothesis, nonvascular plants such as mosses are thought to be intermediates in the progression from green algae to vascular plants (Fig. 23.8). We have no way to decide which hypothesis more accurately approximates the actual origin of terrestrial sporophytes. The presence of *Horneophyton*, *Lyonophyton*, and *Sciadophyton* supports the transformation hypothesis, as does the similarity between *Anthoceros* and *Horneophyton*. But in mosses and liverworts, several species with the most complex capsules, which are parts of the sporophyte phase, seem to be the most derived species. Perhaps mosses and liverworts arose separately from green algae by interpolation. At present, data are insufficient to be certain.

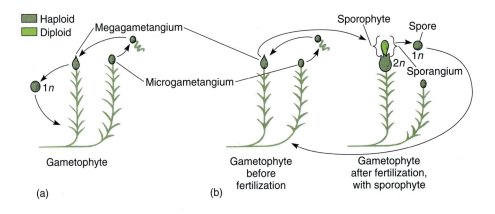

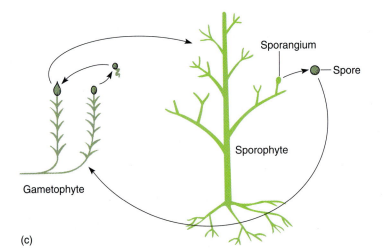

FIGURE 23.8
The interpolation theory. (a) The very earliest land plants were postulated to have no sporophyte; instead the zygote "germinated" by meiosis. (b) At a later stage in evolution, the zygote would germinate mitotically and produce a simple sporophyte that in the early stages would have consisted of a sporangium and perhaps also a foot. (c) With continued evolution, the sporophyte would have become progressively more elaborate while the gametophytes became simpler.

ZOSTEROPHYLLOPHYTES

Another group of early vascular plants are the **zosterophyllophytes,** also called the zoster-ophylls, named after the principal genus *Zosterophyllum* (Fig. 23.9a; Table 23.3). They were small herbs without secondary growth. Many of their features were similar to those of rhyniophytes, but three characteristics make us think they are a distinct group: Their sporangia were lateral, not terminal; the sporangia opened transversely along the top edge (Fig. 23.9b); and their xylem was an exarch protostele, that is, protoxylem on the outer margin and metaxylem in the center (see Fig. 23.5; Table 23.4).

Although these distinctions may seem minor, we are certain that in all the more recent plants with large leaves (ferns, conifers, flowering plants), sporangia are terminal rather than lateral; thus, plants like *Rhynia* may have been the transitions between algae and later seed plants, but *Zosterophyllum* was not. Instead, some of the simplest vascular plants alive today, the lycophytes, have lateral sporangia, and they may represent a line of evolution based on *Zosterophyllum*-like ancestors (Fig. 23.10).

TABLE 23.3	*Classification of Division Zosterophyllophyta**

Order Zosterophyllales
    Family Zosterophyllaceae   *Zosterophyllum, Crenaticaulis, Rebuchia, Sawdonia*

* All species in the group are extinct.

TABLE 23.4	*Characteristics of Rhyniophytes and Zosterophyllophytes*

	Rhyniophytes	**Zosterophyllophytes**
Sporangia		
location	terminal	lateral
dehiscence	along the side	across the top
Protoxylem	endarch	exarch

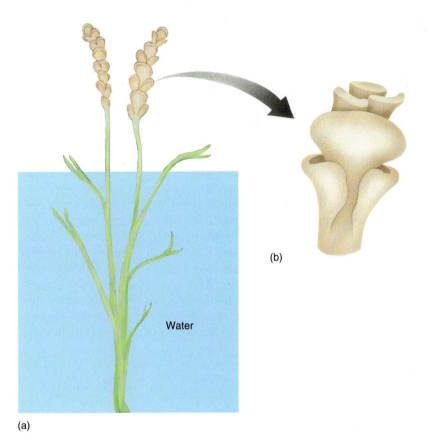

(b)

(a)

FIGURE 23.9
(a) *Zosterophyllum rehenanum* plants were quite similar to those of the rhyniophytes, but the ends of fertile branches bore numerous lateral sporangia, not a single terminal one. Also, sporangia (b) opened by a suture that passed over the top of the sporangium, not up its side.

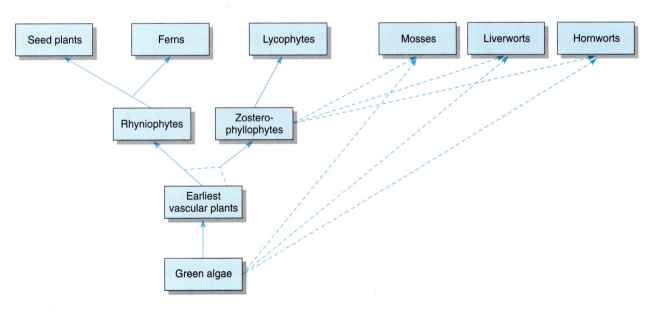

FIGURE 23.10
The very earliest vascular plants—those that immediately preceded *Cooksonia*—are unknown. Out of that group came rhyniophytes and zosterophyllophytes. We are not sure whether they evolved as two completely separate lines or as one line that soon diverged into one group with terminal sporangia and endarch protoxylem and a second line with lateral sporangia and exarch protoxylem. Nor do we know whether the nonvascular plants evolved out of these very earliest vascular plants or directly from green algae.

FIGURE 23.11

This reconstruction of the zosterophyllophyte *Rebuchia ovata* shows several significant advances in body construction. The stems that bore sporangia were specialized, keeping the sporangia close together and elevated into the wind; other stems branched repeatedly, forming a large photosynthetic surface. Some branches were larger than others, so branching was no longer strictly equal dichotomies. These derived features are not in the line that led to ferns and seed plants; analogous evolutionary changes had to occur in that line as well.

Zosterophyllum is a genus of extinct plants that grew as small bunches, only about 15 cm high. They probably grew in swampy, marshy areas: Upper portions of their stems had cuticle, ordinary epidermal cells, and stomata, but lower portions did not, presumably because they grew under water. Stems of *Zosterophyllum* were naked (smooth), branched dichotomously, and contained a small amount of xylem that consisted of tracheids with annular and scalariform secondary walls. Sporangial walls were several layers thick, and all spores were the same size, so they must have been homosporous.

Other genera of zosterophyllophytes show that significant morphological changes evolved quickly. In *Rebuchia* the sporangia occurred together on the ends of specialized branches (Fig. 23.11). In *Crenaticaulis* some branching was pseudomonopodial: Larger, trunklike shoots bore smaller, shorter lateral shoots. These are the morphological bases for producing cones and trunks. Several of the zosterophyllophytes had a smooth surface (see Fig. 23.9), but others had outgrowths of tissue called **enations** that ranged from quite small to long, thin scales. Enations increased the photosynthetic surface area of the plants, and in *Asteroxylon* they contained stomata and a small trace of vascular tissue that ran from the stele through the cortex to the base of the enation (Fig. 23.12).

The zosterophyllophytes and rhyniophytes shared so many characters that we believe they must have evolved from a common ancestor. They represent two lines of evolution, the rhyniophytes giving rise later to the ancestors of seed plants and ferns, and the zosterophyllophytes giving rise to the lycophytes. (See the section on the microphyll line of evolution.)

PSILOTUM

DIVISION PSILOTOPHYTA

This small division (one family with two genera) contains the simplest of all living vascular plants, *Psilotum* and *Tmesipteris* (Fig. 23.13; Table 23.5). *Psilotum* is constructed very much like *Rhynia*, and until the mid-1970s *Rhynia,* even though extinct, was often placed in this division. The resemblance is more than skin deep: *Psilotum* (the "whisk fern") is a small

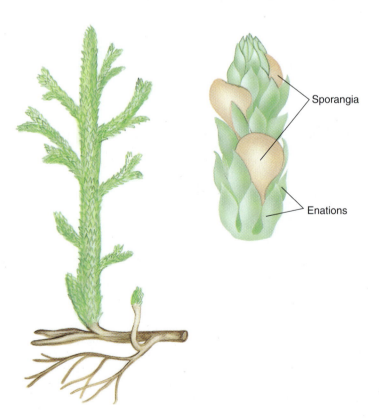

Sporangia

Enations

FIGURE 23.12

Reconstruction of *Asteroxylon,* an early lycophyte, showing the surface covered with enations, small flaps of photosynthetic tissue. In species without enations, stems were round and oriented vertically, not very good for harvesting sunlight.

(a)

(b)

(c)

TABLE 23.5	*Classification of Division Psilotophyta*

Class Psilotopsida
   Order Psilotales
      Family Psilotaceae   *Psilotum, Tmesipteris*

FIGURE 23.13

(a) *Psilotum* growing with mosses and ferns in a Hawaiian rainforest. (b) The growth habit of *Psilotum* strongly resembles that of a rhyniophyte—dichotomous axes with only a few enations. (c) *Tmesipteris* growing in New Zealand. *(W. E. Ferguson)*

plant with prostrate rhizomes and upright stems that branch dichotomously and have an epidermis, cortex, and a simple vascular cylinder with no pith—a protostele (Fig. 23.14). Xylem consists of annularly or helically thickened tracheids. *Psilotum* is unique among living vascular plants in that it has no roots or leaves. The shoot does have occasional small projections of tissues considered to be enations. As in *Rhynia*, sporangia are produced at the ends of branches.

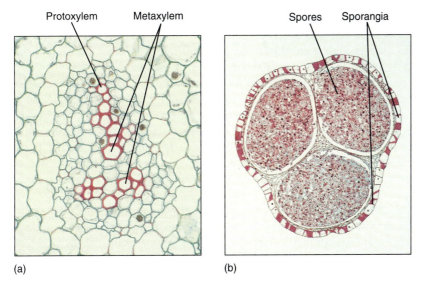

(a)

(b)

FIGURE 23.14

(a) Vascular structure of *Psilotum* is simple: a protostele and phloem surrounding the xylem. Protoxylem is exarch (to the exterior of the metaxylem), which is similar to the zosterophyllophytes rather than the rhyniophytes (× 100). (b) Sporangia of *Psilotum* are actually considered to be three sporangia fused together; each is at the end of an extremely short branch, basically a three-branched shoot subtended by an enation (× 8).

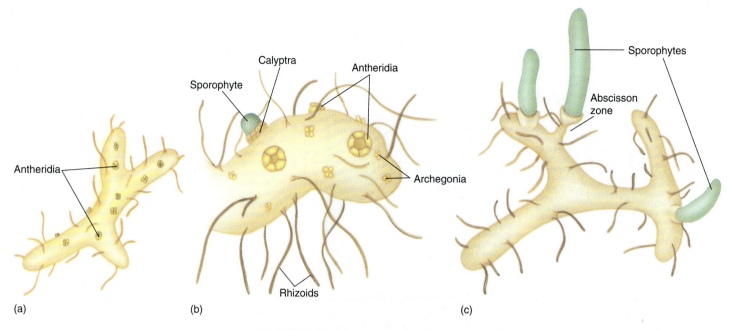

FIGURE 23.15

Gametophytes of *Psilotum* are small cylinders (a) that live heterotrophically, nourished by endophytic soil fungi. (b) Each gametophyte bears both antheridia and archegonia, and because the egg and zygote are retained in the archegonium, the young sporophyte initially is nourished by the parent gametophyte; during early development there is proliferation of gametophyte tissue as a protective calyptra. (c) As the sporophyte grows and enlarges, it forms an abscission zone and detaches from the gametophyte, becoming independent.

Particularly important are the gametophytes, short, branched cylinders less than 2 mm in diameter. Their surface is covered with elongate cells, rhizoids, that act like roots and at least anchor and perhaps also absorb (Fig. 23.15). The gametophytes have no chlorophyll but instead are heterotrophic, forming either a symbiotic or a parasitic association with soil fungi; fungi invade most cells of the gametophyte and provide it with sugars and minerals. The only internal differentiation is that gametophytes contain a small amount of vascular tissue similar to that of sporophytes: A central mass of tracheids is surrounded by phloem and an endodermis. *Psilotum* is the only species whose gametophytes contain vascular tissue. A vascularized gametophyte supports the transformation theory of the origin of sporophytes. Although sporophytes and gametophytes of *Psilotum* are easily distinguishable from each other, they share many fundamental features.

Despite the close resemblance of *Rhynia* and *Psilotum,* fossil evidence suggests that *Rhynia* itself became extinct millions of years ago. No fossils of it have been found after the Middle Devonian Period, 360 million years ago. At present, most morphologists consider *Psilotum* and *Tmesipteris* to be living representatives of ancient lines, surviving remnants of the very first stages in the organization of a vascular plant, but we are not certain how closely related they are to rhyniophytes. An alternative theory postulates that *Psilotum* and *Tmesipteris* are actually highly modified relatives of ferns, and they resemble *Rhynia* only because of convergent evolution.

Psilotum occurs in tropical and subtropical regions. In the United States, it can be found in the Gulf Coast states from Florida to Texas as well as in Hawaii. *Tmesipteris* is limited to Australasia, primarily Australia and other South Pacific islands.

THE MICROPHYLL LINE OF EVOLUTION: DIVISION LYCOPHYTA

Division Lycophyta contains species that represent a distinct line of evolution out of the first land plants. Lycophytes have lateral sporangia and exarch protosteles, so they may have come from a *Zosterophyllum* type of ancestor (Fig. 23.16; Table 23.6).

TABLE 23.6 *Classification of Division Lycophyta*

Order Asteroxylales*
  Family Asteroxylaceae — *Asteroxylon*

Order Protolepidodendrales*
  Family Protolepidodendraceae — *Baragwanathia, Leclercqia, Protolepidodendron*

Order Lycopodiales
  Family Lycopodiaceae — *Lycopodium, Phylloglossum, Lycopodites**

Order Lepidodendrales*
  Family Lepidodendraceae — *Lepidodendron, Lepidophloios, Lepidostrobus, Sigillaria, Stigmaria*

Order Selaginellales
  Family Selaginellaceae — *Selaginella, Selaginellites**

Order Isoetales
  Family Isoetaceae — *Isoetes, Stylites, Isoetites**

Order Pleuromeiales*
  Family Pleuromeiaceae — *Pleuromeia*

* All species in this group are extinct.

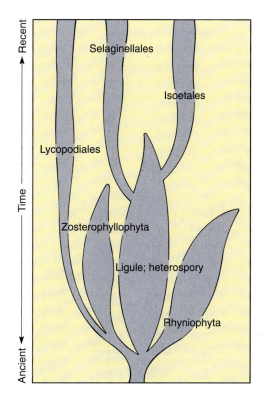

FIGURE 23.16
A proposed phylogeny of division Lycophyta. The lycopods are believed to have originated from zosterophylls and diversified early; then most became extinct and now only the genus *Lycopodium* still exists with living species. The selaginellas and *Isoetes* may have evolved from rhyniophytes that were extremely similar to the zosterophyllophytes and then diversified rapidly into many groups, several of which contained large trees with secondary growth. Most also become extinct, and now only three genera are left.

MORPHOLOGY

The earliest lycophytes were members of the genera *Drepanophycus* and *Baragwanathia* (Fig. 23.17a). They were similar to their presumed ancestors, the zosterophyllophytes, with an important difference: Their enations were large, up to 4 cm long, and they contained a single well-developed trace of vascular tissue. Such enations must have been effective at increasing photosynthesis, and they could be called leaves. However, "leaf" is an ambiguous term, and enations in the division Lycophyta are called **microphylls** for clarity (Fig. 23.17b–e). "Micro-" refers to their evolution from small enations, not to their actual size. In some plants, they were up to 78 cm long. This is not the line of evolution that led to ferns and seed plants; microphylls are not the same as the leaves you are familiar with.

Another important advance over the type of body organization of the *Rhynia-Zosterophyllum* type was the evolution of true roots that allowed sporophytes to anchor firmly, absorb efficiently, and thus to grow to tremendous size.

Many extinct lycophytes such as *Lepidodendron, Sigillaria,* and *Stigmaria* had a vascular cambium and secondary growth (Figs. 23.18 to 23.20). Their wood looked remarkably like the secondary xylem of pines and other living conifers, having a pith, rays, and elongate tracheids. However, the vascular cambium had one major flaw: Its cells apparently could not undergo radial longitudinal division, so new fusiform initials could not be produced. As the wood grew to a larger circumference, the cambial cells became increasingly wider tangentially (Fig. 23.20b). No specimen has ever been found with wood more than about 10 cm thick. After that much secondary growth, the cambial cells may have stretched so much that they could no longer function.

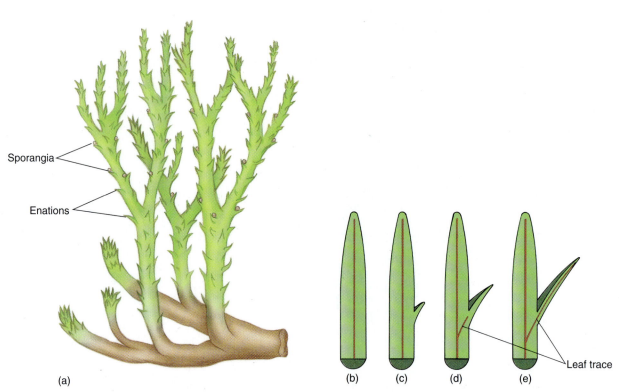

Sporangia

Enations

(a)

(b)   (c)   (d)   (e)

Leaf trace

FIGURE 23.17

(a) *Drepanophycus* was an early lycophyte that was still small and simple. (b to e) Microphylls in lycophytes are believed to have evolved as enations. Originally they were small, simple flaps of photosynthetic tissue (c). Later they became larger (d) and were vascularized (e). See Figure 23.25 for evolution of megaphylls.

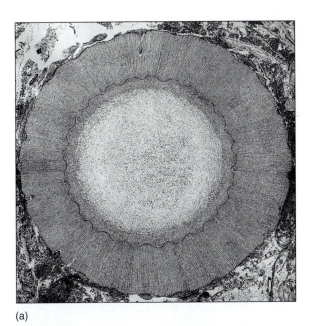

(a)

(b)

FIGURE 23.20

(a) Wood of *Sigillaria* consisted of large tracheids and superficially resembled the wood of modern conifers. However, it was not as derived in its pitting and other features of its secondary wall, and it was formed by a vascular cambium that could not form new fusiform initials (× 80). *(Courtesy of T. Delevoryas, University of Texas)* (b) In the vascular cambia of seed plants (top), as the cambium is pushed outward by accumulation of new secondary xylem, its circumference becomes greater, but fusiform initials divide by radial walls, creating new fusiform initials. An old cambium has many more cells than a young cambium (see Fig. 8.7). In arborescent lycophytes (bottom), cambial cells could not undergo this division, so an old cambium had no more cells than a young one. The older cells became stretched circumferentially until they were no longer functional; secondary growth then ceased.

HETEROSPORY

Reproduction of lycophytes was also more sophisticated than that of rhyniophytes and zosterophyllophytes in that the sporangia were clustered together in compact groups called **cones** or **strobili** (sing.: strobilus), which protected the sporangia (Fig. 23.21). Although many species remained homosporous, others became heterosporous, having microspores and megaspores that germinated to give rise to distinct microgametophytes and megagametophytes, respectively. Heterospory has special significance in land plants, because it is a necessary precondition for the evolution of seeds (see Figs. 23.1 and 23.7). In the lycophytes *Lepidostrobus* and *Lepidophloios,* the megaspore developed into a megagametophyte without enlarging; the megagametophyte existed completely within the wall of the original megaspore, which was up to 10 mm long. Furthermore, the megaspore in some species was retained within the sporophyll, protected by thick-walled cells of the sporangium. This is remarkably similar to ovules and seeds in the seed plants, the most important difference being that in these lycophytes, the sporangium dehisced (much like modern anthers do) and the megaspore wall cracked, exposing the archegonia. Sperm cells could swim to the egg during fertilization.

The lycophytes are remarkable in that they represent an ancient line of evolution distinct from the seed plants but demonstrate convergent evolution in several characters: leaves, roots, secondary growth, and almost seeds. In the Devonian and Carboniferous Periods, this group dominated the swampy areas of Earth with extensive forests of large trees, but most became extinct; at present the entire division contains only five genera, *Lycopodium, Phylloglossum, Selaginella, Isoetes,* and *Stylites.*

FIGURE 23.21

(a) *Lycopodium cernuum* is a common species that has extensive rhizomes, vertical chlorophyllous shoots, and sporangia clustered into cones. The shoots are actually leaning against and being supported by the surrounding grasses and shrubs. (b) *Lycopodium obscurum,* with sporangia clustered into cones at the tips of the branches. *(Ed Reschke)* (c and d) *Lycopodium lucidulum* is one of several lycopod species in which the sporangia are distributed among the leaves rather than in strobili. *(James W. Richardson/VU)* (e) *Selaginella* plants tend to be more delicate than those of *Lycopodium* and are often overlooked or mistaken for mosses. The leaves usually are very small and delicate and sporangia are inconspicuous. *(Dennis Drenner)*

EXTANT GENERA

Lycopodium ("ground pine" or "club moss") is fairly common in forests from tropical regions to the arctic (Fig. 23.21). All living species, about 200, are small herbs with prostrate rhizomes that have true roots and short upright branches. Microphylls are spirally arranged on their stems, and secondary growth never occurs. The sporangia may be arranged in cones or distributed along the shoots (*L. lucidulum*). All *Lycopodium* species are homosporous, a relictual trait. Spores germinate and grow into bisexual gametophytes that produce both antheridia and archegonia. In some species gametophytes are green and photosynthetic; in others they are subterranean and heterotrophic, nourished by fungi.

Selaginella is less common in temperate North America, and its plants are smaller and easily overlooked (Fig. 23.21e). Probably the best-known species is the resurrection plant, *S. lepidophylla,* which curls up, turns brown, and appears dead upon drying but uncurls and regreens when moistened. Unlike *Lycopodium, Selaginella* has the more derived condition of being heterosporous, and the megagametophyte develops inside the megaspore wall (see Figs. 23.1 and 23.22a). The megaspore is not retained on the sporophyte, however,

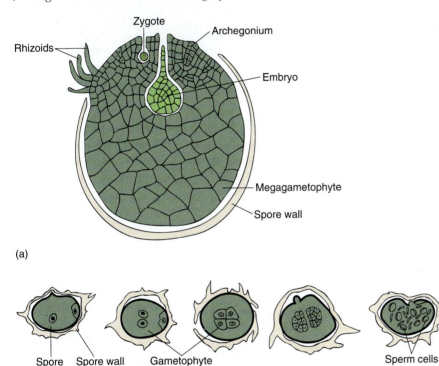

(a)

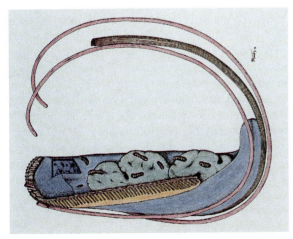

(b)

(c)

FIGURE 23.22

(a) The megagametophyte of some selaginellas develops almost entirely within the megaspore wall. The spore cracks open, exposing the archegonia (megagametangia) and permitting fertilization by swimming sperm cells. (b) The microgametophytes of selaginellas develop within the microspore wall, then liberate motile sperms.
(c) Reconstruction of the organelles of a sperm cell of *Lycopodium obscurum.* These sperms cells are similar to those of mosses (see Fig. 22.16) and also strongly resemble sperm cells of green algae. Color coding is like that of Figure 22.16: Blue = nucleus, green = plastid, red = flagella and basal bodies, brown = mitochondrion, yellow = microtubules that anchor the flagella. *(Courtesy of K. Renzaglia, B. Tackett, Jr., and A. Maden, East Tennessee State University)*

FIGURE 23.23
Although *Isoetes* is a small plant, its short basal cormlike stem has a small amount of secondary growth. (*W. H. Hodges/Peter Arnold, Inc.*)

and is not seedlike. The microgametophytes also develop within the spore wall and consist of a single vegetative cell and an antheridium (Fig. 23.22b). Many flagellate sperms are produced and then released as the spore wall ruptures. The sperms swim to the archegonia through a film of dew or other available moisture

Selaginellas can be distinguished from lycopodiums by the presence of a small flap of tissue, the **ligule,** on the upper surface of *Selaginella* leaves. Although ligules are simple structures and their adaptive advantages are unknown, they are ancient and can be used to distinguish early fossil selaginellas from fossil lycopods.

Isoetes is a genus of about 60 species of small, unusual plants called quillworts which grow in wet, muddy areas that occasionally become dry (Fig. 23.23). Their body consists of a small cormlike stem that has roots attached below and leaves above. *Isoetes* is heterosporous like *Selaginella,* and almost every leaf contains sporangia. Microphylls in this genus also have ligules. Weak cambial growth results in the production of additional cortex parenchyma to the exterior and a type of vascular tissue to the interior; the latter tissue is a mixture of tracheids, sieve elements, and parenchyma. *Stylites* plants are very similar to those of *Isoetes,* and several botanists have suggested that the two species of *Stylites* are really extreme forms of *Isoetes* (Table 23.6).

The anatomy and morphology preserved in fossil lycophytes indicate many instances of convergent evolution with the seed plants. In both lines, elaborate, efficient leaves, wood, bark, and roots evolved. Sporangia became separated from nonreproductive organs and were grouped together into strobili. In both lycophytes and seed plants, heterospory evolved, as did endosporial development of the megagametophyte. In seed plants, the megaspores were retained in the megasporangium and evolved into seeds; this nearly happened in certain extinct lycophytes. We cannot tell from the fossils if convergent evolution of specialized metabolisms also occurred, but *Isoetes* does have CAM photosynthesis. It also has a unique means of obtaining its carbon dioxide—absorption by the roots from soil or mud. Its leaves have a thick cuticle but no stomata. We can only wonder what the metabolism of the extinct treelike lycophytes was like. The extinction of so many of the lycophytes is unfortunate; they were complex, sophisticated plants that had many superb adaptations. They probably underwent interesting responses to changes in season, and they must have been able to resist numerous types of pathogenic bacteria, fungi, and insects. Those aspects of their biology will probably remain unknown to us forever.

BOX 23.1   *Molecular Studies of the Evolution of Early Land Plants*

The evolutionary relationships of the early land plants are still not understood well. Plants like mosses, liverworts, *Rhynia*, and lycopods are so simple that there are not many characters to compare. Furthermore, when two groups do resemble each other, we cannot always be certain if the common characters are homologous (derived from the same ancestral character) or analogous (derived from different ancestral characters and resembling each other only as a result of convergent evolution).

Recently, an analysis of the arrangement of genes in chloroplasts has been used to investigate which early land plants are related. The circular DNA from the chloroplasts was isolated from a variety of plants; then DNA restriction enzymes were used to analyze the order in which genes occur (see Chapter 15). Although only a few groups have been studied, it turns out that mosses, liverworts, and the lycophytes (*Lycopodium, Selaginella,* and *Isoetes*) have one arrangement, whereas ferns, *Psilotum,* and the seed plants (the conifers and flowering plants of Chapters 24 and 25) all have an alternative arrangement.

From numerous studies on many types of plants, it had been concluded that chloroplast genomes are quite stable and this type of rearrangement is rare. It seems unlikely that this type of rearrangement could occur twice, producing similar gene arrangements in three distinct groups. Thus the fact that mosses, liverworts, and lycophytes share a particular gene arrangement supports the hypothesis that these three groups are closely related. Assuming that these three groups are related, which is more relictual, more similar to the ancestral condition? Based on the organization of xylem and sporangia, we are certain that the lycophytes are related to the zosterophyllophytes, which were derived from the rhyniophytes. Consequently, if the lycophytes did not evolve from mosses and liverworts, then the mosses and liverworts must have evolved out of the lycophytes or zosterophyllophytes. This would be consistent with the lack of fossil mosses and liverworts during the early stages of land plant evolution. Furthermore, if these nonvascular plants evolved from the zosterophyllophyte/lycophyte line of evolution, that means that they did not evolve directly from green algae, as hypothesized by the dashed lines in Figure 23.10. If true, then there were at most two origins from algae (hornworts and rhyniophytes) or just one (all land plants).

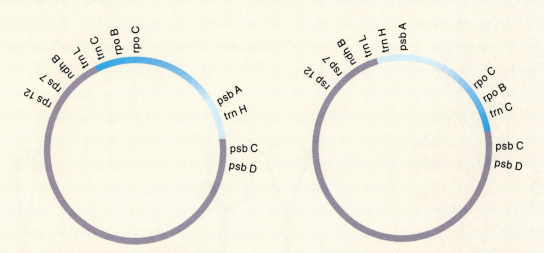

Order of genes on the circular DNA of plastids of liverworts, mosses, and lycophytes (a), and of seed plants (b). Notice that the section shown in blue is reversed in nonvascular plants relative to that in the vascular plants. (*Based on research of L. A. Raubeson, Yale University, and R. K. Jansen, University of Texas*)

THE MEGAPHYLL LINE OF EVOLUTION

DIVISION TRIMEROPHYTOPHYTA

Division Trimerophytophyta was proposed in 1968 for three genera of extinct plants, *Trimerophyton, Psilophyton,* and *Pertica* (Table 23.7). The fossils strongly resemble those of rhyniophytes, having terminal sporangia that dehisced laterally, homospory, dichotomous branching, and an endarch vascular cylinder of tracheids. Trimerophytes, however, are considered a distinct advancement out of the rhyniophytes because of several special features. Most important is the trend of **overtopping.** Rather than stems of equal length as in true dichotomous branching, trimerophytes had an unequal branching in which one stem was more vigorous (see Fig. 23.3b). In later species the inequality caused by overtop-

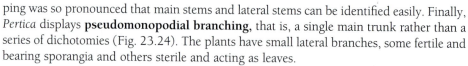

TABLE 23.7 *Classification of Division Trimerophytophyta**

Order Trimerophytales
    Family Trimerophytaceae   *Psilophyton, Pertica, Trimerophyton*

* All species in this group are extinct.

ping was so pronounced that main stems and lateral stems can be identified easily. Finally, *Pertica* displays **pseudomonopodial branching,** that is, a single main trunk rather than a series of dichotomies (Fig. 23.24). The plants have small lateral branches, some fertile and bearing sporangia and others sterile and acting as leaves.

Simultaneously, the positioning of branches became more regular and controlled. In rhyniophytes, the points of dichotomy were irregular and unpredictable, but in some species of trimerophytes, lateral branches were arranged in a regular spiral phyllotaxy. Other types of phyllotaxy that occurred were alternate, decussate (opposite leaves arranged in four rows), distichous (leaves in two rows), tetrastichous (alternate leaves, in four rows), and even whorled. Correlated with the evolution of a pseudomonopodial growth habit and the presence of numerous photosynthetic lateral branches was an increase in the vigor and robustness of the plants. Individuals of *Pertica quadrifaria* had stems 1.5 cm wide and about 1 m tall, whereas those of *P. dalhousii* are estimated to have been as much as 3 m tall, about the height of tall shrubs today. Although a dense stand of *P. dalhousii* would not be called a forest, there would have been vertical stratification of light, air movement, and humidity within the canopy.

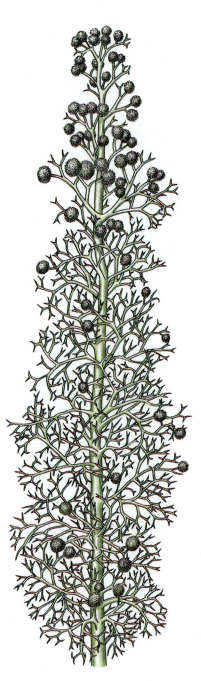

FIGURE 23.24

Pertica quadrifaria had one main trunk from which grew small branches; the smallest twigs still show dichotomous branching, but the larger stems branched pseudomonopodially. This branch pattern also evolved separately in the lycophytes, but here in the trimerophytes, it is the ancestor to the stem structure of seed plants. The globular structures are clusters of sporangia.

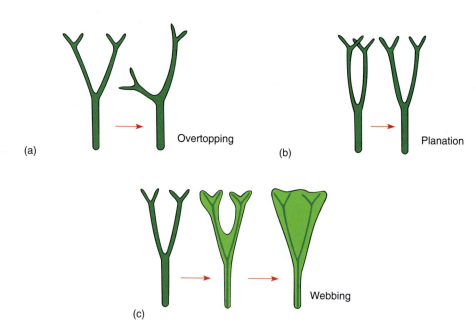

FIGURE 23.25

The leaves of ferns, *Equisetum*, and seed plants are megaphylls that arose by the evolutionary modification of stems. (a) The first step was the elimination of equal dichotomous branching; instead, one branch (the main axis) overtops the other, which remains smaller and lateral. (b) Next, lateral branch systems restricted their branching to just one plane (planation); they stopped producing three-dimensional branch systems. (c) Finally, the spaces between close branches developed a thin sheet of chlorophyll-containing cells in a type of webbing.

The trimerophytes became distinct from the rhyniophytes during the Lower Devonian and existed until the Upper Devonian Period, then came to an end by evolving into the ancestors of modern seed plants, ferns, and the arthrophytes (see Chapter 24).

ORIGIN OF MEGAPHYLLS

At least three distinct types of analogous structures called leaves occur in plants: (1) leaves on the gametophytes of nonvascular plants; (2) enations/microphylls of zosterophyllo-phytes and division Lycophyta; and (3) **megaphylls,** leaves that evolved from branch systems and are present in all seed plants, ferns, and arthrophytes (horsetails, see Fig. 23.26). Megaphyll evolution is summarized by the **telome theory.** Imagine a plant like *Pertica* consisting of a main stem and smooth, cylindrical, dichotomously branching lateral stems (Fig. 23.24). The ultimate twigs, those of the last dichotomy, are known as **telomes.** Now imagine that all subdivisions of a lateral branch become aligned in one plane (**plana-tion;** Fig. 23.25) and that parenchyma develops between telomes and even lower branches (**webbing**). This is not a leaf, but it has suitable characteristics to be the ancestor to leaves (see Fig. 24.6). Also, the leaves we see on the trees and herbs of the flowering plants are the result of about 300 million years of evolutionary refinement. If the branch system involved in this evolution produced sporangia, the resulting structure would not be just a leaf, but rather a **sporophyll,** a sporangium-bearing leaflike structure. For one reason or another, the plants with this organization outcompeted those with microphylls, the lycophytes. This is not to say that the megaphyll type of organization itself was the main reason for their success. But today, megaphyllous plants are by far the more common.

DIVISION ARTHROPHYTA

The division Arthrophyta (also called Sphenophyta) contains several genera of extinct plants and one genus, *Equisetum,* with 15 extant species known as horsetails or scouring rushes (Table 23.8). The living plants are all herbs without any secondary growth, and although certain species may attain a height of up to 10 m, they are usually less than 1 m tall (Fig. 23.26). Their aerial stems have a characteristic jointed structure, with a whorl of fused leaves at the nodes. The leaves are small and have just a single trace of vascular tissue, but they are small megaphylls, not microphylls. If branches are present, they alternate with the leaves rather than being located in the leaf axils. These are important pieces of evidence that arthrophytes are distinct from ferns and seed plants. The vertical, aerial stems arise from deep subterranean rhizomes; in harsh habitats, aerial shoots die during winter, but rhizomes persist and plants can spread vigorously. True roots are present, being produced at the rhizome's nodes. The evolutionary origin of roots is still uncertain.

(a)

(b)

FIGURE 23.26
Equisetum plants are small, having vertical shoots that arise from subterranean, highly branched rhizomes. In some species such as *E. debile* (a), each shoot is both photosynthetic and reproductive; in other species such as *E. telmateia* (b), two distinct types of shoots are produced. *(a, R. and L. Mitchell; b, W. E. Ferguson)*

TABLE 23.8	*Classification of Division Arthrophyta*
Order Hyeniales*	
Family Hyeniaceae	*Hyenia*
Order Calamitales*	
Family Calamitaceae	*Arthropitys, Astero-phyllites, Calamites*
Order Sphenophyllales*	
Family Sphenophyllaceae	*Sphenophyllum*
Order Equisetales	
Family Equisetaceae	*Equisetum, Equisetites**

* All species in this group are extinct.

FIGURE 23.27

The stem anatomy of *Equisetum* is dominated by numerous canals. The center of the pith (this is a siphonostele) is torn apart, forming a central canal or pith canal, and around the remnants of the pith are carinal canals formed by the breakdown of protoxylem. Just exterior to the carinal canals are metaxylem and phloem; vascular bundles are endarch. No secondary growth occurs in any living *Equisetum*. The cortex contains cortical canals that alternate with the carinal canals, each cortical canal being located near a furrow in the stem surface. The adaptive value of these canals is not known, but this highly characteristic structure allows us to identify fossils related to *Equisetum*. (*Jim Solliday/ Biological Photo Service*)

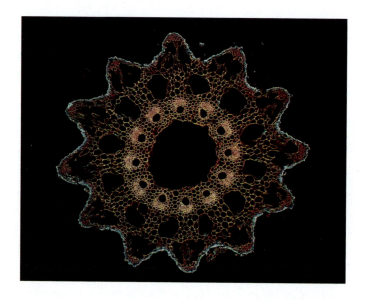

Sporangiophores

FIGURE 23.28

The sporangia of *Equisetum* occur in groups of five to ten clustered together on a sporangiophore. Sporangiophores are grouped into strobili, although in strobili of lycophytes, sporangia occur individually.

Internal structure is as distinctive as external morphology (Fig. 23.27). Stems have a pith, so these are siphonosteles, not protosteles as in all other plants mentioned so far (compare Fig. 23.27 with Fig. 23.5). Protoxylem forms next to the pith, interior to the metaxylem, so it is endarch, a rhyniophyte/trimerophyte trait distinctly different from the exarch protoxylem of zosterophyllophytes. Vessels are rare outside the flowering plants, occurring only in *Selaginella,* four ferns, three genera of gymnosperms, and *Equisetum.* Stem elongation causes some cells to be stretched and torn, forming canals. The cortex of *Equisetum* is composed of large cells, the outer ones chlorophyllous. Air canals may be located in the outer cortex. The epidermis contains stomatal pores and guard cells, as well as such large amounts of silica that the stems are tough and rigid.

Reproductive structures in *Equisetum* are specialized; sporangia never occur singly, but always in groups of five to ten located on an umbrella-shaped **sporangiophore** (Fig. 23.28). This has a short stalk and a flat, shield-shaped head from which the sporangia project, parallel to the sporangiophore stalk. The sporangiophores are always arranged in compact spirals forming a strobilus. *Equisetum* is homosporous, so all plants have only one type of sporangiophore and strobilus, but in some species, the strobilus occurs at the tip of a green photosynthetic shoot, whereas in others it is borne on a special, colorless reproductive shoot.

After spores are released, they germinate on moist soil and develop into small (1 mm across) green gametophytes. They have no vascular tissue at all and no epidermis or stomata; the body is just a small mass of parenchyma. Whereas the gametophytes of *Psilotum* are still somewhat complex, those of *Equisetum,* ferns, and seed plants are all small and simple. The gametophytes of *Equisetum* are either male or bisexual. Antheridia release numerous multiflagellate sperms that swim to the archegonia and eggs. The megagamete and zygote are never released, but rather are retained and nourished by the gametophyte.

The evolutionary line of arthrophytes can be traced back through the trimerophytes to the rhyniophytes (Table 23.8). In two groups of early arthrophytes, the Sphenophyllales and the Calamitales, a vascular cambium produced secondary xylem. Presumably secondary phloem was produced also, but the preservation of the fossils is not good enough to be certain (Figs. 23.29 and 23.30). These plants became large trees up to 30 cm in diameter and more than 20 m tall. They had true monopodial growth, a main trunk, lateral branches, true leaves, and true roots.

The arthrophyte vascular cambium evolved independently from that of the lycophytes, yet the two suffered from the same defect: The fusiform initials could not undergo radial longitudinal division to produce more fusiform initials. As the wood accumulated and pushed the cambium outward, the arthrophyte fusiform initials finally became too large to function, and secondary growth ceased.

(a)

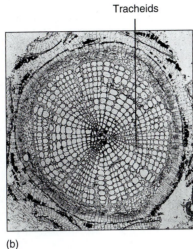

Tracheids

(b)

FIGURE 23.29

(a) This reconstruction of a *Sphenophyllum* shows the whorled leaves characteristic of arthrophytes; a strobilus is also visible. *(B-83047c Field Museum of Natural History, Chicago)* (b) Many extinct arthrophytes were treelike, having a wood of large tracheids. As in the lycophyte vascular cambium, the arthrophyte fusiform initials could not undergo longitudinal radial divisions, so no new fusiform initials could be formed and secondary growth was limited (×100). *(Courtesy of D. A. Eggert, University of Illinois)*

FIGURE 23.30

Calamites was an obvious relative to the living equisetums, having jointed stems with whorls of leaves or branches at each node. A plant like this would have been a small tree about 10 m tall.

DIVISION PTERIDOPHYTA

Aside from the arthrophyte line of evolution out of the trimerophytes, probably two more led to the plants that dominate Earth at present: ferns (division Pteridophyta) and seed plants (the divisions of the next chapters; see Fig. 23.42).

Ferns first appeared in the Devonian Period and then diversified greatly (Figs. 23.31 and 23.32). The division Pteridophyta is still large, with many modern families, genera, and at least 10,000 species (Table 23.9). Ferns can be found in almost any habitat. Moist,

FIGURE 23.31

This reconstruction of a fern ancestor, *Rhacophyton*, shows spirally arranged branch systems that had evolved nearly to the point of being called leaves. Note that some are fertile with sporangia and some are sterile; the former evolved into sporophylls, the latter into foliage leaves.

FIGURE 23.32

A later fern, *Phlebopteris*, from the Triassic Period. Even though the Triassic occurred from 230 to 190 million years ago, plants had evolved so many modern features that this fossil is not an ancestor to ferns in general, but to a particular family of living ferns, the Matoniaceae. *(Courtesy of T. Delevoryas, University of Texas, and R. C. Hope)*

BOX 23.2 *Form Genera*

Although we can be fairly sure that a plant like *Sciadophyton* was the gametophyte of a plant like *Rhynia* or *Cooksonia*, we can probably never know for certain. These fossils are the separate pieces of a life cycle, just as the individual fossils of spores, leaves, fruits, flowers, stems, and roots are the pieces of plants. How do we know which fossil parts came from the same species? As a plant dies, its leaves and flowers usually abscise, so we have numerous individual leaf and flower fossils mixed with the remains of bare stems. It is rare to find, at present or in the past, entire meadows or forests that consist of just a single species; usually many species grow mixed together. We find many types of fossil leaves mixed with many types of fossil stems, flowers, wood, and other plant parts. If we find a branch with leaves or flowers still attached, that is, if there is an **organic connection**, we can establish which organs are part of one plant.

Until we know which parts were portions of the same species, we must use **form genera**, which are created for types of isolated organs, tissues, spores, or pollen. If a fossil leaf is found which appears to be distinct from all others, it is named as a new form genus of leaf. As more fossil leaves are found in rocks of similar age, some appear to be related to the first and are assigned to that form genus. Mixed with the leaves may be twigs or large pieces of wood. We may be almost certain that the leaves have been produced by the twigs, based on this corre-lation of common occurrence at the same time and the same place. However, the leaves cannot be assigned to the form genus of the wood unless they are actually found still connected to a twig. Only when an organic connection has been established can two form genera be combined. For spores and pollen, no organic connection ever occurs, but occasionally an unopened sporangium or anther is found, so certain spores or pollen grains can be associated with those sporangia or anthers. Even when two form genera have been shown to be parts of the same species, it is still often much simpler to continue using the names of the form genera.

TABLE 23.9 *Classification of Division Pteridophyta*

Class Cladoxylopsida*

Class Coenopteridopsida*

Class Ophioglossopsida
  Order Ophioglossales
    Family Ophioglossaceae   *Ophioglossum, Botrychium*

Class Marattiopsida
  Order Marattiales
    Family Marattiaceae   *Marattia, Angiopteris, Psaronius**

Class Filicopsida
  Order Filicales
    Family Schizaeaceae   *Schizaea*
    Family Gleicheniaceae   *Gleichenia*
    Family Osmundaceae   *Osmunda*
    Family Matoniaceae   *Matonia*
    Family Polypodiaceae†   *Adiantum* (maidenhair fern),
        Asplenium (spleenwort),
        Blechnum, Cheilanthes,
        Dryopteris (shield fern),
        Pellaea, Platycerium,
        Polypodium, Polystichum (sword fern),
        Pteridium (bracken fern),
        Woodsia
    Family Cyatheaceae   *Cnemidaria, Cyathea*
    Family Hymenophyllaceae   *Trichomanes*
  Order Marsileales
    Family Marsileaceae   *Marsilea, Regnellidium*
  Order Salviniales
    Family Salviniaceae   *Azolla, Salvinia*

* All species in this group are extinct.

† This family contains almost all the living species of ferns.

FIGURE 23.33

(a) Many ferns grow in hot, dry climates with little rainfall. If transplanted to habitats often described as typical for ferns—cool, moist, and shady—these would die within days. *Azolla* (b) and *Salvinia* (c) do not look like ferns at first glance, but the structure of their leaves and sporangia shows that they are ferns. *(b and c, W. E. Ferguson)* (d) Several genera of ferns such as this *Cyathea* are referred to as tree ferns because they have very long vertical trunks and huge fronds. However, unlike trees, these never have secondary growth and are actually giant herbs. *(Patti Murray/Earth Scenes)*

shady forests and lakesides are often considered "typical" fern habitats, but species of *Woodsia* and *Cheilanthes* occur in dry, hot deserts; *Salvinia* and *Azolla* grow floating on water; and *Ceratopteris* lives submerged below water (Fig. 23.33). Other genera contain epiphytes (*Polypodium*) or vines (*Lygodium*). All ferns are perennial and herbaceous; none is woody, but some do achieve the size of small trees—the tree ferns *Angiopteris, Cyathea, Cnemidaria,* and others. Although called tree ferns, they never have secondary xylem.

The fern sporophyte consists of a single axis, either a vertical shoot or a horizontal rhizome, that bears both true roots and megaphyllous leaves. The vascular system of the stem is an endarch siphonostele, a derived trait similar to that of division Arthrophyta and the seed plants (Fig. 23.34). At each node, a leaf trace diverges from the siphonostele,

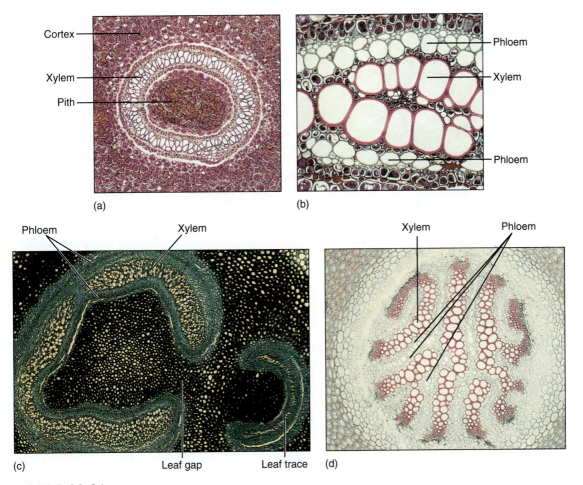

FIGURE 23.34

Anatomy of fern stems is variable—not only from species to species, but also within a single plant. (a) In this *Dennstaedtia* and many other ferns, the xylem forms a complete ring in transverse section. (b) Phloem occurs on both the exterior and interior of the xylem in ferns; phloem interior to xylem is extremely rare in seed plants ($\times$ 20). (c) In this *Adiantum*, the vascular tissue makes almost a complete cylinder. A leaf gap occurs where vascular tissue moves to the leaves. Leaf gaps are characteristic of ferns; they do not occur in the microphyll line of evolution (lycophytes). (d) Members of the microphyll line of evolution, such as this *Lycopodium annotinum,* do not have pith ($\times$ 20).

leaving a small segment of the vascular cylinder as just parenchyma; this region is a **leaf gap.** A vascular cambium has been reported to occur in one fern, *Botrychium.*

Leaves of ferns may be leathery or delicate, only one cell thick in the filmy ferns, but many layers thick in most. They have an upper layer of palisade parenchyma and a lower region of spongy mesophyll. The leaves are small (*Trichomanes*) or up to several meters long (tree ferns), but they are almost always compound with a rachis and leaflets (Fig. 23.35). The staghorn fern (*Platycerium*) and bird's-nest fern (*Asplenium*) have simple leaves. Fern leaf primordia have a distinct apical cell, unlike the leaves of seed plants, and as the primordium grows, it curves inward, producing a young leaf that is tightly coiled and commonly known as a fiddlehead (Fig. 23.36). As the leaf expands, it uncoils and becomes

FIGURE 23.35
Fern leaves (fronds) have a petiole and a lamina, the lamina being almost always pinnately compound. The midrib is a rachis, and the leaflets may also be compound.

FIGURE 23.36
Fern leaves have a highly characteristic development: The young leaves are tightly coiled and uncurl as they expand. This is true not only of the rachis, but of the leaflets as well. This circinate vernation does not occur in nonfern plants. *(Courtesy of S.-C. Hsiao, University of Texas)*

flat. Fern leaves, especially the larger ones, usually contain a considerable amount of vascular tissue. The veins typically branch dichotomously; they are never parallel as in the monocots. Although a type of reticulate venation does occur in a few ferns, it appears more like an imperfect dichotomous venation than the true reticulate venation of dicots.

In some species, for example *Blechnum spicant* and cinnamon fern (*Osmunda cinnamomea*), certain leaves serve only for photosynthesis and others serve only as sporophylls, but in most ferns, each leaf does both (Fig. 23.37). On the underside of the leaf are **sori** (sing.: sorus), clusters of sporangia where meiosis occurs (Fig. 23.38). Most ferns are homosporous; only two groups of water ferns (Marsileales and Salvineales) are heterosporous.

When fern spores germinate, they grow into small, simple heart-shaped or ribbon-shaped photosynthetic gametophytes with unicellular rhizoids on the lower surface but with no vascular tissue and no epidermis (Fig. 23.39). Each usually bears both antheridia and archegonia. Antheridia develop early and can be difficult to detect because they occur among the rhizoids; archegonia develop later, close to the gametophyte apex. When the environment is sufficiently moist, antheridia release motile sperm cells that could easily be mistaken for unicellular green algae except for the absence of plastids. These swim to the archegonia and fertilize the egg.

FIGURE 23.37

(a) This cinnamon fern (*Osmunda cinnamomea*) contains both foliage leaves that never bear sporangia and highly distinct sporophylls that do bear sporangia. (*John N. A. Lott/Biological Photo Service*) (b) Most ferns, such as this *Pellaea ovata*, have only one type of leaf; its broad, thin shape is selectively advantageous for photosynthesis, and it bears sporangia on the underside. (*Robert and Linda Mitchell*) (c) In almost all ferns such as this sword fern (*Polystichum*), the sporangia occur in clusters, sori, covered by an indusium. (*W. E. Ferguson*)

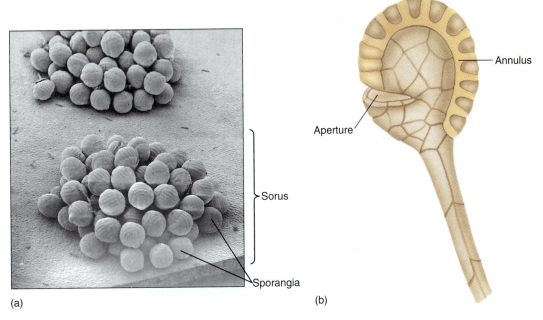

FIGURE 23.38

(a) A sorus contains many sporangia, each of which consists of a stalk and the actual sporangium body where meiosis occurs. (*Biophoto Associates/Photo Researchers, Inc.*) (b) The sporangium opens when specialized cells of the annulus dehydrate, shrink, and suddenly crack open, expanding rapidly and throwing the spores (× 40).

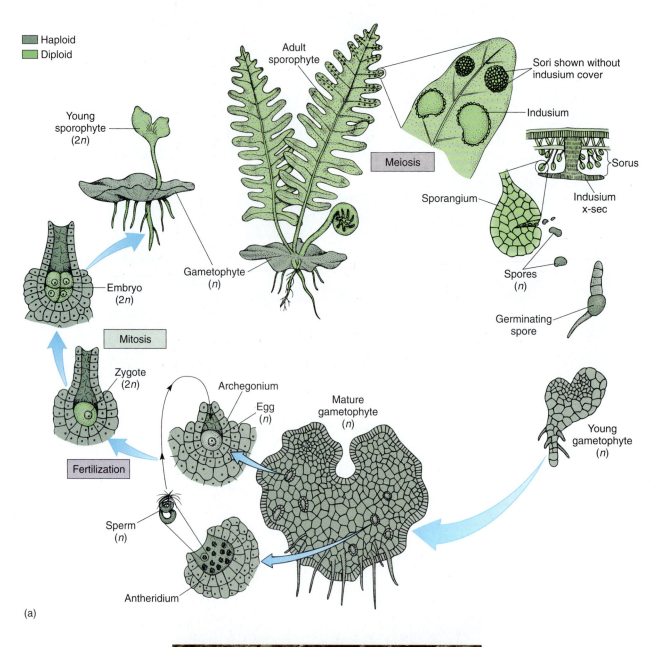

FIGURE 23.39

(a) Life cycle of a fern. Details are given in the text. (b) Fern gametophytes, several with the first leaves of young sporophytes. (*Courtesy of S.-C. Hsiao, University of Texas*)

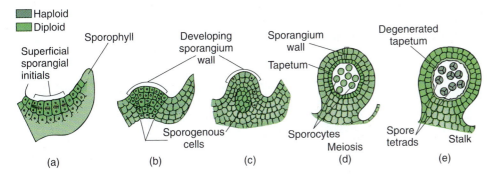

FIGURE 23.40

Development of eusporangia. (a) Surface cells undergo divisions and the inner daughter cells become sporogenous. (b to e) Further divisions result in a large number of spores and a thick sporangium wall.

As in *Equisetum,* the zygote of a fern is retained by the gametophyte that nourishes it. However, the young sporophyte soon produces its first leaves and roots and becomes independent. Its continued growth eventually destroys the small gametophyte. Fern reproduction, with swimming sperms, is still aquatic, but the plants are able to grow in almost any environment as long as there can be at least a temporary high density of gametophytes and a film of dew or rainwater.

Eusporangia and Leptosporangia. Ferns contain two types of sporangia that differ in fundamental aspects of their development. The **eusporangium** is initiated when several surface cells undergo periclinal divisions, resulting in a small multilayered plate of cells (Fig. 23.40). The outer cells develop into the sporangium wall, and the inner cells proliferate into sporogenous tissue. This results in a relatively large sporangium with many spores. Anthers and ovules of flowering plants develop this way as well, and the eusporangium is considered to be the basic type in vascular plants.

Leptosporangia are initiated when a single surface cell divides periclinally and forms a small outward protrusion (Fig. 23.41). This undergoes several more divisions, which result in a small set of sporogenous cells and a thin covering of sterile cells. The leptosporangium produces just a few spores.

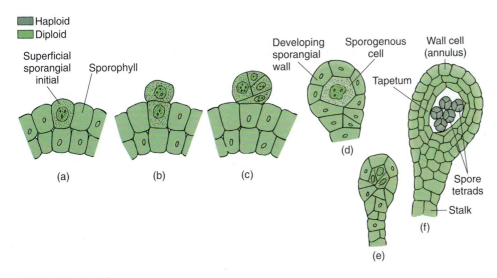

FIGURE 23.41

Development of leptosporangia. (a and b) The sporangium is initiated by division in just a single surface cell. (c to f) After a few cell divisions, the sporangium has a stalk, a thin wall, and only a few spores.

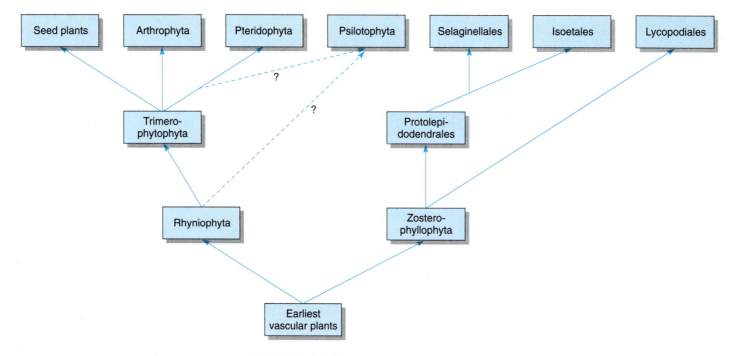

FIGURE 23.42

The lycophytes—lycopods, selaginellas, *Isoetes*—diverged very early from all other living vascular plants. The latter originated by way of the trimerophytes; of this group, the arthrophytes became distinct quite early, as did the ferns. The line labeled "Seed Plants" is discussed in the next two chapters. However, their trimerophyte ancestors were so advanced that seed plants and ferns share many fundamental aspects of biology. The most significant difference is that in the seed plant line, a vascular cambium evolved (for the third time), as did true seeds.

Because most plants have eusporangia whereas leptosporangia occur only in some ferns, we believe that having eusporangia is a relictual condition. Leptosporangia must have evolved as a derived condition after the early ferns had come into existence. At present, only two orders of ferns, Ophioglossales and Marattiales, still retain eusporangia, whereas all other living ferns are leptosporangiate (Table 23.9).

The plants of this chapter—vascular plants without seeds—show great diversity, which at first may seem overwhelming. However, there are only two fundamental lines of evolution out of the rhyniophytes/zosterophyllophytes: the microphyll line (lycophytes) and the megaphyll lines (all the rest except *Psilotum;* Fig. 23.42). The second line originated in rhyniophytes, produced the trimerophytes, then split into arthrophytes, ferns, and seed plants.

SUMMARY

1. Vascular plants first appeared about 420 million years ago, having evolved from green algal ancestors. Necessary modifications would have included methods of water conservation: low surface-to-volume ratio, compact body, cuticle, and coordination of reproduction with presence of water.

2. The presence of plants on land changed the land as a habitat; with crowding, competition for light would have become important and an upright habit selectively advantageous.

3. The evolution of vascular tissue had profound consequences: Relatively strong, vertical stems could be supported, large plant bodies could be integrated, and parts could differentiate and specialize.

4. Early gametophytes were the most elaborate ever, and all subsequent gametophyte evolution has involved simplification and reduction.

5. Vascular cryptogams are embryophytes, not algae, because of their multicellular, jacketed sporangia and gametangia. They have vascular tissue, so they are not mosses, liverworts, or hornworts. Although some have small endosporial megagametophytes, none is retained in the sporangium, so none is a seed.

6. The earliest vascular plants are the rhyniophytes, plants with simple dichotomously branched stems, no leaves, and endarch protoxylem.

7. Zosterophyllophytes were similar to rhyniophytes but had exarch protoxylem and lateral sporangia that opened across their tops. Zosterophyllophytes probably were the ancestors to lycophytes.

8. Division Psilotophyta contains only two living genera; although *Psilotum* strongly resembles *Rhynia,* no fossil evidence exists to link rhyniophytes and psilotophytes.

9. Three types of leaf occur in the plant kingdom: gametophyte leaves of nonvascular plants, microphylls of lycophytes, and megaphylls of all nonlycophyte vascular plants. Microphylls evolved from enations, whereas megaphylls are modified telomes—branch systems.

10. Vascular cambia have evolved at least three times; the two that evolved in ancient lycophytes and arthrophytes suffered from the defect that their cells could not divide with a radial longitudinal wall.

11. Trimerophytes were transitional between rhyniophytes and arthrophytes, ferns, and seed plants. Evolutionary advances in the trimerophytes were pseudomonopodial branching and the first steps in megaphyll evolution.

12. Division Arthrophyta contains only one small genus of living, herbaceous species and several extinct genera of large, complex, treelike plants. All had a characteristic jointed structure and various systems of canals.

13. Division Pteridophyta contains the ferns, with many living species and tremendous diversity. No ferns have secondary growth or seeds, but all have megaphylls.

IMPORTANT TERMS

cone
dichotomous branching
enation
endarch xylem
exarch xylem
leaf gap
leaf trace
megaphyll

microphyll
monopodial branching
overtopping
planation
protostele
rhyniophytes
seed plants
siphonostele

sorus
spermatophytes
sporophyll
strobilus
telome theory
vascular cryptogams
webbing
zosterophyllophytes

REVIEW QUESTIONS

1. What are some of the modifications necessary if an alga is to become evolutionarily adapted to living on land? Is a single modification sufficient, or are several necessary?

2. Why would it be necessary for an evolutionary line to develop stomata and guard cells before it developed an extremely impervious cuticle? Why must vascular tissues precede the evolution of roots and active apical meristems?

3. Draw a plant of *Rhynia*; be certain to include the reproductive parts. Assume that it has a life cycle with an alternation of isomorphic generations and draw a complete life cycle. If *Rhynia* or its contemporaries were the ancestors to the ferns, how did the gametophytes and sporophytes change during evolution?

4. What are the zosterophyllophytes? How did they differ from rhyniophytes? Why do we think they are related to lycophytes but not to ferns and seed plants?

5. Describe the evolution of microphylls and megaphylls. What are enations and telomes?

6. What three groups had a vascular cambium? In which two was there a problem, and what was the problem? Would it be a problem in an environment in which a shrubby body is more advantageous selectively than a large, treelike body?

7. Describe the trimerophytes. From what group did they evolve and what lines of evolution did they produce? Even though all rhyniophytes and trimerophytes are now extinct, would you consider them unsuccessful?

BotanyLinks includes a Directory of Organizations for this chapter.

24

SEED PLANTS I: GYMNOSPERMS

The seeds of conifers are borne in cones, not in fruits. *(D. Cavagnaro)*

CONCEPTS

The life cycle of vascular cryptogams is an alternation of independent, heteromorphic generations. A disadvantage of this life cycle is that the new sporophyte is temporarily dependent upon a tiny gametophyte for its start in life. Consequently, many new sporophytes perish. All the genes of each gametophyte and half the genes of each new sporophyte embryo are identical to those of the maternal sporophyte, so any mutation that improves the survival of the gametophytes or embryos contributes to the reproductive success of the maternal sporophyte. It would be advantageous if the embryo could use the photosynthetic and absorptive capacity of the leaves and roots of the previous sporophyte. In order for this to happen, the megagametophyte and embryo must be retained inside the maternal sporophyte; this is accomplished by retaining the megaspore and allowing the megagametophyte to develop within the sporangium (Fig. 24.1a to d). Such an arrangement requires some alteration in the microgametophyte as well, because retention of the megagametophyte changes its position. Free-living gametophytes develop on the soil, whereas retained ones develop high on the plant in strobili, a position that cannot be reached by swimming sperm cells produced by microgametophytes living at the soil surface (Fig. 24.1e). The modifications that overcame this problem are simple: The megasporophylls with the megagametophytes were arranged in upright conelike structures, allowing microspores to be carried by

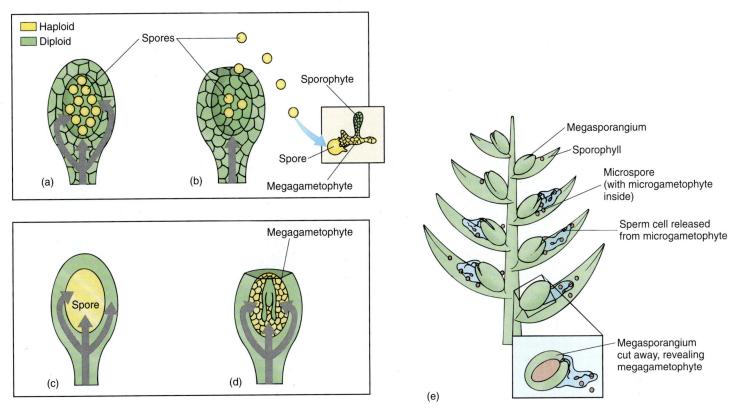

FIGURE 24.1

(a and b) Plants that release megaspores. Only limited amounts of nutrient can be placed in a spore, and after spore release (b) the sporophyte can do nothing more to nourish or protect the gametophytes or new sporophytes. All the genes of each gametophyte and half the genes of each new sporophyte are identical to those of the parental sporophyte, so any mutation that improves the survival of the gametophytes or young progeny sporophytes contributes to the reproductive success of the parental sporophyte. (c and d) Plants that retain megaspores. Mutations that cause a reduction in the number of megaspores and their retention in the sporangium can be selectively advantageous. The sporophyte can nourish and protect its gametophytes and also the subsequent progeny sporophytes, at least for a while. Whereas the free gametophytes in (b) might be killed by a brief drought, those of (d) are provided with water by the root and vascular system of the parent. Also, the sporangium wall may have a sclerenchyma sheath or antiherbivore chemical defenses that protect the spores, gametophytes, and new sporophytes; such protection is not available in free-sporing species once the spores are released. In (d), the embryo is shown embedded in tissue of the megagametophyte; this occurs in gymnosperms—seed plants that are not flowering plants. In flowering plants, the tiny megagametophyte is quickly replaced by endosperm. (e) Megasporangia that are packed together in a cone composed of sporophylls automatically have a means of catching and retaining microspores. These germinate into microgametophytes with antheridia that release sperms within a millimeter or two of the megagametophyte and its archegonia.

wind and dropped into the megasporangiate cone. There they would germinate into a microgametophyte, produce their antheridia and sperm cells, and carry out fertilization. Fertilization itself has changed little, with a moist sporophyll replacing moist soil. These changes produced the first seed plants, the division Pteridospermophyta or seed ferns and the early members of division Coniferophyta, known as class Cordaitles. These two groups are now extinct, but their descendants still exist as the living seed plants. In several lines of evolution, those ending in the cycads and in *Ginkgo,* this method of fertilization still exists, and beautiful swimming sperms are produced. In other lines, those leading to conifers and to flowering plants, sperms have become nonmotile and are carried to the egg by growth of the microgametophyte as a pollen tube.

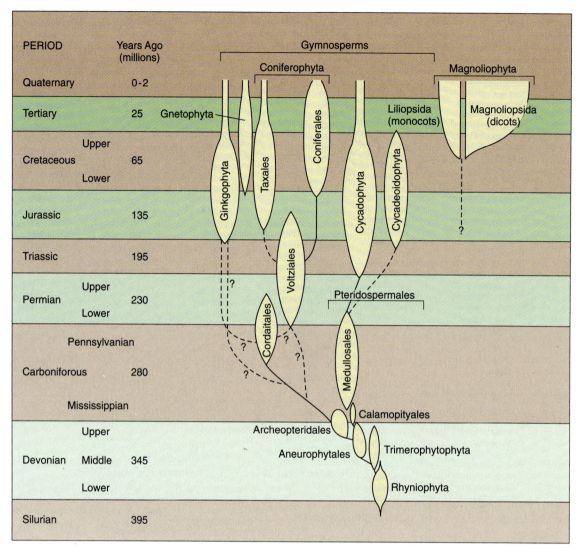

PERIOD		Years Ago (millions)	
Quaternary		0-2	
Tertiary		25	
Cretaceous	Upper	65	
	Lower		
Jurassic		135	
Triassic		195	
Permian	Upper	230	
	Lower		
Carboniforous	Pennsylvanian	280	
	Mississippian		
Devonian	Upper	345	
	Middle		
	Lower		
Silurian		395	

FIGURE 24.2

Proposed phylogeny of seed plants based on the most recent analysis of fossils and living plants. Rhyniophytes are the basic group, and then trimerophytes, which gave rise to several lines, the most important for us being Aneurophytales. An early split occurred into an evolutionary line (conifers and related groups) that would never form flowers and a second line (flowering plants and related groups) that would. The many dashed lines indicate uncertainty; many of these plants have so many similar, derived features that it is not easy to be certain how they are interrelated. Many important evolutionary advances involved subtle changes in metabolism and embryo nutrition, characters not preserved in fossils. (See also *Paleobotany and the Evolution of Plants* by W. N. Stewart and G. W. Rothwell, 1993, Cambridge University Press; and *Paleobotany: An Introduction to Fossil Plant Biology* by T. N. Taylor and E. L. Taylor, 1993, Prentice-Hall.)

An old classification from the 1800s grouped all the seed plants together in a single division, Spermatophyta, with two classes, class Gymnospermae and class Angiospermae. The **gymnosperms** are those plants with "naked ovules," that is, ovules located on flat sporophylls, not enclosed in carpels. **Angiosperms** are the flowering plants, those with carpels, which are believed to be sporophylls that form a tublelike, closed structure. This classification emphasized both the close relationship of all seed plants and the idea that the evolution of the flower was the most significant event in this group. But new fossil evidence indicates that seed plants are a much more ancient group than previously thought, and many evolutionary developments were occurring: the origin of seeds and cones, the origin of a vascular cambium and secondary growth, the origin of vessels, and the origin of many defensive chemical compounds. The different types of gymnosperms are more distinct than we realized before, and some must have evolved to be the ancestors of flowering plants (Fig. 24.2). Consequently, many botanists believe that the old division "Spermatophyta" and classes "Gymnospermae" and "Angiospermae" should be abandoned, and each major group should be elevated to division status. Other botanists believe that these various lines are still rather closely related and form a natural group and should therefore be held together in one division. Either way, the concept of "gymnosperm" is still so useful that it is constantly employed, at least informally, and the word "angiosperm" is still used to refer to flowering plants. The divisions of living seed plants commonly accepted now and used in this book are (1) division Coniferophyta, (2) division Cycadophyta, (3) division Ginkgophyta, (4) division Gnetophyta (these four are gymnosperms), and (5) division Magnoliophyta (the flowering plants). Numerous extinct groups exist, the most significant for us being progymnosperms and seed ferns.

DIVISION PROGYMNOSPERMOPHYTA: PROGYMNOSPERMS

As discussed in the preceding chapter, trimerophytes were an important group of plants that evolved from rhyniophytes (see Table 23.7). One evolutionary trend initiated in trimerophytes was pseudomonopodial branching, in which one branch dominates the plant as a trunk and other branches are smaller and act as lateral branches (see Fig. 23.24). Branch orientation was controlled such that regular spiral, decussate, and distichous branch systems resulted, various species of the genus *Pertica* being excellent examples. As the terminal branches became webbed, the first stages of megaphyll evolution were established. These evolutionary trends were continued in horsetails (division Arthrophyta) and ferns (division Pteridophyta), which evolved out of trimerophytes.

A third group to evolve from trimerophytes was the now-extinct **progymnosperms**, so named because some gave rise later to conifers, cycads, and the other gymnosperms (Table 24.1). Like ferns and horsetails, progymnosperms also developed megaphyllous leaves. However, another feature was just as significant—the evolution of a vascular cambium with unlimited growth potential and capable of producing both xylem and phloem.

Vascular cambia had also arisen in lycophytes and arthrophytes, but in both cases those cambial cells could not undergo radial longitudinal division (see Fig. 23.20b and Box 8.2). As wood accumulated, cambial cells were stretched tangentially and finally stopped functioning while the stem was still relatively narrow, less than 50 cm in diameter. Also, the vascular cambium produced either little or no secondary phloem.

The vascular cambium that evolved in progymnosperms was different. As early as 360 million years ago it was capable of undergoing radial longitudinal divisions, so it could function indefinitely, producing large amounts of both secondary xylem and phloem (Figs. 24.3 and 24.4). The wood was almost indistinguishable from that of many living conifers. It contained elongate tracheids, most with circular bordered pits. It had little or no axial parenchyma. Rays were tall and uniseriate and consisted of procumbent ray tracheids. This wood must have been strong, efficient at conduction, and capable of supporting a large

---

TABLE 24.1 *Classification of Division Progymnospermophyta**

Order Aneurophytales	
Family Aneurophytaceae	*Aneurophyton, Eospermatopteris, Proteokalon, Protopteridium, Tetraxylopteris, Triloboxylon*
Order Archaeopteridales	
Family Archaeopteridaceae	*Archaeopteris, Callixylon*

* All species of this group are extinct.

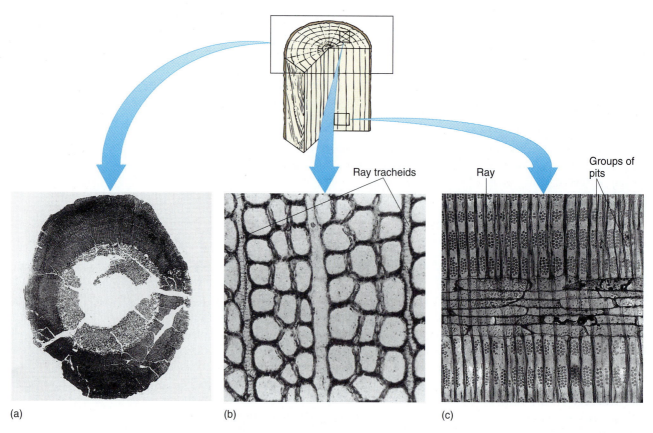

FIGURE 24.3

(a) Unlike the small amounts of wood produced in arborescent lycophytes (see Fig. 23.20a) and arthrophytes (see Fig. 23.29b), the vascular cambium of progymnosperms was able to produce a virtually unlimited amount of both secondary xylem and secondary phloem, as in this *Callixylon brownii.* (b and c) Progymnosperm wood had large vertical tracheids with pitted walls and an abundance of rays with ray tracheids. Ray tracheids do not occur in flowering plants but are common in conifers as horizontal tracheids that permit rapid conduction horizontally in the sapwood. (b) *Callixylon erianum* (× 700). (c) *C. newberryi* (× 375). *(Courtesy of C. B. Beck, University of Michigan)*

mass of leaves and branches. At least some species, *Triloboxylon* and *Proteokalon* for example, had a cork cambium that produced bark. Progymnosperms produced true woody trees: Trunks were up to 1.5 m in diameter and 12 m tall.

Although progymnosperm wood was similar to that of gymnosperms, the two groups must be kept separate because progymnosperms did not have seeds or even ovule precursors. The species that arose later were heterosporous, but megaspores were shed through longitudinal slits in the sporangia; they were not retained in an indehiscent sporangium as occurs in seed plants. Although leaves and wood of progymnosperms were quite advanced, their reproduction was remarkably simple.

ANEUROPHYTALES

The order Aneurophytales contains the more relictual progymnosperms, such as *Aneurophyton, Protopteridium, Proteokalon, Tetraxylopteris, Triloboxylon,* and *Eospermatopteris* (Fig. 24.5). They varied in stature from shrubs (*Protopteridium, Tetraxylopteris*) to large trees, up to 12 m tall. They all had a vascular cambium and secondary growth, but the primary xylem of the stems was a protostele, that is, a solid mass of xylem without pith; rhyniophytes and trimerophytes also had protosteles. The aneurophytes further resembled trimerophytes in having little webbing between their ultimate branches—telomes—and

FIGURE 24.4
Reconstruction of *Archaeopteris,* a small tree about 6 m tall. Although extinct now, *Archaeopteris* flourished about 360 million years ago across northern North America, Canada, and Europe.

(a)

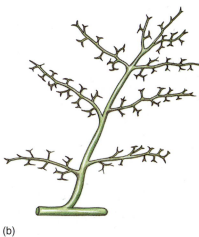

(b)

FIGURE 24.5
(a) *Eospermatopteris* was a member of the Aneurophytales; it had a trunk and what appear to be frondlike leaves, but these were just branch systems; planation and webbing had not yet evolved. (b) A portion of the "leaf" of *Aneurophyton.* Planation had begun, but no webbing or sheet of photosynthetic tissue had yet appeared.

only in *Proteokalon* were telomes arranged in a plane. These could not yet be called leaves. These species were homosporous, all spores being produced in sporangia located at branch tips.

ARCHAEOPTERIDALES

A more derived progymnosperm was *Archaeopteris* in the order Archaeopteridales. These were trees up to 8.4 m tall with abundant wood and secondary phloem (see Fig. 24.4). Unlike aneurophyte stems, the stems of *Archaeopteris* had a siphonostele, pith surrounded by a ring of primary xylem bundles, much like modern conifers and dicots. Although the "fronds" of archaeopterids resembled fern fronds, close examination reveals that they were actually planated branch systems and that the ultimate "pinnules" were really spirally arranged simple leaves. Webbing is only partial in *A. macilenta* (Fig. 24.6), but it is complete in *A. halliana* and *A. obtusa;* these can be considered full-fledged megaphylls. Whether the leaves were annual or perennial is not known; the branches that bore the leaves did not undergo secondary growth but rather remained herbaceous.

Reproduction in archaeopterids was still simple, but it was heterosporous. Megaspores measured up to 300 μm in diameter, microspores only 30 μm in diameter, and each type was produced in its own distinctive sporangium—broad megasporangia and slender microsporangia. The sporangia were terminal on short branches that were mixed with sterile, leaflike branch systems. Megaspores were released from the sporangia, not retained. Seeds were not produced.

Within the progymnosperms the fundamental characters of stems, leaves, and secondary tissues had been established. The current hypothesis is that the Archaeopteridales later gave rise to the ancestors of modern conifers, and the Aneurophytales produced the ancestors of cycads and related plants (see Fig. 24.2).

EVOLUTION OF SEEDS

Investigating the life cycle of extinct plants based on fossil material is difficult, but not impossible. In free-sporing species, spores can be identified with sporophytes only if some spores are trapped in a sporangium that was attached to leaves or wood during fossilization. Spores cannot be identified with gametophytes except when the gametophyte is

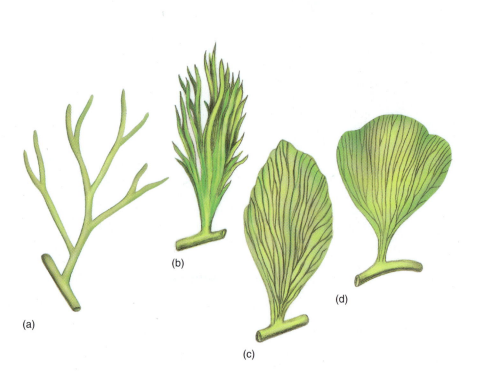

FIGURE 24.6
During the existence of the plants classified in the extinct genus *Archaeopteris,* the final transitions from telomes to leaves occurred. (a) *Archaeopteris fissilis.* (b) *A. macilenta.* (c) *A. halliana.* (d) *A. obtusa.* What appear to be veins in the leaves were actually the shoot axes of the telomes.

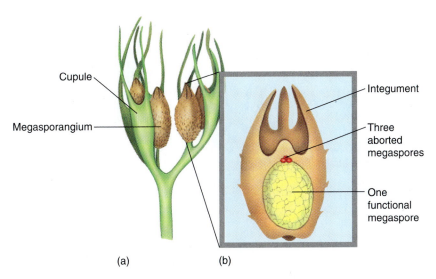

FIGURE 24.7

A reconstruction of the megasporangium (b) and adjacent telomes (a) of *Archaeosperma arnoldii.*
Each sporangium contained only one large megaspore; the other three products of the spore
mother cell meiosis apparently degenerated early. Tissue immediately around the megaspore was
the megasporangium wall. Attached to it and extending upward as finger-like projections was a
layer called an integument. The exact nature of the integument is not known, but it may have
been derived from telomes adjacent to the sporangium (see Fig. 24.8). Above the sporangium was
a space surrounded by integument tissue, a calm, wind-free pollen chamber where pollen or
spores settled. Around each megasporangium and integument was another set of partially fused
telomes.

microscopic and develops within the spore wall. If megaspores are retained in the megasporangium, at least some of the investigation is easier.

At present, the earliest known progymnosperm species with heterospory is *Chauleria*
from the Middle Devonian Period, about 390 million years ago. It appears to have been a
member of the Aneurophytales, and its sporangia contained both small (30 to 40 μm) and
large (60 to 156 μm) spores. The slightly more recent Archaeopteridales (Upper Devonian
Period, 360 million years ago) were heterosporous and had distinct microsporangia and
megasporangia. In *Archaeopteris halliana* the number of megaspores in each megasporangium was low. Another Upper Devonian fossil, *Archaeosperma arnoldii,* was more derived:
The megasporangium produced only one megaspore mother cell, and this produced only
one large, viable megaspore and three small, aborted cells. The megasporangium was
surrounded by a layer of tissue, an **integument,** that projected upward. There was a large
micropyle, a hole in the integument that permitted the sperm cells to swim to the egg after
the megaspore had developed into a megagametophyte and had produced eggs (Fig. 24.7).

In *Archaeosperma* the integument was fused to the sporangium and was solid except at
the tip of the micropyle, where it was lobed. In this condition it is difficult to determine
how the integument arose evolutionarily, but in other fossils, early stages are present. In
Genomosperma kidstoni the megasporangium was closely surrounded by sterile telomes
(Fig. 24.8). In *G. latens* the telomes were fused at the base, and in *Eurystoma angulare* they
were similar to *Archaeosperma,* fused into one structure except at the tip. The sheath of
sterile branches must have been important in trapping windblown microspores, but at first
it would have allowed them to settle anywhere on the megasporangium. As they fused to
each other and to the megasporangium, the space at the top of the megasporangium
became the place where microspores settled, acting as a **pollen chamber** or holding area.

As microspores evolved into pollen grains, changes occurred in the nature of their
wall, in their method of germination, and in the nature of the microgametophyte they
produced. By the Carboniferous Period, just after the Upper Devonian Period (about 340
million years ago), four types of pollen had become common. In some, internal structure is

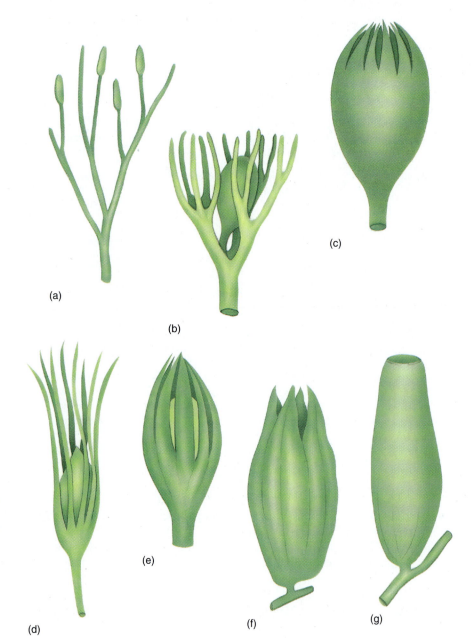

FIGURE 24.8
(a to c) A hypothetical evolution of an integument from telomes. In (a), sporangia are neither clustered nor particularly associated with sterile telomes. (b) One megasporangium is surrounded by sterile telomes that all retain their individuality. (c) The sterile telomes have fused at their bases, forming a protective sheath around the megasporangium. (d to g). Actual fossils that correspond to the hypothesis. (d) *Genomosperma kidstoni*. (e) *G. latens*. (f) *Eurystoma angulare*. (g) *Stamnostoma hyttonense*.

preserved well enough to see that these grains already had internal microgametophytes remarkably similar to those of modern gymnosperms.

As the megasporangia evolved into ovules with integuments, nearby telomes, either on the branches with the sporangia or on adjacent sterile branches, became modified into cupules (see Fig. 24.7a). These tended to be leaflike and to surround the ovules loosely, perhaps giving some protection and helping create wind patterns that would bring in pollen grains. The cupules may have later given rise to the carpel in angiosperms.

DIVISION CONIFEROPHYTA: CONIFERS

CONIFERALES

The conifers are the most diverse (about 50 genera and 550 species) and familiar of the gymnosperms (Fig. 24.9; Table 24.2). They are all trees of moderate to gigantic size; the giant redwoods of California (*Sequoiadendron giganteum*) reach 90 m in height and 10 m in diameter. Conifers are never vines, herbs, or annuals, and they never have bulbs or rhi-

FIGURE 24.9

Division Coniferophyta is large and diverse, containing many familiar plants. (a) Ponderosa pine (*Pinus ponderosa*) *(Runk/Schoenberger from Grant Heilman).* (b) American larch (*Larix laricina*), which is deciduous, losing its leaves in autumn. *(R. Stottlemyer, Michigan Technological University/Biological Photo Service)* (c) Big tree redwood (*Sequoiadendron giganteum*). *(J. N. A. Lott, McMaster University/Biological Photo Service)* (d) Garden juniper (*Juniperus chinensis* var. *procumbens*), which grows as a flat ground cover. *(B. L. Runk from Grant Heilman)*

TABLE 24.2	Classification of Division Coniferophyta
Order Cordaitales*	
Order Voltziales*	
Family Voltziaceae	*Lebachia*
Order Coniferales	
Family Pinaceae	*Abies* (firs), *Cedrus* (cedar), *Larix* (larch), *Picea* (spruce), *Pinus* (pine), *Pseudotsuga* (Douglas fir), *Tsuga* (hemlock)
Family Araucariaceae	*Agathis, Araucaria* (Norfolk Island pine and monkey puzzle)
Family Taxodiaceae	*Metasequoia* (dawn redwood), *Sequoia* (redwood), *Sequoiadendrion* (giant sequoia), *Taxoduim* (bald cypress)
Family Podocarpaceae	*Dacrydium, Podocarpus*
Family Cupressaceae	*Chamaecyparis, Cupressus* (cypress), *Juniperus* (juniper), *Libocedrus* (incense cedar), *Thuja* (arbor-vitae)
Family Cephalotaxaceae	*Cephalotaxus*
Order Taxales	
Family Taxaceae	*Taxus* (yew), *Torreya*

* All species of this group are extinct.

(a)

(b)

(c)

FIGURE 24.10
Conifer leaves are never compound. They are always simple and leathery, needle-like in some, scalelike in others. (a) In eastern or Canadian hemlock *(Tsuga canadensis),* each leaf has two white bands where stomata are located. *(Grant Heilman)* (b) Scalelike leaves of *Chamaecyparis lawsoniana. (William E. Ferguson)* (c) Pines have two types of leaves—small, papery brown scales on the long shoots (visible on the branch) and the long needles borne by the short shoots located in the axils of the papery scale leaves. The long needles were produced last year. White pine *(Pinus strobus). (E. R. Degginger)*

PLANTS & PEOPLE

ECONOMIC IMPORTANCE OF CONIFERS

Conifers are of major economic importance to humans. They occur in extensive forests that supply us with wood used for lumber and paper. In 1972, a peak year for consumption of wood products, we harvested 13.7 billion cubic feet of material from U.S. forests. The wood of conifers, especially pine, is particularly useful for home construction. Lumber from bald cypress is resistant to decay and is commonly used for piers and docks. Redwood is highly prized for its color and durability, so that entire groves are being cut faster than the trees can regrow. Unless demand is reduced, redwoods may finally exist only in a few state and national parks in California.

Chemicals are also obtained from conifers. One important chemical is cellulose itself; after being purified, it can be converted into rayon, cellophane, and other household products. Even more important are the resins extracted from conifer pitch. These are formed in resin canals inside wood and bark; if a tree is cut deeply enough to penetrate to sapwood, pitch flows out and can be harvested. If heated, the more volatile compounds evaporate and then condense into turpentine. Venetian turpentine, extracted from European larch, is an important material for artists. Varnish is also extracted from conifer resin, as is the Canada balsam used for attaching cover glasses to microscope slides of biological specimens. After volatile compounds are removed from resin, the result is a sticky rosin. Applied to the bows of musical instruments such as violins, rosin makes them tacky, increasing the friction between bow and string and resulting in more vibration of the string and therefore more sound. Applied to a pitcher's hand, rosin improves his grip on a baseball.

From prehistoric times until recently, pitch was an important waterproofing agent. It was applied to boats to seal small holes and make the wood more impermeable to water. Ancient Greeks even used pitch to coat the inside of crockery used to store wine. It not only made the clay nonporous but also imparted a unique flavor to the wine.

One of the major uses of wood is as a source of fibers for making paper: The United States consumes as much as 19 million tons of newsprint per year, along with 25 million tons for other types of paper. Virtually all paper was made from flax fibers (still used for linen-based paper) until the 1800s, when a method was discovered for separating fibers of wood from each other. Wood chips are treated with sodium hydroxide or a mixture of sulfurous acid and bisulfites. The treatment dissolves the middle lamella and frees the fibers, which can then be washed and bleached. Many aspects of paper-making damage the environment severely. Trees are either clear-cut from natural forests or are grown on giant plantations of pure pine, from which all other plants are eliminated. The sodium hydroxide, sulfurous acid, and bleaches are significant pollutants. Much of this damage can be avoided easily by recycling used paper: Newspapers, letters, photocopies, and similar papers can be soaked in water to release the fibers, which can be used to make a new batch of paper. No further deforestation, fiber separation, or bleaching is necessary. Certain recycled paper fibers can be made into high-quality paper and others into cardboard that can be used for egg cartons and other packaging, reducing the consumption of styrofoam. Paper recycling reduces forest destruction, acid rain, and the release of chemical pollutants.

---

zomes. As described in Chapter 6, the leaves of conifers are always simple, being either needles or scales. Leaves of most conifers are perennial, persisting for many years (Fig. 24.10); leaves of *Agathis* and *Araucaria* remain even on very old trunks (Fig. 24.11).

The venation of conifer leaves is often simple, with just one or two long veins running down the center of a needle-shaped leaf, or several parallel veins in scale-shaped leaves. Unlike the leaf veins of flowering plants, those of conifers have an endodermis (Fig. 24.12). Furthermore, in addition to the ordinary vascular tissues of the leaf vein, there is also a tissue called transfusion tissue, consisting of transfusion parenchyma cells and transfusion tracheids. The latter are more or less cuboidal and have prominent circular bordered pits. Transfusion parenchyma is intermixed with the tracheids, the two cell types forming a complex three-dimensional pattern that apparently facilitates the transfer of materials between the ordinary vascular tissues and the mesophyll tissue outside the endodermis.

Just as in their progymnosperm ancestor *Archaeopteris*, the wood of modern conifers lacks vessels and their phloem lacks sieve tubes. Tracheids are so narrow that only one or two rows of circular bordered pits can occur on their radial walls (see Chapter 8). All conifers have pollen cones and seed cones, most of which are woody, but in *Juniperus* and *Podocarpus*, seed cones superficially resemble the fruits of flowering plants. The conifers are still a very successful group, forming extensive forests covering over 17 million km²; in many of these forests, flowering plants exist only as herbs, shrubs, or small trees growing in the conifers' shade.

FIGURE 24.11

In almost all species of conifers, each leaf lives for several years, or as in the case of this *Araucaria,* for many years, persisting even as the trunk becomes massive.

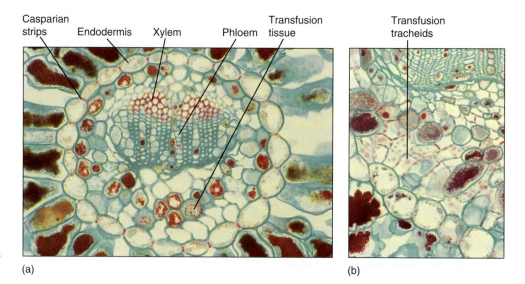

FIGURE 24.12
(a) The vascular bundle of this Douglas fir leaf has an endodermis, xylem, phloem, and transfusion tissue (×80). (b) High magnification of a pine leaf bundle showing secondary phloem and transfusion tracheids. Casparian strips are visible in the endodermis walls (×130).

The pines are good representatives for closer examination. The trees are monopodial, with one main trunk bearing many branches; the wood is composed exclusively of tracheids, but annual rings, spring wood, and summer wood are all visible because large-diameter tracheids are produced in the spring, followed by narrow-diameter tracheids in the summer (Fig. 24.13). Rays are thin and tall and contain both ray parenchyma and ray tracheids. Resin canals, which produce the thick, sticky pitch, run vertically among the tracheids and horizontally in the rays. The wood has almost no axial parenchyma. Phloem contains sieve cells and albuminous cells and also has tall, narrow rays (Fig. 24.14). A cork cambium produces a thick, tough bark that provides excellent protection even from forest fires.

Pines are somewhat atypical in this regard; many other conifers do not have two types of shoot or dimorphic leaves.

Pines, like several other conifers, have two types of shoot, each with a characteristic type of leaf. Tiny papery leaves occur on **long shoots** and in their axils are **short shoots** that produce the familiar long needle leaves (see Figs. 24.10c and 24.15). The leaves have many xeromorphic characters: thick cuticle, sunken stomata, cylindrical shape.

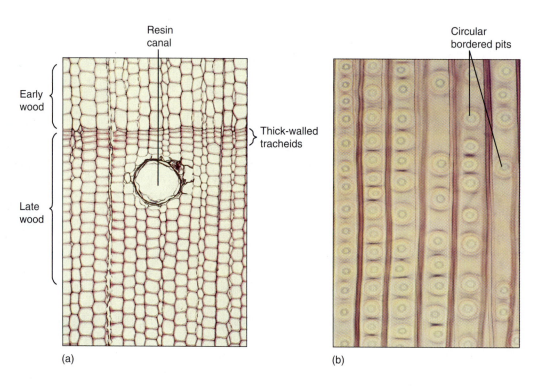

FIGURE 24.13
(a) The wood of pine, like that of all conifers, lacks vessels; it consists almost exclusively of tracheids. Growth rings are visible because late wood contains narrow, thick-walled tracheids, whereas early wood contains wide, thin-walled tracheids (transverse section, ×30).
(b) Pine tracheids are narrow and their circular bordered pits are wide, so only one row of pits fits on a given wall, unlike the dozens of small pits that cover each wall of an angiosperm vessel (see Fig. 5.32a) (×80). Note how similar this wood is to that of progymnosperms (see Fig. 24.3).

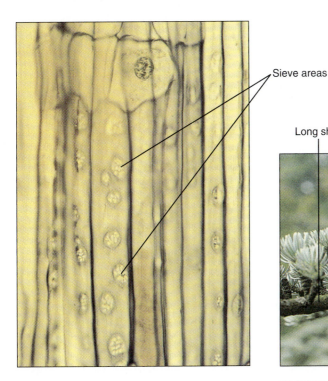

FIGURE 24.14
Phloem of conifers contains sieve cells, not sieve tube members; sieve cells are long and narrow with sieve areas over much of their surface, but they never have horizontal cross walls (sieve plates) with enlarged sieve pores. The sieve areas in this pine are particularly abundant and easy to see (×25).

FIGURE 24.15
In *Cedrus atlantica,* the nature of the short shoot is more obvious because it forms more needles each year and so slowly grows into a visible shoot.

Like all conifers, pines have both pollen cones and seed cones. Pollen cones are **simple cones;** that is, they consist of a single short unbranched axis that bears microsporophylls (Figs. 24.16 and 24.17). Microspore mother cells undergo meiosis and form microspores; then each of these develops endosporially into a small gametophyte with four cells, one of which is a generative cell as in flowering plants (Figs. 24.18 and 24.19). The gametophytes are shed from the tree as pollen and carried by wind; a small percentage land in seed cones, but the great majority land elsewhere and die.

(a)   (b)

FIGURE 24.16
(a) Pollen cones typically occur in clusters near the ends of branches. As the microsporangia dehisce, pollen is liberated to the wind and blown away for distribution. Such a method of gene transfer is inefficient because so few pollen grains land on seed cones; it is successful primarily because conifers grow as dense forests, where each conifer is surrounded by hundreds of potential partners. (b) Pollen cones are simple cones; they have one single stem axis and bear microsporophylls. These are not microsporangia but rather leaflike structures that bear microsporangia (×10). *(Robert and Linda Mitchell)*

685

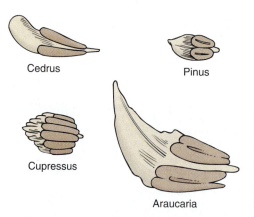

FIGURE 24.17

In genera such as *Araucaria,* each microsporophyll is obviously one sporophyll carrying two sporangia. In *Cupressus* and *Pinus,* the sporophyll has been so simplified that it looks like a lobed sporangium rather than a sporophyll with several sporangia.

Seed cones are more complex than pollen cones: They are **compound cones,** each consisting of a shoot with axillary buds. The short axis bears leaves called **bracts** rather than sporophylls (Fig. 24.20). Each bract has an axillary bud that bears the megasporophylls. In some fossil conifers the individual structures can still be seen, but in all modern conifers, extensive fusion has occurred: The axillary bud is microscopic and its megasporophylls are fused laterally, forming an **ovuliferous scale** (Figs. 24.21 and 24.22). In larch, fir, and Douglas fir (*Pseudotsuga*) the ovuliferous scale is still distinct and can be seen, but in most conifers it is fused to the bract.

Inside each megasporangium, a single large megaspore mother cell undergoes meiosis, with three of the resulting cells degenerating and only one surviving as the megaspore (Fig. 24.23). The megasporangium does not dehisce; the megaspore is retained inside and grows

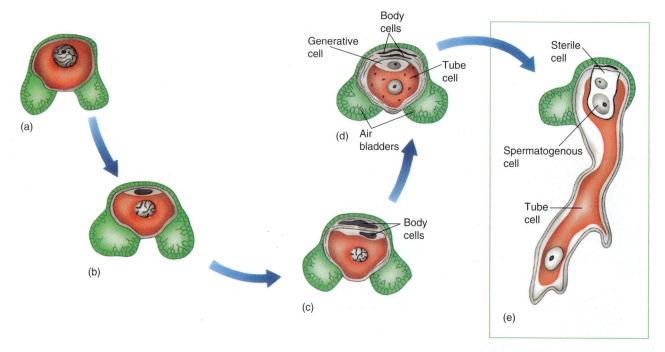

FIGURE 24.18

The microspore (pollen) of pine has one cell and two large air bladders that increase its buoyancy in air. The microspore develops into a small gametophyte in a process similar to that in pollen of flowering plants, except that a few more body cells are formed. First, two mitotic divisions produce two small body cells that degenerate and a large cell that divides, resulting in a generative cell and another body cell, called the tube cell. As in angiosperms, the body cell becomes the pollen tube and the generative cell divides into two nonmotile sperm cells.

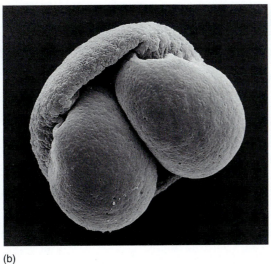

(a)

(b)

FIGURE 24.19

(a) Young pollen grains in which one large body cell and the generative cell are visible (× 200). *(E. R. Degginger/Earth Scenes)* (b) The two air bladders of pine pollen are easily visible in this scanning electron micrograph. The addition of the hollow bladders adds virtually no weight to the pollen but increases its volume so that the pollen's density (weight per volume) is decreased and it does not sink so quickly in air (× 3000). *(R. E. Litchfield/Science Source/Photo Researchers, Inc.)*

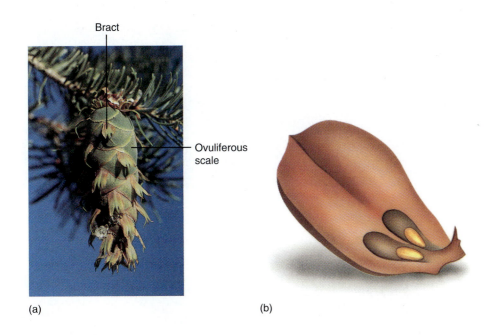

(a)

(b)

FIGURE 24.20

(a) The long three-pointed bracts of this Douglas fir *(Pseudotsuga)* cone are the true leaves of the cone axis. Each one contains an axillary bud whose leaves have fused together side by side into the flat, shieldlike ovuliferous scale just behind each bract. These fused leaves that constitute the ovuliferous scale are actually megasporophylls. (b) If an ovuliferous scale were pulled from the cone and turned over, the two ovules would be visible. The scale obviously does not look like a set of leaves fused together, but look at Figures 24.6 and 24.21.

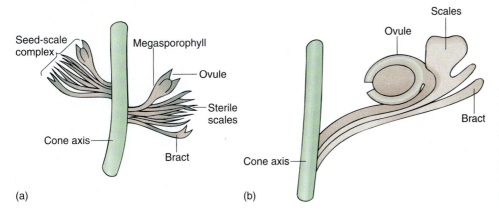

(a)

(b)

FIGURE 24.21

Possible steps in the evolution of seed cones: (a) In *Lebachia,* an early conifer, the bract is visible and the nature of the fertile axillary bud is obvious. Only slight fusion of megasporophylls had occurred. (b) In the Triassic *Voltzia,* the modern condition was essentially complete, with sporophylls fused together.

FIGURE 24.22

(a) In the seed cones of pine, all parts are fused together; even the bract is fused to the ovuliferous scale. When morphologists first began working on pine cones without knowing about fossil structures, they were completely baffled and could not explain such a complex structure. *(William E. Ferguson)* (b) Colorado blue spruce *(Picea pungens). (Runk/Schoenberger from Grant Heilman)* (c) Rock "cedar" *(Juniperus ashei).* (d) Atlantic cedar *(Cedrus atlantica). (R. J. Erwin/ Photo Researchers)*

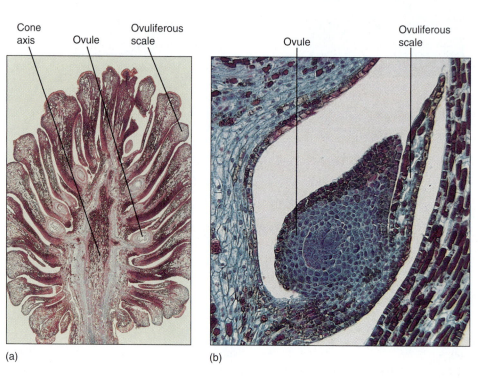

FIGURE 24.23

This section of a pine cone was made just as the megaspore mother cells were about to begin meiosis. Each ovuliferous scale carries two ovules, each of which contains an integument and a nucellus (the actual megasporangium). As in flowering plants, each nucellus usually has only one megasporocyte (megaspore mother cell) and produces only one surviving megaspore. (a) × 2. *(Bruce Iverson)* (b) × 25. *(Robert and Linda Mitchell)*

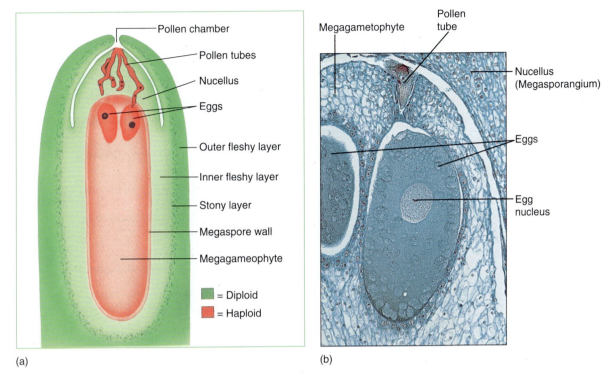

(a)

(b)

FIGURE 24.24

Ovules of pine and other conifers are much larger than those of flowering plants. This cellular megagametophyte is not as large as a moss or liverwort gametophyte, but it is much more plantlike than the megagametophytes of angiosperms (see Fig. 9.16). Above the two eggs is the thick, indehiscent megasporangium (nucellus). Sperms cannot swim to the archegonia and eggs; they must be carried by a pollen tube that digests its way through the megasporangium. Between the megasporangium and the integuments is the pollen chamber (× 20). *(Photograph: James W. Richardson/CBR Images)*

into a large coenocytic megagametophyte by free nuclear divisions and may have as many as 7200 nuclei. Development can take as long as a year, but finally walls form, converting the coenocyte into a cellular megagametophyte. Two or three archegonia form as sets of cells, each surrounding a large egg (Fig. 24.24). Conifer eggs are gigantic cells loaded with carbohydrate and protein. The egg nucleus, although haploid, is swollen to a volume much larger than that typical for entire cells. It is probably filled with DNA synthetases and RNA polymerases, ready for extremely rapid activity once karyogamy occurs.

Unlike pollination in flowering plants, conifer pollen arrives before the egg is mature, and more than a year may pass between pollination and fertilization. The pollen germinates, and a massive pollen tube slowly digests its way toward the megagametophyte as the egg forms. Because the megasporangium does not open, a passageway digested by a pollen tube is necessary. The two or three eggs in one megagametophyte can all be fertilized, but only one zygote develops into an embryo; the other one or two die. A zygote does not immediately form an embryo in conifers; instead, some of the first cells elongate as a **suspensor** that pushes the other cells deep into the megagametophyte (Fig. 24.25). These other cells, called the **proembryo**, develop into the embryo. No double fertilization occurs; rather, the female gametophyte continues to grow and acts as a nutritive tissue similar to endosperm. The mature embryo has the same organization as an angiosperm embryo (radicle, hypocotyl, epicotyl, and cotyledons) but always has many cotyledons, not just one or two (Fig. 24.25d). The seed also resembles that of a flowering plant, but it is borne in a cone, not a fruit. In two conifers the cones become fleshy, fruitlike, and brightly colored — red in *Podocarpus*, blue in *Juniperus* (see Fig. 24.22).

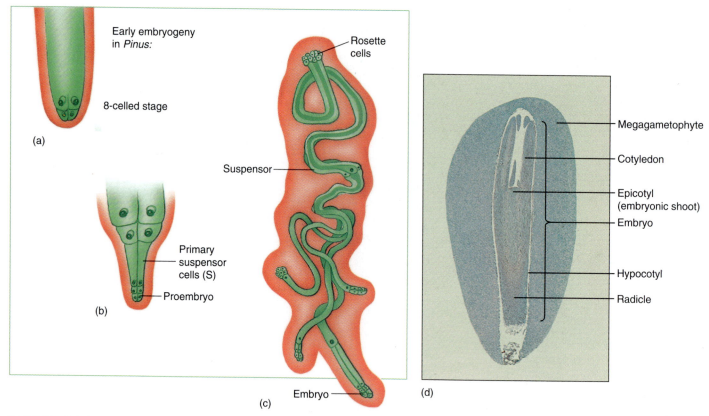

FIGURE 24.25

(a) Just as in angiosperms (see Fig. 9.18), the zygote of conifers does not immediately form an embryo; instead it produces a suspensor that thrusts the proembryo into the megagametophyte (b). (c) In conifers the suspensor can divide and form multiple embryos from each zygote; they do not produce identical twins or quadruplets, however, because only one embryo survives. Also, even though each megagametophyte produces several eggs, only one embryo survives in each ovule (with rare exceptions). (d) A conifer seed looks remarkably like the seed of a flowering plant. It has a seed coat (removed here) derived from the integument, an embryo with several cotyledons, and a nutritive tissue that is actually the megagametophyte, not endosperm. After fertilization, the parental sporophyte transports large amounts of nutrients, supplying everything needed for the embryos, which grow completely heterotrophically. It also fills all cells of the megagametophyte with carbohydrates, proteins, and mineral nutrients; this acts just like the endosperm of flowering plants and is often called endosperm (× 2.5). *(Bruce Iverson)*

ORIGIN AND EVOLUTION OF CONIFERS

As early as 360 million years ago, progymnosperms had a vascular cambium that produced secondary xylem and phloem similar to that in modern conifers. In fact, the only wood characters absent from *Archaeopteris* but present in some modern conifers are resin canals and axial wood parenchyma. The earliest resin canals are found in Lower Cretaceous fossilized wood called *Pityoxylon*, the form genus for wood with characters of the extant genera *Larix, Picea, Pinus,* and *Pseudotsuga*. Axial wood parenchyma may have arisen independently several times. It is found in fossil wood similar to that of Taxodiaceae and Cupressaceae as well as in fossil wood similar to that of *Podocarpus* and *Dacrydium*.

Starting in strata about 300 million years old (late Carboniferous), fossils are found that are considered to be basically true gymnosperms. They are assigned to the groups Cordaitales and Voltziales (See Table 24.2). Cordaitales were small to large trees with gymnosperm wood (Fig. 24.26). Their leaves (form genus *Cordaites*) were strap-shaped and up to 1 m long and 1.5 cm wide. Leaf veins appeared to be parallel because the leaves were so long; the veins actually dichotomized and resembled those of *Archaeopteris macilenta* (see Fig. 24.6b). Voltziales resembled the extant plants known as Norfolk Island pines (*Araucaria excelsa*); they had tall trunks, branches in whorls, and needle-shaped leaves that could have been derived from telomes of *Archaeopteris fissilis*.

The most significant events in the transition of progymnosperms to gymnosperms was the evolutionary modification of reproductive structures. The progymnosperm *Archaeopteris* was heterosporous, but its sporangia were relatively exposed and unprotected; if fossils of the form genus *Archaeosperma* (seeds) were part of *Archaeopteris,* as is suspected, there was a cupule and an integument around the megasporangium. Reproductively, Cordaitales and Voltziales had fewer relictual features than *Archaeopteris* and *Archaeosperma*. Megasporangiate structures of Cordaitales were somewhat similar to modern seed cones of conifers in that they had a primary shoot with sterile bracts; in the bract axils were secondary shoots with fertile and sterile leaves. In some members of Voltziales the secondary shoots had become planar and bilaterally symmetrical (see Fig. 24.21). In *Pseudovoltzia,* the sterile and fertile leaves of the secondary shoots had become fused, like an ovuliferous scale in modern conifers. By the Cretaceous, there was a cone, classified in the form genus *Compsostrobus*, that was almost identical to a pine cone except that its bracts were longer and it was less compact.

In many features, Voltziales were almost identical to modern conifers, and they existed into the Jurassic Period, long after the first modern conifers appeared. Voltziales were so similar to Coniferales and the final transition was so gradual that a distinct division between the two groups is difficult to make.

TAXALES

The order Taxales contains only one family, Taxaceae, with five genera, *Taxus* (yew) and *Torreya* being the most familiar (Table 24.2). Its members strongly resemble species of Coniferales except in some aspects of their reproduction: Rather than having seed cones like the conifers, the taxads bear their seeds terminally on short lateral shoots (Fig. 24.27). They have no cone and after fertilization, as the seed develops, a bright red, fleshy, sweet envelope, the **aril,** grows around the seed. Microsporangia occur in groups of up to nine on flat microsporophylls.

(a)

(b)

FIGURE 24.26
Cordaites was a small tree with long, strap-shaped leaves.

FIGURE 24.27
(a) *Taxus* has individual seeds, each surrounded by a red, fleshy aril. In the young seeds (grey) the aril is only partly expanded. *Taxus cuspidata. (E. R. Degginger)* (b) *Taxus* pollen "cones" are not as tightly clustered as those of conifers but otherwise are not very different. *(George Loun/Visuals Unlimited)*

BOX 24.1   *Tree Breeding Using Molecular Markers*

Many species of conifers are important economically because they produce wood, whereas others produce resins that can be used in the production of varnishes and numerous other chemicals. In many cases, the value of the trees could be increased if they had harder wood or less lignin or if they produced resins with increased amounts of certain chemical compounds. Botanists have been trying to increase the usefulness of conifers by cross-breeding them and subjecting them to artificial selection: saving those with the best allele combinations and eliminating those that have poor qualities. However, this has been an extremely time-consuming process because the trees grow slowly. Also, they might have to be 20 or 30 years old before it is possible to determine whether they have the proper wood hardness or lack of lignin or resin chemical composition.

Recently, R. Sederoff and D. O'Malley, botanists at North Carolina State University, devised a method to speed up the selection process. They have mapped a portion of the genome of loblolly pine, identifying 200 sites on the various chromosomes. It would be best if the mapped sites were each a gene governing some economically important aspect of the pine's phenotype, but we do not yet know much about pine genetics and development. Instead, the mapped sites are simply sites that can be easily identified by DNA tests. Because there are so many, each must be close to some important (if unidentified) gene. If many trees are examined, it is possible to see which mapped sites are associated with particular, useful characters. For example, all trees with low levels of lignin (which would allow for paper production with less pollution) might lack marker X whereas those with abundant lignin might have X. X may not actually be the gene for controlling the amount of lignin production, but it must be so close to the critical gene that crossing-over in meiosis only rarely separates them.

With this type of genome map and with a knowledge of which markers are associated with which characters, it is possible to increase the rate of progress of tree-breeding programs. The usual crosses are made using parents that have desired phenotypes. Once seeds have been produced, they can be planted and then tissue can be tested to see which markers they inherited. By examining the combination of markers present in the DNA of the seedling, it is possible to predict the characters of the tree as an adult. Seedlings that lack the desired character combination could be weeded out, while those that have the proper character combinations would be allowed to grow. By reducing the time required to do an experiment from 20 years to 1 year, our knowledge of conifer growth and development should be greatly improved. Note that this is an example of artificial selection, not natural selection.

Plant Biology Tutor CD-ROM

The initial p is silent; it is pronounced as if spelled teridospermofita.

Members of the Taxales are separated from the Coniferales on the basis of their arillate seed and lack of a typical seed cone. Much of our understanding of gymnosperms is due directly to the extensive and careful work of R. Florin; he concluded that the megasporangiate structure of taxads could not be traced to the Cordaitales or Voltziales but rather must have derived from sporangia similar to those of *Rhynia*. If this is correct, the Taxales are quite distinct phylogenetically from Coniferales. This would mean that megaphylls, roots, a vascular cambium, and secondary growth evolved independently in the taxad line of evolution. But the structure of taxads is so similar to that of conifers that this conclusion is difficult to accept. More recently, it has been suggested that the taxad arillate seed could have evolved from a *Lebachia*-type of reproductive shoot (see Fig. 24.21a) by simple modifications. If so, the Taxales are closely related to the conifers.

DIVISION PTERIDOSPERMOPHYTA: SEED FERNS

Progymnosperms gave rise to another line of gymnospermous plants in addition to the conifers: the cycadophytes (see Fig. 24.2). These are classified as three divisions: Pteridospermophyta (seed ferns, all extinct; Table 24.3), Cycadophyta (cycads, extant), and Cycadeoidophyta (cycadeoids, all extinct).

The earliest seed ferns appeared in the Upper Devonian Period; they were woody plants with fernlike foliage that bore seeds instead of sori on their leaves. Many seed ferns resembled modern tree ferns, but others were vines that either climbed on or scrambled over other plants (Fig. 24.28).

Pteridosperms are thought to have evolved from the Aneurophytales because the earliest seed ferns, such as *Stenomyelon,* had a three-ribbed protostele as in the aneurophytes *Triloboxylon* and *Aneurophyton*. In later species of *Stenomyelon, Calamopitys,* and *Lyginopteris,* most central cells of the stem differentiated as parenchyma, not tracheids (Fig. 24.29a). They had a ring of vascular bundles surrounding a pith; such an arrangement also

TABLE 24.3 *Classification of Division Pteridospermophyta**

Order Lyginopteridales
    Family Lyginopteridaceae   *Lyginopteris*
    Family Callistophytaceae   *Callistophyton*

Order Medullosales
    Family Medullosaceae   *Pachytesta*

Order Caytoniaies

――――――――――――――――――――――――――――――――

* All species in this group are extinct.

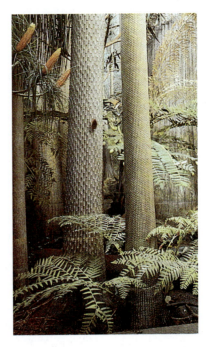

FIGURE 24.28
Reconstruction of a swamp-forest during the Carboniferous Period. The large trees are related to *Lepidodendron,* (*Sigillaria rugosa* on the left, *S. saulli* on the right), and the smaller fernlike plants are seed ferns (*Neuropteris decipiens*), not true ferns. Note the seeds at the ends of some fronds. (*75400c Field Museum of Natural History, Chicago*)

occurs in the stems of all gymnosperms and angiosperms. Seed ferns had a vascular cambium that was long-lived and produced both xylem and phloem; this too is similar to gymnosperms and angiosperms. Although their wood was basically similar to that of their progymnosperm ancestors, there were interesting differences. Tracheids were much longer and wide enough that several rows of circular bordered pits could occur on each radial wall; conifer tracheids are so narrow that only one or two rows can fit on each radial wall. Rays in pteridosperm wood were many cells wide, not just one cell wide, and they were very tall, being large wedges of parenchyma. This wood is much softer and less dense than the wood of conifers and progymnosperms. This type of parenchymatous wood also occurs in cycads and cycadeoids. Around the stem was a thick cortex that contained distinctive radial plates of sclerenchyma just below the epidermis (Fig. 24.29b). The inner cortex contained secretory ducts. In older plants, a cork cambium and bark formed exterior to the secondary phloem; the cortex was shed with the first bark.

The leaves of seed ferns were similar to those of true ferns in overall organization— large, compound, and planar. Unlike ferns, however, the foliage leaves of seed ferns bore

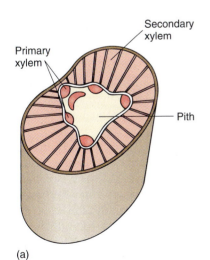

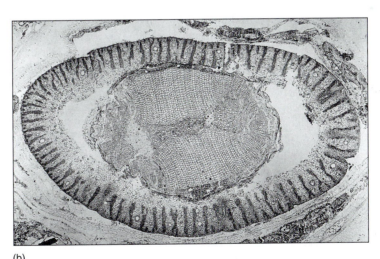

(a)   (b)

FIGURE 24.29
(a) The vascular system of seed plants (a siphonostele) evolved from a protostele as some of the central cells failed to differentiate into tracheids, maturing as parenchyma instead. This is a diagram of *Calamopitys americana*, which represents a stage in which most cells still differentiate into tracheids, but a few form parenchyma (pink region). Normal primary xylem still forms on the edges (red), and then a cambium arises and produces secondary xylem to the exterior (brown area). (b) A transverse section of *Schopfiastrum* stem, showing the beginning of a pith, parenchymatous wood, and the distinctive cortex with plates of sclerenchyma (× 10). (*Courtesy of B. M. Stidd and T. L. Phillips, Western Illinois University*)

FIGURE 24.30
Seed ferns, such as this *Emplectopteris*, bore seeds along their leaves, not in cones. Otherwise, seed fern leaves were remarkably analogous (not homologous) to the leaves of true ferns in terms of venation and tissue structure.

seeds, not sori (Fig. 24.30). Within the seed fern ovule the megasporangium (nucellus) was large and vascularized, with bundles of vascular tissue running into and through it. The integument was attached to the megasporangium only at the base and was vascularized. Seeds could be extremely large, up to 11 cm long and 6 cm in diameter in the now extinct *Pachytesta incrassata;* at maturity they had a thick three-layered seed coat.

DIVISION CYCADOPHYTA: CYCADS

Modern cycads are frequently confused with either ferns or young palm trees because they have stout trunks with pinnately compound leaves (Table 24.4, Fig. 24.31). Most cycads are short plants less than 1 or 2 m tall, but *Macrozamia* can reach heights of 18 m. The trunk is covered with bark and persistent leaf bases that remain on the plant even after the lamina and petiole have abscised. Internally, cycad stems are similar to those of seed ferns—a thick cortex containing secretory ducts surrounds a small amount of parenchymatous wood. Tracheids are long and wide, and rays are massive. Even very old stems have only a small amount of wood; most support is provided by the tough leaf bases. A prominent pith contains secretory canals.

Unlike seed ferns, cycad foliage leaves do not bear ovules. Instead, cycads produce seed cones and pollen cones, each on separate plants; cycads are always dioecious. Pollen cones consist of spirally arranged shield-shaped microsporophylls that bear clusters of microsporangia (Fig. 24.32). Upon germination, pollen grains produce a branched pollen tube and large, multiflagellated sperm cells. Seed cones are variable, with those of *Cycas revoluta* usually considered the most relictual (Fig. 24.33). In this species the seed cone is a large, loose aggregation of leaflike, pinnately compound megasporophylls. Six to eight

TABLE 24.4 *Classification of Division Cycadophyta*	
Order Cycadales	
Family Cycadaceae	*Bowenia, Cycas, Ceratozamia, Dioon, Encephalartos, Lepidozamia, Macrozamia, Zamia*
Family Stangeriaceae	*Stangeria*

FIGURE 24.31

Aspects of the vegetative body of cycads. (a) Cycads tend to have short, stout trunks with tough, leathery, pinnately compound leaves that superficially resemble leaves of ferns or palms (*Cycas revoluta*). *(J. L. Castner)* (b) Cycad leaf veins have transfusion tissue similar to that in conifer leaves (× 50). (c) Transfusion tracheids extend from veins into the leaf blade. Stomata are visible, slightly out of focus, in the lower epidermis (× 32). (d) Some cycads such as this *Dioon spinulosum* do become large, but only if they become extremely old. *(Robert and Linda Mitchell)* (e) Cycads have secondary growth, but only small amounts of wood and phloem are produced, so the trunks do not become very much wider with age. The xylem is of the parenchymatous type (× 32).

(a)

Sporangia

Sporophyll

Cone axis

(b)

(c)

FIGURE 24.32

(a) Pollen cones of cycads tend to be large, more than 30 cm long, like this one of *Cycas circinnalis.* It is a simple cone, with one axis that bears microsporophylls. *(J. L. Castner)* (b) Microsporophylls in cycads may bear many microsporangia. (c) Sperm cells of cycads have hundreds of flagella and must swim to the egg cell. *Zamia* (×165). *(K. Norstog, Fairchild Tropical Gardens)*

FIGURE 24.33

(a) Seed cones of cycads differ greatly from those of conifers; they are simple, they have only one axis, and the megasporophylls are borne on that axis. They have no bracts or axillary buds. *(Robert and Linda Mitchell)* (b) Cycad megasporophylls are not fused structures as is the ovuliferous scale of conifers; instead, the megasporophyll is simple and strongly resembles a foliage leaf, at least in *Cycas circinnalis.* It is not difficult to see how this type of sporophyll could evolve from a seed fern sporophyll (see Fig. 24.30).

(a)

Megasporophyll

Seed

(b)

large ovules occur near the base, but the upper half of the megasporophyll is rather leaflike, similar to the ovule-bearing organs of some seed ferns. Megasporophylls of the cycad *Zamia floridana* are considered more derived. They are shield-shaped and bear only two ovules, and the entire cone is quite compact. Cycad ovules are like those of seed ferns, having a large, vascularized megasporangium and a loosely attached, vascularized integument.

Although Cycadophyta was a much larger group with many more species in earlier times, at present it contains nine or ten genera and about 100 species. Modern cycads are highly prized ornamentals in the warmest parts of the United States; only a few can withstand freezing temperatures in winter. They are almost all tropical with an unusual distribution: Some occur in Cuba and Mexico, others in Australia, still others in southeast Asia or Africa. Cycads are now believed to have been widespread and to have occupied many habitats, but in most areas they have become extinct. The few remaining, widely scattered species are the survivors of a formerly extensive group.

DIVISION CYCADEOIDOPHYTA: CYCADEOIDS

The cycadeoids (all extinct) had vegetative features almost identical to those of cycads (Fig. 24.34; Table 24.5). The two groups differ only in subtle details of the differentiation of stomatal complexes and in leaf trace organization. On such characters alone, cycadeoids would never be considered distinct from cycads. However, cycadeoid reproductive structures were quite different from those of cycads: Individual cones contained both microsporophylls and megasporophylls. Each ovule had a stalk, and the megasporangium was surrounded by an integument that extended out into a long micropyle. Between the ovules were thick, fleshy scales. Microsporophylls were located below the cluster of megasporophylls and curved upward, enveloping the megasporophylls. Each microsporophyll was cup-shaped and contained numerous microsporangia.

TABLE 24.5 *Classification of Division Cycadeoidophyta**

Order Cycadeoidales
    Family Williamsoniaceae
    Family Cycadeoidaceae

___

* All species in this group are extinct.

FIGURE 24.34
Vegetatively, the cycadeoids such as *Cycadeoidea* were very similar to cycads. Internally, like cycads, they had a broad pith, a small amount of parenchymatous xylem, a broad cortex, and a tough outer protective layer formed by persistent leaf bases.

TABLE 24.6 *Classification of Division Ginkgophyta*
Order Ginkgoales
Family Ginkgoaceae *Ginkgo biloba, G. adiantoides**
* This species is extinct.

DIVISION GINKGOPHYTA: MAIDENHAIR TREE

This division contains a single living species, *Ginkgo biloba* (Fig. 24.35; Table 24.6). It may seem unusual to erect an entire division for a single species, but *G. biloba* (the "maidenhair tree") is itself unusual. It looks very much like a large dicot tree with a stout trunk and many branches, but its wood, like that of conifers, lacks axial parenchyma. It has "broad leaves," but they have dichotomously branched veins like seed ferns, not reticulate venation like dicots. Ginkgos have both short shoots, which bear most of the leaves, and long shoots. Reproduction in *Ginkgo* is dioecious and gymnospermous, but cones are not produced. Instead, ovules occur in pairs at the ends of a short stalk and are completely unprotected at maturity (Fig. 24.35c). Pollen is produced in an organ that resembles a catkin, having a stalk and several sporangiophores that each have two microsporangia. Like ovules of pteridosperms and cycads, the ovules of *Ginkgo* are large (1.5 to 2.0 mm in diameter) and develop a three-layered seed coat.

A *Ginkgo* tree itself is beautiful and is a popular ornamental because the leaves turn brilliant yellow in autumn, but the microsporangiate ("male") trees are preferred; when the megasporangiate ("female") trees produce seeds, the outer fleshy layer of the seed emits butyric acid, which has a putrid odor that is difficult to tolerate.

The exact ancestors of ginkgos are not known, but they must have been one of the seed ferns or a closely related group. Ginkgos became abundant during the Mesozoic Era, especially in the mid-Jurassic Period (about 170 million years ago). An abundance of diverse fossilized leaves has been found, but it is difficult to know how many species they represent, because living *G. biloba* trees produce diverse types of leaves depending on the growing conditions. At present, about eight form genera for leaves are recognized. Fossi-

(a)

(b)   Microsporophylls

Megasporophyll   Seed

(c)

FIGURE 24.35

(a) A tree of *Ginkgo biloba,* the only species of this division, could easily be mistaken for a dicot tree. This photo shows its current typical habitat: in cultivation in cities. There may be no natural populations left; it has survived because it has been cultivated as an ornamental. *(Doug Wechsler/ Earth Scenes)* (b) Although *Ginkgo* leaves are broad, they have dichotomous venation like leaves of cycads and seed ferns, not reticulate venation like dicots. Microsporophylls occur in small, conelike clusters, mixed with foliage leaves on short shoots. *(John D. Cunningham/Visuals Unlimited)* (c) Ovules occur in pairs at the end of a stalklike megasporophyll. The ovule itself, the integument, is exposed; no bract or ovuliferous scale protects it. Here, both ovules have developed into seeds. *(Robert Dunne/Photo Researchers)*

lized remains of leaves and wood can be found in almost all areas of the world, especially in high latitudes such as Alaska, Canada, and Siberia near the North Pole, and Patagonia, South Africa, and New Zealand near the South Pole. Ginkgos began to die out early in the Tertiary Period, but two species, *G. biloba* and *G. adiantoides,* persisted. The latter species became extinct during the Pliocene Epoch, about 10 to 12 million years ago.

DIVISION GNETOPHYTA

Division Gnetophyta contains three groups of enigmatic plants, each of which is often placed in its own order: *Gnetum* with 30 species (order Gnetales; Fig. 24.36), *Ephedra* with about 40 species (order Ephedrales; Fig. 24.37), and *Welwitschia mirabilis,* the only species in order Welwitschiales (Fig. 24.38; Table 24.7).

The initial g is silent; it is pronounced as if spelled neatofita.

(a)

(b)

FIGURE 24.36
Gnetum. (a) Plants of *Gnetum* strongly resemble dicots, having broad leaves and woody stems. (*G. Davidse*) (b) Ovules are not borne in cones. (*Robert and Linda Mitchell*)

(a)

(c)

(b)

FIGURE 24.37
Ephedra. (a) Plants of *Ephedra* often occur in dry areas and strongly resemble many types of desert-adapted dicots. Although their reproductive structures are gymnospermous, the microsporangiate cone (b) could be mistaken for a staminate imperfect flower. The naked ovules reveal that they are not angiosperms (c). (*b, William E. Ferguson; c, N. H. (Dan) Cheatham/Photo Researchers, Inc.*)

(a)

(c)

(b)

FIGURE 24.38

Welwitschia mirabilis. (a) Whole plant with torn leaves. *(William E. Ferguson)* (b) Microsporangiate strobili. (c) Megasporangiate strobilus. *(Richard Shiell/Earth Scenes)*

Gnetums are mostly vines or small shrubs with broad leaves similar to those of dicots. They are native to southeast Asia, tropical Africa, and the Amazon Basin. Plants of *Ephedra* are tough shrubs and bushes that inhabit desert regions in northern Mexico and southwestern United States. Their leaves are reduced and scalelike. The few living plants of *Welwitschia* exist only in deserts of South Africa or in cultivation. They have a short wide stem and only two leaves, but the leaves grow perennially from a basal meristem, becoming increasingly longer.

All three genera are unusual in being gymnosperms with vessels in their wood. This had been thought to show that they might be related to primitive angiosperms. However,

TABLE 24.7	*Classification of Division Gnetophyta*
Order Gnetales	
Family Gnetaceae	*Gnetum*
Order Ephedrales	
Family Ephedraceae	*Ephedra*
Order Welwitschiales	
Family Welwitschiaceae	*Welwitschia mirabilis*

their vessel elements evolved from tracheids with circular bordered pits, whereas those of angiosperms derived from scalariform tracheids. Furthermore, angiosperms are thought to have evolved from vessel-less ancestors, the vessels evolving after flowers, not before them.

Unlike the pollen cones of all other gymnosperms, those of gnetophytes are compound and contain small bracts. Seed cones are also compound and contain extra layers of tissue around the ovules; the tissue is variously interpreted as an extra integument, bract, or sporophyll.

The few fossils of gnetophyte organs or tissues are only several million years old, too recent to be of much help in understanding the evolution and ancestry of the group. The pollen is quite distinctive, being spindle-shaped and having narrow ridges. It is easy to recognize, and fossil pollen of this type occurs as far back as the late Triassic Period, but pollen has not helped reveal their origins either.

SUMMARY

1. In a seed, the embryo draws its nutrition from the surrounding megagametophyte (or endosperm), which in turn is supplied with nutrients by the parental sporophyte. The sporangium and associated structures provide protection for the megaspore, megagametophyte, and embryo.

2. In all modern seed plants, the megasporangium (nucellus) does not open, so a pollen tube must deliver the sperm cells to the archegonia. Sperm cells are still motile in all plants except conifers and flowering plants.

3. At present gymnosperms are known to represent several distinct lines of evolution, each of which is classified as a division.

4. Progymnosperms arose from trimerophytes with the evolution of a vascular cambium of unlimited growth potential which produced solid wood with little parenchyma. The evolution of true seeds also began in the progymnosperms, first with retention of the megaspore and then with formation of an integument.

5. One order of progymnosperms, Archaeopteridales, gave rise to conifers; another progymnosperm order, Aneurophytales, evolved into pteridosperms, cycads, cycadeoids, and perhaps angiosperms.

6. Conifers constitute a diverse division of large trees with solid wood, simple pollen cones, and compound seed cones.

7. The members of Taxales produce arillate seeds, not seed cones. Taxales are usually included in division Coniferophyta, but as the most anomalous member.

8. Pteridosperms, seed ferns, gave rise to cycads and cycadeoids. Important evolutionary changes were the formation of pith and of ovules on more or less unspecialized foliage leaves. Their wood contained abundant axial parenchyma.

9. Cycads are modern gymnosperms with compound leaves, parenchymatous wood, and simple seed cones consisting of an axis and a rather leaflike megasporophyll.

10. Cycadeoids are a group of extinct gymnosperms whose vegetative structure was almost indistinguishable from that of cycads. However, their reproductive structures contained both microsporophylls and megasporophylls together.

11. *Ginkgo biloba* is the sole living species of its division. It resembles a dicot tree except that its leaves have dichotomous veins and it bears naked ovules, surrounded by neither a cone nor a carpel.

12. Reproduction in division Gnetophyta is gymnospermous, but plants of *Ephedra* and *Gnetum* resemble dicots. *Welwitschia* has only two leaves, which grow indeterminately from basal meristems. None has any fossil record, and their ancestry is unknown.

IMPORTANT TERMS

angiosperms
aril
compound cone
cone bract
gymnosperms

integument
long shoot
ovuliferous scale
pollen chamber
proembryo

short shoot
simple cone
suspensor

REVIEW QUESTIONS

1. What is the disadvantage of a life cycle in which alternate generations are completely independent of the preceding generation? Are there advantages to this type of life cycle?

2. What is a cone? How are cones of conifers similar to those of lycophytes and cycads? How do they differ?

3. Describe progymnosperms. Why do we believe that they evolved from trimerophytes? What were the significant evolutionary advances that characterized progymnosperms?

4. Name several families and genera of conifers. How many could you recognize on sight? How is their wood similar to that of *Archaeopteris*? How is it different from the wood of flowering plants?

5.   Describe a conifer pollen cone. Why is it a simple cone? Describe a conifer seed cone. Why is it a compound cone?

6.   If you were given an unknown plant and were asked to determine whether it was a conifer or a cycad, what would you look for if given (a) only wood, (b) only leaves, (c) only seed cones? What parts would you need if you had to distinguish between a cycad and a cycadeoid?

7.   Describe the megasporophylls of a cycad seed cone. Why are they evidence that cycads might be related to seed ferns?

8.   Would it have been possible, on a theoretical basis, for an indehiscent megasporangium to evolve before integument-penetrating pollen tubes evolved? Why or why not? What problem would be involved?

BotanyLinks .net Questions

www.jbpub.com/botanylinks

Visit the .net Questions area of BotanyLinks (http://www.jbpub.com/botanylinks/netquestions) to complete this question:

1.   Many conifers are big trees that live for centuries. With all the cutting of forests and loss of habitat, do any forests of old trees still exist? Go to the BotanyLinks home page for more information on this subject.

BotanyLinks includes a Directory of Organizations for this chapter.

SEED PLANTS II: ANGIOSPERMS

25

The seeds of angiosperms develop within the protection of the fruit. *(D. Cavagnaro)*

CONCEPTS

Flowering plants, the most recently evolved and most derived of all plant groups, are classified together in a single division, **division Magnoliophyta**, also known as **division Anthophyta.** The earliest fossils clearly recognizable as parts of flowering plants are preserved pollen grains in rocks more than 130 million years old. We have not found any fossilized leaves or flowers from that time, so we believe the earliest flowering plants must have lived in drier areas where few sites were suitable for fossilizing plant parts, such as large, deep lakes full of fine-grained sediments. The first angiosperms probably were small trees or shrubs, because plants of this type survive well in drier regions and because the most likely ancestors—other seed plants in the gymnosperm groups—are woody trees.

The evolutionary transformation of a gymnosperm into a flowering plant was not a simple process, and it involved many alterations. The most obvious, of course, was the conversion of gymnospermous sporophylls into stamens and carpels, resulting in the formation of flowers. Because gymnosperms have rather flat, leaflike sporophylls, those flowering plants that have similar reproductive parts are believed to be the most relictual. For example, groups such as *Magnolia, Austrobaileya,* and *Degenaria* have flat stamens without distinct filament and anther portions, and the sporogenous tissues, the microsporocytes (microspore mother cells), form relatively large, prominent internal masses (Fig. 25.1).

FIGURE 25.1

Several groups of angiosperms have stamens with what are believed to be relictual features; these features have not undergone much evolutionary change, so they are still similar to the ancestral features. The stamens (microsporophylls) of these four living dicots are much fleshier, flatter, and more leaflike than most other stamens of living species. Also, the sporogenous tissue is rather massive, similar to that of cycads and cycadeoids. *Austrobaileya, Himantandra,* and *Degeneria* are unfamiliar to most students; they are plants of South Pacific islands.

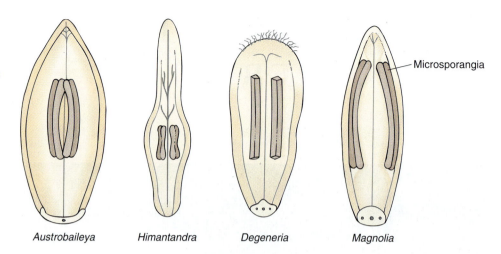

Austrobaileya   *Himantandra*   *Degeneria*   *Magnolia*   — Microsporangia

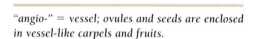

"angio-" = vessel; ovules and seeds are enclosed in vessel-like carpels and fruits.

Carpels in these three genera are similarly leaflike, resembling young leaves whose blades have not yet opened. Instead of a stigma and style, the ovary edges have rows of hairs that function as a stigmatic surface (Fig. 25.2). Whereas ovules in gymnosperms are exposed on the megasporophyll surface, the folded nature of the sporophyll in flowering plants offers the ovules some protection; the carpels are **closed** or **angiospermous,** as opposed to the gymnosperm's open sporophyll. Although botanists often speak of the angiospermous carpel protecting the ovules, probably much more important is the fact that the pollen tube (microgametophyte) must now interact with the stigmatic area in addition to the nucellus, allowing more opportunity for the sporophyte to interact with the pollen and "test" it. Weak pollen or that of the wrong species can be inhibited from germinating, and the sporophyte can permit only healthy pollen of the proper species to pass through the stigma and reach its ovules (Fig. 25.3). Fewer ovules are wasted by exposure to improper or unhealthy pollen.

In the transition from gymnosperms to angiosperms, fertilization evolved such that the second sperm cell of the pollen tube fuses with the polar nuclei of the megagametophyte, producing the endosperm nucleus. This process of **double fertilization** is universal in flowering plants. Within the vegetative body, the major transitions were the evolution of vessel elements and sieve tubes (Fig. 25.4). Leaves became broader and developed reticulate venation, and they became more polymorphic, adaptable to a variety of functions, not just photosynthesis.

FIGURE 25.2

Carpels are megasporophylls, so we expect the most relictual living carpels to be rather leaflike, somewhat resembling megasporophylls of cycads (see Fig. 24.33). Several living angiosperms do have carpels resembling young leaves that have failed to open (a); the two halves of the blade are pressed together, a rather large amount of vascular tissue is present, and a stigma and style are absent. Hairs along the blade margin act like a stigma. If such a carpel produced ovules only at the base and if the upper part elongated, it would begin to resemble a more derived, modern carpel (c).

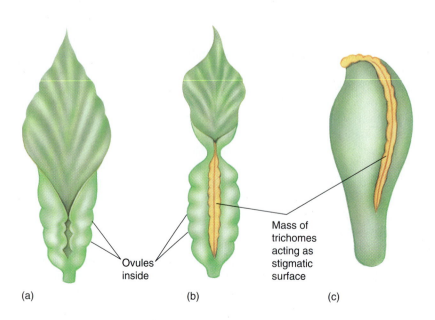

(a)   (b)   (c)   Ovules inside   Mass of trichomes acting as stigmatic surface

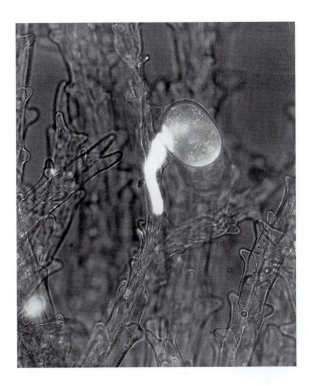

FIGURE 25.3

In many species of angiosperms, complex compatibility interactions occur between either the pollen and stigma or the pollen tube and stigma/style. In some cases, if the stigma and style detect an improper pollen tube, they deposit callose around it and inhibit its growth. The callose block is visible here as a white sheath ($\times$ 60). *(Courtesy of H. L. Mogensen, Northern Arizona University)*

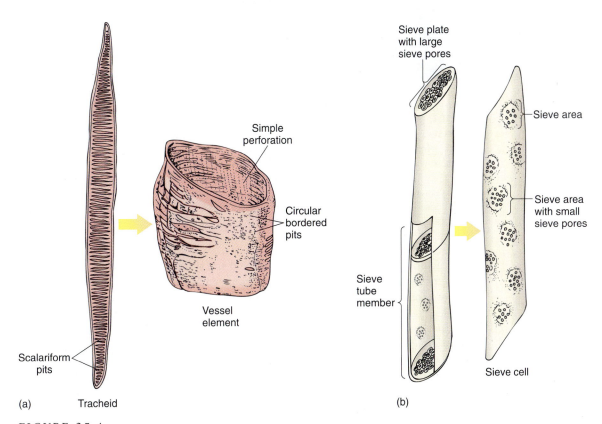

FIGURE 25.4

(a) In living plants, vessel elements that most closely resemble tracheids are long and narrow; short, wide vessel elements are derived. Most long, narrow vessel elements have scalariform perforation plates, so we assume that these vessel elements evolved from tracheids with scalariform pits. (b) Because a sieve element conducts mostly longitudinally, it is most efficient if the sieve areas and sieve pores on its ends are especially large. When this finally occurred, sieve tube members came into being.

(a)

(b)

(c)   Host plant   Tuber

FIGURE 25.5
(a) This dwarf daisy (*Monoptilon*) is by no means the smallest flowering plant. Many species of mustard, often the first flowers of spring, are also tiny. (b and c) *Ombrophytum subterraneum* from the town of Chiu Chiu in northern Chile, the southernmost outpost of the Inca empire. *Ombrophytum* is a parasite on roots of surrounding plants and is reported by people of the area never to emerge above ground. The plant in (c) consists of an irregular, lumpy "tuber" and two inflorescences with tiny pink flowers. The root and stem base of the host are visible. (*Courtesy of Eliana Belmonte, Universidad de Tarapacá*)

Structural modifications are most easily seen and studied, but probably a more fundamental transition was the acquisition of developmental plasticity, the capacity to survive mutations that alter growth and development. Conifers lack developmental plasticity; for example, most people never have trouble recognizing an unfamiliar plant as being a conifer when they see one because all conifers greatly resemble each other. They are all large or giant trees with one or a few main stems and needle- or scale-shaped leaves. They are always green or brown, never brightly colored except for the red seeds of yew and the gray-blue berry-like cones of junipers. This uniformity is even more striking when one considers that gymnosperms dominated Earth for hundreds of millions of years and could have diversified into almost any habitat. On the other hand, angiosperms have tremendous developmental plasticity: Beginning about 125 to 145 million years ago, they have diversified so greatly that not only hundreds of thousands of species, but also thousands of types, now exist. They range from gigantic (*Eucalyptus,* oaks, elms) to tiny (mustards, certain daisies; Fig. 25.5a); woody to herbaceous; perennial to biennial to annual; temperate to tropical; desert habitats to mesic to rainforest to aquatic; autotrophic to partially parasitic to fully parasitic; and epiphytic to subterranean (Fig. 25.5b) to endophytic (living inside another plant, such as *Rafflesia, Tristerix;* see Box 5.1).

The evolutionary changes involved in the conversion of a gymnosperm line into an angiosperm line did not occur instantaneously, nor did they occur in all species. Double fertilization and the ability to form flowers were probably the first transformations, because all flowering plants have these features. Sieve tubes possibly evolved next because only one or two species are thought to lack them. But the flowering plants must have existed and diversified for several million years before vessel elements evolved, because several families do not have vessels; these families must have evolved and become distinct before vessels appeared. All the herbaceous species today do have vessels, so herbaceousness must have developed next, and it probably arose several times in many separate lines. The ability to produce flowers in the first year after seed germination is a necessary trait for annual plants, most of which are herbs. Only much later did characteristics such as succulence, parasitism, and vining shoots develop; only a few isolated families or species have these characters.

Other derived features are the fusion of the petals into one structure (sympetally) and floral zygomorphy, that is, flowers that are bilaterally symmetrical, not radially symmetrical (Table 25.1). By this time the flowering plants had diversified so much that each of these evolutionary innovations involved only a small part of the whole group. It should not be thought that the flowering plants are now "finished," that all their evolution has already happened. Some groups do seem to be changing very little now, but others, especially grasses, composites, bromeliads, and orchids, are still changing and evolving so rapidly that it is difficult to keep up with them. Similarly, we have no reason to expect that the flowering plants will always be the most derived group; lycophytes dominated Earth for millions of years, only to be displaced by gymnospermous seed plants, which in turn were overshadowed by flowering plants.

A special word of caution is necessary about the flowering plants. Angiosperms have evolved in numerous directions and undergone many changes. For example, the *original* condition of early angiosperms is believed to have been large shrubs or small trees; some evolutionary lines appear to have retained this feature, so the shrubby plants of those lines are relictually shrubby. But other lines evolved to be herbaceous or nearly so (a derived

TABLE 25.1	*Relictual and Derived Features in Division Magnoliophyta*	
	Relictual (present in early angiosperms)	**Derived** (present only in some later, more modern angiosperms)
Habit		
Size	large bush, small tree?	large trees, herbs, bulbs, vines, many types
Leaf retention	evergreen	deciduous or leafless
Wood		
Vessels	none	present
Axial parenchyma	none or little	abundant; distribution is important
Rays	all narrow, tall	of several types in one plant: short/wide and narrow/tall
Flowers		
Presence of parts	complete flowers	incomplete/imperfect
Number of each type of part	many	few: sets of 3, 4, 5
Arrangement	spirals	whorls
Symmetry	radial	bilateral
Position of ovary	superior	inferior
Fusion of parts	none	much fusion
Pollination	wind? beetles?	many types
Fruit/seed dispersal	wind?	many types

feature), but then some of the descendants of these herbs became woody and shrubby again. Although they may appear homologous to the relictually shrubby plants, in these lines shrubbiness is actually a derived feature—they are secondarily shrubby. Because of the large amount of evolutionary change that has occurred in the angiosperms, we must always be careful not to mistake analogous features for homologous features.

ORIGIN AND EARLY EVOLUTION OF THE ANGIOSPERMS

Concepts about the nature of early angiosperms have changed as our knowledge of existing and fossil plants has become more complete. In the last century, wind-pollinated members of the subclass Hamamelidae (alders, elms, oaks, plane trees) were considered the most relictual of the living flowering plants and were known as the *Amentiferae* because their inflorescences are aments (an old term for catkin; Fig. 25.6a). At first glance this classification seems reasonable because these species tend to be woody trees, and most gymnosperms such as conifers are also wind-pollinated trees. Once their wood anatomy was studied, however, it became obvious the Hamamelidae could not be relictual plants because their wood contains vessels, fibers, and abundant parenchyma, features not found in gymnosperms. Furthermore, the flowers of Amentiferae are quite specialized: They often lack sepals and petals and are typically unisexual, occurring as either staminate or carpellate flowers, not as perfect flowers. It is difficult to imagine how such simple flowers could evolve into the much more complex flowers of other angiosperms.

About 70 years ago, C. E. Bessey developed the hypothesis of the "ranalean" flower, in which a *Magnolia*-type flower was thought to be relictual (Fig. 25.6b). Such a flower is **generalized**; that is, it has all parts (sepals, petals, stamens, carpels) and these are arranged spirally. Also, carpels occur in a superior position, above the other parts. This ranalean flower is quite different from the wind-pollinated flowers of the Amentiferae. It is easier to postulate the evolution of all the various existing flower types from a ranalean, generalized ancestor than from an amentiferous, specialized type of imperfect flower that lacks sepals and petals. For example, an amentiferous flower could evolve easily and quickly from a generalized, *Magnolia*-type flower by reduction or loss of parts.

This change in the concept of the relictual flower type was dramatic, not only because the differences in structure are so great, but because the ranalean flower is insect-pollinated, whereas gymnosperms are wind-pollinated. However, the magnolias and their rela-

FIGURE 25.6
(a) Catkins of alder *(Alnus rubra)*; each of the three main axes bears many flowers, all with stamens but not carpels. Other catkins on the plant would be carpellate and would lack stamens. *(C. E. Mills, University of Washington/Biological Photo Service)* (b) A flower of *Magnolia* is considered the typical ranalean flower—similar to that of the early angiosperms. All the parts are rather massive and are not specialized for one particular type of pollinator. Very few flowers have so many carpels, almost always fewer than ten; by contrast, seed cones of conifers and cycads have large numbers of megasporophylls. *(Dan Clark from Grant Heilman)*

(a)

Petals   Carpels   Stamens

(b)

tives, with their ranalean flowers, must be considered relictual, because the wood, with all of its numerous characters, is very similar to the wood of gymnosperms. It is now known that the four orders of flower-visiting insects (bees, butterflies, beetles, flies) had evolved at about the time angiosperms were appearing. They could have pollinated the early flowers, and the co-evolution of flowers and pollinators could have begun.

At present, our ideas about the nature of the vegetative body of relictual angiosperms are changing. Considerable doubt exists that plants were heavy, woody trees like either *Magnolia* or members of the Hamamelidae. Instead, they may have been fast-growing woody shrubs of sunny, somewhat dry, open habitats. They may have diversified, some lines becoming larger and more massive, others smaller and herbaceous. More importantly, we now have greater appreciation for the numerous differences between gymnosperms and angiosperms. The transition of some gymnosperms into angiosperms was gradual. We no longer expect to find one specific ultimate ancestor to the angiosperms or to pinpoint the date of origin. Studies now examine the step-by-step evolution of leaves and the development of a broad lamina with reticulate venation and an elongate petiole. In wood, the origin of vessels, scalariform bordered pits, fibers, and parenchyma is being studied. This same approach is being followed with the evolution of stamens, pollen, carpels, and pollinator relationships.

With new discoveries of fossil gymnosperms and with careful analysis of their microscopic details, we now realize that by the Jurassic Period in the middle of the Mesozoic Era (180 million years ago), many of the gymnosperms had features similar to those we associate with angiosperms. In many evolutionary lines various features were advancing at different rates. Some had leaves with reticulate venation; others had the first stages of scalariform bordered pits in their wood. Several had sporophylls that were beginning to look like carpels, and some of these were being visited by beetles. Almost all of this assemblage of gymnosperms with derived features became extinct; only a few produced lines that are still in existence today, such as the gnetophytes and the angiosperms. Our present task is to determine which of the fossil gymnosperms are part of this evolutionary line. We will almost certainly find a series of gymnosperms that resemble angiosperms more closely in a greater number of features but not a "last gymnosperm" or a "first angiosperm."

If the angiosperms evolved from the mid-Mesozoic assemblage of derived gymnosperms, how many lines were involved? Most botanists conclude that there was just one, that angiosperms are monophyletic. Features as complex as double fertilization, flowers, and developmental plasticity probably did not evolve more than once.

At present, many paleobotanists and taxonomists believe that the transition from gymnosperm to angiosperm occurred during the Jurassic and Lower Cretaceous Periods of the Mesozoic Era. The earliest leaf fossils definitely considered to be those of angiosperms are from the Lower Cretaceous Period, about 130 million years ago. They represent both dicot and monocot leaves, for example *Magnoliaephyllum, Ficophyllum* (figlike leaves), *Vitiphyllum* (grapelike leaves), and *Plantaginopsis* (similar to *Plantago*) (Fig. 25.7). These frag-

FIGURE 25.7
A leaf of *Magnoliaephyllum* showing strong similarities to the leaves of living magnolias. However, fine details visible by microscopy show that this is not identical to a true *Magnolia* leaf. (*Courtesy of D. Dilcher, Florida Museum of Natural History, University of Florida*)

ments are rare in such old rocks, constituting only about 2% of all fossils present. In rocks about 50 million years more recent, from the Upper Cretaceous Period, many more types of angiosperm leaf fossils are present, outnumbering the leaf fossils of gymnosperms and ferns. These Upper Cretaceous leaf fossils originally were assigned to extant genera such as *Populus* (poplar), *Quercus* (oak), and *Magnolia* on the basis of overall shape and size. But careful examinations of fine venation, cuticle, and stomata often show that these fossils are not truly part of modern genera.

The original misidentification of these leaves contributed to the erroneous hypothesis that angiosperms must have originated in the Jurassic Period, become abundant enough to leave a few fossils in the Lower Cretaceous Period, but then suddenly, in the short span of 30 to 50 million years, rapidly diversified into most of the major extant families and even many of the modern genera. After this supposed burst, genera such as *Populus, Quercus,* and *Magnolia* would have had to remain relatively unchanged for 80 to 90 million years, up to the present. Such a burst of diversification followed by extreme stability did not seem plausible. With the discovery that the Upper Cretaceous and Tertiary leaf fossils only resemble leaves of modern genera but are not identical to them in subtle features, a much more gradual evolution can be assumed to have occurred. The early angiosperms were well established by the Lower and Mid-Cretaceous Periods, but probably no modern genus was yet in existence, only precursors that greatly resembled modern genera. These would have evolved gradually into true poplars, oaks, magnolias, and so on.

The oldest wood that seems to be from an angiosperm comes from the Aptian Epoch (125 million years ago) of Japan; unequivocal angiosperm wood occurred about 120 million years ago. Flowers and fruits appear for the first time in the Lower Cretaceous Period, which must mean that their forerunners evolved in the Jurassic Period (Fig. 25.8). Often

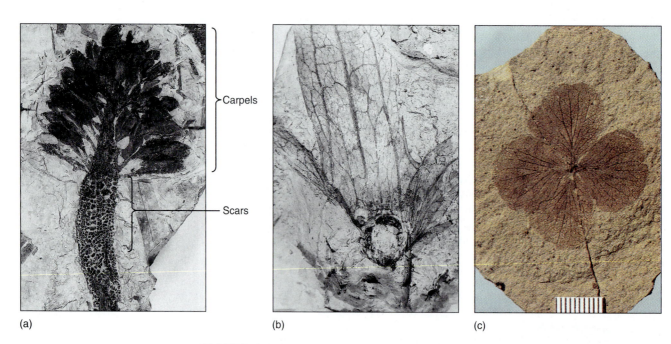

(a)   (b)   (c)

FIGURE 25.8

(a) Part of a flower from the Cretaceous Period; numerous spirally arranged carpels were still attached at the time it died and started to fossilize. The spiral arrangement is similar to that found in flowers of *Magnolia* and water lily. Below the carpels are spirally arranged scars where structures abscised; we do not know if they were sepals, petals, or stamens. (b) Part of a fossil fruit, *Paraoreomunnea puryearensis*, about 60 million years old. It was a winged fruit similar to those produced today by members of Juglandaceae, the family of walnuts and hickories. *(a and b, courtesy of D. Dilcher, Florida Museum of Natural History, University of Florida)* (c) Fossil of a *Hydrangea* flower, 44 million years old. This flower is much more modern and derived than that of Figure 25.8a, having fewer parts. *(Courtesy of S. R. Manchester, Florida Museum of Natural History, University of Florida)*

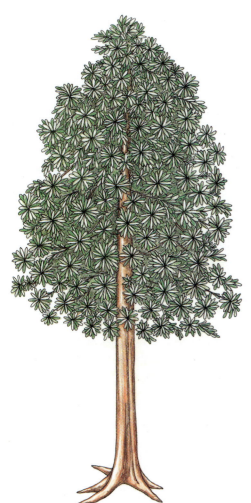

FIGURE 25.9
A reconstruction of a *Glossopteris*, a possible relative to the ancestors of flowering plants. The leaves had a prominent midrib and reticulate venation. The leaves were deciduous and the wood had annual rings, indicating that it lived in a temperate climate.

the fossils consist only of abscised carpels or carpels attached to a receptacle. Unfortunately, sepals, petals, and stamens often abscise even while the flower is still on the plant, so they are rarely present with the carpels. The perianth parts and stamens of modern plants also tend to be delicate and to decay quickly rather than fossilize; the same may have been true of the early flowers. Pollen that could have come from either relictual dicots or monocots is found in Lower Cretaceous strata.

Much attention is now being given to gymnosperms of the Jurassic and Triassic Periods, and the focus is centering on cycadophytes and glossopterids (Fig. 25.9). As more is learned about this group of gymnosperms, we realize how much the various groups had begun to develop angiosperm-like features.

CLASSIFICATION OF FLOWERING PLANTS

The Magnoliophyta is such a large group with so many families, genera, and species that it is rare for an individual taxonomist to attempt to study and classify the entire group. The most recent monograph of the entire division is *An Integrated System of Classification of Flowering Plants* (1981, Columbia University Press) by Dr. Arthur Cronquist of the New York Botanical Garden. Dr. Cronquist's classification is followed here.

Soon after their origin, flowering plants began to follow two distinct lines of evolution, and at present there are two classes: (1) **class Liliopsida,** the monocots, and (2) **class Magnoliopsida,** the dicots (Table 25.2). No single character always distinguishes a monocot from a dicot, and some species would fool most botanists. In general, monocots, as their name implies, have only one cotyledon on each embryo, and other typical characters are

"Dicot" and "monocot" are derived from the old subclass names "Dicotyledonae" and "Monocotyledonae."

TABLE 25.2 *Classification of Division Magnoliophyta*
Class Liliopsida (monocots)
Subclass Alismatidae
Subclass Arecidae
Subclass Commelinidae
Subclass Zingiberidae
Subclass Liliidae
Class Magnoliopsida (dicots)
Subclass Magnoliidae
Subclass Hamamelidae
Subclass Caryophyllidae
Subclass Dilleniidae
Subclass Rosidae
Subclass Asteridae

the following: Their leaves usually have parallel veins because the leaves are elongate and strap-shaped (grasses, lilies, and irises); vascular bundles are distributed throughout the stem, not restricted in one ring; monocots never have ordinary secondary growth and wood. Flowers of monocots have their parts arranged in groups or multiples of three: three sepals, three petals, three stamens, and three carpels (see Fig. 25.21). Dicots are much more diverse and include a greater number of families, genera, and species. Dicots have two cotyledons usually and reticulate venation in the leaves; vascular bundles occur in only one ring in the stem. Dicots can be woody, herbaceous, or succulent or have any of many highly modified forms. Flower parts occur in sets of five most often (see Fig. 25.34), sometimes in sets of four, but only rarely in sets of three.

CLASS LILIOPSIDA

At present, monocots are widely believed to have arisen from early dicots about 80 to 100 million, perhaps even 120 million, years ago. Certain features common to most monocots probably are inherited from their dicot ancestors. Because all monocots lack ordinary secondary growth and wood, the dicot ancestors were probably herbs with either no vascular cambium or little cambial activity. The gynoecia of many monocots are composed of several carpels that are either free of each other or at most only slightly fused together. The perianth is composed of sepals and petals that are usually relatively unspecialized and

Class Liliopsida is known as subclass Monocotyledonae both in older classifications and in several current ones.

FIGURE 25.10
The flowers of water lilies have numerous sepals, stamens, and so on, with all parts arranged in spirals like the leaves of a rosette plant. This is believed to be a relictual feature similar to a gymnosperm cone. Monocot flowers would be derived from an ancestral flower like this by mutations that result in only three or six of each type of floral appendage.

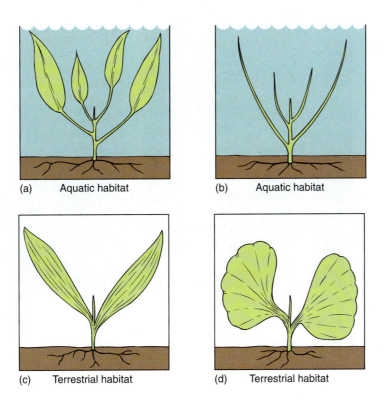

FIGURE 25.11
(a) Monocots are believed to have originated as semiaquatic plants that inhabited swamps and marshes. (b) The leaf may have consisted only of a leaf base and petiole, the entire blade having been lost evolutionarily. (c) A "pseudolamina" evolving from a petiole. (d) A "broad leaf" evolving from a petiole.

have very little fusion. These features must have been present in the group from which the monocots evolved. The dicot order Nymphaeales, which contains water lilies, has similar features. An early set of dicots may have resembled Nymphaeales and may have given rise to both the water lilies and the monocots (Fig. 25.10).

One theory about the parallel venation of monocot leaves postulates that the ancestors had broad leaves and lived as aquatic plants. Gradually the leaves evolved to a more reduced, simple type without a blade, a form that is more adaptive for a submerged leaf (Fig. 25.11a and b). Some of these plants moved into drier habitats where their leaves were not submerged and a broad or long lamina would be advantageous. Mutations that resulted in a basal meristem were selected, resulting in strap-shaped leaves (Fig. 25.11c). In the evolutionary lines of "broadleaf" monocots such as palms, philodendrons, and dieffenbachias, a marginal type of meristem evolved; it was located at the end of the residual leaf, resulting in the formation of what appears to be a petiole and lamina (Figs. 25.11d and 25.20).

Out of these first monocots, five subclasses evolved: (1) Alismatidae, (2) Arecidae, (3) Commelinidae, (4) Zingiberidae, and (5) Liliidae (Fig. 25.12). We do not know which

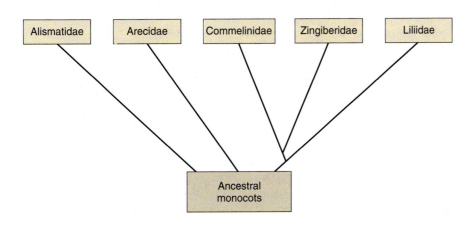

FIGURE 25.12
Diagram of a widely accepted classification of the monocots, class Liliopsida.

(a)   (b)

FIGURE 25.13
Sagittaria, a member of the Alismatidae, is dioecious; plants have either staminate (a) or carpellate (b) flowers.

<table>
<tr><td>

TABLE 25.3 *Classification of Subclass Alismatidae*

Order Alismatales
    Family Alismataceae

Order Hydrocharitales
    Family Hydrocharitaceae

Order Najadales
    Family Najadaceae
    Family Posidoniaceae
    Family Ruppiaceae

</td></tr>
</table>

group evolved first; none seems to be ancestral to any other subclass, so the earliest monocots may have undergone rapid divergent radiation, establishing five evolutionary lines that have been distinct throughout most of the history of monocots. At present the species of subclass Alismatidae retain the greatest number of relictual features, but no other group appears to have arisen from them. Rather, they have changed least through time, while the other groups have evolved more rapidly and more of their characters have become modified by natural selection.

Subclass Alismatidae.   In the Alismatidae are many aquatic herbs such as *Sagittaria* (arrowhead; Fig. 25.13) and many aquarium plants such as *Hydrocharis* and *Naja* (Table 25.3). These plants are most often found in swamps and marshes, partly or entirely submerged. Although many species retain a large number of relictual features, others have become highly modified in response to the unusual selection pressures associated with an aquatic habitat. The plants of many species of the families Hydrocharitaceae, Ruppiaceae, and Najadaceae grow completely submerged in water; they have no transpiration, so mutations that result in the loss of stomata are not selectively disadvantageous. Air chambers make the plants buoyant, so mutations that prevent the formation of lignified fibers are selectively advantageous; the plant does not waste carbon and energy by producing fibers that are not needed for support. Such plants tend to be thin and delicate, all parts of their bodies having very little sclerenchyma and almost no xylem. Tissues have been lost or simplified; the ancestors of Alismatidae were more complex and massive and had more sclerenchyma.

Other members of subclass Alismatidae have part of their bodies emergent from the water, and some are entirely emergent whenever the marsh dries temporarily. These plants are less highly modified, often having large leaves, considerable amounts of fiber, and a thick cuticle on the leaf epidermis. It is necessary for a species to be adapted to the most adverse conditions that it frequently encounters during one plant's lifetime. In the most relictual members of Alismatidae, flowers are large and showy, with three sepals and three petals, but in others they are highly modified, especially in those species in which the flowers do not emerge from the water and pollination occurs below the surface. Such flowers often completely lack a perianth.

Subclass Arecidae. The Arecidae is a small subclass containing only four orders, five families, and about 5600 species (Table 25.4). It is made up of larger plants, either aquatic or terrestrial, very often tropical. The most familiar are palms and aroids such as *Philodendron, Monstera,* and *Dieffenbachia.*

The family Palmae, also known as Arecaceae, contains about 3500 species, over half the species of the subclass. Palms are easily recognizable by their sturdy solitary trunk and large crown of leaves. In a few species, the trunk is prostrate (palmetto palm) or a vine (*Daemonorops*). All species have compound leaves, either pinnately compound (feather palms) or palmately compound (fan palms). The typical coconut is one type of palm fruit. Palm flowers are seldom seen because they are usually tiny, about 5 mm across, and are formed only high up in the tree (Fig. 25.14).

Palms vary somewhat in their body structure, their reproductive biology, and their type of habitat. Consequently they have spread through many tropical and semitropical regions, often being a major aspect of a flora. In most people's minds, palms are synonymous with Florida, Hawaii, and other tropical regions. However, it is instructive to consider palms carefully. The earliest known fossils are 80 million years old, so palms are an ancient group within the monocots. Although in some species the shoot may become massive, they never have any secondary growth. Because the capacity to produce annual, deciduous leaves has not evolved in this family, most palms cannot survive in areas with harsh winters. Because the stems tend to be wide, upward growth is slow, and palms are at a disadvantage when competing for light with trees that can elongate rapidly. As a result, palms are not found in dense forests, but rather in open ones that allow more light to reach the understory plants, such as young palms. Palms are numerous only because they are adapted to a widespread habitat; outside this particular ecological niche, however, they do not thrive.

The Family Araceae (aroids) contains several large orders of familiar houseplants: *Philodendron* (250 species), *Anthurium* (500 species), and *Arisaema* (Jack-in-the-pulpit; 100 species). *Dieffenbachia* (dumb cane) and 1800 other species also belong here (Fig. 25.15). This family is characterized by the evolution of a distinctive inflorescence: tiny flowers, either unisexual or perfect, embedded in a thick stalk, the spadix. Staminate flowers are located near the top of the spadix, carpellate flowers near the base. The spadix is surrounded by a large bract, the spathe (see Fig. 18.7).

TABLE 25.4	*Classification of Subclass Arecidae*

Order Arecales
    Family Arecaceae
Order Cyclanthales
    Family Cyclanthaceae
Order Pandanales
    Family Pandanaceae
Order Arales
    Family Araceae

(a)

(b)

FIGURE 25.14
(a) Palm flowers usually occur in large inflorescences at the top of a plant. *Corypha umbraculifera.* (*Kjell B. Sandved*) (b) Although palm flowers are small, they have typical monocot organization with their parts in sets of three. *Syagrus oleracea.*

FIGURE 25.15
This *Anthurium* (Araceae) is an aroid, one of the monocots that have a dicot type of broad leaf. Many members of this family are adapted to conditions of low light and high humidity; most are members of tropical or subtropical habitats.

Subclass Commelinidae. Subclass Commelinidae contains many familiar groups, such as grasses and sedges (Table 25.5). The evolution of this group has two prominent aspects. First, the flowers have become adapted to wind pollination (Fig. 25.16). Only the flowers of the most archaic members are still pollinated by insects (Fig. 25.17). In almost all other species, pollination is mediated by wind, and structures that would aid in attracting insects, such as nectaries or showy sepals and petals, have been either lost or modified such that they now aid in catching pollen or distributing the fruits after seeds have matured.

The second evolutionary trend is a reduction in body size; most species in the subclass Commelinidae are small herbs. Although they may be fibrous and tough like some grasses and sedges, they are rarely large and massive like palms and philodendrons. The giant bamboos are the only exception.

Members of this subclass fill many ecological niches. They range from tropical species to those that inhabit deserts, grasslands, marshes, forests, and tundra, various species each tolerating extremes of heat, cold, moisture, and drought. Nearly every type of habitat contains at least one member of this subclass.

The spiderwort family, Commelinaceae, contains relatively relictual, showy flowers such as those of *Tradescantia, Zebrina, Commelina,* and *Rhoeo*. They have three large sepals and two or three brightly pigmented petals; although they lack nectaries, they are insect pollinated. The androecium has three stamens, and the gynoecium consists of three carpels fused together into a single compound structure.

The grasses, family Poaceae, have about 8000 species and are much more than just the plants in the lawn. They also include most foods, such as wheat (*Triticum*), barley (*Hordeum*), oats (*Avena*), rye (*Secale*), corn (*Zea*), rice (*Oryza*), and sugar cane (*Saccharum*) as well as a major building material of the tropics, bamboo (the subfamily Bambusoideae). Grasses are extremely abundant in flat, open, dry regions in the central areas of all continents, such as the rangelands of the United States and Africa, the steppes of Russia and Ukraine, and the pampas of Argentina.

All grasses are wind pollinated, so sepals and petals are of little importance and are reduced to bristle-like structures. The three carpels are fused together, but in most grasses the three stigmas remain separate. An "ear" of corn is an inflorescence, and each "silk" is a

TABLE 25.5	*Classification of Subclass Commelinidae*
Order Commelinales	
Family Commelinaceae	
Order Juncales	
Family Juncaceae	
Order Cyperales	
Family Cyperaceae	
Family Poaceae	
Order Typhales	
Family Typhaceae	

FIGURE 25.16
Grass flowers are very reduced, simplified, and wind pollinated. The usual two small, dry scales (called a lemma and a palea) are thought to be remnants of an ancestral bract and two fused tepals; the third tepal does not form. Three anthers with long filaments and three carpels fused together with long feathery styles and stigmas are usually present. Grass flowers occur grouped together in complex, compact inflorescences, and inflorescence characters are important for identifying grasses.

FIGURE 25.17
Flowers of *Commelina* have many relictual features. They are still insect pollinated and have a large, showy perianth. There are three petals, but one is smaller than the other two, causing the flower to be bilaterally symmetrical. It often has six stamens, but three may be very small or suppressed. *(Robert and Linda Mitchell)*

compound style and stigma (see Fig. 9.26). Within the fused ovary of grasses there is just one ovule; once fertilized, it matures into a seed whose seed coat fuses firmly to the developing fruit wall. The objects that we usually call the "seeds" of corn, wheat, and oats are actually single-seeded dry fruits technically known as **caryopses** (singular: caryopsis). Nearly the entire bulk of the seed is endosperm; the embryo is small but well developed.

Closely related to grasses are sedges (Cyperaceae with 4000 species; Fig. 25.18) and rushes (Juncaceae). Other members of this subclass are cattails (Typhaceae).

Subclass Zingiberidae.　This subclass contains some of the most familiar of all houseplants: *Maranta, Calathea*, canna lilies *(Canna)*, gingers *(Zingiber, Hedychium)*, and bromeliads, as well as some that are best known in the warmer southern states—banana *(Musa)*

FIGURE 25.18
Most sedges grow in wet, marshy areas. *Carex stricta. (Pat Lynch/Photo Researchers, Inc.)*

TABLE 25.6   *Classification of Subclass Zingiberidae*
Order Bromeliales
Family Bromeliaceae
Order Zingiberales
Family Cannaceae
Family Heliconiaceae
Family Marantaceae
Family Musaceae
Family Zingiberaceae

and bird-of-paradise (*Strelitzia*) (Table 25.6). Members of subclass Zingiberidae differ from those of the other monocot subclasses discussed so far in that they tend to have large, showy flowers pollinated by insects, birds, or bats. Furthermore, many of flowers have derived features; adjacent sepals are often fused to each other, forming a tube, and the same is often true of the petals. The gynoecium typically consists of three carpels that have fused almost completely, except for small regions that act as nectaries (**septal nectaries**). Very often the gynoecium is inferior, located below the sepals and petals, a condition that is always considered quite derived, the result of considerable evolutionary modification.

Subclass Zingiberidae contains two orders (Bromeliales and Zingiberales), nine families, and almost 3800 species. Members of one subclass of monocots often resemble those of other subclasses; the Zingiberidae show that it can be quite difficult to be certain which groups are related to which. The Bromeliales are often classified with the Commelinidae, and the Zingiberales with the Liliidae. Not everyone would agree that the two orders should be combined and placed in their own subclass, as has been done by Dr. Cronquist. Many characters are being analyzed to understand the phylogeny of these groups. In addition to morphology and anatomy, important features are the nature of crystals, starch grains, silica deposits, protein bodies, chemical characters, and even susceptibility to attack by certain

(a)   (b)

(c)

FIGURE 25.19
(a) The inflorescences of many bromeliads have brilliant pigmentation. One hypothesis postulates that this coloration increases visibility of the flowers in the dim light of a dense rain forest. (*D. Cavagnaro/Visuals Unlimited*) (b) Spanish moss, *Tillandsia usneoides,* is a bromeliad, not a moss. It grows attached to tree trunks as an epiphyte, not a parasite. This species grows in the southeastern United States, where humidity is high and rainfall is frequent. (c) Pineapple plants are terrestrial bromeliads; they grow with their roots in soil. The edible portion of pineapple is a multiple fruit, composed of an entire inflorescence. (*Phil Degginger*)

fungi. It is difficult to know which features are homologous (descent from a common ancestor) and which are analogous (convergent evolution). Until definitive proof of one classification or another can be obtained, we must be aware of the complexity and understand why differences of opinion exist.

Order Bromeliales has only one family, Bromeliaceae, which contains some of the most beautiful tropical epiphytes, their large, brightly colored inflorescences being easily visible even in thick jungle vegetation (Fig. 25.19a). Epiphytic species extend as far northward as the subtropics, Spanish moss and ball moss (both are species of *Tillandsia*) occurring from Florida to Texas (Fig. 25.19b). Other species are terrestrial and usually xerophytic, such as *Puya* of Chile and Peru and pineapples (*Ananas*). These xerophytic terrestrial forms are believed to be relictual, and their capacity to withstand drought may have made it easier for bromeliads to adapt to life as epiphytes. Because bromeliads occur only in the Americas, it is assumed that they evolved after South America separated from Africa about 80 million years ago (see Chapter 27). Had they evolved earlier, numerous species should occur in the African rain forests and coastal deserts.

Order Zingiberales contains eight families; the largest, Zingiberaceae, has about 1000 species. Members of this order are almost all tropical, and most are soft, nonleathery herbs. The leaves are broad and have a petiole, resembling those of dicots, but they are not believed to be homologous with dicot leaves (Figs. 25.11 and 25.20). During the evolution of the Zingiberales, mutations that resulted in the petiole/leaf becoming flattened and broadened were selectively advantageous: The plants grow in the heavy shade of jungle understory where light interception is important.

Subclass Liliidae.   Most botanists consider subclass Liliidae to contain the most derived of all monocots, lilies (family Liliaceae with 4000 species) and orchids (Orchidaceae with over 20,000 species) plus 17 other families (Table 25.7). Most species of Liliidae rely on insect pollination and have large, showy flowers that are often wonderfully aromatic (Figs. 25.21 and 25.22a). Septal or other types of nectaries are usually well developed. A common habit (growth form) is the formation of a tough succulent stem; in the lily family the stem is often subterranean as a corm, rhizome, or bulb. In orchids it is frequently present as an aerial pseudobulb that can persist even during stressful conditions. The leaves and even roots may die, but when conditions become favorable, the pseudobulb produces new leaves, roots, and flowers.

FIGURE 25.20
Some monocots, such as this banana (*Musa*), have broad leaves, but their lamina is not homologous to that of a dicot leaf. In this monocot broad leaf, all vascular bundles run parallel to each other from the midrib to the margin, whereas in a dicot lamina, they diverge and branch into a reticulate pattern.

FIGURE 25.21
Lilies and their relatives are excellent for demonstrating the typical monocot flower: They have three of each type of appendage and all are large enough to be seen easily. The three carpels are fused together, but a transverse section shows three chambers.
(Runk/Schoenberger from Grant Heilman)

TABLE 25.7	*Classification of Subclass Liliidae*
Order Liliales	
Family Agavaceae	
Family Iridaceae	
Family Liliaceae	
Order Orchidales	
Family Orchidaceae	

(a)   (b)

FIGURE 25.22

Members of subclass Liliidae. (a) *Iris.* (b) Like bamboos and palms, yuccas and agaves prove that herbs do not have to be soft and small. The leaves are long and narrowly triangular with parallel venation, and they may be thick at the base. They are arranged as a rosette around a short (agaves) or long (yuccas) stem. Most have a needle-sharp point and razor-like hooked spines along the edges of the leaf. *(Courtesy of David Bogler, University of Texas)*

The lily family contains so many ornamental plants that most people are familiar with them and think of them as the "typical" monocots. Liliaceae contains lilies, day-lilies, tulips, onions, asparagus, and amaryllis. The heavier, coarser agaves and yuccas are sometimes placed in the Liliaceae, sometimes in their own family, the Agavaceae. The lily family is believed to have given rise to the Iridaceae (irises) and Dioscoreaceae (yams).

The Orchidaceae is the largest and most diverse family of all monocots. The most familiar, ornamental ones are epiphytic, but many are terrestrial and one is a subterranean parasite. Orchid flowers are highly modified from the ancestral condition of the angiosperms, being zygomorphic (bilaterally symmetrical) with complex shapes, colors, and fragrances that attract specific pollinators. Orchids are unusual in producing hundreds or thousands of tiny seeds in each fruit. The seeds are dustlike and so undeveloped at "germination" that they must form a symbiosis with certain fungi in order to survive long enough to form roots and leaves.

According to Dr. Cronquist, the orchids must be derived from lilies; the most derived lilies have characters that would be expected to occur in the ancestors of orchids. Virtually all features of the orchids are highly derived and extremely modified. They have little in common with those members of subclass Alismatidae that are thought to retain relictual features of the earliest monocots.

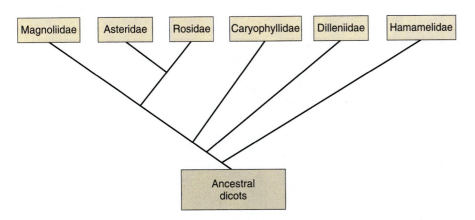

FIGURE 25.23
Diagram of a widely accepted classification of the dicots, class Magnoliopsida.

CLASS MAGNOLIOPSIDA

The dicots, class Magnoliopsida, constitute a much larger group than the monocots and are divided into six subclasses: (1) Magnoliidae, (2) Hamamelidae, (3) Caryophyllidae, (4) Dilleniidae, (5) Rosidae, and (6) Asteridae (Fig. 25.23). They are more difficult to characterize than are the monocots because they are so diverse. Virtually every type of organ, tissue, and metabolism has several or many variations, resulting in hundreds of thousands of species of dicots. A simple description would be both inaccurate and misleading, so each subclass is described below in some detail.

Subclass Magnoliidae. Subclass Magnoliidae is made up of those dicot species that (1) have relictual characters we associate with early angiosperms and (2) are apparently not related to more derived groups (Table 25.8). For example, family Magnoliaceae contains trees with wood similar to that of gymnosperms in that it lacks vessels, fibers, and axial parenchyma; magnolia flowers have many of each organ, and the numerous stamens and carpels are arranged in spirals as leaves usually are (see Fig. 25.6b). The carpels are not fused together into a compound gynoecium. Another important feature is that their pollen grains have only a single germination pore: They are **uniaperturate,** as are monocots. More derived dicots have three germination pores (Fig. 25.24). Other members of the Magnoliidae are the laurels and avocado (Lauraceae), star anise (Illiciaceae), buttercups and anemones (Ranunculaceae), peperomias (Piperaceae), and water lilies (Nymphaeaceae in the order Nymphaeales, which may be related to the ancestors of monocots).

Class Magnoliopsida is known as subclass Dicotyledonae both in older classifications and in several current ones.

TABLE 25.8   *Classification of Subclass Magnoliidae*

Order Magnoliales	Order Nymphaeales
Family Austrobaileyaceae	Family Nelumbonaceae
Family Degeneriaceae	Family Nymphaeaceae
Family Magnoliaceae	Order Ranunculales
Family Winteraceae	Family Berberidaceae
Order Laurales	Family Ranunculaceae
Family Lauraceae	Order Papaverales
Order Piperales	Family Fumariaceae
Family Piperaceae	Family Papaveraceae
Order Illiciales	
Family Illiciaceae	

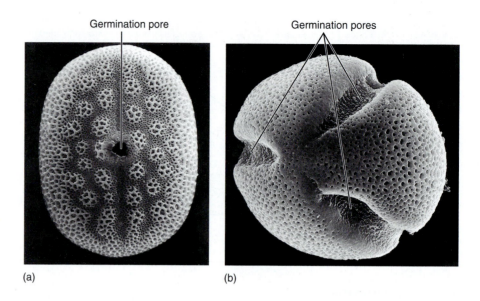

FIGURE 25.24

(a) All early fossil pollen has only one aperture, but almost all living dicots with many derived features have three apertures. Therefore, uniaperturate pollen in living angiosperms, such as this *Siphonoglossa*, is considered a relictual feature. Other features of this genus are derived (× 300). (b) This pollen of *Lophospermum* has three germination pores; the pores are located in grooves, something that does not occur in any early fossil pollen (× 300). *(a, courtesy of R. A. Hilsenbeck, Sul Ross State University; b, W. J. Elisens, University of Oklahoma)*

Several member groups are important medicinally: Papaveraceae (poppies) is the source of a milky latex harvested for opium. Morphine, a strong analgesic (pain killer) that depresses the cerebral cortex, is extracted from opium, as is codeine. *Chondodendron* (Menispermaceae) produces curare, a drug that blocks nerve transmission and causes temporary paralysis; it is used in surgery to relax muscles without the need for deep anesthesia.

Subclass Magnoliidae is unusual in that some of its member families consist of woody tree species (Magnoliaceae, Degeneriaceae, Lauraceae), whereas others are aquatic or semi-aquatic herbs (Nymphaeaceae and many Ranunculaceae). Although the overall aspects of these two types of plant are so different that one would not immediately place magnolias and water lilies together, they have numerous anatomical, chemical, and reproductive similarities. Not long after dicots evolved from gymnosperms to a state somewhat like that of magnolias, they may have divided into two evolutionary lines, one adapting to marshy or aquatic habitats and becoming less woody and more herbaceous. This line may have resulted in water lilies and monocots. The other line would have remained terrestrial and woody, giving rise to the rest of the flowering plants.

Subclass Hamamelidae.   Some of the more common members of the subclass Hamamelidae are the sycamores (Platanaceae); walnut, hickory, and pecan (Juglandaceae); oak, chestnut, and beech (Fagaceae); marijuana and hops (Cannabaceae); alders and birches (Betulaceae); and stinging nettles (Urticaceae) (see Figs. 25.6a and 25.25; Table 25.9). Subclass Hamamelidae consists of an evolutionary line of plants that began to differentiate about 120 million years ago, during the Lower Cretaceous Period, in areas that had alternating wet and dry seasons. In adapting to such a climate, they became deciduous, their leaves abscising each autumn, as opposed to remaining evergreen like members of subclass Magnoliidae. They produce flowers early in spring, before their new leaves expand, so the flowers are exposed to wind; it has been possible for these species to revert to wind pollination. This combination of characters permitted them to migrate out of the tropics into drier, colder temperate regions. At that time, temperate regions were sparsely populated and the members of Hamamelidae expanded their range and diversified rapidly, dominating the regions. However, by the Upper Cretaceous Period, about 80 million years ago, insect-pollinated members of subclasses Rosidae and Dilleniidae began to adapt to temperate climates and to outcompete the Hamamelidae, causing a decline in their numbers and diversity. As insects and Rosidae continue to co-evolve, the Hamamelidae may become even more restricted.

At present, Hamamelidae contains 11 orders, 24 families, and about 3400 species. Many of the families are small and well defined, quite distinct from other families; such characters lead us to suspect that the families are old and relictual, having originated and

FIGURE 25.25
American sycamore (*Platanus occidentalis*) is a member of the Hamamelidae, a wind-pollinated tree.
(*L. Lefever from Grant Heilman*)

TABLE 25.9 *Classification of Subclass Hamamelidae*
Order Trochodendrales
Family Tetracentraceae
Family Trochodendraceae
Order Hamamelidales
Family Hamamelidaceae
Family Platanaceae
Order Urticales
Family Cannabaceae
Family Moraceae
Family Ulmaceae
Family Urticaceae
Order Juglandales
Family Juglandaceae
Order Fagales
Family Betulaceae
Family Fagaceae

TABLE 25.10 *Classification of Subclass Caryophyllidae*

Order Caryophyllales
   Family Aizoaceae
   Family Amaranthaceae
   Family Cactaceae
   Family Caryophyllaceae
   Family Chenopodiaceae
   Family Nyctaginaceae
   Family Phytolaccaceae
   Family Portulacaceae
Order Polygonales
   Family Polygonaceae
Order Plumbaginales
   Family Plumbaginaceae

diversified early. They are not changing rapidly now and are not producing many new species. A large number of species, especially in the orders Fagales, Hamamelidales, and Juglandales, consist of large woody trees, similar to those of Magnoliidae. Only in the order Urticales has there been a marked evolutionary trend toward herbaceousness.

Subclass Caryophyllidae. Examples of Caryophyllidae are pokeweed (Phytolaccaceae), cacti (Cactaceae), iceplant (Aizoaceae), sorel and buckwheat (Polygonaceae), portulaca (Portulacaceae), bougainvillea and four-o'clocks (Nyctaginaceae), spinach, beets, and Russian thistle (Chenopodiaceae), and carnations and chickweed (Caryophyllaceae) (Table 25.10; see Fig. 25.29). The order Caryophyllales, containing 14 families, is held together strongly by many derived characters, but one is especially important. Whereas other flowering plants have anthocyanin pigments in their flowers, almost all Caryophyllales instead produce a group of water-soluble pigments called **betalains** (Fig. 25.26). The ancestral group may have lacked petals in their flowers, a condition common in Ranunculaceae (subclass Magnoliidae) and many Caryophyllidae. Many species of this subclass do have petals now, but their petals are thought not to be homologous with petals of other angiosperms. Instead, in some they appear to be modified sepals, in others modified stamens.

FIGURE 25.26
The order Caryophyllales is characterized by the presence of betalain pigments in flowers and fruits. There are two basic types: betacyanins (red to violet) and betaxanthins (yellow).

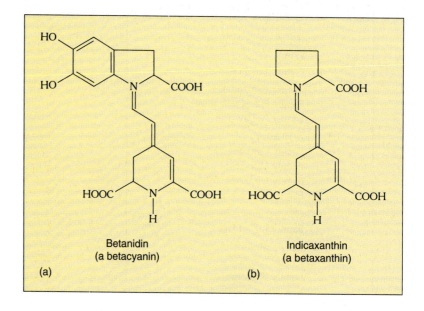

Betanidin
(a betacyanin)

(a)

Indicaxanthin
(a betaxanthin)

(b)

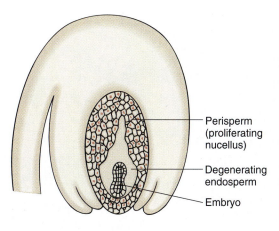

FIGURE 25.27
In Caryophyllales, endosperm develops for only a short while or not at all and is not sufficient to nourish the embryo. Instead, the nucellus (megasporangium) cells proliferate and act like endosperm, forming a tissue called perisperm. Which do you think probably occurred first in evolution—the mutations that caused the lack of formation of endosperm or those that caused the formation of perisperm?

Perisperm (proliferating nucellus)

Degenerating endosperm

Embryo

Another unifying character of the order Caryophyllales is that the endosperm develops little if at all, and a different nutritive tissue (**perisperm**) develops by the proliferation of the nucellus (Fig. 25.27).

A third feature that unites the members of this order and distinguishes them from all other taxa is the nature of the plastids in their sieve tube members. In Caryophyllales, phloem plastids contain deposits of fibrous protein located as a ring just interior to the plastid membrane (Fig. 25.28). In other subclasses, plastids contain either starch or crystalline protein with a different structure.

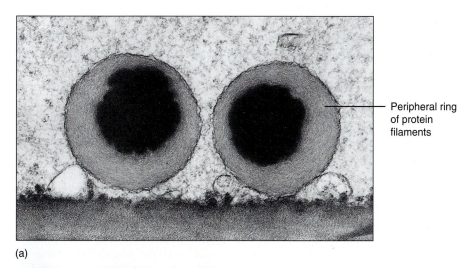

Peripheral ring of protein filaments

(a)

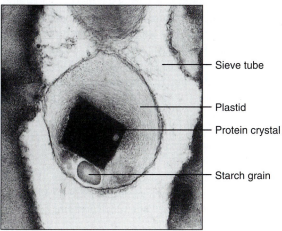

Sieve tube

Plastid

Protein crystal

Starch grain

(b)

FIGURE 25.28
The plastids in sieve tube members are simple, but they may accumulate particles of starch or protein or both. The nature of the accumulated material is highly specific with regard to the family of the organism, and phloem plastid analysis is important in studying the evolution of flowering plants. Sieve tube plastids of *Monococcus* (a) lack a central cubic protein crystal, but those of *Macarthuria* have one (b). Both are members of order Caryophyllales, so both have a peripheral ring of protein filaments. (Both × 40,000) (*Courtesy of H.-Dietmar Behnke, University of Heidelberg*)

(a)   (b)

FIGURE 25.29

Members of subclass Caryophyllidae: (a) Portulaca (*Portulaca grandiflora*), (b) *Carpobrotus.*
(*a, B. Head/Earth Scenes*)

TABLE 25.11	*Classification of Subclass Dilleniidae*

Order Dilleniales
    Family Dilleniaceae
    Family Paeoniaceae

Order Theales
    Family Theaceae

Order Malvales
    Family Bombacaceae
    Family Malvaceae
    Family Sterculiaceae
    Family Tiliaceae

Order Nepenthales
    Family Droseraceae
    Family Nepenthaceae
    Family Sarraceniaceae

Order Violales
    Family Begoniaceae
    Family Cucurbitaceae
    Family Fouquieriaceae
    Family Passifloraceae
    Family Violaceae

Order Salicales
    Family Salicaceae

Order Capparales
    Family Brassicaceae
      (= Cruciferae)
    Family Capparaceae

Order Ericales
    Family Ericaceae
    Family Monotropaceae

Order Primulales
    Family Myrsinaceae
    Family Primulaceae

The Caryophyllidae are postulated to have arisen from ancestors to the family Ranunculaceae or a similar group, about 70 to 80 million years ago. The members of Caryophyllidae are mostly herbaceous, with either no wood, very little wood, or unusual, anomalous wood (see Figs. 8.29 and 25.29). The ancestral group is therefore suspected to have been herbaceous or shrubby, having lost the capacity for extended, massive secondary growth. Those members of Caryophyllidae that are now large trees typically have anomalous secondary growth; because these woody plants evolved from an herbaceous ancestor, their wood is not strictly homologous with that of other angiosperms. All the cell types are the same, but often the organization of the wood is different.

Subclass Dilleniidae.   Examples of Dilleniidae are *Paeonia* (Paeoniaceae), *Camellia* (both the ornamental and the plant from which tea is produced; Theaceae); *Hypericum* (St. John's wort; Clusiaceae); linden tree (Tiliaceae); *Theobroma cacao* (source of **chocolate**; Sterculiaceae); balsa wood (Bombacaceae); cotton, *Hibiscus,* and mallows (Malvaceae); insect-catching pitcher plants (Sarraceniaceae and Nepenthaceae) and sundews (Droseraceae); violets (Violaceae); passion flowers (Passifloraceae); pumpkins, watermelons, and squash (Cucurbitaceae); *Rhododendron,* cranberries, blueberries, heather, and mountain laurel (Ericaceae); and cyclamen and primroses (Primulaceae) (Table 25.11). Subclass Dilleniidae evolved from subclass Magnoliidae. Their most striking evolutionary modifications are fusions of the flower parts and several technical features. Whereas the most relictual flowering plants have numerous flower parts, all separate and free of each other as in *Magnolia,* in Dilleniidae there are strong trends of fusion. Carpels are most commonly fused together into one compound gynoecium, and in about one third of the species, petals are fused together (Fig. 25.30). The most relictual group within the subclass Dilleniidae is the order Dilleniales (which contains peonies), especially the family Dilleniaceae. It does not have many of the characters of the subclass and could be placed rather easily into subclass Magnoliidae; it has been suggested that the Dilleniidae diverged from the Magnoliidae about 75 million years ago.

The central order of Dilleniidae is Theales, which contains camellias, tea, and *Hypericum.* The rest of the subclass seems to have diversified from this group. During the later evolution into the various orders, numerous modifications occurred, leaving few dominant similarities. Instead, this is a diverse group, occupying many habitats and having numerous types of structure, metabolism, and reproductive biology. Dilleniidae contains 13 orders, 78 families, and about 25,000 species.

(a)

(b)

(c)

(d)

(e)

FIGURE 25.30

(a) These flowers of *Cavendishia* in the Ericaceae show the derived feature of sympetaly—petals fused together as one unit, except at their tips. Fused parts are common in recently evolved flowers, rare in ancient ones. (b to e) Members of subclass Dilleniidae: (b) swamp rose-mallow (*Hibiscus palustris*); (c) St. John's wort (*Hypericum spathulatum*); (d) pitcher plant (*Sarracenia alabamense*); (e) *Theobroma cacao*, the source of chocolate. Flowers and fruits are borne on the trunk and large branches, not on twigs. The seeds are harvested, fermented, roasted, and ground into cocoa powder. (*b, S. Rannels from Grant Heilman; c, John Lynch, PHOTO/NATS; d, William E. Ferguson; e, James L. Castner*)

Subclass Rosidae.   Subclass Rosidae is the largest in terms of the number of families it contains (114), and it seems to be a natural group, although it is difficult to give any universal characters (Table 25.12). As a group, the members of Rosidae are more derived than those of subclass Magnoliidae, from which they arose (probably beginning about 115 million years ago), but have none of the specialized characters of the Hamamelidae (wind pollination) or Caryophyllidae (betalains and perisperm). It is difficult to distinguish between the Rosidae and the Dilleniidae, and one theory postulates that they constitute a single group. The Rosidae are especially interesting in that none of them has any of the highly relictual features found in some of the Magnoliidae.

One important character in Rosidae is the presence of pinnately compound leaves. This is believed to have been the ancestral condition for the subclass, the Rosidae having evolved from some member of the Magnoliidae that had pinnately compound leaves. Although some living species have simple leaves, these apparently have arisen from compound leaves, perhaps by suppression of all the leaflets except one. Whereas simple leaves are an early, relictual condition of the division Magnoliophyta, they are a later, derived condition for subclass Rosidae.

TABLE 25.12 *Classification of Subclass Rosidae*

Order Rosales
    Family Crassulaceae
    Family Grossulariaceae
    Family Hydrangeaceae
    Family Pittosporaceae
    Family Rosaceae
    Family Saxifragaceae
Order Fabales
    Family Fabaceae (= Leguminosae)
Order Myrtales
    Family Myrtaceae
    Family Onagraceae
Order Cornales
    Family Cornaceae

Order Santalales
    Family Balanophoraceae
    Family Loranthaceae
    Family Santalaceae
    Family Viscaceae
Order Rafflesiales
    Family Hydnoraceae
    Family Rafflesiaceae
Order Euphorbiales
    Family Euphorbiaceae
Order Rhamnales
    Family Vitaceae

Order Polygalales
    Family Malpighiaceae
    Family Polygalaceae
Order Sapindales
    Family Aceraceae
    Family Burseraceae
    Family Hippocastanaceae
Order Geraniales
    Family Geraniaceae
    Family Oxalidaceae
Order Apiales
    Family Apiaceae (= Umbelliferae)

(a)

(b)

(c)

(d)

FIGURE 25.31
Members of subclass Rosidae: (a) lupine (*Lupinus latifolius*), (b) horse chestnut (*Aesculus hippocastanum*), (c) Mexican plum (*Prunus mexicana*), (d) strawberry (*Fragaria*). (*b, Kent and Donna Dannen/Photo Researchers, Inc.*)

PLANTS & PEOPLE

MAINTAINING GENETIC DIVERSITY

The flowering plants are particularly adept at evolving rapidly; in the short time they have existed, they have diversified greatly. The result is that hundreds of thousands of species exist, each differing from the others in features such as flowers, fruits, vegetative structure, and metabolism. Within this panoply of diverse plants, we and our ancestors have discovered a wealth of foods, medicines, and soul-soothing flowers.

Now, at the end of the 20th century, we have become aware of the horrific ecological destruction occurring on Earth, with vast areas of vegetation being cut or burned or flooded with artificial lakes behind hydroelectric dams. We have also begun to realize that we are causing the extinction of hundreds of species every year, and, in the process, their genetic information is being lost. On a philosophical basis, we can debate whether one species, *Homo sapiens*, has the right to destroy so many other species. But on a more pragmatic basis, we should consider whether we are unwittingly destroying potential sources of new foods and medicines. Consider just two examples: A relatively unknown species, quinoa (*Chenopodium quinoa*), can be grown in relatively dry regions with poor soil, yet it produces an abundance of nutritious, protein-rich seeds. And *Pantadiplandra brazzeana* produces a small protein that is 2000 times sweeter than sugar; this could possibly be used as a completely natural, safe, nutritious, low-calorie sweetener, decreasing the amount of sugar or artificial sweeteners that we consume.

How can botanists preserve the genetic resources of the plants, animals, and microbes that coexist with us? Several types of efforts are needed, and two particularly important ones are habitat preservation (see Plants and People in Chapter 17) and *germ plasm banks* (germ plasm refers to the cells that produce gametes). Germ plasm banks were started several years ago as rather simple seed storage facilities, where bags of many types of seeds were maintained under cool, dry conditions. After several years, while most seeds were still healthy, they would be germinated, grown into mature plants and then their fresh, healthy seeds would be collected and stored. After several years, the process would be repeated.

This is still done for many species, but now the seeds can be kept for extremely long periods, held at very low temperatures or even in liquid nitrogen. Furthermore, it is now possible to maintain some seeds as tissue or cell cultures, either grown slowly in culture tubes or kept as frozen cells. In the foreseeable future, it may be that some plants, or at least some of their genes, will be kept simply as DNA molecules in solution or in cloning vectors.

At present, there are over 100 germ plasm banks throughout the world, preserving at least 3 million types of plant material. Examples are the Svalbard International Seedbank, built in an abandoned coal mine in the permafrost of northern Norway, and the National Seed Storage Laboratory (NSSL) located at Colorado State University and run by the United States. The NSSL was established in 1958 and has more than 232,000 seed types. At present, its size is being quadrupled, and its capacity will be more than 1 million seed samples.

The purpose of germ plasm banks is not merely to store and maintain plant genetic diversity but also to make it available. Botanists can request samples and use them for research and plant breeding programs. If some are found to have useful traits, they can be grown as crops. Of course, it is also hoped that in many cases, we can use these plants to help restore or rebuild ecosystems that we have already damaged.

Although the subclass is named for the Order Rosales, roses should not be considered typical because Rosales is the most relictual order, the one closest to the Magnoliidae. The members of the rest of subclass Rosidae are much more derived and modified (Fig. 25.31). Because Rosidae contains 18 orders, 114 families, and 58,000 species, only a few can be mentioned here. Five of the orders contain almost 75 percent of the species: Fabales (legumes) 14,000 species; Myrtales (*Eucalyptus,* evening primrose) 9000; Euphorbiales (poinsettia, rubber tree) 7600; Rosales (roses, saxifrages) 6600; and Sapindales (maples, horse chestnuts, creosote bush, and the species whose resins are valued as frankincense—*Boswellia*—and myrrh—*Commiphora*) 5400. Members of this subclass include roses, of course, and legumes (peas, beans, peanuts, *Mimosa,* redbud, and clover; Fabaceae = Leguminosae); *Fuchsia,* evening primrose (Onagraceae); dogwood (Cornaceae); mistletoes and some of the other parasites (Santalaceae, Loranthaceae, Viscaceae, and Balanophoraceae); the spurges that look like cacti and often have an extremely poisonous milky latex (Euphorbiaceae); grapes (Vitaceae); maples (Aceraceae); geraniums (Geraniaceae); dill, celery, carrot, parsley, and hemlock (Apiaceae = Umbelliferae).

The rose family is important not only in an evolutionary sense but also economically. Rosaceae contains numerous ornamental genera, including *Rosa, Crategus* (hawthorn), *Spiraea, Cotoneaster, Pyracantha, Photinia, Potentilla, Chaenomeles* (flowering quince), and *Sorbus* (mountain ash). The family also provides most of the fruits that can be grown in temperate climates: *Malus* (apple), *Pyrus* (almond, apricot, cherry, nectarine, peach, plum, prune), *Eriobotrya* (loquat), *Fragaria* (strawberry), and *Rubus* (blackberry, loganberry, and raspberry).

TABLE 25.13 *Classification of Subclass Asteridae*

Order Gentianales
    Family Apocynaceae
    Family Asclepiadaceae
    Family Gentianaceae
Order Solanales
    Family Convolvulaceae
    Family Cuscutaceae
    Family Hydrophyllaceae
    Family Polemoniaceae
    Family Solanaceae
Order Lamiales
    Family Boraginaceae
    Family Lamiaceae
    Family Verbenaceae
Order Plantaginales
    Family Plantaginaceae
Order Scrophulariales
    Family Acanthaceae
    Family Bignoniaceae
    Family Gesneriaceae
    Family Orobanchaceae
    Family Scrophulariaceae
Order Campanulales
    Family Campanulaceae
Order Rubiales
    Family Rubiaceae
Order Dipsacales
    Family Caprifoliaceae
    Family Dipsacaceae
Order Asterales
    Family Asteraceae

Subclass Asteridae.   The most derived subclass of dicots is the Asteridae (Table 25.13), which contains plants such as sunflower, periwinkle, petunia, and morning glory. The Asteridae, having evolved out of the Rosidae perhaps as recently as 60 million years ago, are now far removed from the relictual subclass Magnoliidae. The majority of Asteridae can be easily distinguished from all other flowering plants on the basis of three features: (1) They have sympetalous flowers (their petals are fused together into a tube); (2) they always have just a few stamens, never more than the number of petal lobes; and (3) stamens alternate with petals (Fig. 25.32). The Asteridae exploit very specialized pollinators that recognize complex floral patterns, and such plants could not evolve before derived, sophisticated insects appeared.

Many chemical differences exist between this group and all others. It lacks many specialized chemicals found in other subclasses: betalains present in the Caryophyllidae, benzyl-isoquinoline alkaloids common in Magnoliidae, and ellagic acid (Fig. 25.33a) or proanthocyanins common in Rosidae, Dilleniidae, and Hamamelidae. Instead, many Asteridae have **iridoid compounds** (Fig. 25.33b), which occur only rarely outside this subclass.

At present, this subclass has the greatest number of species (about 60,000), but they are grouped into a small number of large families. One family (sunflowers, daisies; Asteraceae = Compositae) contains fully one third of all the species and is the largest family of dicots. Examples of this subclass are milkweeds (Asclepiadaceae); potato, tomato, red peppers, eggplant, tobacco, deadly nightshade, and petunia (Solanaceae); morning glory (Convolvulaceae); thyme, mints, and lavender (Lamiaceae); and Asteraceae with sunflowers, dandelions, lettuce, *Chrysanthemum,* ragweed, and thistle (Fig. 25.34).

Many members of this subclass are extremely important medicinally: Apocynaceae (oleander family) contains periwinkle, *Vinca,* from which are extracted vinblastine and vincristine, two of our most potent anticancer drugs. Rubiaceae contains, in addition to coffee (*Coffea*), *Cinchona,* from which we derive the antimalaria drug quinine. Another rubiaceous genus, *Cephaelis,* provides us with ipecac, an emetic used to induce vomiting in cases of oral poisoning. Heart disease is treated with cardiac glycosides extracted from *Digitalis* (Scrophulariaceae, the snapdragon family); these compounds make heart muscle beat more slowly and strongly, increasing output of blood from the heart and improving circulation.

The most basal orders of subclass Asteridae, such as Gentianales (gentian and milkweed families) and Solanales (families of potato, morning glory, dodder, phlox), tend to be only moderately specialized compared with the rest of the subclass. They do have the fused petals and reduced number of stamens, but they show no major, distinctive evolutionary trends. They have many body types ranging from herbs to water plants to vines and occasionally small trees.

FIGURE 25.32
Flowers of potato (*Solanum tuberosum*) have stamens that alternate with the fused petals.

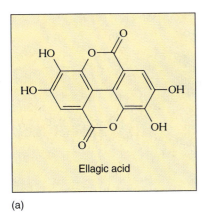

(a)   (b)

FIGURE 25.33

(a) The presence or absence of certain classes of chemicals is an important character in studying the relationships of plants (chemotaxonomy). This is ellagic acid, which is common in three dicot subclasses but absent from one of the most derived, the Asteridae. Its presence deters many insects from eating the plants, but by about 50 million years ago, many types of insects had evolved a tolerance to it. (b) Many Asteridae produce iridoid compounds such as this cornin. They are a relatively new class of chemical defense compounds, so new that few insects tolerate them; iridoid compounds may be partly responsible for the success of the Asteridae by keeping herbivory to a minimum.

(a)   (c)

FIGURE 25.34

Members of subclass Asteridae: (a) potato (*Solanum tuberosum*), (b) milkweed (*Asclepias viridis*), (c) goatsbeard (*Tragopogon* species). Each outer "petal" is a ray floret and each central flower is a disk floret. The dark cylinders are anthers that have fused around styles. The innermost disk florets have not yet opened.

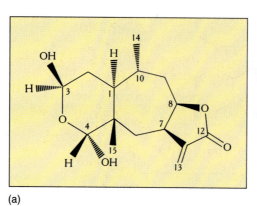

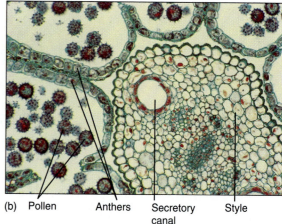

(a)   (b) Pollen   Anthers   Secretory   Style
                                                                   canal

FIGURE 25.35

(a) This is hymenoxon, a sesquiterpene lactone, a member of the chemical arsenal of the Asteridae. This particular compound occurs in bitterweed (*Hymenoxys odorata*) of the southwestern United States; the plant is eaten by sheep only during droughts when no other plants are available. Sesquiterpene lactone causes hemorrhaging of all internal organs of the sheep. (b) Many members of Asteraceae contain secretory canals lined with cells that produce a variety of toxic compounds. This canal is in the style of sunflower (*Helianthus*).

The order Scrophulariales (families of olive, *Penstemon* and snapdragon, gesneriads, acanths, and trumpet-creeper) is more derived in its floral characters; the flowers tend to be bilaterally symmetrical and of sizes and shapes that permit only certain insects to enter and effect pollination. The floral bilateral symmetry further forces the insect to enter in one particular orientation. Pollinators capable of doing this tend to be sophisticated and able to recognize different flowers easily. They learn which flowers provide the greatest rewards, then search for other flowers of identical shape, color pattern, and fragrance, thus providing efficient pollination for the plant species.

The order Asterales consists of the single giant family Asteraceae. It contains 1100 genera and 20,000 species distributed worldwide in almost all habitats except dense, dark forests. They range from important food plants (*Lactuca*—lettuce) to ornamentals to weeds. The characteristic daisy/sunflower type of inflorescence makes them instantly recognizable. Members of Asteraceae have a wide range of unique chemical defenses against herbivores: sesquiterpene lactones, monoterpenes, terpenoids, and latex canals that contain polyacetylene resins (Fig. 25.35). The presence of these chemicals makes composites extremely resistant to animals that eat plants or lay eggs in them; it also causes them to be irritating to human skin, resulting in numerous cases of contact dermatitis. The Asteraceae is a young family, perhaps no more than 36 million years old, which probably originated in the Oligocene Epoch of the Tertiary Period.

SUMMARY

1. All flowering plants, members of division Magnoliophyta, have a closed carpel, a long pollen tube, double fertilization, and, with very few exceptions, sieve tubes and vessels.

2. Division Magnoliophyta is almost universally believed to be monophyletic; all its evolutionary lines can be traced back to a single common ancestor. The earliest fossils of leaves and wood date to about 125 to 130 million years ago, and fossil pollen from earlier periods has been found.

3. The earliest angiosperms may have been weedy shrubs that grew in areas not conducive to fossilization.

4. Early characteristics of angiosperms are believed to be flowers with many sepals, petals, stamens, and carpels; radially symmetrical flowers; insect pollination; wood without vessels, then wood with tracheid-like vessels with scalariform perforation plates; wood with little axial parenchyma; and exclusively woody and perennial plants, none herbaceous or annual.

5. Some of the major evolutionary modifications that have occurred in many groups are reduction in the number of each flower organ; imperfect flowers; bilateral symmetry of flowers; nectaries and fragrances that attract

only specific pollinators; wind pollination; various seed or fruit distribution mechanisms; biennial or annual herbaceous habit; wood with shorter, wider vessel members; larger amounts of wood parenchyma; and numerous types of antipredator compounds—perhaps two or three generations of these.

6. Division Magnoliophyta contains two classes, Liliopsida (monocots) and Magnoliopsida (dicots).

7. Monocots can generally be recognized by long, strap-shaped leaves; flower parts in threes; parallel leaf venation; numerous vascular bundles in the stem, not arranged in a ring; and no ordinary secondary growth.

8. Dicots can generally be recognized by broad, not strap-shaped leaves; reticulate leaf venation; flower parts in fives or fours, but not threes; vascular bundles in the stem arranged in one ring; and woody growth or an annual or biennial herbaceous body.

9. In the classification presented here, class Liliopsida contains five subclasses and class Magnoliopsida six.

IMPORTANT TERMS

Amentiferae
angiospermous carpel
anthocyanin pigments
betalain pigments
class Liliopsida
class Magnoliopsida

closed carpel
dicots
Division Anthophyta
Division Magnoliophyta
double fertilization
generalized flower

iridoid compounds
monocots
perisperm
ranalean flower

REVIEW QUESTIONS

1. What transformations must have occurred if gymnosperms really were the ancestors of angiosperms? Which might have occurred earlier (are present in all relictual angiosperms) and which later (are absent from some of the most relictual angiosperms)?

2. What is meant by a closed carpel as opposed to an open cone scale?

3. Describe what the early angiosperms may have been like with regard to flowers, body, ecology, wood, and pollen. For each of the characters, what living angiosperms still have those characters? In what ways have those characters changed in certain lines of evolution?

4. What are the scientific names of the two classes of angiosperms? What are their common names? What are some of the characters that you would use to identify the class of a specimen you had never seen before?

5. Which dicot subclass is characterized by wind pollination? By the presence of perisperm and betalain pigments? By sympetalous flowers with few stamens and a recently evolved class of antiherbivore compounds?

6. What is a "ranalean" flower? Describe its features. Do any monocots have ranalean flowers? Any dicots? What group was formerly known as the "Amentiferae"? Why were they considered relictual?

7. Why were many modern genera of dicots initially thought to have come suddenly into existence about 130 million years ago? Why do we now think that those leaf fossils do not really belong to the modern genera they were named for?

BotanyLinks .net Questions
www.jbpub.com/botanylinks

Visit the .net Questions area of BotanyLinks (http://www.jbpub.com/botanylinks/netquestions) to complete these questions:

1. Many plants provide us with medicinal chemicals. Go to the BotanyLinks home page to investigate which plants are useful, what chemicals they provide, and the illnesses that are treated.

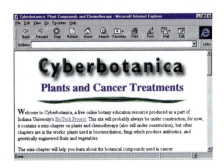

2. Have you ever visited a botanical garden? Are there any close to where you live or are going to school? Go to the BotanyLinks home page to find a botanical garden located in your area.

BotanyLinks includes a Directory of Organizations for this chapter.

Plants do not exist in isolation; they interact with the physical environment, with animals and microbes, and with other plants.

IV
ECOLOGY

The focus of this section is the plant in its environment—plant biology in the broadest scope. An important aspect of the biology of a plant is to view it as a member of a population; population biology differs from the biology of individual plants. If a plant that is a member of a dense, crowded population becomes infested with aphids, the probability is high that other plants in the population will also become infested with the progeny of the aphids from the first plant. But if a plant is a member of a sparse, widely scattered population, it has little likelihood of being found by the aphids from the first plant. The plants we are considering may be absolutely identical, but their fates will differ because of the nature of the populations of which they are a part.

Every plant is also affected by all other aspects of its immediate environment, which in turn is part of a larger environment—a continent or island or perhaps the water of a lake or ocean. Climate must also be considered. None of these factors is uniform in either time or space; there are differences in soil, rainfall patterns, disturbances, and day length. On a larger scale, periods of global warming alternate with ice ages while the continents drift across the surface of the planet. In addition, each organism is affected by at least some of the other organisms around it, and as they undergo their own evolution, their impacts on each other are altered. Earth's surface is composed of a rich, changing mosaic of diverse habitats. Within this milieu, evolution by natural selection has produced hundreds of thousands of species of prokaryotes, protists, algae, fungi, animals, and plants. Their various adaptations for particular aspects of the environment have determined where they will survive and where they will be outcompeted by organisms that are better adapted.

26

POPULATIONS AND ECOSYSTEMS

As the glacier retreats from this valley, hardy pioneer plants colonize the open spaces that become available. *(David Muench)*

CONCEPTS

Ecology is the study of organisms in relation to all aspects of their surroundings. Throughout this book, emphasis has been placed on the importance of analyzing structure, metabolism, and diversity in terms of adaptation and fitness. Mutations result in new alleles that alter the phenotype; then natural selection eliminates those that are less well-adapted but retains those that increase fitness. Most factors have been discussed individually—for example, the effect of trichomes in deterring insects from chewing leaves and the ability of carotenoids to protect chlorophyll from excess sunlight. But most factors that affect a plant's health and survival are present together in the habitat, acting on the plant simultaneously. We must try to understand the plant in relation to its entire habitat, to all components of its surroundings.

    An individual plant never exists in isolation in a habitat; instead, there are other individuals of the same species and together they constitute a population (Fig. 26.1). Individuals of plant populations often do not interact strongly, but **the biology of the population is more than just the sum of all the biologies of individuals.** In a species

FIGURE 26.1
The numerous individuals of one species constitute a population. They depend on each other for gametes during sexual reproduction and also support a healthy population of pollinators and seed dispersers. Because there are several species, the photograph shows a community of several populations. When the soil and atmospheric components are added, an ecosystem results.

that consists of either dioecious or self-sterile individuals, a population can carry out successful sexual reproduction, but an individual cannot. If pollinated by animals, a single individual usually cannot produce enough nectar or pollen to keep even one pollinator alive, but a population of plants can sustain a population of pollinators. On the negative side, a population may be dense enough that spores from a pathogenic fungus are ensured of landing on at least one susceptible individual and then spreading to others, whereas if an individual could exist in isolation far removed from others, it might be safe from pathogens, predators, and natural catastrophes such as fires (Fig. 26.2).

A population also does not exist in isolation, but rather coexists with numerous populations of other plant species as well as populations of animals, fungi, protists, and prokaryotes. All the populations together constitute a **community,** which when considered along with the physical, nonliving environment is an **ecosystem.** These contribute additional levels of interactions and complexity and make it more difficult to be precisely certain of the effects of factors on individuals. The presence of trichomes may deter leaf-eating insects, but those insects might have frightened away more damaging egg-laying insects. Besides, the leaf-eating insects might have had a beneficial role by dropping nutrient-rich fecal pellets to the soil where they might have encouraged the growth of mycorrhizal fungi. The structure, metabolism, and diversity of plants cannot be fully understood without understanding the ecosystem.

Burned area

FIGURE 26.2
Trees in dense populations were killed by fire. Identical trees in sparse populations survived.

FIGURE 26.3

The redwoods here completely dominate the habitat, each tree containing many times the bulk and volume of any single herb. They provide filtered light and protection from wind for the understory plants, which are incapable of tolerating full sunlight. But do the herbs affect the trees in any way? Yes: The redwood seeds germinate only after a low, cool fire fueled by the understory herbs and shrubs.

PLANTS IN RELATION TO THEIR HABITATS

The **habitat** is the set of conditions in which an organism completes its life cycle. For migratory animals, the winter area, summer area, and migration routes are all habitat components. No plant is migratory, but portions of plants are: spores, pollen, fruits, seeds, and vegetative propagules.

How much of the surroundings should be considered part of the habitat is debated. Many factors do not appear to affect certain plants at all. The presence of small herbs on the forest floor does not seem to influence the large trees; their presence or absence has little effect on the mineral nutrition of the trees or on their pollination (Fig. 26.3). An experiment might remove all the small annuals and then examine whether the trees are affected, but it would be difficult to measure the growth of whole trees, especially their roots, and it might take years to see any effect on the trees' metabolism. The small herbs may be important; perhaps they harbor spiders that catch insects that would otherwise kill the trees' seedlings. If so, then removing the herbs might cause an increase in the insect population and decreased survival of tree seedlings; with fewer seedlings of this species, it might be possible for the seedlings of a different species to survive better. After many years, the forest composition would be changed as a result of the removal of the herbs.

On the other hand, we do know that many components impact others directly. Pollinators are critically important aspects of the habitat for the plant species they pollinate, and any disease organisms or predators that prey on those pollinators are also important to the plant. Those aspects of the habitat that definitely affect a plant constitute its **operational habitat,** whereas all the components, whether with known effect or not, are its habitat.

For example, consider the redwood forests in California. It was discovered that there were no redwood seedlings in the national parks, and redwood seeds were not germinating. After a recent forest fire, redwood seeds sprouted and grew vigorously. The fire not only stimulated the seeds to germinate but, by burning the understory plants, released minerals that increased the soil fertility. We now realize that the shrubs and herbs are vital to the success of the giant redwoods: They fuel quick, cool fires that are too low to damage the large redwoods but are necessary for seed germination. The policy of preventing forest fires—a natural factor of this ecosystem—was harming the redwoods.

Habitat components are of two types, **abiotic** and **biotic.** The abiotic components are nonliving and are physical phenomena: climate, soil, latitude, altitude, and disturbances like fires, floods, and avalanches. Biotic components are the living factors: the plant itself, other plant species, and species of animals, fungi, protists, and prokaryotes.

ABIOTIC COMPONENTS OF THE HABITAT

Climate.   Climate is critically important to all organisms; most species are restricted to certain regions primarily because they cannot live in climatic conditions outside those regions (see Chapter 27). Climate itself has many components—temperature, rainfall, relative humidity, and winds being just a few.

The average temperature of a habitat is not as important as its extremes: the lowest winter temperature and the highest summer temperature. Many species of bromeliads, aroids, and orchids are restricted to the tropics because those are frost-free habitats. Rainforests along the west coast of California, Oregon, and Washington receive adequate rain, but the freezing winters prevent them from being suitable habitats for most tropical species. On the other hand, many temperate trees must have a winter dormancy period accompanied by weeks of subfreezing temperature in order to be vernalized and bloom. If cultivated in areas with warm winters, the plants grow and survive well but do not reproduce. The highest or lowest temperatures during a plant's lifetime are important. If a tree species must be ten years old before it can flower but killing temperatures occur every eight or nine years, the species cannot survive there.

The growing season of an area is often determined by the date of the last severe, killing frost in the spring and the first killing frost of autumn. The length of the growing season must be adequate for sufficient photosynthesis, growth, development, and reproduction; if

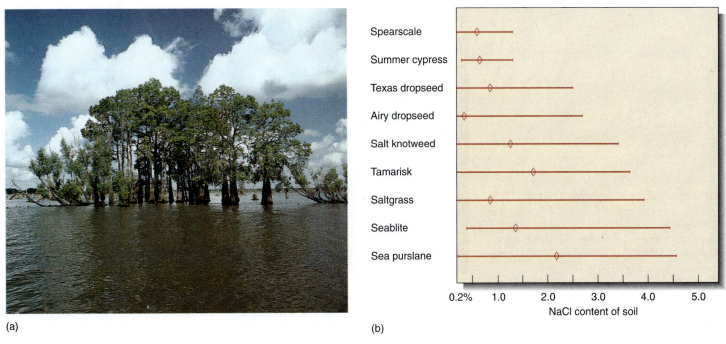

(a)   (b)

FIGURE 26.4

(a) Although most plants would grow better if given more water, the swampy bayous of southern Louisiana contain so much water that most plants cannot grow; their roots drown. Bald cypress (*Taxodium*) is well-adapted, however; its roots form "knees" filled with aerenchyma that allow atmospheric oxygen to diffuse into the submerged roots. (b) Plants vary in their ability to tolerate salty soil. These species occur around salt flats in Oklahoma and Kansas; sea purslane and seablite can survive even in the presence of strong salt concentrations, whereas others tolerate less. By being tolerant of soil salt, these plants can occupy habitats on the edges of salt flats which are free of other plants. Not shown are the many species that have virtually no salt tolerance. Diamond marks indicate the optimal concentrations.

not, a reproductive population cannot survive even if it can tolerate the temperature extremes of winter and summer.

Moisture occurs as rain or snow or as hail that supplies water but also damages leaves, buds, flowers, and animals. Habitats range from extremely dry (deserts) through progressively more moist all the way to marshes, lakes, and rivers that are virtually all water. Just as with temperature, the total amount of precipitation may not be as important as the seasonal extremes or the timing of the precipitation. A constant drizzle that occurs almost year-round supports certain types of ecosystems, whereas the same amount of rain, distributed as just winter snowfall and occasional summer thunderstorms, results in a different type of ecosystem.

Numerous metabolic processes respond proportionally to abiotic factors. Once there is sufficient moisture for marginal survival, increased amounts of water produce increased growth and reproduction. There is usually an upper limit: With too much water, roots drown for lack of soil oxygen (Fig. 26.4a). Between the low and high extremes is the **tolerance range** of the organism. Ranges vary greatly from species to species (Fig. 26.4b). Some are extremely broad: Most temperate plants across the northern United States and Canada, especially in the midwest, tolerate summer highs of over 100°F and winter lows below −40°F. Plants of southern Florida, Puerto Rico, and Hawaii tend to have much narrower tolerance ranges, being killed both by cool temperatures and by hot ones.

Soil Factors.   Soils are formed by the breakdown of rock. Initially the resulting soil is thin and virtually identical to the parent rock in its chemical composition; consequently, young soils are variable in the amounts of macronutrients and micronutrients they have available. Because nitrogen is not a significant component of any type of rock, all young soils are deficient in it.

FIGURE 26.5
This "soil" consists mostly of rather large rock fragments with little water-holding capacity and very few dissolved essential elements. The plants growing here, pioneers, not only tolerate these conditions but actually change them. The acids they release as they decompose greatly accelerate chemical weathering and soil formation.

The first plants that invade a new soil, called **pioneers**, must be able to tolerate severe conditions. The soil is sandy, with relatively large particle size, and most minerals are still locked in the rock matrix. The soil has little water-holding capacity, and the first plants have no neighbors to help moderate the wind, provide transpired humidity, or otherwise temper the environment (Fig. 26.5). Pioneer plants often are associated with nitrogen-fixing prokaryotes; many lichens contain cyanobacteria, and angiosperms have root nodules. As the pioneers live on the soil, they change it significantly; carbon dioxide from root respiration produces carbonic acid and accelerates chemical weathering. Dead plant parts such as leaves, fruits, roots, and bark become substrates for soil organisms, and their decay

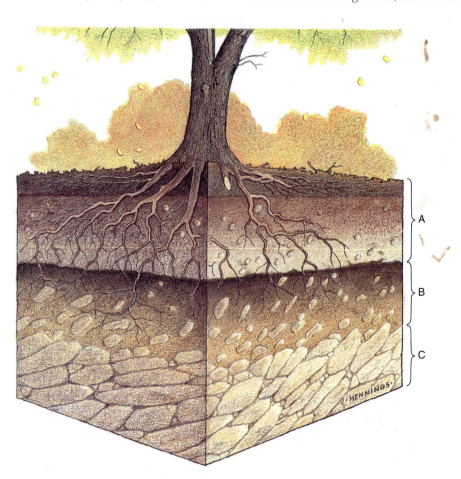

FIGURE 26.6
Most soils show three horizontal layers, or horizons. The relative thicknesses of the A horizon (zone of leaching) and B horizon (zone of deposition) depend on many factors, the abundance of vegetation and humus being especially important.

contributes humus, greatly increasing the soil's water-holding capacity. Roots may penetrate to the bedrock, entering larger cracks and then expanding and breaking the rock physically.

After many years, a thick soil may result that has a distinct soil profile with three layers or horizons (Fig. 26.6). The **A horizon** is uppermost and is sometimes called the zone of leaching; it consists of litter and debris, and as this breaks down, rainwater washes nutrients from it downward into the next layer, the **B horizon**, or zone of deposition. The B horizon is the area where materials from the A horizon accumulate. It is rich in nutrients and contains both humus and clay. Below is the **C horizon**, composed mostly of parent rock and rock fragments.

Whereas young soils differ because of the chemical nature of their parental rock, older, more mature soils are less diverse. As the rock weathers, essential elements are absorbed by roots and become trapped in the plant body; nonessential elements are leached away. As the plants or their parts die, they fall to the ground and decay slowly, releasing the essential elements which re-enter the soil. There they are taken up again. As a result, essential elements cycle repeatedly, alternating between organisms and the A and B horizons, whereas other elements are gradually washed downward into the water table and removed by underground water flow.

Latitude and Altitude.   Latitude contributes many factors to the abiotic environment. At the equator all days are 12 hours long, no seasonal variation occurs, and plants cannot measure season by photoperiod. At progressively higher latitudes to either the north or south, summer days become progressively longer, as do winter nights (Fig. 26.7; Table 26.1). Above the Arctic and Antarctic Circles, summer days are 24 hours long, as are winter nights. At intermediate and higher latitudes, day length is an excellent indicator of season, and some species are sensitive to photoperiod.

The amount of light energy that strikes a given area of Earth's surface each year also varies with latitude; in equatorial regions, the sun is always nearly overhead, and each square meter of surface receives a maximum amount of radiation (Fig. 26.8). At higher latitudes the sun is only rarely overhead, usually only near midsummer. At other times, when the sun is low, light strikes the Earth obliquely and less energy is received per square meter; even at noon on winter days the sun is low in the sky and the days are not bright. At high latitudes, temperatures fluctuate greatly on both a daily and a seasonal basis. Soil formation is slow in the cold latitudes, and often what soil has formed is blown away by

TABLE 26.1	*Average Dates for First Flowering of Plants in Northwestern Ohio**
Hepatica	April 13
Spring beauty	April 17
Bloodroot	April 18
Yellow trout lily	April 23
Pepper root	April 23
Early blue violet	April 25
Dutchman's breeches	April 28
Yellow violet	May 2
Wild blue phlox	May 2
Wake-robin	May 5
Jack-in-the-pulpit	May 14
May apple	May 16

* The time of flowering of these insect-pollinated species must coincide with conditions appropriate for pollinator metabolism as well as flower metabolism. Also, pollinators must become active or migrate into an area when conditions have become appropriate for the flower metabolism that provides them with nectar.

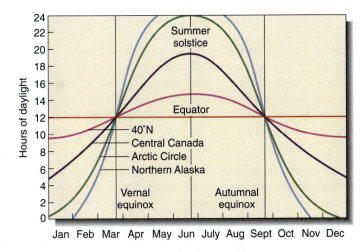

FIGURE 26.7
The line for day length at 40 degrees north, which corresponds to the center of the United States, shows that the longest day (summer solstice, about June 21) is just less than 15 hours long, whereas the shortest day (winter solstice, about December 21) is only about 10 hours long. Closer to the equator, the difference between the longest and shortest days is less, and near the equator the difference is too little for plants and animals to be able to use it as a seasonal indicator (see Table 26.1).

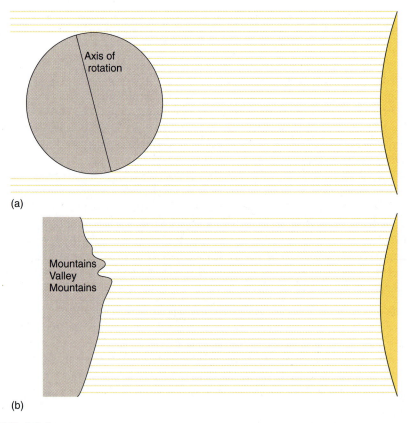

FIGURE 26.8

(a) During winter in the Northern Hemisphere, the North Pole is pointed away from the sun and the amount of energy received per square meter is at a minimum. The South Pole points toward the sun and the Southern Hemisphere receives more direct light, resulting in a large amount of energy per square meter. Six months later the positions are reversed. (b) The variability of energy received per square meter of surface is affected by topography. Away from the equator, the side of a mountain or valley that faces the equator also faces the sun and receives more energy. The sides that face the poles are cooler. Very often, the two sides of a valley running east to west have different vegetation.

strong winds. In much of Alaska and Canada, soil was scraped away during the ice ages 18,000 years ago.

Regions of high altitudes on mountain tops are similar to those at high latitude. There are high winds and poor soil; much or all of the year is cold, and the growing season is short. Water may be present mostly as snow and ice, so physiological drought is frequent. The areas may have varying day lengths, depending on their latitude. An additional stress present in high-altitude habitats is intense ultraviolet light. High altitudes are above much of Earth's atmosphere and thus are not fully shielded by ozone, oxygen, carbon dioxide, and water vapor.

Disturbance. Disturbances are phenomena such as fires, landslides, snow avalanches, and floods; they produce a significant, often radical change in an ecosystem quickly (Fig. 26.9). Disturbances affect the biotic factors directly, often completely eliminating many or all individuals from an area and also altering the soil, but they have little or no impact on other abiotic factors such as climate, latitude, or altitude. The elimination of large numbers of individuals by a disturbance alters species relationships in the ecosystem. Man-made disturbances have been caused by insecticides, herbicides, hunting, and habitat destruction (Fig. 26.10).

Fire is a natural, common component of many dry ecosystems. With little moisture, fallen leaves and twigs decay so slowly that a thick layer of debris builds up. The living plants tend to have waxy cuticles and water-proofing resins that make them especially

(a)

(b)

(c)

(d)

FIGURE 26.9

(a) Fires have a significant impact on the ecosystems. They kill not only plants but also pollinators, herbivores, and pathogenic fungi and bacteria. Fire also releases minerals back to the soil and decreases shading. A quick, cool fire does not damage rhizomes, tubers, seeds, or trees with thick bark. *(S. J. Krasemann/Peter Arnold)* (b) Within a few months after a fire, mosses, grasses, and wildflowers have begun regrowth. (c and d) The eruption of Mt. St. Helens radically altered the surrounding ecosystems, but there too plants, animals, and other organisms are recovering. *(c, M. Shafer/Peter Arnold; d, Larry Nielsen/Peter Arnold)*

flammable. Lightning storms often occur without rainfall, starting fires that burn rapidly and cause great destruction. Many of the plants and animals of such ecosystems have become fire-resistant as a result of natural selection caused by frequent fires. The bark of certain species of pine trees is so thick that a rapidly moving, moderate fire does no damage to the vascular cambium and other living tissues; the lower portions of the trunk have no branches because of self-pruning, so flames cannot reach high enough to ignite needles; only herbs and small shrubs are burned. Furthermore, the cones of lodgepole pine and jack pine open only after being exposed to the heat of fire; this adaptation results in the release of seeds after many competing plants and predatory animals have been killed and the forest

FIGURE 26.10

Humans destroy habitats on a massive scale; once this hydroelectric dam is completed, it will permanently flood the valley upstream and convert a tropical forest to a lake. Downstream, the flooding that had been a natural part of the habitat, important in maintaining the community, will be eliminated. The availability of electrical power will almost certainly result in an immigration of people, causing further land clearing and habitat destruction.

floor is open and sunny. Also the soil is enriched by the minerals in the ash, making it an ideal site for pine seedlings. However, if fires do not occur frequently enough, usually because foresters put them out, understory shrubs and small trees grow tall and large; so much brush and dead wood accumulate that when fire does occur, it is extremely hot. As it burns to the top of the shrubs and understory trees, flames may reach the lowest branches of the pines, igniting the crowns (Fig. 26.11). Once this occurs, the fire can spread rapidly through the canopy of trees, killing them.

Many grasses have adaptations that permit them to benefit from fires. Many prairie grasses in the midwest and the saw grasses of the Florida everglades grow in dense clumps with their shoot tips and leaf primordia at or below ground level, protected from fire by soil

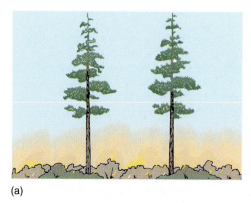
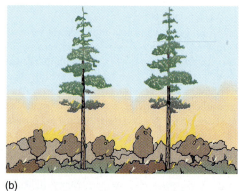

(a)   (b)

FIGURE 26.11

(a) If fire occurs frequently, understory shrubs are burned back before they become tall and there is always a large space between them and the lowest branches of the dominant trees. Fire cannot get high enough to ignite the tree canopy. (b) If fire occurs infrequently, understory shrubs become tall, reaching the lowest branches of the dominant trees. Fire can then burn upward from the shrubs into the canopy, igniting the highly flammable needles, twigs, and cones. Even if the trees are not killed outright, their shoot meristems are destroyed and no further growth is possible.

and the living bases of leaves. Leaf tips may be dead and dry, and when fire occurs, the dead portions burn, releasing their minerals, but the bases of the plants are unharmed.

Annuals and short-lived plants do not survive fires, but their seeds, buried underground, do. Plants with bulbs, rhizomes, tubers, or corms easily survive small fires.

BIOTIC COMPONENTS OF THE HABITAT

The Plant Itself.   An individual itself, just by being in a habitat, modifies the habitat and is a part of it. Habitat modification may be beneficial, detrimental, or neutral to the continued success of that species in the habitat. In the beech/oak forests of the northern United States, the trees modify the habitat by producing a dense canopy that results in a heavily shaded forest floor (Fig. 26.12a). With such low light levels, few seedlings grow well, but two that do are those of oaks and beeches. As a result, mature trees create a habitat that suits their seedlings and aids their own successful reproduction.

Pine forests are more open, but still the forest floor is shaded by mature pines. Pine seedlings, however, need full sunlight and do not grow well below the canopy of older pines. Seedlings of other species flourish in these conditions and crowd out the few pine seedlings that may occur. Pine trees modify their habitat adversely for their continued

(a)

(b)

(c)

FIGURE 26.12

(a) In this forest, the two dominant species, beeches and oaks, alter the habitat so that it is suitable for their own seedlings. Their seeds can germinate and grow in the heavy shade they provide, but the seedlings of many other species cannot survive in such low light. *(Grant Heilman Photography)* (b) Although much more open and sunny than a beech/oak forest, a pine forest is still too shaded for pine seedlings. Pines seem to alter the habitat adversely for their own long-term survival, but the needles they drop are highly flammable and are the main cause of frequent fires that kill oak seedlings but not pine seedlings. *(William E. Ferguson)* (c) For several years, a pine seedling remains short with the shoot apical meristem well protected at ground level, surrounded by moist, nonflammable living leaves. It can survive quick, cool fires and builds an extensive root system. After about 3 to 7 years, the seedling begins to grow upward very rapidly, with most energy going into stem elongation and little to the roots. Within just 2 or 3 years, the tree may be more than 18 feet tall and the shoot apical meristem is out of danger from fires.

FIGURE 26.13

As a glacier retreats, the rubble and sand left behind is extremely poor soil, and of course the climate is usually harsh. However, several species of pioneers can grow here, and their activity enriches the soil, permitting invasion by less hardy species. Mt. Rainier National Park, Washington.

success; only disturbances can create the open habitats needed for pine seedlings (Fig. 26.12b and c).

As glaciers retreat, they leave behind moraines—great mounds of rubble, sand, and boulders. The soil is poor, with no humus and few available nutrients, but pioneer species such as alder, *Dryas* (in the rose family), willow, and fireweed are able to colonize recently exposed moraines (Fig. 26.13). They survive in the open conditions and tolerate low levels of nutrients; the alders and *Dryas* have root nodules containing symbiotic nitrogen-fixing bacteria that supply nitrogen. Within a few years, the decay of their leaves and bark has enriched the soil sufficiently that sitka spruce and western hemlock become established. After this, however, the spruce/hemlock forest creates too much shade and eliminates the pioneers that have altered the habitat to their own detriment by enriching it.

Other Plant Species.   When several individuals, of either just one or several species, occur together, the possibility for interaction is created. If the interaction is basically beneficial for both organisms, it is described as **mutualism,** but if it is disadvantageous, it is **competition.** Competition is a situation in which two populations do not grow as well together as they do separately because they use the same limited supply of resources. Many plants are believed to compete with others for light, soil nutrients, water, and the attention of pollinators and seed dispersers, among other things. If a single plant were allowed to grow by itself, in many cases it might grow more rapidly, become larger, and produce many more gametes than it would if other plants were nearby. Roots of the other plants might grow among its roots and remove water and nutrients. The competitors might grow taller than it does and then intercept sunlight by putting their leaves above its leaves. Their flowers, even if they did not produce more nectar, might still distract its pollinators such that its pollen would be carried to the stigmas of the wrong species and its own stigmas would receive foreign pollen.

The role of competition has been extensively debated. One theory postulates that the result of competition is **competitive exclusion:** Whichever species is less adapted is excluded from the ecosystem by superior competitors. The species that get sunlight and other resources win; those that do not, lose and are eliminated. If this is true, then very little competition occurs in a typical ecosystem; each species is assumed to be adapted to a particular set of conditions, a **niche,** that no other species is adapted to use as efficiently. For example, some species are adapted to full sunlight, others to partial shade (Fig. 26.14). In an ecosystem, the former must be a canopy tree and the latter an understory species.

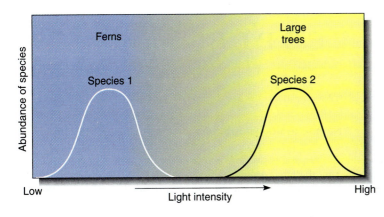

FIGURE 26.14
Both ferns and large trees need light, but many fern species cannot tolerate high light intensity and must have partial shading. Large trees often must have intense light; they do poorly if shaded. Growing together, they do not compete for sunlight; each uses a portion of the resource the other does not use. If either were removed, the other would not grow better.

This theory of little competition predicts that if certain species are removed, the others do not benefit from the unused resources because they are not adapted for them. This is sometimes found to be the case.

The concept of niche is difficult to define exactly; basically it refers to the set of aspects of the habitat that directly affect a species. For example, a particular marsh species occupies a particular semiaquatic niche defined by a range of soil moisture or flooding, a range of seasonal rainfall, a range of temperature, a paucity of root grazing due to lack of swimming herbivores, and the presence of appropriate pollinators. Another species of marsh plant may grow in the same general geographical area, even the same marsh, but occupy a different niche because it may have a different pollinator or may grow in areas of the marsh that are slightly more acidic than the microhabitats occupied by the first species. As long as even one factor differs, they occupy different niches.

As an alternative to the theory of competitive exclusion, a second theory postulates that species overlap in their tolerance ranges (Fig. 26.15a), and, when grown together, each has exclusive use of the portion of the range not used by the other. In the overlap zone

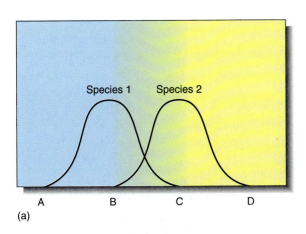

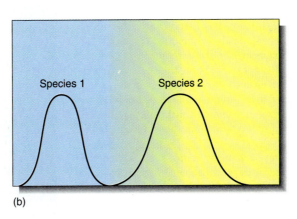

FIGURE 26.15
(a) Two species compete if they can use the same resource. Here, species 1 would have exclusive use of the resource under conditions A to B, whereas species 2 would have exclusive use between C and D. With conditions between B and C, they would compete. If the resource were water availability, species 1 might grow in drier areas and species 2 in wet areas, and both could grow in moist areas. Whether they both actually live in moist areas depends on how well they compete; one may be so efficient as to exclude the other. In areas of competition, one survives better if the other is removed. (b) Over time, mutations that improve the ability of a species to compete may be selected. If such mutations do not occur or if the competitor is too efficient, then mutations may be selected that restrict the efficiency of the species in areas where it does well. In this case, species 1 evolves such that it is specialized in dry habitats and now cannot live in moist conditions. This may increase its total fitness, because the capacity to live in moist environments involves more than just the ability to use abundant water; moist environments also have more insects and fungi that require expensive defense mechanisms.

PLANTS & PEOPLE

NICHES IN THE JET AGE

A niche is defined as a set of conditions in which a particular species can thrive, outcompeting other organisms. Cacti occupy desert niches in North and South America, and water hyacinths proliferate in the waterways of the Amazon rainforest, filling the niche consisting of calm water surfaces in warm areas. But the desert niches occur in many geographical regions other than the western hemisphere, and tropical rivers flow quietly all around the world. Why do cacti and water hyacinths grow only in restricted areas? Basically, it is simply because they have not been able to spread to the other areas because their natural range—the site where they originated evolutionarily—is surrounded by vast regions of inhospitable barriers. Under natural conditions, cacti could establish themselves in the desert niches of Africa, the Middle East, China, and Australia only if migratory birds happened to carry seeds from one continent to another. Such long-distance dispersal is extremely rare.

Or perhaps we should say it *was* extremely rare. Humans have now far surpassed migratory birds as agents of long-distance dispersal. Planes and ships connect all parts of the world, causing an interchange of plant and animal species on an unprecedented scale. Some of the transport is accidental as seeds are caught in clothing of tourists or are attached to hides or other material being shipped. Zebra mussels came to the United States in the bilge water of ships, and insects are carried in fruit and produce and even in the stagnant water trapped in old tires being imported for recycling. Of course even microbes and viruses are transported over long distances, as is the case with HIV, the virus that causes AIDS. Other transport is intentional, as living plants or animals are imported specifically for cultivation as crops or ornamentals. Carp were imported into our rivers as a food source; killer bees were brought to Argentina from Africa for research and then escaped from a broken laboratory cage. The vine kudzu was planted in the southern states as a ground cover to control erosion along road cuts and on canal banks, and water hyacinth apparently was introduced into our waterways simply because it is pretty.

All of these organisms, and many more, share a common feature—they have thrived in the new habitats in which we have placed them. All have found conditions that permit their rapid growth and reproduction. And because we did not bring along their natural predators or pathogens, they tend to be free of the organisms that could limit their expansion. They have all undergone population explosions and are proliferating rapidly. Zebra mussels are clogging waterways through the American northeast and are spreading down the Mississippi River; killer bees have spread throughout South America, Mexico, and Texas and into New Mexico in just ten years. In such cases, it turns out that these organisms are more highly adapted to the conditions here than are our native species. Consequently, the exotic, introduced species outcompete the indigenous species. Carp have crowded out many types of American fish, and kudzu covers thousands of square kilometers of forests, killing the trees by shading them and cutting off their sunlight.

No insecticides or herbicides are specific enough to control only the introduced species, and trying to eliminate plants, insects, and fish by hand is just impossible. Most control efforts center on searching the home habitat to find predators or disease organisms and then introducing those into the new location. But because the original problem was caused by the introduction of an exotic organism, there is of course considerable reluctance to bring in another exotic organism. An ideal solution would be to find a pathogen that preys *only* on the problem species and on no other. Ideally, then, when the original exotic species is eliminated, the second will also die out for lack of food. Caution must be used because the pathogen may be able to attack native species that are related to the exotic species, or a mutation may occur in the pathogen that allows it to attack native species as well.

Already, the transport of species to new habitats where they can proliferate and outcompete native species has been a serious problem. It can only become worse as our jet-age global movement brings about increased travel and transport of material.

Kudzu (*Pueraria lobata*), which was introduced from Japan, has no pathogens or pests here in the United States. It grows rapidly, covering rocks, trees, telephone poles, even buildings. Trees covered by kudzu die from lack of sunlight. (*J. C. Stevenson/Earth Scenes*)

where the habitat is suitable for both species, usage is determined by competition: The one that is more adapted occupies the overlap zone exclusively. The weaker of the two species occupies only part of its potential niche. The two can coexist, but if either is removed, the other then has its full range of resources available and grows better. Examples of this type of competition have been found to occur.

A long-term result of competition should be species modification by natural selection. If two species compete with partial tolerance-range overlap, then mutations are beneficial if they allow each to use more efficiently its exclusive portion of the range. Through time, the species diverge until competition is reduced (Fig. 26.15b).

The geographic ranges of most populations are extensive enough that they contain a diversity of biotic and abiotic factors: Each may include hills, valleys, and plains; rocky soil and rich soil; grasslands and open woodlands. Because of this diversity, a population may be competing for sunlight in one part of its geographic range but competing for water in another and for other factors in still another. The ecosystem diversity causes different subpopulations to specialize for certain features, especially if gene flow is not rapid and thorough. This could be the beginning of divergent speciation, of course, but long before the various subpopulations could be called subspecies, they would be considered **ecotypes,** each specialized in response to particular ecosystem factors at its locality. The various ecotypes resemble each other so strongly that they clearly belong to the same species but have enough differences for ecologists and taxonomists to suspect that separate ecotypes exist. Experiments are necessary to prove that natural selection is in fact altering the genome, because phenotypic differences can result simply from the growing conditions themselves. For example, beans grown in rich soil with adequate water are strikingly different from those grown in poor soil with little water and other stresses. To test if ecotypes really exist, **transplant experiments** are performed: Plants from each site are transplanted to the alternate site and plants from both sites may be grown together in a **common garden** at an intermediate site. If the transplanted individuals take on the phenotype of the naturally occurring plants at that site, then genetic factors were not important, just stresses. But if the transplanted individuals retain their original phenotype or if they die, then genetic divergence had begun.

Also, success in competition depends on factors other than the ones involved in the competition. Species A may grow rapidly and shade species B when water is abundant, but species B may be more drought-tolerant and can outgrow and outcompete species A when water is scarce. Water availability may vary over the geographic range such that A dominates in wet areas and B in dry areas. Also, because rainfall varies from year to year, species A may be more common some years and species B more abundant in other years.

Animal ecology includes a variety of negative interactions in addition to competition; predation—the killing of prey—and parasitism are two examples. Insectivorous plants do trap and kill insects, deriving nitrogenous compounds from them, but no plant kills and digests other plants in a predator-prey relationship. Parasitic plants are not uncommon. Mistletoes are familiar examples, and hundreds of species of parasitic plants in at least 12 families are known.

Numerous animals, fungi, protists, and bacteria are saprotrophs, living on dead organisms; vultures, maggots, wood-rotting fungi, and decay bacteria are examples. No plant species of any type is known to attack and digest dead plant or animal material; roots do absorb the minerals, nitrates, and sulfates that are released from decomposing humus, but the roots do not participate in the enzymatic attack.

Organisms Other Than Plants.   Animals, fungi, and bacteria are obviously important biotic aspects of a plant's habitat. Interrelationships between plants and these other organisms can be either beneficial or detrimental for one or both partners.

Plants and animals have many relationships that are examples of mutualism, in which both species benefit. Most instances of pollination are mutualistic—the animal receives nectar or a portion of the pollen and the plant benefits from pollen transfer. Seed dispersal by fruit-eating animals **(frugivores)** also benefits both species so long as the animal does not chew the seeds and digest the embryos.

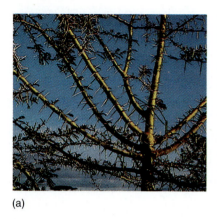

(a)

(b)

(c)

FIGURE 26.16

(a) Ant *Acacia* plant. (b) These thorns on an ant *Acacia* are swollen and hollow and serve as excellent nests for ants. They form as part of the *Acacia's* own normal development; the presence of ants does not induce their formation. (c) The leaftips of ant acacias develop as Beltian bodies (yellow), food bodies rich in glycogen. (d) As long as an acacia is occupied by ants, all other vegetation is kept cleared away, but if the ants are killed with insecticide, the acacias are overgrown quickly. *(a, Zig Leszczynski/Earth Scenes; b, Robert and Linda Mitchell; d, Paul Feeny)*

(d)

FIGURE 26.17

The shoot base of *Hydnophytum* swells and forms hollow chambers that become occupied by ants. There are two types of chambers: Those with smooth walls are used as living quarters, and those with warty walls are used as waste dumps by the ants. The warts on the walls are absorptive cells; the walls absorb nitrogenous compounds as the wastes decompose.

A famous example of mutualism is the association of certain ants and acacias in tropical Central America. Acacias are small trees that have enlarged, hollow thorns at their leaf base. The thorns are used as ready-made, stout, waterproof homes by ants (Fig. 26.16). Because of the large number of thorns on a single plant, the ants benefit from having abundant housing, but in addition, acacias produce nectar and their leaflet tips are modified into golden yellow food bodies (called Beltian bodies) that are filled with glycogen. The ants receive both housing and food. The plant benefits because this species of ant is aggressive; they patrol the plant and attack any animal that touches it, even large mammals. If the leaves of another plant come close to an *Acacia,* the ants destroy the leaf, keeping the *Acacia* unshaded. As a result, the plant is free of pathogenic fungi and insects and it grows in full sunlight. This mutualism is obligate; the plant cannot survive in nature without the ants, and the ants are not found away from the plant. A similar mutualism occurs between ants and *Hydnophytum* in Southeast Asia (Fig. 26.17).

Commensal relationships, in which one species benefits and the other is unaffected, are also common between plants and animals. When birds build nests in trees, the birds benefit and the tree is (usually) unharmed. When sticky fruits or seeds, such as cocklebur, stick to an animal's fur or feathers and then are dispersed, the plant benefits and the animal is unharmed. One-sided negative relationships also occur: Animals trample and kill small plants without being affected. Competition, in which both parties are adversely affected, probably is not common between plants and animals.

Predation is a relationship in which one species benefits and the other is harmed (Fig. 26.18); the species that benefits seeks out the other and uses it specifically for food or some other form of resource. Animals that eat plants are **herbivores** and the process is **herbi-**

vory, but it is often more precisely delimited as **browsing** (eating twigs and leaves of shrubs—deer, giraffes) or **grazing** (eating herbs—sheep, cattle). Insects also lay eggs in plants, then their larvae feed on plant tissues as they tunnel through them. Aphids and spittlebugs suck sap rather than chew tissues. Insects and birds both harvest leaves and twigs for nest construction.

Many interrelationships between plants and fungi or bacteria are harmful to the plant, but the fungi and bacteria are described as being **pathogenic** rather than predatory. Either may cause mild disease or be so virulent that they kill the plants quickly. A large percentage of the microbes are saprotrophs (see Chapter 20), living on dead plant tissues such as leaves, logs, fallen fruit, and sloughed bark. This benefits the fungi and bacteria without harming the living plants, and it can actually help them. This process speeds up the release of mineral nutrients, especially nitrogen compounds, enriching the soil. A case of plants attacking and parasitizing fungi may be known. Orchid seeds are tiny and lack chlorophyll; they remain moribund until invaded by soil fungi, then they turn green and grow well. It had been assumed that the seedling received nutrients and perhaps growth factors from the fungus and in return provided it with carbohydrate. However, tests have not revealed any benefit to the fungus, just to the plant.

Examples of plant-fungus mutualism are well known and recently have been shown to be of much greater importance than ever before suspected, as in the case of mycorrhizae. The mycorrhizal fungus transports phosphate into the plant and receives carbohydrate; both benefit. Many plant species grow only poorly in nature if the soil fungi are killed with a fungicide, and it has been postulated that one reason the plains of the American midwest lack trees is because the type of mycorrhizae beneficial to trees cannot compete well with the mycorrhizae beneficial to grasses. A complex relationship has been discovered with Indian pipe (an achlorophyllous parasitic plant) and a mycorrhizal fungus: Indian pipe parasitizes the fungus, drawing nutrients from it, and the fungus in turn obtains carbohydrates from its other mycorrhizal partners that are chlorophyllous and photosynthetic. Indian pipe basically parasitizes other plants, using the mycorrhiza as a bridge.

The nonplant organisms add a great deal of complexity to a plant's habitat, and numerous types of interrelationships are possible. Only those relationships involving plants have been mentioned here, but the animals are competing with each other, as are the fungi and bacteria. Also, interactions occur between animals, fungi, protists, and bacteria. The ecosystem is extremely complex, and it is virtually impossible to predict how the disruption of one part might affect other parts. Although the operational habitat may be simple, the real habitat contains so many factors linked to so many other factors that we must be careful in our treatment of ecosystems.

"Zorak, you idiot! You've mixed incompatible species in the earth terrarium!"

FIGURE 26.18

The Far Side. Gary Larson (Universal Press Syndicate)

THE STRUCTURE OF POPULATIONS

Populations can be thought of as having many types of structure; their distribution through the habitat is an important one, as is the age structure of the individuals.

GEOGRAPHIC DISTRIBUTION

Boundaries of the Geographic Range. The ability of a plant species to spread throughout a geographic area is a result of its adaptations to the abiotic and biotic components of that area. Although most habitat components act on the plant simultaneously and most should be considered important, at any given time and locality, one factor alone determines the health of the plant. This factor, whatever it may be, is the **limiting factor.** As described for photosynthesis, at a medium level of carbon dioxide, increasing the amount of light causes an increase in photosynthesis, whereas increasing the concentration of carbon dioxide does not; light is the limiting factor. As light intensity is increased, however, a point is reached at which brighter light does not cause more rapid photosynthesis. Then, an increase in the level of carbon dioxide does result in greater photosynthesis, and carbon dioxide becomes the limiting factor.

FIGURE 26.19

On modern American farms, plants are given optimal amounts of fertilizer and water, pesticides and insecticides keep pathogens under control, and a variety of weed-control measures eliminate competition from other species. If the seeds are not planted too close together, each plant has adequate room and grows as rapidly as possible. It is limited only by its own innate capacity for growth. *(Courtesy of G. P. Mauseth)*

The concept of limiting factors applies to all aspects of a plant's interaction with its habitat. In areas of high rainfall, more water probably does not result in better plant growth, but the plants that are growing in the shady part of that ecosystem might benefit from more light, whereas those that are growing in the sunny spots might grow faster if more nitrogen compounds were available in the soil. Still other plants might not respond to extra light or nitrogen but would benefit from decreased herbivory. If each plant received extra amounts of the factor that had been limiting it, its growth would increase until some other factor became limiting. If the shaded plant received extra light, its growth rate would increase to the point where nitrogen perhaps became limiting. If a plant is placed in an optimal environment and given adequate amounts of nutrients, light, water, and freedom from pathogens, growth and reproduction increase greatly, but not infinitely; at some point the plant's innate capacity becomes the limiting factor. Crops on irrigated farms with weed control and pesticides are an example (Fig. 26.19).

Any factor of the ecosystem can act as a limiting factor. Water is important to many species; most cannot live in desert regions because of lack of water and most cannot live in marshes because of excess water. Extreme temperature inhibits plant growth in many regions; even if given adequate water, some plants cannot conduct it as rapidly as it would be transpired at high temperatures. For other species, high temperatures apparently cause enzyme systems to lose synchronization, and metabolism does not function correctly. Lack of warmth in winter is a limiting factor that keeps many species restricted to the tropics; temperate rain forests have adequate moisture and even richer soil than do the natural habitats of most tropical species, but the temperate rain forest habitat has freezing winter temperatures.

Biotic factors are also critical; many desert plants grow much more rapidly if given more water than occurs in their habitat, but their ranges do not extend into moist regions. Another important consideration is that plant species that rely on animals for pollination or seed dispersal cannot reproduce where their animal partners do not exist; the geographic range of these plants may be set by the limiting factors of the animals.

Soil factors often produce abrupt boundaries for the geographic ranges of populations. Both mineral composition and soil texture are important (Fig. 26.20). Soils derived from

(a)   (b) — Alluvial fan

FIGURE 26.20

(a) The distribution of these plants is easy to understand. Most of this granite outcrop is so smooth that all seeds are washed off by rain—nothing can grow on it. But where it has cracked, soil accumulates, seeds germinate, and plants thrive. The plant distribution is controlled by the physics of crystallization and fracturing of large masses of granite. (b) Soil that washes out of valleys accummulates as a dry delta called an alluvial fan. The soil is open and porous and usually has its own types of plants, distinct from those in the hills or in the valley. *(Bertram G. Murray, Jr./ Earth Scenes)*

limestone, sandstone, or serpentine often have characteristic species growing on them. Beaches with loose, porous, sandy soils have species distinct from those on nearby soils that are more compact and contain more humus. The limiting factor for a particular species may be the same factor over its entire geographic range, but often it varies from area to area.

Local Geographic Distribution.   In addition to large-scale geographic distribution of the entire population, small-scale, local distribution of individuals with respect to each other is also important. Individuals have one of three types of local distribution: random, clumped, or uniform. The term **random distribution** is used whenever there is no obvious, identifiable pattern to the position of individuals (Fig. 26.21a). A random pattern is one that has no predictive value; knowing the position of one plant does not let you estimate the position of another plant. In most habitats, many individuals seem to be distributed at random, but that may simply be due to the presence of many small-scale patterns that are difficult to detect or one large pattern that is too complex to see.

Clumped distributions are those in which the spacing between plants is either small or large, but rarely average (Fig. 26.21b). This can result from many factors. The seeds of a plant often fall near the plant, not at uniform or random distances from it. If a bird or other animal eats many fruits and seeds, it will probably "deposit" them all together in a nice neat lump.

Uniform distributions are the types that occur in orchards and tree plantations; all individuals are evenly spaced from their neighbors (Fig. 26.21c). In natural populations, uniform distributions are not extremely common; those that do occur are thought to result from intraspecies competition. The roots of one individual may establish a zone that prevents the germination or growth of others. Zones can also be established, at least theoretically, by the release from the plant of chemicals that inhibit other plants. Such chemicals are called **allelochemics** and the inhibition is **allelopathy.** One example may be the purple salvias of California (*Salvia leucophylla*); they grow in a relatively uniform spacing with a zone of bare soil surrounding each shrub (Fig. 26.22). Several chemicals, particularly terpenes, are given off from these plants and have been shown to inhibit growth of plants in the laboratory; it is known also that these do accumulate in the soil near *Salvia*. It is not certain that these actually are allelochemics, however, because some experiments have

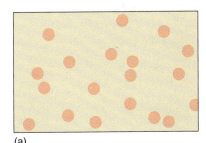

(a)

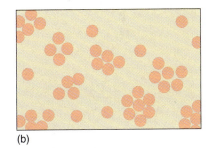

(b)

(c)

FIGURE 26.21

Types of distribution on a small scale. (a) This appears to be a random distribution, but examination of a larger area might have revealed a pattern. (b) Clumped distribution. (c) Uniform distribution.

FIGURE 26.22

Each bush of *Salvia leucophylla* is surrounded by a zone devoid of vegetation. Allelochemics given off by the salvias penetrate the top layers of soil and inhibit seed germination. If a seedling can get its root through the top zone, the plant may grow well, but herbivory by rabbits and mice that live in the protection of the salvias then becomes a problem. *(Roger Del Moral, University of Washington)*

shown that rabbits, mice, and birds that live in the salvia shrubs are mostly responsible for the bare patches; if a wire cage is placed over a bare patch to exclude animals, plants do grow there. Presumably the bare patches are narrow because it is risky for the animals to venture too far from the cover of the salvia. It may be possible that terpenes are allelochemics under certain circumstances but that typically, animal herbivory prevents other plants from growing near salvia.

AGE DISTRIBUTION: DEMOGRAPHY

The manner in which a population responds to various factors in its habitat is affected partly by its **age distribution**, its **demography**—the relative proportions of young, middle-aged, and old individuals. Analysis of age distribution has been applied mostly to animal populations and may be difficult to apply to plants, but the fundamental aspects are important and easy to understand. Imagine a species in which a pair of individuals produces four offspring by the time they die; the four offspring also double their numbers, so there are eight after they die. Future generations would contain 16, 32, 64, 128 individuals, and so on. This population is undergoing an exponential rate of increase (Fig. 26.23a), and there are always greater numbers of young individuals than old ones. This is important in determining whether most members are very young and highly susceptible; moderately young and vigorous; older and well-established; or very old and senescent.

Two factors affect the possible rate of population growth: generation time and intrinsic rate of natural increase. **Generation time**, the length of time from the birth of one individual until the birth of its first offspring, affects the rapidity of population growth: Annuals have a generation time of 1 year or less and can increase rapidly, whereas most conifers and angiosperm trees must be several years old before they produce their first seeds. For easier comparisons, we often measure population increase in terms of generations, not years.

The second factor, **intrinsic rate of natural increase** or **biotic potential**, is the number of offspring produced by an individual that actually live long enough to reproduce under ideal conditions. Even with optimal conditions, a large percentage of seeds do not germinate and many seedlings die before they are old enough to reproduce, so the biotic potential does not equal the number of seeds produced. For many species, biotic potential is a large number, represented in population equations as **r**. Plants that produce a large number of healthy, viable seeds over their lifetimes have a large *r*, a large biotic potential,

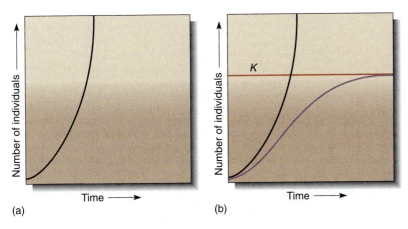

FIGURE 26.23

(a) Curve representing the number of individuals in a population that is growing exponentially. For this to occur indefinitely, growth must be controlled only by the organism's innate capacity for growth and the number of individuals present. In real situations, habitat limitations cause growth to be somewhat slower than the theoretical maximum. (b) Graph showing the more realistic situation in which the carrying capacity, K, is included. Instead of increasing infinitely (black curve), the growth rate (purple curve) decreases as the population size approaches the carrying capacity of the habitat. When population size equals carrying capacity, population increase stops; the death rate equals the birth (germination) rate.

and their populations can potentially increase greatly each generation. A species that produces fewer seeds than another species can reproduce faster than the second if it has a very short generation time, however. Mustard plants are small, live for only 1 year, and produce just a few seeds each, whereas oaks are large trees, each of which produces thousands of seeds in its lifetime. But a mustard population can grow more rapidly than can an oak population.

The biotic potential is measured under ideal conditions, but such conditions do not often occur in nature. Furthermore, once they do occur, even in a laboratory experiment, the very existence of the plants finally disrupts those ideal conditions. Once the population becomes large, the plants must compete for water, nutrients, and space. The number of individuals in each population that can live in a particular ecosystem is limited; that number is the **carrying capacity** and it is symbolized by **K**. Theoretically, a population increases until the number of its individuals (N) becomes close to K; at that time, crowding and competion result in poorer growth, lower reproduction, and decreased chances that seedlings will be in suitable sites (Fig. 26.23b). Birth rate (germination) decreases and death rate increases. These factors continue as the number of individuals continues to approach K; when N and K are equal, population growth stops because death rate equals birth rate.

Many factors cause death rate to increase and birth rate to decrease as population size approaches carrying capacity (N approaches K). A large, densely crowded population is an ideal target for herbivores; most of the progeny of one pair of insects are likely to find suitable plants wherever they go in a dense plant population, so most survive. The large plant population is ideal for the insects, so the insects' population growth rate can increase toward their own biotic potential. The same is true of pathogenic fungi and bacteria. Even without considering predators or pathogens, large populations of plants may alter the environment physically, making it less ideal, as mentioned for the shading of pine seedlings by mature pine trees. A habitat filled with herbivores or pathogens or shade has a lower carrying capacity than the original habitat had.

On the basis of theory alone, we would expect that once a species invades a new habitat, it would undergo exponential growth as in Figure 26.23b, first increasing rapidly, later more slowly, and finally remaining stable with numbers that neither increase nor decrease. In reality, many factors prevent real populations from acting like ideal ones. As

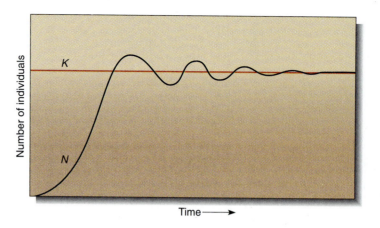

FIGURE 26.24
If a population increases slowly toward the carrying capacity of its environment, it may slow at *K*, as shown in Figure 26.23b. But often growth is too rapid and a large number of extra individuals temporarily survive; they then die off because of limited resources and the population falls below the carrying capacity.

the population approaches K, it may not slow sufficiently and may overshoot the carrying capacity; there would be too many individuals temporarily, followed by a die-off that drops the number far below the carrying capacity (Fig. 26.24). With other species, as N approaches *K*, an explosive increase in the populations of many types of pests may occur, and the numbers of aphids, mites, caterpillars, and fungi may increase. The plant's pests have overshot their own *K* and they kill so many plants that the plant population falls far below its *K*, perhaps almost to zero. From there, it may then increase again, undergoing major cycles. Alternatively, as with pines, as the population increases, they change the environment to one that favors other species; their own reproduction falls to zero and as the adults die, the entire population is lost.

r- AND *K*-SELECTION

As a population increases, theoretically it goes through a young phase in which numbers of individuals are low and resources are plentiful. Population growth is limited by the species' own biotic potential, *r*. Later, conditions are crowded, resources are more scarce, and population growth is governed by the carrying capacity of the ecosystem, *K*. The lifetime of any single individual is typically much shorter than the time required for a population to pass through this full development. Therefore, which is more advantageous to the species, to become adapted to *r* conditions or to *K* conditions? The two are very different and require distinct, often mutually exclusive adaptations.

r-Selection.   A disturbance usually produces *r* conditions. A fire or flood destroys many individuals in the area, and resources are plentiful for the few that remain, whether they are seeds, survivors, or immigrants carried in by wind, animals, or the flood itself. Pioneers that produce many seeds quickly have an advantage in that most seeds find suitable sites; if

TABLE 26.2	*Selectively Advantageous Traits in r-Selected and K-Selected Species*
Adaptive in *r*-Sites	**Adaptive in *K*-Sites**
annual	perennial
early maturity	late maturity
many small seeds	fewer, larger seeds
few mechanical or chemical defenses	many defenses

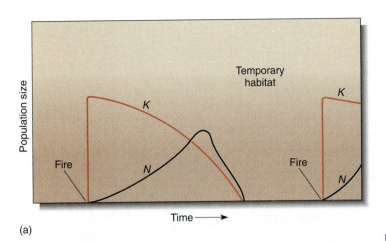

(a)

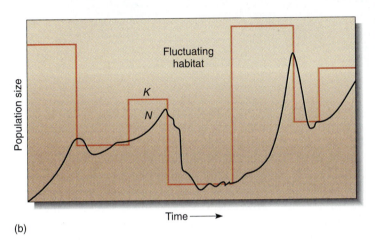

(b)

FIGURE 26.25

(a) Most habitats have no carrying capacity ($K = 0$) for species that are not a natural component of the habitat's ecosystem. But a disturbance such as a fire, flood, landslide, or disease outbreak may alter the habitat, suddenly making it suitable for new species. A good example is a construction site: Where new roads and shopping malls are to be built, the clearing of the site suddenly increases the carrying capacity for many weeds. They had not been present before or were present in very low numbers because they compete poorly with the normal vegetation present before construction. The carrying capacity for the weed species soon begins to fall as other plants invade and the natural vegetation returns to areas not covered with asphalt. (b) A site's carrying capacity may vary less drastically as the populations of competitors, pests, and pathogens rise and fall in their own cycles. If most of the herbivores are suddenly killed by a bacterial disease outbreak, the carrying capacity for the plants suddenly increases. This lasts only until new herbivores move in or some other pest responds to the increased number of uneaten plants.

seeds form too slowly, the sites may be filled by seeds of other invaders (Fig. 26.25). Because population density is low, the spread of predators and pathogens is slow, so the threat from them is not great; having antiherbivore and antifungal defenses is not so important because neither of those two agents is the limiting factor. Actually, the biotic potential is the limiting factor, so mutations that increase r are selectively advantageous (Table 26.2). Plants grow quickly, have few defensive compounds or structures, flower quickly, and produce many small seeds. Because most disturbances are impossible to predict and are widely scattered, seeds also must be adapted for widespread dispersal. r-Selected species typically are annuals or small shrubby perennials because the disturbed habitat gradually changes back into a crowded one that is no longer suitable for the pioneer r species (see Fig. 27.20). As more species of plants, animals, and fungi re-establish themselves in the area, the r-selected species are at a disadvantage; they have few defenses against predators and are too short to compete for sunlight. Only another disturbance can save them at this site; usually their population numbers fall to zero or close to it. The species itself survives because many seeds have emigrated to other sites, at least a few of which are appropriately disturbed areas (Fig. 26.26).

Some types of disturbances are predictable: killing temperatures in temperate winters and lethal hot/dry conditions in desert summers. r-Selected species are ideally suited for these environments. In spring, the habitat becomes suitable and may be almost devoid of plants. Small annuals grow and reproduce quickly, and population growth is extremely rapid until the habitat becomes disturbed by winter or summer climate.

K-Selection.   Conditions in a crowded habitat, where a population is close to its carrying capacity, select for phenotypes very different from those that are beneficial in a disturbed

FIGURE 26.26

Most plants of disturbed habitats, such as these coastal dune plants, are *r*-selected species: Their wind-blown seeds land, germinate and grow quickly, and then produce many more seeds. They can grow so rapidly and reproduce so abundantly because almost no energy or mineral resources are spent on antiherbivore defenses or drought adaptations. Because of periodic storms, freshly exposed, highly disturbed sand dunes are always ready for invasion. Cape Lookout National Seashore. *(C. C. Lockwood/Earth Scenes)*

habitat (Fig. 26.27; Table 26.2). In a disturbed region, virtually every spot is a suitable site for seed germination and growth, but in a *K* habitat, almost every possible site is filled. Once an individual dies, its site may become occupied by the seeds or rhizomes of a different species. It is advantageous to live for a long time, holding on to a site. To survive as a long-lived perennial is difficult, however, and large amounts of carbon and energy must be diverted into antipredator defenses. These resources are then not available for growth or reproduction, both of which are much slower than in an *r*-selected species. Many long-lived conifers such as redwoods, douglas firs, and bristle-cone pines are good examples of *K*-selected species. *K*-Selected species also face intense competition from other plant species, and therefore adaptations that increase the ability to use scarce resources are beneficial. Examples of such adaptation may be the capacity to use low amounts of light or soil strata that are poor in nutrients.

Species that are *r*-selected can occur next to ones that are *K*-selected. Avalanches in dense forests open up small sites suitable for *r*-selected species. The same is true for hurricanes, fires, and floods. The floor of a deciduous forest is temporarily an *r* site during springtime, between the time when temperatures become warm enough for germination and growth and the time when the canopy trees put out their new leaves. For several weeks, the forest floor is sunny, warm, rich in nutrients, and temporarily uncrowded. Small *r*-selected plants that grow and reproduce quickly can complete their life cycles before the larger trees come out of dormancy and block the sunlight.

FIGURE 26.27

In a stable environment with stable climate and an ecosystem that has few disturbances, the carrying capacity does not fluctuate as it does in Figure 26.25. Under such conditions, competition and lack of sites may be the dominant limitations on population increase.

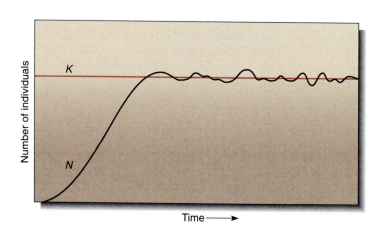

THE STRUCTURE OF ECOSYSTEMS

Many concepts can be considered in the structure of an ecosystem. Different botanists think about ecosystems in ways that are influenced by their own interests. Four of the most commonly mentioned structures are the physiognomic structure, temporal structure, species diversity, and trophic levels.

PHYSIOGNOMIC STRUCTURE

The physical size and shape of the organisms and their distribution in relation to each other and to the physical environment constitute the **physiognomic structure.** Trees, shrubs, and herbs are the three most useful categories, but in addition, a system of **life forms** was defined by C. Raunkiaer in 1934 (Table 26.3). The criterion for classification was the means by which the plant survives stressful seasons, such as by placing buds below ground (geophytes: bulbs, rhizomes) or winterizing aerial buds (phanerophytes: trees, vines). Regions of the world that have similar climatic conditions have similar physiognomic structures unless the soil is particularly unusual (Fig. 26.28).

An almost infinite number of combinations of life forms, vertical structures, and characteristics such as abundance of broadleaf plants, conifers, and sclerophyllous plants are possible (Fig. 26.29). An almost infinite number of types of ecosystems might be expected, but actually only a few basic types exist, as described in Chapter 27. Although differences exist between various types of forests or grasslands or marshes, each category is easily recognized as a common type of ecosystem.

TEMPORAL STRUCTURE

The changes that an ecosystem undergoes with time constitute its **temporal structure;** the time span can be as short as a day or can encompass seasons or decades. For animals, a daily cycle can be especially obvious, with some animals active at night (nocturnal) and others during the day (diurnal). Many plants also have daily rhythms of flower opening and closing.

Plants change dramatically with the season, as do the other organisms. Spring is typically a time of renewed activity, the production of flowers and new leaves. This is not simultaneous for all species; often the understory plants become active earliest, benefitting from the open canopy. Wind-pollinated flowers are usually produced before leaves expand and block the wind. Leafing out and flowering must be coordinated not only with the end of low-temperature stress conditions but also with the habits of pollinators and herbivores. It does a plant no good to produce flowers when temperatures are still too low for its pollinators. Through late spring and early summer, various species flower at distinct times, controlled by plant maturity, photoperiod, or adequate rainfall. In most ecosystems there is

TABLE 26.3 *Life Forms of Raunkiaer*	
Life Form	**Means of Surviving Stress**
Therophytes	Annual life span; survive stress as seeds.
Geophytes	Buds are underground on rhizomes, bulbs, corms.
Hemicryptophytes	Buds are located at surface of soil, protected by leaf and stem bases: many grasses and rosette plants.
Chamaephytes	Buds are located above ground, but low enough to not be exposed to strong winds: small shrubs.
Phanerophytes	Buds are located high, on shoots at least 25 to 30 cm above ground: trees and large shrubs.

(a)

(b)

FIGURE 26.28

(a) This desert scrub vegetation with small trees, large bushes, and saguaro cactus is characteristic of much of the southwestern United States and northern Mexico, where summers are hot with occasional rainstorms and winters are cool and moist but freezes are not severe. *(John Gerlach/Earth Scenes)* (b) The northern part of the central valley in Chile, just north of Santiago, has climatic conditions similar to those in the southwestern United States, and the vegetation of the two regions resemble each other in habit (life forms), distribution, and other features. Only by looking carefully does one notice that the species are very different.

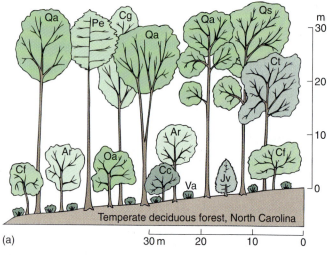

(a)

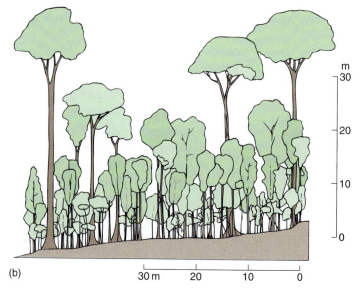

(b)

FIGURE 26.29

The presence of trees, shrubs, and herbs (phanerophytes, chamaephytes, and therophytes) gives an ecosystem vertical structure. The trees form the canopy; the shrubs, short trees, and saplings of tall trees form a middle level understory or subcanopy; the herbs and seedlings are the ground-level plants. Vines can occur in all three levels. (a) A temperate forest. Ar, *Acer rubrum* (red maple); Cc, *Cercis canadensis* (redbud); Cf, *Cornus florida* (dogwood); Cg, *Carya glabra* (pignut hickory); Ct, *Carya tomentosa* (mockernut hickory); Jv, *Juniperus virginiana* (red cedar); Oa, *Oxydendrum arboreum* (sourwood); Pe, *Pinus echinata* (shortleaf pine); Qa, *Quercus alba* (white oak); Qs, *Quercus stellata* (post oak); Va, *Viburnum affine* (arrow-wood). (b) A tropical rain forest has a much more complex structure. For more information about the research represented by these figures, see *Plants, Man, and the Ecosystem* by W. D. Billings and *Tropical Rainforests of the Far East* by T. C. Whitmore.

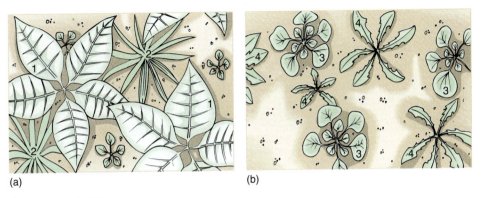

(a)   (b)

FIGURE 26.30

The temporal patchiness of an ecosystem is important; a large perennial plant occupies a site year after year, so the site is unavailable to seeds. But a site is occupied only temporarily by short-lived plants and becomes available for seeds when the plant dies. (a) Species 1 and 2 are winter rosettes that occupy most of the site through winter and early spring. (b) After they flower, fruit, and die, their sites are vacated and species 3 and 4, which germinate in later spring, find many sites available.

no time during the summer when nothing is in bloom. The continual presence of some species in flower provides nectar and pollen for insects throughout the season.

Species that bloom early in spring probably form fruits and mature seeds during summer; species that flower later release their seeds in late summer or autumn. Once seeds are dispersed, some remain dormant until the following spring. Others germinate and grow into a low rosette that survives the winter, even growing slightly on warm winter days. When spring arrives, the seedling is already well-rooted and can begin to grow quickly while the seeds of competitors are just starting to germinate (Fig. 26.30).

Late summer and autumn bring changes that depend on the ecosystem; in the northern United States, herbs die while shrubs and trees develop resting buds. The entire plant enters light dormancy; then the first cool days initiate deep dormancy. Leaves and fruits are abscised, removing the last sources of food for most animals. In the southern and southwestern United States, cooler autumn weather is often more welcome than the first warm days of spring because the summer is so much more severe than the winter. Most gardening is done in autumn rather than spring, and fall wildflowers are abundant and dramatic. The growing season extends at least to December for shrubs and many herbs, and the small rosette plants may never become truly dormant.

In tropical ecosystems, winter and summer do not exist, but an alternation of dry and wet seasons governs ecosystem change. Coastal marsh, wetland, and reef ecosystems may be strongly affected by seasonality; rainy seasons dilute the salt water, whereas in dry seasons rivers deliver less fresh water and mineral-rich silt.

Over long periods—many years—most ecosystems undergo gradual, often dramatic changes. This process of succession is discussed in the next chapter.

SPECIES COMPOSITION

Species composition refers to the number and diversity of species that coexist in an ecosystem, and it depends on whether the climate is mild or stressful, the soil is rich or poor, and the species' tolerance ranges are broad or narrow (Figs. 26.31 and 26.32). Stressful climates with poor soils support a low number of species, because so few species are adapted to such conditions (Fig. 26.33). On the other hand, mild climates and rich soils support an abundance of species because most plants have tolerance ranges that include such climatic and soil conditions (Fig. 26.34). Competition is intense, but apparently natural selection has resulted in habitat partitioning, with each species occupying a narrow portion of the various resource gradients (Fig. 26.32). The presence of a large number of species actually creates more niches that can be filled by new species; the presence of trees makes it possible for epiphytes and parasites to occur in the ecosystem.

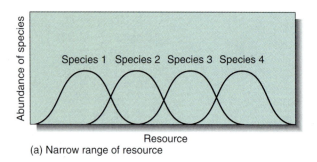
(a) Narrow range of resource

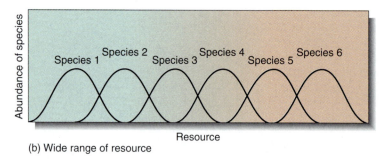
(b) Wide range of resource

FIGURE 26.31
In the ecosystem represented by (a), the range of the resource available is narrow compared to that of the ecosystem in (b); in general, it is able to support fewer species. For example, the resource may be water, with the right ecosystem having a variety of areas that range from dry to moist to lakes or streams, whereas the left ecosystem is just marshy or just desert.

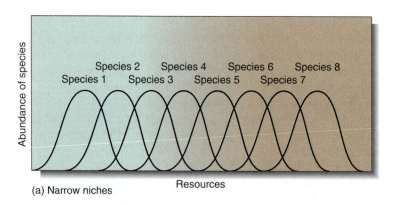
(a) Narrow niches

FIGURE 26.32
Two ecosystems with similar ranges of resources can differ in the number of species they contain. One may be occupied by a large number of very specialized species, each adapted to only a narrow range of the resource, whereas the other may be occupied by a few species of generalists that grow well under a variety of conditions and exclude most competitors.

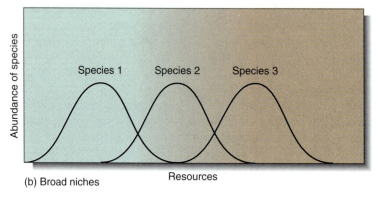
(b) Broad niches

FIGURE 26.33
Adapting to certain harsh conditions is difficult, and only a few species may have succeeded. Consequently, harsh habitats may have many unused resources.

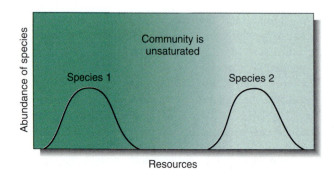

FIGURE 26.34
This climate is nonstressful, the soil is rich, and no toxins are present. Many species live here, and many more could if protected from competition. Species diversity here is high.

TROPHIC LEVELS

Trophic levels are basically feeding levels. Each ecosystem contains some members, autotrophs, that bring energy into the system. Photosynthesis is by far the dominant method, accounting for virtually all energy input. Green vascular plants are most important, but algae and cyanobacteria carry on about one third of all photosynthesis worldwide, and lichens and nonvascular plants are important in cold, high latitudes and altitudes. Chemosynthetic bacteria bring chemical energy into ecosystems, and although this is minor now, it was the only method before photosynthesis evolved (Fig. 26.35).

Autotrophs are known as **primary producers** for obvious reasons, and they are the first step of any food web. They are the energy and nutrient supply (food) for the herbivores, which constitute the **primary consumers** (sometimes called secondary producers). Herbivores are preyed on by carnivores, the **secondary consumers.** Omnivores exist at both trophic levels. **Decomposers** such as fungi and bacteria break down the remains of all types of organisms, even those of other decomposers (Fig. 26.36).

FIGURE 26.35
Recently, colonies of tube worms and clams have been discovered on the deep sea floor near hydrothermal vents, areas similar to Yellowstone's geysers, where hot, mineral-rich water erupts into the ocean. The water contains large amounts of hydrogen sulfide, which is oxidized by chemotrophic bacteria (see Chapter 19). However, the bacteria occur symbiotically within the tissues of the tube worms and clams, and they leak carbohydrates to their symbiotic hosts. Basically, the bacteria act like chemosynthetic "chloroplasts." A further exotic metabolism is that the blood of tube worms is able to carry hydrogen sulfide to the bacteria in their tissues, even though hydrogen sulfide is highly toxic to most animals. This is an unusual ecosystem that is not dependent on sunlight and in which plants are not involved at all. *(J. Frederick Grassle/Woods Hole Oceanographic Institution)*

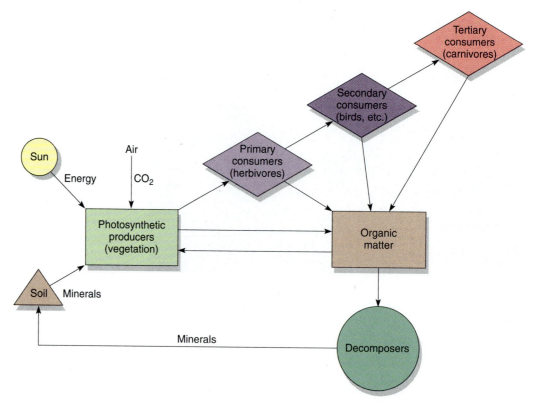

FIGURE 26.36
Energy and materials flow through the ecosystem as a network rather than in a straight line. Each level contributes to the decomposers (they even decompose other decomposers), and omnivores eat both plants and other animals. The primary producers never draw energy from the other stages, but many depend on animals and decomposers to return essential elements to the soil. At each step, respiration returns most of the carbon back to the air, and most of the energy is liberated as heat.

As plants photosynthesize, energy and carbon compounds enter the ecosystem. As the plants are eaten, the energy and the carbon compounds move to the herbivore trophic level, then to the carnivore trophic level, and finally on to the decomposers. This is referred to as the **energy flow** and the **carbon flow** of the ecosystem. At each step much of the food is used in respiration, resulting in the production of ATP, heat, carbon dioxide, and water. The carbon dioxide is released back to the atmosphere, where it can be used in photosynthesis again; the energy temporarily contributes to warming the planet, but it is ultimately radiated into space. The portion of the food that is not respired is available for growth and reproduction. As a very rough approximation, about 90% of an animal's food is respired and 10% is retained as growth or gametes. Therefore, when herbivores eat 1000 kg of grass, 900 kg are respired and 100 kg become herbivore tissues. Similarly, a carnivore that eats the 100 kg of herbivore would only gain 10 kg, and after dying would support just 1 kg of decomposers.

Along with the flow of energy and carbon, minerals flow through the ecosystem. Plants absorb minerals from soil and incorporate them into their bodies as amino acids, ATP, coenzymes, and so on, and these are consumed by herbivores. Many of these compounds are broken down by animals, and the minerals are lost as waste. After organisms die, the minerals that remain in their bodies are released by decomposers and become part of the soil again. In addition, minerals can be completely lost from an ecosystem as they are carried away by rainwater and streams, especially if erosion is occurring or as cities dump mineral-rich human waste into rivers. Such minerals are carried to the oceans and become part of marine ecosystems. Many algae, protists, and marine animals have bodies with

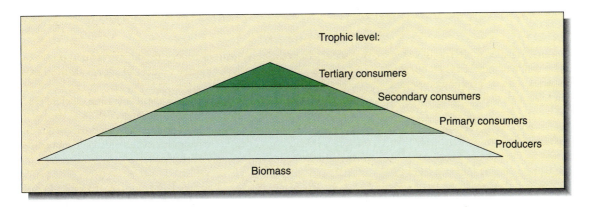

FIGURE 26.37

A pyramid such as this can represent the number of organisms at each level, the weight of all the individuals at each level, the respiration at each level, and so on. Such analysis can be difficult to do, but these relationships are extremely important. If a national park is established to preserve a species of wolf or mountain lion, it must be known how extensive its food web is and how large an area must be set aside to ensure an adequate number of primary consumers.

heavy, mineral-rich shells and bones; as the organisms die, the bodies sink to the depths of the ocean and are unavailable to living organisms of the land or ocean surface.

The movement of energy and biomass from one trophic level to the next is often represented as a pyramid of energy or a pyramid of biomass (Fig. 26.37). Such pyramids are useful for illustrating these principles in textbooks, but in reality it is extremely difficult to measure all the biomass of the primary producers of even a simple ecosystem. It is almost impossible to measure it accurately in a forest ecosystem, and it is impossible to measure the biomass or energy content of the fungi and bacteria that are acting as decomposers. Pyramids of numbers are more easily obtained; often there are fewer individuals in each higher trophic level, but of course millions of aphids can live on one tree.

Pyramids of biomass and numbers have become important as a means of illustrating the impact of the human introduction of herbicides, pesticides, and toxic wastes. Although some of these are biodegradable, many are not and they persist unchanged and toxic in the environment. Imagine a toxic herbicide applied at the low rate of one part per million—1 gm per 1000 kg of pasture grass. Herbivores will eat the 1000 kg of plants, and the pyramid of biomass shows that they will respire 900 kg away, but the 1 gm of herbicide is not oxidized or excreted. Instead it accumulates in the animals, now at a concentration of 1 gm per 100 kg. This is ten times stronger than the application rate. As carnivores eat the herbivores, the concentration rises to 1 gm per 10 kg, now 100 times stronger. Human metabolism is no different; these chemicals move through the food web to us as well. The toxic materials are not distributed uniformly within the bodies of animals and humans but become concentrated in our livers and kidneys as well as in fatty tissues such as our brain and spinal cord.

The role of decomposers in the ecosystem is vitally important; decomposers are so ubiquitous as well as microscopic that we take them for granted. If there were no decomposers in nature, plants could grow only as rapidly as weathering breaks down rock; as soon as minerals were released, they would be absorbed and locked into a plant and would then move through the food web. When the organisms died, the minerals would remain locked in the nondecomposing body. Sterilized (canned) food does not decay even though moist and kept at room temperature because no decomposers are present. Only by the physical and chemical weathering of rock and dead bodies could plant growth continue. But with decomposers present, minerals are recycled rapidly; decomposers release the minerals to the environment through their membranes, and their own bodies are so small and delicate that after death they are broken down immediately by other decomposers or by inorganic chemical weathering.

SUMMARY

1. Ecology is the study of organisms in relation to all aspects of their surroundings. Each plant is part of a population of individuals of the same species, and these co-exist with other species, forming a community. When climate and physical surroundings are also considered, the whole is an ecosystem.

2. The biology of a population is not simply the sum of all the individuals because plants interact positively and negatively with each other.

3. Habitat components may be biotic or abiotic. Abiotic factors include climate, soil, latitude, altitude, and disturbances. Biotic factors include all living organisms and the results of their activities.

4. Disturbances are disruptions of portions of an ecosystem; although they may cause great destruction, they may also be necessary for maintaining the ecosystem such that there is not a succession, a long-term change in species present.

5. By existing in an ecosystem, a plant affects the ecosystem; the plant may produce changes that are detrimental for the survival of itself or its progeny, or it may alter the ecosystem such that it is more suitable for its progeny.

6. The amount of competition varies from species to species and from one ecosystem to another. In general, where climates are mild and soils rich, species diversity is great, and competition is most important in exclud-ing species from the ecosystem. In harsher conditions, fewer species exist and the conditions themselves limit the number of species that can exist successfully.

7. With a large geographic range, a species may face different types of competition or stress in different parts of its range.

8. Plants interact with animals, fungi, protists, and bacteria in numerous ways, but most interactions basically can be divided into mutualistic (beneficial to both) or predatory (detrimental to one, beneficial to the other).

9. Populations have at least two basic types of structure: (1) their geographic range, which is set by limiting factors, and their local distribution (random, clumped, or uniform); (2) the distribution of the ages of all the individuals (demography).

10. The biotic potential of a species (r) is a measure of its reproductive success: the mean number of offspring per parent that survive to reproductive maturity. An r-selected species is one that is adapted to sites temporarily free of competitors and rich in resources.

11. The carrying capacity (K) of an ecosystem is a measure of the number of individuals of a particular species that can be sustained by the ecosystem. A K-selected species is one that is adapted to stable, relatively unchanging sites where competition may be very important.

IMPORTANT TERMS

biotic potential	demography	K-selection	pathogenic	primary producers
carrying capacity	ecosystem	life forms	pioneer	r-selection
commensal relationship	ecotype	limiting factor	population	tolerance range
community	habitat	mutualism	predation	trophic levels
competition	herbivory	niche	primary consumers	

REVIEW QUESTIONS

1. Define each of the following: population, community, habitat, ecosystem. Why is the biology of a population different from the biology of an individual?

2. How much of a plant's environment should be considered its habitat? Would the habitat of a wind-pollinated, wind-dispersed species be simpler than one with animal-mediated pollination and seed dispersal? Why?

3. Why are climate extremes often more important than average climatic conditions? What is the "growing season" of a plant, and what are some conditions that affect its length?

4. What is a pioneer species? Are they more likely to be r-selected or K-selected? Why?

5. Fire is a critically important, natural disturbance in many ecosystems. Describe some plant adaptations that allow plants to survive fires.

6. What is competition between plants? What is competitive exclusion? How does competition affect natural selection and evolution?

7. Describe biotic potential. Describe carrying capacity. How do these two factors affect population growth and the ultimate stable size of a population?

8. Consider the population growth curve of Figure 26.23b. Would a young population (left side of curve) have a greater proportion of young individuals than an old population (right side of curve where it has reached K)? Why? Does it matter if a population has a greater proportion of young, middle-aged, or old individuals? The answer is obvious for humans, but does it matter for plants?

BotanyLinks includes a Directory of Organizations for this chapter.

BIOMES

The grassland prairie biome extends across the entire center of the United States and Canada, covering thousands of square miles.

CONCEPTS

Earth's land surface is covered almost entirely by **biomes,** extensive groupings of many ecosystems characterized by the distinctive aspects of the dominant plants. For example, some of the biomes of North America are the temperate deciduous forests, subalpine and montane coniferous forests, grasslands, and deserts. Plant life is absent only in the harshest deserts (the Atacama in Chile and Peru, the Sahara in Africa, the Gobi in China) and in the land regions covered permanently by ice (most of Antarctica and the tops of high mountains; Fig. 27.1). In all other areas, the rock and soil are at least temporarily free of ice and have some liquid water; primary producers—plants, algae, and cyanobacteria—carry out photosynthesis and support food webs of consumers and decomposers.

Biomes vary from extremely simple, as in tundra, to more complex grasslands, temperate forests, and tropical rain forests. Biome complexity and physiognomy are most strongly influenced by two abiotic factors: climate and soil. A particular type of biome, such as grassland or temperate deciduous forest, may occur in various regions of Earth because the same set of climatic and soil factors occurs in various regions (see Fig. 26.28). At all sites, the physiognomy, the appearance, of a biome is similar, but often the actual species present differ considerably from one area to another. For example, temperate grasslands are easily recognizable in the central plains of the United States, the steppes of Russia, the pampas of Argentina, and the veldt of Africa (Fig. 27.2); all are dominated by grasses and large mammals and are devoid of trees except along rivers. Despite the strong physical

FIGURE 27.1
Antarctica is one of the few places on Earth where plant life simply cannot exist; in the area shown, even if plants were introduced, they would die. Lack of long-distance seed dispersal is not the reason for the absence of plants, as you may have guessed. *(Courtesy of Tom Taylor, Ohio State University)*

(a)

(b)

(c)

FIGURE 27.2
Under the proper conditions of climate and soil, grasslands almost invariably develop, as opposed to rainforest or desert. Wherever those conditions exist, a specific type of vegetation can be expected. Even though the actual species present in one differ from those in the other, the biomes are still recognizable as grasslands because of convergent evolution due to similar environmental conditions. Grasslands in (a) Wyoming, (b) Kenya, (c) Brazil. *(b, Peter J. Bryant/Biological Photo Service; c, Luiz Claudio Marigo/Peter Arnold, Inc.)*

similarity, each has its own set of characteristic species because few plants have long-distance dispersal mechanisms capable of moving pollen or seeds to all continents. Species that become adapted to the grassland niche of one continent do not occur on other continents because no birds, mammals, winds, or other means carry seeds such long distances. Despite the lack of gene flow between widely separated units of a biome, the physiognomic similarity persists as a result of convergent and parallel evolution: Climate, soil, and other habitat factors in each area select for similar phenotypes. If portions of a biome are not too widely separated, some gene flow can also contribute to the uniformity of biome appearance, and various regions do actually have some species in common.

The geographic locations of the Earth's biomes are determined by many factors, but two of overriding importance are (1) world climate and (2) positions of the continents. If either or both of these factors were different, the locations of the biomes would be different. In this regard, it is important to realize that both do change: The climate undergoes cycles of cooling and warming, and the continents shift position because of continental drift. In the past, conditions on Earth's surface were quite different, and so were the distributions of plants and animals. Similarly, the conditions will be altered in the future, and organisms will be affected by that as well.

WORLD CLIMATE

Earth's climatic conditions are the result of its tilted axis of rotation and the presence of the atmosphere and oceans.

EFFECTS OF EARTH'S TILT

Imagine a planet whose axis of rotation is exactly perpendicular to the plane of its orbit. At the equator, the sun would rise exactly in the east every day of the year; it would pass directly overhead and set in the west. There would be no change, no seasons. At all sites not on the equator, 70 degrees N for instance, the sun would never be overhead; every day it would rise in the southeast, pass low in the sky, and set in the southwest. Maximum heating would be at the equator; all other regions would always receive only oblique lighting and would be much cooler. If there were no atmosphere or oceans, heat could not be transferred from the equator to the poles, and there would be a tremendous temperature gradient between those regions.

Earth's axis of rotation is tilted 23.5 degrees away from perpendicular to the orbital plane (see Fig. 26.8). At summer solstice, June 21 or 22, the North Pole points as directly toward the sun as possible, and in the Northern Hemisphere, the sun has reached its highest point in the sky. The sun appears to be overhead at noon for those people located at 23.5 degrees N latitude, the Tropic of Cancer, which runs just south of California, Texas, and Florida. The days are their longest and summer has officially begun. North of the Arctic Circle, the sun is visible even at midnight.

As Earth continues its orbit, the axis of rotation points less toward the sun, which appears to rise and set more to the south and is lower in the southern sky at noon. By September 23, autumnal equinox, Earth has made one-quarter orbit, the sun is directly over the equator, and days are exactly 12 hours long: Autumn begins. After another 3 months, winter solstice, December 21 or 22, the South Pole points as directly as possible toward the sun; the sun is directly over the Tropic of Capricorn, 23.5 degrees S latitude. Summer begins in the Southern Hemisphere, but in the north, winter begins and the days are at their shortest; above the Arctic Circle, the Sun never rises. With another one-quarter orbit, the sun moves back toward the equator, arriving there at March 21, the vernal equinox.

The most intense solar heating is not confined simply to the equator, but rather moves seasonally northward, then southward. All parts of the planet experience seasonality, although in the region between 23.5 degrees N and 23.5 degrees S, the tropics, not as much change occurs between summer and winter as in the temperate regions outside this band.

Plant Biology Tutor CD-ROM

ATMOSPHERIC DISTRIBUTION OF HEAT

The atmosphere and oceans, being fluids, develop convection currents and massive flows when heated in one area and cooled in another. They distribute heat from the tropics to the temperate regions and all the way to the poles.

Much of the tropical zone is occupied by the Pacific, Atlantic, and Indian Oceans (Fig. 27.3). Throughout the year, the water and air receive solar heat, causing tremendous amounts of evaporation into the air. The air warms, expands, then rises high into the atmosphere. As it flows upward, the surrounding air pressure decreases, and the moist rising air expands even more. The expansion causes the air to cool, decreasing its ability to hold moisture. Water vapor condenses into rain and falls back to the surface in torrential storms, producing tropical rainforests in Central America, northern South America, central Africa, and Southeast Asia.

After rising, the air is pushed northward and southward by the continued rising of more tropical air below it (Fig. 27.4). While spreading at high altitude, the air radiates heat to space, cooling even more. By the time it reaches about 30 degrees N or S latitude (the horse latitudes) it has cooled, contracted, and become dense enough to sink. It encounters greater atmospheric pressure at lower altitudes and is compressed and heated. As it warms, its water-holding capacity increases. Because its capacity to hold water is greater but its water content is the same, the air has become drier. But it had already lost much of its moisture while initially rising, so it is now extremely dry. Land areas below this descending air contain the world's hot, dry desert biomes (Fig. 27.3).

Once back at Earth's surface, part of the air spreads toward the equator and part flows toward the poles. Earth's rotation causes the air moving toward the equator to be deflected westward. It moves as a northeast trade wind in the Northern Hemisphere and a southeast trade wind in the Southern Hemisphere. Air spreading toward the poles from the horse latitudes is deflected eastward and blows as a prevailing westerly.

The actual area of descending dry air varies with the season, being farther north when the sun is near the Tropic of Cancer (Northern Hemisphere summer) and farther south when the sun is near the Tropic of Capricorn. The United States, during its winter, is located entirely within the influence of the prevailing westerlies; all winter weather comes from the Pacific and Arctic Oceans, moving eastward across the continent. During summer in the United States, the northeast trade winds have moved northward far enough to influence the states along the Gulf of Mexico; northeast tradewinds bring summer storms westward from the Atlantic onto the east coast and gulf, supplying summer rains.

Continental Climate.   The size of a land mass influences the weather it receives. Larger islands such as the Hawaiian Islands, Guam, and Puerto Rico have mountains that force air to rise as it blows across them. Rising air cools and rain forms. Low-lying, small islands like the Florida Keys and the smallest islands in the Bahamas are too flat to affect the air; these **desert islands** are extremely dry, often with no fresh water streams or lakes (Fig. 27.5).

Continents also cause air to rise, cool, and drop precipitation. If the topography is fairly flat, air rises gradually and rains are distributed over an extensive area. Summer storms from the Atlantic and Gulf of Mexico move through the eastern United States, the Mississippi Valley, and the plains states; the land rises so gradually that rainfall covers half a continent.

If topography is mountainous, as on the United States' west coast, rain is dropped in a narrow area. The prevailing westerlies bring moist air from the Pacific Ocean onto land

Winds are named after their sources: A westerly wind comes from the west and blows to the east.

FIGURE 27.3

Most land on Earth is distributed rather far north at present, and the tropical zone, between the Tropics of Cancer and Capricorn, is mostly water. The greatest amount of solar energy falls on water, causing great evaporation and humidifying the atmosphere but leading to relatively little temperature change. If more land were in the tropics, the soil would heat more, which in turn would heat the air, but the oceans would stay cool and evaporate little moisture to the atmosphere; all regions would receive much less rain.

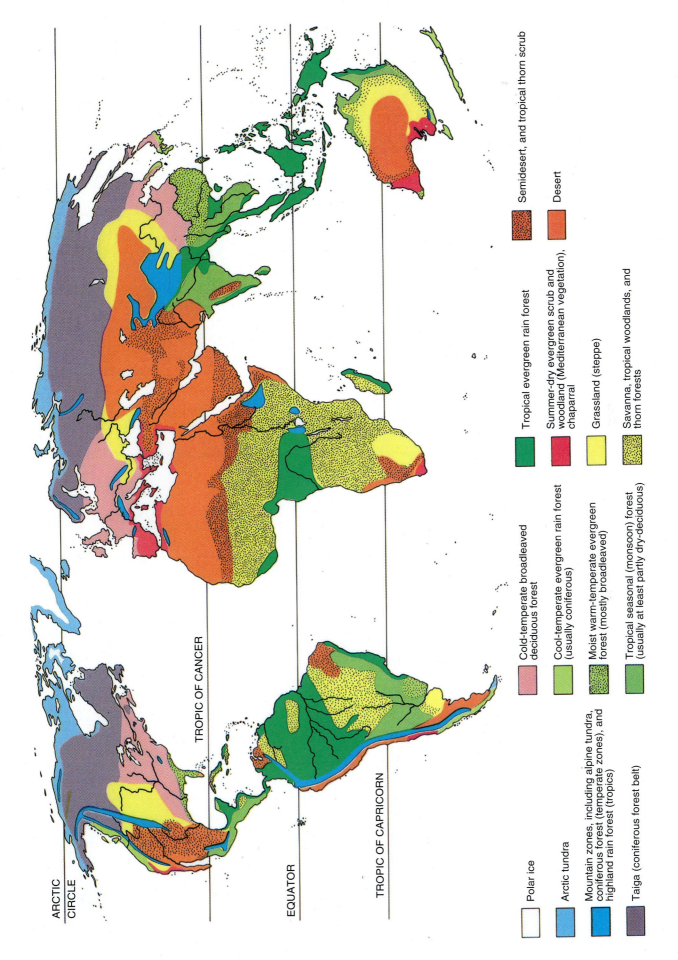

ARCTIC
CIRCLE

TROPIC OF CANCER

EQUATOR

TROPIC OF CAPRICORN

Semidesert, and tropical thorn scrub

Desert

Tropical evergreen rain forest

Summer-dry evergreen scrub and
woodland (Mediterranean vegetation),
chaparral

Grassland (steppe)

Savanna, tropical woodlands, and
thorn forests

Cold-temperate broadleaved
deciduous forest

Cool-temperate evergreen rain forest
(usually coniferous)

Moist warm-temperate evergreen
forest (mostly broadleaved)

Tropical seasonal (monsoon) forest
(usually at least partly dry-deciduous)

Polar ice

Arctic tundra

Mountain zones, including alpine tundra,
coniferous forest (temperate zones), and
highland rain forest (tropics)

Taiga (coniferous forest belt)

771

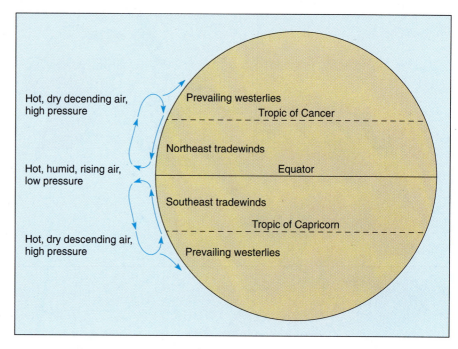

FIGURE 27.4

Air rises above the equator and then spreads northward and southward at high altitudes. By the time it has reached the tropics, it is cool enough to contract, become dense, and sink. At the surface, some flows back toward the equator as tradewinds, and some continues toward the poles as prevailing westerlies. The entire pattern is shifted northward during the Northern Hemisphere summer and southward during the winter.

FIGURE 27.5

(a) When a mountain or other large land mass forces air to rise, the air expands because of the lower pressure at high altitude. If it cools below its dew point, it drops rain or snow. On the lee side of the mountain, descent of the air compresses and heats it, raising its ability to hold water. Rather than bringing rain to the area, it may actually dry out the soil. (b) A low island does not cause the air to rise, so that source of cooling is not present. Air blows across undisturbed, and the island receives little rain even though the air may be extremely humid.

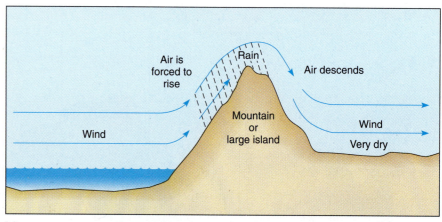

(a)

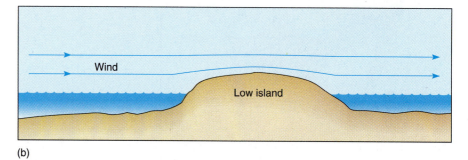

(b)

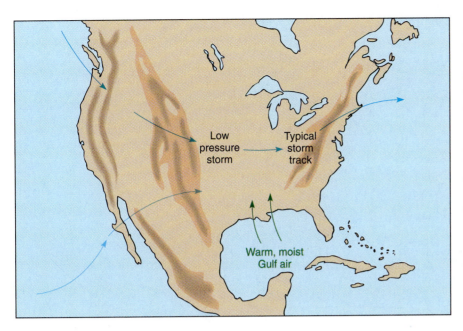

FIGURE 27.6
During our summer, tropical storms sweep west through the Caribbean, then turn to the north or northeast and bring moisture inland as far west as central Texas and north toward the Great Lakes. Fortunately, no southern coastal mountain range blocks this movement or the central and eastern United States would be very dry. During our winter, air circulation patterns move south so much that westerlies dominate the entire continent. Virtually all weather systems come from the west coast or Canada, dropping their moisture in the Rocky Mountains. Snow storms in the plains states drop little snow; along the northeastern United States some moist Atlantic air is drawn over land and mixes with cold western air, resulting in heavy snowstorms. The southeastern United States remains cool but not cold, receiving rain when moist Caribbean air occasionally pushes north over land. The southwestern United States tends to remain dry in winter, receiving some snowfall when Pacific westerlies push slightly northward.

over California, Oregon, Washington, and western Canada. The air is immediately forced to rise by coastal mountain ranges—the Cascades in Oregon, Washington, and Canada and the Coastal Range in California. Strong rains fall on the western slopes, but after the air crosses the summit, it descends and warms and rains cease. The eastern slopes are much drier than the western ones. The decreased rain on the landward side of mountains is called a **rain shadow.** As air continues eastward, it encounters the Sierra Nevadas and the Rocky Mountains; these are higher than the coastal ranges, so air is forced upward again, far enough to cool it even more than before, and more rain falls. Descending east of the Rockies, the air warms, rains cease, and the air moves across the central plains. By now the air is very dry and only rarely provides precipitation; if it were not for the Atlantic moisture moving northward out of the Gulf of Mexico, this area would be a desert instead of a grassland (Fig. 27.6).

The size of a land mass also affects its temperature fluctuation. On a large island, moist air is always rising over the mountains, and clouds are frequent. Temperatures have a narrow range. Along the coasts of a continent, cloudy, rainy weather is frequent and temperature fluctuation is mild. Farther inland, air is dry and clear; lack of clouds exposes the land to full daytime insolation, and heating is rapid and extreme. At night, clear skies allow the land to radiate infrared energy into space, with none reflected by clouds; cooling is also rapid and extreme.

Oceanic Distribution of Heat

Water in the Pacific and Atlantic Oceans flows in giant circular currents driven by the air circulation patterns. The ocean currents distribute heat from the tropics to the poles, lessening the temperature gradient that would otherwise exist. Also, as warm tropical surface water moves to higher latitudes, large amounts of water evaporate into the temperate prevailing westerlies, giving them more humidity than they would have if the oceans did not circulate.

As the trade winds blow across the tropics from east to west, friction between air and water causes equatorial currents to form. During the weeks that a particular mass of water flows along the equator as part of the Atlantic Equatorial currents, it absorbs huge amounts of energy and warms significantly (Fig. 27.7). At the western side of the ocean basin, it is deflected northward by the tip of Brazil, then part is deflected by Florida and enters the Gulf of Mexico as a counterclockwise current. The water's warmth permits high evaporation into the air and keeps much of the gulf coast humid. The rest of the current moves

These are westerly currents; whereas winds are named for their sources, ocean currents are named according to their destinations. Easterly winds produce westerly currents.

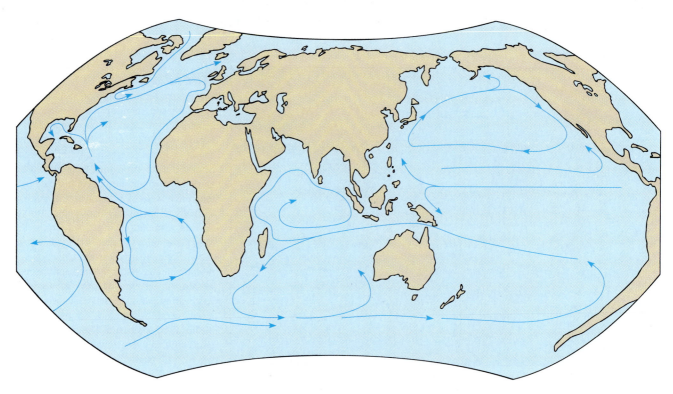

FIGURE 27.7
Ocean water in the two basins, Atlantic and Pacific, circulates in four giant circles, clockwise in the Northern Hemisphere and counterclockwise south of the equator. For the United States, this means that the warm waters that could moderate our climate and give us more precipitation are in the wrong place. The warm Gulf Stream supplies moisture to prevailing westerly winds that carry it across the narrow Atlantic to Europe. The warm current of the Pacific also supplies moisture to prevailing westerlies, but because the Pacific Ocean is so wide, much of the moisture is lost as rain as the air moves across the cooler water in the eastern Pacific.

along the east coast as the Gulf Stream; because this is a latitude dominated by westerly winds, the warm current does not keep the land as warm and wet as one would expect. Near New Jersey and New York, the current turns eastward toward Europe; at the turning point, cold polar water moves south along the east coast of Canada and the northeastern United States.

The Pacific Ocean has a similar pattern with westerly equatorial currents. The Philippines and Indonesia act as a barricade, deflecting water north and south. The turn is abrupt, and much northern water heads eastward as the Kuro Shio current toward Canada and Washington. Once at the west coast of North America, some water turns northward toward Alaska as the Alaska current, but the bulk turns south as the California current. The westerlies absorb huge amounts of moisture as they blow for thousands of miles across the warm Pacific waters, and even though much falls as rain over the ocean, a large amount remains in the air and keeps the coasts wet.

The constant friction of the trade winds not only powers the westerly equatorial currents but also actually causes water to pile up on the west side of an ocean. Directly at the equator the trade winds are relatively weak, and water is able to flow downhill from west to east as an equatorial countercurrent. This extremely warm water is deflected northward and moves along the west coast of Mexico and California, keeping them warm all year.

CONTINENTAL DRIFT

PRESENT POSITION OF THE WORLD'S CONTINENTS

One of the most important factors in determining the climate of a region is that region's latitude. Near the equator in the tropical zone, climate is warm and humid, but a region located either to the north or south is exposed to cooler, drier conditions (see Fig. 27.3).

If the continents occupied different positions, they could cause the whole Earth to have an altered climate. The presence of the large continent of Antarctica at the South Pole allows huge amounts of fresh water to accumulate there as snow fields and glacial ice, lowering sea levels and exposing more land surface. It also increases ocean salinity by trapping fresh water for thousands of years. Increased salinity increases seawater density, affecting oceanic circulation and sea level. If Central America did not exist, water could circulate between the Pacific and Atlantic Oceans, resulting in new oceanic circulation patterns that would affect heat distribution to the poles.

The climate of a land mass is also affected by its size, shape, and presence of mountains. If a continent is small and flat, moist oceanic air can blow across it and bring precipitation to all parts. If it is large and flat, the central regions are too far from the oceans to receive much moisture, most of it having fallen as rain or snow closer to the coastline. The center of the giant continent of Eurasia suffers this fate. Mountains create rain shadows if located on the side of the continent close to the source of wind. The United States would be more moist if its highest mountains were in the east or if it were located farther south where the northeast trade winds could bring moisture in across the low Appalachians and Adirondacks.

PAST POSITIONS OF THE WORLD'S CONTINENTS

Cambrian Period.   During the Cambrian Period, while all life was still aquatic and nothing lived on land, several separate continents were distributed in a vast ocean (Fig. 27.8a and b). Eurasia consisted of either one continent or two that were already close and moving together; the collision of western Europe with eastern Europe and Asia apparently caused the formation of the Ural Mountains. This ancient land mass, whether one or two continents, was also farther south than at present, located in a warmer climate.

In the Southern Hemisphere was a giant continent called **Gondwanaland**; it was composed of South America, Africa, India, Australia, and Antarctica. Gondwanaland was constructed such that South America and Africa were attached, and Brazil and northwest Africa were located at the South Pole. The rest of the continent extended northward toward the tropics, but on the opposite side of the world from North America and Eurasia.

Middle and Late Paleozoic Era.   At about the time life was beginning to move onto land and rhyniophytes were evolving, the continents drifted together (Fig. 27.8b and c). First, during the Silurian Period, North America collided with Eurasia, forming the Appalachian

---

BOX 27.1   *Measuring Ancient Continental Positions and Climates*

Continents move across Earth's surface, propelled by circulation patterns in the mantle. It is possible to discover where the continents have been located in the past by employing several techniques; the most important is studying paleomagnetism. Many rocks contain magnetic minerals, and when these rocks are fluid, as in lava or magma, the minerals become aligned with Earth's magnetic field. Once rock solidifies, its magnetic alignment is fixed. If the magnetic orientation of a rock does not point to the north magnetic pole, we know that there has been a shift in the rock's position since it was formed. Studying the magnetic orientation of extensive blocks of rock of known age reveals the position of the rock at the time of formation.

Another technique important for identifying past positions of continents is the analysis of their past climates. Signs of ancient glaciation (characteristic scouring marks) indicate that the continent must have been near a polar region or at high altitude. The mineral bauxite forms only under moist, warm conditions, so its presence is an excellent indicator of tropical location in the past. Mineral deposits called evaporites are formed by evaporation of water from small, shallow seas located on a continent near a coast, like the Dead Sea today. They indicate that temperatures were high and precipitation low. Carbonic acid in seawater reacts with calcium to form carbonate, a temperature-dependent process; carbonates are found only in warm tropical waters. The presence of fossilized coral also indicates warm, shallow seas, whereas coal fields are derived from the fossilization of plants that lived in warm, rainy conditions.

In many areas, the history of each square mile is known for a period of millions of years, and the drifting has been mapped in great detail. This allows us to consider how continental drift has affected the origin of life on land and the evolution of ecosystems and biomes.

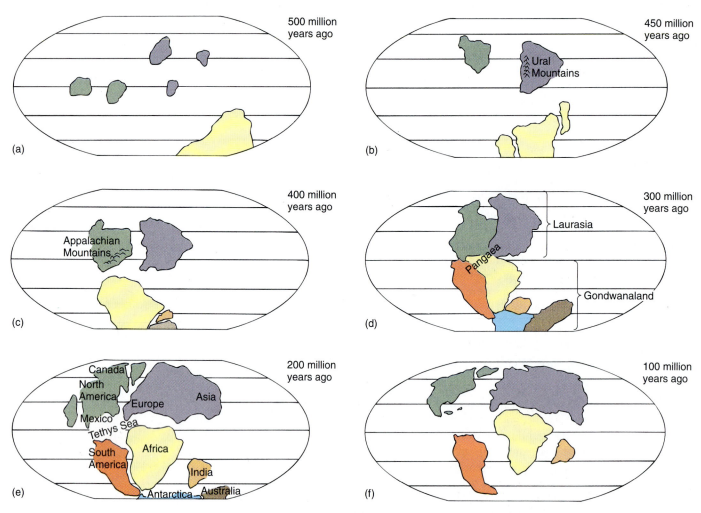

FIGURE 27.8

The continents have changed not only position, but size, shape, and profile as certain areas rise into mountain chains and other areas are covered by ocean when the sea level rises because of polar ice cap melting. Oceanic circulation changes as continents deflect or channel current movement, and this affects both transfer of moisture to the atmosphere and precipitation patterns.

Mountains, which have almost entirely eroded away in the intervening 400 million years. This new continent is known as **Laurasia.** Next, Gondwanaland moved north and ran into Laurasia, pushing up the Alps. Virtually all land on the entire globe was located together as one supercontinent, **Pangaea.** The Laurasian portion was still in the tropics, and much of Gondwanaland was in the southern temperate and polar regions.

Pangaea, which existed for millions of years, through the end of the Paleozoic Era, had a diverse climate. The Laurasian portion was warm, moist, and tropical, whereas southern Gondwanaland was frozen and heavily glaciated. Large regions were swampy lowlands. Initially, Pangaea must have been relatively flat, but the collisions caused the formation of extensive mountain ranges in its center: the Appalachians and the Urals, as just mentioned. These must have caused rain shadows in the interior of Pangaea, and its enormous physical dimensions would have caused the central regions to be dry anyway. It was in the diverse conditions of Pangaea that most major groups of land plants arose: the Aneurophytales, Trimerophytophyta, and Archaeopteridales appeared during the Devonian Period as Pangaea finished forming, and the diversification of the horsetails, lycopods, Medullosales, and Cordaitales occurred during the Carboniferous Period.

Mesozoic Era. In the early Mesozoic Era (the Triassic Period), just after the Permian Period, the climate worldwide may have warmed. There is no evidence of glaciation anywhere, although the polar regions did remain cool; away from the poles, the climate was warm and equable. Many new plant groups evolved at this time: Caytoniales, Cycadeoidophyta, Cycadophyta, and Ginkgophyta. Ferns increased in number, but many of the first major groups started to disappear. As the Triassic Period continued, the climate became more arid; the Triassic Period is the driest of all geological times. By the time of the Jurassic Period, Pangaea had many contrasting climates: the North American and Gondwanaland sections were arid, but the western Eurasia region was less so. Humid zones occurred in Australia and India, as well as in central and eastern Eurasia.

In the Jurassic Period, Pangaea began to break up; the North American segment moved northwestward. Once separation from Eurasia occurred, the north Atlantic Ocean formed, producing a maritime influence on the two new coasts. Eurasia also began moving northward, separating from the southern portion of Pangaea. This resulted in the formation of a waterway, called the Tethys Sea, between the northern and southern continents. This area had alternated between marshy swampland and a shallow coastal sea, but with continental separation, the Tethys Sea became deep, and oceanic circulation between southern and northern continents could occur on a large scale. The northward movement of the North American and Eurasian continents brought them out of the tropics and into the temperate zones, their northern regions extending into the north polar zones. Conifers evolved and spread throughout the regions that were becoming cooler and moister.

At about the same time, the Antarctic-Australian segment separated from the rest of Gondwanaland, and later Australia broke away and moved northward into the south temperate zone (Fig. 27.8e and f). India separated from Africa and moved rapidly over a long distance. At the time of separation, India was located at the edge of the south temperate zone (about 30 degrees S latitude) and had a flora adapted to such a climate. But it migrated rapidly, in just a few tens of millions of years, into the tropics, across the equator, and into the north temperate zone. The rapid fluctuation of climate in less than 70 million years was too much of a disturbance, and massive extinctions occurred. Once it arrived at the Eurasian continent, the two collided, forming the Himalayan mountains. As the two continents neared each other, plants could invade the foreign territory, some migrating northward onto the mainland, others spreading south from Eurasia onto India. This was to the detriment of India; its flora and fauna had not yet had time to become well adapted to the conditions of the north temperate zone, whereas the invaders from the mainland were well adapted.

In the Cretaceous Period, about 110 million years ago, South America separated from Africa, forming the southern Atlantic Ocean and allowing more humid oceanic air to reach west Africa. Until the break-up of Pangaea, Africa had the misfortune of being the dry center of the supercontinent. Its climate was so arid and severe that it had few species of plants or animals. With the formation of the Atlantic Ocean, Africa's climate became much more humid and conducive to plant growth, but even today the continent as a whole has many fewer indigenous species than one would expect if it had had a mild climate for a longer time.

Most continental pieces remained close together long enough for many newly evolved groups to migrate to most of the continents. The break-up of Pangaea occurred just as the flowering plants were becoming established, but as the North and South Atlantic Oceans widened, gene flow between the continents was reduced, and each developed unique floras. For example, the central deserts of Gondwanaland would have been ideal for cacti, but no cacti occur naturally in Africa, only in the Americas. The cactus family did not evolve until after the South Atlantic was wide enough to prevent the dispersal of cactus seeds from South America to Africa.

The latest major event in continental drift was (is) the collision of South America and North America. The two are coming together, and as always happens when continents collide, a mountain range forms. This collision started so recently, about 5 to 13 million years ago, that the mountains are short and are mostly still submerged below sea level. Only their tops protrude and are known as Central America and the Caribbean Islands. The formation of Central America created a continuous land bridge between the two continents, allowing plant species to be interchanged.

Plant Biology Tutor CD-ROM

THE WORLD BIOMES AT PRESENT

As a result of continental drift, the United States is currently located almost exclusively in the north temperate zone and receives much of its weather from the Pacific and Arctic Oceans in winter and from the Pacific Ocean and tropical Atlantic Ocean in summer. Much of Alaska is situated in the polar zone, while Hawaii, Puerto Rico, and Guam lie in tropical waters.

MOIST TEMPERATE BIOMES

Temperate Rainforests.   The northwest coast of the United States is formed by a series of mountain ranges that force the westerly winds from the Pacific Ocean to rise as soon as they come ashore. On the Olympic Peninsula in Washington, rain on the western side of the Olympic Mountains is often above 300 cm/yr. The rains are reliably present through autumn, winter, and spring, with only a brief period of summer dryness. Winters are mild and only rarely is there frost; summers are warm but not hot. These conditions extend from northern California to Alaska, but they end at the summit of the Coastal Range, about 200 km from the coast.

Plant life in the temperate rainforest biome is dominated by giant long-lived conifers. Coastal redwoods of California are up to 100 m tall, and Douglas fir, western hemlock, and western red cedar form a canopy that reaches 60 to 70 meters. The forests can attain old age with little disturbance: Hurricanes and tornadoes do not occur, and fires are rare. The stability of the forest and the daily fog result in a rich growth of epiphytic mosses, liverworts, and ferns; the ground is also covered with shrubs, herbs, and ferns (Fig. 27.9). Temperate rainforests, containing totally different species, also occur in southwest Chile, which has the same climate.

FIGURE 27.9
The Olympic National Park in Washington State encompasses a superb temperate rainforest, one of the *moist temperate biomes.* Abundant rain occurs almost every day, temperatures are never extremely hot or cold, and disturbances are rare or nonexistent. The ecosystems in this biome are very stable and are dominated by *K*-selected species; weedy, *r*-selected species are rare. (*David Muench Photography*)

FIGURE 27.10
This is a drier montane forest (moist temperate biome), which may become warm in July and August and have several weeks without much precipitation. The forest is open and sunny and contains few epiphytes. Pines, firs, and cedars are abundant and dominant in the north; they are joined by oaks in the south.

Drier Montane and Subalpine Forests. As the westerly winds continue inland, they are drier and shed less rain on the Cascade Mountains and the Sierra Nevadas, then even less on the Rocky Mountains; each range feels the rain shadow effect of the preceding mountains. Montane forests occur at the bases of these mountains, subalpine forests at higher elevations (Fig. 27.10). The first inland range in California is the Sierra Nevada; its lower elevations hold a montane forest of ponderosa pine, and some oaks from the valley floor may extend some distance up the slopes. At higher elevations is a mixed conifer forest, one containing many species of conifers: ponderosa pine, Douglas fir, white fir, incense cedar, and sugar pine. The most famous residents are the giant sequoia, the largest organisms in the world, much more massive than whales, being 80 m tall, up to 10 m in circumference, and weighing over 400 metric tons when dry. At the higher elevations are subalpine forests of lodgepole pine, whitebark pine, and mountain hemlock. These form open stands and give way to alpine meadows at their upper boundaries (Fig. 27.11).

This species has been greatly harmed by lumbering and is currently found in only 75 groves.

FIGURE 27.11
High elevations have increased precipitation and winds and decreased temperature and growing season. The montane forest gives way to alpine meadows. Many areas are flat enough that small bogs and marshes form, being rich in sedges. Rocky Mountain National Park. (*David Muench Photography*)

FIGURE 27.12

Much of the dryness of the Rocky Mountains is due to edaphic (soil) conditions as well as to climate. Because these mountains are still very young and steep, soil formation is slow and soil erosion is rapid. The thin, rocky soil is not very effective at holding the moisture that does fall, and in summer, plants often have little soil moisture available.

The Rocky Mountains are our most massive mountain range, extending in a broad band from Alaska south through Canada, Idaho, Montana, Wyoming, Utah, Colorado, Arizona, New Mexico, and Mexico. Being the third range from the coast, it is driest, receiving as little as 40 cm of rain per year; the southern rockies are especially dry and warm. The subalpine forest contains Engelman spruce over the entire length of the range, but in the north other trees are alpine larch and whitebark pine; in the south, varieties of bristlecone pine are part of the subalpine forest. The montane forests typically contain Douglas fir in their higher elevations, ponderosa pine in lower ones. The montane forests on the Rocky Mountains are frequently subjected to fire, as often as every 5 years under natural conditions. Ponderosa pine is well-adapted to fire and survives it easily. Other plants are killed, and fires create open grassy areas around the pines.

In drier montane and subalpine forests, soil is shallow, rocky, and very well drained; it tends to be acidic. Water stress in summer is not uncommon, due both to sparse rain and to rapid runoff through porous soils on steep slopes (Fig. 27.12). Much of the available moisture comes as the melting of winter snow.

In the eastern United States, the Adirondacks and Appalachians are tall enough to support montane and subalpine forests. In the Adirondacks, a spruce/fir subalpine forest extends down to 760 m and contains balsam fir (Fig. 27.13). In the warmer southern climate, the lower limit for subalpine forest is higher: 990 m in the Appalachians and 1400

FIGURE 27.13

In contrast to the Rocky Mountains, the Appalachians are ancient, having arisen during the formation of Laurasia prior to the formation of Pangaea. They have since been eroded extensively so that they are low and gently sloped and have a thick, rich soil. The higher elevations support subalpine forests of conifers, and the lower elevations have montane forests of hardwoods (dicots). *(Jim Tuten/Earth Scenes)*

in the Smoky Mountains. Above these elevations are stands of red and black spruce and Fraser's fir. Below the subalpine conifer forests are montane forests composed of hardwoods, not conifers; these are described in the next section.

Temperate Deciduous Forests. The climate that produces the temperate deciduous forest biome is one with cold winters and warm but not hot summers and relatively high precipitation in all seasons. Whereas the drier montane and subalpine forests of the Rocky Mountains have streams with running water restricted mostly to springtime when snowpack is melting, the northeastern temperate deciduous forest has streams that flow year round. Much of the precipitation in the northeastern region is derived from summer weather systems that move northward out of the central Atlantic Ocean or from winter storms that blow southward from Arctic seas.

Temperate deciduous forest in the United States occupies lower, warmer regions, whereas higher, cooler elevations support the subalpine vegetation of the Appalachians and Adirondacks (Fig. 27.14). Dominant trees vary geographically, but tall, broadleaf decidu-

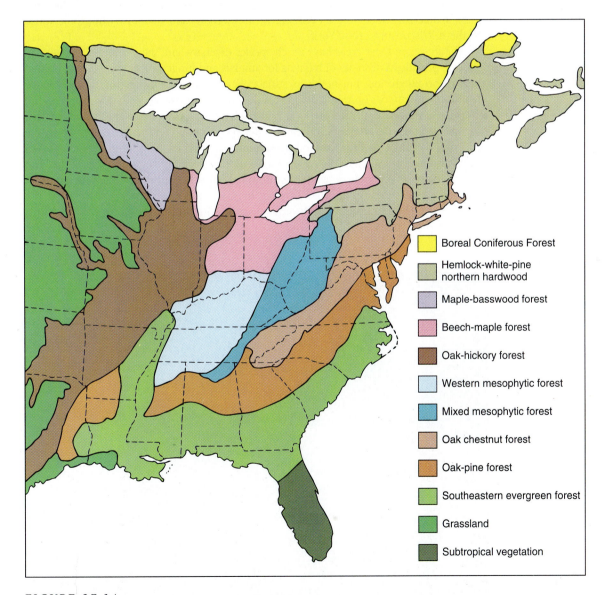

Boreal Coniferous Forest

Hemlock-white-pine northern hardwood

Maple-basswood forest

Beech-maple forest

Oak-hickory forest

Western mesophytic forest

Mixed mesophytic forest

Oak chestnut forest

Oak-pine forest

Southeastern evergreen forest

Grassland

Subtropical vegetation

FIGURE 27.14
The temperate deciduous forest (moist temperate biome) is an extensive and complex biome in the eastern United States consisting of many subdivisions. This is an area of low, nonmountainous topography with cold, wet winters and warm, rainy summers. Soils are deep.

ous trees such as maple and oak are frequent everywhere, maples being more common in the north and oak in the south. Intermixed and forming a subcanopy are dogwood, hop hornbeam, and blue beech. The shrub layer is sparse because of the heavy shade provided by broadleaf dominants, but witch hazel, spicebush, and gooseberry are common. The ground layer is covered with r-selected herbs that thrive during the spring warmth just before trees leaf out. Such a spring sunny period does not occur in the evergreen forests.

The foliage of these angiosperm trees contains fewer defensive chemicals than do the needles and scale leaves of a conifer. There may be 3 to 4 metric tons of broadleaf foliage per hectare in a temperate deciduous forest, and as much as 5% is consumed by herbivores; the rest is abscised in autumn and decays quickly in the humid conditions produced by frequent rain. A thick layer of litter does not accumulate.

The forest is not uniform across its entire breadth. The geographic extent of temperate deciduous forest biome contains numerous soil types, various altitudes, and gradients of temperature and precipitation. As many as nine subdivisions have been recognized, such as oak/hickory forest (Illinois, Missouri, Arkansas), oak/chestnut forest (Pennsylvania, the Virginias), oak/pine forest (east Texas, northwest Louisiana, and northern parts of Mississippi, Alabama, Georgia, and the Carolinas), beech/maple forest (Michigan, Indiana, Ohio), maple/basswood forest (Wisconsin, Minnesota), and hemlock/white pine forest (northern Wisconsin to New York).

The forests in these areas have a rich species diversity, and their considerable vertical structure provides a diversity of habitats for animals and fungi. Much animal life is located above or below ground, but not at its surface. Few birds nest on the ground; most make their nests on branches or in holes in the trunk. Small mammals may burrow, but many are arboreal. As in the coniferous forest at higher elevations here and in the inland western mountains, strong climatic differences exist between summer and winter, accented by the deciduous nature of the broadleaf forest. Autumn leaf fall ends the feeding period for aerial insect leaf-eaters but initiates the season for decomposers and insect herbivores of the litter zone. Many birds emigrate to wintering grounds, but other species may immigrate from farther north. Disturbance is not common; fires are rare and hurricanes that enter the regions usually lose destructive power quickly and change into widespread weaker storms (Fig. 27.15).

FIGURE 27.15
Hurricanes move westward in the Atlantic but turn north-northeastward as they approach land. Many enter land between Texas and Florida, then move up the Mississippi River Valley or just east of it. The winds lose their hurricane force immediately and change into a giant storm system that supplies summer rainfall. The photograph shows Hurricane Elena photographed from space. (NASA)

FIGURE 27.16
Fast, low, cool fires burn frequently through southeastern evergreen forests, one of the *moist temperate biomes*. This kills nonadapted species but leaves pines and cabbage palms: Without fires, seedlings of broadleaf trees would soon overtop the pines and make the forest floor too shaded for pine seedlings; within less than 100 years it could convert to an oak/beech forest. Notice that the flames are too low to reach the needles of the pine trees. (*Dale Jackson/Visuals Unlimited*)

Southeastern Evergreen Forests. This biome shows the powerful effect that disturbance and soil type can exert (see Fig. 26.12a). The southeastern evergreen forest biome occurs at the southern edge of the oak/pine component of the temperate deciduous forest, along the northern portions of the Gulf States, the top of Florida, and the coast of the Carolinas. Its climate is similar to that of the inland region except that winters are warmer; frosts occur but the ground does not freeze. Precipitation is higher than in inland temperate deciduous forest areas, but the soil is so sandy and porous that rainwater percolates downward rapidly and runs off into streams. Shortly after a rain, the soil is dry. The region has a definite dry aspect to it.

In addition to rapid drainage, the biome is shaped by frequent fires; lightning initiates fires that burn rapidly through the litter of fallen pine needles that decompose only slowly (see Figs. 26.12 and 27.16). As a result of fire disturbance and soil conditions, the forest consists almost purely of fire-adapted longleaf pine with occasional oaks. The fact that fire rather than climate is the primary biome determinant has been proven by fire prevention: Repeatedly suppressing fire lets broadleaf seedlings survive and quickly changes the forest from pines to oaks, hickories, beeches, and evergreen magnolias.

DRY TEMPERATE BIOMES

Grasslands. The entire central plains of North America, extending from the Texas coast to and beyond the Canadian border is—or more accurately, was—grassland, often referred to as prairie. This part of North America has no mountains, being too far from the various continental collision zones; it is remarkably flat, with at most low, rolling hills. It is characterized climatically as drier than the forests discussed so far, being located in rain shadows of the prevailing westerlies but out of reach of many Atlantic weather systems. Rain is only about 85 cm/yr. Seasons vary from bitterly cold winters, especially in the north, to very hot summers. Climate and vegetation factors have produced some of the richest soil in existence anywhere. Rainfall is sufficient to promote reasonably rapid weathering of rock, but it is not so great as to leach away valuable elements. Grasses produce abundant foliage that, rather than abscising and falling to the soil, is eaten by herbivores, often large mammals such as cattle or, in the past, buffalo. Vegetable matter is returned to the soil as manure that decomposes rapidly, enriching the soil and increasing its water-holding capacity. Because of the lack of trees, except along rivers, no treetop habitats are available for animals like squirrels. All small mammals burrow and birds nest on the ground.

FIGURE 27.17
Many groups are undertaking efforts to re-establish natural prairies. Native grasses are planted and non-native species weeded out. The animal communities must be re-established also to truly recreate the biome and make it self-perpetuating. Extensive acreage would be required to supply enough primary consumers to support the secondary consumers. *(Pat Armstrong/Visuals Unlimited)*

Because the soil is so rich, almost all the grasslands have been converted to farms; virtually nothing remains of the original biome. Efforts are being made to re-establish grassland prairie by removing all domesticated plants and seeding in grasses and other species known to have occurred originally, such as composites, mints, and legumes (Fig. 27.17). It is still too early to tell how successful these reconstruction efforts will be, but many fear that the plots are too small (a few hectares) and should instead be extensive enough to support herds of buffalo if the full ecosystem is to be restored.

It is not known for certain what the critical factors were that caused this region to be grassland as opposed to forest. The buffalo may have been important; they roamed in unimaginable numbers, and tree seedlings would have been poorly adapted to survive their grazing (Fig. 27.18). Also, Indians set prairies afire periodically, which caused the grasses

FIGURE 27.18
Large mammalian herbivores such as buffalo, deer, moose, and elk are important factors in their ecosystems. Grasses have basal meristems on their leaves and the shoot apical meristems are low, so grasses can be grazed without being killed. Grazing of shrubs and tree saplings destroys the leaves and the shoot apical meristems, so grazing tends to maintain a prairie as a grassland, free of trees. But if too many cattle are fenced into an area, they overgraze it, killing even the grasses by eating leaves so quickly that no photosynthesis can occur. In this overgrazed pasture, the oaks and cacti have a selective advantage because their tough, bitter leaves or spines deter grazing.

FIGURE 27.19
If trees are widely spaced and grass grows between them, it is a woodland biome, a type of *dry temperate biome.* These are often called parklands in the United States, savannas elsewhere. This is a transition biome, representing the interface between a forest and a grassland, but it is not just the mixing of the two. This is especially obvious when animal species are considered: Many birds nest in the trees but feed in the grasslands; these birds cannot live in pure forest or pure grassland. *(David Muench Photography)*

to grow especially luxuriantly the following year, providing better feed for buffalo; seedlings of tree species were killed by the fire. A recent hypothesis is that grasses support a type of mycorrhizal fungus that outcompetes and eliminates the type of fungus necessary for forest trees.

Extensive grasslands occur between the Cascade Mountains and Rocky Mountains, the result of the Cascade's rain shadow. These extend from northern Washington south through Oregon and Nevada. Much of them remains, but they are grazed; large parts have been cleared and used for irrigated farms.

Shrublands and Woodlands. A woodland is similar to a forest except that trees are widely spaced and do not form a closed canopy. If grass grows between the trees, the biome is known as a savanna instead (Fig. 27.19). Shrublands are similar except that trees are replaced by shrubs. Woodlands and shrublands often occur as the transition between a moist forest and a dry grassland or between grassland and desert. Soils often have a high clay content.

The chaparral in California is a well-known shrubland (Fig. 27.20a). Its climate consists of a rainy, mild winter followed by a dry, hot summer; drought occurs every year. Much of the California chaparral is dominated by short shrubs, 1 to 3 m tall, but at higher elevations manzanita, buckthorn, and scrub oak occur. Many plants have dimorphic root systems: Some roots spread extensively just below the soil surface, but a tap root system reaches great depths. The two together allow the plants to gather water from the deep, constant water table and from brief rains that penetrate only a few centimeters into the soil.

Fires occur frequently in California chaparral and are always in the news because of the houses they destroy. Low rainfall allows dry litter to accumulate without decomposing, and dead shrubs persist upright as dry sticks. An area typically burns every 30 to 40 years, with fires most frequent in summer. Winter rains cause flooding, erosion, and mudslides after a fire because no vegetation remains to hold soil in place. Although the shrubs and trees are fire-adapted and resprout quickly, the main new growth is by annual and perennial herbs. These are present before the fire as seeds, bulbs, rhizomes, or other protected structures, and after the burn they grow vigorously, free of shading by charred shrubs; release of minerals from the ash also enriches the soil (Fig. 27.20c). A few years after fire, larger shrubs dominate the biome again and herbs are suppressed, perhaps by allelopathy, perhaps by recovery of the herbivore population.

(a)

(b)

(c)

FIGURE 27.20

(a) This is California chaparral, one of the most famous *shrubland biomes;* such shrubland extends from California across the southwestern United States and northern Mexico into west Texas. Fire is an important disturbance that maintains this biome; it is being invaded by the large mammal *Homo sapiens,* which is protected by fire insurance policies. *(John D. Cunningham/Visuals Unlimited)* (b) The resinous leaves and accumulated dry litter cause frequent fires to be inevitable. *(David J. Cross/ Peter Arnold, Inc.)* (c) Seedlings of *r*-selected species flourish in the spaces opened by the burning of the dominant shrubs. These spots remain open only temporarily because the shrubs are fire-adapted and recover after a few years. By that time, the *r*-selected species will have grown and reproduced abundantly, limited mostly by their biotic potential. Before their population numbers can approach the carrying capacity of the open ecosystem, the growth of the shrubs begins to lower the carrying capacity of these smaller herbs. They are not lost from the ecosystem, however; their seeds either blow to newly burned sites or lie in the ground and sprout after the next fire. If fire were suppressed for many years, the seeds would finally die and these *r*-selected species would be lost from the biome. *(Mike Andrews/Earth Scenes)*

Farther east, drier climates and higher elevations result in pinyon/juniper woodland instead of chaparral shrubland. Rainfall is only 25 to 50 cm, and soils are rocky, shallow, and infertile. The vegetation is a savanna of pinyon pine—small, slow-growing trees with short needles—and juniper trees that may have the stature of large shrubs (Fig. 27.21). In Arizona and New Mexico, oaks may be important. Trees are widely spaced and between them is a grassy vegetation; the species are bunch grasses that grow in clumps, not the mat-forming species of the central plains. Between the bunch grasses is bare soil. Sagebrush and bitterbush occur in the northern parts of the biome.

Desert.   The driest regions of temperate areas are occupied by deserts, where rainfall is less than 25 cm/yr. Deserts are either cold or hot, based on their winter temperatures (Figs. 27.22 and 27.23). A hot desert has warm winter temperatures. In the United States, this climate occurs in rocky, mountainous areas of southern California, Arizona, New Mexico, and west Texas, but in other parts of the world it may occur on flat, sandy plains, as in the Sahara. Three separate and highly distinct deserts actually occur in the southwestern

FIGURE 27.21
East from the California chaparral is the drier pinyon/juniper woodland, shown here in southern Colorado and northern New Mexico. Soils are rocky, thin, and poor, and in many areas chaparral grades into a desert or desert-grassland.

(a)

(b)

FIGURE 27.22
(a) The Great Basin Desert is a cold desert (in the winter only); it extends from near Las Vegas in the south well into Washington state, encompassing much of Nevada, eastern Oregon, and western Idaho. (b) The Great Basin Desert has two dominant plants, sagebrush (*Artemesia tridentata*) and a bunch grass (*Bromus tectorum*). Even the northern parts have cacti (*Pediocactus* and prickly pear), although they tend to be inconspicuous.

(a)

(b)

FIGURE 27.23

(a) For most North Americans, "desert" means Arizona and saguaro cactus *(Carnegiea);* actually, this species of cactus is quite atypical and seems to be poorly adapted; it may be in the process of becoming extinct. (b) Unlike streams and rivers in the eastern temperate deciduous forests, those of the western deserts have water in them only after a thunderstorm. At that time the water flows as a flash flood, moving at tremendous speed and carrying not just silt but entire boulders. Usually no water plants are associated with such streams, but their banks may contain trees such as this palo verde *(Cercidium),* whose deep roots tap the moisture that remains in the stream bed after the surface water has run off. (*b, Stephanie S. Ferguson, William E. Ferguson*)

United States and northern Mexico: the Chihuahuan Desert in west Texas and New Mexico, the Sonoran Desert in Arizona and northwestern Mexico, and the Mohave Desert in southeastern California, southern Nevada, and northwestern Arizona.

Deserts soils are rocky and thin; what little soil occurs may blow away, leaving nothing but pebbles. Brief, intense thundershowers wash soil out of mountains and deposit it in large alluvial fans at valley entrances; fans often have the deepest soil and their own distinct vegetation. The most abundant plants in our hot deserts are creosote bush, bur sage, agaves, and prickly pear. Most perennial plants have one or several defenses against herbivores—chemicals and spines being the most common. Joshua tree, an arborescent lily, grows in the Mohave desert, and numerous other needle-leaf yuccas and agaves are abundant; cacti are ubiquitous. Deserts are highly patchy ecosystems, and slight variations in soil type, drainage, elevation, or covering vegetation can cause abrupt changes in vegetation. In valleys or mountains, slopes that face the equator intercept light almost perpendicularly and so are much warmer and drier than those that face away from the equator and are lighted obliquely (see Fig. 26.8). The hotter side has the richer xerophyte vegetation.

Alpine Tundra.   The biome located above the highest point at which trees survive on a mountain, the timberline, is alpine tundra (Fig. 27.24). In the equatorial region, elevations as high as 4500 m can support tree growth, but in the cooler regions at 40 degrees N, elevations as low as 3500 m are too severe for trees. Alpine tundra is cold much of the year, with a short growing season limited by a late snow melt in spring and early snowfall in autumn. Soils are thin and have undergone little chemical weathering. Summer days can be surprisingly warm and clear, and many plants flourish in the brief summer. Nights are

FIGURE 27.24

Plants in the *Alpine tundra biome* face short, cold growing seasons, and often snow is never gone from the shaded areas below cliffs. Soils are thin and ultraviolet light is intense. Mineral Point, Colorado. (*David Muench Photography*)

generally cold in all seasons, and severe, violent weather can occur at any time. The dominant forms of plant life are grasses, sedges, and herbs such as saxifrages, buttercups, and composites. Dwarf plants growing with densely packed stems and leaves are common and are known as cushion plants. Much of the alpine tundra land occurs as flat meadows with shallow alpine marshes.

It is not known for certain why trees do not grow above the tree line. Those near the highest elevation are short and typically have branches mostly on the side away from the wind; at slightly higher elevation, trees are extremely misshapen and gnarled, known by the German word *krummholz;* the forest is called an elfin forest. It is believed that blowing snow and ice abrade the trees' surfaces, permitting desiccation and death.

In North America, most alpine tundra biomes occur on the tall mountains in the west, but Mt. Washington in New Hampshire and Whiteface Mountain in New York are high enough to have regions of alpine tundra.

Polar Biomes

Arctic Tundra. Polar regions contain few significantly dry areas (Fig. 27.25). Precipitation, usually snow, may be low, but evaporation is also low, so moisture persists. In the extreme northern latitudes, the ground freezes to great depths during winter, and summer is too cool to melt anything more than the top few centimeters. Below this, soil is permanently frozen and is known as **permafrost.**

Like alpine tundra, arctic tundra has a short growing season of 3 months or less, and temperatures are cool, averaging only about 10°C. Freezing temperatures can occur on any day of the year. Arctic soils have a high clay content and are poor in nitrogen because nitrogen-fixing microbes are sparse. Permafrost prevents drainage when the soil surface melts in summer, so soils are waterlogged and marshy. Bogs, ponds, and shallow lakes are common.

Arctic tundra vegetation contains even more grasses and sedges than does alpine tundra, as well as many more mosses and lichens. Almost nothing is taller than 20 cm, even the dwarf willows and birches that occur. Many plants have underground storage tubers, bulbs, or succulent roots; more than 80% of a plant's biomass may be underground even during summer.

FIGURE 27.25
Arctic tundra, one of the *polar biomes,* is marshy and wet when not frozen solid; permafrost prevents water from seeping into the soil, and the area is so flat that runoff is slow. This is true only of the arctic tundra, an area that was on the trailing edge during the formation of Pangaea and has not collided with any other tectonic plate, so it has not undergone mountain building. Antarctic tundra is located on the Andes Mountains where drainage is excellent. Bering Land Bridge National Preserve, Alaska. *(William E. Ferguson)*

Boreal means northern (aurora borealis = northern lights), just as austral means southern (Australia).

Boreal Coniferous Forests.   Just south of arctic tundra is a broad band of forest, the boreal coniferous forest (Fig. 27.26). This forest occurs completely across Alaska and Canada and throughout northern Eurasia in Scandinavia and Russia. The Russian name for this biome is *taiga,* a term frequently used in the West. The boreal forest is an ancient biome, and its formation was strongly influenced by the diversification of division Coniferophyta just as the North American and Eurasian plates were breaking away from Pangaea and their northern parts were leaving the tropical zone and entering the north temperate, subarctic zone.

(a)

FIGURE 27.26
(a) The boreal forest is gigantic, stretching not only along all of Canada and Alaska, but also across Scandinavia and northern Russia. Within any square kilometer, there may be thousands of individuals of the same species, so the wind pollination of conifers is highly efficient. Banff National Park, Canada. *(Ken Cole/Earth Scenes)* (b) Conifers dominate the boreal forest; flowering plants tend to be understory shrubs and herbs. Mt. Rainier National Park.

(b)

Boreal forest is almost exclusively coniferous; conifers appear to be adapted to this climate because they are evergreen and capable of photosynthesizing immediately whenever a sunny day occurs. Deciduous angiosperms would be limited to the short growing season, less than 4 months long. Conifers have drooping branches that shed snow easily; without this architecture, snow loads can easily break off limbs. Although the Coniferophyta contains many species, the boreal forest is not rich in diversity; it is not unusual for only two species to completely dominate thousands of square miles. Black spruce and white cedar may dominate western areas, while white spruce and balsam fir cover much of the eastern area. Shrubs are not abundant but include blueberries, cherries, and gooseberries. Herbs are also sparse. Not much disturbance occurs in the boreal forest; fire may occur in the south and insect plagues in northern parts. This biome does contain many large mammals such as moose, caribou, deer, grizzly bear, and timber wolves. Boreal forest is continuous with and grades into subalpine and montane forests that occur farther south. Many species may occur in both biomes.

TROPICAL BIOMES

Most tropical biomes are characterized by a lack of freezing temperatures. On high mountains in the tropics, cool or cold nights and winters occur, but only at very high elevations is frost encountered. Under conditions of high rainfall, tropical rainforests develop, but the drier areas contain tropical grasslands and savannas.

Tropical Rainforests.   Tropical rainforests occur close to the equator; Hawaii, Puerto Rico, and Guam have extensive rain forests (Fig. 27.27a). Precipitation is high, typically over 200 cm/yr and often as much as 1000 cm/yr (10 m—over 30 feet—of rain). Rains typically occur every day; the morning may be cool and fresh, but clouds develop rapidly

(a)   (b)

FIGURE 27.27
(a) Unlike temperate rainforests, tropical rainforests (one of the *tropical biomes*) are never exposed to freezing conditions. Photosynthesis can occur throughout the year. Because leaves are not deciduous, their useful lifetime is not limited to just several summer months, but rather they can be effective for several years. On mountainous areas, landslides are a frequent disturbance, but in flat plains, ecosystems may be extremely stable. Panama. (b) Millions of acres of tropical rainforests are being cleared for farming every year; after the native plants have been cut and burned, the soil is temporarily rich from the mineral content of the ash. But this quickly leaches away because of the low clay content of the soil; after a year or two, crops fail and the farmers clear more land. Unfortunately, the forest cannot recover the abandoned fields; the trees are not adapted to the mineral-free soil and need to have a thick layer of humus.

FIGURE 27.28
Tropical rainforest is a sea of green; small, inconspicuous flowers would never be noticed by pollinators unless they had strong fragrances. Many tree species undergo massive flowering, involving not only simultaneous opening of all flowers on one tree but also the flowering of all trees within an area. This is not controlled by photoperiod because day length changes too little near the equator. Instead, in some plants it is triggered by changes in air temperature or humidity associated with a rainstorm at the end of the dry season.

and rain almost invariably falls by noon. After a rain, there are large clouds in a clear sky and bright sunlight; the temperature quickly rises and the relative humidity of the air is close to 100%.

High temperatures and moisture cause much more rapid soil transformation here than in other biomes. Many elements are leached from the soil, leaving behind just a matrix of aluminum and iron oxides. Humus decays rapidly, and there is little development of soil horizons. An extensive system of roots and mycorrhizae catch and recycle minerals as litter decays. Almost all available essential elements exist in the organisms, not in the soil (Fig. 27.27b).

The dominant trees are angiosperms; virtually no conifers occur naturally in the tropical rainforest. The canopy is 30 to 40 m above ground level, but numerous large trees emerge above all others. A subcanopy may occur at about 10 to 25 m. Some trees have massive, gigantic trunks, but most are slender. In an undisturbed area where the canopy has remained intact for years, ground vegetation is minimal and it is easy to walk through the forest. Localized disturbances happen when a large tree dies and falls, creating a hole in the canopy. Suddenly, light is available on the forest floor, and herbs and shrubs proliferate. However, a tree soon grows up and fills the canopy, blocking light.

Virtually all small shrubs and herbs occur as epiphytes, located high in the canopy, nearer the light. Orchids, bromeliads, aroids, and cacti are common. Numerous vines are anchored in the soil, but their stems grow to the canopy, then branch and leaf out profusely. Leaves and roots can be separated by a narrow stem 50 m long.

Tropical rainforest is synonymous with species diversity. A single hectare may have well over 40 species of trees, often up to 100, and a single tree may harbor thousands of species of insects, fungi, and epiphytic plants. With such diversity, each hectare may have only one or two individuals of a particular species, especially of trees. No dominants occur. With such low population density for most species, wind pollination is not successful. All plants must be pollinated by animals, and being noticed by a pollinator can be difficult; many subcanopy species have brilliantly colored flowers that are easy to see in the dim light, and scents tend to be so strong that insects and birds find the flowers quickly. Large trees can undergo a massive flowering that no pollinator could possibly ignore (Fig. 27.28).

Tropical Grasslands and Savanna.   Those areas of the tropics with lower rainfall develop as savannas (a few trees; Fig. 27.29) or grasslands (no trees). The best known savannas are in Africa, but they also occur in Brazil (the cerrados), Venezuela (the llanos), and Australia (Fig. 27.30). Under natural conditions, undisturbed by humans, the vegetation consists mostly of bunch grasses up to 1.5 to 2 m tall. Most of the trees are open, flat-topped, widely

FIGURE 27.29

Along coastlines in tropical areas, the low-lying, flat, sandy terrain receives relatively little rainfall, and the sandy soil allows what little rain there is to seep away quickly. Conditions are desert-like, except for very high humidity. There may be a coastal thorn-scrub biome, so named because the trees are short and most bear spines. Although the area is relatively open, walking is almost impossible because of the numerous sharp spines everywhere. Also, the heat and humidity drain away most devotion to plant study.

FIGURE 27.30

Tropical grassland is a *tropical biome* in which rainfall is not abundant enough to support forest or thorn-scrub. Unlike temperate grasslands, they have no cold winter, but rather three seasons: warm and wet, cool and dry, hot and dry. Otherwise, many aspects are similar to those of temperate grasslands. Mauna Kea, Hawaii. *(Sydney Thomson/Earth Scenes)*

scattered legumes. The South American savannas do not have many large grazers, but those of Africa are famous—zebras, wildebeests, and giraffes. An unexpected but very important grazer is the termite; termites are abundant and build giant nests of soil particles and plant debris. They bring in large amounts of plant material, digest it, and add their fecal material to the termite mound. Colonies are so abundant and the mounds so large that termites are a major link in nutrient cycling.

SUMMARY

1. Biomes are groupings of ecosystems; although controlled by diverse factors such as climate and soil, many biomes are distinctive enough to be easily recognizable.

2. Earth's climatic patterns result from several factors, including the tilt of the axis of rotation, the circulation of the atmosphere and oceans, and the positions, shapes, and sizes of the land masses.

3. The warmest and most humid zone worldwide is located between the Tropics of Cancer and Capricorn; the temperate zones are farther from the equator and are cooler and drier and dominated by westerly winds; the polar regions are cold.

4. The climate of a particular land mass is determined by its latitude, its size, the warmth of adjacent ocean currents, humidity and temperature of winds, and rain shadows produced by mountains.

5. The nature of many types of mineral deposits is affected by climatic conditions of heat and humidity; the minerals act as a record of past climatic conditions. Certain fossil life forms (coal, coral) are also good indicators. Paleomagnetism shows the past orientation and position of the continents.

6. In the late Paleozoic Era, just as land plants were appearing, all continents came together, forming Pangaea; during the Jurassic Period, Pangaea fragmented into the present continents. As each drifted, many moved into new latitudes and their climates changed dramatically, affecting their life forms. Altered ocean circulation permitted moist air to cover more of the land surface.

7. Moist temperate biomes have high rainfall, and winters are cool or cold. Several biomes are temperate rainforest, montane and subalpine forest, temperate deciduous forest, and southeastern evergreen forest.

8. Dry temperate biomes have cold winters and moderate rainfall: grasslands, shrublands, woodlands, desert, and alpine tundra.

9. Polar biomes have low light intensity, long summer days, long winter nights, and permafrost. In arctic tundra, much of the biomass is located below ground as bulbs and tubers. The boreal forest (taiga) is an extensive circumpolar forest dominated by conifers.

10. Tropical biomes are influenced strongly by the complete absence of freezing temperatures; where rainfall is high, tropical rainforest develops. If rain is less abundant, tropical grasslands exist.

IMPORTANT TERMS

biome
boreal coniferous forest
chaparral
continental drift
deciduous forest
desert
evergreen forest

Gondwanaland
grasslands
Laurasia
montane forest
Pangaea
permafrost
rain shadow

shrubland
temperate rainforest
tropical rainforest
tundra
woodland

REVIEW QUESTIONS

1. What is the latitude where you live? Which coast is closer to your location? Does it influence your climate significantly?

2. What is the precipitation pattern in your area—mostly in summer or winter? What is typically the longest time between rains—10 days, a month? Does this dry period occur in summer or winter?

3. Are the soils in your area thin, rocky, and immature or thick and rich? What is the nature of the bedrock that produces the soil—igneous, sedimentary, metamorphic?

4. Of the biomes discussed in this chapter, in which do you live? What are the dominant plants in your biome? The dominant animals other than humans? What are the most significant causes of ecological alteration in your area—urban development, logging, agriculture? What have been some of the changes caused by this?

5. Name all the state and national parks or wilderness areas located within one day's drive of your home. How many of these have you visited? Of all the biomes mentioned in this chapter, how many have you traveled through? Did you notice the changes in vegetation?

BotanyLinks .net Questions

www.jbpub.com/botanylinks

Visit the .net Questions area of BotanyLinks (http://www.jbpub.com/botanylinks/netquestions) to complete this question:

1. What resources are available for studying the world's biomes and biodiversity? Go to the BotanyLinks home page to research this subject.

BotanyLinks includes a Directory of Organizations for this chapter.

GLOSSARY

Numbers after definitions are the chapters where the principal discussions occur. Italicized terms are defined elsewhere in the Glossary.

A channel   The groove in the ribosome small subunit in which the free amino acid-carrying tRNA occurs. Alternative: *P channel*. 15

A horizon   The uppermost soil layer, the zone of leaching. 26

abiotic   Refers to things that are not and never have been alive. Compare: *biotic*. 26

abscisic acid   A hormone involved in resistance to stress conditions, stomatal closure, and other processes. 14

abscission zone   The region at the base of an organ, such as a leaf or fruit, in which cells die and tear, permitting the organ to fall cleanly away from the stem with a minimum of damage. 6

absorption spectrum   A graph of the relative ability of a pigment to absorb different wavelengths of light. Compare: *action spectrum*. 10

accessory fruit   A fruit that contains nonovarian tissue. Synonym: false fruit. Alternative: *true fruit*. 9

accessory pigment   A pigment that has an absorption spectrum different from that of chlorophyll *a* and that transfers its absorbed energy to chlorophyll *a*. 10

actinomorphic   Synonym for *regular flower; radially symmetrical*. 9

action spectrum   A graph of the relative rates of reaction of a process as influenced by different wavelengths of light. Compare: *absorption spectrum*. 10

activation energy   The energy needed to overcome the electrical repulsion between two molecules, such that they can react chemically. 2

active transport   The forced pumping of molecules from one side of a membrane to the other by means of molecular pumps located in the membrane. 3, 12

adaptive radiation   Divergent evolution in which a species rapidly diverges into many new species. 17

adenosine triphosphate (ATP)   A cofactor that contributes either energy or a phosphate group or both to a reaction; as it does so, it loses either one or two phosphate groups, becoming either ADP or AMP. 2, 10, 11

adult plant   A plant that is mature enough to flower. Alternative: *juvenile plant*. 14

adventitious   Refers to an organ that forms in an unusual place; refers primarily to roots that form on leaves, nodes, or cuttings rather than on another root. 7

agamospermy   A set of methods of asexual reproduction that involve cells of the ovule and result in seeds and fruit. 9

aggregate fruit   A fruit that develops from the crowding together of several separate carpels of one flower. Alternatives: *simple fruit* and *multiple fruit*. 9

albuminous cell   In gymnosperm phloem, a nurse cell connected to and controlling an enucleate sieve cell. Compare: *companion cell*. 5

albuminous seed   A seed that contains large amounts of endosperm. Alternative: *exalbuminous seed*. 9

all-or-none response   A situation in which an organism either responds to a stimulus or does not respond; the level of response is not correlated with the level of stimulus. Alternative: *dosage-dependent response*. 14

alleles   Versions of a gene that differ from each other in their nucleotide sequences. 16

allelochemic   See *allelopathy*.

allelopathy   The inhibition of germination or growth of one species by chemicals (allelochemics) given off by another species. 26

allopatric speciation   Speciation that occurs when two or more populations of one species are physically separated such that they cannot interbreed. Alternative: *sympatric speciation*. 17

alternation of generations   A type of plant life cycle in which a diploid spore-forming plant gives rise to haploid gamete-forming plants, which in turn give rise to more diploid spore-forming plants. The generations may be similar morphologically (isomorphic) or dissimilar (heteromorphic). 9, 21–24

amino acid   A small molecule containing an amino group and a carboxyl group; the monomers of proteins. 2

amino acid attachment site   In transfer RNA, the 3′ end where the amino acid is carried. 15

amino sugar   A sugar that contains an amino group; a component of the chitin of fungal cell walls. 2, 20

amylase   An enzyme that digests amylose (starch). 2

amylopectin   A simple, branched polysaccharide containing only glucose residues. Much of starch is amylopectin. 2

amyloplast   See *plastid*.

amylose   A simple unbranched polysaccharide containing only glucose residues. Much of starch is amylose. 2

anabolism   Metabolism in which large molecules are constructed from small ones. Alternative: *catabolism*. 10

analogous features   Features that resemble each other but are not based on homologous genes, those related by descent from common ancestral genes. Alternative: *homologous features.* 18

anaphase   The third phase of mitosis; at the metaphase-anaphase transition, centromeres divide and the two chromatids of a chromosome become independent chromosomes. During anaphase, the two are pulled to opposite poles of the spindle by spindle microtubules. 4

anaphase I   The third phase of meiosis I, similar to anaphase of mitosis except that at the metaphase I–anaphase I transition, no division of centromeres occurs. Instead, one homolog is pulled away from the other in each pair, thus reducing the number of sets of chromosomes in each daughter nucleus to the haploid condition from the diploid. 4

anaphase II   A phase of meiosis II, similar to anaphase of mitosis. 4

androecium (pl.: androecia)   A collective term referring to all the stamens of one flower. 9

angiosperm   Informal term for flowering plants, members of division Magnoliophyta; their seeds develop within a fruit. Also called anthophytes. 5, 24, 25

angiospermous sporophyll   The sporophyll of a flowering plant, the carpel that encloses the ovule. 25

anion   An ion carrying a negative charge. 2

anisogamy   A type of sexual reproduction in which the two gametes are only slightly different; usually one is larger and both are motile. Alternatives: *isogamy* and *oogamy.* 21

annual plant   A plant that completes its life cycle in one year or less. Compare: *biennial* and *perennial.* 5, 9

annual ring   In secondary xylem, the set of wood, usually early wood and late wood, produced in one year. 8

annular thickening   A pattern of secondary wall deposition in tracheids and vessel elements; the wall occurs as separate rings. 5

anomalous secondary growth   Any form of secondary growth that does not conform to that typically occurring in gymnosperms and dicots. 8

anoxygenic photosynthesis   Bacterial photosynthesis that does not use water for an electron donor and does not release oxygen as a waste product. 19

anther   The portion of a stamen that contains sporogenous tissue which produces microspores (pollen). 9, 25

antheridiophore   In liverworts, an umbrella-shaped outgrowth of the gametophyte, bearing antheridia. 22

antheridium (pl.: antheridia)   A small structure that produces sperm cells, or the equivalent of sperm cells in ascomycete fungi. 20, 22–25

anthophyte   A rarely used informal term for an angiosperm, a flowering plant. 5, 25

anticlinal wall   A wall perpendicular to a nearby surface, especially the outer surface of the plant. Alternative: *periclinal wall.* 8

anticodon   In transfer RNA, the nucleotide triplet complementary to the codon of mRNA. 15

antipodal cell   One of several (usually three) cells in the angiosperm megagametophyte, located opposite the egg cell and the synergids. 9

apical dominance   The suppression of axillary buds by the growing, active apical bud of a shoot. 14

apoplast   The intercellular spaces and cell walls of a plant; all the volume of a plant that is not occupied by protoplasm (the symplast). 3, 5, 7

archaebacteria   Prokaryotes that have unusual types of metabolism and membrane lipids, perhaps representing extremely archaic forms of life. 19

archegoniophore   In liverworts, an outgrowth of the gametophyte, bearing archegonia and having a stalk with radiating fingers of tissue. 22

archegonium (pl.: archegonia)   Any structure in true plants (not algae) that produces an egg; the megagametangium of true plants. 22, 23

aril   A thick fleshy envelope around some seeds. 24

artificial selection   The process in which humans purposefully alter the gene pool of a species by selective breeding. 17

artificial system of classification   A classification not based on evolutionary, phylogenetic relationships but on other characters. Alternative: *natural system of classification.* 18

ascocarp   Fruiting body of ascomycete fungi. 20

ascogenous hyphae   See *ascogonium.*

ascogonium (pl.: ascogonia)   In ascomycete fungi, a small tubular branch that contacts an antheridium and receives nuclei from it; it then produces heterokaryotic ascogenous hyphae, which contribute to the ascocarp. 20

ascospores   Spores produced by meiosis in asci. 20

ascus (pl.: asci)   In ascomycete fungi, the swollen end of a hypha where karyogamy and meiosis occur. 20

atactostele   The vascular system of monocots, a set of bundles not restricted to forming one ring. Compare: *eustele* and *protostele.* 5

autotroph   An organism that synthesizes its own organic compounds, using only carbon dioxide and mineral nutrients. Compare: *heterotroph.* 10

auxiliary cell   In red algae, a cell that receives the diploid nucleus from the fertilized carpogonium. 21

auxins   Hormones involved in cell elongation, apical dominance, and rooting, among other processes. 14

axial tissue   In a woody stem or root, the tissue derived from fusiform cambium cells. Alternative: *ray.* 8

axillary bud   A bud located in a leaf axil, just above the attachment point of a leaf. May be either a leaf bud or a floral bud. Alternative: *terminal bud.* 5

axoneme   In flagella, the set of 9 + 2 microtubules. 21

B horizon   In soil, the zone of deposition, which receives leached minerals from the A horizon above it. 26

bacteriochlorophyll   Light-harvesting pigment involved in bacterial photosynthesis. 19

bacteriophage   A virus that attacks bacteria. 15

bar   A measure of pressure and of water potential; equals 0.987 atmospheres or 1.02 kg/cm². 12

basal body   See *centrioles.*

basidiocarp   In basidiomycete fungi, a fruiting body such as a mushroom or puffball. 20

basidiospores   Spores produced by meiosis in basidia. 20

basidium (pl.: basidia)   In basidiomycete fungi, the cell in which karyogamy and meiosis occur and which then produces basidiospores. 20

betalains   Water-soluble pigments characteristic of the flowering plant class Caryophyllales. 25

biennial plant   A plant that requires 2 years to complete its life cycle, with cold winter temperature in the first year being necessary for reproduction. 5, 14, 25

biliprotein   A protein in cyanobacteria and red algae which associates with phycobilin, forming a phycobilisome. 19, 21

binary fission   Cell division in prokaryotes; a term occasionally used for cell division in yeasts. 19

binomial classification   A system of providing scientific names to organisms, each name consisting of the genus name and the species epithet. 18

biome   An extensive grouping of ecosystems, characterized by the distinctive aspects of dominant plants. 27

biotic Refers to living things. Compare: *abiotic.* 26

biotic potential The intrinsic rate of natural increase, the number of offspring produced by an individual which live long enough to reproduce. Symbol: *r.* 26

biotroph A fungus that slowly draws nutrients from living hosts, often without killing them. Synonym: *parasite.* 20

biparental inheritance Inheritance of genes from two parents, the most common case for nuclear genes. Alternative: *uniparental inheritance.* 16

blind pit See *pit.*

bloom (algal) A sudden increase in the numbers of algae when environmental conditions become particularly favorable for growth. 19

bordered pit In a xylem-conducting cell, a pit in the secondary wall having a thickened rim (border). Alternative: *simple pit.* 5

bract A small, often thickened and protective leaflike structure; bracts usually protect developing inflorescences. 9

broadleaf plant Informal term for any member of the flowering plant class Magnoliopsida: a dicot. Alternative: *monocot.* 5, 25

browser An herbivore that eats twigs and leaves of shrubs. Compare: *grazer.* 26

bryophyte A term without uniform definition; used by some to refer only to mosses, by others to refer to mosses, liverworts, and hornworts. 22

bud scale A small, specialized leaf, usually waxy or corky, that protects an unopened bud. 5, 6

bulb A short, subterranean, vertical stem that has fleshy, scalelike leaves. Example: onion. 5

bulbil A small bulblike axillary plantlet that serves as a means of vegetative reproduction. 9

bundle sheath A set of cells, which may be parenchyma, collenchyma, or sclerenchyma, that encases some or all of the vascular bundles of a leaf. 6, 10

bundle sheath extension A set of cells, usually fibers, that extends from the bundle sheath to the upper or lower (or both) epidermis of a leaf. 6

C_3 cycle Synonym for Calvin/Benson cycle, one type of stroma reaction. 10

C_4 metabolism (C_4 photosynthesis) A set of metabolic reactions in which carbon dioxide is fixed temporarily into organic acids that are transported to bundle sheaths, where they release the carbon dioxide and C_3 photosynthesis occurs. 10

C horizon The deepest soil layer, composed of parental rocks and rock fragments. 26

callose A long-chain carbohydrate polymer that seals certain regions, e.g., damaged sieve elements (12) or growing pollen tubes (9).

Calvin/Benson cycle Synonym for C_3 cycle, one type of stroma reaction. 10

calyptra (pl.: calyptras) In nonvascular plants, a small sheath of cells, derived from the archegonium, which covers the top of the capsule. 22

calyx (pl.: calyces) A collective term for all the sepals of one flower. 9

capsule In mosses and liverworts, the sporophyte generation. 22

carbohydrates Organic compounds composed of carbon backbones with hydrogens and oxygens attached in a ratio of about 2 : 1; sugars, starch, and cellulose are examples. 2

carbon fixation Photosynthetic conversion of carbon dioxide into an organic molecule, with carbon being reduced in the process. 10

carotenoid A class of lipid-soluble accessory pigments in chloroplasts and chromoplasts. 10

carpel Organ of a flower that contains ovules and is involved in the production of megaspores, seeds, and fruits. See *gynoecium.* 9, 25

carpogonium (pl.: carpogonia) In red algae, an egglike cell that fuses with a spermatium. 21

carpospores In red algae, diploid spores produced by the carposporophyte that grow into tetrasporophytes. 21

carposporophyte In red algae, the mass of diploid cells that arise after the auxiliary cell receives the diploid nucleus. 21

carrying capacity The number of individuals of a population that can live in a particular ecosystem. 26

caryopsis (pl.: caryopses) A single-seeded dry fruit that is fused to the enclosed seed; found in grasses and often mistaken for a seed rather than a fruit and a seed. 25

Casparian strip (band) A layer of impermeable lignin and suberin in the walls of endodermal cells, preventing diffusion of material through that portion of the wall. 7

catabolism Metabolism in which large molecules are broken down into smaller ones. Alternative: *anabolism.* 10

catalyst A material that reduces the activation energy of a reaction, permitting it to occur more rapidly at a lower temperature. 2

cation A positively charged ion. 2

cation exchange In soil, the release of an essential element cation from a soil particle and its replacement by a proton. 13

cavitation In xylem, the breaking of a water column when tension overcomes the cohesive nature of water; an embolism forms. 12

cDNA *Complementary DNA.* 15

cell cycle arrest When cells stop dividing, they undergo cell cycle arrest. The cells may become dormant temporarily (dormant buds, seeds) or they may differentiate and mature. 4

cell plate During cell division, the new cell wall forms inside a large vesicle surrounded by phragmoplast microtubules. The wall, vesicle, and phragmoplast together constitute the cell plate. 4

cellulose A polysaccharide composed only of glucose residues linked by beta-1, 4-glycosidic bonds; it is the major strengthening component of plant cell walls. 2, 3, 4

centimorgan (cM) Synonym for *map unit.* 16

central cell In the megagametophyte in a flower's ovule, the cell that contains two nuclei (usually) and develops into endosperm after fertilization. 9

centrifugal growth Growth outward from a common point. 21

centrioles In animals and some fungi and algae, organelles that act as basal bodies for organizing the microtubules of flagella. 3

centromere Region of a chromosome that holds the two chromatids together prior to anaphase of mitosis or anaphase II of meiosis. Spindle microtubules attach to centromeres and move the chromosomes during division. 4

CF_0–CF_1 complex Part of ATP synthase. CF_0 is a proton channel; CF_1 is a set of enzymes that phosphorylates ADP to ATP. 10

chemical messenger A chemical that, by its presence, carries information from one area to another. 14

chemiosmotic phosphorylation The synthesis of ATP from ADP and phosphate using the energy of an osmotic gradient and a gradient of electrical charge; occurs in chloroplasts and mitochondria. 10

chemoautotroph A bacterium that has the ability to obtain energy from chemical reactions (without photosynthesis) and obtains its carbon from carbon dioxide. Compare: *photoautotroph.* 19

chemosynthetic origin of life Theory that life began through a series of chemical reactions on primitive Earth when conditions were quite different than they are today. 17

chiasma (pl.: chiasmata) During diplotene of prophase I of meiosis, as homologous chromosomes begin to move away from each other, they are held together by chiasmata, thought to be tangles in the chromosomes. During diakinesis, the chiasmata slide to the ends of the chromosomes and disappear. 4

chitin   A polymer in fungal cell walls, composed in part of amino sugars. 2, 20

chlamydospore   In fungi, a segment of hyphae rich in oil or glycogen and having thick, melanized walls; an extremely resistant spore. 20

chlorophyll   Pigment involved in capturing the light energy that drives photosynthesis; found in plants, algae, and cyanobacteria. See *bacteriochlorophyll.* 3, 10

chloroplast   See *plastids.*

chlorosis   A common symptom of mineral deficiency, a yellowing of leaves due to lack of chlorophyll. 13

chromatid   A portion of a chromosome consisting of one DNA double helix and its histones. Before S phase, each chromosome consists of just one chromatid, but after DNA replication in S phase, each chromosome consists of two chromatids. 4

chromatin   The complex formed when histone proteins bind to DNA. 3, 15, 16

chromoplast   See *plastids.*

chromosome   Each nuclear DNA double helix is complexed with histone into a chromosome, which consists structurally of one (pre-S phase) or two (post-S phase) chromatids plus a centromere (4). The circles of DNA found in prokaryotes (19), plastids, and mitochondria are occasionally called chromosomes.

chrysolaminarin   Reserve polysaccharide found in algae of division Chrysophyta (diatoms and their relatives). 21

cilium (pl.: cilia)   Similar to a flagellum, only shorter. 3, 21

circadian rhythm   An endogenous rhythm whose period is approximately 24 hours long. 14

circinate vernation   Refers to the development of a fern leaf in which it must uncoil as it expands. 23

circular bordered pit   A pit that is circular in cross-section and has a thickened rim (border) that slightly overarches the pit chamber. See *pit.* 5

cis-position   In a carbon-carbon double bond, the situation in which two groups of interest lie on the same side of the bond. Alternative: trans-*position.* 2

citric acid cycle   Metabolic pathway in which acetyl-CoA is oxidized to carbon dioxide as reduced electron carriers are generated. Synonyms: Krebs cycle and tricarboxylic acid cycle. 11

cladophyll   A flattened stem that resembles a leaf. 5

clamp connection   In basidiomycete fungi, the connection between two adjacent cells in hyphae; clamp connections are part of a mechanism that ensures heterokaryosis. 20

cleavage furrow   In the cell division of some algae, the inward furrowing of the cell membrane. 21

climacteric fruits   Fruits that undergo a sudden burst of metabolism and ripening (the climacteric) as the last step of maturation. Alternative: *nonclimacteric fruits.* 14

clumped distribution   A distribution of plants in space such that they occur in groups. Alternatives: *uniform* and *random distribution.* 26

codon   In mRNA, a set of three nucleotides that specifies an amino acid to be incorporated into a protein. 15

coenocyte   A cell, usually large, that has many nuclei, up to several thousand. 4, 20, 21

coenzyme   Synonym for *cofactor.*

coenzyme A   A carrier molecule, able to pick up an acetyl group, becoming acetyl-CoA. 2, 11

coevolution   A type of evolution in which two species become increasingly adapted to each other, resulting in a highly specific interaction. 9

cofactor   A small molecule essential to the activity of an enzyme. 2

cohesion-tension hypothesis   Hypothesis that as water is pulled upward by transpiration, its molecules cohere sufficiently to withstand the tension. 12

coleoptile   The outermost sheathing leaf of a grass seedling, providing protection for the shoot within. 14

collenchyma   Collenchyma cells have only primary walls, but these are thickened at the corners of the cell and thin elsewhere. The walls are plastically deformable; if stretched to a new shape, they retain that shape. Alternatives: *parenchyma* and *sclerenchyma.* 5

colony   A group of cells all derived from one recent mother cell and held together by an extracellular matrix, but not closely adhering to each other and not integrated as a single individual. 21

commensal relationship   An interaction of two species in which one benefits and the other is unaffected. 26

community   All the populations of a region. 26

companion cell   In the phloem of angiosperms, a nurse cell that is connected to, and is a sister cell to, an enucleate sieve tube member. Alternative: *albuminous cell.* 5

compartmentalization   The formation of numerous compartments, usually surrounded by semipermeable membranes, such that each compartment has a distinct metabolism. 3

compatibility barrier   Chemical interactions that prevent the fertilization of a gamete by an inappropriate gamete. 9

competition   An interaction of two species which is disadvantageous to one or both. Compare: *mutualism.* 26

competitive exclusion   The inability of a species to grow in part of its range due to competition from another species more adapted to that part of the range. 26

competitive inhibitor   A small molecule that inhibits an enzyme by binding to the active site. Alternative: *noncompetitive inhibitor.* 2

complementary DNA   DNA synthesized with reverse transcription; the DNA is complementary to the RNA substrate and similar to the gene that coded for the RNA. 15

complete dominance   A situation in which the presence of one allele completely masks the presence of the homologous allele. Alternative: *incomplete dominance.* 16

complete flower   A flower having sepals, petals, stamens, and carpels. Alternative: *incomplete flower.* 9

compound cone   A cone with several lateral axes attached to a main axis; conifer seed cones are compound. Compare: *simple cone.* 24

compound leaf   A leaf in which the blade consists of several separate parts (leaflets), all attached to a common petiole. **palmately compound**   All leaflets are attached to the same point. **pinnately compound**   Leaflets are attached to the rachis, an extension of the petiole. Alternative: *simple leaf.* 6

compression wood   See *reaction wood.*

conceptacle   In the brown alga *Fucus,* a small cavity in which sperms and eggs are produced. 21

cone   A compact collection of reproductive structures on a short axis. Synonym: *strobilus.* 23, 24

cone scale   In conifer seed cones, a scale that bears seeds; it is a flattened shoot with fused sporophylls. Synonym: *ovuliferous scale.* 24

conidiophore   In fungi, a hypha that gives rise to conidia. 20

conidium (pl.: conidia)   In fungi, spores formed by the segmentation of the end of a hypha. 20

conjugation   A method of genetic exchange occurring in bacteria and certain green algae; DNA is passed through a tube that joins two adjacent cells. 19, 21

conjugation pilus (pl.: pili)   In bacteria, the narrow tube through which DNA passes during conjugation. 19

conjunctive tissue In monocots, the pithlike region in which vascular bundles are located in stems and roots. 5

continental climate A climate characterized by dry air and great changes of temperature from summer to winter. 27

convergent evolution Evolution of two phenotypically distinct species, organs or metabolisms such that they strongly resemble each other, usually because they are responding to similar selection pressures. Compare: *parallel evolution* and *divergent speciation*. 17

cork cambium (pl.: cambia) A layer of cells that produces the cork cells of bark. Also called phellogen. 8

cork cell A cell in bark that has walls encrusted with suberin; cork prevents loss of water through the bark and prevents entry of pathogens. 8

corm A subterranean, vertical stem that is thick and fleshy and has only thin papery leaves. Example: gladiolus. 5

corolla A collective term for all the petals of a single flower. 9

cortex In stems and roots, the primary tissue located between the epidermis and the phloem. 5

cotyledon In embryos of seed plants, the rather leaflike structures involved in either nutrient storage (most dicots and gymnosperms) or nutrient transfer from the endosperm (most monocots). 9, 24

covalent bond A chemical bond in which electrons are shared between two atoms. 2

cpDNA Plastid DNA; it is in the form of closed circles, without histones. Usually there are many circles per plastid. See *mtDNA*. 3, 21

Crassulacean acid metabolism (CAM) A metabolism in which carbon dioxide is absorbed at night and fixed temporarily into organic acids. During daytime, the acids break down, carbon dioxide is released, and C_3 photosynthesis occurs. 10

crista (pl.: cristae) One of the tubular or vesicular folds of the mitochondrial inner membranes. 3, 11

critical night length The length of darkness that must be exceeded by short-day plants, or not exceeded by long-day plants, for flowering to be initiated. 14

cross-fertilization The fertilization of a gamete by a genetically distinct gamete derived from a different parent: Alternative: self-fertilization; see *cross-pollination*. 16

cross-field pitting In secondary xylem, pit-pairs formed between ray cells and axial tracheids or vessels. 8

crossing over During prophase I of meiosis, after homologous chromosomes have paired and a synaptonemal complex has formed, the DNA of the homologs breaks and the end of each homolog is attached to the other homolog, resulting in two new chromosomes. 4, 16

cross-pollination The pollination of a flower by pollen from a completely different plant. Alternative: *self-pollination*. 9

cuticle A layer of cutin on epidermal cells; the cuticle reduces water loss but also unavoidably restricts the entry of carbon dioxide. 5

cutin A polymer of fatty acids that is water impermeable; it forms a layer (cuticle) on the epidermis. 2, 5

cyanide-resistant respiration Synonym for *thermogenic respiration*. 11

cyanophycin granule A storage particle of nitrogen compounds, found in some cyanobacteria. 19

cyclic electron transport The flow of electrons from P700 back to plastoquinone in photosynthesis, such that there is proton pumping but no synthesis of NADPH. Alternative: *noncyclic electron transport*. 10

cytochromes Small electron carriers that contain iron. 10, 11

cytokinesis Division of the protoplasm of a cell, as opposed to nuclear division, karyokinesis. 4

cytokinins A class of hormones involved in cell division, apical dominance, and embryo development among other things. 14

cytoplasm Protoplasm consists of nucleus, vacuoles, and cytoplasm. 3

cytosol Synonym for *hyaloplasm*.

dark reactions Synonym for *stroma reactions*.

day-neutral plant A plant that is induced to flower by factors other than night length. 14

degenerate code In the genetic code, many amino acids are coded by several codons, not just one each. 15

dehydration reaction A chemical reaction in which a proton is lost from one reactant and a hydroxyl from the other, creating a water molecule. 2

deletion mutation A mutation involving loss of DNA. 16

demography The study of the age distribution of the individuals of a population. 26

denitrification The conversion, by microbes, of nitrate to nitrogen gas, thus making nitrogen no longer available to plants. 19

deoxyribonucleic acid (DNA) The information molecule in nuclei, plastids, mitochondria, and prokaryotes. 2, 14, 15, 16

deoxyribose A five-carbon sugar occurring in DNA. 2

derived features Features present in modern organisms but not in their ancestors. Alternative: *relictual features*. 1, 18

determinate growth Growth that stops at a genetically predetermined size. Typical of leaves and flowers but not of whole shoots and roots. Alternative: *indeterminate growth*. 5

diakinesis see *prophase I*.

diatom Common name for algae of the class Bacillariophyceae; they have two silica shells that fit together like the parts of a Petri dish. 21

diatomaceous earth Oceanic sediments formed by accumulation of the silica shells of diatoms. 21

diatoxanthin A xanthophyll pigment found in brown algae. 21

dibiontic A life cycle with an alternation of generations. Alternative: *monobiontic*. 21

dichotomous branching A forking that results in two nearly equal branches (or leaf veins or secretory ducts). Alternatives: *monopodial* and *sympodial branching*. 23, 24, 25

dicot Informal term for any member of the flowering plant class Magnoliopsida, a broadleaf plant. Alternative: *monocot*. 5

dicotyledon Synonym for *dicot*.

dictyosome A stack of thin vesicles held together in a flat or cup-shaped array; dictyosomes receive vesicles from endoplasmic reticulum along their forming face, then modify the material in the vesicle lumen or synthesize new material. Vesicles swell and are released from the maturing face. See *Golgi apparatus*. 3

differentially (selectively) permeable membrane A membrane that permits the passage of certain types of particles and inhibits the passage of others. 12

diffuse porous wood Wood in which the vessels of late wood are about as numerous and as wide as those of early wood. Alternative: *ring porous wood*. 8

diffusion The random motion of particles from regions of higher concentration to regions of lower concentration. 12

dihybrid cross A cross in which two characters are considered simultaneously. 16

dikaryotic The condition of having two genetically different nuclei present in one cell, especially in ascomycete and basidiomycete fungi. Alternative: *monokaryotic*. 20

dimorphic Having two distinct morphologies, such as in plants with a juvenile and an adult form. 5, 9

dinoflagellate Common name for algae of division Pyrrhophyta. Most have two flagella, each of which lies in a groove. 21

dioecy   The condition in which a species has two types of sporophyte—one with stamens and one with carpels. See *monoecy*. 9

diploid   Refers to two full sets of chromosomes in each nucleus, as typically found in sporophytes and zygotes. See *haploid*. 4, 16

diplotene   See *prophase I*.

disaccharide   A small carbohydrate composed of just two simple sugar residues. 2

distal   Refers to the position of a structure distant from a point of reference; relative to a stem, a leaf blade is distal to the petiole. Alternative: *proximal*. 5

diurnal   Daytime; a diurnal plant opens its flowers at sunrise. Alternative: *Nocturnal*. 14

divergent speciation   The evolution of part of a species into a new species, with the remainder of the species continuing as the original species or evolving into a third, new species. Alternative: *phyletic speciation*. 17

DNA cloning   Producing large numbers of identical copies of DNA, usually by inserting it into bacteria and allowing the bacteria to multiply. 15

DNA denaturation   Synonym for *DNA melting*.

DNA hybridization   The slow cooling of a mixture of short DNA molecules such that complementary strands encounter each other and hydrogen bond into double helices. 15

DNA ligase   An enzyme that can attach two strands of DNA to each other, repairing nicks and linking Okazaki fragments into complete molecules of DNA. 15

DNA melting   Heating a DNA double helix gently until the hydrogen bonds are broken and one molecule separates from the complementary molecule. Synonym: DNA denaturation. 15

DNases (DNAases)   Enzymes that digest DNA. 15

dominant trait   A trait whose phenotype completely masks that of the alternative (recessive) allele in the heterozygous condition. 16

dosage-dependent response   A situation in which the amount of response is correlated with the amount of stimulus. Alternative: *all-or-none response*. 14

double fertilization   The process unique to angiosperms in which one sperm fertilizes the egg (forming a zygote) and the other sperm fertilizes the polar nuclei (forming the primary endosperm nucleus). 9

duplication division   Synonym for *mitosis*.

early wood   Synonym for *spring wood*.

ecosystem   The set of physical nonliving environmental factors of a region, plus the communities of organisms of that region. 26

ecotypes   Races of a species that are each adapted to particular environmental factors in certain parts of the species range. 26

edaphic   Refers to the soil: soil factors are edaphic factors. 26

egg apparatus   A name for the egg cell and the one or two adjacent synergids in an angiosperm megagametophyte. 9

elater   In the sporangia of liverworts and horsetails, small twisted cells that push the spores out of the sporangium. 22, 23

elasticity   A property of sclerenchyma walls; if stretched to a new size or shape, they return to their original size and shape once the deforming force is removed. Alternative: *plasticity*. 5

electron carrier   A cofactor that carries electrons between reactions. Examples: nicotinamide adenine dinucleotide (phosphate), flavin adenine dinucleotide. 2

electron transport chain   A series of electron carriers that transfer electrons from a donor, which becomes oxidized, to a receptor, which becomes reduced. 10, 11, 13, 19

electronegativity   A measure of an atom's or a molecule's tendency to give off or take on electrons. 2

elicitor   Something that provokes a response; in pathogenic fungi, a component of the fungal cell wall that provokes the plant to produce phytoalexins. 20

Embden-Meyerhoff pathway   Synonym for *glycolysis*.

embolism   A packet of water vapor formed in xylem when the water column cavitates (breaks). Often called an air bubble. 12

embryo sac   A common synonym for the megagametophyte of flowering plants. 9

embryophyte   A rarely used term for all plants that are not algae—that is, all that have multicellular reproductive structures with sterile tissue. 21

enation   A small, projecting flap of tissue, thought to have been the ancestor of leaves in the lycophytes. 23–25

end-product inhibition   A type of regulatory mechanism in which an enzyme at the front of a metabolic pathway is inhibited by the end product. Synonym: feedback inhibition. 2

endergonic reaction   A reaction that absorbs energy. 2

endocarp   The innermost layer of the fruit wall, the pericarp. See *exocarp* and *mesocarp*. 9

endocytosis   A process of absorbing material into a cell by forming an invagination in the plasma membrane, then pinching it shut and forming a vesicle. See *exocytosis*. 3

endodermis   A sheath of cells surrounding the vascular tissue of roots (and occasionally horizontal stems); their Casparian strips prevent uncontrolled diffusion between root cortex and root vascular tissue by means of walls and intercellular spaces. 6

endogenous rhythm   A rhythm generated entirely within an organism whose periodicity is not maintained by an external rhythm. 14

endomembrane system   The membranous organelles of a cell. 3

endoplasmic reticulum (ER)   A system of narrow tubes and sheets of membrane that form a network throughout the cytoplasm. If ribosomes are attached, it is rough ER (RER) and is involved in protein synthesis. If no ribosomes are attached, it is smooth ER (SER) and is involved in lipid synthesis. 3

endoreduplication   The repeated synthesis of all nuclear DNA without its partitioning into separate nuclei by division. See *gene amplification*. 4

endosperm   The tissue, usually polyploid, which is formed during double fertilization only in angiosperms and which nourishes the developing embryo and seedling. 9

endosymbiont theory   The theory that postulates that plastids and mitochondria arose as prokaryotes that were living symbiotically within an early eukaryotic cell. 21

endothermic reaction   A reaction that absorbs heat. 2

enhancer elements   Regions of DNA upstream from the structural region of a gene, which increase the ability of RNA polymerases to transcribe the gene. 15

entrainment   The resetting of an endogenous rhythm by an exogenous stimulus. 14

entropy   A measure of disorder in a system. 10

epicotyl   In the embryo of a seed, the embryonic shoot, located above the cotyledons. Synonym: plumule. 9

epidermis   The outermost layer of the plant primary body, covering leaves, flower parts, young stems, and roots. 5

epigynous   See *inferior ovary*.

epiphyte   A plant that grows on another plant, either attached to it or climbing over it, but not parasitizing it. 5, 26

epistasis   The control of the expression of one gene by another, distinct gene, often involving the genes for the various enzymes catalyzing a single metabolic pathway. 16

essential element   An element required for normal growth and reproduction. Major (macro) essential elements are needed in relatively large concentration; minor (micro, trace) essential elements are needed only in low concentrations. 13

essential organs In flowers, the spore-producing organs—stamens and carpels. Alternative: *nonessential organs.* 9

ethylene A hormone involved in fruit ripening, the initiation of aerenchyma in submerged roots and stems, and other aspects of development. 14

eukaryotes Organisms that have true nuclei and membrane-bounded organelles. Eukaryotes are plants, animals, fungi, and protists, but not bacteria. Alternative: *prokaryotes.* 3, 19

eustele The vascular system of the stems of seed plants, composed of bundles around a pith. Compare: *protostele, siphonostele,* and *atactostele.* 23, 24

evolution The change of nucleotide sequences in a species' DNA through natural selection, genetic drift, or accident. 17

exalbuminous seed A seed with little or no endosperm at maturity. Alternative: *albuminous seed.* 9

excited state electron An electron that has absorbed a quantum and moved to a higher orbital; it has more energy than when it is in its ground state. 10

exergonic reaction A reaction that releases energy. 2

exocarp Outermost layer of the fruit wall; in fleshy fruits, the rind or peel. See *mesocarp* and *endocarp.* 9

exocytosis The transfer of vesicle or vacuole-carried material to the outside of a cell by fusion of the plasma membrane with the membrane of the vesicle or vacuole. See *endocytosis.* 3

exons Those portions of the structural region of a gene whose information is actually translated into protein. Alternative: *intron.* 15

exothermic reaction A reaction that releases heat. 2

extrinsic protein Synonym for peripheral protein. Alternative: *intrinsic protein.* 3

eyespot Synomym for *stigma* in phototactic algae.

facilitated diffusion The diffusion of hydrophilic (polar or charged) molecules through a membrane by means of channels composed of hydrophilic, membrane-spanning proteins. 3

facultative aerobe (facultative anaerobe) An organism that can use oxygen if present or survive without it. 11, 19

false fruit See *accessory fruit.*

fascicular cambium The portion of the vascular cambium that develops within a vascular bundle. Alternative: *interfascicular cambium.* 8

fatty acid Long chains of carbon with only hydrogens as side groups and a carboxyl group at one end. If all carbon-carbon bonds are single, the fatty acid is saturated; if it contains any double bonds, it is unsaturated. 2, 3, 11

feedback inhibition Synonym for *end-product inhibition.*

fermentation Synonym for anaerobic respiration. See *respiration.*

ferredoxin An iron-containing proteinaceous electron carrier in photosynthesis. 10

fertilization Fusion of two gametes (or their equivalents in some fungi and algae). Synonym: *syngamy.* 9, 20–25

fiber A sclerenchyma cell that is long and tapered and has pointed ends; provides a tissue with strength and flexibility. Compare: *sclereid.* 5, 8

field capacity The amount of water held by a soil after drainage due to gravity has been completed. 13

filament The stalk of a stamen, it elevates the anther. 9, 25

first filial generation (F_1) The progeny of an experimental cross. 16

first-order reaction A reaction involving only a single molecule, which breaks apart during the reaction. 2

flagellin Protein monomer that makes up the flagella of prokaryotes. 19

flagellum (pl.: flagella) Like a cilium but longer, an organelle of locomotion. Both flagella and cilia have a 9 + 2 arrangement of microtubules. 3, 20, 21

floridean starch A storage polysaccharide found in red algae. 21

fluid mosaic Biological membranes are two-dimensional fluids in which various types of lipids and intrinsic proteins can diffuse laterally. 3

fluorescence The spontaneous emission of a quantum by an excited electron, which allows the electron to return to its ground state. 10

form genus A genus based on a character that is not a reliable indicator of evolutionary relationships. Unrelated species are grouped together. 18

founder The first individual(s) that establishes a population in a new habitat; especially important in adaptive radiation. 17

frameshift error During the translation of mRNA, a misalignment of the ribosome such that the triplets it reads are not true codons. 15

freely permeable membrane A membrane that allows everything to pass through. See *impermeable* and *semipermeable membrane.* 3

fret A set of thylakoid membranes that connect grana in chloroplasts. 10

frond Nontechnical term for the leaf of a fern. 23

frugivore A fruit-eating animal, important in the dissemination of seeds. 9, 29

fruit In angiosperms, the structure that forms from carpels and associated tissues after fertilization. 9

fruiting body In fungi, the spore-producing and spore-disseminating structures such as a mushroom or puffball. 20

fucoxanthin Xanthophyll pigment found in brown algae. 21

funiculus (pl.: funiculi) The stalk of an ovule. 9, 25

fusiform initials In a vascular cambium, the long cells with tapered ends that give rise to axial cells of the secondary xylem and secondary phloem. Alternative: *ray initials.* 8

G_1 Part of interphase of the cell cycle, G_1 (gap 1) is the interval between cell division and the synthesis of DNA in the nucleus. G_1 is often the longest phase, during which the nucleus actively directs cytoplasmic metabolism. 4

G_2 Part of interphase of the cell cycle between the synthesis of DNA and the beginning of nuclear division. 4

gametangium (pl.: gametangia) Any structure that produces gametes. 21–25

gamete A haploid sex cell, such as an egg or sperm. **megagamete** A large, immobile gamete; an egg. **microgamete** A small, often mobile gamete; a sperm. 9, 21–25

gametophore The leafy stem of a moss gametophyte. 22

gametophyte A haploid plant that produces gametes. Alternative: *sporophyte.* **megagametophyte** A gametophyte that produces megagametes (eggs) only. **microgametophyte** A gametophyte that produces microgametes (sperms) only. 9, 21–25

gas vesicle In certain prokaryotes, a conical vesicle of gas, surrounded by a layer of protein; provides buoyancy. 19

gene In DNA, a sequence of nucleotides which contains information necessary for the metabolism and structure of an organism. 15

gene amplification The repeated synthesis of the DNA of just one or a few genes, not the entire genome. See *endoreduplication.* 4, 15

gene flow The movement of alleles within a population by the movement of pollen or seeds. 17

gene pool The total population of all the alleles in all the sex cells of the individuals of a population. 17

generative cell In the pollen grains of seed plants, the cell that gives rise directly to the sperm cells. Alternative: *vegetative cell.* 9, 24

genetic code   The set of nucleotide triplets in DNA that code for amino acids to be inserted during protein synthesis. 15

genetic drift   In a small gene pool, the alteration in allele frequencies mostly by accidents rather than by natural selection. 17

genome   All the alleles of an organism. 4

genotype   The set of alleles present in an organism's genome. Alternative: *phenotype.* 15

genus (pl.: genera)   A group of species closely related by descent from a common ancestor. 18

gibberellic acid   A natural gibberellin. 14

gibberellins   A class of hormones involved in stem elongation, seed germination, and other processes. 14

gluconeogenesis   Formation of glucose from 3-phosphoglyceraldehyde. 10, 11

glucose   One of the most abundant simple sugars, a six-carbon monosaccharide. Component of starch, cellulose, and many metabolic pathways. 2, 10, 11

glycocalyx (pl.: glycocalyces)   A secretion of mucilaginous material that surrounds many prokaryotes. 19

glycolipid   Lipid molecules with sugars attached. See *glycoprotein.* 3

glycolysis   The metabolic pathway by which glucose is broken down to pyruvic acid. Synonym: *Embden-Meyerhoff pathway.* 11

glycoprotein   A protein with sugars attached; often the sugars occur in short chains less than ten sugars long. See *glycolipid.* 3

glyoxysome   See *microbody.*

Golgi apparatus   A collection of interconnected dictyosomes, as many as several thousand, the entire set forming a cup-shaped apparatus. Often treated as a synonym for dictyosome; Golgi apparatuses are rare in plants, common in animals. Synonym: Golgi body. 3

Golgi body   See *Golgi apparatus.*

Gondwanaland   Southern portion of the ancient continent of Pangaea. Compare: *Laurasia.* 27

granum (pl.: grana)   A set of flat vesicles in chloroplasts, involved in chemiosmotic phosphorylation. 3, 11

grazer   An herbivore that eats low herbs such as grasses. Compare: *browser.* 26

ground meristem   A term refering to any expanse of meristematic tissue that produces a somewhat uniform mature tissue. 5

ground state electron   An electron in its most stable orbital, when it contains the least amount of energy. Alternative: *excited state electron.* 10

guard cells   A pair of epidermal cells capable of adjusting their size and shape, causing the stomatal pore to open when they swell and close when they shrink. 5

gymnosperm   Common name for plants with naked seeds; all seed plants that are not angiosperms. Examples: conifers and cycads. 24

gynoecium (pl.: gynoecia)   A collective term referring to all the carpels of a flower. 9

habit   The characteristic shape or appearance of the individuals of a species. 5

habitat   The set of conditions in which an organism completes its life cycle. 26

hairpin loop   A small kink in the end of a messenger RNA molecule that is being transcribed; the kink appears involved in signaling that transcription should stop. 15

half-inferior ovary   An ovary that is partly inferior, such that the sepals, petals, and stamens appear attached to its side; those appendages are perigynous. Alternatives: *superior* and *inferior ovary.* 9

haploid   Refers to one full set of chromosomes per nucleus. Gametes, spores, and gametophytes typically have haploid nuclei. See *diploid.* 4, 16, 21

hardwood   A term applied to both dicot trees and shrubs and to their wood, because in general dicot wood contains fibers. Alternative: *softwood.* 8

haustorium (pl.: haustoria)   The structure by which a parasite enters and draws nutrients from a plant; in fungi, it is a hypha; in mistletoes and similar parasites, it is a modified root. 7, 20

heartwood   The colored, aromatic wood in the center of a trunk or branch; all the wood parenchyma cells have died and no water conduction is occurring. Alternative: *sapwood.* 8

helical thickening   A pattern of secondary wall deposition in tracheids and vessel elements; the wall occurs as one or two helical bands. 5

hemicelluloses   A set of cell wall polysaccharides that crosslink cellulose molecules in plant cell walls. 3

hemiparasite   A parasite that draws water, minerals, and perhaps some organic material from its host but also carries out photosynthesis. Compare: *holoparasite.* 26

herb   A plant that consists only of primary tissues; lacking wood. 5

herbivore   An animal that eats plants. Compare *browser* and *grazer.* 26

heteroblasty   The phenomenon in which an individual plant produces several different types of leaves. 6

heterocyst   In cyanobacteria, specialized cells in which nitrogen fixation occurs. 19

heterokaryosis   In fungi, a condition in which each cell has two or more nuclei of different mating types. Alternative: *homokaryosis.* 20

heteromorphic generations   A dibiontic life cycle in which sporophyte and gametophyte are easily distinguishable morphologically. Alternative: *isomorphic generations.* 21

heterospory   A condition in which the life cycle of a plant contains two types of spores, microspores and megaspores. Alternative: *homospory.* 9, 21–25

heterotroph   An organism that obtains its carbon from organic molecules, not from carbon dioxide. Compare: *autotroph.* 10

heterozygote   A diploid organism with two different alleles for a particular gene. Alternative: *homozygote.* 16

hexose monophosphate shunt   Synonym for *pentose phosphate pathway.*

hilum (pl.: hila)   Scar produced when a seed breaks from the funiculus. 9

histones   A set of basic nuclear proteins that complex together and with DNA, first forming nucleosomes and then complexing further into chromosomes. 3, 4, 15, 16

hnRNA   Heterogeneous nuclear RNA. 15

holdfast   The portion of a nonmotile, attached alga that holds the alga to its substrate. 21

holoparasite   A parasite that draws all its water, minerals, and organic material from its host and is unable to carry out photosynthesis. Compare: *hemiparasite.* 26

homokaryosis   In fungi, a condition in which all nuclei in the mycelium are identical genetically. Alternative: *heterokaryosis.* 20

homologous chromosomes   In a diploid nucleus, each type of chromosome is present as a pair, one inherited paternally, the other maternally. 4

homologous features   Features that are the phenotypic expression of homologous genes, those related by descent from common ancestral genes. Alternative: *analogous features.* 18

homospory   A condition in which the life cycle of a plant contains only one type of spore. Alternative: *heterospory.* 9, 21–25

homozygote   A diploid organism with two identical alleles for a particular gene. Alternative: *heterozygote.* 16

hormone   A chemical that is produced by one part of a plant, often in response to a stimulus, and then is transported to other parts and induces responses in appropriate sites. 14, 15

hyaloplasm   The liquid substance of protoplasm, excluding all the organelles such as nuclei, plastids, ribosomes. Synonym: cytosol. 3

hybrid sterility   A postzygotic isolation mechanism in which a hybrid zygote can grow into an adult but cannot form fertile gametes. 17

hydrogen bonding   The weak attraction between polar molecules. 2, 12

hydroids   In the stems of some mosses, cells that resemble tracheids and are involved in water conduction. 22

hydrolysis   The breaking of a chemical bond by adding water to it, a proton being added to one product and a hydroxyl to the other. 2

hydrophilic   "Water loving"—refers to compounds that are relatively soluble in water and other polar solvents and insoluble in lipids and other nonpolar solvents. Opposite: *hydrophobic.* 2, 3

hydrophobic   "Water fearing"—refers to compounds that are relatively insoluble in water and other polar solvents and soluble in lipids and other nonpolar solvents. Opposite: *hydrophilic.* 2, 3

hydroponics   The growing of plants in a water solution, without the use of soil. 13

hymenium (pl.: hymenia)   In a fungus fruiting body, the layer of fertile cells that produce spores. 20

hypersensitive reaction   An extreme susceptibility to a fungus, such that a plant cell dies before nutrients can be drawn from it. 20

hyphae (sing.: hypha)   The long, narrow filaments, either coenocytic or cellular, that constitute the body of a fungus. 20

hypocotyl   The portion of an embryo axis located between the cotyledons and the radicle. 9

hypogynous   See *superior ovary.* 9

hypothesis (pl.: hypotheses)   A model of a phenomenon constructed from observations of the phenomenon. It must make testable predictions about the outcome of future observations or experiments. 1

immobile essential element   An element that cannot be removed from mature tissues; if a plant becomes deficient in an immobile element, young tissues show symptoms even though older tissues may have extra. Alternative: *mobile essential element.* 13

imperfect flower   A flower lacking either stamens or carpels or both. Alternative: *perfect flower.* 9

imperfect fungi   Fungi that do not undergo sexual reproduction or whose sexual reproduction has never been observed. 20

impermeable   A barrier that allows nothing to pass through. See *freely permeable* and *semipermeable membrane.* 3

incipient plasmolysis   As a plant cell is losing water and shrinking, the point at which the protoplast just begins no longer to exert pressure against the wall. 12

included phloem   In certain types of anomalous secondary growth, patches of secondary phloem may be located within the secondary xylem. 8

incomplete dominance   A situation in which the phenotypes of both alleles of a heterozygote are expressed. Alternative: *complete dominance.* 16

incomplete flower   A flower that is missing one or more of the four basic appendages (sepals, petals, stamens, carpels, or any combination). Alternative: *complete flower.* 9

indehiscent   Remaining closed at maturity, not opening; true of many fruits and the megasporangia of seed plants. 9, 24, 25

independent assortment   In a double heterozygote, the distribution of one allele of one gene during meiosis is not linked to the distribution of either allele of the homologous gene. Alternative: *linked genes.* 16

indeterminate growth   Growth not limited by a plant's own genetic development program: most trees have indeterminate growth. Alternative: *determinate growth.* 5

indole acetic acid (IAA)   An auxin. 14

infection thread   During the invasion of roots by nitrogen-fixing bacteria, the bacteria are encased n an invagination of the plant cell wall, the infection thread. 7

inferior ovary   An ovary located below the sepals, petals, and stamens; those appendages are epigynous. Alternatives: *superior* and *half-inferior ovary.* 9

inflorescence   A discrete group of flowers. 9

inner bark   The innermost, living layer of bark, located between the vascular cambium and the innermost cork cambium. Alternative: *outer bark.* 8

insertion mutation   A mutation involving insertion of the new DNA into a sequence of pre-existing DNA. 16

insertion sequence   A small transposable element that contains only the genes coding for the enzymes necessary for the element's excision and insertion. See *transposon.* 16

integument   In flowers, the covering layer over the nucellus of an ovule. Usually two integuments (inner and outer) are present. 9, 25

intercalary meristem   See *meristem.*

interfascicular cambium   The portion of the vascular cambium that develops from parenchyma cells located between vascular bundles. Alternative: *fascicular cambium.* 8

interkinesis (pl.: interkineses)   The portion of meiosis that occurs between telophase I and prophase II. There is no duplication of DNA. 4

internode   Portion of a stem where there are no leaves; portion between nodes. 5

interphase   The portion of the cell cycle that is not cell division; interphase consists of G_1, S, and G_2. 4

interpolation theory   The theory that vascular plants arose from monobiontic algae and a sporophyte generation was gradually interpolated into the life cycle. Compare: *transformation theory.* 23

intrinsic protein   A protein that is an integral part of a membrane, deeply embedded in it; it cannot be washed out of the membrane easily. Alternative: *peripheral protein.* 3

intron   A portion of the structural region of a gene whose information is not translated into protein; instead, the intron-transcribed portions of the hnRNA are removed and digested. Alternative: *exon.* 15

inversion mutation   A mutation in which a portion of the DNA double helix is excised, turned end for end, and ligated into place with reverse order. 16

ionic bond   Chemical bond between two molecules, one of which is negatively charged, the other positively charged. 2

isogamy   Sexual reproduction in which all gametes are structurally identical; there are no sperms and eggs. Alternatives: *anisogamy* and *oogamy.* 9, 21–23

isolation mechanism   A structure or metabolism that inhibits the movement of substances from one region to another. 12

isomorphic generations   A dibiontic life cycle in which sporophytes and gametophytes are almost indistinguishable. Alternative: *heteromorphic generations.* 21

isopentenyl adenosine   A natural cytokinin. 14

isotype specimen   A specimen obtained from the same plant or clone as the type specimen. 18

juvenile plant   A plant that is too immature to flower, even if otherwise appropriate stimuli are present. Alternative: *adult plant.* 14

K   Symbol for *carrying capacity.* 26

karyogamy   Fusion of the nuclei of two gametes after protoplasmic fusion (plasmogamy). See *syngamy.* 9, 20

karyokinesis (pl.: karyokineses)   Division of a nucleus, as opposed to cell division, cytokinesis. The two types of karyokinesis are mitosis and meiosis. 4, 16

kinetin   An artificial cytokinin. 14

kinetochore   In a chromosome, the kinetochore is the point at which spindle microtubules attach to the centromere. 4

Krebs cycle   Synonym for *citric acid cycle* and tricarboxylic acid cycle.

labyrinthine wall   Synonym for *transfer wall.*

lamina (pl.: laminae or laminas)   The broad, expanded part of a leaf; not the petiole. 6

laminarin   A reserve polysaccharide found in brown algae. 21

late wood   Synonym for *summer wood.*

lateral meristem   A name describing the position of the vascular cambium and cork cambium. 8

lateral veins   The major, large vascular bundles of a leaf, which are attached to the midrib or the petiole. Larger than minor veins. 6

Laurasia   Northern portion of the ancient continent of Pangaea. Compare: *Gondwanaland.* 27

leaf axil   Portion of a node immediately above the attachment point of a leaf. 5

leaf gap   In fern vascular tissue, an area above a leaf trace where there is no conducting tissue. 23

leaf primordium (pl.: primordia)   An extremely early stage in leaf development, when the leaf exists only as a pronounced bulging of the shoot apical meristem. 6

leaf scar   The region on a stem where a leaf was attached prior to abscission. 6

leaf sheath   Almost exclusively in monocots, the basal portion of a leaf, wrapped around the stem above a node. 6

leaf trace   A vascular bundle that extends from the stem vascular bundles through the cortex and enters a leaf. 6

lenticel   In bark, a region of cork cells with intercellular spaces, permitting diffusion of oxygen into inner tissues. 8

leptoids   In the stems of some mosses, elongate cells that resemble sieve cells and are involved in carbohydrate conduction. 22

leptotene   See *prophase I.*

lethal allele   An allele whose expression results in death. 16

leucoplast   See *plastid.*

life forms   A classification of the ways plants are adapted morphologically for surviving stressful seasons. 26

light compensation point   The level of illumination at which photosynthetic fixation of carbon dioxide just matches respiratory loss. 10

light-dependent reactions   In photosynthesis, the set of reactions directly driven by light. Alternative: *stroma reactions.* 10

lignin   A complex compound that impregnates most secondary cell walls, making them stronger, more waterproof, and resistant to attack by fungi, bacteria, and animals. 3, 8, 11

ligule   In selaginellas, a small flap of tissue on the upper surface of a leaf. 23

limiting factor   Of the several factors necessary for a process, the one that is the most scarce and sets the rate of the process. 26

linkage group   A set of genes that do not undergo independent assortment, being part of one chromosome. 16

linked genes   Genes located close to each other on a chromosome undergo crossing over only rarely, so they are linked. Alternative: *independent assortment.* 16

lipid body   A spherical droplet of oil or other lipid, common in the cells of many seeds. 3

lipids   A class of compounds that are hydrophobic and water insoluble. Examples: fats, oils, and waxes. 2

lipopolysaccharide layer   In gram-negative bacteria, a layer of sugar-bearing lipids located just exterior to the peptidoglycan layer. 19

lithotroph   An organism that obtains its energy by metabolizing inorganic compounds, either oxidizing them or reducing them, depending on the compound and the environment. 19

littoral zone   On a sea coast, the region between low tide and high tide; a common habitat for brown algae. 21

locule   The cavity within a structure such as a sporangium, gametangium, or carpel. 25

long-day plant   A plant that is induced to flower by nights shorter than the critical night length. 14

lumen   The interior of any structure such as a vesicle, vacuole, oil chamber, or resin duct. 3

lysis   The bursting of an animal cell or a plant protoplast due to excessive absorption of water. Do not confuse with plasmolysis. 12

lysosome   An organelle that contains digestive enzymes and is involved in recycling the components of worn-out organelles; in plants, the central vacuole acts as the lysosome. 3

macronutrient   Synonym for major *essential element.*

mannitol   A reserve polysaccharide in brown algae. 21

manoxylic wood   Wood that contains significant amounts of axial parenchyma, such as that of cycads. Compare: *pycnoxylic wood.* 24

map unit   A measure of the separation of genes on a chromosome; one map unit equals a 1% probability that crossing over will occur between them. Synonym: centimorgan (cM). 16

maternal inheritance   Uniparental inheritance for genes contributed by the megagamete but not by the microgamete. 16

mating types   Biochemical distinctions such that gametes of identical mating type cannot fuse; only those of compatible mating types can undergo syngamy. 20

matric potential   A component of water potential; a measure of the effect of a matrix on a substance's ability to absorb or release water. 12

matrix   The liquid within a mitochondrion. 3, 11

megagametangium (pl.: megagametangia)   A structure that produces megagametes (eggs). 21

megagamete   In an oogamous species, the larger, immobile gamete. Synonym: egg. 9, 21–25

megagametophyte   See *gametophyte.*

megaphyll   A leaf that has evolved from a branch system. Present in ferns and all seed plants. Alternative: *microphyll.* 23–25

megasporangium (pl.: megasporangia)   A structure that produces megaspores. 9, 21–25

megaspore   A large spore that grows into a megagametophyte that produces egg cells. Alternative: *microspore.* 9, 21–25

megaspore mother cell   Synonym for *megasporocyte.*

megasporocyte   In a heterosporous species, a cell that undergoes meiosis, resulting in the production of a megaspore. Synonym: megaspore mother cell. 9, 23, 24

meiosis   Reduction division, a process in which nuclear chromosomes are duplicated once but divided twice, such that the resulting nuclei each have only one half as many chromosomes as the mother cell. 4, 16

meiosis I   The first division of meiosis, during which the chromosome number per nucleus is reduced. Synapsis and crossing over occur during meiosis I. 4

meiosis II   The second division of meiosis, during which centromeres divide and the two chromatids of one chromosome become independent chromosomes. 4

meristem   A group of cells specialized for the production of new cells. **apical meristem**   Located at the farthest point of the tissue or organ produced. **basal meristem**   Located at the base. **intercalary meristem**   Located between the apex and the base. **lateral meristem**   Located along the side. 5

meristoderm   In brown algae, the outer layer of the body, which is both meristematic and photosynthetic. 21

mesocarp   The middle layer of the fruit wall. See *exocarp* and *endocarp.* 9

mesokaryotes   A term proposed for dinoflagellate algae, because their nuclear and cytoplasmic organization appears to be intermediate between those of prokaryotes and eukaryotes. 21

mesophyll   All tissues of a leaf except the epidermis. 6

metaphase   The second phase of mitosis during which chromosomes move to the center of the spindle, the metaphase plate. 4

metaphase I   The second phase of meiosis I, similar to metaphase of mitosis except that homologous pairs of chromosomes are involved. 4

metaphase II   A phase of meiosis II, similar to metaphase of mitosis. 4

metaphase plate   The region in the center of the spindle where chromosomes become aligned during nuclear division. 4

metaphloem   The part of the primary phloem that differentiates late, after adjacent cells have completed their elongation. Alternative: *protophloem*. 5

metaxylem   The part of the primary xylem that differentiates late, after adjacent cells have completed their elongation. Alternative: *protoxylem*. 5

MPa   Megapascal, a unit of pressure measurement equivalent to about 70 pounds per square inch. 12

microbody   A class of two types of small, vesicle-like organelles. Peroxisomes are involved in photorespiration, the detoxification of harmful products of photosynthesis. Glyoxysomes are involved in respiring stored fatty acids. 3, 10, 11

microfibril   As adjacent molecules of cellulose are synthesized, they crystalize into a microfibril, which may be 10 to 25 nm wide. 3

microfilaments   A structural element composed of actin and believed to be involved in the movement of organelles other than flagella, cilia, or chromosomes. 3

microgametangium (pl.: microgametangia)   A structure that produces microgametes (sperm cells). 21

microgametophyte   See *gametophyte*.

micronutrient   Synonym for minor *essential element*.

microphyll   The type of leaf that evolved from an enation; present in lycophytes. Alternative: *megaphyll*. 23, 24

microplasmodesmata (sing.: microplasmodesma)   In cyanobacteria, small holes in the walls of heterocysts, connecting the heterocysts to adjacent cells. 19

micropyle   In an ovule, the small apical opening created where the integuments do not meet; the pollen tube enters through the micropyle. 9, 25

microsporangium (pl.: microsporangia)   A structure that produces microspores. 9, 21–25

microspore   A small spore that grows into a microgametophyte that produces sperm cells. Alternative: *megaspore*. 25

microspore mother cell   Synonym for *microsporocyte*.

microsporocyte   In a heterosporous species, a cell that undergoes meiosis, resulting in the production of four microspores. Synonym: microspore mother cells. 9

microtubules   A skeletal element in eukaryotic cells, composed of alpha and beta tubulin. Microtubules constitute the mitotic spindle, phragmoplast, and axial component of flagella. 3

middle lamella (pl.: lamellae or lamellas)   The layer of adhesive pectin substances that acts as the glue that holds the cells of a multicellular plant together. 3

midrib   The large, central vascular bundle of a leaf. Alternative: *lateral* and *minor veins*. 6

minor essential element   See *essential element*.

minor veins   The smallest veins of a leaf, branching off lateral veins. There is no criterion that strictly distinguishes between minor veins and lateral veins. 6

mitochondrion (pl.: mitochondria)   The eukaryotic organelle involved in aerobic respiration, particularly the citric acid cycle and respiratory electron transport. 3, 11

mitosis (pl. mitoses)   Duplication division—a type of nuclear division (karyokinesis) in which nuclear chromosomes are first duplicated, then divided in half, one daughter nucleus receiving one set, the other daughter nucleus receiving the other set. Alternative: *meiosis*. 4, 16, 21

mobile essential element   An element that can be removed from mature tissues and transported to young or newly formed tissues. Alternative: *immobile essential element*. 13

molecular pump   An integral membrane protein that forces molecules from one side of a membrane to the other, using energy in the process called active transport. 3, 12

monobiontic   A life cycle with only one free-living generation; there is no alternation of generations. Alternative: *dibiontic*. 21

monocot   Informal term for any member of the flowering plant class Liliopsida. Examples, lily, iris, palm, agave. Alternative: *dicot*. 5

monocotyledon   Synonym for *monocot*.

monoecy   The condition in which a species has imperfect flowers (some staminate, others carpellate), but both are located on the same sporophyte. See *dioecy*. 9

monohybrid cross   A cross in which only a single trait is analyzed, disregarding all other traits. 16

monokaryotic   In fungi, having one type of nucleus per cell. Alternative: *dikaryotic*. 20

monomer   The subunit of a polymer. 2

monopodial branching   Branching in which one shoot is dominant and forms a distinct trunk, all other shoots being significantly different from the trunk. Alternatives: *sympodial* and *dichotomous branching*. 23–25

monosaccharide   Synonym for *simple sugar*; The monomer of polysaccharides. 2

morphogenic (morphogenetic) response   A response in which the quality of the plant changes, such as conversion from a vegetative state to a floral state. 14

motor cells   Cells that swell and shrink in plant organs capable of repeated, reversible movement, such as insect traps and petioles of leaves that undergo sleep movements. 12

mtDNA   Mitochondrial DNA: it is in the form of closed circles, without histones. See *cpDNA*. 3, 29

mucigel   The mucilaginous, slimy material secreted by root caps and root hairs. 7

multiple fruit   A fruit formed by the crowding together of the individual fruits of an entire inflorescence. Alternatives: *simple fruit* and *aggregate fruit*. 9

murein   Synonym for *peptidoglycan*. 19

mutagen   Any chemical or physical force that causes a change in the sequence of nucleotides in DNA. 16

mutation   Any change in the sequence of DNA. 16

mutualism   An interaction of two species in which both species benefit. Compare: *competition*. 26

mycelium (pl.: mycelia)   The diffuse mass of hyphae which constitutes the vegetative body of a fungus. 20

mycorrhizae (sing.: mycorrhiza)   Fungi that form a symbiotic relationship with roots, usually of benefit to plants because they provide phosphorus. **ectomycorrhizae**   A type in which the fungi invade only the outermost cells of the root. **endomycorrhizae**   A type in which the fungi invade all cells of the root cortex. 7, 13

nastic response   A nongrowth response that is stereotyped and not oriented with regard to the stimulus. 14

natural selection   The preferential survival, in natural conditions, of those individuals whose alleles cause them to be more adapted than other individuals with different alleles. 17

natural system of classification   A classification based on evolutionary, phylogenetic relationships. Alternative: *artificial system of classification.* 18

necrotroph   A fungus similar to a parasite which attacks its host virulently, killing it and absorbing the released nutrients. 20

nectary   A gland that secretes a sugary solution that typically attracts pollinators. 9, 25

netted venation   Synonym for *reticulate venation,* typically found in leaves of dicots. 6

niche   The set of conditions exploited best by one species. 26

nitrification   The conversion, by microbes, of ammonia to nitrate. 19

nitrogen assimilation   The incorporation of ammonium into organic compounds within an organism. 13

nitrogen fixation   The conversion of atmospheric nitrogen into any compound that can be used by plants, typically either nitrate or ammonium. 13

nitrogenase   The enzyme responsible for nitrogen fixation. 13

nocturnal   Nighttime; a nocturnal plant opens its flowers at dusk. Alternative: *diurnal.* 13

node   Point on a stem where a leaf is attached. 5

nodule   In roots of plants that form symbiotic associations with nitrogen-fixing bacteria, regions of the root swell, forming nodules whose cells contain the bacteria. 7

nonclimacteric fruits   Fruits that ripen slowly and steadily, without a sudden burst of metabolism (the climacteric) at the end. Alternative: *climacteric fruits.* 14

noncompetitive inhibitor   A small molecule that inhibits an enzyme by attaching to some site other than the active site. Alternative: *competitive inhibitor.* 2

noncyclic electron transport   The flow of electrons from water to NADPH during the light-dependent reactions of photosynthesis. Alternative: *cyclic electron transport.* 10, 19

nondisjunction   During meiosis II, the nonseparation of the two chromatids of a chromosome, such that one daughter cell receives both while the other daughter cell receives none. 16

nonessential organs   In flowers, the sepals and petals. Alternative: *essential organs.* 9

nonpolar molecule   A molecule that does not carry even a partial charge anywhere. 2

nonstoried cambium   A cambium in which the fusiform initials are not aligned horizontally. Alternative: *storied cambium.* 8

nucellus (pl.: nucelli)   The megasporangium of an ovule. 9, 25

nuclear envelope   A set of two membranes, the inner and the outer nuclear envelopes, that surround the nucleus. 3

nuclear pores   Structures in the nuclear envelope that are involved in transport of material between nucleus and cytoplasm. 3

nucleic acid   A polymer of nucleotides. 2, 15, 16

nucleoid   In prokaryotes, the portion of the protoplasm where DNA circles are concentrated. 19

nucleolus (pl.: nucleoli)   Organelles located within the nucleus, nucleoli are areas where ribosomal RNAs are synthesized and assembled into ribosomal subunits. 3, 15

nucleoplasm   The substance located in the nucleus—DNA, histones, RNA, enzymes, nucleic acids, water. 3

nucleosome   A nuclear particle composed of histones with DNA wrapped around them; a basic aspect of chromosome structure. 15

nucleotide   The monomer of nucleic acids; each nucleotide consists of a nitrogenous base, a sugar, and a phosphate group. 2, 15, 16

nucleus (pl.: nuclei)   In eukaryotic cells, the organelle that contains

DNA and is involved in inheritance, metabolism control and ribosome synthesis. 3

obligate aerobe   An organism that must have oxygen to survive. Synonym: strict aerobe. Alternative: *obligate anaerobe.* 11, 19

obligate anaerobe   An organism that is killed by exposure to oxygen. Synonym: strict anaerobe. Alternative: *obligate aerobe.* 11, 19

Okazaki fragments   During DNA replication, the set of short fragments that grow discontinuously along one strand of DNA and must be ligated into one continuous DNA molecule. 16

oligosaccharide   A compound made up of a few simple sugars (monosaccharides). 2

ontogeny   Synonym for development, morphogenesis. 14, 15

oogamy   A type of sexual reproduction in which the two gametes are distinctly different structurally; one is a microgamete (sperm), and the other is a megagamete (egg). Alternatives: *anisogamy* and *isogamy.* 9, 21–25

oogonium   Synonym for megagametangium in certain organisms. 21

operational habitat   Those aspects of a habitat that definitely affect the organism being considered. 26

operculum (pl.: opercula)   In mosses, the lidlike top of a sporangium. 22

organ   A structure composed of a variety of tissues; seed plants are considered to have only three organs: roots, stems, and leaves. 5

organelles   The "little organs" of a cell, such as nuclei, plastids, mitochondria, and ribosomes. Many are membrane-bounded compartments, others are nonmembrane structures composed of protein or protein and RNA. 3

osmosis   Diffusion through a membrane. 12

osmotic potential   A component of water potential; a measure of the effect of solute particles on a substance's ability to absorb or release water. 12

outcrossing   Synonym for cross fertilization. 16

outer bark   The outermost dead layers of bark, from the surface to the innermost cork cambium. Alternative: *inner bark.* 8

ovary   In a flower, the base of the carpel; the region that contains ovules and will develop into a fruit. 9, 25

overtopping   In the evolution of unequal branching, the ability of one shoot to grow for a longer time than the other shoot that resulted from the branching. 23, 24

ovule   The structure in a carpel that contains the megasporangium and will develop into a seed. 9, 25

ovuliferous scale   The scale that bears the ovule in gymnosperm seed cones. Synonym: *cone scale.* 24

oxidation state   A measure of the number of electrons added to or removed from a molecule during an oxidation-reduction reaction. 10, 11

oxidative phosphorylation   The formation of ATP from ADP and phosphate, powered by energy released through respiration. 10, 11

oxidize   To raise the oxidation state of a molecule by removing an electron from it. 10, 11

oxidizing agent   An electron carrier that is not carrying electrons. Alternative: *reducing agent.* 10, 11

P680   The reaction center of photosystem II. 10

P700   The reaction center of photosystem I. 10

P$_{fr}$ See *phytocrome.*

P$_r$ See *phytochrome.*

P channel   The groove in the ribosome small subunit in which the nascent protein-carrying tRNA occurs. Alternative: *A channel.* 15

pachytene   See *prophase I.*

palisade parenchyma Any part of leaf mesophyll in which the cells are elongate and aligned parallel to each other. 6

palmately compound See *compound leaf.*

Pangaea The ancient supercontinent composed of all the world's land, it existed in the late Paleozoic Era and consisted of Laurasia (north) and Gondwanaland (south). 27

parallel evolution The evolution of similar homologous features in two or more groups due to similar selective pressures, the features being derived from a common ancestral feature. Compare: *convergent evolution.* 17

parallel venation Almost exclusively in monocot leaves, a pattern in which all veins run approximately parallel to each other, either from the base of the leaf to its tip or from the midrib to the margin. Alternative: *reticulate venation.* 6

paramylon A storage polysaccharide in euglenoid algae. 21

parasexual cycle In fungi, a condition in which compatible nuclei of a heterokaryotic mycelium fuse, then undergo meiosis and crossing over, even though they are not in a sporangium. 20

parasite See *biotroph.*

parenchyma Cells with only thin primary walls; all other features are highly variable from type to type. Alternatives: *collenchyma* and *sclerenchyma.* 5

parental type chromosome A chromosome that, after meiosis, has not undergone crossing over. Alternative: *recombinant chromosome.* 16

passage cell A cell in the endodermis that has only Casparian strips whereas all surrounding endodermis cells have thickened waterproof walls. 7

pectic substances A set of polysaccharides that constitute the middle lamella and act as the glue that holds together the cells of multicellular plants. 3

pedicel The stalk of an individual flower. Compare: *peduncle.* 9

peduncle The stalk of an inflorescense, a group of flowers. Compare: *pedicel.* 9

pellicle In euglenoid algae, a layer of elastic proteins on the cell surface. 21

pentose phosphate pathway A type of respiration in which glucose is converted either to ribose or erythrose. Synonyms: hexose monophosphate shunt and phosphogluconate pathway. 11

peptide bond The chemical bond that holds the amino acid residues together in a protein. 2

peptidoglycan The polymer that constitutes the main source of strength in the wall of bacteria. 19

perennial plant A plant that lives for more than 2 years. Compare: *annual* and *biennial plant.* 5, 9

perfect flower A flower that has both stamens and carpels. Alternative: *imperfect flower.* 9

perforation In a vessel element, the hole(s) where both primary and secondary walls are missing. Alternative: *pit.* 5

pericarp Technical term for the fruit wall, composed of one or more of the following: exocarp, mesocarp, endocarp. 9

periclinal wall A wall that is parallel to a nearby surface, especially the outer surface of the plant. Alternative: *anticlinal wall.* 8

pericycle An irregular band of cells in the root, located between the endodermis and the vascular tissue. 7

periderm Technical term for bark; it consists of cork, cork cambium, and any enclosed tissues such as secondary phloem. 8

perigynous See *half-inferior ovary.*

peripheral protein A membrane protein that is only weakly associated with the surface of the membrane. Alternative: *intrinsic protein.* 3

perisperm A nutritive tissue in seeds of the dicot order Caryophyllales, formed as nucellus cells proliferate. 9, 25

peristome teeth In a moss capsule, the one or two sets of teeth-like structures around the mouth of the sporangium. 22

permeable membrane A membrane through which materials can pass. Alternative: *impermeable membrane.* Compare: *differentially permeable membrane.* 2, 12

peroxisome See *microbody.*

petals The appendages, usually colored, on a flower, most often involved in attracting pollinators. See also *corolla.* 9, 25

petiole The stalk of a leaf. 6

petiolule The stalk that attaches a leaflet to the rachis of a compound leaf. 6

phage Synonym for *bacteriophage.*

phellem Technical term for cork. 8

phelloderm Parenchyma cells produced to the inside by the cork cambium; usually only a layer or two are formed, and phelloderm is not present in all species. 8

phellogen Synonym for *cork cambium.*

phenotype The physical, observable characteristics of an organism. Alternative: *genotype.* 16, 17

phloem The portion of vascular tissues involved in conducting sugars and other organic compounds, along with some water and minerals. Alternative: *xylem.* 5, 8

phosphogluconate pathway Synonym for *pentose phosphate pathway.*

phospholipid A type of lipid containing two fatty acids and a phosphate group bound to glycerol. 2

phosphorylation The attaching of a phosphate group to a substrate. See *chemiosmotic phosphorylation, photophosphorylation,* and *substrate level phosphorylation.* 2, 10, 11

photic zone In aquatic environments, upper regions that are illuminated sufficiently to allow photosynthesis. 19

photoautotroph An organism that obtains its energy through photosynthesis and its carbon from dioxide. Compare: *chemoautotroph.* 10, 11

photoperiod In reference to cycles of light and darkness, the length of time that uninterrupted light is present. 14

photophosphorylation The formation of ATP from ADP and phosphate by means of light energy; a part of photosynthesis. 10, 19

photorespiration The oxidation of phosphoglycolate produced when RuBP carboxylase adds oxygen, not carbon dioxide, to RuBP. 10

photosynthetic unit A cluster of photosynthetic pigments and electron carriers embedded in the chloroplast membrane. 10

photosystem I The pigments and electron carriers that transfer electrons from P700 to NADPH. 10

photosystem II The pigments and electron carriers that transfer electrons from water to P700 in photosystem I. 10

phototroph An organism that obtains its energy through photosynthesis. 19

phragmoplast During cell division, the phragmoplast is a set of short microtubules oriented parallel to the spindle microtubules; it catches dictyosome vesicles and guides them to the site where the new cell wall (cell plate) is forming. See *phycoplast.* 4, 21

phycobilins The accessory pigments of cyanobacteria and red algae. Phycocyanin absorbs blue light, and phycoerythrin absorbs red light. 19, 21

phycobilisome In cyanobacteria and red algae, a particle involved in photosynthesis and composed of phycobilins and biliproteins. 19, 21

phycocyanin See *phycobilins.*

phycoerythrin See *phycobilins.*

phycoplast In the cell division of some algae, a set of microtubules oriented parallel to the plane of the new cell wall and involved in wall formation. Compare: *phragmoplast.* 21

phyletic speciation The evolution of one species into a new species, such that the original species no longer exists. Alternative: *divergent speciation*. 17

phyllode   Synonym for *cladophyll*.

phyllotaxy   The arrangement of leaves and axillary buds on a stem. 5

phytoalexins   Lipid-like or phenolic compounds produced by plants in response to attacks by fungi. 20

phytochrome   A pigment involved in many aspects of morphogenesis in which the stimulus is red light or the length of a dark period. P_{fr} absorbs far-red light, P_r absorbs red light. 14

phytoferritin   A protein molecule that binds and stores iron; mostly found in plastids. 3

pinna (pl.: pinnas or pinnae)   Technical name for a leaflet of a fern. 23

pinnately compound   See *compound leaf*.

pioneers   The first plants to inhabit an area that previously had no life. 26

pit   In a sclerenchyma cell, an area where there is no secondary wall over the primary wall and material can pass into or out of the cell.
  blind pit   A pit that does not meet another pit in the adjacent cell.
  pit-pair   A set of aligned pits in adjacent sclerenchyma cells. 5

pit connection   In red algae, a large hole in the wall between two cells. 21

pit membrane   The set of two primary walls and middle lamella that occurs between the two pits of a pit-pair. 5

pit plug   In red algae, material that fills the hole (pit connection) between two cells. 21

pith   The region of parenchyma located in the center of most shoots and some roots, surrounded by vascular bundles. 5

placenta (pl.: placentas or placentae)   Tissue in the ovary of a carpel to which the ovules are attached. 9

planation   In the telome theory of the origin of megaphylls, the concept that all branching occurred in one plane, resulting in a flat system. 23

plant growth substance   Term used for any hormone-like compound, whether natural or artificial. 14

plasma membrane   The semipermeable membrane that surrounds the protoplasm of a cell. Synonym: plasmalemma. 3

plasmalemma   Synonym for *plasma membrane*.

plasmid   A small circle of DNA occurring in some bacteria and acting like a bacterial chromosome. 15

plasmodesma (pl.: plasmodesmata)   A narrow hole in a primary wall, containing some cytoplasm, plasma membrane, and a desmotubule; a means of communication between cells. See *symplast*. 3, 5

plasmodium (pl.: plasmodia)   The body of a slime mold, a large mass of protoplasm with hundreds or thousands of nuclei. 20

plasmogamy   The fusion of the cytoplasm of two gametes during sexual reproduction. See *karyogamy*. 9, 20

plasmolysis   The shrinking of a cell due to loss of water. Do not confuse with lysis. 12

plasticity   A property of collenchyma walls; once stretched to a new shape or size, usually by growth, the wall retains that new shape or size. Alternative: *elasticity*. 5

plastids   A family of organelles within plant cells only. **proplastids** Young plastids common in meristematic cells. **chloroplasts** Chlorophyll-rich plastids that carry out photosynthesis. **amyloplasts** Plastids that store starch. **chromoplasts** Plastids that contain red or yellow pigments, located in flowers and fruits. **leucoplasts** Colorless plastids. 3

plastochron   The length of time required for an apical meristem to make the cells of one node and internode. 5

plastocyanin   A copper-containing electron carrier. 10

plastoglobulus (pl.: plastoglobuli)   A droplet of lipid located within a plastid. 3

plastoquinone   A class of lipid-soluble electron carriers. 10

pleiotropic effects   The multiple phenotypic expressions of a single allele whose activity affects various aspects of metabolism. 16

pleomorphic bacteria   Bacteria that lack cell walls and thus do not have a constant shape. 19

plumule   Synonym for *epicotyl*.

plurilocular gametangium (pl.: gametangia)   In brown algae, a multicellular structure in which each cell produces a gamete. Alternative: *unilocular sporangium*. 21

pneumatocyst   Synonym for air bladder, a swollen, hollow structure in brown algae; it increases buoyancy. 21

point mutation   A mutation affecting only a single nucleotide. 16

polar molecule   A molecule that has a partial positive charge at one site and a partial negative charge at another site. 2

polar nuclei (sing.: nucleus)   The two nuclei of the central cell of the megagametophyte in a flowering plant; after fertilization, they become the endosperm nucleus. 9

polar transport   Transport in one direction based on an organ's structure, regardless of its spatial orientation. 14

pollen   In seed plants, the microspores and microgametophytes. 9, 23–25

pollen chamber   In gymnosperms, a cavity just above the nucellus in the ovule, the site where pollen accumulates and germinates. 24

pollen tube   After landing on a compatible stigma or gymnosperm megasporophyll, a pollen grain germinates with a tubelike process that carries the sperm cells to the vicinity of the egg cell. 9, 24

polymer   A large compound composed of a number of subunits, monomers. 2

polymerase chain reaction   A method of copying minute quantities of DNA using bacterial enzymes. 15

polyploid   Refers to a nucleus that contains three or more sets of chromosomes. 16

polysaccharide   A compound made up of many simple sugars (monosaccharides). 2

polysome   The complex formed when numerous ribosomes bind to the same molecule of messenger RNA. 3, 15

population   All the individuals of a species that live in a particular area at the same time and can interact with each other. 17, 26

postzygotic isolation mechanisms   Phenomena that prevent successful interbreeding of two populations but that act after fertilization; the two sets of chromosomes are incompatible and cannot produce a fertile adult. Alternative: *prezygotic isolation mechanism*. 17

P-protein   Phloem-protein, or fibrillar protein that plugs sieve pores and prevents leakage if sieve elements are damaged. 12

predation   A relationship in which one species benefits and the other is harmed; not often used in botany. 26

presentation time   The length of time a stimulus must be present in order for an organism to perceive it. 14

pressure flow hypothesis   The hypothesis that flow in phloem is due to active loading in sources and active unloading in sinks. 12

pressure potential   A component of water potential; a measure of the effect of pressure or tension on a substance's ability to absorb or release water. 12

prezygotic isolation mechanism   Phenomena that prevent successful interbreeding of two populations but that act so early that fertilization is not possible. Alternative: *postzygotic isolation mechanism*. 17

primary cell wall   A cell wall present on all plant cells except some sperm cells; it is formed during cell division and is usually thin, but some may be thick. See *secondary cell wall* and *collenchyma*. 3

primary consumer   In ecology, a synonym for herbivore. 26

primary endosperm nucleus   The nucleus formed by the fusion of one sperm nucleus and two polar nuclei. 9

primary growth   The production of new cells by shoot and root apical meristems and leaf primordia. Alternative: *secondary (woody) growth.* 5, 8

primary phloem   The phloem of the primary body; it differentiates from cells derived from apical meristems or forms in leaves, flowers, and fruits. Compare: *secondary phloem.* 5, 8

primary pit field   An area of a primary cell wall that is especially thin and contains numerous plasmodesmata. See *plasmodesma.* 3, 5

primary plant body   The herbaceous body produced by apical meristem (roots, stems, leaves, flowers, and fruits). Alternative: *secondary (woody) plant body.* 5, 8

primary producer   In ecology, synonym for *autotroph.* 26

primary structure of a protein   The order of the sequence of amino acids in the protein. 2, 15, 16

primary tissues   The tissues derived more or less directly from an apical meristem or leaf primordium; the tissues of the primary plant body. Alternative: *secondary tissues.* 2, 5

primary xylem   The xylem of the primary body; the xylem produced from cells derived from apical meristems or formed in leaves, flowers, and fruits. Compare: *secondary xylem.* 5, 8

primer RNA   During DNA replication, a short piece of RNA that is synthesized against open DNA and from which DNA polymerase can begin building a new molecule of DNA. 16

primordium (pl.: primordia)   A small mass of cells that grows into an organ such as a leaf or petal. 5

procambium (pl.: procambia)   Tissue that matures into primary xylem and primary phloem. 5

prochlorophytes   A group of prokaryotes that have both chlorophyll *a* and *b*; believed to be closely related to the ancestors of plastids in algae and plants. 19, 21

procumbent cell   In rays in secondary xylem, cells that are longer radially than they are tall; they typically have little or no cross-field pitting. Alternative: *upright cell.* 8

proembryo   An early stage of embryo development, usually considered to encompass the stages between the zygote and the initiation of the cotyledon primordia. 9

progymnosperms   A group of extinct plants believed to have been the ancesters of gymnosperms. 2

prokaryotes   Organisms that have no true nucleus or membrane-bounded organelles. Prokaryotes are eubacteria, cyanobacteria, and archaebacteria. Alternative: *eukaryotes.* 3, 19

promoter region   That portion of a gene in which control molecules and RNA polymerases bind during gene activation and transcription. Alternative: *structural region.* 15

propagules   Parts of a plant involved in reproduction and dissemination. Examples are seeds, bulbs, plant pieces that can form roots and grow into a new plant. 9, 26

prophase   The initial phase of mitosis during which the nucleolus and nuclear membrane break down, chromosomes begin to condense, and the spindle begins to form. 4

prophase I   The first phase of meiosis, similar to prophase of mitosis, with the following additional processes: **leptotene**   Initiation of chromosome condensation. **zygotene**   The pairing of homologous chromosomes (synapsis). **pachytene**   Formation of the synaptonemal complex. **diplotene**   Homolog separation and chiasmata become visible. **diakinesis**   Complete separation of homologs and terminalization of chiasmata. 4

proplastid   See *plastids.*

protease   An enzyme that digests protein. 2

protoderm   A term that refers to any immature epidermal cell. 5

protonema (pl.: protonemata)   In nonvascular plants, the mass of alga-like cells that grow from the spore during germination. 22

protophloem   The part of the primary phloem that differentiates early, while adjacent cells are still elongating. Alternative: *metaphloem.* 5

protoplasm   All the substance of a cell, usually considered not to include the cell wall. The protoplasm of a single cell is a protoplast. See *cytoplasm* and *hyaloplasm.* 3

protostele   A vascular cylinder that has no pith; common in roots and early vascular plants. Alternative: *siphonostele.* 23–25

protoxylem   The part of the primary xylem that differentiates early, while adjacent cells are still elongating. Alternative: *metaxylem.* 5, 23–25

provascular tissue   Cells in the primary plant body that later differentiate into xylem, phloem, or vascular cambium. Synonym: *procambium.* 5

proximal   Refers to the position of a structure near a point of reference; relative to a stem, a petiole is proximal to a leaf blade. Alternative: *distal.* 5

pseudomonopodial branching   A type of sympodial branching that strongly resembles monopodial branching, having what appears to be one main shoot. 23

pulvinus (pl.: pulvini)   A jointlike region of a petiole where motor cells are located and flexion occurs during nastic responses. 14

purine   One of the two types of nitrogenous bases occurring in nucleotides; purines have two ring structures. Compare: *pyrimidine.* 2

pycnoxylic wood   Wood with little or no axial parenchyma, such as that of gymnosperms and progymnosperms. Compare: *manoxylic wood.* 24

pyrenoid   In many algae and hornworts, a region of the choroplast believed to be involved in polymerization of polysaccharides. 21, 22

pyrimidine   One of the two types of nitrogenous bases occurring in nucleotides; pyrimidines have only a single ring structure. Compare: *purine.* 2

quantum (pl.: quanta)   A particle of electromagnetic energy. Synonym: photon. 10

quaternary structure   The association of several proteins into a large structure such as a microtubule or synthetic complex. 2

quiescent center   Portion of the root apical meristem in which cell division does not occur. 7

r   Symbol for *biotic potential.*

rachis (pl.: rachises)   The extension of the petiole in a compound leaf; all leaflets are attached to the rachis. 6

radially symmetrical   Divisible into two equal halves by any median longitudinal section. Synonym: *actinomorphic.* 9

radicle   The main root of a seed; it is the direct continuation of the embryonic stem. 7, 9

rain shadow   The diminished amount of rainfall on the leeward side of a mountain compared to the side that faces an ocean. 27

random distribution   In ecology, the distribution of individuals in the habitat with no obvious, identifiable pattern. Compare: *clumped* and *uniform distribution.* 26

raphe   In seeds, a ridge caused by the fusion of the funiculus to the side of the ovule. In diatoms, a groove in the shell. 21

raphide   A long, narrow, needle-like crystal, occurring in clusters in specialized cells. 3

ray   In secondary xylem and phloem, a radial series of cells produced by ray initials. Alternative: *axial tissue.* 8

ray initials   In a vascular cambium, the short cells that give rise to the rays of the secondary xylem and phloem. Alternative: *fusiform initials.* 8

ray tracheid   Horizontal tracheids in the secondary xylem rays of gymnosperms. 8

reaction center   A special chlorophyll *a* molecule actually involved in the transfer of electrons in photosynthesis. 10

reaction wood   Wood formed in response to mechanical stress. **tension wood**   The reaction wood of dicots, formed on the upper side of a branch. **compression wood**   The reaction wood of gymnosperms, formed on the lower side of a branch. 8

reannealing   Synonym for DNA hybridization by slowly cooling a mixture of single-stranded DNA molecules. 15

receptacle   The stem (axis) of a flower, to which all the other parts are attached. 8 In the brown alga *Fucus,* the ends of the branches where conceptacles are located. 21

recessive trait   A trait whose phenotype is completely masked by that of the alternative (dominant) allele in the heterozygous condition. 16

recombinant chromosome   A chromosome that results from crossing-over in meiosis, being composed of parts of the paternal and maternal homologs. Alternative: *parental type chromosome.* 16

recombinant DNA   DNA constructed from pieces of DNA from several sources, either through crossing-over in meiosis or laboratory manipulation. 15

red tide   A bloom of algae in which the cells become so numerous as to give the water a reddish tint. 21

redox potential   The tendency of a molecule to accept or donate electrons during a chemical reaction. 10

reduce   To lower the oxidation state of a molecule by adding an electron to it. 10, 11

reducing agent   An electron carrier that is carrying electrons. Alternative: *oxidizing agent.* 10, 11

reducing power   The ability of an electron carrier to force electrons onto another compound. 10, 11

reduction division   Synonym for *meiosis.*

regular flower   A radially symmetrical flower. Alternative: *zygomorphic flower.* 9

relictual feature   A feature that occurs in a modern organism and was inherited relatively unchanged from an ancient ancestor. Alternative: *derived feature.* 1, 18

replication fork   In DNA replication, the point at which the double helix opens and formation of new DNA occurs. 16

replicon   During DNA replication, a short segment of DNA that has opened and where replication is occurring. 16

reproductive barrier   Any physical or metabolic phenomenon that prevents two members of a species from interbreeding. 17

reproductive isolation   The inability of some members of a species to interbreed with other members through either biotic or abiotic reproductive barriers. 17

respiration   The breakdown of molecules such that part of their energy is used to make ATP. If oxygen is required as an electron acceptor, the process is aerobic respiration; if not, it is anaerobic respiration (fermentation). 11, 19

respiratory quotient   An indicator of the type of substrate being respired, RQ = (carbon dioxide liberated)/(oxygen consumed). 11

resting phase   Old synonym for *interphase.*

restriction endonucleases   Enzymes that recognize specific sites in DNA double helices, then cut the two strands in complementary sites. 15

restriction map   A map of a DNA molecule made by exposing it to restriction endonucleases, thus showing the number of cleavage sites and the number of bases between sites. 15

reticulate thickening   A pattern of secondary wall deposition in tracheids and vessel elements; the wall is deposited as a series of intersecting bands that have a netlike appearance. 5

reticulate venation   A netlike pattern of veins in a leaf, found primarily in leaves of dicots (broadleaf plants). Synonym: netted venation. Alternative: *parallel venation.* 6

retrovirus   A virus whose genetic material is single-stranded RNA; the most common type of plant virus. 15

reverse transcriptase   An enzyme that catalyzes the synthesis of DNA, using RNA as a template. 15

rhizoids   In certain fungi, algae, and nonvascular plants, cells or parts of cells that project into the substrate and anchor the organism. 20–22

rhizome   A fleshy, horizontal, subterranean stem involved in allowing the plant to migrate laterally. Examples: bamboo, iris. 5

rhyniophytes   The common name for *Rhynia* and its close relatives, the earliest vascular plans. 23

ribonucleic acid (RNA)   A polymer of ribose-containing nucleotides; there are three classes: messenger RNA, transfer RNA, and ribosomal RNA. 2, 15, 16

ribophorin   An integral membrane protein that binds ribosomes to the membrane, usually in rough endoplasmic reticulum. 3

ribose   A five-carbon sugar occurring in ribonucleic acid, among other things. 2

ribosome   An organelle responsible for protein synthesis; ribosomes consist of a large subunit and a small subunit, both made of proteins and ribosomal RNA (rRNA). 3, 15

ring porous wood   Wood in which the early wood has more numerous and larger vessels than the late wood. Alternative: *diffuse porous wood.* 8

RNA polymerase I   The enzyme that transcribes ribosomal genes into ribosomal RNA. 15

RNA polymerase II   The enzyme responsible for transcribing genes into messenger RNA. 15

RNA polymerase III   The enzyme responsible for transcribing the 5S RNA gene into 5S RNA. 15

root cap   A layer of parenchyma cells that cover and protect the root apex. 7

root hair zone   Region of a root tip, just proximal to the zone of elongation, where epidermal cells grow out as root hairs. 7

root pressure   Water in root xylem is under pressure due to the active accumulation of salts by the endodermis and the accompanying influx of water. 7, 12

rough ER (RER)   See *endoplasmic reticulum.*

rubisco   Synonym for *RuBP carboxylase.*

RuBP carboxylase   The enzyme in photosynthesis that carboxylates RuBP, thus bringing carbon into the plant's metabolism. Synonym: rubisco. 10

S phase   The synthesis phase of the cell cycle, during which nuclear DNA is replicated (synthesized). 4, 16

saprobe   Synonym for *saprotroph.*

saprotroph   A fungus that attacks an organism that has died from other causes. 20

sapwood   The light-colored, light-scented outermost wood of a trunk or branch; conduction is still occurring and many wood parenchyma cells are alive. Alternative: *heartwood.* 8

scalariform thickening   A pattern of secondary wall deposition in tracheids and vessel elements; the wall is interrupted by broad, short pits that cause the wall to have a ladder-like appearance. 5

scientific method   A means of analyzing the physical universe. Observations are used as the basis for constructing a hypothesis that predicts the outcome of future observations or experiments. Anything that can never be verified cannot be accepted as part of a scientific hypothesis. 1

scientific name　The binomial name of a species, consisting of the genus name and the species epithet. 18

sclereid　A sclerenchyma cell that is rather cubical, not long like a fiber. Masses of sclereids provide a tissue with strength and rigidity. Alternative: *fiber*. 5

sclerenchyma　Sclerenchyma cells have both a primary wall and an elastic secondary wall; if stretched to a new size or shape, the wall returns to its original size and shape after the deforming force is removed. Alternatives: *parenchyma* and *collenchyma*. 5

sclerotium (pl.: sclerotia)　In fungi, a mass of tightly adhering, resistant hyphae able to survive harsh conditions for years. 20

scutellum (pl.: scutella)　In grass seeds, the single cotyledon, which is shield-shaped and digests and absorbs the endosperm during germination. 9

secondary cell wall　A cell wall present only in certain cells (sclerenchyma) and formed only after cell division has been completed. When present, the secondary wall is located interior to the primary wall and is typically impregnated with lignin. See *primary cell wall, sclerenchyma,* and *xylem*. 3, 5

second-order reaction　A reaction involving the collision of two particles. 2

secondary consumer　In ecology, a synonym for carnivores and omnivores. 26

secondary (woody) growth　Growth that occurs by means of either the vascular cambium or the cork cambium. It results in wood and bark, the secondary tissues. Alternative: *primary growth*. 5, 8

secondary phloem　Phloem derived from the vascular cambium. Compare: *primary phloem*. 5, 8

secondary (woody) plant body　The wood and bark produced by the vascular cambium and cork cambium. Alternative: *primary plant body*. 5, 8

secondary structure　Short sequences of regular helix or regular pleating in a protein. 2

secondary tissues　The tissues of the secondary plant body—those produced by the vascular cambium and the cork cambium. Alternative: *primary tissues*. 8

secondary xylem　Xylem derived from the vascular cambium. Compare: *primary xylem*. 5, 8

seed coat　The protective layer on a seed; the seed coat develops from one or both integuments. Synonym: *testa*. 9

selectively permeable membrane　Synonym for *differentially permeable membrane*.

self-assembly　The automatic assembly of a larger structure solely due to interaction of charges and hydrophobic/hydrophilic regions on the molecules. 2

selfing　Pollinating a plant's stigma with pollen from the same plant or a plant of identical genotype. 16

self-pollination　The pollination of a flower by pollen from the same flower or another flower on the same plant. Alternative: *cross-pollination*. 9

semiconservative replication　Refers to the fact that during DNA replication, one new molecule is paired with one original molecule such that in every new chromosome, half the DNA is conserved from the pre-existing chromosome. 16

semipermeable membrane　A membrane that is relatively permeable to some substances and relatively impermeable to others. Synonym: *differentially permeable*. See *impermeable* and *freely permeable*. 3, 12

sepal　In flowers, the outermost of the fundamental appendages, most often providing protection of the flower during its development. See *calyx*. 9

septum (pl.: septa)　Synonym for cross wall, especially with regard to fungi and algae. 20

sessile　Refers to an organ that has no stalk but rather is attached directly to the stem or other underlying organ. For leaves, alternative is petiolate. 6

seta (pl.: setae)　In mosses and liverworts the stalk of the capsule, located between the foot and the sporangium. 22

sexual reproduction　Reproduction in which genomes of two individuals are brought together in one nucleus followed by meiosis with crossing-over. 9, 16, 20

short-day plant　A plant that is induced to flower by nights longer than the critical night length. 14

sieve area　In phloem, an area on a sieve element wall in which numerous sieve pores occur. 5

sieve cell　The phloem conducting cell in nonangiosperms; sieve cells are long and tapered with small sieve areas over much of their surfaces. Alternative: *sieve tube member*. 5

sieve element　Refers to either or both types of phloem-conducting cells: sieve cells and sieve tube members. 5

sieve plate　In phloem, the end walls of sieve tube members, bearing one or several large sieve areas with large sieve pores. 5

sieve pore　In sieve elements, the holes (enlarged plasmodesmata) in the primary walls; sieve pores permit movement of phloem sap from one sieve element to another. 5

sieve tube　In the phloem of angiosperms, a column of sieve tube members interconnected by large sieve areas and sieve pores. 5

simple cone　A cone with just one axis, bearing only sporophylls. Most cones are simple cones. Alternative: *compound cone*. 24

simple fruit　A fruit that develops from a single carpel or the fused carpels of a single flower. Alternatives: *aggregate fruit* and *multiple fruit*. 9

simple leaf　A leaf in which the blade consists of just one part. Alternative: *compound leaf*. 6

simple pit　In sclerenchyma cells, a pit with no border. Alternative: *bordered pit*. 6

simple sugar　A sugar that is not composed of smaller sugar molecules, the monomer for polysaccharides. Synonym: *monosaccharide*. 2

sink　In phloem transport, any organ or tissue that receives material transported by the phloem. Alternative: *source*. 12

siphonostele　A vascular cylinder that contains pith; common in stems but absent in early vascular plants. Compare: *protostele*. 23–25

siphonous　In algae, a synonym for coenocyte—a long, tubular cell with many nuclei and few or no cross walls. 21

smooth ER (SER)　See *endoplasmic reticulum*.

softwood　A term applied to both gymnosperms and their wood, because few gymnosperms have any fibers in their wood. Alternative: *hardwood*. 8

somatic mutation　A mutation in a cell that is not a gamete and does not give rise to gametes. 16

sorus (pl.: sori)　In ferns, a cluster of sporangia on the underside of leaves. 23

source　In phloem transport, any organ or tissue that supplies material to be transported. Alternative: *sink*. 12

speciation　The conversion of one species or population of a species into a new species. 17

species　A set of individuals that are closely related by descent from a common ancestor and can reproduce with each other but not with members of any other species. 18

spermatium (pl.: spermatia)　Generally, a synonym for sperm cell, but in botany it usually refers only to the nonmobile sperms of red algae. 21

spermatophytes   Plants that produce seeds. Alternative: *vascular crypto-gams.* 23

spherosome   See *lipid body.*

spindle   The framework of microtubules that pulls the chromosomes from the center of the cell to the poles during nuclear division. 4

split genome virus   A type of virus in which the genome occurs as two or more separate double helices of DNA, each packaged in separate viral particles. 15

spongy mesophyll   Any part of leaf mesophyll in which the cells are not aligned parallel to each other and are separated by large intercellular spaces. 6

sporangiophore   In the arthrophytes (sphenophytes), a stalked, um-brella-like structure that bears sporangia. 23

sporangiospores   In fungi, spores that form inside the swollen tip of a hypha. 20

sporangium (pl.: sporangia)   A structure that produces spores. 9, 20–25

spore   A single cell that is a means of asexual reproduction; it can grow into a new organism but cannot fuse like a gamete. 9, 19–25

sporophyll   A leaf that bears sporangia. 23–25

sporophyte   A diploid plant that produces spores. Alternative: *gameto-phyte.* 9, 21–25

spring wood   In secondary xylem, the wood formed early in the season, usually with an abundance of vessels in angiosperms or with wide tracheids in gymnosperms. Also called early wood. Alternative: *sum-mer wood.* 8

stamens   The organs of a flower involved in producing microspores (pollen). See also *androecium, anther,* and *filament.* 9, 25

start codon   In messenger RNA, a codon (set of three nucleotides) that indicates the beginning of information for protein synthesis. 15

statocytes   Cells within the root cap that detect the direction of gravity. 14

stele   The set of vascular tissues in a root or stem (but not in a leaf). 5, 23, 24

sterigma (pl.: sterigmata)   In basidiomycete fungi, the narrow tube that connects basidiospores to basidia. 20

stigma (pl.: stigmas)   In the carpel of a flower, the receptive tissue to which pollen adheres. 9, 25 In algae, an eyespot, a set of pigment droplets involved in detecting light direction. 21

stipe   The stalk of certain organisms—fern leaves, kelps, mushrooms. 20, 21, 23

stipules   Small flaps of tissue located at the base of a leaf, near its at-tachment to the stem. Stipules may range from quite leaflike to small and inconspicuous. 6

stolon   An aerial stem with elongate internodes; it establishes plantlets periodically when it contacts soil. Example: strawberry. 5

stoma (pl.: stomata)   A word sometimes used to mean "stomatal pore," the intercellular space between guard cells through which carbon dioxide and water are exchanged, and sometimes used to mean "sto-matal complex," the stomatal pore plus guard cells plus associated cells. 5

stomatal pore   The intercellular space between two guard cells; carbon dioxide is absorbed through the pore and water is lost. 5

stop codon   In messenger RNA, a codon (set of three nucleotides) that indicates the end of information for protein synthesis. 15

storied cambium   A vascular cambium in which fusiform initials are aligned horizontally. Alternative: *nonstoried cambium.* 8

strict aerobe   Synonym for *obligate aerobe.*

strict anaerobe   Synonym for *obligate anaerobe.*

strobilus (pl.: strobili)   Synonym for *cone.*

stroma reactions   In plant photosynthesis, the set of reactions that occur in the stroma and are not directly powered by light. Synonym: dark reactions. Alternative: *light-dependent reactions.* 10

stromatolite   Large stonelike growths of cyanobacteria, formed in shal-low, warm sea water. Some are 2.7 billion years old. 19

structural region   The portion of a gene consisting of nucleotide triplets that specify which amino acids are to be incorporated into protein. Alternative: *promotor region.* 15

style   In the carpel, the tissue that elevates the stigma above the ovary. 9, 25

subapical meristem   The region of a shoot or root just proximal to the apical meristem. 5

suberin   Lipid material that causes the hydrophobic properties of cork cell walls and the Casparian strip of the endodermis. 7, 8

substrate   The reactant acted upon by an enzyme. 2

substrate level phosphorylation   The formation of ATP from ADP by having a phosphate group transferred to it from a substrate mole-cule. 10, 11

substrate specificity   The ability of an enzyme to distinguish one sub-strate from similar substrates. 2

summer wood   In secondary xylem, the wood formed late in the season, usually with few or no vessels in angiosperms, or with narrow tra-cheids in gymnosperms. Also called late wood. Alternative: *spring wood.* 8

superior ovary   An ovary located above the sepals, petals, and stamens; those appendages are hypogenous. Alternatives: *inferior* and *half-infe-rior ovary.* 9

suspensor   In seed plant embryos, the stalk of cells that pushes the em-bryo into the endosperm. 9

symbiotic relationship   A relationship in which two or more organisms live closely together. 21

sympatric speciation   Speciation that occurs within a limited geo-graphic range; populations are separated by biotic reproductive barriers, not by physical differences. Alternative: *allopatric speciation.* 17

sympetally   The condition of having the petals of a flower fused to-gether into a tube. 25

symplast   The protoplasm of all the cells in a plant are interconnected by plasmodesmata; the entire mass is the symplast or symplasm. See *apoplast.* 3, 5, 7

sympodial branching   A branching pattern in which what appears to be one main shoot (the trunk) is actually a series of lateral branches, each of which displaces the apex of the shoot that bears it. Alterna-tives: *monopodial* and *pseudomonopodial branching.* 23–25

synapsis (pl.: synapses)   The pairing of homologous chromosomes during zygotene of prophase I of meiosis. Synapsis precedes crossing-over. 4

synaptonemal complex   In prophase I of meiosis, after homologous chromosomes have paired (undergone synapsis), a protein complex, the synoptonemal complex, holds them together. It is composed of a central element connected by fine fibers to the lateral elements that actually connect to DNA. 4

synergid   In the egg apparatus of an angiosperm megagametophyte, there is an egg and one or two adjacent cells, synergids; the pollen tube enters one of the synergids. 9

syngamy   The fusion of a sperm and an egg. 9

TATA box   In the promotor region of many genes, a region rich in thy-mine- and adenine-containing nucleotides, believed important for RNA polymerase binding. 15

taxis   A response in which a cell swims toward or away from a stimu-lus. 14

taxon   A term that refers to any taxonomic group such as species, genus, family, and so on. 18

telome   In a plant with dichotomous branching, the last two twigs produced by the last bifurcation. 23

telome theory   The theory that leaves (megaphylls) of arthrophytes, ferns, and seed plants evolved from branch systems (telomes) by overtopping, planation, and webbing. 23

telophase   The fourth and last phase of mitosis, during which the chromosomes decondense, the nucleolus and nuclear envelope reform, the spindle depolymerizes, and the phragmoplast appears. 4

telophase I   The fourth phase of meiosis I, similar to telophase of mitosis. However, in many organisms, telophase I and prophase II are often shortened or eliminated, and full nuclei are not formed between meiosis I and II. 4

telophase II   A phase of meiosis II, similar to telophase of mitosis. 4

temperate virus   A virus whose genome has been incorporated into the host's genome, being replicated simultaneously with host DNA; the virus produces few or no symptoms. 15

tendril   An organ that attaches a vine to a support by wrapping around it. It may be a modified leaf, leaflet, or shoot. Example: grape. 5, 6

tension wood   See *reaction wood.*

tepal   Refers to members of a perianth when it is not certain if they are really sepals or petals. 9

terminal bud   A bud located at the extreme apex of a shoot; usually present only in winter as a dormant bud. Alternative: *axillary bud.* 5

tertiary structure of a protein   The overall three-dimensional shape of an entire protein molecule. 2

test cross   A cross involving one parent known to be homozygous recessive for the trait being considered. 16

testa (pl.: testas)   Synonym for *seed coat.*

tetrads   During meiosis I, after homologs have paired and condensed sufficiently, the four chromatids are visible as a tetrad. 4

tetraploid   Refers to four full sets of chromosomes within a single nucleus. 4, 16

tetraspores   In red algae, the spores produced by meiosis in the tetrasporophyte. 21

tetrasporophyte   In red algae, the diploid generation equivalent to a sporophyte generation in other algae. 21

thallophyte   An old, rarely used term to distinguish organisms that are not embryophytes; algae and fungi. 20, 21

theory   After a hypothesis has been confirmed by numerous observations or experiments, it is considered to be a theory. 1

thermogenic respiration   Respiration in which electron transport is uncoupled from ATP synthesis, so heat is generated. Synonym: cyanide-resistant respiration. 11

thigmotropic response   A tropic response with touch as the stimulus. 14

threshold   The level or intensity of stimulus that must be present during the presentation time in order for an organism to perceive it. 14

thylakoids   The photosynthetic membranes of chloroplasts. 3, 10

ti plasmid   A plasmid from the bacterium *Agrobacterium tumefaciens;* this is a commonly used vector for recombinant DNA studies in plants. 15

tinsel flagellum   A flagellum covered with many minute hairlike projections. Alternative: *whiplash flagellum.* 20, 21

tolerance range   The range of environmental conditions in which an organism can live and reproduce. 26

tonoplast   The vacuolar membrane. 3

trace element   See *essential element.*

tracheary element   A term referring to either or both types of xylem-conducting cell: tracheids and vessel elements. 5

tracheid   A xylem-conducting cell; tracheids tend to be long and tapered, and they never have a perforation—a complete hole in the primary wall—as vessel elements do. 5, 12, 25

transamination   The transferral of an amino group from one molecule to another; important in the synthesis of amino acids. 13

transcription   The "reading" of DNA by RNA polymerase with the simultaneous production of RNA. 15

transduction   In a receptive tissue after the perception of a stimulus, transduction is the change that allows the tissue to communicate that the stimulus has occurred. 14 In bacteria, a method of genetic exchange occurring when a bacterium incorporates DNA carried in by a virus. 19

transfer cells   Cells involved in rapid short-distance transfer of material; they have transfer (labyrinthine) walls. 12

transfer wall   In transfer cells, walls whose inner surface is highly convoluted, thus increasing the surface area of the plasma membrane and the number of molecular pumps present. Synonym: *labyrinthine wall.* 12

transformation   In bacteria, a method of genetic exchange occurring when a bacterium incorporates a piece of DNA from the environment. 19

transformation theory   The theory that vascular plants arose from algae that had an alternation of isomorphic generations, each of which was gradually transformed into the types of sporophytes and gametophytes present today. Compare: *interpolation theory.* 23

translation   In protein synthesis, the utilization of mRNA to guide the incorporation of amino acids into protein. 15

translocation   Long-distance transport of water and nutrients by xylem and phloem. 12

transpiraton   Loss of water vapor through the epidermis. **transcuticular transpiration**   Loss through the cuticle. **transstomatal transpiration**   Loss through stomata. 12

transposable element   A region of DNA that codes for enzymes that catalyze the release of the element and its insertion into a different site in the DNA. See also *transposon* and *insertion sequence.* 16

***trans*-position**   In a carbon-carbon double bond, the situation in which two groups of interest lie on opposite sides of the bond. Alternative: *cis*-position. 2

transposon   A large transposable element that carries, in addition to the insertion sequence, other genes that code for proteins not directly associated with transposition. 16

tricarboxylic acid cycle   Synonym for *citric acid cycle* and Krebs cycle. 11

trichogyne   In ascomycete fungi, a narrow tube that transfers nuclei from an antheridium to an ascogonium. 20

trichome   A plant hair; often restricted to structures that contain only cells derived from the epidermis. 5

triglyceride   A type of lipid consisting of three fatty acids bound to one molecule of glycerol. 2

triploid   A nucleus that contains three sets of chromosomes. 16

tropic response   A growth response oriented with regard to the stimulus. Synonym: tropism. 14

true fruit   A fruit that developed only from carpel tissue, not containing any other tissue. Alternative: *accessory fruit.* 9

trumpet hyphae   In some brown algae, the phloem-like cells that conduct photosynthate through the alga. 21

tuber   A short, fleshy, horizontal stem, involved in storing nutrients but not in migrating laterally. Example: potato. 5

tubulin   See *microtubules.*

turgid   Filled with water to such a degree that the surface of the cell or plant is firm. 12

turgor pressure   The pressure with which a protoplast presses against the cell wall when a cell is turgid. 12

tylosis (pl.: tyloses)   After a vessel stops conducting because of cavitation, adjacent cells may push cytoplasm into the vessel through pits, plugging the vessel. 8

type specimen   A single specimen that is the absolute standard for the species and its scientific name. 18

uniform distribution   In ecology, the distribution of individuals in the habitat such that they are evenly spaced, as in an orchard, rather than being clumped or occurring at random. 26

unilocular sporangium   In brown algae, a multinucleate, coenocytic sporangium where meiosis occurs. Alternative: *plurilocular gametangia*. 21

uniparental inheritance   Inheritance of genes from just one parent, the most common case for plastid and mitochondrial genes; sperm cells typically do not contribute these organelles to the zygote. Alternative: *biparental inheritance*. Synonym: *maternal inheritance*. 16

uniseriate   Consisting of just one row; often used for things that consist of one layer. 5

upright cell   In rays in secondary xylem, cells that are taller than long; they typically have extensive cross-field pitting. Alternative: *procumbent cell*. 8

vacuole   A membrane-bounded (tonoplast) space larger than a vesicle which stores material, either dissolved in water or as a crystalline or flocculent mass. 3

vacuole membrane   Synonym for *tonoplast*.

valence electrons   The electrons that actually participate during a chemical reaction. 2

variegation   A pattern of spots, stripes, or patches in leaves or other organs, caused by plastid mutations. 16

vascular bundle   A column of vascular tissue, typically both xylem and phloem together, but in leaves sometimes consisting of only one or the other. 5

vascular cambium   The meristem that produces secondary vascular tissues—secondary xylem (wood) and secondary phloem (inner bark). 8

vascular cryptogams   The vascular plants that do not produce seeds, such as lycopods, horsetails, and ferns. Alternative: *spermatophytes*. 23

vector   Pieces of DNA that are used to insert experimental, recombinant DNA into bacteria or eukaryotes. 15

vegetative   Refers to phenomena or parts of a plant not involved in sexual reproduction. 9

vegetative cell   In the pollen grain of seed plants, the cell or cells that do not give rise to the sperm cells; the cell that is not the generative cell. 9, 24

venation   The pattern of veins in a tissue or organ. 5, 6

vernalization   The cold treatment necessary for biennials to initiate flowering. 14

vesicle   A small space enclosed by a single membrane. Vesicles are similar to vacuoles, but smaller, generally being unresolvable by light microscopy. 3

vessel   In xylem, a column of vessel elements interconnected by pairs of perforations. 5

vessel element (vessel member)   A xylem conducting cell that has one or two perforations—a complete large hole in the primary wall that permits water to flow easily from one vessel member to another. 5, 25

violaxanthin   A xanthophyll pigment in brown algae. 21

water potential   The chemical potential of water; a measure of the ability of a substance to absorb or release water relative to another substance. Components: *osmotic potential, pressure potential, matric potential*. 12

wax   An extremely hydrophobic polymer of long-chain fatty acids; wax contributes to the water-retaining capacity of the epidermis. 2

webbing   In the telome theory of the origin of megaphylls, the concept that the lamina originated by the production of parenchyma cells between the telomes. 23

whiplash flagellum   A flagellum that has a smooth surface. Alternative: *tinsel flagellum*. 20, 21

whorl   A set of leaves or flower parts, all attached to the stem or receptable at the same level. 5, 6

wood   Secondary xylem. 8

woody plant   A plant that undergoes secondary growth by means of a vascular cambium which produces secondary xylem (wood) and secondary phloem. Alternative is an herbaceous plant. 5, 8

xerophyte   A plant adapted to desert conditions. 5, 6, 27

xylem   The water- and mineral-conducting portion of vascular tissues, containing either tracheids or vessel elements or both; parenchyma, fibers, and sclereids are also frequent components of xylem. Alternative: *phloem*. 5

zeatin   A natural cytokinin. 14

zone of elongation   Region of a root tip, just proximal to the root apical meristem, where cells undergo pronounced elongation. 7

zoospore   A spore capable of swimming. 21

zygomorphic flower   A bilaterally symmetrical flower. Alternative: *regular flower* (actinomorphic). 9, 25

zygosporangium   In zygomycete fungi, a large multinucleate sporangium formed by fusion of two compatible hyphae. 20

zygote   The diploid cell formed as the result of the fusion of two gametes. 4, 9, 21, 22

zygotene   See *prophase I*.

Note: Page numbers in *italics* refer to illustrations; page numbers followed by (t) refer to tables.

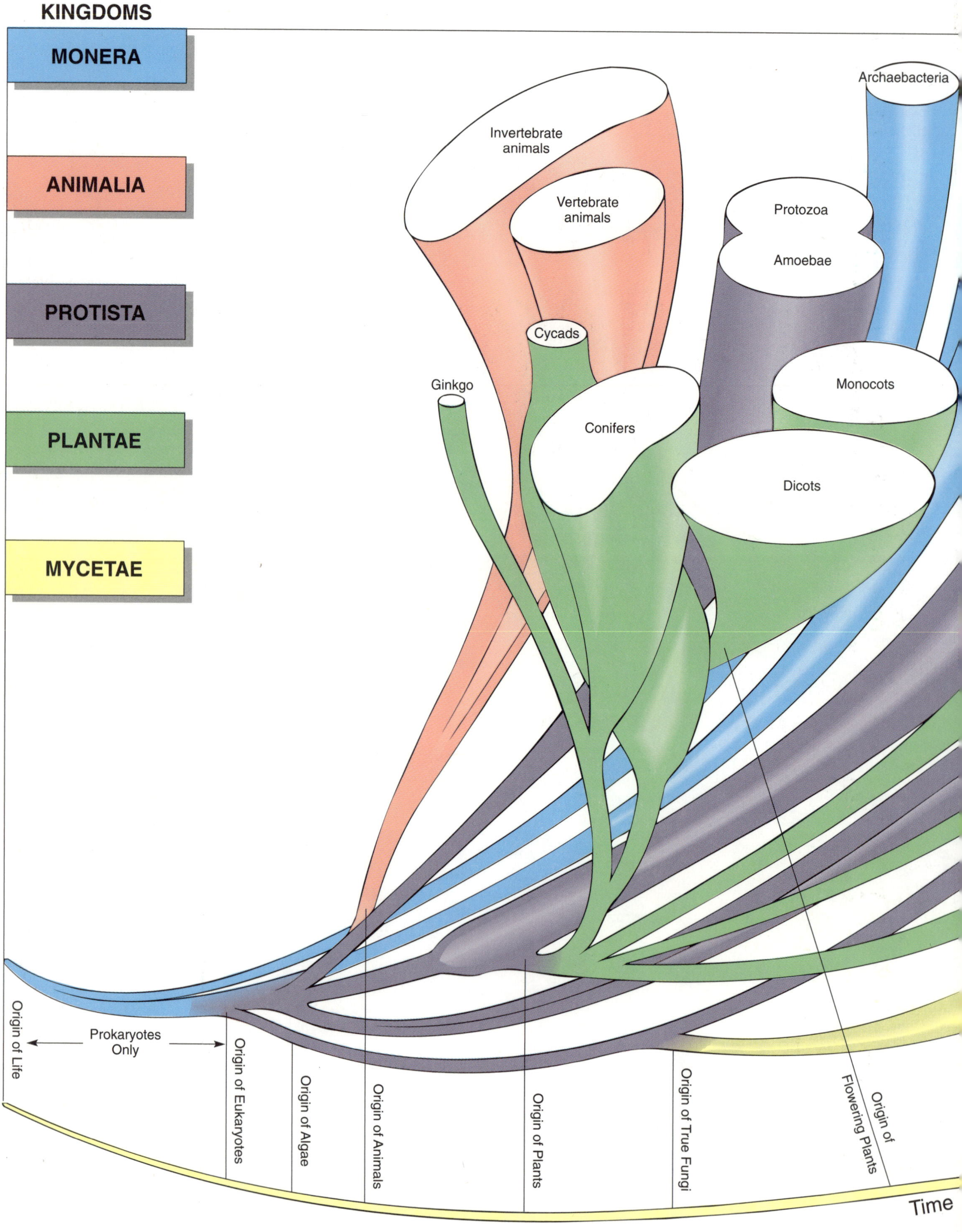

KINGDOMS

MONERA

ANIMALIA

PROTISTA

PLANTAE

MYCETAE

Archaebacteria

Invertebrate animals

Vertebrate animals

Protozoa

Amoebae

Cycads

Monocots

Ginkgo

Conifers

Dicots

Origin of Life

Prokaryotes Only

Origin of Eukaryotes

Origin of Algae

Origin of Animals

Origin of Plants

Origin of True Fungi

Origin of Flowering Plants

Time